Electronic Circuits with MATLAB®, PSpice®, and Smith Chart

Electronic Circuits with MATLAB®, PSpice®, and Smith Chart

Won Y. Yang, Jaekwon Kim, Kyung W. Park, Donghyun Baek, Sungjoon Lim, Jingon Joung, Suhyun Park, Han L. Lee, Woo June Choi, and Taeho Im

WILEY

This edition first published 2020

Registered Office
John Wiley & Sons, Inc., 111 River Street, Hoboken, NJ 07030, USA

Editorial Office
111 River Street, Hoboken, NJ 07030, USA

For details of our global editorial offices, customer services, and more information about Wiley products visit us at www.wiley.com.

Wiley also publishes its books in a variety of electronic formats and by print-on-demand. Some content that appears in standard print versions of this book may not be available in other formats.

Library of Congress Cataloging-in-Publication data applied for

ISBN: 9781119598923

Cover design: Wiley
Cover image: © green_01/Shutterstock

Set in 10/12pt Warnock by SPi Global, Pondicherry, India

Printed in the United States of America

V10015854_112719

To our parents and families
who love and support us
and
to our teachers and students
who enriched our knowledge

Contents

Preface

The aim of this book is not to let the readers drowned into a sea of computations. More hopefully, it aims to inspire the readers with mind and strength to make full use of the MATLAB and PSpice softwares so that they can feel comfortable with mathematical equations without caring about how to solve them and further can enjoy developing their ability to analyze/design electronic circuits. It aims also to present the readers with a steppingstone to radio frequency (RF) circuit design from junior–senior level to senior-graduate level by demonstrating how MATLAB can be used for the design and implementation of microstrip filters. The features of this book can be summarized as follows:

1) For representative examples of designing/analyzing electronic circuits, the analytical solutions are presented together with the results of MATLAB design and analysis (based on the theory) and PSpice simulation (similar to the experiment) in the form of trinity. This approach gives the readers not only information about the state of the art, but also confidence in the legitimacy of the solution as long as the solutions obtained by using the two software tools agree with each other.
2) For representative examples of impedance matching and filter design, the solution using MATLAB and that using Smith chart have been presented for comparison/crosscheck. This approach is expected to give the readers not only confidence in the legitimacy of the solution, but also deeper understanding of the solution.
3) The purposes of the two softwares, MATLAB and PSpice, seem to be overlapped and it is partly true. However, they can be differentiated since MATLAB is mainly used to design circuits and perform a preliminary analysis of (designed) circuits while PSpice is mainly used for detailed and almost real-world simulation of (designed) circuits.
4) Especially, it presents how to use MATLAB and PSpice not only for designing/analyzing electronic and RF circuits but also for understanding the underlying processes and related equations without having to struggle with time-consuming/error-prone computations.

The contents of this book are derived from the works of many (known or unknown) great scientists, scholars, and researchers, all of whom are deeply appreciated. We would like to thank the reviewers for their valuable comments and suggestions, which contribute to enriching this book.

We also thank the people of the School of Electronic and Electrical Engineering, Chung-Ang University for giving us an academic environment. Without affections and supports of our families and friends, this book could not be written. We gratefully acknowledge the editorial, Brett Kurzman and production staff of John Wiley & Sons, Inc. including Project Editor Antony Sami and Production Editor Viniprammia Premkumar for their kind, efficient, and encouraging guide.

Program files can be downloaded from https://wyyang53.wixsite.com/mysite/publications. Any questions, comments, and suggestions regarding this book are welcome and they should be mailed to wyyang53@hanmail.net.

Won Y. Yang et al.

About the Companion Website

Do not forget to visit the companion website for this book:

www.wiley.com/go/yang/electroniccircuits

Scan this QR code to visit the companion website.

There you will find valuable material designed to enhance your learning, including the following:

- Learning Outcomes for all chapters
- Exercises for all chapters
- References for all chapters
- Further reading for all chapters
- Figures for Chapters 16, 22, and 30

1

Load Line Analysis and Fourier Series

CHAPTER OUTLINE

1.1 Load Line Analysis

The v-i characteristic of a nonlinear resistor such as a diode or a transistor is often described by a curve on the v-i plane rather than by a mathematical relation. The v-i characteristic curve can be obtained by using a curve tracer for nonlinear resistors. To analyze circuits containing a nonlinear resistor, we should use the *load line analysis*. To grasp the concept of the load line, consider the graphical analysis of the circuit in Figure 1.1(a), which consists of a linear resistor R_1, a nonlinear resistor R_2, a DC voltage source V_s, and an AC voltage source of small amplitude $v_\delta \ll V_s$. Kirchhoff's voltage law (KVL) can be applied around the mesh to yield the mesh equation as

Electronic Circuits with MATLAB®, PSpice®, and Smith Chart, First Edition. Won Y. Yang, Jaekwon Kim, Kyung W. Park, Donghyun Baek, Sungjoon Lim, Jingon Joung, Suhyun Park, Han L. Lee, Woo June Choi, and Taeho Im.

Companion website: www.wiley.com/go/yang/electroniccircuits

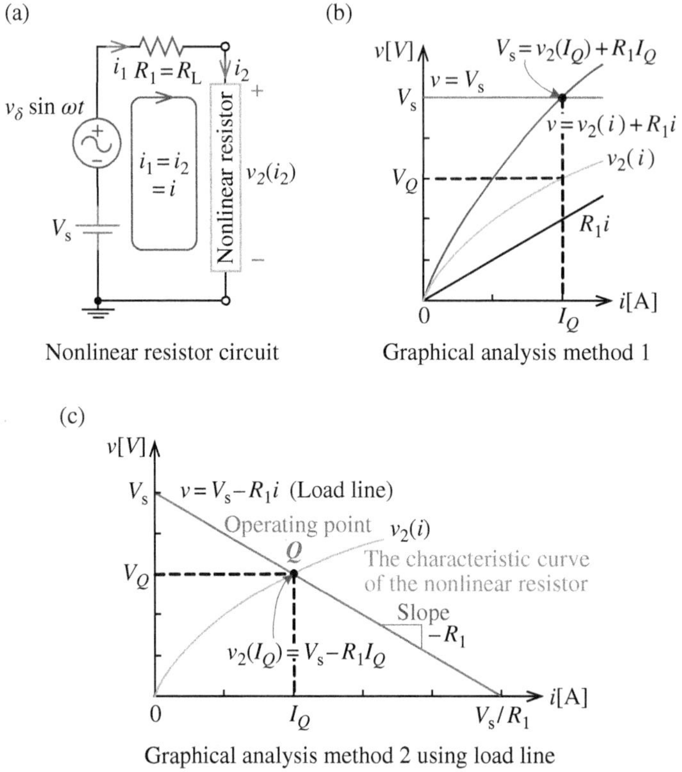

Figure 1.1 Graphical analysis of a linear/nonlinear resistor circuit.

$$R_1\, i + v_2(i) = V_s \tag{1.1.1}$$

where the *v*-*i* relationship of R_2 is denoted by $v_2(i)$ and represented by the characteristic curve in Figure 1.1(b). We will consider a graphical method, which yields the *quiescent, operating,* or *bias point* Q = (I_Q, V_Q)*, that is, a pair of the current through and the voltage across* R_2 *for* $v_\delta = 0$.

Since no specific mathematical expression of $v_2(i)$ is given, we cannot use any analytical method to solve this equation and that is why we are going to resort to a graphical method. First, we may think of plotting the graph for the LHS (left-hand side) of Eq. (1.1.1) and finding its intersection with a horizontal line for the RHS (right-hand side), that is, $v = V_s$ as depicted in Figure 1.1(b). Another way is to leave only the nonlinear term on the LHS and move the other term(s) into the RHS to rewrite Eq. (1.1.1) as

$$v_2(i) = V_s - R_1 i \tag{1.1.2}$$

and find the intersection, called the *operating point* and denoted by Q (quiescent point), of the graphs for both sides as depicted in Figure 1.1(c). The straight line with the slope of $-R_1$ is called the *load line*. This graphical method is better than the first one in the aspect that it does not require us to plot a new curve for $v_2(i) + R_1 i$. That is why it is widely used to analyze nonlinear resistor circuits in the name of 'load line analysis'. Note the following:

- Most resistors appearing in this book are linear in the sense that their voltages are linearly proportional to their currents so that their voltage-current relationships (VCRs) are described by Ohm's law

$$v = R\,i \tag{1.1.3}$$

  and consequently, their v-i characteristics are described by straight lines passing through the origin with the slopes corresponding to their resistances on the i-v plane. However, they may have been modeled or approximated to be linear just for simplicity and convenience, because all physical resistors more or less exhibit some nonlinear characteristic. The problem is whether or not the modeling is valid in the range of practical operation so that it may yield the solution with sufficient accuracy to serve the objective of analysis and design.
- A curve tracer is an instrument that displays the v-i characteristic curve of an electric element on a cathode-ray tube (CRT) when the element is inserted into an appropriate receptacle.

1.1.1 Load Line Analysis of a Nonlinear Resistor Circuit

Consider the circuit in Figure 1.1(a), where a linear load resistor $R_1 = R_L$ and a nonlinear resistor R connected in series are driven by a DC voltage source V_s in series with a small-amplitude AC voltage source producing the virtual voltage as

$$v_s(t) = V_s + v_\delta \sin\omega t \tag{1.1.4}$$

The VCR $v(i)$ of the nonlinear resistor R is described by the characteristic curve in Figure 1.2.

As depicted in Figure 1.2, the upper/lower limits as well as the equilibrium value of the current i through the circuit can be obtained from the three operating points, that is, the intersections (Q_1, Q, and Q_2) of the characteristic curve with the following three load lines.

$$v = V_s + v_\delta - R_L i \tag{1.1.5a}$$

$$v = V_s - R_L i \tag{1.1.5b}$$

$$v = V_s - v_\delta - R_L i \tag{1.1.5c}$$

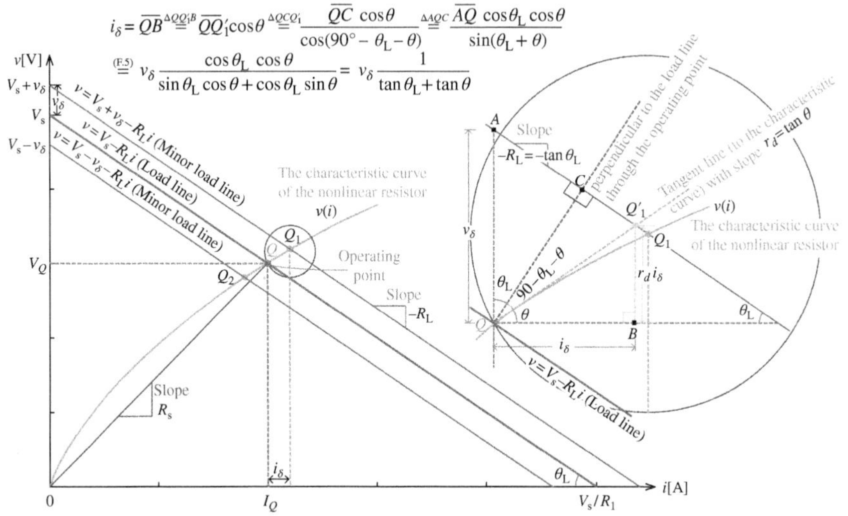

Figure 1.2 Variation of the voltage and current of a nonlinear resistor around the operating point **Q**.

Although this approach gives the exact solution, we gain no insight into the solution from it. Instead, we take a rather approximate approach, which consists of the following two steps.

- Find the equilibrium (I_Q, V_Q) at the major operating point Q, which is the intersection of the characteristic curve with the DC load line (1.1.5b).
- Find the two approximate minor operating points Q_1' and Q_2' from the intersections of the tangent to the characteristic curve at Q with the two minor load lines (1.1.5a) and (1.1.5c).

Then we will have the current as

$$i(t) = I_Q + i_\delta \sin\omega t \tag{1.1.6}$$

With the *dynamic, small-signal,* or *AC resistance* r_d defined to be the slope of the tangent to the characteristic curve at Q as

$$r_d = \left.\frac{dv}{di}\right|_Q \tag{1.1.7}$$

let us find the analytical expressions of I_Q and i_δ in terms of V_s and v_δ, respectively. Referring to the encircled area around the operating point in Figure 1.2, we can express i_δ in terms of v_δ as

$$i_\delta = \overline{QB} \overset{\Delta QQ_1'B}{=} \overline{QQ_1'}\cos\theta \overset{\Delta QCQ_1'}{=} \frac{\overline{QC}\cos\theta}{\cos(90^\circ - \theta_L - \theta)} \overset{\Delta AQC}{=} \frac{\overline{AQ}\cos\theta_L\cos\theta}{\sin(\theta_L + \theta)}$$

$$\overset{(F.5)}{=} v_\delta \frac{\cos\theta_L\cos\theta}{\sin\theta_L\cos\theta + \cos\theta_L\sin\theta} = v_\delta \frac{1}{\tan\theta_L + \tan\theta} \tag{1.1.8}$$

This corresponds to approximating the characteristic curve in the operation range by its tangent at the operating point. Noting that

- the load line and the tangent to the characteristic curve at Q are at angles of $(180^\circ - \theta_L)$ and θ to the positive i-axis,
- the slope of the load line is $\tan(180^\circ - \theta_L) = -\tan\theta_L$ and it must be $-R_L$, which is the proportionality coefficient in i of the load line Eq. (1.1.2); $\tan\theta_L = R_L$, and
- the slope of the tangent to the characteristic curve at Q is the dynamic resistance r_d defined by Eq. (1.1.7); $\tan\theta = r_d$,

we can write Eq. (1.1.8) as

$$i_\delta = \frac{v_\delta}{R_L + r_d} \tag{1.1.9}$$

Now we define the *static* or *DC resistance* of the nonlinear resistor R to be the ratio of the voltage V_Q to the current I_Q at the operating point Q as

$$R_s = \frac{V_Q}{I_Q} = \frac{V_s - R_L I_Q}{I_Q} \tag{1.1.10}$$

so that the DC component of the current, I_Q, can be written as

$$I_Q = \frac{V_s}{R_L + R_s} \tag{1.1.11}$$

Finally, we combine the above results to write the current through and the voltage across the nonlinear resistor R as follows.

$$i(t) = I_Q + i_\delta \sin\omega t = \frac{V_s}{R_L + R_s} + \frac{v_\delta}{R_L + r_d}\sin\omega t \tag{1.1.12}$$

$$v(t) = R_s I_Q + r_d i_\delta \sin\omega t = \frac{R_s}{R_L + R_s}V_s + \frac{r_d}{R_L + r_d}v_\delta \sin\omega t \tag{1.1.13}$$

This result implies that the nonlinear resistor exhibits twofold resistance, that is, the *static resistance* R_s to a DC input and the *dynamic resistance* r_d to an AC input of small amplitude. That is why r_d is also called the (*small-signal*) *AC resistance*, while R_s is called the *DC resistance*.

Remark 1.1 Operating Point and Static/Dynamic Resistances of a Nonlinear Resistor

1) For a nonlinear resistor R_2 connected with linear resistors in a circuit excited by a DC source and a small-amplitude AC source, its *operating point* $Q = (V_Q, I_Q)$ is the intersection of its characteristic curve $v(i)$ and the load line.
2) The v-intercept of the load line ($v = V_s - R_L i$) is determined by the DC component (V_s) of the voltage source. The slope of the load line is determined by the equivalent resistance (R_L) of the linear part seen from the pair of terminals of the nonlinear resistor (see Problem 1.2).
3) The *static* or *DC resistance* (R_s) is the ratio of the voltage V_Q to the current I_Q at the operating point Q.
4) The *dynamic*, *small-signal*, *AC*, or *incremental resistance* (r_d) is the slope of the tangent to the characteristic curve at Q.
5) Once we have R_L, R_s, and r_d, we can use Formulas (1.1.12) and (1.1.13) to find approximate expressions for the voltage and current of the nonlinear resistor.
6) As for linear resistors, we do not say the static or dynamic resistance, since they are identical.
7) The relationship between the AC (small-signal) components of voltage across and current through the nonlinear resistor can be attributed to the Taylor series expansion of its VCR $v(i)$ up to the first-order term around the operating point $Q = (V_Q, I_Q)$.

$$v(i) \approx V_Q + \left.\frac{dv}{di}\right|_Q (i - I_Q) = V_Q + r_d i_\delta \text{ with } r_d = \left.\frac{dv}{di}\right|_Q \tag{1.1.14}$$

Remark 1.2 DC Analysis and Small-Signal (AC) Analysis

1) The procedure to analyze a circuit (which contains nonlinear resistors like a diode or a transistor and is driven by a high DC voltage source V_s [for biasing] and a low AC voltage source $v_\delta \sin \omega t$ [for amplification]) consists of two steps. The first step, called DC analysis, is to remove the AC voltage source $v_\delta \sin \omega t$ and find the operating point $Q = (V_Q, I_Q)$ of the nonlinear resistor, which corresponds to the load line analysis. The second step, called small-signal (AC) analysis, is to find the dynamic resistance r_d of the nonlinear resistor (from the slope of its i-v characteristic curve or the derivative of its VCR equation at Q-point), remove the DC voltage source V_s, regard the nonlinear resistor as a linear resistor r_d (corresponding to a linear

approximation of the characteristic curve), and find the AC components (i_δ sin ωt, $r_d i_\delta$ sin ωt) of the current through and voltage across the nonlinear resistor. The DC solution and AC solution can be added up to yield the complete solution.

2) As the magnitude v_δ of the AC voltage becomes large, the large-signal model (see Section 2.1.1 for a diode) or the characteristic curve itself might have to be used for analysis since the nonlinear behavior of the nonlinear resistor may become conspicuous, leading to unignorable distortion of the voltage/current waveforms obtained using the small-signal analysis.

1.1.2 Load Line Analysis of a Nonlinear *RL* circuit

As an example of applying the load line analysis for a nonlinear first-order circuit, consider the circuit of Figure 1.3.1(a), which consists of a nonlinear resistor, a linear resistor $R = 2\,\Omega$, and an inductor $L = 14$ H, and is driven by a DC voltage source of $V_s = 12$ V and an AC voltage source $v_\delta \sin \omega t = 2.8 \sin t$[V]. The v-i relationship of the nonlinear resistor is $v(i) = i^3$ and described by the characteristic curve in Figure 1.3.1(b). Applying KVL yields the following mesh equation:

$$L\frac{di(t)}{dt} + Ri(t) + i^3(t) = V_s + v_\delta \sin t;$$

$$14\frac{di(t)}{dt} + 2i(t) + i^3(t) = 12 + 2.8\sin t \tag{1.1.15}$$

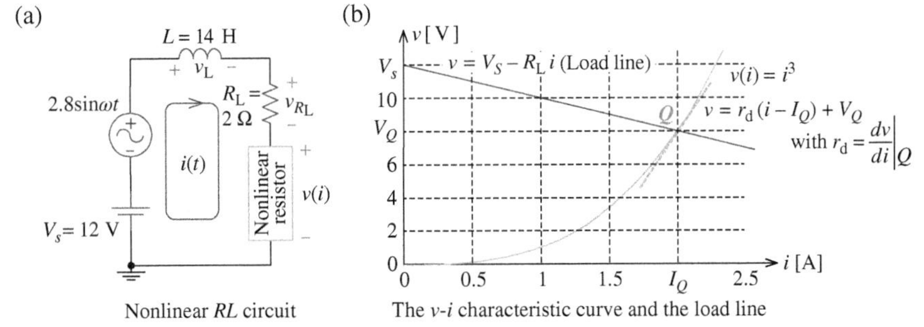

Figure 1.3.1 A nonlinear *RL* circuit and its load line analysis.

First, we can draw the load line on the graph of Figure 1.3.1(b) to find the operating point Q from the intersection of the load line and the v-i characteristic curve of the nonlinear resistor. If the function $v(i) = i^3$ of the characteristic curve is available, we can also find the operating point Q as the DC solution to Eq. (1.1.15) by removing the AC source and the time derivative term to write

$$2I_Q + I_Q^3 = 12 \tag{1.1.16}$$

and solving it as

$$I_Q = 2\,\text{A}, V_Q = v_2(I_Q) = I_Q^3 = 8\,\text{V} \rightarrow Q = (I_Q, V_Q) = (2\,\text{A}, 8\,\text{V}) \tag{1.1.17}$$

Eq. (1.1.16) can be solved by running the following MATLAB statements:

```
>>eq_dc=@(i)2*i+i.^3-12; I0=0; IQ=fsolve(eq_dc,I0)
```

Then, as a preparation for analytical approach, we linearize the nonlinear differential equation (1.1.15) around the operating point Q by substituting $i = I_Q + \delta i = 2 + \delta i$ into it and neglecting the second or higher degree terms in δi as

$$\begin{aligned} &14\frac{d(2+\delta i)}{dt} + 2(2+\delta i) + (2+\delta i)^3 = 12 + 2.8\sin t \\ &;\frac{d}{dt}\delta i(t) = -\delta i(t) + 0.2\sin t \end{aligned} \tag{1.1.18}$$

Note that we can set the slope of the characteristic curve as the dynamic resistance r_d of the nonlinear resistor:

$$r_d = \left.\frac{dv}{di}\right|_Q = 3I_Q^2 = 12\,\Omega \tag{1.1.19}$$

and apply KVL to the circuit with the DC source V_s removed and the nonlinear resistor replaced by r_d to write the same linearized equation as Eq. (1.1.18):

$$\begin{aligned} &14\frac{di(t)}{dt} + 2i(t) + i^3(t) = V_s + 2.8\sin t \\ &\underset{V_s\rightarrow 0}{\overset{i^3(t)\rightarrow r_d i(t)}{\longrightarrow}} 14\frac{di(t)}{dt} + (2+12)i(t) = 2.8\sin\omega t \end{aligned} \tag{1.1.20}$$

We solve the first-order linear differential equation with zero initial condition $\delta i(0) = 0$ to get $\delta i(t)$ by running the following MATLAB statements:

```
>>syms s
  dIs=2.8/(s^2+1)/14/(s+1); dit_linearized=ilaplace(dIs)
  dit1_linearized=dsolve('Dx=-x+0.2*sin(t)','x(0)=0') % Alternatively
```

This yields

```
dit_linearized = exp(-t)/10 - cos(t)/10 + sin(t)/10
```

which means

$$\delta i(t) = \frac{1}{10}e^{-t} - \frac{1}{10}\cos t + \frac{1}{10}\sin t \tag{1.1.21}$$

We add this AC solution to the DC solution I_Q to write the approximate analytical solution for $i(t)$ as

$$i(t) = I_Q + \delta i(t) \overset{(1.1.17)}{\underset{(1.1.21)}{=}} 2 + 0.1(e^{-t} - \cos t + \sin t) \tag{1.1.22}$$

Now, referring to Appendix D, we use the MATLAB numerical differential equation (DE) solver 'ode45()' to solve the first-order nonlinear differential equation (1.1.15) by defining it as an anonymous function handle:

```
>>di=@(t,i)(12+2.8*sin(t)-2*i-i.^3)/14; % Eq.(1.1.15)
```

and then running the following MATLAB statements:

```
>>i0=IQ; tspan=[0 10]; % Initial value and Time span
  [t,i_numerical]=ode45(di,tspan,i0); % Numerical solution
```

We can also plot the numerical solution together with the analytical solution as black and red lines, respectively, by running the following MATLAB statements:

```
>>i_linearized=eval(IQ+dit_linearized); % Analytical solution (1.1.22)
  plot(t,i_numerical,'k', t,i_linearized,'r')
```

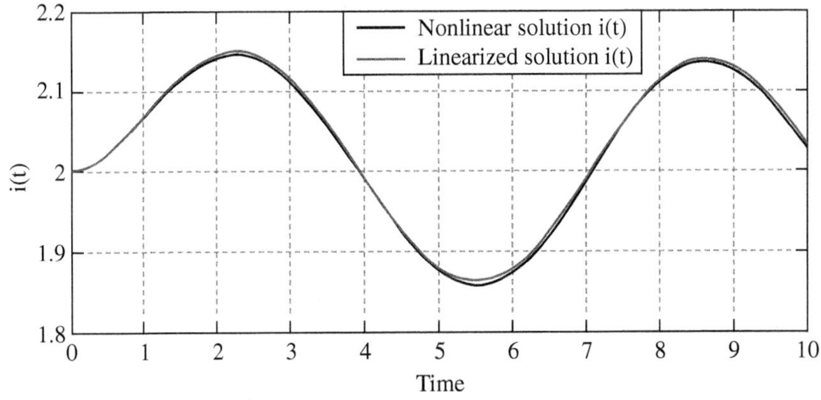

Figure 1.3.2 The linearized solution and nonlinear solution for the nonlinear *RL* circuit of Figure 1.3.1.

```
%elec01f03.m - for the analysis of a nonlinear RL Circuit
clear, clf
global RL L Vs vd
N=1000; i_step=0.003; i=[0:N]*i_step; % Range on the current axis
RL=2; L=14; Vs=12; vd=2.8;
eq_dc=@(i,RL,Vs)Vs-RL*i-i.^3; % Eq.(1.1.16): DC part of Eq.(1.1.20)
IQ=fsolve(eq_dc,0,optimset('fsolve'),RL,Vs) % Current at operating point Q
VQ=IQ^3; % Voltage at the Q-point
v= Vs - RL*i; % Load line
v2= i.^3; % Characteristic curve
i1=1.7:i_step:2.3; % Range on which to plot the tangent line
v4 = 3*IQ^2*(i1-IQ)+VQ; % Tangent line to the characteristic curve at Q
subplot(211), plot(i,v,'k', i,v2,'b', i1,v4,'r', IQ,VQ,'mo')
% Use the Laplace transform to solve the linearized differential eq.
syms s
dIs=0.2/(s^2+1)/(s+1); dit_linearized=ilaplace(dIs)
% This yields 0.1*(exp(-t) -cos(t) +sin(t)) +2; Eq.(1.1.22).
% Alternatively, use the symbolic differential solver dsolve() as
dit1_linearized = dsolve('Dx=-x+0.2*sin(t)','x(0)=0')
% Use nonlinear ODE solver ode45() to solve Eq.(1.1.15).
di=@(t,i)(12+2.8*sin(t)-2*i-i.^3)/14; % Eq.(1.1.15)
[t,i]=ode45(di,[0 10],IQ); % Numerical sol to Eq.(1.1.15)
i_linearized=eval(IQ+dit_linearized); % Eq.(1.1.22) for time range t
% Plot the Analytical (linearized) and Numerical (nonlinear) solutions.
subplot(212), plot(t,i,'k', t,i_linearized,'r'), ylabel('i(t)')
legend('Nonlinear solution i(t)','Linearized solution i(t)')
title('Analytical (linearized) and Numerical (nonlinear) solutions')
```

This will yield the plots of the numerical solution $i(t)$ and the approximate analytical solution (1.1.22) for the time interval [0,10 s] as depicted in Figure 1.3.2.

Overall, we can run the above MATLAB script "elec01f03.m" to get Figure 1.3.1(b) (the load line together with the characteristic curve) and Figure 1.3.2 (the numerical nonlinear solution together with the analytical linearized solution) together.

1.2 Voltage-Current Source Transformation

Two electric circuits are said to be *externally equivalent* with respect to a pair of terminals if their terminal voltage-current relationships are identical so that they are indistinguishable from outside. The *source transformation* refers to

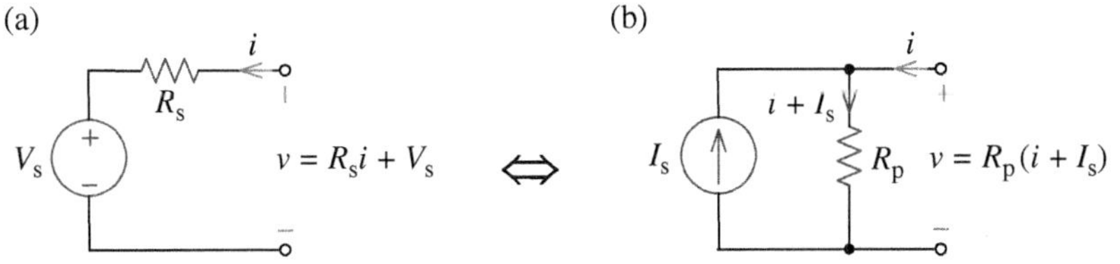

Voltage source with a series resistor   Current source with a parallel resistor

Figure 1.4 Equivalence of voltage and current sources.

the conversion of a voltage source in series with an element like a resistor (Figure 1.4(a)) to a current source in parallel with the element (Figure 1.4(b)), or vice versa in such a way that the two circuits are (externally) equivalent w.r.t. their terminal characteristics. What is the relationship among the values of the voltage source V_s, the current source I_s, the series resistor R_s, and the parallel resistor R_p required for the external equivalence of the two source models? To find it out, we write the VCR of each circuit as

$$v = R_s i + V_s \tag{1.2.1a}$$

$$v = R_p(i + I_s) = R_p i + R_p I_s \tag{1.2.1b}$$

where the current through R_p is found to be $(i + I_s)$ by applying KCL at the top node of the resistor R_p in Figure 1.4(b). In order for these two polynomial equations (in i) to be identical for any value of v and i, their coefficients (including the constant term) should be the same:

$$R_p = R_s = R \tag{1.2.2a}$$

$$V_s = R_p I_s \text{ or, equivalently, } I_s = \frac{V_s}{R_s} \tag{1.2.2b}$$

This *source equivalence condition* is used for voltage-to-current or current-to-voltage source transformation.

1.3 Thevenin/Norton Equivalent Circuits

Thevenin's theorem says that any network consisting of linear elements and independent/dependent sources as shown in Figure 1.5(a) may be replaced at a pair of its terminals (nodes) by the Thevenin equivalent circuit, which consists of a single element of impedance Z_{Th} in series with a single independent voltage source V_{Th} (see Figure 1.5(b)), where the values of Z_{Th} and V_{Th} are determined as follows:

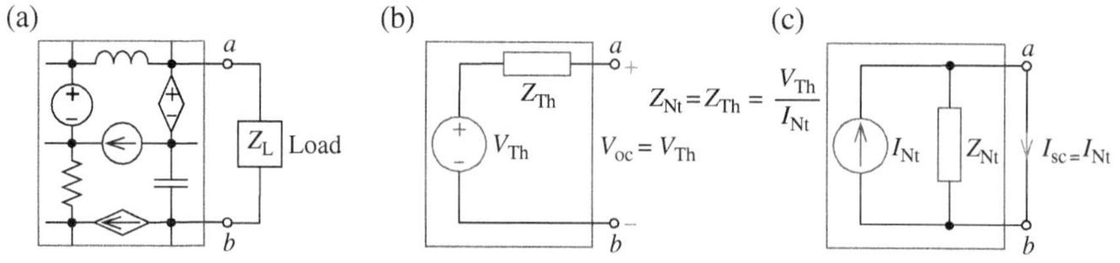

Arbitrary network with a load at its terminals *a-b*

Thevenin equivalent and open-circuit voltage with $Z_L = \infty$

Norton equivalent and short-circuit current with $Z_L = 0$

Figure 1.5 Thevenin/Norton equivalents of an arbitrary circuit seen from terminals *a* to *b*.

T1. Thevenin eqavuivalent voltage source V_{Th}:
The *open-circuit voltage* across the terminals, that is, the voltage across the open-circuited terminals *a-b* (with $Z_L = \infty$).

T2. Thevenin equivalent impedance Z_{Th}:
The equivalent impedance of the circuit (with all the independent sources removed) seen from the terminals *a-b*, where an impedance is a 'generalized' resistance.

Norton's theorem says that any linear network may be replaced at a pair of its terminals by the Norton equivalent circuit, which consists of a single element of impedance Z_{Nt} in parallel with a single independent current source I_{Nt} (see Figure 1.5(c)), where the values of Z_{Nt} and I_{Nt} are determined as follows:

N1. Norton equivalent current source I_{Nt}:
The *short-circuit current* through the terminals, that is, the current through the short-circuited terminals *a-b* (with $Z_L = 0$).

N2. Norton equivalent impedance Z_{Nt}:
The equivalent impedance of the circuit (with all the independent sources removed) seen from the terminals *a-b*

Since Thevenin and Norton equivalents are equivalent in representing a linear circuit seen from a pair of two terminals, one can be obtained from the other by using the source transformation introduced in Section 1.2. This suggests another formula for finding the equivalent impedance as

$$Z_{Th} = Z_{Nt} = \frac{V_{Th}}{I_{Nt}} = \frac{V_{oc}(\text{the open-circuit voltage})}{I_{sc}(\text{the short-circuit current})} \tag{1.3.1}$$

Note that we should find the equivalent impedance after removing every independent source, that is, with every voltage/current source short-/open-circuited. For networks having no dependent source, the series/parallel combination and Δ-Y/Y-Δ conversion formulas often suffice for the purpose of

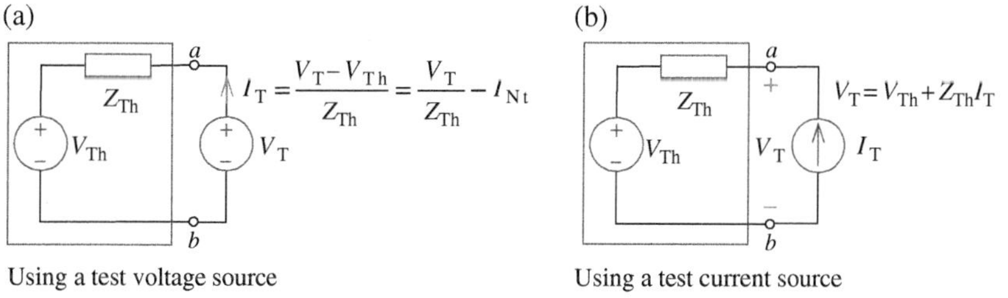

Figure 1.6 One-shot method for obtaining Thevenin equivalent.

finding the equivalent impedance. For networks having dependent sources, we should apply a test voltage source V_T across the terminals, measure the current I_T through the terminals (see Figure 1.6(a)), and write down the relationship between V_T and I_T as

$$I_T = \frac{V_T - V_{Th}}{Z_{Th}} \quad \rightarrow \quad V_T = Z_{Th} I_T + V_{Th} \tag{1.3.2}$$

In this relationship, we find the value of the equivalent impedance Z_{Th} from the proportionality coefficient of the term in I_T, and the equivalent source V_{Th} from the constant term independent of I_T. This suggests a *one-shot method* of finding the values of the equivalent impedance Z_{Th} and the equivalent source V_{Th} or I_{Nt} at a time. This one-shot method of finding the equivalent may be applied for any linear networks with or without independent/dependent sources. Another way to find the equivalent circuit is to apply a test current source I_T through the terminals (see Figure 1.6(b)), measure the voltage V_T across the terminals, and write down the relationship between V_T and I_T as Eq. (1.3.2).

Example 1.1 Thevenin Equivalent of Bridge Network

a) Find the Thevenin equivalent circuit of the bridge network in Figure 1.7(a1) seen from terminals 2-3.

Since the voltages at nodes 2 and 3 are determined by the voltage divider rule, we can get the open-circuit voltage across the terminals 2-3 as

$$V_{Th} = v_2 - v_3 = \frac{R_2}{R_1 + R_2} V_s - \frac{R_4}{R_3 + R_4} V_s = \frac{R_2 R_3 - R_1 R_4}{(R_1 + R_2)(R_3 + R_4)} V_s \tag{E1.1.1}$$

which is the value of the Thevenin equivalent voltage source.

To find the equivalent impedance, we remove (deactivate) the voltage source by short-circuiting it as depicted in Figure 1.7(b1). Then we use the parallel/series combination formulas to get the resistance between the two terminals 2 and 3 as

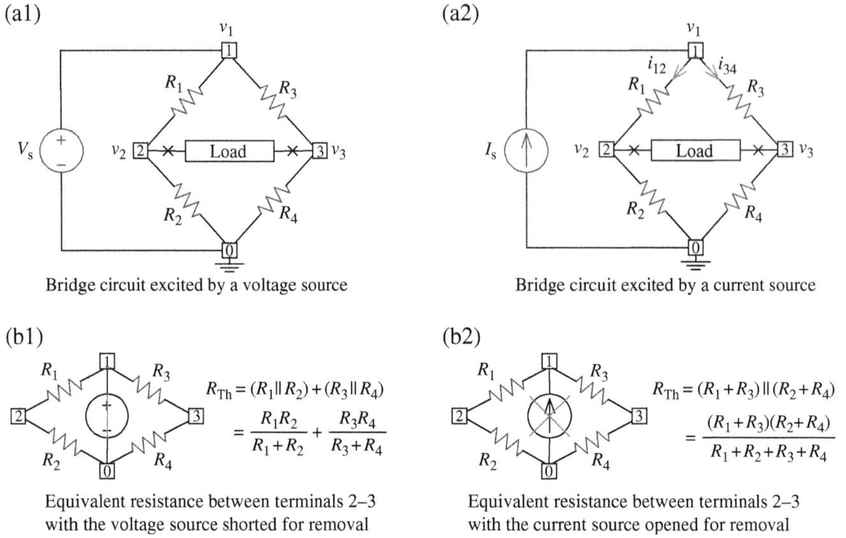

Figure 1.7 Thevenin equivalents of bridge circuits for Example 1.1.

$$R_{Th} = (R_1 \| R_2) + (R_3 \| R_4) = \frac{R_1R_2}{R_1+R_2} + \frac{R_3R_4}{R_3+R_4} \tag{E1.1.2}$$

b) Find the Thevenin equivalent circuit of the bridge network in Figure 1.7(a2) seen from terminals 2-3.

Since the currents through R_1-R_2 and R_3-R_4 are determined by the current divider rule, we can get the open-circuit voltage across terminals 2-3 as

$$\begin{aligned} V_{Th} = v_2 - v_3 = R_2 i_{12} - R_4 i_{34} &= \frac{R_2(R_3+R_4)I_s - R_4(R_1+R_2)I_s}{(R_1+R_2)+(R_3+R_4)} \\ &= \frac{R_2R_3 - R_1R_4}{(R_1+R_2)+(R_3+R_4)} I_s \end{aligned} \tag{E1.1.3}$$

which is the value of the Thevenin equivalent voltage source.

To find the equivalent impedance, we remove (deactivate) the current source by open-circuiting it as depicted in Figure 1.7(b2). Then we use the series/parallel combination formulas to get the resistance between the two terminals 2 and 3 as

$$R_{Th} = (R_1+R_3) \| (R_2+R_4) = \frac{(R_1+R_3)(R_2+R_4)}{R_1+R_2+R_3+R_4} \tag{E1.1.4}$$

Example 1.2 Thevenin Equivalent of a Network Having Dependent Source
Let us find the Thevenin equivalent of the network of Figure 1.8(a) seen from the terminals *a-b* two times, once by using the test voltage source method and once by using the test current source method.

a) Test Voltage Source Method
With the test voltage source V_T applied across the terminals *a-b* as depicted in Figure 1.8(a), we may well use the mesh analysis. We transform the βi_B[A]-source in parallel with R_E into a $\beta R_E i_B$[V]-source in series with R_E as depicted in Figure 1.8(b). Then, noting that the controlling variable i_B is the same as the mesh current, we label the mesh current i_B and write the mesh equation as

$$(R_B + R_E)i_B = V_T - \beta R_E i_B \tag{E1.2.1}$$

$$; V_T = (R_B + R_E + \beta R_E)i_B = \{R_B + (\beta + 1)R_E\}I_T \tag{E1.2.2}$$

Matching this relation with Eq. (1.3.2) yields

$$Z_{Th} = R_B + (\beta + 1)R_E \tag{E1.2.3}$$

b) Test Current Source Method
With the test current source I_T applied through the terminals *a-b* as depicted in Figure 1.8(c), we may well use the node analysis. Noting that the controlling variable i_B is the same as I_T, we set up the node equation and solve it as

$$\begin{bmatrix} 1/R_B & -1/R_B \\ -1/R_B & 1/R_B + 1/R_E \end{bmatrix} \begin{bmatrix} v_1 \\ v_2 \end{bmatrix} = \begin{bmatrix} I_T \\ \beta i_B \end{bmatrix} = \begin{bmatrix} I_T \\ \beta I_T \end{bmatrix} \tag{E1.2.4}$$

$$; \begin{bmatrix} v_1 \\ v_2 \end{bmatrix} = \frac{I_T}{1/R_B R_E} \begin{bmatrix} 1/R_B + 1/R_E & 1/R_B \\ 1/R_B & 1/R_B \end{bmatrix} \begin{bmatrix} 1 \\ \beta \end{bmatrix} = \begin{bmatrix} R_B + (\beta + 1)R_E \\ (\beta + 1)R_E \end{bmatrix} I_T \tag{E1.2.5}$$

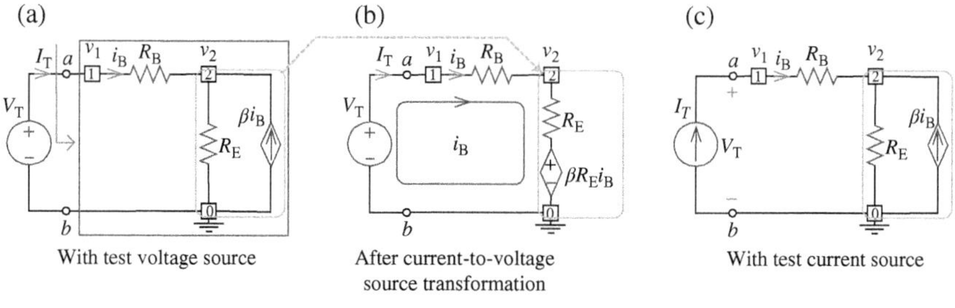

Figure 1.8 To find the equivalent circuit for Example 1.2.

This solution with $v_1 = V_T$ yields the same result as obtained in (a).

$$V_T = v_1 = (R_B + (\beta + 1)R_E)I_T \tag{E1.2.6}$$

Example 1.3 (Thevenin) Equivalent Resistance of a Network Having Dependent Source

Find the Thevenin equivalent of a common gate (CG) amplifier in Figure 1.9 seen from s-0(GND).

Noting that the circuit has a (dependent) current source of $g_m v_{gs} = (v_g - v_s)/r_{gs}$ where $g_m = 1/r_{gs}$ and $v_g = 0$, we apply a test current source of I_T at node s as shown in Figure 1.9 to write node equations as

$$\begin{bmatrix} 1/r_{gs} + 1/r_o & -1/r_o \\ -1/r_o & 1/r_o + 1/R_L \end{bmatrix} \begin{bmatrix} v_s \\ v_d \end{bmatrix} = \begin{bmatrix} I_T \\ -g_m v_{gs} \end{bmatrix}$$

$$\underset{v_g = 0}{\overset{v_{gs} = v_g - v_s = -v_s}{\longrightarrow}} \begin{bmatrix} 1/r_{gs} + 1/r_o & -1/r_o \\ -1/r_o - g_m & 1/r_o + 1/R_L \end{bmatrix} \begin{bmatrix} v_s \\ v_d \end{bmatrix} = \begin{bmatrix} I_T \\ 0 \end{bmatrix}$$

We can solve this equation to get

$$\begin{bmatrix} v_s \\ v_d \end{bmatrix} = \begin{bmatrix} r_{gs} + r_o & -r_{gs} \\ -(r_{gs} + r_o)R_L & (r_o + R_L)r_{gs} \end{bmatrix}^{-1} \begin{bmatrix} r_{gs} r_o I_T \\ 0 \end{bmatrix}$$
$$= \frac{1}{r_{gs} + r_o} \begin{bmatrix} (r_o + R_L) r_{gs} & \text{x} \\ (r_{gs} + r_o)R_L & \text{x} \end{bmatrix} \begin{bmatrix} I_T \\ 0 \end{bmatrix} \tag{E1.3.1}$$

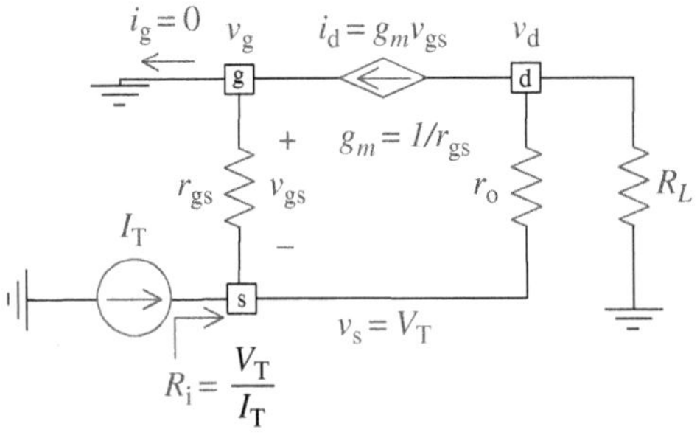

Figure 1.9 To find the equivalent resistance seen from s-0 for Example 1.3.

This solution with $v_s = V_T$ yields the equivalent resistance seen from s-0 as

$$R_i = \frac{V_T}{I_T} = \frac{(r_o + R_L)\, r_{gs}}{r_{gs} + r_o} = \frac{r_o + R_L}{1 + g_m r_o} \tag{E1.3.2}$$

Example 1.4 Input Resistance, Output Resistance, and Voltage Gain
Consider the circuit in Figure 1.10.

a) Find the input resistance R_i (considering the load resistance R_L).

$$R_i = R_1 + R_L \tag{E1.4.1}$$

b) Find the output resistance R_o (considering the source resistance R_s).

$$R_o = R_s + R_1 \tag{E1.4.2}$$

c) Find the output resistance R_{o1} without considering the source resistance R_s, i.e. considering $R_s = 0$.

$$R_{o1} = R_1 \tag{E1.4.3}$$

d) Find the voltage gain $A_v = v_o/v_i$ and express it in terms of R_{o1} and R_L.

$$A_v = \frac{v_o}{v_i} = \frac{R_L}{R_1 + R_L} = \frac{R_L}{R_{o1} + R_L} \tag{E1.4.4}$$

e) Express the global voltage gain $G_v = v_o/v_s$ in terms of R_o and R_L. Also express it in terms of R_s, R_i, R_{o1}, and R_L.

$$G_v = \frac{v_o}{v_s} = \frac{R_L}{R_s + R_1 + R_L} = \frac{R_L}{R_o + R_L} = \frac{R_i}{R_s + R_i}\frac{R_L}{R_{o1} + R_L} = \frac{R_i}{R_s + R_i} A_v \tag{E1.4.5}$$

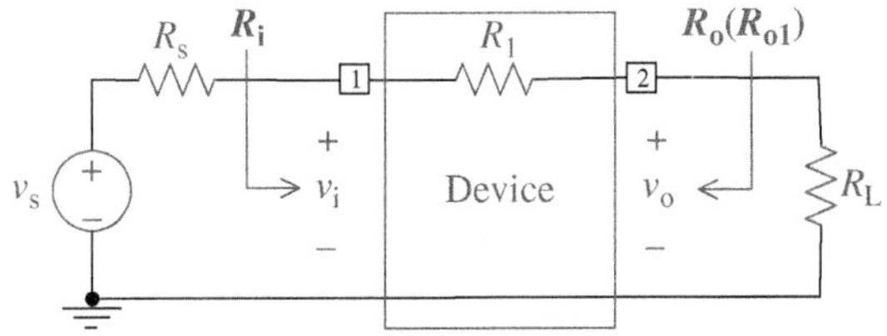

Figure 1.10 Input resistance, output resistance, and voltage gain for Example 1.4

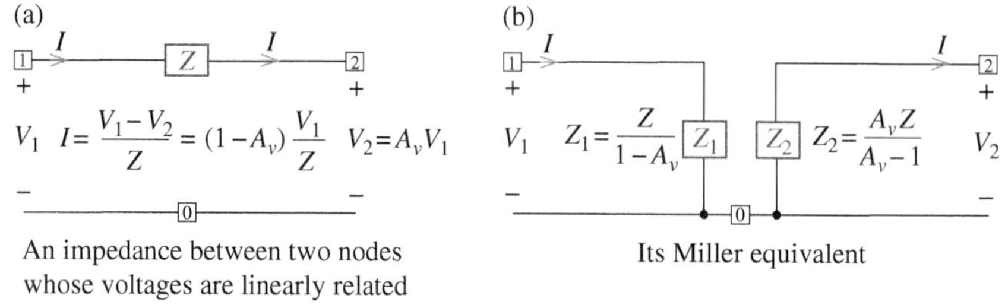

Figure 1.11 Miller's theorem.

1.4 Miller's Theorem

Miller's theorem states that an element of impedance Z connected between two nodes 1 and 2 as shown in Figure 1.11(a), each with node voltage V_1 and $V_2 = A_v V_1$, respectively, can be replaced by an equivalent consisting of two impedances Z_1 and Z_2:

$$Z_1 = \frac{V_1}{I} = \frac{Z}{1 - A_v} \text{ and } Z_2 = \frac{V_2 = A_v V_1}{-I} = \frac{A_v Z}{A_v - 1} \tag{1.4.1}$$

each of which is connected between the nodes 1/2 and a common node 0 as shown in Figure 1.11(b). It can easily be shown that the two circuits of Figure 1.11(a) and (b) are equivalent both from terminals 1 to 0 and 2 to 0. If the element is a capacitor of impedance $Z = 1/sC$, the equivalent capacitances seen from terminals 1 to 0 and 2 to 0 will be

$$C_1 = C(1 - A_v) > C \text{ and } C_2 = \frac{C(A_v - 1)}{A_v} > C \text{ if } A_v < 0 \tag{1.4.2}$$

Here, C_1 is referred to as the *Miller (effective) capacitance* where $(1 - A_v)$ is called the *Miller multiplier* and the increase in the equivalent input capacitance is the *Miller effect.*

1.5 Fourier Series

A function $x(t)$ of t is said to be periodic with period T if $x(t) = x(t + T)\ \forall\ t$, where T[s] and $\omega_0 = 2\pi/T$[rad/s] are referred to as the *fundamental period* and *fundamental (angular) frequency*, respectively, if T is the smallest positive real number to satisfy the equation for periodicity. According to the theory on Fourier series, any practical periodic function $x(t)$ can be represented as a sum of (infinitely) many trigonometric (cosine/sine) functions or complex exponential functions, which is called the *Fourier series representation*. There are four forms of Fourier series representation as follows:

<Trigonometric form>

$$x(t) = a_0 + \sum_{k=1}^{\infty} a_k \cos k\omega_0 t + \sum_{k=1}^{\infty} b_k \sin k\omega_0 t \text{ with } \omega_0 = \frac{2\pi}{T} \; (T : \text{the period of } x(t)) \tag{1.5.1a}$$

where the Fourier coefficients a_0, a_k, and b_k are

$$a_0 = \frac{1}{T}\int_T x(t)\,dt \text{ (the integral over one period } T)$$

$$a_k = \frac{2}{T}\int_T x(t)\cos k\omega_0 t\,dt \tag{1.5.1b}$$

$$b_k = \frac{2}{T}\int_T x(t)\sin k\omega_0 t\,dt$$

<Magnitude-and-Phase form>

$$x(t) = d_0 + \sum_{k=1}^{\infty} d_k \cos(k\omega_0 t + \phi_k) \tag{1.5.2a}$$

where the Fourier coefficients are

$$d_0 = a_0, \;\; d_k = \sqrt{a_k^2 + b_k^2}, \;\; \phi_k = \tan^{-1}\left(\frac{-b_k}{a_k}\right) \tag{1.5.2b}$$

<Sine-and-Phase form>

$$x(t) = d_0' + \sum_{k=1}^{\infty} d_k' \sin(k\omega_0 t + \phi_k') \tag{1.5.3a}$$

where the Fourier coefficients are

$$d_0' = a_0 = d_0, \;\; d_k' = \sqrt{a_k^2 + b_k^2} = d_k, \;\; \phi_k' = \tan^{-1}\left(\frac{a_k}{b_k}\right) = \phi_k + \frac{\pi}{2} \tag{1.5.3b}$$

<Complex Exponential form>

$$x(t) = \frac{1}{T}\sum_{k=-\infty}^{\infty} c_k e^{jk\omega_0 t} \tag{1.5.4a}$$

where the Fourier coefficients are

$$c_k = \int_T x(t)e^{-jk\omega_0 t}\,dt \text{ (the integral over one period } T) \tag{1.5.4b}$$

Here, the kth frequency $k\omega_0$ $(k > 1)$ with fundamental frequency $\omega_0 = 2\pi/T = 2\pi f_0$[rad/s] ($T$: period) is referred to as the kth *harmonic*. The above four forms of Fourier representation are equivalent and their Fourier coefficients are related with each other as follows:

$$c_0 = \int_T x(t)\,dt = Td_0 = Ta_0$$

$$c_k = \int_T x(t)\,e^{-jk\omega_0 t}\,dt = \int_T x(t)\,(\cos k\omega_0 t - j\sin k\omega_0 t)\,dt = \frac{T}{2}(a_k - jb_k) = \frac{T}{2}d_k\angle\phi_k$$

$$c_{-k} = \int_T x(t)\,e^{jk\omega_0 t}\,dt = \int_T x(t)\,(\cos k\omega_0 t + j\sin k\omega_0 t)\,dt = \frac{T}{2}(a_k + jb_k)$$

$$= \frac{T}{2}d_k\angle-\phi_k = c_k^*$$

$$a_0 = \frac{c_0}{T},\ a_k = \frac{c_k + c_{-k}}{T} = \frac{2\mathrm{Re}\{c_k\}}{T},\ b_k = \frac{c_{-k} - c_k}{jT} = -\frac{2\mathrm{Im}\{c_k\}}{T}$$

The plot of the Fourier coefficients (1.5.2b), (1.5.3b), or (1.5.4b) versus frequency $k\omega_0$ is referred to as the *spectrum*, which can be used to describe the spectral contents of a signal, that is, depict what frequency components are contained in the signal and how they are distributed over the low-/medium-/high-frequency range. The Fourier analysis performed by PSpice yields the Fourier coefficients (1.5.3b) and the corresponding spectra of chosen variables (see Example 1.5).

<Effects of Vertical/Horizontal Translations of x(t) on the Fourier coefficients>

Translating $x(t)$ along the vertical axis by $\pm A$ (+: upward, −: downward) causes only the change of Fourier coefficient $d_0 = a_0$ for $k = 0$ (DC component or average value) by $\pm A$. Translating $x(t)$ along the horizontal (time) axis by $\pm t_1$ (+: rightward, −: leftward) causes only the change of phases (ϕ_k's) by $\mp k\omega_0 t_1$, not affecting the magnitudes d_k or d_k' or $|c_k|$.

$$c_k' \overset{(1.5.4b)}{=} \int_T x(t-t_1)\,e^{-jk\omega_0 t}dt = \int_T x(t-t_1)\,e^{-jk\omega_0(t-t_1+t_1)}dt$$

$$= e^{-jk\omega_0 t_1}\int_T x(t-t_1)\,e^{-jk\omega_0(t-t_1)}dt \overset{(1.5.4b)}{=} c_k e^{-jk\omega_0 t_1}$$

$$= |c_k|\angle(\phi_k - k\omega_0 t_1) \tag{1.5.5}$$

Note that $x(t-t_1)$ is obtained by translating $x(t)$ by t_1 in the positive (rightward) direction for $t_1 > 0$ and by $-t_1$ in the negative (leftward) direction for $t_1 < 0$ along the horizontal (time) axis.

1.5.1 Computation of Fourier Coefficients Using Symmetry

If a function $x(t)$ has one of the following three symmetries, we can make use of it to reduce the computation for computing Fourier coefficients, as summarized in Table 1.1:

Table 1.1 Reduced computation of Fourier coefficients exploiting symmetries of $x(t)$.

	Symmetry of a periodic function with period T		
Fourier coefficients	**Even symmetry:** $x(t) = x(-t)$	**Odd symmetry:** $x(t) = -x(-t)$	**Half-wave symmetry:** $x(t) = -x(t + T/2)$
a_0(DC term)	$a_0 = \frac{2}{T}\int_0^{T/2} x(t)\,dt$	0	0
$a_k(k = \text{odd})$	$a_k = \frac{4}{T}\int_0^{T/2} x(t)\cos k\omega_0 t dt$	0	$a_k = \frac{4}{T}\int_0^{T/2} x(t)\cos k\omega_0 t dt$
$a_k(k = \text{even})$	$a_k = \frac{4}{T}\int_0^{T/2} x(t)\cos k\omega_0 t dt$	0	0
$b_k(k = \text{odd})$	0	$b_k = \frac{4}{T}\int_0^{T/2} x(t)\sin k\omega_0 t\,dt$	$b_k = \frac{4}{T}\int_0^{T/2} x(t)\sin k\omega_0 t\,dt$
$b_k(k = \text{even})$	0	$b_k = \frac{4}{T}\int_0^{T/2} x(t)\sin k\omega_0 t\,dt$	0

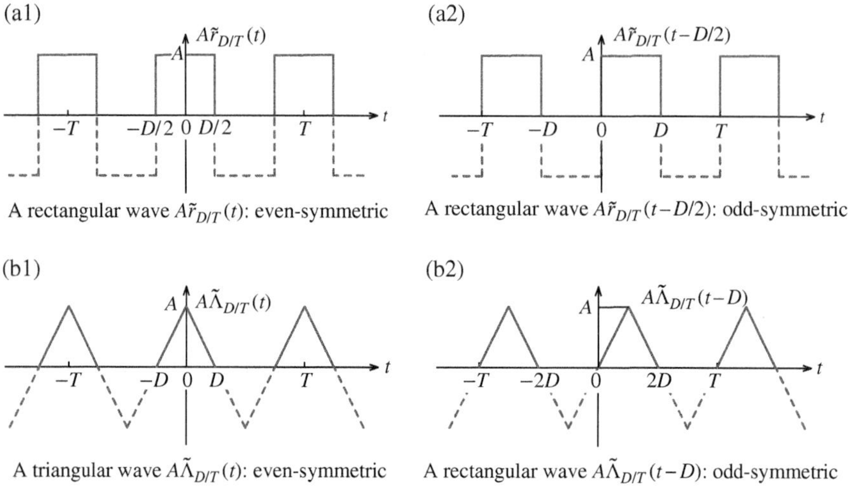

(a1)
A
−T
−D/2 0 D/2
T
t
A rectangular wave $A\tilde{r}_{D/T}(t)$: even-symmetric
(a2)
$A\tilde{r}_{D/T}(t-D/2)$
A
−T
−D
0
D
T
t
A rectangular wave $A\tilde{r}_{D/T}(t-D/2)$: odd-symmetric
(b1)
A
$A\tilde{\Lambda}_{D/T}(t)$
−T
−D 0 D
T
t
A triangular wave $A\tilde{\Lambda}_{D/T}(t)$: even-symmetric
(b2)
A
$A\tilde{\Lambda}_{D/T}(t-D)$
−T
−2D
0
2D
T
t
A rectangular wave $A\tilde{\Lambda}_{D/T}(t-D)$: odd-symmetric

Figure 1.12 Examples of rectangular and triangular waves.

1) Even symmetry: $x(t) = x(-t)$
2) Odd symmetry: $x(t) = -x(-t)$
3) Half-wave symmetry: $x(t) = -x\left(t + \frac{T}{2}\right)$ (T : the period)

Note that an even (symmetric) function can become an odd (symmetric) one just by translation and vice versa as illustrated in Figure 1.12.

Example 1.5 Fourier Spectra of a Rectangular (Square) Wave and a Triangular Wave

a) Spectrum of a Rectangular (Square) Wave (Figure 1.13(a1))

Consider the even rectangular wave $x(t)$ with height A, duration D, and period T:

$$x(t) = A\,\tilde{r}_{D/T}(t) \text{ where } \tilde{r}_{D/T}(t) = \begin{cases} 1 & \text{for } |t - mT| \le D/2 \ (m: \text{an integer}) \\ 0 & \text{elsewhere} \end{cases} \tag{E1.5.1}$$

We use Eq. (1.5.1b) to compute the Fourier coefficients as

$$a_0 = \frac{1}{T}\int_{-T/2}^{T/2} A\,\tilde{r}_{D/T}(t)\,dt = \frac{A}{T}\int_{-D/2}^{D/2} 1\,dt = \frac{AD}{T} \tag{E1.5.2}$$

$$a_k \overset{\text{even } x(t)}{=} \frac{4}{T}\int_0^{D/2} A r_{D/T}(t) \cos k\omega_0 t\, dt \overset{(\text{F.34})}{=} \frac{4A}{T}\frac{\sin k\omega_0 t}{k\omega_0}\bigg|_0^{D/2} = \frac{2AD}{T}\frac{\sin(k\pi D/T)}{k\pi D/T} \qquad \text{(E1.5.3)}$$

$$b_k \overset{\text{even } x(t)}{=} \frac{2}{T}\int_{-T/2}^{T/2} x(t) \sin k\omega_0 t\, dt \overset{\text{odd}}{=} 0 \text{ with } \omega_0 = \frac{2\pi}{T} \qquad \text{(E1.5.4)}$$

Thus, we can write the Fourier series representation of the rectangular wave $x(t)$ as

$$A\tilde{r}_{D/T}(t) \overset{(1.5.1\text{a})}{=} \frac{AD}{T} + \sum_{k=1}^{\infty} \frac{2AD}{T}\frac{\sin(k\pi D/T)}{k\pi D/T} \cos k\omega_0 t \qquad \text{(E1.5.5)}$$

In the case of $T=2D$ as depicted in Figure 1.13(a1), we have the Fourier coefficients as

$$a_k = \begin{cases} A/2 & \text{for } k = 0 \\ (-1)^m 2A/k\pi & \text{for } k = 2m+1 \\ 0 & \text{for } k = 2m > 0 \text{ (positive even number)} \end{cases} \quad \text{and } b_k = 0 \qquad \text{(E1.5.6)}$$

and the corresponding (magnitude) spectrum is depicted in Figure 1.13(b1).

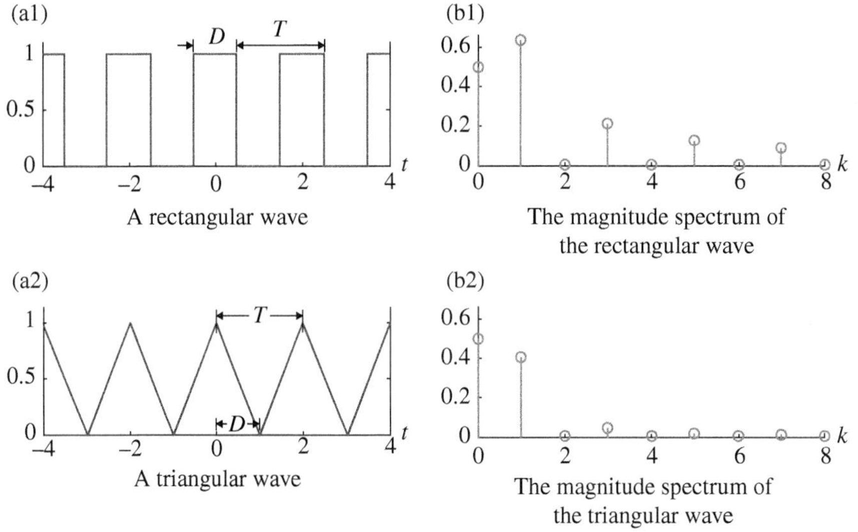

Figure 1.13 Rectangular/triangular waves and their magnitude spectra.

b) Spectrum of a Triangular Wave (Figure 1.13(a2))

Consider the even triangular wave $x(t)$ with maximum height A, duration $2D$, and period T:

$$x(t) = A\tilde{\Lambda}_{D/T}(t) \text{ where } \tilde{\Lambda}_{D/T}(t) = \begin{cases} 1 - t/D \text{ for } |t - mT| \le D(m: \text{an integer}) \\ 0 \qquad \text{elsewhere} \end{cases} \tag{E1.5.7}$$

We use Eq. (1.5.1b) to obtain the Fourier coefficients as

$$a_0 = \frac{1}{T}\int_{-T/2}^{T/2} A\tilde{\Lambda}_{D/T}(t)\,dt \overset{\text{even}}{=} \frac{2A}{T}\int_0^D \left(1 - \frac{t}{D}\right)dt \overset{(\text{F.32})}{=} \frac{2A}{T}\left(t - \frac{t^2}{2D}\right)\bigg|_0^D = \frac{AD}{T} \tag{E1.5.8}$$

$$a_k \overset{\text{even}}{=} \frac{4}{T}\int_0^D A\left(1 - \frac{t}{D}\right)\cos k\omega_0 t\,dt \text{ with } \omega_0 = \frac{2\pi}{T}$$

$$\overset{(\text{F.34}),(\text{F.36})}{=} \frac{4A}{T}\left(\frac{\sin k\omega_0 t}{k\omega_0}\bigg|_0^D - \frac{1}{D}\left(\frac{\sin k\omega_0 t}{k\omega_0}t\bigg|_0^D - \int_0^D \frac{\sin k\omega_0 t}{k\omega_0}dt\right)\right)$$

$$\overset{(\text{F.35})}{=} -\frac{4A\cos k\omega_0 t}{TD\,(k\omega_0)^2}\bigg|_0^D = \frac{4A}{TD}\frac{(1 - \cos k\omega_0 D)}{(k\omega_0)^2}$$

$$\overset{(\text{F.14})}{=} \frac{2AD}{T}\frac{\sin^2(k\omega_0 D/2)}{(k\omega_0 D/2)^2} = \frac{2AD}{T}\frac{\sin^2(k\pi D/T)}{(k\pi D/T)^2} \tag{E1.5.9}$$

$$b_k \overset{\text{even}\,x(t)}{=} \frac{2}{T}\int_{-T/2}^{T/2} x(t)\sin k\omega_0 t\,dt \overset{\text{odd}}{=} 0 \tag{E1.5.10}$$

Thus, we can write the Fourier series representation of the triangular wave $x(t)$ as

$$A\tilde{\Lambda}_{D/T}(t) \underset{(\text{E1.5.8,9,10})}{\overset{(1.5.1\text{a})}{=}} \frac{AD}{T} + \sum_{k=1}^{\infty} \frac{2AD}{T}\frac{\sin^2(k\pi D/T)}{(k\pi D/T)^2}\cos k\omega_0 t \tag{E1.5.11}$$

In the case of $T = 2D$ as depicted in Figure 1.13(a2), we have the Fourier coefficients as

$$a_k = \begin{cases} A/2 & \text{for } k = 0 \\ 4A/(k\pi)^2 & \text{for } k = 2m+1 \\ 0 & \text{for } k = 2m > 0 \text{ (positive even number)} \end{cases} \tag{E1.5.12}$$

$$b_k = 0 \tag{E1.5.13}$$

and the corresponding (magnitude) spectrum is depicted in Figure 1.13(b2).

Example 1.6 Fourier Spectra of Rectified Cosine Waves

a) Spectrum of a Cosine Wave (Figure 1.14(a1))

Consider a cosine wave $x(t) = A\cos\omega_0 t$ with $\omega_0 = 2\pi/T$. In the case of single tone periodic signal, the time function is the very Fourier series representation with a single nonzero coefficient.

$$a_1 \overset{(1.5.1b)}{=} \frac{2A}{T}\int_{-T/2}^{T/2} \cos\omega_0 t \cos\omega_0 t\,dt \overset{(F.15)}{=} \frac{2A}{T}\int_{-T/2}^{T/2} \frac{1+\cos 2\omega_0 t}{2}\,dt = A \quad \text{(E1.6.1)}$$

The (magnitude) spectrum is depicted in Figure 1.14(b1).

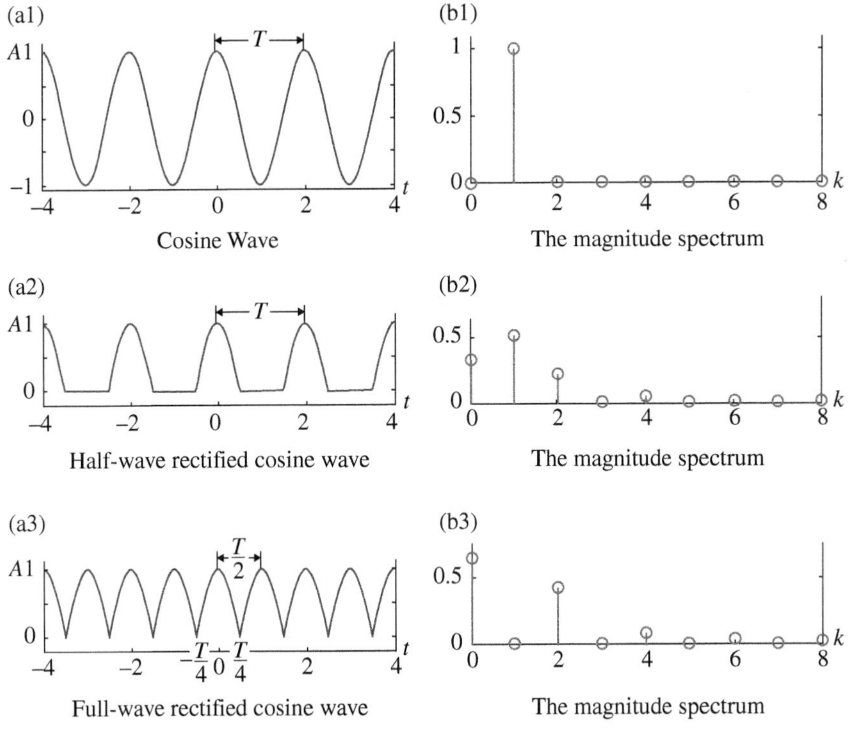

Figure 1.14 Half/full-wave rectified cosine waves and their magnitude spectra

b) Spectrum of a Half-Wave Rectified Cosine Wave (Figure 1.14(a2))

Consider a half-wave rectified cosine wave $x_h(t) = \max(A\cos\omega_0 t, 0)$. We use Eq. (1.5.1b) to obtain the Fourier coefficients as

$$a_0 = \frac{1}{T}\int_{-T/4}^{T/4} A\cos\omega_0 t\,dt \overset{\text{even}}{=} \frac{2A}{T}\int_0^{T/4}\cos\omega_0 t\,dt \overset{\text{(F.34)}}{=} \frac{2A}{\omega_0 T}\sin\omega_0 t\big|_0^{T/4} = \frac{A}{\pi} \quad \text{(E1.6.2)}$$

$$a_k \overset{\text{even}}{=} \frac{4}{T}\int_0^{T/4} A\cos\omega_0 t\cos k\omega_0 t\,dt \text{ with } \omega_0 = \frac{2\pi}{T}$$

$$\overset{\text{(F.11)}}{=} \frac{2A}{T}\int_0^{T/4}(\cos(k+1)\omega_0 t + \cos(k-1)\omega_0 t)\,dt$$

$$\overset{\text{(F.34)}}{=} \frac{2A}{(k+1)\omega_0 T}\sin(k+1)\omega_0 t\big|_0^{T/4} + \frac{2A}{(k-1)\omega_0 T}\sin(k-1)\omega_0 t\big|_0^{T/4}$$

$$\overset{k\neq 1}{=} \frac{A}{(k+1)\pi}\sin(k+1)\frac{\pi}{2} + \frac{A}{(k-1)\pi}\sin(k-1)\frac{\pi}{2}$$

$$\overset{k=2m\text{(even)}}{=} (-1)^m\frac{A}{\pi}\left(\frac{1}{k+1} - \frac{1}{k-1}\right) = \frac{(-1)^{m+1}2A}{(k^2-1)\pi} \quad \text{(E1.6.3)}$$

$$a_1 \overset{\text{even}}{=} \frac{4}{T}\int_0^{T/4} A\cos^2\omega_0 t\,dt \overset{\text{(F.15)}}{=} \frac{2A}{T}\int_0^{T/4}(1+\cos 2\omega_0 t)\,dt = \frac{A}{2} \quad \text{(E1.6.4)}$$

$$b_k \overset{\text{even } x(t)}{=} \frac{2}{T}\int_{-T/4}^{T/4} A\cos\omega_0 t\sin k\omega_0 t\,dt \overset{\text{odd}}{=} 0 \quad \text{(E1.6.5)}$$

Thus, we can write the Fourier series representation of the half-wave rectified cosine wave as

$$x_h(t) \underset{\text{(E1.6.2,3,4,5)}}{\overset{\text{(1.5.1a)}}{=}} \frac{A}{\pi} + \frac{A}{2}\cos\omega_0 t - \sum_{k=2m}^{\infty}\frac{(-1)^m 2A}{(k^2-1)\pi}\cos k\omega_0 t \quad \text{(E1.6.6)}$$

The corresponding (magnitude) spectrum is depicted in Figure 1.14(b2).

c) Spectrum of a Full-Wave Rectified Cosine Wave (Figure 1.14(a3))

Consider a full-wave rectified cosine wave $x_f(t) = |A\cos\omega_0 t|$, which can be regarded as the sum of a half-wave rectified cosine wave $x_h(t)$ and its $T/2$-shifted version $x_h(t-T/2)$. Thus, instead of using Eq. (1.5.1b), we use Eq. (E1.6.6) to write its Fourier series representation as

$$x_f(t) = x_h(t) + x_h\left(t - \frac{T}{2}\right) \overset{(E1.6.6)}{=} \frac{A}{\pi} + \frac{A}{2}\cos\omega_0 t - \sum_{k=2m}^{\infty} \frac{(-1)^m 2A}{(k-1)^2\pi}\cos k\omega_0 t$$

$$+\frac{A}{\pi} + \frac{A}{2}\cos\omega_0\left(t - \frac{T}{2}\right) - \sum_{k=2m}^{\infty} \frac{(-1)^m 2A}{(k-1)^2\pi}\cos k\omega_0\left(t - \frac{T}{2}\right) \quad \text{with } \omega_0 T = 2\pi$$

$$\overset{(F.4)}{=} \frac{2A}{\pi} - \sum_{k=2m}^{\infty} \frac{(-1)^m 4A}{(k^2-1)\pi}\cos k\omega_0 t \overset{k=2m}{=} \frac{2A}{\pi} - \sum_{m=1}^{\infty} \frac{(-1)^m 4A}{(4m^2-1)\pi}\cos 2m\omega_0 t \quad \text{(E1.6.7)}$$

The corresponding (magnitude) spectrum is depicted in Figure 1.14(b3).
(cf.) In fact, the fundamental frequency of a full-wave rectified cosine wave is $2\omega_0 = 4\pi/T$, which is two times that of the original cosine wave.

d) MATLAB program to get the Fourier spectrum
Once you have saved a periodic function as an M-file, you can use the MATLAB function 'CTFS_trigonometric()' to find its trigonometric Fourier series coefficients as illustrated below. Interested readers are invited to run the above script 'elec01e05.m' to get the Fourier coefficients and the spectra for a pure cosine wave and half/full-wave rectified cosine waves in Figure 1.14.

```
function [a0,a,b]=CTFS_trigonometric(x,T,N)
%Find the Fourier coefficients a0, ak, bk for k=1:N
%x: A periodic function with period T
%T: Period, N: Maximum order of Fourier coefficients
w0=2*pi/T; % Fundamental frequency [rad/s]
xcoskw0t=[x '(t).*cos(k*w0*t)'];
  x_coskw0t=inline(xcoskw0t,'t','k','w0');
xsinkw0t=[x '(t).*sin(k*w0*t)'];
  x_sinkw0t=inline(xsinkw0t,'t','k','w0');
tol=1e-6; % Tolerance on numerical error
a0= quadl(x_coskw0t,-T/2,T/2,tol,[],0,w0)/T; % Eq.(1.5.1a)
for k=1:N
  a(k)= 2/T*quadl(x_coskw0t,-T/2,T/2,tol,[],k,w0); % Eq.(1.5.1a)
  b(k)= 2/T*quadl(x_sinkw0t,-T/2,T/2,tol,[],k,w0); % Eq.(1.5.1a)
end
```

```
%elec01e05.m : to plot Fig.1.14
% plots the CTFS spectra of (half/full-rectified) cosine waves
clear, clf
global T, T=2; w0=2*pi/T; % Period and Fundamental frequency
N=8; kk=1:N; % Frequency indices
tt=[-400:400]*T/200; % Time interval of 4 periods
for i=1:3
  if i==1
    x ='cos_wave'; % Fourier coefficients from analytical results
    a0_t= 0; a_t= zeros(1,N); a_t(1)=1; b_t= zeros(1,N); % (E1.6.1)
  elseif i==2
    x ='cos_wave_half_rectified';
    a0_t= 1/pi; a_t= [1/2 zeros(1,N-1)]; b_t= zeros(1,N); % (E1.6.3)
    for m=1:floor(N/2), a_t(2*m)= -(-1)^m*2/(4*m^2-1)/pi; end
  else
    x ='cos_wave_full_rectified';
    a0_t= 2/pi; a_t= zeros(1,N); b_t= zeros(1,N); % (E1.6.7)
    for m=1:floor(N/2), a_t(2*m)= -(-1)^m*4/(4*m^2-1)/pi; end
  end
  xt = feval(x,tt); % Original signal
  [a0,a,b] = CTFS_trigonometric(x,T,N);
  discrepancy_between_numeric_and_analytic= [a0 a b]-[a0_t a_t b_t]
  xht = a0; % starting from the DC term
  % reconstruct x(t) from its Fourier series representation (1.5.1a)
  for k=1:N
     xht = xht + a(k)*cos(k*w0*tt) + b(k)*sin(k*w0*tt);
  end
  subplot(319+i*2), plot(tt,xt,'k-', tt,xht,'r:')
  d1 = [a0 sqrt(a.^2+b.^2)];
  phi1k = [0 atan2(a,b)]; % Eq.(1.5.3b)
  subplot(320+i*2)
  stem([0:length(d1)-1], d1)
end
```

```
function y=cos_wave(t)
global T, y= cos(2*pi/T*t);
```

```
function y=cos_wave_half_rectified(t)
global T, y= max(cos(2*pi/T*t),0);
```

```
function y=cos_wave_full_rectified(t)
global T, y= abs(cos(2*pi/T*t));
```

1.5.2 Circuit Analysis Using Fourier Series

In the previous section, we discussed how to represent a periodic function in Fourier series. In this section, we study how to analyze a linear circuit excited by a source whose value can be described as a periodic function of time, which is not necessarily a sinusoidal function. The procedure which we take to use Fourier series for analyzing circuits is as follows:

1) Find the Fourier series representation of the periodic input function.
2) Based on the fact that a (linear) system having transfer function $G(s)$ responds to a sinusoidal input of frequency $k\omega_0$ with the frequency response $G(jk\omega_0)$, apply the phasor method to find the individual responses to usually a few main terms, not necessarily every harmonic term, of the input represented in Fourier series.
3) If necessary, add the DC and the fundamental frequency and harmonic components of the output based on the superposition principle to find the total response.
   (cf.) Note that the superposition principle holds only for linear systems.

Example 1.7 ***RC*** **Circuit Excited by a Rectangular (Square) Wave Voltage Source**

Figure 1.15(a) shows the PSpice schematic of an *RC* circuit excited by a rectangular (square) wave voltage source of height $\pm V_{\text{m}} = \pm\pi$, period $T = 2$ [s], and duration (pulse width) $D = 1$ [s], where the voltage across the capacitor is taken as the output. Figure 1.15 (b) shows the input and output waveforms obtained from the PSpice simulation. Figure 1.15(c) shows the Fourier spectra of the input and output obtained by clicking the fast Fourier transform (FFT) button on the toolbar in the PSpice A/D (Probe) window. Figure 1.15(d) shows how to fill in the Simulation Settings dialog box to get the Fourier analysis results (for chosen variables) printed in the output file. Figure 1.15(e) shows the contents of the output file that can be viewed by clicking PSpice>View_Output_File on the top menu bar and pull-down menu in the Capture window or View/Output_File on the top menu bar and pull-down menu in the Probe window.

a) Let us find the three leading frequency components of the input $v_{\text{i}}(t)$ and output $v_{\text{o}}(t)$. Rather than applying Eq. (1.5.1b) to get the Fourier coefficients of the rectangular wave input $v_{\text{i}}(t)$, we will make use of the result obtained in Example 1.5. Note the following:

   - A rectangular wave with height $A = 2\pi$, period $T = 2$, and duration $D = 1$ can be expressed by the Fourier series representation as follows:

$$x(t) \overset{(\text{E1.4.5})}{=} \frac{AD}{T} + \sum_{k=1}^{\infty} \frac{2AD}{T} \frac{\sin(k\pi D/T)}{k\pi D/T} \cos k\omega_0 t$$

$$= \pi + \sum_{k=1}^{\infty} 2\pi \frac{\sin(k\pi/2)}{k\pi/2} \cos k\pi t \qquad \text{(E1.7.1)}$$

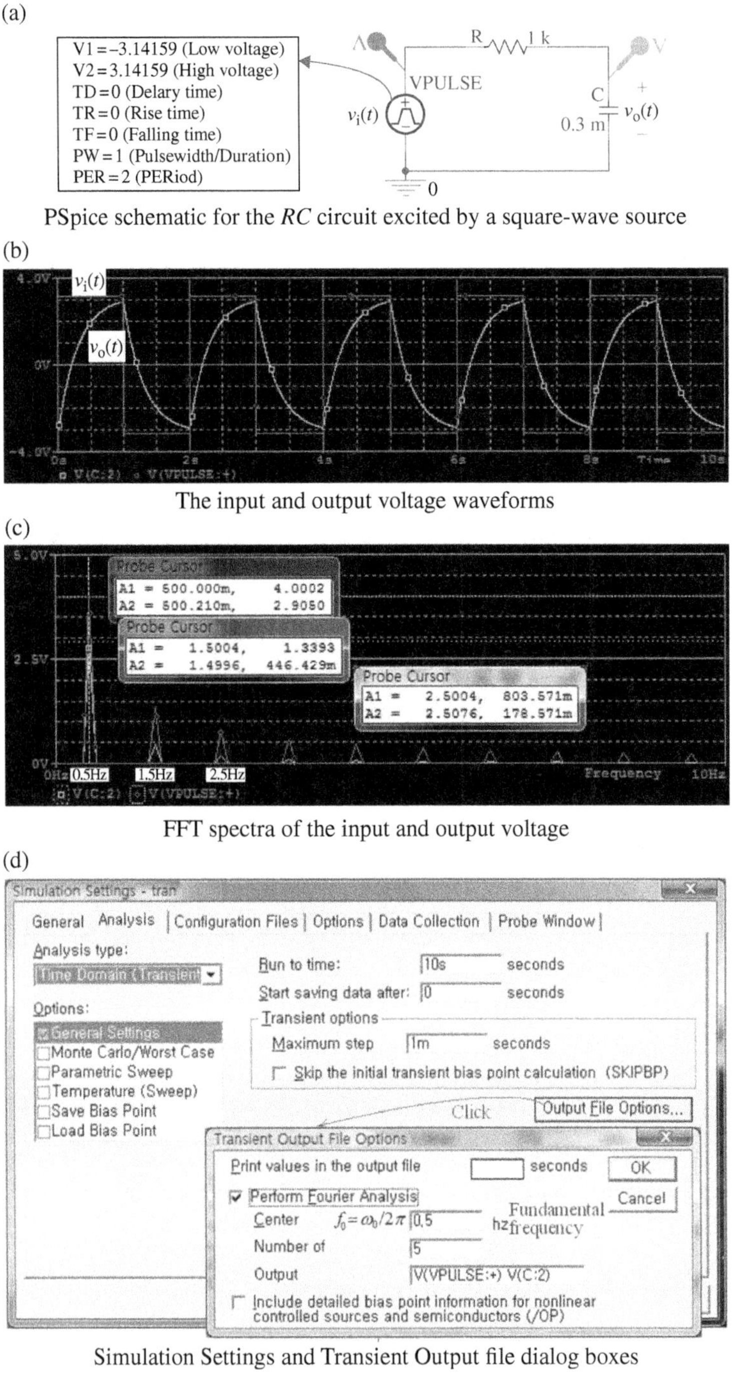

Figure 1.15 PSpice simulation and analysis for Example 1.7 ("RC_excited_by_square_wave.opj").

(e)

```
C:#OrCAD#Yang#Elec#RC_excited_by_square_wave-PSpiceFiles#SCHEMATIC1#tran#tran.out
*Profile: "SCHEMATIC1-tran" [C:\..\OrCAD\Yang\Elec\RC_excited_by_square_wave-PS
**********************************************************************
*Analysis directives:
.TRAN  0 10s 0 1m
.FOUR 0.5 5 V([N01762]) V([N01780])

**** INCLUDING SCHEMATIC1.net ****
* source RC_EXCITED_BY_SQUARE_WAVE
R_R          N01762   N01780   1k
C_C          0        N01780   0.3m
V_VPULSE     N01762   0
+PULSE -3.14159 3.14159 0 1m 1m 1 2
```

Contents of the netlist file

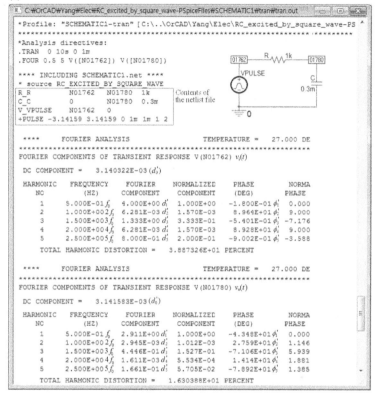

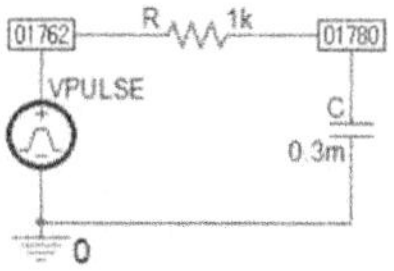


```
 ****     FOURIER ANALYSIS                    TEMPERATURE =   27.000 DE
******************************************************************
FOURIER COMPONENTS OF TRANSIENT RESPONSE V(N01762) vi(t)

 DC COMPONENT =   3.140322E-03 (d0')

 HARMONIC   FREQUENCY      FOURIER        NORMALIZED     PHASE              NORMA
    NO        (HZ)         COMPONENT      COMPONENT      (DEG)              PHASE
    1     5.000E-01 f0     4.000E+00 d1'  1.000E+00   -1.800E-01 phi1'   0.000
    2     1.000E+00 2f0    6.281E-03 d2'  1.570E-03    8.964E+01 phi2'   9.000
    3     1.500E+00 3f0    1.333E+00 d3'  3.333E-01   -5.401E-01 phi3'  -7.176
    4     2.000E+00 4f0    6.281E-03 d4'  1.570E-03    8.928E+01 phi4'   9.000
    5     2.500E+00 5f0    8.000E-01 d5'  2.000E-01   -9.002E-01 phi5'  -3.588
     TOTAL HARMONIC DISTORTION =   3.887326E+01 PERCENT

 ****     FOURIER ANALYSIS                    TEMPERATURE =   27.000 DE
******************************************************************
FOURIER COMPONENTS OF TRANSIENT RESPONSE V(N01780) vo(t)

 DC COMPONENT =   3.141583E-03 (d0')

 HARMONIC   FREQUENCY      FOURIER        NORMALIZED     PHASE              NORMA
    NO        (HZ)         COMPONENT      COMPONENT      (DEG)              PHASE
    1     5.000E-01 f0     2.911E+00 d1'  1.000E+00   -4.348E+01 phi1'   0.000
    2     1.000E+00 2f0    2.945E-03 d2'  1.012E-03    2.759E+01 phi2'   1.146
    3     1.500E+00 3f0    4.446E-01 d3'  1.527E-01   -7.106E+01 phi3'   5.939
    4     2.000E+00 4f0    1.611E-03 d4'  5.534E-04    1.414E+01 phi4'   1.881
    5     2.500E+00 5f0    1.661E-01 d5'  5.705E-02   -7.892E+01 phi5'   1.385
     TOTAL HARMONIC DISTORTION =   1.630388E+01 PERCENT
```

PSpice output file containing the FFT analysis result

Figure 1.15 (Continued)

- The rectangular wave input $v_i(t)$ (in Figure 1.15(b)) can be obtained by translating this rectangular wave downward by π and rightward by $T/4 = 0.5$.

Thus, we can express $v_i(t)$ as

$$v_i(t) = x(t-0.5) - \pi = \sum_{k=1}^{\infty} 2\pi \frac{\sin(k\pi/2)}{k\pi/2} \cos k\pi(t-0.5)$$

$$= \sum_{k=\text{odd}}^{\infty} \frac{4}{k} \sin k\pi t = 4\sin \pi t + \frac{4}{3}\sin 3\pi t + \frac{4}{5}\sin 5\pi t + \cdots \quad \text{(E1.7.2)}$$

Since the RC circuit has the frequency response

$$G(s) = \frac{1/sC}{R + 1/sC} = \frac{1}{1 + sRC};$$

$$G(j\omega) = \frac{1}{1 + j\omega RC} = \frac{1}{\sqrt{1 + (\omega RC)^2}} \angle -\tan^{-1}\omega RC, \quad \text{(E1.7.3)}$$

its output to the input $v_i(t)$ (described by Eq. (E1.7.2)) can be written as

$$v_o(t) = \sum_{k=2m+1}^{\infty} \frac{4}{k\sqrt{1+(k\omega_0 RC)^2}} \sin\left(k\omega_0 t - \tan^{-1} k\omega_0 RC\right). \quad \text{(E1.7.4)}$$

The three leading frequency components of the output phasor are

$$\mathbf{V}_o^{(1)} \overset{k=1}{=} G(j\omega_0)\mathbf{V}_i^{(1)} = \frac{4}{\sqrt{1+(\omega_0 RC)^2}} \angle -\tan^{-1}\omega_0 RC$$

$$\overset{RC=0.3}{=} \frac{4}{\sqrt{1+(0.3\pi)^2}} \angle -\tan^{-1} 0.3\pi = 2.91 \angle -43.3^\circ \quad \text{(E1.7.5)}$$

$$\mathbf{V}_o^{(3)} \overset{k=3}{=} G(j3\omega_0)\mathbf{V}_i^{(3)} = \frac{4}{3\sqrt{1+(3\omega_0 RC)^2}} \angle -\tan^{-1} 3\omega_0 RC$$

$$\overset{RC=0.3}{=} \frac{4}{3\sqrt{1+(0.9\pi)^2}} \angle -\tan^{-1} 0.9\pi = 0.4446 \angle -70.5^\circ \quad \text{(E1.7.6)}$$

$$\mathbf{V}_o^{(5)} \overset{k=5}{=} G(j5\omega_0)\mathbf{V}_i^{(5)} = \frac{4}{5\sqrt{1+(5\omega_0 RC)^2}} \angle -\tan^{-1} 5\omega_0 RC$$

$$\overset{RC=0.3}{=} \frac{4}{5\sqrt{1+(1.5\pi)^2}} \angle -\tan^{-1} 1.5\pi = 0.166 \angle -78^\circ \quad \text{(E1.7.7)}$$

where the phases are taken with that of sin ωt as a reference as in Eq. (1.5.3b) (called the sine-and-phase form of Fourier series representation) in order to match this Fourier analysis result with that seen in the PSpice simulation output file (Figure 1.15(e)). The magnitude ratios among the leading three frequency components of the input and output are

$$\text{input}: \ |\mathbf{V}_i^{(1)}| : |\mathbf{V}_i^{(3)}| : |\mathbf{V}_i^{(5)}| = 4 : \frac{4}{3} : \frac{5}{3} = 15 : 5 : 3$$

$$\text{output}: \ |\mathbf{V}_o^{(1)}| : |\mathbf{V}_o^{(3)}| : |\mathbf{V}_o^{(5)}| = 2.91 : 0.4446 : 0.166 = 15 : 2.3 : 0.86$$

This implies that the relative magnitudes of high-frequency components to low ones become smaller after the input signal passes through the filter, resulting in a considerable decrease of the total harmonic distortion (THD) from 38.9 to 16.3% as can be seen in the output file (Figure 1.15(e)). This is a piece of evidence

that the RC circuit with the capacitor voltage taken as the output functions as a low-pass filter.
(cf.) See Section 1.5.3 for the definition of total harmonic distortion (THD).

b) Let us determine the minimum value of RC required to keep the relative magnitude of the second leading frequency ($3\omega_0$) component to the first one (ω_0) not larger than, say, 12.5%.

$$\frac{4}{3\sqrt{1+(3\omega_0 RC)^2}} \le \frac{4}{\sqrt{1+(\omega_0 RC)^2}} \times 0.125$$

$$;RC \ge \frac{\sqrt{3.2353}}{\omega_0} = \frac{1.8}{\pi} = 0.5725 \tag{E1.7.8}$$

```
%solve_eqE178.m : to solve Eq.(E1.7.8)
eq=@(w0RC) sqrt(1+w0RC^2)-0.375*sqrt(1+9*w0RC^2);
w0RC0=1; w0RC=fsolve(eq,w0RC0); w0=pi; RC=w0RC/w0
```

c) We can run the MATLAB script "elec01e07.m" (listed below) to get the following Fourier analysis results:

```
>>elec01e07
  Magnitude_and_phase_of_input_spectrum =
              0    0.0000         0
         1.0000    4.0000    0.0000
         2.0000    0.0000   89.9999
         3.0000    1.3333    0.0000
         4.0000    0.0000   90.0001
         5.0000    0.8000    0.0000
  THD of input=3.8873e-001 and THD of output=1.6304e-001
  Magnitude_and_phase_of_output_spectrum =
              0    0.0000         0
         1.0000    2.9109  -43.3038
         2.0000    0.0000   27.9466
         3.0000    0.4446  -70.5224
         4.0000    0.0000   14.8561
         5.0000    0.1661  -78.0192
```

```
%elec01e07.m : to solve Example 1.7
%Perform Fourier analysis to solve the RC circuit excited by a square wave
global T D Vm
T=2; w0=2*pi/T; D=1; Vm=pi; % Period, Frequency, Duration, Amplitude
N=5; kk=0:N; % Frequency indices to analyze using Fourier analysis
tt=[-300:300]*T/200; % Time interval of 3 periods
x='elec01e05_f'; % Bipolar square wave input function defined in an M-file
RC=0.3; n=1; d=[RC 1]; % Numerator/Denominator of xfer function (E1.7.3)
[Y,X,THDy,THDx]=Fourier_analysis(n,d,x,T,N);
Mag_and_phase_of_input_spectrum = [kk; abs(X); angle(X)*180/pi].'
fprintf('THD of input=%10.4e and THD of output=%10.4e',THDx,THDy)
Mag_and_phase_of_output_spectrum = [kk; abs(Y); angle(Y)*180/pi].'
xt = feval(x,tt); % Input signal for tt
yt= Y(1); % DC component of the output signal
for k=1:N % Fourier series representation of the output signal
  yt = yt + abs(Y(k+1))*sin(k*w0*tt + angle(Y(k+1))); % Eq. (1.5.3a)
end
subplot(221), plot(tt,xt, tt,yt,'r') %plot input/output signal waveform
subplot(222), stem(kk,abs(X)), hold on, stem(kk,abs(Y),'r') % their spectra
```

```
function y=elec01e07_f(t)
% defines a bipolar square wave with period T, duration D, and amplitude Vm
global T D Vm
t= mod(t,T); y=((t<=D)-(t>D))*Vm;
```

```
function [Y,X,THDy,THDx]=Fourier_analysis(n,d,x,T,N)
%Input:   n = Numerator polynomial of transfer function G(s)
%         d = Denominator polynomial of transfer function G(s)
%         x = Input periodic function
%         T = Period of the input function
%         N = Highest frequency index of Fourier analysis
%Output:  Y = Fourier coefficients [Y0,Y1,Y2,...] of the output
%         X = Fourier coefficients [X0,X1,X2,...] of the input
%      THDy = Total Harmonic Distortion factor of the output
%      THDx = Total Harmonic Distortion factor of the input
% Copyleft: Won Y. Yang, wyyang53@hanmail.net, CAU for academic use only
if nargin<5, N=10; end
w0=2*pi/T; kk=0:N; % Fundamental frequency and Frequency index vector
[a0,a,b] = CTFS_trigonometric(x,T,N); % trigonometric Fourier coefficients
Xmag = [a0 sqrt(a.^2+b.^2)]; Xph = [0 atan2(a,b)]; % Eq. (1.5.3b)
X= Xmag.*exp(j*Xph); % Input spectrum
Gw= freqs(n,d,kk*w0); % Frequency response
Y= X.*Gw; % Output spectrum
%Total Harmonic Distortion factors of the input and output
THDx= sqrt(Xmag(3:end)*Xmag(3:end).')/Xmag(2); % Eq. (1.5.11)
Ymag= abs(Y); % the magnitude of the output spectrum
THDy= sqrt(Ymag(3:end)*Ymag(3:end).')/Ymag(2);
```

Note that the MATLAB script "elec01e06.m" uses the function '`Fourier_analysis()`' (listed below), which takes the numerator (n) and denominator (d) of the transfer function $G(s)$, the (periodic) input function (x) defined for at least one period [−T/2,+T/2], the period (T), and the order (N) of Fourier analysis as input arguments and produces the output (yt) for one period, the sine-and-phase form of Fourier coefficients Y and X of the output and input (for k =0, … , N), and the THDs of the output and input as output arguments.

1.5.3 RMS Value and Distortion Factor of a Non-Sinusoidal Periodic Signal

Let a non-sinusoidal periodic signal have the magnitude-and-phase or sine-and-phase form of Fourier series representation as

$$x(t) \overset{(1.5.2a)}{=} X_0 + \sum_{k=1}^{\infty} X_{km} \cos(k\omega_0 t + \phi_k) = X_0 + \sum_{k=1}^{\infty} X_{km} \sin(k\omega_0 t + \phi_k') \qquad (1.5.8)$$

Then its *rms (root-mean-square)* or *effective value* can be computed as

$$X_{\rm rms} = \sqrt{\frac{1}{T}\int_T x^2(t)\,dt} = \sqrt{\frac{1}{T}\int_T \left(X_0^2 + \sum_{k=1}^{\infty} X_{km}^2 \cos^2(k\omega_0 t + \phi_k)\right) dt}$$

$$= \sqrt{X_0^2 + \sum_{k=1}^{\infty} \frac{1}{T}\int_T X_{km}^2 \cos^2(k\omega_0 t + \phi_k)\,dt} \overset{(F.15)}{=} \sqrt{X_0^2 + \sum_{k=1}^{\infty} X_k^2} \qquad (1.5.9)$$

which is the square root of the sum of the squared DC (average) value (X_0^2) and the squared rms values ($X_k^2 = X_{km}^2/2$) of every (fundamental and harmonic) frequency component contained in the signal. In this derivation, we have made use of the fact that the integral of a product of two sinusoids of different frequencies over their common period is zero:

$$\int_T \cos(k\omega_0 t + \phi_k)\cos(m\omega_0 t + \phi_m)\,dt \overset{(F.11)}{=} \frac{1}{2}\int_T (\cos((k+m)\omega_0 t + \phi_k + \phi_m)$$
$$+ \cos((k-m)\omega_0 t + \phi_k - \phi_m))\,dt = 0 \qquad (1.5.10)$$

As a measure of how far a periodic signal is from its sinusoid of fundamental frequency, the *THD* (*total harmonic distortion*), also referred to as the *distortion factor*, is defined to be the ratio of the rms value of harmonic components to that of fundamental frequency component as follows:

$$THD = \sqrt{\frac{\sum_{k=2}^{\infty} X_k^2}{X_1}} = \sqrt{\frac{\sum_{k=2}^{\infty} X_{km}^2}{X_{1m}}} \qquad (1.5.11)$$

Problems

1.1 Load Line on the v-i Characteristic Curve Instead of the i-v Characteristic Curve

In Section 1.1, we write the equation for the load line to plot on the i-v characteristic curve as

$$v_2(i) \overset{(1.1.2)}{=} V_s - R_1\, i \tag{P1.1.1}$$

(a) How do you write the equation for the load line to plot on the v-i characteristic curve?

(b) What is the expression for the slope of the tangent line to the v-i characteristic curve of a diode at the operating point $Q = (V_Q, I_Q)$? Note that it is called the *dynamic, small-signal, AC,* or *incremental conductance.*

1.2 Load Line Analysis of a Nonlinear Resistor Circuit

Consider the circuit of Figure P1.2(a), which contains a nonlinear resistor whose voltage-current characteristic curve is depicted in Figure P1.2(b).

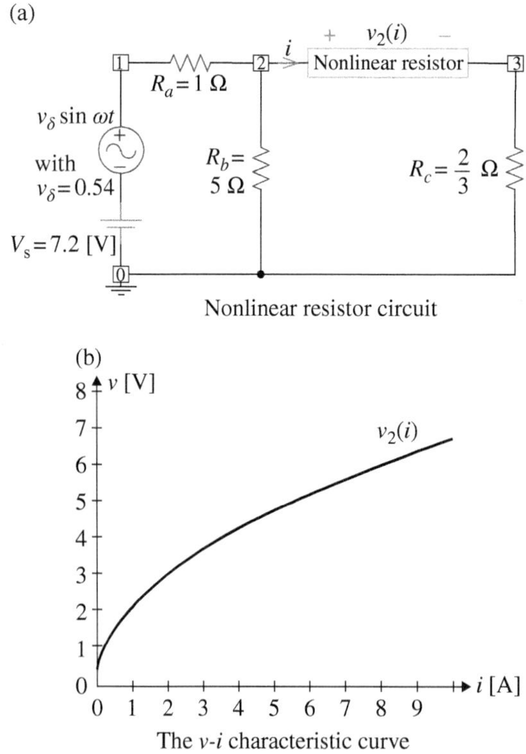

Figure P1.2 Graphic analysis of a nonlinear resistor circuit.

(a) Find the Thevenin equivalent of the circuit (with the nonlinear resistor open-circuited) seen from nodes 2-3.
(b) Draw the load line on the characteristic curve of the nonlinear resistor to find the operating point (I_Q, V_Q).
(c) Noting that $v_2(i) = 1.5\sqrt{2i}$, find the static and dynamic resistances of the nonlinear resistor at the operating point Q by using Eqs. (1.1.10) and (1.1.7), respectively.
(d) Using Eqs. (1.1.12) and (1.1.13), express the current through and the voltage across the nonlinear resistor in terms of the DC/AC components of the input voltage V_s and v_δ.

1.3 A Nonlinear (First-Order) *RC* Circuit Driven by a Sinusoidal Source
Consider the circuit of Figure P1.3(a) that consists of a nonlinear resistor, a linear resistor $R = 500$ kΩ, and a capacitor $C = 14$ [μF] and is driven by a DC voltage source of $V_s = 6$ [V] and an AC voltage source $v_\delta \sin t = 1.4 \sin t$ [V]. The v-i relationship of the nonlinear resistor is $i_2(v) = v^3$[μA] and described by the characteristic curve in Figure P1.3(b). We can apply KCL at node 1 to write the following node equation:

(a)

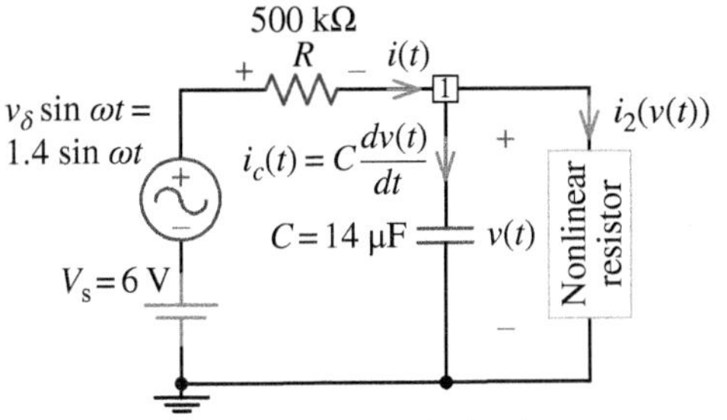

Nonlinear *RC* circuit

(b)

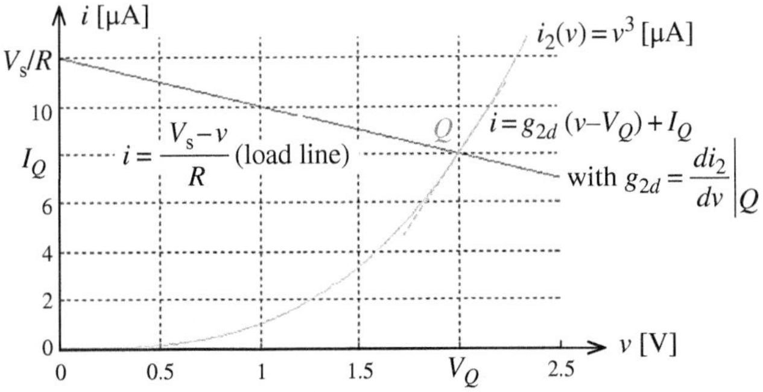

The *i*-*v* characteristic curve and the load line

Figure P1.3

$$C\frac{dv(t)}{dt} + 10^{-6}v^3(t) = \frac{1}{R}(V_s + v_\delta \sin t - v(t))$$

$$;\frac{dv(t)}{dt} = -\frac{10^{-6}}{C}v^3(t) + \frac{1}{RC}(V_s + v_\delta \sin t - v(t)) \quad \text{(P1.3.1)}$$

(a) Verify that the equation for the operating point Q in DC steady state is obtained by removing the AC source and the time derivative term and it can be solved as

$$2V_Q + V_Q^3 = 12;\ V_Q = 2\text{V},\ I_Q = i_2(V_Q) = V_Q^3 = 8\mu A$$
$$\rightarrow Q = (V_Q, I_Q) = (2\text{V}, 8\mu A) \quad \text{(P1.3.2)}$$

Also, draw the load line on the graph of Figure P1.3(b) to find the operating point Q from the intersection of the load line and the v–i characteristic curve of the nonlinear resistor.

(b) Use a MATLAB differential equation solver like 'ode45()' to solve the nonlinear differential equation (P1.3.1) as done in Section 1.1.2. Also, try an iterative (dynamic) method to solve the nonlinear differential equation as coded in the following script "elec01p03b.m," where the capacitor voltage or the voltage at node 1 is computed as

$$v(t) = v(0) + \frac{1}{C}\int_0^t i_C(\tau)\,d\tau$$

$$\text{with } i_C(\tau) = i(\tau) - i_2(\tau) = \frac{V_s + v_\delta \sin\tau}{R} - v^3(\tau) \times 10^{-6} \quad \text{(P1.3.3)}$$

$$\xrightarrow[\text{trapezoidal integration rule}]{\text{descretize with}} \quad v[n+1] = v[n] + \frac{1}{2C}(i_C[n] + i_C[n-1])T \quad \text{(P1.3.4)}$$

Plot the two solutions together for comparison.

```
%elec01p03b.m
% On-line iterative solution of Eq.(P1.3.1)
R=5e5; C=14e-6; Vs=6; vd=1.4;
ftn=@(v,R,Vs)(Vs-v)/R*1e6-v.^3;
VQ=fsolve(ftn,0,optimset('fsolve'),R,Vs), IQ=VQ^3/1e6;
t=[0:0.01:10]; % Time vector
v(1)=VQ; % To let v(t) begin with VQ
% To start the trapezoidal rule with the right-side integration rule
iC(1)=(Vs+vd*sin(t(1))-v(1))/R-v(1)^3/1e6; % Capacitor current
v(2)=v(1)+iC(1)/C*(t(2)-t(1)); % Voltage at node 1
for n=2:length(t)-1
   iC(n)=(Vs+vd*sin(t(n))-v(n))/R-v(n)^3/1e6; % Eq.(P1.3.3)
   v(n+1)=v(n)+(iC(n)+iC(n-1))/C/2*(t(n+1)-t(n)); % Eq.(P1.3.4)
end
plot(t,v)
```

1.4 (Thevenin) Equivalent Resistance of a Network Having Dependent Source
Find the Thevenin equivalent of a CG amplifier in Figure P1.4 seen from d-0 (GND), which has a (dependent) current source of $g_m v_{gs} = (v_g - v_s)/r_{gs}$ where $g_m = 1/r_{gs}$ and $v_g = 0$.

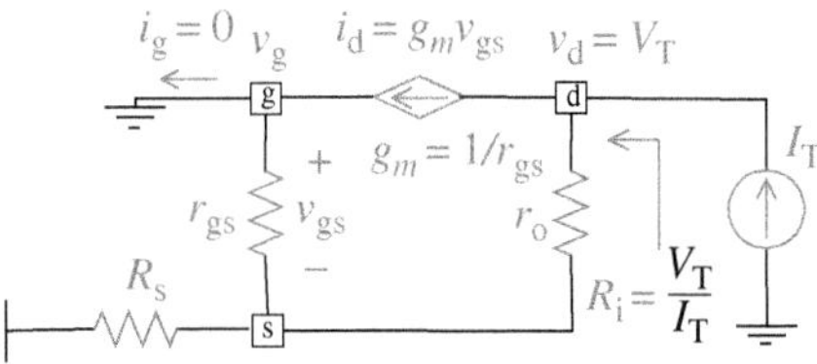

Figure P1.4 To find the equivalent resistance seen from d-0 for Problem 1.4.

1.5 Lowpass Filtering of a Full-Wave Rectified Cosine Wave
Figure P1.5(a) shows the PSpice schematic of a rectifier circuit in which a full-wave rectified voltage $v_i(t) = |10\cos(2\pi 60t)|$ is made smooth by an *LCR* low-pass filter as can be observed from the input and output signals and their spectra in Figure 2.11(b) and (c). Note that in the Simulation Settings dialog box, the PSpice simulation parameters for Time Domain (Transient) Analysis are set as

Run_to_time = 100 ms and Maximum_Step_size = 100 us

Note also that in the Transient Output File Options dialog box (opened by clicking on Output File Options button in the Simulation Settings dialog box), the parameters for FFT analysis are set as in the following:

Center (fundamental) frequency: 60 Hz
Number of harmonics: 5
Output variables: V(L:2) V(L:1)

Figure P1.5(b) shows the FFT analysis results for the input and output voltage waveforms of the *LRC* filter as listed in the PSpice simulation output file (opened by selecting PSpice>View_Output_File from the top menu bar of the schematic window).

(a) Noting that the transfer function and frequency response of the LRC filter are

$$G(s) = \frac{R\|(1/sC)}{sL + R\|(1/sC)} = \frac{R/sC}{sL(R+1/sC)+R/sC} = \frac{1}{s^2LC + sL/R + 1} \tag{P1.5.1}$$

$$G(j\omega) = \frac{1}{1-\omega^2LC + j\omega L/R} = \frac{1}{\sqrt{(1-\omega^2LC)^2 + (\omega L/R)^2}} \angle -\tan^{-1}\frac{\omega L/R}{1-\omega^2LC} \tag{P1.5.2}$$

(a)

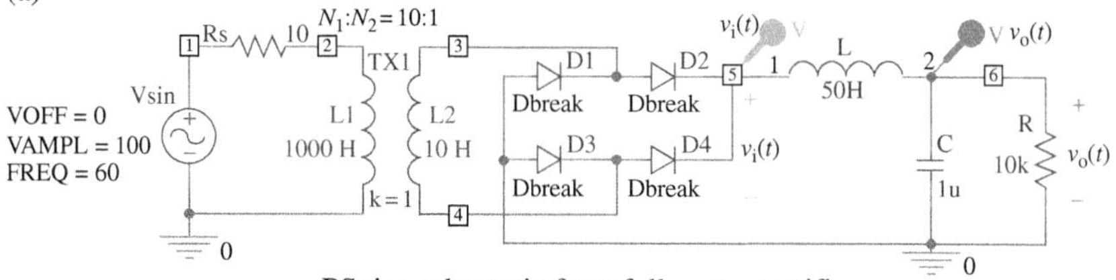

PSpice schematic for a full-water rectifier

(b)

```
**** INCLUDING SCHEMATIC1.net ****
* source ELEC01P04
V_Vsin    N01      0
  +SIN  0  100  60   0   0   0
R_Rs      N01      N02       10
X_TX2     N02      0         N03     N04     SCHEMATIC1_TX2
D_D1      0        N03       Dbreak
D_D2      N03      N05       Dbreak
D_D3      0        N04       Dbreak
D_D4      N04      N05       Dbreak
C_C       0        N06       1u
L_L       N05      N06       50H
R_R       0        N06       10k

.subckt  SCHEMATIC1_TX2  1  2  3  4
K_TX2      L1_TX2 L2_TX2 1
L1_TX2     1 2 1000H
L2_TX2     3 4 10H
.ends SCHEMATIC1_TX2
```

Contents of the netlist file

```
****        FOURIER ANALYSIS                    TEMPERATURE =    27.000 DEG C
******************************************************************************
```

FOURIER COMPONENTS OF TRANSIENT RESPONSE V(N05) $v_i(t)$

DC COMPONENT = 5.091168E+00

HARMONIC NO	FREQUENCY (HZ)	FOURIER COMPONENT	NORMALIZED COMPONENT	PHASE (DEG)	NORMALIZED PHASE (DEG)
1	6.000E+01	1.275E-03	1.000E+00	-6.193E+01	0.000E+00
2	1.200E+02	4.244E+00	3.328E+03	-8.983E+01	3.403E+01
3	1.800E+02	8.031E-04	6.298E-01	-9.255E+01	9.325E+01
4	2.400E+02	8.500E-01	6.666E+02	-8.990E+01	1.578E+02
5	3.000E+02	8.913E-04	6.990E-01	-9.413E+01	2.155E+02

FOURIER COMPONENTS OF TRANSIENT RESPONSE V(N06) $v_o(t)$

DC COMPONENT = 4.585037E+00

HARMONIC NO	FREQUENCY (HZ)	FOURIER COMPONENT	NORMALIZED COMPONENT	PHASE (DEG)	NORMALIZED PHASE (DEG)
1	6.000E+01	2.089E-02	1.000E+00	4.202E+01	0.000E+00
2	1.200E+02	1.556E-01	7.449E+00	9.476E+01	1.071E+01
3	1.800E+02	5.884E-03	2.817E-01	1.660E+01	-1.095E+02
4	2.400E+02	9.324E-03	4.463E-01	6.653E+01	-1.016E+02
5	3.000E+02	3.483E-03	1.668E-01	1.346E+01	-1.966E+02

FFT analysis result from PSpice simulation (listed in the PSpice simulation output file)

Figure P1.5 PSpice simulation of a rectifier for Problem 1.5 ("elec01p05.opj").

and the Fourier series representation of the full-wave rectified cosine input is

$$v_i(t) \overset{(E1.4.7)}{=} \frac{2A}{\pi} - \sum_{m=1}^{\infty} \frac{(-1)^m 4A}{(4m^2-1)\pi} \cos 2m\omega_0 t$$

$$\underset{\omega_0 = 2\pi 60 = 377}{\overset{A=10}{=}} \frac{20}{\pi} + \frac{40}{3\pi}\cos 754t - \frac{40}{15\pi}\cos 1508t + \cdots, \qquad \text{(P1.5.3)}$$

verify that the first two major frequency components of the output can be expressed as

$$|\mathbf{V}_o^{(0)}| = |G(j0)||\mathbf{V}_i^{(0)}|$$

$$= \left.\frac{1}{\sqrt{(1-\omega^2 LC)^2 + (\omega L/R)^2}}\right|_{\substack{\omega=0\\ R=10\,000}} \frac{20}{\pi} = \frac{20}{\pi} \qquad \text{(P1.5.4)}$$

$$|\mathbf{V}_o^{(2)}| = |G(j2\omega_0)||\mathbf{V}_i^{(2)}|$$

$$= \left.\frac{1}{\sqrt{(1-(2\omega_0)^2 LC)^2 + (2\omega_0 L/R)^2}}\right|_{\substack{2\omega_0=754\\ R=10\,000}} \frac{40}{3\pi} \qquad \text{(P1.5.5)}$$

Write the condition for the relative magnitude of the major harmonic to the DC component to be less than r_{max}. Run the MATLAB script "elec02f12.m" (Section 2.2.6) together with MATLAB function '`elec02f12_f()`' to plot the region of (C, L) satisfying the condition as shown in Figure 2.12. Is the point (C = 1 μF, L = 50 H) in the 5%-admissible region?

(b) Modify the MATLAB script "elec01e06.m" (for Example 1.6) so that it can use the MATLAB function '`Fourier_analysis()`' to perform the spectral analysis of the input and output voltage waveforms for the full-wave rectifier in Figure P1.5a and run it. Fill up Table P1.5 with the magnitudes of each frequency component obtained from the MATLAB analysis and PSpice simulation (up to four significant digits) for comparison. Is the relative magnitude of the major harmonic to the DC component less than 5%?

Table P1.5 Fourier analysis results obtained using MATLAB and PSpice.

	Components	PSpice	MATLAB
Input $v_i(t)$	DC component ($k = 0$)	5.091	
	Fundamental (ω_0)	1.275×10^{-3}	0
	Second harmonic ($2\omega_0$)		
Output $v_o(t)$	DC component ($k = 0$)		
	Fundamental (ω_0)	2.089×10^{-2}	0
	Second harmonic ($2\omega_0$)		

2

Diode Circuits

CHAPTER OUTLINE

2.1 The *v-i* Characteristic of Diodes

Figure 2.1(a) shows the symbol for a diode, which allows an electric current (as a low resistance) in the forward direction, while blocking current (as a large resistance) in the reverse direction. Figure 2.1(b) shows the *v-i* characteristic of a practical diode, which can be approximated by the *Shockley diode equation* (named after transistor coinventor William Bradford Shockley) or the *diode law*:

$$i_D = I_S\left(e^{v_D/\eta V_T} - 1\right); \quad v_D = \eta V_T \ln(i_D/I_S + 1) \tag{2.1.1a,b}$$

Electronic Circuits with MATLAB®, PSpice®, and Smith Chart, First Edition. Won Y. Yang, Jaekwon Kim, Kyung W. Park, Donghyun Baek, Sungjoon Lim, Jingon Joung, Suhyun Park, Han L. Lee, Woo June Choi, and Taeho Im.

Companion website: www.wiley.com/go/yang/electroniccircuits

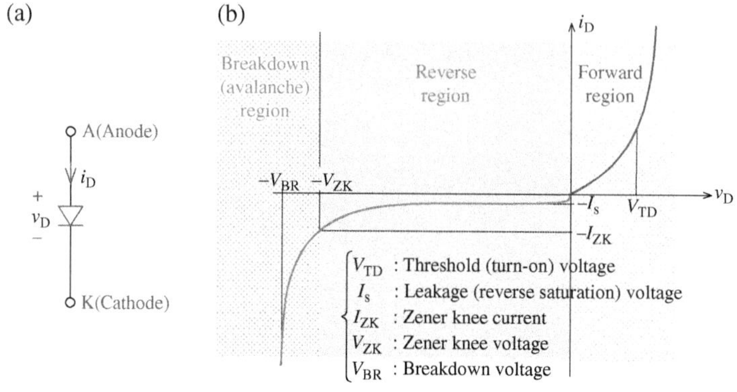

Symbol for a diode — The v-i characteristic of a diode

Figure 2.1 Symbol and *v-i* characteristic of a diode.

where I_S, η, and V_T are the *leakage* (*reverse saturation*) *current*, the empirical constant (called the *emission coefficient* or *ideality factor*) between 1 and 2, and the thermal voltage, respectively.

Note that the *thermal voltage* V_T is given by

$$V_T = \frac{T_K\,[°K]}{11605} = \frac{T_C\,[°C] + 273}{11605} \approx 25.9\,[\text{mV}]\,\text{at}\,27\,°C\,(T:\text{temperature}) \quad (2.1.2)$$

2.1.1 Large-Signal Diode Model for Switching Operations

When the signal applied to such electronic devices as diodes and transistors is large in comparison with the bias level, they show ON-OFF behavior, functioning like a switch.

The *v-i* characteristic curve of a diode in terms of its static behavior can be approximated by a solid/dotted piecewise linear (PWL) line for the forward-/reverse-bias mode as depicted in Figure 2.2(a). According to the approximation, the operation of a diode in the forward-/reverse-bias mode is represented by the equivalent model depicted in Figure 2.2(b).

2.1.2 Small-Signal Diode Model for Amplifying Operations

Figure 2.3(a) and (b) shows the high-frequency AC models of forward-/reverse-biased diodes, respectively. Note that the *junction* (or *depletion* or *transition*) *capacitance* defined as the ratio of the incremental change (Δq_j) in the charge (in the depletion layer) to that (Δv_D) in the anode-to-cathode bias voltage v_D can be expressed as

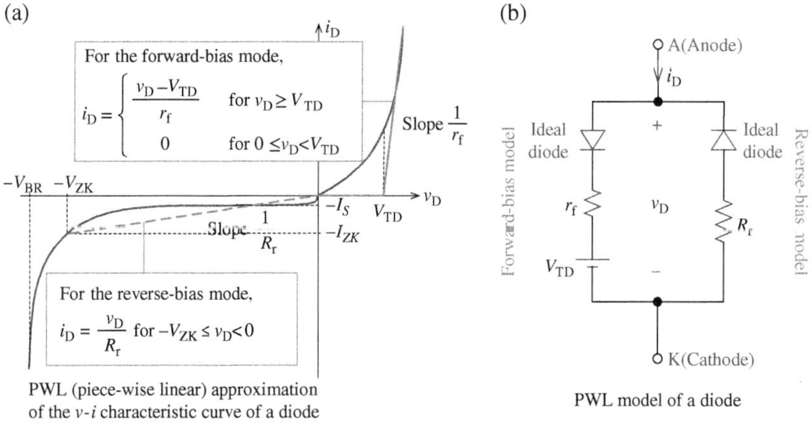

Figure 2.2 PWL approximation of the *v-i* characteristic curve of a diode and the corresponding model.

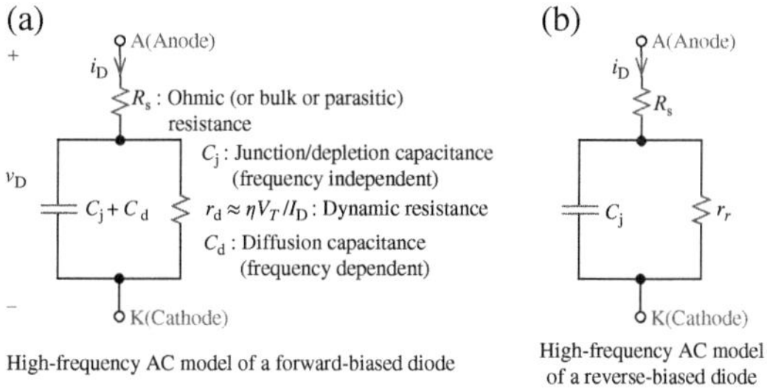

Figure 2.3 High-frequency AC (small-signal) model of a diode.

$$C_j = \frac{dq_j}{dv_D} = \frac{C_{j0}}{(1 - v_D/V_j)^M} \tag{2.1.3}$$

where M: *junction gradient coefficient,* V_j: (*built-in*) *junction potential* with the value of 0.5~0.9 V for a Si (silicon) diode and 0.2~0.6 V for a Ge (germanium) diode, and C_{j0}: *zero-bias junction capacitance*

Note also that the *diffusion* (or *transit time*) *capacitance* due to the diffusion of carriers from anode to cathode in the forward-bias mode can be expressed as

$$C_d = \left.\frac{dq_D}{dv_D}\right|_Q = \left.\frac{t_T di_D}{dv_D}\right|_Q \overset{(2.1.1a)}{=} \frac{t_T I_Q}{\eta V_T} \tag{2.1.4}$$

where t_T: *transit* or *storage time* taken for the charge to cross the diode, I_Q: diode current at the operating point Q, η: *emission coefficient* or *ideality factor,* and,

$$V_T \overset{(2.1.2)}{=} \frac{T_K[°K]}{11605} = \frac{T_C[°C]+273}{11605} \approx 25.9[mV] \text{ for } T_C = 27°C : \textit{thermal voltage}$$

Note also that the dynamic resistance r_d can be approximated as

$$r_d = \left.\frac{dv_D}{di_D}\right|_Q = \left.\frac{1}{di_D/dv_D}\right|_Q \overset{(2.1.1a)}{=} \left.\frac{1}{(I_S/\eta V_T)\exp(v_D/\eta V_T)}\right|_Q \overset{(2.1.1a)}{=} \left.\frac{1}{(I_S+i_D)/\eta V_T}\right|_Q$$

$$= \left.\frac{\eta V_T}{I_S+i_D}\right|_Q \underset{i_D=I_Q}{\overset{I_S<<I_Q}{\approx}} \frac{\eta V_T}{I_Q} \tag{2.1.5}$$

2.2 Analysis/Simulation of Diode Circuits

The procedure of performing a large-signal analysis for diode circuits can be summarized:

1) Assume (guess) the ON/OFF state of each diode.
2) For ON/OFF state, replace the diode by the forward-/reverse-bias model of each diode like the ones depicted in Figure 2.2(b).
3) Solve the circuit and verify the result and the assumptions:
   - For a diode assumed to be ON (with $v_D \approx V_{TD}$), the initial guess is justified if $i_D > 0$; otherwise, i.e. if $i_D < 0$, resume the analysis with the assumption that the diode is OFF.
   - For a diode assumed to be OFF (with $i_D = 0$), the initial guess is justified if $v_D < V_{TD}$; otherwise, i.e. if $v_D \geq V_{TD}$, resume the analysis with the assumption that the diode is ON.

2.2.1 Examples of Diode Circuits

See the following examples.

Example 2.1 Analysis of an AND Gate Using PWL Model

For the circuit of Figure 2.4(a), find the output voltage v_o to different values of input voltages $\{v_1, v_2\}$ where the PWL diode model parameters in Figure 2.2(b) are $r_f = 30$ Ω, $V_{TD} = 0.685$ V, $R_r = \infty\ \Omega$, and $I_s = 1 \times 10^{-15}$ A. Note that all voltages are measured w.r.t. the reference node, which is grounded.

a) For $v_1 = v_2 = 5$ [V], we assume that D_1 and D_2 are OFF so that $v_o = 5$ V(High) as described by the equivalent in Figure 2.4(b1). We check the validity of this assumption:

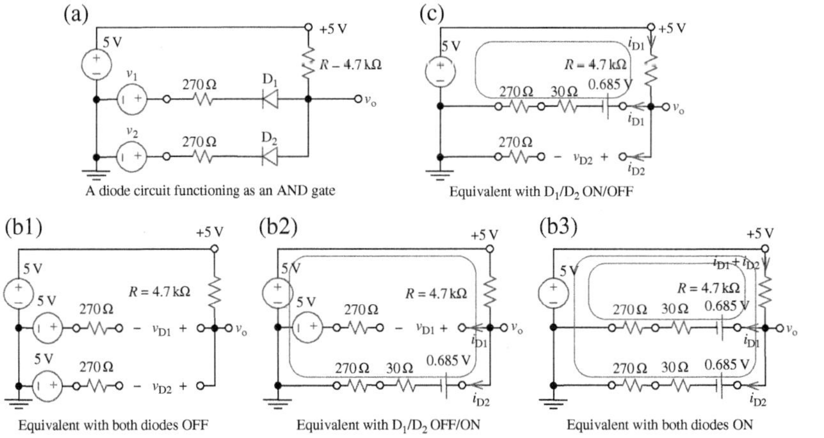

Figure 2.4 A diode circuit functioning as an AND gate and its equivalents for different input values.

$$\left.\begin{aligned} v_{D_1} &= 5-5=0\,\text{V} \underset{D_1:\text{OFF}}{\overset{\text{O.K.}}{<}} 0.685 = V_{TD} \\ v_{D_2} &= 5-5=0\,\text{V} \underset{D_2:\text{OFF}}{\overset{\text{O.K.}}{<}} 0.685 = V_{TD} \end{aligned}\right\} \overset{\text{Valid}}{\rightarrow} v_o = 5\,\text{V} : \text{High} \qquad \text{(E2.1.1)}$$

b) For $v_1 = 5$ [V] and $v_2 = 0$ [V], we assume that D_1 is OFF and D_2 is ON as described by the equivalent in Figure 2.4(b2) so that we can get

$$i_{D_1} = 0\,\text{A},\ i_{D_2} \underset{\text{to the outer loop}}{\overset{\text{KVL}}{=}} \frac{5-0.685}{4700+270+30} = 0.863\ [\text{mA}] > 0$$

$$\rightarrow v_o = 5 - 4700(i_{D_1} + i_{D_2}) = 5 - 4700 \times 0.863 \times 10^{-3} = 0.944\ [\text{V}] \qquad \text{(E2.1.2)}$$

We check the validity of this assumption:

$$\left.\begin{aligned} v_{D_1} = v_o - 5 = 0.944 - 5 = -4.056\,\text{V} \underset{D_1:\text{OFF}}{\overset{\text{O.K.}}{<}} 0.685 = V_{TD} \\ i_{D_2} = 0.863\ [\text{mA}] \underset{D_2:\text{ON}}{\overset{\text{O.K.}}{>}} 0 \end{aligned}\right\} \overset{\text{Valid}}{\rightarrow} v_o = 0.944\,\text{V} : \text{Low} \qquad \text{(E2.1.3)}$$

c) For $v_1 = v_2 = 0$ [V], we assume that D_1 and D_2 are ON as described by the equivalent in Figure 2.4(b3) so that we can apply Kirchhoff's Voltage Law (KVL) to the inner/outer loops with $i_{D_1} = i_{D_2} = i_D$ to obtain

$$4700 \times 2i_D + (270+30)i_D = 5 - 0.685;\ i_D = 0.445\ [\text{mA}] > 0$$
$$\rightarrow v_o = 5 - 4700(i_{D_1} + i_{D_2}) = 5 - 4700 \times 0.89 \times 10^{-3} = 0.817\ [\text{V}] \qquad \text{(E2.1.4)}$$

We check the validity of this assumption:

$$\left.\begin{array}{l} i_{D_1} = 0.445\ [\text{mA}] \underset{D_1:\text{ON}}{\overset{\text{O.K.}}{>}} 0 \\ i_{D_2} = 0.445\ [\text{mA}] \underset{D_2:\text{ON}}{\overset{\text{O.K.}}{>}} 0 \end{array}\right\} \overset{\text{Valid}}{\rightarrow} v_o = 0.817\ \text{V : Low} \tag{E2.1.5}$$

Are we lucky that all the assumptions have turned out to be true? Then let us try with another assumption for the last case. For $v_1 = v_2 = 0$ [V], we assume that D_1 is ON and D_2 is OFF as described by the equivalent in Figure 2.4(c) so that we can apply KVL to the inner loop to get

$$i_{D_1} \underset{\text{to the inner loop}}{\overset{\text{KVL}}{=}} \frac{5-0.685}{4700+270+30} = 0.863\ [\text{mA}] > 0,\ i_{D_2} = 0\ \text{A}$$

$$\rightarrow v_o = 5 - 4700 i_{D_1} = 5 - 4700 \times 0.863 \times 10^{-3} = 0.944\ [\text{V}] \tag{E2.1.6}$$

We check the validity of this assumption:

$$\left.\begin{array}{l} i_{D_1} = 0.863\ [\text{mA}] \underset{D_1:\text{ON}}{\overset{\text{O.K.}}{>}} 0 \\ v_{D_2} = 0.944\ [\text{V}] \underset{D_2:\text{OFF}}{\overset{\text{Not O.K.}}{<}} 0.685 = V_{TD} \end{array}\right\} \begin{array}{l}\text{Not valid because the assumption of} \\ D_2\ \text{OFF has turned to be wrong.}\end{array} \tag{E2.1.7}$$

If we use the (nonlinear) exponential model (2.1.1) for a more exact analysis, the KVL equation in $i_{D_1} = i_{D_2} = i_D$ can be set up as

$$270 i_D + \eta V_T \ln(i_D/I_S + 1) + 4700 \times 2 i_D = 5 \tag{E2.1.8}$$

since the current through $R = 4.7\ \text{k}\Omega$ is $i_{D_1} + i_{D_2} = 2i_D$. Here, Eq. (2.1.1b) has been used to express the diode voltage v_D in terms of the diode current i_D. To solve this nonlinear equation, the MATLAB function 'fsolve()' can be used as listed in the following script "elec02e01.m":

```
%elec02e01.m
nVT=0.0259; Is=1e-15; % Thermal voltage, Reverse saturation current
rf=30; VTD=0.685; % Diode PWL model parameters
R = 4700;
vDi = @(iD)nVT*log(iD/Is+1); % Eq.(2.1.1b)
eq=@(i)vDi(i)+(2*R+270)*i-5; % Eq.(E2.1.8) with exponential model
%eq = @(i)VTD+(rf+2*R+270)*i-5; % KVL equation with PWL model
iD = fsolve(eq,1e-3) % Diode current
vD = vDi(iD) % Diode voltage
vo = 5 - R*2*iD % Output voltage
```

Running this script yields

```
iD = 4.4522e-04,   vD = 0.6947,   vo = 0.8149
```

which is close to the above result (Eq. (E2.1.5)) of analysis using the PWL model.

Example 2.2 Analysis of a Two-Diode Circuit Using CVD (Constant Voltage Drop) Model

For the circuit of Figure 2.5(a), replace the diodes (assumed to be ON) by the *CVD model* (that is, the PWL model with $r_f = 0\,\Omega$ and $V_{TD} = 0.7$ V) and find the voltages v_1 and v_2. Compare them and those obtained using the exponential model (Eq. 2.1.1) with $I_s = 10 \times 10^{-15}$ A and $\eta V_T = 25.9$ mV.

Note that there are four possible states for the two diodes D_1 and D_2: ON-ON, ON-OFF, OFF-ON, and OFF-OFF. First, assuming that both D_1 and D_2 are ON, we replace them by the Constant Voltage Drop (CVD) model to draw the equivalent as shown in Figure 2.5(b). Then the voltages v_1 and v_2 can easily be found as

$$v_1 = 0.7\,\text{V},\ v_2 = v_1 - 0.7 = 0\,\text{V} \tag{E2.2.1}$$

yielding

$$i = \frac{V_{s1} - v_1}{R_1} = \frac{10 - 0.7}{10\text{k}} = 0.93\,\text{mA}$$

$$;\, i_{D2} = \frac{v_2 - V_{s2}}{R_2} = \frac{0 - (-6)}{5\text{k}} = 1.2\,\text{mA}, \tag{E2.2.2}$$

$$i_{D1} = i - i_{D2} = 0.93 - 1.2 = -0.27\,\text{mA}$$

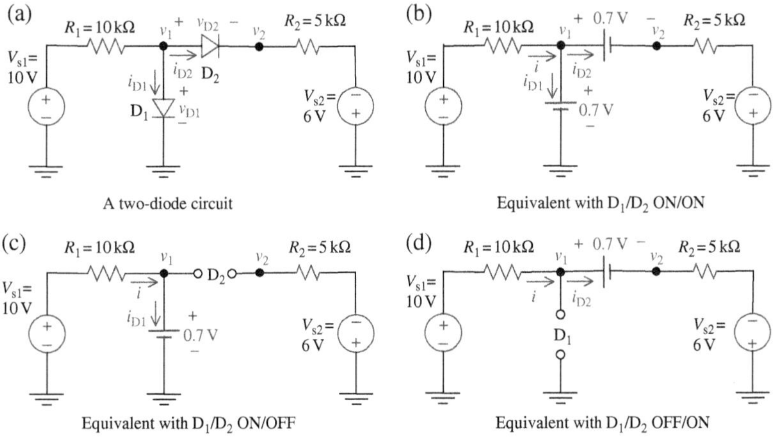

Figure 2.5 A two-diode circuit for Example 2.2.

However, $i_{D_1} = -0.27$ mA < 0 contradicts the assumption that D_1 is ON. That is why we make another assumption that D_1 and D_2 are ON and OFF, respectively, draw the corresponding equivalent as depicted in Figure 2.5(c) and get

$$i = \frac{V_{s1} - v_1}{R_1} = \frac{10 - 0.7}{10\text{k}} = 0.93\,\text{mA},\ v_{D2} = v_1 - v_2 = 0.7 - (-6) = 6.7\,\text{V} \tag{E2.2.3}$$

Still, $v_{D_2} = 6.7$ V > 0 contradicts the assumption that D_2 is OFF. That is why we make another assumption that D_1 and D_2 are OFF and ON, respectively, draw the corresponding equivalent as depicted in Figure 2.5(d) and get

$$i = i_{D2} = \frac{V_{s1} - 0.7 - V_{s2}}{R_1 + R_2} = \frac{10 - 0.7 - (-6)}{10\text{k} + 5\text{k}} = 1.02\,\text{mA},$$

$$v_1 = V_{s1} - R_1 i = 10 - 10 \times 1.02 = -0.2\,\text{V} \tag{E2.2.4}$$

This yields $v_{D_1} = v_1 = -0.2$ V < 0 and $i_{D_2} = i = 1.02$ mA > 0, suiting the assumption that D_1 and D_2 are OFF and ON, respectively. Therefore, the solution is $v_1 = -0.2$ V and $v_2 = v_1 - 0.7 = -0.9$ V.

If we use the (nonlinear) exponential model (2.1.1) for a more exact analysis, KCL can be applied at nodes 1 and 2 to yield a set of two node equations in v_1 and v_2 as

$$\begin{aligned} &\frac{V_{s1} - v_1}{R_1} - I_S\left(e^{v_1/\eta V_T} - 1\right) - I_S\left(e^{(v_1 - v_2)/\eta V_T} - 1\right) = 0 \\ &I_S\left(e^{(v_1 - v_2)/\eta V_T} - 1\right) - \frac{v_2 - V_{s2}}{R_2} = 0 \end{aligned} \tag{E2.2.5}$$

where Eq. (2.1.1a) has been used to express diode current i_D in terms of diode voltage v_D. To solve this set of nonlinear equations, the MATLAB function '`fsolve()`' can be used as listed in the following script "elec02e02.m":

```
%elec02e02.m
Is=10e-15; nVT=0.0259; % Diode exponential model parameters
Vs1=10; Vs2=-6; R1=1e4; R2=5e3; % Circuit parameter values
iD = @(vD) Is*(exp(vD/nVT)-1); % Eq.(2.2.1a)
eqs = @(v) [(Vs1-v(1))/R1-iD(v(1))-iD(v(1)-v(2));
            iD(v(1)-v(2))-(v(2)-Vs2)/R2]; % Eq.(E2.2.5)
v = fsolve(eqs,[1 1]), ID2=iD(v(1)-v(2))
```

Running this script yields the following, which is close to the above result with the CVD model:

```
v = -0.2289   -0.8855,   ID2 = 0.0010
```

2.2.2 Clipper/Clamper Circuits

Figure 2.6(a1) and (a2) respectively show *clipper* circuits for clipping the upper and lower portion of the input signal above/below the reference level of $(V_1 + V_D)/(-V_2 - V_D)$, which is determined by the DC voltage source connected in series with the diode. Figure 2.6(b1) and (b2) show their input and output voltage waveforms obtained from PSpice simulation. Running the following MATLAB script yields a similar result.

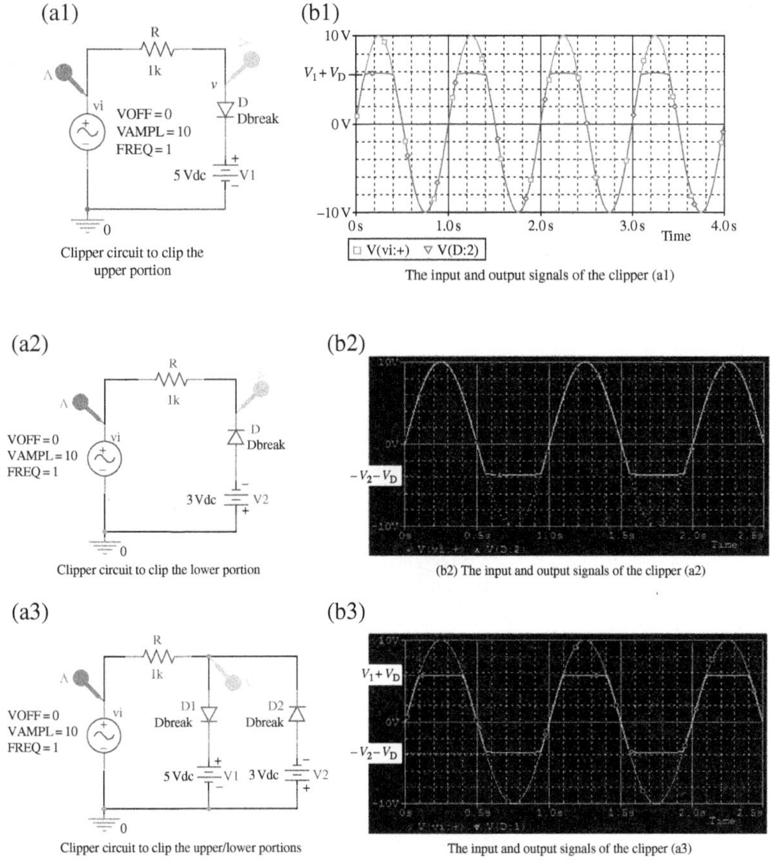

Figure 2.6 Clipper circuits to clip the upper/lower portion of the input signal and their input/output signals.

```
%elec02f06_1_clipper.m
Is=10e-15; nVT=0.0259; VD=0.7;
iD=@(vD)Is*(exp(vD/nVT)-1); % Eq.(2.2.1a)
t=[0:0.01:2.5]; vi=10*sin(2*pi*t); R=1e3; V1=5;
v10=0; Nt=length(t);
options=optimoptions ('fsolve','Display','none');
for n=1:Nt
  eq = @(v)vi(n)-v-R*iD(v-V1); % KCL equation in v
  if n>1, v10=v(n-1); end; v(n)=fsolve(eq,v10,options);
end
plot(t,vi,'r:', t,v,'g', t,V1+VD,'k:')
```

Figure 2.6(a3) shows a two-level clipper circuit, which combines two clipper circuits for clipping the upper and lower portions of the input signal so that the output voltage can be kept within the range of $[-V_2 - V_D, V_1 + V_D]$. Figure 2.6(b3) shows its input and output voltage waveforms.

Figure 2.7(a1) and (a2) shows positive/negative clamper circuits (called clamped capacitors or DC restorers), which push the input signal (within $[V_{i,min}, V_{i,max}]$) upward/downward by the positive/negative capacitor voltage

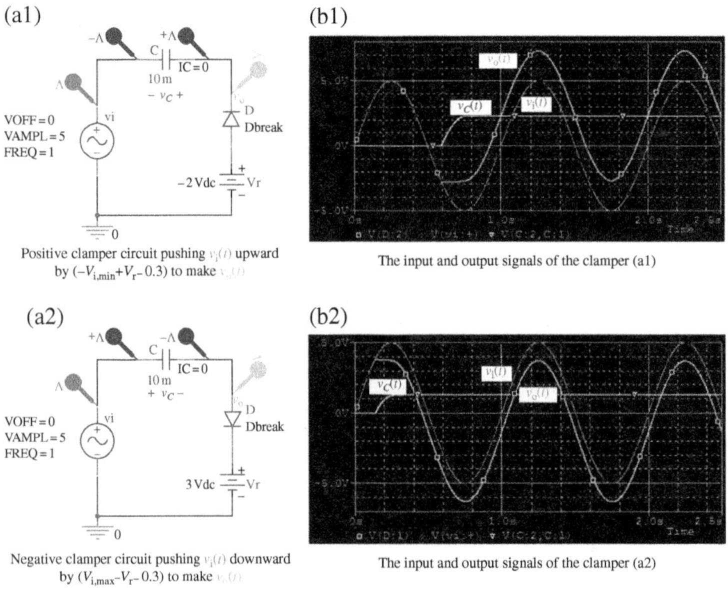

Figure 2.7 Positive/negative clamper circuits and their input/output signal waveforms ("clipper_clamper.opj").

charged (when $v_i = V_{i,min}/V_{i,max}$), i.e. $V_C = (-V_{i,min} + V_r - V_D)/(V_{i,max} - V_r - V_D)$ so that their outputs are related with their inputs as

$$v_o \approx v_i - V_{i,min} + V_r - V_D \tag{2.2.1a}$$

$$v_o \approx v_i - V_{i,max} + V_r + V_D \tag{2.2.1b}$$

Note that the trough/peak level of the positive/negative clamper outputs will be $(V_r - V_D)/(V_r + V_D)$, respectively, where V_D depends on the capacitance C and how long C has been charged.

2.2.3 Half-wave Rectifier

The diode in the circuit of Figure 2.8(a) can be represented by the CVD model as shown in Figure 2.8(b), which is valid in the forward mode of the diode, i.e. while $v_i \geq V_{os}$ where the offset or cut-in voltage V_{os} (slightly less than the threshold voltage V_{TD}) is the diode voltage at which the diode starts to turn on. Here, we can find the cut-in or ignition angle ϕ at which the diode starts to turn on:

$$V_m \sin\phi = V_{os}; \qquad \phi = \sin^{-1}\frac{V_{os}}{V_m} \tag{2.2.2}$$

Similarly, the extinction angle at the end of the (first) positive half-cycle is $\pi - \phi$ (see Figure 2.8(c)). Note that the peak inverse voltage (PIV) of the diode is V_m.

2.2.4 Half-wave Rectifier with Capacitor – Peak Rectifier

Figure 2.9.1(a) shows the PSpice schematic of a half-wave rectifier composed of a diode, a capacitor, and a resistor, called a *peak rectifier*, which conducts in the forward direction for $v_D \geq V_{TD} = 0.65$ V. Figure 2.9.1(b) and (c) shows the equivalent circuits of the rectifier for $v_i \geq v_o + V_{TD}$ and $v_i < v_o + V_{TD}$, respectively. Figure 2.9.1(d) shows the PSpice simulation result obtained from the Transient Analysis with Run_to_time of 40 ms. From this PSpice simulation result, we find

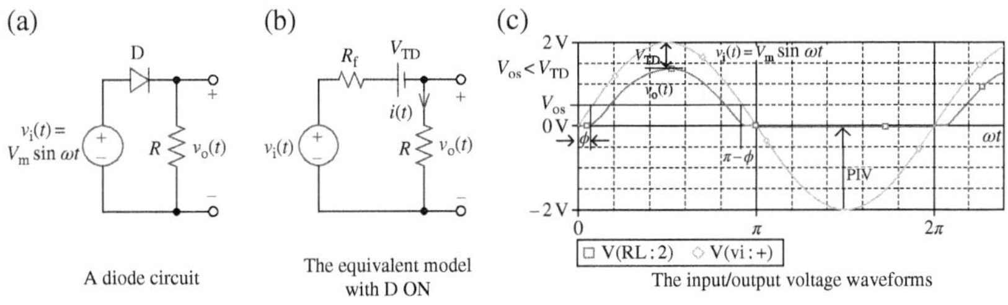

Figure 2.8 A half-wave rectifier and its input/output voltage waveforms.

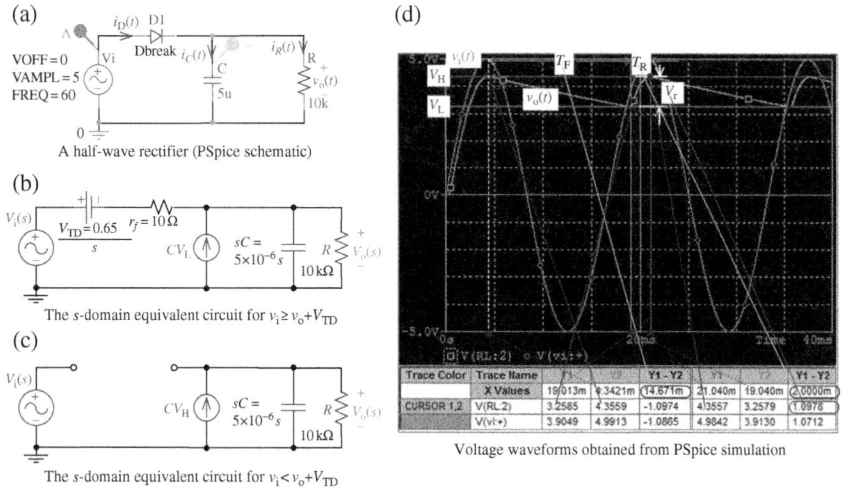

Figure 2.9.1 Equivalent circuits and input/output voltage waveforms of a half-wave rectifier ("rectifier_halfwave.opj").

the upper/lower limit V_H/V_L of the output voltage $v_o(t)$ and the rising/falling period T_R/T_F as

$$V_H = 4.36\,[\mathrm{V}],\quad V_L = 3.26\,[\mathrm{V}],\quad T_R = 2.0184\,[\mathrm{ms}],\quad T_F = 14.757\,[\mathrm{ms}]$$

To obtain these parameters, after getting the output voltage waveform in the PSpice A/D (Probe) window, click the Toggle Cursor button on the toolbar to activate the two cross-type cursors on the graph. Then use the left/right mouse button and/or arrow/shift-arrow key or click the appropriate toolbar button to move them to the peak/trough and read their coordinates from the Probe Cursor box. If you have two or more waveforms on the Probe window, you can choose one which you want to take a close look at by clicking the name of the corresponding variable under the graph.

To get the upper/lower limit V_H/V_L of the output voltage $v_o(t)$ and the rising/falling period T_R/T_F via an analytical approach using MATLAB, we set up the following equations:

$$V_H e^{-T_F/RC} - V_L = 0 \quad \text{(for falling period)} \tag{2.2.3a}$$

$$V_H - V_L - V_m(1 - \cos\omega T_R) = 0 \; \text{(for rising period)} \tag{2.2.3b}$$

$$\omega(T_R + T_F) = 2\pi;\; T_R + T_F - 1/f = 0 \tag{2.2.3c}$$

Noting that V_H is already known as $V_H = V_m - V_{TD} = 5 - 0.65 = 4.35$, we solve this set of equations to find $V_L = 3.23$, $T_R = 0.0018$, and $T_F = 0.0149$ by saving these equations into an M-file named, say, 'halfwave_rectifier_eq.m' and running the following MATLAB script "do_halfwave_rectifier.m."

```
%do_halfwave_rectifier.m
clear
global Vm f VTD
Vm=5; f=60; VTD=0.65;
VH=Vm-VTD; % Local maximum (High Voltage) of vo(t)
R=1e4; C=5e-6;
x0=[0 0 0]; % Initial guess of [VL TR TF]
% VL: Low Voltage (Local Min) of vo(t), TR: Rise Time, TF: Falling Time
x=fsolve('halfwave_rectifier_eq',x0,optimoptions ('fsolve'),C,R)
VL=x(1); TR=x(2); TF=x(3);
fprintf('\n VH=%8.4f, VL=%8.4f, TR=%8.4f, TF=%8.4f\n', VH,VL,TR,TF)
```

```
function y=halfwave_rectifier_eq(x,C,R)
global Vm f VTD
VH=Vm-VTD; w=2*pi*f; T=1/f;
VL=x(1); TR=x(2); TF=x(3);
y= [VH*exp(-TF/R/C)-VL; % Eq. (2.2.3a)
    VH-VL-Vm*(1-cos(w*TR)); % Eq. (2.2.3b)
    TR+TF-T]; % Eq. (2.2.3c)
```

Now, to perform the MATLAB simulation of the rectifier circuit with capacitor filter in Figure 2.9.1(a), we discretize the integro-differential equation for the voltage-current relationship (VCR) of the capacitor as

$$v_o(t) = \frac{1}{C}\int i_C(\tau)\,d\tau \underset{\text{with sampling period } T}{\overset{\text{Discretization}}{\longrightarrow}} \begin{cases} v_o[n+1] \overset{\text{Left-side rule}}{=} v_o[n] + \frac{1}{C} i_C[n]T \\ \text{or} \\ v_o[n+1] \overset{\text{Trapezoidal rule}}{=} v_o[n] + \frac{1}{2C}(i_C[n-1] + i_C[n])T \end{cases} \tag{2.2.4a}$$

where

$$i_C(t) = i_D(t) - i_R(t) \tag{2.2.4b}$$

$$\text{with } i_D(t) \overset{(2.1.1a)}{=} I_S\left(e^{v_D(t)/\eta V_T} - 1\right) \overset{v_D = v_i - v_o}{=} I_S\left(e^{(v_i(t) - v_o(t))/\eta V_T} - 1\right)$$

The numerical solution process to find the output voltage of a half-wave rectifier as shown in Figure 2.9.1(a) has been cast into the following MATLAB function '`rectifier_RC()`'. The following MATLAB script "elec02f09.m" uses this function to find $v_o(t)$ of the half-wave rectifier (in Figure 2.9.1(a)) as Figure 2.9.2 together with the values of $V_H = 4.36$, $V_L = 3.26$, $T_R = 0.002$, and $T_F = 0.0147$.

```
function [vo,vD,iD,iR]=rectifier_RC(vit,R,C,Is,nVT)
% vit = [vi; t] : 2-row matrix consisting of the input signal and time
% R,C = Resistance and Capacitance
% Copyleft: Won Y. Yang, wyyang53@hanmail.net, CAU for academic use only
if nargin<5, nVT=(273+27)/11605; end % Thermal voltage
iDv = @(vD) Is*(exp(vD/nVT)-1); % Eq. (2.1.1a)
[Nr,Nc]=size(vit);
if Nc==2, vi=vit(:,1); ts=vit(:,2); N=Nr;
 elseif Nr==2, vi=vit(1,:); ts=vit(2,:); N=Nc;
end
dt=ts(2)-ts(1); % Sampling interval
vo(1) = 0; % Initial value of the output voltage vo(t)
for n=1:N
    vD(n) = vi(n)-vo(n); iD(n) = iDv(vD(n)); iR(n) = vo(n)/R;
    iC(n) = iD(n) - iR(n); % Eq. (2.2.4b)
    vo(n+1) = vo(n) + iC(n)/C*dt; % Eq. (2.2.4a)
end
vo = vo(2:end); % To make the size of vo equal to that of vi
```

```
%elec02f09.m
% To simulate the rectifier in Fig. 2.9.1(a)
clear, clf
Vm=5; f=60; w=2*pi*f; P=1/f; % Amplitude, Frequency, Period of vi(t)
R=1e4; C=5e-6; % Values of R and C of the RC filter
dt=1e-5; t=0:dt:0.04; vi=Vm*sin(w*t); % Time range, Input voltage vi(t)
Is=1e-14; % Saturation current
vo=rectifier_RC([vi; t],R,C,Is);
subplot(313), plot(t,vi,'g', t,vo,'r')
N1=floor(P/dt); nn0=1:N1;
[VH,imax]=max(vo(nn0)); VH % The 1st peak value of vo(t)
tH1=t(imax); % The 1st peak time
N2=2*N1+10; nn1=N1:N2;
[VL,imin]=min(abs(vo(nn1))); VL
tL1=t(N1-1+imin); % The 1st valley time
[emin,imin]=min(abs(vo(nn1)-VH));
tH2=t(N1-1+imin); % The 2nd peak time
TF=tL1-tH1, TR=tH2-tL1 % Falling/Rising period
```

```
>>elec02f09
   VH = 4.3567 % Upper limit (High value) of the output voltage
   VL = 3.2586 % Lower limit (Low value) of the output voltage
   TF = 0.0147 % Falling time
   TR = 0.0020 % Rising time
```

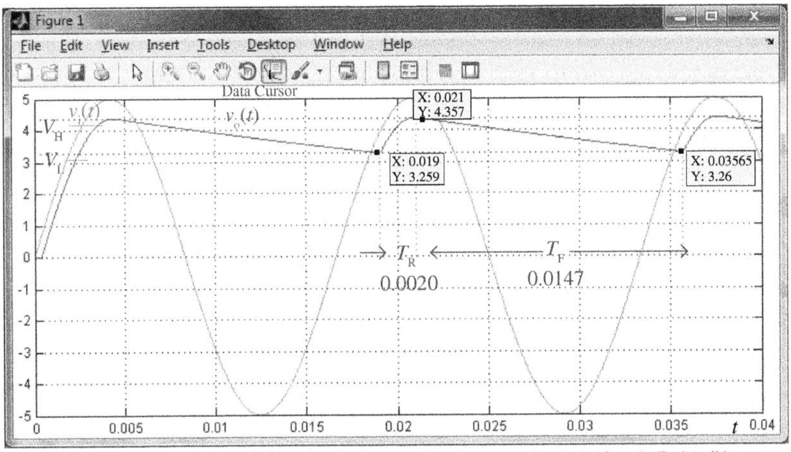

The three special points have been marked by using the data cursor function that can be selected from the Tools pulldown menu.

Figure 2.9.2 Input/output voltage waveforms of the half-wave rectifier.

These results are close to those obtained via an analytical approach or the PSpice simulation results depicted in Figure 2.9.1(d).

One observation about the behavior of the output voltage $v_o(t)$ made from Figures 2.9.1(d) or 2.9.2 is that $v_o(t)$ follows the input voltage $v_i(t) = V_m \sin(\omega t) = 5 \sin(2\pi f t)$ ($f = 60$ Hz) promptly when rising up, but very lazily when falling down, which is helpful for making the rectifier output $v_o(t)$ smooth with a small ripple. Why are the behaviors of the circuit different for the two cases of the capacitor being charged and discharged? It is because the time constant of the circuit with the capacitor being charged (via the diode) from the source is much shorter than that with the diode off and the capacitor being discharged, as can be seen from the equivalent circuits with the diode ON/OFF depicted in Figure 2.9.1(b) and (c):

$$\text{The charging time constant: } T_c = (r_f \parallel R)C \approx 10 \times 5 \times 10^{-6} = 5 \times 10^{-5}\,\text{s}$$

$$\text{The discharging time constant: } T_D = RC \approx 10 \times 10^3 \times 5 \times 10^{-6} = 5 \times 10^{-2}\,\text{s}$$

(cf.) This kind of circuits can be used not only for rectifying and smoothing an AC voltage into a DC voltage in power supplies, but also for demodulating a conventional amplitude modulated (AM) signal to restore the message signal in communication systems.

2.2.5 Full-wave Rectifier

Figure 2.10(a1) and (a2) shows a full-wave rectifier using a center-tapped transformer and another full-wave rectifier using a diode bridge, respectively.

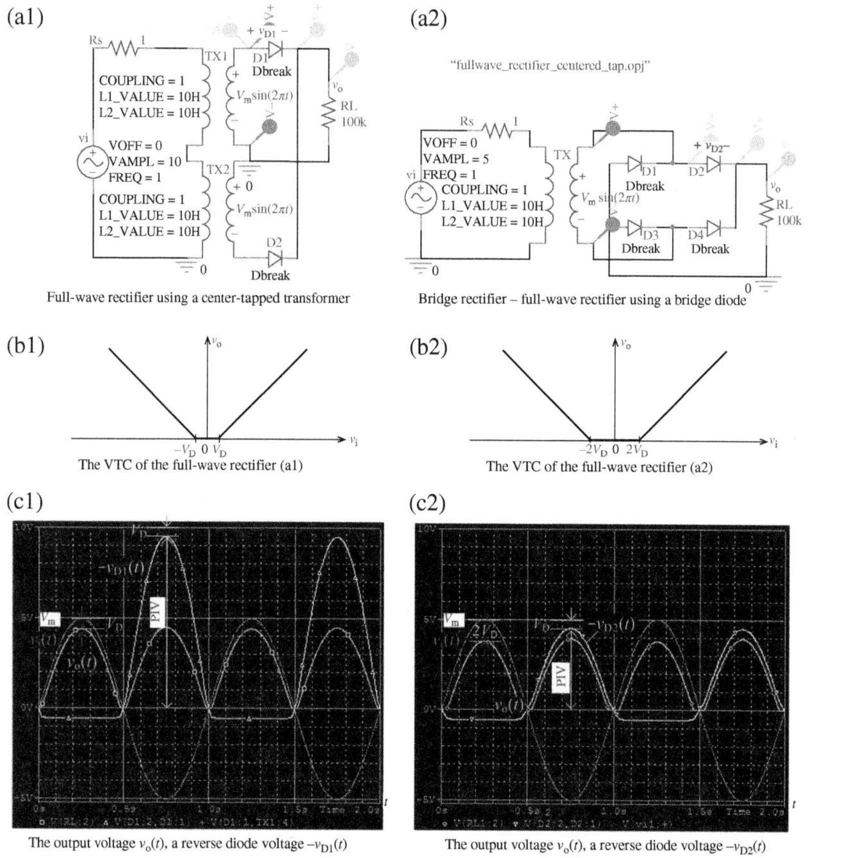

(a1)
(a2)
"fullwave_rectifier_centered_tap.opj"
Full-wave rectifier using a center-tapped transformer
Bridge rectifier – full-wave rectifier using a bridge diode
(b1)
(b2)
The VTC of the full-wave rectifier (a1)
The VTC of the full-wave rectifier (a2)
(c1)
(c2)
The output voltage $v_o(t)$, a reverse diode voltage $-v_{D1}(t)$
The output voltage $v_o(t)$, a reverse diode voltage $-v_{D2}(t)$

Figure 2.10 Two full-wave rectifiers and their voltage transfer characteristics (VTCs), output voltage waveforms, and peak inverse voltages (PIVs).

Figure 2.10(b1) and (b2) shows their input-output relationships, called voltage transfer characteristic (VTC):

$$v_o = \begin{cases} 0 & \text{if } v_i < V_D \\ |v_i - V_D| & \text{if } v_i \ge V_D \end{cases} \tag{2.2.5a}$$

$$v_o = \begin{cases} 0 & \text{if } v_i < 2V_D \\ |v_i - 2V_D| & \text{if } v_i \ge 2V_D \end{cases} \tag{2.2.5b}$$

Figure 2.10(c1) and (c2) shows their PSpice simulation results (obtained from the Transient analysis with $v_i(t) = V_m \sin(2\pi t)$ ($V_m = 5$ V)) for the output voltage $v_o(t)$ and a reversed diode voltage $-v_D(t)$, from which we see their PIVs for a diode as

$$\text{PIV} = 2V_m - V_D \tag{2.2.6a}$$

$$\text{PIV} = V_m - V_D \tag{2.2.6b}$$

2.2.6 Full-wave Rectifier with *LC* Filter

Figure 2.11(a) shows the PSpice schematic of a rectifier circuit in which a full-wave rectified voltage $v_i(t) = |10 \sin(2\pi 60t)|$ is made smooth by an LC low-pass filter as can be observed from the input and output signals and their spectra in Figure 2.11(b) and (c), respectively. We are going to find the condition on LC (with $R = 10\,\text{k}\Omega$) that should be satisfied to keep the relative magnitude of the major harmonic component to the DC component less than r_{max}, say, 5%. To this end, we will perform the Fourier analysis to find the two leading frequency components (including the DC term) of the input $v_i(t)$ and output $v_o(t)$.

Since the input voltage $v_i(t)$ to the LC filter is a full-wave rectified cosine wave, we can use (E1.5.7) (in Example 1.5) to write its Fourier series representation as

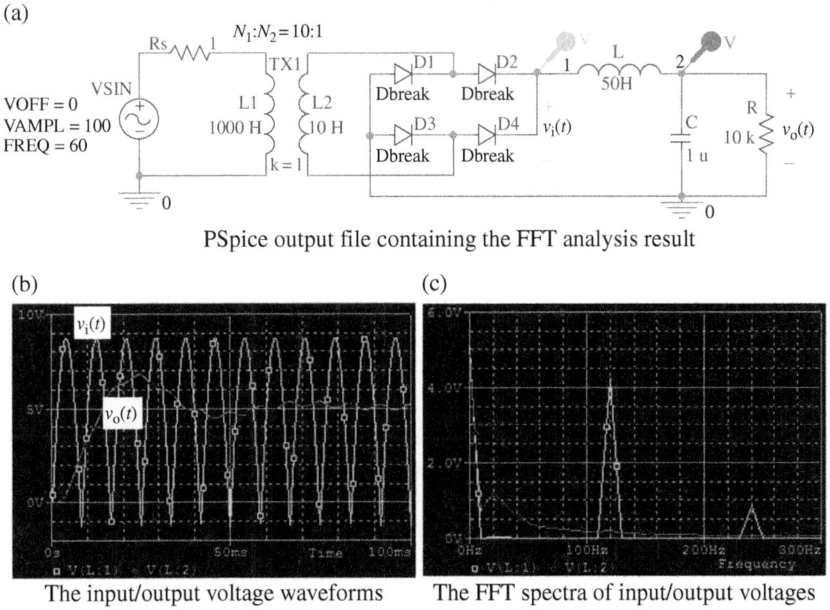

Figure 2.11 PSpice simulation of a full-wave rectifier ("rectifier_fullwave.opj").

$$v_i(t) \overset{(E1.5.7)}{=} \frac{2A}{\pi} - \sum_{m=1}^{\infty} \frac{(-1)^m 4A}{(4m^2-1)\pi} \cos 2m\omega_0 t$$

$$\underset{\omega_0 = 2\pi 60 = 377}{\overset{A=10}{=}} \frac{20}{\pi} + \frac{40}{3\pi}\cos 754t - \frac{40}{15\pi}\cos 1508t + \cdots \tag{2.2.7}$$

The transfer function and frequency response of the LCR filter are

$$G(s) = \frac{R\|(1/sC)}{sL + R\|(1/sC)} = \frac{R/sC}{sL(R+1/sC)+R/sC} = \frac{1}{s^2LC + sL/R + 1}$$

$$;G(j\omega) = \frac{1}{1-\omega^2LC + j\omega L/R} = \frac{1}{\sqrt{(1-\omega^2LC)^2 + (\omega L/R)^2}} \angle -\tan^{-1}\frac{\omega L/R}{1-\omega^2LC} \tag{2.2.8}$$

Thus, the magnitudes of the DC component and the first harmonic in the output are

$$|\mathbf{V}_o^{(0)}| = |G(j0)||\mathbf{V}_i^{(0)}| = \left.\frac{1}{\sqrt{(1-\omega^2LC)^2 + (\omega L/R)^2}}\right|_{\substack{\omega=0\\ R=10000}} \frac{20}{\pi} = \frac{20}{\pi} \tag{2.2.9}$$

$$|\mathbf{V}_o^{(2)}| = |G(j2\omega_0)||\mathbf{V}_i^{(2)}| = \left.\frac{1}{\sqrt{(1-(2\omega_0)^2LC)^2 + (2\omega_0 L/R)^2}}\right|_{\substack{2\omega_0=754\\ R=10000}} \frac{40}{3\pi} \tag{2.2.10}$$

Consequently, we can write the condition on LC (with $R = 10\ \text{k}\Omega$) for the relative magnitude of the major harmonic to the DC component to be less than r_{max} as

$$|\mathbf{V}_o^{(2)}| \le r_{max}\,|\mathbf{V}_o^{(0)}|\ \left.\frac{1}{\sqrt{(1-(2\omega_0)^2LC)^2 + (2\omega_0 L/R)^2}}\right|_{\substack{2\omega_0=754\\ R=10000}} \frac{40}{3\pi} \le \frac{20}{\pi} r_{max}$$

$$;(1-(2\omega_0)^2LC)^2 + \left(\frac{2\omega_0 L}{R}\right)^2 = (1-754^2LC)^2 + \left(\frac{754L}{10000}\right)^2 \ge \left(\frac{2}{3r_{max}}\right)^2 \tag{2.2.11}$$

It seems to be good to determine the admissible region for (C, L) satisfying the design specification on filtering off harmonics with $r_{max} = 5$ and 10%. This can be done by saving Eq. (2.2.11) in an M-file named 'elec02f12_f.m' and running

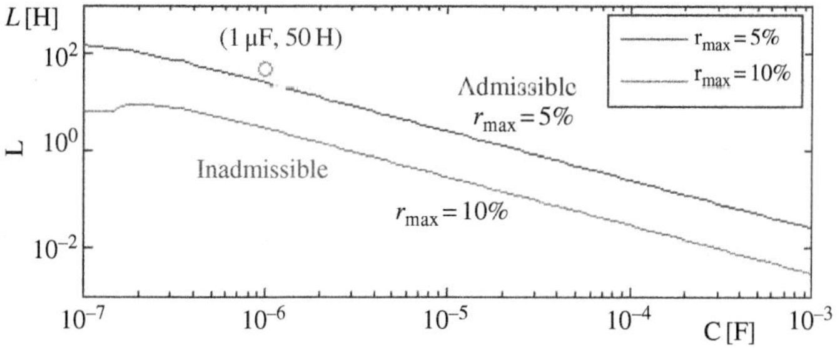

Figure 2.12 Admissible region for (*C*, *L*) to satisfy the design specification on the relative magnitude of the major harmonic to the DC component.

the following MATLAB script "elec02f12.m" that uses the nonlinear equation solver '`fsolve()`' to solve Eq. (2.2.11) for L at different values of C in some range and plot the C-L curves for r_{max} = 5 and 10% as shown in Figure 2.12. Note that the region below the curves in Figure 2.12 is not admissible in the sense that no positive values of (C, L) in the region can satisfy the above inequality constraint (2.2.11).

```
%elec02f12.m
clear, clf
global R w0
R=1e4; w0=2*pi*60;
CC=logspace(-7,-3,100); % Range of C to 100 points on [10^-7,10^-3]
ftn='elec02f12_f'; gss='brgm'; % The function to solve
ratioms=[0.05 1];
for i=1:numel(ratioms)
  for m=1:length(CC)
    C= CC(m); % Given the value of C
    if m<2, L0=10; else L0=LL(m-1); end % Initial guess of L
    LL(m) = fsolve(ftn,L0,optimset('Display','off'),C,ratioms(i));
  end
  loglog(CC,LL,gss(i)), hold on
end
plot(1e-6,50,'mo'); axis([CC([1 end]) 1e-3 1e3]);
xlabel('C'); ylabel('L')
```

```
function y=elec02f12_f(L,C,r_max)
global R w0
w=2*w0; y= (1-w^2*L*C)^2 + (w*L/R)^2 - (2/3/r_max)^2; % Eq.(2.2.11)
```

2.2.7 Precision Rectifiers

Figure 2.13(a) shows a basic precision half-wave rectifier circuit, which consists of an OP amp (having a diode in its negative-feedback path) and a resistor where the OP amp output voltage v_{o0} is basically determined as $A(v_+ - v_-) = A(v_i - v_-)$ (A: the open-loop gain) unless it exceeds the saturation voltage $\pm V_{om}$ (Eq. (5.1.1) in Section 5.1). How will it work? Let us assume that the diode is off so that no current flows through R and thus $v_- = 0$.

Case 1 When $v_i = v_+ > 0$,

v_{o0} will go positive to make $v_D = v_{o0} - v_- > 0$ so that the forward-biased diode will conduct. This will make a closed feedback path between the output and the negative input terminal of the OP amp so that by the virtual short principle (Remark 5.1),

$$v_o \approx v_i \quad \text{if} \quad v_i > 0 \tag{2.2.12a}$$

Case 2 When $v_i = v_+ < 0$,

v_{o0} will go negative to make $v_D = v_{o0} - v_- < 0$ so that the reverse-biased diode will be cut off. Thus, no current will flow through R, causing

$$v_o = 0 \quad \text{if} \quad v_i < 0 \tag{2.2.12b}$$

where the negative feedback is broken up so that the output voltage v_{o0} of the OP amp operating in the open-loop mode will go to the negative saturation voltage $-V_{om}$.

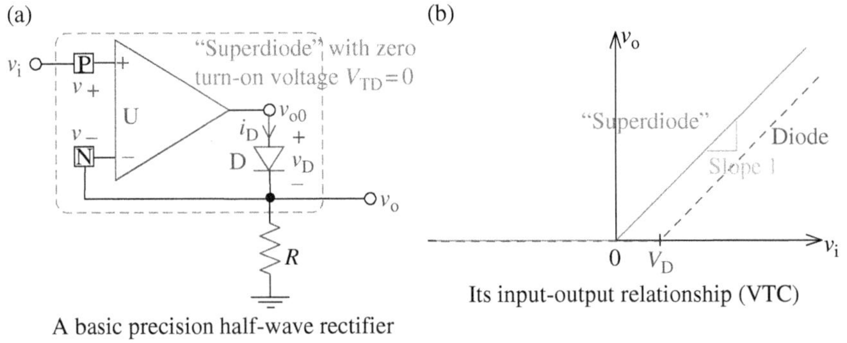

Figure 2.13 The basic precision half-wave rectifier, called a "superdiode."

Figure 2.13(b) shows the input-output relationship (solid line), called VTC, of the half-wave rectifier circuit based on Eq. (2.2.12a,b). Note that it does not have the threshold voltage V_D unlike that (dotted line) of the simple half-wave rectifier circuit in Figure 2.8(a) and that is why such a combination of a diode with an OP amp may well be called a "superdiode" with $V_D = 0$. Isn't it marvelous? Now, let us see some improved versions of this basic precision rectifier.

2.2.7.1 Improved Precision Half-wave Rectifier

Figure 2.14(a) shows an improved precision half-wave rectifier circuit. Let us see how it works.

Case 1 When $v_i > 0$,

v_{o0} (desiring to be $A(v_+ - v_-) \sim -Av_i$) will go negative to make $v_{D_1} = v_- - v_{o0} > 0$ so that the forward-biased diode D_1 will conduct. This will make a closed feedback path via D_1 between the output and the negative input terminal of the OP amp so that by the virtual short principle, $v_- \approx v_+ = 0$ V (virtual ground) and thus $v_{o0} = v_- - v_{D_1} = v_- - V_D = -V_D$ (clamped at one diode drop below zero), keeping D_2 off. Then no current flows through R_2 so that

$$v_o = v_- - R_2 i_{R2} \approx 0 \quad \text{if} \quad v_i > 0 \tag{2.2.13a}$$

Case 2 When $v_i < 0$,

v_{o0} will be positive to forward bias D_2 (through R_2-R_1-v_i) and reverse bias D_1 (with $v_{D_1} = v_- - v_{o0} < 0$) so that D_1/D_2 will be off/on, respectively. Then a negative feedback path is established through D_2-R_2 so that by the virtual short principle, $v_- \approx v_+ = 0$ V and thus

$$v_o = v_- - R_2 i_{R2} = 0 - R_2 i_{R1} = -\frac{R_2}{R_1} v_i \overset{R_1 = R_2}{=} -v_i \quad \text{if} \quad v_i < 0 \tag{2.2.13b}$$

This causes $v_{D_1} = v_- - v_{o0} = 0 - (v_o + V_D) < 0$ so that the reverse-biased D_1 can be kept off. Note that whether $v_i > 0$ or $v_i < 0$, a negative feedback path is maintained so that time required to bring the OP amp out of saturation (when the sign of v_i changes from – to +) can be saved.

Figure 2.14(b) shows the input-output relationship, called VTC, of the half-wave rectifier circuit based on Eqs. (2.2.13a) and (2.2.13b). The PSpice simulation results (obtained from the Transient analysis with $v_i(t) = V_m \sin(2\pi t)$ ($V_m = 5$ V)) of the two precision half-wave rectifiers in Figures 2.13(a) and 2.14(a) for the

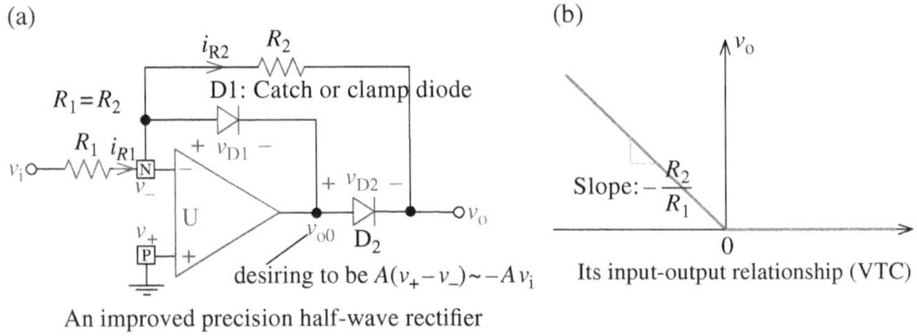

Figure 2.14 An improved (fast) precision half-wave rectifier.

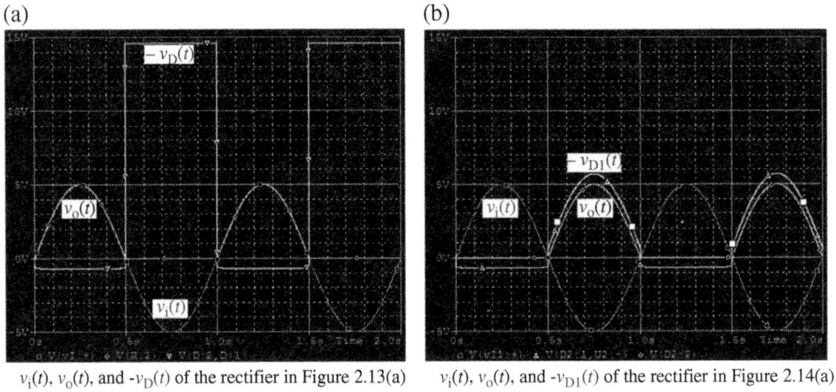

Figure 2.15 Simulation results of two precision half-wave rectifiers ("elec02f15a.opj," "elec02f15b.opj").

output voltage $v_o(t)$ and a reversed diode voltage $-v_D(t)$ are shown in Figure 2.15 (a) and (b). From the graphs, we see their PIVs for a diode as

$$\text{PIV} = \max\{-v_D(t)\} = V_{om} = 14.6\,\text{V} \quad \text{for the precision half-wave rectifier in Figure 2.13a} \tag{2.2.14a}$$

$$\text{PIV} = \max\{-v_{D1}(t)\} = V_m + V_D \quad \text{for the precision half-wave rectifier in Figure 2.14a} \tag{2.2.14b}$$

where V_m and V_{om} are the maxima of the input voltage $v_i(t)$ and OP amp output voltage $v_{o0}(t)$, respectively.

2.2.7.2 Precision Full-wave Rectifier

Figure 2.16.1(a) shows a precision full-wave rectifier circuit, which consists of a precision half wave rectifier, called a "superdiode" (Figure 2.13(a)), in the upper part and another precision half-wave rectifier with an inverting OP amp circuit (Figure 2.14(a)) in the lower part. When $v_i = v_{+1} > 0$, the upper part lets the input v_i pass and when $v_i = v_{-2} < 0$, the lower part inverts the input v_i with gain $-R_2/R_1$ (see Eq. (6.2.8)). Therefore, as shown in Figure 2.16.1(b), the VTC of this rectifier is

$$v_o = \begin{cases} v_i & \text{when } v_i \geq 0 \\ -(R_2/R_1)v_i & \text{when } v_i < 0 \end{cases} \tag{2.2.15}$$

Figure 2.16.2(a) shows another precision full-wave rectifier circuit, which consists of a "superdiode" (Figure 2.13(a)) (with an additional capacitor C in parallel with diode D_1) in the left part and an inverting OP amp circuit (with diode D_1 connected to its positive input terminal) in the right part. Note that the two OP amps U_1/U_2 have negative feedback through C/R_2 so that by the virtual short principle, we always have $v_{-1} \approx v_{+1} = v_i$ and $v_{-2} \approx v_{+2}$. Let us see how it works:

Case 1 When $v_i = v_{+1} > 0$,

v_{o1} (desiring to be $A(v_{+1} - v_{-1})$) will be high enough to make D_1 off (with $v_{D_1} = v_{-1} - v_{o1} < V_{TD}$) and D_2 on (with $v_{D_2} = v_{o1} - v_{+2} = V_D$) so that v_o will be

$$v_o = v_{-2} \left(\underset{\text{by negative feedback}}{\overset{\text{virtual short}}{\approx}} v_{+2} \right) = v_{-1} \underset{\text{by negative feedback}}{\overset{\text{virtual short}}{\approx}} v_{+1} = v_i \text{ if } v_i > 0 \tag{2.2.16a}$$

where no current flows through R_1-R_2 since C as well as D_1 will not conduct once it has instantly been charged to $v_C = v_{-1} - v_{o1} = v_{+2} - v_{o1} = -v_{D_2} = -V_D$.

(a)

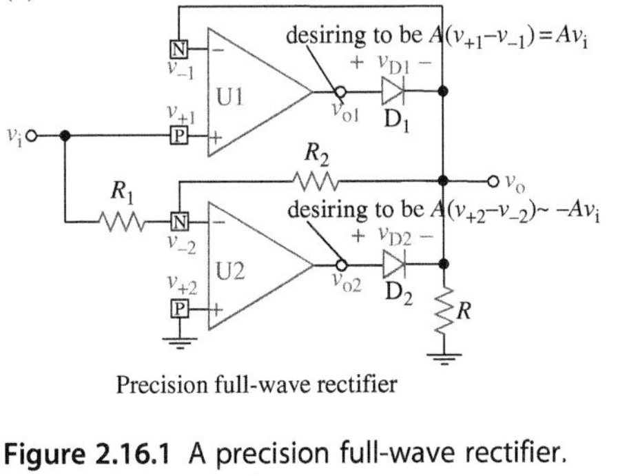

Precision full-wave rectifier

(b)

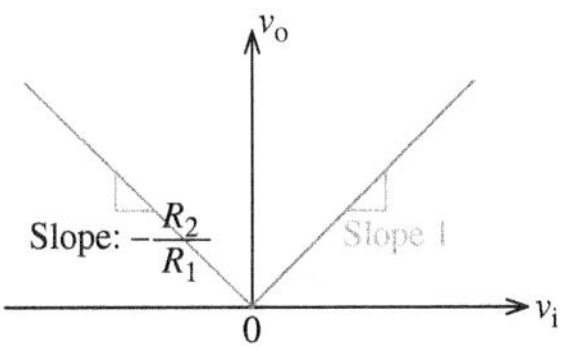

Its input-output relationship (VTC)

Figure 2.16.1 A precision full-wave rectifier.

Case 2 When $v_i = v_{+1} \approx v_{-1} < 0$,

v_{o1} will be low enough to make D_1 on (with $v_{D_1} = v_{-1} - v_{o1} = V_D$) and $\underline{D_2 \text{ off}}$ (with $v_{D_2} = v_{o1} - v_{+2} < V_{TD}$) so that $i_{R3} = 0$ and thus $v_{-2} \approx v_{+2} = R_3 i_{R3} = 0$. Then the OP amp U_2 together with R_1 and R_2 functions as an inverting amplifier with gain $-R_2/R_1$ (see Eq. (5.2.8)) so that

$$v_o = v_{-2} - R_2 i_{R2} = 0 - R_2 i_{R1} = -R_2 \frac{v_{-1} - v_{-2}}{R_1} = -\frac{R_2}{R_1} v_i \overset{R_1 = R_2}{=} -v_i \text{ if } v_i < 0 \qquad (2.2.16b)$$

where C as well as D_2 will not conduct once it has instantly been charged to $v_C = v_{D_1} = V_D$.

Therefore, the VTC of this rectifier can also be described by Eq. (2.2.15) or Figure 2.16.2(b), which is identical to Figure 2.16.1(b).

How do the two precision full-wave rectifiers compare with? First, the input impedance (seen from v_i) of the latter is much larger than that (R_1) of the former. Second, the PIVs of D_1 and D_2 in the latter (where the OP amps never enter the saturation region) are $V_m + V_D$ in common, while those in the former (where the OP amps enter the negative saturation region when the diode in their negative feedback paths is off) are as high as $\max\{-v_{Di}(t)\} = V_m + V_{om}$ in common where V_m and V_{om} are the maxima of the input voltage $v_i(t)$ and OP amp output voltage $v_{oo}(t)$, respectively. From the PSpice simulation results shown in Figure 2.17, we see that the PIVs of the diodes in the two rectifiers are

$$\text{PIV} = \max\{-v_{D1}(t)\} = V_m + V_{om} = 19.6\,\text{V} \qquad (2.2.17a)$$
for precision full-wave rectifier in Figure 2.16.1a

$$\text{PIV} = \max\{-v_{D2}(t)\} = V_m + V_D = 5.7\,\text{V} \qquad (2.2.17b)$$
for precision full-wave rectifier in Figure 2.16.2a

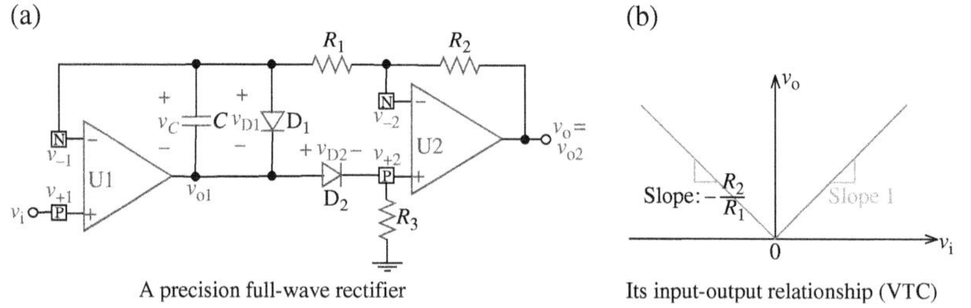

Figure 2.16.2 A precision full-wave rectifier with high input impedance [E-2, J-2, W-8].

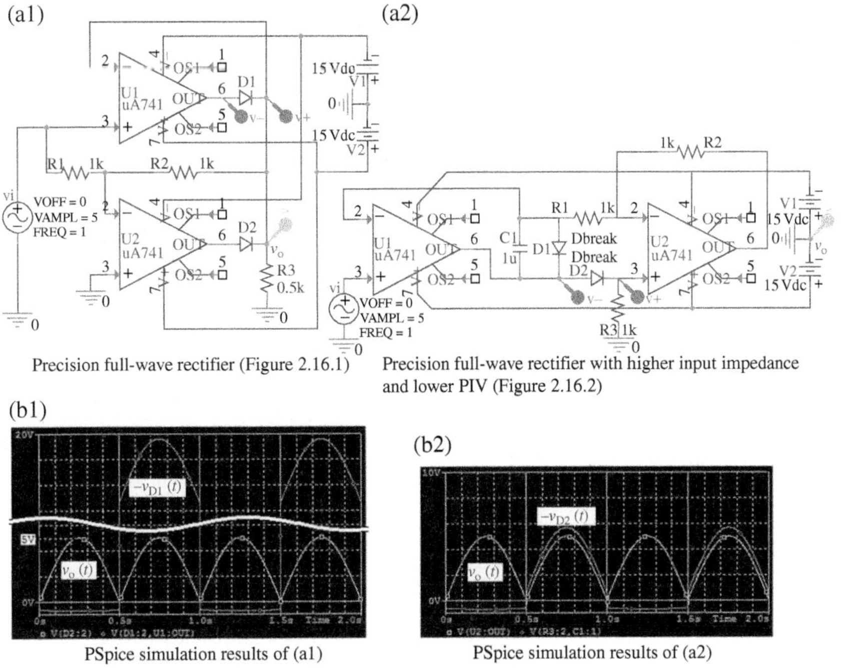

Figure 2.17 PSpice simulation of the two precision full-wave rectifiers ("elec02f17_1.opj," "elec02f17_2.opj").

2.2.8 Small-Signal (AC) Analysis of Diode Circuits

Consider the circuit with the PSpice schematic of Figure 2.18.1(a) where the PSpice Model for the diode D1N4148 (opened by selecting the diode and clicking on Edit>PSpice Model from the top menu bar) is shown in Figure 2.18.1(b). For PSpice simulation with the schematic, we do the following:

- Fill in the Simulation Settings dialog box (for Transient or Bias_Point analysis) as shown in Figure 2.18.1(c) and click OK to close the dialog box.
- Place a Current Marker to measure the diode current $i_D(t)$ and click Run to get the waveform of $i_D(t)$ as shown in Figure 2.18.1(d) or the bias point analysis on the schematic or in the output file.

To get the *v*-*i* characteristic of the diode and draw the load lines in the PSpice A/D Window, we do the following:

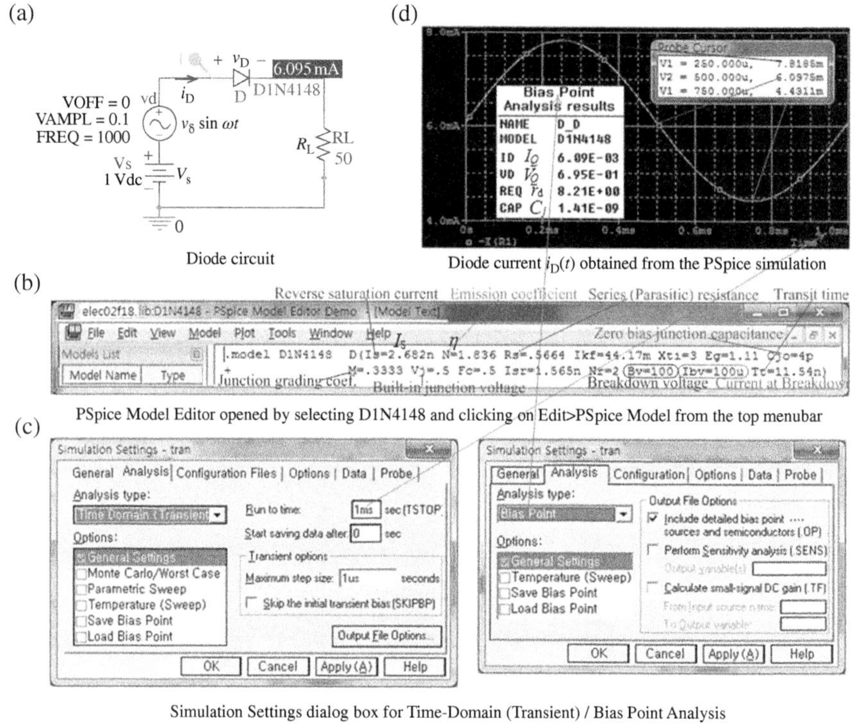

Figure 2.18.1 Simulation of a diode circuit driven by a DC source and an AC source ("e lec02f18.opj").

- Construct the PSpice schematic as Figure 2.18.2(a)
- Fill in the Simulation Settings dialog box for DC Sweep analysis (Figure 2.18.2 (b)) and click Run.
- Click Trace>Add Trace from the top menu bar to open the Add Traces dialog box and fill in the box as Figure 2.18.2(e), which yields the load lines (Figure 2.18.2(c)).

(Q) Can the waveform of $i_D(t)$ (Figure 2.18.1(d)) be predicted from the Q-points in Figure 2.18.2(c)?

To determine the diode constants of the diode D1N1418, we can use the curve fitting toolbox 'cftool' (in MATLAB) to fit the data points (v_D, i_D) of the v-i characteristic curve (Figure 2.18.2(c)) to the Shockley diode equation. Noting that curve fitting works better for linear functions than for nonlinear functions, let us approximate the Shockley diode equation and take the logarithms of both sides to linearize as

(a)

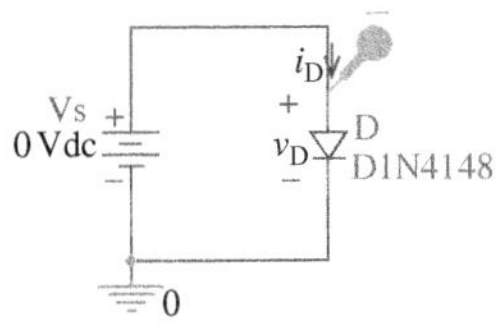

PSpice schematic to get the $v_D - i_D$ characteristic of diode D1N4148

(b)

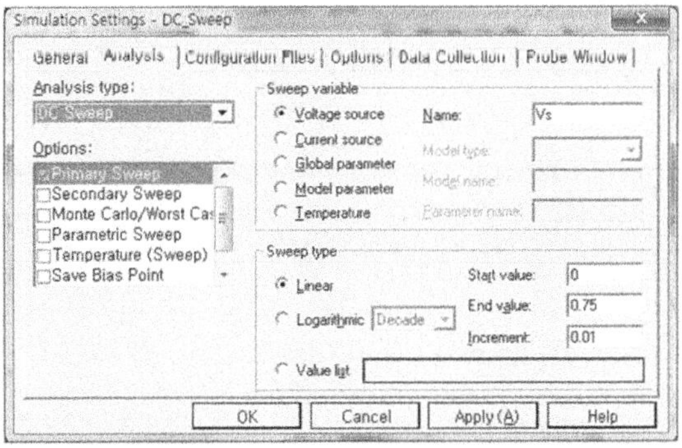

Simulation Settings dialog box for DC Sweep Analysis with v_D varying

(c)

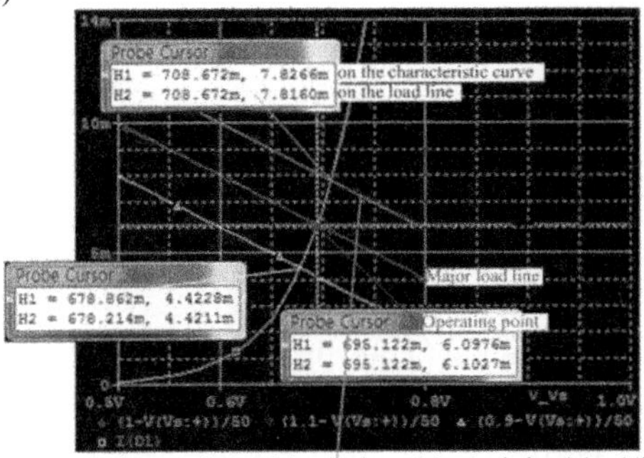

$v_D - i_D$ characteristic of the diode and (major/minor) load lines

(d)

V_Vs(v_D)	I(D1) (i_D)
0	1.19902057879578e-014
0.2	2.40015790026797e-007
0.3	1.86364150960873e-006
0.4	1.43180879474607e-005
0.5	0.0001102225816466
0.6	0.0009080914716930
0.65	0.0025275134909428
0.69	0.0055208189362651
0.70	0.0066569022663167
0.71	0.0079934574760804
0.75	0.0158058113219442

Data file obtained by clicking on I(D1) and copying/pasting onto the notepad

(e)

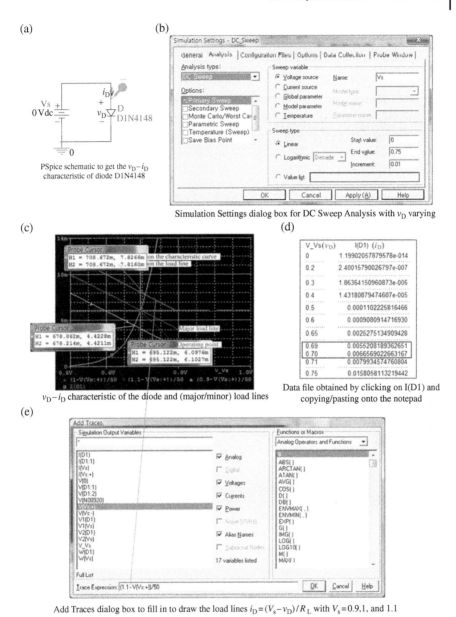

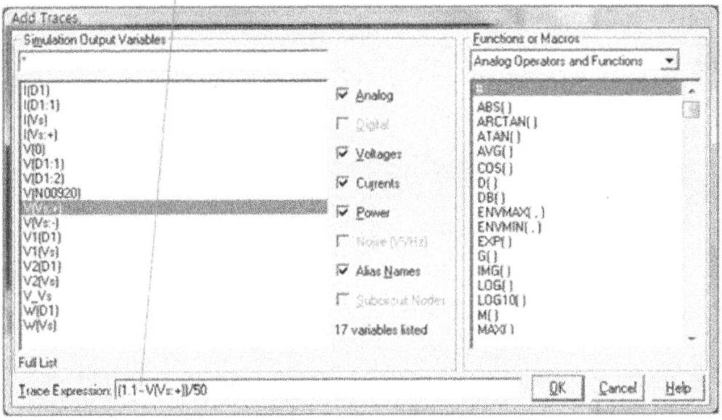

Add Traces dialog box to fill in to draw the load lines $i_D = (V_s - v_D)/R_L$ with $V_s = 0.9, 1$, and 1.1

Figure 2.18.2 PSpice simulation to get the *v*-*i* characteristic and operating point(s) for ("elec02f18b.opj").

$$i_D \overset{(2.1.1)}{=} I_S\left(e^{v_D/\eta V_T}-1\right) \xrightarrow{I_S \text{ negligibly small}} i_D \approx I_S e^{v_D/\eta V_T}$$
$$\underset{\text{of both sides}}{\overset{\text{Taking logarithms}}{\longrightarrow}} \ln i_D = \frac{1}{\eta V_T} v_D + \ln I_S \tag{2.2.18}$$

To determine the parameters $1/\eta V_T$ and $\ln I_S$ of Eq. (2.2.18) which fit the PSpice simulation data, do the following:

- Select I(D1) below the current waveform graph (Figure 2.18.2(c)), press 'CTRL+c' (copy), and then press 'CTRL+v' (paste) into a notepad to create a data file named 'vD_iD.dat' (Figure 2.18.2(d)) where the first line containing the variable descriptions should be deleted.
- Run the following MATLAB statements:

```
>>load vD_iD.dat, vD=vD_iD(:,1); iD=vD_iD(:,2);
  ln_iD=log(iD); % Take natural logarithm to linearize fitting ftn
  cftool % To start the curve fitting toolbox
```

  This will open the CFTOOL window as shown in Figure 2.18.3(a).

- Click Data button to open the Data dialog box (Figure 2.18.3(b)).
- In the Data dialog box, put 'vD' and 'ln_iD' into the fields of X Data and Y Data, respectively, and click Create data set button to create a new data set named 'ln_iD vs. vD'.
- Click Fitting button to open the Fitting dialog box (Figure 2.18.3(c)).
- In the Fitting dialog box, select 'ln_iD vs. vD' and 'Custom Equations' in the fields of Data set and Type of fit, respectively.
- Click New button inside Custom Equations panel to open a New Custom Equation dialog box.
- In the New Custom Equation dialog box, click the tab General Equations and construct the model equation corresponding Eq. (2.2.18) as

```
y = a*x +ln_Is
```

  where a and ln_Is stand for $1/\eta V_T$ and $\ln I_S$, respectively. Then click OK button to close the New Custom Equation dialog box.

- In the Fitting dialog box, click Apply button to make the following results appear in the box below Result:

(a)

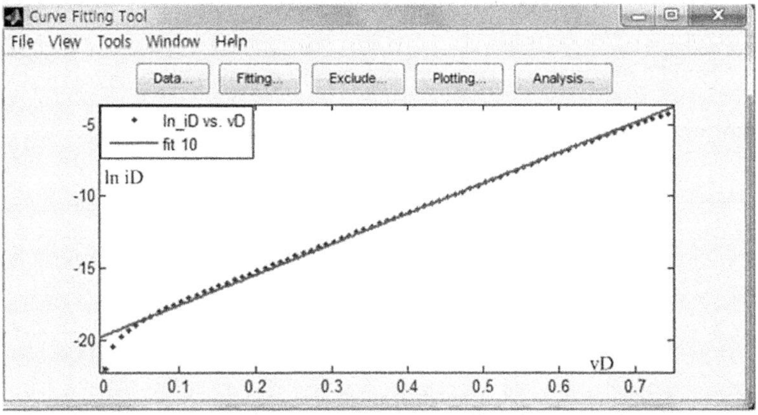

Curve Fitting Tool window opened by typing 'cftool' at the MATLAB prompt

(b)

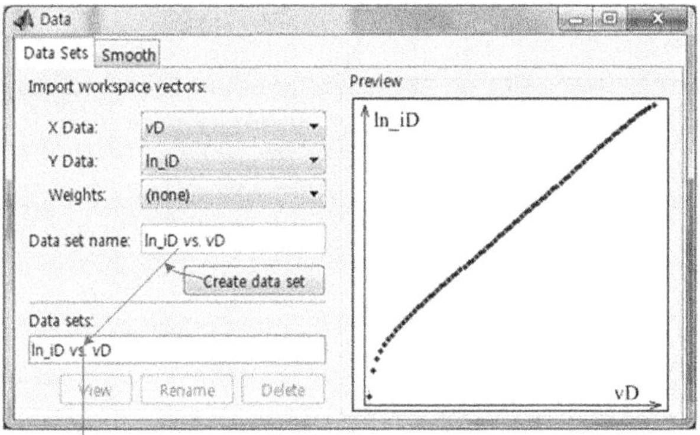

Data dialog box opened by clicking Data button in the CFTOOL window

(c)

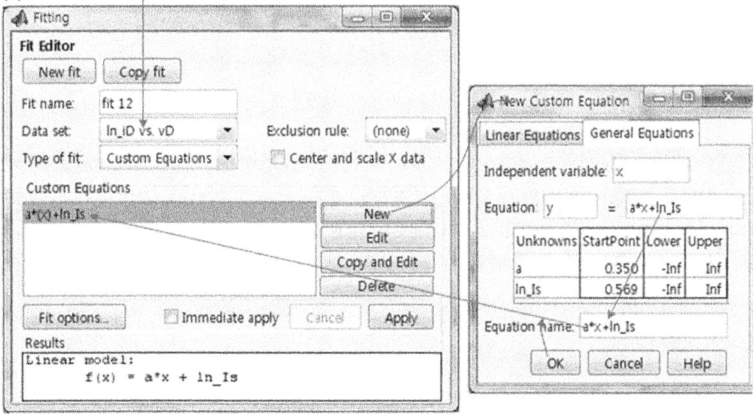

Fitting dialog box opened by clicking Fitting button in the CFTOOL window

Figure 2.18.3 Using cftool for curve fitting to determine the diode constants.

```
Coefficients (with 95% confidence bounds):
    a = 21.24 (20.87, 21.6)
    ln_Is = -19.72 (-19.88, -19.56)
```

This implies that the parameters ηV_T and I_S have been determined as

$$\eta V_T = 1/a = 1/21.24 = 0.0471 \text{ and } I_S = e^{\ln I_s} = e^{-19.72} = 2.7272 \times 10^{-9} \quad (2.2.19)$$

where the true values of ηV_T and I_S can be found from the PSpice Model of the diode (Figure 2.18.1(b)) as

$$\eta V_T \underset{\text{Fig.2.18.1(b)}}{\overset{(2.1.2)}{=}} 1.836 \times \frac{27+273}{11605} = 0.0475 \text{ and } I_S = \ 2.682 \times 10^{-9} \quad (2.2.20)$$

(Q) Are the curve fitting results fine? To be honest with our readers, the accuracy of curve fitting depends on not only the number of measured data points but also the data range.

With the diode constants found in the corresponding PSpice Model Editor (Figure 2.18.1(b)) or obtained from the curve fitting (Figure 2.18.1(c)), let us make a theoretical analysis of the diode circuit (Figure 2.18.1(a)) by using the nonlinear equation solver '`fsolve()`' of MATLAB to solve the (nonlinear) KCL equation

$$i_D(t) = \frac{v_{R_L}(t)}{R_L} = \frac{v_s(t) - v_D(t)}{R_L}$$

$$;I_S\left(e^{v_D(t)/\eta V_T} - 1\right) = \frac{v_s(t) - v_D(t)}{R_L} \text{ with } v_s(t) = V_s + v_\delta \sin \omega t \quad (2.2.21)$$

for $v_D(t)$ or to solve the (nonlinear) KVL equation

$$v_s(t) - v_D(t) = R_L i_D(t)$$

$$;v_s(t) - \eta V_T \ln\left(\frac{i_D(t)}{I_S} + 1\right) = R_L i_D(t) \text{ with } v_s(t) = V_s + v_\delta \sin \omega t \quad (2.2.22)$$

for $i_D(t)$. To do this job, we compose the MATLAB script "elec02f1804.m" as below and run it to get Figure 2.18.4, which shows the diode current waveform (plotted as blue line) together with those obtained from the PSpice simulation (Figure 2.18.1(d)) and the theoretical analysis.

(Q) How would you copy the PSpice simulation data (Figure 2.18.1(d)) into a MATLAB graph?

```
%elec02f1804.m
Vs=1; vd=0.1; % Amplitudes of DC/AC voltage sources
RL=50; % Load resistance
f=1000; w=2*pi*f; % Frequency of the AC source
Is=2.682e-9; nVT=0.0475; % Diode constants Eq.(2.2.20)
dt=1e-6; t=[0:1000]*dt; % Time range for solution
vst = Vs + vd*sin(w*t); % Source voltage waveform
% To solve Eq.(2.2.21) for vD and use Eq.(2.1.1) to find iD(t) from vD(t)
eq_2221=@(vD,Is,nVT,RL,vs)RL*Is*(exp(vD/nVT)-1)-vs+vD; % Eq.(2.2.21)
options=optimset('Display','off');
for n=1:length(t)
    vD0=0.7; vD(n)=fsolve(eq_2221,vD0,options,Is,nVT,RL,vst(n));
    iD(n)=Is*(exp(vD(n)/nVT)-1);
end
% To solve Eq.(2.2.22) for iD(t)
eq_2222=@(iD,Is,nVT,RL,vs)vs-nVT*log(iD/Is+1)-RL*iD; % Eq.(2.2.22)
for n=1:length(t)
    iD0=0.006; iD1(n)=fsolve(eq_2222,iD0,options,Is,nVT,RL,vst(n));
end
load t_iD.dat % iD(t) data from PSpice
t_PSpice=t_iD(:,1); iD_PSpice=t_iD(:,2);
VQ=695e-3; IQ=6.1e-3; % Operating point from Fig.2.18.2(c)
% To find the dynamic distance rd
dvD=0.7-0.69; diD=0.00665690237656236-0.00552081875503063;
rd1=dvD/diD % Eq.(2.2.23a) using PSpice data in data file 'vD_iD.dat'
rd2=nVT/(Is+IQ) % Eq.(2.2.23b) using the Shockley diode equation
rd0=8.21; % from the output file obtained from Bias Point analysis
% To find iD(t) using the analytical expression Eq.(1.1.12)
iD2 = IQ+vd/(RL+rd0)*sin(w*t); % Eq.(1.1.12)
plot(t,iD,t,iD1,'m:',t_PSpice,iD_PSpice,'g',t,iD2,'r'), grid on, shg
legend('fsolve vD-iD','fsolve iD-vD','PSpice','Eq.(1.1.12) with rd0')
```

Now, to get an overview of the input-output relationship of the diode circuit, let us perform a (theoretical) small-signal analysis by using Eqs. (1.1.12) and (1.1.13). First, we get the operating point Q as $(V_Q, I_Q) = (0.695, 0.0061)$ from the intersection of the (major) load line and the v-i characteristic curve in Figure 2.18.2(c). Then, we find the dynamic resistance r_d by reading REQ = 8.21 from the Bias Point analysis result shown in the output file (Figure 2.18.1(d)) or by using the PSpice data (stored in the data file 'vD_iD.dat') or Eq. (2.1.5) as follows:

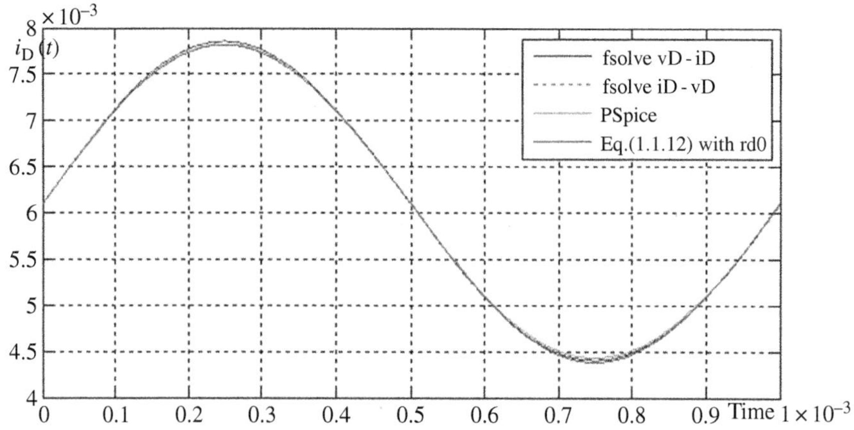

Figure 2.18.4 Diode current waveforms obtained in various ways.

$$r_d \overset{(2.1.5)}{=} \frac{\Delta v_D}{\Delta i_D} = \frac{0.7 - 0.69}{0.006657 - 0.005521} = 8.80\ \Omega \tag{2.2.23a}$$

$$r_d \overset{(2.1.5)}{=} \left.\frac{\eta V_T}{i_D}\right|_{i_D = I_Q} \underset{\text{Figure 2.18.1b}}{\overset{(2.2.19)}{=}} \frac{1.836 \times (273 + 27)/11605}{0.00609} = 7.79\ \Omega \tag{2.2.23b}$$

Which one of the three values [8.21 (from PSpice Bias Point analysis), 8.80 (from PSpice simulation data), 7.79 (from the theoretical formula)] should we use for Eqs. (1.1.12) and (1.1.13)? The authors have no idea. How about 8.21 close to their average? Thus, we use Eqs. (1.1.12) and (11.13) with $r_d = 8.21$ to write the theoretical equation for the diode currents/voltages as

$$i_D(t) \overset{(1.1.12)}{=} I_Q + i_\delta \sin \omega t = I_Q + \frac{v_\delta}{R_L + r_d} \sin \omega t = 0.00609 + \frac{0.1}{50 + 8.21} \sin \omega t \tag{2.2.24}$$

$$v_D(t) \overset{(1.1.13)}{=} V_Q + r_d i_\delta \sin \omega t = 0.695 + \frac{8.21 \times 0.1}{50 + 8.21} \sin \omega t \tag{2.2.25}$$

This diode current waveform with $r_d = 8.21$ is depicted together with the ones obtained in other ways in Figure 2.18.4. The diode current waveforms are expected to become closer to each other if the magnitude v_δ of AC voltage is smaller so that the approximation of the v-i characteristic curve by its tangent line at Q-point can become more accurate.

2.3 Zender Diodes

Figure 2.19(a) and (b) shows the symbol and the v-i characteristic for a special kind of diode, called a Zener diode where $v_Z = -v_D$ and $i_Z = -i_D$. A *Zener diode* behaves as other normal diodes in the forward-bias mode but, in the reverse-bias mode, withstands the reverse diode current (up to $I_{Z,max}$) while keeping $v_Z = -v_D$ close to the *Zener* (*breakdown*) *voltage* V_Z (even with some variation in its current i_D) as long as $i_Z = -i_D$ remains between $I_{Z,min}$ and $I_{Z,max}$. The maximum reverse current $I_{Z,max}$ that the diode can endure is $P_{Z,max}/V_Z$ where $P_{Z,max}$ is the maximum power dissipation. The minimum reverse current $I_{Z,min}$ needed to keep the diode in the reverse breakdown mode is slightly below the *Zener knee current* I_{ZK}, at which the diode exhibits the reverse breakdown. The v-i characteristic curve of a Zener diode in terms of its static behavior can be approximated by a red/brown PWL line for the forward-/reverse-bias mode as depicted in Figure 2.19(b) where the diode resistance r_f (in the forward-bias mode) and the *Zener resistance* r_z (in the reverse-bias mode) are defined as

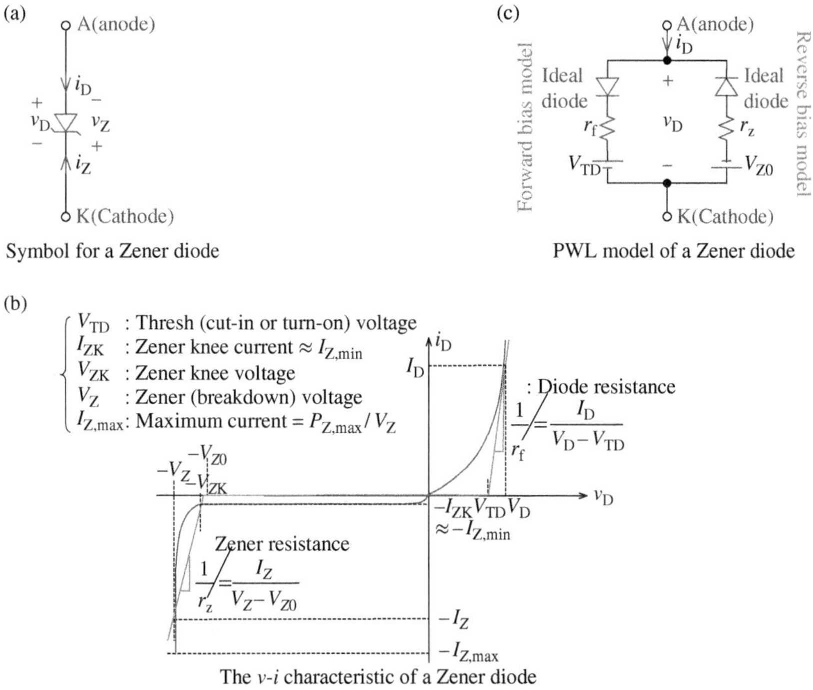

Figure 2.19 Symbol, v-i characteristic, and PWL model of a Zener diode.

$$r_f = \frac{V_D - V_{TD}}{I_D} \text{ at operating point } (V_D, I_D) \tag{2.3.1a}$$

$$r_z = \frac{V_Z - V_{Z0}}{I_Z} \text{ at operating point } (V_Z, I_Z) \tag{2.3.1b}$$

Here, $-V_{Z0}$ is the diode voltage at the intersection of the straight line having slope $1/r_z$ with the voltage axis and it is almost equal to the Zener knee voltage $-V_{ZK}$. According to the approximation, the v-i characteristic of a Zener diode in the forward-/reverse-bias mode is represented by the equivalent model depicted in Figure 2.19(c).

Figure 2.20(a) shows a Zener (shunt) regulator, which is supposed to maintain an almost constant output voltage for all the variation of the DC voltage source v_s and the load resistor R_L. In order for the regulator to function properly in the breakdown region, the Zener diode current i_Z should be bounded as

$$I_{Z,\min} \le \overbrace{\frac{v_{s,\min} - V_Z}{R_s} - \frac{V_Z}{R_{L,\min}}}^{\text{Maximum load condition}} \le i_Z \le \overbrace{\frac{v_{s,\max} - V_Z}{R_s}}^{\text{No load condition}} \le I_{Z,\max} \tag{2.3.2}$$

Thus, the source resistance R_s should satisfy the following inequality:

$$\frac{v_{s,\max} - V_Z}{I_{Z,\max}} \le R_s \le \frac{v_{s,\min} - V_Z}{(V_Z/R_{L,\min}) + I_{Z,\min}} \tag{2.3.3}$$

If $I_{Z,\max} = P_{Z,\max}/V_Z$ has not yet been specified, R_s can be set to some value slightly less than the upper bound of Inequality (2.3.3):

$$R_s \overset{\text{slightly less than}}{\approx (\le)} \frac{v_{s,\min} - V_Z}{(V_Z/R_{L,\min}) + I_{Z,\min}} \tag{2.3.4}$$

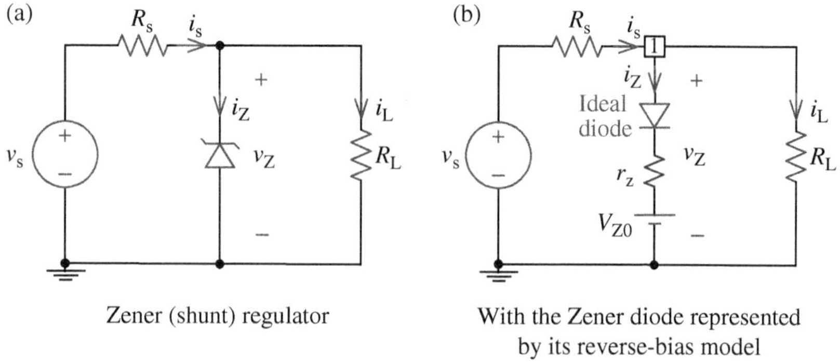

Figure 2.20 Zener (shunt) regulator.

If R_s is fixed, the minimum load resistance can be determined by substituting $V_Z = V_{ZK}$ and $i_Z = I_{ZK}$ into Inequality (2.3.3) as

$$R_{L,\min} = \frac{V_{ZK}}{(v_{s,\min} - V_{ZK}/R_s) - I_{ZK}} \tag{2.3.5}$$

Then the power rating of the Zener diode should be determined so that the following inequality can be satisfied:

$$P_{Z,\max} = V_Z I_{Z,\max} \ge V_Z \frac{v_{s,\max} - V_Z}{R_s} \tag{2.3.6}$$

Figure 2.20(b) shows the Zener regulator with the Zener diode represented by the reverse bias model (Figure 2.19(c)), for which we can apply KCL at node 1 to write the node equation in v_Z as

$$\frac{v_Z - v_S}{R_s} + \frac{v_Z - V_{Z0}}{r_z} + \frac{v_Z}{R_L} = 0 \tag{2.3.7}$$

This can be solved for v_Z to yield

$$v_Z = \frac{v_s/R_s + V_{Z0}/r_z}{1/R_s + 1/r_z + 1/R_L} \tag{2.3.8}$$

According to this (approximate) expression of v_Z, its variations w.r.t. v_s, R_L, and I_L are computed as

$$\text{Line regulation}: \frac{dv_Z}{dv_s} \overset{(2.3.8)}{=} \frac{1/R_s}{1/R_s + 1/r_z + 1/R_L} \tag{2.3.9a}$$

$$\frac{dv_Z}{dR_L} \overset{(2.3.8)}{=} \frac{(v_S/R_s + V_{Z0}/r_z)/R_L^2}{(1/R_s + 1/r_z + 1/R_L)^2} \tag{2.3.9b}$$

$$\text{Load regulation}: \frac{dv_Z}{dI_L} = \frac{dR_L}{dI_L}\frac{dv_Z}{dR_L} \underset{(2.3.9b)}{\overset{I_L = V_Z/R_L}{=}} -\frac{R_L^2}{V_Z}\frac{(v_S/R_s + V_{Z0}/r_z)/R_L^2}{(1/R_s + 1/r_z + 1/R_L)^2}$$

$$= -\frac{v_S/R_s + V_{Z0}/r_z}{V_Z(1/R_s + 1/r_z + 1/R_L)^2} \tag{2.3.9c}$$

This implies that the sensitivities of v_Z w.r.t. v_s/R_L become better (smaller) as R_s/R_L increases, respectively.

Note that if $v_Z < V_{Z0}$ so that $i_Z = (v_Z - V_{Z0})/r_z < 0$, we set $i_Z = 0(I_{ZK})$ and use the voltage divider rule to determine v_Z as if there were no Zener diode branch:

$$v_Z = \frac{R_L}{R_s + R_L} v_s \tag{2.3.10}$$

The following MATLAB function 'Zener_regulator()', given the values of V_Z, I_Z, r_Z, I_{ZK}, R_L, R_s, v_s, and dv_s (the variation of v_s) for a Zener regulator (shown in Figure 2.20(a)), returns the (output) voltage v_Z (across the Zener diode or load resistor) and the current i_Z through the Zener diode, together with the sensitivities of v_Z w.r.t. v_s, R_L, and I_L.

```
function [vZ,iZ,dvZdvs,dvZdRL,dvZdIL,Rsmax,RLmin,VZ0] =
                      Zener_regulator(VZ,IZ,rz,IZK,RL,Rs,vs,dvs)
% Input:  VZ = (Nominal) voltage across Z at the operating point
%         IZ = (Nominal) current through Z at the operating point
%         rz = Zener resistance
%        IZK = Zener knee current
%         RL = Load resistance
%         Rs = Source resistance
%         vs = Source voltage
%        dvs = Absolute variation of vs
% Output: vZ = Output voltage across Z||RL
%         iZ = Current through the Zener diode Z
% Copyleft: Won Y. Yang, wyyang53@hanmail.net, CAU for academic use only
if nargin<8, dvs = 0; end
Rsmax = (vs-dvs-VZ)./(VZ./RL+IZK); % Eq.(2.3.4)
VZ0 = VZ - rz*IZ; % Almost equal to the Zener knee voltage VZK
VZK = VZ0+rz*IZK; % Zener knee voltage
RLmin = VZK./((vs-dvs-VZK)/Rs-IZK); % Eq.(2.3.5)
N_RL=length(RL); N_vs=length(vs);
if N_RL<N_vs, RL=repmat(RL,1,N_vs); end
if N_vs<N_RL, vs=repmat(vs,1,N_RL); end
numerator = vs/Rs+VZ0/rz;
denominator = 1/Rs+1/rz+1./RL;
vZ = numerator./denominator; % Eq.(2.3.8)
iZ = (vZ-VZ0)/rz;
for n=1:length(iZ)
   if iZ(n)<0 % As if there were no Zener diode branch
     iZ(n) = 0; vZ(n) = RL(n)/(Rs+RL(n))*vs(n); % Eq.(2.3.10)
   end
end
dvZdvs = 1/Rs./denominator; % Eq.(2.3.9a)
dvZdRL = numerator./RL.^2./denominator.^2; % Eq.(2.3.9b)
dvZdIL = -dvZdRL.*RL.^2/VZ; % Eq.(2.3.9c)
```

Example 2.3 A Zener (Shunt) Regulator

Consider the Zener regulator shown in Figure 2.20(a), whose purpose is to keep the output voltage across R_L close to 4.7 V where the source voltage V_s and the load resistance R_L vary between 12 ± 2 [V] and between 200 and 1000 Ω, respectively. Let the values of the device parameters of the Zener diode Z (1N750) be

$$V_Z = 4.7\,\text{V},\ I_Z = 20\,\text{mA},\ I_{ZK} = 1\,\text{mA},\ r_z = 1.5\ \Omega \tag{E2.3.1}$$

where the maximum and minimum currents for the Zener diode to function properly in the breakdown region are $I_{Z,max} = P_{Z,max}/V_Z = 352.5[\text{mW}]/4.7[\text{V}] = 75$ [mA] and $I_{Z,min} = 5[\text{mA}]$, respectively.

a) Use Eq. (2.3.3) to fix the value of the source resistance R_s:

$$124 = \frac{14-4.7}{0.075} = \frac{v_{s,max} - V_Z}{I_{Z,max}} \le R_s \le \frac{v_{s,min} - V_Z}{(V_Z/R_{L,min}) + I_{Z,min}} = \frac{10-4.7}{4.7/200 + 0.005} = 186 \tag{E2.3.2}$$

Let us fix R_s as 180 Ω.

b) Use Eq. (2.3.1b) together with Eq. (E2.3.1) to determine V_{Z0} for the PWL model (Figure 2.19(c)) of the Zener diode Z:

$$V_Z \overset{(2.3.1b)}{=} V_{Z0} + r_z I_Z;\ V_{Z0} = V_Z - r_z I_Z \underset{r_z = 1.5\,\Omega}{\overset{V_Z = 4.7\text{V},\ I_Z = 20\,\text{mA}}{=}} 4.7 - 1.5 \times 0.02 = 4.67\text{V} \tag{E2.3.3}$$

c) Use the equivalent circuit (Figure 2.20(b)) with the Zener diode Z replaced by its PWL model to find the output voltage $v_o = v_Z$ (across $R_L = 200\ \Omega$ in parallel with the Zener diode) to the source voltage $v_s = 14$, 12, and 8.5V.

We can use Eq. (2.3.8) to find v_o:

$$v_o = v_Z \overset{(2.3.8)}{=} \frac{v_s/R_s + V_{Z0}/r_z}{1/R_s + 1/r_z + 1/R_L} = \frac{v_s/180 + 4.67/1.5}{1/180 + 1/1.5 + 1/200} = \begin{cases} 4.712\text{V} & \text{for } v_s = 14\text{V} \\ 4.696\text{V} & \text{for } v_s = 12\text{V} \\ 4.667\text{V} & \text{for } v_s = 8.5\text{V} \end{cases} \tag{E2.3.4}$$

However, $v_Z = 4.667 < V_{Z0} = 4.67$ [V] is not possible as long as the Zener diode Z is ON. Thus, we have to recalculate v_Z for $v_s = 8.5$V on the assumption that Z is OFF so that $i_Z = 0$ A. Then, as if there were no Zener diode branch, the voltage divider rule can be used to find v_Z as

$$v_Z = \frac{R_L}{R_s + R_L} v_S = \frac{200}{180 + 200} 8.5 = 4.474\text{V} \tag{E2.3.5}$$

Here, we can use the above MATLAB function '`Zener_regulator()`' as

```
VZ=4.7; IZ=0.02; rz=1.5; IZK=1e-3;
Rs=180; RL=200; vss=[14 12 8.5]; dvs=14-12;
[vos,iZ,dvodvs,dvodRL]=...
Zener_regulator(VZ,IZ,rz,IZK,RL,Rs,vss,dvs)
```

Running this block of MATLAB statements yields

```
vos =     4.7120   4.6957   4.4737
iZ  =     0.0280   0.0171        0
dvodvs =  0.0082   0.0082   0.0082
```

(Q1) Are these values of the output voltage v_o close to the above results with the PWL model?

(Q2) Is the change of the output voltage v_o (from 4.696 to 4.712 V) due to that of the source voltage v_s by $\Delta v_s = 2$ V (from 12 to 14 V) predictable from the line regulation $dv_o/dv_s = 0.0082$?

d) Use the equivalent circuit (Figure 2.20(b)) with $v_s = 12$ V to find the output voltage $v_o = v_Z$ for $R_L = 200$, 600, and 1000 Ω.

We can use Eq. (2.3.8) to find v_o:

$$v_o = v_Z \overset{(2.3.8)}{=} \frac{v_s/R_s + V_{ZO}/r_z}{1/R_s + 1/r_z + 1/R_L} = \frac{12/180 + 4.67/1.5}{1/180 + 1/1.5 + 1/R_L} = \begin{cases} 4.696\,\text{V} & \text{for } R_L = 200\ \Omega \\ 4.719\,\text{V} & \text{for } R_L = 600\ \Omega \\ 4.724\,\text{V} & \text{for } R_L = 1\ k\Omega \end{cases} \tag{E2.3.6}$$

Here, we can use the above MATLAB function '`Zener_regulator()`' as

```
VZ=4.7; IZ=0.02; rz=1.5; IZK=1e-3;
Rs=180; RLs=[200 600 1000]; vs=12;
[vos,iZ,dvodvs,dvodRL]=Zener_regulator(VZ,IZ,rz,IZK,RLs,Rs,vs)
```

Running this block of MATLAB statements yields

```
vos =     4.6957    4.7189   4.7236
iZ  =     0.0171    0.0326   0.0357
dvodRL =  1.0e-03 * 0.1733   0.0195   0.0070
```

(Q1) Are these values of the output voltage v_o close to the above results with the PWL model?

(Q2) Is the change of the output voltage v_o (from 4.719 to 4.724 V) due to that of the load resistance R_L by $\Delta R_L = 400$ Ω (from 600 Ω to 1 kΩ) predictable from $dv_o/dR_L = 1.95 \times 10^{-5}$? If not, why is that? How about Δv_o for a 10% change of R_L like $\Delta R_L = 60$ Ω from $R_L = 600$ Ω?

e) PSpice Simulation

Figure 2.21(a) shows the PSpice schematic of the Zener regulator with $V_s = 12$ V and $R_s = 180\ \Omega$. Referring to Section II.5.4, do the DC Sweep and Parametric Sweep Analyses to get the plots of v_o versus v_s for different values {200, 400, ..., 1000} of R_L as follows:

1) Click the value of R_L (to be varied) and set it to {Rvar} (including the curly brackets) in the schematic (Figure 2.21(a)).
2) Click the Place Part button on the tool palette in the Schematic window to open the Place Part dialog box, get PARAM (contained in the library 'special.olb'), and click OK to place it somewhere in the schematic (Figure 2.21(a)).
3) Double-click the PARAMETERS: placed in the schematic to open the Property Editor spreadsheet (Figure 2.21(b)) and click the New Column button to open the Add New Column box (Figure 2.21(c)), in which you can type Rvar and 100 into the Name and Value fields, respectively, where the numerical value entered as 100 here does not matter.
4) Click the New Simulation Profile button to create a new simulation profile named, say, 'DC_sweep', click the Edit Simulation Profile button to open the Simulation Settings dialog box, and fill it out as depicted in Figure 2.21 (d), where the menu of Options is selected as Parametric Sweep, the Sweep variable is chosen to be Global parameter named 'Rvar', and the Sweep type is set to Linear with Start value 200, End value 1000, and Increment 200. If the values of parameters do not form an arithmetic progression (with constant difference), you may type the list of the parameter values (inside brackets) into the Value list field as illustrated in Figure 2.21(d). Then click OK to close the Simulation Settings dialog box.
5) Click the Voltage/Level Marker button to put the voltage probe pin at the output node, click the Run button to perform the simulation, and see the multiple curves of v_o versus v_s appearing in the PSpice A/D (Probe) Window as shown in Figure 2.21(e).

Now, to check if the (maximum) Zener diode current obtained with the largest value of RL = 1000 exceeds $I_{Z,max} = 75$ [mA], do the following:

6) Click the Current Marker button to put the current probe pin at the K(Cathode) node of the Zener diode, and see the multiple curves of i_Z versus v_s.
7) Click the rightmost one among the symbols for –I(D) to select the curve of i_Z for RL = 1000.
8) Click the Toggle Cursor button on the tool bar in the Probe window to have the cross-type cursor appear, move it to the right end of the current waveform corresponding to $v_s = 14$ V, and read the value of $i_Z = 46.664$ [mA], which turns out to be less than $I_{Z,max} = 75$ [mA].

(a)

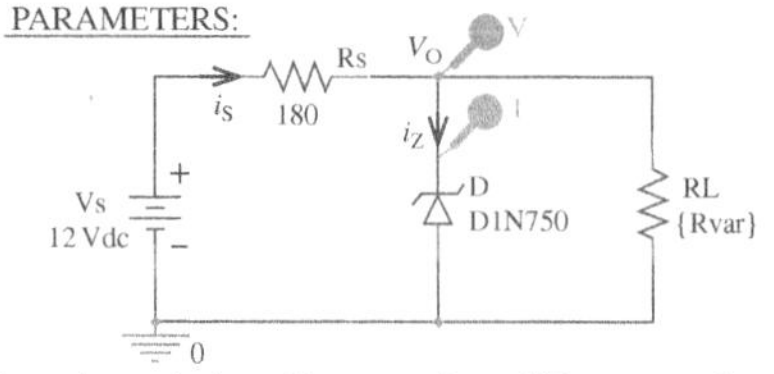

PSpice schematic for a Zener regulator ("Zener_regulator.opj")

(b)

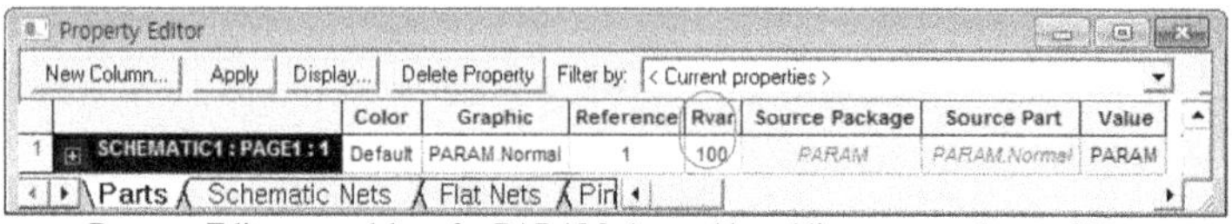

Property Editor spreadsheet for PARAM changed by adding a new column of Rvar

(c)

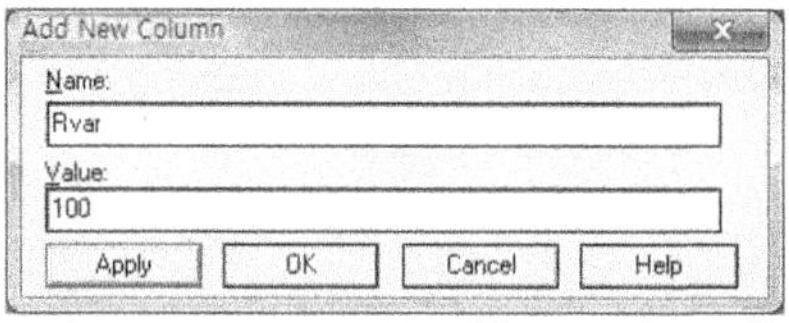

Add New Column dialog box filled to add a new column of Rvar1

(d)

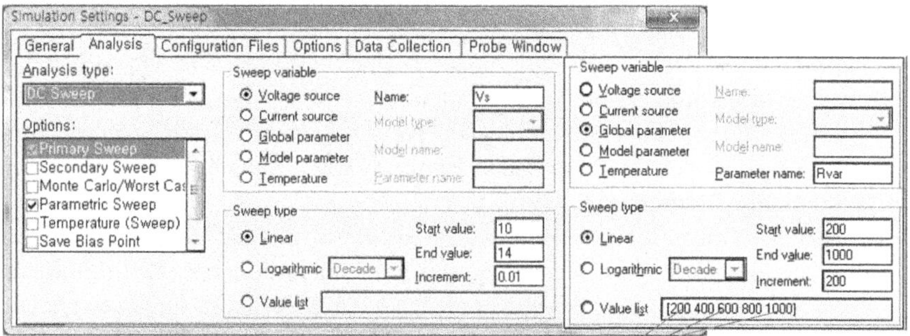

Simulation Settings dialog box for DC Sweep and Parametric Sweep Analyses

(e)

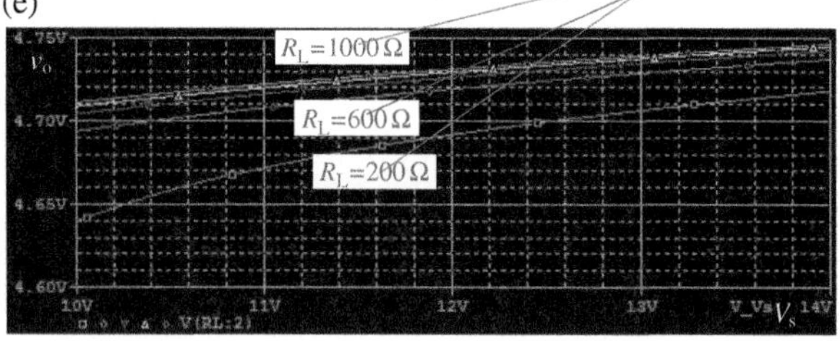

Output voltage versus source voltage for different values of RL

Figure 2.21 From PSpice schematic to simulation result for a Zener (shunt) regulator.

f) To determine the power rating of R_s, find the maximum power that can be dissipated by R_s.

$$P_{R_s,\max} = \frac{(v_{s,\max} - V_Z)^2}{R_s} = \frac{(14-4.7)^2}{180} = 481\,[\mathrm{mW}] \tag{E2.3.7}$$

Figure 2.22(a1) and (a2) show a symmetrical *limiter* using two Zener diodes and one using normal diodes, respectively, where Figure 2.22(b1) and b2 shows their input-output relationships called VTCs. Note that the upper/lower limits of the Zener limiter output are $(V_{TD} + V_Z)/(\_V_{TD} - V_Z)$ (Figure 2.22(b1)) and those of the diode limiter output are $(V_{TD} + V_1)/(-V_{TD} - V_2)$ (Figure 2.22(b2)).

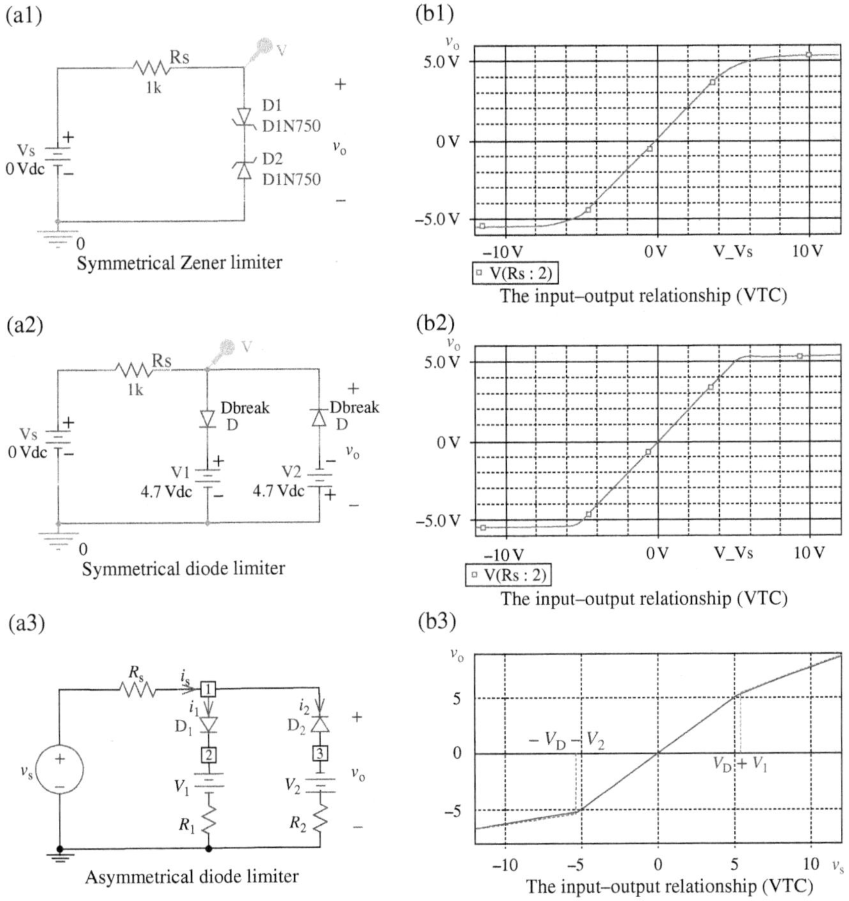

Figure 2.22 A limiter using Zener diodes and two ones using normal diodes ("Zener_limiter.opj," "Diode_limiters.m").

Figure 2.22(a3) and (b3) shows an asymmetrical *limiter* using two D_i-V_i-R_i paths (i = 1, 2) in parallel, and its input-output relationship, respectively. Note that the slopes of the input-output relationship above the point (V_D+V_1, V_D+V_1) and below the point $(-V_D-V_2, -V_D-V_2)$ are determined as the voltage divider gain of the circuit with the diodes and voltage sources removed by short-circuiting:

$$\frac{R_1}{R_s+R_1} \quad \text{and} \quad \frac{R_2}{R_s+R_2} \tag{2.3.11}$$

To analyze this limiter based on the exponential model (Eq. (2.1.1)) for the diodes, we can apply KCL at nodes 1, 2, and 3 to write a set of node equations:

$$\begin{aligned} &\frac{v_s-v_1}{R_s}-i_D(v_1-v_2)+i_D(v_3-v_1)=0 \\ &i_D(v_1-v_2)-\frac{v_2-V_1}{R_1}=0 \\ &i_D(v_3-v_1)+\frac{v_3+V_2}{R_2}=0 \end{aligned} \tag{2.3.12}$$

where $i_D(v_D)=I_S\left(e^{v_D/\eta V_T}-1\right)$ and use the MATLAB function '`fsolve()`' to solve it as listed in the following MATLAB script "diode_limiters.m."

```
%diode_limiters.m
% Copyleft: Won Y. Yang, wyyang53@hanmail.net, CAU for academic use only
clear, clf
Is=10e-15; % Saturation current
VT=(273+27)/11605; % Thermal voltage
iD = @(vD) Is*(exp(vD/VT)-1); % Eq.(2.1.1a)
options = optimset('TolFun',1e-10, 'Display','off');
vs = [-12:0.05:12]; % Range of the source voltage signal
Rs=1e4; R1=1e4; R2=2500; V1=4.7; V2=4.7;
v=zeros(length(vs),3); % Initialize the voltage values to zero
for n=1:length(vs)
  eq = @(v) [(vs(n)-v(1))/Rs-iD(v(1)-v(2))+iD(v(3)-v(1));
            iD(v(1)-v(2))-(v(2)-V1)/R1;
            iD(v(3)-v(1))+(v(3)+V2)/R2]; %Eq.(2.3.12)
  if n<2, v0 = [0 0 0]; else v0 = v(n-1,:); end
  v(n,:) = fsolve(eq,v0,options);
end
Vsm=12; VD=0.7; VD1=VD+V1; VD2=VD+V2;
plot(vs,v(:,1), [VD1 VD1 Vsm], [0 VD1 VD1+(Vsm-VD1)*R1/(Rs+R1)], 'm:', ...
  [-Vsm -VD2 -VD2], [-VD2-(Vsm-VD2)*R2/(Rs+R2) -VD2 0], 'm:') %Eq.(2.3.11)
```

Problems

2.1 Diode Circuits

(a) Consider the diode circuit of Figure P2.1.1(a1) where there are four possible states for the two diodes D_1 and D_2: ON-ON, ON-OFF, OFF-ON, and OFF-OFF.

- First, assuming that both D_1 and D_2 are ON, replace them by the Constant Voltage Drop (CVD) model (with $V_D = 0.7$ V) to draw the equivalent as shown in Figure P2.1.1(a2), find the currents I_{D_1} and I_{D_2}, and check the validity of the solution.
- For a more exact analysis using the (nonlinear) exponential model (2.1.1) with $I_s = 1 \times 10^{-14}$[A] and $V_T = (27 + 273)/11605$[V], apply Kirchhoff's current law (KCL) at nodes 1 and 2 to write two node equations as

$$\begin{aligned} &i_{D1}(0 - v_1) + i_{D2}(v_2 - v_1) - (v_1 + 10)/R_1 = 0 \quad \text{at node 1} \\ &i_{D2}(v_2 - v_1) - (10 - v_2)/R_2 = 0 \quad \text{at node 2} \end{aligned} \tag{P2.1.1}$$

and use the MATLAB function '`fsolve()`' to solve them for v_1 and v_2. To this end, complete the following MATLAB script "elec02p01a.m" and run it to find v_1, v_2, I_{D_1}, and I_{D_2}.

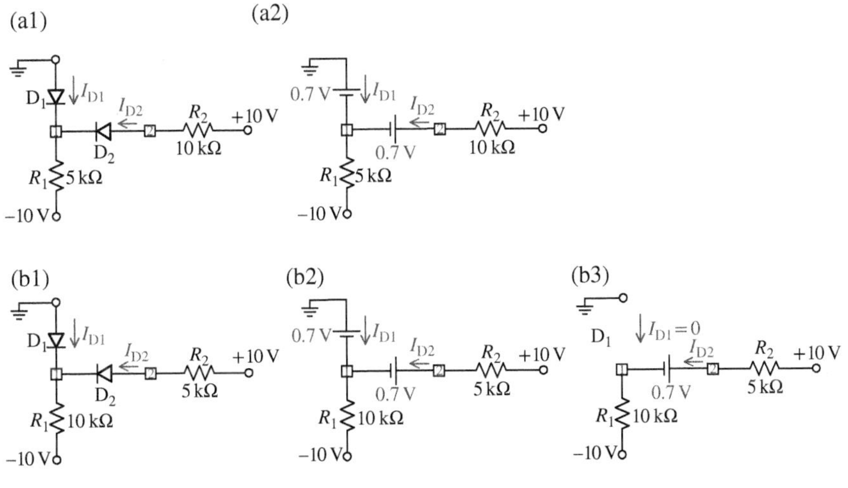

Figure P2.1.1 Diode circuits for Problem 2.1.

```
%elec02p01a.m
clear, clf
T=27; nVT=(T+273)/11605; % Temperature[Celsius] and Thermal voltage
Is=1e-14; % Leakage current of the diodes
% Diode exponential model parameters
iD = @(vD)Is*(exp(vD/VT)-1); % Eq.(2.2.1a)
R1=5e3; R2=10e3; % Circuit parameter values
eq = @(v) [iD(-v(?))+iD(v(2)-v(?))-(v(?)+10)/R1; ...
         iD(v(?)-v(1))-(10-v(?))/R2];
v0 = [0; 0]; % Initial guess of [v1; v2]
options=optimset('Display','off', 'TolX',1e-10, 'TolFun',1e-10);
v = fsolve(eq,v0,options); % Solving the set of nonlinear eqs
ID1 = iD(-v(1)); ID2 = iD(v(?)-v(?));
fprintf(' ID1=%8.3fmA, ID2=%8.3fmA, V1=%8.4fV, V2=%8.4fV\n',
ID1*1e3,ID2*1e3,v(1),v(2));
```

(b) Consider the diode circuit of Figure P2.1.1(b1).
- First, assuming that both D_1 and D_2 are ON, replace them by the CVD model (with $V_D = 0.7$ V) to draw the equivalent as shown in Figure P2.1.1(b2) and find the currents I_{D_1} and I_{D_2}. If the solution turns out to be invalid, try with another assumption.
- For a more exact analysis using the (nonlinear) exponential model (2.1.1) with $I_S = 6 \times 10^{-16}$ A, make a slight modification of the above MATLAB script "elec02p01a.m" and run it to find v_1, v_2, I_{D_1}, and I_{D_2}.

(c) Perform the PSpice simulation of the diode circuit of Figure P2.1.1(a1) with the Analysis type of Bias Point by taking the following steps:
1) Compose the PSpice schematic as shown in Figure P2.1.2(a).
2) Create a simulation profile (with the analysis type of Bias Point) by selecting 'Bias Point' as the analysis type in the Simulation Settings dialog box as shown in Figure P2.1.2(b).
3) Run the PSpice schematic and click on the 'Enable Bias Voltage Display'/'Enable Bias Current Display' button in the toolbar of the Capture CIS Window to see the bias-point analysis results on voltages/currents at/through each node/branch.

2.2 Bridge Rectifier Circuit

Consider the bridge rectifier circuit of Figure P2.2(a) consisting of four diodes. Based on the (nonlinear) exponential model (2.1.1) with $I_s = 1 \times 10^{-14}$[A] and

(a)

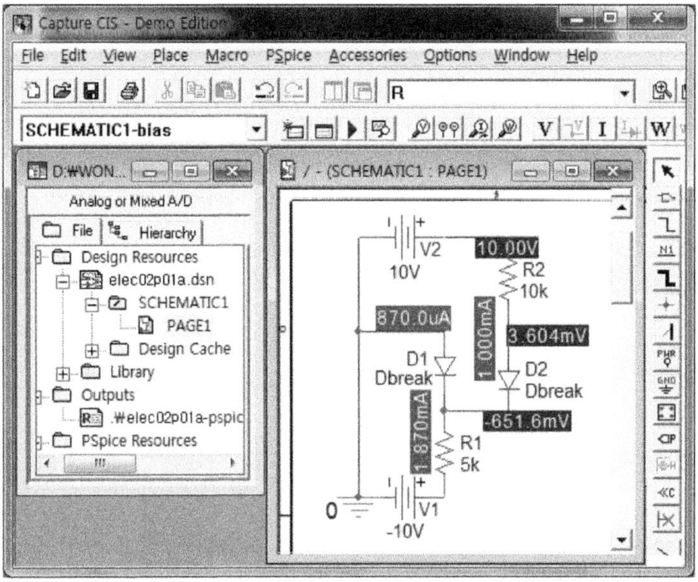

PSpice schematic for the circuit of Figure P2.1.1(a)

(b)

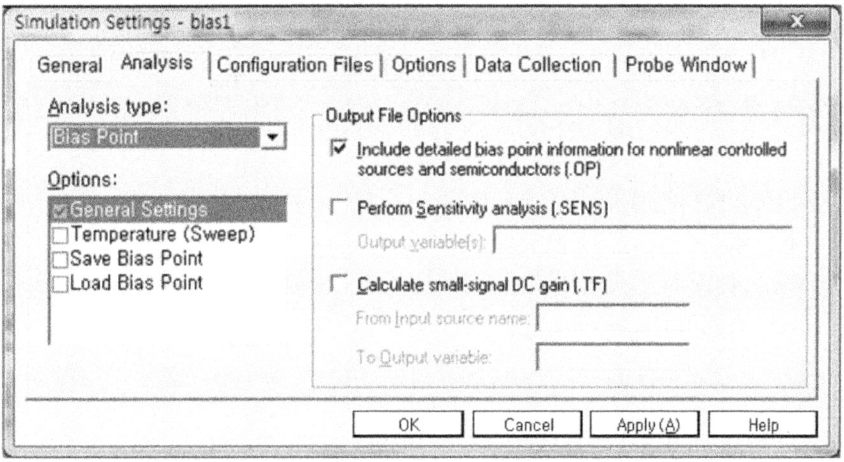

Simulation Settings dialog box for Bias Point analysis

Figure P2.1.2 PSpice simulation for the circuit of Figure P2.1.1(a1) – "elec02p01.opj."

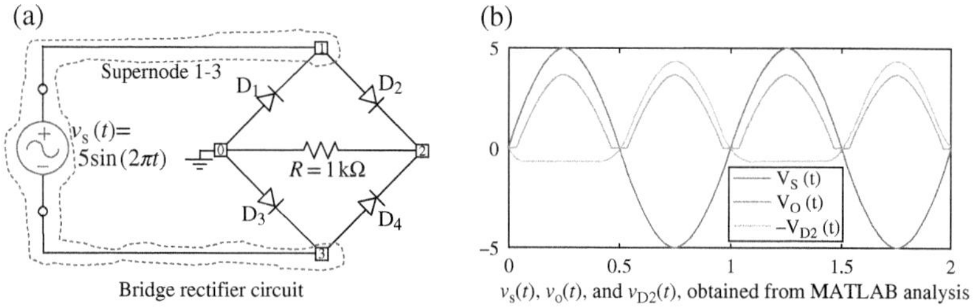

Figure P2.2 A bridge rectifier circuit and its analysis results for Problem 2.2.

$V_T = (27+273)/11605$[V], apply KCL at nodes 1 and 2 to write two node equations as

$$\begin{aligned} &i_{D2}(v_1-v_2)+i_{D4}(v_1-v_s-v_2)-i_{D1}(0-v_1)-i_{D3}(0+v_s)=0 \quad \text{at supernode 1-3} \\ &i_{D2}(v_1-v_2)+i_{D4}(v_1-v_s-v_2)-v_2/R_2=0 \quad \text{at node 2} \end{aligned} \tag{P2.2.1}$$

and use the MATLAB function '`fsolve()`' to solve them for v_1 and v_2 where $v_s(t) = 5\sin(2\pi t)$ [V]. To this end, complete the following MATLAB script "elec02p02.m" and run it to plot $v_s(t)$, $v_o(t) = v_2(t)$, and $-v_{D_2}(t)$ (the reverse

```
%elec02p02.m
Is=1e-14; T=27; VT=(T+273)/11605; % Device parameters
R=1e3; Vm=5; f=1; w=2*pi*f; % Circuit parameters
t=[0:1e-3:2]; vs=Vm*sin(w*t); vst=[vs; t];
[v1s,vos]=bridge_rectifier(vst,R,Is,VT);
plot(t,vs, t,vos, t,vos-v1s,'m')
```

```
function [v1s,v2s]=bridge_rectifier(vst,R,Is,VT)
vs=vst(1,:); t=vst(2,:);
iD=@(vD) Is*(exp(vD/VT)-1);
options=optimoptions('fsolve','Display','none');
for n=1:numel(t)
  vsn=vs(n);
  eq=@(v) [iD(v(1)-v(?))+iD(v(?)-vsn-v(2))-iD(-v(?))-iD(vsn-v(1));
           iD(v(?)-v(2))+iD(v(1)-vsn-v(?))-v(2)/R]*1e6;
  if n<2, v0=[0.7 0]; else v0=v; end
  v=fsolve(eq,v0,options);
  v1s(n)=v(1); v2s(n)=v(2);
end
```

voltage of diode D_2). Compare the waveforms with the PSpice simulation results shown in Figure 2.10c2.

2.3 Diode Clipping Circuits – Diode Limiters

```
%clippers.m
% Copyleft: Won Y. Yang, wyyang53@hanmail.net, CAU for academic use only
Is=10e-15; VT=25e-3; % Saturation current, Thermal voltage
iDvD = @(vD) Is*(exp(vD/VT)-1).*(vD>0);
iZvZ = @(vD,VZ0,rz) Is*(exp(vD/VT)-1).*(vD>0)-max(-VZ0-vD,0)/rz;
vZiZ = @(iD,VZ0,rz,IZK) VT*???(iD/Is+1).*(iD>0) +... % Eq.(P2.3.1)
        (VZ0/???)*iD.*(-IZK<=iD&iD<0)-(iD<-IZK).*(VZ0-??*iD);
options = optimset('TolFun',1e-10, 'Display','off');
% Symmetrical diode limiter
vi=[-10:0.05:10]; % Range of the input signal
Rs=1e3; V1=2; V2=-2;
vo=zeros(size(vi)); % Initialize the output voltage values to zero
for n=1:length(vi)
  eq = @(vo) (vi(n)-??)/Rs-iDvD(??-V1)+iDvD(V2-??);
  if n<2, vo0=0; else vo0=vo(n-1); end
  vo(n) = fsolve(eq,vo0,options);
end
subplot(331), plot(vi,vo)
% Symmetrical Zener limiter
vi=[-10:0.05:10]; % Range of the input signal
VZ0=4.67; IZK=4e-12; VZ=4.7; IZ=2e-3; rz=(VZ-VZ0)/IZ % Eq.(2.3.1b)
vo=zeros(size(vi)); % Initialize the output voltage values to zero
for n=1:length(vi)
  eq=@(i,VZ0,rz,IZK) vZiZ(?,VZ0,rz,IZK)-vZiZ(-?,VZ0, rz,IZK)
                                        +Rs*?-vi(n);...
  if n<2, i0=-1; else i0=i(n-1); end
  i(n) = fsolve(eq,i0,options,VZ0,rz,IZK); vo(n) = vi(n)-Rs*i(n);
end
subplot(332), plot(vi,vo)
% To plot the v-i characteristic curve of a Zener diode
i=[-5:0.01:1]*1e-3; vz=vZiZ(i,VZ0,rz,IZK);
subplot(333), plot(vz,i, [-5 1],[0 0],'k', [0 0], i([1 end]),'k')
```

Consider the two diode clipping circuits (called limiters or clippers) each in Figure P2.3.1(a1) and (a2) where $V_T = 25$ mV, the reverse saturation current of the diodes is $I_s = 10^{-14}$ A, and the Zener diode has $V_Z = 4.67/4.7$ V at $I_Z = 0/2$mA and $I_{ZK} = 4 \times 10^{-12}$ A. Note that Figure P2.3.1(b1) and (b2) (obtained from PSpice simulation with Transient analysis) show their output voltage waveforms to sinusoidal input voltage of frequency 1 Hz and amplitude 5 V and 10 V, respectively, and Figure P2.3.1(c1) and c2 (obtained from PSpice simulation with DC sweep analysis and voltage source VSIN replaced by VDC) show their input-output relationships called voltage transfer characteristics (VTCs).

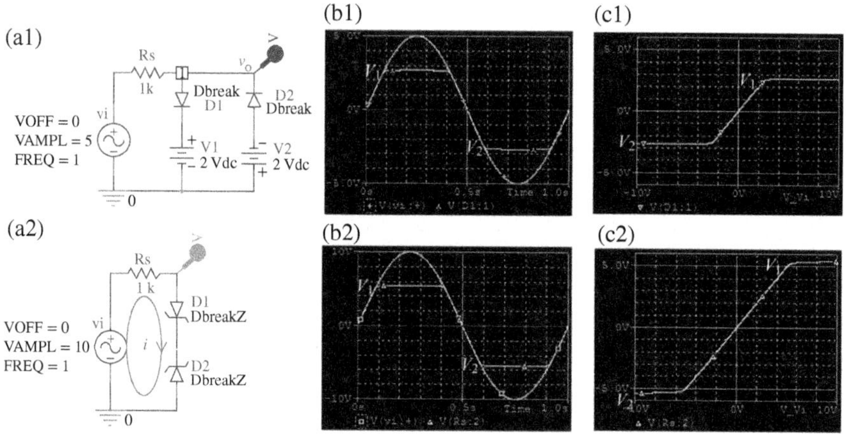

Figure P2.3.1 Diode clipping circuits (diode limiters) for Problem 2.3.

(a) Referring to Figure 2.19, complete the following equation describing the typical v-i characteristic of a Zener diode, which has been piecewise linear (PWL) approximated in the reverse-biased region:

$$v_Z(i_Z) = \begin{cases} V_T \ln(i_Z/I_s + 1) & \text{if } i_Z \geq 0 \\ (V_{Z0}/\square) i_Z & \text{if } -I_{ZK} \leq i_Z < 0 \\ V_{Z0} - \square i_Z & \text{if } i_Z < -I_{ZK} \end{cases} \tag{P2.3.1}$$

(b) Complete the sixth to seventh lines of the above MATLAB script "clippers.m" to create a MATLAB function handle defining the function $v_Z(i_Z)$ with three parameters V_{Z0}, r_Z, and I_{ZK} where `(iZ>0)`, `(-IZK<=iZ&iZ<0)`, and `(iZ<-IZK)` are logical expressions, each of which becomes 1 or 0 depending on whether it is true or not. Then noting that $V_{Z0} = 4.67$ V, use Eq. (2.3.1b) to determine the Zener resistance r_Z and plot the v-i

characteristic of the Zener diode, i.e. i_Z versus v_Z for $i_Z = (-5{:}0.01{:}1) \times 10^{-3}$ [A] as shown in Figure P2.3.2(a) by running the last two lines of "clippers.m."

(c) To plot the v-i characteristic (Figure P2.3.2(c)) of the Zener diode through PSpice simulation with DC sweep analysis, do the following:

1) Create a PSpice schematic shown in Figure P2.3.2(b1) where the device parameters of the Zener diode part 'DbreakZ' are set in the PSpice Model Editor (opened by clicking on that part so that it will be highlighted in pink and selecting the menu Edit>PSpice Model) as shown in Figure P2.3.2(b2).

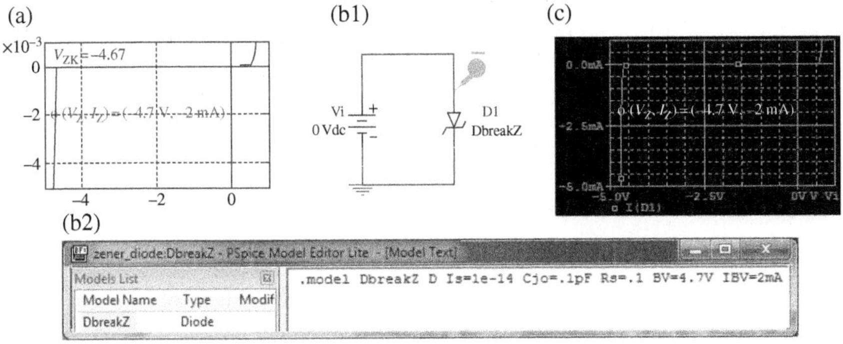

Setting the device parameters of Zener diode 'DbreakZ' in the Model Editor

Figure P2.3.2 The v-i characteristic curve of the Zener diodes in Figure P2.3.1(a2).

2) Click on New Simulation Profile button to open the Simulation Setting dialog box where you are supposed to set the Analysis type (DC Sweep), Sweep variable (Voltage source: Vi), and Sweep type (Linear: Start/End value = −4.8/0.8, Increment = 0.01) appropriately.

3) Place a current marker at the anode of the Zener diode and run the PSpice schematic to see the VTC as shown in Figure P2.3.2(c).

(d) Write the KCL equation in v_o for node 1 in the clipper of Figure P2.3.1(a1) and the KVL equation in i for the mesh in the clipper of Figure P2.3.1(a2). With those equations, complete the above MATLAB script "clippers.m" so that it can plot the VTC curves of the two clippers. Run it to plot the VTCs and see if they are close to those shown in Figure P2.3.1(c1) and (c2).

2.4 Diode Clampers – DC Restorers

Consider the two diode clampers in Figure P2.4(a1) and (a2) that are driven by a PWL voltage source generating a square wave with magnitude ±5 V and period 2 s. Perform PSpice simulations of the clampers (with the SKIPBP box checked in the Simulation Settings dialog box) to get their input/output voltage waveforms as shown in Figure P2.4(b1) and (b2). Do their input-output relationships conform with Eqs. (2.2.1a) and (2.2.1b)?

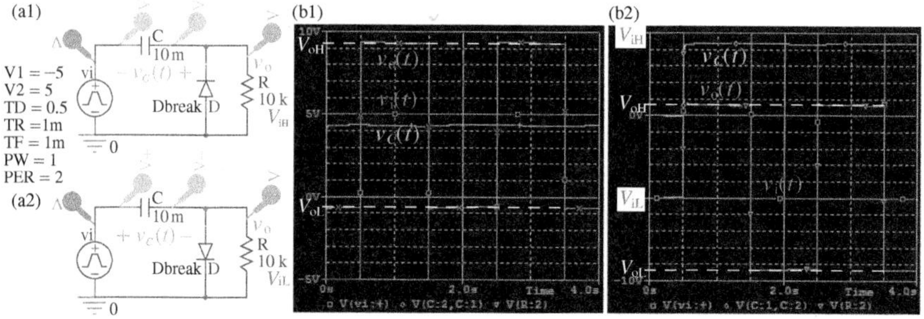

Figure P2.4 Two clamper circuits and their PSpice simulation results ("dc_restorer.opj").

2.5 MATLAB and PSpice Simulations of Voltage Doubler and Quadrupler

```
%voltage_doubler.m
Is=2.682e-9; T=27; nVT=1.836*(273+T)/11605; % Diode constants
iDv=@(vD)Is*(exp(vD/nVT)-1);
dt=0.001; tf=10; ts=0:dt:tf;
Vm=5; f=2; w=2*pi*f; vs=Vm*sin(w*ts);
C1=1e-6; C2=1e-6; vC1(1)=0; vC2(1)=0;
for n=1:length(ts)
  vD1(n)=v?(n)-vC?(n); iD1(n)=iDv(vD1(n));
  vD2(n)=-vC?(n)-vD?(n); iD2(n)=iDv(vD2(n));
  iC1(n)=iD?(n)-iD?(n);
  vC1(n+1) = vC1(n) + iC?(n)/C1*dt;
  vC2(n+1) = vC2(n) + sgn*iD?(n)/C2*dt;
end
plot(ts,vs,'r', ts,-vC2(1:end-1),'g');
grid on; legend('vs(t)','vo(t)')
```

Consider the voltage doubler/quadrupler circuits driven by a sinusoidal voltage source of amplitude 5 V and frequency 2 Hz, whose PSpice schematics are shown in Figures P2.5.1(a) and P2.5.2(a), respectively.

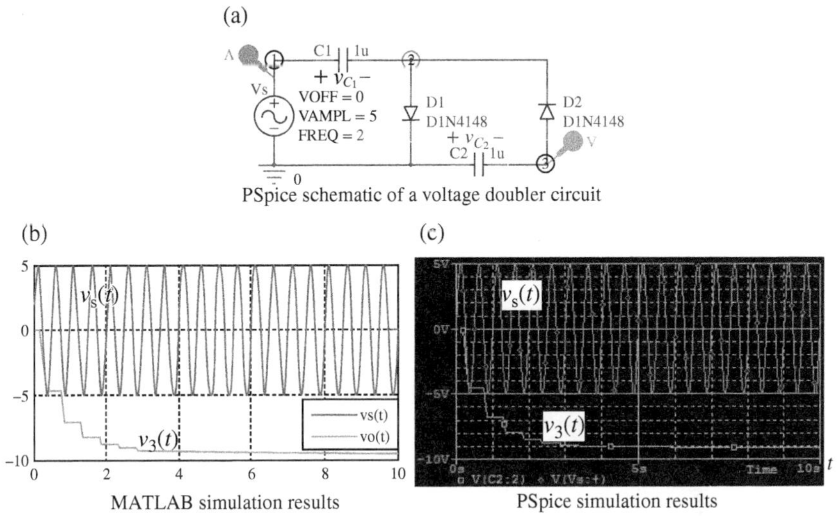

Figure P2.5.1 MATLAB and PSpice simulations of a voltage doubler circuit ("voltage_doubler.opj").

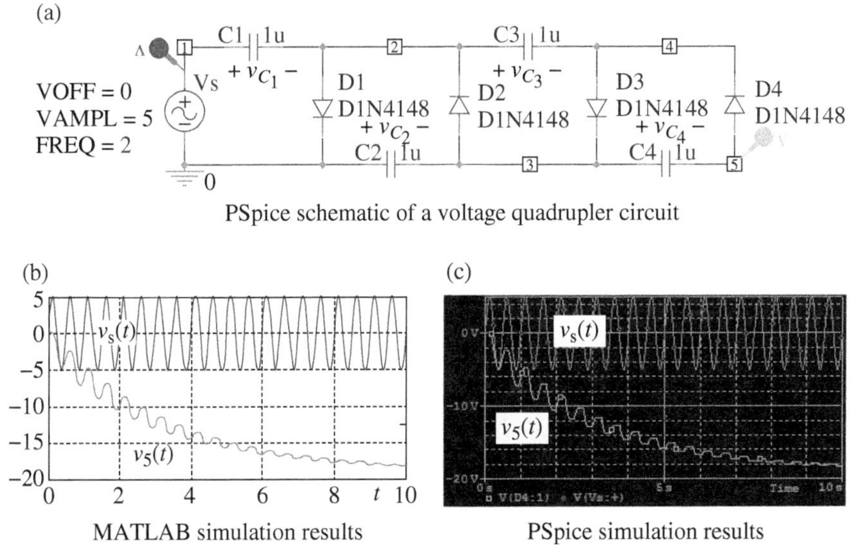

Figure P2.5.2 PSpice and MATLAB simulations of a voltage quadrupler circuit ("voltage_quadrupler.opj").

(a) Complete the above MATLAB script "voltage_doubler.m" to simulate the voltage doubler for 10s to get its input and output waveforms as shown in Figure P2.5.1(b). Also perform PSpice simulation of the circuit to get its input and output waveforms as shown in Figure P2.5.1(c).

(b) Complete the following MATLAB script "voltage_quadrupler.m" to simulate the voltage quadrupler for 10 s to get its input and output waveforms as shown in Figure P2.5.2(b). Also perform PSpice simulation of the circuit to get its input and output waveforms as shown in Figure P2.5.2(c).

```
%voltage_quadrupler.m
clear
Is=2.682e-9; T=27; nVT=1.836*(273+T)/11605; % Diode constants
iDv = @(vD)Is*(exp(vD/nVT)-1);
dt=0.001; tf=10; ts=0:dt:tf; % Time vector
Vm=5; f=2; w=2*pi*f; vs=Vm*sin(w*ts); % Input source voltage
C1=1e-6; C2=1e-6; C3=1e-6; C4=1e-6;
vC1(1)=0; vC2(1)=0; vC3(1)=0; vC4(1)=0;
v2(1)=0; v3(1)=0; v4(1)=0; v5(1)=0;
for n=1:length(ts)
  vD1 = v?(n); iD1 = iDv(vD1);
  vD2 = v?(n)-v?(n); iD2 = iDv(vD2);
  vD3 = v?(n)-v?(n); iD3 = iDv(vD3);
  vD4 = v?(n)-v?(n); iD4 = iDv(vD4);
  vC1(n+1) = vC1(n) + (iD1?iD2?iD3?iD4)/C1*dt;
  vC2(n+1) = vC2(n) + (iD2?iD3?iD4)/C2*dt;
  vC3(n+1) = vC3(n) + (iD3?iD4)/C3*dt;
  vC4(n+1) = vC4(n) + iD?/C4*dt;
  v2(n+1) = vs(n) - vC?(n+1);
  v3(n+1) = -vC?(n+1);
  v4(n+1) = v2(n+1) - vC?(n+1);
  v5(n+1) = v3(n+1) - vC?(n+1);
end
plot(ts,vs, ts,v5(1:end-1),'r'), grid on
```

2.6 Half-wave Rectifier Circuit Fed by a Triangular Voltage Source

Consider the half-wave rectifier circuit of Figure P2.6(a) where the voltage source v_i generates a triangular voltage waveform (with amplitude 10 V and frequency $f = 1$ kHz) shown in Figure P2.6(c).

(a)

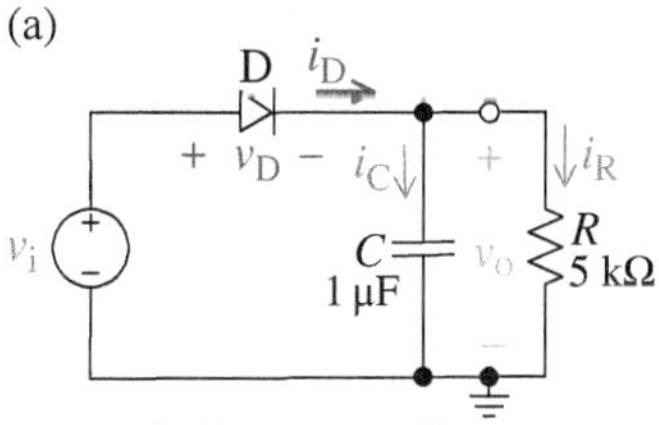

A half-wave rectifier

(b)

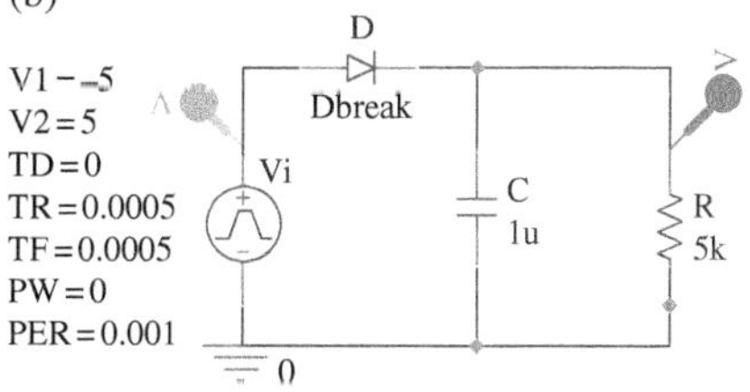

The PSpice schematic for the half-wave rectifier

(c)

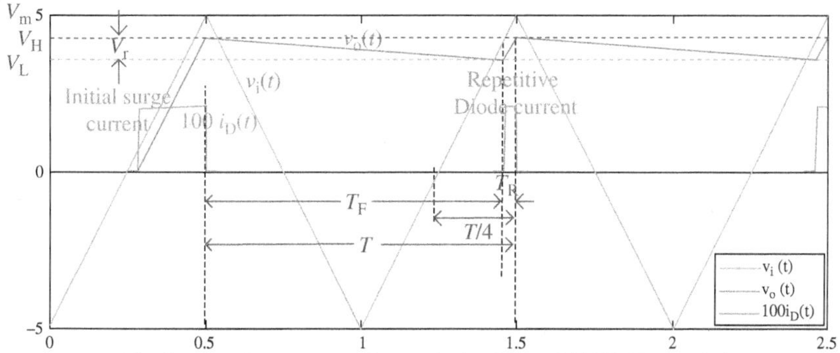

The input and output waveforms obtained from MATLAB simulation

(d)

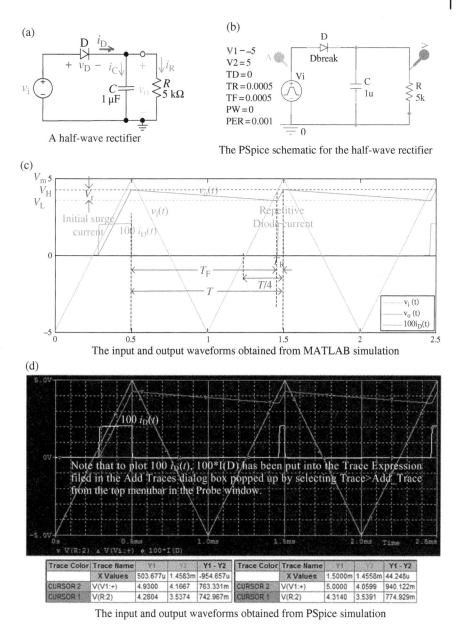

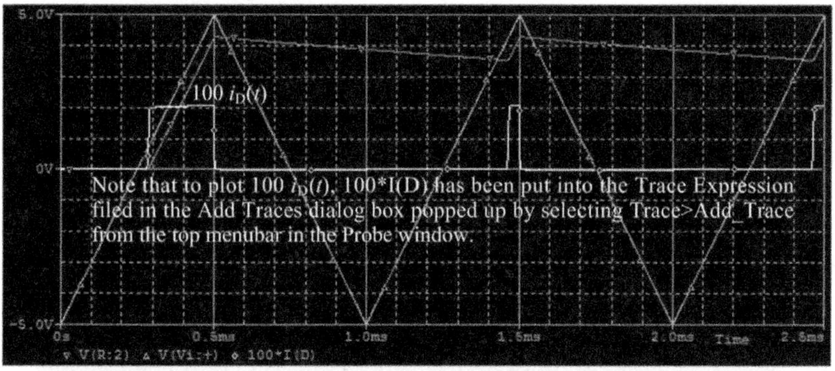

Trace Color	Trace Name	Y1	Y2	Y1 - Y2
	X Values	503.677u	1.4583m	-954.657u
CURSOR 2	V(V1:+)	4.9300	4.1667	763.331m
CURSOR 1	V(R:2)	4.2804	3.5374	742.967m

Trace Color	Trace Name	Y1	Y2	Y1 - Y2
	X Values	1.5000m	1.4558m	44.248u
CURSOR 2	V(V1:+)	5.0000	4.0599	940.122m
CURSOR 1	V(R:2)	4.3140	3.5391	774.929m

The input and output waveforms obtained from PSpice simulation

Figure P2.6 Half-wave rectifier circuit fed by a triangular voltage source.

(a) To get the upper/lower limit V_H/V_L of the output voltage $v_o(t)$ and the rising/falling period T_R/T_F via an analytical approach using MATLAB, we set up the following equations like Eq. (2.2.3) referring to Figure P2.6(c):

$$V_H e^{-T_F/RC} - V_L = 0 \quad \text{(for the falling period)} \tag{P2.6.1a}$$

$$V_H - V_L - V_m T_R/(T/4) = 0 \text{ (for the rising period)} \tag{P2.6.1b}$$

$$T_R + T_F - 1/f = 0 \tag{P2.6.1c}$$

Noting that V_H is already known as $V_H = V_m - V_{TD} = 5 - 0.65 = 4.35$, we can solve this set of equations to find $V_L = 3.94$, $T_R = 0.02$ ms, and $T_F =$ 0.98ms by saving these equations into an M-file named, say, 'elec02p06_f.m' and running the following MATLAB script "elec02p06a.m."

```
%elec02p06a.m
global Vm f VTD T D
Vm=5; f=1000; T=1/f; D=T/2; VTD=0.65;
VH=Vm-VTD; % Local maximum (High Voltage) of vo(t)
R=5e3; C=1e-6;
x_0=[0 0 0]; % Initial guess of [VL TR TF]
x=fsolve('elec02p06_f',x_0,optimset('fsolve'),C,R)
VL=x(1); TR=x(2); TF=x(3);
fprintf('\n VH=%8.4f, VL=%8.4f, TR=%8.4fms, TF=%8.4fms',
                                          VH,VL,TR*1000,TF*1000)
Vr=VH-VL, VH*T/R/C % Ripple Eq.(P2.6.2)
IC_avg=C*Vr/TR; % Eq.(P2.6.4)
Vo_avg=(VH?VL)/2 % Eq.(P2.6.5)
IR_avg=Vo_avg/R % Eq.(P2.6.6)
ID_avg=IC_avg?IR_avg % Eq.(P2.6.7)
```

```
function y=elec02p06_f(x,C,R)
global Vm f VTD
VH=Vm-VTD; w=2*pi*f; T=1/f; VL=x(1); TR=x(2); TF=x(3);
y=[VH*exp(-TF/R/C)-VL;
VH-VL-4*Vm*TR/T; TR+TF-T]; % Eqs.(P2.6.1)
```

(b) To simulate the half-wave rectifier by using MATLAB, complete and run the following script "elec02p06b.m" (after "elec02p06a.m") to get the simulation result as shown in Figure P2.6(c).

```
%elec02p06b.m
tf=2.5*T; dt=tf/5000; t=0:dt:tf; % Time range
% Input voltage
tri_wave_=@(t,D,T)mod(t,T)/D.*(mod(t,T)<D)+ ...
        (1-mod((t-D),T)/D).*(mod(t,T)>=D).*(mod(t,T)<2*D);
vi = 2*Vm*tri_wave_(t,D,T)-Vm; % Input voltage waveform
% Rectification
Is = 1e-14; % Saturation current
[vo,vD,iD,iR] = rectifier_RC([vi; t],R,?,Is);
plot(t,vi,'g', t,vo,'r', t,100*iD,'m', t([1 end]),[0 0],'k')
legend('v_i (t)','v_o (t)','100*i_D (t)')
N1=floor(T/dt); nn0=1:N1;
[VH,imin]=max(vo(nn0)); VH % Te 1st peak value of vo(t)
tH1=t(imin); % The 1st peak time
N2=2*N1+10; nn1=N1:N2;
[VL,imin]=min(abs(vo(nn1))); VL
tL1=t(N1-1+imin); % The 1st valley time
[emin,imin]=min(abs(vo(nn1)-VH));
tH2=t(N1-1+imin); % The 2nd peak time
TF=tL1?tH1 % Falling period
TR=tH?-tL1 % Rising period
```

(c) Assuming that $T_R \ll T \ll RC$ so that $T_F = T - T_R \approx T \ll RC$, show that the ripple voltage $V_r = V_H - V_L$ is approximately

$$V_r = V_H - V_L = V_H - V_H e^{-T_F/RC} \approx V_H - V_H\left(1 - \frac{T_F}{RC}\right)$$

$$\approx V_H - V_H\left(1 - \frac{T}{RC}\right) = V_H\left(\frac{T}{RC}\right) \tag{P2.6.2}$$

Noting that the capacitor voltage charged during the rising period T_R is equal to the ripple voltage V_r and is related with the capacitor current i_C or its average $I_{C,avg}$ is as

$$V_r = \frac{1}{C}\int_{T-T_R}^{T} i_C(t)\,dt \approx \frac{1}{C} I_{C,avg} T_R, \tag{P2.6.3}$$

find the average of the capacitor current through C:

$$I_{C,avg} \overset{(P2.6.3)}{=} \frac{C}{T_R} V_r \tag{P2.6.4}$$

Also, noting that the average of the output voltage (across $R||C$) is

$$V_{o,avg} = \frac{1}{2}(V_H + V_L), \tag{P2.6.5}$$

find the average of the resistor current through R:

$$I_{R,\mathrm{avg}} = \frac{V_{\mathrm{o,avg}}}{R} \overset{(\mathrm{P2.6.5})}{=} \frac{V_{\mathrm{H}} + V_{\mathrm{L}}}{2R} \tag{P2.6.6}$$

Also, find the average of the diode current through D:

$$I_{\mathrm{D,avg}} = I_{C,\mathrm{avg}} + I_{R,\mathrm{avg}} \overset{(\mathrm{P2.6.4,6})}{=} \frac{C}{T_{\mathrm{R}}} V_{\mathrm{r}} + \frac{V_{\mathrm{H}} + V_{\mathrm{L}}}{2R} \tag{P2.6.7}$$

(d) To simulate the half-wave rectifier by using PSpice, draw the schematic as Figure P2.6(b) and perform the Transient analysis for 2.5 ms to get the input and output voltages and diode current as shown in Figure P2.6 (d). You can activate or deactivate two cross cursors by clicking on the toggle cursor button in the PSpice A/D (Probe) window and move them to any two points by left-/right-clicking or pressing the (left or right) arrow/shift-arrow to read their coordinates simultaneously where you can left-/right-click on the symbol of the variable (below the waveforms) which you want to read.

(e) Does the value of $I_{\mathrm{D,avg}}$ agree with the (approximate) average of the diode current $i_{\mathrm{D}}(t)$ (during the ON period) shown in Figure P2.6(c) or (d)?

2.7 Precision Full-wave Rectifier

Consider the precision full-wave rectifier in Figure P2.7(a) where the maximum (saturation) output voltage and maximum (short-circuit) output current of the OP Amp μA741 are $V_{\mathrm{om}} = 14.6$ V and $I_{\mathrm{os}} = 40$ mA, respectively, for a bipolar power supply of ±15 V.

(a) Noting that the OP Amp U1 has always a negative feedback path (via R_2) so that $v_{-1} = v_{+1}$ by the virtual short principle, let us see how the rectifier works. If $v_{\mathrm{i}} > 0$, $v_{\mathrm{o}0}$ becomes (low, high) so that D_1 can be (off, on) and thus $i_{\mathrm{o}0} = 0$, $i_{R_3} = 0$, and $v_{-1} = v_{+1} = v_{\mathrm{i}}$. Then $i_{R_1} = i_{R_2} = \square$ so that $v_o = v_{-1} = \square$. If $v_{\mathrm{i}} < 0$, $v_{\mathrm{o}0}$ becomes (low, high) so that D_1 can be (off, on) to activate the negative feedback path for the OP Amp U2, resulting in $v_{-1} = v_{+1} = v_{-2} = v_{+2} = \square$ (virtual ground). Then the OP Amp U1 functions as an inverting amplifier with gain $-R_2/R_1$ so that $v_{\mathrm{o}} = \square = \square$.

(b) Doesn't the value of R_3 seem to matter? To check your idea, resimulate the rectifier with $R_3 = 100\ \Omega$ to get $v_{\mathrm{o}}(t)$, $v_{-2}(t)$, and $10i_{\mathrm{D}}(t)$ as shown in Figure P2.7(c) where $v_{\mathrm{o}}(t)$ has been severely distorted in the middle of the negative cycle of $v_{\mathrm{i}}(t)$. Why is that? It is because $v_{-1} = v_{+1} = v_{-2} \neq v_{+2} = 0$, i.e. the virtual short principle does not hold despite the negative feedback path via D_1 with $i_{\mathrm{D}} = i_{\mathrm{o}0} > 0$. What caused this situation starting from just when v_{i} is about to decrease below −4 V so that $-i_{R_3} = i_{\mathrm{o}0}$ is going to increase over $(v_{+1} - v_{\mathrm{i}})/R_3 = (v_{-2} - v_{\mathrm{i}})/R_3 = 40$ mA $= I_{\mathrm{os}}$? What is the minimum value of R_3 to avoid such a situation?

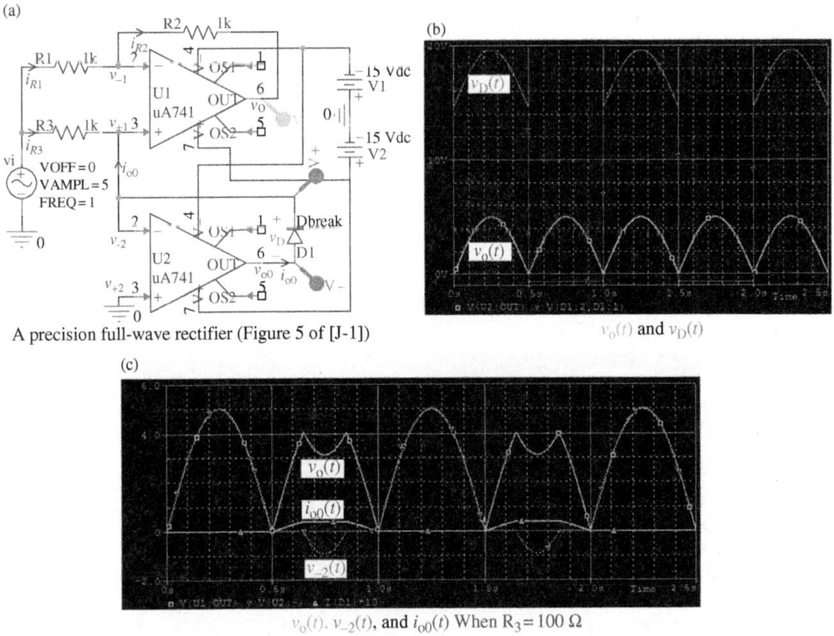

Figure P2.7 A precision full-wave rectifier.

2.8 Precision Full-wave Rectifier

Consider the precision full-wave rectifier with the PSpice schematic shown in Figure P2.8(a) where $R_1 = R_2 = R_3 = R_4 = R_5 = 1\,\text{k}\Omega$ and the maximum (saturation) output voltage of the OP Amp μA741 is $V_{om} = 14.6\,\text{V}$ for a bipolar power supply of ±15 V.

(a) Noting that the OP Amp U1 as well as U2 has always a negative feedback path via R_2-D_1 (when $v_i > 0$) or via D_2-R_5 (when $v_i < 0$) so that $v_{-1} = v_{+1} = 0$ and $v_{-2} = v_{+2}$ by the virtual short principle, let us see how the rectifier works. When $v_i > 0$, v_{o0} becomes (low, high) so that D_1/D_2 can be (on/off, off/on) and thus $i_{R_5} = \square$, $v_{-2} = v_{+2} = v_{-1} = v_{+1} = \square$ where both U1 and U2 work as linear amplifiers each with gain $-R_2/R_1$ and $-R_4/R_3$, respectively. Then

$$v_o = \boxed{} \qquad \text{(P2.8.1)}$$

When $v_i < 0$, v_{o0} becomes (low, high) so that D_1/D_2 can be (on/off, off/on). KCL at node 1 yields a node equation with v_3 as an unknown variable:

$$\frac{v_i - 0}{R_1} + \frac{v_3 - 0}{R_5} + \frac{v_3 - 0}{\square} = 0;\ v_3 = \boxed{} \qquad \text{(P2.8.2)}$$

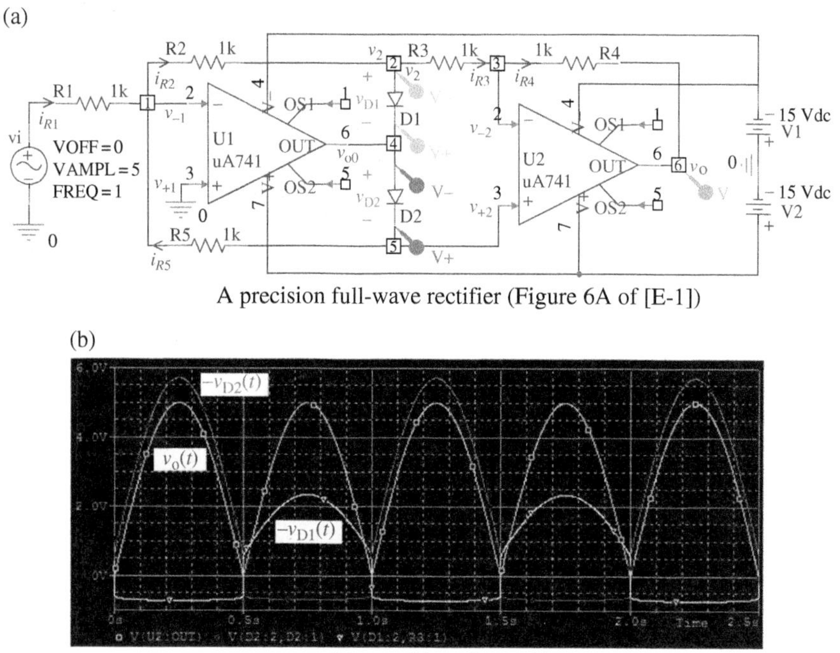

A precision full-wave rectifier (Figure 6A of [E-1])

$v_o(t)$, $-v_{D1}(t)$, and $-v_{D2}(t)$

Figure P2.8 A precision full-wave rectifier.

Then, regarding R_2-R_3-R_4 as a voltage divider, we can find the output voltage v_o as

$$v_o = \frac{R_2 + R_3 + R_4}{\square} v_3 = \square \tag{P2.8.3}$$

(b) Find the peak inverse voltages (PIVs) of the diodes D_1 and D_2.

When $v_i = V_m = 5$ V, v_{o0} becomes low so that D_1 is on with $v_{D_1} = V_{D_1} = 0.7$ V and D_2 is reverse-biased with

$$v_{D_2} = v_4 - v_5 = (v_2 - \square) - 0 = \square$$

When $v_i = -V_m = -5$ V, v_{o0} becomes high so that D_2 is on with $v_{D_2} = V_{D_2} = 0.7$ V and D_1 is reverse-biased with

$$v_{D_1} = v_2 - v_4 = v_2 - (v_5 + \square) = v_2 - (v_3 + \square)$$
$$= \frac{R_2}{\square} V_m - \frac{\square}{R_2 + R_3 + R_4} V_m - \square = \square \tag{P2.8.4}$$

These PIVs can also be seen from the simulation result in Figure P2.8(b).

2.9 Zener (Shunt) Regulator

Consider the Zener regulator shown in Figure 2.20(a) where the Zener diode has $V_Z = 6.8$ V at $I_Z = 5$ mA, $r_Z = 20\ \Omega$, and $I_{ZK} = 0.2$ mA, and the supply voltage v_S varies by ±1 V around its nominal value 10 V.

(a) Determine V_{Z0} of the linear i_Z-v_Z relation (2.3.1b) describing the almost-straight line part of the i-v characteristic curve of the Zener diode.

Substituting $(I_Z, V_Z) = (5\text{ mA}, 6.8\text{ V})$ and $r_z = 20\ \Omega$ into Eq. (2.3.1b) yields

$$v_Z \overset{(2.3.1b)}{=} V_{Z0} + r_z i_Z \rightarrow V_{Z0} = \boxed{} \tag{P2.9.1}$$

(b) Use the equivalent circuit (Figure 2.20(b)) with the Zener diode Z replaced by its PWL model to find the output voltage $v_o = v_Z$ (across $R_L = 2\text{ k}\Omega$ in parallel with the Zener diode) to the source voltage $v_s = 10, 9, 8.3$ V.

We can use Eq. (2.3.8) to get

$$v_Z|_{R_L=2k\Omega} \overset{(2.3.8)}{=} \left.\frac{v_S/R_s + V_{Z0}/r_z}{1/R_s + 1/r_z + 1/R_L}\right|_{R_L=2k\Omega} = \boxed{} \begin{array}{l}\text{for } v_s = 10\text{ V}\\ \text{for } v_s = 9\text{ V}\\ \text{for } v_s = 8.3\text{ V}\end{array} \tag{P2.9.2}$$

```
>>([10 9 8.3]/500+6.7/20)/(1/500+1/20+1/2000)
```

However, for the third value $v_Z = 6.697\text{ V} < V_{Z0}$, we set $i_Z = 0$ (I_{ZK}) and use the voltage divider rule to redetermine v_Z as if there were no Zener diode branch:

$$v_Z = \boxed{} \tag{P2.9.3}$$

(c) Make use of Eq. (2.3.9a) to guess how v_Z (the voltage across the Zener diode or $R_L = 2\text{ k}\Omega$) will be changed by the ±1V-change in v_s.

$$\Delta v_Z|_{R_L=2\text{ k}\Omega} = \Delta v_S \left.\frac{dv_Z}{dv_S}\right|_{R_L=2\text{ k}\Omega} \overset{(2.3.9a)}{=} \boxed{} \approx \boxed{} \tag{P2.9.4}$$

You can run the following MATLAB statements to get a help in finding the answers to the above questions:

```
>>VZ=6.8; IZ=5e-3; rz=20; IZK=0.2e-3;
  vss=[10 9 8.3]; Rs=500; RL=2e3;
```

```
[vos,iZ,dvodvs,dvodRL,dvodIL,Rsmax,RLmin,VZ0] = ...
  Zener_regulator(VZ,IZ,rz,IZK,RL,Rs,vss)
```

Is your guess close to the real change in v_Z (across $R_L = 2\ \text{k}\Omega$ to $v_s = 10\text{V}$) caused by the -1V-change in v_s, which can be observed in Eq. (P2.9.2)?

(d) Use the equivalent circuit (Figure 2.20(b)) to find the output voltage $v_o = v_Z$ (across $R_L = 2, 1.9, 1.4\ \text{k}\Omega$ in parallel with the Zener diode) to the source voltage $v_s = 9\ \text{V}$.

We can use Eq. (2.3.8) to get

$$v_Z\big|_{v_s=9\ \text{V}} \overset{(2.3.8)}{=} \left.\frac{v_s/R_s + V_{Z0}/r_z}{1/R_s + 1/r_z + 1/R_L}\right|_{v_s=9\text{V}} = \frac{\boxed{}}{\boxed{} + 1/R_L} = \begin{cases} \boxed{} & \text{for } R_L = 2\ \text{k}\Omega \\ \boxed{} & \text{for } R_L = 1.9\ \text{k}\Omega \\ \boxed{} & \text{for } R_L = 1.4\ \text{k}\Omega \end{cases} \quad \text{(P2.9.5)}$$

```
>>(9/500+6.7/20)./(1/500+1/20+1./[2 1.9 1.4]/1e3)
```

However, for the third value $v_Z = 6.696\ \text{V} < V_{Z0}$, we set $i_Z = 0$ (I_{ZK}) and use the voltage divider rule to redetermine v_Z as if there were no Zener diode branch:

$$v_Z = \boxed{} \quad \text{(P2.9.6)}$$

(e) Make use of Eq. (2.3.9b) to guess how v_Z (to $v_s = 9\ \text{V}$) will be changed by the $\pm 100\ \Omega$-change in R_L.

$$\Delta v_Z\big|_{v_s=9\text{V}} = \Delta R_L \left.\frac{dv_Z}{dR_L}\right|_{v_s=9\text{V}} \overset{(2.3.9b)}{=} \pm 100 \left.\frac{\boxed{}}{\left(\boxed{} + 1/R_L\right)^2}\right|_{v_s=9\text{V}}$$

$$= \pm 100 \frac{\boxed{}}{\left(\boxed{} + 1/R_L\right)^2} = \boxed{} \quad \text{(P2.9.7)}$$

You can run the following MATLAB statements to get a help in finding the answers to the above questions:

```
>>vs=9; Rs=500; RLs=[2e3 1.9e3 1.4e3];
  [vos,iZ,dvodvs,dvodRL,dvodIL,Rsmax,RLmin,VZ0]=...
  Zener_regulator(VZ,IZ,rz,IZK,RLs,Rs,vs); vos, dvodRL
```

Is your guess close to the real change in v_Z (across $R_L = 2\ \text{k}\Omega$ to $v_s = 9\text{V}$) caused by the $-100\ \Omega$-change in R_L, which can be observed in Eq. (P2.9.5)?

(f) To see that for $v_s \le 8.3\text{V}$, v_Z gets distinctly away from $V_Z = 6.8\text{V}$ at $R_L = 2\ \text{k}\Omega$, perform a PSpice simulation by taking the following steps:

1) Draw the PSpice schematic including a DC voltage source Vdc (10V), a sinusoidal voltage source VSIN (with amplitude 2 V and frequency 1 Hz), a Zener diode DbreakZ, two resistors $R_s = 500\ \Omega$ and $R_L = 2\ \text{k}\Omega$, and a Ground, as shown in Figure P2.9.1(a). If DbreakZ is not found in the Part

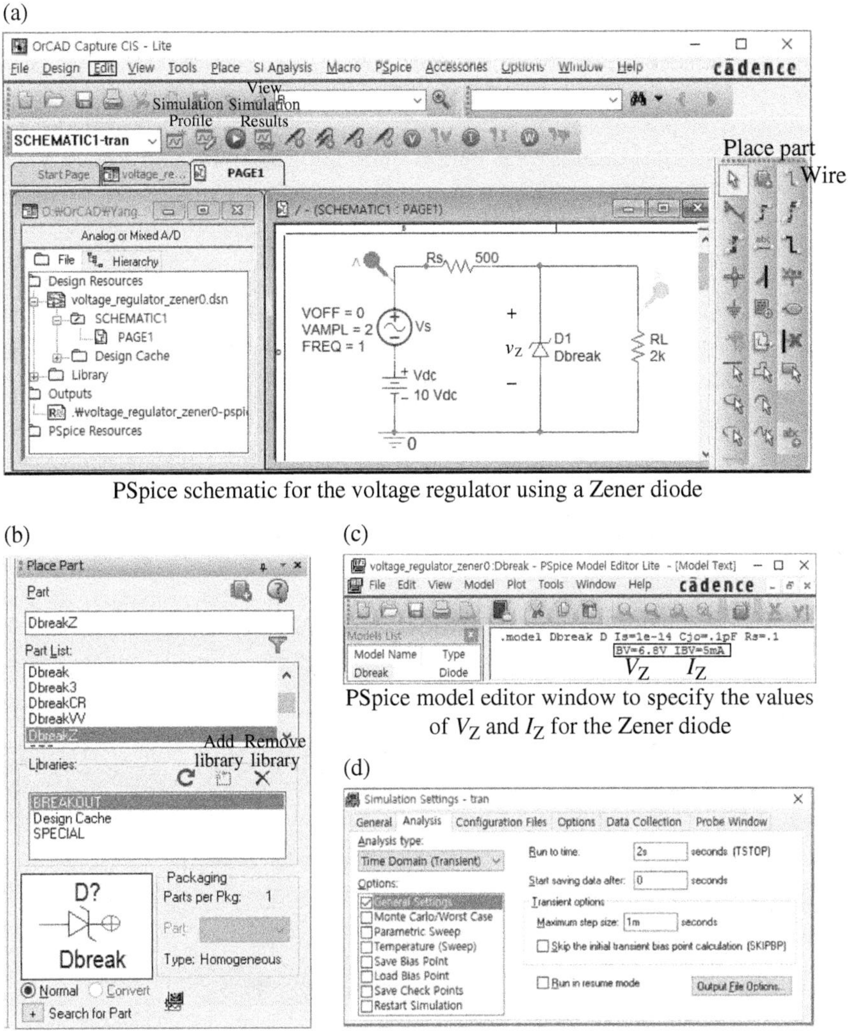

PSpice schematic for the voltage regulator using a Zener diode

PSpice model editor window to specify the values of V_Z and I_Z for the Zener diode

Place Part dialog box to fetch and place a Zener diode 'DbreakZ'

Simulation Settings dialog box

Figure P2.9.1 PSpice simulation for the Zener regulator in Problem 2.9 ("voltage_regulator_zener0.opj").

List (Part Browser dialog box) when you type 'DbreakZ' in the Part text box of the Place Part dialog box (Figure P2.9.1(b)) opened by Place part button on the tool palette (Figure P2.9.1(a)), you should insert the BEAKOUT library into the Libraries list by clicking on the Add library button and selecting Cadence > ... > tools > capture > library > pspice > breakout.olb.

2) Select the symbol of DbreakZ by clicking on it, select 'Edit > PSpice Model' (from the menu bar) to open the PSpice Model Editor window, and type 'BV = 6.8V IBV = 5mA' to set $V_Z = 6.8\text{V}$ and $I_Z = 5\,\text{mA}$ (Figure P2.9.1(c)).
3) Set the Analysis type, Run to time, and Maximum step size as 'Time Domain (Transient)', 2 s, and 1 ms, respectively, in the Simulation Settings dialog box opened by clicking on the New/Edit Simulation Profile button from the toolbar (Figure P2.9.1(d)).
4) Place the voltage markers at the upper terminals of the voltage source Vs and resistor RL as shown in Figure P2.9.1(a).
5) Click Run to start the simulation and get the input/output voltage waveforms as shown in Figure P2.9.2.

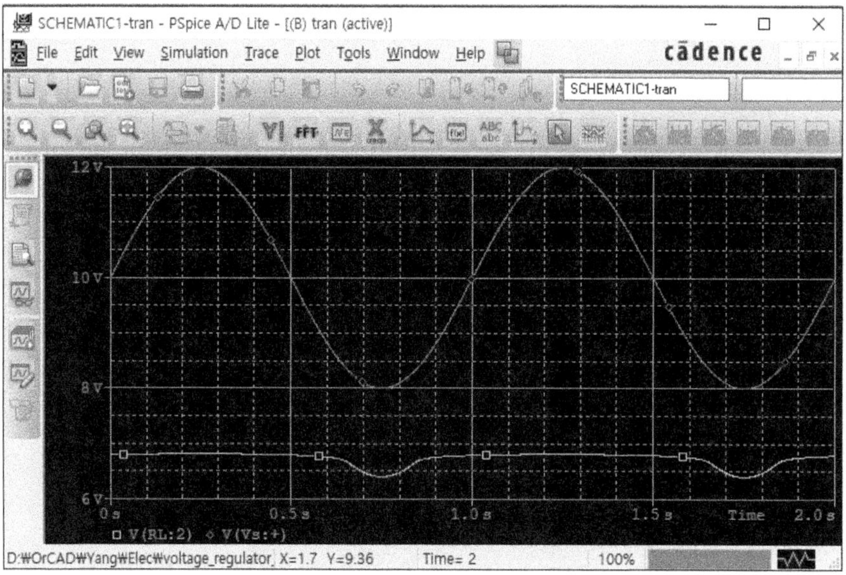

Figure P2.9.2 PSpice simulation results for Problem 2.9.

Does the output voltage v_Z drop conspicuously below $V_Z = 6.8\text{V}$ as the source voltage v_s becomes lower than 8.3V? How can the value of v_Z to the input source voltage $v_s = 8\text{V}$ be estimated?

3

BJT Circuits

CHAPTER OUTLINE

Electronic Circuits with MATLAB®, PSpice®, and Smith Chart, First Edition. Won Y. Yang, Jaekwon Kim, Kyung W. Park, Donghyun Baek, Sungjoon Lim, Jingon Joung, Suhyun Park, Han L. Lee, Woo June Choi, and Taeho Im.

Companion website: www.wiley.com/go/yang/electroniccircuits

- BJT (Bipolar Junction Transistor)
  - NPN-BJT
  - PNP-BJT
- FET (Field Effect Transistor)
  - JFET (Junction FET)
    - N-channel
    - P-channel
  - MOSFET (Metal-Oxide-Semiconductor FET)
    - Enhancement mode
      - N-channel
      - P-channel
    - Depletion mode
      - N-channel
      - P-channel

Figure 3.1 Classification of transistors.

Transistor stands for '*tran*sfer' + 're*sistor*', meaning that the basic operation of a transistor is to transfer an input signal from a resistor to another resistor. Transistors are broadly classified into two groups, i.e. Bipolar Junction Transistors (BJTs) and Field Effect Transistors (FETs). The two groups of transistors are further classified as depicted in Figure 3.1.

3.1 BJT (Bipolar Junction Transistor)

The (NPN/PNP) BJT is a three-terminal device formed from two P-N junctions (like diodes) which share a common (P/N-type) semiconductor and is widely used for various purposes including amplification and switching.

3.1.1 Ebers-Moll Representation of BJT

Figure 3.2(a1)/(a2), (b1)/(b2), and (c1)/(c2) shows the symbols, the basic structures, and the Ebers-Moll models for NPN/PNP types of BJT, respectively. According to the Ebers-Moll models, the emitter and collector currents of NPN/PNP-BJTs are described as

$$i_E = \mp i_{ED} \pm \alpha_R i_{CD} \overset{(2.1.1)}{\underset{\eta=1}{=}} \mp I_{SE}\left(e^{\pm v_{BE}/V_T} - 1\right) \pm \alpha_R I_{SC}\left(e^{\pm v_{BC}/V_T} - 1\right) \tag{3.1.1a}$$

$$i_C = \pm \alpha_F i_{ED} \mp i_{CD} \overset{(2.1.1)}{\underset{\eta=1}{=}} \pm \alpha_F I_{SE}\left(e^{\pm v_{BE}/V_T} - 1\right) \mp I_{SC}\left(e^{\pm v_{BC}/V_T} - 1\right) \tag{3.1.1b}$$

(Upper/lower one of each double sign applies to NPN/PNP-BJT)

where the typical values of I_{SE} (*reverse saturation current of B-E junction*) and $I_{SC} = \alpha_F I_{SE}/\alpha_R$ (*reverse saturation current of B-C junction*) are in the order of $10^{-15} \sim 10^{-12}$ and those of α_F and α_R are as follows:

$$0.98 \le \alpha_F \le 0.998, \quad 0.1 - 0.4 \le \alpha_R \le 0.8 \tag{3.1.2}$$

Note that the forward and reverse saturation leakage currents are the same as each other and the transistor *saturation current* I_S:

$$\alpha_R I_{SC} = \alpha_F I_{SE} = I_S \text{ (reciprocity condition)} \tag{3.1.3}$$

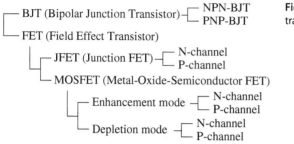

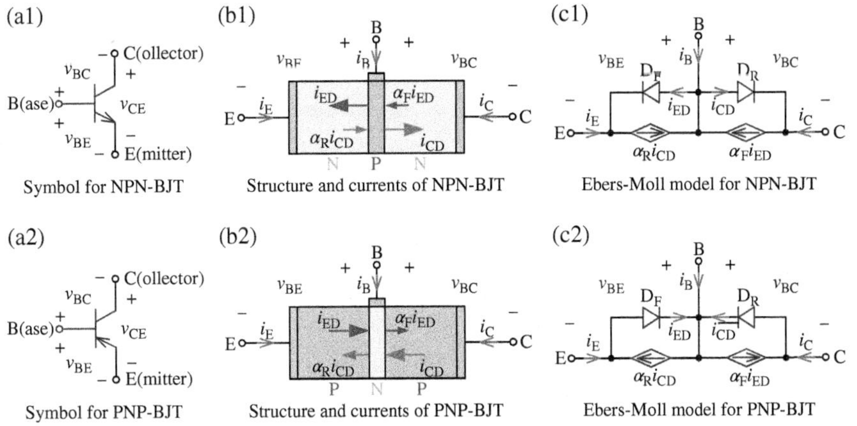

Figure 3.2 Symbols, basic structures, and models for NPN/PNP-type Bipolar Junction Transistors (BJTs).

We can use these equations to write the relations between the emitter current i_E and the collector current i_C as

$$\begin{aligned} i_E &\overset{(3.1.1a)}{=} \mp i_{ED} \pm \alpha_R i_{CD} \overset{(3.1.1b)}{=} \mp i_{ED} \pm \alpha_R(\mp i_C + \alpha_F i_{ED}) \\ &= \mp(1-\alpha_F\alpha_R) i_{ED} - \alpha_R i_C = \mp(1-\alpha_F\alpha_R) I_{ES}\left(e^{\pm v_{BE}/V_T} - 1\right) - \alpha_R i_C \\ &= \mp I_{EO}\left(e^{\pm v_{BE}/V_T} - 1\right) - \alpha_R i_C \end{aligned} \tag{3.1.4a}$$

$$\begin{aligned} i_C &\overset{(3.1.1b)}{=} \pm \alpha_F i_{ED} \mp i_{CD} \overset{(3.1.1a)}{=} \pm \alpha_F(\mp i_E + \alpha_R i_{CD}) \mp i_{CD} \\ &= -\alpha_F i_E \mp (1-\alpha_F\alpha_R) i_{CD} - \alpha_R i_C = -\alpha_F i_E \mp (1-\alpha_F\alpha_R) I_{CS}(e^{\pm v_{BC}/V_T} - 1) \\ &= -\alpha_F i_E \mp I_{CO}(e^{\pm v_{BC}/V_T} - 1) \end{aligned} \tag{3.1.4b}$$

where $I_{EO} = (1 - \alpha_F\alpha_R)I_{SE}/I_{CO} = (1 - \alpha_F\alpha_R)I_{SC}$ is the *reverse emitter/collector current*, which flows through the B-E/B-C junction when the junction is highly reverse biased and the collector/emitter terminal is open so that $i_C = 0/i_E = 0$, respectively.

Also, we can apply Kirchhoff's Current Law (KCL) to the BJT or the closed surface including its three terminals to write

$$i_C + i_E + i_B = 0 \tag{3.1.5}$$

Noting that the reverse emitter/collector currents I_{EO}/I_{CO} are negligibly small, the relationships among the three currents i_C, i_E, and i_B of NPN-BJT with $v_{BE} \geq 0.7$ (forward biased) and $v_{BC} < 0.4$ (not forward biased enough) can be written as

$$i_C \underset{I_{CO}\approx 0,\, v_{BC}<0.4}{\overset{(3.1.4b)}{=}} -\alpha_F i_E \tag{3.1.6a}$$

$$i_B \overset{(3.1.5)}{=} -i_E - i_C \overset{(3.1.6a)}{=} -(1-\alpha_F)i_E;\ i_E = -\frac{1}{1-\alpha_F} i_B \tag{3.1.6b}$$

$$i_C \overset{(3.1.6a)}{=} -\alpha_F i_E \overset{(3.1.6b)}{=} \frac{\alpha_F}{1-\alpha_F} i_B = \beta_F\ i_B \text{ with } \alpha_F \overset{(3.1.9b)}{=} \frac{\beta_F}{\beta_F+1} \tag{3.1.6c}$$

$$i_E \overset{(3.1.5)}{=} -i_C - i_B \overset{(3.1.6c)}{=} -(\beta_F+1)i_B \tag{3.1.6d}$$

According to [H-2], the collector and base currents of an NPN-BJT can be expressed as

$$i_C \underset{\text{[H-2]}}{\overset{(8.6.6)}{=}} \left\{ I_S\left(e^{v_{BE}/V_T}-1\right) - I_{SC}\left(e^{v_{BC}/V_T}-1\right)\right\}\left(1+\frac{v_{CE}}{V_A}\right) \tag{3.1.7a}$$

$$i_B \underset{\text{[H-2]}}{\overset{(8.6.7)}{=}} \frac{I_S}{\beta_F}\left(e^{v_{BE}/V_T}-1\right) + \frac{I_{SC}}{\beta_R+1}\left(e^{v_{BC}/V_T}-1\right) \tag{3.1.7b}$$

where $I_S = \alpha_F I_{ES}$, $I_{SC} = I_S/\alpha_R$ (typically within $10I_S$), and V_A is the *Early voltage*, whose typical value is 10~100 V. Here, compared with Eq. (3.1.1), the additional term proportional to v_{CE}/V_A has been included to account for the Early effect (also called the base-width modulation effect) that I_S increases as increasing v_{CE} results in a decrease in the effective base width of BJT. Noting that based on these equations, the collector current i_C can be expressed in terms of v_{CE}, v_{BE}, and i_B as

$$i_C \overset{(3.1.7a,b)}{=} \left\{\beta_F i_B - \frac{\beta_F}{\beta_R+1} I_{SC}\, e^{(v_{BE}-v_{CE})/V_T}\right\}\left(1+\frac{v_{CE}}{V_A}\right) \tag{3.1.8}$$

we can run the following MATLAB script "plot_iC_vs_vCE.m" to plot i_C versus v_{CE} (with v_{BE} = 0.7 V) for several values of i_B as shown in Figure 3.3 where the (dotted) extrapolation of every i_C curve intersects the v_{CE}-axis at common point $v_{CE} = -V_A$.

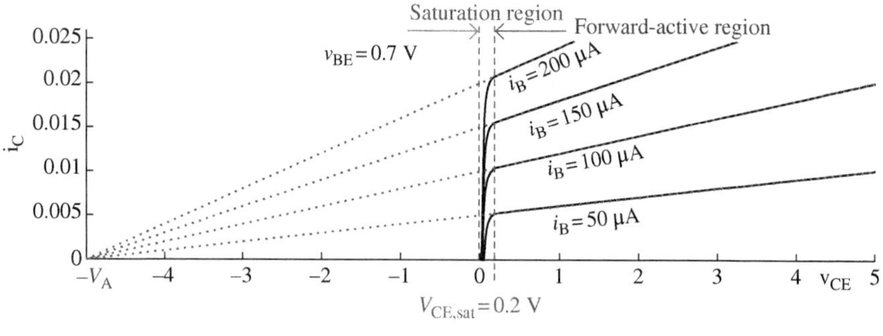

Figure 3.3 The i_C-v_{CE} characteristic curves for different values of constant i_B (with v_{BE} = 0.7V).

```
%plot_iC_vs_vCE.m
clear, clf
% BJT parameters (Saturation current, Current gain, Early voltage)
Is=1e-16; betaF=100; alphaR=0.1; betaR=alphaR/(1-alphaR); VA=5;
Isc=Is/alphaR; % CB(Collector-Base) saturation current
VT=(273+27)/11605; % Thermal voltage
subplot(311), hold on
iC_in_vCE_iB = @(vCE,iB,vBE) (betaF*iB-(betaF/(betaR+1)+1)*Isc* exp
                  ((vBE-vCE)/VT)).*(1+vCE/VA); % Eq.(3.1.8)
% To plot the iC-vCE curve for several values of iB (with vBE=0.7V)
dvCE=0.01; vCEs=[0:dvCE:5]; vBE=0.7;
for n=1:4
   iB=n*5e-5; plot(vCEs,iC_in_vCE_iB(vCEs,iB,vBE))
   plot([-VA vCEs(end)],[0 iC_in_vCE_iB(vCEs(end),iB,vBE)],'r:')
end
plot([0.2 0.2],[0 0.1],'m:') % vCE=VCE,sat=0.2V
xlabel('v_{CE}'), ylabel('i_C')
axis([-VA vCEs(end) 0 0.026])
```

3.1.2 Operation Modes (Regions) of BJT

Table 3.1 shows the four operation modes (regions) of NPN-BJT(Si) determined by the bias conditions of B-E and B-C junctions. Note the following:

- When $v_{BE} = 0.7$ V and $v_{CE} > 0.3$ V so that $v_{BC} = v_{BE} - v_{CE} < 0.4$, the BJT operates in the forward-active region.
- v_{CE} is 0.3 V at the edge of saturation and becomes $v_{CE,sat} = 0.2$ V in the 'deep' saturation mode.
- In the forward-active mode where $v_{BC} < 0.4$, the first terms (proportional to I_S) are dominant over the second terms (proportional to I_{SC}) in Eqs. (3.1.7a) and (3.1.7b) so that $i_C \approx \beta_F i_B$.
- In the reverse-active mode where $v_{BC} > 0.5$, the second terms (proportional to I_{SC}) are dominant over the first terms (proportional to I_S) in Eqs. (3.1.7a) and (3.1.7b) so that $i_C \approx -(\beta_R + 1)i_B$. This corresponds to Eq. (3.1.6d) where the roles of the collector and emitter terminals have been switched.

3.1.3 Parameters of BJT

To analyze BJT circuits, we need to define the following parameters for the forward-/reverse-active mode:

Table 3.1 Operation modes (regions) of NPN–BJT (Si) with $V_{TD} = 0.5\,V$.

Operation mode		Forward-active	Cut-off	Saturation	Reverse-active
Bias condition	B-E	Forward ($v_{BE} \geq V_{TD}$)	Reverse ($v_{BE} < V_{TD}$)	Forward ($v_{BE} \geq V_{TD}$)	Reverse ($v_{BE} < 0.4$)
	B-C	Reverse ($v_{BC} < 0.4$)	Reverse ($v_{BC} < 0$)	Forward ($v_{BC} \geq 0.4$)	Forward ($v_{BC} \geq V_{TD}$)
Functions		Current-controlled current source $i_C = -\alpha_F i_E$, $v_{BE} = 0.7\,V$	Open switch	Closed switch $v_{BE} = 0.7\,V$ $v_{CE} = 0.2\,V$	Roles of E and C terminals switched $i_E = -\alpha_R i_C$

1) **Forward-active mode**

<CE (Common-Emitter), *forward large-signal (DC) current gain*>

$$h_{FE} = \beta_F = \frac{\alpha_F}{1-\alpha_F} \overset{(3.1.6c)}{=} \left.\frac{i_C}{i_B}\right|_{v_{CE}=V_{CE,Q}} \quad \text{: normally, } 50-500 \tag{3.1.9a}$$

<CB (Common-Base), *forward large-signal (DC) current gain*>

$$h_{FB} = \alpha_F = \frac{\beta_F}{\beta_F+1} \underset{(3.1.6d)}{\overset{(3.1.6c)}{=}} \left.\frac{i_C}{i_E}\right|_{v_{CE}=V_{CE,Q}} \quad \text{: normally, } 0.98-0.998 \tag{3.1.9b}$$

<CE (Common-Emitter), *forward small-signal (AC or incremental) current gain*>

$$h_{fe} = \beta = \left.\frac{di_C}{di_B}\right|_{v_{CE}=V_{CE,Q}} = \left.\frac{i_c}{i_b}\right|_{vce=0} \tag{3.1.10a}$$

<CB (Common-Base), *forward small-signal (AC or incremental) current gain*>

$$h_{fb} = \alpha = \frac{\beta}{\beta+1} = -\left.\frac{di_C}{di_E}\right|_{v_{CE}=V_{CE,Q}} = -\left.\frac{i_c}{i_e}\right|_{vce=0} \tag{3.1.10b}$$

2) **Reverse-active mode**

<CE (Common-Emitter), *reverse large-signal (DC) current gain*>

$$h_{RE} = \beta_R = \frac{\alpha_R}{1-\alpha_R} = \left.\frac{i_E}{i_B}\right|_{v_{CE}=V_{CE,Q}} \quad \text{: normally, } 1 \sim 5 \tag{3.1.11a}$$

<CB (Common-Base), *reverse large-signal (DC) current gain*>

$$h_{RB} = \alpha_R = \frac{\beta_R}{\beta_R+1} = -\left.\frac{i_E}{i_C}\right|_{v_{CE}=V_{CE,Q}} \quad \text{: normally, } 0.16 \sim 0.5 \tag{3.1.11b}$$

3.1.4 Common-Base Configuration

Figure 3.4(a) shows a common-base (CB) NPN-BJT circuit in which the input is applied to the BEJ (B-E junction) and the output is available from the BCJ (B-C junction) so that the base terminal is common between the input and the output. Note that since the BEJ/BCJ are forward/reverse biased, we can expect the BJT to operate in the forward-active mode (Table 3.1). Also, note that since the BEJ and BCJ are nonlinear resistors, we may have to apply the load line analysis for both the B-E loop and the B-C loop.

To perform a comparatively exact analysis considering the nonlinearity of the circuit, we draw the load line on the B-E characteristic curve (Figure 3.4(b)) where the load line equation can be obtained by applying KVL around the B-E loop:

$$-i_E = \frac{V_{EE} - v_{BE}}{R_E = 1\,\text{k}\Omega} = 0.001(10 - v_{BE}) \tag{3.1.12}$$

The intersection of the load line with the B-E characteristic curve is the *operating (bias) point* Q_E:

$$v_{BE} = 0.7\,\text{V}, i_E = -9.3\,\text{mA} \tag{3.1.13}$$

where it does not matter which one of many B-E characteristic curves with different values of v_{CB} is used to determine the operating point because B-E characteristic curve varies little with v_{CB}. Then we draw the load line on the B-C characteristic curve (Figure 3.4(c)) where the load line equation can be obtained by applying KVL around the B-C loop:

$$i_C = \frac{V_{CC} - v_{CB}}{R_C = 0.5\,\text{k}\Omega} = 0.002(10 - v_{CB}) \tag{3.1.14}$$

The intersection of the load line with the B-C emitter characteristic curve for $-i_E = 9.3$ mA gives the operating point Q_C:

$$v_{CB} = 5.35\,\text{V}, \quad i_E = 9.3\,\text{mA} \tag{3.1.15}$$

To be strict, unless the B-E characteristic curve varies little with v_{CB}, we should relocate Q_C with the B-E characteristic curve for $v_{CB} = 5.35$ V and repeat the same process iteratively. Then this theoretical analysis becomes time-consuming even for such a simple circuit. However, as a practical means, when the BEJ (B-E junction) is surely forward biased, we often set the BEJ voltage as

$$V_{BE} = 0.7\,\text{V} \tag{3.1.16}$$

(a)

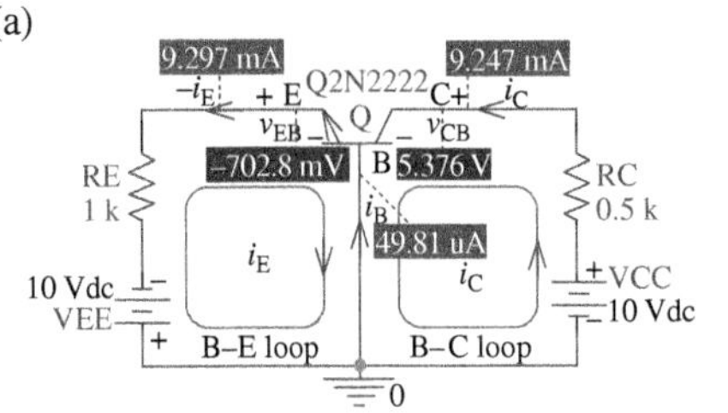

PSpice schematic for a CB BJT circuit ("CB_BJT.opj")

(b)

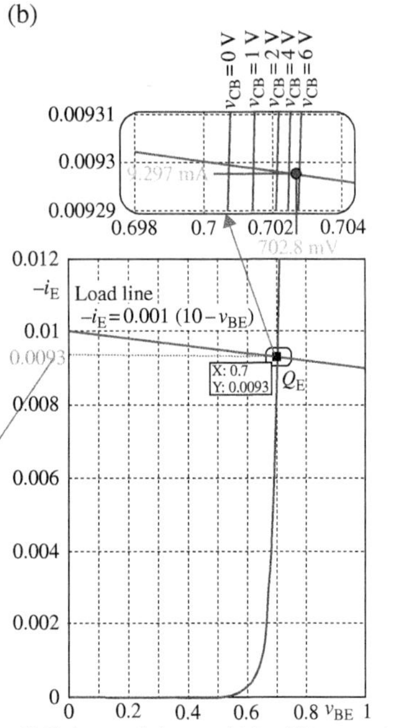

B-E characteristic curve(s) for different v_{CB}'s

(c)

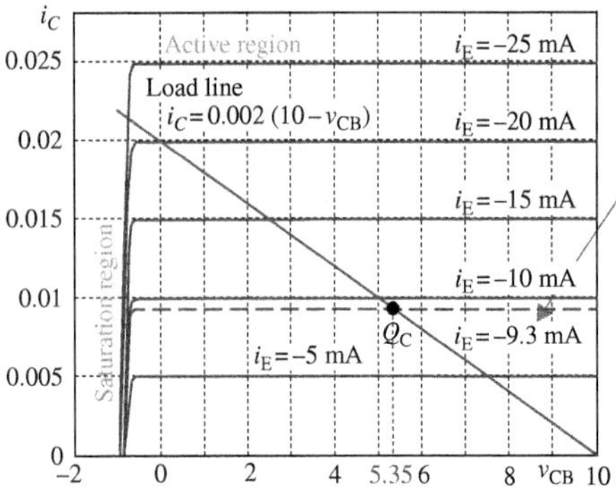

B-C characteristic curve(s) for different i_E's

(d)

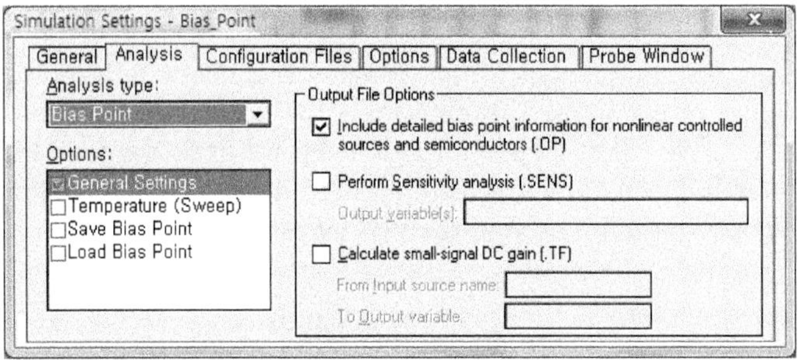

Simulation Settings dialog box for PSpice schematic for a CB BJT circuit in (a) ("CB_BJT.opj")

(e)

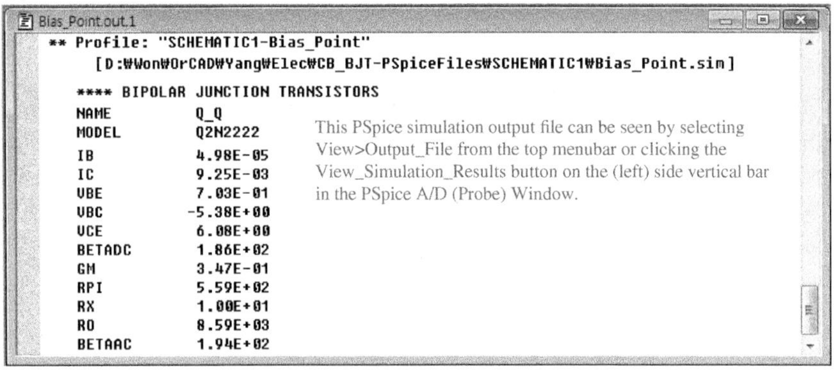
```
Bias_Point.out.1
** Profile: "SCHEMATIC1-Bias_Point"
   [D:₩Won₩OrCAD₩Yang₩Elec₩CB_BJT-PSpiceFiles₩SCHEMATIC1₩Bias_Point.sim]

**** BIPOLAR JUNCTION TRANSISTORS
NAME        Q_Q
MODEL       Q2N2222
IB          4.98E-05
IC          9.25E-03
VBE         7.03E-01
VBC        -5.38E+00
VCE         6.08E+00
BETADC      1.86E+02
GM          3.47E-01
RPI         5.59E+02
RX          1.00E+01
RO          8.59E+03
BETAAC      1.94E+02
```
This PSpice simulation output file can be seen by selecting View>Output_File from the top menubar or clicking the View_Simulation_Results button on the (left) side vertical bar in the PSpice A/D (Probe) Window.

PSpice simulation output file containing the Bias Point analysis result

Figure 3.4 A Common-Base (CB) BJT circuit and related *v-i* characteristic curves.

and instead of performing the load line analysis for the B-E loop, use Eqs. (3.1.12), (3.1.6c), (3.1.9b), and (3.1.14) together with $\beta_F = 186$ (BETADC from the PSpice simulation output file or databook) to obtain the following:

$$-i_E \overset{(3.1.12)}{=} 0.001(10 - v_{BE}) = 0.0093\,\text{A} = 9.3\,\text{mA} \tag{3.1.17a}$$

$$i_C \overset{(3.1.6c)}{=} -\alpha_F i_E \overset{(3.1.9b)}{=} \frac{\beta_F}{\beta_F + 1}(-i_E) \underset{\text{output file}}{\overset{\text{PSpice}}{=}} \frac{186}{186+1} 9.3 \approx 9.3\,\text{mA} \tag{3.1.17b}$$

$$v_{CB} \overset{(3.1.14)}{=} V_{CC} - R_C i_C$$

$$\overset{(3.1.17b)}{=} 10 - 500 \times 0.0093 = 5.35\,\text{V (BCJ reverse biased)} \tag{3.1.17c}$$

Now, to perform the PSpice simulation, we create an OrCAD/PSpice project named, say, "CB_BJT.opj," compose the schematic as depicted in Figure 3.4(a), make a Simulation Settings dialog box (with Bias Point analysis type) as depicted in Figure 3.4(d), and run it to get the PSpice simulation output some part of which is shown in Figure 3.4(e). The Bias Point analysis result can also be made seen in the schematic (Figure 3.4(a)) by clicking on the 'Enable Bias Voltage Display' and 'Enable Bias Current Display' buttons in the tool bar above the Schematic Editor window.

3.1.5 Common-Emitter Configuration

Figure 3.5(a) shows a common-emitter (CE) NPN-BJT circuit in which the input is applied to the BEJ (B-E junction) and the output is available from the CEJ (C-E junction) so that the emitter terminal is common between the input and the output. Note that we can expect the BJT to operate in the forward-active mode since the BEJ/BCJ are forward/reverse biased (Table 3.1).

To perform a comparatively exact analysis considering the nonlinearity of the circuit, we do the following:

- Setting the BEJ voltage to $v_{BE} = 0.7$ V, apply KVL around the B-E loop to find the base current i_B as

$$R_B i_B + v_{BE} = V_{BB};\ i_B = \frac{V_{BB} - v_{BE}}{R_B} = \frac{5 - 0.7}{86\,\text{k}\Omega} = 50\,\mu\text{A} \tag{3.1.18}$$

- Draw the load line on the C-E characteristic curve(s) (Figure 3.5(b)) where the load line equation can be obtained by applying KVL around the C-E loop:

$$i_C = \frac{V_{CC} - v_{CE}}{R_C = 0.5\,\text{k}\Omega} = 0.002(10 - v_{CE}) \tag{3.1.19}$$

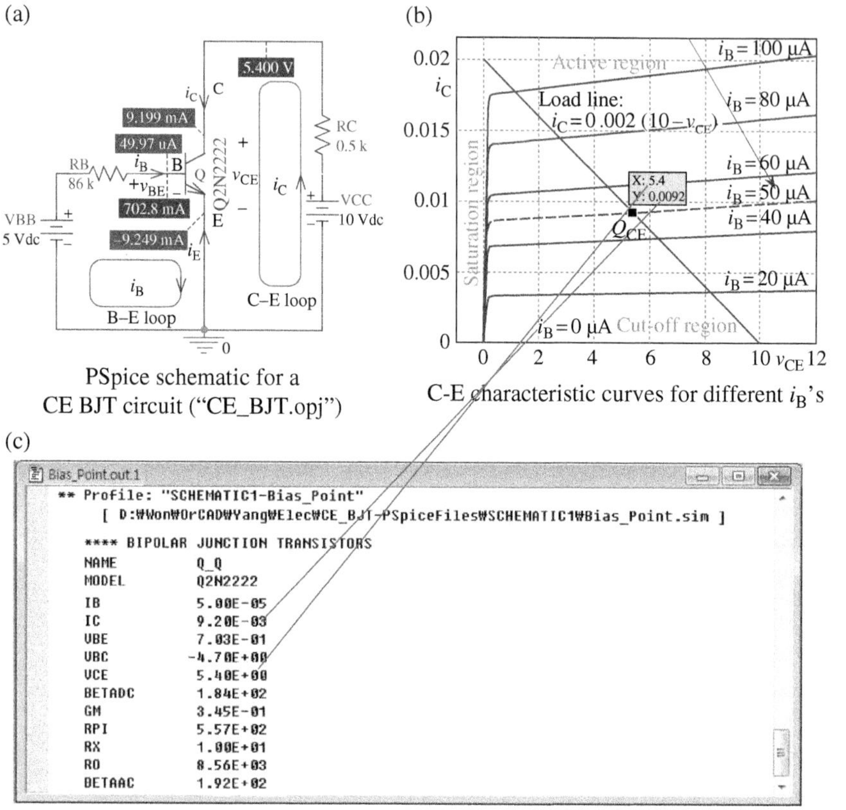

Figure 3.5 A Common-Emitter (CE) BJT circuit and related *v-i* characteristic curves.

- The intersection of the load line with the C-E characteristic curve for $i_B = 50\ \mu A$ gives the operating point Q_{CE}:

$$v_{CE} = 5.4\,\text{V},\ i_C = 9.2\,\text{mA} \tag{3.1.20}$$

Most often, as a more practical means instead of the load line analysis, we use Eqs. (3.1.6c), (3.1.18), and (3.1.19) together with $\beta_F = 184$ (BETADC from the PSpice simulation output file or databook) to obtain the following:

$$i_C \overset{(3.1.6c)}{=} \beta_F i_B \underset{\text{PSpice output file}}{\overset{(3.1.18)}{=}} 184 \times 50\ \mu\text{A} = 9.2\ \text{mA} \tag{3.1.21a}$$

$$v_{CE} \underset{\text{KVL around the C-E loop}}{\overset{(3.1.19)}{=}} V_{CC} - R_C i_C$$

$$\overset{(3.1.21a)}{=} 10 - 500 \times 0.0092 = 5.4\,\text{V} > 0.3\,\text{V} \tag{3.1.21b}$$

Whichever of the graphical or analytical methods we may use, we need to check if $v_{CE} > 0.3$ V so that the BJT will not enter the saturation mode. If v_{CE} turned out to be not greater than 0.3 V, then we would have to set $v_{CE} = v_{CE,sat} = 0.2$ V and use Eq. (3.1.19) to find the collector current i_C.

Now, to perform the PSpice simulation, we create an OrCAD/PSpice project named, say, "CE_BJT.opj," compose the schematic as depicted in Figure 3.5(a), make a Simulation Settings dialog box (with Bias Point analysis type) as depicted in Figure 3.4(d), and run it to get the PSpice simulation result as shown in Figure 3.5(c), which is a part of the PSpice simulation output file that can be seen by selecting View>Output_File from the top menu bar. The Bias Point analysis result can also be made seen in the schematic (Figure 3.5(a)) by clicking on the 'Enable Bias Voltage Display' and 'Enable Bias Current Display' buttons in the tool bar above the Schematic Editor window.

3.1.6 Large-Signal (DC) Model of BJT

Figure 3.6(a)/(b)/(c)/(d) shows the large-signal (DC) models of an NPN-BJT for the forward-active/saturation/reverse-active/cut-off modes, respectively.

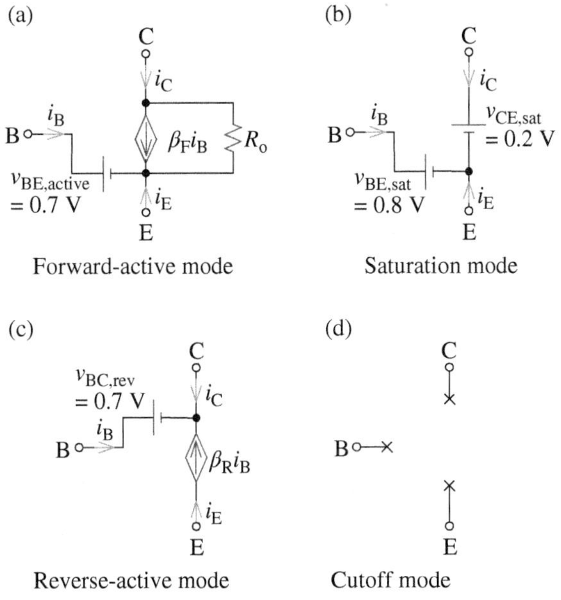

Figure 3.6 Large-signal models of NPN-BJT in different operation modes (regions).

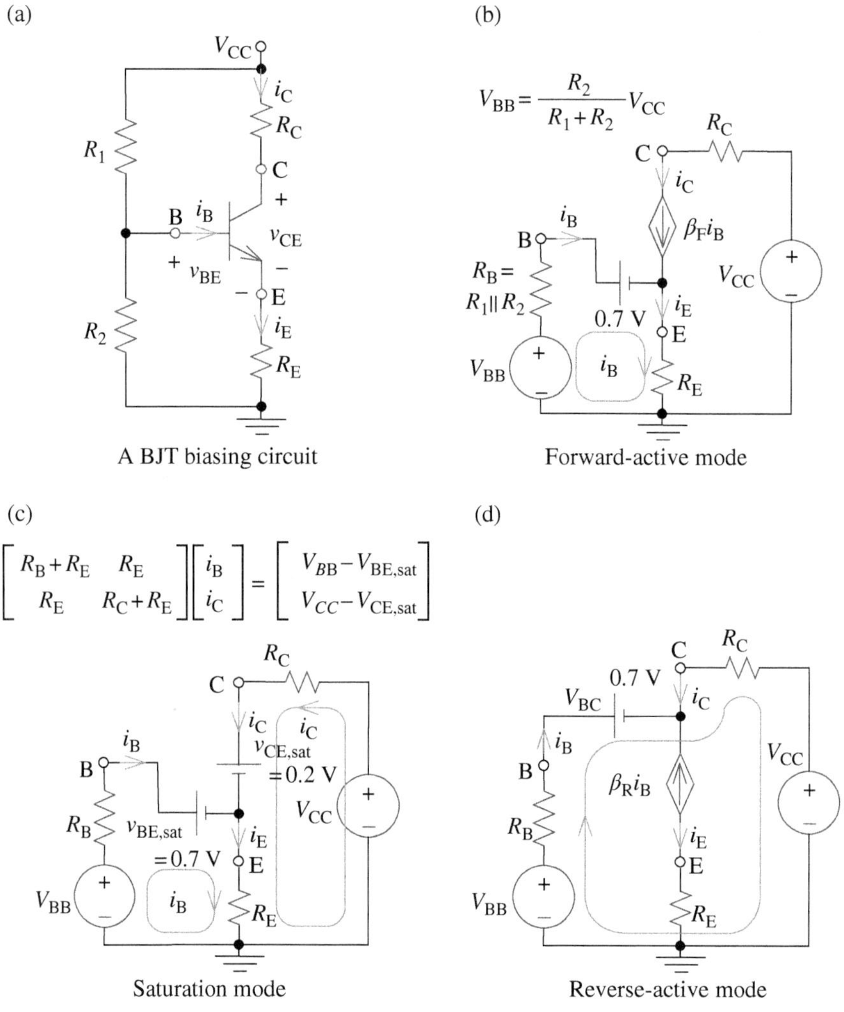

Figure 3.7 A BJT biasing circuit and its equivalents in different operation modes (regions) of BJT.

Figure 3.7(a) shows a typical (DC driven) BJT biasing circuit. Figure 3.7(b)/(c)/(d) shows its equivalents with the BEJ biasing side replaced by its Thevenin equivalent and with the BJT replaced by its model in the forward-active/saturation/reverse-active mode, respectively.

```
function [VBQ,VEQ,VCQ,IBQ,IEQ,ICQ,mode]=...
     BJT_DC_analysis(VCC,R1,R2,RC,RE,beta)
% Copyleft: Won Y. Yang, wyyang53@hanmail.net, CAU for academic use only
% VCC is supposed to be given as [VCC VEE].
l_beta=length(beta);
if l_beta<2, betaR=1; betaF=beta; else betaR=beta(2); betaF=beta(1); end
if length(RE)>1, RE=sum(RE); end
if length(VCC)>1, VEE=VCC(2); VCC=VCC(1); else VEE=0; end
if R1>20, VBB=VEE*R2/(R1+R2); RB=parallel_comb([R1 R2]);
 else VBB=R1; RB=R2;
end
VBEQ=0.7; IBQ=(VBB-VEE-VBEQ)/(RB+(1+betaF)*RE);
if IBQ>0 % Forward-active or Saturation mode
  mode='forward-active'; % Fig. 3.7(b)
  ICQ=betaF*IBQ; IEQ=IBQ+ICQ; VEQ=VEE+RE*IEQ; VBQ=VEQ+VBEQ;
  VCQ=VCC-RC*ICQ;
  if VCQ-VEQ<0.3, % Saturation mode: Fig. 3.7(c)
    mode='saturation'; VCEQ=0.2; VBEQ=0.7;
    x=[RB+RE RE; RE RC+RE]\[VBB-VEE-VBEQ; VCC-VEE-VCEQ];
    IBQ=x(1); ICQ=x(2); IEQ=IBQ+ICQ;
    VEQ=RE*IEQ; VBQ=VEQ+VBEQ; VCQ=VEQ+VCEQ;
  end
 else % Cut-off mode or Reverse-active mode
  mode='cut-off'; VEQ=0; VBQ=VBB; VCQ=VCC; IBQ=0; ICQ=0; IEQ=0;
  if VBQ-VCQ>0 % Reverse-active mode: Fig. 3.7(d)
   mode='reverse-active'; VBCQ=0.7;
   IBQ=(VBB-VBCQ-VCC)/(RB+(1+betaR)*RC); IEQ=-betaR*IBQ; ICQ=-IBQ+IEQ;
   VCQ=VCC-RC*ICQ; VBQ=VCQ+VBCQ; VEQ=RE*IEQ;
   if VCQ-VEQ>-0.2, % Reverse-saturation mode
     mode='reverse-saturation'; VCEQ=-0.2; VBCQ=0.7;
     x=[RB+RC -RC; -RC RE+RC]\[VBB-VBCQ-VCC; VCC-VEE-VCEQ];
     IBQ=x(1); IEQ=x(2); ICQ=-IBQ+IEQ;
     VEQ=RE*IEQ; VCQ=VEQ+VCEQ; VBQ=VCQ+VBCQ;
  end
 end
end
disp(' VCC VEE VBB VBQ VEQ VCQ IBQ IEQ ICQ')
fprintf('%5.2f %5.2f %5.2f %5.2f %5.2f %5.2f %4.2e %4.2e %4.2e',
                         VCC,VEE,VBB,VBQ,VEQ,VCQ,IBQ,IEQ,ICQ)
fprintf('\n in the %s mode with VCE,Q=%6.2f[V]\n', mode,VCQ-VEQ);
```

```
function Zp=parallel_comb(Zs)
Zp = 1./sum(1./Zs.');
```

The above MATLAB function '`BJT_DC_analysis()`' can be used to analyze NPN-BJT biasing circuits (driven by DC sources) and find the values of $V_{B,Q}$, $V_{E,Q}$, $V_{C,Q}$, $I_{B,Q}$, $I_{E,Q}$, and $I_{C,Q}$ (voltages/currents at/through the base, emitter, and collector terminals) at the operating point. Note the following about its usage:

```
function [VBQ,VEQ,VCQ,IBQ,IEQ,ICQ,mode]=...
    BJT_PNP_DC_analysis(VEE,R1,R2,RE,RC,beta)
% Copyleft: Won Y. Yang, wyyang53@hanmail.net, CAU for academic use only
% VEE is supposed to be given as [VEE VCC].
l_beta=length(beta);
if l_beta<2, betaF=beta; betaR=1; else betaR=beta(2); betaF=beta(1); end
if length(RE)>1, RE=sum(RE); end
if length(VEE)>1, VCC=VEE(2); VEE=VEE(1); else VCC=0; end
if R1>20, VBB=VEE*R2/(R1+R2); RB=parallel_comb([R1 R2]);
 else VBB=R1; RB=R2;
end
VEBQ=0.7; IBQ=(VEE-VBB-VEBQ)/(RB+(1+betaF)*RE);
if IBQ>0 % Forward-active or Saturation mode
  mode='forward-active';
  ICQ=betaF*IBQ; IEQ=IBQ+ICQ;
  VEQ=VEE-RE*IEQ; VBQ=VEQ-VEBQ; VCQ=VCC+RC*ICQ;
  if VEQ-VCQ<0.3, % Saturation mode
   mode='saturation';
   VECQ=0.2; VEBQ=0.7;
   x=[RB+RE RE; RE RC+RE]\[VEE-VBB-VEBQ; VEE-VCC-VECQ]; % KVL
   IBQ=x(1); ICQ=x(2); IEQ=IBQ+ICQ;
   VEQ=VEE-RE*IEQ; VBQ=VEQ-VEBQ; VCQ=VCC+RC*ICQ;
  end
else % Cut-off mode or Reverse-active mode
 mode='cut-off';
 VEQ=VEE; VBQ=VBB; VCQ=0; IBQ=0; ICQ=0; IEQ=0;
 if VBQ-VCQ<0 %Reverse-active mode Ex.11.4
  mode='reverse-active';
  VCBQ=0.7; IBQ=(-VBB-VCBQ)/(RB+(1+betaR)*RC);
  IEQ=-betaR*IBQ; ICQ=-IBQ+IEQ; VEQ=VEE-RE*IEQ; VBQ=VCQ-VCBQ; VCQ=RC*ICQ;
  if VEQ-VCQ>-0.2, %Reverse-saturation mode
   mode='reverse-saturation';
   VECQ=-0.2; VCBQ=0.7;
   x=[RB+RC -RC; -RC RE+RC]\[VCC-VBB-VCBQ; VEE-VCC-VECQ];
   IBQ=x(1); IEQ=x(2); ICQ=-IBQ+IEQ;
   VEQ=VEE-RE*IEQ; VCQ=VEQ-VECQ; VBQ=VCQ-VCBQ;
  end
 end
end
disp(' VEE  VCC  VBB  VBQ  VEQ  VCQ   IBQ    IEQ    ICQ')
fprintf('%5.2f %5.2f %5.2f %5.2f %5.2f %5.2f %9.2e %9.2e %9.2e',
                       VEE,VCC,VBB,VBQ,VEQ,VCQ,IBQ,IEQ,ICQ)
fprintf('\n in the %s mode with VEC,Q=%6.2f[V]\n', mode,VEQ-VCQ);
```

- If the emitter terminal is connected (via R_E) to another voltage source V_{EE}, the first input argument should be a two-dimensional vector $[V_{CC}\ V_{EE}]$.
- If the base terminal is connected (via R_B) to another voltage source V_{BB}, the second and third input arguments should be V_{BB} and R_B, respectively.

Likewise, the above MATLAB function '`BJT_PNP_DC_analysis()`' has been composed to analyze typical (DC driven) PNP-BJT biasing circuits.

Instead of the (linear) large-signal model as in Figure 3.6, the exponential model of an NPN-BJT based on Eq. (3.1.7) (with $V_A = \infty$ to exclude the Early effect) can be used to write KVL equations in v_{BE} and v_{BC} along the two paths V_{CC}-R_C-CBJ-BEJ-R_E-V_{EE} and V_{CC}-R_C-CBJ-R_B-V_{BB} (for the NPN-BJT circuit in Figure 3.8(a)) as

$$V_{CC} - V_{EE} - R_C i_C(v_{BE}, v_{BC}) + v_{BC} - v_{BE} - R_E\{i_C(v_{BE}, v_{BC}) + i_B(v_{BE}, v_{BC})\} = 0 \tag{3.1.22a}$$

$$V_{CC} - V_{BB} - R_C i_C(v_{BE}, v_{BC}) + v_{BC} - R_B i_B(v_{BE}, v_{BC}) = 0 \tag{3.1.22b}$$

where

$$i_C(v_{BE}, v_{BC}) \underset{V_A=\infty}{\overset{(3.1.7a)}{=}} I_S e^{v_{BE}/V_T} - I_{SC} e^{v_{BC}/V_T} \text{ with } V_T = \frac{273 + T[°C]}{11\,605}[V] \tag{3.1.23a}$$

$$i_B(v_{BE}, v_{BC}) \underset{V_A=\infty}{\overset{(3.1.7b)}{=}} \frac{I_S}{\beta_F} e^{v_{BE}/V_T} + \frac{I_{SC}}{\beta_R + 1} e^{v_{BC}/V_T} \tag{3.1.23b}$$

Also, KVL equations in v_{EB} and v_{CB} can be written along the two paths V_{EE}-R_E-EBJ-BCJ-R_C-V_{CC} and V_{BB}-R_B-BCJ-R_C-V_{CC} (for the PNP-BJT circuit in Figure 3.8(b)) as

$$V_{EE} - V_{CC} - R_E\{i_C(v_{EB}, v_{CB}) + i_B(v_{EB}, v_{CB})\} - v_{EB} + v_{CB} - R_C i_C(v_{EB}, v_{CB}) = 0 \tag{3.1.24a}$$

$$V_{BB} - V_{CC} - R_B i_B(v_{EB}, v_{CB}) + v_{CB} - R_C i_C(v_{EB}, v_{CB}) = 0 \tag{3.1.24b}$$

where

$$i_C(v_{EB}, v_{CB}) \underset{V_A=\infty}{\overset{(3.1.7a)}{=}} I_S\, e^{v_{EB}/V_T} - I_{SC}\, e^{v_{CB}/V_T} \tag{3.1.25a}$$

$$i_B(v_{EB}, v_{CB}) \underset{V_A=\infty}{\overset{(3.1.7b)}{=}} \frac{I_S}{\beta_F} e^{v_{EB}/V_T} + \frac{I_{SC}}{\beta_R + 1} e^{v_{CB}/V_T} \tag{3.1.25b}$$

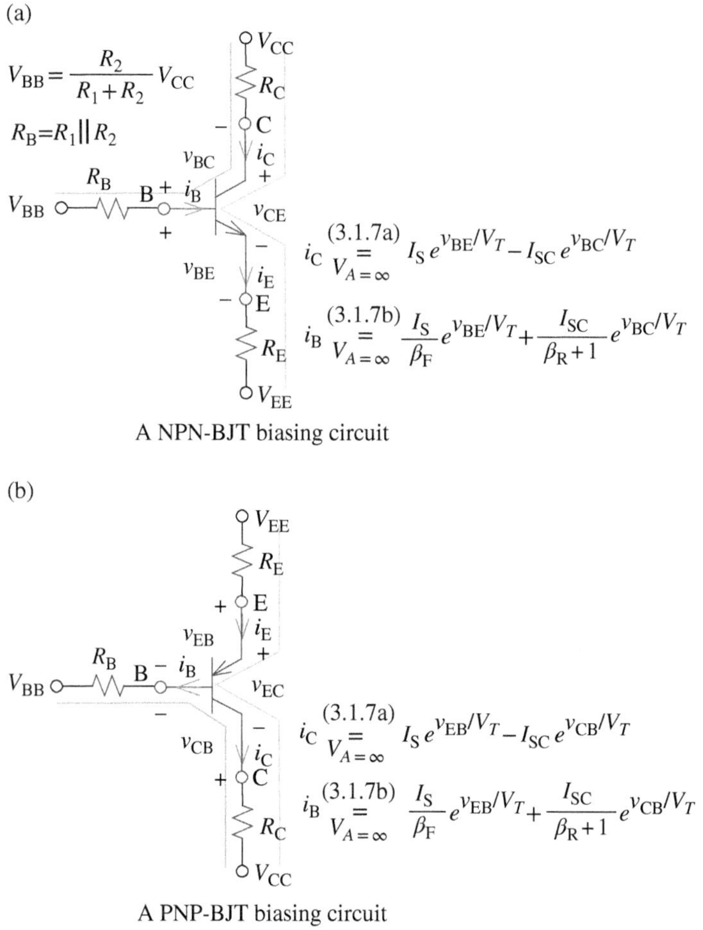

Figure 3.8 NPN/PNP BJT biasing circuits and their *i-v* relations.

The following MATLAB function '`BJT_DC_analysis_exp()`' can be used to analyze an NPN-BJT biasing circuit (based on the exponential model) and find the values of $V_{B,Q}$, $V_{E,Q}$, $V_{C,Q}$, $I_{B,Q}$, $I_{E,Q}$, and $I_{C,Q}$ (voltages/currents at/through the base, emitter, and collector terminals) at the operating point. Note the following about '`BJT_DC_analysis_exp()`':

- If you want to use it for analyzing a PNP-BJT circuit, attach the minus sign to the sixth input argument `beta`.
- The sixth input argument `beta` is expected to be given as $[\pm\beta_F \ \beta_R \ I_S]$.
- In this 'nonlinear' approach, active-or-saturated is not clear-cut but only a matter of degree.

```
function [VBQ,VEQ,VCQ,IBQ,IEQ,ICQ,mode]=...
    BJT_DC_analysis_exp(VCC, R1,R2,RC,RE,beta,VA,T)
% For a PNP-BJT, betaF(1) should be negative.
% Copyleft: Won Y. Yang, wyyang53@hanmail.net, CAU for academic use only
if nargin<8, T=27; end; VT = (273+T)/11605; % Thermal voltage
if nargin<7, VA=1e4; end % Early voltage
l_beta=length(beta);
sgn = 2*(beta(1)>0)-1; bF=abs(beta); % sgn = 1/-1 for NPN/PNP-BJT
if l_beta<2, bR=1; Is=1e-14;
 elseif l_beta<3, bR=bF(2); bF=bF(1); Is=1e-14;
 else Is=bF(3); bR=bF(2); bF=bF(1);
end
if length(RE)>1, RE=sum(RE); end
if length(VCC)>1, VEE=VCC(2); VCC=VCC(1); else VEE=0; end
if R1>20
  VBB=(VCC*(sgn>0)+VEE*(sgn<0))*R2/(R1+R2); RB=parallel_comb([R1 R2]);
  else VBB=R1; RB=R2;
end
alphaR=bR/(bR+1); Isc=Is/alphaR;
v0 = [0.7; 0.4]; % Initial guess for v=[vBE vBC]/[vEB vCB]
iC = @(v) Is*exp(v(1)/VT)-Isc*exp(v(2)/VT);
iB = @(v) Is/bF*exp(v(1)/VT)+Isc/(bR+1)*exp(v(2)/VT);
eq = @(v) [sgn*(VCC-VEE)-v(1)+v(2)-RC*iC(v)-RE*(iC(v)+iB(v));
          sgn*(VCC-VBB)+v(2)-RC*iC(v)+RB*iB(v)]; %Eq.(3.1.22)/(3.1.24)
options=optimoptions('fsolve','Display','off','Diagnostics','off');
v = fsolve(eq,v0,options);
ICQ=iC(v); IBQ=iB(v); IEQ=IBQ+ICQ;
VBE=v(1); VBC=v(2);
VCQ=VCC-sgn*RC*ICQ; VBQ=VCQ+sgn*VBC; VEQ=VBQ-sgn*VBE;
if ICQ>=1e-6, mode='forward-active';
 if ICQ/IBQ<bF-0.01, mode='saturation'; end
elseif ICQ<-1e-6, mode='reverse-active';
 if -IEQ/IBQ<bR-0.01, mode='reverse-saturation'; end
else %Cutoff mode
 mode='cut-off'; VEQ=VEE; VBQ=VBB; VCQ=VCC; IBQ=0; ICQ=0; IEQ=0;
end
if nargout<1
 fprintf('\nResults of analysis using the exponential model')
 fprintf('\n with betaF=%4.0f, betaR=%4.1f, VA=%8.2f[V], Is=%8.4f[fA]
                                        \n', bF,bR,VA,Is*1e15);
 if R1>20, fprintf(' and R1=%8.2f[kOhm], R2=%8.2f[kOhm], RC=%6.0f
                [Ohm], RE=%6.0f[Ohm]\n', R1/1e3,R2/1e3,RC,RE);
  else fprintf(' and VBB=%5.2f[V], RB=%8.2f[kOhm], RC=%6.0f[Ohm],
                        RE=%6.0f[Ohm]\n', VBB,RB/1e3,RC,RE);
 end
 disp(' VCC    VEE    VBB    VBQ    VEQ    VCQ    IBQ    IEQ    ICQ')
 fprintf('%5.2f %5.2f %5.2f %5.2f %5.2f %5.2f %9.2e %9.2e %9.2e',...
       VCC,VEE,VBB,VBQ,VEQ,VCQ,IBQ,IEQ,ICQ)
 if sgn>0, vce='VCE'; else vce='VEC'; end % NPN or PNP
 if ICQ>0, icq='ICQ'; iCQ=ICQ; beta=bF; % Forward or Reverse
  else icq='IEQ'; iCQ=IEQ; beta=bR; sgn=-sgn;
 end
 fprintf(['\n in the %s mode with ' vce ',Q=%6.2f[V]'],mode,sgn*(VCQ-VEQ));
 fprintf(['\nwhere beta_forced = ' icq '/IBQ= %6.2f with beta= %6.2f\n'],
                                                  iCQ/(IBQ+eps),beta)
end
```

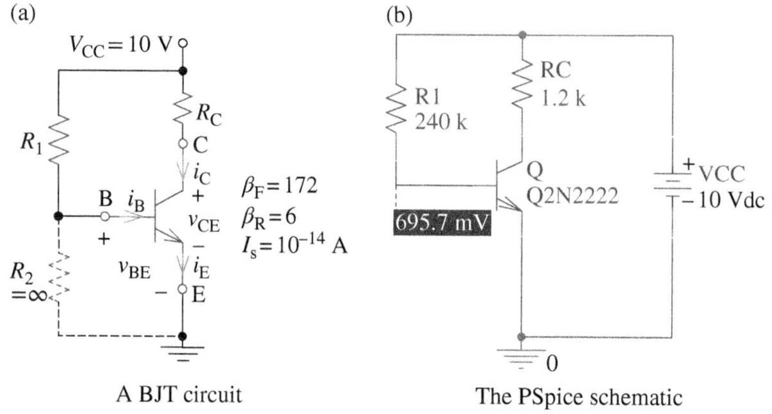

Figure 3.9 A BJT circuit and its PSpice schematic for Example 3.1.

Example 3.1 A BJT Circuit

Perform MATLAB analysis and PSpice simulation of the BJT circuit in Figure 3.9(a) for the following values of R_1 and R_C where the values of forward/reverse DC current gains β_F and β_R are assumed to be 172 and 6, respectively:

(a) R_1 = 350 kΩ, R_C = 1.5 kΩ	(b) R_1 = 350 kΩ, R_C = 2.25 kΩ
(c) R_1 = 240 kΩ, R_C = 1.5 kΩ	(d) R_1 = 240 kΩ, R_C = 1.2 kΩ

We can do a MATLAB analysis for the case of (a) by running the following statements:

```
>>betaF=172; betaR=6;
  VCC=10; VBB=10; R1=350e3; R2=1e10; RC=1500; RE=0;
  BJT_DC_analysis(VCC,R1,R2,RC,RE,[betaF betaR]);
  BJT_DC_analysis(VCC,VBB,R1,RC,RE,[betaF betaR]); % Alternatively,
  % considering B as connected to another voltage source (via R1)
```

This yields the following results:

```
 VCC   VEE   VBB   VBQ   VEQ   VCQ   IBQ      IEQ      ICQ
10.00 0.00 10.00  0.70  0.00  3.14  2.66e-05 4.60e-03 4.57e-03
      in the forward-active mode with VCE,Q= 3.14 [V]
```

Together with the results of PSpice simulation (for Bias Point analysis) using the schematic in Figure 3.9(b), the MATLAB analysis results are listed in the following table. Note that the MATLAB analysis and PSpice simulation results differ because the values of v_{BE} and β_F varying with operating point are not adaptively considered in the MATLAB analysis.

(Q) Would you increase/decrease the values of R_1/R_C to keep the BJT from entering the saturation mode?

		$I_{B,Q}$ (μA)	$I_{C,Q}$ (mA)	$V_{BE,Q}$ (V)	$V_{BC,Q}$ (V)	$V_{CE,Q}$ (V)	Mode
(a) R_1 = 350 kΩ	MATLAB	26.6	4.57	0.7	−2.44	3.14	Forward-active
R_C = 1.5 kΩ	PSpice	26.6	4.59	0.685	−2.43	3.12	Forward-active
(b) R_1 = 350 kΩ	MATLAB	26.6	4.36	0.7	0.5	0.2	Saturation
R_C = 2.25 kΩ	PSpice	26.6	4.35	0.685	0.48	0.204	Saturation
(c) R_1 = 240 kΩ	MATLAB	38.7	6.53	0.7	0.5	0.2	Saturation
R_C = 1.5 kΩ	PSpice	38.8	6.52	0.695	0.479	0.217	Saturation
(d) R_1 = 240 kΩ	MATLAB	38.7	6.66	0.7	−1.30	2.00	Forward-active
R_C = 1.2 kΩ	PSpice	38.8	6.73	0.696	−1.22	1.92	Forward-active

Example 3.2 A BJT Circuit

Perform MATLAB analysis and PSpice simulation of the BJT circuit in Figure 3.10 (a1) where $\beta_F = 172$ and $\beta_R = 6$.

To use the MATLAB function '`BJT_DC_analysis()`' for analyzing the BJT circuit, we should consider it as a standard BJT biasing circuit with possibly different value(s) of resistances and voltage sources as depicted in Figure 3.10(a2) and run the following MATLAB statements:

```
>>betaF=172; betaR=6;
  VCC=10; R1=3e5; R2=1e10; RC=1500; RE=500;
  BJT_DC_analysis(VCC,R1,R2,RC,RE, [betaF betaR]);
```

This yields the following results:

```
VCC    VEE  VBB   VBQ  VEQ   VCQ    IBQ      IEQ       ICQ
10.00  0.00 10.00 2.78 2.08 3.79 2.41e-05 4.16e-03 4.14e-03
               in the forward-active mode with VCE,Q= 1.71 [V]
```

The results of PSpice simulation (for Bias Point analysis) using the schematic in Figure 3.10(b1) or (b2) are listed in the PSpice schematic and also can be seen in Simulation Output file (opened by selecting PSpice>View_Output_File in the top menu bar of the Capture CIS Window):

$$V_{B,Q} = 2.73\,\text{V}, V_{E,Q} = 2.05\,\text{V}, V_{C,Q} = 3.88\,\text{V}, I_{B,Q} = 24.22\,\mu\text{A}, I_{C,Q} = 4.08\,\text{mA}$$

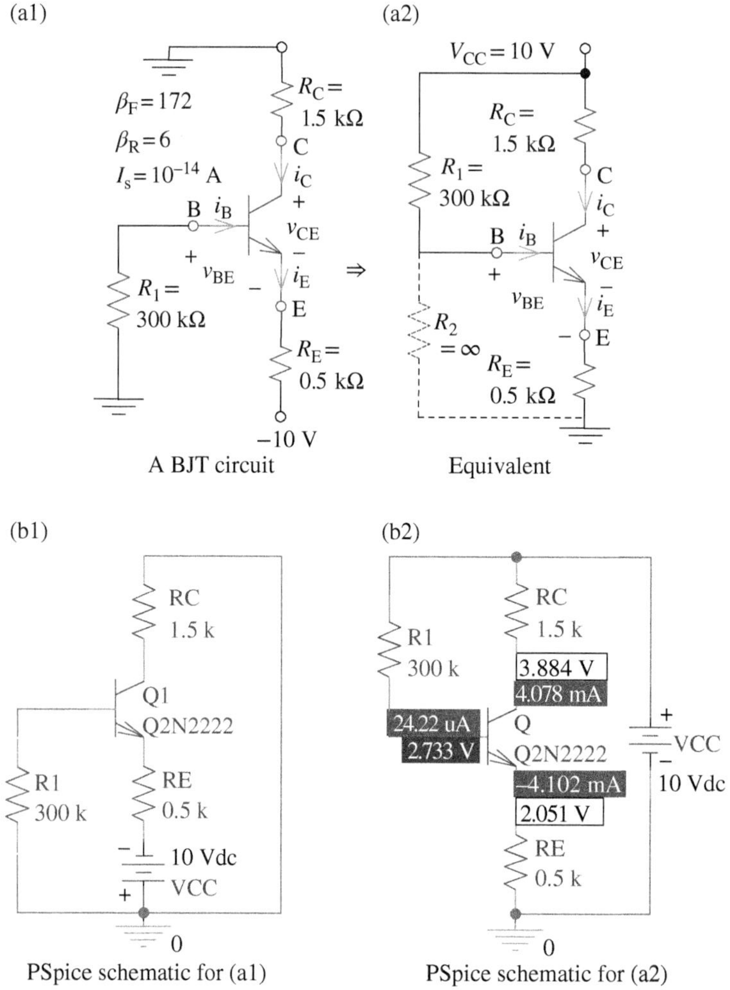

Figure 3.10 A BJT circuit, its equivalent, and their PSpice schematics for Example 3.2.

Example 3.3 A BJT Circuit

Perform MATLAB analysis and PSpice simulation of the BJT circuit in Figure 3.11(a).

We can use the MATLAB function '`BJT_DC_analysis()`' or '`BJT_DC_analysis_exp()`' to analyze the BJT circuit by running the following MATLAB statements:

```
>>betaF=172; betaR=6; Is=1e-14;
  VCC=10; R1=1e5; R2=1e5; RC=500; RE=1e3;
  BJT_DC_analysis(VCC,R1,R2,RC,RE, [betaF betaR]);
  BJT_DC_analysis_exp(VCC,R1,R2,RC,RE, [betaF betaR Is]);
  % More exact analysis using the exponential model
```

This yields the following results:

```
 VCC   VEE   VBB   VBQ  VEQ  VCQ   IBQ       IEQ       ICQ
10.00 0.00 5.00 4.04 3.34 8.34 1.93e-05 3.34e-03 3.32e-03
in the forward-active mode with VCE,Q= 5.01 [V]
 VCC   VEE VBB   VBQ  VEQ  VCQ   IBQ       IEQ       ICQ
10.00 0.00 5.00 4.03 3.35 8.34 1.93e-05 3.35e-03 3.33e-03
```

The results of PSpice simulation (for Bias Point analysis) using the schematic in Figure 3.11(b) are listed in the Simulation Output file as follows:

$$V_{BE,Q} = 6.76E-01, V_{BC,Q} = -4.30E+00, V_{CE,Q} = 4.98E+00,$$
$$I_{B,Q} = 1.94E-05, I_{C,Q} = 3.34E-03$$

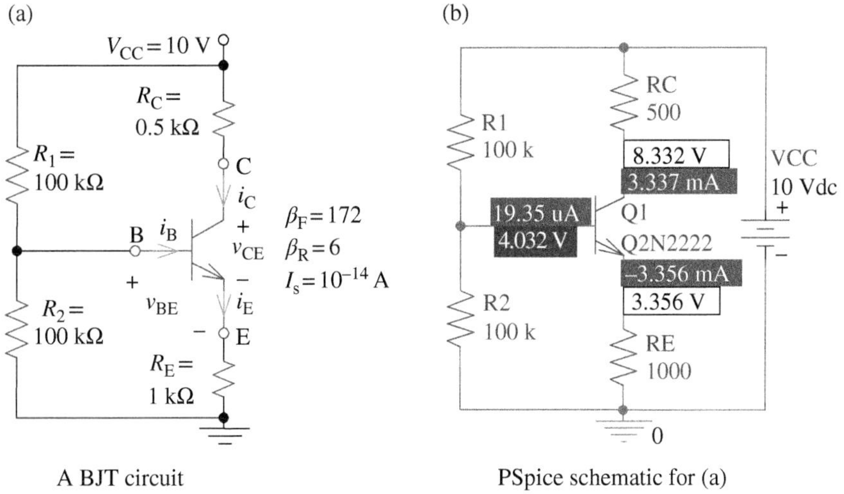

Figure 3.11 A BJT circuit and its PSpice schematic for Example 3.3.

Example 3.4 A BJT Circuit (in Reverse-Active Mode)
Perform MATLAB analysis and PSpice simulation of the BJT circuit in Figure 3.12(a).

We can use the MATLAB function ‘`BJT_DC_analysis()`’ to analyze the BJT circuit by running the following statements:

```
>>betaF=172; betaR=6;
  VCC=-15; VBB=0; R1=1e10; R2=15e3; RC=2e3; RE=2e3;
  BJT_DC_analysis(VCC,R1,R2,RC,RE, [betaF betaR]);
  BJT_DC_analysis(VCC,VBB,R2,RC,RE, [betaF betaR]); % Alternatively,
  % considering B as connected to another voltage source VBB via R1
```

This yields the following results:

```
 VCC    VEE    VBB    VBQ     VEQ    VCQ     IBQ         IEQ         ICQ
-15.00 0.00 -0.00 -7.40 -5.92 -8.10 4.93e-04 -2.96e-03 -3.45e-03
in the reverse-active mode with VCE,Q= -2.18 [V]
```

The results of PSpice simulation (for Bias Point analysis) using the schematic in Figure 3.12(b) are listed in the Simulation Output file as follows:

$$V_{BE,Q} = -1.44E+00, V_{BC,Q} = 6.83E-01, V_{CE,Q} = -2.12E+00,$$
$$I_{B,Q} = 4.92E-04, I_{C,Q} = -3.47E-03$$

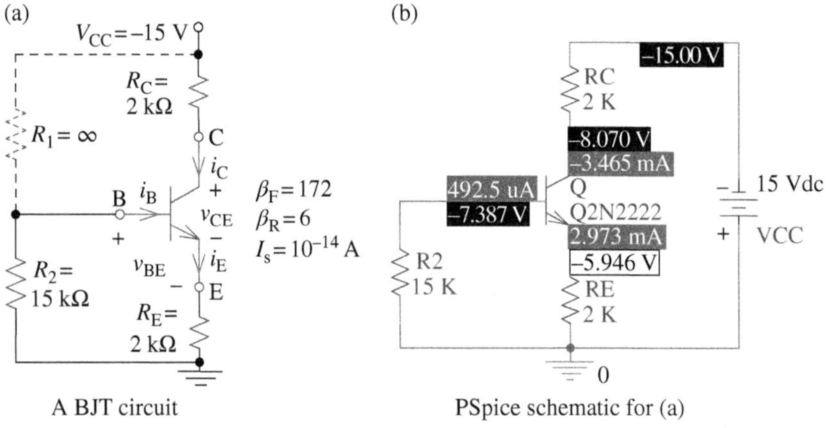

Figure 3.12 A BJT circuit and its PSpice schematic for Example 3.4.

Example 3.5 A BJT Circuit

Consider the BJT circuit in Figure 3.13(a) where the device parameters of the BJT Q_1 are $\beta_F = 100$, $\beta_R = 1$, and $I_s = 10^{-14}$ A.

a) Determine the value of V_{BB} such that the BJT operates at the edge of saturation with $v_{CE} = 0.3$ V.

With $v_{CE} = 0.3$ V, we can get i_C and i_B as

$$i_C = \frac{V_{CC} - V_C}{R_C} = \frac{V_{CC} - v_{CE}}{R_C} = \frac{10 - 0.3}{1} = 9.7\,[\text{mA}], \tag{E3.5.1}$$

$$i_B = \frac{i_C}{\beta_F} = \frac{9.7\,\text{m}}{172} = 56.4\,[\mu\text{A}] \tag{E3.5.2}$$

To make $i_B = 56.4\ \mu$A, the value of V_{BB} should be

$$V_{BB} = V_{BE} + R_B i_B = 0.7 + 10\,\text{k} \times 0.0564\,\text{m} = 1.264\,[\text{V}] \tag{E3.5.3}$$

b) With $V_{BB} = 1.26$ V, make a MATLAB analysis to find v_{CE}.

```
>>betaF=172; betaR=6;
  VCC=10;  VBB=1.26; RB=10e3; RC=1e3; RE=0;
  BJT_DC_analysis(VCC,VBB,RB,RC,RE, [betaF betaR]);
```

This yields the following results:

```
 VCC     VEE    VBB    VBQ    VEQ    VCQ     IBQ          IEQ        ICQ
10.00   0.00  1.26   0.70   0.00   0.37   5.60e-05   9.69e-03   9.63e-03
in the forward-active mode with VCE,Q= 0.37 [V]
```

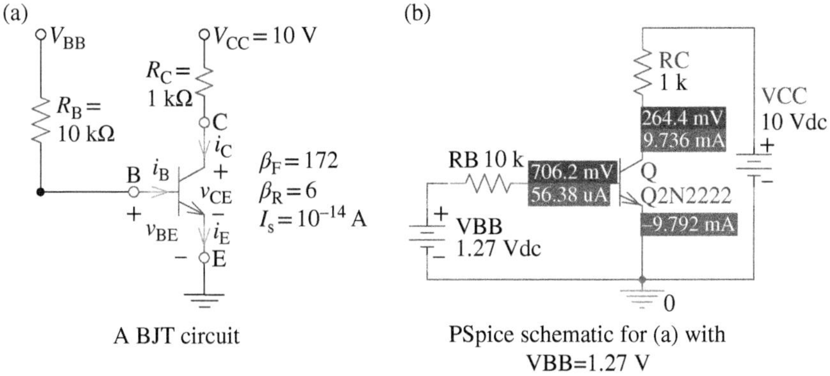

Figure 3.13 A BJT circuit and its PSpice schematic for Example 3.5.

c) With V_{BB} = 1.27 V, make a MATLAB analysis to find v_{CE}.

```
>>betaF=172; betaR=6;
  VCC=10;  VBB=1.27;  RB=10e3;  RC=1e3;  RE=0;
  BJT_DC_analysis(VCC,VBB,RB,RC,RE,[betaF betaR]);
```

This yields the following results:

```
 VCC   VEE   VBB  VBQ   VEQ   VCQ      IBQ        IEQ        ICQ
10.00 0.00 1.26 0.70 0.00 0.20 5.70e-05 9.86e-03 9.80e-03
in the saturation mode with VCE,Q= 0.20[V]
```

(Note) This example illustrates the usage of '`BJT_DC_analysis()`' for analyzing a circuit of BJT biased using two power supplies V_{CC} and V_{BB}.

Example 3.6 A PNP-BJT Circuit (in Saturation)

Consider the PNP-BJT circuit in Figure 3.14(a1) where the device parameters of the BJT Q_1 are $\beta_F = 100$, $\beta_R = 1$, and $I_s = 10^{-14}$ A.

a) Analyze the circuit to find v_{CE} and i_C and determine the operation mode of Q_1.

Noting that the EBJ is forward biased by V_{EE} = 5 V, we assume that Q_1 is in the forward-active mode so that v_{EB} = 0.7 V and apply KVL to the V_{EE}-v_{EB}-R_B loop to write

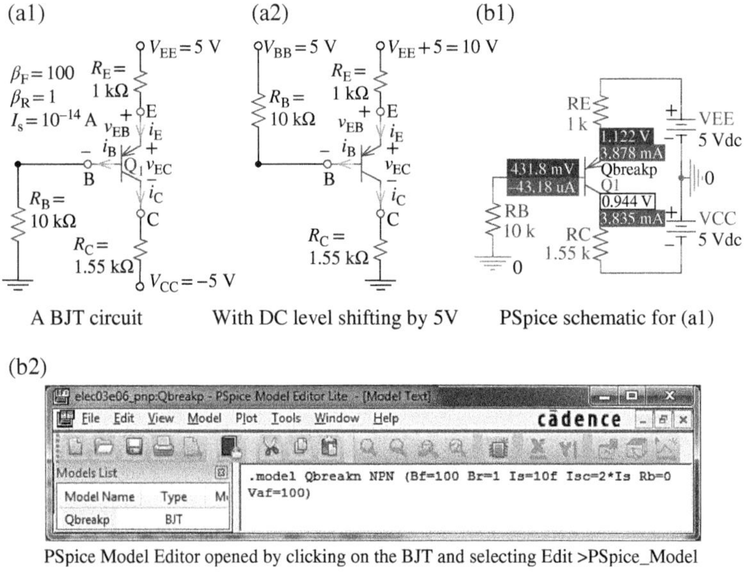
(a1)
(a2)
(b1)
A BJT circuit
With DC level shifting by 5V
PSpice schematic for (a1)
(b2)
.model Qbreakn NPN (Bf=100 Br=1 Is=10f Isc=2*Is Rb=0 Vaf=100)
PSpice Model Editor opened by clicking on the BJT and selecting Edit >PSpice_Model

Figure 3.14 A BJT circuit and its PSpice schematic for Example 3.6 ("elec03e06_pnp.opj").

$$R_B I_B + v_{EB} + (\beta+1)R_E I_B = V_{EE};$$
$$I_B = \frac{V_{EE} - v_{EB}}{R_B + (\beta+1)R_E} = \frac{5-0.7}{10+101\times 1} = 0.039\,[\text{mA}] \quad \text{(E3.6.1)}$$

$$I_E = \frac{V_{EE} - V_E}{R_E} = \frac{V_{EE} - (R_B I_B + v_{EB})}{R_E}$$
$$= \frac{5-(10\times 0.039+0.7)}{1} = 3.91\,[\text{mA}] \quad \text{(E3.6.2)}$$

$$V_C = V_{CC} + R_C I_C = V_{CC} + R_C(I_E - I_B)$$
$$= -5 + 1.55(3.91 - 0.039) = 1\,[\text{V}] \quad \text{(E3.6.3)}$$

$$V_E = V_{EE} - R_E I_E = 5 - 1\times 3.91 = 1.09\,[\text{V}] \quad \text{(E3.6.4)}$$

To see if Q_1 is really in the forward-active region, we should check

$$v_{EC} = V_E - V_C = 1.09 - 1 = 0.09\,[\text{V}] \overset{?}{\geq} 0.3\,\text{V} : \text{No} \quad \text{(E3.6.5)}$$

Now, assuming that Q_1 is in saturation so that $v_{EB} = 0.7$ V and $v_{EC} = 0.2$ V, we apply KCL at Q_1 to write

$$I_E = I_B + I_C; \quad \frac{V_{EE} - (V_C + v_{EC})}{R_E} = \frac{V_C + v_{EC} - v_{EB} - V_{BB}}{R_B} + \frac{V_C - V_{CC}}{R_C} \quad \text{(E3.6.6)}$$

and solve it to find

$$V_C = \frac{(V_{EE} - 0.2)/R_E + (V_{BB} + 0.5)/R_B + V_{CC}/R_C}{1/R_E + 1/R_B + 1/R_C}$$
$$= \frac{(5-0.2)/1 + 0.5/10 - 5/1.55}{1/1 + 1/10 + 1/1.55} = 0.93\,\text{V} \quad \text{(E3.6.7)}$$

Then we can get

$$I_B = \frac{V_C - 0.5}{R_B} = \frac{0.93 - 0.5}{10} = 0.043\,[\text{mA}] > 0, \quad \text{(E3.6.8)}$$
$$I_C = \frac{V_C - V_{CC}}{R_C} = \frac{0.93 + 5}{1.55} = 3.83\,[\text{mA}] > 0 \quad \text{(E3.6.9)}$$

To see if Q_1 is really in the saturation region, we should check

$$\beta_{\text{forced}} = \left.\frac{i_C}{i_B}\right|_{\text{saturation}} = \frac{I_C}{I_B} = \frac{3.83}{0.043} \approx 89.1 \leq \beta_F = 100 : \text{OK} \quad \text{(E3.6.10)}$$

Thus, Q_1 is indeed in saturation.

(Q) To get Q_1 out of the saturation region, would you increase or decrease $R_B/R_E/R_C$?

b) Use the MATLAB function 'BJT_PNP_DC_analysis()' or 'BJT_DC_analysis_exp()' to analyze the circuit.

We run the following MATLAB statements:

```
>>betaF=100;  betaR=1;
  VEE=5;  VCC=-5;  VBB=0;  RB=10e3;  RE=1e3;  RC=1.55e3;
  BJT_PNP_DC_analysis([VEE VCC],VBB,RB,RE,RC,[betaF betaR]);
```

to get

```
VEE    VCC    VBB    VBQ    VEQ    VCQ    IBQ       IEQ       ICQ
5.00 -5.00   0.00   0.43   1.13   0.93  4.31e-05  3.87e-03  3.83e-03
       in the saturation mode with VEC,Q= 0.20
```

Alternatively, we can modify the circuit of Figure 3.14(a1) as Figure 3.14(a2) (with a DC-level shifting by 5 V) and run the following MATLAB statements to get the same result, but with DC shifted by +5 V:

```
>>BJT_PNP_DC_analysis(VEE+5,VBB+5,RB,RE,RC,[betaF betaR]);
```

to get

```
  VEE    VCC    VBB    VBQ    VEQ    VCQ    IBQ       IEQ       ICQ
 10.00   0.00   5.00   5.43   6.13   5.93  4.31e-05  3.87e-03  3.83e-03
       in the saturation mode with VEC,Q= 0.20
```

As another alternative, we can use the MATLAB function 'BJT_DC_analysis_exp()' with negative-signed value of betaF(β_F) to get a similar (hopefully, more accurate) result:

```
>>Is=1e-14;
BJT_DC_analysis_exp([VCC VEE],VBB,RB,RC,RE,[-betaF betaR Is]);
```

to get a similar (hopefully, more accurate) result:

```
VCC    VEE    VBB    VBQ    VEQ    VCQ    IBQ       IEQ       ICQ
-5.00  5.00   0.00   0.43   1.12   0.95  4.31e-05  3.88e-03  3.84e-03
       in the saturation mode with VEC,Q= 0.17[V]
       where beta_forced = ICQ/IBQ = 89.05 with beta = 100.00
```

c) Perform a PSpice simulation of the circuit for the bias-point analysis.

We draw the PSpice schematic (Figure 3.14(b1)), click on the part: PNP-BJT Q_1 (QbreakP belonging to the PSpice library 'breakout.olb') to select it, select the menu Edit>PSpice_Model to open the PSpice Model Editor window

(Figure 3.14(b2)) and set the values of Bf(β_F) = 100 and Is = 10^{-14} A, click the New Simulation Profile button to open the Simulation Settings dialog box and set the Analysis type to Bias Point, click the Run button, and click the Enable Bias Voltage/Current Display button or PSpice>View_Output_File to read the values of voltage/current obtained from the simulation as shown in Figure 3.14(b1).

Example 3.7 Two-BJT Circuit (in Forward-Active Mode)

Consider the two-BJT circuit in Figure 3.15(a) where the device parameters of the BJTs Q_1 and Q_2 are β_F=100, β_R=1, and I_s=10^{-14} A in common.

a) Analyze the circuit to find V_{E1}, V_{C1}, V_{E2}, and V_{C2}.

First, we replace the voltage divider (consisting of V_{CC1}, R_1, and R_2) by its Thevenin equivalent (consisting of V_{BB1}=$V_{CC1}R_2/(R_1+R_2)$= 5 V and R_{B1}=$R_1||R_2$ = 63.5 kΩ) as shown in Figure 3.15(b). Then we assume that both BJTs operate in the forward-active region (with v_{BE1} = 0.7 V and v_{EB2} =0.7 V) and write the KVL equation (in I_{B1}) along the path V_{BB1}-R_{B1}-v_{BE1}-R_{E1} as

$$R_{B1}I_{B1} + v_{BE1} + (\beta+1)R_{E1}I_{B1} = V_{BB1}, \tag{E3.7.1}$$

which can be solved for I_{B1} as

$$I_{B1} = \frac{V_{BB1} - v_{BE1}}{R_{B1} + (\beta+1)R_{E1}} = \frac{5-0.7}{63.5 + 101\times 1.5} = 0.02\,[\text{mA}] \tag{E3.7.2}$$

Thus, we can find

$$I_{C1} = \beta I_{C1} = 100\times 0.02 = 2\,[\text{mA}],\quad I_{E1} = I_{B1} + I_{C1} = 2.02\,[\text{mA}] \tag{E3.7.3,4}$$

$$V_{E1} = R_{E1}I_{E1} = 1.5\times 2.02 = 3.03\,[\text{V}] \tag{E3.7.5}$$

Now, noting that the current through R_{C1} is not I_{C1} but (I_{C1}−I_{B2}), we write the KVL equation (in I_{B2}) along the mesh V_{EE2}-R_{E2}-v_{EB2}-R_{C1}-V_{CC1} as

$$R_{E2}(\beta+1)I_{B2} + v_{EB2} - R_{C1}(I_{C1} - I_{B2}) = V_{EE2} - V_{CC1} \tag{E3.7.6}$$

which can be solved for I_{B2} as

$$I_{B2} = \frac{V_{EE2} - V_{CC1} - v_{EB2} + R_{C1}I_{C1}}{R_{C1} + (\beta+1)R_{E2}}$$

$$\overset{(E3.7.3)}{=} \frac{10-10-0.7+1\times 100\times 0.02}{1+101\times 1} = 0.0127\,[\text{mA}] \tag{E3.7.7}$$

Thus, we can find

$$I_{C2} = \beta I_{B2} \overset{(E3.7.7)}{=} 100\times 0.0127 = 1.27\,[\text{mA}] \tag{E3.7.8}$$

$$I_{E2} \underset{(E3.7.8)}{\overset{(E3.7.7)}{=}} I_{B2} + I_{C2} = 1.29\,[\text{mA}] \tag{E3.7.9}$$

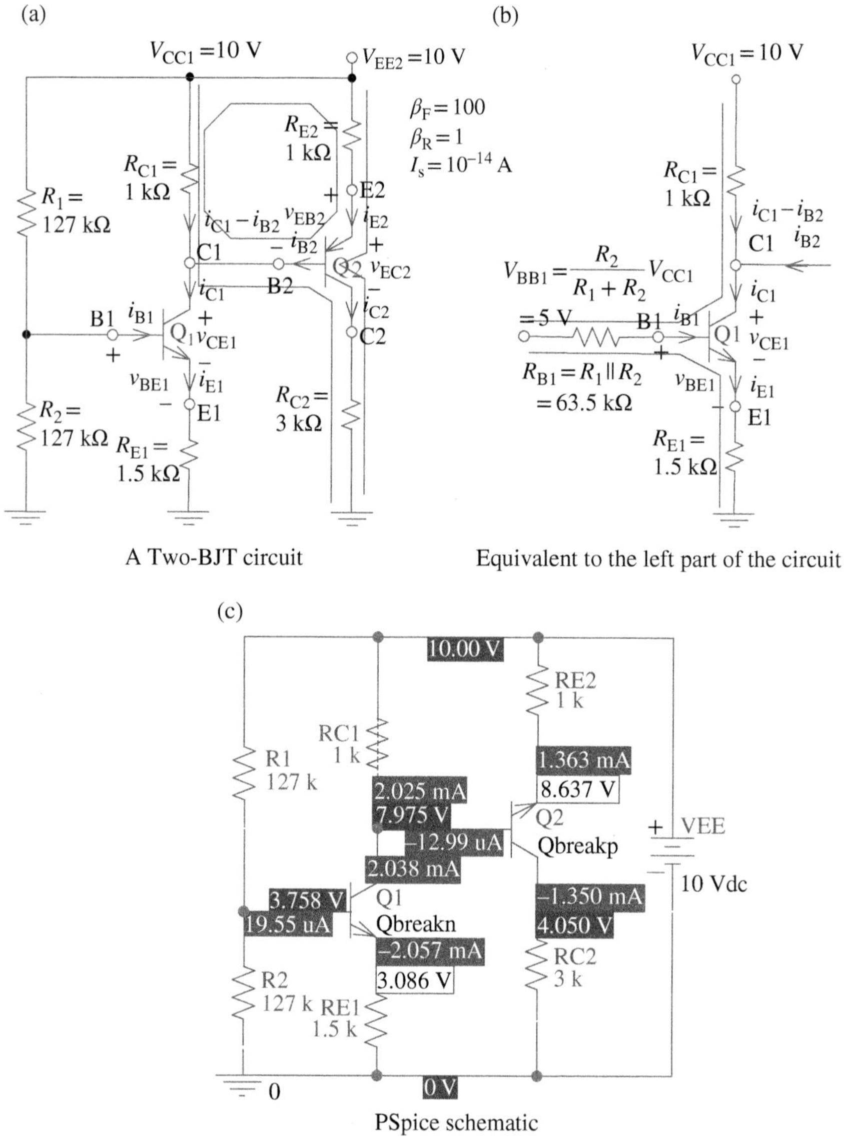

Figure 3.15 A two-BJT BJT circuit for Example 3.7.

$$V_{E2} = V_{EE2} - R_{E2} I_{E2} \overset{(E3.7.9)}{=} 10 - 1.5 \times 1.29 = 8.71\ [V] \tag{E3.7.10}$$

$$V_{C2} = R_{C2} I_{C2} \overset{(E3.7.8)}{=} 3 \times 1.27 = 3.81\ [V] \tag{E3.7.11}$$

$$V_{C1} = V_{E2} - v_{EB2} \overset{(E3.7.10)}{=} 8.71 - 0.7 = 8.01\ [V] \tag{E3.7.12}$$

Now we check if Q_1 and Q_2 operate in the forward-active region:

$$v_{CE1} = V_{C1} - V_{E1} \underset{(E3.7.5)}{\overset{(E3.7.12)}{=}} 8.01 - 3.03 = 4.98\,[V] \overset{?}{\geq} 0.3\,V : OK \tag{E3.7.13a}$$

$$v_{EC2} = V_{E2} - V_{C2} \underset{(E3.7.11)}{\overset{(E3.7.10)}{=}} 8.71 - 3.81 = 4.90\,[V] \overset{?}{\geq} 0.3\,V : OK \tag{E3.7.13b}$$

This implies that Q_1 and Q_2 operate in the forward-active region.

b) Use the MATLAB function 'BJT_DC_analysis()'/'BJT_PNP_DC_analysis()' to analyze the Q_1/Q_2 part, respectively.

First, we run the following MATLAB statements for the analysis of the Q_1 (NPN-BJT) part:

```
>>betaF=100; betaR=1; VCC1=10; VEE2=10; beta=[betaF betaR];
  R1=127e3; R2=127e3; RC1=1e3; RE1=1.5e3; RE2=1e3; RC2=3e3;
  [VB1,VE1,VC1,IB1,IE1,IC1,mode1]= ...
  BJT_DC_analysis(VCC1,R1,R2,RC1,RE1,beta);
```

to get the following analysis result for the Q_1 part:

```
 VCC    VEE    VBB   VBQ    VEQ    VCQ     IBQ        IEQ        ICQ
10.00  0.00  5.00 3.73  3.03  8.00  2.00e-05  2.02e-03  2.00e-03
in the forward-active mode with VCE,Q= 4.97[V]
```

Then, noting that $V_{C1} = V_{CC1} - R_{C1}I_{C1} = 8$ has been obtained with Q_2 unconnected, i.e. $I_{B2} = 0$, we run the following MATLAB statements for the analysis of the Q_2 (PNP-BJT) part with the Thevenin equivalent consisting of a voltage source $V_{C1} = 8$ and a resistor $R_{B2} = R_{C1}$ at B_2:

```
>>VBB2=VC1; RB2=RC1;
BJT_PNP_DC_analysis(VEE2,VBB2,RB2,RE2,RC2,beta);
```

to get the following analysis result for the Q_2 part:

```
 VEE    VCC    VBB    VBQ    VEQ    VCQ     IBQ        IEQ        ICQ
10.00  0.00  8.00  8.01  8.71  3.82  1.27e-05  1.29e-03 1.27e-03
in the forward-active mode with VEC,Q= 4.89[V]
```

Q) V_{C1} should be equal to V_{B2}. However, we have VC1 = 8.00 and VB2 = 8.01. Which one is wrong and how should it be corrected?

Alternatively, we can use the exponential model to set up the KVL equations (in v_{BE1}, v_{BC1}, v_{EB2}, and v_{CB2}) along the four paths V_{CC1}-R_{C1}-v_{CB1}-R_{B1}-V_{BB1}, V_{BB1}-R_{B1}-v_{BE1}-R_{E1}, V_{CC1}-R_{C1}-v_{BC2}-R_{C2}, and V_{EE2}-R_{E2}-v_{EB2}-v_{BC2}-R_{C2} as

$$V_{CC1} - R_{C1}\{i_C(v_{BE1}, v_{BC1}) - i_B(v_{EB2}, v_{CB2})\} + v_{BC1} + R_{B1}i_B(v_{BE1}, v_{BC1}) - V_{BB1} = 0 \tag{E3.7.14a}$$

```
%elec03e07.m
clear
betaF=100; betaR=1; alphaR=betaR/(betaR+1); beta=[betaF betaR];
Is=1e-14; Isc=Is/alphaR; VT=25e-3; % BJT parameters
VCC1=10; VEE2=10; R1=127e3; R2=127e3;
RC1=1e3; RE1=15e2; RE2=1e3; RC2=3e3;
% To find the equivalent of the voltage divider biasing circuit
VBB1=VCC1*R2/(R1+R2); RB1=parallel_comb([R1 R2]);
% PWL model based approach
vBE1=0.7; vEB2=0.7;
IB1 = (VBB1-vBE1)/(RB1+(betaF+1)*RE1) % Eq.(E3.7.2)
IC1 = betaF*IB1, IE1 = IB1+IC1 % Eq.(E3.7.3,4)
VE1 = RE1*IE1 % Eq.(E3.7.5)
IB2 = (VEE2-VCC1-vEB2+RC1*IC1)/(RC1+(betaF+1)*RE2) % Eq.(E3.7.7)
IC2 = betaF*IB2, IE2 = IB2+IC2 % Eq.(E3.7.8,9)
VE2 = VEE2-RE2*IE2, VC2 = RC2*IC2 % Eq.(E3.7.10,11)
VC1 = VE2-vEB2 % Eq.(E3.7.12)
vCE1 = VC1-VE1, vEC2 = VE2-VC2 % Eq.(E3.7.13a,b)
% Using BJT_DC_analysis() and BJT_PNP_DC_analysis()
[VB1_,VE1_,VC1_,IB1_,IE1_,IC1_,mode1]= ...
                 BJT_DC_analysis(VCC1,R1,R2,RC1,RE1,beta);
VBB2=VEE2-RC1*IC1_; RB2=RC1;
[VB2_,VE2_,VC2_,IB2_,IE2_,IC2_,mode2]= ...
                 BJT_PNP_DC_analysis(VEE2,VBB2,RB2,RE2,RC2,beta);
VC1_=VC1_+RC1*IB2_, VB2_ % VC1 should be compensated or set equal to VB2
% Exponential model based approach
iC=@(v)Is*exp(v(1)/VT)-Isc*exp(v(2)/VT); % Eq.(3.1.23a)
iB=@(v)Is/betaF*exp(v(1)/VT)+Isc/(betaR+1)*exp(v(2)/VT); % Eq.(3.1.23b)
% Eq.(E3.7.14) with v=[vBE1 vBC1 vEB2 vCB2]
eq=@(v)[VCC1-VBB1+v(2)-RC1*(iC(v(1:2))-iB(v(3:4)))+RB1*iB(v(1:2));
          VBB1-RB1*iB(v(1:2))-v(1)-RE1*(iC(v(1:2))+iB(v(1:2)));
          VCC1+v(4)-RC1*(iC(v(1:2))-iB(v(3:4)))-RC2*iC(v(3:4));
          VEE2-v(3)+v(4)-RE2*(iC(v(3:4))+iB(v(3:4)))-RC2*iC(v(3:4))];
options=optimset('Display','off','Diagnostics','off');
v0 = [0.7; 0.4; 0.7; 0.4]; % Initial guess for v=[vBE1 vBC1 vEB2 vCB2]
v = fsolve(eq,v0,options);
VBE1=v(1), VBC1=v(2), VEB2=v(3), VCB2=v(4)
IB1Q = iB(v(1:2)), IC1Q = iC(v(1:2))
IB2Q = iB(v(3:4)), IC2Q = iC(v(3:4))
VC1Q = VCC1-RC1*(IC1Q-IB2Q), VB1Q=VC1Q+VBC1, VE1Q=VB1Q-VBE1
VE2Q = VEE2-RE2*(IC2Q+IB2Q), VB2Q=VE2Q-VEB2, VC2Q=VB2Q+VCB2
VCE1Q = VC1Q-VE1Q, VEC2Q=VE2Q-VC2Q
```

$$V_{BB1} - R_{B1} i_B(v_{BE1}, v_{BC1}) - v_{BE1} - R_{E1}\{i_C(v_{BE1}, v_{BC1}) + i_B(v_{BE1}, v_{BC1})\} = 0 \quad \text{(E3.7.14b)}$$

$$V_{CC1} - R_{C1}\{i_C(v_{BE1}, v_{BC1}) - i_B(v_{EB2}, v_{CB2})\} + v_{CB2} - R_{C2} i_C(v_{EB2}, v_{CB2}) = 0 \quad \text{(E3.7.14c)}$$

$$V_{EE2} - R_{E2}\{i_C(v_{EB2}, v_{CB2}) + i_B(v_{EB2}, v_{CB2})\} - v_{EB2} + v_{CB2} - R_{C2} i_C(v_{EB2}, v_{CB2}) = 0 \quad \text{(E3.7.14d)}$$

where

$$i_C(v_{BE}, v_{BC}) \underset{V_A=\infty}{\overset{(3.1.7a)}{=}} I_S\, e^{v_{BE}/V_T} - I_{SC}\, e^{v_{BC}/V_T} \quad \text{(E3.7.15a)}$$

$$i_B(v_{BE}, v_{BC}) \underset{V_A=\infty}{\overset{(3.1.7b)}{=}} \frac{I_S}{\beta_F} e^{v_{BE}/V_T} + \frac{I_{SC}}{\beta_R + 1} e^{v_{BC}/V_T} \quad \text{(E3.7.15b)}$$

This set of nonlinear equations in a unknown vector $\mathbf{v} = [v_{BE1}\ v_{BC1}\ v_{EB2}\ v_{CB2}]$ can be solved by using the MATLAB function '`fsolve()`' as shown in the above script "elec03e07.m":

A part of the results obtained from running this script is as follows:

```
IC1Q=0.0020, IC2Q=0.001355
VC1Q=VB2Q=7.9907, VB1Q=3.7155, VE1Q=3.0647, VE2Q=8.6315,
VC2Q=4.0649
```

(Q) Which one is closer to the PSpice simulation results (shown in Figure 3.15 (c)), the PWL-model-based solution or the exponential-model-based solution?

c) Use the PSpice software to simulate the circuit.

We draw the PSpice schematic (Figure 3.15(c)) where the device parameters of the NPN-BJT Q_1 (QbreakN) and PNP-BJT Q_2 (QbreakP) picked up from the PSpice library 'breakout.olb' have been set in the PSpice Model Editor window as shown in Figure 3.14(b2). The simulation results obtained by performing Bias Point analysis are shown in the schematic.

Example 3.8 Three-BJT Circuit

Consider the three-BJT circuit in Figure 3.16(a) where the device parameters of the BJTs Q_1, Q_2, and Q_3 are $\beta_F = 100$, $\beta_R = 1$, and $I_s = 10^{-14}$ A in common.

a) Analyze the circuit to find V_{E1}, V_{C1}, V_{E2}, V_{C2}, V_{E3}, and V_{C3}.

First, noting that the Q_1-Q_2 part of the circuit is identical to the circuit of Figure 3.15(a), we copy all the analysis results of Example 3.7 except for V_{C2}, because the current through R_{C2} decreases from I_{C2} by I_{B3}. To find I_{B3}, we assume that Q_3 also operates in the forward-active mode (with $v_{BE3} = 0.7$ V) and write the KVL equation (in I_{B3}) along the path R_{C2}-v_{BE3}-R_{E3} as

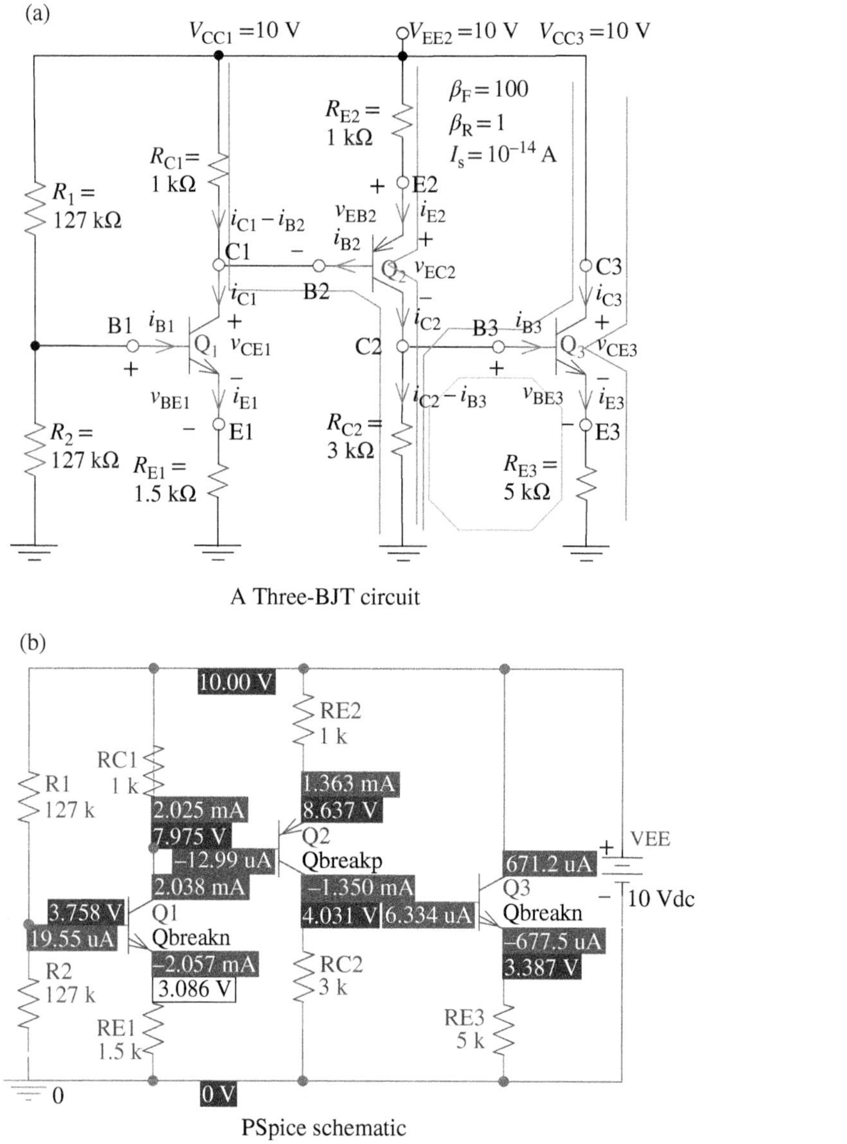

Figure 3.16 A three-BJT BJT circuit for Example 3.8.

$$-R_{C2}(I_{C2} - I_{B3}) + v_{BE3} + R_{E3}(\beta_3 + 1)I_{B3} = 0 \tag{E3.8.1}$$

which can be solved for I_{B3} as

$$I_{B3} = \frac{R_{C2}I_{C2} - v_{BE3}}{R_{C2} + (\beta_3 + 1)R_{E3}} = \frac{3 \times 1.27 - 0.7}{3 + 101 \times 5} = 0.00612\,[\text{mA}] \tag{E3.8.2}$$

Thus, we can find

$$I_{C3} = \beta I_{B3} = 100 \times 0.00612 = 0.612\,[\mathrm{mA}] \tag{E3.8.3}$$

$$I_{E3} = I_{B3} + I_{C3} = 0.618\,[\mathrm{mA}] \tag{E3.8.4}$$

$$V_{E3} = R_{E3} I_{E3} = 5 \times 0.618 = 3.09\,[\mathrm{V}] \tag{E3.8.5}$$

$$V_{C2} = V_{B3} = V_{E3} + V_{BE3} = 3.09 + 0.7 = 3.79\,[\mathrm{V}] \tag{E3.8.6}$$

Now we check if Q_2 and Q_3 operate in the forward-active region:

$$v_{EC2} = V_{E2} - V_{C2} \underset{(E3.8.6)}{\overset{(E3.7.10)}{=}} 8.71 - 3.79 = 4.92\,[\mathrm{V}] \overset{?}{\geq} 0.3\,\mathrm{V}:\ \mathrm{OK} \tag{E3.8.7a}$$

$$v_{CE3} = V_{C3} - V_{E3} \overset{(E3.8.5)}{=} 10 - 3.09 = 6.91\,[\mathrm{V}] \overset{?}{\geq} 0.3\mathrm{V}:\mathrm{OK} \tag{E3.8.7b}$$

This implies that Q_2 and Q_3 operate in the forward-active region.

b) With the analysis result of the Q_1/Q_2 part obtained in Example 3.7, go on to use the MATLAB function '`BJT_DC_analysis()`' for the analysis of the Q_3 part.

Regarding Q_3 as driven by the Thevenin equivalent consisting of $V_{BB3} = V_{C20} = R_{C2} I_{C2}$ (with Q_3 unconnected) and $R_{B3} = R_{C2}$, we run the following MATLAB statements for the analysis of the Q_3 (NPN-BJT) part:

```
>>VCC3=10; RC2=3e3; IC2=1.27e-03; % From the solution of Example 3.7
  VBB3=RC2*IC2; RB3=RC2; RC3=0; RE3=5e3; betaF=100; betaR=1;
  BJT_DC_analysis(VCC3,VBB3,RB3,RC3,RE3, [betaF betaR]);
```

to get the following analysis result for the Q_3 part:

```
 VCC    VEE    VBB    VBQ    VEQ     VCQ      IBQ        IEQ        ICQ
10.00  0.00   3.81   3.79   3.09   10.00   6.12e-06   6.18e-04   6.12e-04
in the forward-active mode with VCE,Q= 6.91 [V],
```

Alternatively, we can use the exponential model to set up the KVL equations (in v_{BE1}, v_{BC1}, v_{EB2}, v_{CB2}, v_{BE3}, and v_{BC3}) along the six paths V_{CC1}-R_{C1}-v_{CB1}-R_{B1}-V_{BB1}, V_{BB1}-R_{B1}-v_{BE1}-R_{E1}, V_{CC1}-R_{C1}-v_{BC2}-R_{C2}, V_{EE2}-R_{E2}-v_{EB2}-v_{BC2}-R_{C2}, V_{CC3}-v_{CB3}-R_{C2}, and V_{CC3}-v_{CB3}-v_{BE3}-R_{E3} as

$$V_{CC1} - R_{C1}\{i_C(v_{BE1}, v_{BC1}) - i_B(v_{EB2}, v_{CB2})\} + v_{BC1} + R_{B1} i_B(v_{BE1}, v_{BC1}) - V_{BB1} = 0 \tag{E3.8.8a}$$

$$V_{BB1} - R_{B1} i_B(v_{BE1}, v_{BC1}) - v_{BE1} - R_{E1}\{i_C(v_{BE1}, v_{BC1}) + i_B(v_{EB2}, v_{CB2})\} = 0 \tag{E3.8.8b}$$

$$V_{CC1} - R_{C1}\{i_C(v_{BE1}, v_{BC1}) - i_B(v_{EB2}, v_{CB2})\} + v_{CB2} \\ - R_{C2}\{i_C(v_{EB2}, v_{CB2}) - i_B(v_{BE3}, v_{BC3})\} = 0 \tag{E3.8.8c}$$

$$V_{EE2} - R_{E2}\{i_C(v_{EB2}, v_{CB2}) + i_B(v_{EB2}, v_{CB2})\} - v_{EB2} + v_{CB2} \\ - R_{C2}\{i_C(v_{EB2}, v_{CB2}) - i_B(v_{BE3}, v_{BC3})\} = 0 \tag{E3.8.8d}$$

$$V_{CC3} + v_{BC3} - R_{C2}\{i_C(v_{EB2}, v_{CB2}) - i_B(v_{BE3}, v_{BC3})\} = 0 \tag{E3.8.8e}$$

$$V_{CC3} + v_{BC3} - v_{BE3} - R_{E3}\{i_C(v_{BE3}, v_{BC3}) + i_B(v_{BE3}, v_{BC3})\} = 0 \tag{E3.8.8f}$$

This set of nonlinear equations in an unknown vector $\mathbf{v} = [v_{BE1}\ v_{BC1}\ v_{EB2}\ v_{CB2}\ v_{BE3}\ v_{BC3}]$ can be solved by using the MATLAB function '`fsolve()`' as shown in the following script "elec03e08.m":

```
%elec03e08.m
betaF=100; betaR=1; alphaR=betaR/(betaR+1);
Is=1e-14; Isc=Is/alphaR; VT=25e-3; % BJT parameters
VCC1=10; VEE2=10; VCC3=10;
R1=127e3; R2=127e3; RC1=1e3; RE1=1.5e3; RE2=1e3; RC2=3e3; RC3=0; RE3=5e3;
VBB1=VCC1*R2/(R1+R2); RB1=parallel_comb([R1 R2]);
% Exponential model based approach
iC=@(v)Is*exp(v(1)/VT)-Isc*exp(v(2)/VT);
iB=@(v)Is/betaF*exp(v(1)/VT)+Isc/(betaR+1)*exp(v(2)/VT);
% Eq.(E3.8.8) with v=[vBE1 vBC1 vEB2 vCB2 vBE3 vBC3]
eq=@(v)[VCC1-VBB1+v(2)-RC1*(iC(v(1:2))-iB(v(3:4)))+RB1*iB(v(1:2));
        VBB1-RB1*iB(v(1:2))-v(1)-RE1*(iC(v(1:2))+iB(v(1:2)));
VCC1+v(4)-RC1*(iC(v(1:2))-iB(v(3:4)))- RC2*(iC(v(3:4))-iB(v(5:6)));
VEE2-v(3)+v(4)-RE2*(iC(v(3:4))+iB(v(3:4)))-RC2*(iC(v(3:4))-iB(v(5:6)));
VCC3+v(6)-RC2*(iC(v(3:4))-iB(v(5:6)));
VCC3+v(6)-v(5)-RE3*(iC(v(5:6))+iB(v(5:6)))];
options=optimset('Display','off','Diagnostics','off');
v0 = [0.7; 0.4; 0.7; 0.4; 0.7; 0.4]; % Initial guess for v
v = fsolve(eq,v0,options);
VBE1=v(1), VBC1=v(2), VEB2=v(3), VCB2=v(4), VBE3=v(5), VBC3=v(6)
IB1Q = iB(v(1:2)), IC1Q = iC(v(1:2))
IB2Q = iB(v(3:4)), IC2Q = iC(v(3:4))
IB3Q = iB(v(5:6)), IC3Q = iC(v(5:6))
VC1Q = VCC1-RC1*(IC1Q-IB2Q), VB1Q=VC1Q+VBC1, VE1Q=VB1Q-VBE1
VE2Q = VEE2-RE2*(IC2Q+IB2Q), VB2Q=VE2Q-VEB2, VC2Q=VB2Q+VCB2
VC3Q = VCC3-RC3*IC3Q, VB3Q=VC3Q+VBC3, VE3Q=VB3Q-VBE3
```

A part of the results obtained from running this script is as follows:

```
IC1Q=0.0020, IC2Q=0.0014, IC3Q=6.7745e-4
VC1Q=VB2Q=7.9907, VB1Q=3.7155, VE1Q=3.0647
VC2Q=VB3Q=4.0446, VE2Q=8.6315
VC3Q=10, VE3Q=3.4211
```

c) Use the PSpice software to simulate the circuit.
We draw the PSpice schematic (Figure 3.16(b)) where the device parameters of the NPN-BJTs Q_1, Q_3 (QbreakN), and PNP-BJT Q_2 (QbreakP) have been set in the PSpice Model Editor window as shown in Figure 3.14(b2). The simulation results obtained by performing Bias Point analysis are shown in the schematic.

Example 3.9 Two-BJT Circuit (in Saturation Mode)
Consider the two-BJT circuit in Figure 3.17(a) where the device parameters of the BJTs Q_1 and Q_2 are $\beta_F = 100$, $\beta_R = 1$, and $I_s = 10^{-14}$ A in common.

a) Analyze the circuit to find V_{E1}, V_{C1}, V_{E2}, and V_{C2}.
First, as done in Example 3.7, we assume that Q_1 operates in the forward-active region (with $v_{BE1} = 0.7$ V) and try to find I_{B1}, I_{C1}, I_{E1}, V_{E1}, V_{C1}, etc. as

$$I_{B1} \overset{(E3.7.2)}{=} \frac{V_{BB1} - v_{BE1}}{R_{B1} + (\beta+1)R_{E1}} = \frac{5-0.7}{5+101\times 1.32} = 0.0311\,[\text{mA}] \tag{E3.9.1}$$

$$I_{C1} = \beta I_{B1} = 100\times 3.14\,[\text{mA}],\quad I_{E1} = I_{B1} + I_{C1} = 3.14\,[\text{mA}] \tag{E3.9.2,3}$$

$$V_{E1} = R_{E1}I_{E1} = 1.32\times 3.14 = 4.14\,[\text{V}] \tag{E3.9.4}$$

$$V_{C1} = V_{CC1} - R_{C1}I_{C1} = 10 - 1.8\times 3.11 = 4.4\,[\text{V}] \tag{E3.9.5}$$

$$v_{CE1} = V_{C1} - V_{E1} \underset{(E3.9.5)}{\overset{(E3.9.4)}{=}} 4.40 - 4.14 = 0.26\,[\text{V}] \overset{?}{\geq} 0.3\,\text{V} : \text{No.} \tag{E3.9.6}$$

Since this implies that Q_1 may be in the saturation mode, we assume $v_{CE1} = 0.2$ V (referring to Example 3.6), write the KCL equation (in V_{C1}) at Q_1, and solve it to find

$$\begin{aligned} V_{C1} &= \frac{(V_{BB1} - 0.5)/R_{B1} + V_{CC1}/R_{C1} + v_{CE1}/R_{E1}}{1/R_{B1} + 1/R_{C1} + 1/R_{E1}} \\ &= \frac{(5-0.5)/5 + 10/1.8 + 0.2/1.32}{1/5 + 1/1.8 + 1/1..32} = 4.37\,[\text{V}] \end{aligned} \tag{E3.9.7}$$

where I_{B2} has been assumed to be small enough to make $i_{RC1} \approx I_{C1}$.

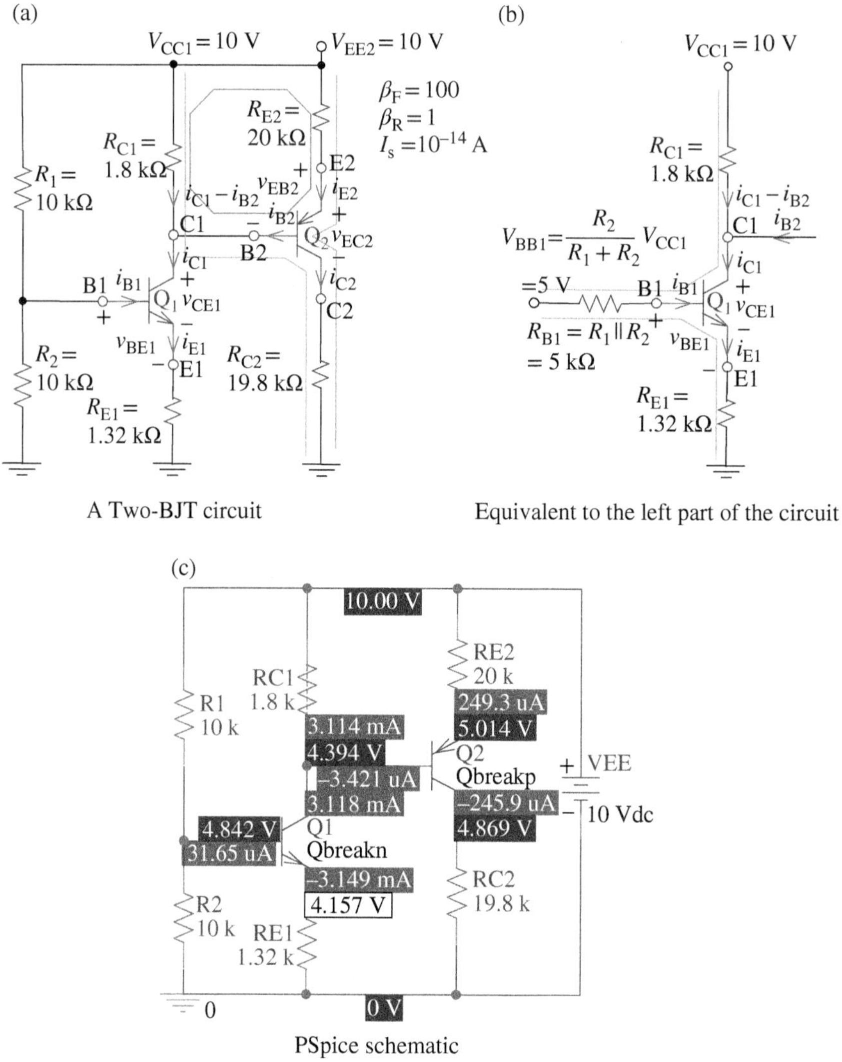

Figure 3.17 A two-BJT BJT circuit for Example 3.9.

Although we can proceed further in this way to find V_{C2}, why not use the MATLAB functions '`BJT_DC_analysis()`'/'`BJT_PNP_DC_analysis()`' to analyze the Q_1/Q_2 part, respectively? As done in Example 3.7b, we run the following MATLAB statements:

```
>>betaF=100; betaR=1; VCC1=10; VEE2=10;
  R1=10e3; R2=10e3; RC1=1.8e3; RE1=1.32e3; RE2=20e3; RC2=19.8e3;
```

```
[VB1,VE1,VC1,IB1,IE1,IC1,mode1]=...
  BJT_DC_analysis(VCC1,R1,R2,RC1,RE1,[betaF betaR]);
VBB2=VEE2-RC1*IC1; RB2=RC1;
[VB2,VE2,VC2,IB2,IE2,IC2,mode2]=...
  BJT_PNP_DC_analysis(VEE2,VBB2,RB2,RE2,RC2,[betaF betaR]);
```

to get

```
  VCC     VEE    VBB   VBQ    VEQ    VCQ       IBQ        IEQ        ICQ
 10.00   0.00   5.00  4.87   4.17   4.37   2.67e-05   3.16e-03   3.13e-03
in the saturation mode with VCE,Q= 0.20 [V]
 VEE     VCC    VBB    VBQ    VEQ    VCQ       IBQ        IEQ        ICQ
 10.00   0.00   4.37   4.37   5.07   4.87   7.56e-07   2.47e-04   2.46e-04
in the saturation mode with VEC,Q= 0.20 [V]
```

Alternatively, as done in Example 3.7b, we can use the exponential model to set up a set of KVL equations (in v_{BE1}, v_{BC1}, v_{EB2}, and v_{CB2}) and solve it by running the following MATLAB script:

```
%elec03e09.m
clear
betaF=100; betaR=1; alphaR=betaR/(betaR+1);
Is=1e-14; Isc=Is/alphaR; VT=25e-3; % BJT parameters
VCC1=10; VEE2=10;
R1=10e3; R2=10e3; RC1=1.8e3; RE1=1.32e3; RE2=20e3; RC2=19.8e3;
VBB1=VCC1*R2/(R1+R2); RB1=parallel_comb([R1 R2]);
iC = @(v) Is*exp(v(1)/VT)-Isc*exp(v(2)/VT);
iB = @(v) Is/betaF*exp(v(1)/VT)+Isc/(betaR+1)*exp(v(2)/VT);
eq=@(v) [VCC1-VBB1+v(2)-RC1*(iC(v(1:2))-iB(v(3:4)))+RB1*iB(v(1:2));
VBB1-RB1*iB(v(1:2))-v(1)-RE1*(iC(v(1:2))+iB(v(1:2)));
VCC1+v(4)-RC1*(iC(v(1:2))-iB(v(3:4)))-RC2*iC(v(3:4));
VEE2-v(3)+v(4)-RE2*(iC(v(3:4))+iB(v(3:4)))-RC2*iC(v(3:4))];
options=optimset('Display','off','Diagnostics','off');
v0 = [0.7; 0.4; 0.7; 0.4]; %Initial guess for v=[vBE1 vBC1 vEB2 vCB2]
v = fsolve(eq,v0,options);
VBE1=v(1), VBC1=v(2), VEB2=v(3), VCB2=v(4)
IB1Q = iB(v(1:2)), IC1Q = iC(v(1:2))
IB2Q = iB(v(3:4)), IC2Q = iC(v(3:4))
VC1Q=VCC1-RC1*(IC1Q-IB2Q), VB1Q=VC1Q+VBC1, VE1Q=VB1Q-VBE1
VE2Q = VEE2-RE2*(IC2Q+IB2Q), VB2Q=VE2Q-VEB2, VC2Q=VB2Q+VCB2
vCE1Q = VC1Q-VE1Q, vEC2Q = VE2Q-VC2Q
```

A part of the results obtained from running this script is as follows:

```
IC1Q=0.0031, IC2Q=2.4564e-04
VC1Q=VB2Q=4.3742, VB1Q=4.8377, VE1Q=4.1759, VE2Q=4.9730, VC2Q=4.8637
vCE1Q = 0.1983,  vEC2Q = 0.1093
```

b) Use the PSpice software to simulate the circuit.

We draw the PSpice schematic as Figure 3.17(c) and perform the Bias Point analysis to get the simulation results shown in the schematic.

3.1.7 Small-Signal (AC) Model of BJT

Figure 3.18(a)/(b) shows the high/low frequency small-signal (AC) models of an NPN-BJT for the forward-active mode, respectively, where

g_m: *transconductance (gain)*

$$g_m = \left.\frac{\partial i_C}{\partial v_{BE}}\right|_Q \overset{(3.1.1b)}{\approx} \frac{\partial}{\partial v_{BE}}\left\{\pm \alpha_F I_{ES}\left(e^{\pm v_{BE}/V_T}-1\right)\right\}\Bigg|_Q = \frac{1}{V_T}\alpha_F I_{ES} e^{\pm v_{BE,Q}/V_T} \overset{(3.1.1b)}{\approx} \frac{|I_{C,Q}|}{V_T} \tag{3.1.26}$$

$r_{be} = r_\pi = \beta / g_m$: incremental resistance of forward-biased B-E junction (3.1.27)
(a few hundred to several thousand Ω)

r_b: base-spreading resistance (40∼400 Ω)

$$r_o = \frac{1}{\partial i_C / \partial v_{CE}|_Q} = \frac{V_A(\text{Early voltage})}{I_{C,Q}} : \text{output resistance} \tag{3.1.28}$$

(tens ∼ hundreds of kΩ) due to Early effect

r_{bc}: incremental resistance of B-C junction (several MΩ)
C_{be} (C_D, C_π, CJE): diffusion capacitance of B-E junction (tens to hundreds of pF)
C_{bc} (C_T, C_μ, CJC): transition/depletion capacitance of reverse-biased B-C junction (0.1∼100 pF)

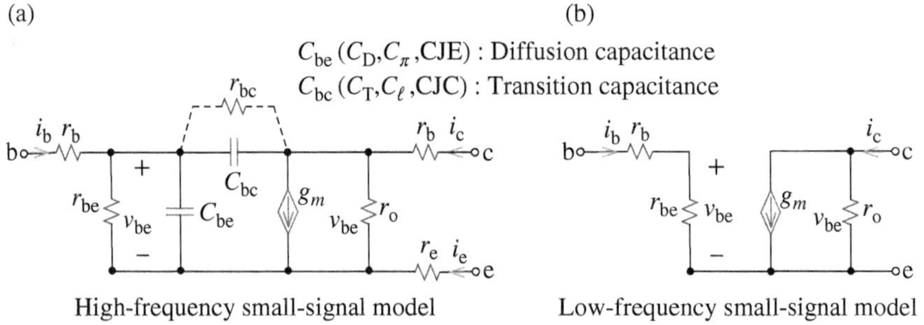

Figure 3.18 Hybrid-π small-signal models of NPN-BJT with or without frequency dependence.

Note that compared with the high-frequency model in Figure 3.18(a), the low-frequency model in Figure 3.18(b) has no capacitance because the magnitudes of impedance or reactance of C_{be} and C_{bc} are large enough to be regarded as virtually open:

$$C_{be} = 50\,\text{pF} = 50 \times 10^{-12}\,[\text{F}] \rightarrow \frac{1}{\omega C_{be}} \overset{\omega=10^4}{=} \frac{1}{10^4 \times 50 \times 10^{-12}} = 2\,\text{M}\Omega$$

$$C_{bc} = 2\,\text{pF} = 2 \times 10^{-12}\,[\text{F}] \rightarrow \frac{1}{\omega C_{bc}} \overset{\omega=10^4}{=} \frac{1}{10^4 \times 2 \times 10^{-12}} = 50\,\text{M}\Omega$$

Note also that referring to the low-frequency model in Figure 3.18(b), the trans-conductance, g_m, is related with the CE, small-signal (AC) current gain $\beta \overset{(3.1.8a)}{=} i_c/i_b$ as

$$g_m = \frac{i_c}{v_{be}} = \frac{i_c}{r_{be} i_b} = \frac{\beta}{r_{be}} \left[\frac{1}{\Omega} = \frac{\text{A}}{\text{V}} = \text{Siemens}\right] \tag{3.1.29}$$

3.1.8 Analysis of BJT Circuits

For the analysis of BJT circuits, the following three steps are taken where Table 3.2 shows the notations representing the DC/AC components and total solutions:

1) DC Analysis
   - Remove every AC source (by open/short-circuiting the current/voltage sources, respectively) and open/short-circuit every capacitor/inductor, respectively, to find the DC equivalent circuit.
   - Replace the BJTs by their large-signal models (Figure 3.6) if necessary.
   - Find the DC voltage/currents $\{V_{BE,Q}, V_{CE,Q}, I_{B,Q}, I_{E,Q}, I_{C,Q}, \ldots\}$ corresponding to the operating point Q.
2) AC Analysis
   - Determine the small-signal parameters such as g_m, β, and $r_{be} = r_\pi$, r_o, ...
   - Remove every DC source (by open/short-circuiting the current/voltage sources) and short/open-circuit every capacitor/inductor to find the AC equivalent circuit.
   - Replace the BJTs by their small-signal models (Figure 3.18) if necessary.
   - Find the AC voltage/currents $\{v_{be}, v_{ce}, i_b, i_e, i_c, \ldots\}$.
3) Superposition
   Add the DC voltages/currents and AC voltages/currents to find the complete solution: $v_{BE} = V_{BE,Q} + v_{be}$, $v_{CE} = V_{CE,Q} + v_{ce}$, $i_B = I_{B,Q} + i_b$, $i_E = I_{E,Q} + i_e$, $i_C = I_{C,Q} + i_c$

Table 3.2 Symbols representing DC and AC variables

	DC components at operating point Q	AC components: Instantaneous values	AC components: r.m.s. values	DC + AC components: Instantaneous values	DC + AC components: r.m.s. values
Base current	$I_{B,Q}$	i_b	I_b	i_B	I_B
Voltage across C-E junction	$V_{CE,Q}$	v_{ce}	V_{ce}	v_{CE}	V_{CE}

To see how the above procedure can be applied, let us consider the BJT circuit in Figure 3.19.1(a) where the roles of the three capacitors are as follows:

- C_s is used for injecting (coupling) the AC input to the base terminal of the BJT and also for blocking the DC source to keep the bias conditions undisturbed.
- C_L is used for extracting the AC output signal from the collector terminal of the BJT without disturbing the DC Q-point.
- C_E is used to make the AC signal bypass R_{E2} so that the emitter resistance should be regarded as R_{E1}+R_{E2} for setting the DC bias conditions and R_{E1} for producing AC output signal.

Note that C_s and C_L are called coupling or blocking capacitors while C_E is called a bypass capacitor. Whatever they are called, all of the capacitors are commonly supposed to provide a very large/small impedance (or reactance $X_C = 1/\omega C$) for DC($\omega = 0$)/AC($\omega > 0$) signals like being virtually open ($X_C = \infty$)/short($X_C = 0$)-circuited where ω represents the frequency of the input signal.

Now, along the procedure listed in the above box, we take the following steps:

1) **DC Analysis**

   - Remove every AC source (by open/short-circuiting current/voltage sources) and open/short-circuit every capacitor/inductor to find the DC equivalent circuit as shown in Figure 3.19.1(b).
   - Redraw Figure 3.19.1(b) as Figure 3.19.1(c) by replacing the BEJ biasing side with its Thevenin equivalent and also replacing the BJT with its large-signal model (Figure 3.6(b)).
   - For the circuit in Figure 3.19.1(c), find the DC voltage/currents corresponding to the operating point Q where the CE, forward large-signal (DC) current gain βF of the BJT is assumed to be 180.

$$V_{BE,Q} = 0.7\,V \tag{3.1.30}$$

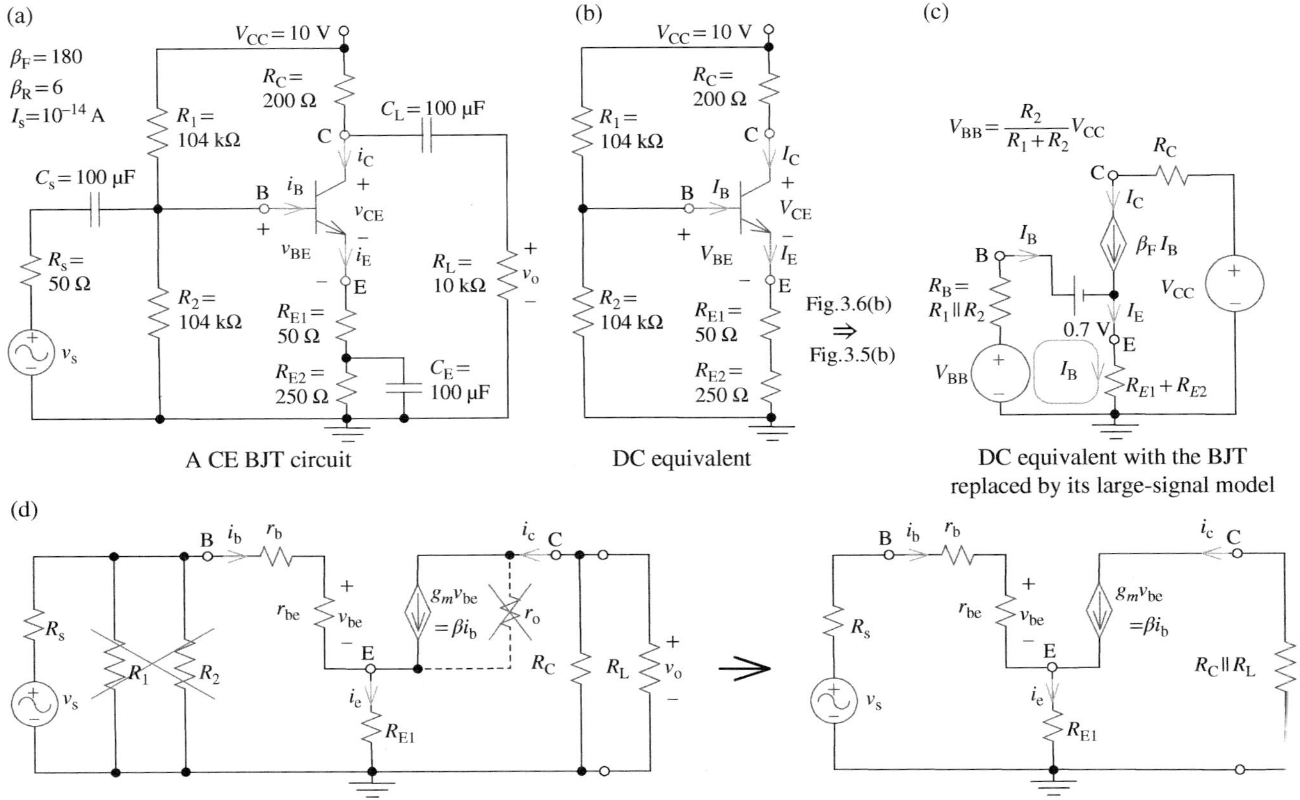

Low-frequency AC equivalent with the BJT replaced by its small-signal model and its simplification for node analysis

Figure 3.19.1 A CE BJT circuit and its DC/AC equivalents.

$$I_{B,Q} = \frac{V_{BB} - V_{BE,Q}}{R_B + (1+\beta_F)(R_{E1}+R_{E2})} = \frac{10 \times 104/(104+104) - 0.7}{(104\|104) + (1+180)(0.05+0.25)}$$
$$= 0.0405\,\text{mA} \tag{3.1.31}$$

$$I_{C,Q} = \beta_F I_{B,Q} = 180 \times 0.0405\,\text{mA} = 7.28\,\text{mA} \tag{3.1.32}$$

$$V_{CE,Q} = V_{CC} - R_C I_{C,Q} - (R_{E1}+R_{E2}) I_{E,Q} = 10 - 0.2 \times 7.28 - 0.3 \times (7.28 + 0.04)$$
$$= 6.35\,\text{V} \overset{\text{OK}}{\geq} 0.3\,\text{V} \tag{3.1.33}$$

The DC analysis can be done by running the following statements:

```
>>VCC=10; betaF=180; betaR=6;
  R1=104000; R2=104000; RC=200; RE=[50 250];
  BJT_DC_analysis(VCC,R1,R2,RC,RE,[betaF,betaR]);
```

This yields the following results that conform with the above hand-calculated results:

```
 VCC    VEE    VBB    VBQ    VEQ    VCQ       IBQ          IEQ          ICQ
10.00  0.00   5.00   2.90   2.20   8.54  4.05e-005  7.32e-003  7.28e-003
     in the forward-active mode with VCE,Q= 6.35
```

2) **AC Analysis**

- Determine the small-signal parameters such as g_m, β, and $r_{be} = r_\pi$, r_o, ...

$$g_m \overset{(3.1.26)}{\approx} \left.\frac{|I_{C,Q}|}{V_T}\right|_{T=27°C+273=300°K} \overset{(3.1.32)}{\underset{(2.1.2)}{=}} \frac{7.28\,\text{mA}}{300/11\,605\,\text{V}} \tag{3.1.34}$$
$$= 0.282\,[\text{S}]\ (\text{GM} = 0.278 \text{ from PSpice})$$

$$\beta = \left.\frac{di_C}{di_B}\right|_{v_{CE}=V_{CE,Q}} = 194\,(\text{BETAAC} = 194 \text{ from PSpice}) \tag{3.1.35}$$

$$r_{be} \overset{(3.1.27)}{=} \frac{\beta}{g_m} = \frac{194}{0.282} = 688\,\Omega\ (\text{RPI} = 698 \text{ from PSpice}) \tag{3.1.36}$$

$$r_b = 10\,\Omega\,(\text{RX} = 10 \text{ from Pspice}) \tag{3.1.37}$$

$$r_o = V_A(\text{early voltage})/I_{C,Q} = 100/0.0728$$
$$= 13.7\,\text{k}\Omega\,(\text{RO} = 10\,800 \text{ from Pspice}) \tag{3.1.38}$$

- Remove every DC source (by open/short-circuiting the current/voltage sources) and short/open-circuit every (large) capacitor/inductor to find the AC equivalent circuit as Figure 3.19.1(d) where the BJT is replaced by its low-frequency small-signal model (Figure 3.18(b)).

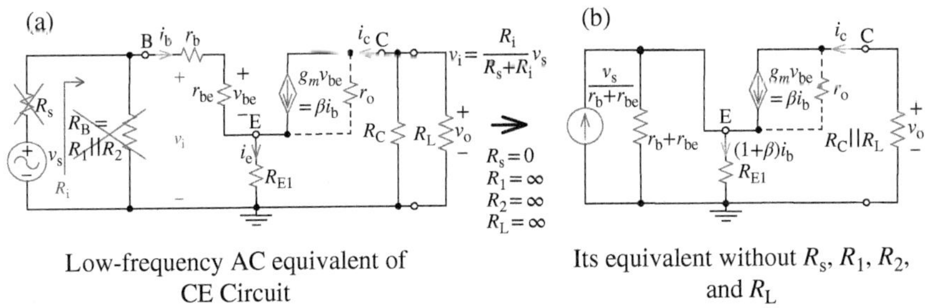

(a)
Low-frequency AC equivalent of CE Circuit
(b)
Its equivalent without R_s, R_1, R_2, and R_L

Figure 3.19.2 Low-frequency AC equivalent without R_s, R_1, R_2, and R_L to find the open-loop gain $A_{vo} = v_o/v_i$.

- To find the voltage gain $A_v = v_o/v_i$, we remove R_s and $R_B = R_1 \| R_2$ and then make a V-to-I source transformation for node analysis to get the circuit as shown in Figure 3.19.2(b).
- Then we can set up the node equation as

$$\begin{bmatrix} \frac{1}{r_b + r_{be}} + \frac{1}{R_{E1}} + \frac{1}{r_o} & -\frac{1}{r_o} \\ -\frac{1}{r_o} & \frac{1}{r_o} + \frac{1}{R_C \| R_L} \end{bmatrix} \begin{bmatrix} v_e \\ v_o \end{bmatrix} = \begin{bmatrix} \frac{v_s}{r_b + r_{be}} + g_m (v_i - v_e) \frac{r_{be}}{r_b + r_{be}} \\ -g_m (v_i - v_e) \frac{r_{be}}{r_b + r_{be}} \end{bmatrix}$$

$$; \begin{bmatrix} \frac{1 + g_m r_{be}}{r_b + r_{be}} + \frac{1}{R_{E1}} + \frac{1}{r_o} & -\frac{1}{r_o} \\ -\frac{1}{r_o} - \frac{g_m r_{be}}{r_b + r_{be}} & \frac{1}{r_o} + \frac{1}{R_C \| R_L} \end{bmatrix} \begin{bmatrix} v_e \\ v_o \end{bmatrix} = \begin{bmatrix} \frac{1 + g_m r_{be}}{r_b + r_{be}} v_i \\ -\frac{g_m r_{be}}{r_b + r_{be}} v_i \end{bmatrix} \tag{3.1.39a}$$

In case $R_{E1} = 0$, we have just one-node circuit and get the voltage gain as

$$\frac{1}{R_C \| R_L \| r_o} v_o = g_m v_{be} = g_m \frac{r_{be}}{r_b + r_{be}} v_i;\ A_v = \frac{v_o}{v_i} = g_m \frac{r_{be}}{r_b + r_{be}} (R_C \| R_L \| r_o) \tag{3.1.39b}$$

- Once we have got the open-loop voltage gain A_{vo} (with $R_L = \infty$) and the input/output resistances R_i/R_o (see Eq. (3.2.1)/(3.2.4)), we can easily get the overall voltage gain (considering R_s) as

$$G_v = \frac{R_i}{R_s + R_i} A_v = \frac{R_i}{R_s + R_i} A_{vo} \frac{R_L}{R_o + R_L} \tag{3.1.40}$$

```
function [VBQ,VEQ,VCQ,IBQ,IEQ,ICQ,Av,Ai,Ri,Ro,gm,rbe,ro]=
              BJT_CE_analysis(VCC,rb,Rs,R1,R2,RC,RE,RL,beta,Vsm,VA,T)
% Vsm = Amplitude of the AC input signal applied to base terminal through Cs
% Copyleft: Won Y. Yang, wyyang53@hanmail.net, CAU for academic use only
if nargin<12, T=27; end % Ambient temperature
if nargin<11, VA=1e4; end % Early voltage
l_beta=length(beta); % bF may have been given as [betaF betaR betaAC Is]
if l_beta<3, betaAC=beta(1); bF=beta; % bF=[betaF betaR]
 else betaAC=beta(3); bF=beta(1:2); if l_beta>3, bF=[bF beta(4)]; end
end
[VBQ,VEQ,VCQ,IBQ,IEQ,ICQ,mode]=BJT_DC_analysis(VCC,R1,R2,RC,RE,bF);
RE1=RE(1);
if R1>20, RB=parallel_comb([R1 R2]); else RB=R2; end % R1=VBB?
[gm,rbe,re,ro]=gmrbero_BJT(ICQ,betaAC,VA,T);
if Rs==0, Rs=0.01; end
RCL= parallel_comb([RC RL]);
if RE1>=1
 Y=[(1+gm*rbe)/(rb+rbe)+1/RE1+1/ro -1/ro; % gm*rbe=betaAC
 -1/ro-gm*rbe/(rb+rbe) 1/ro+1/RCL];
  vec = Y\[1+gm*rbe;-gm*rbe]/(rb+rbe); % Eq.(3.1.39a)
  Av=vec(2);
 else
  Av=-gm*rbe/(rb+rbe)*parallel_comb([ro RCL]); % Eq.(3.1.39b)
end
RbeE = rb+rbe+(betaAC+1)*RE1;
Ri = parallel_comb([RB RbeE]); % Eq.(3.2.1)
Ro = parallel_comb([RC ro]); % Eq.(3.2.4)
Ai = -betaAC*RC/(RC+RL); % Eq.(3.2.2)
Gv=Av/(Rs/Ri+1); % Eq.(3.1.40)
if nargin>9 & abs(Vsm)>0
 Ibm = Vsm/(Rs+RbeE);
 if Ibm>=IBQ, fprintf(' Possibly crash into the cutoff region since
              Ibm(%6.2fuA)>IBQ(%6.2fuA)\n',Ibm*1e6,IBQ*1e6); end
 if abs(Gv*Vsm-(1+betaAC)*Ibm*RE1)>=VCQ-VEQ-0.3 % Eq.(3.1.47)
   fprintf(' Possibly violate the saturation region since |Av*Vsm-
     (1+betaAC)*Ibm*RE1(%5.2f)|>VCQ-VEQ-0.3(%5.2f)\n',
     abs(Av*Vsm-(1+betaAC)*Ibm*RE1), VCQ-VEQ-0.3);
 end
end
fprintf(' Ri=%9.3f[kOhm], Ro=%8.0f[Ohm]\n Gv=Ri/(Rs+Ri)xAv = %6.3fx%
                    8.2f =%8.2f\n', Ri/1e3,Ro,1/(Rs/Ri+1),Av,Gv);
```

```
function [gm,rbe,re,ro]=gmrbero_BJT(ICQ,beta,VA,VT)
if VT>0.1, VT=(273+VT)/11605; end % considering VT as T
gm=abs(ICQ)/VT; rbe=beta/gm; re=rbe/(beta+1); ro=VA/abs(ICQ); %(3.1.26-28)
```

If we assume that R_1, R_2, and r_o are so large (compared with other resistors) as to be negligible as parallel elements, we can approximate the AC equivalent as Figure 3.19.1(d) so that we can write the (small-signal) base current i_b, collector current i_c, and output voltage v_o as

$$i_b \approx \frac{v_s}{R_s + r_b + r_{be} + (1+\beta)R_{E1}}; \quad i_c = \beta i_b \approx \frac{\beta v_s}{R_s + r_b + r_{be} + (1+\beta)R_{E1}} \tag{3.1.41}$$

$$v_o = -(R_C \| R_L) i_c \approx \frac{-\beta(R_C \| R_L)}{R_s + r_b + r_{be} + (1+\beta)R_{E1}} v_s$$

$$\overset{(3.1.35\sim37)}{=} \frac{-194(200 \| 10\,000)}{50 + 10 + 688 + (1+194)50} v_s = -3.62 v_s \tag{3.1.42}$$

The DC/AC analysis procedure, which has been cast into the MATLAB function '`BJT_CE_analysis()`' listed above, can be carried out by running the following statements:

```
>>VCC=10; Vsm=0.02; rb=10; betaF=180; betaR=6;   betaAC=194;
  Rs=50; R1=104000; R2=104000; RC=200; RE=[50 250];   RL=10000;
  BJT_CE_analysis(VCC,rb,Rs,R1,R2,RC,RE,RL,[betaF betaR betaAC],Vsm);
```

This yields the following result that conforms with that obtained above:

```
gm= 281.664[mS], rbe=  689[Ohm], ro= 1373[kOhm]
Gv=Ri/(Rs+Ri)xAv = 0.994 x -3.64 =  -3.62
```

Figure 3.20(a), (b), and (c) shows the PSpice schematic of the BJT circuit in Figure 3.19.1(a), its simulation result of the input/output voltage waveforms,

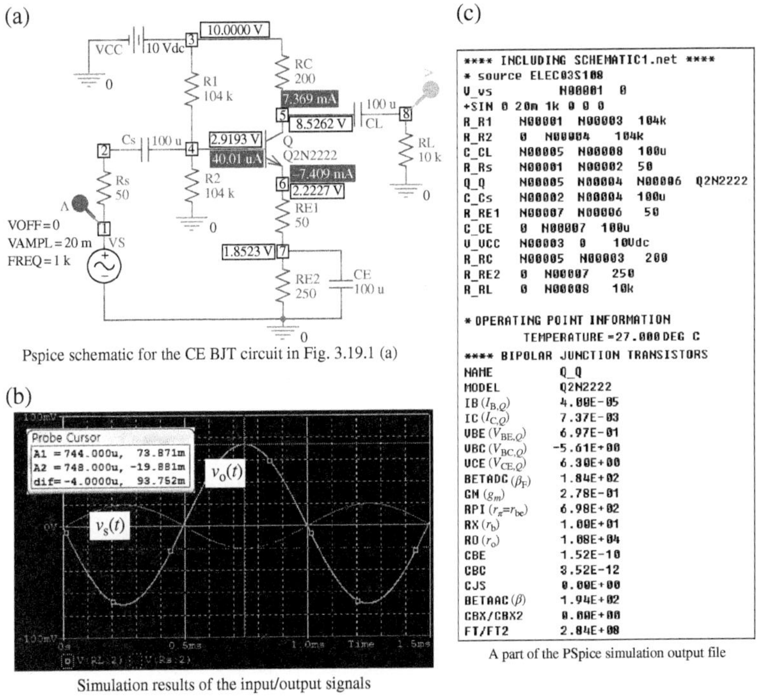

Figure 3.20 PSpice simulation of the CE BJT circuit in Figure 3.19.1a ("elec03f20.opj").

and a part of simulation output file containing the netlist and the bias (operating) point information, respectively. Note that the voltage gain can be computed from the ratio between the negative/positive peak values of input/output signals shown in the Probe Cursor box as

$$A_v = \frac{v_{om}}{v_{sm}} = \frac{73.871}{-19.881} = -3.72 \tag{3.1.43}$$

where the negative sign indicates a phase shift of 180° between the input and the output.

Now, consider the following question:

> Won't the voltage gain and/or the linear input-output relationship be changed when the amplitude (v_{sm}) of the AC input voltage v_s increases?

To find out the answer to this question, let us make a soft experiment of increasing v_{sm} to 0.3 V, 1 V, and 1.5 V. The PSpice simulation results are depicted in Figure 3.21(a), (b), and (c), which show that the upper part of the output voltage waveform is distorted for v_{sm} = 1 V and both the upper and lower parts of the output voltage waveform are distorted for v_{sm} = 1.5 V. (This explains the meaning of 'small-signal', illustrating that the small-signal analysis result based on the linear approximation is valid only within a certain range of the input signal.) Why is that? To understand why the upper and/or lower parts of the output voltage waveform to large inputs are distorted, we should use the load line analysis by drawing the DC load line to determine the operating point Q and drawing the AC load line at the operating point Q (see Figure 3.22(a)) if there exists a (bypass) capacitor (like C_E) connected in parallel with the emitter resistor R_{E2} (see Figure 3.19.1(a) or 3.20(a)).

Note that the equations of the DC/AC load lines are obtained by applying KVL through V_{CC}-R_C-v_{CE}-R_E as

$$\text{DC load line: } i_C \approx \frac{1}{R_C + R_{E1} + R_{E2}}(V_{CC} - v_{CE}) = \frac{1}{0.2 + 0.05 + 0.25 = 0.5\,\text{k}\Omega}(10 - v_{CE}) \tag{3.1.44}$$

$$\text{AC load line: } i_C = \frac{1}{(R_C \| R_L) + R_{E1}}(V_{CC} - v_{CE}) \approx \frac{1}{(0.2\|10) + 0.05 \approx 0.25\,\text{k}\Omega}(10 - v_{CE}) \tag{3.1.45}$$

where the bypass capacitor C_E is assumed to have a negligibly small reactance $1/\omega C_E$ like a short-circuit for the input signal frequency ω and the (negative) emitter current $-i_E$ is assumed to be (almost) equal to the collector current i_C since the base current is negligibly small compared with the collector/emitter currents.

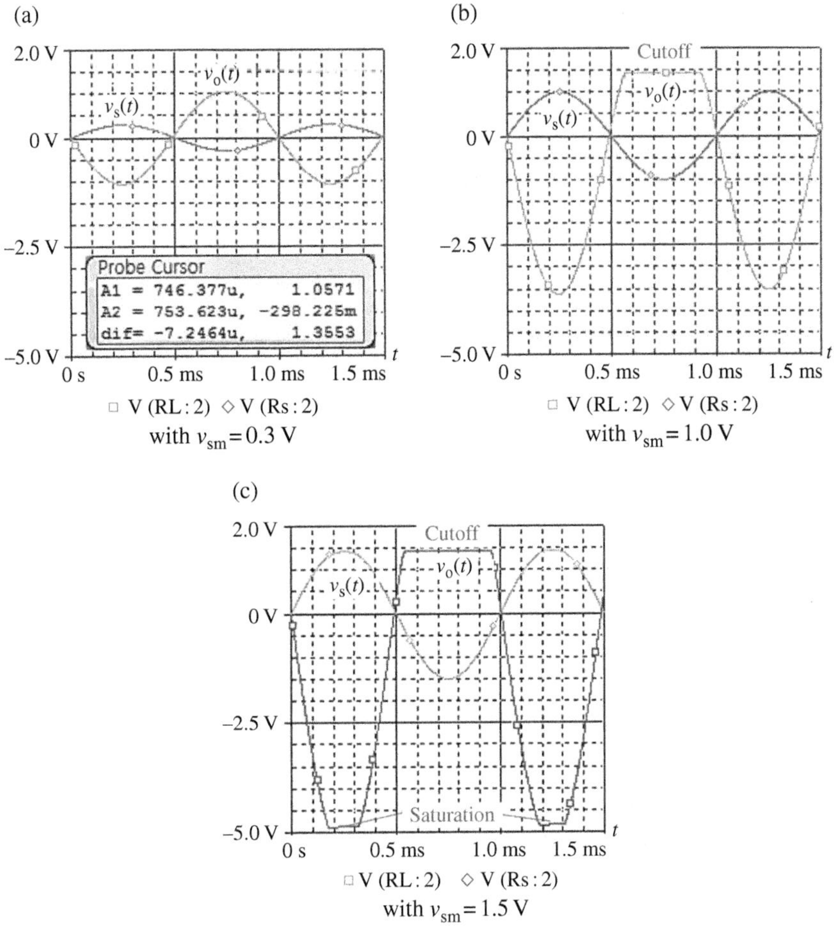

Figure 3.21 Distortion in the output voltage of the circuit of Figure 3.20a due to a large amplitude of input.

Note also that the AC input moves the instantaneous operating point along the AC load line around the quiescent operating point Q, i.e. the intersection of the DC load line and the CE characteristic curve corresponding to the base current i_B determined by the DC biasing circuit. With this background knowledge, Figure 3.22(a) together with b-d shows how i_C (the collector current) and v_{CE} (the collector-to-emitter voltage) vary with the variation of i_B due to the AC signal. About Figure 3.22, there are several observations to make:

- A BJT crashes into the cutoff region if the amplitude of i_b (the AC component of base current i_B computed roughly by Eq. (3.1.41)) exceeds $I_{B,Q}$ so that the total base current $i_B = I_{B,Q} + i_b$ may become zero where

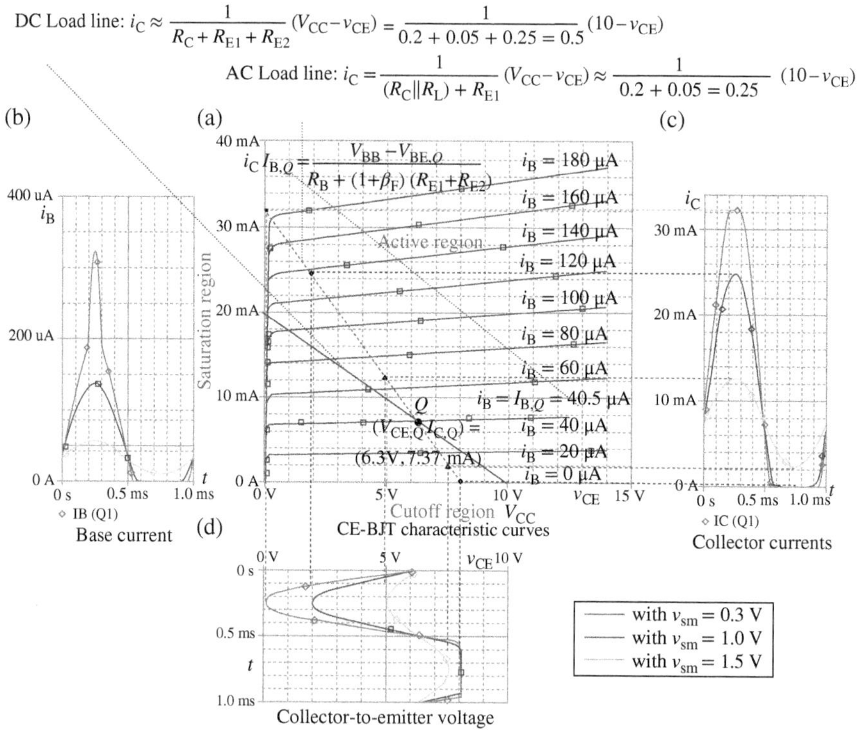

Figure 3.22 DC and AC load lines for the CE BJT circuit in Figure 3.19.1a.

$$i_{bm} \overset{(3.1.41)}{\approx} \frac{v_{sm}}{R_s + r_b + r_{be} + (1+\beta)R_{E1}} \tag{3.1.46}$$

- A BJT trespasses on the saturation region if the maximum variation of v_{ce} (the AC component of collector-to-emitter voltage) exceeds $V_{CE,Q} - V_{CE,sat}$(0.2 V) where

$$v_{ce,m} = v_{om} - (1+\beta)R_{E1} i_{bm} \underset{(3.1.46)}{\overset{(3.1.42)}{=}} \frac{\beta(R_C \| R_L) - (1+\beta)R_{E1}}{R_s + r_b + r_{be} + (1+\beta)R_{E1}} v_{sm} \tag{3.1.47}$$

- Increasing R_{E1} (with the voltage gain A_v [Eq. (3.1.49)] decreased) reduces the DC base current

$$I_{B,Q} \overset{(3.1.31)}{=} \frac{V_{BB} - V_{BE,Q}}{R_B + (1+\beta_F)(R_{E1} + R_{E2})} = \frac{V_{CC}R_2/(R_1 + R_2) - V_{BE,Q}}{R_B + (1+\beta_F)(R_{E1} + R_{E2})} \tag{3.1.48}$$

and decreases the slope $1/(R_C + R_{E1} + R_{E2})$ of the DC load line so that the operating point Q can move left (toward the saturation region) downwards (toward the cutoff region). However, the possibility of the BJT to trespass on the saturation region and/or crash into the cutoff region decreases because the maximum variations of v_{ce} (Eq. (3.1.47)) and $i_c = (\beta + 1)i_b$ decrease more abruptly than the operating point Q moves left downwards.

	V_{sm} (V)	V_{CC} (V)	R_1 (kΩ)	R_2 (kΩ)	R_C (Ω)	R_{E1} (Ω)	R_{E2} (Ω)	$I_{B,Q}$ (μA)	$I_{C,Q}$ (mA)	$V_{CE,Q}$ (V)	Remark
(1)	1.5	10	104	104	200	50	250	40	7.37	6.3	Cutoff, Sat
(2)	1.5	10	104	104	200	200	250	32.1	5.83	6.19	Cutoff
(3)	1.5	10	104	104	200	200	100	40	7.37	6.3	Normal
(4)	1.5	10	104	104	400	200	100	40.4	7.31	4.87	Normal

- Decreasing R_{E2} (with the voltage gain A_v (Eq. (3.1.49)) unaffected) increases the DC base current $I_{B,Q}$ (Eq. (3.1.48)) and the slope $1/(R_C + R_{E1} + R_{E2})$ of the DC load line so that the operating point Q can move right (away from the saturation region) upwards (away from the cutoff region). Thus, the possibility of the BJT to be saturated and/or cut off decreases.
- Decreasing R_C decreases the maximum variation of v_{ce} (Eq. (3.1.47)) to reduce the possibility of the BJT to trespass on the saturation region, but it reduces the voltage gain

$$A_v = \frac{v_o}{v_s} \overset{(3.1.42)}{=} \frac{\beta(R_C \| R_L)}{R_s + r_b + r_{be} + (1+\beta)R_{E1}} \tag{3.1.49}$$

- Increasing VCC (with the voltage gain A_v unaffected) increases the DC base current $I_{B,Q}$ (3.1.48) and pushes the v_{CE}-intercept rightwards so that the operating point Q can move right upwards. Thus, the possibility of the BJT to be saturated and/or cut off decreases.

The PSpice simulation results of the circuit in Figure 3.19.1(a) or 3.20(a) are depicted in Figure 3.23, which supports the observations stated above. Figure 3.23(a1)/(b1) shows the simulation results and load line analysis for the circuit with $R_C = 200\,\Omega$, $R_{E1} = 50\,\Omega$, and $R_{E2} = 250\,\Omega$, respectively. Figure 3.23(a2)/(b2) shows the simulation results and load line analysis for the circuit with $R_C = 200\,\Omega$, $R_{E1} = 200\,\Omega$, and $R_{E2} = 250\,\Omega$, respectively, supporting the observation that increasing R_{E1} (with the voltage gain A_v decreased) will move the operating point Q left downwards, but will decrease the maximum variations of v_{ce} and i_c more abruptly so that the possibility of the BJT to trespass

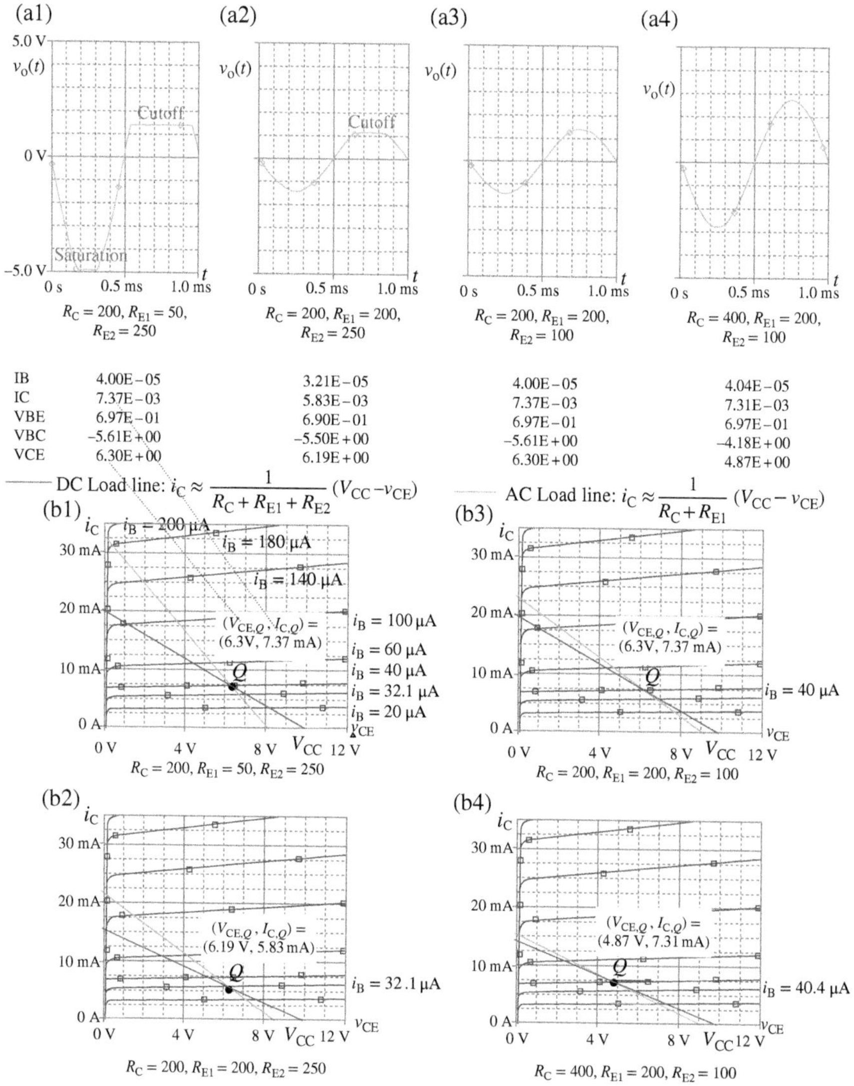

Figure 3.23 DC and AC load line for the CE BJT circuit (in Figure 3.19.1a) with different resistor values.

on the saturation region and/or crash into the cutoff region can decrease. Figure 3.23(a3)/(b3) shows the simulation results and load line analysis for the circuit with $R_C = 200\ \Omega$, $R_{E1} = 200\ \Omega$, and $R_{E2} = 100\ \Omega$, respectively, supporting the observation that decreasing R_{E2} (with the voltage gain A_v unaffected) will move the operating point Q right upwards so that the possibility of the BJT to trespass on the saturation region and/or crash into the cutoff region can decrease. Figure 3.23(a4)/(b4) shows the simulation results and load line analysis for the circuit with $R_C = 400\ \Omega$, $R_{E1} = 200\ \Omega$, and $R_{E2} = 100\ \Omega$, respectively, implying that increasing R_C may increase the voltage gain A_v without trespassing on the saturation region or crashing into the cutoff region thanks to the increase of linearity margin secured by increasing R_{E1} and decreasing R_{E2}.

Can the possibility of a BJT to trespass on the saturation region or crash into the cutoff region be predicted by the MATLAB function 'BJT_CE_analysis()'? Let us try it for the above four cases:

```
>>VCC=10; Vsm=1.5; rb=10; betaF=180; betaR=6; betaAC=194;
Rs=50; R1=104e3; R2=104e3; RC=200; RE=[50 250]; RL=1e4; % (1)
BJT_CE_analysis(VCC,rb,Rs,R1,R2,RC,RE,RL,[betaF betaR betaAC],Vsm);

 VCC    VEE    VBB    VBQ    VEQ    VCQ    IBQ        IEQ        ICQ
 10.00  0.00   5.00   2.90   2.20   8.54   4.05e-05   7.32e-03   7.28e-03
 in the forward-active mode with VCE,Q= 6.35
 gm= 281.664[mS], rbe= 689[Ohm], ro= 1373[kOhm]
 Possibly crash into the cutoff region
 Possibly violate the saturation region
 Gv=Ri/(Rs+Ri)xAv = 0.994 x -3.64 = -3.62

>>Rs=50; R1=104e3; R2=104e3; RC=200; RE=[200 250]; RL=1e4; % (2)
BJT_CE_analysis(VCC,rb,Rs,R1,R2,RC,RE,RL,[betaF betaR betaAC],Vsm);

 VCC    VEE    VBB    VBQ    VEQ    VCQ    IBQ        IEQ        ICQ
 10.00  0.00   5.00   3.32   2.62   8.84   3.22e-05   5.83e-03   5.80e-03
 in the forward-active mode with VCE,Q= 6.22
 gm= 224.360[mS], rbe= 865[Ohm], ro= 1724[kOhm]
 Possibly crash into the cutoff region
 Gv=Ri/(Rs+Ri)xAvo = 0.998 x -0.95 = -0.95

>>Rs=50; R1=104e3; R2=104e3; RC=200; RE=[200 100]; RL=1e4; % (3)
BJT_CE_analysis(VCC,rb,Rs,R1,R2,RC,RE,RL,[betaF betaR betaAC],Vsm);

 VCC    VEE    VBB    VBQ    VEQ    VCQ    IBQ        IEQ        ICQ
 10.00  0.00   5.00   2.90   2.20   8.54   4.05e-05   7.32e-03   7.28e-03
 in the forward-active mode with VCE,Q= 6.35
 gm= 281.664[mS], rbe= 689[Ohm], ro= 1373[kOhm]
 Gv=Ri/(Rs+Ri)xAv = 0.998 x -0.96 = -0.96

>>Rs=50; R1=104e3; R2=104e3; RC=400; RE=[200 100]; RL=1e4; % (4)
BJT_CE_analysis(VCC,rb,Rs,R1,R2,RC,RE,RL,[betaF betaR betaAC],Vsm);

 VCC    VEE    VBB    VBQ    VEQ    VCQ    IBQ        IEQ        ICQ
 10.00  0.00   5.00   2.90   2.20   7.09   4.05e-05   7.32e-03   7.28e-03
 in the forward-active mode with VCE,Q= 4.89
 gm= 281.664[mS], rbe= 689[Ohm], ro= 1373[kOhm]
 Gv=Ri/(Rs+Ri)xAv = 0.998 x -1.88 = -1.88
```

For the four cases, the MATLAB function 'BJT_CE_analysis()' seems to have worked well in terms of its prediction about the possibility of the BJT to be saturated or cut off.

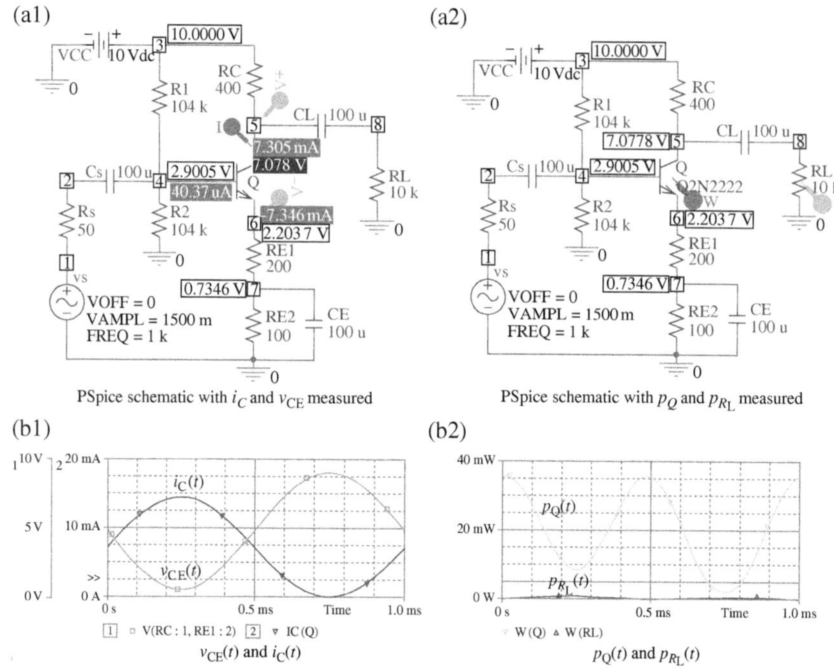

Figure 3.24 Measurement of BJT power in PSpice ("elec03f24.opj").

Now, to find the DC power of the BJT for the last case by using MATLAB, run the following MATLAB statements:

```
>> [VBQ,VEQ,VCQ,IBQ,IEQ,ICQ,Av]=BJT_CE_analysis(VCC,rb,Rs,R1,R2,RC,RE,
                                   RL, [betaF betaR betaAC],Vsm);
>>PQ_DC=(VCQ-VEQ)*ICQ+(VBQ-VEQ)*IBQ % DC power of BJT Q
   PQ_DC= 0.0356 % 35.6mW
```

The instantaneous (DC + AC) power of the BJT can easily be found as depicted in Figure 3.24(b2) by running the PSpice schematic (Figure 3.24(a2)) with a power (W) marker placed at the center of the device. Why is the instantaneous power $p_Q(t)$ always below the DC power $P_{Q,DC}$ =35.6 mW of BJT? It is because $v_{ce}\, i_c < 0$ (Figure 3.24(b1)) so that the AC power of the BJT is negative, implying that the BJT Q is acting as an AC source (active element) supplying an AC power to the other parts.

3.1.9 BJT Current Mirror

Consider the circuit of Figure 3.25(a) where the BJT Q_1 is said to be diode-connected' or 'connected in diode configuration' since its collector and base terminals are short-circuited so that it behaves like a diode. Why is a BJT used as a diode? It is for efficiency of fabricating integrated circuit (IC) with matched devices. For proper operation of the circuit, the two BJTs Q_1 and Q_2 must be matched in the sense that they have identical current gains (α_F,β_F) and characteristic curves.

Let us analyze the circuit of Figure 3.25(a), which is called a *current mirror* because the currents of the two matched BJTs sharing the same v_{BE} are equal. Assume that the voltage sources V1 and V2 forward-bias the B-E junctions and reverse-bias the B-C junctions ($v_{BC} < 0.4$) of both BJTs to let them operate in the forward-active mode so that $v_{BE1} = v_{BE2} = 0.7$ V and

$$i_{C1} \underset{\text{NPN}}{\overset{(3.1.1b)}{=}} \alpha_F I_{ES}(e^{v_{BE1}/V_T}-1)-I_{CS}(e^{v_{BC1}/V_T}-1) \approx \alpha_F I_{ES}(e^{v_{BE1}/V_T}-1)$$
$$= \alpha_F I_{ES}(e^{v_{BE2}/V_T}-1) \approx i_{C2} \tag{3.1.50}$$

$$i_{B1} \overset{(3.1.6c)}{=} \frac{1}{\beta_F} i_{C1} \approx \frac{1}{\beta_F} i_{C2} \overset{(3.1.6c)}{=} i_{B2} \tag{3.1.51}$$

Noting that the voltage at node 1 (the lump of terminals C1-B1-B2) is $v_{BE1} = 0.7$ V, we apply KCL at the node to write

$$I_R = \frac{V_1 - v_{BE1}}{R} = i_{C1} + i_{B1} + i_{B2} \underset{(3.1.51)}{\overset{(3.1.6c)}{\approx}} \left(1+\frac{2}{\beta_F}\right) i_{C1} \overset{(3.1.50)}{\approx} \left(1+\frac{2}{\beta_F}\right) i_{C2} \tag{3.1.52}$$

which yields the output current as

$$i_o = i_{C2} \overset{(3.1.52)}{\approx} \frac{I_R}{1+2/\beta_F} \approx \frac{V_1 - v_{BE1}}{R} = \frac{15-0.7}{10000} = 0.0014\,\text{A} = 1.4\,\text{mA} \tag{3.1.53}$$

This is supported by the PSpice simulation result (with Bias Point analysis) listed in Figure 3.25(a), which shows that the current i_{C2} supplied by the current mirror is constant as about 1.4 mA for different values of V2 and, therefore, the current mirror can be used as a current source.

Let us analyze the circuit of Figure 3.25(b), which is also called a *circuit mirror* because the currents of the two matched BJTs Q_1 and Q_2 are equal. Assume that the voltage sources V1 and V2 forward-bias the B-E junctions and reverse-bias the B-C junctions ($v_{BC} < 0.4$) of the three BJTs to let them operate in the forward-active mode so that $v_{BE1} = v_{BE2} = v_{BE3} = 0.7$ V and

$$i_{C1} \underset{\text{NPN}}{\overset{(3.1.1b)}{\approx}} \alpha_F I_{ES}\left(e^{v_{BE1}/V_T}-1\right) = \alpha_F I_{ES}\left(e^{v_{BE2}/V_T}-1\right) \approx i_{C2} = i_C \tag{3.1.54}$$

$$i_{B1} \overset{(3.1.6c)}{=} \frac{1}{\beta_F} i_{C1} \approx \frac{1}{\beta_F} i_{C2} \overset{(3.1.6c)}{=} i_{B2};\ i_{B1} \approx i_{B2} = i_B \tag{3.1.55}$$

Noting that the voltage at node 1 (the lump of terminals E3-B1-B2) is $v_{BE1} = 0.7$ V and the voltage at node 2 (the lump of terminals C1-B3) is $v_{BE1} + v_{BE3} = 1.4$ V, we can write

$$\text{KCL at node 1:}\ i_{E3} = i_{B1} + i_{B2} = 2\,i_{B2} \overset{(3.1.6c)}{=} \frac{2}{\beta_F} i_{C2} \tag{3.1.56}$$

(a)

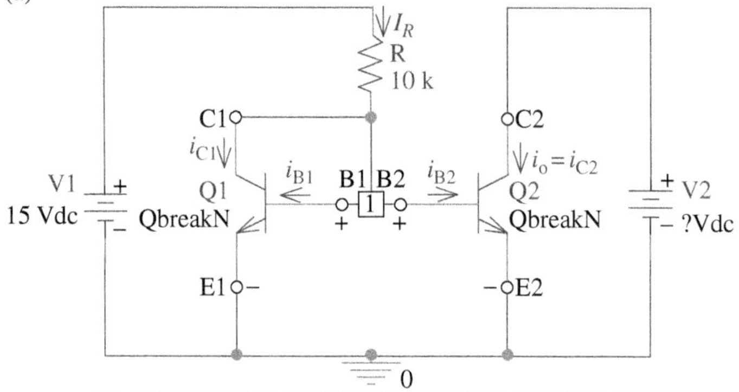

V_2	1 V	5 V	10 V	20 V	40 V
I_{C2}	1.39 mA	1.39 mA	1.39 mA	1.39 mA	1.39 mA

Current mirror using two BJTs

(b)

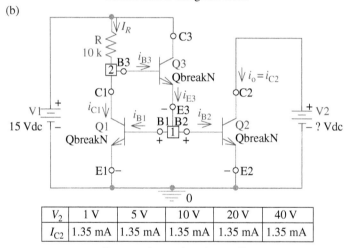

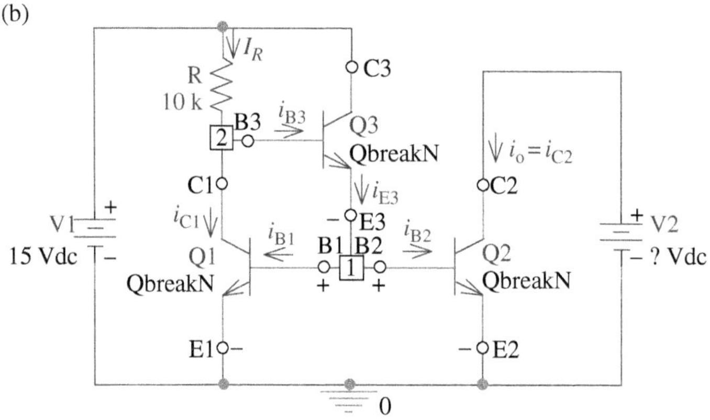

V_2	1 V	5 V	10 V	20 V	40 V
I_{C2}	1.35 mA	1.35 mA	1.35 mA	1.35 mA	1.35 mA

Current mirror using three BJTs

Figure 3.25 Current mirrors using BJTs ("elec03f25.opj").

$$\text{Eq.(3.1.6d) for } Q_3:\ i_{B3} \overset{(3.1.6d)}{=} \frac{1}{1+\beta_F} i_{E3} \overset{(3.1.56)}{=} \frac{2}{\beta_F(1+\beta_F)} i_{C2} \tag{3.1.57}$$

$$\text{KCL at node 2:}\ I_R = i_{C1} + i_{B3} \overset{(3.1.57)}{=} \left(1 + \frac{2}{\beta_F(1+\beta_F)}\right) i_{C2} \tag{3.1.58}$$

This yields the output current as

$$\begin{aligned} i_o = i_{C2} &\overset{(3.1.58)}{\approx} \frac{I_R}{1 + 2/\beta_F(1+\beta_F)} \\ &\approx \frac{V_1 - v_{BE1} - v_{BE3}}{R} = \frac{15-1.4}{10\,000} = 0.00136\,\text{A} = 1.36\,\text{mA} \end{aligned} \tag{3.1.59}$$

This is supported by the PSpice simulation result (with Bias Point analysis) listed in Figure 3.25(b), which shows that the current i_{C2} supplied by the current mirror is constant as about 1.35 mA for different values of V2 and, therefore, the current mirror can be used as a current source.

Let us compare the sensitivities of i_{C2} w.r.t. β_F for the two current sources:

$$(a):\ i_C \overset{(3.1.53)}{\approx} \frac{I_R}{1+2/\beta_F} \quad \rightarrow S_{\beta_F}^{i_C} = \frac{\partial i_C}{\partial \beta_F} = \frac{2I_R}{(1+2/\beta_F)^2\beta_F^2} = \frac{2I_R}{(\beta_F+2)^2} \tag{3.1.60}$$

$$(b):\ i_C \overset{(3.1.59)}{\approx} \frac{I_R}{1+2/\beta_F(1+\beta_F)} \rightarrow S_{\beta_F}^{i_C} = \frac{2(2\beta_F+1)I_R}{(1+2/\beta_F(1+\beta_F))^2\beta_F^2(1+\beta_F)^2} = \frac{2(2\beta_F+1)I_R}{(\beta_F(1+\beta_F)+2)^2} \tag{3.1.61}$$

This implies that the current mirror (b) has smaller sensitivity of the output current i_{C2} w.r.t. β_F (roughly proportional to $1/\beta_F^3$) compared with that of the current mirror (a) (roughly proportional to $1/\beta_F^2$).

Now, to analyze the 3-BJT current mirror (Figure 3.26(a)) and that with R replaced by a current source I (Figure 3.26(b)) by using the exponential model, we apply KCL at nodes 1 and 2 to write

$$i_{E3}\left(\begin{bmatrix} v_{BE3}=v_2-v_1 \\ v_{BC3}=v_2-V_1 \end{bmatrix}\right) - i_{B1}\left(\begin{bmatrix} v_{BE1}=v_1 \\ v_{BC1}=v_1-v_2 \end{bmatrix}\right) - i_{B2}\left(\begin{bmatrix} v_{BE2}=v_1 \\ v_{BC2}=v_1-V_2 \end{bmatrix}\right) = 0 \quad \text{at node 1 of (a), (b)} \tag{3.1.62a}$$

$$\frac{V_1-v_2}{R} - i_{B3}\left(\begin{bmatrix} v_{BE3}=v_2-v_1 \\ v_{BC3}=v_2-V_1 \end{bmatrix}\right) - i_{C1}\left(\begin{bmatrix} v_{BE1}=v_1 \\ v_{BC1}=v_1-v_2 \end{bmatrix}\right) = 0 \text{ at node 2 of (a)} \tag{3.1.62b}$$

$$I - i_{B3}\left(\begin{bmatrix} v_{BE3}=v_2-v_1 \\ v_{BC3}=v_2-V_1 \end{bmatrix}\right) - i_{C1}\left(\begin{bmatrix} v_{BE1}=v_1 \\ v_{BC1}=v_1-v_2 \end{bmatrix}\right) = 0 \text{ at node 2 of (b)} \tag{3.1.62c}$$

where $i_{Ck}(v_{BEk}, v_{BCk})$ and $i_{Bk}(v_{BEk}, v_{BCk})$ are defined by Eqs. (3.1.23a,b), and $i_{Ek}(\cdot,\cdot) = i_{Bk}(\cdot,\cdot) + i_{Ck}(\cdot,\cdot)$ for all k. The following MATLAB function '`BJT3_current_mirror()`' solves the set of Eqs. (3.1.62a,b) for circuit (a) or Eqs. (3.1.62a,c) for circuit (b) depending on whether the value of the third input argument R is greater than or equal to 1 or not. It returns the output current $i_o = i_{C2}$ and $\mathbf{v} = [v_1\ v_2]$ for possibly several values of V_2 (given as the second to last elements of the fourth input argument V12). For instance, we can solve the circuit of Figure 3.25(b) to get i_o for V1 = 15 V and V2 = {1, 5, 10, 20, 40} by running the following MATLAB statements:

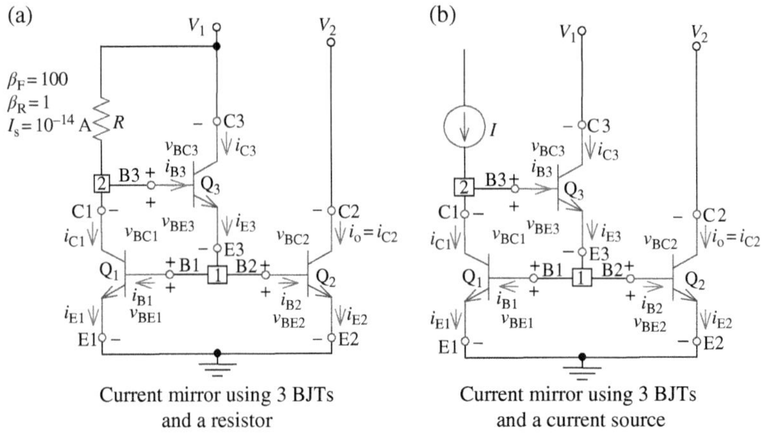

Figure 3.26 Current mirrors using a BJT as a diode.

```
>>R=1e4; Is=1e-14; io=BJT3_current_mirror([100 1],Is,R,[15 1 5 10 20 40])
```

```
function [io,vs]=BJT3_current_mirror(betaF,Is,R,V12,VT)
% Analyze a current mirror consisting of 3 BJTs and an R or I-source
% If R<1, it will be regarded as a current source I=R.
% Copyleft: Won Y. Yang, wyyang53@hanmail.net, CAU for academic use only
if nargin<5, VT=(273+27)/11605; end % Thermal voltage
if length(betaF)<2, betaR=1; else betaR=betaF(2); betaF=betaF(1); end
if length(Is)<2, VA=inf; else VA=Is(2); Is=Is(1); end
V1=V12(1); if length(V12)>1, V2s=V12(2:end); else V2s=V12; end
alphaR=betaR/(betaR+1);
Isc=Is/alphaR; % Collector-Base saturation current
options=optimset('Display','off','Diagnostics','off');
iC=@(v) Is*exp(v(1)/VT)-Isc*exp(v(2)/VT); % Eq. (3.1.23a) with v=[vBE vBC]
iB=@(v) Is/betaF*exp(v(1)/VT)+Isc/(betaR+1)*exp(v(2)/VT); % Eq. (3.1.23b)
iE=@(v) Is*(1+1/betaF)*exp(v(1)/VT)-Isc*betaR/(betaR+1)*exp(v(2)/VT);
for n=1:length(V2s)
  V2=V2s(n);
  if R>=1 % Eq. (3.1.62a,b) with a resistor
   eq=@(v) [iE([v(2)-v(1) v(1)-V1])-iB([v(1) v(1)-v(2)])-iB([v(1) v(1)-V2]);
           V1-v(2)-R*(iB([v(2)-v(1) v(2)-V1])+iC([v(1) v(1)-v(2)]))];
  else I=R; % Eq. (3.1.62a,c) with a current source
   eq=@(v) [iE([v(2)-v(1) v(2)-V1])-iB([v(1) v(1)-v(2)])-iB([v(1) v(1)-V2]);
            I-iB([v(2)-v(1) v(2)-V1])-iC([v(1) v(1)-v(2)])]*1e3;
   end
   v0=[0.7 0.4]; % Initial guess
   v=fsolve(eq,v0,options); vs(n,:)=v; io(n)=iC([v(2) v(2)-V2]);
end
```

3.1.10 BJT Inverter/Switch

Figure 3.27(a1)/(a2) shows the PSpice schematics of BJT inverter for Transient/DC_Sweep analysis. Figure 3.27(b1) shows the input and output voltage waveforms of the inverter (obtained from the Transient analysis) where the input 1(high)/0 (low) drives the BJT into the saturation/cutoff mode so that the output $v_o = v_{CE}$ can go to logic 0(low)/1(high) with high-to-low/low-to-high propagation delay t_{pHL}/t_{pLH} that are defined as the times between the 50% input and 50% output.

Note that in order for the BJT to go back and forth between the saturation and cutoff modes, the collector current $i_{C,sat}$ in the saturation mode should be less than β_F times the base current i_B:

$$i_{C,sat} = \frac{V_{CC} - V_{CE,sat}}{R_C} < \beta_F\, i_B = \beta_F \frac{v_i - V_{BE,sat}}{R_B} \tag{3.1.63}$$

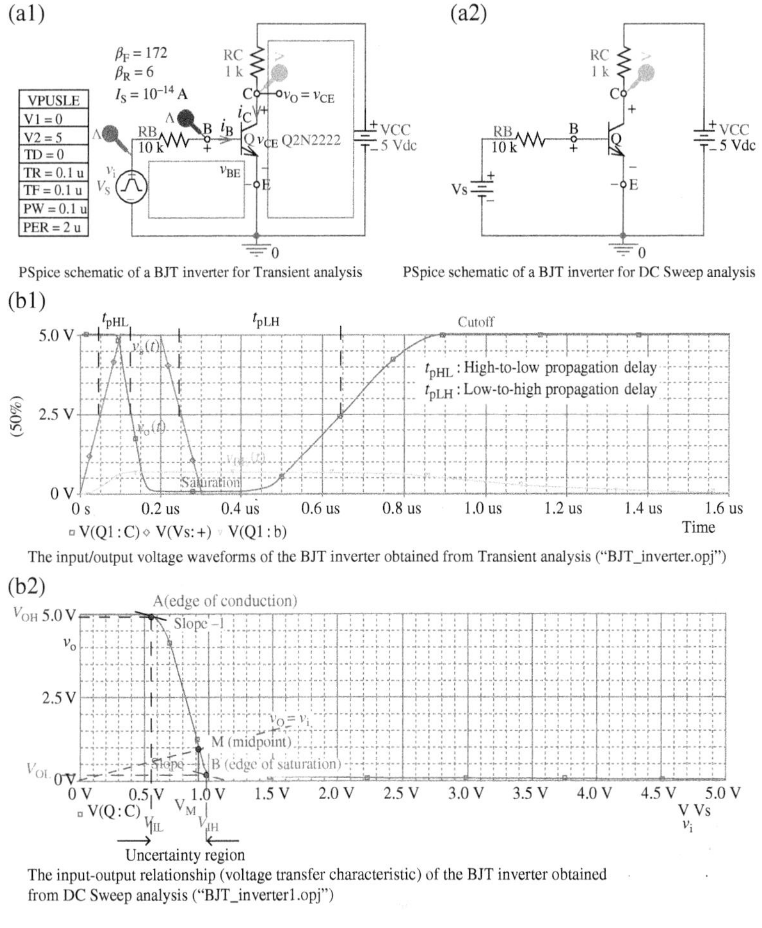

Figure 3.27 BJT inverter.

This condition can easily be satisfied by taking a small R_B and a large R_C. But the following should be noted:

- A small R_B makes the input impedance small so that the fan-in of the gate can be decreased where *fan-in* is the number of logic gates that can be connected to its input without deteriorating the input signal or producing an undefined or incorrect output.
- A large R_C makes the output impedance and loading effect (due to it) large so that the (output-high) fan-out of the gate can be decreased where *fan-out* is the number of logic gates that can be connected to its output (as loads) without producing an undefined or incorrect output.

Figure 3.27(b2) shows the input(v_i)-output(v_o) relationship, called the *VTC* (*voltage transfer characteristic*), of the inverter (obtained from the DC Sweep analysis). In the VTC, the output low/high levels V_{OH}/V_{OL} are defined as the minimum/maximum values of output v_o corresponding to logic 1/0, respectively and V_{IH}/V_{IL} are defined as the minimum/maximum values of input v_s that can be interpreted as logic 1/0, respectively, where V_{IL}/V_{OH} are the input/output at point A with slope of −1 and V_{IH}/V_{OL} are the input/output at point B with slope of −1. Also, we define the midpoint M as the intersection of the VTC and line $v_o = v_i$, which can be thought of as the boundary at which the inverter switches its output from one state to the other.

Figure 3.28(a) shows a practical VTC together with an ideal VTC. As measures of how much the gate can tolerate the variation of signal levels without causing any erroneous logical state, Figure 3.28(b1)/(b2) shows the high and low noise margins for practical/ideal VTCs where the *high* and *low noise margins* are defined as

$$NM_H = V_{OH} - V_{IH}, NM_L = V_{IL} - V_{OL} \tag{3.1.64}$$

What are the physical meanings of the high/low noise margins? As can be seen in Figure 3.28(b1) or (b2), the high/low noise margin means how much the high (1)/low(0) signal can decrease/increase without being mistaken for a low(0)/high(1) signal by the next-stage (load) gate (the same kind of inverter), i.e. without misleading (mistakenly driving) the load inverter into the cutoff/saturation mode like a low/high voltage. The *absolute noise margin* is defined as the smaller of the two noise margins:

$$NM = \min\{NM_H, NM_L\} \tag{3.1.65}$$

Note that the noise immunity measured by the absolute noise margin is maximized by the ideal VTC with abrupt switching at $V_{IL} = V_{IH} = (V_{OL} + V_{OH})/2$, which has maximum logic swing from V(0) to V(1), but no transition region.

To analyze the BJT inverter circuit in Figure 3.27(a1) by using the exponential model, we can apply KVL around the two meshes to write

$$\begin{aligned} v_i - v_{BE} - R_B i_B(v_{BE}, v_{BC}) &= 0 \\ V_{CC} - R_C i_C(v_{BE}, v_{BC}) + v_{BC} - v_{BE} &= 0 \end{aligned} \tag{3.1.66}$$

where $i_C(v_{BE}, v_{BC})$ and $i_B(v_{BE}, v_{BC})$ are defined by Eqs. (3.1.23a,b).

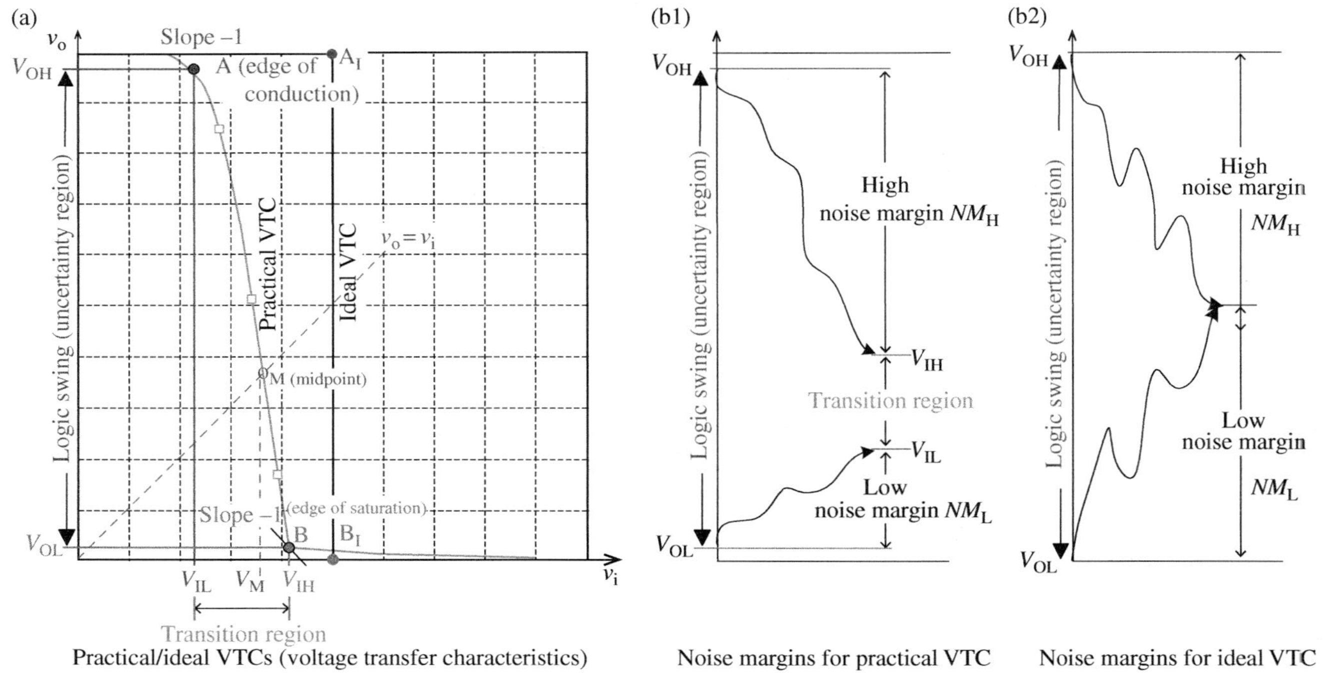

Figure 3.28 Practical/ideal voltage transfer characteristics (VTCs) and the corresponding noise margins.

```
function [VIL,VIH,VOL,VOH,VM,NML,NMH,Pavg,vo1,iC1]=...
BJT_inverter(betaF,Is,RB,RC,VCC,vi1,VT)
% Plot the VTC of a BJT inverter (with no output argument)
% which consists of an NPN-BJT and resistors RB/RC between vi/VCC and B/C.
% Copyleft: Won Y. Yang, wyyang53@hanmail.net, CAU for academic use only
if nargin<7, VT=(273+27)/11605; end % Thermal voltage
if length(betaF)<2, betaR=1; else betaR=betaF(2); betaF=betaF(1); end
alphaR=betaR/(betaR+1); Isc=Is/alphaR; % CB saturation current
dvi=5e-3; vis=[0:dvi:VCC]; vCE_sat=0.2;
options=optimset('Display','off','Diagnostics','off');
iC = @(v) Is*exp(v(1)/VT)-Isc*exp(v(2)/VT); % Eq.(3.1.23a) with v=[vBE vBC]
iB = @(v) Is/betaF*exp(v(1)/VT)+Isc/(betaR+1)*exp(v(2)/VT); % Eq.(3.1.23b)
for n=1:length(vis)
  vi = vis(n);
  if vi>=0.4 % v = [vBE vBC]
   eq = @(v) [vi-v(1)-RB*iB(v); VCC-RC*iC(v)+v(2)-v(1)]; % Eq.(3.1.66)
   if ~exist('v'), v0=[0.7 0.4]; else v0=v; end % Initial guess
   v = fsolve(eq,v0,options);
   vos(n)=v(1)-v(2); iCs(n)=(VCC-vos(n))/RC; % Eqs.(3.1.67),(3.1.66b)
  else vos(n)=VCC; iCs(n)=0; % Cutoff mode
  end
end
[VIL,VIH,VOL,VOH,VM,NML,NMH,VL,Pavg]=...
find_pars_of_inverter(vis, vos,iCs,VCC);
if nargout==0|nargin<6
   plot(vis,vos,[VIL VM VIH],[VOH VM VOL],'ro');
end
if nargin>5 % If you want vo1 for vi1
    for i=1:length(vi1)
      [dmin,imin]=min(abs(vis-vi1(i)));
      vo1(i)=vos(imin); iC1(i)=iCs(imin);
    end
 else vo1=vos; iC1=iCs;
end
```

```
function [VIL,VIH,VOL,VOH,VM,NML,NMH,VL,Pavg]=...
                    find_pars_of_inverter(vis,vos,is,Vs,VH)
if nargin<5, VH=Vs; end % Highest output voltage
dvs=abs(diff(vis)); dv=min(dvs(find(dvs>1e-6)));
[pks,locs]=findpeaks(1./abs(diff([vos(1) vos])+dv));
[pks,inds]=sort(pks,'descend'); inds1=locs(inds(1:2));
[VLH,inds2]=sort(vis(inds1)); % Points with slope=-1
VIL=VLH(1); VIH=VLH(2); VOH=vos(inds1(inds2(1))); VOL=vos(inds1(inds2(2)));
NML = VIL-VOL; NMH = VOH-VIH; % Eq.(3.1.64)
[em,imin]=min(abs(vis-vos)); VM=vis(imin); % Midpoint
[em,imin]=min(abs(vis-VH)); VL=vos(imin); % Virtual lowest output
Pavg=Vs*mean([max(is) min(is)]); % Average power for on-off periods
```

Once we solve this set of equations to find v_{BE} and v_{BC} for a given value of the input voltage v_i, we can find the output voltage v_o as

$$v_o = v_{CE} = v_{BE} - v_{BC} \tag{3.1.67}$$

The process of solving Eq. (3.1.66) to find v_o for $v_i = 0 \sim V_{CC}$, finding $V_{IL}, V_{IH}, V_{OL}, V_{OH}$, and V_M, and plotting v_o versus v_i has been cast into the above MATLAB function 'BJT_inverter()'. We can run the following script "plot_VTC_BJT_inverter.m" (which uses 'BJT_inverter()')

```
%plot_VTC_BJT_inverter.m
VCC=5; RB=1e4; RC=1e3;
betaF=180; betaR=6; Is=1e-14;
[VIL,VIH,VOL,VOH,VM,NML,NMH,Pavg]=...
          BJT_inverter([betaF betaR],Is,RB,RC,VCC)
```

to get the VTC as shown in Figure 3.28(a) and the inverter parameter values as

```
VIL = 0.565, VIH= 0.990, VOL= 0.151, VOH= 4.971, VM= 0.920
NML = 0.414, NMH= 3.981, VL = 0.031
Pavg= 12.423[mW]
```

3.1.11 Emitter-Coupled Differential Pair

Figure 3.29(a) shows an *emitter-coupled* (in the sense that the emitter terminals of the two BJTs are connected) or *differential* (in the sense that its output varies with the differential input $v_d = v_{BE1} - v_{BE2}$) pair. To analyze this circuit, we assume that both BJTs operate in the forward-active mode so that we can use Eq. (3.1.1b) to write their approximate collector/emitter currents as

$$i_{C_1} \underset{\text{NPN}}{\overset{(3.1.1b)}{\approx}} \alpha_F I_{ES}(e^{v_{BE1}/V_T} - 1) \approx \alpha_F I_{ES} e^{v_{BE1}/V_T}$$

$$; i_{E_1} \overset{(3.1.6a)}{=} \frac{1}{\alpha_F} i_{C_1} \approx I_{ES} e^{v_{BE1}/V_T} \tag{3.1.68a}$$

$$i_{C_2} \underset{\text{NPN}}{\overset{(3.1.1b)}{\approx}} \alpha_F I_{ES}(e^{v_{BE2}/V_T} - 1) \approx \alpha_F I_{ES} e^{v_{BE2}/V_T}$$

$$; i_{E2} \overset{(3.1.6a)}{=} \frac{1}{\alpha_F} i_{C_2} \approx I_{ES} e^{v_{BE2}/V_T} \tag{3.1.68b}$$

Then their ratios can be approximately written as

$$\frac{i_{C_2}}{i_{C_1}} = \frac{i_{E_2}}{i_{E_1}} \underset{(3.1.68a)}{\overset{(3.1.68b)}{\approx}} e^{(v_{BE2} - v_{BE1})/V_T} = e^{-v_d/V_T} \quad \text{with } v_d = v_{BE1} - v_{BE2} \tag{3.1.69}$$

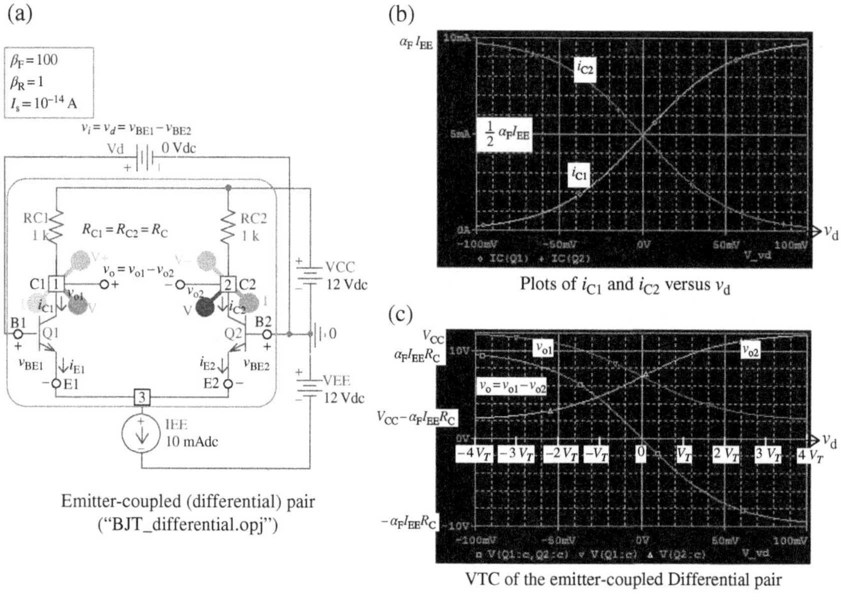

Emitter-coupled (differential) pair ("BJT_differential.opj")
Plots of i_{C1} and i_{C2} versus v_d
VTC of the emitter-coupled Differential pair

Figure 3.29 Emitter-coupled (differential) pair and its VTC.

Also, we apply KCL at the node E1-E2 to write

$$i_{E1} + i_{E2} \overset{(3.1.6a)}{=} \frac{i_{C_1}}{\alpha_F} + \frac{i_{C_2}}{\alpha_F} \overset{KCL}{=} I_{EE}; \quad i_{C1} + i_{C2} = \alpha_F I_{EE} \tag{3.1.70}$$

Combining Eqs. (3.1.69) and (3.1.70) yields the expression of each collector current as

$$i_{C_1} \approx \frac{\alpha_F I_{EE}}{1 + e^{-v_d/V_T}} \to \begin{cases} \alpha_F I_{EE} \approx I_{EE} (\text{closed switch}) & \text{if } v_d > 4V_T \\ 0 \quad (\text{open switch}) & \text{if } v_d < -4V_T \end{cases} \tag{3.1.71a}$$

$$i_{C2} \approx \frac{e^{-v_d/V_T} \alpha_F I_{EE}}{1 + e^{-v_d/V_T}} = \frac{\alpha_F I_{EE}}{1 + e^{v_d/V_T}} \to \begin{cases} 0 \ (\text{open switch}) \text{ if } v_d > 4V_T \\ \alpha_F I_{EE} (\text{closed switch}) \text{ if } v_d < -4V_T \end{cases} \tag{3.1.71b}$$

where these collector currents are depicted in Figure 3.29(b). Then we can write the (differential) output voltage as

$$\begin{aligned} v_o &= v_{o1} - v_{o2} = v_{C1} - v_{C2} = (V_{CC} - R_C i_{C1}) - (V_{CC} - R_C i_{C2}) \\ &= R_C (i_{C2} - i_{C1}) = \frac{e^{-v_d/V_T} - 1}{1 + e^{-v_d/V_T}} \alpha_F I_{EE} R_C \end{aligned} \tag{3.1.72}$$

This differential output voltage v_o, together with v_{o1} and v_{o2}, is shown in Figure 3.29(c). From Figure 3.29 and Eqs. (3.1.71a,b) and (3.1.72), note the following:

- If $-1.5V_T < v_d < 1.5V_T$, the differential output voltage v_o and other signals vary almost linearly with the differential input v_d, allowing the emitter-coupled pair to be used as an amplifier with a voltage gain of

$$\left.\frac{v_o}{v_d}\right|_{v_d=1.5V_T} \overset{(3.1.72)}{=} \frac{(e^{-1.5}-1)}{1.5(1+e^{-1.5})V_T}\alpha_F I_{EE} R_C \underset{V_T=25\text{mV}}{\overset{I_{EE}=10\text{mA},R_C=1\text{k}\Omega}{\approx}} 170 \tag{3.1.73}$$

- If $v_d > 4V_T$, we have

$$i_{C1} \approx \alpha_F I_{EE}, i_{C2} \approx 0,\ v_{o1} \approx V_{CC} - \alpha_F I_{EE} R_C \underset{\text{of } I_{EE} \text{ and } R_C}{\overset{\text{Proper choice}}{\rightarrow}} 0, \text{and } v_{o2} \approx V_{CC} \tag{3.1.74a}$$

On the other hand, if $v_d < -4V_T$, we have

$$i_{C1} \approx 0,\ i_{C2} \approx \alpha_F I_{EE}, v_{o1} \approx V_{CC}, \text{ and } v_{o2} \approx V_{CC} - \alpha_F I_{EE} R_C \underset{\text{of } I_{EE} \text{ and } R_C}{\overset{\text{Proper choice}}{\rightarrow}} 0 \tag{3.1.74b}$$

It is implied that a large swing of the differential input $v_d = \pm 4V_T$ makes the two BJTs Q_1/Q_2 operate as closed/open or open/closed switches, producing two distinct levels of differential output v_o depending on whether $v_d = 4V_T$ or $v_d = -4V_T$.

The amplifying/switching properties are extensively used in analog/digital circuits, respectively. That is why the emitter-coupled or differential pair is one of the most important configurations employed in ICs.

To analyze the BJT differential pair circuit in Figure 3.29(a), we can apply KCL at nodes 1, 2, and 3 to write

$$\begin{aligned} &V_{CC} - v_1 - R_C i_{C1}(v_d - v_3, v_d - v_1) = 0 \text{ at node 1} \\ &V_{CC} - v_2 - R_C i_{C2}(0 - v_3, 0 - v_2) = 0 \text{ at node 2} \\ &i_{E1}(v_d - v_3, v_d - v_1) + i_{E2}(0 - v_3, 0 - v_2) - I_{EE} = 0 \text{ at node 3} \end{aligned} \tag{3.1.75}$$

where $i_{Ck}(v_{BEk}, v_{BCk})$ and $i_{Bk}(v_{BEk}, v_{BCk})$ are defined by Eqs. (3.1.23a,b). The process of solving this set of equations to find $\mathbf{v} = [v_1\ v_2\ v_3]$ for $v_d = -V_{dm} \sim V_{dm}$ and plotting $v_o = v_1 - v_2$, i_{C1}, i_{C2} (together with their analytic values computed by Eqs. (3.1.72,74) versus v_d has been cast into the following MATLAB function '`BJT_differential()`'. We can run

```
>>betaF=100; betaR=1; Is=1e-14; IEE=10e-3; RC=1e3; VCC=12;
  BJT_differential([betaF betaR],Is,IEE,RC,VCC);
```

to get the graphs of i_{C1}, i_{C2}, and v_o as shown in Figure 3.29(b) and (c).

```
function [vo1s,vo2s,iC1s,iC2s]=BJT_differential(betaF,Is,IEE,RC,VCC,Vdm)
% Analyze an NPN-BJT differential (emitter-coupled) pair
% (to find the outputs vo1/vo2 to a range of differential input vd=-Vdm~Vdm
%   and plot its VTC in case of no output argument)
% which consists of 2 NPN-BJTs, 2 resistors RC1=RC2=RC, and an I-src IEE.
% Copyleft: Won Y. Yang, wyyang53@hanmail.net, CAU for academic use only
VT=(273+27)/11605; % Thermal voltage
if nargin<6, Vdm=4*VT; end
dvd=Vdm/100; vds=[-Vdm:dvd:Vdm]; % Differential input
if length(betaF)<2, betaR=1; else betaR=betaF(2); betaF=betaF(1); end
alphaF=betaF/(1+betaF); alphaR=betaR/(betaR+1);
Isc=Is/alphaR; % Collector-Base saturation current
options=optimset('Display','off','Diagnostics','off');
iC = @(v) Is*exp(v(1)/VT)-Isc*exp(v(2)/VT); % Eq.(3.1.23a) with v=[vBE vBC]
iE = @(v) Is*(1+1/betaF)*exp(v(1)/VT)-Isc*betaR/(betaR+1)*exp(v(2)/VT);
for n=1:length(vds)
  vd = vds(n);
  eq=@(v) [VCC-v(1)-RC*iC([vd-v(3) vd-v(1)])
           VCC-v(2)-RC*iC([0-v(3) 0-v(2)])
           iE([vd-v(3) vd-v(1)])+iE([0-v(3) 0-v(2)])-IEE]; % Eq.(3.1.75)
  if ~exist('v'), v0=[VCC VCC/2 -0.7]; else v0=v; end % Initial guess
  v = fsolve(eq,v0,options);
  vo1s(n)=v(1); vo2s(n)=v(2); vos(n)=v(1)-v(2); % Eq.(3.1.72)
  iC1s(n)=iC([vd-v(3) vd-v(1)]); iC2s(n)=iC([-v(3) -v(2)]);
end
iC1s_a=alphaF*IEE./(1+(exp(-vds/VT))); % Eq.(3.1.71a)
iC2s_a=alphaF*IEE./(1+(exp(vds/VT))); % Eq.(3.1.71b)
if nargout==0
  subplot(211)
  plot(vds,iC1s,'g', vds,iC2s,'r', vds,iC1s_a,'k:', vds,iC2s_a,'b:')
  legend('iC1','iC2','iC1_a','iC2_a'); xlabel('v_d'); ylabel('i_C');
  vos_a=(exp(-vds/VT)-1)./(1+exp(-vds/VT))*alphaF*IEE*RC; % Eq.(3.1.72)
  subplot(212)
  plot(vds,vos,'g', vds,vo1s,'r', vds,vo2s, vds,vos_a,'k:')
  legend('vo','vo1','vo2','vo_a'); xlabel('v_d'); ylabel('v_o');
end
```

3.2 BJT Amplifier Circuits

This section deals with several configurations of BJT amplifier, i.e. the CE (common-emitter) amplifier, the CC (common-collector) amplifier (called an emitter follower), the CB (common-base) amplifier, and cascaded or compound multistage amplifier.

3.2.1 Common-Emitter (CE) Amplifier

Figure 3.30 shows a CE amplifier and its low-frequency AC equivalent (which is the same as Figure 3.19.1(d)) where the BJT has been replaced by the equivalent in Figure 3.18(b), and the biasing resistances $R_1 \| R_2$ and BJT output resistance r_o are assumed to be so large as to be negligible as parallel resistors. Let us find the input resistance, current gain, voltage gain, and output resistance.

1) **Input Resistance** R_i

   To find the input resistance, i.e. the equivalent resistance seen from the source side, we apply KVL for the left mesh (denoted in a gray closed curve) with $R_B = R_1 \| R_2$ neglected to write

$$R_s i_b + r_b i_b + r_{be} i_b + R_{E1}(\beta + 1) i_b = v_s; \{r_b + r_{be} + (\beta + 1) R_{E1}\} i_b = v_s - R_s i_b = v_b$$

(a)

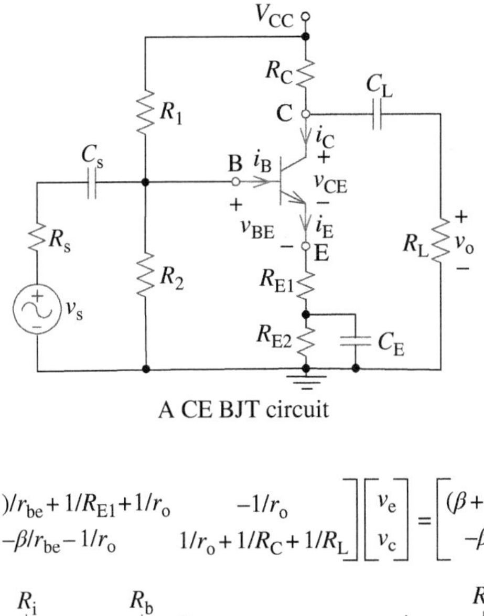

A CE BJT circuit

(b)

$$\begin{bmatrix} (\beta+1)/r_{be} + 1/R_{E1} + 1/r_o & -1/r_o \\ -\beta/r_{be} - 1/r_o & 1/r_o + 1/R_C + 1/R_L \end{bmatrix} \begin{bmatrix} v_e \\ v_c \end{bmatrix} = \begin{bmatrix} (\beta+1)/r_{be} \\ -\beta/r_{be} \end{bmatrix}$$

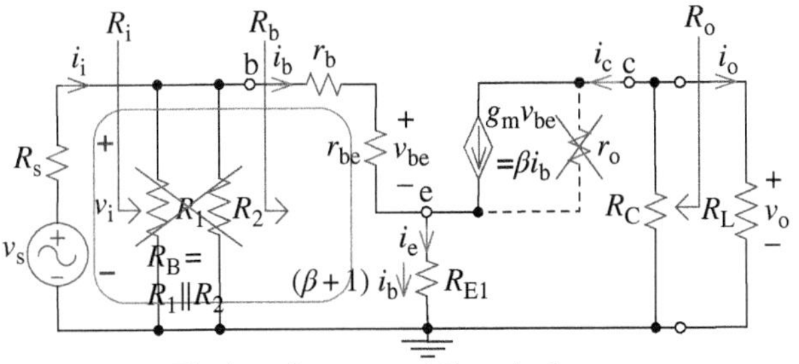

The low-frequency AC equivalent

Figure 3.30 A CE (common-emitter) BJT circuit and its low-frequency AC equivalent.

This yields the input resistance as

$$R_b = \frac{v_b}{i_b} = r_b + r_{be} + (\beta + 1)R_{E1}$$

$$\overset{\text{With } R_B = R_1 \| R_2}{\longrightarrow} R_i = \frac{v_i}{i_i} = R_B \| R_b = R_B \| \{r_b + r_{be} + (\beta + 1)R_{E1}\} \tag{3.2.1}$$

2) **Current Gain A_i**

The output current i_o through the load resistor R_L can be expressed as

$$i_o = -\frac{R_C}{R_C + R_L} i_c = -\frac{R_C}{R_C + R_L} \beta i_b$$

Thus, the current gain, i.e. the ratio of the output current i_o to the input current $i_i = i_b$ is

$$A_i = \frac{i_o}{i_i = i_b} = -\frac{\beta R_C}{R_C + R_L} \tag{3.2.2}$$

3) **Voltage Gains G_v and A_v**

The overall voltage gain, i.e. the ratio of the output voltage v_o to the source voltage v_s is

$$G_v = \frac{v_o}{v_s} = \frac{v_i}{v_s}\frac{v_o}{v_i} = \frac{R_i}{R_s + R_i} A_v$$

$$= -\frac{R_i}{R_s + R_i} \frac{\beta (R_C \| R_L)}{r_b + r_{be} + (\beta + 1)R_{E1}} \quad \text{with } R_i \text{ given by Eq.(3.2.1)} \tag{3.2.3}$$

4) **Output Resistance R_o**

To find the (Thevenin) equivalent resistance seen from the load side, we need to remove the (independent) voltage source v_s by short-circuiting it. Then no current flows of itself so that we have $i_b = 0$, $v_{be} = 0$, and $i_c = 0$ even if a test voltage or current source is applied to the output port. Therefore, the output resistance turns out to be

$$R_o = R_C \| r_o \approx R_C \tag{3.2.4}$$

This AC analysis process to find R_i, R_o, A_i, and A_v has been included in the MATLAB function '`BJT_CE_analysis()`' presented in Section 3.1.8. If a current source supplying a BJT with its DC emitter current $I_{E,Q}$ is given instead of the biasing circuit as depicted in Figure 3.31(a), the following MATLAB function '`BJT_CE_analysis_I()`' can be used for the AC analysis.

```
function [Av,Ai,Ri,Ro,gm,rbe,ro,vsm,vom]=...
    BJT_CE_analysis_I(Rs,RB,RC,RE,RL,IEQ,beta,VA,T,vbem)
% To analyze CE amplifier biased by IEQ
% beta = [betaAC rb] if a nonzero base resistance is given.
% If the 10th input argument vbem (maximum vbe such that the BJT operated
% in the linear region) is given, the following outputs will be returned.
% vsm = Max amplitude of vs ensuring the linear operation of BJT.
% vom = Max amplitude of vo when the BJT operates in the linear operation.
% Copyleft: Won Y. Yang, wyyang53@hanmail.net, CAU for academic use only
if nargin<9, T=27; end % Ambient temperature
if nargin<8, VA=1e4; end % Early voltage
if length(beta)>1, rb=beta(2); beta=beta(1); else rb=0; end
if length(RB)>1, RB=parallel_comb(RB); end
if sum([Rs RB RC RE RL rb]<0), error('Resistance must be positive!'); end
ICQ = IEQ*beta/(beta+1); % Collector current at the bias point
[gm,rbe,re,ro]=gmrbero_BJT(ICQ,beta,VA,T);
RbeE = rbe + (beta+1)*RE(1);
Ri = parallel_comb([RB RbeE]); % Eq.(3.2.1)
Ro = parallel_comb([RC ro]); % Eq.(3.2.4)
Ai = -beta/(1+RL/RC); % Eq.(3.2.2)
Avo = -beta*Ro/RbeE; % Eq.(3.2.3) with neither RL nor ro
Av = Avo/(Ro/RL+1); % Eq.(3.2.3) with RL, but without considering ro
Gv = Av/(Rs/Ri+1); % Taking Rs into consideration
if nargout<1
 fprintf('\n gm=ICQ/VT=%8.3f[mS], re=1/gm=%6.0f[Ohm], rbe=beta/gm=
  %6.0f [Ohm], ro=VA/ICQ=%7.2f[kOhm]\n', gm*1e3,re,rbe,ro/1e3);
 fprintf(' Ri=RB||RbeE=%8.2f kOhm, Ro=RC||ro=%6.0f Ohm', Ri/1e3,Ro);
 fprintf('\n Av=-beta*Ro/RbeE x RL/(Ro+RL) =%8.2f x%5.2f =%8.2f\n
  Gv=Ri/(Rs+Ri)xAv =%5.3f xAv =%8.2f\n', Avo,RL/(Ro+RL),Av,Ri/(Rs+Ri),Gv);
end
if nargin<10, vbem = NaN; end
vsm=(Rs/Ri+1)*RbeE/rbe*vbem; vom=abs(Gv)*vsm;
```

Example 3.10 AC Analysis of CE Amplifier

a) Consider the CE circuit in Figure 3.19.1(a) or 3.30(a) where $V_{CC} = 10\,\text{V}$, $r_b = 10\,\Omega$, $R_s = 50\,\Omega$, $R_1 = 104\,\text{k}\Omega$, $R_2 = 104\,\text{k}\Omega$, $R_C = 200\,\Omega$, $R_{E1} = 50\,\Omega$, $R_{E2} = 250\,\Omega$, $R_L = 10\text{k}\,\Omega$, $\beta_F = 180$, $\beta_R = 6$, $\beta_{AC} = 194$, and $V_A = 10^4$ V. Find R_i, R_o, $A_v = v_o/v_i$, and $G_v = v_o/v_s$.

We can use the values of β, r_b, r_{be}, and r_o obtained in Section 3.1.8 (Eqs. (3.1.35-38)) for the AC analysis as follows:

$$R_i = \frac{v_i}{i_i} \overset{(3.2.1)}{=} R_B||(r_b + r_{be} + (\beta+1)R_{E1})$$

$$= 52\,\text{k}||(10 + 688 + 195 \times 50) = 8.7\,\text{k}\Omega \tag{E3.10.1}$$

$$R_o \overset{(3.2.4)}{=} R_C||r_o \approx R_C = 200\,\Omega \tag{E3.10.2}$$

$$A_v \overset{(3.2.3)}{=} -\frac{\beta(R_C\|R_L)}{r_b + r_{be} + (\beta+1)R_{E1}}$$

$$\underset{(3.1.37)}{\overset{(3.1.36)}{=}} -\frac{194\times(200\|10\,000)}{10+688+195\times 50} = -3.64 \tag{E3.10.3}$$

$$G_v = \frac{v_o}{v_s} = \frac{v_i}{v_s}\frac{v_o}{v_i} = \frac{R_i}{R_s+R_i}A_v$$

$$\underset{(E3.10.3)}{\overset{(E3.10.1)}{=}} -\frac{8.7}{0.05+8.7}3.64 = -0.994\times 3.64 = -3.62 \tag{E3.10.4}$$

This AC analysis can also be done by using the MATLAB function 'BJT_CE_analysis()' although the results will be slightly different since r_o is taken into consideration in the MATLAB function:

```
>>VCC=10;  Vsm=0.02;  rb=10;  betaF=180;  betaR=6; betaAC=194;
  Rs=50; R1=104000; R2=104000; RC=200; RE=[50 250]; RL=10000;
  BJT_CE_analysis(VCC,rb,Rs,R1,R2,RC,RE,RL,[betaF betaR betaAC],Vsm);
```

Running these MATLAB statements yields

```
Results of analysis using the PWL model
with betaF= 180, betaR= 6.0
  and R1= 104.00[kOhm], R2= 104.00[kOhm], RC= 200[Ohm], RE=300[Ohm]
 VCC    VEE    VBB    VBQ   VEQ    VCQ      IBQ         IEQ        ICQ
10.00  0.00  5.00  2.90  2.20  8.54 4.05e-05  7.32e-03  7.28e-03
  in the forward-active mode with VCE,Q= 6.35[V]
where beta_forced = ICQ/IBQ = 180.00 with beta = 180.00
  gm= 281.664[mS], rbe=  689[Ohm], ro= 1373.39[kOhm]
Ri=  8.701 kOhm, Ro= 200 Ohm
Gv=Ri/(Rs+Ri)xAvoxRL/(Ro+RL) = 0.994 x -3.71 x 0.980 = -3.62
```

b) Consider the CE circuit in Figure 3.31(b) that is biased by a current source of $I = 7.32$ mA, which is equal to the value of $I_{E,Q}$ obtained in (a).

   Since the circuit of Figure 3.31(b) is the same as that of Figure 3.19.1(a) or 3.30(a) dealt with in (a) (except for the current source I) and the value of the current source for biasing the BJT is equal to the DC emitter current $I_{E,Q}$ = 7.32 mA (obtained in Section 3.1.8), we will get the same AC analysis results as in part (a) by using the MATLAB function 'BJT_CE_analysis_I()' (without having to do the DC analysis):

```
>>VCC=10; Vsm=0.02; rb=10; betaAC=194;
  Rs=50;  RB=52000;  RC=200;  RE=[50 250];  RL=10000;  IEQ=7.32e-3;
  BJT_CE_analysis_I(Rs,RB,RC,RE,RL,IEQ,[betaAC rb]);
```

Running these MATLAB statements yields

```
gm=ICQ/VT= 281.710[mS], re=1/gm= 4[Ohm],
rbe=beta/gm= 689[Ohm], ro=VA/ICQ= 1373.16[kOhm]
Ri=RB||RbeE= 8.700 kOhm, Ro=RC||ro= 200 Ohm
Av=-beta*Ro/RbeE x RL/(Ro+RL) =  -3.71 x 0.98 =  -3.64
Gv=Ri/(Rs+Ri)xAv =0.994 xAv =  -3.62
```

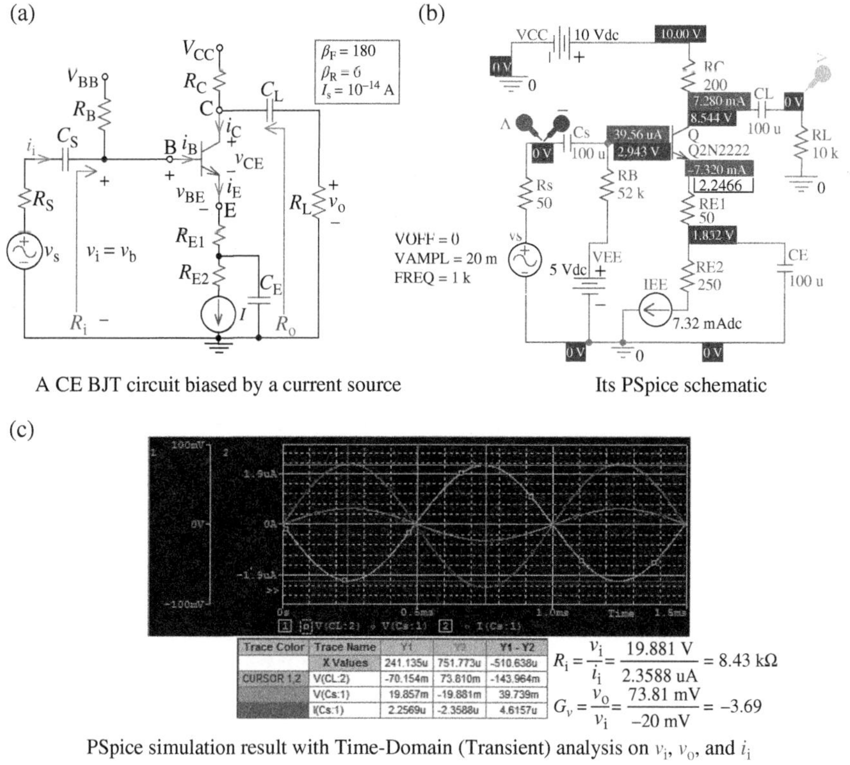

Figure 3.31 A CE circuit and its PSpice simulation results ("elec03e10a.opj"/"elec03e10b.opj").

The theoretical values of R_i, R_o, and A_v are close to not only those for the circuit of Figure 3.19.1(a) calculated in (a) but also those obtained from the PSpice simulation as shown in Figure 3.31(c):

$$R_i = \frac{v_i}{i_i} = \frac{19.881\,\text{V}}{2.3588\,\mu\text{A}} = 8.43\,\text{k}\Omega \tag{E3.10.5}$$

$$G_v = \frac{v_o}{v_i} = \frac{73.81\,\text{mV}}{-20\,\text{mV}} = -3.69 \tag{E3.10.6}$$

3.2.2 Common-Collector (CC) Amplifier (Emitter Follower)

Figure 3.32 shows a CC amplifier and its low-frequency AC equivalent where the BJT has been replaced by the equivalent in Figure 3.18(b) and the BJT output resistance r_o is assumed to be so large as to be negligible as a parallel

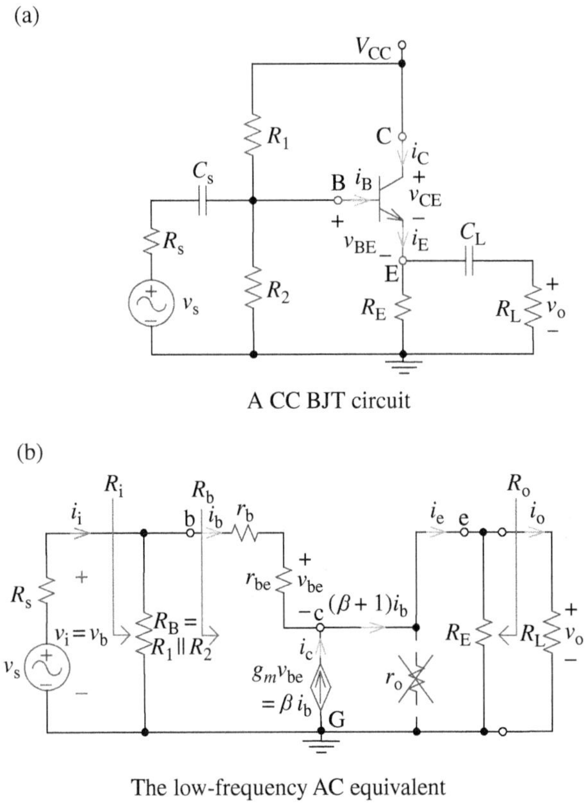

Figure 3.32 A CC (common-collector) BJT circuit and its low-frequency AC equivalent.

resistor. Let us find the input resistance, current gain, voltage gain, and output resistance.

1) **Input Resistance** R_i

To find the input resistance from the relationship between $v_i = v_b$ and i_i, we express the voltages at nodes e and b in terms of i_b as

$$v_e = (R_E \| R_L)i_e = (\beta+1)(R_E \| R_L)i_b; v_b = v_e + (r_b + r_{be})i_b$$
$$= (\beta+1)(R_E \| R_L)i_b + (r_b + r_{be})i_b = \{(\beta+1)(R_E \| R_L) + r_b + r_{be}\}i_b$$

This yields the equivalent resistance R_b seen from terminals b-G as

$$R_b = \frac{v_b}{i_b} = (\beta+1)(R_E \| R_L) + r_b + r_{be}$$

so that we can write the input resistance (including $R_B = R_1 \| R_2$) as

$$R_i = \frac{v_i = v_b}{i_i} = R_B \| \{(\beta + 1)(R_E \| R_L) + r_b + r_{be}\} \quad (3.2.5)$$

2) **Current Gain A_i**

The output current i_o through the load resistor R_L can be expressed as

$$i_o = \frac{R_E}{R_E + R_L} i_e = \frac{R_E}{R_E + R_L} (\beta + 1) i_b$$

Thus, the current gain, i.e. the ratio of the output current i_o to the input current $i_i = i_b$ is

$$A_i = \frac{i_o}{i_i} = \frac{i_b}{i_i}\frac{i_o}{i_b} = \frac{R_B}{R_B + r_b + r_{be} + (\beta + 1)(R_E \| R_L)} \frac{(\beta + 1) R_E}{R_E + R_L}$$

$$= \frac{(\beta + 1) R_B R_E}{(R_B + r_b + r_{be})(R_E + R_L) + (\beta + 1) R_E R_L} \quad (3.2.6)$$

3) **Voltage Gains G_v and A_v**

The voltage gain (with $R_s = 0$) is

$$A_v = \frac{v_o}{v_i} = \frac{(\beta + 1)(R_E \| R_L)}{r_b + r_{be} + (\beta + 1)(R_E \| R_L)} \overset{r_b + r_{be} \ll (\beta + 1)(R_E \| R_L)}{\longrightarrow} \frac{(\beta + 1)(R_E \| R_L)}{(\beta + 1)(R_E \| R_L)} = 1 \quad (3.2.7a)$$

The overall voltage gain, i.e. the ratio of the output voltage v_o to the source voltage v_s is

$$G_v = \frac{v_o}{v_s} = \frac{v_i}{v_s}\frac{v_o}{v_i} = \frac{R_i}{R_s + R_i} A_v \quad (3.2.7b)$$

where R_i is given by Eq. (3.2.5). This implies that if $R_s \ll R_i$ and $r_b + r_{be} \ll (\beta + 1)(R_E \| R_L)$, the output voltage is almost equal to the source voltage and that is why the CC amplifier is called an *emitter follower* or *buffer amplifier*.

```
function [VBQ,VEQ,VCQ,IBQ,IEQ,ICQ,Av,Ai,Ri,Ro,gm,rbe,ro]=...
  BJT_CC_analysis(VCC,rb,Rs,R1,R2,RC,RE,RL,beta,Vsm,VA,T)
% R1,R2 can be replaced by VBB,RB if VBB-RB is connected to the base.
% If beta=[betaF betaR betaAC], large-signal model-based analysis
% If beta=[betaF betaR betaAC Is], the exponential model-based analysis
% Copyleft: Won Y. Yang, wyyang53@hanmail.net, CAU for academic use only
if nargin<12, T=27; end % Ambient temperature
if nargin<11, VA=1e4; end % Early voltage
if sum([rb Rs R1 R2 RC RE RL]<0), error('Resistance must be positive!'); end
l_beta=length(beta);
if l_beta<3, betaAC=beta(1); bF=beta; % bF=[betaF betaR]
 else betaAC=beta(3); bF=beta(1:2); % bF=[betaF betaR betaAC ..]
  if l_beta>3, bF=[bF beta(4)]; end % bF=[betaF betaR betaAC Is]
end
[VBQ,VEQ,VCQ,IBQ,IEQ,ICQ,mode]=BJT_DC_analysis(VCC,R1,R2,RC,RE,bF);
if strcmp(mode,'forward-active')==0
  fprintf('\n AC analysis is impossible in BJT_CC_analysis()\n
           because the BJT is in the saturation mode!\n');
  Av=NaN; Ai=NaN; Ri=NaN; Ro=NaN; gm=NaN; rbe=NaN; ro=NaN;
  return;
end
RE1=RE(1);
if R1>20, RB=parallel_comb([R1 R2]); else RB=R2; end % R1=VBB?
if RE1<1, error('RE1 of CC amplifier (e-follower) may not be zero!'); end
[gm,rbe,re,ro]=gmrbero_BJT(ICQ,betaAC,VA,T);
RsB=parallel_comb([Rs RB]);
RELo=parallel_comb([RE RL ro]);
betaRELo=(betaAC+1)*RELo; rbebetaRELo=rb+rbe+betaRELo;
Av=betaRELo/rbebetaRELo; % Eq.(3.2.7a)
RbeEL = rb+rbe+(betaAC+1)*REL;
REpRL = RE1*RL_Av/(RE1+RL_Av);
if nargin>9 & abs(Vsm)>0
  Ibm = Vsm/Rs/rbebetaRELo/(1/Rs+1/RB+1/rbebetaRELo);
  if Ibm>=IBQ, fprintf('\n Possibly crash into the cutoff region\n'); end
  if abs(Gv*Vsm)>=VCQ-VEQ-0.2 % Eq.(3.1.47)
    fprintf('\n Possibly violate the saturation region\n');
  end
end
Ri=parallel_comb([RB RbeEL]); % Eq.(3.2.5)
Ro=parallel_comb([RE1 (RsB+rb+rbe)/(betaAC+1)]); % Eq.(3.2.8)
Ai=(betaAC+1)*RB*RE1/((RB+rb+rbe)*(RE1+RL)+(betaAC+1)*RE1*RL);
                                                       % Eq.(3.2.6)
Gv=Av/(Rs/Ri+1); % Eq.(3.2.7b) considering Rs
fprintf('\t gm=%8.3f[mS], rbe=%6.0f[Ohm], ro=%7.2f[kOhm]\n',
             gm*1e3,rbe,ro/1e3)
fprintf(' Ri=%9.3f[kOhm], Ro=%6.0f[Ohm]\n Gv=Ri/(Rs+Ri)xAv =%5.3f x
         %8.3f =%8.3f\n', Ri/1e3,Ro,1/(Rs/Ri+1),Av,Gv);
```

4) **Output Resistance R_o**

To find the (Thevenin) equivalent resistance seen from the load side, we remove the (independent) voltage source v_s by short-circuiting it and apply a test voltage source V_T to the output port. Then, the base current i_b and the test current I_T through V_T are computed as

$$i_b = -\frac{V_T}{(R_s \| R_B) + r_b + r_{be}}$$

$$I_T = -i_e + i_{R_E} = (\beta + 1)i_b + i_{R_E} = \frac{(\beta + 1)V_T}{(R_s \| R_B) + r_b + r_{be}} + \frac{V_T}{R_E}$$

Thus, we find the output resistance as

$$R_o = \frac{V_T}{I_T} = \frac{1}{(\beta + 1)/((R_s \| R_B) + r_b + r_{be}) + 1/R_E} = R_E \| \frac{(R_s \| R_B) + r_b + r_{be}}{\beta + 1} \quad (3.2.8)$$

The emitter follower has a very low output resistance (3.2.8), which enables the circuit to provide its load with much current without paying much attention to the loading effect. It also has a very high input resistance (3.2.5), which enables the circuit to save the current provided by its source (driver). In short words, the emitter follower is modest enough not to burden its source as well as generous to its load. (Isn't the emitter follower praiseworthy? Who can blame such a nice guy for not amplifying the voltage?) That is the main feature of emitter follower with an almost unity voltage gain.

Example 3.11 AC Analysis of CC Amplifier (Emitter Follower)
Consider the CC circuit in Figure 3.33(a) where $V_{CC} = 15$ V, $r_b = 0\ \Omega$, $R_s = 20$ kΩ, $R_1 = 345$ kΩ, $R_2 = 476$ kΩ, $R_C = 4$ kΩ, $R_E = 5940\ \Omega$, $R_L = 1$ kΩ, $\beta_F = 100$, $\beta_R = 1$, $\beta_{AC} = 100$, and $V_A = 100$ V. Find R_i, R_o, $A_v = v_o/v_i$, and $G_v = v_o/v_s$ and compare the value of G_v with that obtained from PSpice simulation.

We can use the values of $r_{be} = 2586\ \Omega$ and $r_o = 10^4$ kΩ (obtained by using the MATLAB function '`BJT_DC_analysis()`' and '`gmrbero_BJT()`' in '`BJT_CC_analysis()`') for the AC analysis as follows:

$$R_i = \frac{v_i}{i_i} \overset{(3.2.5)}{=} R_B \| \{(\beta + 1)(R_E \| R_L) + r_b + r_{be}\}$$
$$= 345 \| 476 \| \{101(5.94 \| 1) + 2.586\} = 61.61\ \text{k}\Omega \quad (E3.11.1)$$

$$R_o \overset{(3.2.8)}{=} R_E \| \frac{(R_s \| R_B) + r_b + r_{be}}{\beta + 1}$$
$$= 5940 \| \frac{(20 \| 345 \| 476) + 2586}{101} = 198.7\ \Omega \quad (E3.11.2)$$

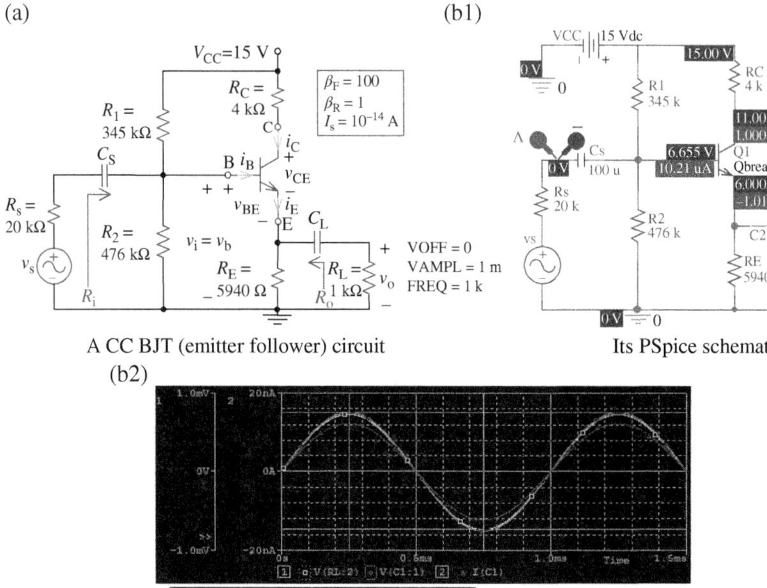

Trace Color	Trace Name	Y1	Y2	Y1 - Y2
	X Values	242.286u	752.297u	-510.012u
CURSOR 1,2	V(RL:2)	728.495u	-733.053u	1.4616m
	V(Cs:1)	752.561u	-752.691u	1.5053m
	I(Cs)	12.313n	-12.360n	24.673n

$$R_i=\frac{v_{i,pp}}{i_{i,pp}}=\frac{752.6+752.7\ \mu V}{12.31+12.36\ nA}=61\ k\Omega$$

$$G_v=\frac{v_{o,pp}}{v_{s,pp}}=\frac{728.5+733.1\ \mu V}{2\ mV}=0.731$$

Figure 3.33 A CC amplifier (emitter follower) and its PSpice simulation results ("elec03e11.opj").

$$A_v \overset{(3.2.7a)}{=} \frac{(\beta+1)(R_E\|R_L)}{r_b+r_{be}+(\beta+1)(R_E\|R_L)} = \frac{101\times(5940\|1000)}{2586+101\times(5940\|1000)} = 0.971 \tag{E3.11.3}$$

$$G_v = \frac{v_o}{v_s} = \frac{v_i}{v_s}\frac{v_o}{v_i} = \frac{R_i}{R_s+R_i}A_v \underset{(E3.11.3)}{\overset{(E3.11.1)}{=}} \frac{61.61}{20+61.61}0.971 = 0.733 \tag{E3.11.4}$$

This AC analysis can also be done by using the above MATLAB function '`BJT_CC_analysis()`' although the results will be slightly different since r_o is taken into consideration:

```
>>VCC=15; Vsm=0.001;  rb=0;  betaF=100;  betaR=1;  betaAC=betaF;
  Rs=2e4; R1=345e3; R2=476e3; RC=4e3; RE=5.94e3; RL=1e3;
  BJT_CC_analysis(VCC,rb,Rs,R1,R2,RC,RE,RL,[betaF,betaR,betaAC],Vsm);
```

If the CC circuit is biased by a current source of IEQ = 1 mA, run the MATLAB function '`BJT_CC_analysis_I()`' as

```
>>IEQ=0.001; RB=parallel_comb([R1 R2]);
  BJT_CC_analysis_I(Rs,RB,RC,RE,RL,IEQ,[betaAC rb]); % Alternative
```

Running the above MATLAB statements yields

```
Results of analysis using the PWL model
with betaF= 100, betaR= 1.0
and R1= 345.00[kOhm], R2= 476.00[kOhm], RC= 4000[Ohm], RE= 5940[Ohm]
 VCC     VEE     VBB     VBQ     VEQ     VCQ      IBQ        IEQ        ICQ
15.00   0.00    8.70    6.70    6.00    11.00  1.00e-05  1.01e-03  1.00e-03
 in the forward-active mode with VCE,Q= 5.00[V]
where beta_forced = ICQ/IBQ = 100.00 with beta = 100.00
 gm= 38.669[mS], rbe= 2586[Ohm], ro= 10003.67[kOhm]
Ri= 61.606 kOhm, Ro= 199 Ohm
Gv=Ri/(Rs+Ri)xAv= 0.755 x  0.97 =  0.73
```

```
function [Av,Ai,Ri,Ro,gm,rbe,ro,vsm,vom]=...
   BJT_CC_analysis_I(Rs,RB,RC,RE,RL,IEQ,beta,VA,T,vbem)
% To analyze CC amplifier given IEQ
% beta = [beta rb] if a nonzero base resistance is given.
% If the 10th input argument vbem (maximum vbe such that the BJT operated
% in the linear region) is given, the following outputs will be returned.
% vsm = Max amplitude of vs ensuring the linear operation of BJT.
% vom = Max amplitude of vo when the BJT operates in the linear operation.
% Copyleft: Won Y. Yang, wyyang53@hanmail.net, CAU for academic use only
if nargin<9, T=27; end % Ambient temperature
if nargin<8, VA=1e4; end % Early voltage
if length(RB)>1, RB=parallel_comb(RB); end
if length(beta)>1, rb=beta(2); beta=beta(1); else rb=0; end
% If there is no RB/RE, they must be given as very large values like 1e10.
if sum([Rs RB RC RE RL rb]<0), error('Resistance must be positive!'); end
ICQ = IEQ*beta/(beta+1); % Collector current at the bias point
[gm,rbe,re,ro]=gmrbero_BJT(ICQ,beta,VA,T);
RsB=parallel_comb([Rs RB]); REL=parallel_comb([RE RL]);
RbeEL = rb+rbe+(beta+1)*REL;
Ri=parallel_comb([RB RbeEL]); % Eq.(3.2.5)
Ro=parallel_comb([RE (RsB+rbe)/(beta+1)]); % Eq.(3.2.8)
Ai = (beta+1)*RB*RE/((RB+rbe)*(RE+RL)+(beta+1)*RE*RL); % Eq.(3.2.6)
Av = REL/(re+REL); % Eq.(3.2.7)
Gv = Av/(Rs/Ri+1); % Considering Rs
if nargout<1
  fprintf('\n gm=ICQ/VT=%8.3f[mS], re=1/gm=%6.0f[Ohm], rbe=beta/gm=
       %6.0f[Ohm], ro=VA/ICQ=%7.2f[kOhm]\n', gm*1e3,re,rbe,ro/1e3);
  fprintf(' Ri=RB||{rbe+(beta+1)(RE||RL)}=%9.2f kOhm,
          Ro=RE||[{(Rs||RB)+rbe}/(beta+1)]=%6.0f Ohm', Ri/1e3,Ro);
  fprintf('\n Av=(RE||RL)/(re+(RE||RL))=%8.3f, Gv=Ri/(Rs+Ri)xAv =
                          %5.3f xAv =%8.3f\n', Av,Ri/(Rs+Ri),Gv);
end
if nargin<10, vbem = NaN; end
vsm = (Rs/Ri+1)*RbeEL/rbe*vbem; vom = Gv*vsm;
```

These values of R_i and G_v are close to the PSpice simulation results shown in Figure 3.33(b2):

$$R_i = \frac{v_{i,pp}}{i_{i,pp}} = \frac{752.6 + 752.7\,\mu V}{12.31 + 12.36\,nA} = 61\,k\Omega \tag{E3.11.5}$$

$$G_v = \frac{v_{o,pp}}{v_{s,pp}} = \frac{728.5 + 733.1\,\mu V}{2\,mV} = 0.731 \tag{E3.11.6}$$

As listed above, we have the MATLAB function '`BJT_CC_analysis_I()`', which can be used for the AC analysis of a CC circuit biased by a DC current source $I_{E,Q}$.

3.2.3 Common-Base (CB) Amplifier

Figure 3.34.1 shows a CB amplifier and its low-frequency AC equivalent where the BJT has been replaced by the equivalent in Figure 3.18(b) and the BJT output resistance r_o is assumed to be so large as to be negligible as a parallel resistor. Let us find the input resistance, current gain, voltage gain, and output resistance.

1) **Input Resistance R_i**

To find the input resistance from the relationship between $v_i = v_e$ and i_i, we apply KCL at node c to write

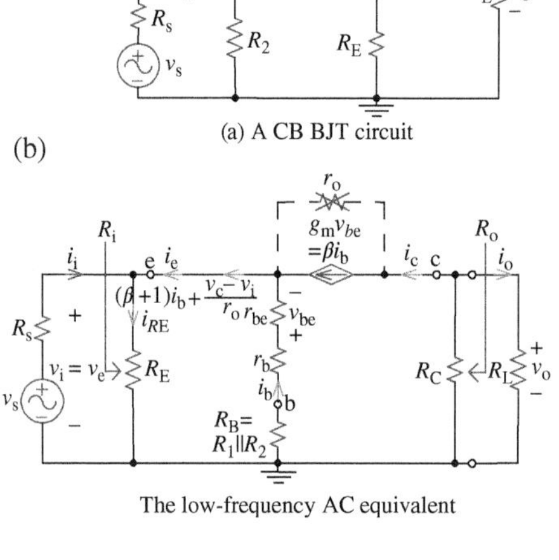
(a)
(a) A CB BJT circuit
(b)
The low-frequency AC equivalent

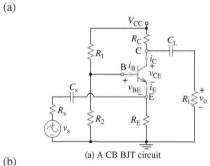

Figure 3.34.1 A CB (common-base) BJT circuit and its low-frequency AC equivalent.

$$\left(\frac{1}{r_o}+\frac{1}{R_C\|R_L}\right)v_c=\frac{v_i}{r_o}-\beta i_b;$$

$$v_c=\frac{(R_C\|R_L)\{1+\beta r_o/(r_b+r_{be}+R_B)\}}{r_o+(R_C\|R_L)}v_i \text{ with } i_b=-\frac{v_i}{r_b+r_{be}+R_B}$$

Thus, we can find the input resistance as

$$R_i=\frac{v_i}{i_i}=\frac{v_i}{v_i/R_E-(\beta+1)i_b-(v_c-v_i)/r_o}=R_E\|\frac{(R_L+r_o)(r_b+r_{be}+R_B)}{R_L+r_b+r_{be}+R_B+(\beta+1)r_o}$$

$$\overset{r_o=\infty}{\longrightarrow} R_E\|\frac{r_b+r_{be}+R_B}{\beta+1}. \tag{3.2.9}$$

This input resistance is very small compared with that (Eq. (3.2.1)) of CE amplifier and that (Eq. (3.2.5)) of CC amplifier.

2) **Current Gain A_i**

The current i_{R_E} through the emitter resistor R_E can be expressed in terms of the base current i_b through r_{be}-r_b-R_B (connected in parallel with R_E) as

$$i_{R_E}=-\frac{r_b+r_{be}+R_B}{R_E}i_b$$

Applying KCL at node e yields the expression of the input current i_i in terms of i_b as

$$i_i=i_{R_E}-(\beta+1)i_b=-\left(\frac{r_b+r_{be}+R_B}{R_E}+\beta+1\right)i_b$$

The output current i_o through the load resistor R_L can be expressed as

$$i_o=-\frac{R_C}{R_C+R_L}i_c=-\frac{R_C}{R_C+R_L}\beta i_b$$

Thus, the current gain, i.e. the ratio of the output current i_o to the input current i_i is

$$A_i=\frac{i_o}{i_i}=\frac{\frac{R_C}{R_C+R_L}\beta}{\frac{r_b+r_{be}+R_B}{R_E}+\beta+1}=\frac{R_C R_E}{R_C+R_L}\frac{\beta}{r_b+r_{be}+R_B+(\beta+1)R_E} \tag{3.2.10}$$

3) **Voltage Gains G_v and A_v**

To find the voltage gain $A_v=v_o/v_i$ (with $R_s=0$), we apply KCL at node c (of the circuit in Figure 3.34.1(b)) to write the node equation and solve it as

$$\frac{v_o-v_i}{r_o}+\frac{v_o}{R_C\|R_L}=-\beta i_b\,;\ \frac{v_o}{r_o}+\frac{v_o}{R_C\|R_L}=\frac{v_i}{r_o}+\beta\frac{v_i}{r_b+r_{be}+R_B}$$

$$;A_v=\frac{v_o}{v_i}=\frac{1/r_o+\beta/(r_b+r_{be}+R_B)}{1/r_o+1/(R_C\|R_L)} \tag{3.2.11a}$$

The overall voltage gain, i.e. the ratio of the output voltage v_o to the source voltage v_s is

$$G_v = \frac{v_o}{v_s} = \frac{v_i}{v_s}\frac{v_o}{v_i} = \frac{R_i}{R_s + R_i}A_v \tag{3.2.11b}$$

where R_i is given by Eq. (3.2.9).

```
function [VBQ,VEQ,VCQ,IBQ,IEQ,ICQ,Av,Ai,Ri,Ro,gm,rbe,ro]=...
    BJT_CB_analysis(VCC,rb,Rs,R1,R2,RC,RE,RL,beta,Vsm,VA,T)
% R1,R2 can be replaced by VBB,RB if VBB-RB is connected to the base.
% If beta=[betaF betaR betaAC], large-signal model-based analysis
% If beta=[betaF betaR betaAC Is], the exponential model-based analysis
% Copyleft: Won Y. Yang, wyyang53@hanmail.net, CAU for academic use only
if nargin<12, T=27; end % Ambient temperature
if nargin<11, VA=1e4; end % Early voltage
if sum([rb Rs R1 R2 RC RE RL]<0), error('Resistance must be positive!'); end
l_beta=length(beta);
if l_beta<3, betaAC=bF(1); bF=beta; % beta=[betaF betaR]
 else betaAC=beta(3); bF=beta(1:2); % beta=[betaF betaR betaAC ..]
      if l_beta>3, bF=[bF beta(4)]; end % beta=[betaF betaR betaAC Is]
end
[VBQ,VEQ,VCQ,IBQ,IEQ,ICQ,mode]=BJT_DC_analysis(VCC,R1,R2,RC,RE,bF);
if strcmp(mode,'forward-active')==0
   fprintf('\n AC analysis is impossible in BJT_CB_analysis()\n
                              because the BJT is in the saturation mode!\n');
   Av=NaN; Ai=NaN; Ri=NaN; Ro=NaN; gm=NaN; rbe=NaN; ro=NaN; return;
end
RE1=RE(1);
if R1>20, RB=parallel_comb([R1 R2]); else RB=R2; end % R1=VBB?
[gm,rbe,re,ro]=gmrbero_BJT(ICQ,betaAC,VA,T);
RCL=parallel_comb([RC RL]); RbeB=rb+rbe+RB;
Av=(1/ro+betaAC/RbeB)/(1/ro+1/RCL); % Eq.(3.2.11a)
Ri = parallel_comb([RE1 RbeB/(betaAC+1)]); % Eq.(3.2.9)
ro1=ro+parallel_comb([Rs RE1 RbeB])*(1+betaAC/RbeB*ro);
Ro = parallel_comb([RC ro1]); % Eq.(3.2.12)
Ai = betaAC*RC*RE1/(RC+RL)/(RbeB+(betaAC+1)*RE1); % Eq.(3.2.10)
Gv = Av/(Rs/Ri+1); % Eq.(3.2.11b)
if nargin>9 & abs(Vsm)>0 % Detailed AC analysis
 Ibm = Vsm/Rs/RbeB/(1/Rs+1/RE1+(betaAC+1)/RbeB); % Eq.(3.1.38)
  if Ibm>=IBQ, fprintf('\n Possibly crash into the cutoff region\n');
    end
 if abs(Gv*Vsm+RbeB*Ibm)>=VCQ-VEQ-0.2 % Eq.(3.1.39)
   fprintf('\n Possibly violate the saturation region\n');
 end
end
fprintf('\t gm=%8.3f[mS], rbe=%6.0f[Ohm], ro=%7.2f[kOhm]\n',
                                                    gm*1e3,rbe, ro/1e3)
fprintf(' Ri=%9.3f[kOhm], Ro=%6.0f[Ohm]\n Gv=Ri/(Rs+Ri)xAv=%7.3f x%8.2f
                              =%8.2f\n', Ri/1e3,Ro,1/(Rs/Ri+1),Av,Gv);
```

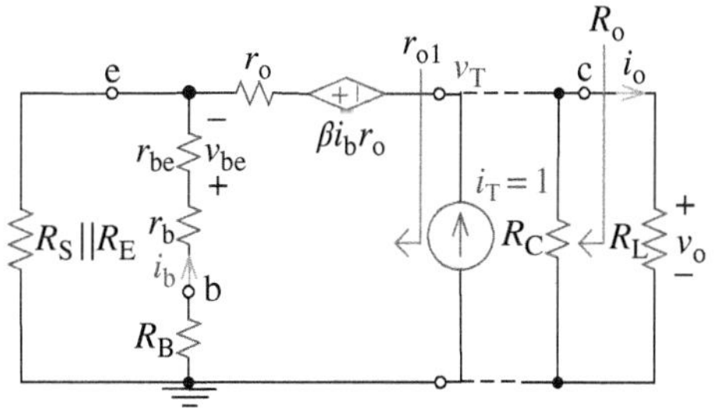

Figure 3.34.2 To find output resistance R_o of the CB circuit.

4) Output Resistance R_o

To find the equivalent resistance seen from the load side, we remove the (independent) voltage source v_s by short-circuiting it, make a I-to-V source transformation of the dependent current source βi_b into the voltage source $\beta i_b r_o$ in series with r_o as shown in Figure 3.34.2. Then we apply a test current source of $i_T = 1$ A and find the voltage across it:

$$R_o = R_C||r_{o1} \text{ with } r_{o1} = r_o + \{R_s||R_E||(r_{be} + r_b + R_B)\}\left(1 + \frac{\beta r_o}{r_{be} + r_b + R_B}\right) \quad (3.2.12)$$

This process for analyzing a CB amplifier to find their input/output resistances and voltage/current gains has been cast into the above MATLAB function 'BJT_CB_analysis()' and the following one 'BJT_CB_analysis_I()' for the case where the amplifier is excited by current source.

```
function [Av,Ai,Ri,Ro,gm,rbe,ro,vsm,vom]=...
              BJT_CB_analysis_I(Rs,RB,RC,RE,RL,IEQ,beta,VA,T,vbem)
% To analyze CB amplifier given IEQ
% beta = [beta rb] if a nonzero base resistance is given.
% If the 10th input argument vbem (maximum vbe such that the BJT operated
% in the linear region) is given, the following outputs will be returned.
% vsm = Max amplitude of vs ensuring the linear operation of BJT.
% vom = Max amplitude of vo when the BJT operates in linear operation.
% Copyleft: Won Y. Yang, wyyang53@hanmail.net, CAU for academic use
if nargin<9, T=27; end % Ambient temperature
if nargin<8, VA=1e4; end % Early voltage
if length(RB)>1, RB=parallel_comb(RB); end
if length(beta)>1, rb=beta(2); beta=beta(1); else rb=0; end
% If there is no RB/RE, they must be given as very large values like 1e10.
if sum([Rs RB RC RE RL rb]<0), error('Resistance must be positive!'); end
ICQ = IEQ*beta/(beta+1); % Collector current at the bias point
[gm,rbe,re,ro]=gmrbero_BJT(ICQ,beta,VA,T);
```

```
RbeB=rb+rbe+RB; RE1=RE(1);
Ri = parallel_comb([RE1 RbeB/(beta+1)]); % Eq.(3.2.9)
%Ro = parallel_comb([RC ro]); % Eq.(3.2.12)
ro1= ro+parallel_comb([Rs RE RbeB])*(1+beta/RbeB*ro);
Ro = parallel_comb([RC ro1]); % Eq.(3.2.12)
Ai = beta*RC*RE/(RC+RL)/(RB+rbe+(beta+1)*RE1); % Eq.(3.2.10)
Avo= beta*Ro/RbeB; % Eq.(3.2.11a) with no RL
Av = Avo/(Ro/RL+1); % Eq.(3.2.3) considering RL
Gv = Av/(Rs/Ri+1); % Eq.(3.2.11b) Considering Rs
if nargout<1
  fprintf('\n gm=ICQ/VT=%8.3f[mS],re=1/gm=%6.0f[Ohm],rbe=beta/gm=
    %6.0f[Ohm], ro=VA/ICQ=%7.2f[kOhm]\n', gm*1e3,re,rbe,ro/1e3);
  fprintf(' Ri=RE||{(RB+rbe)/(beta+1)}=%9.3f kOhm, Ro=RC||ro=%6.0f
   Ohm', Ri/1e3,Ro);
 fprintf('\n Av=-alpha*Ro/(re+RE)xRL/(Ro+RL)=%8.2f x%5.2f =%8.2f,
   Gv=Ri/(Rs+Ri)xAv =%5.3f xAv =%8.2f\n', vo,RL/(Ro+RL),Av,
   Ri/(Rs+Ri),Gv);
end
if nargin>9, vsm=(Rs+Ri)/re*vbem; vom=Gv*vsm;
else vsm=NaN; vom=NaN; end
```

Example 3.12 AC Analysis of CB Amplifier
Consider the CB circuit in Figure 3.35(a) where $V_{CC} = 15$ V, $r_b = 0\ \Omega$, $R_s = 1$ kΩ, $R_1 = 26.2$ kΩ, $R_2 = 16.2$ kΩ, $R_C = 10$ kΩ, $R_E = 19.8\ \Omega$, $R_L = 10$ kΩ, $\beta_F = 100$, $\beta_R = 1$, $\beta_{AC} = 100$, and $V_A = 10^4$ V. Find R_i, R_o, $A_v = v_o/v_i$, and $G_v = v_o/v_s$, and compare their values with those obtained from PSpice simulation.

We can use the values of $r_{be} = 10\,327\ \Omega$ and $r_o = 399.47$ kΩ (obtained by using the MATLAB function '`BJT_DC_analysis()`' and '`gmrbero_BJT()`') for the AC analysis as follows:

$$R_i = \frac{v_i}{i_i} \overset{(3.2.9)}{=} R_E \| \frac{r_b + r_{be} + R_B}{\beta + 1} = 19.8 \| \frac{10.327 + (26.2\|16.2)}{101}$$
$$= 19.8 \| \frac{20.337}{101} = 199\,\Omega \tag{E3.12.1}$$

$$R_o \underset{r_o=\infty}{\overset{(3.2.12)}{=}} R_C = 10\,\text{k}\Omega \tag{E3.12.2}$$

$$A_v \underset{r_o=\infty}{\overset{(3.2.11a)}{=}} \frac{v_o}{v_i} = \frac{\beta(R_C\|R_L)}{r_b + r_{be} + R_B} = \frac{100 \times (10\|10)}{20.337} = 24.6 \tag{E3.12.3}$$

$$G_v = \frac{v_o}{v_s} = \frac{v_i}{v_s}\frac{v_o}{v_i} = \frac{R_i}{R_s + R_i}A_v \underset{(E3.12.3)}{\overset{(E3.12.1)}{=}} \frac{0.199}{1 + 0.199} 24.6 = 4.083 \tag{E3.12.4}$$

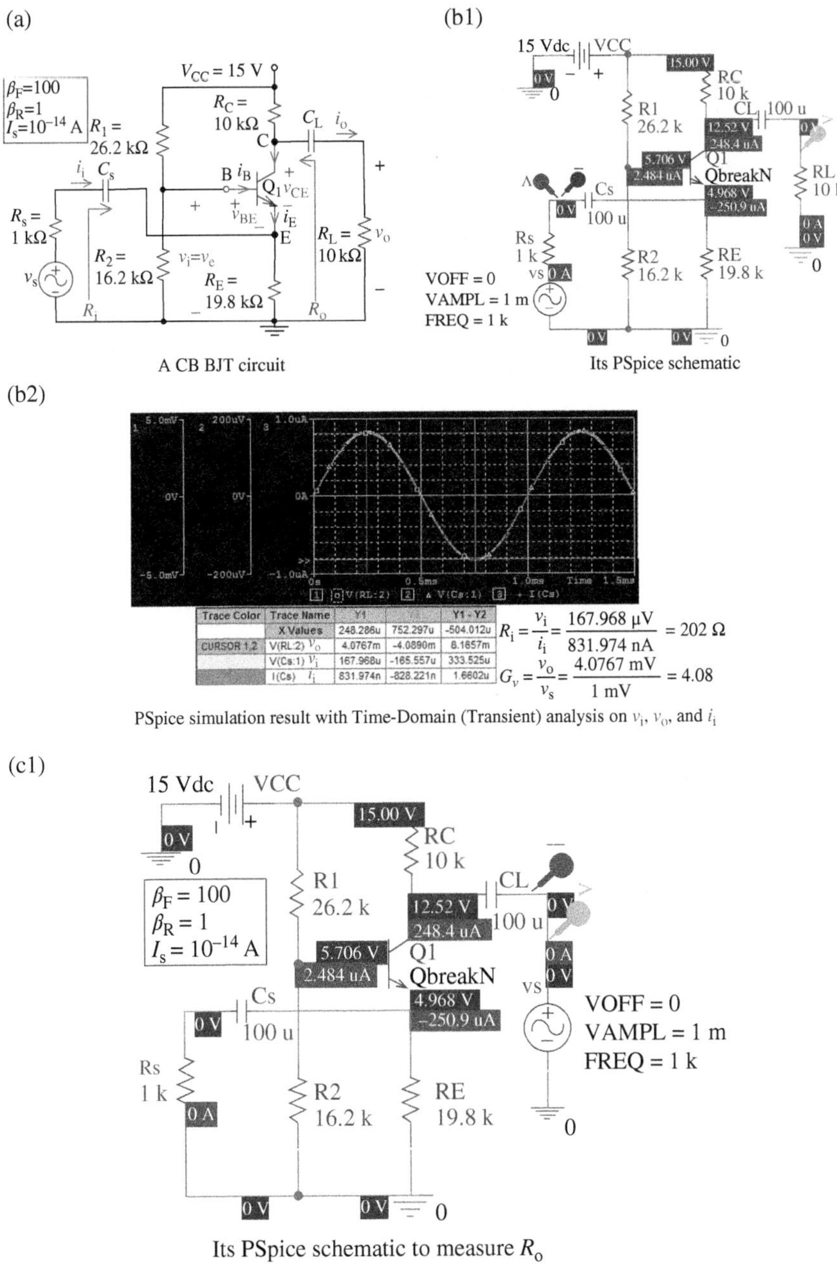

Trace Color	Trace Name	Y1	[illegible]	Y1 - Y2
	X Values	248.286u	752.297u	-504.012u
CURSOR 1,2	V(RL:2) v_o	4.0767m	-4.0890m	8.1657m
	V(Cs:1) v_i	167.968u	-165.557u	333.525u
	I(Cs) i_i	831.974n	-828.221n	1.6602u

$$R_i = \frac{v_i}{i_i} = \frac{167.968\ \mu V}{831.974\ nA} = 202\ \Omega$$

$$G_v = \frac{v_o}{v_s} = \frac{4.0767\ mV}{1\ mV} = 4.08$$

Figure 3.35 A CB circuit and its PSpice simulation results ("elec03e12.opj").

(c2)

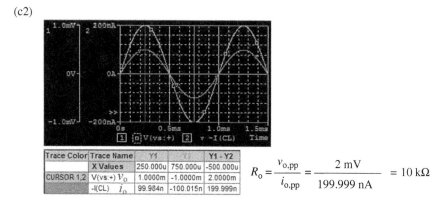

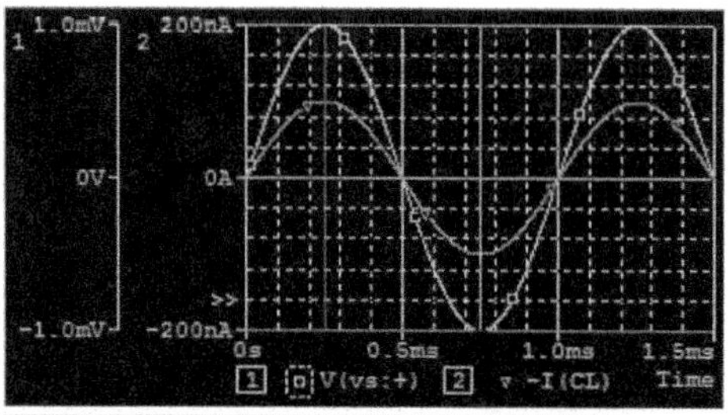

Trace Color	Trace Name	Y1	Y2	Y1 - Y2
	X Values	250.000u	750.000u	-500.000u
CURSOR 1,2	V(vs:+) v_o	1.0000m	-1.0000m	2.0000m
	-I(CL) i_o	99.984n	-100.015n	199.999n

$$R_o = \frac{v_{o,pp}}{i_{o,pp}} = \frac{2\ \text{mV}}{199.999\ \text{nA}} = 10\ \text{k}\Omega$$

Voltage/current waveforms applied/measured from the output

Figure 3.35 (Continued)

This AC analysis can also be done by using the above MATLAB function '`BJT_CB_analysis()`' although the results will be slightly different since r_o is taken into consideration:

```
>>VCC=15; Vsm=0.001; rb=0; betaF=100; betaR=1; betaAC=betaF;
  Rs=1e3; R1=26.2e3; R2=16.2e3; RC=10e3; RE=19.8e3; RL=1e4;
  BJT_CB_analysis(VCC,rb,Rs,R1,R2,RC,RE,RL, [betaF,betaR,betaAC],Vsm);
```

If the CB circuit is biased by a current source of IEQ = 251 μA, run the MATLAB function '`BJT_CB_analysis()`' with `IEQ` and `RB=`$R_1||R_2$ in place of `R1` and `R2` or '`BJT_CB_analysis_I()`' as follows:

```
>>IEQ=253e-6; RB=parallel_comb([R1 R2]); % Alternative
BJT_CB_analysis_I(Rs,RB,RC,RE,RL,IEQ, [betaAC rb]); % Another alternative
```

Running these MATLAB statements yields

```
Results of analysis using the PWL model
with betaF= 100, betaR= 1.0
and R1= 26.20[kOhm], R2= 16.20[kOhm], RC= 10000[Ohm], RE= 19800[Ohm]
 VCC     VEE     VBB     VBQ     VEQ     VCQ        IBQ         IEQ          ICQ
15.00   0.00    5.73    5.71    5.01    12.50   2.50e-06   2.53e-04    2.50e-04
in the forward-active mode with VCE,Q= 7.49[V]
where beta_forced = ICQ/IBQ = 100.00 with beta = 100.00
      gm= 9.684[mS], rbe= 10327[Ohm], ro= 39947.48[kOhm]
Ri=  0.199kOhm, Ro= 10000Ohm
Gv=Ri/(Rs+Ri)xAv= 0.166 x 24.58 =  4.09
```

These values of R_i, R_o, and G_v are close to the PSpice simulation results shown in Figure 3.35(b2) and (c2):

$$R_i = \frac{v_i}{i_i} = \frac{167.968\,\mu\text{V}}{831.974\,\text{nA}} = 202\,\Omega \tag{E3.12.5}$$

$$R_o = \frac{v_{o,pp}}{i_{o,pp}} = \frac{2\,\text{mV}}{199.999\,\text{nA}} = 10\,\text{k}\Omega \tag{E3.12.5}$$

$$G_v = \frac{v_o}{v_s} = \frac{4.0767\,\text{mV}}{1\,\text{mV}} = 4.08 \tag{E3.12.6}$$

where in Figure 3.35(b2), an AC (sine-wave) voltage source of 1 mV is applied at the output terminal to find the output resistance R_o from the voltage-current relationship:

The next section will show how the MATLAB functions presented above can be used to analyze a multistage amplifier.

3.2.4 Multistage Cascaded BJT Amplifier

Table 3.3 lists the formulas for finding the input/output resistances, voltage gain, and current gain of the CE/CC/CB amplifiers.

Note that to find the input/output resistance of a CC configuration requires the input/output resistance of the next/previous stage corresponding to its load/source resistance R_L/R_s as implied by Eq. (3.2.5)/(3.2.8). That is why, for a systematic analysis of a multistage amplifier containing one or more CC configurations, we should find the input/output resistance of each stage, starting from the last/first stage backwards/forwards to the first/last stage where the load

Table 3.3 Characteristics of Common-Emitter/Common-Collector/Common-Base (CE/CC/CB) amplifiers.

	CE	CC	CB
R_i	$R_B \| \{r_b + r_{be} + (\beta + 1)R_{E1}\}$ (3.2.1)	$R_B \| \{r_b + r_{be} + (\beta+1)(R_E \| R_L)\}$ (3.2.5)	$R_E \| \dfrac{r_b + r_{be} + R_B}{\beta + 1}$ (3.2.9)
R_o	$R_C \| r_o \approx R_C$ (3.2.4)	$R_E \| \dfrac{(R_s \| R_B) + r_b + r_{be}}{\beta + 1}$ (3.2.8)	$R_C \| r_{o1}$ (3.2.12)
A_v	$-\dfrac{\beta(R_C \| R_L)}{r_b + r_{be} + (\beta+1)R_{E1}}$ (3.2.3)	$\dfrac{(\beta + 1)(R_E \| R_L)}{r_b + r_{be} + (\beta + 1)(R_E \| R_L)}$ (3.2.7)	$\dfrac{\beta(R_C \| R_L)}{r_b + r_{be} + R_B}$ (3.2.11)
A_i	$-\dfrac{\beta R_C}{R_C + R_L}$ (3.2.2)	$\dfrac{(\beta+1)R_B R_E}{(R_B+r_b+r_{be})(R_E+R_L) + (\beta+1)R_E R_L}$ (3.2.6)	$\dfrac{R_C R_E}{R_C+R_L} \dfrac{\beta}{r_b+r_{be}+R_B+(\beta+1)R_E}$ (3.2.10)

resistance to each stage except the last one is the input resistance of the next stage and the source resistance to each stage except the first one is the output resistance of the previous stage.

```
function [Av,Avo,Gv,Ri,Ro]=Av_CE(ro_,RE_,RC_)
% Put 1 as the 1st input argument ro_ if ro~=inf.
% Put 0 as the 2nd input argument RE_ if RE=0.
% Put 0 as the 3rd input argument RC_ if RC=inf.
syms b rb rbe ro Rs RB RC RE RL
if nargin>1&RE_==0, RE=0; end
if nargin>2&RC_==0, RC=inf; end
RCL=parallel_comb([RC RL]); vi=1;
if nargin>0&ro_>0,
 if RE==0
  Ri=parallel_comb([RB rbe]); Av=-b/rbe*parallel_comb([ro RCL]);
 else
  Y=[(b+1)/rbe+1/RE+1/ro -1/ro;
    -1/ro-b/rbe 1/ro+1/RCL];
  vec=Y\[(b+1)/rbe; -b/rbe]; ve=vec(1); vc=vec(2);
  ib=(vi-ve)/rbe; Ri=parallel_comb([RB vi/ib]); Av=vc;
 end
else
 ro=inf; rbeRE=rbe+(b+1)*RE;
 Ri=parallel_comb([RB rbeRE]); % Eq.(3.2.1)
 Av=-b*RCL/rbeRE; % Eq.(3.2.3)
end
Ro=parallel_comb([RC ro]); % Eq.(3.2.4)
Avi=1/(Rs/Ri+1); AvL=1/(Ro/RL+1);
Avo=Av/AvL; %Avo=-b*RC/(rbe+(b+1)*RE);
if nargout<1, fprintf('\n Av =\n'); pretty(simplify(Av)); end
```

```
function Ri=Ri_CE(ro_,RE_,RC_)
% Put 1 as the 1st input argument ro_ if ro~=inf.
% Put 0 as the 2nd input argument RE_ if RE=0.
% Put 0 as the 3rd input argument RC_ if RC=inf.
if nargin>2, [Av,Avo,Gv,Ri]=Av_CE(ro_,RE_,RC_);
  elseif nargin>1, [Av,Avo,Gv,Ri]=Av_CE(ro_,RE_);
  elseif nargin>0, [Av,Avo,Gv,Ri]=Av_CE(ro_);
  else [Av,Avo,Gv,Ri]=Av_CE;
end
```

```
function Ro=Ro_CE(ro_)
% Put 1 as the 1st input argument ro_ if ro~=inf.
syms ro RC
if nargin<1|ro_==0, ro=inf; end
Ro=parallel_comb([RC ro]); % Eq.(3.2.4)
if nargout<1, fprintf('\n Ro =\n'); pretty(simplify(Ro)); end
```

```
function [Av,Avo,Gv,Ri,Ro]=Av_CC(ro_,RE_)
% Put 0 as the 1st input argument ro_ if ro=inf.
% Put 0 as the 2nd input argument RE_ if RE=inf.
syms b rbe ro Rs RB RE RL
if nargin>0&ro_==0, ro=inf; end
if nargin>1&RE_==0, RE=inf; end
REL=parallel_comb([ro RE RL]);
b1REL=(b+1)*REL; RsB=parallel_comb([Rs RB]);
Ri=parallel_comb([RB rbe+b1REL]); % Eq.(3.2.5)
Av=b1REL/(rbe+b1REL); % from Eq.(3.2.7)
Avi=1/(Rs/Ri+1); Avo=(b+1)*RE/(rbe+(b+1)*RE); %Eq.(3.2.7):Open-loopgain
Gv=Avi*Av;
```

```
function Ri=Ri_CC(ro_,RE_)
syms b rbe RB RE RL
if nargin>1, [Av,Avo,Gv,Ri]=Av_CC(ro_,RE_);
 elseif nargin>0, [Av,Avo,Gv,Ri]=Av_CC(ro_);
 else [Av,Avo,Gv,Ri]=Av_CC;
end
```

```
function Ro=Ro_CC()
syms b rbe ro Rs RB RE RL
RsB=parallel_comb([Rs RB]);
Ro=parallel_comb([RE (RsB+rbe)/(b+1)]); % Eq.(3.2.8)
if nargout<1, fprintf('\n Ro =\n'); pretty(simplify(Ro)); end
```

```
function [Av,Avo,Gv,Ri,Ro]=Av_CB(ro_,RE_)
% Put 1 as the 1st input argument ro_ if ro~=inf.
% Put 0 as the 2nd input argument RE_ if RE=inf.
syms b rbe ro Rs RB RC RE RL
if nargin>1&RE_==0, RE=inf; end
RCL=parallel_comb([RC RL]);
if nargin>0&ro_>0
 RLroRB=(RL+ro)*(rbe+RB)/(RL+rbe+RB+(b+1)*ro);
 Ri=parallel_comb([RE RLroRB]); % Eq.(3.2.9)
 ro1=ro+parallel_comb([Rs RE rbe+RB])*(1+b*ro/(rbe+RB));
 Av=(1+b*ro/(rbe+RB))/(1+ro/RCL); % Eq.(3.2.11)
else ro=inf; ro1=inf;
 Ri=parallel_comb([RE (rbe+RB)/(b+1)]); % Eq.(3.2.9)
 Av=b*RCL/(rbe+RB); % Eq.(3.2.11)
end
Ro=parallel_comb([RC ro1]); % Eq.(3.2.12)
Avi=1/(Rs/Ri+1); AvL=1/(Ro/RL+1); Avo=Av/AvL; % Eq.(3.2.11)
Gv=Avi*Av;
```

```
function Ri=Ri_CB(ro_,RE_)
% Put 1 as the 1st input argument ro_ if ro~=inf.
% Put 0 as the 2nd input argument RE_ if RE=inf.
syms b rbe ro Rs RB RE RC RL
if nargin>1&RE_==0, RE=inf; end
if nargin<1|ro_==0, ro=inf; end
RCL=parallel_comb([RC RL]);
VT=1; ib=-VT/(rbe+RB);
vc=(VT/ro-b*ib)/(1/ro+1/RCL);
IT=-(b+1)*ib-(vc-VT)/ro;
Ri=parallel_comb([RE VT/IT]);
if nargout<1, fprintf('\n Ri =\n'); pretty(simplify(Ri)); end
```

```
function Ro=Ro_CB(ro_,RE_)
% Put 0 as the 1st input argument ro_ if ro=inf.
% Put 0 as the 2nd input argument RE_ if RE=inf.
syms b rbe ro Rs RB RC RE
if nargin>1&RE_==0, RE=inf; end
if nargin>0&ro_==0, ro1=inf;
 else ro1=ro+parallel_comb([Rs RE rbe+RB])*(1+b*ro/(rbe+RB));
end
Ro=parallel_comb([RC ro1]); % Eq.(3.2.12)
if nargout<1, fprintf('\n Ro =\n'); pretty(simplify(Ro)); end
```

Each of the formulas listed in Table 3.3 has been coded in MATLAB as above so that they can be called individually as symbolic expressions whenever and wherever needed.

Let us consider the CE amplifier of Figure 3.36(a1) where the device parameters of the NPN-BJT Q_1 are $\beta_F = 100$, $\beta_R = 1$, $\beta_{AC} = 100$, $V_A = 10^4$ V, and $r_b = 0\ \Omega$. Its PSpice simulation result is shown in Figure 3.36(b1) where the overall voltage gain turns out to be

$$G_{v,s} = \frac{v_o}{v_s} = -\frac{(43.9 + 45.2)\,\text{mV}}{2 \times 10\,\text{mV}} = -4.46 \tag{3.2.13}$$

Note that the theoretical value of the overall voltage gain is

$$G_v = \frac{v_o}{v_s} = \frac{v_i}{v_s}\frac{v_o}{v_i} = \frac{R_{i1}}{R_s + R_{i1}}A_v = \frac{4.61}{10 + 4.61}(-14.1) = -4.45 \tag{3.2.14}$$

This can be obtained by running the following MATLAB statements:

```
>>rb=0; betaF=100; betaR=1; betaAC=100;
  VCC=10; Vsm=0.01; beta=[betaF betaR betaAC];
  Rs=1e4; RL=1e3; R11=7e4; R12=3e4; RC1=5e3; RE1=[0 5e3]; Rs1=Rs; RL1=RL;
  BJT_CE_analysis(VCC,rb,Rs1,R11,R12,RC1,RE1,RL1,beta,Vsm);
```

which yields

```
 VCC   VEE   VBB   VBQ   VEQ   VCQ     IBQ        IEQ        ICQ
10.00 0.00 3.00 2.91 2.21 7.81 4.37e-06 4.42e-04 4.37e-04
in the forward-active mode with VCE,Q= 5.61[V]
   gm= 16.915[mS], rbe= 5912[Ohm], ro= 22869.57[kOhm]
Ri= 4.613 kOhm, Ro= 4999 Ohm
Gv=Ri/(Rs+Ri)xAv= 0.316 x -14.10 =  -4.45
```

Figure 3.36(a2) and (b2) shows a CE-CC amplifier and its PSpice simulation result where the p-p (peak-to-peak) value of the overall AC output voltage has turned out to be 20.5 times that of the AC input voltage. To analyze this multistage amplifier (containing a stage of CC configuration), we first find the input resistance of each stage, starting from the last stage backwards to the first stage:

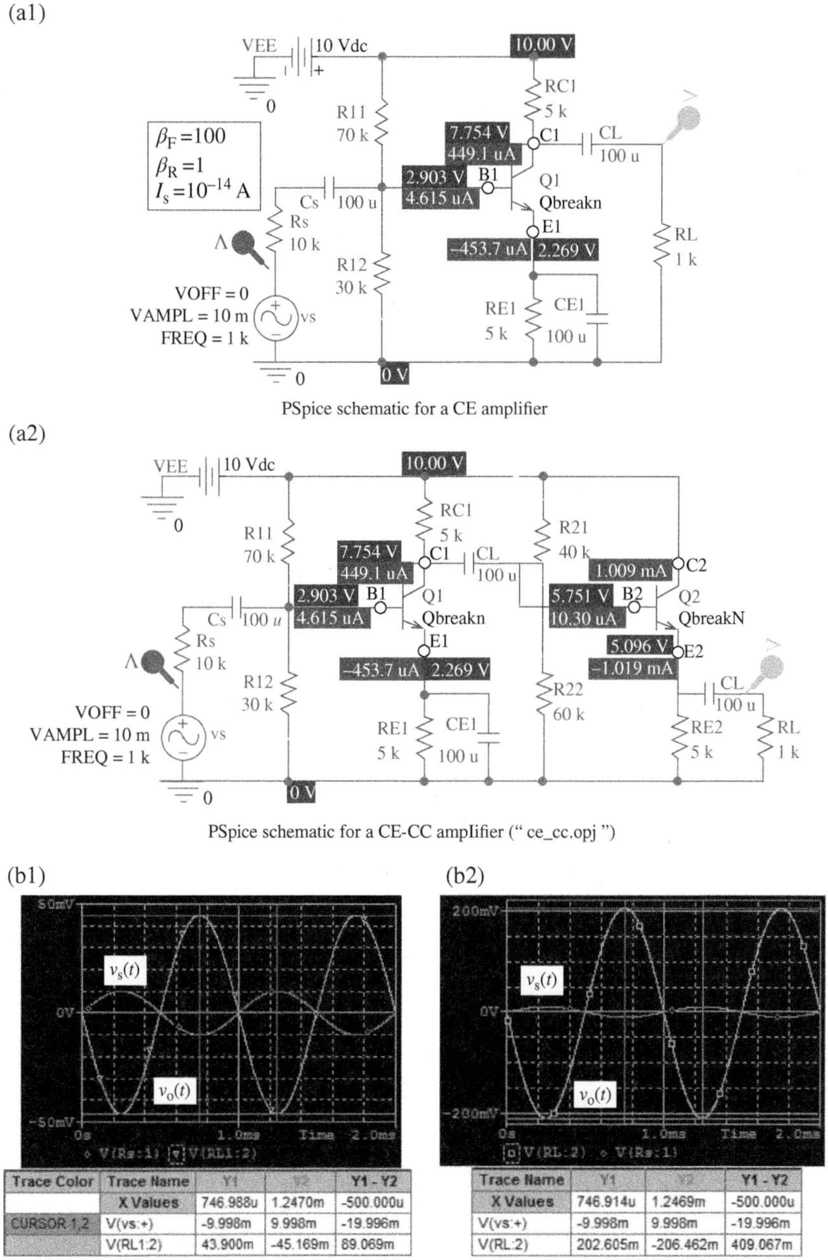

Trace Color	Trace Name	Y1	Y2	Y1 - Y2
	X Values	746.988u	1.2470m	-500.000u
CURSOR 1,2	V(vs:+)	-9.998m	9.998m	-19.996m
	V(RL1:2)	43.900m	-45.169m	89.069m

Its simulation result

Trace Name	Y1	Y2	Y1 - Y2
X Values	746.914u	1.2469m	-500.000u
V(vs:+)	-9.998m	9.998m	-19.996m
V(RL:2)	202.605m	-206.462m	409.067m

Its simulation result

Figure 3.36 A single-stage amplifier of CE configuration and a two-stage amplifier of CE-CC configurations.

```
>>rb=0; betaF=100; betaR=1; betaAC=100;
VCC=10; Vsm=0.01; Rs=1e4; RL=1e3; beta=[betaF betaR betaAC];
R21=4e4; R22=6e4; RC2=0; RE2=5e3;
% Find Ri, Av, and Ai of Stage 2/1 starting from last one
Rs2=0; RL2=RL; [VBQ,VEQ,VCQ,IBQ,IEQ,ICQ,Av2,Ai2,Ri2,Ro2_0]= ...
 BJT_CC_analysis(VCC,rb,Rs2,R21,R22,RC2,RE2,RL2,beta);
R11=7e4; R12=3e4; RC1=5e3; RE1=[0 5e3]; Rs1=Rs; RL1=Ri2; Vsm0=Vsm;
[VBQ,VEQ,VCQ,IBQ,IEQ,ICQ,Av1,Ai1,Ri1,Ro1]= ...
  BJT_CE_analysis(VCC,rb,Rs1,R11,R12,RC1,RE1,RL1,beta,Vsm0);
% Now, analyze each stage forwards from the 2nd one
Rs2=Ro1; Vsm1=Ri1/(Rs+Ri1)*Av1*Vsm0
[VBQ,VEQ,VCQ,IBQ,IEQ,ICQ,Av2,Ai2,Ri2,Ro2]= ...
  BJT_CC_analysis(VCC,rb,Rs2,R21,R22,RC2,RE2,RL2,beta,Vsm1);
Vom=Av2*Vsm1, Gv = Vom/Vsm,  Ri1/(Rs+Ri1)*Av1*Av2
```

Running these statements yields

```
 VCC      VEE     VBB    VBQ   VEQ     VCQ        IBQ          IEQ          ICQ
10.00    0.00    3.00   2.91  2.21   7.81    4.37e-06    4.42e-04   4.37e-04
  gm= 16.915[mS], rbe= 5912[Ohm], ro= 22869.57[kOhm] % Stage 1 of CE
  Ri= 4.613kOhm, Ro= 5912Ohm, Gv=Ri/(Rs+Ri)xAv= 0.316 x -66.79 = -21.09
  Vsm1 =  -0.2109
 VCC     VEE     VBB     VBQ     VEQ     VCQ        IBQ          IEQ          ICQ
10.00   0.00   6.00    5.76   5.06  10.00   1.00e-05   1.01e-03   1.00e-03
  gm= 38.756[mS], rbe= 2580[Ohm], ro= 9981.13[kOhm] % Stage 2 of CC
  Ri= 18.799kOhm, Ro= 66 Ohm, Gv=Ri/(Rs+Ri)xAv= 0.79 x 0.97 =  0.77

  Gv = -20.4588 %Overall voltage gain of the CE-CC amplifier
```

This implies that the overall voltage gain of the CE-CC stage is −20.5 (as confirmed by the PSpice simulation result $G_{v,s}$=−409.1 mV/20 mV=−20.5 in Figure 3.36(b2)), which is much greater than that (−4.45) of the CE stage (Eq. (3.2.14)) despite the additional CC stage whose voltage gain is less than one by itself.

(Q) Why is that?

Figure 3.37(a) and (b) shows a three-stage BJT amplifier consisting of CE-CE-CC configurations and its PSpice simulation result where the overall voltage gain has turned out to be $G_{v,s}$ = 6.64. To analyze this multistage amplifier (containing a CC configuration), we first find the input resistance of each stage, starting from the last stage backwards to the first stage:

```
>>VCC=20; Vsm=5e-3; rb=0; betaF=100; betaR=1; betaAC=100; Is=1e-16;
Rs=100; RL=1e4; % Source resistance and Load resistance
R31=5e4; R32=5e4; RC3=0; RE3=200; beta=[betaF,betaR,betaAC,Is];
R21=1e5; R22=1e5; RC2=200; RE2=100;
R11=1e5; R12=1e5; RC1=1e3; RE1=[250 50];
Rs3=0; RL3=RL; [VBQ,VEQ,VCQ,IBQ,IEQ,ICQ,Av3,Ai3,Ri3,Ro3_0]= ...
    BJT_CC_analysis(VCC,rb,Rs3,R31,R32,RC3,RE3,RL3,beta);
Rs2=0; RL2=Ri3; [VBQ,VEQ,VCQ,IBQ,IEQ,ICQ,Av2,Ai2,Ri2,Ro2_0]= ...
   BJT_CE_analysis(VCC,rb,Rs2,R21,R22,RC2,RE2,RL2,beta);
Rs1=Rs; RL1=Ri2; Vsm0=Vsm;
[VBQ,VEQ,VCQ,IBQ,IEQ,ICQ,Av1,Ai1,Ri1,Ro1]= ...
  BJT_CE_analysis(VCC,rb,Rs1,R11,R12,RC1,RE1,RL1,beta,Vsm0);
```

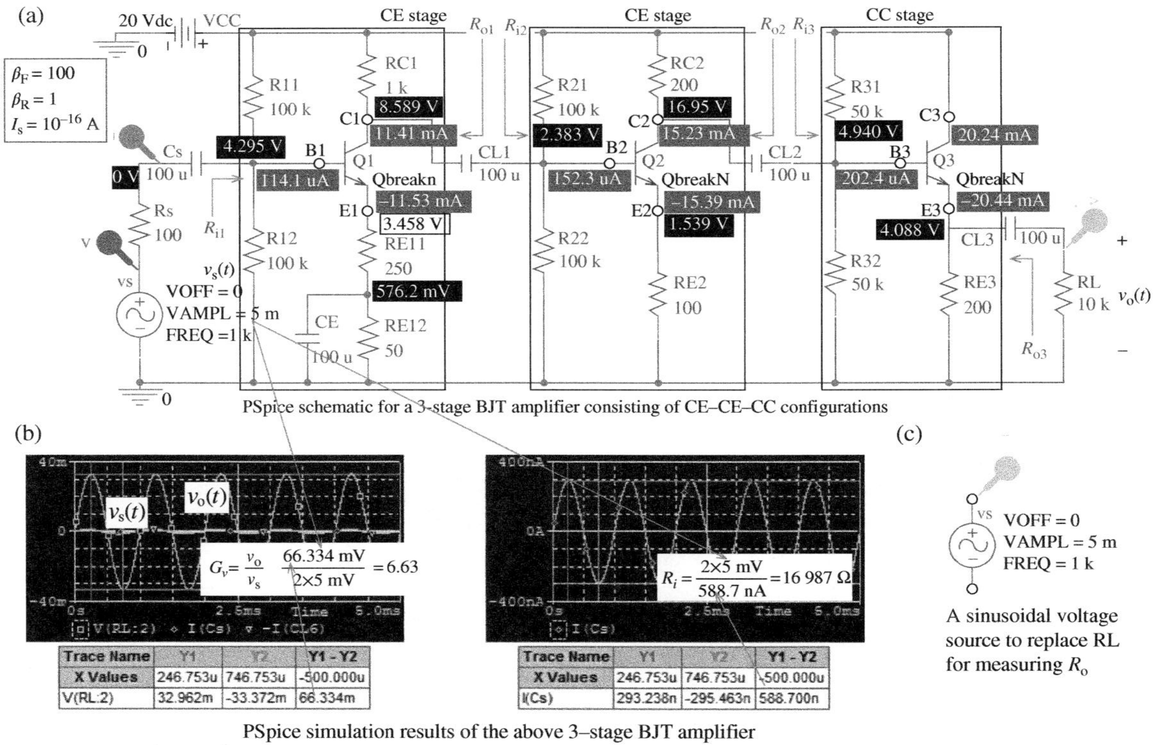

Figure 3.37 A three-stage cascaded BJT amplifier and its PSpice simulation ("ce_ce_cc.opj").

where the load resistance R_L and the input resistances R_{i3}/R_{i2} of stage 3/2 have been put as the load resistances of stage 3 and 2/1, successively and respectively. Note that 0 has been put as the third input argument (corresponding to Rs3/Rs2) of 'BJT_CC_analysis()'/'BJT_CE_analysis()' for stage 3/2 because their source or input resistances are not yet known. That is why the output resistance of the CC stage (to be computed by Eq. (3.2.8) depending on R_s) is not expected to have been found properly. However, the source resistance R_s has properly been put as the third input argument of 'BJT_CE_analysis()' for stage 1. Running the above MATLAB statements yields the following:

```
 Result of provisional analysis for Stage 3
  VCC    VEE     VBB     VBQ     VEQ     VCQ       IBQ         IEQ        ICQ
20.00  0.00  10.00   4.94   4.09  20.00  2.02e-04  2.04e-02  2.02e-02
  in the forward-active mode with VCE,Q= 15.91[V]
 where beta_forced = ICQ/IBQ = 100.00 where beta = 100.00
     gm= 782.947[mS], rbe=  128[Ohm], ro=  494.07[kOhm]
  Ri=  11.088[kOhm], Ro=   1[Ohm]
  Gv=Ri/(Rs+Ri)xAv= 1.000x   0.99 =   0.99 % Not yet meaningful
 Result of provisional analysis for Stage 2
  VCC    VEE     VBB     VBQ     VEQ     VCQ       IBQ         IEQ        ICQ
20.00  0.00  10.00   2.38   1.54  16.95  1.52e-04  1.54e-02  1.52e-02
  in the forward-active mode with VCE,Q= 15.41[V]
 where beta_forced = ICQ/IBQ = 100.00 where beta = 100.00
     gm= 589.311[mS], rbe=  170[Ohm], ro=  656.42[kOhm]
  Ri=   8.520[kOhm], Ro=  200[Ohm]
  Gv=Ri/(Rs+Ri)xAv= 1.000x  -1.91 =  -1.91 % Not yet meaningful
 Results of analysis for Stage 1
  VCC    VEE     VBB     VBQ     VEQ     VCQ       IBQ         IEQ        ICQ
20.00  0.00  10.00   4.29   3.46   8.59  1.14e-04  1.15e-02  1.14e-02
  in the forward-active mode with VCE,Q= 5.13[V]
 where beta_forced = ICQ/IBQ = 100.00 where beta = 100.00
     gm= 441.426[mS], rbe=  227[Ohm], ro=  876.33[kOhm]
  Ri=  16.877[kOhm], Ro=  999[Ohm]
  Gv=Ri/(Rs+Ri)xAv= 0.994x  -3.51 =  -3.49
```

Then, to find the overall voltage gain, we multiply the product of the voltage gains of every stage with the voltage gain of the front voltage divider as

$$G_v = \frac{R_{i1}}{R_s + R_{i1}} A_{v1} A_{v2} A_{v3} = 0.994 \times (-3.51) \times (-1.91) \times 0.99 = 6.60 \tag{3.2.15}$$

```
>>Gv=Ri1/(Rs+Ri1)*Av1*Av2*Av3
    ans = 6.6374
```

How close this is to the PSpice simulation result (6.63) shown in Figure 3.37(b)!

Now, to find the output resistance of the last stage of CC, starting from the first stage forwards to the last stage, we use 'BJT_CE_analysis()' (with Rs1 = Rs and RL1 = Ri2), 'BJT_CE_analysis()' (with Rs2 = Ro1 and RL2 = Ri3), and 'BJT_CC_analysis()e' (with Rs3 = Ro2 and RL3 = RL) for stage 1, 2, and 3, respectively. Here, the analysis of stage 1 does not have to be repeated since it has already been taken care of above.

```
>>Rs2=Ro1; RL2=Ri3; Vsm1=Av1*Vsm;
  [VBQ,VEQ,VCQ,IBQ,IEQ,ICQ,Av2,Ai2,Ri2,Ro2]= ...
  BJT_CE_analysis(VCC,rb,Rs2,R21,R22,RC2,RE2,RL2,beta,Vsm1);
  Rs3=Ro2; RL3=RL; Vsm2=Av2*Vsm1;
  BJT_CC_analysis(VCC,rb,Rs3,R31,R32,RC3,RE3,RL3,beta,Vsm2);
```

These MATLAB statements can be run to yield the following:

```
Results of analysis for Stage 2
Ri=  8.520[kOhm], Ro=  200[Ohm]
 Results of analysis for Stage 3
Ri=  11.088 kOhm, Ro=    3 Ohm
```

All the above MATLAB statements have been put into the MATLAB function 'CE_CE_CC()' so that it can be run by typing the following at the MATLAB prompt:

```
>>Rs=100; RL=1e4; Vsm=5e-3; VCC=20;
  [Gv,Avs,Ais,Ris,Ros]=CE_CE_CC(Rs,RL,Vsm,VCC)
```

Note the following about it:

- The BJT parameters such as r_b(rb), β_F(betaF), β_R(betaR), β_{AC}(betaAC), and I_s have been set to the default values of Qbreak that can be read from the PSpice simulation output file.
- First, to find the input resistances of stage 3(CC), stage 2(CE), and stage 1 (CE) (that will be load resistance to their previous stages), 'BJT_CC_analysis()', 'BJT_CE_analysis()', and 'BJT_CE_analysis()' with their third/eigth input arguments Rs3=0/RL3=RL, Rs2=0/RL2=Ri3, and Rs1=Rs/RL1=Ri2, respectively, have been run backwards starting from the last stage. Then the overall voltage gain can be computed as above.
- To get the proper values of Ro2 and Ro3, each stage (starting from the second one) is analyzed forwards by running the corresponding analysis function with the third/eigth input argument Rs2=Ro1/RL2=Ri3 and Rs3=Ro2/RL3=RL, respectively.

```
function [Gv,Avs,Ais,Ris,Ros,Vom]=CE_CE_CC(Rs,RL,Vsm,VCC)
rb=0; betaF=100; betaR=1; betaAC=100; VA=1e4; Is=1e-16;
R11=1e5; R12=1e5; RC1=1000; RE1=[250 50]; R21=1e5; R22=1e5; RC2=200; RE2=100;
R31=5e4; R32=5e4; RC3=0; RE3=200; beta=[betaF,betaR,betaAC,Is];
% Find Ri, Av, and Ai of Stage 3/2/1 starting from last one.
Rs3=0; RL3=RL; [VBQ,VEQ,VCQ,IBQ,IEQ,ICQ,Av3,Ai3,Ri3,Ro3_0]= ...
      BJT_CC_analysis(VCC,rb,Rs3,R31,R32,RC3,RE3,RL3,beta);
Rs2=0; RL2=Ri3; [VBQ,VEQ,VCQ,IBQ,IEQ,ICQ,Av2,Ai2,Ri2,Ro2_0]= ...
      BJT_CE_analysis(VCC,rb,Rs2,R21,R22,RC2,RE2,RL2,beta);
Rs1=Rs; RL1=Ri2; Vsm0=Vsm;
[VBQ,VEQ,VCQ,IBQ,IEQ,ICQ,Av1,Ai1,Ri1,Ro1]= ...
   BJT_CE_analysis(VCC,rb,Rs1,R11,R12,RC1,RE1,RL1,beta,Vsm0,VA);
Gv=Ri1/(Rs+Ri1)*Av1*Av2*Av3; Avs=[Av1 Av2 Av3]; Ais=[Ai1 Ai2 Ai3];
fprintf('Gv=Ri/(Rs+Ri)*Av1*Av2*Av3=%4.2fx%8.3fx%8.3fx%8.3f=%9.3f\n', ...
     Ri1/(Rs+Ri1),Av1,Av2,Av3,Gv)
% You don't have to go further unless you want to get Ro of each stage
% because Avs of all stages have already and properly been obtained.
Rs2=Ro1; RL2=Ri3; Vsm1=Ri1/(Rs+Ri1)*Av1*Vsm;
[VBQ,VEQ,VCQ,IBQ,IEQ,ICQ,Av2,Ai2,Ri2,Ro2]= ...
     BJT_CE_analysis(VCC,rb,Rs2,R21,R22,RC2,RE2,RL2,beta,Vsm1,VA);
Rs3=Ro2; RL3=RL; Vsm2=Av2*Vsm1;
[VBQ,VEQ,VCQ,IBQ,IEQ,ICQ,Av3,Ai3,Ri3,Ro3]= ...
   BJT_CC_analysis(VCC,rb,Rs3,R31,R32,RC3,RE3,RL3,beta,Vsm2,VA);
Vom=Av3*Vsm2; Ris=[Ri1 Ri2 Ri3]; Ros=[Ro1 Ro2 Ro3];
```

The hand calculations to find the input/output resistances can be done as follows:

$$R_{i1} \overset{(3.2.1)}{=} R_{11}\|R_{12}\|(r_{b1}+r_{be1}+(\beta_1+1)R_{E11}) = 100\text{k}\|100\text{k}\|(223+101\times 250) = 16877\,\Omega$$

$$R_{o1} \overset{(3.2.4)}{=} R_{C1} = 1000\,\Omega$$

$$R_{i2} \overset{(3.2.1)}{=} R_{21}\|R_{22}\|(r_{b2}+r_{be2}+(\beta_2+1)R_{E2}) = 100\text{k}\|100\text{k}\|(167+101\times 100) = 8518\,\Omega \quad (3.2.16)$$

$$R_{o2} \overset{(3.2.4)}{=} R_{C2} = 200\,\Omega$$

$$R_{i3} \overset{(3.2.5)}{\approx} R_{31}\|R_{32}\|\{r_{b3}+r_{be3}+(\beta_3+1)(R_{E3}\|R_L)\} = 50\text{k}\|50\text{k}\|\{126+101\times(200\|10^4)\} = 11089\,\Omega$$

$$R_{o3} \overset{(3.2.8)}{=} R_{E3}\|\frac{(R_{o2}\|R_{B3})+r_{b3}+r_{be3}}{\beta_3+1} = 200\|\frac{(200\|50\,000)+126}{101} = 3.17\,\Omega$$

Figure 3.38 shows the PSpice simulation result for measuring the overall output resistance R_{o3}, which yields the measured value of R_{o3} as 10/2.7951 = 3.58 Ω.

Now, let us consider what will happen if we remove the third stage of CC (emitter follower) configuration to make a two-stage amplifier as depicted in Figure 3.39(a). Then the output voltage across R_L = 100 kΩ will become a bit higher as depicted in Figure 3.39(b1), which can be thought of as a natural result from omitting the CC stage with voltage gain Av3 = 0.98 < 1. However, will we

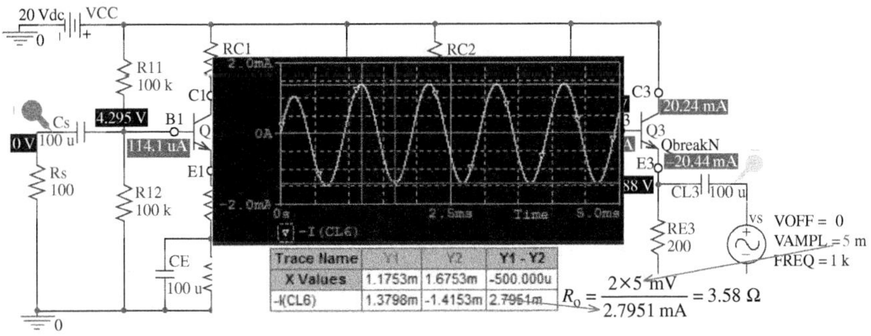

Figure 3.38 PSpice schematic and its simulation for measuring the overall output resistance.

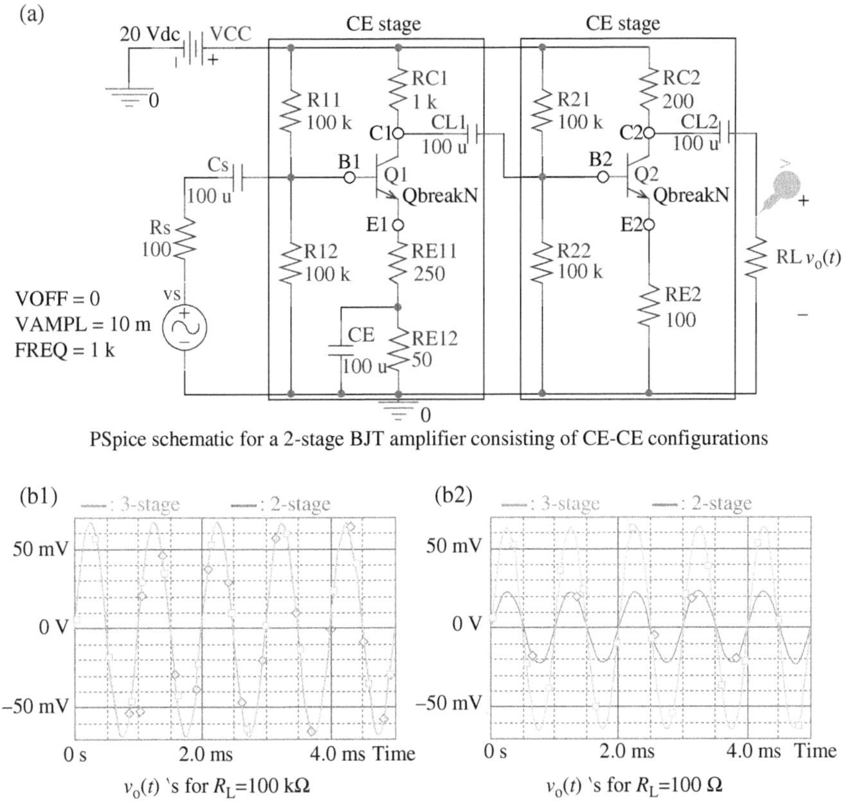

Figure 3.39 A two-stage BJT amplifier and its PSpice simulation results ("ce_ce.opj").

have a similar result even for a smaller load resistor like R_L = 100 Ω? To our surprise, the PSpice simulation result for R_L = 100 Ω depicted in Figure 3.39 (b2) shows that the output voltage of the two-stage amplifier has become much lower than that of the three-stage amplifier, which reveals the potential value of the CC stage. What is the strength of the CC stage to reduce the loading effect so that the voltage drop due to a larger load (with smaller resistance) can be very small? It is the large input resistance and the small output resistance of CC stage compared with those of CE stage as mentioned in Section 3.2.2. That is why we are willing to pay the extra cost of equipping an amplifier with a CC stage as the last one even if it may lower the output voltage a bit for a normal load.

To use MATLAB for showing the strength of CC stage, we can compose the following MATLAB function 'CE_CE()' with R_L as an input argument, which analyzes the two-stage amplifier consisting of CE-CE stages. Then we run it and the above MATLAB function 'CE_CE_CC()' with RL = 100:

```
>>Rs=100; RL=100; Vsm=0.01; VCC=20;
  Gv1=CE_CE(Rs,RL,Vsm,VCC), Gv2=CE_CE_CC(Rs,RL,Vsm,VCC)
```

which yields

```
Gv1 =  2.2669  Gv2 =  6.4350
```

This shows that as the load becomes larger, i.e. as R_L becomes smaller, the role of a CC stage to reduce the loading effect becomes more remarkable.

```
function [Gv,Avs,Ais,Ris,Ros,Vom]=CE_CE(Rs,RL,Vsm,VCC)
rb=0; betaF=100; betaR=1; betaAC=100; VA=1e4; Is=1e-16;
beta=[betaF,betaR,betaAC,Is]
R21=1e5; R22=1e5; RC2=200; RE2=100;
R11=1e5; R12=1e5; RC1=1000; RE1=[250 50];
% Find Ri, Av, and Ai of Stage 2/1 starting from last one.
Rs2=0; RL2=RL; [VBQ,VEQ,VCQ,IBQ,IEQ,ICQ,Av2,Ai2,Ri2,Ro2_0]= ...
  BJT_CE_analysis(VCC,rb,Rs2,R21,R22,RC2,RE2,RL2,beta);
Rs1=Rs; RL1=Ri2; Vsm0=Vsm;
[VBQ,VEQ,VCQ,IBQ,IEQ,ICQ,Av1,Ai1,Ri1,Ro1]= ...
BJT_CE_analysis(VCC,rb,Rs1,R11,R12,RC1,RE1,RL1,beta,Vsm0,VA);
Gv=Ri1/(Rs+Ri1)*Av1*Av2; Avs=[Av1 Av2]; Ais=[Ai1 Ai2];
fprintf(' Gv=Ri/(Rs+Ri)*Av1*Av2=%4.2fx%8.3fx%8.3f=%9.3f\n', ...
  Ri1/(Rs+Ri1),Av1,Av2,Gv)
% Now, analyze each stage forwards from the 2nd one to find their Ros.
Rs2=Ro1; RL2=RL; Vsm1=Ri1/(Rs+Ri1)*Av1*Vsm;
[VBQ,VEQ,VCQ,IBQ,IEQ,ICQ,Av2,Ai2,Ri2,Ro2]= ...
  BJT_CE_analysis(VCC,rb,Rs2,R21,R22,RC2,RE2,RL2,beta,Vsm1,VA);
Vom=Av2*Vsm1; Ris=[Ri1 Ri2]; Ros=[Ro1 Ro2];
```

3.2.5 Composite/Compound Multi-Stage BJT Amplifier

1) **CC-CE (Darlington) Amplifier**

Figure 3.40(a)/(b) shows CC-CE (Darlington) amplifiers using two NPN/PNP BJTs, respectively. The amplifiers can be used to achieve a large current gain since the overall output current is the sum of the two BJT collector currents:

$$\begin{aligned} i_c &= i_{c1} + i_{c2} = \beta_1 i_{b1} + \beta_2 i_{b2} = \beta_1 i_{b1} + \beta_2 i_{e1} \\ &= \beta_1 i_{b1} + \beta_2(\beta_1 + 1) i_{b1} = (\beta_1 + \beta_2 + \beta_1\beta_2) i_{b1} \end{aligned} \tag{3.2.17}$$

2) **CC-CC (Darlington) Amplifier**

Figure 3.40(c) shows a CC-CC (Darlington) amplifier using two NPN BJTs. This amplifier can also be used to achieve a large current gain since the overall output current is the emitter current of the second BJT:

$$i_o = i_{e2} = (\beta_2 + 1) i_{b2} = (\beta_2 + 1) i_{e1} = (\beta_2 + 1)(\beta_1 + 1) i_{b1} \tag{3.2.18}$$

3) **CE-CB (Cascode) Amplifier**

Figure 3.40(d) shows a CE-CB (cascode) amplifier using two NPN BJTs. Its output current and output voltage across the collector resistor R_C are

$$i_o = -i_{c2} = -\frac{\beta_2}{\beta_2+1} i_{e2} = -\frac{\beta_2}{\beta_2+1} i_{c1} = -\frac{\beta_2\beta_1}{\beta_2+1} i_{b1} = -\frac{\beta_2\beta_1}{\beta_2+1}\frac{v_s}{R_s + r_{be1}} \tag{3.2.19}$$

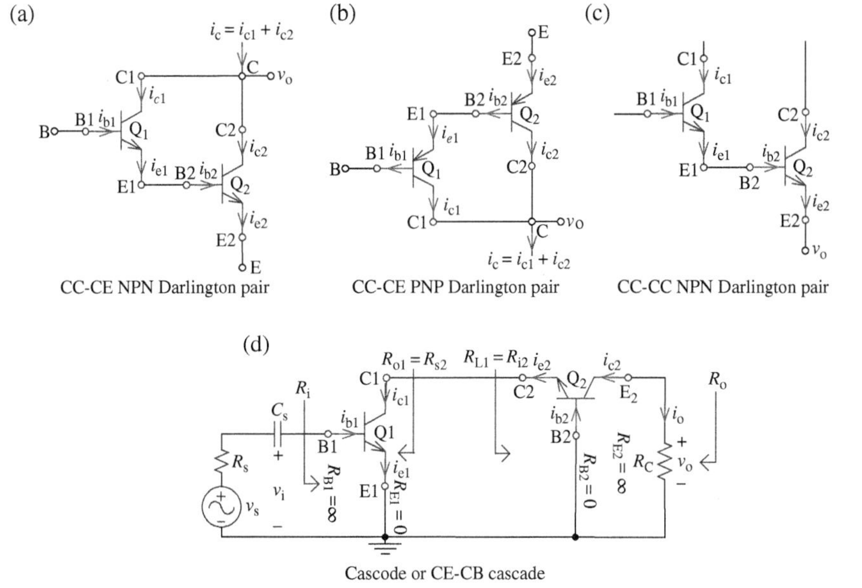
(a)
(b)
(c)
CC-CE NPN Darlington pair
CC-CE PNP Darlington pair
CC-CC NPN Darlington pair
(d)
Cascode or CE-CB cascade

Figure 3.40 Various compound (composite) BJT amplifier circuits.

$$v_o = -R_C i_{c2} = \frac{\beta_2 \beta_1}{\beta_2 + 1} \frac{R_C}{R_s + r_{be1}} v_s \tag{3.2.20}$$

Thus, the current/voltage gains are

$$A_i = \frac{i_o}{i_s = i_{b1}} \overset{(3.2.19)}{=} -\frac{\beta_2 \beta_1}{\beta_2 + 1} \overset{\beta_2 \gg 1}{\approx} -\beta_1 \tag{3.2.21}$$

$$G_v = \frac{v_o}{v_s} \overset{(3.2.20)}{=} \frac{\beta_2 \beta_1}{\beta_2 + 1} \frac{R_C}{R_s + r_{be1}} \overset{\beta_2 \gg 1}{\approx} \beta_1 \frac{R_C}{R_s + r_{be1}} \tag{3.2.22}$$

Not only the voltage/current gains but also the input/output resistances are close to those of the CE amplifier (with $R_E = 0$, $R_B = \infty$, and R_C in place of $R_C \| R_L$) discussed in Section 3.2.1. Then what is the additional BJT for? Compared with the single-stage CE amplifier, the load resistance of the first CE stage is the input resistance of the second CB stage, which is so small that the possibility of Q_1 to enter the saturation region can be reduced. That is one of the advantages that we gain from the additional (second) BJT.

It may be interesting and convenient to use the MATLAB functions listed above for deriving, say, Eq. (3.2.22) (the voltage gain of the CE-CB (cascode) amplifier shown in Figure 3.40(d)) as follows:

```
>>syms b b1 b2 ro ro1 ro2 rbe rbe1 rbe2 Rs RB RC RE RL
  Ri2=subs(Ri_CB,{b,rbe,RB,RE},{b2,rbe2,0,inf})
         %Input resistance from last
  Ri1=subs(Ri_CE,{rbe,RB,RE},{rbe1,inf,0})
  Ro1=subs(Ro_CE,RC,inf); % Output resistance from the first stage
  Ro2=subs(Ro_CB,{b,rbe,Rs,RB,RE},{b2,rbe2,Ro1,0,inf})
  Av1=subs(Av_CE,{b,rbe,RC,RE,RL},{b1,rbe1,inf,0,Ri2})
  Av2=subs(Av_CB,{b,rbe,RB,RE,RL},{b2,rbe2,0,inf,inf})
  Ri=Ri1; Ro=Ro2; % Overall input and output resistances
  Gv=Ri/(Rs+Ri1)*Av1*Av2
```

where the output resistance R_{o1} of stage 1 (CE) has been put as the source resistance R_{s2} of stage 2 (CB) to find the output resistance R_{o2} of stage 2 and the input resistance R_{i2} of stage 2 has been put as the load resistance R_{L1} of stage 1 to find the voltage gain A_{v1} of stage 1. Running these statements yields the overall input/output resistances and voltage gain as

```
Ri1 = rbe1 % The input resistance of the first stage
Ro2 = RC % The output resistance of the last stage
Gv = -(RC*b1*b2)/((Rs + rbe1)*(b2 + 1)) % see Eq.(3.2.22)
```

This result conforms with Eq. (3.2.22). On the other hand, if we consider the (internal) output resistance r_o (due to the Early effect) of each BJT, we can run the following statements:

```
>>Ri2=subs(Ri_CB(1),{b,ro,rbe,RB,RE,RL},{b2,ro2,rbe2,0,inf,inf})
Ri=limit(limit(limit(subs(Ri_CE(1),rbe,rbe1),RB,inf ),RC,inf ),RE,0)
Ro1=subs(Ro_CE(1),{ro,RC},{ro1,inf});
Ro=subs(Ro_CB(1),{b,ro,rbe,Rs,RB,RC,RE},{b2,ro2,rbe2,Ro1,0,inf,inf})
Av1=limit(limit(subs(Av_CE(1),{b,ro,rbe},{b1,ro1,rbe1}),RL,Ri2),RE,0)
Av2=subs(Av_CB(1),{b,ro,rbe,RB,RE,RL},{b2,ro2,rbe2,0,inf,inf})
Gv=Av1*Av2; % Overall voltage gain with Rs=0
Gvo=limit(Gv,RC,inf ); % Overall onen-loop voltage gain with RC=inf
pretty(simplify(Gvo))
```

to get the overall input/output resistances and voltage gain of the CE-CB amplifier with $R_s = 0$ and $R_C = \infty$ (open) as

```
Ri = rbe1
Ro = ro2 + ((b2*ro2)/rbe2 + 1)/(1/rbe2 + 1/ro1)
b1 ro1 (rbe2 + b2 ro2)
----------------% comparable to the results in Sec. 7.5.6 of [S-2]
rbe1 (rbe2 + ro1)
```

This implies

$$R_i = r_{be1}$$

$$R_o = r_{o2} + \frac{\beta_2 r_{o2}/r_{be2} + 1}{1/r_{be2} + 1/r_{o1}} = r_{o2} + \{\beta_2 (r_{o2}/r_{be2}) + 1\}(r_{be2} \| r_{o1})$$
$$= r_{o2} + (g_{m2} r_{o2} + 1)(r_{be2} \| r_{o1})$$

$$G_v = \frac{v_o}{v_i} = \frac{\beta_1 r_{o1}(r_{be2} + \beta_2 r_{o2})}{r_{be1}(r_{be2} + r_{o1})} \overset{r_{be2} \gg \beta_2 r_{o2}}{\approx} g_{m1} r_{o2} \frac{\beta_2 r_{o1}}{r_{be2} + r_{o1}}$$
$$= g_{m1} r_{o1} g_{m2} (r_{be2} \| r_{o1}) \text{ with } g_m = \frac{\beta}{r_{be}}$$

Note that 1 has been put as the first input argument of the MATLAB functions such as '`Ri_CB()`' to include the effect of r_o. Note also that the MATLAB function '`limit()`' is more useful than '`subs()`' for substituting a zero or an infinity into a complicated MATLAB expression.

Example 3.13 AC Analysis of a Cascode (CE-CB) Amplifier
Consider the cascode (CE-CB) amplifier in Figure 3.41(a) where the device parameters of the BJTs Q_1 and Q_2 are $r_b = 0\ \Omega$, $\beta_F = 100$, $\beta_R = 1$, $\beta_{AC} = 100$, $V_A = 10^4$ V, and $V_T = 25.9$ mV. Find the overall voltage gain $G_v = v_o/v_s$ and see how close it is to that obtained from PSpice simulation.

Assuming that Q_1 and Q_2 operate in the forward-active mode so that $I_{C1} = \beta_F I_{B1} = I_{E2} = (\beta_F + 1)I_{B2}$, $I_{B2} = \alpha_F I_{B1} = \beta_F I_{B1}/(\beta_F + 1)$, we write a set of two KVL equations (in two unknowns I_{B1} and I_{R2}) along the path R_1-R_2-R_3 and the mesh R_3-BEJ$_1$-R_E and solve it as

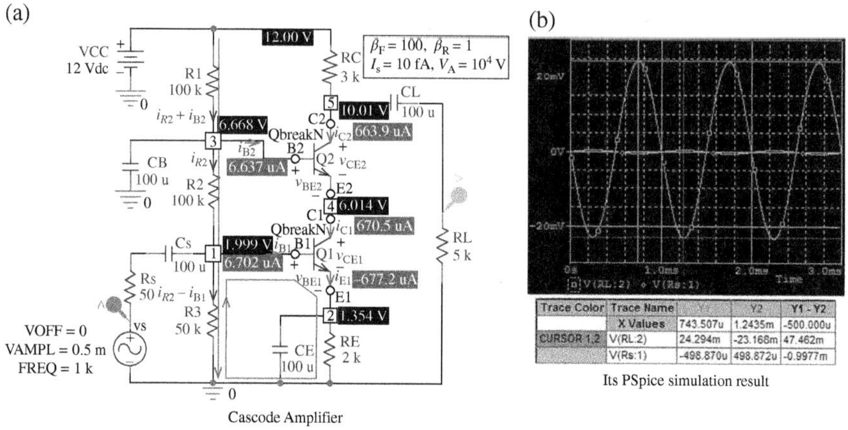

Figure 3.41 A cascode (CE-CB) amplifier circuit and its PSpice simulation result ("elec03e13_cascode.opj").

$$
\begin{aligned}
&V_{CC}-R_1(I_{R2}+\alpha_F I_{B1})-R_2 I_{R2}-R_3(I_{R2}-I_{B1})=0 \\
&R_3(I_{R2}-I_{B1})-V_{BE1}-R_E(\beta_F+1)I_{B1}=0
\end{aligned}
;\quad
\begin{bmatrix}\alpha_F R_1-R_3 & R_1+R_2+R_3\\ -(\beta_F+1)R_E-R_3 & R_3\end{bmatrix}
\begin{bmatrix}I_{B1}\\ I_{R2}\end{bmatrix}=\begin{bmatrix}V_{CC}\\ V_{BE1}\end{bmatrix}
$$

$$
;\begin{bmatrix}0.99\times 100-50 & 100+100+50\\ -101\times 2-50 & 50\end{bmatrix}
\begin{bmatrix}I_{B1}\\ I_{R2}\end{bmatrix}=\begin{bmatrix}12\\ 0.7\end{bmatrix};\;
\begin{bmatrix}I_{B1}\\ I_{R2}\end{bmatrix}=\begin{bmatrix}49 & 250\\ -252 & 50\end{bmatrix}^{-1}\begin{bmatrix}12\\ 0.7\end{bmatrix}=\begin{bmatrix}0.00649\\ 0.0467\end{bmatrix}\text{mA}
\tag{E3.13.1}
$$

Thus, we have $I_{C1} = \beta_F I_{B1} = 0.649$ mA and $I_{C2} = \alpha_F I_{C1} = 0.99 \times 0.649 = 0.643$ mA so that

$$
r_{be1} \overset{(3.1.27)}{=} \frac{\beta}{g_{m1}} \overset{(3.1.26)}{=} \frac{\beta V_T}{I_{C1}} = \frac{100\times 25.9\,\text{mV}}{0.649\,\text{mA}} = 3.99\,\text{k}\Omega
\tag{E3.13.2}
$$

Then we use Eq. (3.2.22) to find the overall voltage gain as

$$
\begin{aligned}
G_v &= \frac{v_o}{v_s} \overset{(3.2.22)}{=} \frac{\beta_2\beta_1}{\beta_2+1}\frac{R_C\|R_L}{R_s+r_{be1}} \overset{(E3.13.2)}{=} \frac{100\times 100}{101}\frac{3000\|5000}{50+3990} \\
&= 46 \approx \frac{47.462\,\text{mV}}{1\,\text{mV}}\,(\text{PSpice})
\end{aligned}
\tag{E3.13.3}
$$

All these calculations can be done by using MATLAB as follows:

```
>>betaF=100; betaR=1; betaAC=100; T=27; VT=(27+273)/11605;
  VA=1e4; alphaF=betaF/(betaF+1); VBE1=0.7; VBE2=0.7;
  VCC=12; Rs=50; R1=100e3; R2=100e3; R3=50e3; RC=3e3; RE=2e3; RL=5e3;
  Z = [R1*alphaF-R3 R1+R2+R3; -(betaF+1)*RE-R3  R3];
  I = Z\[VCC; VBE1]; % Eq.(E3.13.1)
  IB1=I(1); IC1=betaF*IB1, IE1=IB1+IC1; IC2=alphaF*IC1, IB2=IC2/betaF;
  [gm1,rbe1,re1,ro1]=gmrbero_BJT(IC1,betaF,VA,VT); gm1, rbe1
  Av = betaF*alphaF*parallel_comb([RC RL])/(Rs+rbe1) % Eq.(E3.13.3)
```

Example 3.14 AC Analysis of a CC-CC Darlington Amplifier

For the amplifier whose AC equivalent is shown in Figure 3.42(a), find the input/output resistances and the overall voltage gain $G_v = v_o/v_s$ in terms of β_1, β_2, r_{be1}, r_{be2}, R_s, and R_E where $g_{mk} = \beta_k/r_{bek}$.

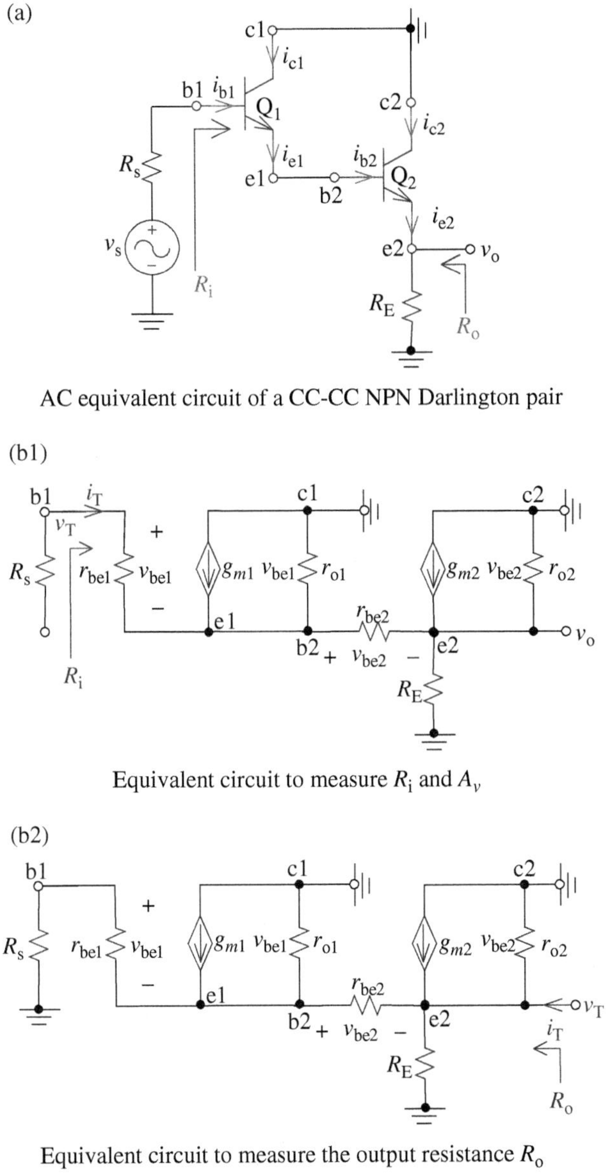
(a)
AC equivalent circuit of a CC-CC NPN Darlington pair
(b1)
Equivalent circuit to measure R_i and A_v
(b2)
Equivalent circuit to measure the output resistance R_o

Figure 3.42 A CC-CC Darlington amplifier circuit and its equivalents to find R_i, R_o, and A_v.

To find the input resistance and the voltage gain, we draw the equivalent (with the BJTs replaced by the model in Figure 3.18(b)) as depicted in Figure 3.42(b1) and write a set of two node equations (in two unknowns v_{e1} and v_o) as

$$\begin{bmatrix} 1/r_{be1}+1/r_{o1}+1/r_{be2} & -1/r_{be2} \\ -1/r_{be2} & 1/r_{be2}+1/R_E+1/r_{o2} \end{bmatrix} \begin{bmatrix} v_{e1} \\ v_o \end{bmatrix} = \begin{bmatrix} v_T/r_{be1}+g_{m1}(v_T-v_{e1}) \\ g_{m2}(v_{e1}-v_o) \end{bmatrix}$$

$$; \begin{bmatrix} g_{m1}+1/r_{be1}+1/r_{o1}+1/r_{be2} & -1/r_{be2} \\ -g_{m2}-1/r_{be2} & g_{m2}+1/r_{be2}+1/R_E+1/r_{o2} \end{bmatrix} \begin{bmatrix} v_{e1} \\ v_o \end{bmatrix} = \begin{bmatrix} v_T(1/r_{be1}+g_{m1}) \\ 0 \end{bmatrix} \quad \text{(E3.14.1)}$$

```
%elec03e14.m
syms b b1 b2 rbe rbe1 rbe2 gm1 gm2 ro ro1 ro2 Rs RB  RC  RE RL
% To find the input resistance and voltage gain
vT=1; % Test input voltage
Y= [gm1+1/rbe1+1/ro1+1/rbe2 -1/rbe2;
   -gm2-1/rbe2 gm2+1/rbe2+1/RE+1/ro2];
v=Y\[vT*(1/rbe1+gm1); 0]; % solve Eq. (E3.14.1)
ve1=v(1); vo=v(2); iT=(vT-ve1)/rbe1;
% Input resistance
Ri=vT/iT; Ri=subs(Ri,{gm1,gm2},{b1/rbe1,b2/rbe2});
% Input resistance (starting from the last stage possibly with RL)
Ri2=subs(Ri_CC(1),{b,rbe,ro,RB,RE,RL},{b2,rbe2,ro2,inf,RE,inf});
Ria=subs(Ri_CC(1),{b,rbe,ro,RB,RE,RL},{b1,rbe1,ro1,inf,inf,Ri2});
simplify(Ri-Ria)
Rib=limit(limit(Ri,ro1,inf ),ro2,inf ); % With ro1=inf and ro2=inf
fprintf('\n Ri=\n'); pretty(simplify(Rib))
% Voltage gain
Av=vo/vT;
Av=subs(Av,{gm1,gm2},{b1/rbe1,b2/rbe2});
Gv=simplify(Ri/(Rs+Ri)*Av); % Overall voltage gain
Av1=subs(Av_CC(1),{b,rbe,ro,RB,RE,RL},{b1,rbe1,ro1,inf,inf,Ri2});
Av2=subs(Av_CC(1),{b,rbe,ro,RB,RE,RL},{b2,rbe2,ro2,inf,RE,inf});
Gva=Ria/(Rs+Ria)*Av1*Av2; % Overall voltage gain
simplify(Gv-Gva)
Avb=limit(limit(Av,ro1,inf ),ro2,inf ); % With ro1=inf and ro2=inf
Gvb=simplify(Rib/(Rs+Rib)*Avb);
fprintf('\n Gv=\n'); pretty(simplify(Gvb))

% Output resistance
ve1=vT/rbe2/((1+gm1*rbe1)/(Rs+rbe1)+1/ro1+1/rbe2);
iT=vT*(1/ro2+1/RE)+(vT-ve1)*(1/rbe2+gm2);
Ro=vT/iT;
Ro=subs(Ro,{gm1,gm2},{b1/rbe1,b2/rbe2});
% Output resistance (starting from the 1st stage with Rs)
Ro1=subs(Ro_CC(1),{b,rbe,ro,Rs,RB,RE},{b1,rbe1,ro1,Rs,inf,inf});
Roa=subs(Ro_CC(1),{b,rbe,ro,Rs,RB,RE},{b2,rbe2,ro2,Ro1,inf,RE});
simplify(Ro-Roa)
Rob=limit(limit(Ro,ro1,inf ),ro2,inf );
fprintf('\n Ro=\n'); pretty(simplify(Rob))
```

We can solve this set of equations to find v_{e1}, v_o, $i_T = (v_T - v_{e1})/r_{be1}$, $R_i = v_T/i_T$, and $A_v = v_o/v_T$. If we assume that $r_{o1} = \infty$ and $r_{o2} = \infty$, we can obtain

$$R_i = r_{be1} + (\beta_1 + 1)\{r_{be2} + (\beta_2 + 1)R_E\} \tag{E3.14.2}$$

$$G_v = \frac{(\beta_1 + 1)(\beta_2 + 1)R_E}{R_s + (1 + \beta_1 + \beta_2 + \beta_1\beta_2)R_E + r_{be1} + (\beta_1 + 1)r_{be2}} \tag{E3.14.3}$$

On the other hand, to find the output resistance, we draw the equivalent as depicted in Figure 3.42(b2) and write a node equation (in unknown v_{e1}) as

$$\left(\frac{1}{R_s + r_{be1}} + \frac{1}{r_{o1}} + \frac{1}{r_{be2}}\right)v_{e1} = g_{m1}v_{be1} + \frac{v_T}{r_{be2}} = g_{m1}\left(-\frac{r_{be1}}{R_s + r_{be1}}v_{e1}\right) + \frac{v_T}{r_{be2}}$$

$$;\left(\frac{1 + g_{m1}r_{be1}}{R_s + r_{be1}} + \frac{1}{r_{o1}} + \frac{1}{r_{be2}}\right)v_{e1} = \frac{v_T}{r_{be2}};\ i_T = \frac{v_T}{r_{o2}} + \frac{v_T}{R_E} + (v_T - v_{e1})\left(\frac{1}{r_{be2}} + g_{m2}\right) \tag{E3.14.4}$$

We can solve this equation to find v_{e1} and $R_o = v_T/i_T$. If we assume that $r_{o1} = \infty$ and $r_{o2} = \infty$, we can obtain

$$R_o = R_E \| \left[\frac{(\beta_1 + 1)r_{be2} + r_{be1} + R_s}{(\beta_1 + 1)(\beta_2 + 1)}\right] \tag{E3.14.5}$$

Alternatively, we can use the MATLAB functions '`Ri_CC()`', '`Av_CC()`', and '`Ro_CC()`' to obtain the same results by running the above MATLAB script "elec03e14.m":

```
Ri= RE + rbe1 + rbe2 + RE b1 + RE b2 + b1 rbe2 + RE b1 b2

  Gv=                       RE (b1 + 1) (b2 + 1)
         ------------------------------------------------------------
         RE + Rs + rbe1 + rbe2 + RE b1 + RE b2 + b1 rbe2 + RE b1 b2

  Ro=              RE (Rs + rbe1 + rbe2 + b1 rbe2)
         ------------------------------------------------------------
         RE + Rs + rbe1 + rbe2 + RE b1 + RE b2 + b1 rbe2 + RE b1 b2
```

Example 3.15 AC Analysis of a CC-CC (Darlington) Amplifier

Consider the CC-CE amplifier in Figure 3.43 where the device parameters of the BJTs Q_1 and Q_2 are $r_b = 0\ \Omega$, $\beta_F = 100$, $\beta_R = 1$, $\beta_{AC} = 100$, $V_A = 10^8$ V, and $V_T = 25.9$ mV. Find the overall input/output resistances and voltage gain $G_v = v_o/v_s$ and see how close they are to those obtained from PSpice simulation.

The exponential-model-based DC analysis and the AC analysis can be performed by running the following MATLAB script "elec03e15.m":

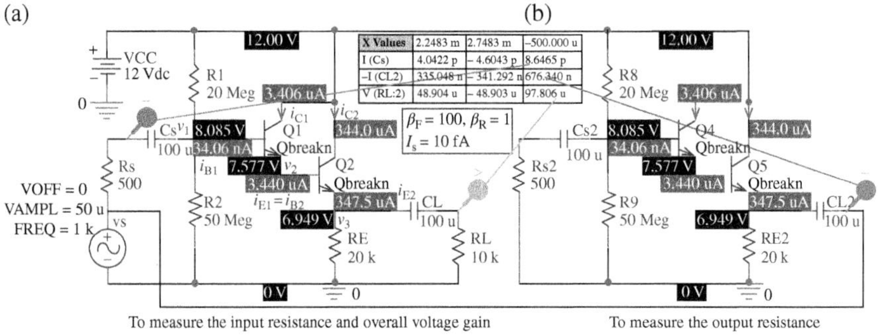

Figure 3.43 A CC-CC amplifier circuit and its PSpice simulation result ("elec03e15.opj").

```
%elec03e15.m
bF=100; bR=1; aR=bR/(bR+1); Is=1e-14; Isc=Is/aR;
T=27; VT=(T+273)/11605; VA=1e8;
VCC=12; R1=20e6; R2=50e6; VBB=VCC*R2/(R1+R2); RE=20e3;
% Exponential-model-based DC analysis
iC=@(v)Is*exp(v(1)/VT)-Isc*exp(v(2)/VT); % Eq.(3.1.23a)
iB=@(v)Is/bF*exp(v(1)/VT)+Isc/(bR+1)*exp(v(2)/VT); % Eq.(3.1.23b)
iE=@(v)iB(v)+iC(v);
eq=@(v)[VCC-v(1)-R1*(v(1)/R2+iB([v(1)-v(2) v(1)-VCC]));
        (iE([v(1)-v(2) v(1)-VCC])-iB([v(2)-v(3) v(2)-VCC]))*1e6;
        v(3)-RE*iE([v(2)-v(3) v(2)-VCC])]; % Node equations in v1,v2,v3
v0 = VBB-[0 0.7 1.4]; % Initial guess for v=[v1 v2 v3]
v = fsolve(eq,v0) % Solution to the set of node equations
VBE1=v(1)-v(2); VBC1=v(1)-VCC; VBE2=v(2)-v(3); VBC2=v(2)-VCC;
IC1Q=iC([VBE1 VBC1]), IC2Q=iC([VBE2 VBC2])
% AC analysis
[gm1,rbe1_,re1_,ro1_]=gmrbero_BJT(IC1Q,bF,VA,VT)
[gm2,rbe2_,re2_,ro2_]=gmrbero_BJT(IC2Q,bF,VA,VT)
Rs_=500; RB_=parallel_comb([R1 R2]); RE_=20e3; RL_=10e3; b1_=bF; b2_=bF;
syms b b1 b2 rbe rbe1 rbe2 ro ro1 ro2 Rs RB RC RE RL
% Input resistance (starting from the last stage with RL)
Ri2=subs(Ri_CC,{b,rbe,RB,RE,RL},{b2_,rbe2_,inf,RE_,RL_});
Ri2=eval(Ri2);
Ri1=subs(Ri_CC,{b,rbe,RB,RE,RL},{b1_,rbe1_,RB_,inf,Ri2});
Ri=eval(Ri1)
% Output resistance (starting from the 1st stage with Rs)
Ro1=subs(Ro_CC,{b,rbe,Rs,RB,RE},{b1_,rbe1_,Rs_,RB_,inf});
Ro1=eval(Ro1);
```

```
Ro2=subs(Ro_CC,{b,rbe,Rs,RB,RE},{b2_,rbe2_,Ro1,inf,RE_});
Ro=eval(Ro2)
% Voltage gain
Av1=subs(Av_CC,{b,rbe,RB,RE,RL},{b1_,rbe1_,RB_,inf,Ri2});
Av1=eval(Av1);
Av2=subs(Av_CC,{b,rbe,RB,RE,RL},{b2_,rbe2_,inf,RE_,RL_});
Av2=eval(Av2);
Gv=Ri/(Rs_+Ri)*Av1*Av2
% PSpice simulation results
Ri_pspice=100e-6/8.6465e-12-Rs_, Ro_pspice=100e-6/676.34e-9
Gv_pspice=97.806e-6/100e-6
```

Example 3.16 AC Analysis of a CC-CE Amplifier and a CC-CB Amplifier

For the two amplifiers whose AC equivalents are shown in Figure 3.44, find the input/output resistances (R_i/R_o) and the overall voltage gain $G_v = v_o/v_s$ in terms of β_1, β_2, r_{be1}, r_{be2}, r_{o1}, r_{o2}, R_s, and R_L where $g_{mk} = \beta_k/r_{bek}$, but assume $r_{o1} = r_{o2} = \infty$ for R_i and G_v, and assume $r_{o1} = r_{o2} \gg r_{be1} = r_{be2} \gg R_s$ and $\beta_1 = \beta_2 \gg 1$ for R_o.

For the amplifier in Figure 3.44(a), we can run the following MATLAB script "elec03e16a.m" to get

$$R_i = r_{be1} + (\beta_1+1)r_{be2},\ G_v = -\frac{(\beta_1+1)\beta_2 R_L}{R_s + r_{be1} + (\beta_1+1)r_{be2}},\ \text{and } R_o = r_{o2} \tag{E3.16.1}$$

```
%elec03e16a.m
syms b b1 b2 rbe rbe1 rbe2 ro ro1 ro2 Rs RB RC RE RL
Ri2=subs(Ri_CE,{b,rbe,RB,RC,RE},{b2,rbe2,inf,inf,0});
Ri=subs(Ri_CC,{b,rbe,RB,RC,RE,RL},{b1,rbe1,inf,0,inf,Ri2})
Av1=subs(Av_CC,{b,rbe,RB,RC,RE,RL},{b1,rbe1,inf,0,inf,Ri2});
Av2=subs(Av_CE,{b,rbe,RB,RC,RE},{b2,rbe2,inf,inf,0});
Gv=eval(Ri/(Rs+Ri)*Av1*Av2); % Overall gain
pretty(simplify(Gv))
Ro1=subs(Ro_CC(1),{b,rbe,ro,RB,RE},{b1,rbe1,ro1,inf,inf})
Ro=subs(Ro_CE(1),{b,rbe,ro,Rs,RB,RC,RE},{b2,rbe2,ro2,Ro1,inf,inf,0})
```

For the amplifier in Figure 3.44(b), we can run the following MATLAB script "elec03e16b.m" to get

$$R_i = r_{be1} + \frac{\beta_1+1}{\beta_2+1}r_{be2},\ G_v = -\frac{(\beta_1+1)\beta_2 R_L}{(\beta_2+1)(R_s+r_{be1}) + (\beta_1+1)r_{be2}},\ \text{and } R_o = 2r_{o2} \tag{E3.16.2}$$

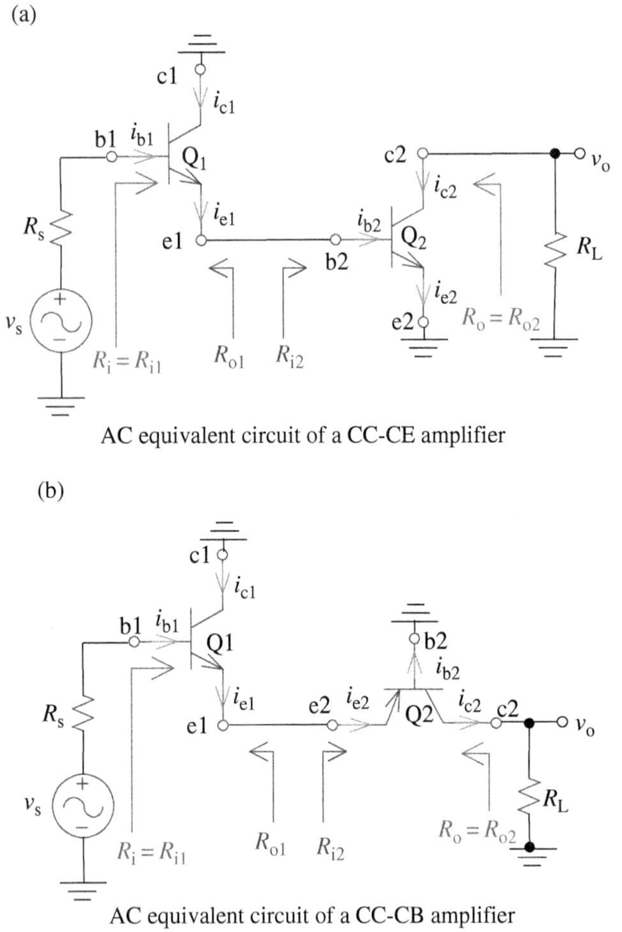

Figure 3.44 A CC-CE amplifier and a CC-CB amplifier.

```
%elec03e16b.m
syms b b1 b2 rbe rbe1 rbe2 ro ro1 ro2 Rs RB RC RE RL
Ri2=subs(Ri_CB,{b,rbe,RB,RC,RE},{b2,rbe2,0,inf,inf});
Ri=subs(Ri_CC,{b,rbe,RB,RC,RE,RL},{b1,rbe1,inf,0,inf,Ri2})
Av1=subs(Av_CC,{b,rbe,RB,RC,RE,RL},{b1,rbe1,inf,0,inf,Ri2});
Av2=subs(Av_CB,{b,rbe,RB,RC,RE},{b2,rbe2,0,inf,inf});
Gv=Ri/(Rs+Ri)*Av1*Av2; % Overall gain
pretty(simplify(Gv))
Ro1=subs(Ro_CC(1),{b,rbe,ro,RB,RE},{b1,rbe1,ro1,inf,inf})
Ro=subs(Ro_CB(1),{b,rbe,ro,Rs,RB,RC,RE},{b2,rbe2,ro2,Ro1,0,inf,inf})
```

Note that to remove a resistance in the formulas for R_i, R_o, A_v, ... due to its non-existence in a circuit, it is enough to set its value to 0/∞ if it is series-/parallel-combined with other resistance(s) without having to compare the circuit with the corresponding model in Fig. 3.30/3.32/3.34.1.

3.3 Logic Gates Using Diodes/Transistors[C-3, M-1]

This section will discuss the DTL (Diode-Transistor Logic) NAND gate, TTL (Transistor-Transistor Logic) NAND gate, and ECL (Emitter-Coupled Logic) OR/NOR gate.

3.3.1 DTL NAND Gate

Figure 3.45 shows a basic DTL NAND gate consisting of a diode AND gate cascaded with a BJT inverter where the binary inputs v_{i1} and v_{i2} are supposed to be one of the two voltage levels corresponding to low (logic 0) and high (logic 1):

$$V(0) = V_{CE,sat} = 0.2\,[\mathrm{V}]\,\text{and}\,V(1) = V_{CC} = 5\,[\mathrm{V}] \tag{3.3.1}$$

Let us look over several aspects of the DTL NAND gate.

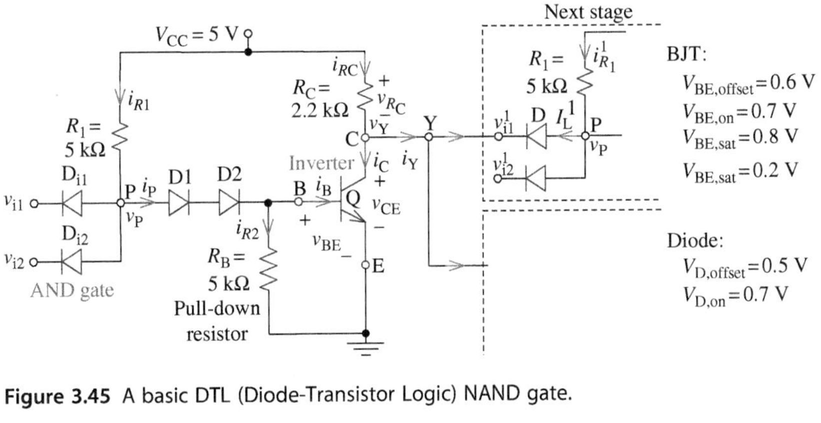

Figure 3.45 A basic DTL (Diode-Transistor Logic) NAND gate.

1) **Logic Function**

To check if the circuit operates as a NAND gate, let us consider the following two cases:

- At least one of the two inputs v_{i1} and v_{i2} is low as $V(0)=V_{CE,sat}=0.2$ V.
- Both of the two inputs v_{i1} and v_{i2} are high as $V(1)=V_{CC}=5$ V.

Case I At Least One of the Two Inputs v_{i1} and v_{i2} is Low as $V(0) = V_{CE,sat} = 0.2$ V

In this case, the input diode connected to the low input will be forward-biased through R_1 by the voltage source V_{CC} so that its voltage can be $V_{D,on} = 0.7$ V. Then the voltage at node P will be

$$v_P = V(0) + V_{D,on} = 0.2 + 0.7 = 0.9\,[\text{V}] \tag{3.3.2}$$

Since this voltage is not high enough to turn on the two series-connected diodes D1-D2 and the B-E junction of Q, the two diodes will be off so that $i_{R2} = 0$, $v_{BE} = 0$, and $i_B = 0$, which forces the BJT Q to be in the cutoff mode. Then the output voltage will be

$$v_Y = V_{CC} = 5\,[\text{V}]\,(\text{high}) \tag{3.3.3}$$

since $i_{RC} = 0$ so that $v_{RC} = R_C i_{RC} = 0$ as long as no output current flows into the next-stage (load) gate(s). This state of high output voltage is most probably true if the next-stage gates are the same gates, since the current through the reverse-biased diodes in the next-stage gate will not be much. However, if the number of next-stage (load) gates exceeds a certain number (called the output-high fan-out), then the state of high output voltage may be jeopardized. Here are a couple of QAs:

(Q1) What if we remove one of the two series-connected diodes D1 and D2?

(A1) With only one diode, say, D1, we still have $v_Y = 5$ V (high) since $v_P = 0.9$ V is not high enough to turn on the diodes D1 and the B-E junction of Q. However, if $v_P \geq 0.9 + 0.3$ [V] due to some noise, then the result will differ. Another diode D2 has been inserted additionally to increase the high (logic 1) noise margin so that the high output can be less affected by the noise.

(Q2) How much current will flow into the previous stage when one of the inputs is low, say, $v_{i1} = 0.2$ V due to the situation where the BJT Q^{-1} of the previous stage is saturated?
(Note that superscripts 1 and −1 stand for the previous and next stages, respectively.)

(A2) Since the series-connected diodes D1-D2 are off, the current through R_1

$$i_{R_1} = \frac{V_{CC} - v_P}{R_1} = \frac{5 - 0.9\,\text{V}}{5\,\text{k}\Omega} = 0.82\,[\text{mA}] \tag{3.3.4}$$

will flow back into the previous stage, which is supposed to be in the saturation mode. This current will be the basis for determining the output-low fan-out (of the previous gate) because it may increase the collector current of BJT Q^{-1} (of the previous stage) so that Q^{-1} can get out of the saturation mode to enter the forward-active region. Note that the output-high/low fan-outs of a gate are the maximum numbers of load gates of similar design that can be connected to the output of the (driver) gate without affecting its high/low output, respectively.

Case II All the Inputs v_{i1} and v_{i2} are High as $v_Y = V_{CC} = 5\,\text{V}$

In this case, all the input diodes will be off so that the voltage source V_{CC} can turn on the two series-connected diodes D1-D2 and make the B-E junction of BJT Q forward-biased. Then the BJT Q will be in the forward-active or saturation mode. Suppose Q is in the forward-active mode. Then the voltages at nodes B and P will be

$$v_B = V_{BE,on} = 0.7\,[\text{V}] \text{ and } v_P = v_B + 2V_{D,on} = 0.7 + 2 \times 0.7 = 2.1\,[\text{V}] \tag{3.3.5}$$

respectively so that the current through R_1 and the current through R_B are

$$i_{R_1} = \frac{V_{CC} - v_P}{R_1} = \frac{5 - 2.1\,\text{V}}{5\,\text{k}\Omega} = 0.58\,[\text{mA}] \text{ and } i_{R_B} = \frac{v_B}{R_B} = \frac{0.7\,\text{V}}{5\,\text{k}\Omega} = 0.14\,[\text{mA}] \tag{3.3.6}$$

respectively. Thus, we can subtract i_{RB} from i_{R1} to get the base current of Q as

$$i_B = i_{R_1} - i_{R_B} = 0.58 - 0.14 = 0.44\,[\text{mA}] \tag{3.3.7}$$

and, accordingly, find the collector current of Q and the collector-to-emitter voltage (equal to the output voltage) as

$$i_C = \beta_F i_B = 100 \times 0.44 = 44\,[\text{mA}] \tag{3.3.8}$$

$$v_Y = v_{CE} = V_{CC} - R_C i_C = 5 - 2.2 \times 44 \overset{\text{No}}{<} 0.3\,[\text{V}] \tag{3.3.9}$$

However, this implies that Q cannot be in the forward-active mode. Therefore, we suppose that Q is in the saturation mode so that the collector-to-emitter voltage is

$$v_Y = V_{CE,sat} = 0.2\,[V]\,(\text{low}) \tag{3.3.10}$$

Then, on the assumption that no current flows from the next stage into Q, the collector current will be

$$i_C = i_{R_C} = \frac{V_{CC} - v_Y}{R_C} = \frac{5 - 0.2\,V}{2.2\,k\Omega} = 2.182\,[mA] \tag{3.3.11}$$

The voltages at nodes B and P will be

$$v_B = V_{BE,sat} = 0.8\,[V]\,\text{and}\, v_P = v_B + 2V_{D,on} = 0.8 + 2 \times 0.7 = 2.2\,[V] \tag{3.3.12}$$

respectively, so that the current through R_1 and the current through R_2 are

$$i_{R_1} = \frac{V_{CC} - v_P}{R_1} = \frac{5 - 2.2\,V}{5\,k\Omega} = 0.56\,[mA] \text{ and } i_{R_B} = \frac{v_B}{R_2} = \frac{0.8\,V}{5\,k\Omega} = 0.16\,[mA] \tag{3.3.13}$$

respectively. Thus, we can subtract i_{RB} from i_{R1} to get the base current of Q as

$$i_B = i_{R_1} - i_{R_B} = 0.58 - 0.16 = 0.4\,[mA] \tag{3.3.14}$$

Here is a QA:

(Q3) What is the minimum value of the CE large-signal forward-active current gain β_F of the BJT Q to guarantee that it is in the saturation mode when all the inputs are high?

(A3) It is equal to the ratio of the collector current to the base current:

$$\beta_{F,min} = \frac{i_C}{i_B} \overset{(3.3.11)}{\underset{(3.3.14)}{=}} \frac{2.182\,[mA]}{0.4\,[mA]} = 5.46 \tag{3.3.15}$$

2) **Fan-out**

When the output voltage is low, i.e. $v_Y = V_{CE,sat} = 0.2$ V with the BJT Q (in the current stage) saturated, it can let the input diode of a load gate (connected to the output node Y) forward-biased so that the current through R_1 of a load gate (in the next stage)

$$i_{R_1} \overset{(3.3.4)}{=} \frac{V_{CC} - v_P}{R_1} = \frac{5 - 0.9\,V}{5\,k\Omega} = 0.82\,[mA] \tag{3.3.16}$$

can flow back (sink) into the BJT Q of the current stage in addition to the existing collector current (Eq. (3.3.11)) coming through R_C. Therefore, if the number of similar load gates connected to the output node Y is N, the maximum collector current of Q will be

$$i_{C,max} = i_{R_C} + Ni_{R_1} \overset{(3.3.11)}{\underset{(3.3.16)}{=}} 2.182\,[\text{mA}] + 0.82N < \beta_F i_B \qquad (3.3.17)$$

From the condition that this maximum collector current should be less than $\beta_F i_B$ in order to keep Q saturated, we can determine the *output-low fan-out* of the basic NAND gate as the minimum integer satisfying the above inequality (3.3.17):

$$N < \frac{\beta_F i_B - i_{R_C}}{i_{R_1}} = \frac{0.4\beta_F - 2.182}{0.82} \rightarrow \text{floor}\left(\frac{0.4\beta_F - 2.182}{0.82}\right) \qquad (3.3.18)$$

How about the output-high fan-out of the DTL gate? When the output is high, it can let the input diodes of the load gates reverse-biased and then the additional current flowing through R_C and going into the next stage is just the sum of the reverse leakage currents that is not so large as to put a limitation on the fan-out.

3) **Role of the Pull-down Resistor R_B**

If there were no connection through R_B between node B and ground so that $i_{RB} = 0$, the base current $i_B = i_{R1} - i_{RB}$ would be larger so that the fan-out determined by Eq. (3.3.18) could be increased. Then, what is the pull-down resistor R_B for? Its role is to decrease the turn-off time (saturation-to-cutoff switching time) by providing another path (in parallel with the BE junction of Q) for the reverse base current so that excess minority carriers can be removed quickly from the base while Q enters the cutoff mode from the saturation mode. Note that a smaller value of the pull-down resistor R_B makes the turn-off time shorter, but on the other hand, it decreases the base current $i_B = i_{R1} - i_{RB}$, reducing the driving capability (measured by fan-out).

4) **Voltage Transfer Characteristic (VTC) and Noise Margin**

Table 3.4 and Figure 3.46 show the (piecewise linear) VTC of the DTL NAND gate depicted in Figure 3.45. Figure 3.46 also shows the low/high noise margins NM_L/NM_H defined as the maximum widths of the range in which the input voltage can vary without changing the high/low output voltage. Note that the (dotted) VTC for the gate using only one diode between nodes P and B implies that saving one diode results in a considerable reduction of the low noise margin NM_L.

Table 3.4 Voltage transfer characteristic (VTC) of a DTL NAND gate.

Region	Range of v_i [V]	v_P [V]	v_B [V]	D1-D2	Q	v_Y [V]
1	$0\sim0.5 = V_{BE}^{0} + 2V_{D,offset}^{0.6} - V_{D,on}^{0.7}$	$v_i + V_{D,\,on}$ $0.7\sim1.2$	0	OFF	Cutoff	5.0
2	$0.5\sim1.3 = V_{BE,offset}^{0.6} + 2V_{D,on}^{0.7} - V_{D,on}^{0.7}$	$v_i + V_{D,\,on}$ $1.2\sim2.0$	$0\sim0.5$	OFF	Cutoff	5.0
3	$1.3\sim1.7 = V_{BE,sat}^{0.8} + 2V_{D,on}^{0.7} - V_{D,offset}^{0.5}$	$v_i + V_{D,\,on}\sim V_{D,\,offset}$ $2.0\sim2.2$	$0.5\sim0.8$	ON	Forward-active	5.0–0.2
4	$1.7\sim5.0$	$V_{BE,\,sat} + 2V_{D,\,on}$ 2.2 V	0.8	ON	Saturated	0.2

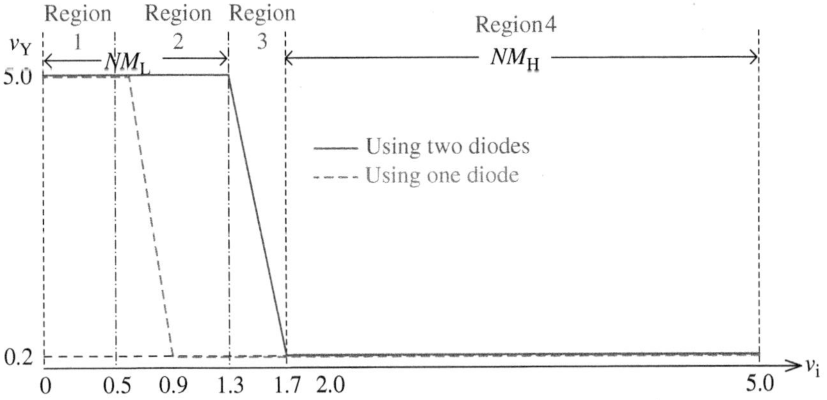

Figure 3.46 Voltage transfer characteristic of a DTL NAND gate.

3.3.2 TTL NAND Gate

3.3.2.1 Basic TTL NAND Gate Using Two BJTs

Figure 3.47 shows a basic TTL NAND gate using two BJTs where the multiple BEJs (BE junctions) of BJT Q_1 replace the input diodes and the BCJ (BC junction) of BJT Q_1 replaces the diode D1 of the DTL NAND gate. Note that the input clamping diodes are placed to keep the input voltages from going below −0.7 V so that Q_1 can be protected from any large negative input voltage.

To check if the circuit operates as a NAND gate, let the forward/reverse DC current gains be

$$\beta_F = 20 \text{ and } \beta_R = 0.1 \tag{3.3.19}$$

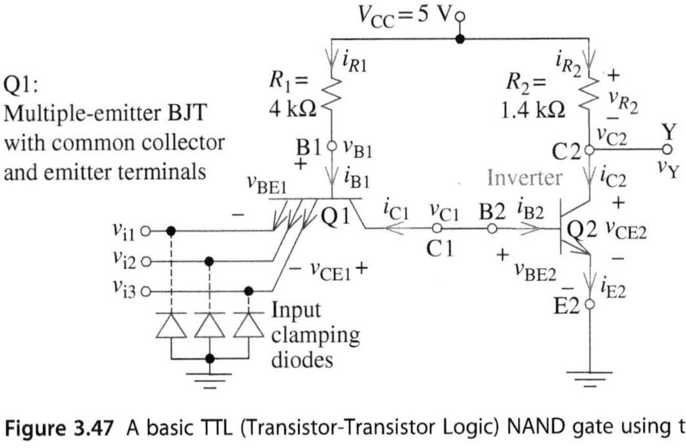

Figure 3.47 A basic TTL (Transistor-Transistor Logic) NAND gate using two BJTs.

respectively, and consider the following two cases:

- At least one of the inputs is low as $V(0) = V_{CE,sat} = 0.2$ V.
- All the inputs are high as $V(1) = V_{CC} = 5$ V.

Case I At Least One of the Three Inputs v_{i1}, v_{i2}, and v_{i3} is Low as $V(0) = V_{CE,sat} = 0.2$ V

In this case, the BEJ of Q_1 connected to the low input will be forward-biased through R_1 by the voltage source V_{CC}, but Q_1 will be not in the forward-active mode but in the saturation mode because its collector current cannot be as large as $\beta_F i_{B1} = \beta_F(V_{CC} - V_{BE1,on})/R_1$ due to the back-to-back connection of the BCJ of Q_1 and BEJ of Q_2. Then we have $v_{CE1} = V_{CE,sat} = 0.2$ V so that

$$v_{B2} = v_{C1} = v_i + V_{CE1,sat} = 0.2 + 0.2 = 0.4\,[\text{V}] \tag{3.3.20}$$

Since this voltage is not high enough to forward-bias the BEJ of Q_2, the BJT Q_2 will be in the cutoff mode. Then the output voltage will be

$$v_Y = V_{CC} = 5\,[\text{V}]\,(\text{high}) \tag{3.3.21}$$

since $i_{R2} = 0$ so that $v_{R2} = R_2 i_{R2} = 0$ as long as no output current flows into the load gate(s) (in the next stage). This state of high output voltage is most probably true if the load gates are the same kind of gates, since the current through the reverse-biased BEJ of Q_1^1 in the load gate will not be much. Then the current through R_1

$$i_{R_1} = i_{B_1} = \frac{V_{CC} - v_{B1}}{R_1} = \frac{V_{CC} - (v_i + V_{BE1,sat})}{R_1} = \frac{5 - (0.2 + 0.8)}{4\,\text{k}\Omega} = 1\,[\text{mA}] \tag{3.3.22}$$

will flow back into the previous stage, possibly annoying its BJT Q_2^{-1}, which is supposed to be in the saturation mode.

Case II All the Inputs v_{i1}, v_{i2}, and v_{i3} are High as $v_Y = V_{CC} = 5$ V

In this case, the BEJ/BCJ of Q_1 are reverse-/forward-biased so that the BJT Q_1 can be in the reverse-active mode where the roles of emitter and collector are switched. Then the collector current of Q_1 will be

$$i_{C1} = -(\beta_R + 1)i_{B1} = -i_{B2} \tag{3.3.23}$$

Assume that the BJT Q_2 is saturated by accepting this current as its base current. Then we have

$$\begin{aligned} v_{B1} = v_{B2} + V_{BC1,\text{reverse-active}} &= V_{BE2,sat} + V_{BC1,\text{reverse-active}} \\ &= 0.8 + 0.7 = 1.5\,[\text{V}] \end{aligned} \tag{3.3.24}$$

$$i_{R_1} = i_{B_1} = \frac{V_{CC} - v_{B1}}{R_1} \overset{(3.3.24)}{=} \frac{5-1.5}{4\,\text{k}\Omega} = 0.875\,[\text{mA}] \tag{3.3.25}$$

$$i_{C1} = -\,i_{B2} \overset{(3.3.23)}{=} -\,(\beta_R + 1)i_{B1} \overset{(3.3.25)}{=} -1.1 \times 0.875 = -0.963\,[\text{mA}] \tag{3.3.26}$$

If no load current flows back from the next stage into Q_2, the collector current of Q_2 will be

$$i_{C2} = i_{R_2} = \frac{V_{CC} - v_{C2}}{R_2} = \frac{V_{CC} - V_{CE2,sat}}{R_2} = \frac{5-0.2}{1.4\,\text{k}\Omega} \approx 3.43\,[\text{mA}] \tag{3.3.27}$$

which turns out to be less than $\beta_F i_{B2} \overset{(3.3.26)}{=} 20 \times 0.963 = 19.26\,[\text{mA}]$. This justifies our assumption that Q_2 is saturated. Thus, we can rest assured that the output voltage is low:

$$v_Y = v_{C2} = V_{CE2} = 0.2\,[\text{V}]\ (\text{low}) \tag{3.3.28}$$

Example 3.17 A Logic Inverter Consisting of Two BJTs

Consider the BJT circuit in Figure 3.48(a) where the device parameters of the BJTs Q_1 and Q_2 are $\beta_F = 100$, $\beta_R = 1$, and $I_s = 10^{-15}$ in common.

We can run the following MATLAB script "elec03e17.m" to plot v_{BE1}, $i_{C1} = -i_{B2}$, and $v_o = v_{CE2}$ for the input voltage $v_i = 0{\sim}2$ [V] as shown in Figure 3.48(b). Also, we can perform the PSpice simulation to plot v_{BE1}, $i_{C1} = -i_{B2}$, and $v_o = v_{CE2}$ for $v_i = 0{\sim}2$ [V].

```
%elec03e17.m
betaF=100; betaR=1; alphaR=betaR/(betaR+1);
Is=1e-15; Isc=Is/alphaR; T=27; VT=(T+273)/11605; % BJT parameters
VCC=5; R1=4e3; R2=1.4e3; vis=[0:0.01:2];
iC=@(v)Is*exp(v(1)/VT)-Isc*exp(v(2)/VT); % Eq.(3.1.23a)
iB=@(v)Is/betaF*exp(v(1)/VT)+Isc/(betaR+1)*exp(v(2)/VT); %Eq.(3.1.23b)
options=optimoptions('fsolve','TolX',1e-6,'Display','off');
for n=1:numel(vis)
  vi=vis(n); % A set of node equations in v=[vB1 vC1=vB2 vC2]
  eq=@(v,vi)[VCC-R1*iB([v(1)-vi v(1)-v(2)])-v(1); % KCL at node 1
    (iC([v(1)-vi v(1)-v(2)])+iB([v(2) v(2)-v(3)]))*1e6; % KCL at node 2
     VCC-R2*iC([v(2) v(2)-v(3)])-v(3)]; % KCL atnode 3
  if n<2, v0=vi+[0.7 1 2]; else v0=v; end % Initial guess for v
  [v,err(n,:)]=fsolve(eq,v0,options,vi);
  vos(n)=v(3); vBE1s(n)=v(1)-vi; iCs(n)=iC([v(1)-vi v(1)-v(2)]);
end
subplot(221), plot(vis,iCs), grid on
subplot(223), plot(vis,vos,'r', vis,vBE1s), grid on
```

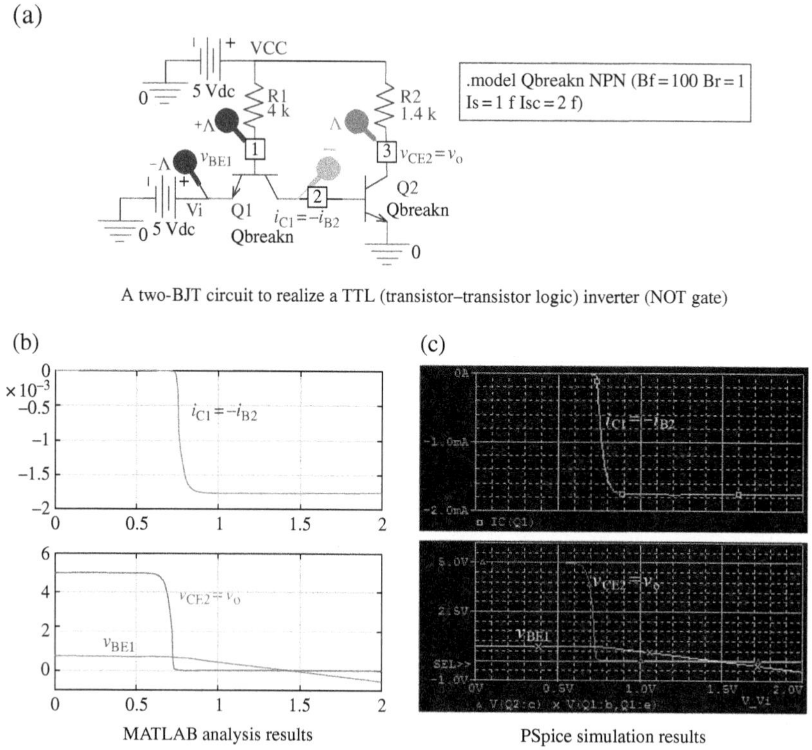

Figure 3.48 A logic inverter consisting of two BJTs and its MATLAB analysis & PSpice simulation.

3.3.2.2 TTL NAND Gate Using Three BJTs

Figure 3.49 shows a TTL NAND gate using three BJTs where compared with the basic TTL NAND gate of Figure 3.47, Q_3 increases not only the fan-out (by raising the overall current gain) but also the noise margin since the BEJ of Q_3 provides an additional diode offset voltage (like D2 in the TTL NAND gate). Let us look over several aspects of the TTL NAND gate.

1) **Logic Function**

To check if the gate operates as a NAND gate, let us consider the following two cases:

- At least one of the inputs is low as $V(0) = V_{CE,sat} = 0.2$ V.
- All the inputs are high as $V(1) = V_{CC} = 5$ V.

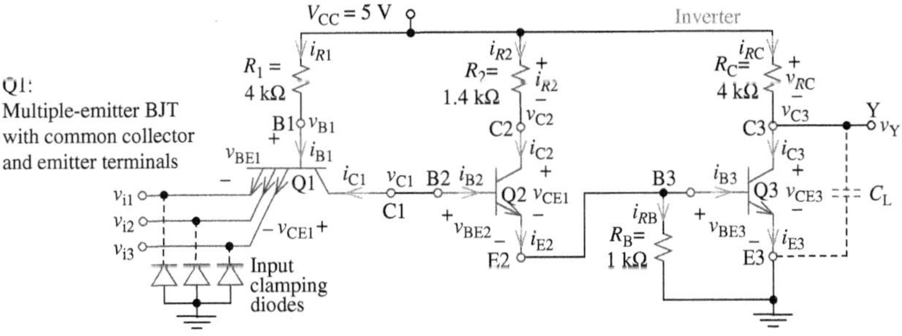

Figure 3.49 A TTL NAND gate using three BJTs.

Case I At Least One of the Inputs v_{i1}, v_{i2}, and v_{i3} is Low as $V(0) = V_{CE,sat} = 0.2$ V

In this case, the BEJ of Q_1 connected to the low input will be forward-biased through R_1 by the voltage source V_{CC}, but Q_1 will be not in the forward-active mode but in the saturation mode because its collector current cannot be as large as $\beta_F i_{B1} = \beta_F(V_{CC} - V_{BE1,on})/R_1$ due to the back-to-back connection of the BCJ of Q_1 and BEJ of Q_2. Then we have $v_{CE1} = V_{CE,sat} = 0.2$ V so that

$$v_{B2} = v_{C1} = v_i + V_{CE1,sat} = 0.2 + 0.2 = 0.4\,[\text{V}] \tag{3.3.29}$$

Since this voltage is not high enough to forward-bias the BEJs of Q_2 and Q_3, the two BJTs will be in the cutoff mode. Then the output voltage will be

$$v_Y = V_{CC} = 5\,[\text{V}]\,(\text{high}) \tag{3.3.30}$$

since $i_{RC} = 0$ so that $v_{RC} = R_C i_{RC} = 0$ as long as no output current flows into the load gate(s) (in the next stage). This state of high output voltage is most probably true if the load gates are the same gates, since the current through the reverse-biased BEJ of Q_1^1 (of the next stage) in the load gate will not be much. Then the current through R_1

$$i_{R_1} = i_{B_1} = \frac{V_{CC} - v_{B1}}{R_1} = \frac{V_{CC} - (v_i + V_{BE1,sat})}{R_1} = \frac{5 - (0.2 + 0.8)}{4\text{k}\Omega} = 1\,[\text{mA}] \tag{3.3.31}$$

will flow back into the driver gate (in the previous stage), possibly annoying its BJT Q_2^{-1}, which is supposed to be in the saturation mode.

Case II All the Inputs v_{i1}, v_{i2}, and v_{i3} are High as $v_Y = V_{CC} = 5\,\text{V}$

In this case, the BEJ/BCJ of Q_1 are reverse-/forward-biased so that the BJT Q_1 can be in the reverse-active mode where the roles of emitter and collector are switched. Then the collector current of Q_1 will be

$$i_{C1} = -(\beta_R + 1)i_{B1} = -i_{B2} \tag{3.3.32}$$

Assume that Q_2 and Q_3 are saturated by accepting $i_{B2} = -i_{C1}$ and $i_{B3} = i_{E2} - i_{RB}$, respectively, as their base currents. Then we have

$$v_{B1} = V_{BE2,\text{sat}} + V_{BE3,\text{sat}} + V_{BC1,\text{reverse-active}} = 0.8 + 0.8 + 0.7 = 2.3\,[\text{V}] \tag{3.3.33}$$

$$i_{B_1} = i_{R_1} = \frac{V_{CC} - v_{B1}}{R_1} \overset{(3.3.33)}{=} \frac{5 - 2.3}{4\,\text{k}\Omega} = 0.675\,[\text{mA}] \tag{3.3.34}$$

$$i_{B2} = -i_{C1} \overset{(3.3.32)}{=} (\beta_R + 1)i_{B1} \overset{(3.3.34)}{=} 1.1 \times 0.675 = 0.7425\,[\text{mA}] \tag{3.3.35}$$

$$v_{C2} = V_{CE2,\text{sat}} + V_{BE3,\text{sat}} = 0.2 + 0.8 = 1\,[\text{V}] \tag{3.3.36}$$

$$i_{C2} = i_{R_2} = \frac{V_{CC} - v_{C2}}{R_2} \overset{(3.3.36)}{=} \frac{5 - 1}{1.4\text{k}\Omega} \approx 2.86\,[\text{mA}] \overset{\text{OK}}{<} \beta_F i_{B2} \tag{3.3.37}$$

$$i_{E2} = i_{B2} + i_{C2} = 0.7425 + 2.86 = 3.6025\,[\text{mA}] \tag{3.3.38}$$

$$i_{R_B} = \frac{V_{BE3,\text{sat}}}{R_B} = \frac{0.8}{1\,\text{k}\Omega} = 0.8\,[\text{mA}] \tag{3.3.39}$$

$$i_{B3} = i_{E2} - i_{R_B} \underset{(3.3.39)}{\overset{(3.3.38)}{=}} 3.6025 - 0.8 = 2.8025\,[\text{mA}] \tag{3.3.40}$$

If no load current flows back from the next stage into Q_3, the collector current of Q_3 will be

$$i_{C3} = i_{R_C} = \frac{V_{CC} - v_{C3}}{R_C} = \frac{V_{CC} - V_{CE3,\text{sat}}}{R_C} = \frac{5 - 0.2}{4\text{k}\Omega} = 1.2\,[\text{mA}] \overset{\text{OK}}{<} \beta_F i_{B3} \tag{3.3.41}$$

Since $i_{C2} < \beta_F i_{B2}$ and $i_{C3} < \beta_F i_{B3}$, Q_2 and Q_3 turn out to be saturated (as we assumed) so that we can rest assured of the low output voltage:

$$v_Y = v_{C3} = V_{CE3,\text{sat}} = 0.2\,[\text{V}]\,(\text{low}) \tag{3.3.42}$$

2) **Output-high Fan-out**

In Figure 3.50, suppose the output voltage of the driver gate is high, i.e. $v_Y = V_{CC} = 5\,\text{V}$ with Q_2 and Q_3 cutoff, which can drive the BJTs Q_1^1

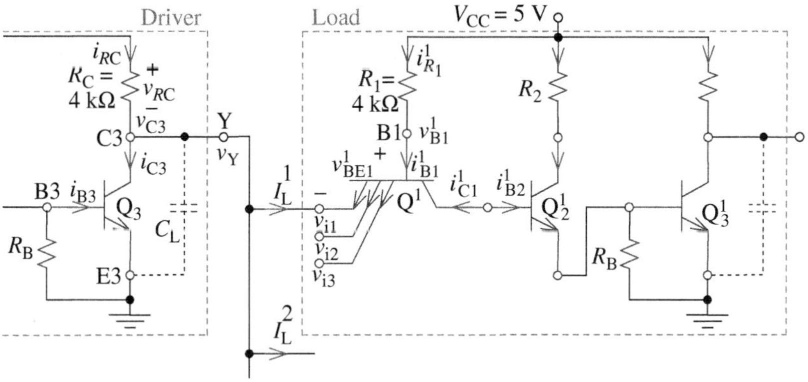

Figure 3.50 A TTL NAND gate as a driver with multiple loads.

(connected to the output node Y), Q^1_2, and Q^1_3 of the load gate (in the next stage) into the reverse-active, saturation, and saturation modes, respectively. Then the load current (flowing into the load gate) equal to the emitter current of Q^1_1 can be obtained as

$$v^1_{B1} = V_{BE2,sat} + V_{BE3,sat} + V_{BC1,\text{reverse-active}} = 0.8 + 0.8 + 0.7 = 2.3[\text{V}]$$

$$i^1_{B1} = i^1_{R_1} = \frac{V_{CC} - v^1_{B1}}{R^1_1} = \frac{5 - 2.3}{4\text{k}\Omega} = 0.675\,[\text{mA}]$$

$$I^1_L = i^1_{E1} = \beta_R i^1_{B1} = 0.1 \times 0.675 = 0.0675\,[\text{mA}] \tag{3.3.43}$$

If the number of load gates connected to the output node Y is N, the current through R_C of the driver gate is $N I^1_L$, which will decrease the output voltage as

$$v_Y = V_{CC} - R_C i_{R_C} = V_{CC} - R_C N I^1_L \tag{3.3.44}$$

In order for the output voltage v_Y not to be lower than the minimum output voltage 3 V (corresponding to logic 1) for all the voltage drop due to the loading effect, the following condition should be met:

$$v_Y = V_{CC} - R_C i_{R_C} = V_{CC} - R_C N I^1_L \overset{?}{\geq} 3\text{V}$$
$$;N \leq \frac{V_{CC} - 3}{R_C I^1_L} \overset{(3.3.43)}{=} \frac{5-3}{4 \times 0.0675} = \frac{2}{0.27} = 7.4 \rightarrow 7 \tag{3.3.45}$$

Therefore the output-high fan-out of the TTL NAND gate of Figure 3.49 is 7.

3) **Output-low Fan-out**

In Figure 3.50, suppose the output voltage of the driver gate is low, i.e. $v_Y = V_{CE,sat} = 0.2$ V with the BJT Q saturated, which can drive the BJTs Q_1^1 (connected to the output node Y), Q_2^1, and Q_3^1 of the load gate (in the next stage) into the saturation, cutoff, and cutoff modes, respectively. Then the load current (flowing from the load gate) equal to the (negative) emitter current of Q_1^1 can be obtained as

$$I_L^1 = i_{E1}^1 \overset{(3.1.5)}{=} -i_{B1}^1 - i_{C1}^1 = -i_{B1}^1 + i_{B2}^1 \overset{Q_2^1\text{ off}}{=} -i_{B1}^1 = -i_{R_1}^1$$
$$= -\frac{V_{CC} - v_{B1}^1}{R_1^1} = -\frac{V_{CC} - (v_Y + V_{BE1,sat})}{R_1^1} = -\frac{5-(0.2+0.8)}{4\,\mathrm{k\Omega}} = -1\,[\mathrm{mA}] \tag{3.3.46}$$

If the number of load gates connected to the output node Y is N, the collector current i_{C3} of Q_3 in the driver gate will be

$$i_{C3} = i_{R_C} - NI_L^1 \underset{(3.3.46)}{\overset{(3.3.41)}{=}} 1.2 + N\,[\mathrm{mA}] \overset{?}{<} \beta_F i_{B3} \tag{3.3.47}$$

In order for Q_3 not to exit the saturation mode, the following condition should be met:

$$i_{C3} \overset{(3.3.47)}{=} 1.2 + N\,[\mathrm{mA}] < \beta_F i_{B3} \overset{(3.3.40)}{=} 20 \times 2.8025 = 56.05\,[\mathrm{mA}];\quad N \le 54 \tag{3.3.48}$$

This implies that the output-low fan-out of the TTL NAND gate of Figure 3.49 is 54. Therefore, the fan-out of the TTL NAND gate is 9, which is the lower of the output-low and output-high fan-outs.

3.3.2.3 Totem-Pole Output Stage

Consider again the TTL NAND gate of Figure 3.49 or 3.50 where C_L denotes the capacitive load consisting of parasitic capacitances of wires and reverse-biased diodes (of the load gates). The capacitive load C_L may cause a long low-to-high transition time (as can be seen from Figure 3.27(b1)) since it must be charged from $V_{CE,sat} = 0.2$ to $V_{CC} = 5.0$ by the current i_{R_C} through R_C. A smaller R_C reduces the output delay, but also results in a more power dissipation of $(V_{CC} - V_{CE,sat})^2/R_C$ when the output is low, i.e. $v_Y = V_{CE,sat}$.

To resolve this dilemma, the (passive) pull-up resistor R_C is made into an active pull-up circuit by inserting a BJT Q_4 (together with a diode D) between R_C and Q_3 as depicted in Figure 3.51. The circuit is called a TTL NAND gate with a *totem-pole output stage* where totem poles are ancient traditional sculptures that were carved as the emblem of a family or clan by the Northwest American Indian tribes. Since its logic function is the same with the previous NAND gates, let us focus on the role of the totem-pole output stage consisting of Q_4-D-Q_3, especially during the low-to-high transition of the output voltage v_Y.

First, let all the inputs of the gate be high. Then the BJTs Q_1, Q_2, and Q_3 will be in the reverse-active, saturation, and saturation modes, respectively, so that the output voltage can be

$$v_Y = v_{C3} = V_{CE3,sat} = 0.2\,[V]\,(\text{low}) \tag{3.3.49}$$

What difference does the additional BJT(Q_4)-diode(D) pair make in comparison with R_C alone? It is expected to cut off the current i_{RC} so that R_C can dissipate no power during the low state of the output. Such an expectation comes true because the voltage difference between B4 and C3 (or Y)

$$v_{B4} - v_{C3} = v_{C2} - v_{C3} = V_{BE3,sat} + V_{CE2,sat} - V_{CE3,sat} = 0.8 + 0.2 - 0.2 = 0.8\,[V] \tag{3.3.50}$$

is not high enough to turn on Q_4-D. (It would be not the case without D.)

Now, suppose that at least one of the inputs becomes low. Then the BJTs Q_1, Q_2, and Q_3 operate in the saturation, cutoff, and cutoff modes, respectively, so that the voltage $v_{B4} = v_{C2}$ can be pulled up high via R_2 enough to turn on (saturate) Q_4-D where the output voltage (across C_L) will remain at 0.2 V for the moment since the capacitor voltage cannot change instantaneously. Then C_L will be charged by the emitter current of Q_4 (saturated)

$$\begin{aligned} i_{E4} &= i_{B4} + i_{C4} = i_{R2} + i_{RC} = \frac{V_{CC} - (v_{C2} = v_{B4})}{R_2} + \frac{V_{CC} - v_{C4}}{R_C} \\ &= \frac{V_{CC} - (v_Y + V_{D,on} + V_{BE4,sat})}{R_2} + \frac{V_{CC} - (v_Y + V_{D,on} + V_{CE4,sat})}{R_C} \\ &= \frac{5 - 1.7}{1.4\,k\Omega} + \frac{5 - 1.1}{0.1\,k\Omega} = 2.36 + 39 = 41.36\,[mA] \end{aligned} \tag{3.3.51}$$

$$(i_{C4} < \beta_F i_{B4} \text{ ensures that } Q_4 \text{ is saturated in the meantime})$$

till i_{E4} becomes almost zero and accordingly, Q_4 and D are just at the cut-in condition with v_{BE4}=$V_{BE,offset}$=0.6 and v_D=$V_{D,offset}$=0.5 so that the output will reach

$$v_Y = V_{CC} - V_{BE4,offset} - V_{D,offset} = 5 - 0.6 - 0.5 = 3.9\,[V]\,(\text{high}) \tag{3.3.52}$$

(Q4) What is the role of the BJT Q_4 placed atop Q_3?

(A4) • Static aspect: when the output voltage is high, Q_4 can afford more load current with a smaller current through and voltage drop across R_2 (with a much smaller output resistance) so that the output-high fan-out can be increased.

- Dynamic aspect: when the output changes from low (with Q_3 saturated) to high (with Q_3 cutoff), Q_4 transfers from cutoff to saturation one jump ahead of the transfer of Q_3 so that it can supply current to C_L (charged to 0.2 V) as a source, reducing the low-to-high switching time.

(Q5) What is the role of the diode D placed between the two BJTs Q_3 and Q_4?
(A5) Without D, the voltage $v_{B4} - v_{C3} = 0.8$ V (Eq. (3.3.50)) will turn on Q_4 alone when the output voltage is low as $v_Y = v_{C3} = V_{CE3,sat} = 0.2$ V so that the collector current i_{C4} flowing through R_C may result in a power dissipation.

(Q6) What is the role of R_C?
(A6) With $R_2 = 0$, the current determined by Eq. (3.3.51) would increase to reduce the low-to-high switch time. However, when Q_4 turns on before Q_3 turns off, the supply voltage V_{CC} would be short-circuited through C4-E4-D-C3-E3, possibly damaging Q_4, D, and Q_3. Therefore, R_C is needed to limit such current spikes.

(Q7) Is there any disadvantage of the totem-pole output stage?
(A7) Yes. The disadvantage is a lower output voltage (Eq. (3.3.52)) corresponding to logic 1.

1) Output-high Fan-out

In Figure 3.51, suppose the output voltage of the driver gate is high, i.e. $v_Y = 3.9$ V with $Q_2/Q_3/Q_4$ (in the current stage) cutoff/cutoff/saturated~cut-in, which lets the load current of $i_L = NI_L^1 = N \times 0.0675\,\text{mA}$ (Eq. (3.3.43)) flow into N load gates in the next stage. The BJT Q_4 is supposed to use its emitter current $i_{E4} = \beta_F i_{B4}$ to supply this load current while the output voltage obtained by subtracting the voltage drops $R_2 i_{R2} = R_2 i_{B4}$, $V_{BE4,sat}$, and $V_{D,on}$ from V_{CC} should be higher than the minimum output voltage 3 V (corresponding to logic 1):

$$v_Y = V_{CC} - R_2 i_{R_2} - V_{BE4,act} - V_{D,on} = 5 - 1.4 i_{B4} - 0.7 - 0.7 = 3.6 - 1.4\frac{i_{E4}}{\beta_F + 1}$$

$$= 3.6 - 1.4\frac{i_L}{\beta_F + 1} \overset{(3.3.43)}{=} 3.6 - 1.4\frac{0.0675N}{21} = 3.6 - 0.0045N > 3\,[\text{V}]$$

$$;N < \frac{3.6 - 3}{0.0045} = 133.3 \rightarrow 133 \tag{3.3.53}$$

Therefore, the output-high fan-out of the TTL NAND gate with a totem-pole output stage of Figure 3.51 is 133. Compare this with Eq. (3.3.45) for the NAND gate without the totem-pole output stage.

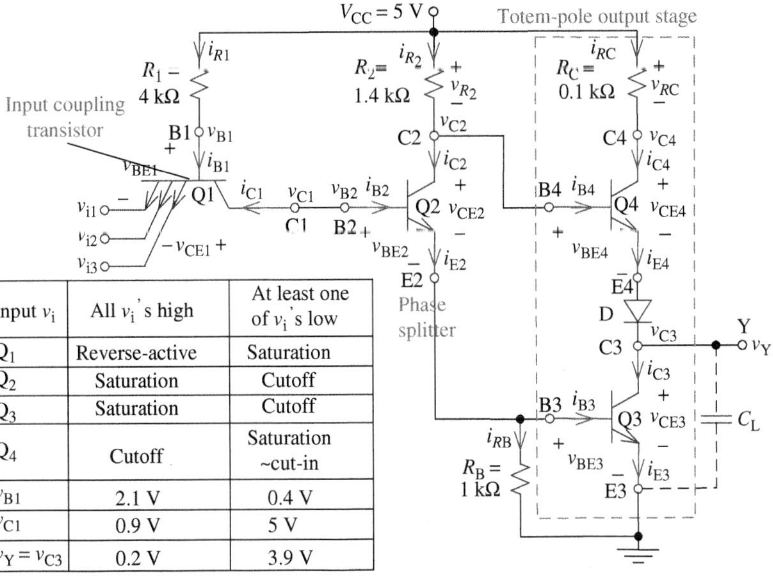

Input v_i	All v_i's high	At least one of v_i's low
Q_1	Reverse-active	Saturation
Q_2	Saturation	Cutoff
Q_3	Saturation	Cutoff
Q_4	Cutoff	Saturation ~cut-in
v_{B1}	2.1 V	0.4 V
v_{C1}	0.9 V	5 V
$v_Y = v_{C3}$	0.2 V	3.9 V

Figure 3.51 A TTL NAND gate with a totem-pole output stage.

2) Output-low Fan-out

In Figure 3.51, suppose the output voltage of the driver gate is low, i.e. $v_Y = V_{CE,sat} = 0.2$ V with $Q_2/Q_3/Q_4$ (in the current stage) saturated/saturated/cutoff. In order for the saturation mode of Q_3 not to be disturbed by the load current (flowing from the load gate), the condition described by Eq. (3.3.47) or (3.3.48) with $i_{RC} = 0$ should be satisfied:

$$i_{C3} = i_{R_C} - NI_L^1 \overset{i_{RC}=0}{\underset{(3.3.46)}{=}} N\,[\text{mA}] \overset{?}{<} \beta_F i_{B3} \overset{(3.3.40)}{=} 20 \times 2.8025 = 56.05\,[\text{mA}]$$
$$;N \le 56 \qquad (3.3.54)$$

Compare this with Eq. (3.3.48) for the NAND gate without the totem-pole output stage.

3) PSpice Simulation of the TTL NAND Gate with a Totem-Pole Output Stage

Figure 3.52(a)/(b) shows the PSpice schematic and its simulation result (for Transient Analysis with maximum stepsize 1 ns) of the TTL NAND gate with a totem-pole output stage depicted in Figure 3.51. Here are several observations about the simulation result (Figure 3.52(b)) where one of the

(a)

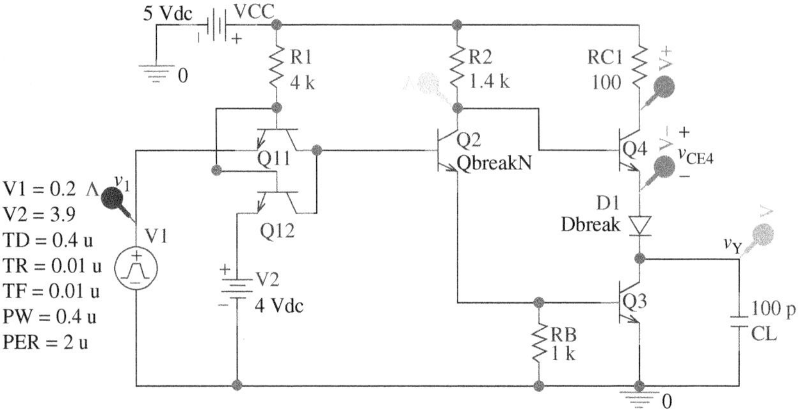

PSpice schematic for the TTL NAND gate with a totem-pole output stage ("nand_ttl_totem.opj")

(b)

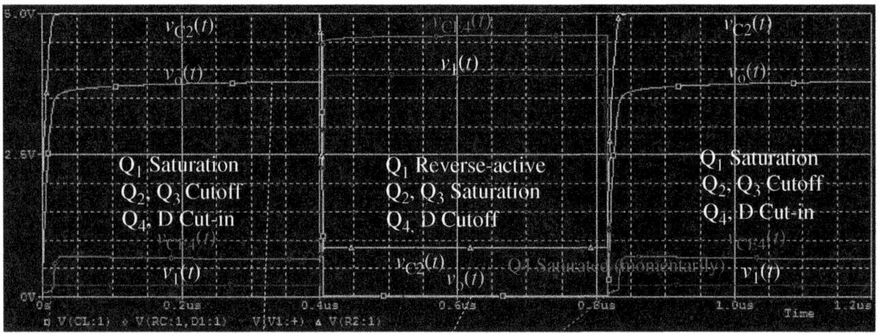

PSpice simulation result

Figure 3.52 PSpice simulation of the TTL NAND gate with a totem-pole output stage depicted in Figure 3.51

two input voltages is fixed as 4 V(HIGH) and the other $v_1(t)$ is a rectangular pulse plotted as a green line:

- When the input $v_1(t)$ is LOW/HIGH, the output $v_Y(t)$ is HIGH/LOW. This implies that the circuit works fine as a NAND gate.
- When at least one of the inputs, say, $v_1(t)$ plotted as a blue line is LOW (0.2 V), $Q_1/Q_2/Q_3/Q_4$/D are in the saturation/cutoff/cutoff/cut-in/cut-in modes, respectively, but Q_4 and D momentarily enter the saturation mode with $v_{CE4} \leq 0.2$ V and $v_D \geq 0.75$ V, respectively, (before cut-in) to charge C_L

during the LOW-to-HIGH transition time of $v_Y(t)$. After transients, Q_4 and D are in the cut-in mode where $v_{CE4} = 0.66$ V, $v_{BE4} = 0.66$ V, $v_D = 0.55$ V, and $i_D - 15\ \mu A$.

- When all the inputs are HIGH (3.9 V), $Q_1/Q_2/Q_3/Q_4$/D are in the reverse-active/saturation/saturation/cutoff/cutoff modes, respectively, where $v_{BC1} = 0.75$ V, $v_{BE2} = 0.82$ V, $v_{CE2} = 0.038$ V, $v_{BE3} = 0.82$ V, $v_{CE3} = 0.018$ V, $v_{BE4} \le 0.5$ V, $v_{CE4} = 4.6$ V, and $v_D \le 0.4$ V.
- The high output voltage is 3.8 V, being close to 3.9 V predicted by Eq. (3.3.52), but the low output voltage is 0.018 V, being lower than $V_{CE3,sat} = 0.2$ V (predicted by Eq. (3.3.49)).

3.3.2.4 Open-Collector Output and Tristate Output

The output impedance of the TTL NAND gate (with totem-pole output stage) is very low irrespective of whether its output is HIGH or LOW. In most cases, the low output impedance is desired because it contributes towards improving the fan-out capability by reducing the loading effect. However, in the case of bus contention where different gates attempt to drive a wired-OR output (as depicted in Figure 3.53(a)) into different logic states, the low output impedance is not good because it may cause an excessive current to flow from HIGH-output gates to LOW-output gates. Against such a happening, open-collector outputs (as depicted in Figure 3.53(b)) can be used where each one of the gates with an open-collector output drives the wired-OR output LOW if it wants a LOW output; otherwise it lets its output float (leaving up to other gates' decision) so that the wired-OR output can be pulled up HIGH (via an external pull-up resistor connected to VCC) only when no gate pulls down the output by asserting LOW.

Another measure against bus contention is to use the tristate output illustrated in Figure 3.53(c) where if the (low-active) Disable input $\overline{\text{Dis}}$ is 0.2 V (LOW), the BJTs Q_1, Q_2, and Q_3 are in the saturation, cutoff, and cutoff modes, respectively, and the diode D is ON so that $v_{B4}=v_{Dis}+v_{D,ON}=0.2+0.7=0.9$[V]. This voltage will turn on just Q_4-R_4 so that $v_{B5}=v_{B4}-0.7=0.2$[V]. This voltage is not sufficient to turn on Q_5. Thus, both Q_3 and Q_5 are OFF so that the gate can let its output float with a very high output resistance when $\overline{\text{Dis}}$ is low. Otherwise, i.e. if $\overline{\text{Dis}}$ is high, the gate performs a usual NAND function. This gate is said to have a tristate or three-state output because it presents three outputs, i.e. HIGH, LOW, and FLOAT (high impedance) states. The $\overline{\text{Dis}}$ input is used to select only one gate among the gates that are wired-OR connected. Note that the tristate outputs must be pulled up or down to keep the output from being floated, i.e. indeterminate or in the high impedance (Hi-Z) state when all the drivers are disabled.

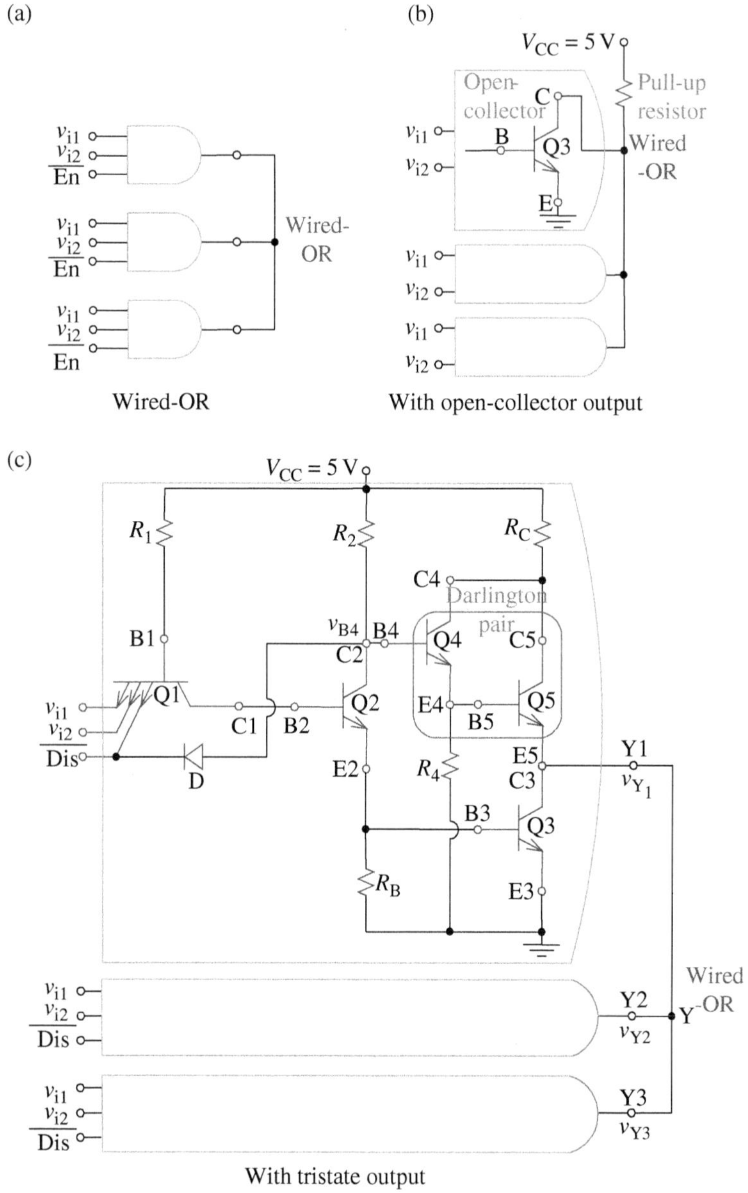

Figure 3.53 Tristate output and open-collector output for wired-OR connection

3.3.3 ECL (Emitter-Coupled Logic) OR/NOR Gate

Figure 3.54 shows an ECL (Emitter-Coupled Logic) OR/NOR gate, which consists of a differential amplifier using emitter-coupled pair (Section 3.1.11), a reference voltage supplier, and a level shifter. Compared with the TTL family, the ECL family achieves very fast switching and short propagation delays by keeping the BJTs in the forward-active mode (away from the saturation mode) so that there can be no excessive stored charge to remove quickly. However, it exhibits more power consumption, reduced noise margin, smaller voltage swing, and higher fan-out capability.

Here is a rough analysis of the ECL circuit in Figure 3.54, which is designed to have all the BJTs operate in the forward-active mode although the two input BJTs Q_1 and Q_2 may be saturated when the corresponding input voltage v_{i1}/v_{i2} is much higher than $V_{ref} = 2.25$ [V] supplied to the base B3 of Q_3. The reference voltage supplier consisting of R_{C4}-Q_4-R_4-Q_8-R_8 supplies the nodes B3 and B7 with

$$v_{B4} = V_{CC} - \frac{R_{C4}}{R_{C4} + R_4 + R_8}(V_{CC} - v_{BE4} - v_{BE8}) = 3.3 - \frac{2.1}{2.1 + 1.5}(3.3 - 1.5)$$
$$= 2.25\,[\text{V}] = V_{ref} \qquad (3.3.55a)$$

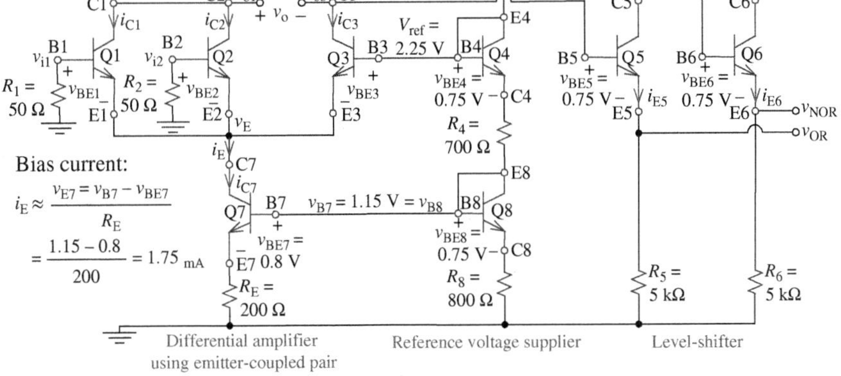

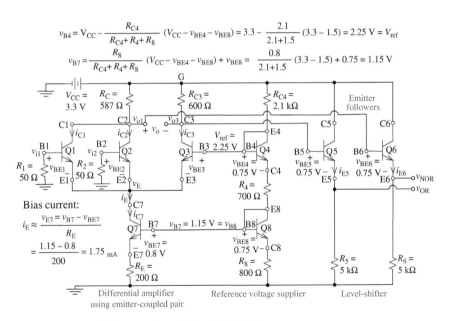

Figure 3.54 An ECL (Emitter-Coupled Logic) OR/NOR gate.

$$v_{B7} = \frac{R_8}{R_{C4}+R_4+R_8}(V_{CC}-v_{BE4}-v_{BE8}) + v_{BE8} = \frac{0.8}{2.1+1.5}(3.3-1.5)+0.75$$
$$= 1.15\,[\mathrm{V}] \tag{3.3.55b}$$

The BJT Q_7, given $v_{B7} = 1.15$ [V] at its base terminal B7, produces its emitter current

$$i_{R_E} = \frac{v_{B4}-v_{BE4}}{R_E} = \frac{1.15-0.8}{0.2} = 1.75\,[\mathrm{mA}] \tag{3.3.56}$$

so that Q_7 makes the current $i_{C7} = \alpha_F i_{RE}/(1+\alpha_F) \approx 1.75$ [mA] through $Q_1|Q_2|$ Q_3-Q_7-R_E like a current source of 1.75 [mA]. This current $i_{C7} \approx 1.75$ [mA] will flow through R_C (with Q_1 and/or Q_2 in the forward-active mode and Q_3 in the cutoff mode) if any one of the two inputs v_{i1} and v_{i2} is higher than $v_{B4} = 2.25$[V] $= V_{ref}$; otherwise, i.e. none of the inputs is higher than $v_{B4} = 2.25$[V] $= V_{ref}$, the current $i_{C7} \approx 1.75$ [mA] will flow through R_{C3} (with Q_1 and Q_2 in the cutoff mode and Q_3 in the forward-active mode). In the former case, the emitter voltages of Q_5 and Q_6 will be

$$v_{OR} = V_{CC} - v_{BE5} \approx 3.3-0.75 = 2.55\,[\mathrm{V}] \tag{3.3.57a}$$
$$v_{NOR} = V_{CC} - R_C i_E - v_{BE6} \approx 3.3-0.587\times 1.75-0.75 = 1.52\,[\mathrm{V}] \tag{3.3.57b}$$

In the latter case, the emitter voltages of Q_5 and Q_6 will be

$$v_{OR} = V_{CC} - R_{C3} i_E - v_{BE5} \approx 3.3-0.6\times 1.75-0.75 = 1.5\,[\mathrm{V}] \tag{3.3.58a}$$
$$v_{NOR} = V_{CC} - v_{BE6} \approx 3.3-0.75 = 2.55\,[\mathrm{V}] \tag{3.3.58b}$$

As an exception, when the input voltage v_{i1}/v_{i2} is much higher than $V_{ref} =$ 2.25[V] (supplied to the base B3 of Q_3), the corresponding BJT $Q_1|Q_2$ will be saturated with $v_{CE} \le 0.2$[V] so that v_{NOR} will rise with the input voltage v_i:

$$v_{NOR} = v_i - v_{BEi} + v_{CEi} - v_{BE6}$$
$$= v_i - 0.8 + (0-0.2) - 0.75 = v_i - (1.35-1.55)\,[\mathrm{V}] \tag{3.3.59}$$

This rough analysis results are supported by the VTC that is obtained from the PSpice simulation (with DC Sweep analysis type) and depicted in Figure 3.55 (b1). Note that the input voltages v_{i1}/v_{i2} applied to the ECL gates should never be much higher than $V_{ref} = 2.25$ [V] to keep the BJTs $Q_1|Q_2$ from being saturated.

(Q8) What is the role of R_1/R_2 each connected in parallel with the input v_{i1}/v_{i2}?

(A8) It provides a matching termination impedance for a 50Ω-transmission line feeding the input. It will also pull down an input terminal which is connected to no gate so that indeterminate logic levels and noise can be prevented.

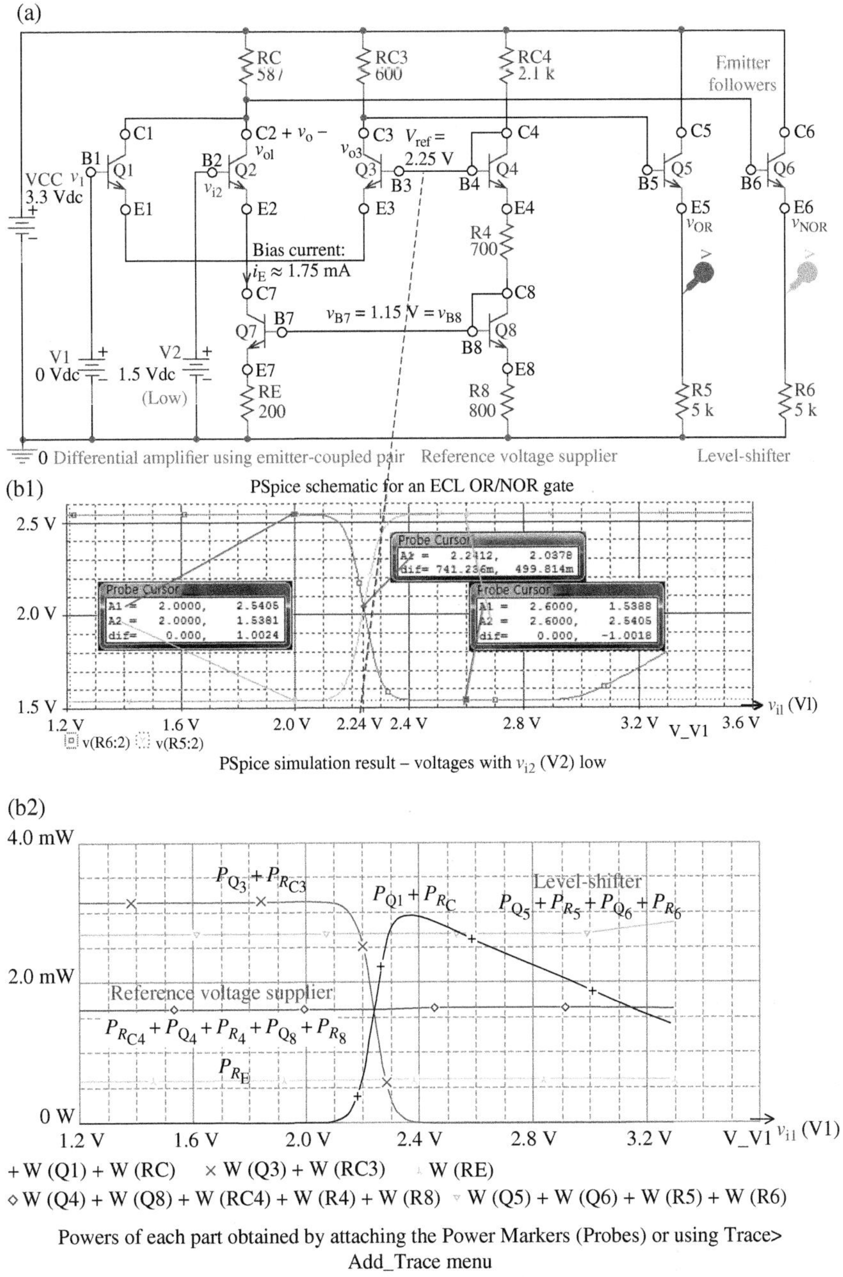

Figure 3.55 PSpice simulation of an ECL OR/NOR gate ("or_ecl3.opj").

(Q9) Why is the VTC for v_{OR}=v_{C3} − 0.75 symmetric while that for v_{NOR} = v_{C2} − 0.75 is asymmetric?

(A9) Because Q_3 involved in producing v_{C3} is driven by the fixed reference voltage V_{ref} = 2.25[V], while $Q_1|Q_2$ involved in producing v_{C2} are driven by the varying input voltage v_i, which may saturate the corresponding BJT $Q_1|Q_2$ if it becomes much higher than V_{ref} = 2.25[V] (supplied to the base B3 of Q_3).

(Q10) Why is R_C = 587 Ω (connected with $Q_1|Q_2$) slightly smaller than R_{C3} = 600 Ω (connected with Q_3)?

(A10) R_C has been made slightly smaller than R_{C3} to have $R_C i_{E,2}$ (with $v_i > V_{ref} \rightarrow$ $Q_1|Q_2$ ON and Q_3 OFF $\rightarrow v_{NOR} = V_{CC} - R_C i_{E,2} - v_{BE6}$) close to $R_{C3} i_{E,3}$ (with $v_i < V_{ref} \rightarrow Q_1|Q_2$ OFF and Q_3 ON $\rightarrow v_{OR} = V_{CC} - R_{C3} i_{E,3} - v_{BE5}$) where i_{E2} of $Q_1|Q_2$ driven by $v_i > V_{ref}$ is slightly larger than i_{E3} of Q_3 driven by V_{ref} .

Figure 3.55(b2) shows the power dissipations of each part that are obtained by attaching the Power Markers (Probes) to the parts or using the Trace>Add_Trace menu. For example, to see the sum of the powers dissipated by the reference voltage supplier (consisting of R_{C4}-Q_4-R_4-Q_3-R_{C4}), you should type the following expression into the Trace Expression field in the lower part of the Add Traces dialog box opened by selecting the Trace>Add Trace menu in the PSpice A/D (Probe) Window (see Appendix D.2.8):

```
W(RC4)+W(Q4)+W(R4)+W(Q8)+W(R8)
```

The power curve of $Q_1|Q_2$ plotted in gray tells us that $Q_1|Q_2$ will dissipate less power if they transit from the forward-active mode into the saturation mode as the corresponding input voltages applied to their base terminals become much higher than V_{ref} = 2.25[V].

3.4 Design of BJT Amplifier

This section will show how a CE amplifier (Figure 3.30(a)) with a desired voltage gain $A_{v,d}$ or CC amplifier (Figure 3.32(a)) with a desired input resistance $R_{i,d}$ can be designed, i.e. how the values of resistors constituting the circuits can be determined to satisfy the design specification.

3.4.1 Design of CE Amplifier with Specified Voltage Gain

To maximize the AC swing of output voltage v_o along the AC load line (see Figure 3.56.2) of CE amplifier (Figure 3.56.1(a)), it may be good to set the

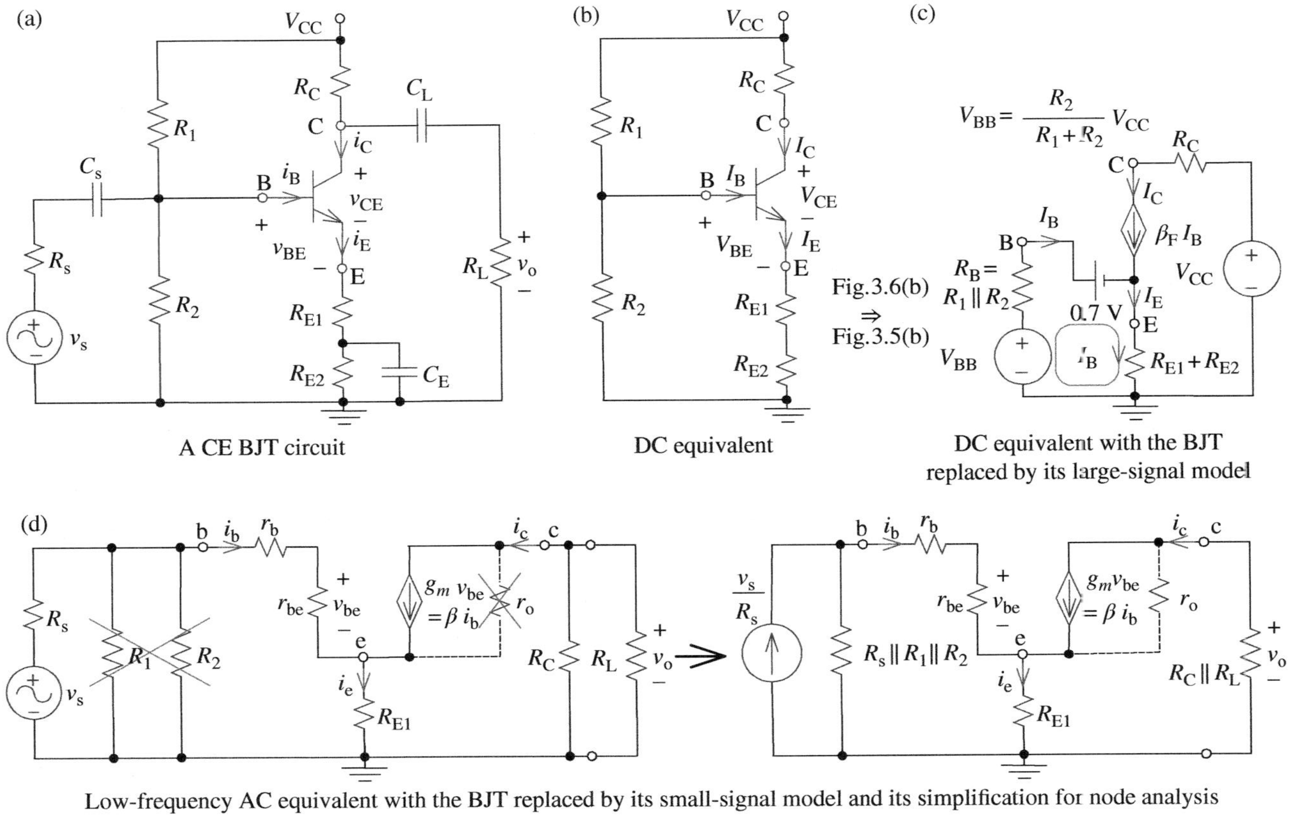
(a)
A CE BJT circuit
(b)
DC equivalent
(c)
Fig.3.6(b) ⇒ Fig.3.5(b)
DC equivalent with the BJT replaced by its large-signal model
(d)
Low-frequency AC equivalent with the BJT replaced by its small-signal model and its simplification for node analysis

Figure 3.56.1 A CE (common-emitter) amplifier and its DC/AC equivalents (see Figure 3.19.1).

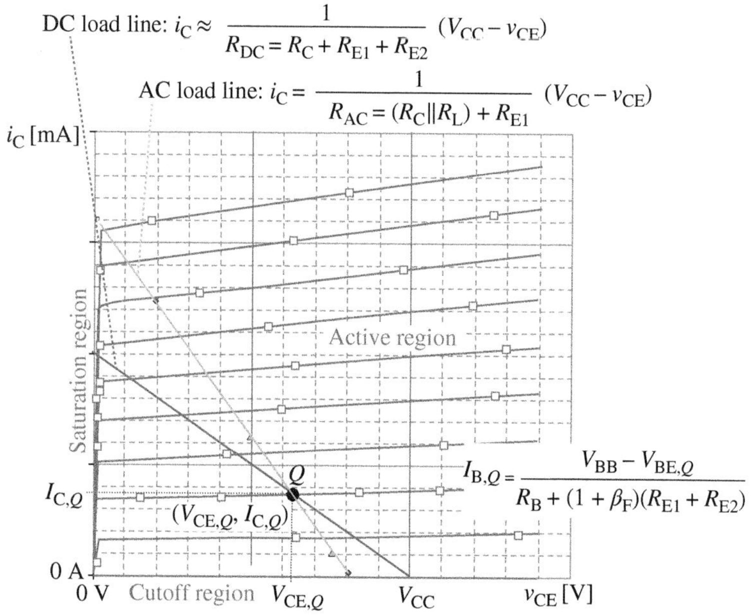

Figure 3.56.2 DC and AC load lines for the CE (common-emitter) amplifier in Figure 3.56.1(a) (see Figure 3.22).

collector current $I_{C,Q}$ and collector-to-emitter voltage $V_{CE,Q}$ of BJT at the operating point Q as half the maximum collector current and about one third of V_{CC}, respectively, and also to determine such a value of R_C that the i_C intercept of the AC load line can be about $2I_{C,Q}$:

$$I_{C,Q} = \frac{I_{C,\max}}{2},\ V_{CE,Q} = K_C V_{CC} \text{ with } K_C \approx \frac{1}{3} (\text{design constant}) \tag{3.4.1}$$

$$I_{C,Q} + \frac{V_{CE,Q}}{R_{AC} \approx (R_C \| R_L)} = 2I_{C,Q};\ R_{AC} \approx \frac{1}{1/R_C + 1/R_L} = \frac{V_{CE,Q}}{I_{C,Q}}$$
$$;\ R_C = \frac{1}{I_{C,Q}/V_{CE,Q} - 1/R_L} \tag{3.4.2}$$

Here, R_{E1}, which is a part of the AC resistance $R_{AC} \overset{(3.1.45)}{=} R_{E1} + (R_C \| R_L)$, has been neglected because it is presumably much less than $(R_C \| R_L)$ as well as has not yet been determined.

Then, in the DC equivalent circuit, the emitter and collector resistors are supposed to share the voltage drop $V_{CC} - V_{CE,Q} = (1 - K_C) V_{CC}$:

$$R_E I_{E,Q} + R_C I_{C,Q} \overset{I_{E,Q}\approx I_{C,Q}}{\approx} (R_E + R_C) I_{C,Q} = (1-K_C)V_{CC}$$
$$; \ R_E = (1-K_C)\frac{V_{CC}}{I_{C,Q}} - R_C \tag{3.4.3}$$

where

$$I_{E,Q} \overset{(3.1.6c,d)}{=} \frac{\beta_F + 1}{\beta_F} I_{C,Q}, \ I_{B,Q} \overset{(3.1.6c)}{=} \frac{1}{\beta_F} I_{C,Q} \tag{3.4.4}$$

Now, to determine the resistances R_1 and R_2 of DC biasing circuit (Figure 3.56.1(b)), we let the equivalent base resistance R_B=$R_1||R_2$ be 1/10 times the equivalent emitter resistance $(\beta_F$+$1)R_E$:

$$R_B = R_1 || R_2 \approx \frac{1}{10}(\beta_F + 1)R_E \tag{3.4.5}$$

Then, the equivalent base voltage source V_{BB} and the biasing resistances R_1 and R_2 are determined as follows:

$$V_{BB} = V_{E,Q} + V_{BE,Q} + R_B I_{B,Q} = R_E I_{E,Q} + 0.7 + R_B I_{C,Q}/\beta_F \tag{3.4.6}$$

$$R_1 = \frac{R_1 R_2/(R_1+R_2)}{R_2/(R_1+R_2)} = \frac{R_B}{V_{BB}/V_{CC}} \tag{3.4.7a}$$

$$R_2 = \frac{1}{1/R_B - 1/R_1} = \frac{R_B}{(V_{CC} - V_{BB})/V_{CC}} \tag{3.4.7b}$$

Now, with the transconductance g_m and base-emitter resistance r_{be}

$$g_m \overset{(3.1.26)}{=} \left.\frac{|I_{C,Q}|}{V_T}\right|_{T=273+27^\circ K} = \frac{|I_{C,Q}|}{300/11600} = \frac{|I_{C,Q}|}{25.9\,\text{mV}} \text{ and } r_{be} \overset{(3.1.27)}{=} \frac{\beta}{g_m} \tag{3.4.8}$$

Eq. (3.2.3) is used to determine the dual emitter resistance $[R_{E1}, R_{E2}]$ so that the desired voltage gain $A_{v,d}$ can be achieved:

$$A_v = \frac{v_o}{v_s} = -\frac{R_i}{R_s + R_i}\frac{\beta(R_C||R_L)}{r_b + r_{be} + (\beta+1)R_{E1}} \overset{R_s=0}{=} -\frac{\beta(R_C||R_L)}{r_b + r_{be} + (\beta+1)R_{E1}} = A_{v,d}$$

$$; R_{E1} = \frac{1}{\beta+1}\left\{\frac{\beta(R_C||R_L)}{|A_{v,d}|} - r_b - r_{be}\right\}, \ R_{E2} = \overset{(3.4.3)}{R_E} - R_{E1} \tag{3.4.9}$$

Note that the minimum power ratings of R_1, R_2, R_C, R_{E1}, and R_{E2} should be

$$P_{R_1}=\frac{(V_{CC}-V_{BB})^2}{R_1}, P_{R_2}=\frac{V_{BB}^2}{R_2}, P_{R_C}=R_C I_{C,Q}^2, \text{ and } P_{R_{Ei}}=R_{Ei} I_{E,Q}^2 \tag{3.4.10}$$

This procedure of designing a CE amplifier with a specified voltage gain $A_{v,d}$ gain has been cast into the following MATLAB function 'BJT_CE_design()' where the default values of design constant Kc and ambient temperature T are set to 1/3 and 27 [°C], respectively.

```
function [R1,R2,RC,RE1,RE2,PRs]= ...
  BJT_CE_design(VCC,beta,rb,AvdRB,ICQ,RL,T,Kc)
% Design a CE amp with given Avd (possibly & RB) at Q=(VCEQ=KC*VCC,ICQ).
% Avd_RB=Avd or [Avd RB]
% Output: PRs=[PR1 PR2 PRC PRE1 PRE2]: Power ratings of R1,R2,RC,RE1,RE2
% Copyleft: Won Y. Yang, wyyang53@hanmail.net, CAU for academic use only
if nargin<8, Kc=1/3; end % Design constant s.t. VCQ=Kc*VCC;
if nargin<7, T=27; end % Ambient temperature
betaF=beta(1);
if numel(beta)>1, betaAC=beta(2); else betaAC=betaF; end
VCEQ=Kc*VCC; % Eq.(3.4.1)
RC=max(1/(ICQ/VCEQ-1/RL),10); RE=(1-Kc)*VCC/ICQ-RC; % Eq.(3.4.2,3)
IEQ=(betaF+1)*ICQ/betaF; IBQ=ICQ/betaF; % Eq.(3.4.4)
Avd=AvdRB(1);
if numel(AvdRB)>1, RB=AvdRB(2);
else RB=(betaF+1)*RE/10; end % Eq.(3.4.5)
VBEQ=0.7; VBB=RE*IEQ+VBEQ+RB*IBQ; % Eq.(3.4.6)
R1=RB/VBB*VCC; R2=RB/(VCC-VBB)*VCC; % Eq.(3.4.7a,b)
gm=ICQ*11605/(273+T); rbe=betaF/gm; % Eq.(3.4.8)
RE1=max((betaAC*parallel_comb([RC RL])/abs(Avd)-rb-rbe)/(betaAC+1),0);
RE2=RE-RE1; % Eq.(3.4.9)
PR1=(VCC-VBB)^2/R1; PR2=VBB^2/R2; PRC=RC*ICQ^2;
PRE1=RE1*IEQ^2; PRE2=RE2*IEQ^2; % Eq.(3.4.10)
PRs=[PR1 PR2 PRC PRE1 PRE2]; % Power ratings of R1,R2,RC,RE1, and RE2
if RE2<10
 if RC>10&(abs(Avd)<betaAC*parallel_comb([RC RL])/(rb+rbe))
  disp('Try again with smaller/larger values of ICQ/VCC')
  [R1,R2,RC,RE1,RE2,PRs]= ...
         BJT_CE_design(VCC,beta,rb,AvdRB,0.9*ICQ,RL,T,Kc);
 else error('Try with a higher VCC or another TR having a larger beta.')
 end
else
 disp('Design Results')
 disp('     R1       R2       RC       RE1        RE2         Avd')
 fprintf('%8.0f %8.0f %8.0f %8.0f %8.0f %8.0f\n', R1,R2,RC,RE1,RE2,Avd)
end
```

```
%design_CE_BJT.m
betaF=189; betaR=6; betaAC=189; rb=10; VA=100;
VCC=18; Vsm=0.1; Rs=50; RL=10000;
Avd=20; ICQ=0.02; % Design parameters
[R1,R2,RC,RE1,RE2]=BJT_CE_design(VCC,[betaF betaAC],rb,Avd,ICQ,RL);
[VBQ,VEQ,VCQ,IBQ,IEQ,ICQ,Av]=BJT_CE_analysis(VCC,rb,Rs,R1,R2,RC,...
  [RE1 RE2],RL,[betaF,betaR,betaAC],Vsm,VA);
```

To use the function for designing a CE BJT (Q2N2222) amplifier with voltage gain $A_{v,\mathrm{d}} = -20$, we run the above MATLAB script "design_CE_BJT.m," which yields

```
>>design_CE_BJT
 Design Results
    R1          R2        RC        RE1        RE2        Avd
   13945       9147      309        14         277         20
 Results of analysis using the PWL model
  VCC    VEE   VBB   VBQ   VEQ  VCQ       IBQ        IEQ        ICQ
  18.00  0.00  7.13  6.55  5.85 11.81 1.06e-04  2.01e-02  2.00e-02
  gm= 773.667[mS], rbe=  244[Ohm], ro=  5.00[kOhm]
  Gv=Ri/(Rs+Ri)xAv= 0.974 x -19.88 = -19.37
```

These results mean the following values of the resistances of designed CE amplifier:

$$R_1 = 13945\,\Omega, R_2 = 9147\,\Omega(\text{open}), R_C = 309\,\Omega(\text{short}), R_{E1} = 14\,\Omega, \text{and}\, R_{E2} = 277\,\Omega \quad (3.4.11)$$

where the BJT parameters, the source/load resistances, and V_{CC} are given as

$$\beta_F = 189, \beta_R = 6, \beta_{AC} = 189, r_b = 10\,\Omega, R_s = 50\,\Omega, R_L = 10\,\mathrm{k}\Omega, \text{and}\, V_{CC} = 18\,\mathrm{V} \quad (3.4.12)$$

The results of using 'BJT_CE_analysis()' (see Section 3.2.1) to analyze the designed circuit are

$$V_{B,Q} = 6.55\,\mathrm{V}, V_{C,Q} = 11.81\,\mathrm{V}, V_{E,Q} = 5.85\,\mathrm{V}, I_{B,Q} = 106\,\mu\mathrm{A}, I_{E,Q} = 20.1\,\mathrm{mA}, \text{and}\, I_{C,Q} = 20\,\mathrm{mA}$$

with the expected voltage gain −19.37, transconductance $g_m = 0.774$, and BEJ resistance $r_{be} = 244\,\Omega$.

Figure 3.57(a) and (b) shows the PSpice schematic of the designed CE amplifier and its simulation results. Although the resulting voltage gain $A_{v,\mathrm{PSpice}} = -3.7172/0.2 = -18.6$ is somewhat smaller than $A_{v,\mathrm{d}} = -20$ required by the design specification, the difference is not so big as to damage the reliability of the MATLAB design and analysis functions.

(a)

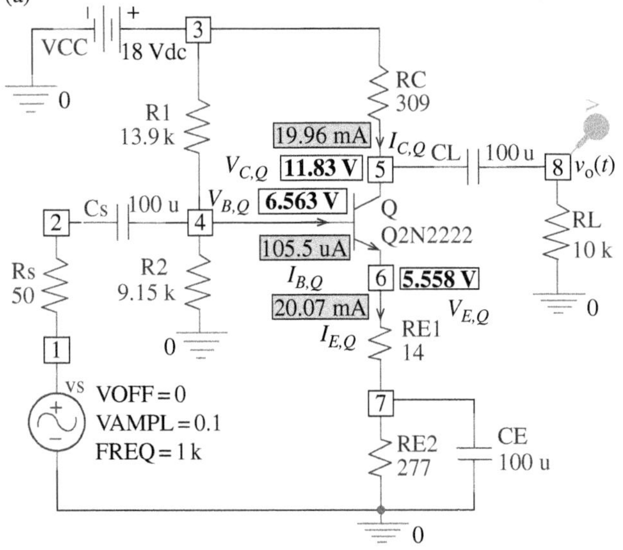

PSpice schematic for a CE BJT circuit

(b)

Cursor box

Trace Name	Y1	Y2	Y1–Y2
X Values	1.7363 m	2.2308 m	–494.500 u
V(CL:2)	1.8940	–1.8232	3.7172

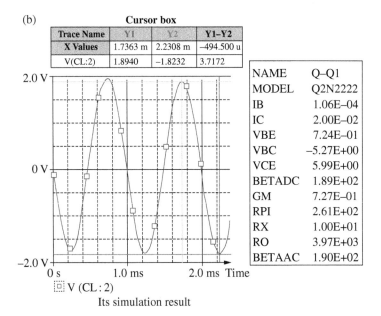

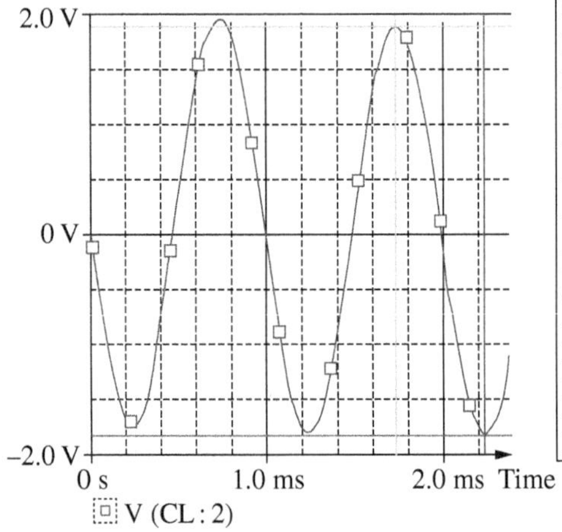

NAME	Q–Q1
MODEL	Q2N2222
IB	1.06E–04
IC	2.00E–02
VBE	7.24E–01
VBC	–5.27E+00
VCE	5.99E+00
BETADC	1.89E+02
GM	7.27E–01
RPI	2.61E+02
RX	1.00E+01
RO	3.97E+03
BETAAC	1.90E+02

Its simulation result

Figure 3.57 A CE circuit and its PSpice simulation ("elec03f53.opj").

3.4.2 Design of CC Amplifier (Emitter Follower) with Specified Input Resistance

Let us consider how to determine the values of resistors constituting the CC amplifier (Figure 3.58(a)) such that the input resistance (Eq. (3.2.5)) has a given value $R_{i,d}$.

$$R_i \overset{(3.2.5)}{=} \frac{1}{1/R_B + 1/\{(\beta+1)(R_E\|R_L) + r_b + r_{be}\}}$$
$$= R_B\|\{(\beta+1)(R_E\|R_L) + r_b + r_{be}\} = R_{i,d} \tag{3.4.14}$$

To maximize the AC swing of output voltage v_o, it may be good to let the C-E junction and R_E share the applied voltage V_{CC} half and half (at the operating point Q) in the DC equivalent circuit (Figure 3.58(b) or (c)) so that

$$V_{CE,Q} = \frac{V_{CC}}{2} \text{ and } V_{E,Q} = R_E I_{E,Q} \approx R_E I_{C,Q} = \frac{V_{CC}}{2} \tag{3.4.15a,b}$$

KVL around the V_{BB}-R_B-0.7 V-V_E loop yields

$$V_{BB} - R_B I_{B,Q} - 0.7 = V_{E,Q} \overset{(3.4.15b)}{=} \frac{V_{CC}}{2} \tag{3.4.16}$$

From Eqs. (3.4.15b) and (3.4.16), we can have the expressions of R_E and R_B in terms of $I_{C,Q}$ as

```
function [R1,R2,RE,PRs]=BJT_CC_design(VCC,beta,rb,RidVBB,RL,T,ICmax)
%Design a CC amp with given Rid (possibly & VBB) at Q=(VCEQ=KC*VCC,ICQ).
% Output: PRs=[PR1 PR2 PRE] : Power ratings of R1, R2, RE
% Copyleft: Won Y. Yang, wyyang53@hanmail.net, CAU for academic use only
if nargin<7, ICmax=0.8; end % Maximum collector current
if nargin<6, T=27; end % Ambient temperature
betaF=beta(1);
if numel(beta)>1, betaAC=beta(2); else betaAC=betaF; end
Rid=RidVBB(1); if numel(RidVBB)>1, VBB=RidVBB(2); else VBB=VCC; end
IC=[1/100:1/1000:1]*ICmax; % Range of collector current
REs=VCC/2./IC; %Eq.(3.4.17a)
RBs=betaF*(VCC/2-0.7)./IC; % Eq.(3.4.17b)
gms=IC*11605/(273+T); rbes=betaF./gms; % Eq.(3.4.8)
Ris=(betaAC+1)*REs*RL./(REs+RL)+rb+rbes; % Eq.(3.2.5)
[tmp,i]=min(abs(RBs.*Ris./(RBs+Ris)-Rid)); % Eq.(3.4.14)
ICQ=IC(i); IBQ=ICQ/betaF; IEQ=(betaF+1)*IBQ;
RE=REs(i); RB=RBs(i); Ri=Ris(i); VB=VBB-RB*IBQ;
R1=min(RB/VBB*VCC,1e9);
R2=min(RB/(VCC-VBB)*VCC,1e9); % Eq.(3.4.7a,b)
PRs=[(VCC-VB)^2/R1 VB^2/R2 RE*IEQ^2]; % Power ratings of R1 and RE
if i==1 % ICQ=ICmax/100
 disp('Try again with another BJT with larger current gain beta!')
 else disp('Design Results'), disp(' R1 RE Rid ICQ')
   fprintf('%8.0f %8.0f %8.0f %8.5f\n', R1,RE,Rid,ICQ)
end
```

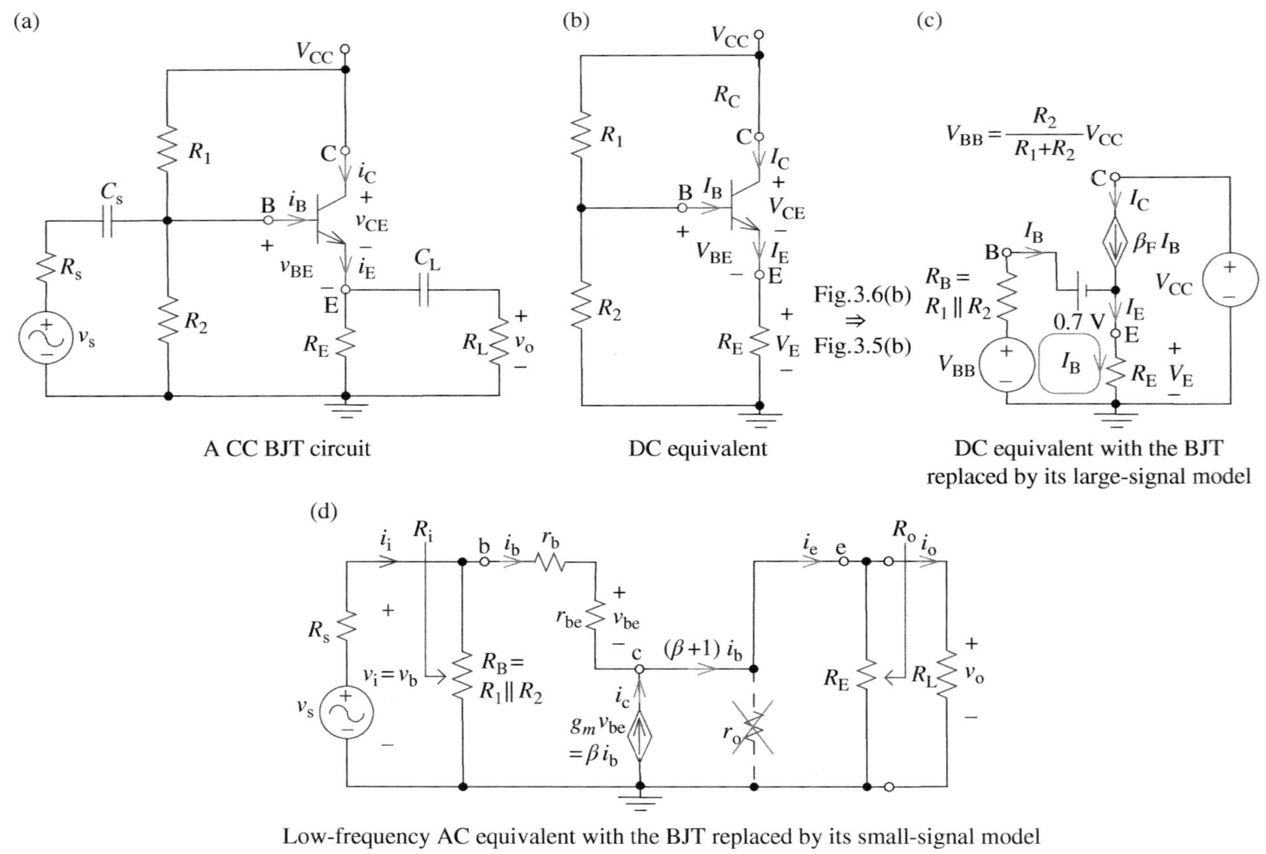
(a)
(b)
(c)
A CC BJT circuit
DC equivalent
Fig.3.6(b) ⇒ Fig.3.5(b)
DC equivalent with the BJT replaced by its large-signal model
(d)
Low-frequency AC equivalent with the BJT replaced by its small-signal model

Figure 3.58 A CC (common-collector) or emitter follower circuit and its DC/AC equivalents (see Figure 3.19.1).

$$R_E \overset{(3.4.15b)}{=} \frac{V_{CC}}{2I_{C,Q}} \tag{3.4.17a}$$

$$R_B \overset{(3.4.16)}{=} \frac{V_{BB}-0.7-V_{CC}/2}{I_{B,Q}} = \frac{\beta_F(V_{BB}-0.7-V_{CC}/2)}{I_{C,Q}}$$

$$\underset{R_1=R_B,R_2=\infty}{\overset{V_{BB}=V_{CC}}{\longrightarrow}} \frac{\beta_F(V_{CC}/2-0.7)}{I_{C,Q}} \tag{3.4.17b}$$

where R_2 has been assumed to be open-circuited to maximize the range of R_B. Then it is key to choose $I_{C,Q}$ from the range ($I_{C,max}/100$, $I_{C,max}/2$) such that Eq. (3.4.14) with Eqs. (3.4.17a,b) can be satisfied. Once such a value of $I_{C,Q}$ is determined, the resistances R_E, R_1=R_B, the resulting input resistance R_i, and the output resistance R_o can be computed using Eqs. (3.4.17a), (3.4.17b), (3.4.14), and (3.2.8) where Eqs. (3.4.14) (or Eq. (3.2.5)) and (3.2.8) have been coded in the MATLAB function 'BJT_CC_analysis()'. However, if $I_{C,Q}$ turns out to be small enough to endanger the swing of AC collector current, i.e.

$$I_{C,Q} < \beta i_b = \beta\frac{v_b}{R_i} = \beta\frac{v_s}{R_s+R_i} \tag{3.4.18}$$

then another BJT with larger current gain β_F should be used. This procedure of designing a CC amplifier with a specified input resistance $R_{i,d}$ has been cast into the above MATLAB function 'BJT_CC_design()' where the default values of $I_{C,max}$ and ambient temperature T are set to 0.8 and 27[°C], respectively.

```
%design_CC_BJT.m
betaF=200; betaR=6; betaAC=200; rb=10; VA=100;
VCC=18; Vsm=0.01; Rs=50; RC=0; RL=5000;
Rid=5e4; RidVBB=[Rid VCC]; % Desired input resistance wnd VBB=VCC
[R1,R2,RE]=BJT_CC_design(VCC,[betaF betaAC],rb,RidVBB,RL);
[VBQ,VEQ,VCQ,IBQ,IEQ,ICQ,Av,Ai,Ri,Ro]=BJT_CC_analysis(VCC,rb,Rs,
     R1,R2, RC,RE,RL, [betaF,betaR,betaAC],Vsm,VA);
```

To use the function for designing a CC BJT (Q2N2222) amplifier with input resistance, $R_{i,d}$ = 50 kΩ, we run the above MATLAB script "design_CC_BJT.m," which yields

```
>>design_CC_BJT
  Design Results
   R1            RE        Rid        ICQ
  98810          536      50000    0.01680
 Analysis Results
  VCC    VBQ    VEQ    VCQ        IBQ          IEQ          ICQ
 18.00  9.72  9.02 18.00  8.38e-005  1.68e-002  1.68e-002
  in the forward-active mode with VCE,Q= 8.98
 Ri= 47.175[kOhm], Ro= 2[Ohm]
 Gv=Ri/(Rs+Ri)xAv = 0.999 x 1.00 = 1.00
```

These results mean the following values of the resistances of the designed CC amplifier:

$$R_1 = 98\,810\,\Omega,\ R_2 = \infty\,\Omega(\text{open}),\ R_C = 0\,\Omega(\text{short}),\ \text{and}\ R_E = 536\,\Omega \qquad (3.4.19)$$

where the BJT parameters, the source/load resistances, and V_{CC} are given as

$$\beta_F = 200, \beta_R = 6, \beta_{AC} = 200, r_b = 10\,\Omega \qquad (3.4.20a)$$

$$R_s = 50\,\Omega, R_L = 5\,\text{k}\Omega, \text{and}\ V_{CC} = 18\,\text{V} \qquad (3.4.20b)$$

The results of using '`BJT_CC_analysis()`' (see Section 3.2.2) to analyze the designed circuit are

$$R_i = 47.175\,\text{k}\Omega, R_o = 2\,\Omega(1.8\,\Omega), \text{and}\, A_v = 1.00 \qquad (3.4.21)$$

'Is the input resistance $R_i = 47.175$ kΩ of designed CC amplifier close to $R_{i,d} = 50$ kΩ?

Figure 3.59(a) and (b) shows the PSpice schematics and their simulation results for measuring the input/output resistances of the designed circuit. The PSpice measured input resistance $R_{i,PSpice} = V_s/I_s - R_s = 0.1\ \text{V}/1.984\ \mu\text{A} - 50\ \Omega = 50.35$ kΩ is very close to $R_{i,d} = 50$ kΩ required by the design specification. The PSpice measured output resistance $R_{o,PSpice} = 0.1\ \text{V}/62.793\ \text{mA} = 1.59\ \Omega$ is close to $R_o = 1.8\ \Omega$ predicted by the MATLAB analysis result.

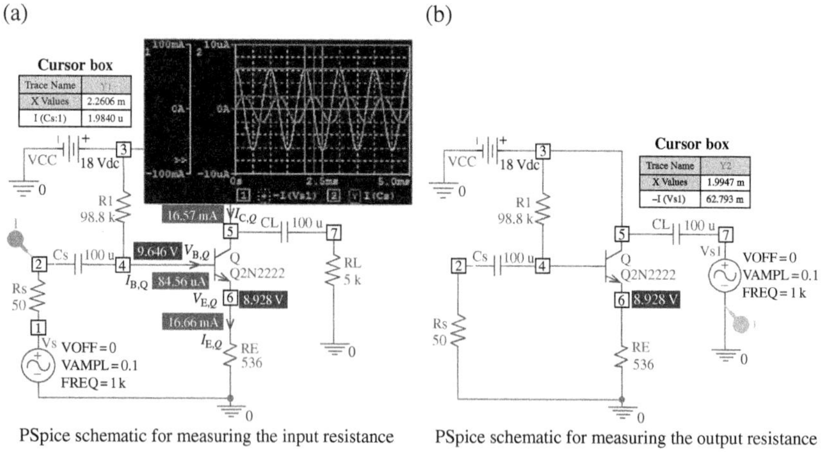

Figure 3.59 A CC circuit and its PSpice simulation ("elec03f55.opj").

3.5 BJT Amplifier Frequency Response

In this section, we will find the transfer function $G(s) = V_o(s)/V_i(s)$ and frequency response $G(j\omega)$ for a CE amplifier, a CC amplifier, and a CB amplifier with the BJT replaced by the high-frequency small-signal model shown in Figure 3.18(a).

3.5.1 CE Amplifier

Figure 3.60(a) and (b) shows a CE amplifier circuit and its high-frequency small-signal equivalent, respectively, where one more load capacitor C_{LL}, in addition to the output capacitor C_L, is connected in parallel with the load resistor R_L. For the equivalent circuit shown in Figure 3.60(b), a set of three node equations in V_1, $V_2 = V_c$, and $V_3 = V_e$ can be set up as

$$\begin{bmatrix} Y_B + 1/r_{be} + sC_{be} + sC_{bc} & -sC_{bc} & -1/r_{be} - sC_{be} \\ -sC_{bc} & sC_{bc} + 1/r_o + Y_C & -1/r_o \\ -1/r_{be} - sC_{be} & -1/r_o & 1/r_{be} + sC_{be} + 1/r_o + Y_E \end{bmatrix} \begin{bmatrix} V_1 \\ V_2 \\ V_3 \end{bmatrix} = \begin{bmatrix} I_{Nt} \\ -g_m v_{be} = -g_m(V_1 - V_3) \\ g_m v_{be} = g_m(V_1 - V_3) \end{bmatrix} \tag{3.5.1}$$

where

$$I_{Nt} = \frac{R_B Y_B}{Z_s + R_B} V_s,\ Y_B = \frac{1}{Z_B = \dfrac{1}{Y_s + G_B} + r_b},\ Y_s = \frac{1}{Z_s} = \frac{1}{R_s + 1/sC_s}$$
$$Y_C = \frac{1}{R_C \| \{1/sC_L + (R_L \| 1/sC_{LL})\}},\ Y_E = \frac{1}{R_{E1} + (R_{E2} \| 1/sC_E)},\ G_B = \frac{1}{R_B} \tag{3.5.2}$$

(Q) How about if $R_{E1} = 0$?

Here, I_{Nt} and Y_B are the values of Norton current source and admittance looking back into the source part from terminals 1 to 0 as shown in Figure 3.60(c1) and (c2). This equation can be rearranged into a solvable form with all the unknown terms on the LHS as

$$\begin{bmatrix} Y_B + 1/r_{be} + sC_{be} + sC_{bc} & -sC_{bc} & -1/r_{be} - sC_{be} \\ -sC_{bc} + g_m & sC_{bc} + 1/r_o + Y_C & -1/r_o - g_m \\ -1/r_{be} - sC_{be} - g_m & -1/r_o & 1/r_{be} + sC_{be} + 1/r_o + Y_E + g_m \end{bmatrix} \begin{bmatrix} V_1 \\ V_2 \\ V_3 \end{bmatrix} = \begin{bmatrix} I_{Nt} \\ 0 \\ 0 \end{bmatrix} \tag{3.5.3}$$

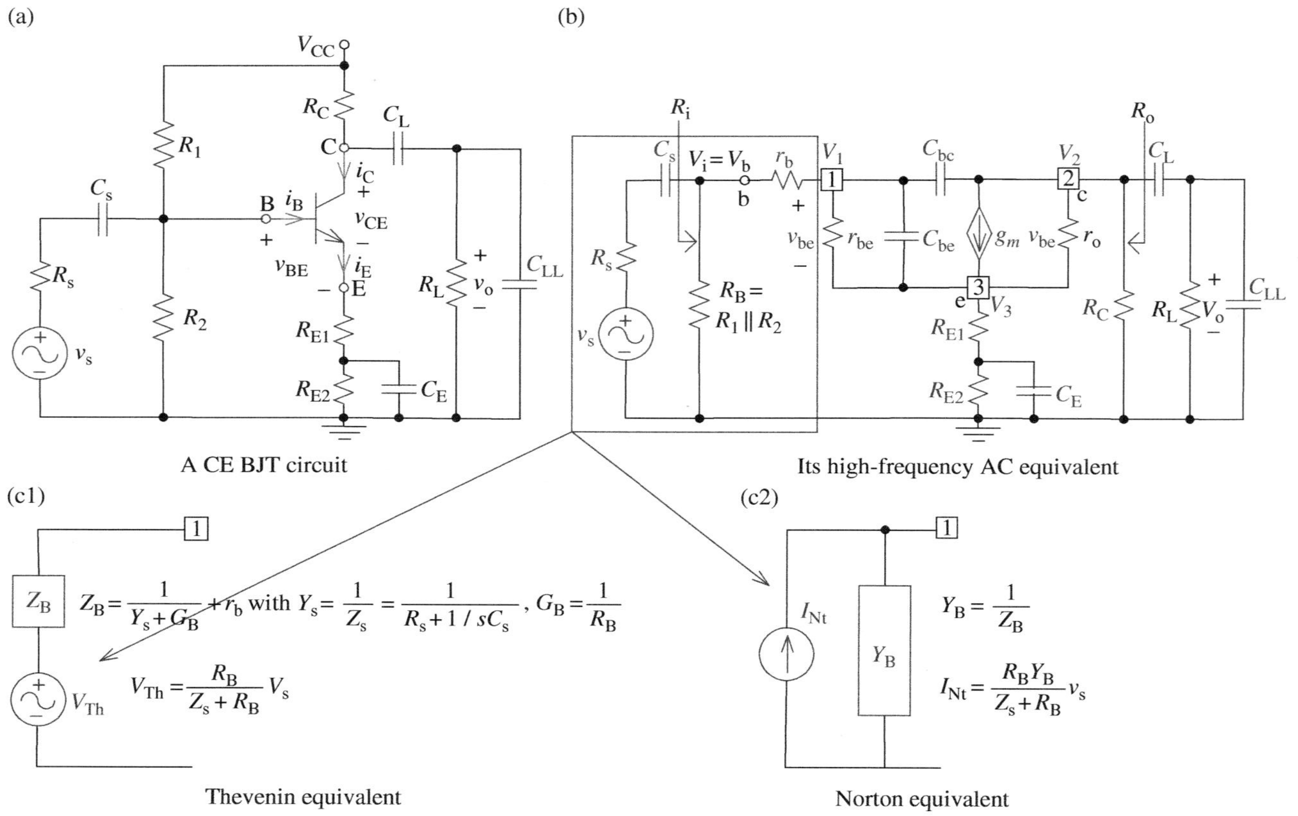

Figure 3.60 A CE (common-emitter) BJT circuit and its high-frequency AC equivalent.

and solved for V_1, V_2, and V_3. Then the transfer function and frequency response can be found as

$$G(s) = \frac{V_o(s)}{V_s(s)} \text{ and } G(j\omega) \text{ with } V_o(s) = \frac{R_L \| (1/sC_{LL})}{1/sC_L + (R_L \| 1/sC_{LL})} V_2(s) \quad (3.5.4)$$

This process to find the transfer function and frequency response has been cast into the following MATLAB function 'BJT_CE_xfer_ftn()'.

```
function [Gs,Av,Ri,Ro,Gw,Cbe,Cbc]=...
  BJT_CE_xfer_ftn(VCC,Rs,Cs,R1,R2,RC,RECE,CL,RLCLL,beta,CVbm,rb,VA,T,w)
%To find the transfer function/frequency response of a CE amplifier
% RECE=[RE1 RE2 CE], RLCLL=[RL CLL] with RL and CLL in parallel
% beta=[betaF betaR betaAC]
% CVbm=[Cbe0,Cbc0,Vb,m] : BEJ|BCJ zero-base capacitance, Junction potential
% Copyleft: Won Y. Yang, wyyang53@hanmail.net, CAU for academic use only
if nargin<14, T=27; end % Ambient temperature
if nargin<13, VA=1e4; end
if nargin<, rb=0; end
Cbe0=CVbm(1); Cbc0=CVbm(2);
if numel(CVbm)<3, Vb=0.7; m=0.5; else Vb=CVbm(3); m=CVbm(4); end
RL=RLCLL(1); if numel(RLCLL)<2, CLL=0; else CLL=RLCLL(2); end
if length(RECE)==1, RE1=0; RE2=RECE; CE=0;
 elseif length(RECE)==2, RE1=RECE(1); RE2=RECE(2); CE=0;
 else RE1=RECE(1); RE2=RECE(2); CE=RECE(3);
end
syms s; ZL=parallel_comb([RL 1/s/CLL]); RE=[RE1 RE2]; Vsm=0.001;
[VBQ,VEQ,VCQ,IBQ,IEQ,ICQ,Av,Ai,Ri,Ro,gm,rbe,ro]=...
  BJT_CE_analysis(VCC,rb,Rs,R1,R2,RC,RE,RL,beta,Vsm,VA,T);
Cbe=Cbe0/(1-(VBQ-VEQ)/Vb)^m;
Cbc=Cbc0/(1-(VBQ-VCQ)/Vb)^m; % Eq.(3.5.5)
sCs=s*Cs; sCE=s*CE; sCL=s*CL; sCbe=s*Cbe; sCbc=s*Cbc;
Zs=Rs+1/sCs; Ys=1/Zs; % Eq.(3.5.2c)
RB=parallel_comb([R1 R2]); GB=1/RB;
YC=1/parallel_comb([RC ZL+1/sCL]); % Admittance at terminal C (Eq.(3.5.2d))
YB=1/(1/(Ys+GB)+rb); % Admittance at terminal B
if RE1+RE2>0
 YE=1/(RE1+1/(1/RE2+sCE)); % Admittance at terminal E (Eq.(3.5.2e))
 Y=[YB+1/rbe+sCbe+sCbc    -sCbc          -1/rbe-sCbe;
      -sCbc+gm          sCbc+1/ro+YC   -1/ro-gm;
     -1/rbe-sCbe-gm          -1/ro      1/rbe+sCbe+1/ro+YE+gm];
 V=Y\[RB/(Zs+RB)*YB; 0; 0]; % Eq.(3.5.3) with Eq.(3.5.2a)
else % without RE
 Y=[YB+1/rbe+sCbe+sCbc   -sCbc;   -sCbc+gm   sCbc+1/ro+YC];
 V=Y\[RB/(Zs+RB)*YB; 0];
end
Gs=V(2)*ZL/(1/sCL+ZL); % Transfer function Vo(s)/Vs(s) Eq.(3.5.4)
if nargin>14, Gw=subs(Gs,'s',j*w); else Gw=0; end % Frequency response
```

Note the following about the internal capacitances of a BJT, the base-to-emitter capacitance C_{be}, the base-to-collector capacitance C_{bc}, and the collector-to-emitter capacitance C_{ce} [W-7]:

- C_{ce} can usually be ignored since $C_{ce} \ll C_{be}$.
- C_{be} and C_{bc} can be modeled as voltage-dependent capacitors with values determined by

$$C_{be} = \frac{C_{be0}}{\left(1 - V_{BE,Q}/\phi_{be}\right)^{m_{be}}}, \quad C_{bc} = \frac{C_{bc0}}{\left(1 - V_{BC,Q}/\phi_{bc}\right)^{m_{bc}}} \tag{3.5.5}$$

where C_{be0}(CJE)/C_{bc0}(CJC): zero-bias B-E/B-C junction capacitances, $V_{BE,Q}$/$V_{BC,Q}$: quiescent B-E/B-C voltages[V], m_{be}(MJE)/m_{bc}(MJC): B-E/B-C grading coefficient, and ϕ_{be}(VJE)/ϕ_{bc}(VJC): B-E/B-C built-in potential[V].

Referring to Figure 3.61, four break (pole or corner) frequencies determining roughly the frequency response magnitude can be determined as [R-2]

$$\omega_{c1} \underset{(8.40)}{\overset{[R\text{-}1]}{=}} \frac{1}{C_s(R_s + R_i)}, \quad \omega_{c2} \underset{(8.41)}{\overset{[R\text{-}1]}{=}} \frac{1}{C_L(R_o + R_L)} \tag{3.5.6a,b}$$

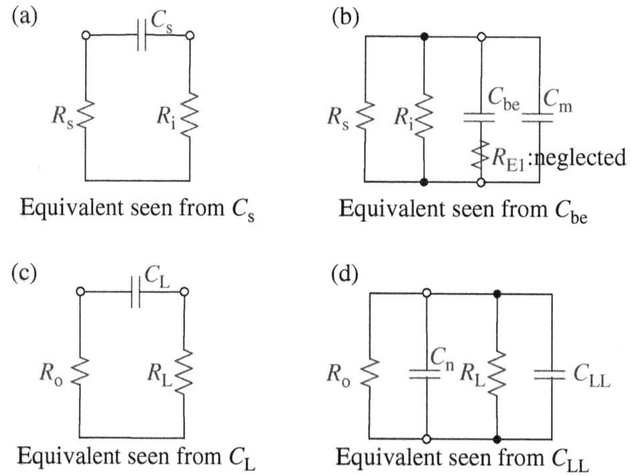

Figure 3.61 (Approximate) equivalent circuits for finding the equivalent resistance seen from each capacitor.

$$\omega_{c3} \underset{(8.57)}{\overset{[R\text{-}1]}{=}} \frac{1}{(C_{be}+C_m)(R_s\|R_i)}, \quad \omega_{c4} \underset{(8.58)}{\overset{[R\text{-}1]}{=}} \frac{1}{(C_n+C_{LL})(R_o\|R_L)} \tag{3.5.6c,d}$$

Note the following about these four break frequencies:

- ω_{C1}, related with the coupling capacitor C_s, is obtained as the reciprocal of the product of C_s and the equivalent resistance as seen from C_s (see Figure 3.61(a)).
- ω_{C2}, related with the output capacitor C_L, is obtained as the reciprocal of the product of C_L and the equivalent resistance as seen from C_L (see Figure 3.61(c)).
- ω_{C3}, related with the internal capacitors $C_{be}|C_{bc}$, is obtained as the reciprocal of the product of (C_{be}+C_m) and the equivalent resistance as seen from C_{be} (see Figure 3.61(b)).
- ω_{C4}, related with the internal capacitors $C_{bc}|C_{LL}$, is obtained as the reciprocal of the product of (C_n+C_{LL}) and the equivalent resistance as seen from C_{LL} (see Figure 3.61(d)).
- The passband of the CE amplifier will be approximately lower-/upper-bounded by the second/third highest break frequencies where not only the values but also the order of the frequencies varies with the values of the related capacitances and resistances.

Note also that R_i/R_o are the input/output resistances of the CE amplifier, respectively, and C_m/C_n are the Miller equivalent capacitances for capacitor C_{bc} seen from the input/output sides, respectively (see Eq. (1.4.2)):

$$C_m = C_{bc}(1-A_v) \text{ and } C_n = C_{bc}(A_v-1)/A_v \quad \text{with } A_v = V_2/V_1 \text{ in Figure 3.60(b)} \tag{3.5.7}$$

The following MATLAB function '`break_freqs_of_CE()`' uses Eqs. (3.5.6-a,b,c,d) to find the four break frequencies:

```
function fc=break_freqs_of_CE(Rs,Cs,CL,RL,CLL,Cbc,Cbe,Av,Ri,Ro)
% To find the 4 break frequencies of a CE amplifier
RsRi=parallel_comb([Rs Ri]); RoRL=parallel_comb([Ro RL]);
Cm=Cbc*(1-Av); % Eq.(3.5.7a)
Cn=Cbc*(1-1/Av); % Eq.(3.5.7b)
fc(1)=1/2/pi/Cs/(Rs+Ri); % Eq.(3.5.6a)
fc(2)=1/2/pi/CL/(Ro+RL); % Eq.(3.5.6b)
fc(3)=1/2/pi/(Cbe+Cm)/RsRi; % Eq.(3.5.6c)
fc(4)=1/2/pi/(CLL+Cn)/RoRL; % Eq.(3.5.6d)
```

Example 3.18 Frequency Response of a CE Amplifier
Consider the CE circuit in Figure 3.62(a) where V_{CC} = 12 V, R_s = 1 kΩ, C_s = 1 μF, R_1 = 300 kΩ, R_2 = 160 kΩ, R_C = 22 kΩ, R_{E1} = 3 kΩ, R_{E2} = 10 kΩ, C_E = 10 μF, C_L = 1 μF, R_L = 100 kΩ, C_{LL} = 1 nF, and the BJT parameters are β_F=100, β_R = 1, β_{AC}=100, V_A=10^4 V, I_s = 10^{-14} A, r_b = 0 Ω, $C_{be}(C_{je})$=10 pF, $C_{bc}(C_{jc})$=1pF, V_b(Vj) = 0.7 V, and m(Mj) = 0.5. Plot the frequency response for f = 1∼100 MHz and see how close it is to the PSpice simulation result. Also estimate the lower and upper 3 dB frequencies.

We can use the MATLAB function 'BJT_CE_xfer_ftn()' to find the frequency response and plot its magnitude curve (as Figure 3.62(c)) by running the following MATLAB statements:

```
>>VCC=12; Rs=1e3; R1=3e5; R2=160e3; RC=22e3; RE=[3e3 10e3]; RL=1e5;
  rb=0; Cs=1e-6; CE=1e-5; CL=1e-6; CLL=1e-8;
  Is=1e-14; betaF=100; betaR=1; betaAC=betaF;
  Cbe0=1e-11; Cbc0=1e-12; Vb=0.7; m=0.5; VA=1e4; T=27;
  beta=[betaF betaR betaAC Is]; CVbm=[Cbe0 Cbc0 Vb m];
  RECE=[RE CE]; RLCLL=[RL CLL]; % CL|CLL=Inf|0; without CL|CLL

   f=logspace(0,8,801); w=2*pi*f; % Frequency range
   [Gs,Av,Ri,Ro,Gw,Cbe,Cbc]=...
    BJT_CE_xfer_ftn(VCC,Rs,Cs,R1,R2,RC,RECE,CL,RLCLL,beta,CVbm,rb,VA,T,w);
  GmagdB=20*log10(abs(Gw)+1e-10); % Frequency response magnitude in dB
  semilogx(f,GmagdB), hold on; Gmax=max(GmagdB);
  fc=break_freqs_of_CE(Rs,Cs,CL,RL,CLL,Cbc,Cbe,Av,Ri,Ro);
  fprintf('\n fc1=%12.3e, fc2=%12.3e, fc3=%12.3e, and fc4=%12.3e\n', fc);
  semilogx(fc(1)*[1 1],[0 Gmax-3],'b:', fc(2)*[1 1],[0 Gmax-3],'g:',
      fc(3)*[1 1],[0 Gmax-3],'r:', fc(4)*[1 1],[0 Gmax-3],'m:')
```

This yields Figure 3.62(c) and the values of the four break frequencies as follows:

```
fc1= 2.007e+00, fc2= 1.305e+00, fc3= 4.956e+06, and fc4= 8.829e+02
```

from which we can get rough estimates of the lower/upper 3 dB frequencies as the second/third highest frequencies f_{cl} = 2 Hz and f_{cu} = 883 Hz, respectively.

3.5.2 CC Amplifier (Emitter Follower)

Figure 3.63(a) and (b) shows a CC amplifier circuit and its high-frequency small-signal equivalent, respectively, where one more load capacitor C_{LL}, in addition to the output capacitor C_L, is connected in parallel with the load resistor R_L. For the equivalent circuit shown in Figure 3.63(b), a set of three node equations in V_1, V_2 = V_e, and V_3 = V_c can be set up as

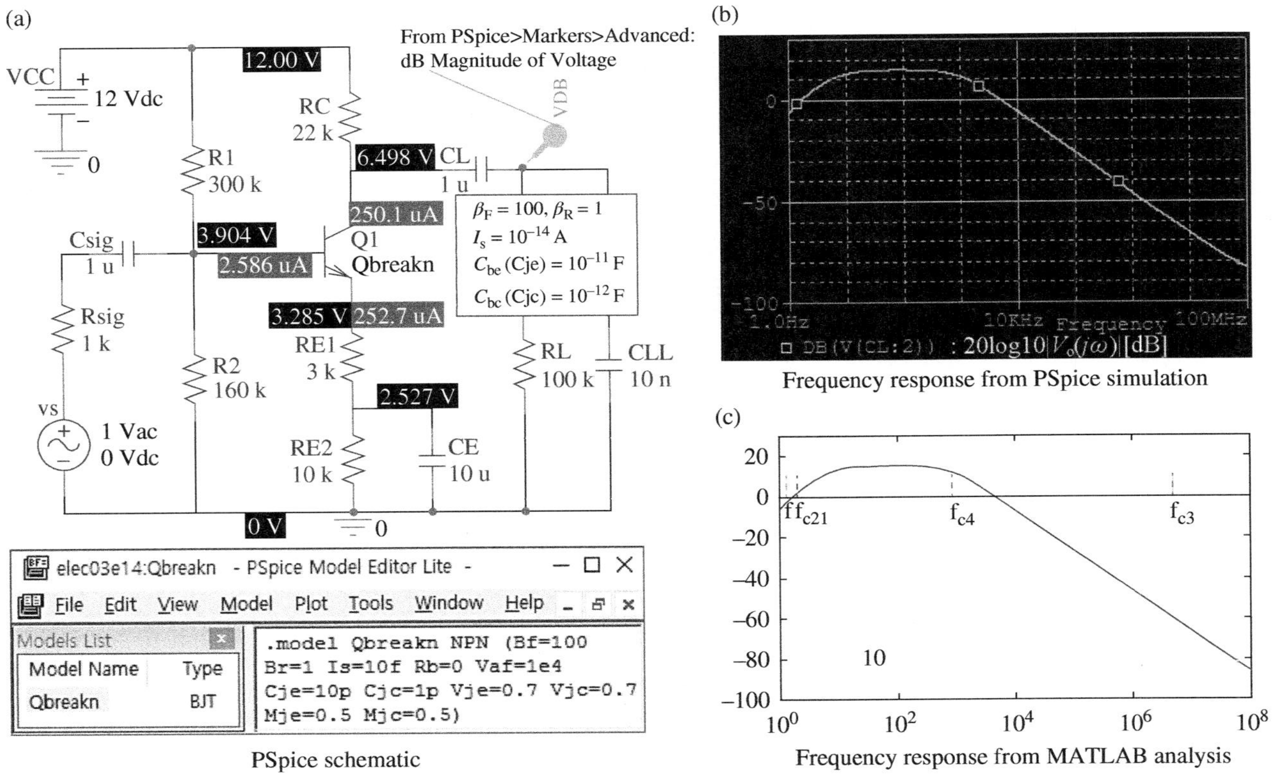

Figure 3.62 A CE amplifier circuit and its frequency response from PSpice and MATLAB ("elec03e18.opj").

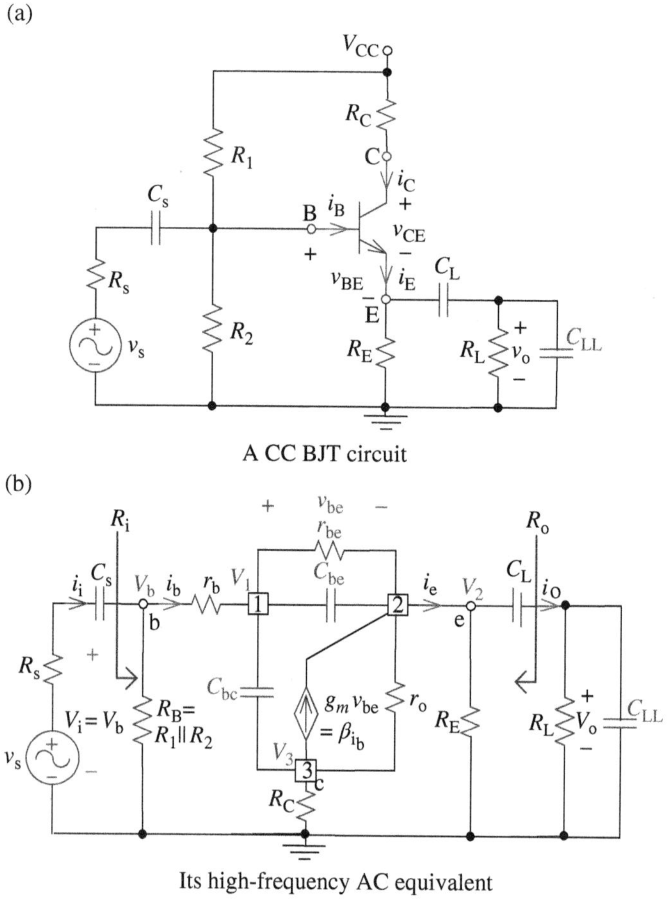
(a)
A CC BJT circuit
(b)
Its high-frequency AC equivalent

Figure 3.63 A CC (common-collector) BJT circuit and its high-frequency AC equivalent.

$$\begin{bmatrix} Y_B + 1/r_{be} + sC_{be} + sC_{bc} & -1/r_{be} - sC_{be} & -sC_{bc} \\ -1/r_{be} - sC_{be} - g_m & 1/r_{be} + sC_{be} + 1/r_o + Y_E + g_m & -1/r_o \\ -sC_{bc} + g_m & -1/r_o - g_m & sC_{bc} + 1/r_o + 1/R_C \end{bmatrix} \begin{bmatrix} V_1 \\ V_2 \\ V_3 \end{bmatrix} = \begin{bmatrix} I_{Nt} \\ 0 \\ 0 \end{bmatrix} \tag{3.5.8}$$

where

$$I_{Nt} = \frac{R_B Y_B}{Z_s + R_B} V_s, \; Y_B = \frac{1}{Z_B = \dfrac{1}{Y_s + G_B} + r_b}, \; Y_s = \frac{1}{Z_s} = \frac{1}{R_s + 1/sC_s}$$

$$Y_E = \frac{1}{R_E \| \{1/sC_L + (R_L \| 1/sC_{LL})\}}, \; G_B = \frac{1}{R_B} \tag{3.5.9}$$

(Q) How about if $R_C = 0$?

```
function [Gs,Av,Ri,Ro,Gw,Cbe,Cbc]=...
  BJT_CC_xfer_ftn(VCC,Rs,Cs,R1,R2,RC,RE,CL,RLCLL,beta,CVbm,rb,VA,T,w)
%To find the transfer function/frequency response of a CC amplifier
% RLCLL=[RL CLL] with RL and CLL in parallel
% CVbm=[Cbe0,Cbc0,Vb,m] : BEJ|BCJ zero-base capacitances, Junction potential
% Copyleft: Won Y. Yang, wyyang53@hanmail.net, CAU for academic use only
if nargin<14, T=27; end % Ambient temperature
if nargin<13, VA=1e4; end
if nargin<12, rb=0; end
Cbe0=CVbm(1); Cbc0=CVbm(2);
if numel(CVbm)<3, Vb=0.7; m=0.5; else Vb=CVbm(3); m=CVbm(4); end
RL=RLCLL(1); if numel(RLCLL)<2, CLL=0; else CLL=RLCLL(2); end
syms s; ZL=parallel_comb([RL 1/s/CLL]); Vsm=0.001;
[VBQ,VEQ,VCQ,IBQ,IEQ,ICQ,Av,Ai,Ri,Ro,gm,rbe,ro]=...
    BJT_CC_analysis(VCC,rb,Rs,R1,R2,RC,RE,RL,beta,Vsm,VA,T);
Cbe=Cbe0/(1-(VBQ-VEQ)/Vb)^m; Cgd=Cgd0/(1-(VBQ-VCQ)/Vb)^m; % Eq.(3.5.5)
sCs=s*Cs; sCL=s*CL; sCbe=s*Cbe; sCbc=s*Cbc;
Zs=Rs+1/sCs; Ys=1/Zs; % Eq.(3.5.8c)
RB=parallel_comb([R1 R2]); GB=1/RB;
YE=1/parallel_comb([RE ZL+1/sCL]); % Admittance at terminal C (Eq.(3.5.9d))
YB=1/(1/(Ys+GB)+rb); % Admittance at terminal B
if RC>0
 Y=[YB+1/rbe+sCbe+sCbc -1/rbe-sCbe          -sCbc;
      -1/rbe-sCbe-gm 1/rbe+sCbe+1/ro+YE+gm -1/ro;
      -sCbc+gm-1/ro-gm                      sCbc+1/ro+1/RC];
 V=Y\[RB/(Zs+RB)*YB; 0; 0]; % Eq.(3.5.8) with Eq.(3.5.9a)
else % without RC
 Y=[YB+1/rbe+sCbe+sCbc        -1/rbe-sCbe;
   -1/rbe-sCbe-gm          1/rbe+sCbe+1/ro+YE+gm];
 V=Y\[RB/(Zs+RB)*YB; 0];
end
Gs=V(2)*ZL/(ZL+1/sCL); % Transfer function Vo(s)/Vs(s)
if nargin>14, Gw=subs(Gs,'s',j*w); else Gw=0; end % Frequency response
```

Solving this set of equations, we can find the transfer function and frequency response as

$$G(s)=\frac{V_o(s)}{V_s(s)} \text{ and } G(j\omega) \text{ with } V_o(s)=\frac{R_L||(1/sC_{LL})}{1/sC_L+(R_L||1/sC_{LL})}V_2(s) \quad (3.5.10)$$

This process to find the transfer function and frequency response has been cast into the above MATLAB function '`BJT_CC_xfer_ftn()`'.

There are four break (pole or corner) frequencies determining roughly the frequency response magnitude:

$$\omega_{c1} \overset{[R-1]}{\underset{(8.40)}{=}} \frac{1}{C_s(R_s+R_i)},\quad \omega_{c2} \overset{[R-1]}{\underset{(8.41)}{=}} \frac{1}{C_L(R_o+R_L)} \quad (3.5.11a,b)$$

$$\omega_{c3} \overset{[R-1]}{\underset{(8.57)}{=}} \frac{1}{(C_{bc}+C_m)(R_s||R_i)},\quad \omega_{c4} \overset{[R-1]}{\underset{(8.58)}{=}} \frac{1}{(C_n+C_{LL})(R_o||R_L)} \quad (3.5.11c,d)$$

where R_i/R_o are the input/output resistances of the CC amplifier, respectively, and C_m/C_n are the Miller equivalent capacitances for capacitor C_{be} seen from the input/output side, respectively (see Eq. (1.4.2)):

$$C_m = C_{be}(1-A_v) \text{ and } C_n = C_{be}(A_v-1)/A_v \quad \text{with } A_v = V_2/V_1 \text{ in Figure 3.63(b)} \quad (3.5.12)$$

Note that Eq. (3.5.11) is just like Eq. (3.5.6) with C_{bc} and C_{be} switched.

The following MATLAB function '`break_freqs_of_CC()`' uses Eqs. (3.5.11a,b,c,d) to find the four break frequencies:

```
function fc=break_freqs_of_CC(Rs,Cs,CL,RL,CLL,Cbc,Cbe,Av,Ri,Ro)
% To find the 4 break frequencies of a CC amplifier
RsRi=parallel_comb([Rs Ri]); RoRL=parallel_comb([Ro RL]);
Cm=Cbe*(1-Av); % Eq.(3.5.12a)
Cn=Cbe*(1-1/Av); % Eq.(3.5.12b)
fc(1)=1/2/pi/Cs/(Rs+Ri); % Eq.(3.5.11a)
fc(2)=1/2/pi/CL/(Ro+RL); % Eq.(3.5.11b)
fc(3)=1/2/pi/(Cbc+Cm)/RsRi; % Eq.(3.5.11c)
fc(4)=1/2/pi/(CLL+Cn)/RoRL; % Eq.(3.5.11d)
```

Example 3.19 Frequency Response of a CC Amplifier
Consider the CE circuit in Figure 3.60(a) where V_{CC} = 12 V, R_s = 1 kΩ, C_s = 1 μF, R_1 = 300 kΩ, R_2 = 160 kΩ, R_C = 0 kΩ, R_E = 13 kΩ, C_L = 1 μF, R_L = 100 kΩ, C_{LL} = 1 nF, and the BJT parameters are $\beta_F = 100$, $\beta_R = 1$, $\beta_{AC} = 100$, $V_A = 10^4$ V, $I_s = 10^{-14}$ A, r_b = 0 Ω, $C_{be}(C_{je})$ = 10 pF, $C_{bc}(C_{jc})$ = 1 pF, V_b(Vj) = 0.7 V, and m(Mj) = 0.5. Plot the frequency response for f = 1∼100 MHz and estimate the lower and upper 3 dB frequencies.

We can use the MATLAB function '`BJT_CC_xfer_ftn()`' to find the frequency response and plot its magnitude curve (as Figure 3.64(c)) by running the following MATLAB statements:

```
>>VCC=12; Rs=1e3; R1=3e5; R2=160e3; RC=0; RE=13e3; RL=1e5;
  Cs=1e-6; CL=1e-6; CLL=1e-8;
  betaF=100; betaR=1; betaAC=betaF; Is=1e-14;
  rb=0; Cbe0=1e-11; Cbc0=1e-12; Vb=0.7; m=0; VA=1e4; T=27;
  beta=[betaF betaR betaAC Is]; CVbm=[Cbe0 Cbc0 Vb m];
  RLCLL=[RL CLL]; % CL|CLL=Inf|0; without CL|CLL

f=logspace(0,8,801); w=2*pi*f; % Frequency range
[Gs,Av,Ri,Ro,Gw,Cbe,Cbc]= BJT_CC_xfer_ftn...
(VCC,Rs,Cs,R1,R2,RC,RE,CL,RLCLL,beta,CVbm,rb,VA,T,w);
GmagdB=20*log10(abs(Gw)+1e-10); % Frequency response magnitude in dB
semilogx(f,GmagdB), hold on; Gmax=max(GmagdB);
fc=break_freqs_of_CC(Rs,Cs,CL,RL,CLL,Cbc,Cbe,Av,Ri,Ro);
fprintf('\n fc1=%12.3e, fc2=%12.3e, fc3=%12.3e, and fc4=%12.3e\n',fc);
semilogx(fc(1)*[1 1],[0 Gmax-3],'b:', fc(2)*[1 1],[0 Gmax-3],'g:',
         fc(3)*[1 1],[0 Gmax-3],'r:', fc(4)*[1 1],[0 Gmax-3],'m:')
```

This yields Figure 3.64(c) and the values of the four break frequencies as follows:

```
fc1= 1.644e+00, fc2= 1.590e+00, fc3= 2.974e+08, and fc4= 1.437e+05
```

from which we can get rough estimates of the lower/upper 3 dB frequencies as the second/third highest frequencies f_{cl}=1.644 Hz and f_{cu}=143.7 kHz, respectively.

3.5.3 CB Amplifier

Figure 3.65(a) and (b) shows a CB amplifier circuit and its high-frequency small-signal equivalent, respectively, where one more load capacitor C_{LL}, in addition to the output capacitor C_L, is connected in parallel with the load resistor R_L. Note that if the terminal B(ase) is not AC grounded via a capacitor C_B, we should let $C_B = 0$ in Figure 3.65(b). For the equivalent circuit shown in Figure 3.65(b), a set of three node equations in $V_1 = V_e$, $V_2 = V_c$, and $V_3 = V_b$ can be set up as

$$\begin{bmatrix} Y_E + sC_{be} + 1/r_{be} + 1/r_o + g_m & -1/r_o & -sC_{be} - 1/r_{be} - g_m \\ -1/r_o - g_m & 1/r_o + sC_{bc} + Y_C & -sC_{bc} + g_m \\ -sC_{be} - 1/r_{be} & -sC_{bc} & sC_{be} + 1/r_{be} + 1/(r_b + Z_B) + sC_{bc} \end{bmatrix} \begin{bmatrix} V_1 \\ V_2 \\ V_3 \end{bmatrix} = \begin{bmatrix} I_{Nt} \\ 0 \\ 0 \end{bmatrix} \tag{3.5.13}$$

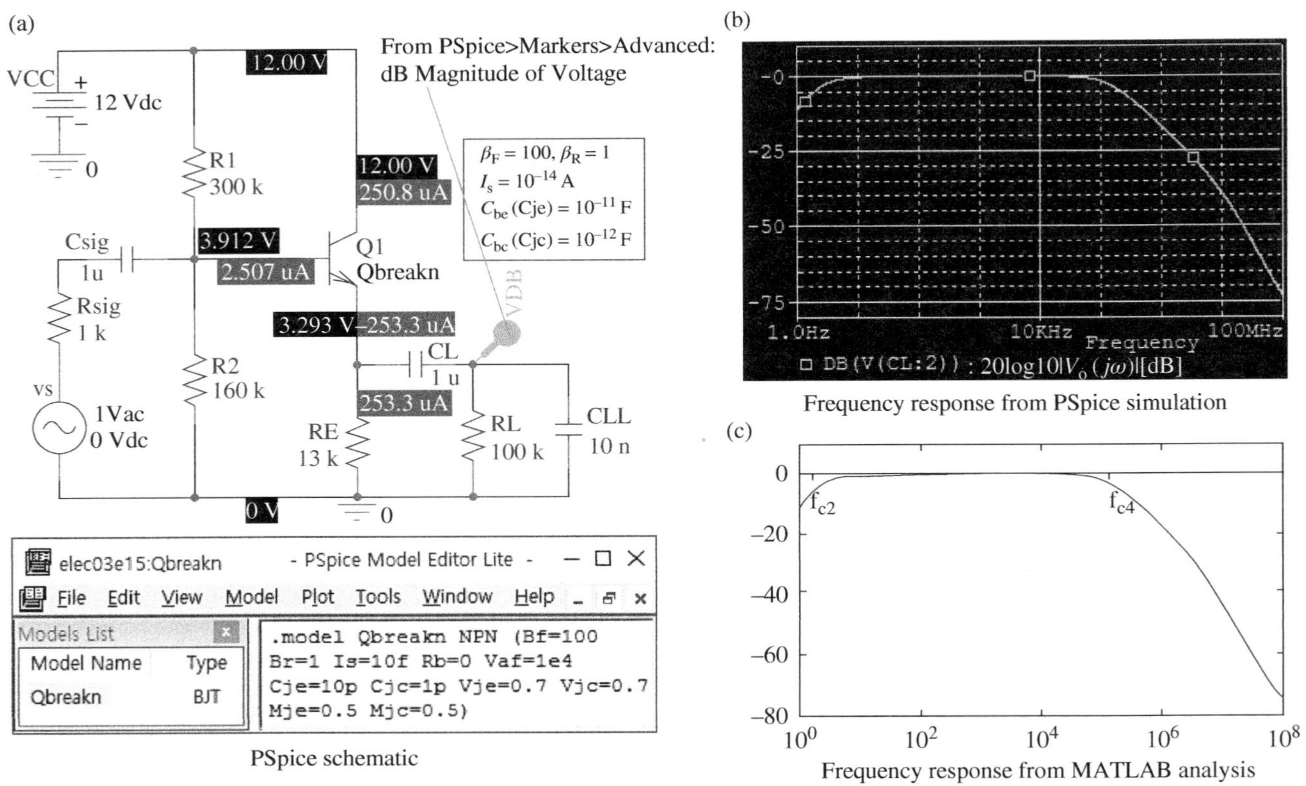

Figure 3.64 A CC amplifier circuit and its frequency response from PSpice and MATLAB ("elec03e19.opj").

where

$$I_{Nt} = Y_s V_s,\quad Y_s = \frac{1}{Z_s} = \frac{1}{R_s + 1/sC_s},\quad Y_E = Y_s + \frac{1}{R_E} \tag{3.5.14a,b,c}$$

$$Z_B = R_B \| (1/sC_B),\ \text{and}\ Y_C = \frac{1}{R_C \| \{1/sC_L + (R_L \| 1/sC_{LL})\}} \tag{3.5.14d,e}$$

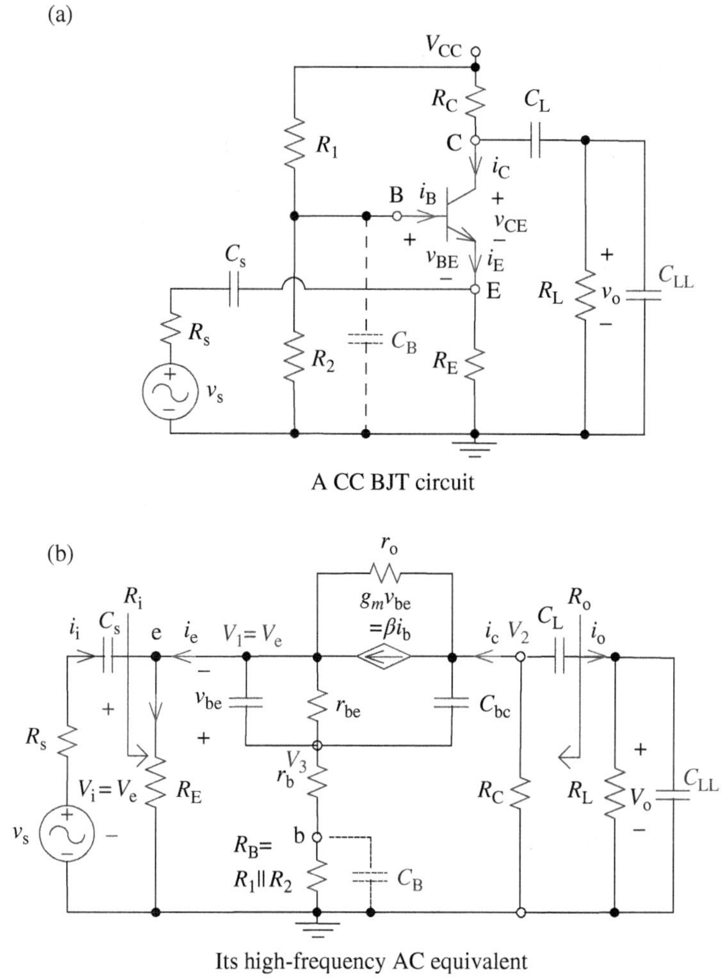

Figure 3.65 A CB (common-base) BJT circuit and its high-frequency AC equivalent.

(Q) How about if $r_b = 0$ and $R_B = 0$ or C_B is so large that R_B can be regarded as AC-shorted?

```
function [Gs,Av,Ri,Ro,Gw,Cbe,Cbc,gm,rbe,ro]=...
 BJT_CB_xfer_ftn(VCC,Rs,Cs,R1,R2,CB,RC,RE,CL,RLCLL,beta,CVbm,rb,VA,T,w)
%To find the transfer function/frequency response of a CB amplifier
% CB must be set to 0 if B is not AC grounded via a capactor CB.
% RLCLL=[RL CLL] with RL and CLL in parallel
% CVbm=[Cbe0,Cbc0,Vb,m]: BEJ|BCJ zero-base capacitances, Junction potential
% Copyleft: Won Y. Yang, wyyang53@hanmail.net, CAU for academic use only
if nargin<15, T=27; end % Ambient temperature
if nargin<14, VA=1e4; end
if nargin<13, rb=0; end
Cbe0=CVbm(1); Cbc0=CVbm(2);
if numel(CVbm)<3, Vb=0.7; m=0.5; else Vb=CVbm(3); m=CVbm(4); end
RL=RLCLL(1); if numel(RLCLL)<2, CLL=0; else CLL=RLCLL(2); end
syms s; ZL=parallel_comb([RL 1/s/CLL]); Vsm=0.001;
[VBQ,VEQ,VCQ,IBQ,IEQ,ICQ,Av,Ai,Ri,Ro,gm,rbe,ro]=...
  BJT_CB_analysis(VCC,rb,Rs,R1,R2,RC,RE,RL,beta,Vsm,VA,T);
Cbe=Cbe0/(1-(VBQ-VEQ)/Vb)^m; Cgd=Cgd0/(1-(VBQ-VCQ)/Vb)^m; %Eq.(3.5.5)
sCs=s*Cs; sCL=s*CL; sCbe=s*Cbe; sCbc=s*Cbc; sCB=s*CB;
Zs=Rs+1/sCs; Ys=1/Zs;
RB=parallel_comb([R1 R2]); ZB=parallel_comb([RB 1/sCB]);
GE=1/RE; YE=Ys+GE;
YC=1/parallel_comb([RC 1/sCL+ZL]); % Impedance at terminal C
if rb+RB>0&CB<0.1
 Y=[YE+sCbe+1/rbe+1/ro+gm     -1/ro             -sCbe-1/rbe-gm;
      -1/ro-gm             1/ro+sCbc+YC           -sCbc+gm;
      -sCbe-1/rbe                -sCbc        sCbe+1/rbe+1/(rb+ZB)+sCbc];
 V=Y\[Ys; 0; 0]; % Eq.(3.5.13)
else % If rb=0 and (RB=0 or CB is so large that RB can be AC-shorted)
 Y=[YE+sCbe+1/rbe+1/ro+gm     -1/ro;     -1/ro-gm    1/ro+sCbc+YC];
 V=Y\[Ys; 0];
end
Gs=V(2)*ZL/(ZL+1/sCL); % Transfer function Vo(s)/Vs(s)
if nargin>15, Gw=subs(Gs,'s',j*w); else Gw=0; end % Frequency response
```

Solving this set of equations, we can find the transfer function and frequency response as

$$G(s)=\frac{V_o(s)}{V_s(s)} \text{ and } G(j\omega) \text{ with } V_o(s)=\frac{R_L\|(1/sC_{LL})}{1/sC_L+(R_L\|1/sC_{LL})}V_2(s) \quad (3.5.15)$$

This process to find the transfer function and frequency response has been cast into the above MATLAB function '`BJT_CB_xfer_ftn()`'.

There are four break (pole or corner) frequencies determining roughly the frequency response magnitude:

$$\omega_{c1} \underset{(8.40)}{\overset{[R-1]}{=}} \frac{1}{C_s(R_s + R_i)}, \qquad \omega_{c2} \underset{(8.41)}{\overset{[R-1]}{=}} \frac{1}{C_L(R_o + R_L)} \tag{3.5.16a,b}$$

$$\omega_{c3} = \frac{1}{C_{be}R_{be}}, \qquad \omega_{c4} = \frac{1}{C_{LL}(R_o \| R_L)} \tag{3.5.16c,d}$$

where R_i/R_o are the input/output resistances of the CB amplifier, respectively, and

$$R_{be} = (R_E + r_b + R_B) \| r_{be} \tag{3.5.17}$$

The following MATLAB function '`break_freqs_of_CB()`' uses Eqs. (3.5.16a,b,c,d) to find the four break frequencies:

```
function fc=break_freqs_of_CB(Rs,Cs,CL,RL,CLL,Cbc,Cbe,Av,Ri,Ro,Rbe)
% To find the 4 break frequencies of a CB amplifier
% Rbe = parallel_comb([RE+rb+RB rbe]) by Eq.(3.5.17)
RsRi=parallel_comb([Rs Ri]); RoRL=parallel_comb([Ro RL]);
fc(1)=1/2/pi/Cs/(Rs+Ri); fc(2)=1/2/pi/CL/(Ro+RL); % Eq.(3.5.16a,b)
fc(3)=1/2/pi/Cbe/Rbe; fc(4)=1/2/pi/CLL/RoRL; % Eq.(3.5.16c,d)
```

Example 3.20 Frequency Response of a CB Amplifier

Consider the CB circuit in Figure 3.66(a) where V_{CC} = 12 V, R_s = 1 kΩ, C_s = 1 μF, R_1 = 300 kΩ, R_2 = 160 kΩ, R_C = 22 kΩ, R_E = 13 kΩ, C_L = 1 μF, R_L = 100 kΩ, C_{LL}=1 nF, and the BJT parameters are β_F=100, β_R=1, β_{AC}=100, V_A=10^4 V, I_s = 10^{-14} A, r_b = 0 Ω, $C_{be}(C_{je})$ = 10 pF, $C_{bc}(C_{jc})$ = 1 pF, V_b(Vj) = 0.7 V, and m(Mj) = 0.5.

a) Plot the frequency response for f = 1~100 MHz and estimate the lower and upper 3 dB frequencies.

   We can use the MATLAB function '`BJT_CB_xfer_ftn()`' to find the frequency response and plot its magnitude curve (as Figure 3.66(c)) by running the following MATLAB statements:

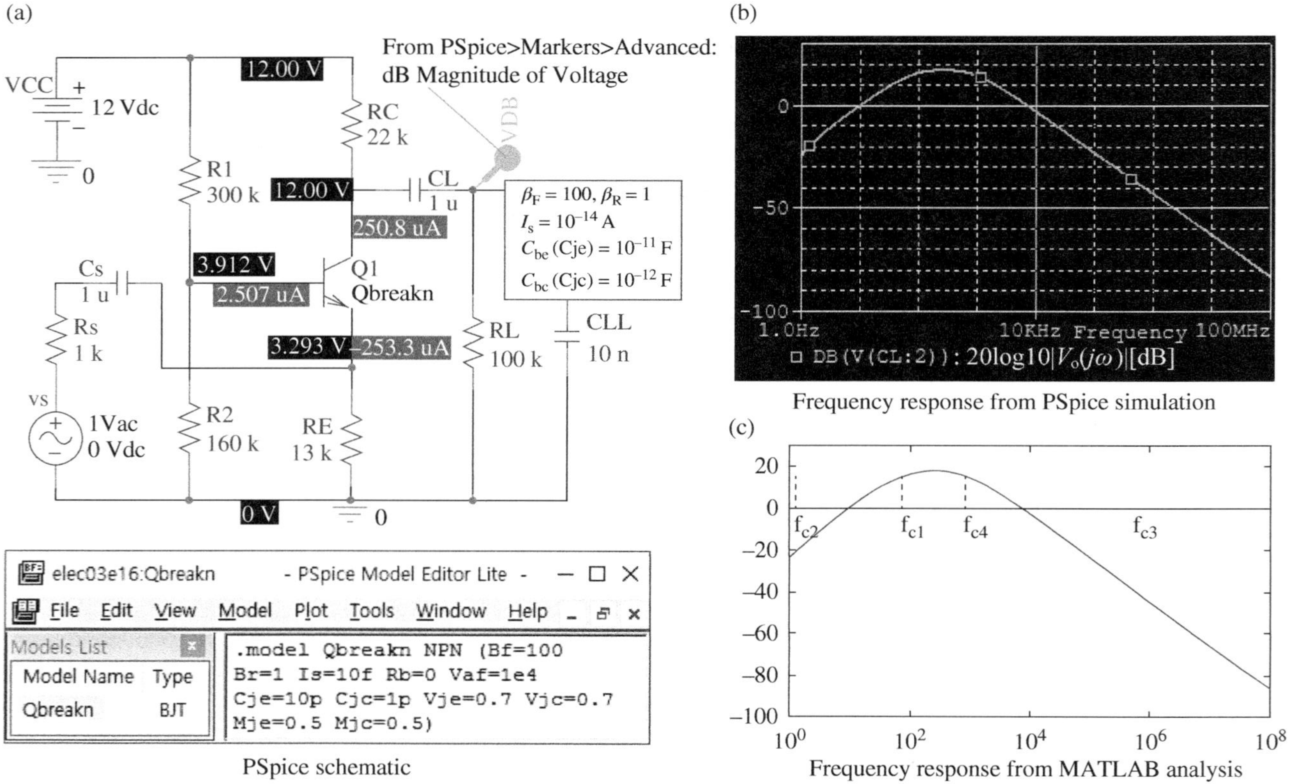

Figure 3.66 A CB amplifier circuit and its frequency response from PSpice and MATLAB ("elec03e20.opj").

```
>>VCC=12; Rs=1e3; R1=3e5; R2=160e3; RC=22e3; RE=13e3; RL=1e5;
  Cs=1e-6; CL=1e-6; CLL=1e-8; RLCLL=[RL CLL]; CB=0;
  % CL|CLL=Inf|0; without CL|CLL
  rb=0; betaF=100; betaR=1; betaAC=betaF; Is=1e-14;
  Cbe0=1e-11; Cbc0=1e-12; Vb=0.7; m=0.5; VA=1e4; T=27;
  beta=[betaF betaR betaAC Is]; CVbm=[Cbe0 Cbc0 Vb m];
  f=logspace(0,8,801); w=2*pi*f; % Frequency range
  [Gs,Av,Ri,Ro,Gw,Cbe,Cbc]= BJT_CB_xfer_ftn ...
    (VCC,Rs,Cs,R1,R2,CB,RC,RE,CL,RLCLL,beta,CVbm,rb,VA,T,w);
  GmagdB=20*log10(abs(Gw)+1e-10); % Frequency response magnitude [dB]
  semilogx(f,GmagdB), hold on; Gmax=max(GmagdB);
```

This yields Figure 3.66(c). Then, to estimate the lower and upper 3 dB frequencies, run the following MATLAB statements:

```
>>RB=parallel_comb([R1 R2]);
  Rbe=parallel_comb([RE+rb+RB*(CB<1) rbe]); % Eq.(3.5.17)
  % If CB does not exist, its value should be set to 1 or larger.
  fc=break_freqs_of_CB(Rs,Cs,CL,RL,CLL,Cbc,Cbe,Av,Ri,Ro,Rbe);
  fprintf('\nfc1=%12.3e, fc2=%12.3e, fc3=%12.3e, and fc4=%12.3e\n',
  fc); semilogx(fc(1)*[1 1],[0 Gmax-3],'b:', fc(2)*[1 1],[0 Gmax-3],'g:',
              fc(3)*[1 1],[0 Gmax-3],'r:', fc(4)*[1 1],[0 Gmax-3],'m:')
```

This yields the four break frequencies as

```
fc1= 7.786e+01, fc2= 1.305e+00, fc3= 5.714e+05, and fc4= 8.830e+02
```

From these, we can get rough estimates of the lower/upper 3 dB frequencies as the second/third highest frequencies f_{cl} = 77.86 Hz and f_{cu} = 883 Hz, respectively.

b) Which one of the capacitances C_s, C_L, C_{be}, and C_{LL} is (the most closely) related with the upper 3 dB frequency?

   Looking into the MATLAB function '`break_freqs_of_CB()`', we can see that f_{c4} stems from C_{LL}. Interested readers are recommended to change C_{LL} into 0.1 nF and see how the upper 3 dB frequency is changed.

3.6 BJT Inverter Time Response

In this section, let us see the time response of a basic BJT inverter (as shown in Figure 3.27(a1) or 3.67(b)) to logic low-to-high and high-to-low input transitions. About Figure 3.27(b1) illustrating the low-to-high and high-to-low propagation delays, one might wonder what makes the circuit seemingly not having any dynamic elements such as capacitors or inductors exhibit a dynamic time response to the piece wisely constant input. It stems from internal parasitic capacitances between the terminals of the BJT, whose effect becomes conspicuous as the frequency or slope of the input increases, while negligible for a DC or low-frequency input.

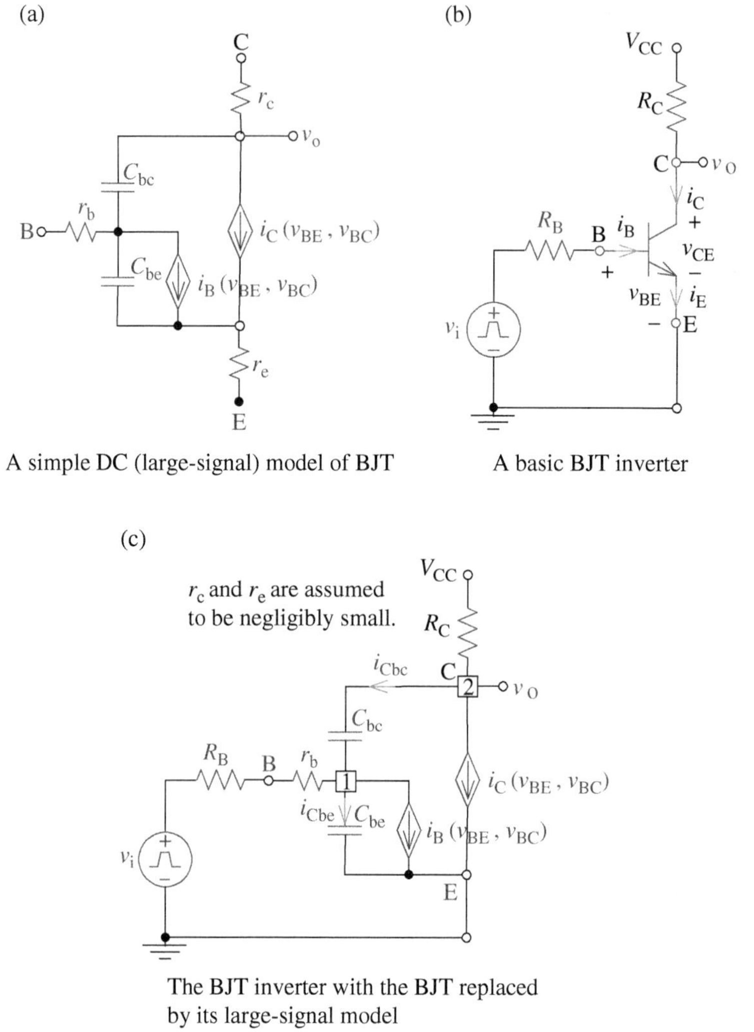

Figure 3.67 A simple large-signal Spice model of BJT and the equivalent of a BJT inverter adopting the model.

Figure 3.67(a) shows a simple DC or large-signal Spice model of a BJT where $i_C(v_{BE},v_{BC})$ and $i_B(v_{BE},v_{BC})$ are defined as Eqs. (3.1.23a,b), respectively, and

r_b: base resistance, r_e: emitter resistance, r_c: collector resistance,
C_{bc}: base-collector junction capacitance, and
C_{be}: base-emitter junction capacitance.

```
function [vo,vCbc,vCbe,iC,iB]=...
BJT_inverter_dynamic(beta,Is,CC,RBb,RC,VCC,vi,dt,VT)
% Analyze an NPN-BJT inverter to find the output vo to an input vi
% which consists of an NPN-BJT and resistors RB/RC between vi/VCC and B/C.
% beta= [betaF betaR]; CC= [Cbe Cbc]; RBb= [RB rb]
% vo: Output voltage(s) to the input voltage(s) vi
betaF=beta(1); if numel(betaF)<2, betaR=1; else betaR=beta(2); end
Cbe=CC(1); Cbc=CC(2);
RB=RBb(1); if numel(RBb)<2, rb=0; else rb=RBb(2); end
alphaR=betaR/(betaR+1); Isc=Is/alphaR; % Collector-Base saturation current
iCv=@(vBE,vBC) Is*exp(vBE/VT)-Isc*exp(vBC/VT); % Eq.(3.1.23a)
iBv=@(vBE,vBC) Is/betaF*exp(vBE/VT)...
+Isc/(betaR+1)*exp(vBC/VT); %Eq.(3.1.23b)
vCbc(1)=VCC; vCbe(1)=0; vB(1)=vi(1); % Initialize
for n=1:length(vi)
  v1(n)=vCbe(n); vo(n)=v1(n)+vCbc(n); vBC(n)=vB(n)-vo(n); %Eq.(3.6.2a,b,d)
  iC(n)=iCv(vB(n),vBC(n)); iB(n)=iBv(vB(n),vBC(n));
  vB(n+1)=v1(n)+rb*iB(n); % Eq.(3.6.2c)
  iCbc(n)=(VCC-vo(n))/RC-iC(n); % Eq.(3.6.1a)
  vCbc(n+1)=vCbc(n)+iCbc(n)*dt/Cbc; % Eq.(3.6.3a)
  iCbe(n)=(vi(n)-v1(n))/(RB+rb)+iCbc(n)-iB(n); % Eq.(3.6.1b)
  vCbe(n+1)=vCbe(n)+iCbe(n)*dt/Cbe; % Eq.(3.6.3b)
end
vCbc=vCbc(1:n); vCbe=vCbe(1:n); %make the size of vCbc|vCbe the same as vi
```

Figure 3.67(c) shows a DC equivalent of the inverter (Figure 3.67(b)) with the BJT replaced by its large-signal model (Figure 3.67(a)). To solve the circuit in a numerical way, we should know the following:

- From the KCL equations at nodes 1 and 2, the currents i_{Cbc} and i_{Cbe}, each through C_{bc} and C_{be}, respectively, can be expressed as

$$i_{Cbc} = \frac{V_{CC} - v_o}{R_C} - i_C(v_{BE}, v_{BC}) \text{ and } i_{Cbe} = \frac{v_i - v_1}{R_B + r_b} + i_{Cbc} - i_B(v_{BE}, v_{BC}) \qquad (3.6.1)$$

where

$$v_1 = v_{Cbe},\ v_o = v_1 + v_{Cbc},\ v_{BE} = v_B = v_1 + r_b\, i_B, \text{ and } v_{BC} = v_B - v_o \qquad (3.6.2a,b,c,d)$$

- These currents charge each one of the capacitors, C_{bc} and C_{be}, respectively, as

$$v_{Cbc} = \frac{1}{C_{bc}} \int i_{Cbc}(t)dt \rightarrow v_{Cbc}[n+1] = v_{Cbc}[n] + \frac{1}{C_{bc}} i_{Cbc}[n] \Delta t \qquad (3.6.3a)$$

$$v_{Cbe} = \frac{1}{C_{be}} \int i_{Cbe}(t)dt \rightarrow v_{Cbe}[n+1] = v_{Cbe}[n] + \frac{1}{C_{bc}} i_{Cbe}[n] \Delta t \qquad (3.6.3b)$$

A process of solving the inverter to find the output $v_o(t)$ to an input $v_i(t)$ (based on these equations) has been cast into the above MATLAB function '`BJT_inverter_dynamic()`'. Note the following about the dynamic behavior of the inverter:

- The collector resistor R_C and the (internal) base-collector capacitance C_{bc} affect how long it takes for C_{bc} to be charged up to V_{CC}, via the time constant during the rising period of $v_o(t)$.
- The base resistor R_B and the (internal) base-emitter capacitance C_{be} affect how long it takes for C_{be} to be charged up to V_{BE} = 0.7 V, via the delay time during the falling period of $v_o(t)$.

Note that the time constant of an RC circuit, i.e. the time taken for its transient response to reach 63.2% of its final value, is RC where R/C is the equivalent resistance/capacitance seen from C/R.

Example 3.21 Time Response of a BJT Inverter

Consider the BJT inverter in Figure 3.68(a) where V_{CC} = 12 V, R_B = 1 kΩ, R_C = 10 kΩ, and the BJT parameters are β_F = 100, β_R = 1, $I_s = 10^{-14}$ A, r_b = 0 Ω, V_A = ∞ V, $C_{be}(C_{je})$ = 10 pF, and $C_{bc}(C_{jc})$ = 1 pF. Use the above MATLAB function '`BJT_inverter_dynamic()`' to find the output $v_o(t)$ to an input $v_i(t)$ plotted as the dotted line in Figure 3.68(b) or (c) and plot it for t = 0~0.2 μs.

To this end, we can complete the following MATLAB script "elec03e21.m" and run it to see the plot of $v_i(t)$ and $v_o(t)$ as shown in Figure 3.68(c). The time constant during the rising period of $v_o(t)$ is estimated as

$$R_C C_{bc} = 10\text{k} \times 10\text{p} = 10^4 \times 10^{-12} = 10^{-8}\,[\text{s}] \qquad \text{(E3.17.1)}$$

```
%elec03e21.m
% To find the dynamic response of a BJT amplifier
VCC=5; RB=1e3; RC=10e3; % Circuit parameters
betaF=100; betaR=1; Is=1e-14; rb=0; % Device parameters
Cbe=1e-11; Cbc=1e-12; beta=[betaF betaR]; CC=[Cbe Cbc];
dt=1e-12; t=[0:200000]*dt; % Time range
ts=[0 2 3 5 6 12 13 15 16 20]*1e4*dt;
vis=[0 0 VCC VCC 0 0 VCC VCC 0 0];
vi=interp1(ts,vis,t); % PWL (piecewise linear) input
[vo,vCbc,vCbe,iC,iB]= ...
   BJT_inverter_dynamic(beta,Is,CC,[RB rb],RC,VCC,vi,dt);
plot(t,vi,'r', t,vo)
```

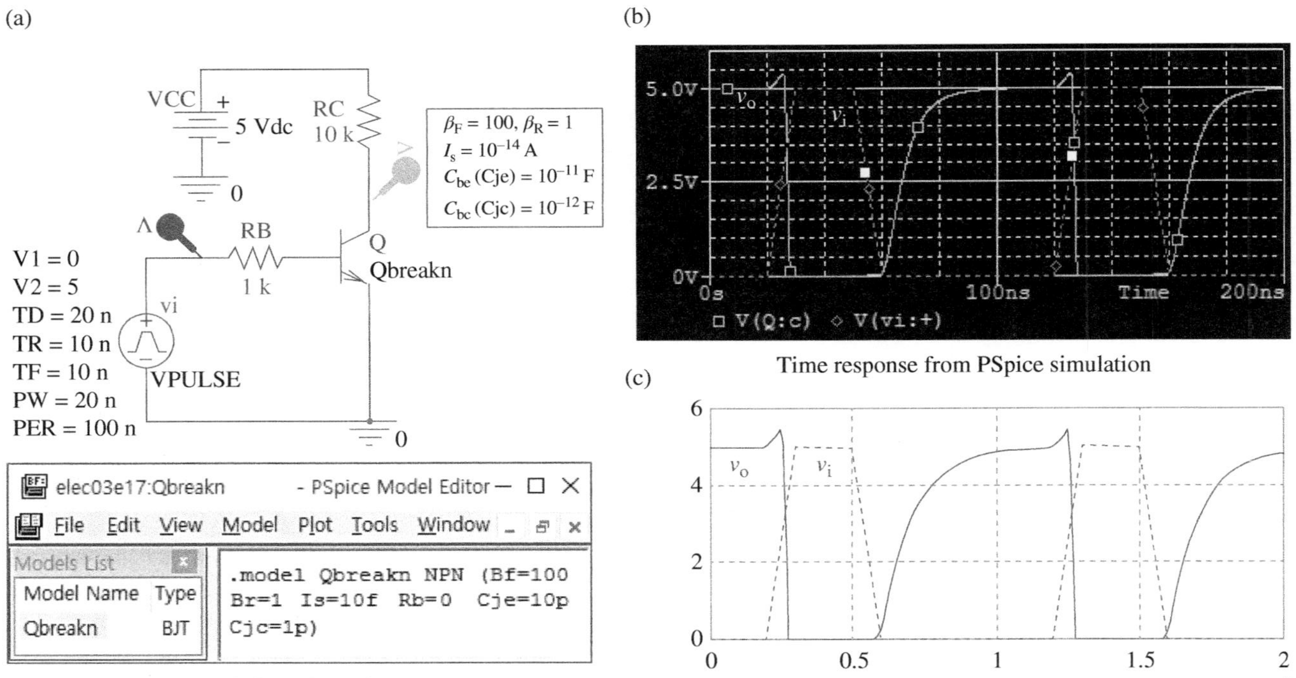

(a)

$V1 = 0$
$V2 = 5$
$TD = 20\,n$
$TR = 10\,n$
$TF = 10\,n$
$PW = 20\,n$
$PER = 100\,n$

$\beta_F = 100, \beta_R = 1$
$I_s = 10^{-14}\,A$
$C_{be}\,(Cje) = 10^{-11}\,F$
$C_{bc}\,(Cjc) = 10^{-12}\,F$

VCC
5 Vdc
RC
10 k
RB
1 k
vi
VPULSE
Qbreakn
Q

(b) Time response from PSpice simulation

(c) Time response from MATLAB simulation

v_o

v_i

PSpice schematic

elec03e17:Qbreakn — PSpice Model Editor — □ ×

File Edit View Model Plot Tools Window – ☐ ×

Models List

Model Name	Type
Qbreakn	BJT

.model Qbreakn NPN (Bf=100
Br=1 Is=10f Rb=0 Cje=10p
Cjc=1p)

Figure 3.68 A BJT inverter and its time response from PSpice and MATLAB ("elec03e21.opj").

(Q1) What makes the salient part of $v_o(t)$ in the beginning of the rising period?

(A1) It is because the BJT Q does not turn on to make v_o low until v_{be} becomes higher than the turn-on voltage 0.5 V and in the meantime, v_{cb} (across the internal capacitance C_{bc} as shown in Figure 3.67(c)) stays at 5 V (due to the continuity of capacitor voltage) so that $v_o = v_{ce} = v_{cb} + v_{be} = 5.5$ V.

(Q2) As R_B and C_{be} increase, does the delay time during the falling period of $v_o(t)$ become longer or shorter?

(Q3) During the falling period of $v_o(t)$, why does it decrease, not exponentially with any time constant, but almost linearly as $v_i(t)$ increases?

(A3) As $v_i(t)$, and accordingly $v_{BE}(t) = v_B(t)$, increases over 0.5 V so that the BJT Q begins to turn on, C_{bc} gets discharged, not via any resistor, but only through the CEJ of Q so that $v_o = v_{Cbc}$ can decrease toward 0 V.

Figure 3.69(a) shows three BJT inverters, one with no load, one with a purely capacitive load, and one with an *RC* load. About Figure 3.69(b) showing their input and outputs, note the following:

(a)

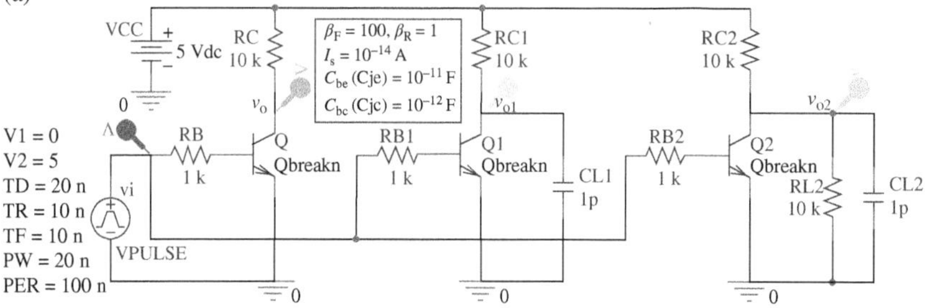

PSpice schematics for BJT inverters with or without capacitive loads

(b)

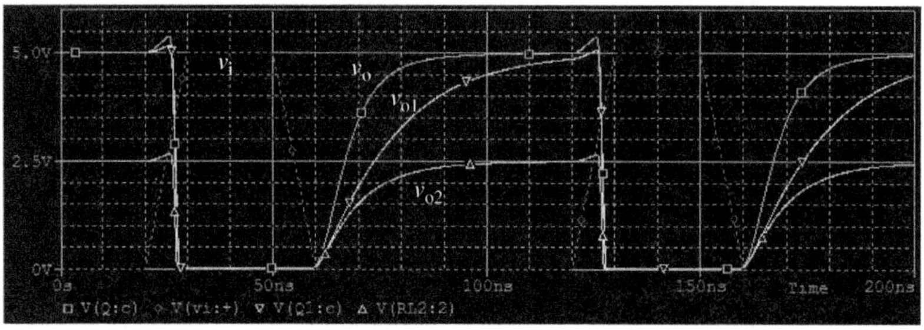

The input and output waveforms of the BJT inverters from PSpice simulation

Figure 3.69 BJT inverter with a capacitive load and its time response from PSpice ("elec03e21.opj").

- The time constant during the rising period of $v_{o1}(t)$ is about twice as long as that of $v_o(t)$ because the equivalent capacitance seen from R_C has increased from $C_{bc} = 1\,\text{pF}$ to $C_{bc}\|C_{L1} = C_{bc} + C_{L1} = 1 + 1 = 2\,\text{pF}$ so that the time constant can be doubled.
- The time constant during the rising period of $v_{o2}(t)$ is about half as long as that of $v_{o1}(t)$ because the equivalent resistance seen from $C_{bc}\|C_{L2}$ has decreased from $R_C = 10\,\text{k}\Omega$ to $R_{C2}\|R_{L2} = 10\,\text{k}\|10\,\text{k} = 5\,\text{k}\Omega$ so that the time constant can halve.

(Q4) During the rising period, why does $v_{o2}(t)$ reach just 2.5 V, which is just half of $V_{DD} = 5\,\text{V}$ unlike $v_o(t)$ or $v_{o1}(t)$?

(Q5) During the falling period, the outputs of all the FET inverters decrease almost linearly as $v_i(t)$ increases. Why is that?

Problems

3.1 Circuit with an NPN-BJT Forward Biased

Consider the BJT biasing circuit in Figure P3.1.1(a) where the device parameters of the BJT Q_1 are $\beta_F = 154$, $\beta_R = 6.092$, and I_s(saturation current) $= 14.34^{-15}$ A.

(a) Referring to Figure P3.1.1(b), find i_C and v_{CE} of the circuit with $R_C = 1$ kΩ, $R_E = 1$ kΩ, and $V_{CC} = 5$ V by hand. Then do the same thing two times, once (based on the piecewise linear [PWL] model) by using

(a)

A BJT biasing circuit

(b)

$$V_{BB} = \frac{R_2}{R_1 + R_2} V_{CC}$$

$$R_B = R_1 \| R_2$$

Forward-active mode

(c)

$$\begin{bmatrix} R_B + R_E & R_E \\ R_E & R_C + R_E \end{bmatrix} \begin{bmatrix} i_B \\ i_C \end{bmatrix} = \begin{bmatrix} V_{BB} - V_{BE,sat} \\ V_{CC} - V_{CE,sat} \end{bmatrix}$$

Saturation mode

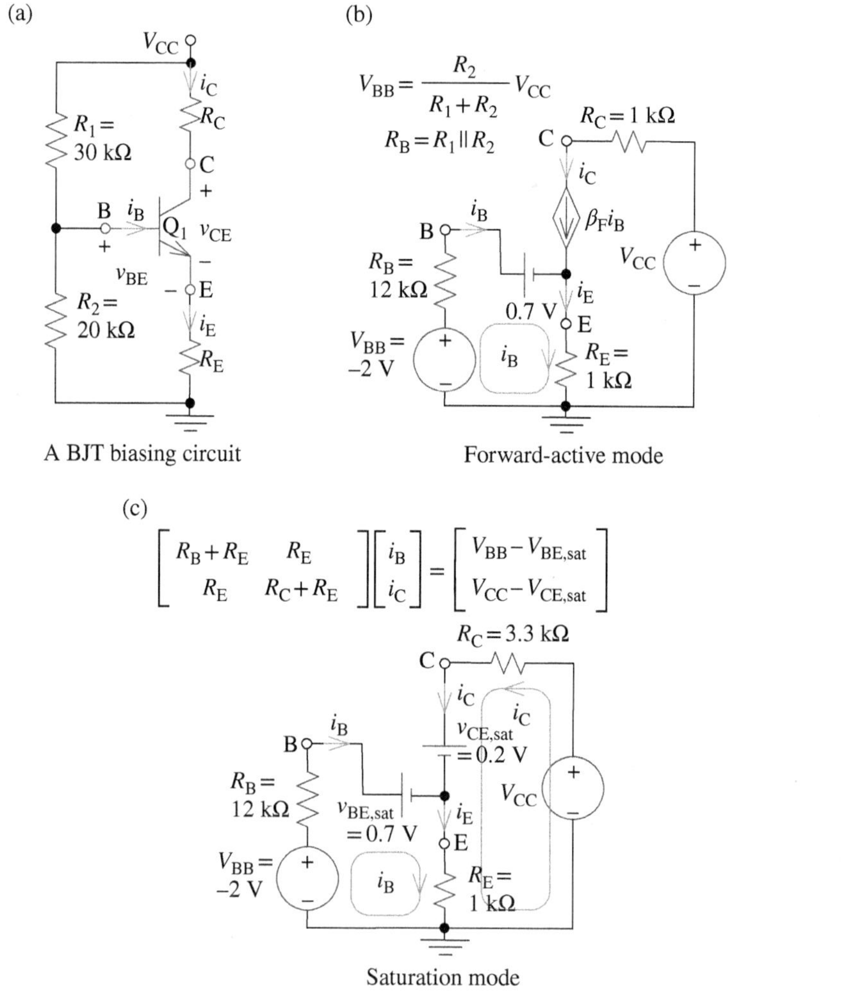

Figure P3.1.1 A BJT biasing circuit and its equivalents in different operation modes (regions) of BJT.

(a)

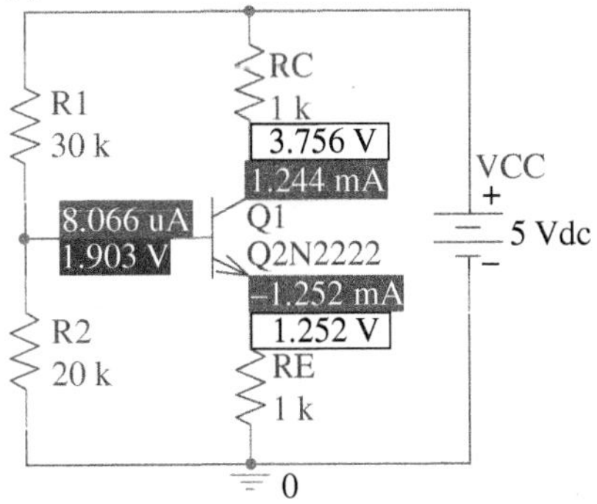

PSpice schematic for Problem 3.1(a)

(b)

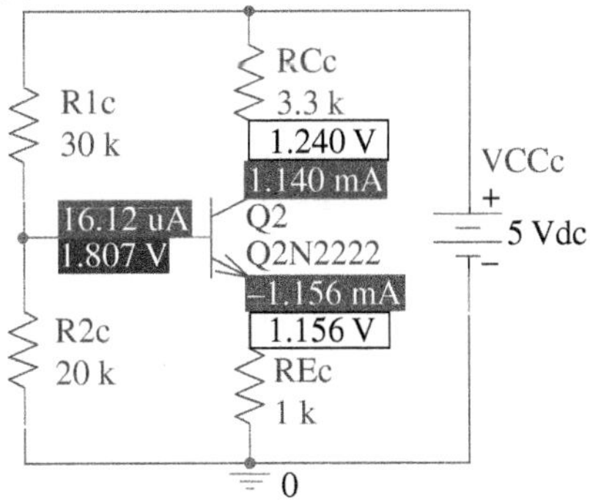

PSpice schematic for Problem 3.1(c)

(c)

** BIPOLAR JUNCTION TRANSISTORS

NAME	Q_Q1	Q_Q2
MODEL	Q2N2222	Q2N2222
IB	8.07E−06	1.61E−05
IC	1.24E−03	1.14E−03
VBE	6.51E−01	6.51E−01
VBC	−1.85E+00	5.67E−01
VCE	2.50E+00	8.39E−02
BETADC	1.54E+02	7.07E+01
GM	4.79E−02	4.42E−02
RPI	3.55E+03	3.57E+03
RX	1.00E+01	1.00E+01
RO	6.10E+04	5.26E+02
BETAAC	1.70E+02	1.58E+02

A part of PSpice output file

Figure P3.1.2 PSpice simulation results.

```
BJT_DC_analysis(VCC,R1,R2,RC,RE,[betaF,betaR]);
```

and once (based on the exponential model) by using

```
BJT_DC_analysis(VCC,R1,R2,RC,RE,[betaF,betaR,Is]);
```

(b) Noting that $v_{CE} = 0.3$ V at the edge of saturation, find such a value of R_C that Q_1 is just barely saturated where $R_E = 1$ kΩ and $V_{CC} = 5$ V.

(c) Referring to Figure P3.1.1(c), find i_B, i_C, and v_{CE} of the circuit with $R_C =$ 3.3 kΩ, $R_E = 1$ kΩ, and $V_{CC} = 5$ V by hand. Then do the same thing two times, once (based on the PWL model) by using

```
BJT_DC_analysis(VCC,R1,R2,RC,RE,[betaF,betaR]);
```

and once (based on the exponential model) by using

```
BJT_DC_analysis(VCC,R1,R2,RC,RE,[betaF,betaR,Is]);
```

(d) Referring to Figure P3.1.2, perform the PSpice simulation to see if the above two analysis results are fine and compare the results with the corresponding MATLAB analysis results.

3.2 Circuit with an NPN-BJT Reverse Biased

Consider the BJT biasing circuit in Figure P3.2.1(a) where the device parameters of the BJT Q_1 are $\beta_F = 154$, $\beta_R = 6.092$, and I_s(saturation

(a)

A BJT biasing circuit

(b)

$$V_{BB} = \frac{R_2}{R_1+R_2} V_{CC}$$

$$R_B = R_1 \| R_2 = 12 \text{ k}\Omega$$

Reverse-active mode

(c)

$$\begin{bmatrix} R_B+R_E & R_E \\ R_E & R_C+R_E \end{bmatrix} \begin{bmatrix} i_B \\ i_C \end{bmatrix} = \begin{bmatrix} V_{BB}-V_{BC,sat}+V_{EC,sat} \\ V_{CC}+V_{EC,sat} \end{bmatrix}$$

Reverse-saturation mode

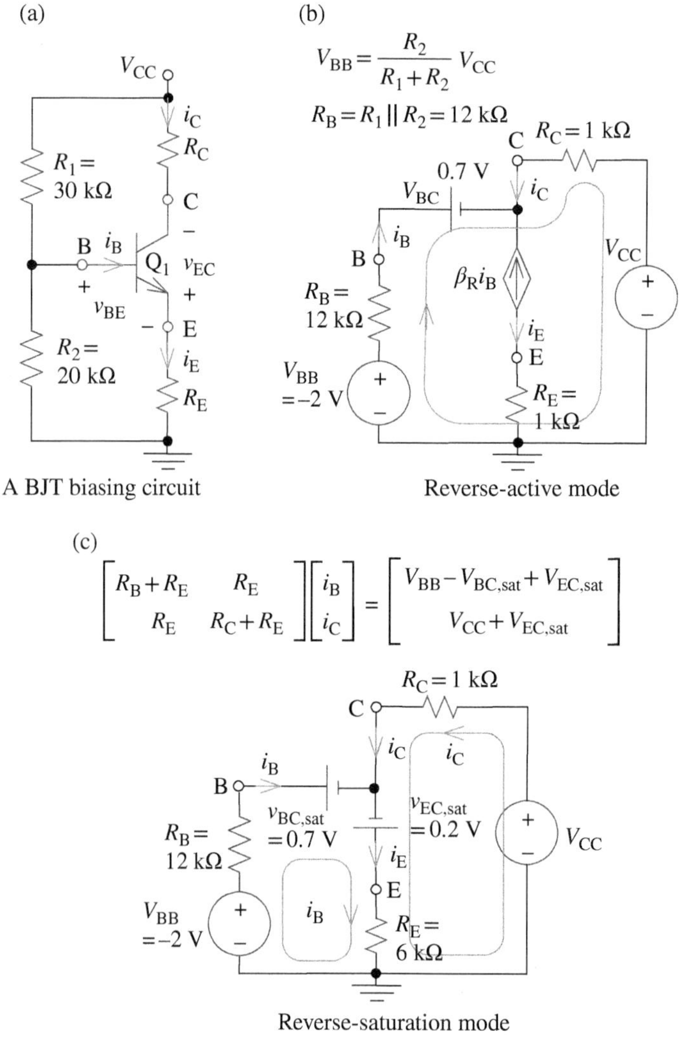

Figure P3.2.1 A BJT biasing circuit and its equivalents in different operation modes (regions) of BJT.

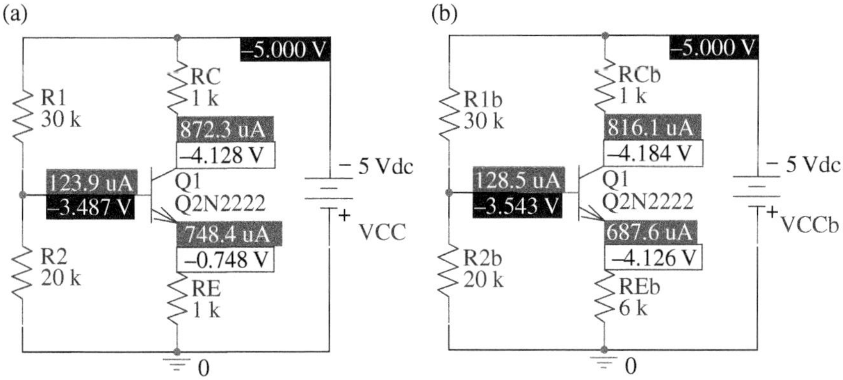

PSpice schematic for Problem 3.2(a)

PSpice schematic for Problem 3.2(c)

(c)

** BIPOLAR	JUNCTION	TRANSISTORS
NAME	Q_Q1	Q_Q2
MODEL	Q2N2222	Q2N2222
IB	1.24E-04	1.29E-04
IC	-8.72E-04	-8.16E-04
VBE	-2.74E+00	5.83E-01
VBC	6.41E-01	6.41E-01
VCE	-3.38E+00	-5.84E-02
BETADC	-7.04E+00	-6.35E+00
GM	-2.89E-02	-2.66E-02
RPI	2.56E+14	3.95E+04
RX	1.00E+01	1.00E+01
RO	3.46E+01	3.35E+01
BETAAC	-7.40E+12	-1.05E+03

A part of PSpice output file

Figure P3.2.2 PSpice simulation results.

current) = 14.34^{-15} A. Note that the roles of the collector and emitter terminals of a BJT are switched when the BJT is reverse biased.

(a) Referring to Figure P3.2.1(b), find i_C and v_{EC} of the circuit with $R_C = 1\ \text{k}\Omega$, $R_E = 1\ \text{k}\Omega$, and $V_{CC} = -5$ V by hand. Then do the same thing two times, once (based on the PWL model) by using

```
BJT_DC_analysis(VCC,R1,R2,RC,RE,[betaF,betaR]);
```

or

```
BJT_DC_analysis(VCC,VBB,RB,RC,RE,[betaF,betaR]);
```

and once (based on the exponential model) by using

```
BJT_DC_analysis_exp(VCC,R1,R2,RC,RE,[betaF,betaR,Is]);
```

(b) Referring to Figure P3.2.1(c), find i_C and v_{EC} of the circuit with $R_C = 1\ k\Omega$, $R_E = 6\ k\Omega$, and $V_{CC} = -5\ V$ by hand. Then do the same thing two times, once (based on the PWL model) by using

```
BJT_DC_analysis(VCC,R1,R2,RC,RE,[betaF,betaR]);
```

and once (based on the exponential model) by using

```
BJT_DC_analysis(VCC,R1,R2,RC,RE,[betaF,betaR,Is]);
```

(c) Referring to Figure P3.2.2, perform the PSpice simulation to see if the above two analysis results are fine and compare the results with the corresponding MATLAB analysis results.

3.3 Circuit with PNP-BJT

Consider the PNP-BJT biasing circuit in Figure P3.3.1(a) where the device parameters of the PNP-BJT Q_1 are $\beta_F = 100$, $\beta_R = 1$, and I_s(saturation current) $= 10 \times 10^{-15}$ A, respectively.

(a) Referring to Figure P3.3.1(b), find i_C and v_{EC} of the circuit with $R_C = 1\ k\Omega$, $R_E = 1\ k\Omega$, and $V_{EE} = 5\ V$ by hand. Then do the same thing two times, once (based on the PWL model) by using

```
BJT_PNP_DC_analysis(VEE,R1,R2,RE,RC,[betaF,betaR]);
```

and once (based on the exponential model) by using

```
BJT_DC_analysis_exp([VCC VEE],R1,R2,RC,RE,
                    -[betaF, betaR,Is]);
```

(b) Noting that $v_{EC} = 0.3\ V$ at the edge of saturation, find such a value of R_C that Q_1 is just barely saturated where $R_E = 1\ k\Omega$ and $V_{EE} = 5\ V$.

(c) Referring to Figure P3.3.1(c), find i_C and v_{EC} of the circuit with $R_C = 3.2\ k\Omega$, $R_E = 1\ k\Omega$, and $V_{EE} = 5\ V$ by hand. Then do the same thing two times, once (based on the PWL model) by using

```
BJT_PNP_DC_analysis(VEE,R1,R2,RE,RC,[betaF,betaR]);
```

and once (based on the exponential model) by using

```
BJT_DC_analysis_exp([VCC VEE],R1,R2,RC,RE,
                    -[betaF, betaR,Is]);
```

(d) Referring to Figure P3.3.1(d), find i_C and v_{CE} of the circuit with $R_C = 1\ k\Omega$, $R_E = 1\ k\Omega$, and $V_{EE} = -5\ V$ by hand. Then do the same thing two times, once (based on the PWL model) by using

```
BJT_PNP_DC_analysis(VEE,R1,R2,RE,RC,[betaF,betaR]);
```

and once (based on the exponential model) by using

```
BJT_DC_analysis_exp([VCC VEE],R1,R2,RC,RE,
                    [betaF,betaR,Is]);
```

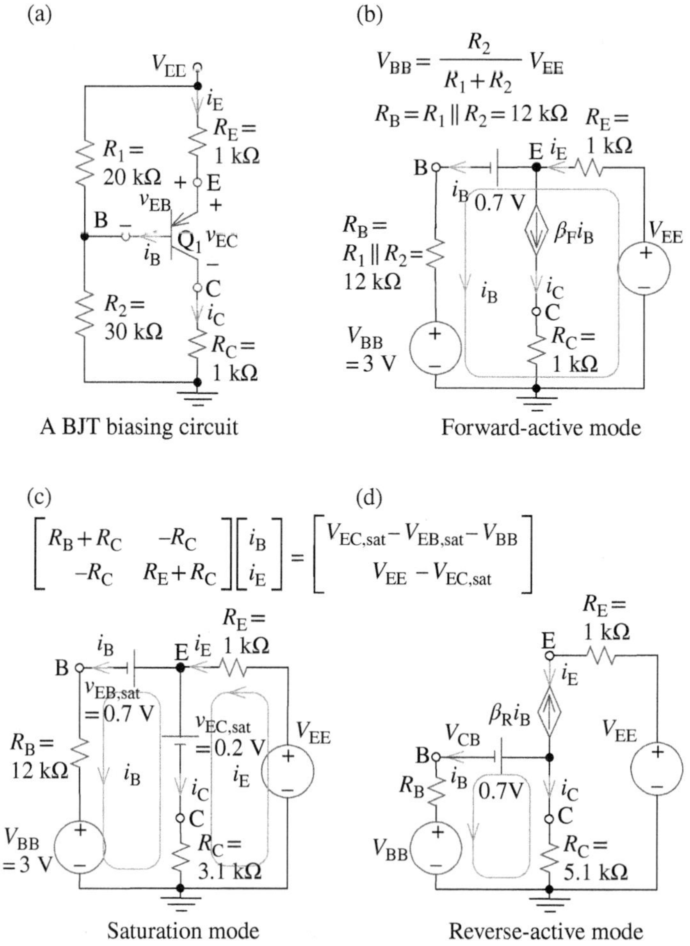

Figure P3.3.1 A PNP-BJT biasing circuit and its equivalents in different operation modes (regions) of BJT.

(e) Referring to Figure P3.3.2, perform the PSpice simulation to see if the above three analysis results are fine by comparing the results with the corresponding MATLAB analysis results obtained above. Note that the device parameters of the NPN-BJT Q_1 (QbreakN) and PNP-BJT Q_2 (QbreakP) picked up from the PSpice library 'breakout.olb' should be set in the PSpice Model Editor window as shown in Figure 3.14(b2).

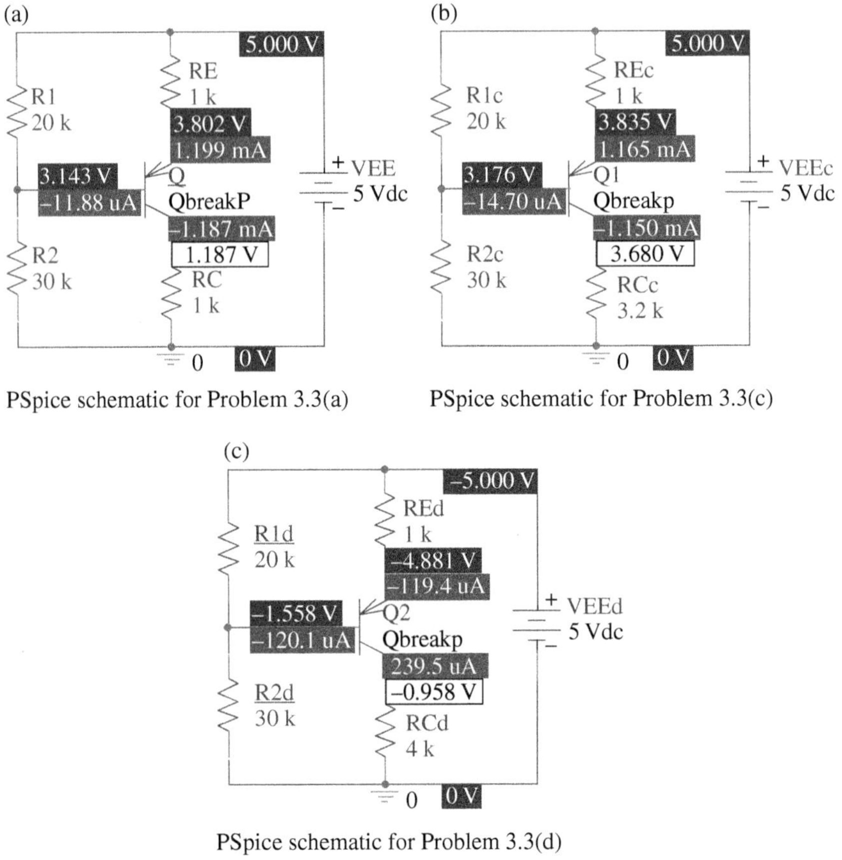

PSpice schematic for Problem 3.3(a)

PSpice schematic for Problem 3.3(c)

PSpice schematic for Problem 3.3(d)

Figure P3.3.2 PSpice simulation results.

3.4 Two-BJT Circuit

Consider the two-BJT circuit in Figure P3.4(a) where the device parameters of the BJTs Q_1 and Q_2 are $\beta_F = 100$, $\beta_R = 1$, and $I_s = 10^{-14}$ A in common.

(a) Referring to Example 3.7, analyze the circuit to find I_{C1}, V_{EC1}, I_{C2}, and V_{CE2}.

(b) Use the MATLAB function '`BJT_PNP_DC_analysis()`'/'`BJT_DC_analysis()`' to analyze the Q_1/Q_2 part, respectively. To this end, run the following MATLAB statements:

```
%elec03p04.m
clear
betaF=100; betaR=1; alphaR=betaR/(betaR+1);
Is=1e-14; Isc=Is/alphaR; VT=25e-3; % BJT parameters
VEE1=10; VCC2=10; R1=3e5; R2=2e5; RE1=55e2; RC1=5e3; RC2=4e3; RE2=4e3;
% To find the equivalent of the voltage divider biasing circuit
VBB1=VEE1*R2/(R1+R2); RB1=parallel_comb([R1 R2]);
% Exponential model based approach
iC=@(v) Is*exp(v(1)/VT)-Isc*exp(v(2)/VT);
iB=@(v) Is/betaF*exp(v(1)/VT)+Isc/(betaR+1)*exp(v(2)/VT);
% Eq. (S3.4.15) with v=[vEB1 vCB1 vBE2 vBC2]
eq=@(v) [VEE1-VBB1-v(?)-RE1*(iC(v(1:2))+iB(v(1:2)))-RB1*iB(v(1:2));
    VBB1+RB1*iB(v(1:2))+v(?)-RC1*(iC(v(1:2))-iB(v(?:?)));
    VCC2+v(?)-RC2*iC(v(3:4))-RC1*(iC(v(?:?))-iB(v(3:4)));
    VCC2+v(?)-v(3)-RC2*iC(v(3:4))-RE2*(iC(v(3:4))+iB(v(3:4)))];
options=optimset('Display','off','Diagnostics','off');
v0 = [0.7; 0.4; 0.7; 0.4]; % Initial guess for v=[vEB1 vCB1 vBE2 vBC2]
v = fsolve(eq,v0,options);
VEB1=v(1), VCB1=v(2), VBE2=v(3), VBC2=v(4)
IB1Q = iB(v(1:2)), IC1Q = iC(v(1:2))
IB2Q = iB(v(3:4)), IC2Q = iC(v(3:4))
VE1Q = VEE1-RE1*(IC1Q+IB1Q), VB1Q=VE1Q-VEB1, VC1Q=VB1Q+VCB1
VC2Q = VCC2-RC2*IC2Q, VB2Q=VC2Q+VBC2, VE2Q=VB2Q-VBE2
VEC1Q = VE1Q-VC1Q, VCE2Q=VC2Q-VE2Q
```

```
>>betaF=100; betaR=1; Is=1e-14; beta=[betaF betaR];
 VEE1=10; VCC2=10; R1=3e5; R2=2e5;
 RE1=55e2; RC1=5e3; RC2=4e3; RE2=4e3;
 [VB1,VE1,VC1,IB1,IE1,IC1,model]= ...
         BJT_PNP_DC_analysis(VEE1,R1,R2,RE1,RC1,beta);
VBB2=???; RB2=???; % Thevenin equivalent of Q1 seen from B2
BJT_DC_analysis(VCC2,VBB2,RB2,RC2,RE2,beta);
```

Also, referring to the MATLAB script "elec03e07.m," complete the above script "elec03p04.m" and run it to do the exponential-model-based analysis.

(c) Referring to Figure P3.4(c), use the PSpice software to simulate the circuit where the device parameters of the NPN-BJT Q_1 (QbreakN) and PNP-BJT Q_2 (QbreakP) picked up from the PSpice library 'breakout.olb' should be set in the PSpice Model Editor window as shown in Figure 3.14(b2). Fill Table P3.4 with the PSpice simulation results and the MATLAB analysis results (obtained in (b)). Which one is closer to the PSpice simulation results, the PWL-model-based solution or the exponential-model-based solution?

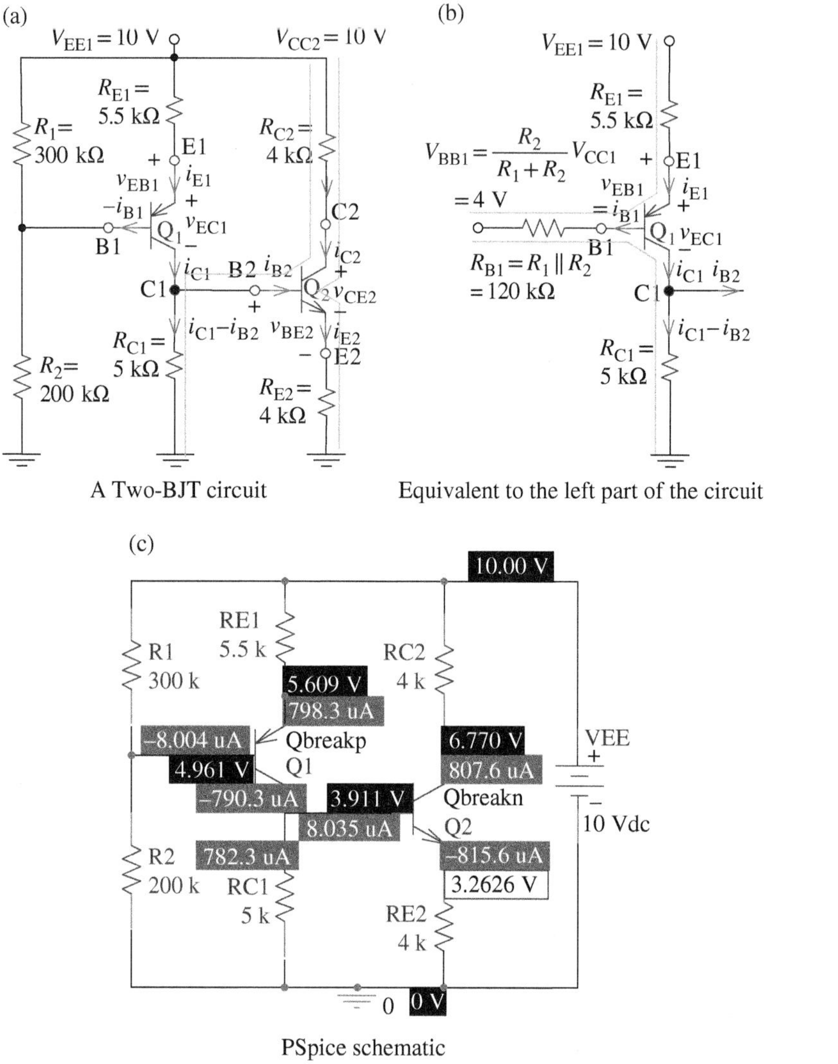

Figure P3.4 A two-BJT circuit for Problem 3.4.

Table P3.4 MATLAB analysis and PSpice simulation results of BJT circuit in Figure P3.4(a).

	I_{C1}	V_{EC1}	I_{C2}	V_{CE2}
PWL model based analysis	0.785 mA			
Exponential-model-based analysis		1.65 V		3.42 V
PSpice simulation			0.808 mA	

3.5 Three-BJT Circuit

Consider the three-BJT circuit in Figure P3.5(a) where the device parameters of the BJTs Q_1, Q_2, and Q_3 are $\beta_F = 100$, $\beta_R = 1$, and $I_s = 10^{-14}$ A in common.

(a) Analyze the circuit to find I_{C1}, V_{EC1}, I_{C2}, V_{CE2}, I_{C3}, and V_{EC3}. Noting that the Q_1-Q_2 part of the circuit is identical to the circuit of Figure P3.4(a), you can copy all the analysis results (except for V_{C2}) of Problem 3.4.

(a)

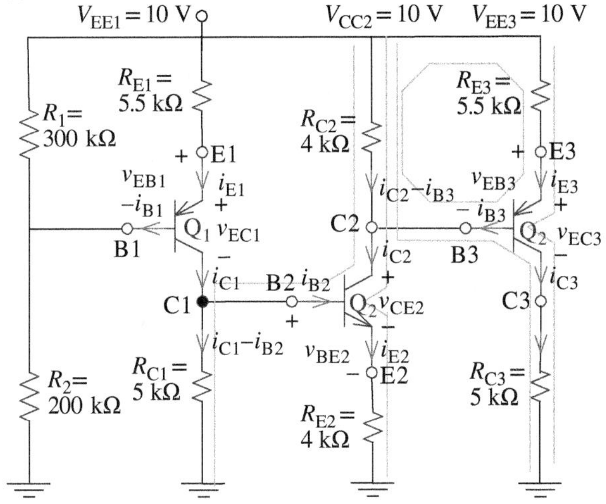

A Three-BJT circuit

(b)

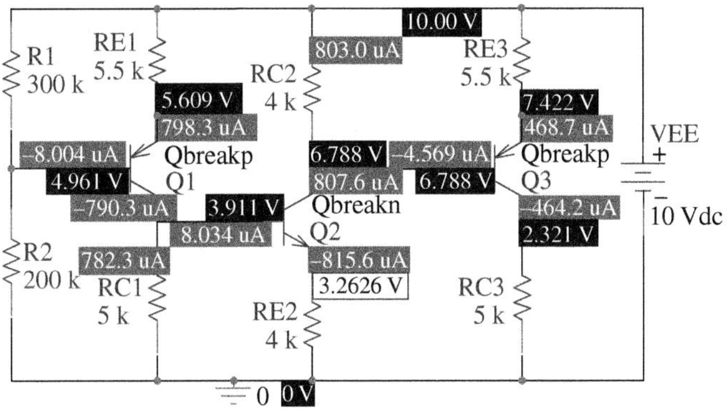

PSpice schematic

Figure P3.5 A three-BJT circuit for Problem 3.5.

Table P3.5 MATLAB analysis and PSpice simulation results of the BJT circuit in Figure P3.5(a).

	I_{C1} (mA)	V_{EC1} (V)	I_{C2} (mA)	V_{CE2} (V)	I_{C3} (mA)	V_{CE3} (V)
PWL-model-based analysis	0.785	1.76	0.788	3.69	0.438	5.37
Exponential-model-based analysis	0.795	1.65	0.819	3.42	0.475	4.98
PSpice simulation	0.790	1.70	0.808	3.51	0.464	5.10

(b) With the analysis result of the Q_1/Q_2 part (except for V_{C2}) obtained in Problem 3.4, go on to use the MATLAB function '`BJT_PNP_DC_analysis()`' for the analysis of the Q_3 part.

(c) Referring to Figure P3.5(b), use the PSpice software to simulate the circuit. Fill Table P3.5 with the PSpice simulation results and the MATLAB analysis results (obtained in (b)).

3.6 Complementary BJT Pair Circuit

```
function [vs,iis,ios,iC1s,iC2s,iB1s,iB2s] = ...
      BJT2_complementary(vis, Ri,RL,VCC,beta,Is)
% To analyze a complementary BJT pair circuit which consists of
% NPN/PNP-BJTs and resistors Ri/RL between vi/E and B/GND.
VT=25e-3; betaF=beta(1); betaR=beta(2);
alphaR=betaR/(betaR+1); Isc=Is/alphaR; % CB saturation current
V1=VCC(1); V2=VCC(2);
iC=@(v)Is*exp(v(1)/VT)-Isc*exp(v(2)/VT); % Eq.(3.1.23a)
iB=@(v)Is/betaF*exp(v(1)/VT)+ ...
Isc/(betaR+1)*exp(v(2)/VT);%Eq.(3.1.23b)
iE = @(v)Is*(1+1/betaF)*exp(v(1)/VT)- ...
Isc*betaR/(betaR+1)*exp(v(2)/VT);
options=optimset('Display','off','Diagnostics','off');
for n=1:length(vis)
  vi=vis(n); v0=[vi/4 vi/8]; % Initial guess
  eq = @(v)[vi-Ri*(iB([v(?)-v(2),v(1)-V1]) ...
                -iB([v(?)-v(1),V2-v(1)]))-v(1);
             v(2)-RL*(iE([v(1)-v(2),v(?)-V1]) ...
                -iE([v(2)-v(1),V2-v(?)]))]; % Eq.(P3.6.2)
  [v,fe] = fsolve(eq,v0,options); vs(n,:)=v;
  iB1s(n)=iB([v(1)-v(2),v(1)-V1]); iC1s(n)=iC([v(1)-v(2),v(1)-V1]);
  iB2s(n)=iB([v(2)-v(1),V2-v(1)]); iC2s(n)=iC([v(2)-v(1),V2-v(1)]);
  iis(n)=(vi-v(?))/Ri; ios(n)=v(?)/RL;
end
```

Consider the complementary BJT pair circuit in Figure P3.6.1(a) where the device parameters of the NPN-/PNP-BJTs Q_1/Q_2 are $\beta_F = 100$, $\beta_R = 1$, and I_s (saturation current) = 10×10^{-15} A. Note that the Kirchhoff's voltage law (KVL) along the mesh B1-E1-E2-B2 yields

$$v_{BE1} + v_{EB2} = 0 \tag{P3.6.1}$$

This implies that both v_{BE1} and v_{EB2} cannot be positive so that Q_1 and Q_2 cannot be simultaneously conducting.

(a) For v_i = 10 V, find i_i, i_o, and v_o.
(b) For v_i = 5 V, find i_i, i_o, and v_o.

(a)

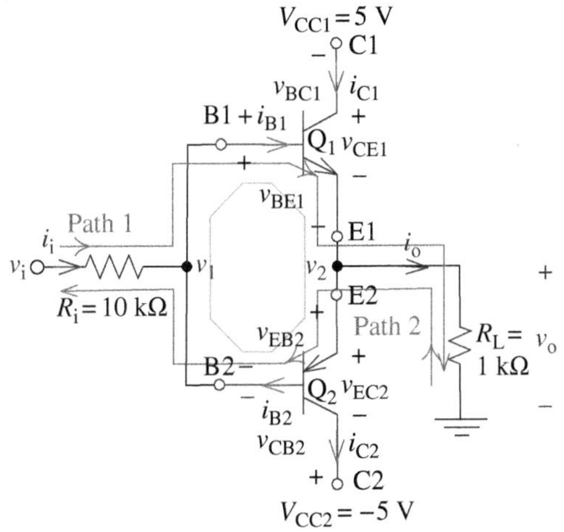

A Complementary BJT pair circuit

(b)

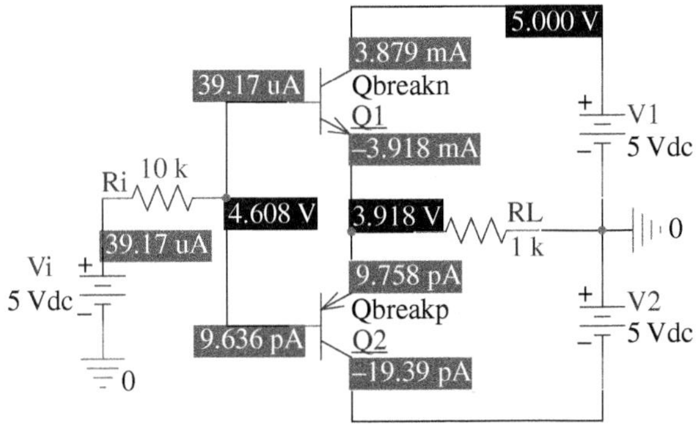

Its PSpice schematic

Figure P3.6.1 A BJT biasing circuit and its PSpice schematic.

(c) For $v_i = 0$ V, find i_i, i_o, and v_o.
(d) For $v_i = -5$ V, find i_i, i_o, and v_o.
(e) For $v_i = -10$ V, find i_i, i_o, and v_o.
(f) Complete the above MATLAB function 'BJT2_complementary()' to analyze the circuit and use it to get i_i, i_o, and v_o for $v_i = \{10, 5, 0, -5, -10\}$ by running the following MATLAB statements:

```
>>Ri=1e4; RL=1e3; betaF=100; betaR=1;
  Is=1e-14; beta=[betaF betaR];
  VCC1=5; VCC2=-5; VCC=[VCC1 VCC2]; vis=[10 5 0 -5 -10];
  [vs,iis,ios]=BJT2_complementary(vis,Ri,RL,VCC,beta,Is);
  fprintf(' vi v1 vo ii io\n')
  disp([vis' vs iis' ios'])
```

(g) Referring to Figure P3.6.1(b), perform the PSpice simulation with the Analysis type of Bias Point for $v_i = 5$ V.
(h) Perform the PSpice simulation (with the Analysis type of DC Sweep and Sweep variable Vi) and the MATLAB analysis (using the MATLAB function 'BJT2_complementary()') for $v_i = [-10:0.02:10]$V to get the two graphs shown in Figure P3.6.2, respectively.

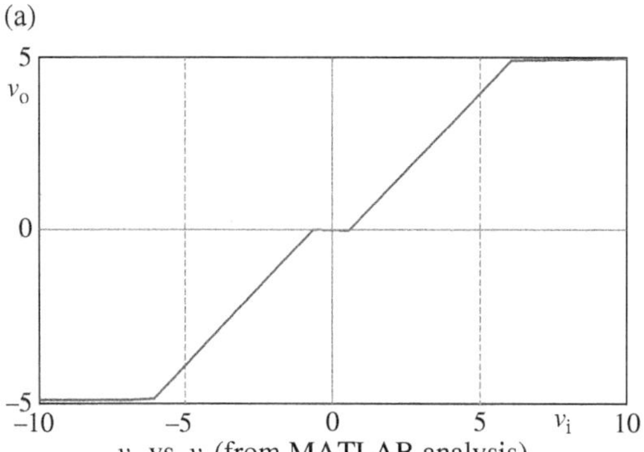

v_o vs. v_i (from MATLAB analysis)

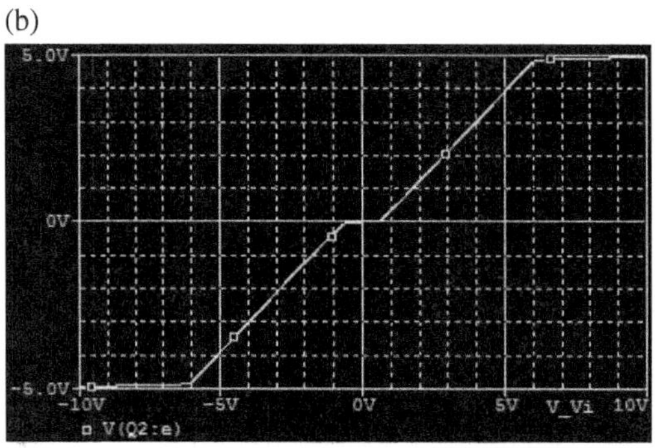

$v_o$vs. v_i (from PSpice simulation)

Figure P3.6.2 v_o versus v_i of the complementary BJT circuit in Figure P3.6.1.

3.7 Current Mirror Consisting of Two BJTs

(a) Consider the current mirror in Figure 3.25(a) where the device parameters of the NPN-BJTs Q_1 and Q_2 are $\beta_F = 100$, $\beta_R = 1$, and I_s (saturation current) $= 10 \times 10^{-15}$ A in common. Referring to the MATLAB function '`BJT3_current_mirror()`' presented in Section 3.1.9, compose a MATLAB function '`BJT2_current_mirror()`' and use it to get $i_o = i_{C2}$ for V1 = 15 V and V2 = {1, 5, 10, 20, 40} by running the following MATLAB statements:

```
>>betaF=100; betaR=1; Is=1e-14; beta=[betaF betaR];
  R=1e4; V1=15; V2=[1 5 10 20 40];
  io=BJT2_current_mirror(beta,Is,R,[V1 V2])
```

(b) Consider the current mirror in Figure P3.7(a) where the device parameters of the NPN-BJTs Q_1, Q_2, and Q_3 are $\beta_F = 100$, $\beta_R = 1$, and I_s

(a)

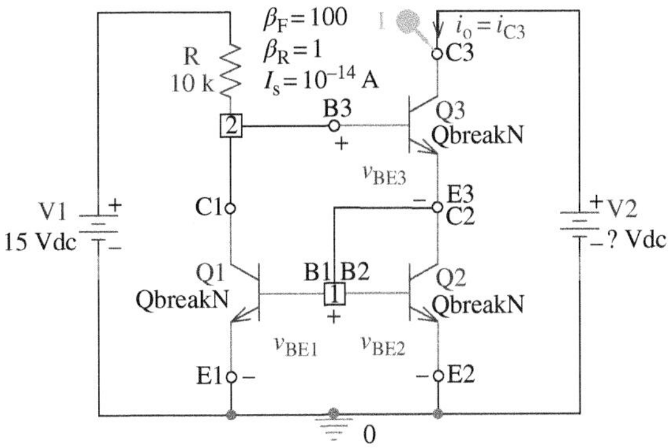

Wilson current mirror

(b)

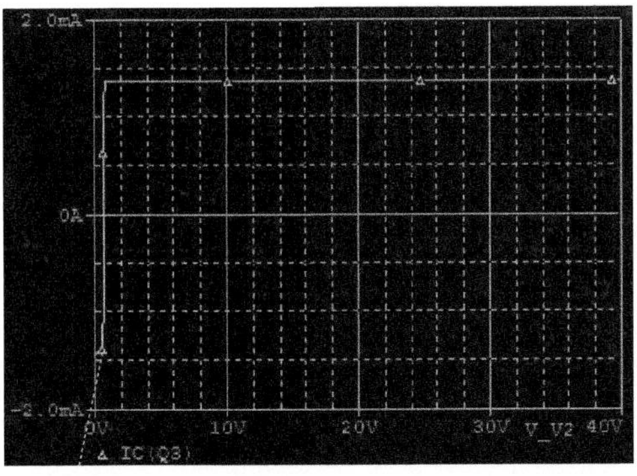

Its PSpice simulation result with DC Sweep analysis ("elec03p07.opj")

Figure P3.7 Wilson current mirror.

(saturation current) = 10×10^{-15} A in common. Show that the output current $i_o = i_{C3}$ is determined as

$$i_o = i_{C3} = \beta_F i_{B3} = \frac{\beta_F + 2}{\beta_F + 1} i_{C2} = \frac{\beta_F + 2}{\beta_F + 1} \frac{\beta_F(\beta_F + 1) I_R}{\beta_F(\beta_F + 1) + \beta_F + 2} \approx \frac{V_1 - v_{BE1} - v_{BE3}}{R} = 1.36\,\text{mA} \tag{P3.7.1}$$

This is supported by the PSpice simulation result (with DC Sweep analysis) shown in Figure P3.7(b). However, why is the output current negative for V2 < 0.66 V?

(c) Referring to the MATLAB function '`BJT3_current_mirror()`' presented in Section 3.1.9, compose a MATLAB function '`BJT3_current_mirror_Wilson()`' to analyze the current mirror of Figure P3.7(a) and use it to get the plot of $i_o = i_{C3}$ versus V2 = {0:0.01:40} for V1 = 15 V.

```
function [io,v]=BJT3_current_mirror_Wilson(betaF,Is,R,V12,VT)
% Analyze a Wilson current mirror consisting of 3 BJTs and an R or I-source
% If R<1, it will be regarded as a current source.
% Copyleft: Won Y. Yang, wyyang53@hanmail.net, CAU for academic use only
if nargin<5, VT=(273+27)/11605; end % Thermal voltage
if length(betaF)<2, betaR=1; else betaR=betaF(2); betaF=betaF(1); end
if length(Is)<2, VA=inf; else VA=Is(2); Is=Is(1); end
V1=V12(1); if length(V12)>1, V2s=V12(2:end); else V2s=V12; end
alphaR=betaR/(betaR+1); Isc=Is/alphaR; % CB saturation current
options=optimset('Display','off','Diagnostics','off');
iC = @(v) Is*exp(v(1)/VT)-Isc*exp(v(2)/VT);
% Eq.(3.1.23a) with v=[vBE vBC]
iB = @(v) Is/betaF*exp(v(1)/VT)+Isc/(betaR+1)*exp(v(2)/VT);
% Eq.(3.1.23b)
iE = @(v) Is*(1+1/betaF)*exp(v(1)/VT)- ...
Isc*betaR/(betaR+1)*exp(v(2)/VT);
for n=1:length(V2s)
  V2=V2s(n);
  if R>=1 % Eq.(3.1.62a,b) with a resistor
  eq=@(v) [iE([v(2)-v(1) v(2)-V2])-iB([v(1) v(1)-v(2)])-iE([v(1) 0]);
        V1-v(2)-R*(iB([v(2)-v(1) v(2)-V2])+iC([v(1) v(1)-v(2)]))];
  else I=R; % Eq.(3.1.62a,c) with a current source
  eq=@(v) [iE([v(2)-v(1) v(2)-V2])-iB([v(1) v(1)-v(2)])-iE([v(1) 0]);
        I-iB([v(2)-v(1) v(2)-V2])-iC([v(1) v(1)-v(2)])]*1e5;
  end
  if n<2, if V2<0.7, v0=[0 0.7]; else v0=[0.7 1.4]; end
   else v0=v; % Initial guess
  end
  v=fsolve(eq,v0,options); vs(n,:)=v; io(n)=iC([v(2)-v(1) v(2)-V2]);
end
```

3.8 BJT Inverter

Consider the BJT inverter in Figure P3.8(a) or 3.27(a) where the device parameters of thc NPN BJT Q arc $\beta_F = 100$, $\beta_R - 1$, and I_s (saturation current) = 10×10^{-15} A.

(a) Complete the following approximate v_i-v_o relationship and find the approximate slope of the VTC of the inverter with the BJT Q_1 operating in the forward-active mode:

$$v_o = V_{CC} - R_C i_C = V_{CC} - R_C \beta_F i_B \approx V_{CC} - \boxed{} (v_i - V_{BE}) \text{ with } V_{BE} = 0.65\,\text{V} \tag{P3.8.1}$$

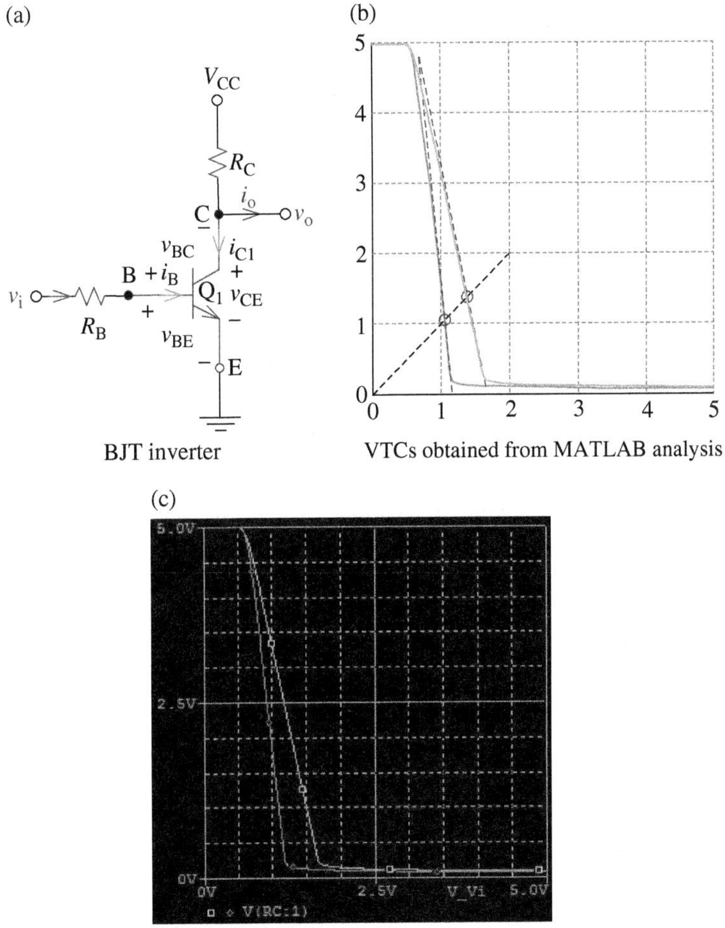

Figure P3.8 A BJT inverter and its voltage transfer characteristic curves (VTCs) obtained using MATLAB/PSpice.

(b) Complete the following MATLAB script "elec03p08b.m," which uses the MATLAB function '`BJT_inverter()`' presented in Section 3.1.10 to plot the VTCs and their linear approximations for R_B = 10 kΩ/20 kΩ where R_C = 1 kΩ and V_{CC} = 5 V as shown in Figure P3.8(b).

```
%elec03p08b.m
clear, clf
VCC=5; RC=1e3;
betaF=100; betaR=1; Is=1e-14;
RBs=[1e4 2e4]; gss='rgb';
dvi=5e-3; vis=[0:dvi:VCC];
VBE=0.65;
for n=1:length(RBs)
  RB=RBs(n);
  [VIL,VIH,VOL,VOH,VM,NML,NMH,PCavg,vos]= ...
     BJT_inverter([betaF betaR],Is,RB,RC,VCC,vis);
  vis1=[VBE VIH];
  vos1 = VCC-betaF*R?/R?*(vis1-VBE); %Eq.(P3.8.1)
  hold on, plot(vis,???,gss(n), vis1,????,':')
end
```

(c) Perform the PSpice simulation (with the Analysis type of DC Sweep and Sweep variable Vi) to plot the VTCs for R_B = 10 kΩ/20 kΩ where R_C = 1 kΩ and V_{CC} = 5 V as shown in Figure P3.8(c). Note that to have multiple waveforms (obtained from different simulations) plotted in a single graph in the Probe (PSpice A/D) window, select File>Append_Waveform(.DAT) on the menu bar to pick up the existing data file(s). If you want to save a data file obtained from the current simulation, say, to plot it together with another data (to be obtained afterward), rename the data file appropriately, which can be located by selecting the menu File>Append_Waveform(.DAT).

3.9 ECL (Emitter-Coupled Logic) – Differential Pair

Consider the two BJT differential pairs, called emitter-coupled logic (ECL), each in Figure P3.9(a1) and (a2) where the device parameters of all the NPN-BJTs are β_F = 100, β_R = 1, and I_s (saturation current) = 10×10^{-15}A in common.

(a) Referring to Eq. (3.1.72), which shows v_{o1} and v_{o2} of the ECL in Figure 3.29, and assuming that both BJTs Q_1 and Q_2 operate in the forward active region, complete the following expressions (in terms of the differential input voltage v_i) for v_{o1} and v_{o2} of the ECL in Figure P3.9(a1):

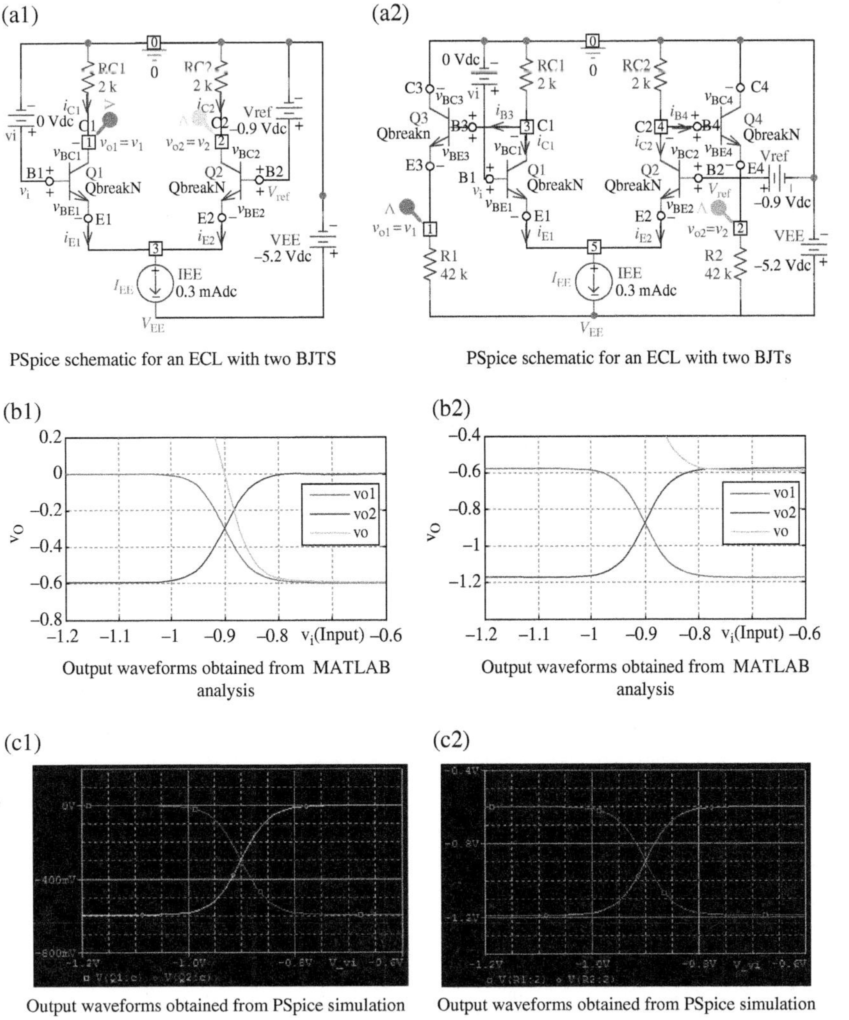
(a1)
PSpice schematic for an ECL with two BJTS
(a2)
PSpice schematic for an ECL with two BJTs
(b1)
Output waveforms obtained from MATLAB analysis
(b2)
Output waveforms obtained from MATLAB analysis
(c1)
Output waveforms obtained from PSpice simulation
(c2)
Output waveforms obtained from PSpice simulation

Figure P3.9 Two ECLs, one with two BJTs and another with four BJTs, and their output waveforms.

$$v_{o1} = -R_{C1}i_{C1} \overset{(3.1.71a)}{=} -\frac{\alpha_F I_{EE} R_{C1}}{1+\boxed{}}, \tag{P3.9.1a}$$

$$v_{o2} = -R_{C2}i_{C2} \overset{(3.1.71b)}{=} -\frac{\alpha_F I_{EE} R_{C1}}{1+\boxed{}} \tag{P3.9.1b}$$

(b) Compared with the ECL in Figure P3.9(a1), the ECL in Figure P3.9(a2) has two more BJTs Q_3 and Q_4 that are expected to shift v_{o1} and v_{o2} (up/down) by one base-emitter drop $V_{BE3} \approx 0.6$ V and $V_{BE4} \approx 0.6$ V, respectively, as can be seen by comparing the output waveforms of Figure P3.9(b1) and (b2) (MATLAB analysis results) or c1 and c2 (PSpice simulation results).

```
function [vo1s,vo2s,iC1s,iC2s]=ECL1(betaF,Is,RC,IEE,Vref,VEE,vis)
% Analyze an ECL (Fig.P3.9(a1)) consisting of 2 NPN-BJTs
% Copyleft: Won Y. Yang, wyyang53@hanmail.net, CAU for academic use only
VT=(273+27)/11605; VBE=0.7; % Thermal voltage
if nargin<7, dvi=VT/10; vis=Vref+[-12*VT:dvi:12*VT]; end % Input
if length(betaF)<2, betaR=1; else betaR=betaF(2); betaF=betaF(1); end
alphaF=betaF/(1+betaF); alphaR=betaR/(betaR+1);
Isc=Is/alphaR; % CB(Collector-Base) saturation current
option=optimset('Display','off','Diagnostics','off');
iC = @(v) Is*exp(v(1)/VT)-Isc*exp(v(2)/VT);
                                    % Eq.(3.1.23a) with v=[vBE vBC]
iB = @(v) Is/betaF*exp(v(1)/VT)+Isc/(betaR+1)*exp(v(2)/VT);
                                    % Eq.(E3.1.23b)
iE = @(v) Is*(1+1/betaF)*exp(v(1)/VT)- ...
                              Isc*betaR/(betaR+1)*exp(v(2)/VT);
for n=1:length(vis)
  vi = vis(n);
  eq=@(v) [-v(?)-RC*iC([vi-v(3) vi-v(?)]); % Eq.(P3.9.2)
          -v(?)-RC*iC([Vref-v(3) Vref-v(?)]);
          (iE([vi-v(?) vi-v(1)])+iE([Vref-v(?) Vref-v(2)])-IEE)*1e4];
  if ~exist('v') % Initial guess
    if vi>Vref, v12=-RC*[alphaF 1-alphaF]*IEE; v3=vi-VBE;
      else v12=-RC*[1-alphaF alphaF]*IEE; v3=Vref-VBE;
     end
     v0=[v12 v3];
    else v0=v;
  end
  v = fsolve(eq,v0,option);
  vo1s(n)=v(1); vo2s(n)=v(2);
  vos(n)=v(1)-v(2); % Eq.(3.1.72)
  iC1s(n)=iC([vi-v(3) vi-v(1)]);
  iC2s(n)=iC([Vref-v(3) Vref-v(2)]);
end
if nargout==0|nargin<7
 subplot(221)
 vo1s_a=-alphaF*IEE*RC./(1+exp(-(vis-Vref)/VT)); % Eq.(3.1.72)
 vo2s_a=-alphaF*IEE*RC./(1+exp((vis-Vref)/VT)); % Eq.(3.1.72)
 plot(vis,vo1s,'r', vis,vo2s, vis,vo1s_a,'k:', vis,vo2s_a,'m:')
 legend('vo1','vo2');
 xlabel('v_i(Input)'); ylabel('v_o'); grid on
end
```

(c) To analyze the ECL circuit of Figure P3.9(a1) using the exponential model of BJT, apply KVL across each of R_{C1} and R_{C2} and apply KCL at node 3 to write the following equations:

```
function [vo1s,vo2s,iC1s,iC2s] = ...
                  ECL2(betaF,Is,RC,R,IEE,Vref,VEE,vis)
% Analyze an ECL (Fig.P3.9(a2)) consisting of 4 NPN-BJTs
% Copyleft: Won Y. Yang, wyyang53@hanmail.net, CAU for academic use only
VT=(273+27)/11605; VBE=0.7; VBE1=0.6; % Thermal voltage
if nargin<8, dvi=VT/10; vis=Vref+[-12*VT:dvi:12*VT]; end % Input
if length(betaF)<2, betaR=1; else betaR=betaF(2); betaF=betaF(1); end
alphaF=betaF/(1+betaF); alphaR=betaR/(betaR+1);
Isc=Is/alphaR; % CB(Collector-Base) saturation current
option=optimset('Display','off','Diagnostics','off');
iC = @(v) Is*exp(v(1)/VT)-Isc*exp(v(2)/VT);
                                  % Eq.(3.1.23a) with v=[vBE vBC]
iB = @(v) Is/betaF*exp(v(1)/VT)+Isc/(betaR+1)*exp(v(2)/VT);
                       % Eq.(3.1.23b)
iE = @(v) Is*(1+1/betaF)*exp(v(1)/VT)-...
         Isc*betaR/(betaR+1)*exp(v(2)/VT);
for n=1:length(vis)
  vi = vis(n);
  eq=@(v) [v(?)-VEE-R*iE([v(3)-v(?) v(3)]);
        v(?)-VEE-R*iE([v(4)-v(?) v(4)]); % Eq.(P3.9.3)
        -v(?)-RC*(iC([vi-v(5) vi-v(?)])+iB([v(?)-v(1) v(?)]));
        -v(?)-RC*(iC([Vref-v(5) Vref-v(?)])+iB([v(4)-v(2) v(4)]));
        (iE([vi-v(?) vi-v(3)])+iE([Vref-v(?) Vref-v(4)])-IEE)*1e4];
  if ~exist('v') % Initial guess
   if vi>Vref, v34=-RC*[alphaF 1-alphaF]*IEE; v12=v34-VBE; v5=vi-VBE;
    else v34=-RC*[1-alphaF alphaF]*IEE; v12=v34-VBE; v5=Vref-VBE;
   end
   v0=[v12 v34 v5];
  else v0=v;
  end
  [v,fe] = fsolve(eq,v0,option);
  vo1s(n)=v(1); vo2s(n)=v(2); vos(n)=v(1)-v(2); % Eq.(3.1.72)
  iC1s(n)=iC([vi-v(5) vi-v(3)]); iC2s(n)=iC([Vref-v(5) Vref-v(4)]);
end
if nargout==0|nargin<8
 subplot(222)
 vo1s_a=-alphaF*IEE*RC./(1+exp(-(vis-Vref)/VT))-VBE1; % Eq.(3.1.72)
 vo2s_a=-alphaF*IEE*RC./(1+exp((vis-Vref)/VT))-VBE1; % Eq.(3.1.72)
 plot(vis,vo1s,'r', vis,vo2s, vis,vo1s_a,'k:', vis,vo2s_a,'m:')
 legend('vo1','vo2'); xlabel('v_i(Input)'); ylabel('v_o');
end
```

$$-v_1 - R_{C1}i_{C1}(v_{BE1}, v_{BC1}) = 0 \tag{P3.9.2a}$$

$$-v_2 - R_{C2}i_{C2}(v_{BE2}, v_{BC2}) = 0 \tag{P3.9.2b}$$

$$i_{E1}(v_{BE1}, v_{BC1}) + i_{E2}(v_{BE2}, v_{BC2}) - I_{EE} = 0 \tag{P3.9.2c}$$

where $i_{Ck}(v_{BEk}, v_{BCk})$ and $i_{Bk}(v_{BEk}, v_{BCk})$ are defined by Eqs. (3.1.23a,b), and $i_{Ek}(\cdot,\cdot) = i_{Bk}(\cdot,\cdot) + i_{Ck}(\cdot,\cdot)$ for all k. Then complete the above MATLAB function '`ECL1()`' so that it can analyze the ECL and run the following statements to plot the output waveforms as shown in Figure P3.9(b1).

```
>>betaF=100; betaR=1; Is=1e-14; RC=2e3; IEE=3e-4;
  Vref=-0.9; VEE=-5.2;
  ECL1([betaF betaR],Is,RC,IEE,Vref,VEE);
```

(d) To analyze the ECL circuit of Figure P3.9(a2) using the exponential model of BJT, apply KVL across each of R_{C1}, R_{C2}, R_1, and R_2 and apply KCL at node 5 to write the following equations:

$$v_1 - V_{EE} - R_1 i_{E3}(v_{BE3}, v_{BC3}) = 0 \tag{P3.9.3c}$$

$$v_2 - V_{EE} - R_2 i_{E4}(v_{BE4}, v_{BC4}) = 0 \tag{P3.9.3d}$$

$$-v_3 - R_{C1}\{i_{C1}(v_{BE1}, v_{BC1}) + i_{B3}(v_{BE3}, v_{BC3})\} = 0 \tag{P3.9.3a}$$

$$-v_4 - R_{C2}\{i_{C2}(v_{BE2}, v_{BC2}) + i_{B4}(v_{BE4}, v_{BC4})\} = 0 \tag{P3.9.3b}$$

$$i_{E1}(v_{BE1}, v_{BC1}) + i_{E2}(v_{BE2}, v_{BC2}) - I_{EE} = 0 \tag{P3.9.3e}$$

where $i_{Ck}(v_{BEk}, v_{BCk})$ and $i_{Bk}(v_{BEk}, v_{BCk})$ are defined by Eqs. (3.1.23a,b), and $i_{Ek}(+,+) = i_{Bk}(+,+) + i_{Ck}(+,+)$ for all k. Then complete the above MATLAB function '`ECL2()`' so that it can analyze the ECL and run the following statements to plot the output waveforms as shown in Figure P3.9(b2).

```
>>betaF=100; betaR=1; Is=1e-14; RC=2e3; IEE=3e-4;
  Vref=-0.9; VEE=-5.2;
  RC=2e3; R=42e3; IEE=3e-4; Vref=-0.9; VEE=-5.2;
  ECL2([betaF betaR],Is,RC,R,IEE,Vref,VEE);
```

3.10 CB Amplifier

Consider the CB amplifier in Figure P3.10(a1) or (a2) where the device parameters of the NPN-BJT are $\beta_F = 100$, $\beta_R = 1$, $\beta_{AC} = 100$, $r_b = 0\ \Omega$, and $I_s = 10 \times 10^{-15}$ A.

(a) Complete the following MATLAB script "elec03p10.m" and run it to find the overall voltage gain G_v and input/output resistances R_i/R_o. Is the result of DC analysis using the PWL model of BJT closer to the PSpice simulation result shown in the schematic (Figure P3.10(a1) or (a2)) than that using the exponential model?

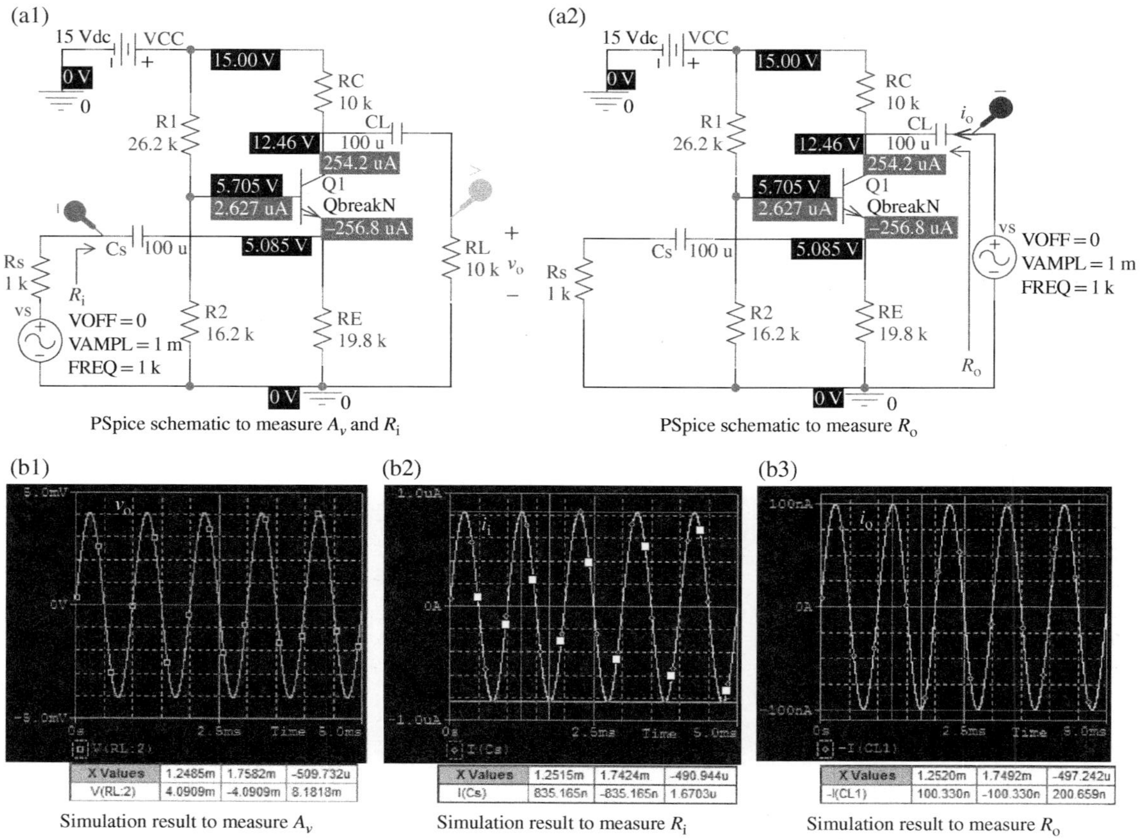

Figure P3.10 A CB amplifier and its PSpice simulation result ("elec03p10.opj").

```
%elec03p10.m
clear
BETAF=100; BETAR=1; betaAC=BETAF; rb=0; VA=1e4; % BJT parameters
beta=[betaF,betaR,betaAC];
VCC=15; Vsm=1e-3; Rs=1e3; RL=1e4; % Source/load resistances
R1=26.2e3; R2=16.2e3; RC=10e3; RE=19.8e3; % Bias circuit
% For DC analysis using the PWL model of BJT
[VBQ,VEQ,VCQ,IBQ,IEQ,ICQ,Av,Ai,Ri,Ro,gm,rbe,ro]=...
  BJT_CB_analysis(VCC,rb,Rs,R1,R2,RC,RE,RL,beta,Vsm,VA);
% For DC analysis using the exponential model of BJT
[VBQ,VEQ,VCQ,IBQ,IEQ,ICQ,Av,Ai,Ri,Ro,gm,rbe,ro]=...
  BJT_CB_analysis(VCC,rb,Rs,R1,R2,RC,RE,RL, [beta Is],Vsm,VA);
```

(b) Are the values of G_v, R_i, and R_o close to those obtained from the PSpice simulations shown in Figure P3.10(b1), (b2), and (b3)?

3.11 CC Amplifier

Consider the CC amplifier in Figure P3.11(a1) or (a2) where the device parameters of the NPN-BJT are $\beta_F = 100$, $\beta_R = 1$, $\beta_{AC} = 100$, $r_b = 0\ \Omega$, and $I_s = 10 \times 10^{-15}$ A.

(a) Complete the following MATLAB script "elec03p11.m" and run it to find the overall voltage gain G_v and input/output resistances R_i/R_o. Is the result of DC analysis using the PWL model of BJT close to that using the exponential model?

```
%elec03p11.m
clear
betaF=100; betaR=1; betaAC=100; rb=0; Is=1e-14; VA=1e4;
beta=[betaF,betaR,betaAC]; % BJT parameters
VCC=15; Vsm=1e-3; Rs=2e4; RL=1e3;
R1=345e3; R2=476e3; RC=4e3; RE=5.94e3;% Bias circuit
% For DC analysis using the PWL model of BJT
[VBQ,VEQ,VCQ,IBQ,IEQ,ICQ,Av,Ai,Ri,Ro,gm,rbe,ro]=...
  BJT_CC_analysis(VCC,rb,Rs,R1,R2,RC,RE,RL,beta,Vsm,VA);
% For DC analysis using the exponential model of BJT
[VBQ,VEQ,VCQ,IBQ,IEQ,ICQ,Av,Ai,Ri,Ro,gm,rbe,ro]=...
BJT_CC_analysis(VCC,rb,Rs,R1,R2,RC,RE,RL, [beta Is],Vsm,VA);
```

(b) Are the values of G_v, R_i, and R_o close to those obtained from the PSpice simulation shown in Figure P3.11(b1), (b2), and (b3)?

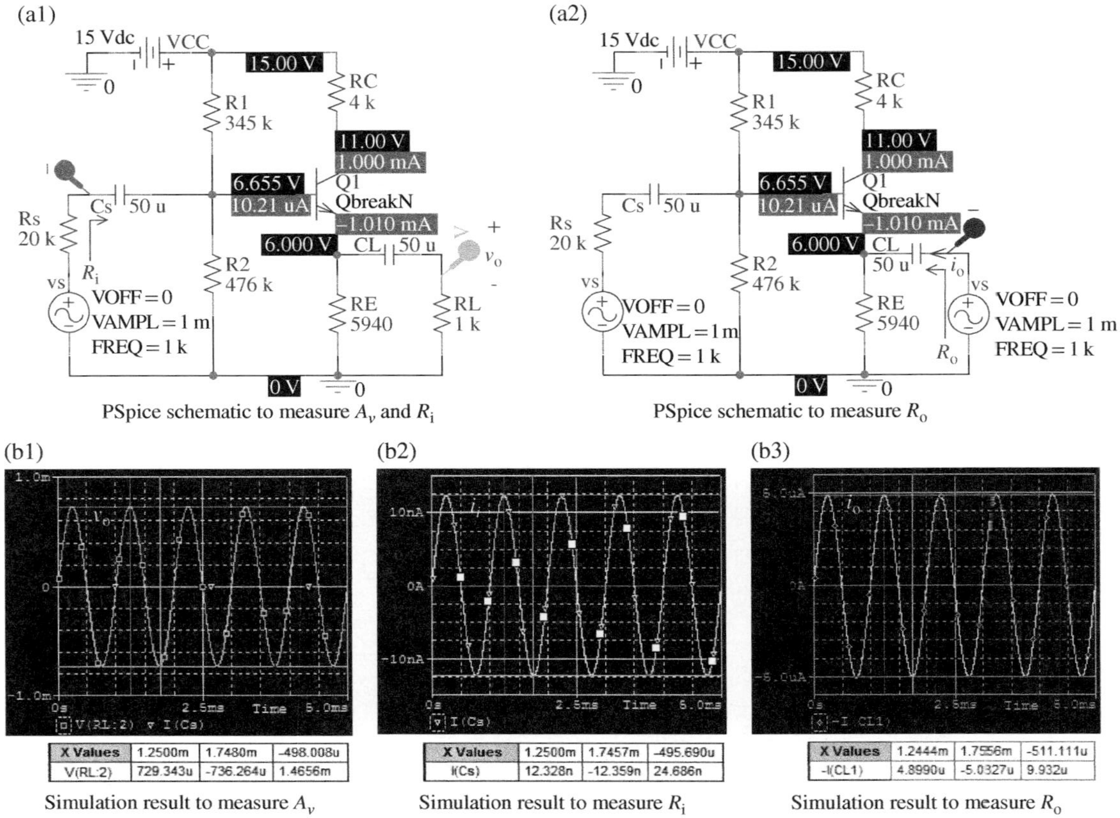

Figure P3.11 A CC amplifier and its PSpice simulation result ("elec03p11.opj").

3.12 Two-Stage BJT Amplifier

Consider the two-stage BJT amplifier in Figure P3.12(a) where the device parameters of the NPN-BJT Q are $\beta_F = 100$, $\beta_R = 1$, $\beta_{AC} = 100$, and $r_b = 0\,\Omega$. Complete the following MATLAB script "elec03p12.m" and run it to find the overall voltage gain G_v and output resistance R_{o2}.

(Note) As long as no CC amplifier (whose input/output resistances are affected by the input/output resistances of the next/previous stage, respectively) is involved, it will be enough to analyze each stage starting from the last one with the source resistance of every stage (except for the first one having the true value of R_s) set to 0 and with the load resistance of every stage (except for the last one having the true value of R_L) set to the input resistance of the next stage.

```
%elec03p12.m
rb=0; betaF=100; betaR=1; betaAC=100; VA=1e4;
beta=[betaF,betaR,betaAC];
VCC=15; Vsm=1e-5; Rs=5e3; RL=2e3; % Source/load resistances
R11=1e5; R12=5e4; RC1=6800; RE1=[0 3925];
R21=1e5; R22=5e4; RC2=6800; RE2=[0 3925];
% Find Ri, Av, and Ai of Stage 2/1 starting from last one.
Rs2=?; RL2=??;
[VBQ,VEQ,VCQ,IBQ,IEQ,ICQ,Avs(2),Ai2,Ri2,Ro2]= ...
  BJT_CE_analysis(VCC,rb,Rs2,R21,R22,RC2,RE2,RL2,beta);
fprintf('\n1st stage of CE')
Rs1=??; RL1=???; Vsm0=Avs(2)*Vsm;
[VBQ,VEQ,VCQ,IBQ,IEQ,ICQ,Avs(1),Ai1,Ri1,Ro1]= ...
   BJT_CE_analysis(VCC,rb,Rs1,R11,R12,RC1,RE1,RL1,beta,Vsm0,VA);
fprintf('\n Overall')
Gv=Ri1/(Rs+???)*prod(???); Vom=Gv*Vsm
fprintf(' Gv=Ri/(Rs+Ri)*Av1*Av2=%4.2fx%8.3fx%8.3f ...
          =%9.3f\n', Ri1/(Rs+Ri1),Avs(1),Avs(2),Gv)
```

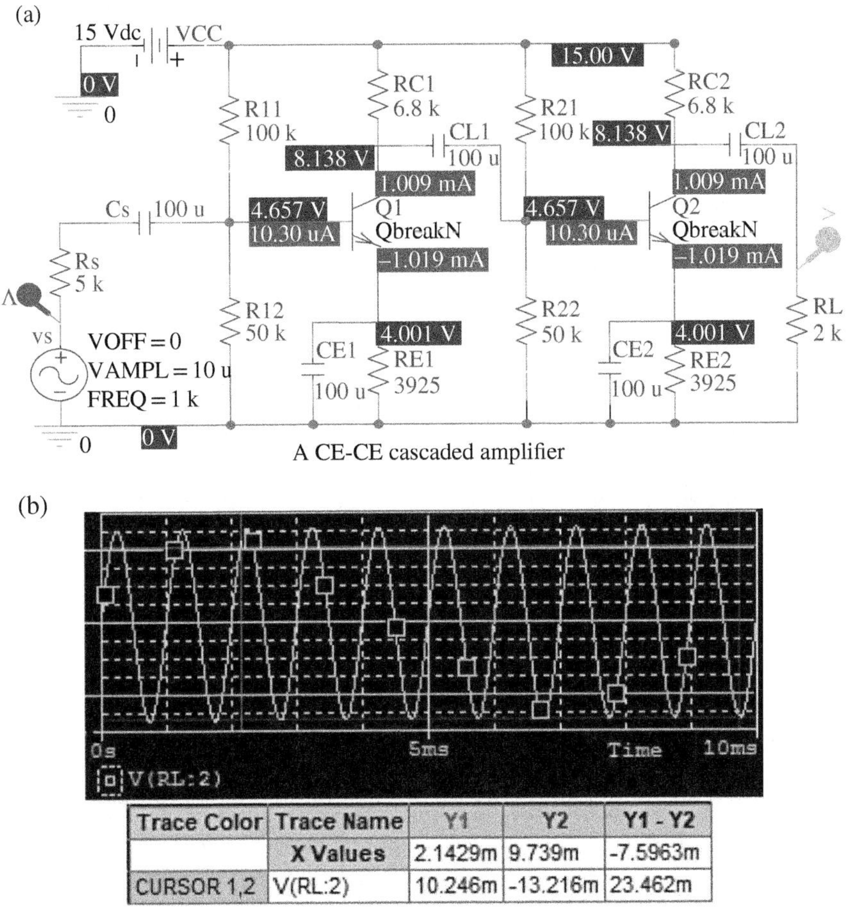

Trace Color	Trace Name	Y1	Y2	Y1 - Y2
	X Values	2.1429m	9.739m	-7.5963m
CURSOR 1,2	V(RL:2)	10.246m	-13.216m	23.462m

PSpice simulation result of the amplifier

Figure P3.12 A two-stage BJT amplifier and its PSpice simulation result ("elec03p12.opj").

3.13 Three-Stage BJT Amplifier

Consider the three-stage BJT amplifier in Figure P3.13.1(a) where the device parameters of all the NPN-BJTs are $\beta_F = 100$, $\beta_R = 1$, $\beta_{AC} = 100$, $r_b = 0\ \Omega$, and $I_s = 1 \times 10^{-16}$ A in common. Note that the first two stages are identical to those of the amplifier in Figure P3.13(a), whose overall voltage gain and output resistance are $G_{v12} = 1329.8$ and $R_{o2} = 6795\ \Omega$, respectively.

```
%elec03p13.m
rb=0; betaF=100; betaR=1; betaAC=100; Is=1e-16; VA=1e4;
beta=[betaF,betaR,betaAC]; % beta=[betaF,betaR,betaAC,Is];
VCC=15; Vsm=1e-5; Rs=5e3; RL=2e3; % Source/load resistances
R11=1e5; R12=5e4; RC1=6800; RE1=[0 3925];
R21=1e5; R22=5e4; RC2=6800; RE2=[0 3925];
R31=5e4; R32=5e4; RC3=0; RE3=200;
Rs3=0; RL3=??; [VBQ,VEQ,VCQ,IBQ,IEQ,ICQ,Av3,Ai3,Ri3,Ro3_0]= ...
 BJT_CC_analysis(VCC,rb,Rs3,R31,R32,RC3,RE3,RL3,beta);
Rs2=0; RL2=???; [VBQ,VEQ,VCQ,IBQ,IEQ,ICQ,Av2,Ai2,Ri2,Ro2_0]= ...
 BJT_CE_analysis(VCC,rb,Rs2,R21,R22,RC2,RE2,RL2,beta);
fprintf('\n1st stage of CE')
Rs1=??; RL1=???; Vsm0=Vsm;
[VBQ,VEQ,VCQ,IBQ,IEQ,ICQ,Av1,Ai1,Ri1,Ro1]= ...
  BJT_CE_analysis(VCC,rb,Rs1,R11,R12,RC1,RE1,RL1,beta,Vsm0,VA);
fprintf('\n2nd stage of CE')
Rs2=???; RL2=???; Vsm1=Ri1/(Rs+Ri1)*Av1*Vsm;
[VBQ,VEQ,VCQ,IBQ,IEQ,ICQ,Av2,Ai2,Ri2,Ro2]= ...
  BJT_CE_analysis(VCC,rb,Rs2,R21,R22,RC2,RE2,RL2,beta,Vsm1,VA);
fprintf('\n3rd stage of CC')
Rs3=???; RL3=??; Vsm2=Av2*Vsm1;
[VBQ,VEQ,VCQ,IBQ,IEQ,ICQ,Av3,Ai3,Ri3,Ro3]= ...
  BJT_CC_analysis(VCC,rb,Rs3,R31,R32,RC3,RE3,RL3,beta,Vsm2,VA);
Vom=Av3*Vsm2, Gv=Ri1/(Rs+Ri1)*Av1*Av2*Av3;
fprintf(' Gv=Ri/(Rs+Ri)*Av1*Av2*Av3=%4.2fx%8.3fx%8.3fx%8.3f...
        =%9.3f\n', Ri1/(Rs+Ri1),Av1,Av2,Av3,Gv)
```

(a) Complete the above MATLAB script "elec03p13.m" and run it to find the overall voltage gain G_v, and the voltage gains (A_{vk}'s) and input/output resistances (R_{ik}/R_{ok}'s) of each stage k. Is the value of G_v close to that obtained from the PSpice simulation shown in Figure P3.13.1(b1)? If you are not satisfied, try with the exponential model of BJT.

(b) Determine the values of the open-loop voltage gains $\{A_{v1\_o}, A_{v2\_o}, A_{v3\_o}\}$ (with $R_{i2} = \infty$, $R_{i3} = \infty$, and $R_L = \infty$ to exclude its next stage) and the output resistances $\{R_{o1\_0}, R_{o2\_0}, R_{o3\_0}\}$ (with $R_s = 0$, $R_{s2} = 0$, and $R_{s3} = 0$ to exclude its previous stage) of each stage. Check if the following relations hold between the voltage gains $\{A_{v1}, A_{v2}, A_{v3}\}$ (considering the next stage) and the open-loop voltage gains $\{A_{v1\_o}, A_{v2\_o}, A_{v3\_o}\}$:

$$A_{v1} = A_{v1_o}\frac{R_{i2}}{R_{o1_0} + R_{i2}}, \quad A_{v2} = A_{v2_o}\frac{R_{i3}}{R_{o2_0} + R_{i3}}, \quad A_{v3} = A_{v3_o}\frac{R_L}{R_{o3_0} + R_L} \tag{P3.13.1}$$

Do you see that $R_{ok\_0} \neq R_{ok}$ for any stage k? Why is that?

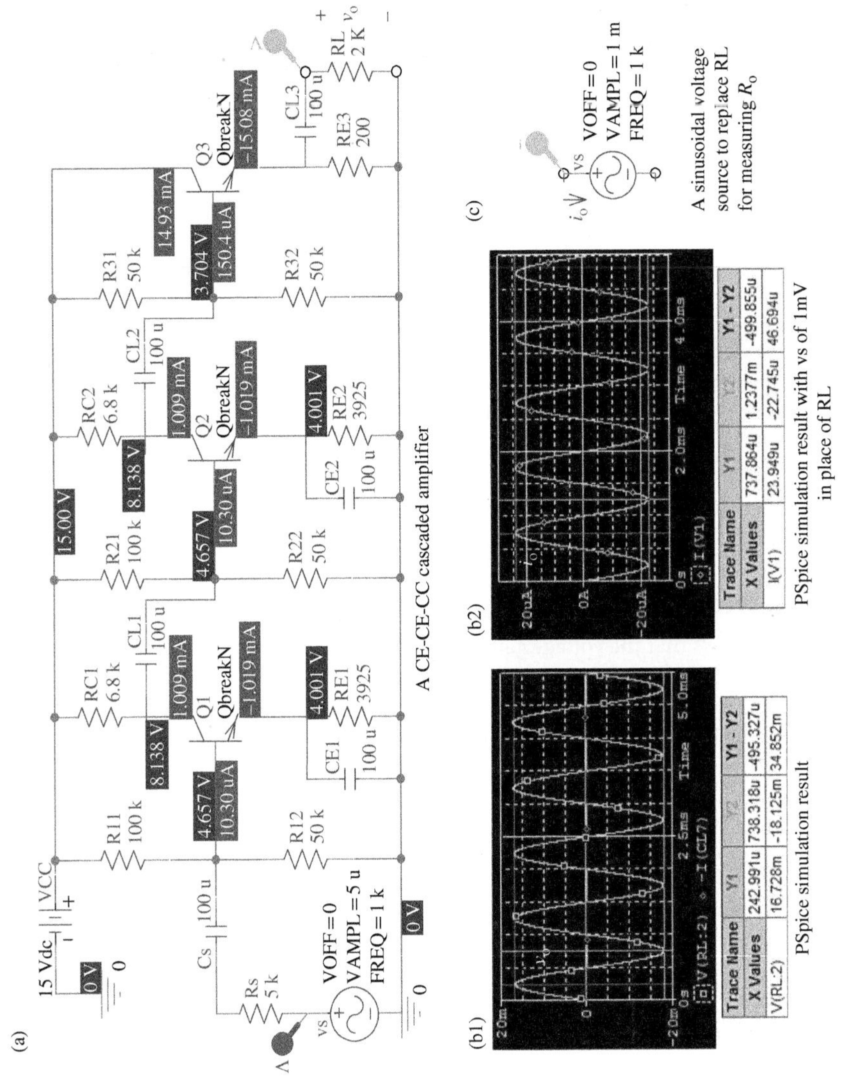

Trace Name	Y1	Y2	Y1 - Y2
X Values	242.991u	738.318u	-495.327u
V(RL:2)	16.728m	-18.125m	34.852m

Trace Name	Y1	Y2	Y1 - Y2
X Values	737.864u	1.2377m	-499.855u
I(V1)	23.949u	-22.745u	46.694u

Figure P3.13.1 A three-stage BJT amplifier and its PSpice simulation result ("elec03p13.opj").

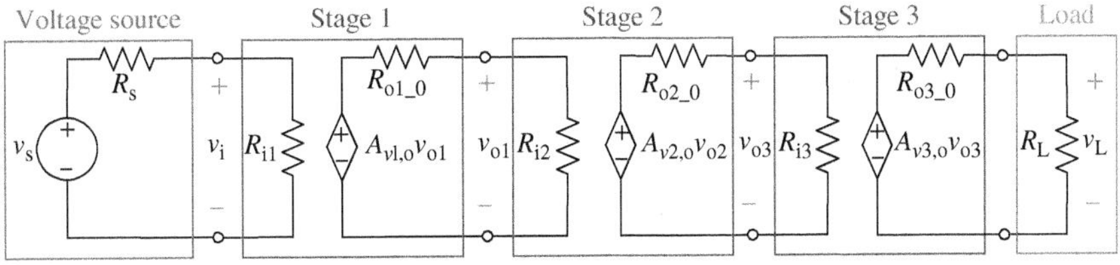

Figure P3.13.2 A voltage source model for the CE-CE-CC amplifier of Figure P3.13.1.

(c) With the values of R_{ok_o}'s and R_{ik}'s together with the open-loop voltage gain A_{vk_o}'s for k = 1, 2, and 3 (obtained in (b)), complete the voltage source model of the three-stage amplifier (Figure P3.13.2) and find the overall voltage gain.

(d) Use the voltage source model to determine the overall voltage gain G_{v12} of the CE-CE stage (with the CC stage removed). Is it smaller than G_v in spite of not having a CC stage with voltage gain $A_{v3} = 0.991$ smaller than 1? Why is that?

(e) Figure P3.13.1(b2) shows another PSpice simulation result on the current i_o for the PSpice schematic with R_L replaced by a sinusoidal voltage source of 1 mV (Figure P3.13.1(c)) and the original voltage source short-circuited for removal. From the simulation result, find the output resistance R_o and see if it is close to the theoretical value obtained in (a).

3.14 Design of CE Amplifier Circuit

Consider the CE BJT amplifier circuit of Figure 3.52.1(a) where the forward DC/AC current gains, reverse DC current gain, and base resistance of the BJT are $\beta_F = 162/\beta_{AC} = 176$, $\beta_R = 6$, and $r_b = 10\ \Omega$, respectively.

(a) Considering the load resistance $R_L = 50\ \text{k}\Omega$, determine such resistor values that the voltage gain $A_{v,d} = -20$ can be achieved with the operating point located near $Q_1 = (V_{CE,Q}, I_{C,Q}) = (3\ \text{V}, 2\ \text{mA})$. Analyze the designed circuit with source/load resistance $R_s = 50\ \Omega/R_L = 50\ \text{k}\Omega$ to find the voltage gain A_v. Is it close to $A_{v,d} = -20$?

(Hint) How about running the following MATLAB statements?

```
>>bF=162; bR=6; bAC=176; rb=10; T=27; % Device constants
  VCC=12; Rs=50; RL=5e4; % Circuit constants
  Avd=20; VCEQd=3; KC=VCEQd/VCC;
  ICQd=2e-3; % Design constants [R1,R2,RC,RE1,RE2]=...
  BJT_CE_design(VCC, [bF bAC],rb,Avd,ICQd,RL,T,KC);
 Vsm=0.1; beta=[bF bR bAC];
 [VBQ,VEQ,VCQ,IBQ,IEQ,ICQ,Av]=...
  BJT_CE_analysis(VCC,rb,Rs,R1,R2,RC, [RE1 RE2],RL,beta,Vsm)
```

(b) Referring to Figure P3.14(a), perform the PSpice simulation of the designed CE amplifier with Time-Domain Analysis to see the output voltage $v_o(t)$ to an AC input $v_s(t) = V_{sm}\sin(2000\pi t)$ (with $V_{sm} = 0.05$ V) for 2 ms (with maximum step size 1 us). Is the voltage gain $V_{om,p\text{-}p}/2V_{sm}$ close to A_v obtained in (a)? Also, on the collector characteristic curves of BJT Q2N2222 (used in the CE amplifier), draw the load line to find the operating point ($V_{CE,Q}$, $I_{C,Q}$) and check if it agrees with the analysis result (obtained in (a)) and PSpice simulation result.

(a)

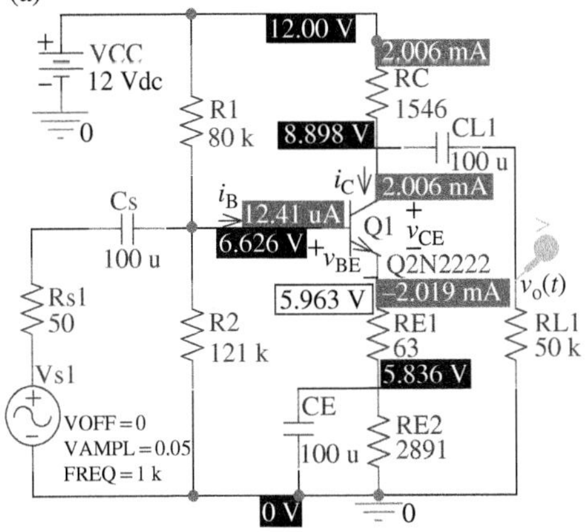

NAME	Q_Q1
MODEL	Q2N2222
IB	1.24E–05
IC	2.01E–03
VBE	6.63E–01
VBC	2.27E+00
VCE	2.94E+00
BETADC	1.62E+02
GM	7.70E–02
RPI	2.29E+03
RX	1.00E+01
RO	3.80E+04
CBE	6.85E–11
CBC	4.54E–12
CJS	0.00E+00
BETAAC	1.76E+02

PSpice schematic for a CE BJT circuit

(b)

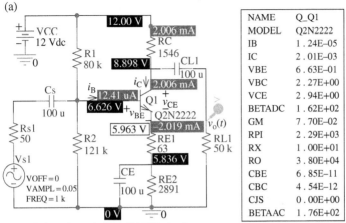

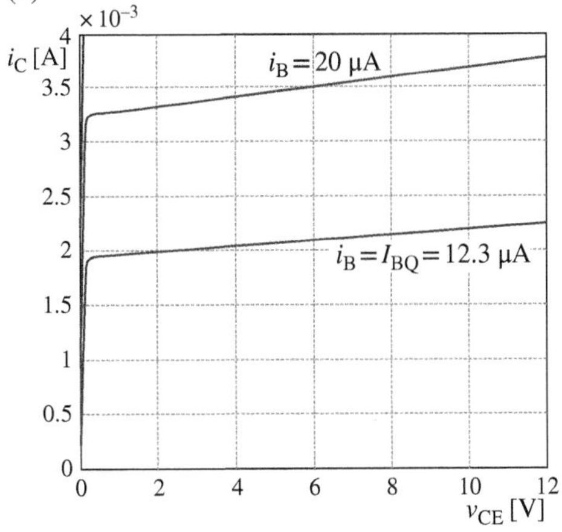

Collector characteristic curves of BJT Q2N2222

Figure P3.14 A CE BJT circuit and the collector characteristic curves of BJT Q2N2222 ("elec03p14.opj").

(c) Perform the PSpice simulation of the designed CE amplifier twice more to see the output voltage $v_o(t)$ to an AC input $v_s(t) = V_{sm} \sin(2000\pi t)$ with larger small-signal input amplitudes $V_{sm} = 0.15$ V and 0.2 V. If the output voltage waveforms are distorted on the upper/lower part, discuss the reason of the distortion in connection with the possibility of the (dynamic) operating point (moved by the small-signal input) to enter the cutoff/saturation region. Does the MATLAB function '`BJT_CE_analysis()`' (with the input voltage amplitude V_{sm} = 0.15/0.2 given as the 10th input argument) give you appropriate warning messages about the possibility of saturation or cutoff regions?

(d) To reduce the distortion of $v_o(t)$, change the resistor values so that the operating point can be located at $Q_2 = (V_{CE,Q}, I_{C,Q}) = (4\text{ V}, 2\text{ mA})$ and $Q_3 = (V_{CE,Q}, I_{C,Q}) = (5\text{ V}, 4\text{ mA})$. Perform the PSpice simulation of the designed CE amplifiers to see the output voltage $v_o(t)$ to an AC input $v_s(t) = V_{sm}\sin(2000\pi t)$ with $V_{sm} = 0.2$ V. Discuss the relationship between the static operating point location and the possibility of distortion.

Remark 3.1 About Fixing the Operating Point of CE BJT Amplifier

1) It seems to be good for reducing the possibility of distortion to locate $V_{CE,Q}$ about halfway between 0 and V_{CC} so that the operating point can stay away from both the saturation and cutoff regions. However, in view of Eq. (3.4.2)/(3.4.3), it should not be so high as to result in negative values of R_C/R_E.
2) The collector current $I_{C,Q}$ at the operating point should not be so small/large as to yield a negative value of R_C/R_E.

$$R_C \overset{(3.4.2)}{=} \frac{1}{I_{C,Q}/V_{CE,Q} - 1/R_L}, \tag{P3.14.1}$$

$$R_E \overset{(3.4.3)}{=} (1-K_C)\frac{V_{CC}}{I_{C,Q}} - R_C = \frac{V_{CC} - V_{CE,Q}}{I_{C,Q}} - R_C \tag{P3.14.2}$$

3.15 Frequency Response of a BJT Cascode Amplifier

Consider the BJT cascode amplifier with PSpice schematic and simulation result on the frequency response in Figure P3.15(a) and (b), respectively, where $V_{CC} = 12$ V, $R_s = 50\ \Omega$, $C_s = 100\ \mu$F, $R_1 = 100$ kΩ, $R_2 = 100$ kΩ, $R_3 = 50$ kΩ, $R_E = 2$ kΩ, $C_E = 100\ \mu$F, $C_B = 100\ \mu$F, $R_C = 3$ kΩ, $C_L = 1\ \mu$F, $R_L = 100$ kΩ, $C_{LL} = 1$ nF, and the values of the BJT parameters of the (high-frequency) small-signal equivalent in Figure P3.15(d) (based on the model in Figure 3.18) are $\beta_F = 100$, $\beta_R = 1$, $\beta_{AC} = 100$, $V_A = 10^4$ V, $I_s = 10^{-14}$ A, $r_b = 0\ \Omega$, $C_{be}(C_{je}) = 10$ pF, and $C_{bc}(C_{jc}) = 1$ pF.

(a) Complete the following MATLAB function '`BJT_cascode_DC_analysis()`', which performs the DC analysis of the BJT cascade amplifier to determine the node voltages and base/collector currents at the operating point.

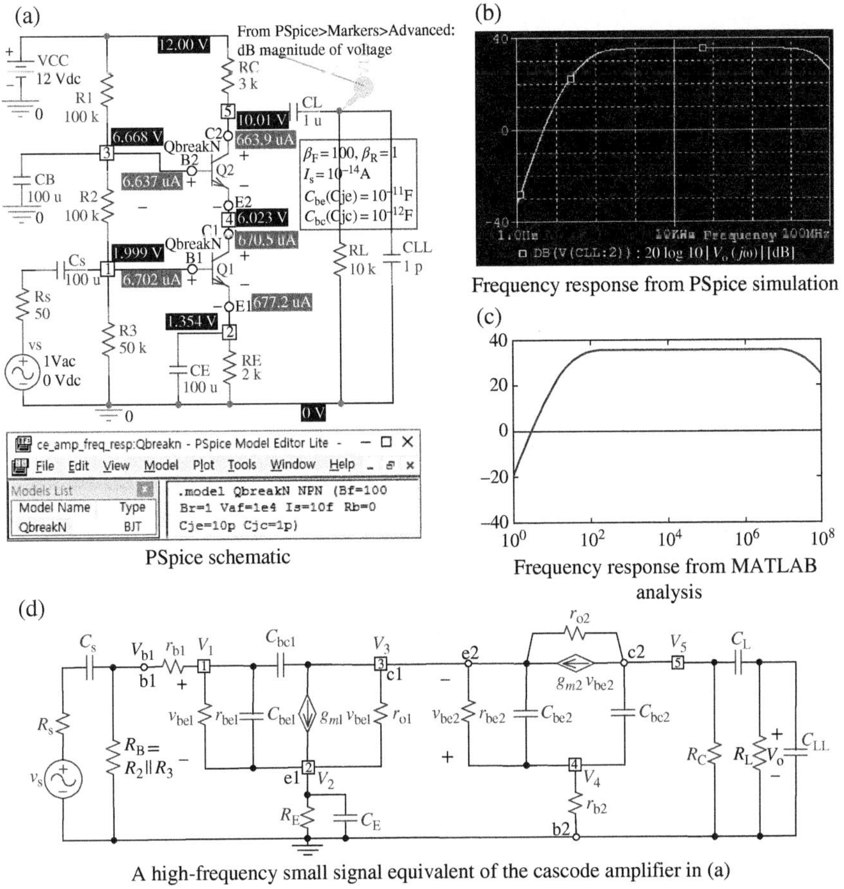
(a)
From PSpice>Markers>Advanced: dB magnitude of voltage
PSpice schematic
(b)
Frequency response from PSpice simulation
(c)
Frequency response from MATLAB analysis
(d)
A high-frequency small signal equivalent of the cascode amplifier in (a)

Figure P3.15 A cascode amplifier circuit and its frequency response from PSpice and MATLAB ("elec03p15.opj").

```
function [v,IB1Q,IC1Q,IB2Q,IC2Q]=...
   BJT_cascode_DC_analysis(VCC,R123,RC,RE,beta,T)
%To do DC analysis of a Cascode (CE-CB) amplifier in Fig. 3.41(a)
% VCC = [VCC VEE], R123=[R1 R2 R3]
% beta=[betaF1 betaR1 Is1; betaF2 betaR2 Is2]
% Copyleft: Won Y. Yang, wyyang53@hanmail.net, CAU for academic use only
if nargin<6, T=27; end; VT=(273+T)/11605;
if 0<T&T<0.1, VT=T; end
if numel(VCC)<2, VEE=0; else VEE=VCC(2); VCC=VCC(1); end
bF1=beta(1,1); bR1=beta(1,2); Is1=beta(1,3);
bF2=beta(end,1); bR2=beta(end,2); Is2=beta(end,3);
aR1=bR1/(bR1+1); Isc1=Is1/aR1; % Collector-Base saturation current
aR2=bR2/(bR2+1); Isc2=Is2/aR2;
R1=R123(1); R2=R123(2); R3=R123(3);
```

```
iC1=@(v)Is1*exp(v(1)/VT)-Isc1*exp(v(2)/VT);
                                     % Eq.(3.1.23a) with v=[vBE vBC]
iB1=@(v)Is1/bF1*exp(v(1)/VT)+Isc1/(bR1+1)*exp(v(2)/VT);
                                                        % Eq.(3.1.23b)
iE1=@(v)Is1*(1+1/bF1)*exp(v(1)/VT)-Isc1*bR1/(bR1+1)*exp(v(2)/VT);
iC2=@(v)Is2*exp(v(1)/VT)-Isc2*exp(v(2)/VT);
% Eq.(3.1.23a) with v=[vBE vBC]
iB2=@(v)Is2/bF2*exp(v(1)/VT)+Isc2/(bR2+1)*exp(v(2)/VT);
                                                        % Eq.(3.1.23b)
iE2=@(v)Is2*(1+1/bF2)*exp(v(1)/VT)-Isc2*bR2/(bR2+1)*exp(v(2)/VT);
% A set of (nonlinear) node equations in v=[v1 v2 v3 v4 v5]:
eq=@(v)[(v(3)-v(?))/R2-(v(?)-VEE)/R3-...iB1([v(?)-v(2) v(?)-v(4)]);
                                                              % node 1
    iE1([v(1)-v(?) v(1)-v(4)])-(v(?)-VEE)/RE; % node 2
    iE2([v(?)-v(4) v(?)-v(5)])-iC1([v(1)-v(2) v(1)-v(4)]); % node 3
   (VCC-v(3))/R1-(v(3)-v(1))/R2-iB2([v(3)-v(?) v(3)-v(5)]); % node 4
    (VCC-v(?))/RC-iC2([v(3)-v(4) v(3)-v(?)])]*1e3; % node 5
% Initial guess for the node voltage vector v=[v1 v2 v3 v4 v5]:
R_123=R1+R2+R3; v10=VCC*R3/R_123; v30=VCC*(R2+R3)/R_123;
v0=[v10 v10-0.7 v30 v30-0.7 (VCC+v30)/2];
% Solve the set of (nonlinear) node equations in v.
options=optimset('Display','off','Diagnostics','off');
v=fsolve(eq,v0,options); % Node voltages at the operating point
% Base/collector currents for Q1 and Q2:
IB1Q=iB1([v(1)-v(2) v(1)-v(4)]); IB2Q=iB2([v(3)-v(4) v(3)-v(5)]);
IC1Q=iC1([v(1)-v(2) v(1)-v(4)]); IC2Q=iC2([v(3)-v(4) v(3)-v(5)]);
```

```
%elec03p15.m
% To find the frequency response of a BJT Cascode (CE-CB) amplifier
% Circuit parameters
VCC=12; Rs=50; R1=100e3; R2=100e3; R3=50e3; RC=3e3; RE=2e3; RL=1e4;
Cs=1e-6; CE=1e-4; CE=inf; CB=1e-4; CL=1e-6; CLL=1e-12; R123=[R1 R2 R3];
% Device parameters for the two BJTs Q1 and Q2
Is=1e-14; betaF=100; betaR=1; betaAC=betaF;
beta=[betaF betaR betaAC Is];
rb1=0; rb2=0; Cbe1=10e-12; Cbe2=10e-12; Cbc1=1e-12; Cbc2=1e-12;
rb=[rb1; rb2]; CC=[Cbe1 Cbc1; Cbe2 Cbc2]; RLCLL=[RL CLL]; VA=1e4; T=27;
f=logspace(0,8,801); w=2*pi*f; % Frequency range
[Gs,fc]=BJT_cascode_xfer_ftn(VCC,Rs,Cs,R123,RC,RE,...
CE,CB,CL,RLCLL,beta,CC,rb,VA,T);
Gw=subs(Gs,s,?*w); % Frequency response = Transfer function with s=jw
GmagdB=20*log10(abs(Gw)+1e-10); Gmax=max(GmagdB);
semilogx(f,GmagdB), hold on
semilogx(fc(1)*[1 1],[0 Gmax-3],'b:', ... fc(6)*[1 1],[0 Gmax-3],'k:')
text(fc(1),-10,'f_{c1}'); ... text(fc(6),-10,'f_{c6}');
```

```
function [Gs,fc,gm1,rbe1,ro1,gm2,rbe2,ro2]=BJT_cascode_xfer_ftn...
(VCC,Rs,Cs,R123,RC,RE,CE,CB,CL,RLCLL,beta,CC,rb,VA,T)
%To find the transfer function of Cascode (CE-CB) amplifier in Fig.P3.15a
% R123=[R1 R2 R3]
% RLCLL=[RL CLL] with RL and CLL in parallel
% beta=[betaF1 betaR1 betaAC1 Is1; betaF2 betaR2 betaAC2 Is2]
% CC=[Cbe1 Cbc1; Cbe2 Cbc2]: BEJ|BCJ capacitances
% Copyleft: Won Y. Yang, wyyang53@hanmail.net, CAU for academic use only
if nargin<15, T=27; end
if T>=0.1; VT=(273+T)/11605; else VT=T; end % Thermal voltage
if nargin<14, VA=1e4; end; VA1=VA(1); VA2=VA(end); ;
if nargin<13, rb=0; end; rb1=rb(1); rb2=rb(end);
if nargin<12, CC=[0 0]; end
Cbe1=CC(1,1); Cbe2=CC(end,1); Cbc1=CC(1,2); Cbc2=CC(end,2);
if size(beta,2)<3
 error('The 11th input argument beta must have [betaF betaR
                                                  betaAC Is]!');
 elseif size(beta,2)==3, beta(:,4)=beta(:,3); beta(:,3)=beta(:,1);
end
bF1=beta(1,1); bR1=beta(1,2); bAC1=beta(1,3); Is1=beta(1,4);
bF2=beta(end,1); bR2=beta(end,2); bAC2=beta(end,3);
Is2=beta(end,4);
RL=RLCLL(1); if numel(RLCLL)<2, CLL=0; else CLL=RLCLL(2); end
[VQs,IB1Q,IC1Q,IB2Q,IC2Q]=...
  BJT_cascode_DC_analysis(VCC,R123,RC,RE,beta(:,[1 2 4]),T);
[gm1,rbe1,re1,ro1]=gmrbero_BJT(IC1Q,bF1,VA1,VT);
[gm2,rbe2,re2,ro2]=gmrbero_BJT(IC2Q,bF2,VA2,VT);
syms s; sCs=s*Cs; sCE=s*CE; sCL=s*CL;
sCbe1=s*Cbe1; sCbc1=s*Cbc1; sCbe2=s*Cbe2; sCbc2=s*Cbc2;
Zs=Rs+1/sCs; Ys=1/Zs; RB=parallel_comb([R2 R3]); GB=1/RB;
ZL=parallel_comb([RL 1/s/CLL]); % Load impedance
YC=1/parallel_comb([RC 1/sCL+ZL]); % Admittance at terminal C
YB=1/(1/(Ys+GB)+rb1); % Admittance at terminal B
YE=(1/RE+sCE)*(CE<inf); % Admittance at terminal E
% Node equations for the high-freq equivalent circuit in Fig. P3.15(d)
if rb2>0
 Y=[YB+1/rbe?+sCbe?+sCbc? -YB-1/rbe?-sCbe? -sCbc? 0 0;
    -YB-1/rbe?-sCbe?-gm? YB+1/rbe?+sCbe?+1/ro1+YE+gm? -1/ro? 0 0;
    -sCbc?+gm? -1/ro?-gm? sCbc?+1/ro?+1/rbe?+sCbe?+1/ro?+gm? ...
                     -1/rbe?-sCbe?-gm? -1/ro2;
      0    0   -1/rbe?-sCbe?  1/rbe?+sCbe?+1/rb?+sCbc? -sCbc?;
      0    0    1/ro?-gm?      -sCbc?+gm? sCbc?+1/ro?+YC];
 V=Y\[RB/(Zs+RB)*YB; 0; 0; 0; 0];
else
 Y=[YB+1/rbe?+sCbe?+sCbc? -YB-1/rbe?-sCbe? -sCbc? 0;
    -YB-1/rbe?-sCbe?-gm? YB+1/rbe?+sCbe?+1/ro?+YE+gm? -1/ro? 0;
    -sCbc1+gm? -1/ro?-gm? sCbc?+1/ro?+1/rbe?+sCbe?+1/ro?+gm? -1/ro?;
      0     0      1/ro?-gm?        sCbc?+1/ro?+YC];
 V=Y\[RB/(Zs+RB)*YB; 0; 0; 0];
end
Gs=V(end)*??/(ZL+1/sCL); % Transfer ftn G(s)=Vo(s)/Vs(s) with Vs(s)=1
Av(1)=-gm1*ro1; Ri(1)=parallel_comb([RB rb1+rbe1]); Ro(1)=ro1;
RCL=parallel_comb([RC RL]); Av(2)=(1+gm2*ro2)*RCL/(ro2+RCL);
Ri(2)=parallel_comb([rbe2 (ro2+RC)/(1+gm2*ro2)]);
Ro(2)=parallel_comb([ro2+rbe2+rb2 RC]);
Cbc=[Cbc1 Cbc2]; Cbe=[Cbe1 Cbe2]; ro=[ro1 ro2];
fc=break_freqs_of_BJT_cascode(Rs,Cs,CL,RL,CLL,Cbc,Cbe,Av,Ri,Ro);
```

```
function fc=...
  break_freqs_of_BJT_cascode(Rs,Cs,CL,RL,CLL,Cbc,Cbe,Av,Ri,Ro)
% To find the 6 break frequencies of a BJT Cascade amplifier
RsRi1=parallel_comb([Rs Ri(1)]);
Ro1Ri2=parallel_comb([Ro(1) Ri(2)]);
Ro2RL=parallel_comb([Ro(2) RL]);
Cm1=Cbc(1)*(1-Av(1)); % Eq.(3.5.6a)
Cn1=Cbc(1)*(1-1/Av(1)); % Eq.(3.5.6b)
fc(1)=1/2/pi/Cs/(Rs+Ri(1));
fc(2)=1/2/pi/CL/(Ro(2)+RL);
fc(3)=1/2/pi/(Cbe(1)+Cm1)/RsRi1;
fc(4)=1/2/pi/Cbe(1)/Ro1Ri2;
fc(5)=1/2/pi/Cbc(2)/Ro(2);
fc(6)=1/2/pi/CLL/Ro2RL;
```

(b) Complete the above MATLAB function '`BJT_cascode_xfer_ftn()`', which solves a set of four or five node equations for the high-frequency small-signal equivalent (Figure P3.15(d)) (depending on r_{be}=0 or not) to find the transfer function $G(s) = V_o(s)/V_s(s)$ of the cascade amplifier.

(c) Complete the above MATLAB script "elec03p15.m" and run it to perform the DC analysis (using '`BJT_cascode_DC_analysis()`'), then based on the DC analysis result, use '`BJT_cascode_xfer_ftn()`' to find the transfer function $G(s)$, and plot the frequency response magnitude $20\log_{10}|G(j\omega)|$ [dB] of the cascade amplifier versus f = 1~100 MHz as shown in Figure P3.15(c).

(d) Perform the PSpice simulation (with AC Sweep analysis) to get the frequency response magnitude curve as shown in Figure P3.15(b). Is the Bias Point analysis result (obtained as a by-product) close to the DC analysis result obtained by using '`BJT_cascode_DC_analysis()`' in (c)?

3.16 Time Response of a BJT Inverter with another Inverter as a Load

Consider the BJT inverter in Figure P3.16(a) where V_{CC} = 12 V, R_B = 1 kΩ, R_C = 5 kΩ, R_{C1} = 5 kΩ, and the BJT parameters are

β_F = 100, β_R = 1, I_s = 10^{-14} A, r_b = 0 Ω, V_A = ∞V, $C_{be1}(C_{je})$ = 10 pF, and $C_{bc1}(C_{jc})$ = 1 pF or 10 pF for Q_1,

and

β_F = 100, β_R = 1, I_s = 10^{-14} A, r_b = 0 Ω, V_A = ∞V, $C_{be2}(C_{je})$ = 10 pF, and $C_{bc2}(C_{jc})$ = 1 pF for Q_2.

Figure P3.16(b1) and (b2) shows the input/output voltage waveforms obtained from PSpice simulations, each with C_{bc1}=1 pF and 10 pF for Q_1, respectively. Referring to Section 3.6 and Figure P3.16(b1-b2), answer the following questions:

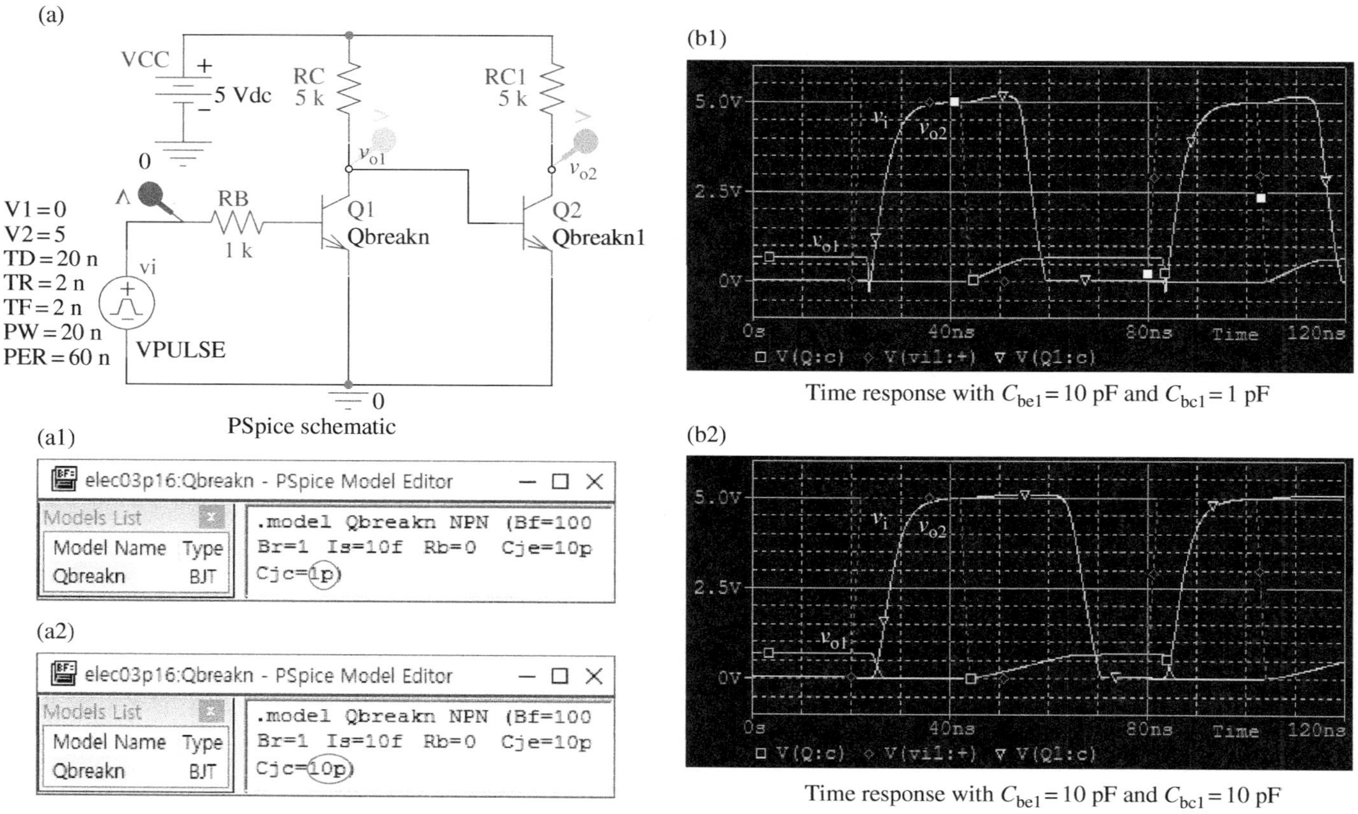

Figure P3.16 BJT inverter with another inverter as a load and its time responses from PSpice ("elec03p16.opj").

(a) During the rising period, why does v_{o1} reach just 0.7 V while v_{o2} reaches V_{CC} = 5 V?
(b) The output voltage $v_{o2}(t)$ of the second inverter shows a small negative spike in the beginning of its rising period in Figure P3.16(b1), but not in Figure P3.16(b2). Explain how such a difference is made by the difference in C_{bc1} = 1 pF or 10 pF.
(c) Explain how a larger value of C_{bc1} makes a longer time constant of v_{o1} during its rising period.

4

FET Circuits

CHAPTER OUTLINE

4.1 Field-Effect Transistor (FET)

As the Bipolar Junction Transistor (BJT) with three terminals, each called the base B, collector C, and emitter E, the Field-Effect Transistor (FET) is also a

Electronic Circuits with MATLAB®, PSpice®, and Smith Chart, First Edition. Won Y. Yang, Jaekwon Kim, Kyung W. Park, Donghyun Baek, Sungjoon Lim, Jingon Joung, Suhyun Park, Han L. Lee, Woo June Choi, and Taeho Im.

Companion website: www.wiley.com/go/yang/electroniccircuits

Table 4.1 BJT versus FET.

	FET (Field Effect Transistor)	BJT (Bipolar Junction Transistor)
	Voltage-operated device	Current-operated device
Input resistance	Large	Small in CE/CB configurations
Output resistance	Large	Small
Power dissipation	Low	High
Noise	Low	Medium
Switching time	Very fast	Fast
Robustness to static charge	Low (more destructible)	High
Thermal stability	Better	
Fabricability	FET can be more easily fabricated with higher density in a smaller space. It can also be connected as *R* or *C*, which makes possible the design of systems consisting of only FETs with no other elements.	

semiconductor device with three terminals, each called the gate G, drain D, and source S. In contrast with the BJT that operates with both types of charge carriers, holes and electrons, the FET is a 'unipolar' device that works with only one type of carriers, holes or electrons. While the BJT can basically be modeled as a current-controlled current source (in the forward-active region) since its collector current i_C depends on its base current i_B, the FET can basically be modeled as a voltage(field)-controlled current source (in the saturation region) since its drain current i_D depends on its gate-to-source voltage v_{GS}. Table 4.1 shows a rough comparison between BJTs and FETs.

Two types of FET are most widely used, junction-gate device called JFET (Junction FET) and insulated-gate device called MOSFET (Metal-Oxide-Semiconductor FET).

4.1.1 JFET (Junction FET)

There are two basic configurations of junction field effect transistor, the *n*-channel JFET and the *p*-channel JFET. As shown in Figures 4.1.1 and 4.1.2, an *n*/*p*-channel JFET consists of a lightly doped *n*/*p*-type semiconductor channel and a highly doped $P(P^+)/N(N^+)$-type semiconductor gate. Figure 4.2 (a) and (b) show typical drain (output) and transfer characteristic curves of a JFET, respectively. There are three regions (operation modes) besides the breakdown region:

1) **Ohmic (Triode) region**

   When $0 < v_{DS} \leq v_{GS} - V_t$ with $V_t < v_{GS} < 0$ (for an *n*-channel JFET), the drain current will be

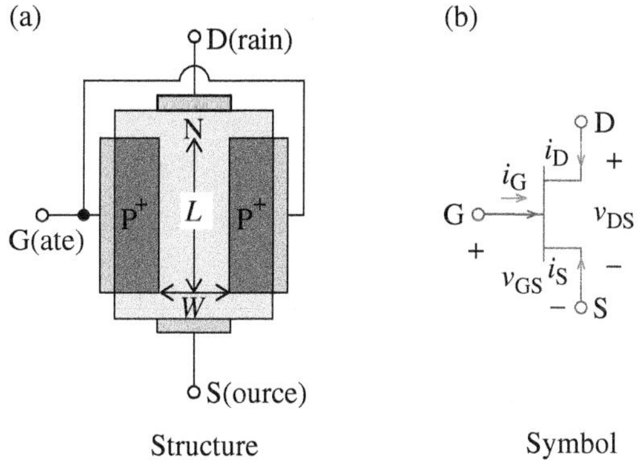

Figure 4.1.1 Structure and symbol of an *n*-channel

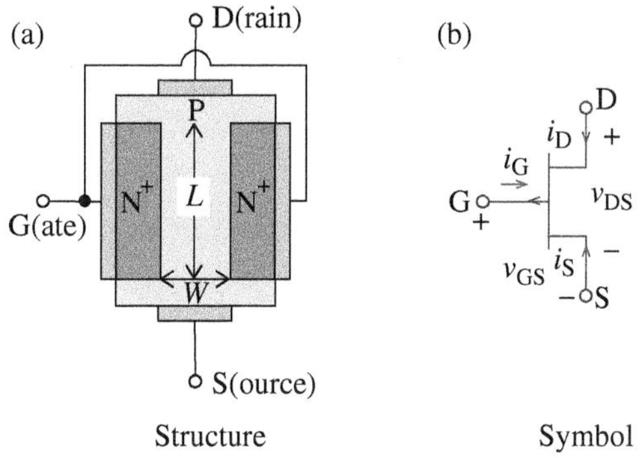

Figure 4.1.2 Structure and symbol of a *p*-channel JFET.

$$i_D = K_p\left\{(v_{GS} - V_t)v_{DS} - \frac{1}{2}v_{DS}^2\right\}(1 + \lambda v_{DS}) \quad \left(I_{DSS} = i_D\Big|_{\substack{v_{GS}=0 \\ v_{DS}=-V_t}}\right) \tag{4.1.1}$$

with $K_p = 2I_{DSS}/V_t^2$ [A/V^2]: *transconductance coefficient* or *conduction parameter*

where V_t [V], called the *threshold, pinch-off,* or *pinch-down voltage,* is such a value of v_{GS} that the drain current i_D drops to zero if $v_{GS} \le V_t$ (for an *n*-channel JFET) or $v_{GS} \ge V_t$ (for a *p*-channel JFET). λ [V^{-1}] is the *channel length modulation* (*CLM*) *parameter,* typically ranging from 0.005 to 0.02 V^{-1}, which accounts for the variation of transconductance coefficient with v_{DS}. I_{DSS} [A], called the *drain-to-source saturation current* or *zero bias (gate voltage) drain current,* is the value of the drain current at $v_{DS} = -V_t$ (with $v_{GS} = 0$).

Why is this region named 'ohmic region'? Because the JFET in the region acts like a voltage-controlled resistor whose resistance is determined by v_{GS}. Since i_D can be large depending on v_{GS} even with a very low v_{DS}, the JFET operating in this region can be viewed as a closed switch with the on-drain resistance $r_{DS,ON}$.

2) **Saturation (Pinch-off) or Constant current region**

When $0 \le v_{GS} - V_t \le v_{DS} < V_{breakdown}$ (for an n-channel MOSFET), the drain current is determined as

$$i_D = \frac{1}{2}K_p(v_{GS} - V_t)^2(1 + \lambda v_{DS}) = I_{DSS}\left(1 - \frac{v_{GS}}{V_t}\right)^2(1 + \lambda v_{DS}) \tag{4.1.2}$$

Note that the boundary between the ohmic and saturation regions plotted as a line of triangle (▿) symbols is described by

$$i_D = \frac{1}{2}K_p v_{DS}^2 \text{ with } K_p = 2I_{DSS}/V_t^2 \text{ and } I_{DSS} = i_D\Big|_{\substack{v_{GS}=0 \\ v_{DS}=-V_t}} \tag{4.1.3}$$

3) **Cutoff region**

If $v_{GS} \le V_t$ (for an n-channel JFET) or $v_{GS} \ge V_t$ (for a p-channel JFET), the leakage drain current $I_{DS,OFF}$ and the gate cutoff current I_{GSS} flow between the drain/gate and the source of an FET. However, the currents are about 1 pA or less than that even under high voltage, the FET can be viewed as an open switch.

4) **Breakdown region**

If the magnitude of the voltage across any two terminals exceeds a certain value, it may cause avalanche breakdown across the gate junction. As can be seen from the drain (output) characteristics in Figure 4.2(a), the breakdown

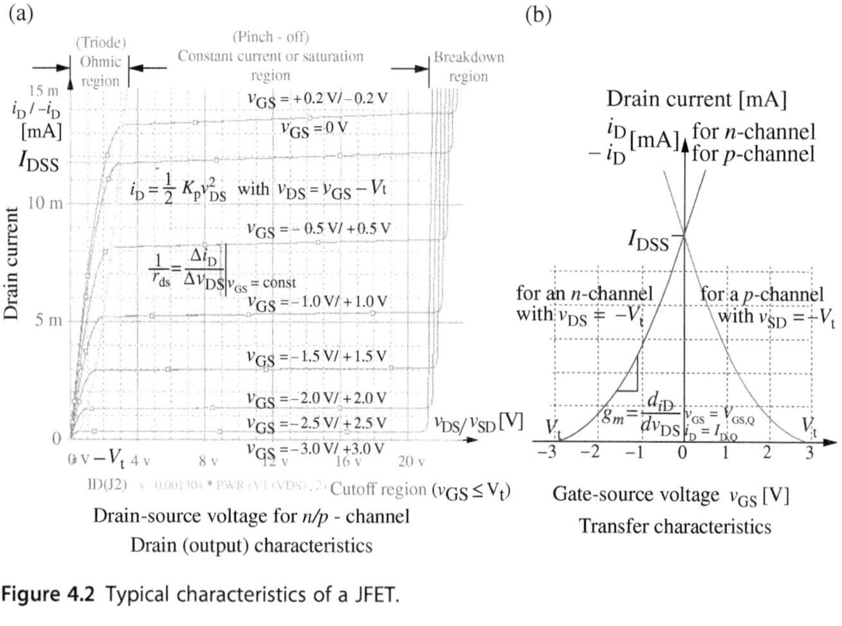

Figure 4.2 Typical characteristics of a JFET.

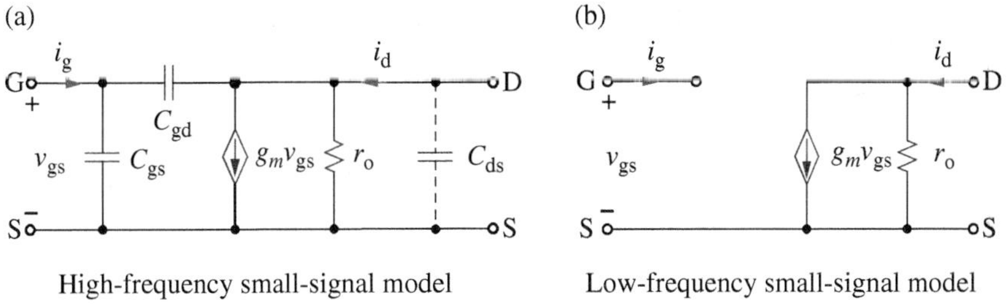

Figure 4.3 Small-signal models of FET.

voltage between drain and source becomes lower as the reverse-bias gate voltage ($|v_{GS}|$) increases in magnitude where the breakdown voltage BV_{DSS} with $v_{GS} = 0$ is specified in the manufacturers' datasheets.

Figure 4.3(a) and (b) shows the high/low frequency small-signal (AC) models of an FET operating in the saturation region, respectively, where $r_o[\Omega]$: incremental resistance between D and S (called the *output resistance*) is the reciprocal of the slope of the output characteristic (Figure 4.2(a)) at the operating (bias) point Q:

$$r_o = \frac{1}{\partial i_D/\partial v_{DS}}\bigg|_{v_{GS}=\text{const}} \overset{(4.1.2)}{\approx} \frac{2}{\lambda K_p(v_{GS}-V_t)^2}\bigg|_Q \overset{(4.1.2)}{=} \frac{1+\lambda I_{D,Q}}{\lambda V_{DS,Q}} \approx \frac{1}{\lambda I_{D,Q}} = \frac{|V_A|}{I_{D,Q}} \tag{4.1.4a}$$

g_m[S]: *transconductance* (*gain*) is the slope of the transfer characteristic (Figure 4.2(b)) at the operating point Q:

$$g_m = \frac{\partial i_D}{\partial v_{GS}}\bigg|_Q \overset{(4.1.2)}{=} \frac{\partial}{\partial v_{GS}}\left\{\frac{1}{2}K_p(v_{GS}-V_t)^2(1+\lambda v_{DS})\right\}\bigg|_Q \approx K_p\left(V_{GS,Q}-V_t\right) \overset{(4.1.2)}{\approx} \frac{2I_{D,Q}}{V_{GS,Q}-V_t} \tag{4.1.4b}$$

$$\lambda = \frac{1}{|V_A|}: \textit{CLM (channel length modulation) parameter} \tag{4.1.4c}$$

normally $0.005 \sim 0.05\,\text{V}^{-1}$

V_A[V]: *channel modulation voltage* or *Early voltage* normally in the range of 20 ~ 200 V

Note that the parameters KP, lambda, Vto, Cgd, and Cgs of the PSpice model for a JFET represent K_p, λ, V_t, C_{gd}, and C_{gs}, respectively, as can be seen from the PSpice Model Editor window (opened by selecting the device and clicking on

Edit > PSpice_Model from the top menu bar in the Schematic Window) or the PSpice simulation output file. Also, the operating point information in the PSpice simulation output file obtained from the Bias Point analysis shows the parameters GM and GDS, each representing g_m and $1/r_o$.

Let us consider the JFET circuit of Figure 4.4(a1) where the voltage appearing across R_S is fed back to bias the JFET. Note that R_S should be chosen large enough to make $v_{GS} < 0$ so that the gate-channel(source) junction can be reverse-biased for normal operation. To analyze this circuit, let us apply Kirchhoff's voltage law (KVL) to the G-S loop to write the load line equation and solve it to find i_D as

$$v_{GS} + R_S i_D = v_G - V_{SS}; \quad i_D = \frac{1}{R_S}(v_G - V_{SS} - v_{GS}) \tag{4.1.5}$$

Then, assuming that the FET will be in the saturation region, we equate this to the drain current Eq. (4.1.2) with $\lambda \approx 0$:

$$\frac{1}{2}K_p(v_{GS} - V_t)^2 \underset{\text{with } \lambda\approx 0}{\overset{(4.1.2)\,\&\,(4.1.5)}{=}} \frac{1}{R_S}(v_G - V_{SS} - v_{GS})$$

$$;\frac{1}{2}K_p v_{GS}^2 + \left(\frac{1}{R_S} - K_p V_t\right) v_{GS} + \frac{1}{2}K_p V_t^2 + \frac{1}{R_S}(V_{SS} - v_G) = 0 \tag{4.1.6}$$

where $v_G = 0$, $R_D = 1\,\text{k}\Omega$, $R_S = 500\,\Omega$, $V_{DD} = 10\,\text{V}$, $V_{SS} = -1\,\text{V}$, $K_p = 2.608 \times 10^{-3}$ A/V^2, $V_t = -3\,\text{V}$, and $I_{DSS} = K_p V_t^2/2 = 11.72\,\text{mA}$. We solve Eq. (4.1.6) to find $V_{GS,Q} = -1.17\,\text{V}$ and substitute it for v_{GS} into Eq. (4.1.5) to get $I_{D,Q}$ = 4.35 mA, which is supported by the operating point $Q = (V_{GS,Q}, I_{D,Q})$ = (-1.17 V, 4.35 mA) obtained from the load line analysis shown in Figure 4.4(b). We can substitute $I_{D,Q} = 4.35\,\text{mA}$ into the KVL equation for the D-S loop to find $V_{DS,Q}$:

$$V_{DD} - R_D i_D - v_{DS} - R_S i_D - V_{SS} = 0$$

$$;V_{DS,Q} = V_{DD} - V_{SS} - (R_D + R_S)I_{D,Q} = 4.48\,\text{V} \tag{4.1.7}$$

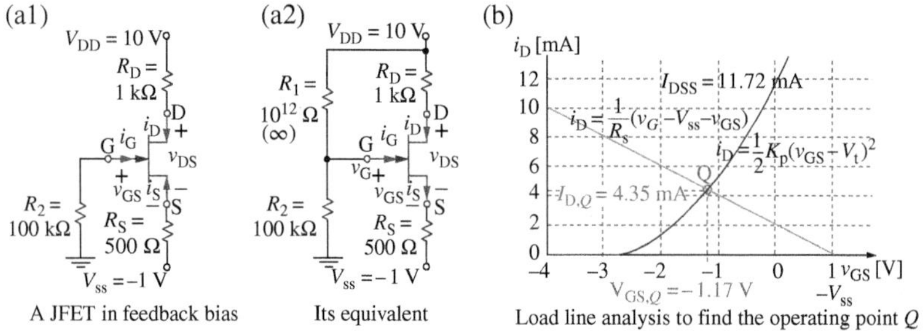

Figure 4.4 DC analysis of a JFET circuit

These computations can be done by running the following MATLAB statements:

```
>>vG=0; RD=1000; RS=500; VDD=10; VSS=-1; Kp=2.608e-3;
    Vt=-3; IDSS=Kp/2*Vt^2
>>x=roots([Kp/2 1/RS-Kp*Vt Kp/2*Vt^2+(VSS-vG)/RS]); % Eq.(4.1.6)
>>VGSQ=x(find((Vt<x&x<vG-VSS)|(Vt>x&x>vG-VSS)))
    VGSQ = -1.1740
>>IDQ=(vG-VSS-VGSQ)/RS % Eq.(4.1.5)
    IDQ = 0.0043
>>VDSQ=VDD-VSS-(RD+RS)*IDQ % Eq.(4.1.7)
    VDSQ = 4.4780
```

As an alternative, the MATLAB function '`FET_DC_analysis()`' (that was made to analyze the standard FET biasing circuits using a voltage divider as depicted in Figure 4.4(a2)) can be used as follows:

```
>>R1=1e12; R2=1e5; [VGQ,VSQ,VDQ,VGSQ,VDSQ,IDQ,mode]=...
     FET_DC_analysis([VDD VSS],R1,R2,RD,RS,Kp,Vt);
  Analysis Results
  VDD    VGQ  VSQ  VDQ  IDQ
  10.00 0.00 1.17 5.65 4.35e-003
  in the saturation mode
```

Here, note the following:

- One of the two roots of the quadratic Eq. (4.1.6) between V_t and $v_G - V_{SS}$ should be selected (see Figure 4.2(b)):

$$V_t \le V_{GS,Q} \le v_G - V_{SS} \text{ (for an } n\text{-channel JFET)} \tag{4.1.8a}$$

$$v_G - V_{SS} \le V_{GS,Q} \le V_t \text{ (for a } p\text{-channel JFET)} \tag{4.1.8b}$$

  If neither of them turns out to be inside the interval $[V_t, v_G - V_{SS}]$, the JFET must be in the cutoff region.
- In order for the above analysis result to be valid, the operating point should be in the saturation region since we have used Eq. (4.1.2) on the assumption that the JFET is saturated. This condition can be assured by checking if the following inequality is satisfied:

$$V_{DS,Q} \ge V_{GS,Q} - V_t \text{ (for an } n\text{-channel JFET)} \tag{4.1.9a}$$

$$V_{DS,Q} \le V_{GS,Q} - V_t \text{ (for a } p\text{-channel JFET)} \tag{4.1.9b}$$

Now, let us consider a question: "What if neither of the two roots of Eq. (4.1.6) satisfies this inequality?" In this case, the JFET must operate in the ohmic region where the circuit can be analyzed by finding the solution of Eq. (4.1.1) with the relations Eqs. (4.1.5) and (4.1.7) substituted for v_{GS} and v_{DS}, respectively:

$$\frac{1}{2}K_p\{2(v_{GS}-V_t)v_{DS}-v_{DS}^2\} \overset{(4.1.1)}{=} i_D$$

$$\downarrow v_{GS} \overset{(4.1.5)}{=} v_G - V_{SS} - R_S i_D \text{ and } v_{DS} \overset{(4.1.7)}{=} V_{DD} - V_{SS} - (R_D + R_S)i_D$$

$$K_p(v_G - V_{SS} - R_S i_D - V_t)\{V_{DD} - V_{SS} - (R_D + R_S)i_D\}$$
$$-\frac{1}{2}K_p\{V_{DD} - V_{SS} - (R_D + R_S)i_D\}^2 = i_D$$

$$;\frac{1}{2}K_p(R_S^2 - R_D^2)i_D^2 + \{K_p((R_D + R_S)(-v_G + V_t) + R_D V_{DD} + R_S V_{SS}) - 1\}i_D$$
$$-\frac{1}{2}K_p(V_{DD} - V_{SS})(V_{DD} + V_{SS} - 2v_G + 2V_t) = 0 \quad (4.1.10)$$

Since this equation is also a quadratic equation, the larger one of its two roots should be taken where it is supposed to satisfy

$$|(R_D + R_S)i_D| < |V_{DD} - V_{SS}| \quad (4.1.11)$$

```
function [VGQ,VSQ,VDQ,VGSQ,VDSQ,IDQ,mode]= ...
             FET_DC_analysis0(vG,VDD,VSS,RD,RS,Kp,Vt,lambda)
% Kp=up*Cox*W/L: Conduction parameter with +/- sign for n/p-channel
% Copyleft: Won Y. Yang, wyyang53@hanmail.net, CAU for academic use only
NP=sign(Kp); Kp=abs(Kp);
if isnumeric(vG), vG=vG(1);
 if RS>1e-3, x=roots([Kp/2 1/RS-Kp*Vt Kp/2*Vt^2+(VSS-vG)/RS]);
  else x=roots([Kp/2*RS 1-Kp*Vt*RS Kp/2*Vt^2*RS+VSS-vG]); % Eq.(4.1.6)
 end
 VGSQ=x(find((Vt<x&x<=vG-VSS)|(vG-VSS<=sgn*x&sgn*x<Vt))); % Eq.(4.1.8)
 if isempty(VGSQ), mode='cut-off'; IDQ=0; VGSQ=vG-VSS; VDSQ=VDD-VSS;
  else IDQ=Kp/2*(VGSQ-Vt)^2; % (vG-VGSQ-VSS)/RS; % Eq.(4.1.5)
   mode='saturation'; VDSQ=VDD-VSS-(RD+RS)*IDQ; % Eq.(4.1.7)
   if VDSQ<VGSQ-Vt % Eq.(4.1.9)
     mode='ohmic'; RDS=RD+RS; RSD=RS-RD; vGp=vG-Vt;
     A=Kp/2*RDS*RSD; B=Kp*(RD*VDD+RS*VSS-RDS*vGp)-1;
     C=-Kp/2*(VDD-VSS)*(VDD+VSS-2*vGp); %Coeffs of Eq.(4.1.10)
     IDQs=roots([A B C]); IDQs=IDQs(find(IDQs*sign(IDQ)>0));
     IDQ=IDQs(find(abs(IDQs)==max(abs(IDQs))));
     VGSQ=vG-VSS-RS*IDQ; VDSQ=VDD-VSS-(RD+RS)*IDQ; % Eq.(4.1.7)
  end
 end
else % When G is connected to D via some RG or directly,
 eq=@(v)VDD-VSS-v-(RD+RS)*Kp/2*(v-Vt).^2.*(1+lambda*v);
 v0=(VDD-VSS)/2; v=fsolve(eq,v0); VDSQ=v; VGSQ=v;
end
[IDQ,mode]=iD_NMOS_at_vDS_vGS(VDSQ,VGSQ,Kp,Vt,lambda);
if NP>0, VSQ=VSS+RS*IDQ; VGQ=VSQ+VGSQ; VDQ=VDD-RD*IDQ;
 else VSQ=VDD-RS*IDQ; VGQ=VSQ-VGSQ; VDQ=VSS+RD*IDQ;
end
```

```
function [VGQ,VSQ,VDQ,VGSQ,VDSQ,IDQ,mode,gm,ro] = ...
          FET_DC_analysis(VDS,R1,R2,RD,RS,Kp,Vt,lambda)
% Normally called like FET_DC_analysis(VDS,R1,R2,RD,RS,Kp,Vt,lambda);
% May be called like FET_DC_analysis(VDS,vG,RD,RS,Kp,Vt,lambda);
% May be called like FET_DC_analysis(VDS,'RG',RD,RS,Kp,Vt,lambda);
% Input: VDS=[VDD VSS] or VDD if VSS=0
%      Kp=kp'(W/L)=up*Cox*W/L or 2*IDS/Vt^2: Conduction parameter
%      Vt                                   : Threshold voltage
% Copyleft: Won Y. Yang, wyyang53@hanmail.net, CAU for academic use only
if length(VDS)>1, VSS=VDS(2); VDD=VDS(1); else VDD=VDS; VSS=0; end
RS=sum(RS); RD=sum(RD); R2=sum(R2);
if RS>=1|RS<eps % When this function is normally called
 if nargin<8, lambda=0; end
 if R2==Inf, vG=VDD; else vG=VDD*R2/(R1+R2); end
 fprintf('\nAnalysis Results with R1=%8.2f[kOhm], R2=%8.2f[kOhm],
    RD=%6.0f[Ohm], RS=%6.0f[Ohm]', R1/1e3,R2/1e3,RD,RS);
else % If RS<1, RS and R1 must be Kp and 'RG' or vG itself, respectively.
  % When R1='RG' (with D-G connected directly or via RG so that vG=vD),
 if nargin>6, lambda=Vt; else lambda=0; end
 Vt=Kp; Kp=RS; RS=RD; RD=R2; vG=R1;
 if isnumeric(R1), fprintf('\nAnalysis Results with vG=%6.2f[V],
                          RD=%6.0f[Ohm], RS=%6.0f[Ohm]', vG,RD,RS);
  else fprintf('\nAnalysis Results with RG between D and G,
                     RD=%6.0f[Ohm], RS=%6.0f[Ohm]',RD,RS);
 end
end
if lambda>1, lambda=1/lambda; end % VA=1/lambda must have been given
[VGQ,VSQ,VDQ,VGSQ,VDSQ,IDQ,mode] = ...
     FET_DC_analysis0(vG,VDD,VSS,RD,RS,Kp,Vt,lambda);
% Transconductance and Output resistance (ro) Eqs.(4.1.4b,a)
gm=Kp*(VGSQ-Vt)*(1+lambda*VDSQ); ro=1/lambda/abs(IDQ);
disp(' VDD  VGQ  VSQ  VDQ  IDQ')
fprintf('%5.2f %5.2f %5.2f %5.2f %4.2e', VDD,VGQ,VSQ,VDQ,IDQ)
fprintf('\n in %s mode with VGD,Q=%6.2f[V]<=Vt=%4.2f,\n',
                                              mode,VGQ-VDQ,Vt)
```

The whole computational process of analyzing the standard *n*-channel FET biasing circuit shown in Figure 4.4(a2) has been cast into the above MATLAB function ‘`FET_DC_analysis()`’. Likewise, the following MATLAB function ‘`FET_PMOS_DC_analysis()`’ has been composed to analyze typical (DC driven) *p*-channel FET biasing circuits.

```
function [VGQ,VSQ,VDQ,VSGQ,VSDQ,IDQ,mode]= ...
             FET_PMOS_DC_analysis(VSD,R1,R2,RS,RD,Kp,Vt)
% Input: VSD= [VSS VDD] or VSS if VDD=0
%     Kp=kp'(W/L)=up*Cox*W/L or 2*IDS/Vt^2: Conduction parameter
%     Vt                                  : Threshold voltage
% Copyleft: Won Y. Yang, wyyang53@hanmail.net, CAU for academic use only
if length(VSD)>1, VSS=VSD(1); VDD=VSD(2); else VSS=VSD; VDD=0; end
if length(RS)>1, RS=sum(RS); end
vG=VDD*R2/(R1+R2)+VSS*R1/(R1+R2);;
[VGQ,VSQ,VDQ,VSGQ,VSDQ,IDQ,mode]= ...
FET_DC_analysis0(vG,VSS,VDD,RD,RS,-Kp,-Vt);
disp(' VSS  VGQ  VSQ  VDQ  IDQ')
fprintf('%5.2f %5.2f %5.2f %5.2f %4.2e', VSS,VGQ,VSQ,VDQ,IDQ)
fprintf('\n in the %s mode with VDG,Q=%6.2f[V] (|Vt|=%4.2f)\n', ...
    mode,VDQ-VGQ,abs(Vt))
```

```
function [VGQ,VSQ,VDQ,VGSQ,VDSQ,IDQ,mode]= ...
           FET_DC_analysis0(vG,VDD,VSS,RD,RS,Kp,Vt,lambda)
% Copyleft: Won Y. Yang, wyyang53@hanmail.net, CAU for academic use only
if nargin<8, lambda=0; end
NP=sign(Kp); Kp=abs(Kp);
if length(vG)<2 % Eq.(4.1.6)
 if RS>1e-3, x=roots([Kp/2 1/RS-Kp*Vt Kp/2*Vt^2+(VSS-vG)/RS]);
  else x=roots([Kp/2*RS 1-Kp*Vt*RS Kp/2*Vt^2*RS+VSS-vG]);
 end
 VGSQ=x(find((Vt<x&x<=vG-VSS)|(vG-VSS<=sgn*x&sgn*x<Vt))); % Eq.(4.1.8)
 else % If vG is given as [vG vGSQ],
 VGSQ=vG(2); vG=vG(1);
end
if isempty(VGSQ), mode='cut-off'; IDQ=0; VGSQ=vG-VSS; VDSQ=VDD-VSS;
 else IDQ=Kp/2*(VGSQ-Vt)^2; % (vG-VGSQ-VSS)/RS; % Eq.(4.1.5)
  mode='saturation'; VDSQ=VDD-VSS-(RD+RS)*IDQ; % Eq.(4.1.7)
  if VDSQ<VGSQ-Vt % Eq.(4.1.9)
   mode='ohmic'; RDS=RD+RS; RSD=RS-RD; vGp=vG-Vt;
   A=Kp/2*RDS*RSD; B=Kp*(RD*VDD+RS*VSS-RDS*vGp)-1;
   C=-Kp/2*(VDD-VSS)*(VDD+VSS-2*vGp);
   IDQs=roots([A B C]); IDQs=IDQs(find(IDQs*sign(IDQ)>0));
   IDQ=IDQs(find(abs(IDQs)==max(abs(IDQs))));
   VGSQ=vG-VSS-RS*IDQ;
   VDSQ=VDD-VSS-(RD+RS)*IDQ; % Eq.(4.1.7)
  end
  if lambda>1e-8 % Eqs.(4.1.12a,b)
   eq=@(v) [vG-v(2)-VSS-RS*iD_NMOS_at_vDS_vGS(v(1),v(2),Kp,Vt,lambda);
    VDD-VSS-v(1)-(RD+RS)*iD_NMOS_at_vDS_vGS(v(1),v(2),Kp,Vt,lambda)];
   v0=[VDSQ VGSQ]; % Initial guess for v=[vDS vGS]
   options=optimset('Display','off','Diagnostics','off');
   v=fsolve(eq,v0,options); VDSQ=v(1); VGSQ=v(2);
   IDQ=iD_NMOS_at_vDS_vGS(VDSQ,VGSQ,Kp,Vt,lambda);
  end
end
% .. ... .
```

```
function iD=iD_NMOS_at_vDS_vGS(vDS,vGS,Kp,Vt,lambda)
if nargin<5, lambda=0; end
vGD=vGS-vDS; ON=(vGS>Vt); SAT=(vGD<=Vt)&ON; TRI=(vGD>Vt)&ON;
iD=Kp*(1+lambda*vDS).*(((vGS-Vt).*vDS-vDS.^2/2).*TRI ...
        +(vGS-Vt).^2/2.*SAT); % Eqs.(4.1.1 and 4.1.2)
```

Before ending this section, let us consider how the FET analysis considering the effect of the CLM parameter λ should be performed when λ is not negligibly small where the value of the voltage v_G at node G is assumed to be given in the circuit of Figure 4.4(a2). In such a case, a set of two KVL equations in v_{GS} and v_{DS}, one for the path GSJ-R_S-V_{SS} and the other for the path V_{DD}-R_D-DSJ-R_S-V_{SS}, should be solved:

$$V_G - v_{GS} - R_S i_D(v_{DS}, v_{GS}) - V_{SS} = 0 \quad (4.1.12a)$$

$$V_{DD} - v_{DS} - (R_D + R_S) i_D(v_{DS}, v_{GS}) - V_{SS} = 0 \quad (4.1.12b)$$

where $i_D(v_{DS}, v_{GS})$ is determined by Eqs. (4.1.1) or (4.1.2) depending on the operation mode of the FET and implemented by the following MATLAB function '`iD_NMOS_at_vDS_vGS()`'. This set of nonlinear equations can be solved by using the MATLAB function '`fsolve()`' as implemented in the above modified version of '`FET_DC_analysis0()`'.

4.1.2 MOSFET (Metal-Oxide-Semiconductor FET)

A MOSFET, with the gate electrode insulated by an oxidized metal layer, can be fabricated to operate as either a depletion-mode or enhancement-mode FET. The MOSFET has a very high input resistance owing to the insulated gate. Due to the very thin gate, MOS devices can easily be damaged by ESD (electrostatic discharge), requiring special precautions.

Figures 4.5.1-4.5.5 show the basic structures and symbols of n/p-channel enhancement and depletion types of MOSFET. Figure 4.6(a) shows typical drain (output) characteristic curves of a MOSFET, where the variables of the horizontal and vertical axes are v_{DS}/v_{SD} and $i_D/-i_D$, respectively, for n/p-channel MOSFETs. Figure 4.6(b) shows typical transfer characteristic curves of n/p-channel enhancement and depletion types of MOSFET, where the sign of the *threshold (gate) voltage* V_t for *n/p*-channel is +/- or -/+ depending on whether the type of MOSFET is enhancement or depletion. Note that the voltage polarities and current directions for an *n*-channel MOSFET (NMOS) and a *p*-channel MOSFET (PMOS) are opposite to each other.

W: device width, L: channel length

L/W = aspect ratio with L and W of the channel

μ_n [cm^2/V/s] : mobility of electrons or positive holes in the channel

C_{ox}[F/cm^2] : gate-to-channel capacitance per unit area due to the gate oxide

$K_p = \mu_n C_{ox} W/L$ [A/V^2] : conduction constant

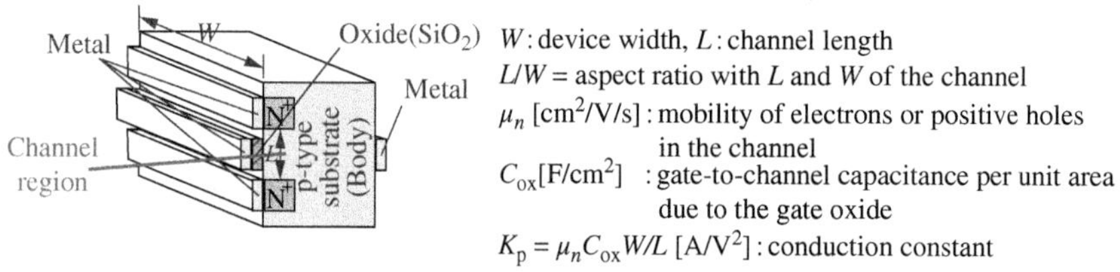

Figure 4.5.1 Structure and NMOS (*n*-channel MOSFET).

Structure   Symbol

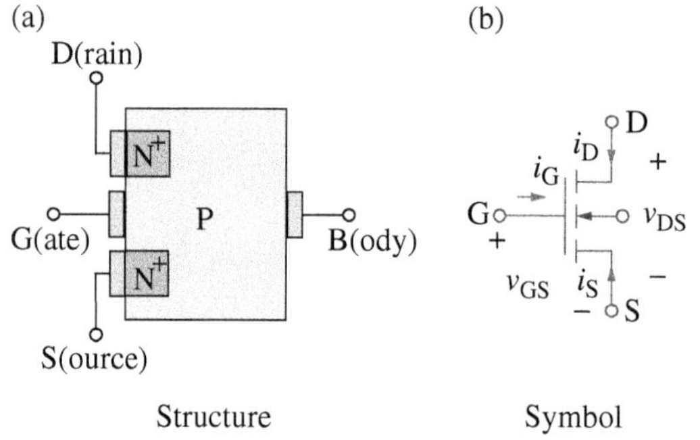

Figure 4.5.2 Structure and symbol of an *n*-channel enhancement MOSFET (NMOS).

Structure   Symbol

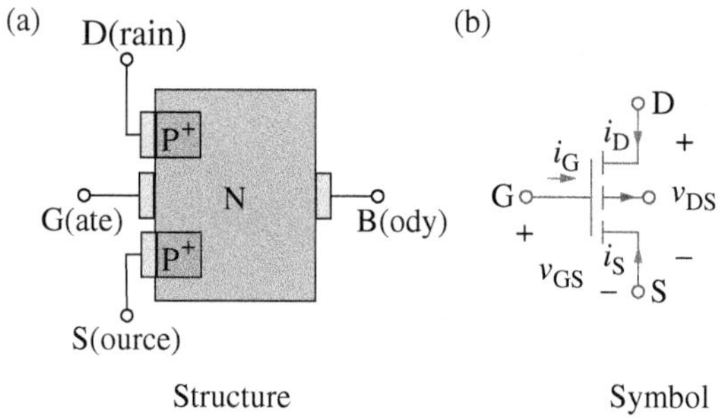

Figure 4.5.3 Structure and symbol of a *p*-channel enhancement MOSFET (PMOS).

Figure 4.5.4 Structure and symbol of an n-channel depletion MOSFET (d-NMOS).

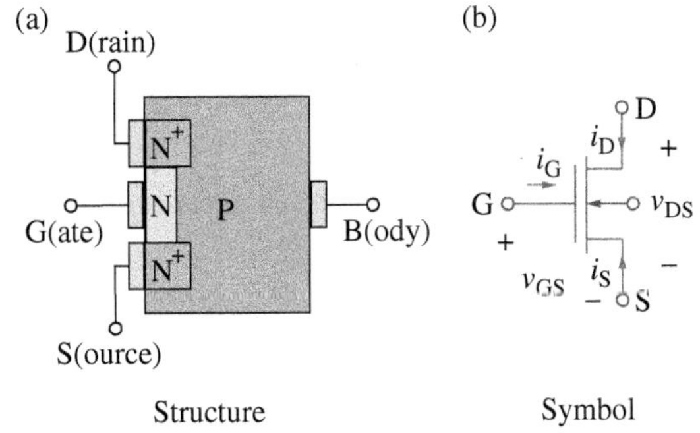

Figure 4.5.5 Structure and symbol of a p-channel depletion MOSFET (d-PMOS).

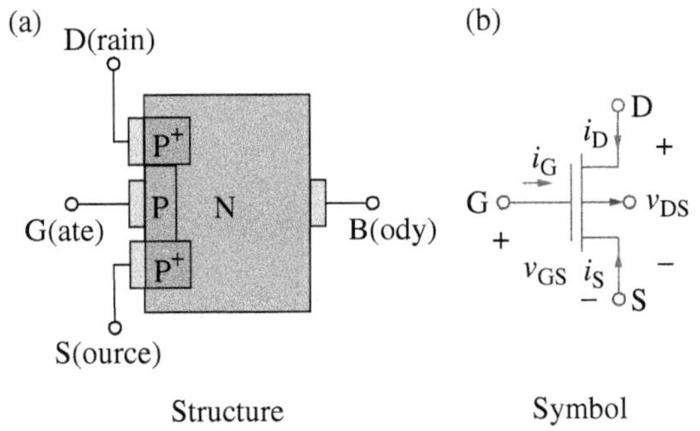

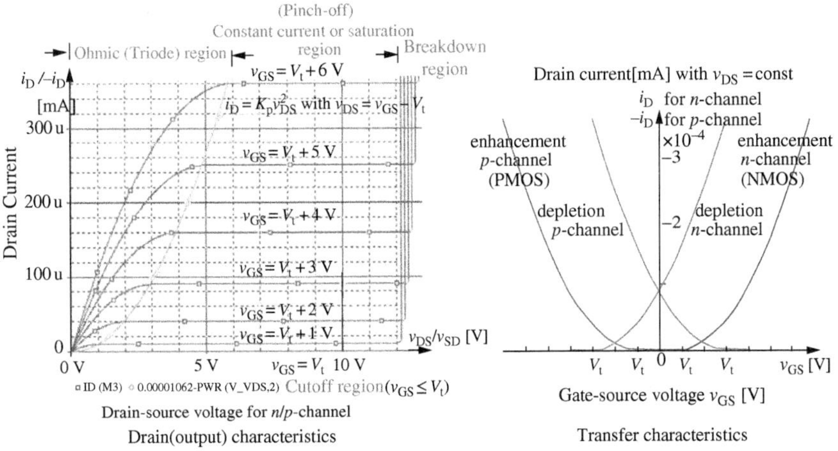

Figure 4.6 Typical characteristics of a MOSFET.

As depicted in Figure 4.6(a), there are three regions (operation modes) besides the breakdown region. Since the operational characteristics of depletion-type MOSFETs (d-MOSFETs) are the same with JFETs, we are going to look over the characteristics of enhancement-mode MOSFETs (e-MOSFETs) only:

1) **Ohmic (Triode) region**
   When $0 < v_{DS} \le v_{GS} - V_t$ with $|V_t| < |v_{GS}|$ (for an NMOS), the drain current is determined as

$$i_D = K_p\left\{(v_{GS} - V_t)v_{DS} - \frac{1}{2}v_{DS}^2\right\}(1 + \lambda v_{DS}) \text{ with } K_p = \mu_n C_{OX}\frac{W}{L}\ [\mathrm{A/V^2}] \tag{4.1.13a}$$

   where $K_p[\mathrm{A/V^2}]$: conduction parameter or transconductance coefficient, $\mu_n[\mathrm{m^2/V/s}]$: mobility of electrons or positive holes in the channel, C_{OX} $[\mathrm{F/m^2}]$: gate-to-channel capacitance per unit area due to the gate oxide, L/W: aspect ratio with length L and width W of the channel, λ $[\mathrm{V^{-1}}]$: *channel length modulation (CLM) parameter*

2) **Saturation (Pinch-off) or Constant current region**
   When $0 \le v_{GS} - V_t \le v_{DS} < V_{\text{breakdown}}$ (for an NMOS), the drain current is determined as

$$i_D = \frac{1}{2}K_p(v_{GS} - V_t)^2(1 + \lambda v_{DS}) \tag{4.1.13b}$$

   The boundary between the ohmic and saturation regions (with $v_{DS} = v_{GS} - V_t$) is described by

$$i_D = \frac{1}{2}K_p v_{DS}^2 \tag{4.1.14}$$

3) **Cutoff region**
   If $v_{GS} \le V_t$ (for an NMOS) or $v_{GS} \ge V_t$ (for a PMOS), the MOSFET will be cut off like an open switch.

Note that Eqs. (4.1.1) and (4.1.2) and Eqs. (4.1.13a) and (4.1.13b), each describing the *i*-*v* relationships of JFET and MOSFET, are identical except for the definition of the conduction parameter K_p and that is why the above MATLAB function '`FET_DC_analysis()`' can be used for the DC analysis of FET circuits whether the FET is a JFET or a MOSFET.

Table 4.2 summarizes the circuit symbols and *i*-*v* relationships of JFET and MOSFET.

Table 4.2 Circuit symbols and *i-v* relationships of JFET and MOSFET.

FET type	*n*-Channel			*p*-Channel		
	JFET	**Enhancement MOSFET**	**Depletion MOSFET**	**JFET**	**Enhancement MOSFET**	**Depletion MOSFET**
Circuit symbols	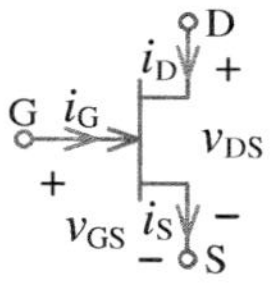	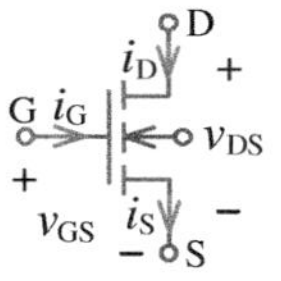	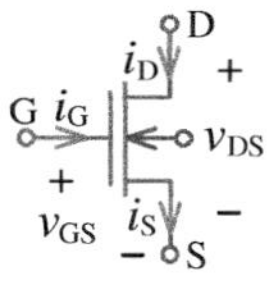	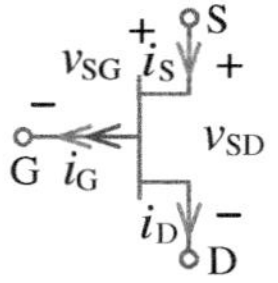	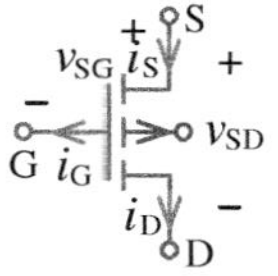	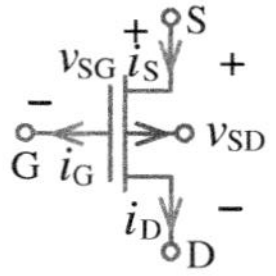
Threshold voltage V_t	–	+	–	+	–	–
Conduction constant K_p	$\frac{2I_{DSS}}{V_t^2}$	Process conduction parameter $K_p = \boxed{\mu_n C_{OX}} \frac{W}{L}$		$\frac{2I_{DSS}}{V_t^2}$	Process conduction parameter $K_p = \boxed{\mu_p C_{OX}} \frac{W}{L}$	
Turn-on condition	$v_{GS} > V_t$ and $v_{DS} > 0$			$v_{SG} > \|V_t\|$ and $v_{SD} > 0$		
Triode region (Ohmic mode)	$v_{GD} = v_G - v_D > V_t > 0$			$v_{DG} = v_D - v_G > \|V_t\|$		
	$i_D \cong K_p\{(v_{GS} - V_t)v_{DS} - v_{DS}^2/2\}$ with overdrive voltage $v_{OV} = v_{GS} - V_t$			$i_D \cong K_p\{(v_{SG} - \|V_t\|)v_{SD} - v_{SD}^2/2\}$ (4.1.13a) with overdrive voltage $v_{OV} = v_{SG} - \|V_t\|$		
Saturation region (Pinch-off mode)	$v_{GD} = v_G - v_D \leq V_t$			$v_{DG} = v_D - v_G \leq \|V_t\|$		
	$i_D \cong K_p(v_{GS} - V_t)^2/2$			$i_D \cong K_p\,(v_{SG} - \|V_t\|)^2/2$ (4.1.13b)		

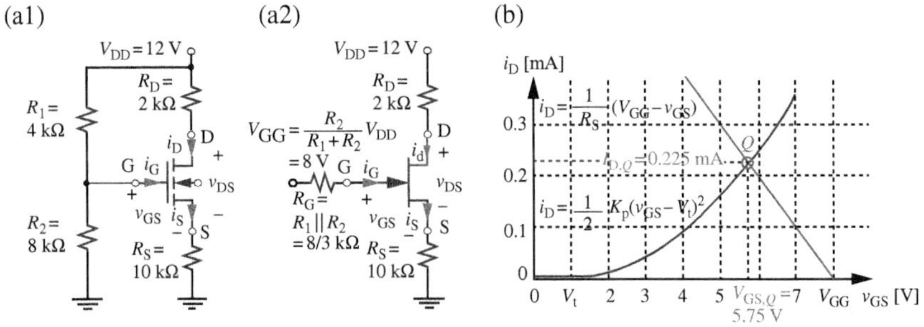

A MOSFET biasing circuit　　Its equivalent　　Load line analysis to find the operating point Q

Figure 4.7 DC analysis of a MOSFET circuit.

Let us consider the standard FET biasing circuit of Figure 4.7(a1) where the NMOS is biased by the voltage divider consisting of V_{DD}-R_1-R_2. To analyze the circuit, we replace the voltage divider by its Thevenin equivalent as shown in Figure 4.7(a2) and write the following equation by equating the two expressions for the drain current, i.e. one from the KVL equation for the V_{GG}-GSJ-R_S loop and the other from Eq. (4.1.13b) with $\lambda = 0$ as

$$\frac{1}{2}K_p(v_{GS} - V_t)^2 = \frac{1}{R_S}(V_{GG} - v_{GS})$$

$$; \frac{1}{2}K_p v_{GS}^2 + \left(\frac{1}{R_S} - K_p V_t\right) v_{GS} + \frac{1}{2}K_p V_t^2 - \frac{1}{R_S}V_{GG} = 0 \tag{4.1.15}$$

where $V_{DD} = 12$ V, $R_1 = 4$ kΩ, $R_2 = 8$ kΩ, $R_D = 2$ kΩ, $R_s = 10$ kΩ, $V_{GG} = V_{DD}R_2/(R_1 + R_2) = 8$ V, $K_p = 2 \times 10^{-5}$ A/V^2, and $V_t = 1$ V. We solve this equation to get $V_{GS,Q} = 5.75$ V and $I_{D,Q} = 0.225$ mA, which conforms with the operating point $Q = (V_{GS,Q}, I_{D,Q}) = (5.75$ V, 0.225 mA$)$ obtained from the load line analysis shown in Figure 4.7(b). Then we substitute $I_{D,Q} = 0.225$ mA for i_D into the KVL equation for the D-S loop to find $V_{DS,Q}$:

$$V_{DD} - R_D i_D - v_{DS} - R_S i_D = 0\ ;\ V_{DS,Q} = V_{DD} - (R_D + R_S)I_{D,Q} = 9.3\text{V} \tag{4.1.16}$$

These hand calculations can be done by running the following MATLAB statements:

```
>>R1=4000; R2=8000; RD=2e3; RS=1e4; VDD=12; Kp=2e-5; Vt=1;
  FET_DC_analysis(VDD,R1,R2,RD,RS,Kp,Vt);
```

which yields

```
VDD    VGQ   VSQ  VDQ    IDQ
12.00 8.00 2.25 11.55 2.25e-004
in the saturation mode with VGD,Q= -3.55[V]<=Vt=1.00
```

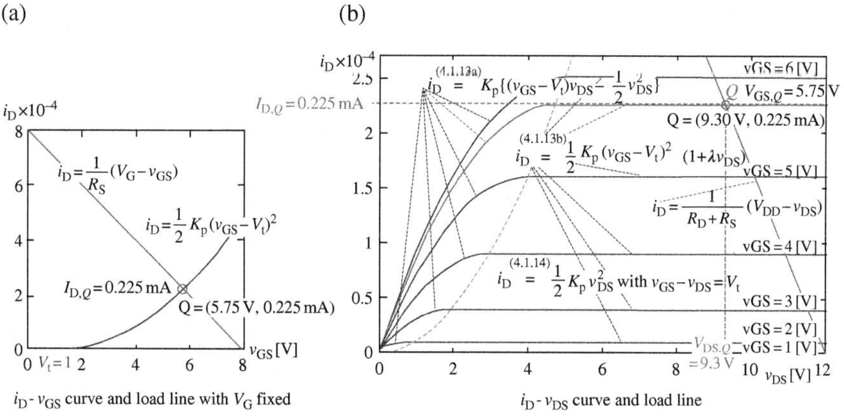

Figure 4.8 Operating point on the i_D-v_{DS} characteristic curve of the NMOS in Figure 4.7(a1).

If you want to locate the operating point $Q = (V_{GS,Q}, I_{D,Q})$ on the i_D-v_{DS} characteristic curve of the NMOS M1 from the viewpoint of load line analysis, run the following MATLAB script "elec04f08.m" to get Figure 4.8.

```
%elec04f08.m
VDD=12; R1=4e3; R2=8e3; RD=2e3; RS=10e3; % Circuit parameters
Kp=2e-5; Vt=1; % FET parameters
VGG=VDD*R2/(R1+R2);
vGSs=[0:1e-3:VGG]; % Range of vGS for which iD and load line are plotted.
iDs=Kp/2*(vGSs-Vt).^2; % iD-vGS characteristic in the saturation mode
iDs_load_line = (VGG-vGSs)/RS; % Load line
[vmin,imin] = min(abs(iDs-iDs_load_line)); % Find the intersection Q
IDQ=iDs(imin), VGSQ=vGSs(imin), VDSQ=VDD-(RD+RS)*IDQ % Q-point
subplot(221)
plot(vGSs,iDs, vGSs,iDs_load_line,'r', VGSQ,IDQ,'mo')
subplot(222)
vDSs=[0:1e-3:VDD]; vOVs=0:5;
for i=1:length(vOVs)
  vGS = Vt + vOVs(i);
  iDs = iD_NMOS_at_vDS_vGS(vDSs,vGS,Kp,Vt); % iD-vDS characteristic
  plot(vDSs,iDs), hold on
  text(0.85*VDD,max(iDs)+5e-6,['vGS=' num2str(vGS) '[V]'])
end
% Boundary between ohmic/saturation regions by Eq.(4.1.14)
plot(vDSs,Kp/2*vDSs.^2,'g:')
plot(vDSs,iD_NMOS_at_vDS_vGS(vDSs,VGSQ,Kp,Vt),'m', VDSQ,IDQ,'mo')
plot([0 VDD],(VDD-[0 VDD])/(RD+RS),'m'), % Load line
axis([0 VDD 0 27e-5])
```

These results can be supported by the Bias Point analysis using the PSpice. After running the OrCAD/PSpice project 'elec04f09.opj' with the schematic (Figure 4.9(a)), you can click PSpice > View_Output_File on the top menu bar of the Schematic window to see the simulation results shown in Figure 4.9(b). The parameters of PSpice parts MbreakN (enhancement *n*-channel)/MbreakND (depletion *n*-channel) (referred to as 'NMOS'/'NMOS_d') can be edited as shown in the PSpice Model Editor (Figure 4.9(c1)) opened by selecting the part (to let it pink-colored) and then clicking Edit > PSpice_Model on the top menu bar of the Schematic window.

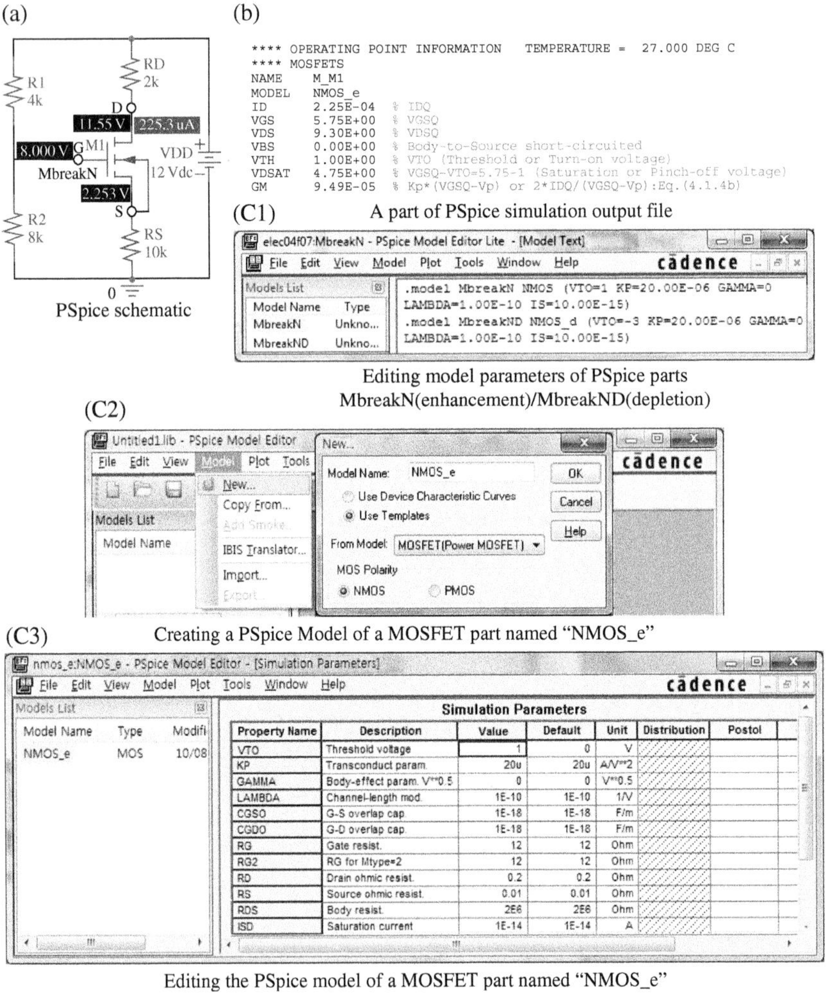

Figure 4.9 PSpice simulation for the FET circuit of Figure 4.7(a1) with creating/editing a PSpice model.

Alternatively, you can create a new PSpice model for FET in the following steps:

- Click File > New > PSpice_Library to open the PSpice Model Editor (Figure 4.9(c2)).
- Click Model/New to open the New Model dialog box, set it as Figure 4.9(c2) and click OK to open the model list and parameter list.
- Set the model parameters as Figure 4.9(c3) where most of them are left as their default values.
- Click File/Save to save the created model into a library file in your own directory.

There is one thing to note about the transconductance gain g_m, which is one of the AC model parameters for an FET. The PSpice Simulation output file (Figure 4.9(b)) shows that the value of g_m(GM) is 9.49×10^{-5}, as can be computed by using Eq. (4.1.4b):

$$g_m \overset{(4.1.4b)}{\approx} K_p\left(V_{GS,Q} - V_t\right) = 2 \times 10^{-5}(5.75 - 1) = 9.49 \times 10^{-5}$$
$$\text{or } g_m \overset{(4.1.4b)}{\approx} \frac{2I_{D,Q}}{V_{GS,Q} - V_t} = \frac{2 \times 2.25 \times 10^{-4}}{5.75 - 1} = 9.47 \times 10^{-5} \quad (4.1.17)$$

Example 4.1 An NMOS Circuit

Consider the NMOS circuit shown in Figure 4.10(a) where the device parameters of the enhancement NMOS (e-NMOS) are K_p=1[mA/V^2], V_t= 1[V], and $\lambda = 0$ or 0.01 [V^{-1}].

a) Assuming that $\lambda = 0\,\mathrm{V}^{-1}$, find i_D, V_D, and V_S.

First, V_G is determined by the voltage divider consisting of R_1 and R_2:

$$V_G = \frac{R_2}{R_1 + R_2} V_{DD} = \frac{10}{10 + 10} 10 = 5\ [\mathrm{V}] \quad (E4.1.1)$$

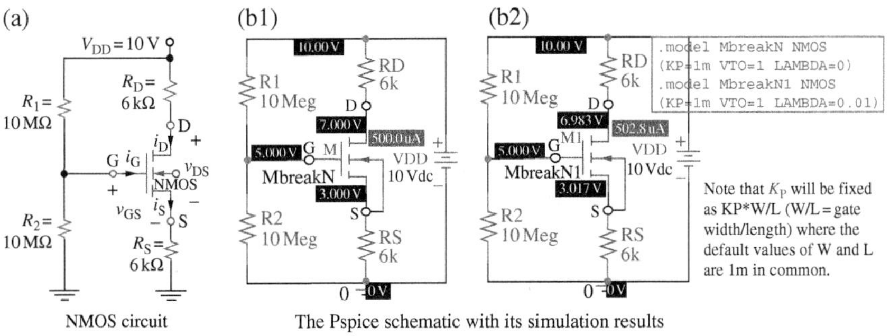

Figure 4.10 NMOS circuit for Example 4.1 ("elec04e01.opj").

Note that this value of $V_G = 5$ V is likely to make $v_{GS} > V_t$ so that the NMOS will be turned on. Then, to tell whether it operates in the saturation or triode mode, we assume that it operates in the <u>triode</u> mode where

$$i_D \overset{(4.1.13a)}{=} K_p\left\{(v_{GS} - V_t)v_{DS} - \frac{1}{2}v_{DS}^2\right\} \text{ [A]} \tag{E4.1.2}$$

We substitute $v_{GS} = V_G - R_S i_D = 5 - 6i_D$ and $v_{DS} = V_{DD} - R_D i_D - R_S i_D = 10 - 12i_D$ into this equation and solve it as

$$i_D = 1 \times \left\{(5 - 6i_D - 1)(10 - 12i_D) - \frac{1}{2}(10 - 12i_D)^2\right\}$$
$$; 11\, i_D - 10 = 0;\ i_D = \frac{10}{11} \text{ [mA]} \tag{E4.1.3}$$

Is this a valid solution? To check the validity, we should see if it satisfies the condition for the triode mode: $v_{GD} > V_t$ and $v_{GS} > 0$:

$$V_D = V_{DD} - R_D I_D = 10 - 6\frac{10}{11} = \frac{50}{11} \text{ [V]}$$
$$\rightarrow V_{GD} = V_G - V_D = 10 - \frac{50}{11} = \frac{5}{11} \text{ [V]} \overset{?}{>} V_t : \text{No} \tag{E4.1.4}$$

$$V_S = R_S I_D = 6 \times \frac{10}{11} = \frac{60}{11} \text{ [V]} \rightarrow V_{GS} = V_G - V_S = 5 - \frac{60}{11} \text{ [V]} \overset{?}{>} V_t : \text{No} \tag{E4.1.5}$$

This means that far from the triode condition, even the turn-on condition is not satisfied, contradicting the assumption of triode mode. Then, we assume the <u>saturation</u> mode where

$$i_D \underset{\lambda=0}{\overset{(4.1.13b)}{=}} \frac{1}{2}K_p(v_{GS} - V_t)^2 \tag{E4.1.6}$$

We substitute $v_{GS} = V_G - R_S i_D = 5 - 6i_D$ into this equation and solve it as

$$i_D = \frac{1}{2}K_p(V_G - V_S - V_t)^2 = \frac{1}{2}(5 - R_S i_D - 1)^2 = \frac{1}{2}(5 - 6i_D - 1)^2 \text{ [mA]}$$
$$;\ \ 18I_D^2 - 25I_D + 8 = 0 \rightarrow I_D = \frac{8}{9} \text{ or } \frac{1}{2} \text{ mA };\ i_D = \frac{1}{2} \text{ [mA]} \tag{E4.1.7}$$

Here, the first root $I_D = (8/9)$ mA should be dumped because it does not satisfy even the turn-on condition: $v_{GS} = V_G - R_S i_D < 0$. How about the second one $I_D = (1/2)$ mA?

$$V_S = R_S I_D = 6 \times \frac{1}{2} = 3 \text{ [V]}$$
$$\rightarrow v_{GS} = V_G - V_S = 5 - 3 = 2 \text{ [V]} \overset{?}{>} V_t : \text{OK (turn-on)} \tag{E4.1.8}$$

$$V_D = V_{DD} - R_D I_D = 10 - 6\frac{1}{2} = 7\ [V]$$
$$\rightarrow v_{GD} = 5 - V_D = 5 - 7 = -2\ [V] \leq^? V_t : \text{OK (saturation)} \tag{E4.1.9}$$

This implies that the FET surely operates in the saturation mode with $v_{DS} = 7 - 3 = 4$ V.

As an alternative to these hand calculations, the MATLAB function '`FET_DC_analysis()`' can be used as follows:

```
>>VDD=10; R1=10e6; R2=10e6; RD=6e3; RS=6e3; Kp=1e-3; Vt=1;
  FET_DC_analysis(VDD,R1,R2,RD,RS,Kp,Vt);
```

This yields the following results:

```
  VDD  VGQ  VSQ  VDQ    IDQ
 10.00 5.00 3.00 7.00 5.00e-04
 in the saturation mode with VGD,Q= -2.00 [V] (Vt=1.00)
```

This analysis result conforms with the PSpice simulation result (with Bias Point analysis) shown in Figure 4.10(b1).

b) Find such a value of R_D that the NMOS operates at the boundary between the saturation and triode regions where $v_{GD} = V_G - V_D = V_t$.

Since i_D has been determined as 0.5 mA regardless of R_D, we have only to find the value of R_D satisfying $v_{GD} = V_G - V_D = V_G - (V_{DD} - R_D i_D) = V_t$:

$$v_{GD} = V_G - V_D = V_G - (V_{DD} - R_D i_D) = 5 - (10 - 0.5R_D) = V_t = 1\,; R_D = 12[k\Omega] \tag{E4.1.10}$$

c) With $\lambda = 0.01\ V^{-1}$, find i_D, V_D, and V_S.

Since it is too difficult to solve Eqs. (4.1.12a,b) by hand, we use the MATLAB function '`FET_DC_analysis()`' as follows:

```
>>lambda=0.01;
  FET_DC_analysis(VDD,R1,R2,RD,RS,Kp,Vt,lambda);
```

This yields the following result, which conforms with the PSpice simulation result shown in Figure 4.10(b2).

```
VDD    VGQ   VSQ   VDQ    IDQ
10.00  5.00  3.02  6.98  5.03e-04
in the saturation mode with VGD,Q= -1.98 [V] (Vt=1.00)
```

Example 4.2 An NMOS Circuit

Consider the NMOS circuit shown in Figure 4.11(a) where the device parameters of the enhancement NMOS (e-NMOS) are $K_p = 25\ [\mu A/V^2]$, $V_t = 1\ [V]$, and $\lambda = 0$ or $0.01\ [V^{-1}]$.

a) Assuming that $\lambda = 0\ V^{-1}$, find i_D, V_D, and V_S.

Noting that R_G has no effect, we use the MATLAB function 'FET_DC_analysis()' as follows:

```
>>VDD=5; VSS=-5; VDS=[VDD VSS]; VG=1; RD=18e3; RS=22e3;
  Kp=25e-6; Vt=1; FET_DC_analysis(VDS,VG,RD,RS,Kp,Vt);
```

This yields the following results:

```
VDD   VGQ    VSQ   VDQ    IDQ
5.00 1.00 -2.82  3.21  9.92e-05
in the saturation mode with VGD,Q= -2.21[V] (Vt=1.00)
```

This analysis result conforms with the PSpice simulation result (with Bias Point analysis) shown in Figure 4.11(b1).

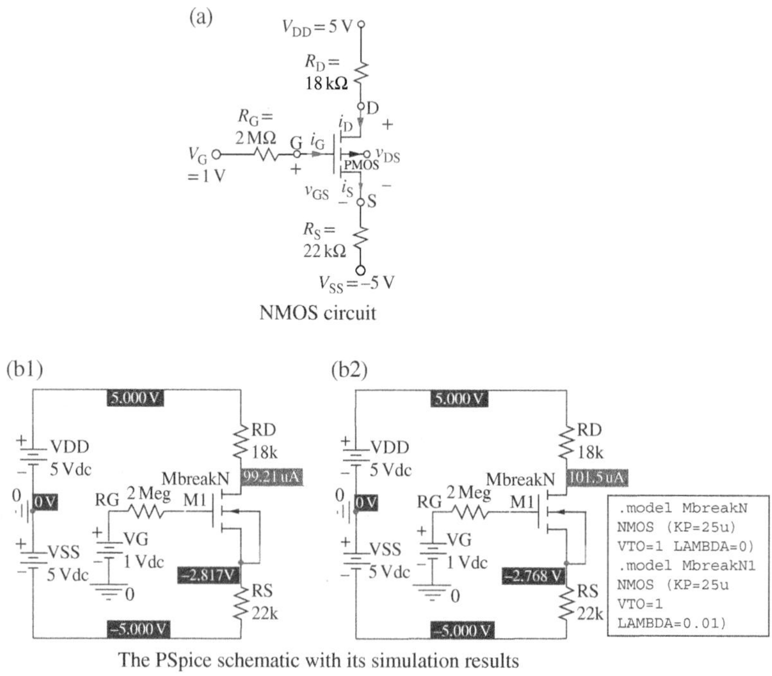

Figure 4.11 NMOS circuit for Example 4.2 ("elec04e02.opj").

b) With $\lambda = 0.01\,\mathrm{V}^{-1}$, find i_D, V_D, and V_S.
To consider the effect of λ, we use the MATLAB function '`FET_DC_analysis()`' as follows:

```
>>lambda=0.01;
  FET_DC_analysis(VDS,VG,RD,RS,Kp,Vt,lambda);
```

This yields the following results:

```
VDD   VGQ   VSQ    VDQ   IDQ
5.00  1.00 -2.77  3.17  1.01e-04
in the saturation mode with VGD,Q= -2.17[V] (Vt=1.00)
```

This analysis result conforms with the PSpice simulation result (with Bias Point analysis) shown in Figure 4.11(b2).

Example 4.3 A PMOS Circuit

Consider the PMOS circuit shown in Figure 4.12(a) where the device parameters of the enhancement PMOS (e-PMOS) are $K_p = 1[\mathrm{mA/V^2}]$, $V_t = -1[\mathrm{V}]$, and $\lambda = 0$ or $0.01[\mathrm{V}^{-1}]$.

a) Assuming that $\lambda = 0\,\mathrm{V}^{-1}$, find i_D, V_D, and V_S.
First, V_G is determined by the voltage divider consisting of R_1 and R_2 :

$$V_G = \frac{R_2}{R_1 + R_2} V_{DD} = \frac{3}{2+3} 5 = 3\ [\mathrm{V}] \tag{E4.3.1}$$

Since $v_{SG} > 0$, the PMOS is supposed to be turned on. Then, to tell whether it operates in the saturation or triode mode, we assume that it operates in the triode mode where

$$i_D \overset{(4.1.13a)}{=} K_p \left\{ (v_{SG} - |V_t|) v_{SD} - \frac{1}{2} v_{SD}^2 \right\} [\mathrm{A}]. \tag{E4.3.2}$$

We substitute $v_{SG} = V_S - V_G = 5 - 3 = 2\,\mathrm{V}$ and $v_{SD} = V_{SS} - R_D i_D = 5 - 6i_D$ into this equation and solve it as

$$i_D = 1 \times \left\{ (2-1)(5-6i_D) - \frac{1}{2}(5-6i_D)^2 \right\}$$

$$;36i_D^2 - 46\,i_D + 15 = 0;\ D = b^2 - 4ac < 0 \tag{E4.3.3}$$

However, this equation has no real solution, contradicting the assumption of triode mode. Then, we assume the saturation mode where

$$i_D \underset{\lambda=0}{\overset{(4.1.13b)}{=}} \frac{1}{2} K_p (v_{SG} - |V_t|)^2 \tag{E4.3.4}$$

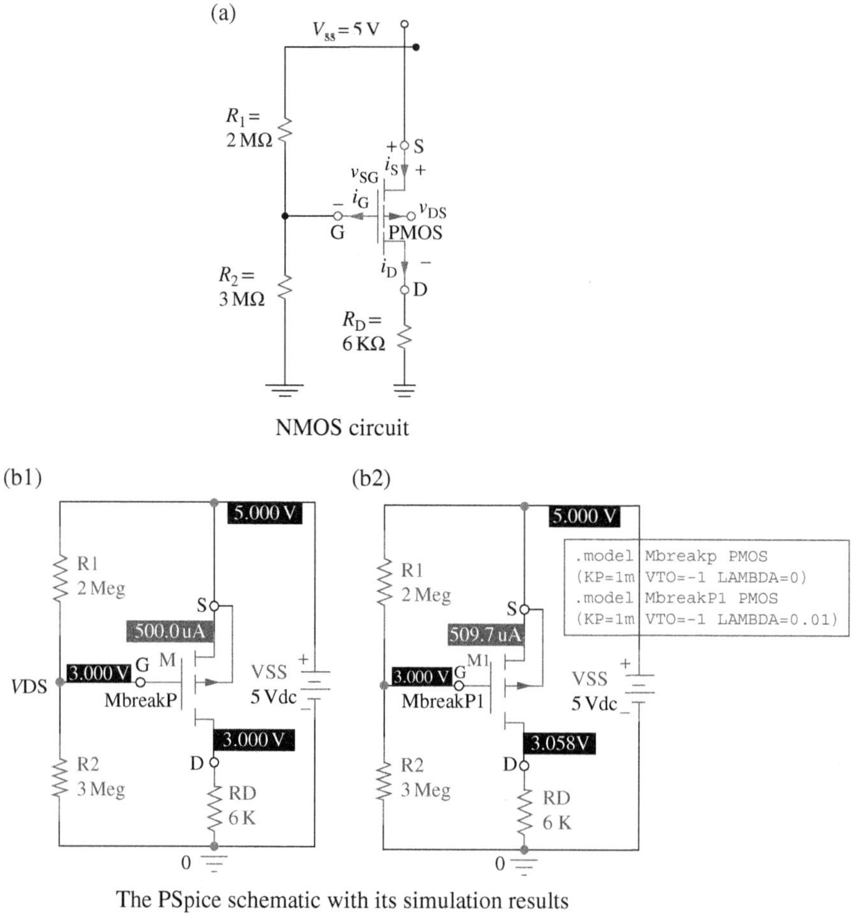

Figure 4.12 PMOS circuit for Example 4.3 ("elec04e03.opj").

We substitute $v_{SG} = 2$ V into this equation to get

$$i_D = \frac{1}{2}1\,(2-1)^2 = 0.5\ [\text{mA}] > 0 \tag{E4.3.5}$$

We should check if this satisfies the saturation condition:

$$v_{DG} = R_D i_D - V_G = 6 \times 0.5 - 3 = 0\ [\text{V}] \overset{?}{\leq} |\,V_t\,|: \text{OK(saturation)} \tag{E4.3.6}$$

This implies that the PMOS surely operates in the saturation mode with $v_{SD} = 5 - 3 = 2$ V.

(Q) What is the value of R_D that the PMOS operates at the boundary between the saturation and triode regions?

b) With $\lambda = 0.01\ \mathrm{V}^{-1}$, find i_D, V_D, and V_S.

Since it is too difficult to solve Eqs. (4.1.12a,b) by hand, we use the MATLAB function '`FET_PMOS_DC_analysis()`' as follows:

```
>>VSS=5; R1=2e6; R2=3e6; RS=0; RD=6e3; Kp=1e-3; Vt=-1; lambda=0.01;
  FET_PMOS_DC_analysis(VSS,R1,R2,RS,RD,Kp,Vt,lambda);
```

This yields the following results:

```
VSS    VGQ    VSQ    VDQ      IDQ
5.00  3.00  5.00  3.06  5.10e-04
in the saturation mode with VDG,Q= 0.06[V] (|Vt|=1.00)
```

This analysis result and that obtained in (a) conform with the PSpice simulation results (with Bias Point analysis) shown in Figure 4.12(b1) and (b2), respectively.

4.1.3 MOSFET Used as a Resistor

Figure 4.13(a) shows a diode-connected e-NMOS with its drain and gate short-circuited. Its i_D-(v_{DS}=v_{GS}) characteristic can be obtained graphically as the dotted line shown in Figure 4.13(c1) from the locus of points with $v_{DS} = v_{GS}$ and also mathematically by substituting $v_{GS} = v_{DS}$ into Eq. (4.1.13b) (for the saturation mode):

$$i_D \overset{(4.1.13b)}{=} \frac{1}{2}K_p(v_{GS} - V_t)^2 = \frac{1}{2}K_p(v_{DS} - V_t)^2 \text{ where } v_{GD} = 0 < V_t \tag{4.1.18}$$

Note that with $v_{DS} = v_{GS}$, Eq. (4.1.4a) and the reciprocal of Eq. (4.1.4b) conform with each other so that a diode-connected e-NMOS can be regarded as just a nonlinear resistor with the small-signal (incremental or dynamic) resistance

$$r_o = \frac{1}{g_m} \text{ since } r_o \overset{(4.1.4a)}{=} \left.\frac{\partial v_{DS}}{\partial i_D}\right|_Q = \left.\frac{\partial v_{GS}}{\partial i_D}\right|_Q \overset{(4.1.4b)}{=} \frac{1}{g_m} \tag{4.1.19}$$

Figure 4.13(a) also shows a diode-connected d-NMOS with its source and gate short-circuited, whose i_D-v_{DS} characteristic can be obtained as the dotted line shown in Figure 4.13(c2) from the locus of points with $v_{GS} = 0$ where its current is limited unlike an e-NMOS. These examples imply that a MOSFET can be used as a nonlinear resistor. The theoretical i_D-v_{DS} characteristics of the e-NMOS/d-NMOS circuits turn out to agree with the PSpice simulation results shown in Figure 4.13(b).

(a)

(b)

```
.model MbreakN NMOS (KP=20u VTO=1)
.model MbreakND NMOS (KP=20u VTO=-3)
```

$V_{TO} = V_t$
= 1 V

V_{GS1} i_{D1}
= V_{DS1} D1

$V_{TO} = Vt$
= −3 V
i_{D2}
D2

MbreakND
M2

+ VDD

G1 ○
M1
Mbreak N

G2 ○

v_{GS2}
= 0

12 Vdc

S1 ○

○ S2

0

Enhancement/depletion NMOSs with
the drain/source and gate short-circuited

PSpice simulation result obtained from the DC Sweep
analysis with VDD as a variable parameter

(c1)

[mA] i_{D1}
0.35
0.3
0.25
0.2
0.15
0.1
0.05
0

$i_D |v_{GS} = 7$ V
$i_D |v_{GS} = 6$ V
i_{D1}
$i_D |v_{GS} = 5$ V
$i_D |v_{GS} = 4$ V
$i_D |v_{GS} = 3$ V
$i_D |v_{GS} = 2$ V

0 1 2 3 4 5 6 V_{DS} [V]

Cutoff ($v_{GS} ≤ V_t$)

The nonlinear resistor characteristic of
the e-NMOS circuit in (a)

(c2)

[mA] i_{D2}
0.35
0.3
0.25
0.2
0.15
0.1
0.05
0

$i_D |v_{GS} = 3$ V
$i_D |v_{GS} = 2$ V
$i_D |v_{GS} = 1$ V
i_{D2}
$i_D |v_{GS} = 0$ V
$i_D |v_{GS} = −1$ V
$i_D |v_{GS} = −2$ V

0 1 2 3 4 5 6 v_{DS} [V]

Cutoff ($v_{GS} ≤ V_t$)

The nonlinear resistor characteristic of
the d-NMOS circuit in (a)

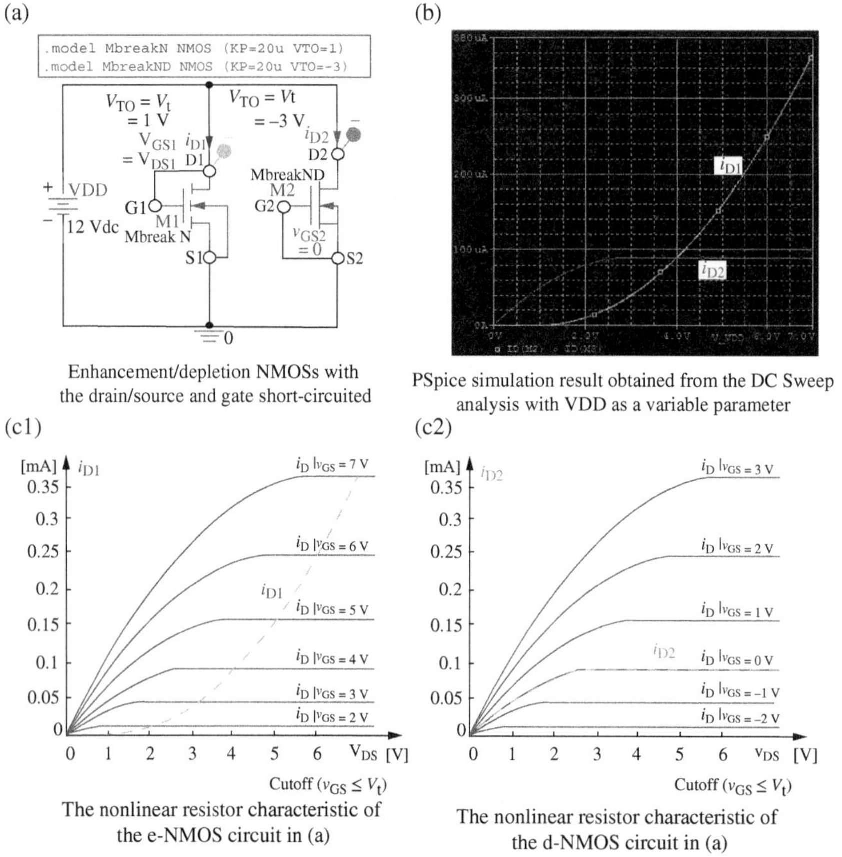

Figure 4.13 Enhancement/depletion type MOSFETs connected as (nonlinear) resistors
("elec04f13.opj").

4.1.4 FET Current Mirror

Consider the circuit of Figure 4.14(a) where the FET M_1 is said to be '*diode-connected*' or 'connected in diode configuration' since its source and gate terminals
are short-circuited so that it behaves like a diode. For proper operation of the
circuit, the two FETs M_1 and M_2 must be matched in the sense that they have
identical conduction parameters K_p, threshold voltages V_t, and CLM parameters λ. Let us analyze the circuit, which is called a *current mirror* because
the currents of the two matched FETs sharing the same v_{GS} are equal. Noting

that since $v_{GD1} = 0 \le V_t$, M_1 (with $v_{GS1} > 0$) operates always in the saturation mode, its drain current can be determined by solving

$$\left.\begin{aligned} v_{DS1} &= V_1 - Ri_{D1} \\ i_{D1} &\overset{(4.1.13b)}{=} \frac{1}{2}K_{p1}(v_{GS1} - V_{t1})^2 = \frac{1}{2}K_{p1}(v_{DS1} - V_{t1})^2 \end{aligned}\right\}$$

$$\rightarrow i_{D1} = \frac{1}{2}K_{p1}(V_1 - Ri_{D1} - V_{t1})^2 \tag{4.1.20}$$

This yields

$$I_{D1} = \frac{-B \pm \sqrt{B^2 - 4AC}}{2A} = \frac{18 \pm \sqrt{18^2 - 4 \times 1 \times 64}}{2} = \cancel{13.123}, \boxed{4.877\,\text{mA}} \tag{4.1.21}$$

where $A = R^2 = 1$, $B = -2\{R(V_1 - V_{t1}) + 1/K_{p1}\} = -18$, and $C = (V_1 - V_{t1})^2 = 64$. Here, I_{D1}=13.123 mA has been dumped because it does not satisfy even the turn-on condition $v_{GS1} = V_1 - RI_{D1} = 9 - 13.123 > V_{t1} = 1$[V]. Thus, since $v_{GS1} = v_{GS2}$ and $V_{t1} = V_{t2}$, the output current will be

$$i_{Oa} = i_{D2} = \frac{1}{2}K_{p2}(v_{GS2} - V_{t2})^2$$

$$= \frac{1}{2}\frac{K_{p2}}{K_{p1}}K_{p1}(v_{GS1} - V_{t1})^2 = \overbrace{\frac{K_{p2}}{K_{p1}}}^{\text{Mirror ratio}}I_{D1} = \frac{(W/L)_2}{(W/L)_1}I_{D1} \tag{4.1.22}$$

This analysis result conforms with the PSpice simulation result (with DC Sweep analysis) shown in Figure 4.14(d). Note that considering the effect of the CLM parameter, the current transfer ratio or current gain, also called the *mirror ratio*, can be written as

$$\frac{i_O}{I_{D1}} = \frac{K_{p2}(1 + \lambda v_{DS2})}{K_{p1}(1 + \lambda v_{DS1})} = \frac{(W/L)_2(1 + \lambda v_{DS2})}{(W/L)_1(1 + \lambda v_{DS1})} \tag{4.1.23}$$

Let us analyze the circuit of Figure 4.14(b) where the three matched FETs have identical threshold voltages V_t and CLM parameters λ, but possibly different conduction parameters $\{K_{p1}, K_{p2}, K_{p3}\}$. Since M_1 and M_3 are diode connected so that $v_{GDk} = 0 \le V_t$ for $k = 1$ and 3, they operate always in the saturation mode (with $v_{GSk} > V_t$ by V_1) and thus their drain currents can be expressed as

$$i_{D1} \overset{(4.1.13b)}{=} \frac{1}{2}K_{p1}(v_{GS1} - V_t)^2 = \frac{1}{2}K_{p1}(v_1 - V_t)^2 \tag{4.1.24a}$$

$$i_{D3} \overset{(4.1.13b)}{=} \frac{1}{2}K_{p3}(v_{GS3} - V_t)^2 = \frac{1}{2}K_{p3}(v_2 - v_1 - V_t)^2 \tag{4.1.24b}$$

```
function [io,ID1,v,Ro,Vomin]=FET3_current_mirror(Kp,Vt,lambda,R,V12)
% To analyze a 3-FET current mirror like Fig.4.14(b)
% with R (resistor) or I (current source)
% If 0<R<1, R will be regarded as a current source I=R.
% Copyleft: Won Y. Yang, wyyang53@hanmail.net, CAU for academic use only
V1=V12(1); if length(V12)>1, V2s=V12(2:end); else V2s=V12; end
if length(Kp)<3, Kp=repmat(Kp(1),1,3); end
iDvGS_GDs=@(vGS,Kp,Vt) Kp/2*(vGS-Vt).^2.*(1+lambda*vGS); % with GD shorted
options=optimset('Display','off','Diagnostics','off'); n0=1;
if R>=1|R==0 % When R is given together with a voltage source V1
  eq=@(v) [(iDvGS_GDs(v(1),Kp(1),Vt)-iDvGS_GDs(v(2)-v(1),Kp(3),Vt))*1e4;
           V1-v(2)-R*iDvGS_GDs(v(2)-v(1),Kp(3),Vt)]; % Eq.(4.1.25)
  v0=V1/3*[1 2]; % Initial guess for v=[v(1) v(2)]
 else I=R; % When a current source I is given
  eq=@(v) [iDvGS_GDs(v(1),Kp(1),Vt)-iDvGS_GDs(v(2)-v(1),Kp(3),Vt);
           I-iDvGS_GDs(v(2)-v(1),Kp(3),Vt);]*1e4;
  v0=Vt+sqrt(2/Kp(1)*I)*[1 2]; % Initial guess for v=[v(1) v(2)]
end
[v,fe]=fsolve(eq,v0,options);
ID1=iDvGS_GDs(v(1),Kp(1),Vt); % Drain current of M1 (reference current)
for n=1:length(V2s) % Output currents for each value of V2
   io(n)=iD_NMOS_at_vDS_vGS(V2s(n),v(1),Kp(2),Vt,lambda);
   if n>1&(V2s(n)-V1)*(V2s(n-1)-V1)<=0, n0=n; end %Index of V2 closest to V1
end
Io=io(n0); ro2=1/lambda/Io; Ro=ro2; % Output current/resistance (4.1.26)
Vomin=v(1)-Vt; % Minimum output voltage
if nargout<1, fprintf('Io=%8.4f[mA], ID1=%8.4f[mA], Ro=%9.0f[kOhm]\n',
                Io*1e3,ID1*1e3,Ro/1e3); end
```

We can apply KCL at nodes 1 and 2 to write

$$i_{D1} = i_{D3} = i_D \tag{4.1.25a}$$

$$\frac{V_1 - v_2}{R} - i_D = 0; \; V_1 - v_2 - Ri_D = 0 \tag{4.1.25b}$$

Once we have solved this set of two nonlinear Eqs. (4.1.24a,b) for v_1 and v_2 (by using the MATLAB function '`fsolve()`'), we can use Eqs. (4.1.22), (4.1.23), or (4.1.13b) to find the output current $i_{Ob} = i_{D2}$ as long as $v_{GD2} = v_1 - V_2 < V_t$ so that M_2 is saturated. If $v_{GD2} = v_1 - V_2 \geq V_t$ (so that M_2 is in the tride region), Eq. (4.1.13a) should be used to find i_{Ob}, as implemented by the MATLAB function '`iD_NMOS_at_vDS_vGS()`'. This solution process for the current

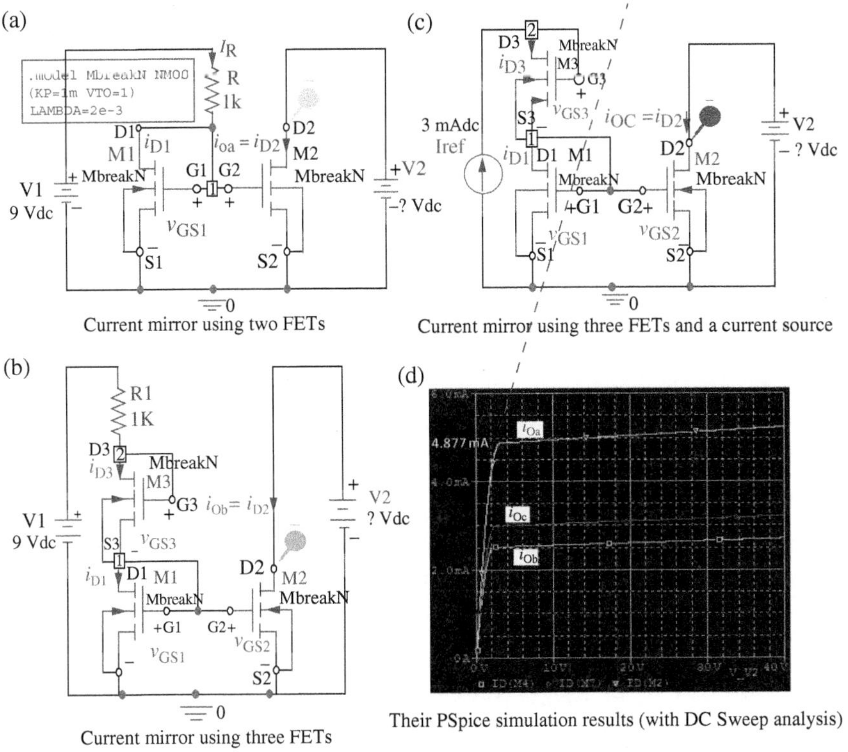

Figure 4.14 Current mirrors using FETs ("elec04f14.opj").

mirror of Figure 4.14(b) has been cast into the above MATLAB function 'FET3_current_mirror()'.

The two current mirrors of Figure 4.14(a) and (b) have the same output resistance:

$$R_o = r_{o2} \overset{(4.1.4a)}{=} \frac{1}{\lambda I_o} \tag{4.1.26}$$

Their minimum output voltage, called the *compliance voltage*, required to keep M_2 in saturation is

$$V_{o,\min} = v_1 - V_t = V_{G2} - V_t = V_{GS2} - V_t \tag{4.1.27}$$

To use the above function 'FET3_current_mirror()' for analyzing the current mirror of Figure 4.14(b), we can run the following statements:

```
>>Kp=1e-3; Vt=1; lambda=2e-3; R=1e3; V1=9;
  V2s=[0:0.01:40]; V12=[V1 V2s];
  [io,ID1,v,Ro,Vomin]= ...
  FET3_current_mirror(Kp,Vt,lambda,R,V12);
  plot(V2s,io, Vomin*[1 1],[0 io(end)],'r:')
```

to get the graph for the output current i_{Ob} versus V2 = 0 ~ 40 V like the corresponding PSpice simulation result shown in Figure 4.14(d).

To analyze the current mirror of Figure 4.14(c) with a current source I, we run

```
>>I=3e-3;
  [io,ID1,v,Ro,Vomin]= ...
  FET3_current_mirror(Kp,Vt,lambda,I,V12);
  plot(V2s,io, Vomin*[1 1],[0 io(end)],'r:')
```

to get the graph for the output current i_{Oc} versus V2 = 0 ~ 40 V like the corresponding PSpice simulation result shown in Figure 4.14(d).

If $R = 0$, it is not so difficult to derive the analytical expression of v_1 even if $K_{p1} \neq K_{p3}$. In this case, we solve Eq. (4.1.25a) for v_1 with $v_2 = V_1$ and Eqs. (4.1.24a,b):

$$\frac{1}{2}K_{p1}(v_1 - V_t)^2 \underset{(4.1.24a,b)}{\overset{(4.1.25a)}{=}} \frac{1}{2}K_{p3}(V_1 - v_1 - V_t)^2 \tag{4.1.28}$$

to get

$$v_1 = \frac{\sqrt{K_{p3}/K_{p1}}}{1 + \sqrt{K_{p3}/K_{p1}}} V_1 + \frac{1 - \sqrt{K_{p3}/K_{p1}}}{1 + \sqrt{K_{p3}/K_{p1}}} V_t \tag{4.1.29}$$

The output current i_{Ob} can be determined from Eqs. (4.1.13b) or (4.1.13a) (with $v_{GS} = v_1$ and $v_{DS} = V_2$) depending on whether $v_{GD} = v_1 - V_2 < V_t$ or $v_{GD} = v_1 - V_2 \geq V_t$ so that M_2 operates in the saturation or triode region.

(Q) What advantage does the current mirror of Figure 4.14(b) get from the additional FET M_3 over that of Figure 4.14(a)?

(A) An advantage is a considerable flexibility in the design of current source since K_{p3}/K_{p1} can be fixed at the will of designer by adjusting the width-to-length ratios (W_n/L_n) of FETs.

The following MATLAB script "elec04f14.m" can be run to get the plot of the output currents i_{Oa}, i_{Ob}, and i_{Oc} versus V2 = 0 ~ 40 V for the three current mirrors shown in Figure 4.14(a-c), as depicted in Figure 4.14(d).

```
%elec04f14.m
V1=9; V2s=[0:0.1:40]; R=1e3; V12=[V1 V2s];
Kp1=1e-3; Kp2=1e-3; Kp3=1e-3; Vt=1; lambda=2e-3;
[ioa,ID1a,va,Roa]=FET2_current_mirror([Kp1 Kp2],Vt,lambda,R,V12);
KC=1;
[ioa_,ID1a_,va_,Roa_]=FET2_current_mirror([Kp1 Kp2],Vt,lambda,R,V12,KC);
Kps=[Kp1 Kp2 Kp3];
[iob,ID1b,vb,Rob]=FET3_current_mirror(Kps,Vt,lambda,R,V12);
R=0;
[iob1,ID1b1,vb1,Rob1]=FET3_current_mirror(Kps,Vt,lambda,R,V12);
sqKpr=sqrt(Kp3/Kp1);
vb1=[vb1(1) (sqKpr*V1+(1-sqKpr)*Vt)/(1+sqKpr)] % conforms with Eq.(4.1.29)
I=3e-3;
[ioc,ID1c,vc,Roc,Vocmin]=FET3_current_mirror(Kps,Vt,lambda,I,V12,1);
subplot(221), plot(V2s,ioa, V2s,ioa_,'m:', V2s,iob,'g', V2s,ioc,'r')
hold on, plot(Vocmin*[1 1],[0 ioc(end)],'r:')
legend('io_a','io_a(another method)','io_b','io_c','Vomin')
```

```
function [io,ID1,v,Ro,Vomin]=FET2_current_mirror(Kp,Vt,lambda,R,V12)
V1=V12(1); if length(V12)>1, V2s=V12(2:end); else V2s=V12; end
if length(Kp)<2, Kp=repmat(Kp(1),1,2); end
iDvGS_GDs=...
  @(vGS,Kp,Vt)Kp/2*(vGS-Vt).^2.*(1+lambda*vGS).*(vGS>Vt); % GD shorted
if nargin<5|KC<1 % Solve the KCL equation as a quadratic polynomial eq
  A=R^2; B=-2*(R*(V1-Vt)+1/Kp(1)); C=(V1-Vt)^2; % Eq.(4.1.21)
  if A~=0, ID1=(A~=0)*(-B-sqrt(B^2-4*A*C))/2/A; else ID1=-C/B; end
  v=V1-R*ID1; % VDS1=VGS1
else % Solve the KCL equation in v as a nonlinear eq
  options=optimset('Display','off','Diagnostics','off');
  eq=@(v)V1-v-R*iDvGS_GDs(v,Kp(1),Vt); % KCL at node 1
  v0=V1/2; % Initial guess for v
  [v,fe]=fsolve(eq,v0,options);
  ID1=iDvGS_GDs(v,Kp(1),Vt);
end
for n=1:length(V2s)
  io(n)=iD_NMOS_at_vDS_vGS(V2s(n),v,Kp(2),Vt,lambda); %than Eq.(4.1.22)
end
Io=io(end); gm2=2*Io/(v(1)-Vt); ro2=1/lambda/Io;
Ro=ro2;
Vomin=v-Vt; % Minimum output voltage
```

Example 4.4 Design of an NMOS Current Source

Consider the current mirror consisting of three NMOSs as depicted in Figure 4.14 (b) where $V_1 = 5$ V, $R = 0\ \Omega$, and the device parameters of the NMOSs are $V_t = 1$ V and $\lambda = 0$. Determine the values of the conduction parameters K_{p1}, K_{p2}, and K_{p3} of the NMOSs so that the reference current, the output current, and the minimum output voltage are $I_{D1} = 0.2$ mA, $I_o = 0.1$ mA, and $V_{o,min} = 0.8$ V, respectively.

First, we use Eq. (4.1.27) to fix such a value of $v_1 = V_{GS1} = V_{GS2}$ that $V_{o,min} = 0.8$ V:

$$v_1 = v_{GS2} = v_{GS1} \overset{(4.1.27)}{=} V_{o,min} + V_t = 1.8\text{V} \tag{E4.4.1}$$

Then we use Eq. (4.1.13b) to determine K_{p2} so that $i_o = i_{D2} = I_o = 0.1$ mA when M_2 is in saturation:

$$K_{p2} \overset{(4.1.13b)}{=} \frac{2I_o}{(v_1 - V_t)^2} = \frac{2 \times 0.1\text{m}}{(1.8-1)^2} = 0.3125\,\text{mS} \tag{E4.4.2}$$

We also use Eq. (4.1.13b) to determine K_{p1} so that $i_{D1} = I_{D1} = 0.2$ mA when M_1 is in saturation:

$$K_{p1} \overset{(4.1.13b)}{=} \frac{2I_{D1}}{(v_1 - V_t)^2} = \frac{2 \times 0.2\text{m}}{(1.8-1)^2} = 0.625\,\text{mS} \tag{E4.4.3}$$

Noting that $v_{GS1} + v_{GS3} = V_1 = 5$ V, the gate-to-source voltage of M_1 can be found as

$$v_{GS3} = V_1 - V_{GS1} \overset{(E4.4.1)}{=} 5 - 1.8 = 3.2\text{V} \tag{E4.4.4}$$

Then we again use Eq. (4.1.13b) to determine K_{p3} so that $i_{D3} = I_{D1} = 0.2$ mA when M_3 is in saturation:

$$K_{p3} \overset{(4.1.13b)}{=} \frac{2I_{D1}}{(v_{GS3} - V_t)^2} = \frac{2 \times 0.2\text{m}}{(3.2-1)^2} = 0.082645\,\text{mS} \tag{E4.4.5}$$

To see if this design works, run the following MATLAB statements:

```
>>V1=5; V2=0.8; R=0; V12=[V1 V2];
  Kp1=0.625e-3; Kp2=0.3125e-3; Kp3=0.082645e-3; Vt=1; lambda=0;
  Kps=[Kp1 Kp2 Kp3];
  [io,ID1,v,Ro,Vomin]=FET3_current_mirror(Kps,Vt,lambda,R,V12)
```

This yields

```
io = 1.0000e-04
ID1 = 2.0000e-04
v =  1.8000   5.0000
Ro =  Inf
Vomin = 0.8
```

Now, let us consider the *double Wilson current mirror* shown in Figure 4.15(a) where the four matched FETs have identical threshold voltages V_t and CLM parameters λ, but possibly different conduction parameters $\{K_{pk}, k = 1, 2, 3, 4\}$. Since

M_2 and M_3 are diode connected so that $v_{GDk} = 0 \leq V_t$ for $k = 2$ and 3, they operate always in the saturation mode and thus their drain currents can be expressed as

$$i_{D2} \overset{(4.1.13b)}{=} \frac{1}{2}K_{p2}(v_{GS2} - V_t)^2 = \frac{1}{2}K_{p2}(v_1 - V_t)^2 \tag{4.1.30a}$$

$$i_{D3} \overset{(4.1.13b)}{=} \frac{1}{2}K_{p3}(v_{GS3} - V_t)^2 = \frac{1}{2}K_{p3}(v_3 - v_2 - V_t)^2 \tag{4.1.30b}$$

We can apply KCL at nodes 2, 1, and 3 to write the node equations:

$$i_{D3}(v_3, v_2) = i_{D1}(v_2, v_1) = i_D, \quad i_{D4}(v_3, v_1) = i_{D2}(v_1) = i_o \tag{4.1.31a,b}$$

$$\frac{V_1 - v_3}{R} - i_D = 0; \; V_1 - v_3 - Ri_D = 0 \tag{4.1.31c}$$

The process of solving this set of equations for v_1, v_2, and v_3 has been cast into the MATLAB function '`FET4_current_mirror_Wilson()`' given below.

The output resistance of this circuit can be determined from its small-signal equivalent shown in Figure 4.15(c):

$$v_T = (i_T - g_{m4}v_{gs4})r_{o4} + v_1 \tag{4.1.32a}$$

$$\text{with } v_{gs4} = v_{g4} - v_{s4} = -g_{m1}v_1\frac{r_{o1}}{r_{o1} + R + 1/g_{m3}}R - v_1 \text{ and } v_1 = \frac{1}{g_{m2}}i_T \tag{4.1.32b}$$

$$R_o = \frac{v_T}{i_T} = r_{o4} + \frac{1}{g_{m2}} + g_{m4}r_{o4}\left(\frac{g_{m1}}{g_{m2}}\frac{r_{o1}R}{r_{o1} + R + 1/g_{m3}} + \frac{1}{g_{m2}}\right)$$

$$\underset{g_{mk} = g_{mk},\, r_{ok} = r_o \forall k}{\overset{1/g_m << r_o << R}{\longrightarrow}} R_o \approx g_m r_o^2 >> r_o \tag{4.1.33}$$

Note that if a reference current source, in place of the voltage source V1-R, is connected into node 3, the value of R should be regarded as ∞. The minimum output voltage required to keep M_4 in saturation is

$$V_{o,\min} = v_3 - V_t = v_{G4} - V_t = v_{G2} + v_{GS4} - V_t \tag{4.1.34}$$

(Q) What advantage/disadvantage does the current mirror of Figure 4.15(a) have due to the additional FET M_4 over that of Figure 4.14(b)?

(A) The output resistance increases (compare Eqs. (4.1.26) and (4.1.33)) and the difference between v_{DS1} and v_{DS2} decreases so that the accuracy of the current transfer ratio (Eq. (4.1.23)) can be expected to get better. However, the minimum output voltage (compare Eqs. (4.1.27) and (4.1.34)) increases so that the output swing, i.e. the range of output voltage v_o for which the output resistance R_o is large, will be decreased. Compare Figures 4.14(d) and 4.15(b1).

```
function [io,ID1,v,Ro,Vomin]= ...
         FET4_current_mirror_Wilson(Kp,Vt,lambda,R,V12)
% To analyze a 4-FET (double) Wilson current mirror like Fig.4.15
%    with R (resistor) or I (current source)
% If 0<R<1, R will be regarded as a current source I=R with with R=inf.
% Copyleft: Won Y. Yang, wyyang53@hanmail.net, CAU for academic use only
V1=V12(1); if length(V12)>1, V2s=V12(2:end); else V2s=V12; end
if length(Kp)<4, Kp=repmat(Kp(1),1,4); end
iDvGS_GDs=@(vGS,Kp,Vt)Kp/2*(vGS-Vt).^2.*(1+lambda*vGS); %GD shorted
options=optimset('Display','off','Diagnostics','off');
n0=1; % To determine the index of V2s closest to V1
for n=1:length(V2s)
  V2=V2s(n); if n>1&(V2-V1)*(V2s(n-1)-V1)<=0, n0=n; end
  if R>=1|R==0, RL=R;
   eq=@(v) [(iDvGS_GDs(v(3)-v(2),Kp(3),Vt)-...
     iD_NMOS_at_vDS_vGS(v(2),v(1),Kp(1),Vt,lambda))*1e4; % Eq.(4.1.31a)
       (iD_NMOS_at_vDS_vGS(V2-v(1),v(3)-v(1),Kp(4),Vt,lambda)- ...
         iDvGS_GDs(v(1),Kp(2),Vt))*1e4; % Eq.(4.1.31b)
       V1-v(3)-R*iDvGS_GDs(v(3)-v(2),Kp(3),Vt)]; % Eq.(4.1.31c)
  else I=R; RL=inf;
   eq=@(v) [iDvGS_GDs(v(3)-v(2),Kp(3),Vt)- ...
         iD_NMOS_at_vDS_vGS(v(2),v(1),Kp(1),Vt,lambda); %Eq.(4.1.31a)
         iD_NMOS_at_vDS_vGS(V2-v(1),v(3)-v(1),Kp(4),Vt,lambda)- ...
             iDvGS_GDs(v(1),Kp(2),Vt); % Eq.(4.1.31b)
         I-iDvGS_GDs(v(3)-v(2),Kp(3),Vt)]*1e4; % Eq.(4.1.31c)
 end
 v0=[V1/4 V1/2 2/3*V1]; % Initial guess for v=[v(1) v(2) v(3)]
 v=fsolve(eq,v0,options); vs(n,:)=v;
 ID1=iDvGS_GDs(v(3)-v(2),Kp(3),Vt);
 io(n)=iD_NMOS_at_vDS_vGS(V2-v(1),v(3)-v(1),Kp(4),Vt,lambda);
end
Io=io(n0); V2=V2s(n0); v=vs(n0,:); % Values of io, V2, vs when V2=V1
gm1=2*ID1/(v(1)-Vt); ro1=1/lambda/ID1;
gm2=2*Io/(v(1)-Vt); ro2=1/lambda/Io;
gm3=2*Io/(v(3)-v(2)-Vt); ro3=1/lambda/ID1;
gm4=2*Io/(v(3)-v(1)-Vt); ro4=1/lambda/Io;
if RL==inf, Ro=ro4+1/gm2+gm4*ro4*(gm1/gm2*ro1+1/gm2);
 else Ro=ro4+(1+gm4*ro4*(gm1*ro1*RL/(ro1+RL+1/gm3)+1))/gm2;
                                             % Eq.(4.1.33)
end
Vomin=v(3)-Vt; % Minimum output voltage by Eq.(4.1.34)
```

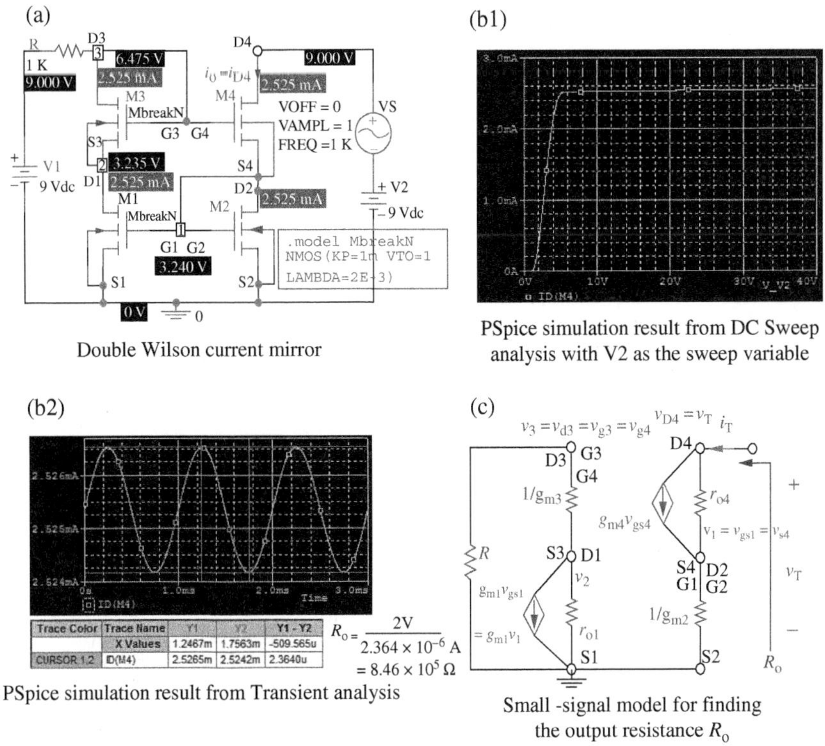
(a)
Double Wilson current mirror
(b1)
PSpice simulation result from DC Sweep analysis with V2 as the sweep variable
(b2)
PSpice simulation result from Transient analysis
$R_o = \frac{2V}{2.364 \times 10^{-6}\,A} = 8.46 \times 10^5\,\Omega$
(c)
Small -signal model for finding the output resistance R_o

Figure 4.15 Double Wilson current mirror using FETs ("elec04f15.opj").

We can use the above MATLAB function 'FET4_current_mirror_Wilson()' to get the plot of the output current of the double Wilson current mirror (Figure 4.15(a)) by running the following MATLAB statements:

```
>>V1=9; V2s=[0:0.1:40]; R=1e3; V12=[V1 V2s];
  Kp=1e-3; Vt=1; lambda=2e-3;
  [io,ID1,vs,Ro,Vomin]= ...
  FET4_current_mirror_Wilson(Kp,Vt,lambda,R,V12); ID1,vs,Ro
  plot(V2s,io, Vomin*[1 1],[0 io(end)],'g:'), legend('io','Vomin');
```

This yields a plot of i_O versus V2 = 0 ~ 40 V as depicted in Figure 4.15(b1) and the following analysis results that are close to the PSpice simulation results displayed on the schematic in Figure 4.15(a):

```
ID1 = 0.0026        % Reference current through M1 and M3
vs = 3.2402  3.2344  6.4746  % v1, v2, v3
Ro = 8.4730e+05  % very close to the PSpice simulation result
```

4.1.5 MOSFET Inverter

Just like the BJT inverter of Figure 3.27 introduced in Section 3.1.10, let us consider an FET inverter (shown in Figure 4.16(a)) using an FET/resistor as a driver/load, respectively. To get its voltage transfer characteristic (VTC), suppose that the input voltage v_i increases from 0 to V_{DD}. While $v_i = v_{GS} \le V_t$, the NMOS is cut off with $i_D = 0$ so that $v_o = V_{DD}$. As $v_i = v_{GS} > V_t$, the NMOS enters the saturation region where the output voltage is determined depending on the input voltage as

$$v_{DS} = V_{DD} - R_D i_D \overset{(4.1.13b)}{\underset{\lambda=0}{=}} V_{DD} - \frac{1}{2} K_p R_D (v_{GS} - V_t)^2 \text{ in the sat region.} \tag{4.1.35}$$

When $v_i = v_{GS}$ increases over $v_{DS} + V_t$ so that $v_{GD} = v_{GS} - v_{DS} > V_t$, the NMOS enters the triode region where the input–output relationship is

$$v_{DS} = V_{DD} - R_D i_D \overset{(4.1.13a)}{=} V_{DD} - R_D K_p \left\{ (v_{GS} - V_t) v_{DS} - \frac{1}{2} v_{DS}^2 \right\} \tag{4.1.36}$$

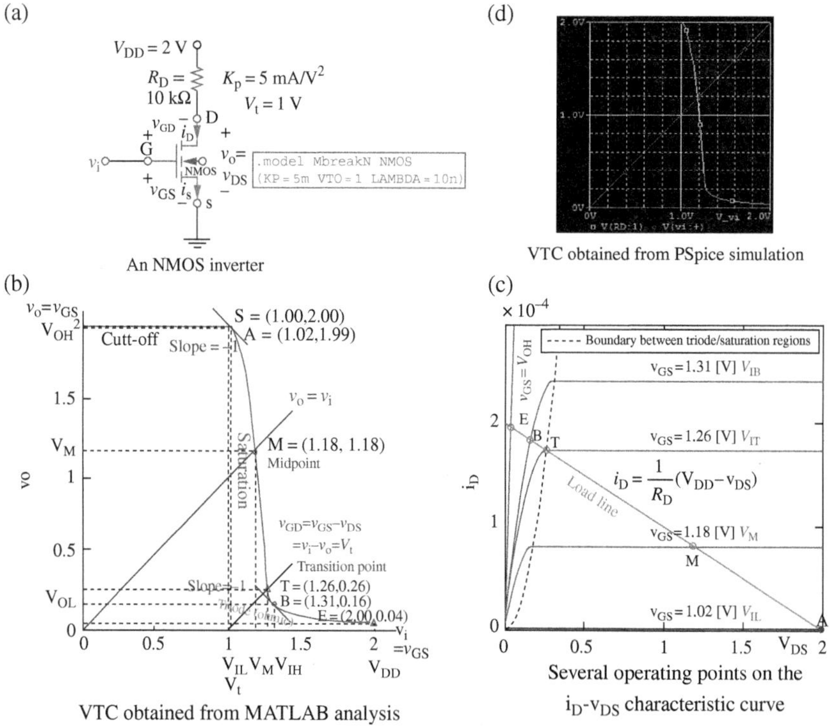

Figure 4.16 An NMOS inverter circuit and its VTC ("elec04f16.opj").

$$; \frac{1}{2}K_pR_Dv_{DS}^2 - \{K_pR_D(v_{GS} - V_t) + 1\}v_{DS} + V_{DD} = 0$$

$$; v_{GS} = V_t + \frac{K_pR_D\, v_{DS}^2/2 + V_{DD} - v_{DS}}{K_pR_Dv_{DS}} \text{ in the triode region.} \tag{4.1.37}$$

The VTC curve in Figure 4.16(b) is based on these two input–output relationships, each for saturation/triode regions. The *transition point* T between the saturation and triode segments on the VTC can be determined from $v_{GD} = v_{GS} - v_{DS} = v_i - v_o = V_t$ together with Eq. (4.1.35). That is, we can substitute $v_{GS} = V_t + v_{DS}$ into Eq. (4.1.35) to write

$$v_{DS} \underset{v_{GS}-V_t=v_{DS}}{\overset{(4.1.35)}{=}} V_{DD} - \frac{1}{2}K_pR_Dv_{DS}^2; \ \frac{1}{2}K_pR_Dv_{DS}^2 + v_{DS} - V_{DD} = 0 \tag{4.1.38}$$

and use the quadratic formula to solve this quadratic equation for the point $T = (V_{IT}, V_{OT})$ as

$$V_{OT} = \frac{\sqrt{2K_pR_DV_{DD} + 1} - 1}{K_pR_D} \tag{4.1.39a}$$

$$V_{IT} = V_{OT} + V_t : \text{Transition point between saturation and triode regions} \tag{4.1.39b}$$

The *switching threshold voltage* V_M, also called the *midpoint voltage*, can be found by substituting $v_{GS} = V_M$ and $v_{DS} = V_M$ into the input-output relationship Eq. (4.1.35) (for the saturation region) and solving it for V_M as

$$V_M \underset{v_i=v_o=V_M}{\overset{(4.1.35)}{=}} V_{DD} - \frac{1}{2}K_pR_D(V_M - V_t)^2$$

$$; \frac{1}{2}K_pR_D(V_M - V_t)^2 + (V_M - V_t) - (V_{DD} - V_t) = 0$$

$$; V_M = \frac{\sqrt{2K_pR_D(V_{DD} - V_t) + 1} - 1}{K_pR_D} + V_t : \text{Switching threshold or midpoint voltage} \tag{4.1.40}$$

Also, V_{IL} (the maximum input voltage that can be interpreted as "0") (at point A in Figure 4.16(b)) and the corresponding output voltage V_{OH} (the minimum output voltage that can be interpreted as "1") can be determined by setting the derivative of Eq. (4.1.35) w.r.t. v_{GS} to -1:

$$v_{DS} \overset{(4.1.35)}{=} V_{DD} - \frac{1}{2}K_pR_D(v_{GS} - V_t)^2; \ \frac{dv_{DS}}{dv_{GS}} = -K_pR_D(v_{GS} - V_t) = -1$$

$$; V_{IL} = V_t + \frac{1}{K_pR_D} \tag{4.1.41a}$$

$$V_{OH} \overset{(4.1.35)}{=} V_{DD} - \frac{1}{2}K_pR_D(v_{GS} - V_t)^2 \overset{v_{GS}=V_{IL}}{=} V_{DD} - \frac{1}{2K_pR_D} \tag{4.1.41b}$$

V_{IH} (the minimum input voltage that can be interpreted as "1") (at point B in Figure 4.16(b)) and the corresponding output voltage V_{OL} (the maximum output voltage that can be interpreted as "0") can be determined by setting the derivative of Eq. (4.1.37) to -1:

$$v_{GS} \overset{(4.1.28)}{=} V_t + \frac{K_p R_D v_{DS}^2/2 + V_{DD} - v_{DS}}{K_p R_D v_{DS}}; \frac{dv_{GS}}{dv_{DS}} = \frac{1}{2} - \frac{V_{DD}}{K_p R_D v_{DS}^2} = -1$$

$$; V_{OL} = \sqrt{\frac{2V_{DD}}{3K_p R_D}} \tag{4.1.42a}$$

$$V_{IH} \overset{(4.1.28)}{=} V_t + \frac{K_p R_D v_{DS}^2/2 + V_{DD} - v_{DS}}{K_p R_D v_{DS}} \overset{v_{DS}=V_{OL}}{=} V_t + \sqrt{\frac{8V_{DD}}{3K_p R_D}} - \frac{1}{K_p R_D} \tag{4.1.42b}$$

```
function [VIL,VIH,VOL,VOH,VM,VIT,VOT,VLH,NML,NMH,PDavg,vo1,iD1]=...
                    NMOS_inverter(Kp,Vt,RD,VDD,vi1)
% Analyze an NMOS inverter consisting of an NMOS and RD between VDD and D
% to find the output vo1 to an input vi1 and plot its VTC
% Vt = Threshold (Pinch-off ) voltage, Kp = mu*Cox*(W/L)
% vo1: Output voltage(s) for input voltage(s) vi=vi1
% iD1: Drain current(s) for input voltage(s) vi=vi1
% Copyleft: Won Y. Yang, wyyang53@hanmail.net, CAU for academic use only
if nargin<5, vi1=0; end
dvi=1e-3; vis=[0:dvi:VDD]; % Full range of the input vi
VOT=(sqrt(2*Kp*RD*VDD+1)-1)/Kp/RD; % Boundary between sat/triode
VIT=VOT+Vt; % Eq.(4.1.30)
[vos,iDs]=vo_iD_NMOS_inverter(Kp,Vt,RD,VIT,VDD,vis);
[VIL,VIH,VOL,VOH,VM,NML,NMH,VL,PDavg] = ...
              find_pars_of_inverter(vis,vos,iDs,VDD);
VH=VDD; % The highest output voltage
if nargin>4&sum(vi1>0&vi1<=VDD)>0 % If you want vo for vi1
    for i=1:length(vi1)
       [dmin,imin] = min(abs(vis-vi1(i)));
       vo1(i)=vos(imin); iD1(i)=iDs(imin);
    end
  else
    vo1=vos; iD1 = iDs;
end
fprintf('\n VIL=%6.3f, VIH=%6.3f, VOL=%6.3f, VOH=%6.3f, VM=%6.3f,
  VIT=%6.3f, VOT=%6.3f, VOE=%6.3f', VIL,VIH,VOL,VOH,VM,VIT,VOT,VOE);
fprintf('\n Noise Margin: NM_L=%6.3f and NM_H=%6.3f', NML,NMH);
fprintf('\n Output signal swing: VOL(%6.2f)~VOH(%6.2f) = %6.2f[V]',
  VOL,VOH,VOH-VOL);
fprintf('\n Average power dissipated=%10.3e[mW]\n', PDavg*1e3);
% Plot the VTC curve
plot(vis,vos, [VIL VM VIH], [VOH VM VOL],'ro')
hold on, plot([Vt VIT VH],[VH VOT VL],'r^')
```

```
function [vo,iD]=vo_iD_NMOS_inverter(Kp,Vt,RD,VIT,VDD,vi)
for n=1:length(vi)
  if vi(n)<=Vt, vo(n)=VDD; iD(n)=0; % Cutoff region
    elseif vi(n)<=VIT % Saturation region
       iD(n)=Kp/2*(vi(n)-Vt).^2; vo(n)=VDD-RD*iD(n); % Eq.(4.1.35)
    else % Triode region
      a=Kp*RD/2; b=-(Kp*RD*(vi(n)-Vt)+1); c=VDD;
      vo(n) = (-b-sqrt(b^2-4*a*c))/2/a; % Eq.(4.1.36)
      iD(n) = (VDD-vo(n))/RD;
  end
end
```

Note that the *high*/*low* and *absolute noise margins* are defined by Eqs. (3.1.64) and (3.1.65) as

$$NM_H = V_{OH} - V_{IH} \tag{4.1.43a}$$

$$NM_L = V_{IL} - V_{OL} \tag{4.1.43b}$$

$$NM = \min\{NM_H, NM_L\} \tag{4.1.44}$$

It may be useful for establishing your overview of the inverter analysis to see that the slope of the VTC (for the saturation region) is $-K_pR_D\,(v_{GS} - V_t)$ and also that the points A = (V_{IL}, V_{OH}), M = (V_M, V_M), T = (V_{IT}, V_{OT}), B = (V_{IH}, V_{OL}), and E = (V_{OH}, V_{OE}) on the VTC can be located on the (pink) load line intersecting the corresponding v_{DS}-i_D characteristic curves of the NMOS in Figure 4.16(c).

The above MATLAB function '`NMOS_inverter()`' uses another function '`vDS_vGS_ NMOS_inverter()`' (implementing Eqs. (4.1.35) and (4.1.36)) to get the set of output voltage v_o of an NMOS inverter for $v_i = 0 \sim V_{DD}$, uses another function '`find_pars_of_inverter()`' (Section 3.1.10) to find the inverter parameters, and then plots the VTC of the inverter.

Example 4.5 Design of an NMOS Inverter Loaded by a Resistor

Determine the values of R_D and K_p for the e-NMOS circuit of Figure 4.16(a) so that it dissipates a power of P_{max} = 0.4 mW at a low output voltage of $v_{o,min}$ = 0.04 V where V_{DD} = 2 V and the device parameters of the NMOS are V_t = 1 V and λ = 0. Plot the VTC of the designed inverter.

First, from the power specification, the drain current for v_o = 0.04 V can be determined as

$$I_{DD} = \frac{P}{V_{DD}} = \frac{0.4\text{m}}{2} = 0.2\ [\text{mA}] \tag{E4.5.1}$$

Then, the value of R_D is determined as

$$R_D = \frac{V_{DD} - v_{o,min}}{I_{DD}} = \frac{2-0.04}{0.2\text{m}} = 9.8 \approx 10\ [\text{k}\Omega] \tag{E4.5.2}$$

Now, noting that the input voltage must be as high as V_{DD} to result in such a low output voltage as 0.04 V (in the triode region), we use Eq. (4.1.13a) to determine the value of K_p so that the drain current can be 0.2 mA for $v_o = v_{DS} = 0.04$ V and $v_i = v_{GS} = V_{DD} = 2$ V:

$$i_D \overset{(4.1.13a)}{=} K_p\left\{(v_{GS} - V_t)v_{DS} - \frac{1}{2}v_{DS}^2\right\}; 0.2\text{m} = K_p\left\{(2-1)0.04 - \frac{1}{2}0.04^2\right\} \quad \text{(E4.5.3)}$$

$$; K_p = \frac{0.2\text{ m}}{0.04 - 0.0016/2} \approx 5\text{ mA/V}^2 \quad \text{(E4.5.4)}$$

You can run the following MATLAB script "elec04e05.m" to plot the VTC and get the parameters of the designed inverter.

```
%elec04e05.m
% To solve Example 4.5
VDD=2; RD=1e4; Kp=5e-3; Vt=1; % Circuit and NMOS device parameters
[VIL,VIH,VOL,VOH,VM,VIT,VOT,VLH,NML,NMH,PDavg]=...
            NMOS_inverter(Kp,Vt,RD,VDD)
```

This yields the VTC as depicted in Figure 4.16(b) and the following results:

```
VIL= 1.020, VIH= 1.307, VOL= 0.163, VOH= 1.990, VM= 1.181
NML= 0.857, NMH= 0.683, VOE= 0.040, Pavg=  0.196[mW]
```

Why is the value of Pavg about half of the power specification, $P_{max} = 0.4$ mW? Because it is the average value of the power for on/off periods.

4.1.5.1 NMOS Inverter Using an Enhancement NMOS as a Load

Figure 4.17(a) shows an e-NMOS inverter, that is an NMOS inverter using a diode-connected enhancement NMOS M_2 (with its gate-drain short-circuited) as a load resistor where $v_{GD2} = 0 < V_{t2}$ so that M_2 is always in saturation with the drain current (common to the two NMOSs):

$$\begin{aligned} i_D &\underset{\lambda=0}{\overset{(4.1.13b)}{=}} \frac{1}{2}K_{p2}(v_{GS2} - V_{t2})^2 = \frac{1}{2}K_{p2}(V_{DD} - v_{DS1} - V_{t2})^2 \\ &= \frac{1}{2}K_{p2}(V_{DD} - v_o - V_{t2})^2 \end{aligned} \quad (4.1.45)$$

Note that we can write a KVL equation in i_D (for path V_{DD}-DS2-DS1) or a KCL equation in v_{DS2} (at node S2-D1) as

$$V_{DD} - v_{DS2}(i_D) = v_{DS1}(i_D) \text{ where } i_{D1} = i_{D2} = i_D \quad (4.1.46a)$$

$$\text{or } i_D(v_{DS2}) = i_D(v_{DS1}) \text{ where } v_{DS2} = V_{DD} - v_{DS1} \quad (4.1.46b)$$

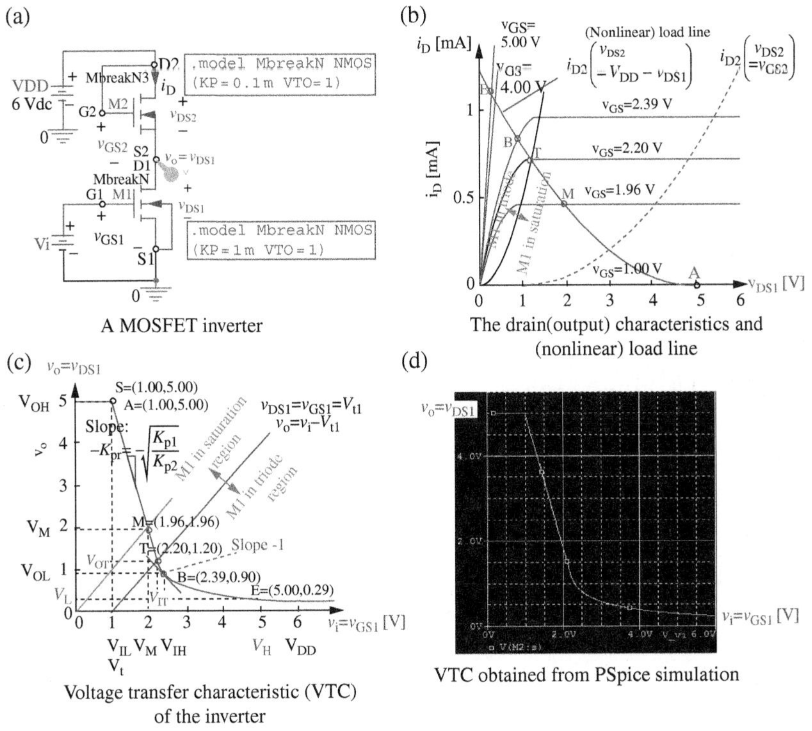

Figure 4.17 An NMOS inverter using an e-NMOS as a load and its VTC ("elec04f17.opj").

where the LHS of each equation corresponds to the (nonlinear) load curve (of M_2 with $V_{t2} = 1$ [V]), which is plotted as the red line together with the i_D-v_{DS1} characteristic curves of M_1 with $V_{t1} = 1$ [V] in Figure 4.17(b).

To get its transfer characteristic, suppose that the input voltage v_i increases from 0 to V_{DD}. While $v_i = v_{GS1} \le V_{t1}$, M_1 is cut off with $i_D = 0$ so that $v_o = V_{DD} - v_{GS2} = V_{DD} - V_{t2}$ since v_{GS2} stays at V_{t2} as long as $i_D = 0$, as described by point A in the VTC shown in Figure 4.17(c). As $v_i = v_{GS} > V_{t1}$, M_1 enters the saturation region where the drain current can be expressed as

$$i_D \underset{\lambda=0}{\overset{(4.1.13b)}{=}} \frac{1}{2}K_{p1}(v_{GS1} - V_{t1})^2 = \frac{1}{2}K_{p1}(v_i - V_{t1})^2 \tag{4.1.47}$$

and the output voltage v_o can be determined in terms of v_i by equating this with Eq. (4.1.45) for i_{D2}:

$$K_{p2}(V_{DD} - v_o - V_{t2})^2 = K_{p1}(v_i - V_{t1})^2$$

$$; v_o = V_{DD} - V_{t2} - K_{pr}(v_i - V_{t1}) \text{ with } K_{pr} = \sqrt{\frac{K_{p1}}{K_{p2}}} \tag{4.1.48}$$

This linear saturation mode (with slope $-K_{pr}$) continues till v_o becomes low enough to make $v_{GD1} = v_i - v_o \geq V_{t1}$ so that M_1 knocks on the door to the triode region, as described by point T:

$$v_i - v_o \overset{(4.1.48)}{=} v_i - V_{DD} + V_{t2} + K_{pr}(v_i - V_{t1}) = V_{t1} \tag{4.1.49}$$

$$;\ v_i = V_{IT} = \frac{V_{DD} + (1 + K_{pr})V_{t1} - V_{t2}}{1 + K_{pr}},\quad V_{OT} = V_{IT} - V_{t1} = \frac{V_{DD} - V_{t2}}{1 + K_{pr}} \tag{4.1.50a,b}$$

If v_i increases over V_{IT}, M_1 enters the triode region where the drain current can be expressed as

$$i_D \overset{(4.1.13a)}{=} K_{p1}\left\{(v_{GS1} - V_{t1})v_{DS1} - \frac{1}{2}v_{DS1}^2\right\} = K_{p1}\left\{(v_i - V_{t1})v_o - \frac{1}{2}v_o^2\right\} \tag{4.1.51}$$

and the output voltage v_o can be determined in terms of v_i by equating Eqs. (4.1.45) and (4.1.51):

$$i_{D2} = \frac{1}{2}K_{p2}(V_{DD} - v_o - V_{t2})^2 = K_{p1}\left\{(v_i - V_{t1})v_o - \frac{1}{2}v_o^2\right\} = i_{D1}$$

$$;\ (K_{p1} + K_{p2})v_o^2 - 2\{K_{p1}(v_i - V_{t1}) + K_{p2}(V_{DD} - V_{t2})\}v_o + K_{p2}(V_{DD} - V_{t2})^2 = 0 \tag{4.1.52}$$

$$;\ v_o = \frac{-B - \sqrt{B^2 - 4AC}}{2A} \text{ with } A = K_{p1} + K_{p2}, C = K_{p2}(V_{DD} - V_{t2})^2, \text{ and } B = -2\{K_{p1}(v_i - V_{t1}) + K_{p2}(V_{DD} - V_{t2})\} \tag{4.1.53}$$

Based on these two input-output relationships (each for the saturation/triode region), we can plot the VTC as depicted in Figure 4.17(c). The VTC can be regarded as having been obtained by taking the values of (v_{GS1}, v_{DS1}) at the operating points, i.e. the intersection points of the load curve of M_2 with the drain characteristic curves of M_1 for each value of $v_i = v_{GS1}$. Figure 4.17(d) shows the VTC obtained from the PSpice simulation, which is quite similar to that (Figure 4.17(c)) obtained from the theoretical analysis.

The *switching threshold voltage* V_M, also called the *midpoint voltage*, can be found by substituting $v_i = V_M$ and $v_i = V_M$ into the input–output relationship Eq. (4.1.48) (for the saturation region) and solving it for V_M as

$$V_M \underset{v_i = v_o = V_M}{\overset{(4.1.48)}{=}} V_{DD} - V_{t2} - K_{pr}(V_M - V_{t1})$$

$$;\ V_M = \frac{V_{DD} - V_{t2} + K_{pr}V_{t1}}{1 + K_{pr}} \text{ with } K_{pr} = \sqrt{\frac{K_{p1}}{K_{p2}}} \tag{4.1.54}$$

How about V_{IL} (the maximum input voltage that can be interpreted as "0") and the corresponding output voltage V_{OH} (the minimum output voltage that can be interpreted as "1")? Unlike the case of the NMOS inverter with a resistive load, it can't be determined as a point of slope -1 because the slope of the VTC abruptly changes from zero to $-K_{pr}$. Instead, we determine (V_{IL}, V_{OH}) at the turn-on point A in Figure 4.17(c):

$$V_{IL} = V_{t1}, \quad V_{OH} = V_{DD} - V_{t2} \tag{4.1.55a,b}$$

V_{IH} (the minimum input voltage that can be interpreted as "1") (at point B in Figure 4.17(c)) and the corresponding output voltage V_{OL} (the maximum output voltage that can be interpreted as "0") can be determined by setting the derivative of dv_i/dv_o (rather than dv_o/dv_i for a technical reason) to -1:

$$\frac{1}{2}K_{p2}(V_{DD} - v_o - V_{t2})^2 = K_{p1}\left\{(v_i - V_{t1})v_o - \frac{1}{2}v_o^2\right\}$$

$$; v_i = \frac{(V_{DD} - v_o - V_{t2})^2}{2K_{pr}^2 v_o} + \frac{1}{2}v_o + V_{t1}$$

$$; \frac{dv_i}{dv_o} = \frac{-2(V_{DD} - v_o - V_{t2})K_{pr}^2 v_o - (V_{DD} - v_o - V_{t2})^2 K_{pr}^2}{2\left(K_{pr}^2 v_o\right)^2} + \frac{1}{2} = -1$$

$$; \frac{(V_{DD} - v_o - V_{t2})(V_{DD} + v_o - V_{t2})}{2K_{pr}^2 v_o^2} = \frac{3}{2}; \; (V_{DD} - V_{t2})^2 - v_o^2 = 3K_{pr}^2 v_o^2$$

$$; V_{OL} = \frac{V_{DD} - V_{t2}}{\sqrt{1 + 3K_{pr}^2}} \tag{4.1.56a}$$

$$V_{IH} = \left.\frac{(V_{DD} - v_o - V_{t2})^2}{2K_{pr}^2 v_o}\right|_{v_o = V_{OL}} + \frac{1}{2}V_{OL} + V_{t1} \text{ with } K_{pr} = \sqrt{\frac{K_{p1}}{K_{p2}}} \tag{4.1.56b}$$

The processes of using these formulas to find the parameters and plotting the VTC have been cast into the following MATLAB function '`NMOS2e_inverter()`'. To analyze the NMOS inverter of Figure 4.17(a), all you need to do is to run the following MATLAB statements:

```
>>VDD=6; Vt1=1; Vt2=1; Kp1=1e-3; Kp2=0.1e-3;
  Kp12=[Kp1 Kp2]; Vt12=[Vt1 Vt2]; NMOS2e_inverter(Kp12,Vt12,VDD);
```

This will yield the following analysis result and the VTC as depicted in Figure 4.17(c), which conforms with that (in Figure 4.17(d)) obtained from PSpice simulation:

```
VIL= 1.00, VIH= 2.39, VOL= 0.90, VOH= 5.00, VM= 1.96, VIT= 2.20, VOT= 1.20
Noise Margin: NML= 0.10 and NM= 2.61, Average power = 3.330e+00 [mW]
```

```
function [VIL,VIH,VOL,VOH,VM,VIT,VOT,VLH,NML,NMH,PDavg,vios]=...
          NMOS2e_inverter(Kp12,Vt12,VDD)
% To find the parameters of an NMOS inverter with another e-NMOS as load
% Kp12=[Kp1 Kp2], Vt12=[Vt1 Vt2]: Threshold (Pinch-off) voltage
% Copyleft: Won Y. Yang, wyyang53@hanmail.net, CAU for academic use only
if numel(Vt12)>1, Vt1=Vt12(1); Vt2=Vt12(2);
  else Vt1=Vt12; Vt2=Vt12; end
if numel(Kp12)>1, Kp1=Kp12(1); Kp2=Kp12(2);
  else Kp1=Kp12; Kp2=Kp12; end
Kpr2=Kp1/Kp2; Kpr=sqrt(Kpr2);
% Transition point (VIT,VOT)
VOT=(VDD-Vt2)/(1+Kpr); VIT=VOT+Vt1; % Eq.(4.1.50)
VIL=Vt1; VOH=VDD-Vt2; % Eq.(4.1.55a,b)
VH=VOH; % Just let VH=VOH (The highest output voltage)
% Cutoff Segment of VTC
dv=0.001; vi_cutoff=0:dv:Vt1; vo_cutoff = VH*ones(size(vi_cutoff));
Nc=numel(vi_cutoff);
% Saturation Segment of VTC
vi_sat = VIL:dv:VIT; % Saturation region
vo_sat = VDD-Vt2-Kpr*(vi_sat-Vt1); % Eq.(4.1.48)
% Alternative to find vo_sat for saturation segment of VTC
a1=1; b1=-2*(VDD-Vt2); c1=(VDD-Vt2)^2-Kpr^2*(vi_sat-Vt1).^2;
vo_sat0 = -b1/2-sqrt((b1/2)^2-a1*c1); % Eq.(4.1.48)
discrepancy_sat=norm(vo_sat0-vo_sat)/norm(vo_sat)
% Triode Segment of VTC
vi_tri=VIT:dv:VDD+1; % Triode region
a2=1+Kpr2; b2=-2*((VDD-Vt2)+Kpr2*(vi_tri-Vt1)); c2=(VDD-Vt2)^2;
vo_tri = (-b2-sqrt(b2.^2-4*a2*c2))/2/a2; % Eq.(4.1.53)
% Put all the segments together
vis=[vi_cutoff vi_sat vi_tri]; vos=[vo_cutoff vo_sat vo_tri];
vios = [vis; vos];
% To find the inverter parameters
VM = (VDD-Vt2+Kpr*Vt1)/(1+Kpr); % Eq.(4.1.54)
VOL = (VDD-Vt2)/sqrt(1+3*Kpr2); % Eq.(4.1.56a)
VIH = (VDD-VOL-Vt2)^2/2/Kpr2/VOL + VOL/2 +Vt1; % Eq.(4.1.56b)
VL = vos(ceil((VH-vis(1))/dv)); % vo to vi=VH (Virtual lowest output)
NMH = VOH-VIH; NML = VIL-VOL; % Eq.(4.1.43a,b)
IDD = Kp2/2*(VDD-VL-Vt2)^2;
PDavg = VDD*IDD/2; % Average power for on-off periods
fprintf('\n VIL=%6.3f, VIH=%6.3f, VOL=%6.3f, VOH=%6.3f, VM=%6.3f,
  VIT=%6.3f, VOT=%6.3f', VIL,VIH,VOL,VOH,VM,VIT,VOT);
fprintf('\n Noise Margin: NM_L=%6.3f and NM_H=%6.3f', NML,NMH);
fprintf('\n Average power = %10.3e[mW]\n', PDavg*1e3);
% To plot the VTC
plot(vis,vos, [VIL VM VIH], [VOH VM VOL], 'ro')
hold on, plot([Vt1 VIT VH], [VH VOT VL], 'r^')
```

4.1.5.2 NMOS Inverter Using a Depletion NMOS as a Load

Figure 4.18(a) shows a d-NMOS inverter, that is an NMOS inverter using a depletion NMOS M_2 (with its gate-source short-circuited) as a load resistor where $v_{GS2}=0>V_{t2}$ so that $i_{D2}=i_{D1}=i_D$ is determined by $v_{DS2}=V_{DD}-v_{DS1}=V_{DD}-v_o$ as

$$i_D \underset{(4.1.13a)}{\overset{(4.1.13b)}{=}} \begin{cases} K_{p2}\left(\overset{0}{v_{GS2}}-V_{t2}\right)^2/2 & \text{if } v_{GD2}=v_{SD2}<V_{t2},\text{i.e.,}v_{DS1}<V_{DD}+V_{t2} \\ K_{p2}\left\{\left(\overset{0}{v_{GS2}}-V_{t2}\right)v_{DS2}-v_{DS2}^2/2\right\} & \text{if } v_{GD2}=v_{SD2}>V_{t2},\text{i.e.,}v_{DS1}>V_{DD}+V_{t2} \end{cases} \tag{4.1.57}$$

The (nonlinear) load curve (of M_2 with $V_{t2}=-1$ [V]) is plotted as the pink line together with the i_D-v_{DS1} characteristic curves of M_1 with $V_{t1}=1$ [V] in (b). Note that the load curve and the i_D-v_{DS1} characteristic curve of M_1 for $v_{GS1}=V_{t1}-\sqrt{K_{p2}/K_{p1}}V_{t2}=2$ are symmetric about $v_{DS1}=V_{DD}/2$ because they can be switched to each other by substituting $v_{DS2}=V_{DD}-v_{DS1}$.

To get its transfer characteristic, suppose that the input voltage v_i increases from 0 to V_{DD}. While $v_i=v_{GS1}\le V_{t1}$, M_1 is cut off with $i_D=0$ while M_2 is turned on, but with $v_{DS2}=0$ (by Eq. (4.1.57b)) so that $v_o=V_{DD}-v_{DS2}=V_{DD}$. As $v_i=v_{GS1}>V_{t1}$, M_2/M_1 enter the triode/saturation region where the drain current can be expressed as

$$i_D \underset{\lambda=0}{\overset{(4.1.13b)}{=}} \frac{1}{2}K_{p1}(v_{GS1}-V_{t1})^2=\frac{1}{2}K_{p1}(v_i-V_{t1})^2 \tag{4.1.58}$$

and the output voltage v_o can be determined in terms of v_i by equating this with Eq. (4.1.57b):

$$I_{D1}=\frac{1}{2}K_{p1}(v_i-V_{t1})^2=K_{p2}\left\{\left(\overset{0}{v_{GS2}}-V_{t2}\right)v_{DS2}-\frac{1}{2}v_{DS2}^2\right\} \tag{4.1.59a}$$

$$;v_o=V_{DD}-v_{DS2}$$

$$=V_{DD}+\frac{B+\sqrt{B^2-4AC}}{2A} \quad \text{with } A=K_{p2},\ B=2K_{p2}V_{t2},\text{ and } C=2I_{D1} \tag{4.1.59b}$$

From Figure 4.18(b), we can see that during this mode, the operating point moves upward from point S (via point A) along the semiparabolic load curve. When will this mode stop and M_2 enter the saturation mode? It is when $v_{GD2}=v_{SD2}=v_{DS1}-V_{DD}=V_{t2}$ so that $v_o=v_{DS1}=V_{DD}+V_{t2}=V_{OT2}$. At this point T_2, we have

$$I_{D1}=\frac{1}{2}K_{p1}(v_i-V_{t1})^2=\frac{1}{2}K_{p2}\left(\overset{0}{v_{GS2}}-V_{t2}\right)^2$$

$$;v_i=V_{IT2}=V_{t1}+\sqrt{K_{p2}/K_{p1}}\,|V_{t2}| \tag{4.1.60}$$

When will this mode stop and M_1 enter the triode mode? It is when $v_{GD1} = v_{GS1} - v_{DS1} = v_i - v_o = V_{t1}$. The output voltage v_o at this point T can be determined by substituting $v_i = v_o + V_{t1}$ into Eq. (4.1.51) as

$$\frac{1}{2}K_{p1}v_o^2 = \frac{1}{2}K_{p2}V_{t2}^2; \quad v_o = V_{OT} = \sqrt{K_{p2}/K_{p1}}\,|V_{t2}| \tag{4.1.61}$$

```
function [VIL,VIH,VOL,VOH,VM,VIT,VOT,VLH,NML,NMH,PDavg,vios]=...
          NMOS2d_inverter(Kp12,Vt12,VDD)
% To find all the parameters of an NMOS inverter with a d-NMOS as load
% Kp12=[Kp1 Kp2], Vt12=[Vt1 Vt2]: Threshold (Pinch-off) voltage
% Copyleft: Won Y. Yang, wyyang53@hanmail.net, CAU for academic use only
Kp1=Kp12(1); Kp2=Kp12(2); Vt1=Vt12(1); Vt2=Vt12(2);
Kpr2=Kp1/Kp2; Kpr=sqrt(Kpr2);
% Transition point (VIT,VOT)
VIT=Vt1+abs(Vt2)/Kpr; IDT=Kp2*Vt2^2/2; % Eq.(4.1.60)
dv=0.001; vis=[0:dv:VIT VIT:dv:VDD];
for n=1:length(vis)
  vi=vis(n);
  if vi<Vt1, iDs(n)=0; vos(n)=VDD;
    elseif vi<VIT % M1/M2: saturation/triode - % Eq.(4.1.59)
     iDs(n)=Kp1/2*(vi-Vt1)^2; A=Kp2; B=2*Kp2*Vt2; C=2*iDs(n);
     vDS2=(-B-sqrt(B^2-4*A*C))/2/A; %vDS2 = min(roots([A B C]));
     vos(n)=VDD-vDS2;
    elseif vi>VIT % M1/M2: triode/saturation - Eq.(4.1.63)
     iDs(n)=Kp2/2*Vt2^2; A=Kp1; B=2*Kp1*(Vt1-vi); C=2*iDs(n);
     vos(n)=(-B-sqrt(B^2-4*A*C))/2/A; %vos(n) = min(roots([A B C]));
    else % M1/M2: saturation/saturation - % Eqs.(4.1.60,61)
    iDs(n)=IDT;
    VOT=sqrt(2/Kp1*IDT); % Transition point of M1 between SAT/Triode
    VOT2=VDD+Vt2; % Transition point of M2 between Triode/SAT
    vos(n) = (VOT+VOT2)/2;
  end
end
vios=[vis; vos]; [VIL,VIH,VOL,VOH,VM,NML,NMH,VL,PDavg]= ...
                  find_pars_of_inverter(vis,vos,iDs,VDD);
VH=VDD; VLH=[VL VH]; % The lowest/highest output voltages
% To plot the VTC
plot(vis,vos, [VIL VM VIH],[VOH VM VOL],'ro')
hold on, plot([Vt1 VIT VH],[VH VOT VL],'r^')
```

Noting that the input voltage v_i at this point is the same as V_{IT} (Eq. (4.1.60)) despite the different values of v_o, we can see that when $v_i = V_{IT}$, v_o may change abruptly between V_{OT} and $V_{OT2} = V_{DD} + V_{t2}$. The midpoint is determined as the constant value of v_i along this SAT-SAT segment:

$$V_M = V_{IT} = V_{t1} + \sqrt{K_{p2}/K_{p1}}\,|V_{t2}| \tag{4.1.62}$$

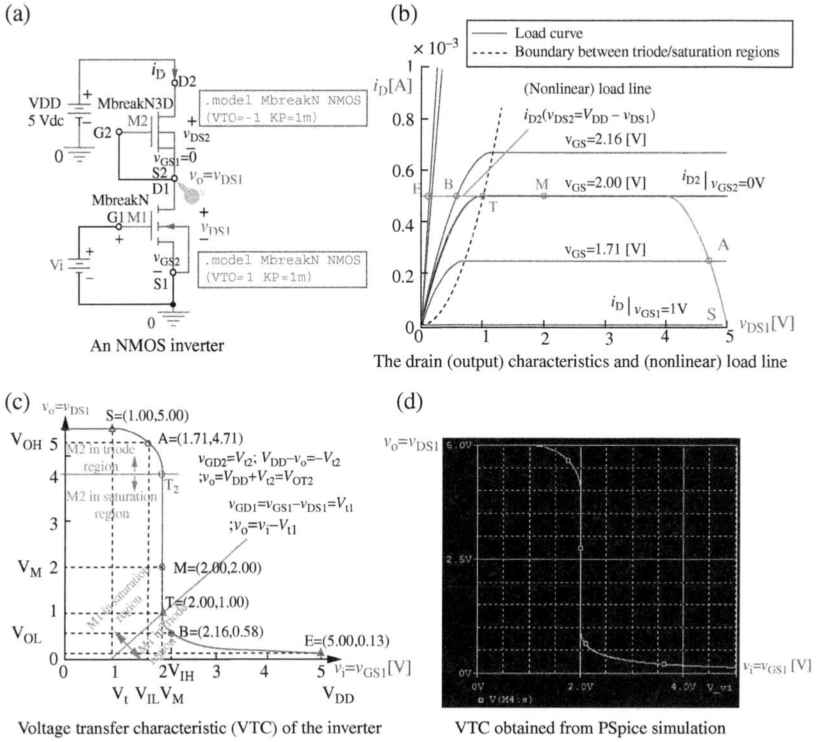

Figure 4.18 An NMOS inverter using a d-NMOS as a load and its VTC ("elec04f18.opj").

As $v_i = v_{GS1}$ increases over V_{IT}, M_1 enters the triode region while M_2 continues to be in saturation so that their drain currents can be expressed as

$$I_{D2} \overset{(4.1.13b)}{\underset{\lambda=0}{=}} \frac{1}{2}K_{p2}\left(\overset{0}{v_{GS2}} - V_{t2}\right)^2 = \frac{1}{2}K_{p2}V_{t2}^2$$

$$i_{D1} \overset{(4.1.13a)}{=} K_{p1}\left\{(v_{GS1} - V_{t1})v_{DS1} - \frac{1}{2}v_{DS1}^2\right\} = K_{p1}\left\{(v_i - V_{t1})v_o - \frac{1}{2}v_o^2\right\}$$

and the output voltage v_o can be determined in terms of v_i by equating these two equations:

$$i_{D1} = K_{p1}\left\{(v_i - V_{t1})v_o - \frac{1}{2}v_o^2\right\} = \frac{1}{2}K_{p2}V_{t2}^2 = I_{D2} \tag{4.1.63a}$$

$$;v_o = \frac{-B - \sqrt{B^2 - 4AC}}{2A} \text{ with } A = K_{p1}, B = 2K_{p1}(V_{t1} - v_i), C = 2I_{D2} \tag{4.1.63b}$$

Based on the two input-output relationships (Eqs. (4.1.59) and (4.1.63)), we can plot the VTC as depicted in Figure 4.18(c). The process of using these formulas to plot the VTC has been cast into the above MATLAB function '`NMOS2d_inverter()`', which uses the MATLAB function '`find_pars_of_inverter()`' (Section 3.1.10) to find the values of the inverter parameters such as V_{IL}/V_{OH}, V_{IH}/V_{OL} (at the two points with the slope of the VTC equal to -1), and V_M (at the midpoint).

To analyze the NMOS inverter of Figure 4.18(a), all you need to do is to run the following MATLAB statements:

```
>>VDD=5; Vt1=1; Vt2=-1; Kp1=1e-3; Kp2=1e-3;
  Kp12=[Kp1 Kp2]; Vt12=[Vt1 Vt2]; NMOS2d_inverter(Kp12,Vt12,VDD);
```

This will yield the following analysis result and the VTC as depicted in Figure 4.18(c), which conforms with that (in Figure 4.18(d)) obtained from PSpice simulation:

```
VIL= 1.708, VIH= 2.155, VOL= 0.577, VOH= 4.706, VM= 2.000
NML= 1.131, NMH= 2.551, VL= 0.127, Pavg=  1.250[mW]
```

Comparing Figures 4.17(c) and 4.18(c), note that the VTC of a d-NMOS inverter has a steeper transition region and, accordingly, higher noise margins than an e-NMOS inverter.

4.1.5.3 CMOS Inverter

Figure 4.19(a) shows a CMOS (Complementary MOSFET) inverter using NMOS and PMOS fabricated on the same chip where both gates/drains are tied to the input/output, respectively, and the sources of Mp(PMOS)/Mn(NMOS) are connected to VDD/GND, respectively. Figure 4.19(b) shows the (blue-lined) i_D-v_{DSn} characteristic curves of Mn depending on v_{GSn} and the (red-lined) $i_D - (v_{SDp} = V_{DD} - v_{DSn})$ characteristic curves of Mp depending on $v_{SGp} = V_{DD} - v_i$. A rough VTC of the inverter can be plotted by connecting the operating points ($v_{SGn} = v_i$, $v_{DSn} = v_o$) {S, A, T_2, T_1, ...}, i.e. the intersection points of the characteristic curve of Mn (for v_{GSn} = 0~1, 1.5, 2, 2.5, 3, 3.5, and 4 ~ 5 V) with that of Mp (for $v_{SGp} = V_{DD} - v_i$ = 5~4, 3.5, 3, 2.5, 2, 1.5, and 1 ~ 0 V), as illustrated in Figure 4.19(b) and (c). Figure 4.19(d) shows the VTC obtained from the PSpice simulation.

Noting from Figure 4.19(b) and (c) that the operating point moves around the five regions I, II, III, IV, and V as the input voltage v_i (applied to the common gate) varies from zero to V_{DD}, let us analyze the inverter to find out the analytical expression of the VTC together with the transition points from a region to another region where the conduction parameters and threshold (turn-on)

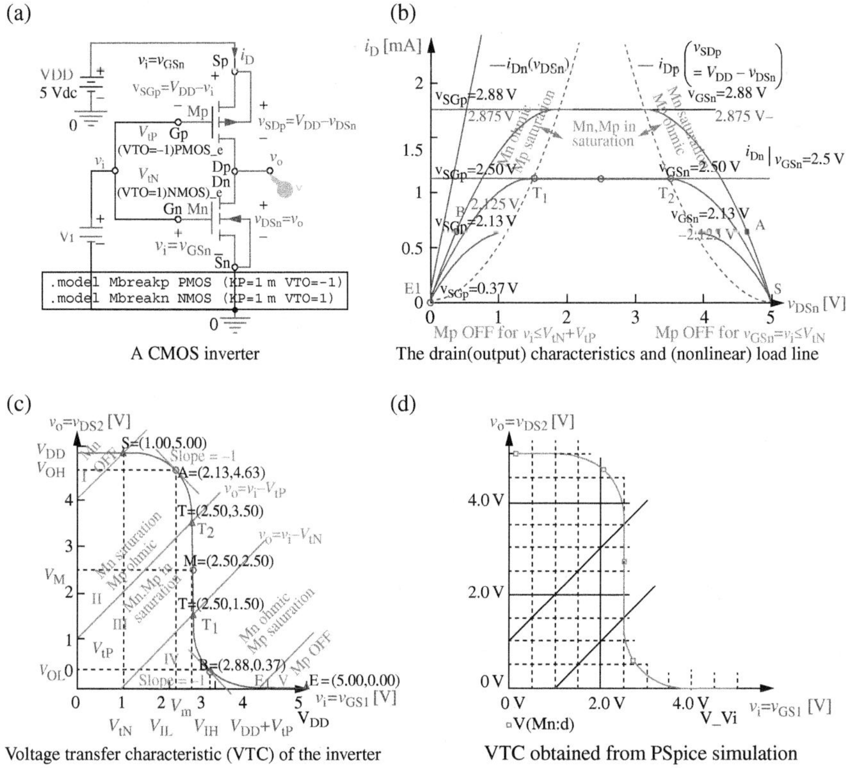

Figure 4.19 A Complementary MOSFET (CMOS) inverter and its voltage transfer characteristic (VTC) ("elec04f19.opj").

voltages of the PMOS/NMOS are $K_{\text{pP}}/K_{\text{pN}}$ and $V_{\text{tP}}/V_{\text{tN}}$, respectively. To this aim, we start from computing the value V_{m} of the input voltage v_{i} (belonging to region III) at which both Mp and Mn operate in the saturation region by equating the drain currents of Mp and Mn described by Eq. (4.1.13b) (with $\lambda = 0$ for simplicity):

$$K_{\text{pP}}(v_{\text{SGP}} + V_{\text{tP}})^2 \underset{\lambda=0}{\overset{(4.1.13\text{b})}{=}} K_{\text{pN}}(v_{\text{GSN}} - V_{\text{tN}})^2$$

$$\underset{v_{\text{SGP}} = V_{\text{DD}} - v_{\text{i}}}{\overset{v_{\text{GSN}} = v_{\text{i}}}{\longrightarrow}} \quad K_{\text{pP}}(V_{\text{DD}} - v_{\text{i}} + V_{\text{tP}})^2 = K_{\text{pN}}(v_{\text{i}} - V_{\text{tN}})^2$$

$$\to \left(1 - \frac{K_{\text{pN}}}{K_{\text{pP}}}\right) v_{\text{i}}^2 + 2\left\{\frac{K_{\text{pN}}}{K_{\text{pP}}} V_{\text{tN}} - (V_{\text{DD}} + V_{\text{tP}})\right\} v_{\text{i}} + (V_{\text{DD}} + V_{\text{tP}})^2 - \frac{K_{\text{pN}}}{K_{\text{pP}}} V_{\text{tN}}^2 = 0 \tag{4.1.64}$$

$$V_m = \frac{(V_{DD}+V_{tP})-KV_{tN} \pm \sqrt{((V_{DD}+V_{tP})-KV_{tN})^2+(K-1)((V_{DD}+V_{tP})^2-KV_{tN}^2)}}{1-K}$$

$$= \frac{(V_{DD}+V_{tP})-KV_{tN} \pm \sqrt{K(V_{DD}+V_{tP})^2-2KV_{tN}(V_{DD}+V_{tP})+KV_{tN}^2}}{1-K}$$

$$= \frac{(V_{DD}+V_{tP})-KV_{tN}-\sqrt{K}(V_{DD}+V_{tP}-V_{tN})}{1-K} = \frac{(1-\sqrt{K})(V_{DD}+V_{tP})+(1-\sqrt{K})\sqrt{K}V_{tN}}{1-K}$$

$$; V_m = \frac{V_{DD}+V_{tP}+\sqrt{K}V_{tN}}{1+\sqrt{K}} \text{ with } K = \frac{K_{pN}}{K_{pP}}: \textit{geometry ratio} \tag{4.1.65}$$

This will be used as the boundary values of v_i between regions II and III or III and IV (see Figure 4.19(c)).

<Region I>

When $0 \le v_i = v_{GSn} \le V_{tN}$, Mn remains OFF while Mp conducts not in the saturation region but in the ohmic (triode) region because the off-transistor Mn keeps Mp from conducting as much as it could with $v_{SGp} = V_{DD} - v_i \ge |V_{tP}|$. Since the resistance of the off-transistor Mn is much greater than that of the on-transistor Mp, almost the whole V_{DD} applies to Mn so that the output voltage is $v_o \approx V_{DD}$.

<Region II>

When $V_{tN} < v_i \le V_m$, Mn/Mp are assumed to be in the saturation/ohmic regions, respectively (see Figure 4.19(b)) and the operating point is computed by equating the drain currents of Mp and Mn that are described by Eqs. (4.1.13b) (with $\lambda = 0$ for simplicity) and (4.1.13a), respectively:

$$\overset{(4.1.13b)\text{ with }\lambda=0}{K_{pN}(v_{GSn}-V_{tN})^2} = \overset{(4.1.13a)}{K_{pP}\left\{2(v_{SGp}+V_{tP})v_{SDp}-v_{SDp}^2\right\}} \underset{v_{SGp}=V_{DD}-v_i,\, v_{SDp}=V_{DD}-v_o}{\overset{vGS_n=v_i}{\longrightarrow}}$$

$$K_{pN}(v_i-V_{tN})^2 = K_{pP}\{2(V_{DD}-v_i+V_{tP})(V_{DD}-v_o)-(V_{DD}-v_o)^2\}$$

$$\rightarrow v_o^2+2(V_{tP}-v_i)v_o+\frac{K_{pN}}{K_{pP}}(v_i-V_{tN})^2+V_{DD}(2(v_i-V_{tP})-V_{DD})=0 \tag{4.1.66}$$

If the (larger) root of this quadratic equation satisfies $v_o > v_i - V_{tP}$, the assumption is right. Otherwise, i.e. if

$$v_o \le v_i - V_{tP} \underset{v_i=V_{DD}-v_{SGp}}{\overset{v_o=V_{DD}-v_{SDp}}{\longrightarrow}} V_{DD}-v_{SDp} \le V_{DD}-v_{SGp}-V_{tP} \rightarrow v_{SDp} \ge v_{SGp}+V_{tP} \tag{4.1.67}$$

then Mp enters the saturation region and this is region III where both Mp and Mn operate in the saturation mode. The boundary between regions II and III is denoted by the pink line corresponding to $v_o = v_i - V_{tP}$ in Figure 4.19(c).

<Region III>

The inverter stays in Region III (with Mp and Mn saturated) as long as the saturation condition of Mn is satisfied, i.e. if

$$v_o \geq v_i - V_{tN} \underset{v_i = v_{GSn}}{\overset{v_o = v_{DSn}}{\longrightarrow}} v_{DSn} \geq v_{GSn} - V_{tN} \tag{4.1.68}$$

```
function [VIL,VIH,VOL,VOH,Vm,VIT1,VOT1,VIT2,VOT2,VLH,NML,NMH,PDavg]=...
         CMOS_inverter(KNP,VtNP,VDD,Kg)
% To find all the parameters of a CMOS inverter.
% KNP = [KN'(WN/LN) KP'(WP/LP)]
% VtNP = [VtN VtP]: Threshold (Pinch-off) voltages of NMOS/PMOS
% Copyleft: Won Y. Yang, wyyang53@hanmail.net, CAU for academic use only
if nargin<4, Kg=0; end
if numel(KNP)>1, KN=KNP(1); KP=KNP(2); else KN=KNP; KP=KNP; end
if numel(VtNP)>1, VtN=VtNP(1); VtP=VtNP(2); else VtN=VtNP;
VtP=-VtNP; end
Kr2=KN/KP; Kr=sqrt(Kr2); % Ratio between conduction parameters of Mn and Mp
VOL=0; VOH=VDD; VH=VOH; % The lowest/highest output voltage
Vm=(VDD+VtP+Kr*VtN)/(1+Kr); % Eq.(4.1.65)
VIT2=Vm; VOT2=Vm-VtP; VIT1=Vm; VOT1=Vm-VtN; % Boundary points of Region III
dv=0.001; vis=[0:dv:VDD]; % Input voltage range
for n=1:length(vis)
  vi=vis(n);
  if vi<VtN, iDs(n)=0; vos(n)=VDD; % Mn OFF, Mp in ohmic region (Region I)
    elseif vi<=Vm % Mn/Mp are saturation/ohmic region (Region II)
      B=2*(VtP-vi); C=Kr2*(vi-VtN)^2+VDD*(2*(vi-VtP)-VDD); %Eq.(4.1.66)
      vos(n)=(-B+sqrt(max(B^2-4*C,0)))/2; iDs(n)=KN/2*(vi-VtN)^2;
    elseif vi<=VDD+VtP % Mp/Mn=saturation/triode (Region IV)
      B=2*(VtN-vi); C=(VDD-vi+VtP)^2/Kr2; % Eq.(4.1.69)
      vos(n)=(-B-sqrt(max(B^2-4*C,0)))/2; iDs(n)=KP/2*(VDD-vi+VtP)^2;
   else iDs(n)=0; vos(n)=0; % Mn in triode region, Mp OFF (Region V)
  end
  % Here, Region III (Mp/Mn: sat) doesn't have to be taken care of
  % since at most only a single value of vi=Vm can belong to that region.
end
[VIL,VIH,VOL,VOH,VM,NML,NMH,VL,PDavg]= ...
            find_pars_of_inverter(vis,vos,iDs,VDD);
if Kg>0
   subplot(221)
   plot(vis,vos), hold on
   plot([VIL VM VIH],[VOH VM VOL],'ro', [VtN VH],[VH VL],'r^')
   title('VTC of a CMOS inverter')
   subplot(222)
   plot(vis,iDs), title('Drain current of the CMOS inverter'), grid on
end
```

```
%elec04f19.m
VDD=5; VtN=1; VtP=-1; KN=1e-3; KP=1e-3;
KNP=[KN KP]; VtNP=[VtN VtP];
Kg=2; % To plot the graphs
[VIL,VIH,VOL,VOH,Vm,VIT1,VOT1,VIT2,VOT2,VLH,NML,NMH,PDavg]=...
          CMOS_inverter(KNP,VtNP,VDD,Kg);
```

<Region IV>

If v_i increases further enough to break the above condition, i.e. $v_o < v_i - V_{tN}$, Mn enters the ohmic region (see Figure 4.19(b)) and the operating point is computed by equating the drain currents of Mp (in the saturation region) and Mn (in the ohmic region) that are described by Eqs. (4.1.13b) (with $\lambda = 0$ for simplicity) and (4.1.13a), respectively:

$$\overset{(4.1.13b)\text{ with }\lambda=0}{K_{pP}(v_{SGp}+V_{tP})^2} = \overset{(4.1.13a)}{K_{pN}\{2(v_{GSn}-V_{tN})v_{DSn}-v_{DSn}^2\}}$$

$$\underset{v_{GSn}=v_i,\,v_{DSn}=v_o}{\overset{v_{SGp}=V_{DD}-v_i}{\longrightarrow}} K_{pP}(V_{DD}-v_i+V_{tP})^2 = K_{pN}(2(v_i-V_{tN})v_o-v_o^2)$$

$$\rightarrow v_o^2 + 2(V_{tN}-v_i)v_o + \frac{K_{pP}}{K_{pN}}(V_{DD}-v_i+V_{tP})^2 = 0 \tag{4.1.69}$$

The (smaller) root of this quadratic equation will be v_o as long as $v_i \le v_{DD}+V_{tP}$.

<Region V>

While Mn is still in the ohmic region, Mp will be OFF if

$$v_i > V_{DD} + V_{tP} \overset{v_i = V_{DD}-v_{SGP}}{\longrightarrow} v_{SGP} < V_{tP} \tag{4.1.70}$$

Then almost the whole V_{DD} applies to Mp (with much larger resistance) so that $v_o \approx 0$.

The process of finding the output voltage of the CMOS inverter to the whole range of the input has been cast into the above MATLAB function '`CMOS_inverter()`'. We have run the above MATLAB script "elec04f19.m" to get not only the following values of the inverter parameters

```
VIL = 2.1250, VIH = 2.8750, VOL = 0.3750, VOH = 4.6250,
VIT2 = 2.5000, VOT2 = 3.5000, Vm = 2.5000, VIT1 = 2.5000, VOT1 = 1.5000,
NML = 1.7500, NMH = 1.7500, PDavg = 0.0028
```

but also the VTC and the drain current of the inverter as depicted in Figures 4.19(c) and 4.20(a), respectively. As can be seen from the VTC in Figure 4.19(c), the lowest/highest output voltage levels are $0/V_{DD}$ so that the CMOS inverter has the maximum output swing. This, together with the symmetry of the VTC, is the reason why a CMOS inverter can have wide noise margins.

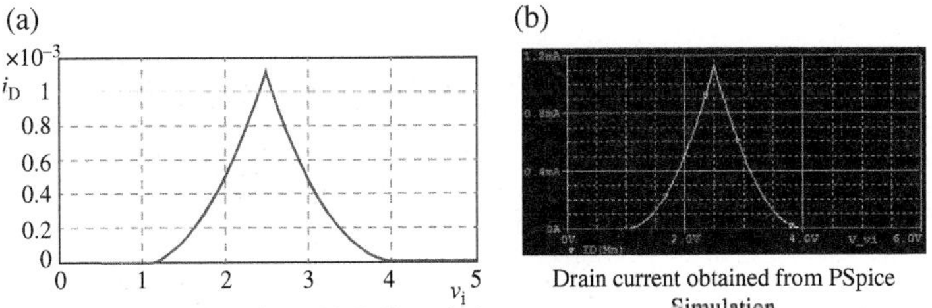

(a) Drain current obtained from MATLAB analysis (b) Drain current obtained from PSpice Simulation

Figure 4.20 Drain currents of the CMOS inverter obtained from MATLAB analysis and PSpice simulation.

Figure 4.20(b) shows the drain current of the inverter obtained from the PSpice simulation, which is quite similar to Figure 4.20(a). What is implied by the current curve(s) in Figure 4.20? The current (drawn from the supply voltage source V_{DD}) is zero when the CMOS inverter is in its output-high/low state so that the CMOS gate can power down in its static condition, which is one of the major advantages of CMOS configuration.

One more thing to note before closing this section is that a formula to determine the geometry ratio K from the given values of V_{DD}, V_m (the boundary value of the input voltage between regions II and III or regions III and IV), V_{tP} (the threshold voltage of PMOS), and V_{tN} (the threshold voltage of NMOS) can be derived from Eq. (4.1.65) as

$$K = \frac{K_{pN}}{K_{pP}} = \left(\frac{V_{DD} + V_{tP} - V_m}{V_m - V_{tN}}\right)^2 \tag{4.1.71}$$

4.1.6 Source-Coupled Differential Pair

Figure 4.21(a) and (b) shows a *source-coupled* (in the sense that the source terminals of the two FETs are connected) or *differential* (in the sense that its output varies with the differential input $v_d = v_{GS1} - v_{GS2}$) pair. To analyze this circuit, we assume that the two FETs have identical conduction parameters K_p and threshold voltages V_t, and both of them operate in the saturation mode so that we can use Eq. (4.1.13b) (with $\lambda = 0$ for ignoring the CLM effect) to write their approximate drain currents as

$$i_{D1} \underset{\lambda=0}{\overset{(4.1.13b)}{=}} \frac{1}{2}K_p(v_{GS1} - V_t)^2 \tag{4.1.72a}$$

$$i_{D2} \underset{\lambda=0}{\overset{(4.1.13b)}{=}} \frac{1}{2}K_p(v_{GS2} - V_t)^2 \tag{4.1.72b}$$

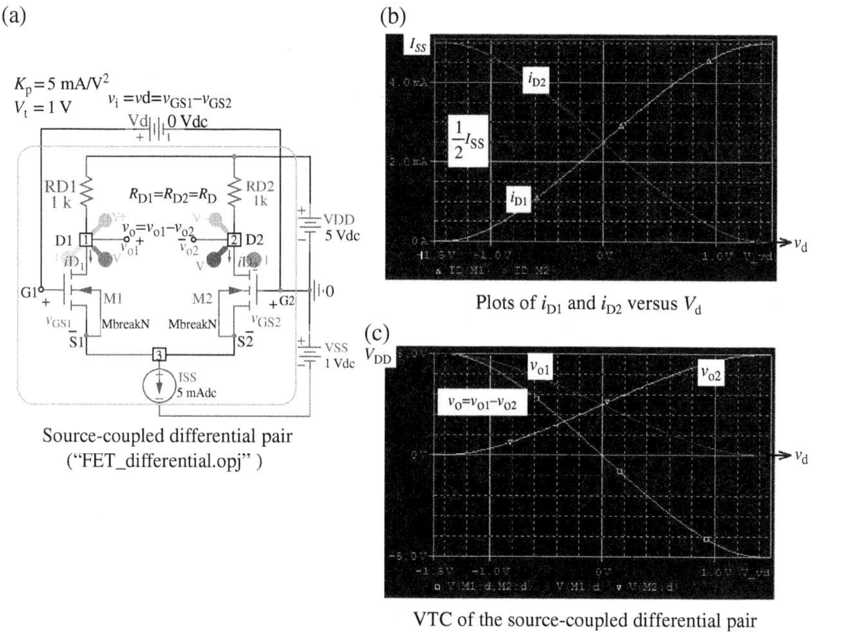

Figure 4.21 Source-coupled (differential) pair and its VTC.

Thus, the difference between their square roots can be written as

$$\sqrt{i_{D1}} - \sqrt{i_{D2}} = \sqrt{K_p/2}\,(v_{GS1} - v_{GS2}) = \sqrt{K_p/2}\,v_d \text{ with } v_d = V_{GS1} - V_{GS2} \quad (4.1.73)$$

Also, we apply KCL at node 3 (S1-S2) to write

$$i_{D1} + i_{D2} \overset{\text{KCL}}{=} I_{SS} = B \quad (4.1.74)$$

Subtracting the square of Eq. (4.1.73) from Eq. (4.1.74) yields

$$\sqrt{i_{D1}}\sqrt{i_{D2}} = \frac{1}{2}\left(I_{SS} - K_p v_d^2/2\right) \rightarrow i_{D1} i_{D2} = \frac{1}{4}\left(I_{SS} - K_p v_d^2/2\right)^2 = C \quad (4.1.75)$$

Thus, i_{D1} and i_{D2} can be found from the two roots of the quadratic equation:

$$x^2 - Bx + C = 0$$

$$;\, i_{D1}, i_{D2} = \frac{B \pm \sqrt{B^2 - 4C}}{2} = \frac{I_{SS} \pm \sqrt{I_{SS}^2 - \left(I_{SS} - K_p v_d^2/2\right)^2}}{2}$$

$$= \frac{I_{SS} \pm v_d\sqrt{K_p I_{SS} - K_p^2 v_d^2/4}}{2} \quad (4.1.76)$$

where these drain currents are depicted in Figure 4.21(b). Then we can write the (differential) output voltage as

$$v_O = v_{O1} - v_{O2} = v_{D1} - v_{D2} = \left(V_{DD} - R_D i_{D1}\right) - \left(VDD - R_D i_{D2}\right) \overset{(4.1.76)}{=} -v_d\sqrt{K_p I_{SS} - K_p^2 v_d^2/4}R_D \quad (4.1.77)$$

This differential output voltage v_o, together with v_{o1} and v_{o2}, is shown in Figure 4.21(c). From Figure 4.21 and Eqs. (4.1.76) and (4.1.77), note the following:

- If $|v_d| < 0.2\sqrt{I_{SS}/K_p}$, the differential output voltage v_o and other signals vary almost linearly with the differential input v_d, allowing the source-coupled pair to be used as an amplifier with a voltage gain of

$$\left.\frac{v_o}{v_d}\right|_{v_d = 0.4\sqrt{I_{SS}/K_p}} \overset{(4.1.77)}{=} -\sqrt{K_p I_{SS} - K_p^2 v_d^2/4}R_D \underset{K_p = 5\text{mA/V}^2}{\overset{I_{SS} = 5\text{mA}, R_D = 1\text{k}\Omega}{\approx}} -5 \quad (4.1.78)$$

- If $v_d > 1.5\sqrt{I_{SS}/K_p}$, we have

$$i_{D1} \approx I_{SS}, i_{D2} \approx 0, v_{o1} \approx V_{DD} - I_{SS}R_D \underset{\text{of } I_{SS} \text{ and } R_D}{\overset{\text{Proper choice}}{\longrightarrow}} 0, \text{ and } v_{o2} \approx V_{DD} \quad (4.1.79a)$$

  On the other hand, if $v_d < -1.5\sqrt{I_{SS}/K_p}$, we have

$$i_{D1} \approx 0,\ i_{D2} \approx I_{SS},\ v_{o1} \approx V_{DD}, \text{ and } v_{o2} \approx V_{DD} - I_{SS}R_D \underset{\text{of } I_{SS} \text{ and } R_D}{\overset{\text{Proper choice}}{\longrightarrow}} 0 \quad (4.1.79b)$$

  It is implied that a large swing of the differential input $v_d = \pm 1.5\sqrt{I_{SS}/K_p}$ makes the two FETs M_1/M_2 operate as a closed or an open switch, producing two distinct levels of differential output v_o depending on whether $v_d = 1.5\sqrt{I_{SS}/K_p}$ or $v_d = -1.5\sqrt{I_{SS}/K_p}$.

The small-signal input-output relationship of the differential pair can be derived by substituting Eqs. (4.1.72a,b) directly into Eq. (4.1.77):

$$\begin{aligned} v_o &\overset{(4.1.77)}{=} R_D(i_{d2} - i_{d1}) \underset{\lambda=0}{\overset{(4.1.13b)}{=}} \frac{1}{2}K_p\{(v_{GS2} - V_t)^2 - (v_{GS1} - V_t)^2\}R_D \\ &= \frac{1}{2}K_p\{(v_{GS2} - V_t)^2 - (v_{GS2} + v_d - V_t)^2\}R_D \\ &\overset{v_d << v_{GS2} - V_t}{\approx} -K_p(v_{GS2} - V_t)R_D v_d \underset{v_{GS2} \approx V_{GS,Q}}{\overset{(4.1.4b)}{=}} -g_m R_D v_d \end{aligned} \quad (4.1.80)$$

Note that the differential (mode) voltage gain $-g_m R_D$ can be regarded as the voltage gain (Eq. (4.2.6b)) of a common-source (CS) amplifier with $R_{S1} = 0$ and $R_L = \infty$. Note also that if v_{o1} or v_{o2}, instead of v_o, is taken as the output, the differential voltage gain $g_m R_D$ is halved:

$$v_{o1} = -\frac{1}{2} g_m R_D v_d, \quad v_{o2} = \frac{1}{2} g_m R_D v_d \tag{4.1.81}$$

The amplifying/switching properties are extensively used in analog/digital circuits, respectively. That is why the source-coupled differential pair is one of the most important configurations employed in ICs.

To analyze the FET differential pair circuit in Figure 4.21(a), we can apply KCL at nodes 1, 2, and 3 to write

$$\begin{aligned} &V_{DD} - v_1 - R_D i_{D1}(v_d - v_3, v_1 - v_3) = 0 \ \text{at node 1} \\ &V_{DD} - v_2 - R_{DC} i_{D2}(0 - v_3, v_2 - v_3) = 0 \ \text{at node 2} \\ &i_{D1}(v_d - v_3, v_1 - v_3) + i_{D2}(0 - v_3, v_2 - v_3) - I_{SS} = 0 \ \text{at node 3} \end{aligned} \tag{4.1.82}$$

where i_{Dk} (v_{GSk}, v_{DSk}) is defined by Eqs. (4.1.13a,b) as

$$i_{Dk} \underset{(4.1.13b)}{\overset{(4.1.13a)}{=}} \begin{cases} K_{pk}\{(v_{GSk} - V_{tk})v_{DSk} - v_{DSk}^2/2\} & \text{if } v_{GDk} = v_{GSk} - v_{DSk} \geq V_{tk} \\ K_{pk}(v_{GSk} - V_{tk})^2/2 & \text{if } v_{GDk} = v_{GSk} - v_{DSk} < V_{tk} \end{cases} \tag{4.1.83}$$

The process of solving this set of equations to find $\mathbf{v} = [v_1 \ v_2 \ v_3]$ for $v_d = -V_{dm} \sim +V_{dm}$ and plotting $v_o = v_1 - v_2$, i_{D1}, i_{D2} (together with their analytic values computed by Eqs. (4.1.79) and (4.1.83)) versus v_d has been cast into the following MATLAB function '`FET_differential()`'. We can run

```
>>Kp=5e-3; Vt=1; lambda=0; ISS=5e-3; RD=1e3; VDD=5; Vdm=1.5;
  dvd=Vdm/1000;
  vds=[-Vdm:dvd:Vdm];
  FET_differential(Kp,Vt,lambda,ISS,RD,VDD,vds);
```

to get the graphs of i_{D1}, i_{D2}, and v_o as depicted in Figure 4.21(b) and (c).

```
function [vo1s,vo2s,iD1s,iD2s]= ...
        FET_differential(Kp,Vt,lambda,ISS,RD,VDD,vds)
% Analyze an FET differential (source-coupled) pair (to find
% the outputs vo1/vo2 to a range of differential input vd=-Vdm~Vdm
% and plot its VTC in case of no output argument)
% which consists of 2 FETs, 2 resistors RD1=RD2=RD, and an I-src ISS.
% vds : Range of differential input vd
% Copyleft: Won Y. Yang, wyyang53@hanmail.net, CAU for academic use only
if nargin<6,
  Vdm=1.5*sqrt(ISS/Kp); dvd=Vdm/1000; vds=[-Vdm:dvd:Vdm]; end
options=optimset('Display','off','Diagnostics','off');
```

```
iD=@(vGS,vDS)Kp*(1+lambda*vDS).*((vGS-Vt).^2/2.*(vGS>=Vt).*(vGS-vDS<Vt)
 +((vGS-Vt).*vDS-vDS.^2/2).*(vGS>=Vt).*(vGS-vDS>=Vt)); % Eq.(4.1.83)
id2 = vds.*sqrt(max(Kp*(ISS-Kp*vds.^2/4),0)); % Eq.(4.1.76): iD1-iD2
idms=find(id2==min(id2)); idm1=idms(1); id2(1:idm1)=min(id2);
idms=find(id2==max(id2)); idm2=idms(end); id2(idm2:end)=max(id2);
iD1s_a=(ISS+id2)/2; iD2s_a=(ISS-id2)/2; % Eq.(4.1.76a,b)
v1_a=VDD-RD*iD1s_a; v2_a=VDD-RD*iD2s_a; vo_a=v1_a-v2_a;
for n=1:length(vds)
  vd = vds(n);
  eq=@(v) [VDD-v(1)-RD*iD(vd-v(3),v(1)-v(3))
           VDD-v(2)-RD*iD(-v(3),v(2)-v(3)) % Eq.(4.1.82)
           (iD(vd-v(3),v(1)-v(3))+iD(-v(3),v(2)-v(3))-ISS)*1e3];
  if ~exist('v'), v0=[v1_a(1) v2_a(1) v3_a(1)]; % Initial guess
   else v0=v;
  end
  %Here, the first values of analytical solutions are used as the initial
  % guess for the numerical solution of Eq. (4.1.80) with vd(1).
  [v,fe]=fsolve(eq,v0,options);
  vo1s(n)=v(1); vo2s(n)=v(2); v3s(n)=v(3);
  vos(n)=v(1)-v(2); % Eq.(4.1.77)
  iD1s(n)=iD(vd-v(3),v(1)-v(3));
  iD2s(n)=iD(-v(3),v(2)-v(3));
end
```

4.1.7 CMOS Logic Circuits

Figure 4.22(a)-(c) shows a two-input CMOS NOR gate, a two-input CMOS NAND gate, and a CMOS (bidirectional) transmission gate together with their truth tables where the substrates of the PMOS(Mp) and NMOS(Mn) are connected to the most positive/negative potentials (that are VDD/GND), respectively. The transmission gate makes possible the tristate output or wired-or connection in CMOS circuits and can be used as building blocks for logic circuitry such as a D flip-flop. Figure 4.23 shows a 256 × 8 bit (256 bytes) NMOS ROM chip where each one of the 256 bytes stored in the memory can be selected by the word line, which is set to High depending on the 8-bit address code input to the 8-input 128-output decoder.

Here are some comparisons of CMOS and NMOS gates:

- CMOS gates have extremely high input impedance, but they are susceptible to damage.
- CMOS gates consume very little power.
- CMOS gates can achieve noise immunity amounting to 45% of the full logic swing.
- Since a CMOS gate requires more transistors than an NMOS gate having the same logic function, not only its size but also its capacitance is larger.

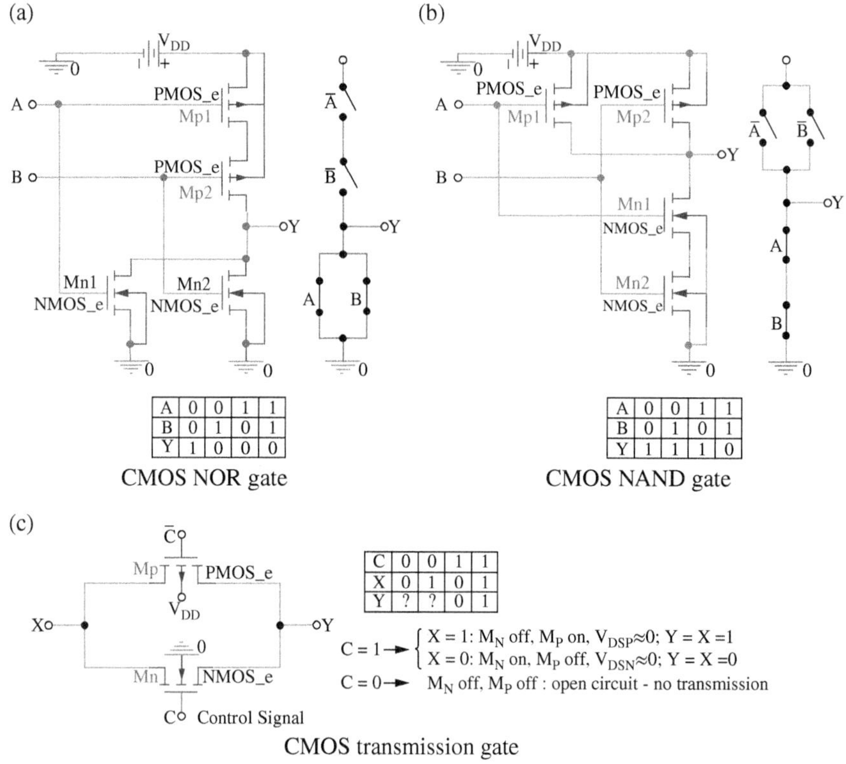

Figure 4.22 Various CMOS logic circuits.

4.2 FET Amplifer

If an FET is used for the purpose of small-signal amplification, it is supposed to be operating in the saturation region (for the range of input signal $v_{GS} = V_{GS,Q} + v_{gs}$ around $V_{GS,Q}$ at the bias point Q) where its drain current is expressed by

$$i_D \underset{\lambda=0}{\overset{(4.1.13b)}{=}} \frac{1}{2}K_p(v_{GS} - V_t)^2 = \frac{1}{2}K_p\left(V_{GS,Q} + v_{gs} - V_t\right)^2 \tag{4.2.1}$$

and additionally, its small-signal component i_d can be approximated linearly in terms of v_{gs}:

$$i_D = I_{D,Q} + i_d \overset{(4.2.1)}{=} \frac{1}{2}K_p(V_{GS,Q} - V_t)^2 + K_p(V_{GS,Q} - V_t)v_{gs} + \frac{1}{2}K_p v_{gs}^2 \tag{4.2.2}$$

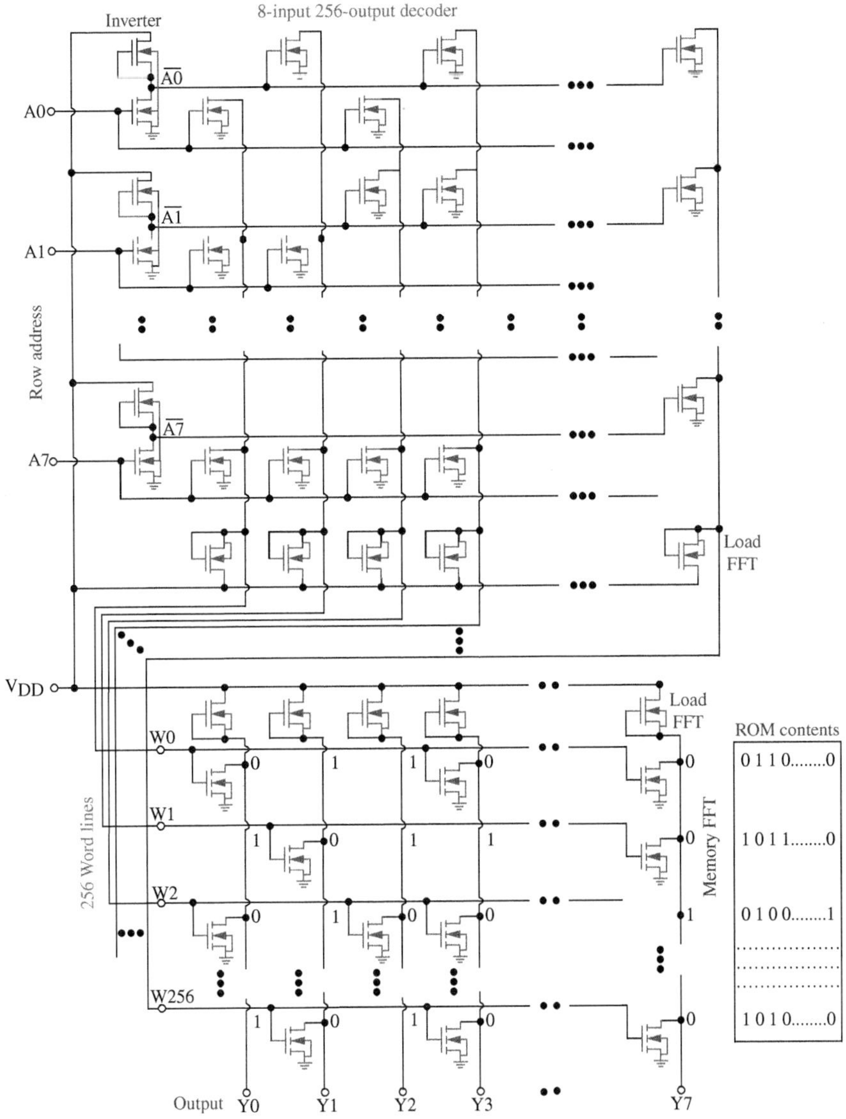

Figure 4.23 256 × 8 bit (256 bytes) NMOS ROM chip.

In order for this linear approximation to be valid, the last term of degree 2 must be negligibly small:

$$\frac{1}{2}K_{\rm p}v_{\rm gs}^2 << K_{\rm p}\left(V_{{\rm GS},Q}-V_{\rm t}\right)v_{\rm gs}$$

$$;\ v_{\rm gs} << 2\left(V_{{\rm GS},Q}-V_{\rm t}\right),\ \text{say, } v_{\rm gs} \le 0.2\left(V_{{\rm GS},Q}-V_{\rm t}\right) \tag{4.2.3}$$

so that

$$i_{\rm d} \approx K_{\rm p}\left(V_{{\rm GS},Q}-V_{\rm t}\right)v_{\rm gs} \tag{4.2.4}$$

This section deals with three configurations of FET amplifier, i.e. the CS (common-source) amplifier, the CD (common-drain) amplifier (called a source follower), and the CG (common-gate) amplifier.

4.2.1 Common-Source (CS) Amplifier

Figure 4.24 shows a CS FET amplifier and its low-frequency AC equivalent where the FET has been replaced by the equivalent in Figure 4.3(b). Let us find the input resistance, voltage gain, and output resistance. The formulas have been implemented in the MATLAB function '`FET_CS_analysis()`'.

1) **Input Resistance $R_{\rm i}$**

   Since the gate current of the FET is almost zero, the equivalent resistance seen from the input side (see Figure 4.24(b1)) is

$$R_{\rm i} = R_1 || R_2 \tag{4.2.5}$$

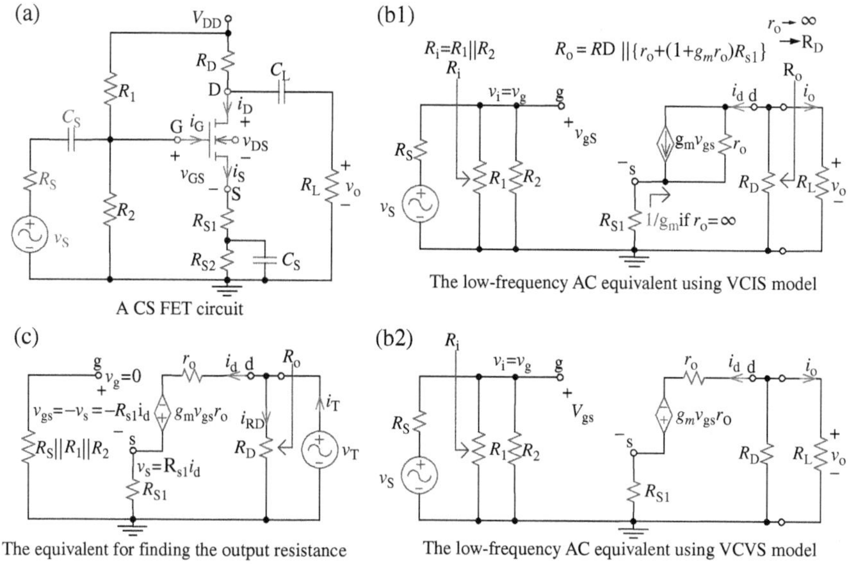

Figure 4.24 A CS (common-source) FET circuit and its low-frequency AC equivalent.

2) **Voltage Gains G_v and A_v**
To find the voltage gain, we write the node equation for the circuit of Figure 4.24(b1) as

$$\begin{bmatrix} 1/R_{S1}+1/r_o & -1/r_o \\ -1/r_o & 1/r_o+1/R_D+1/R_L \end{bmatrix}\begin{bmatrix} v_s \\ v_d=v_o \end{bmatrix}=\begin{bmatrix} g_m v_{gs} \\ -g_m v_{gs} \end{bmatrix}=\begin{bmatrix} g_m(v_g-v_s) \\ -g_m(v_g-v_s) \end{bmatrix}$$

$$;\begin{bmatrix} v_s \\ v_o \end{bmatrix}=\frac{g_m v_g}{(1/R_{S1})(1/r_o)+(1/R_{S1}+1/r_o+g_m)/(R_D\|R_L)}\begin{bmatrix} 1/r_o+1/(R_D\|R_L) & 1/r_o \\ 1/r_o+g_m & 1/R_{S1}+1/r_o+g_m \end{bmatrix}\begin{bmatrix} 1 \\ -1 \end{bmatrix}$$

$$;v_o=\frac{-g_m/R_{S1}}{(1/R_{S1})(1/r_o)+(1/R_{S1}+1/r_o+g_m)/(R_D\|R_L)}v_g$$

$$=\frac{-g_m r_o(R_D\|R_L)}{(R_D\|R_L)+r_o+(1+g_m r_o)R_{S1}}v_g$$

Noting that $v_i = v_g = R_i\, v_s/(R_s + R_i)$, we can write the voltage gains, i.e. the ratios of the output voltage v_o to the source voltage v_s and the input voltage v_i as

$$G_v=\frac{v_o}{v_s}=\frac{v_i}{v_s}\frac{v_o}{v_i}=\frac{R_i}{R_s+R_i}A_v \text{ with } R_i \text{ given by Eq. (4.2.5)} \tag{4.2.6a}$$

$$\text{and } A_v=\frac{v_o}{v_i}=\frac{-g_m r_o(R_D\|R_L)}{(R_D\|R_L)+r_o+(1+g_m r_o)R_{S1}}\overset{\text{if } r_o \gg 1/g_m R_{S1}}{\approx}\frac{-g_m(R_D\|R_L)}{1+g_m R_{S1}} \tag{4.2.6b}$$

3) **Output Resistance R_o**
To find the (Thevenin) equivalent resistance seen from the load side, we need to remove the (independent) voltage source v_s by short-circuiting it as depicted in Figure 4.24(c). Then, with the test voltage v_T applied across R_D, we write the drain current i_d as

$$v_T=(r_o+R_{S1})i_d-g_m r_o v_{gs}$$

$$=(r_o+R_{S1})i_d-g_m r_o(v_g-v_s)\overset{v_g=0}{=}(r_o+R_{S1})i_d+g_m r_o R_{S1} i_d$$

$$;i_d=\frac{1}{r_o+(1+g_m r_o)R_{S1}}v_T$$

Since the test current i_T flowing out of v_T is the sum of i_d and $i_{R_D}=v_T/R_D$:

$$i_T=i_{R_D}+i_d=\frac{1}{R_D}v_T+\frac{1}{r_o+(1+g_m r_o)R_{S1}}v_T$$

Therefore, the output resistance turns out to be

$$R_o = \frac{v_T}{i_T} = \frac{1}{1/R_D + 1/\{r_o + (1 + g_m r_o)R_{S1}\}} = R_D \| \{r_o + (1 + g_m r_o)R_{S1}\} \quad (4.2.7)$$

```
function [VGQ,VSQ,VDQ,IDQ,Av,Ri,Ro,gm,ro,Vom,vsmax]= ...
         FET_CS_analysis(VDD,Rs,R1,R2,RD,RS,RL,Kp,Vt,lambda,Vsm)
% Copyleft: Won Y. Yang, wyyang53@hanmail.net, CAU for academic use only
if nargin<11, Vsm=0.01; end
if nargin<10, lambda=0; end % CLM parameter
if length(RS)==1, RS1=0; RS2=RS; else RS1=RS(1); RS2=RS(2); end
[VGQ,VSQ,VDQ,VGSQ,VDSQ,IDQ,mode,gm,ro]= ...
        FET_DC_analysis(VDD,R1,R2,RD,RS,Kp,Vt,lambda);
if R1<=30, Ri=R2; else Ri=parallel_comb([R1 R2]); end % Eq.(4.2.5)
RDL=parallel_comb([RD RL]); mug=gm*ro;
if ro<inf, ro_mug1_RS1=ro+(mug+1)*RS1; Av=-mug*RDL/(RDL+ro_mug1_RS1);
 else ro_mug1_RS1=inf; Av=-gm*RDL;
end
Ro = parallel_comb([ro_mug1_RS1 RD]); % Eq.(4.2.7)
Avi=1/(Rs/Ri+1); Gv=Avi*Av; % Eq.(4.2.6)
fprintf(' Ri=%8.2f[kOhm], Ro=%6.0f[Ohm]\n Gv=Ri/(Rs+Ri)xAv =%5.2f x%8.2f
    =%8.2f with Kp=%5.2f[mA/V^2], gm=%8.3f[mS], ro=%8.3f[kOhm]\n', ...
       Ri/1e3,Ro,Avi,Av,Gv,Kp*1e3,gm*1e3,ro/1e3)
Vom=Gv*Vsm; vsmax=0.2*(1+gm*RS1)*(VGSQ-Vt)/Avi; % Eq.(4.2.9)
if strcmp(mode(1:3),'sat')~=1
   fprintf('\n This AC analysis is invalid since the FET is not
   in the saturation region\n'); return;
end
if abs(Vom/RDL)>=0.99*IDQ
   fprintf(' Possibly crash into cutoff region since |Av*Vsm/RDL(%9.4fmA)|
   >=|IDQ|(%9.4fmA)\n', abs(Av*Vsm/RDL)*1e3,abs(IDQ)*1e3);
end
if abs(VDSQ)-abs(Vom)<abs(VGSQ-Vt) % Eq.(4.1.9) Sat condition not met
   fprintf(' Possibly violate the ohmic region since |VDSQ|-|Av*Vsm|
   (%6.3f)<|VGSQ-Vt|(%6.3f)\n',abs(VDSQ)-abs(Av*Vsm),abs(VGSQ-Vt));
end
```

4) **Maximum Small-Signal Input $v_{s,max}$ for Linear Amplification**

Noting that if r_o is very large in Figure 4.24(b1),

$$v_g = (1 + g_m R_{S1}) v_{gs} \quad (4.2.8)$$

the maximum small-signal input satisfying the condition (4.2.3) for linear amplification can be determined as

$$\begin{aligned} v_{i,max} &= v_{g,max} = (1 + g_m R_{S1}) v_{gs,max} \\ &= 0.2\,(1 + g_m R_{S1})\left(V_{GS,Q} - V_t\right), v_{s,max} = \frac{R_s + R_i}{R_i} v_{i,max} \end{aligned} \quad (4.2.9)$$

Example 4.6   Analysis of Two CS FET Amplifiers

Consider the CS FET amplifiers in Figure 4.25(a1) and (a2) where $V_{DD} = 12$ V, $R_s = 1$ kΩ, $R_1 = 2.2$ MΩ, $R_2 = 1.5$ MΩ, $R_D = 22$ kΩ, $R_{S1} = 6$ kΩ, $R_{S2} = 6$ kΩ, $R_L = 100$ kΩ, $K_p = 0.5$ mA/V^2, $V_t = 1$ V, and $\lambda = 0.01$ V^{-1}.

```
%elec04e06.m
VDD=12; Rs=1e3; R1=22e5; R2=15e5; RD=22e3; RL=1e5;
Kp=0.5e-3; Vt=1; lambda=0.01; RS=[6e3 6e3]; RS2=[0 sum(RS)];
[VGQ,VSQ,VDQ,IDQ,Av1,Ri1,Ro1,gm1,ro1,Vom1,vsmax1] = ...
   FET_CS_analysis(VDD,Rs,R1,R2,RD,RS,RL,Kp,Vt,lambda) % for (a1)
[VGQ2,VSQ2,VDQ2,IDQ2,Av2,Ri2,Ro2,gm2,ro2,Vom2,vsmax2] = ...
   FET_CS_analysis(VDD,Rs,R1,R2,RD,RS2,RL,Kp,Vt,lambda) % for (a2)
gmRS1=[1+gm1*RS(1) 1+gm2*RS2(1)], vsmax=[vsmax1 vsmax2]
```

(a1)

A CS FET amplifier circuit

(a2)

Another CS FET amplifier Circuit with bypassed source

(b)

DC/AC load lines of the two circuits

(c)

PSpice simulation results on v_{o1} and v_{o2}

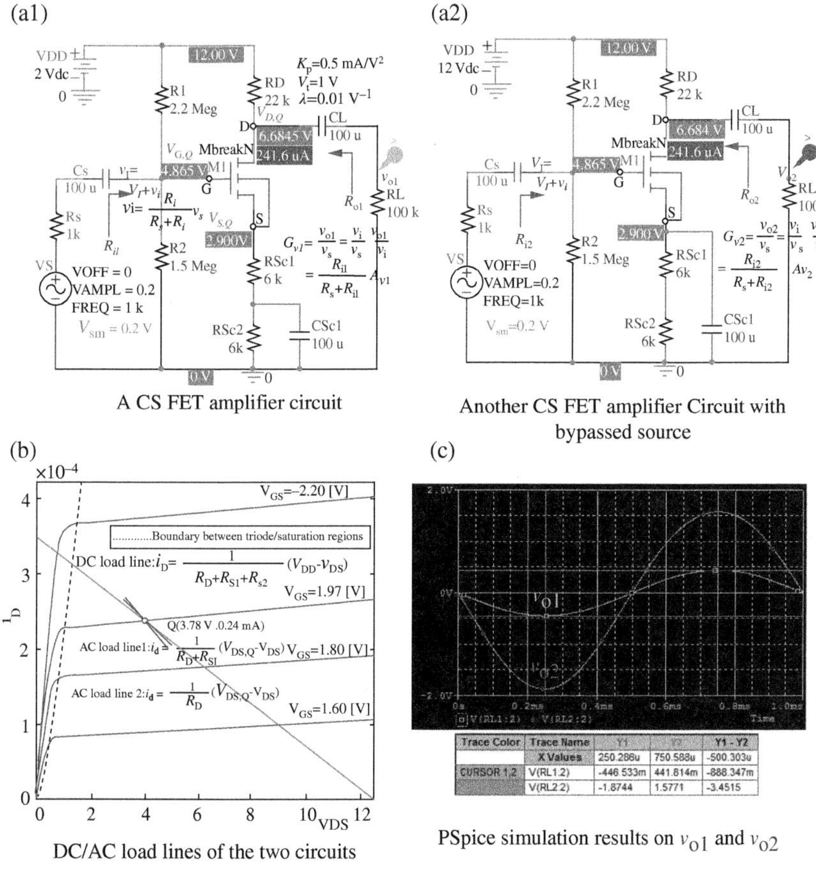

Figure 4.25 Two CS FET amplifiers and their PSpice simulation results ("elec04e06.opj").

To find the values of the amplifier parameters, R_i, R_o, $A_v = v_o/v_i$, and $G_v = v_o/v_s$, and determine the maximum small-signal input $v_{s,max}$ allowing linear amplification for the two circuits, we run the above MATLAB script "elec04e06.m" to get

```
VDD    VGQ   VSQ   VDQ      IDQ
12.00 4.86 2.90 6.68 2.42e-04
 in the saturation mode with VGD,Q= -1.82 [V] (Vt=1.00)
Ri= 891.89 kOhm, Ro= 21723 Ohm % for (a1)
 Gv=Ri/(Rs+Ri)xAv= 1.00 x -2.22 = -2.22 with gm=0.501[mS], ro=429kOhm
Ri= 891.89 kOhm, Ro= 20928 Ohm % for (a2)
 Gv=Ri/(Rs+Ri)xAv= 1.00 x -8.67 = -8.66 with gm=0.501[mS], ro=429kOhm
gmRS1 = [4.0048   1.0000], vsmax = [0.7738   0.1932]
```

These DC + AC analysis results are supported by the PSpice simulation results shown in Figure 4.25(a1), (a2), and (c). As can be guessed from Eqs. (4.2.6b) (A_v inversely proportional to $1 + g_m R_{S1}$) and (4.2.9) ($v_{s,max} \approx v_{i,max}$ proportional to $1 + g_m R_{S1}$), the voltage gain of (a2) is about 4 times that of (a1) while $v_{s,max}$ of (a2) is about 1/4 times that of (a1). Thus, the two amplifiers (a1) and (a2) will not differ much in the sweep range of their output voltage v_o that can be amplified linearly without distortion.

4.2.2 CD Amplifier (Source Follower)

Figure 4.26 shows a CD FET amplifier and its low-frequency AC equivalent where the FET has been replaced by the equivalent in Figure 4.3(b). Let us find the input resistance, voltage gain, and output resistance. The formulas have been implemented in the MATLAB function '`FET_CD_analysis()`'.

1) **Input Resistance R_i**

   Since the gate current of the FET is almost zero, the input resistance is

$$R_i = R_1 \| R_2 \tag{4.2.10}$$

2) **Voltage Gains G_v and A_v**

   To find the voltage gain $A_v = v_o / v_i$, we write the node equation for the circuit of Figure 4.26(b1) as

$$\left(\frac{1}{r_o} + \frac{1}{R_S} + \frac{1}{R_L}\right) v_o = g_m v_{gs} = g_m (v_g - v_s) \underset{v_s = v_o}{=} g_m (v_i - v_s)$$

$$; \left(\frac{1}{r_o} + \frac{1}{R_S} + \frac{1}{R_L} + g_m\right) v_o = g_m v_i \, ; \, \left(\frac{1}{r_o \| R_S \| R_L} + g_m\right) v_o = g_m v_i$$

   Thus, we have the voltage gains as

$$G_v = \frac{v_o}{v_s} = \frac{v_i}{v_s}\frac{v_o}{v_i} = \frac{R_i}{R_s + R_i} A_v \text{ with } R_i \text{ given by Eq.(4.2.10)} \tag{4.2.11a}$$

$$\text{and } A_v = \frac{v_o}{v_i} = \frac{g_m}{g_m + 1/(r_o\|R_S\|R_L)} = \frac{g_m(r_o\|R_S\|R_L)}{1+g_m(r_o\|R_S\|R_L)} \underset{R_s << Ri}{\overset{g_m(r_o\|R_S\|R_L)\gg 1}{\longrightarrow}} = 1 \tag{4.2.11b}$$

```
function [VGQ,VSQ,VDQ,IDQ,Av,Ri,Ro,gm,ro,Vom,vsmax] =
        FET_CD_analysis(VDD,Rs,R1,R2,RD,RS,RL,Kp,Vt,lambda,Vsm)
% If D with RD~=0 is AC grounded, RD should be given as [0 RD].
% Copyleft: Won Y. Yang, wyyang53@hanmail.net, CAU for academic use only
RD1=RD(1); RS1=RS(1);
[VGQ,VSQ,VDQ,VGSQ,VDSQ,IDQ,mode,gm,ro]= ...
         FET_DC_analysis(VDD,R1,R2,RD,RS,Kp,Vt,lambda);
if R1<=30, Ri=R2;
 else Ri=parallel_comb([R1 R2]); % Eq.(4.2.10)
end
gm1=gm/(1+RD1/ro); ro1=ro+RD1; % Against the case where RD is nonzero.
RSL=parallel_comb([RS1 RL]); RdsSL=parallel_comb([ro1 RSL]);
Ro=parallel_comb([1/gm1 RS1 ro1]); % Eq.(4.2.12)
Av=gm1*RdsSL/(1+gm1*RdsSL); Avi=1/(Rs/Ri+1);
Gv=Avi*Av; % Eq.(4.2.11)
Vom=Gv*Vsm;
vsmax=0.2*(1+gm1*RdsSL)*(VGSQ-Vt)/Avi; % Eq.(4.2.13)
fprintf(' Ri=%8.2f kOhm, Ro=%6.0f Ohm\n Gv=Ri/(Rs+Ri)xAv =%7.4f x%8.4f
         =%8.4f',Ri/1e3,Ro,Avi,Av,Gv)
if strcmp(mode(1:3),'sat')~=1
 fprintf('\nThis AC analysis is invalid since the FET isn't saturated\n');
 return;
end
if abs(Av*Vsm/RSL)>=IDQ,
fprintf('\n Possibly crash into cutoff\n'); end
if abs(VDSQ)-Av*Vsm<abs(VGSQ-Vt) % Eq.(4.1.9) is not met
 fprintf('\n Possibly violate the ohmic region\n');
end
```

This implies that if $R_s \ll R_i$ and $g_m\,(r_o\|R_s\|R_L) \gg 1$, the output voltage is almost equal to the input and source voltages and that is why the CD amplifier is called a *voltage follower.*

3) **Output Resistance R_o**

To find the output resistance, we remove the (independent) voltage source, v_i, by short-circuiting it as depicted in Figure 4.26(c). Then, with the test current i_T applied to the output port, we can get

$$v_T = (r_o\|R_S)(g_m v_{gs} + i_T) = (r_o\|R_S)\{g_m(0 - v_T) + i_T\}$$

$$; \quad R_o = \frac{v_T}{i_T} = \frac{r_o\|R_S}{1+g_m(r_o\|R_S)} = \frac{1}{g_m}\|r_o\|R_S \tag{4.2.12}$$

4) **Maximum Small-Signal Input $v_{s,max}$ for Linear Amplification**

Noting that if r_o is very large in Figure 4.26(b1),

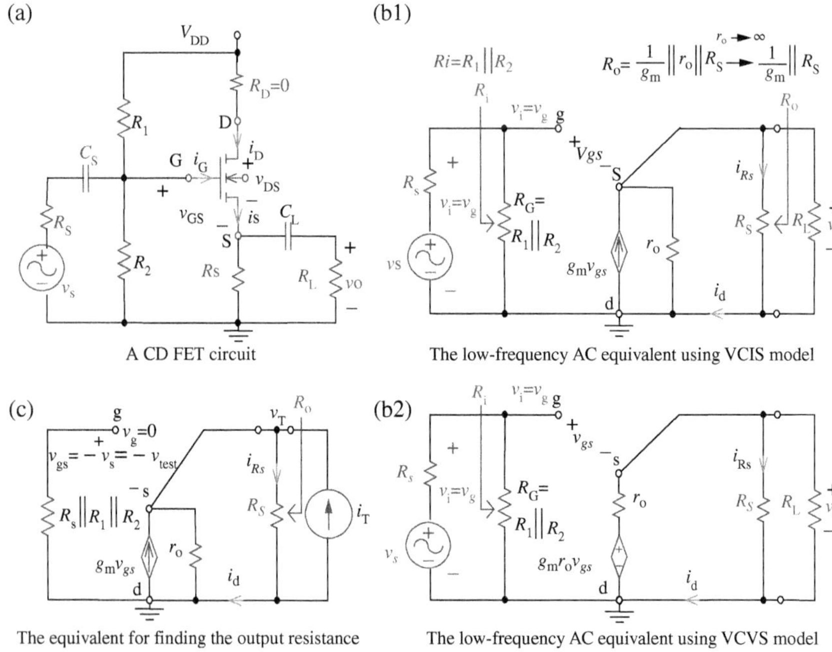

Figure 4.26 A CD (common-drain) FET circuit and its low-frequency AC equivalent.

$$v_g = \{1 + g_m(R_S \| R_L)\} v_{gs} \tag{4.2.13}$$

we can write the maximum small-signal input satisfying the condition (4.2.3) for linear amplification as

$$v_{i,\max} = v_{g,\max} = \{1 + g_m(R_S \| R_L)\} v_{gs,\max} \overset{(4.2.3)}{=} 0.2\{1 + g_m(R_S \| R_L)\} \left(V_{GS,Q} - V_t\right)$$

$$v_{s,\max} = \frac{R_s + R_i}{R_i} v_{i,\max} \tag{4.2.14}$$

Example 4.7 Analysis of a CD FET Amplifier

Consider the CD FET amplifier in Figure 4.27(a) where $V_{DD} = 12\text{ V}$, $R_s = 1\text{ k}\Omega$, $R_1 = 2.2\text{ M}\Omega$, $R_2 = 1.5\text{ M}\Omega$, $R_D = 0\text{ k}\Omega$, $R_S = 12\text{ k}\Omega$, $R_L = 100\text{ k}\Omega$, $K_p = 0.5\text{ mA/V}^2$, $V_t = 1\text{ V}$, and $\lambda = 0.01\text{ V}^{-1}$.

```
%elec04e07.m
VDD=12; Rs=1e3; R1=22e5; R2=15e5; RD=0; RS=12e3;
RL=1e5; Vsm=0.2;
Kp=0.5e-3; Vt=1; lambda=0.01;
[VGQ,VSQ,VDQ,IDQ,Av1,Ri,Ro,gm,ro,Vom,vsmax] = ...
   FET_CD_analysis(VDD,Rs,R1,R2,RD,RS,RL,Kp,Vt,lambda);
fprintf('vsmax =%10.3e[V] \n',vsmax);
```

To find the values of the amplifier parameters R_i, R_o, $A_v = v_o/v_i$, and $G_v = v_o/v_s$, and determine the maximum small-signal input $v_{s,max}$ allowing linear amplification for the circuit, we run the above MATLAB script "clcc04c07.m" to get

```
VDD  VGQ  VSQ  VDQ    IDQ
12.00 4.86 2.92 12.00 2.43e-04
 in the saturation mode with VGD,Q= -7.14 [V] (Vt=1.00)
          gm= 0.515 [mS], ro= 448.241 [kOhm]
 Ri= 891.89 kOhm, Ro= 1665 Ohm
 Gv=Ri/(Rs+Ri)xAv = 0.9989 x 0.8435 = 0.8426
 vsmax = 1.209e+00 [V]
```

These DC + AC analysis results are supported by the PSpice simulation results shown in Figure 4.27(c1) and (c2). As can be guessed from Eqs. (4.2.11) and (4.2.12), the voltage gain, less than 1, may increase toward 1 and the output resistance R_o, smaller than $1/g_m$, may become smaller as the transconductance g_m (Eq. (4.1.4b)) (proportional to $I_{D,Q}$) increases.

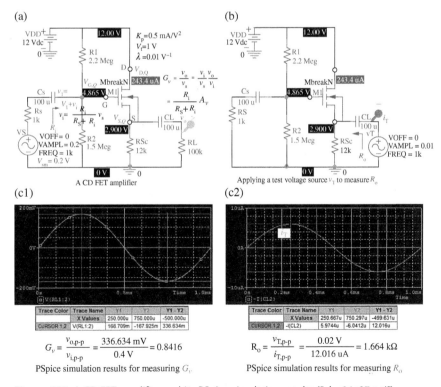

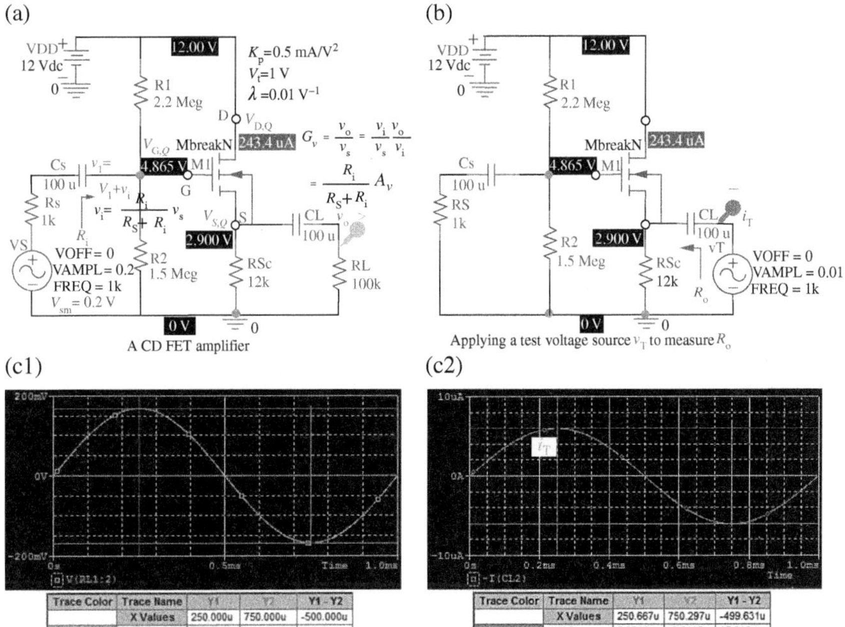

Trace Color	Trace Name	Y1	Y2	Y1 - Y2
	X Values	250.000u	750.000u	-500.000u
CURSOR 1,2	V(RL1:2)	168.709m	-167.925m	336.634m

$$G_v = \frac{v_{o,p\text{-}p}}{v_{i,p\text{-}p}} = \frac{336.634\ \text{mV}}{0.4\ \text{V}} = 0.8416$$

PSpice simulation results for measuring G_v

Trace Color	Trace Name	Y1	Y2	Y1 - Y2
	X Values	250.667u	750.297u	-499.631u
CURSOR 1,2	-I(CL2)	5.9744u	-6.0412u	12.016u

$$R_o = \frac{v_{T,p\text{-}p}}{i_{T,p\text{-}p}} = \frac{0.02\ \text{V}}{12.016\ \text{uA}} = 1.664\ \text{k}\Omega$$

PSpice simulation results for measuring R_o

Figure 4.27 A CD FET amplifier and its PSpice simulation results ("elec04e07.opj").

4.2.3 Common-Gate (CG) Amplifier

Figure 4.28 shows a CG amplifier and its low-frequency AC equivalent where the FET has been replaced by the equivalent in Figure 4.3(b). Let us find the input resistance, voltage gain, and output resistance. The formulas have been implemented in the MATLAB function '`FET_CG_analysis()`'.

1) **Input Resistance** R_i

To find the input resistance, we apply a test voltage source v_s between node s and GND (as depicted in Figure 4.28(c)) and then write the relationship of v_s and i_s as

$$v_s = g_m r_o v_{gs} + \{r_o + (R_D||R_L)\}i_s = -g_m r_o v_s + \{r_o + (R_D||R_L)\}i_s$$
$$;(1 + g_m r_o)v_s = \{r_o + (R_D||R_L)\}i_s$$
$$;v_s = R_{sg} i_s \text{ with } R_{sg} = \frac{r_o + (R_D||R_L)}{1 + g_m r_o} \overset{\text{if } r_o \gg \max\{1/g_m, (R_D||R_L)\}}{\longrightarrow} \frac{1}{g_m}$$

This implies that the equivalent resistance of the FET between node s and GND is R_{sg}. Thus, the input resistance is the parallel combination of R_S and R_{sg}:

$$R_i = R_S||R_{sg} = R_S||\frac{r_o + (R_D||R_L)}{1 + g_m r_o} \overset{\text{if } r_o \gg \max\{1/g_m, (R_D||R_L)\}}{\longrightarrow} R_S||\frac{1}{g_m} \quad (4.2.15)$$

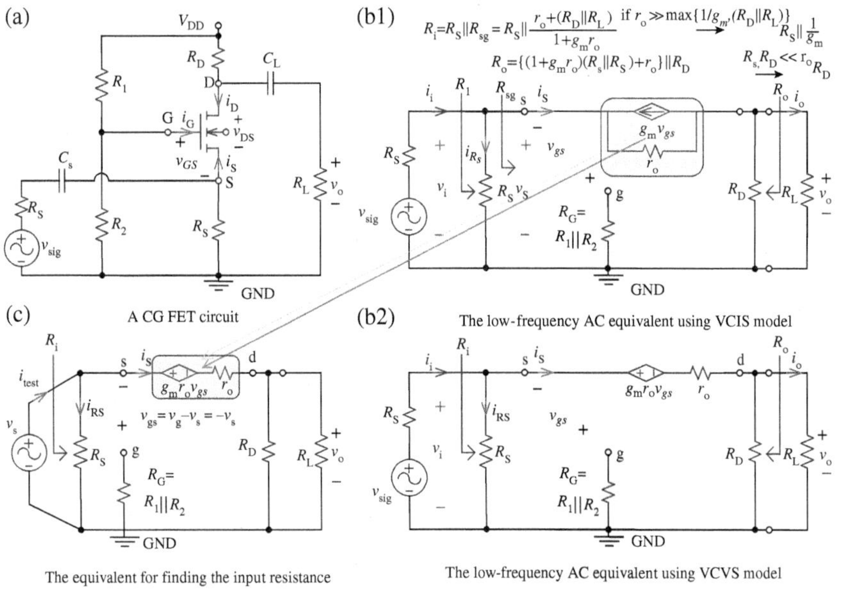

Figure 4.28 A CG (common-gate) FET circuit and its low-frequency AC equivalent.

2) **Voltage Gains G_v and A_v**
Since (during the process of computing R_i) the current through $R_D||R_L$ caused by v_s was found to be

$$i_s = \frac{1 + g_m r_o}{r_o + (R_D||R_L)} v_s$$

the output voltage (across $R_D||R_L$) can be expressed as

$$v_o = (R_D||R_L) i_s = \frac{(1 + g_m r_o)(R_D||R_L)}{r_o + (R_D||R_L)} v_s$$

Thus, taking account of $v_i = R_i\, v_{sig}/(R_s + R_i)$, we can write the voltage gains, i.e. the ratios of the output voltage v_o to the source voltage v_{sig} and the input voltage v_i as

$$G_v = \frac{v_o}{v_{sig}} = \frac{v_i}{v_{sig}} \frac{v_o}{v_i} = \frac{R_i}{R_s + R_i} A_v \text{ with } R_i \text{ given by Eq.(4.2.15)} \quad (4.2.16a)$$

$$\text{and } A_v = \frac{v_o}{v_i} = \frac{(1 + g_m r_o)(R_D||R_L)}{r_o + (R_D||R_L)} \overset{\text{if } r_o \gg \max\{1/g_m, (R_D||R_L)\}}{\approx} g_m (R_D||R_L) \quad (4.2.16b)$$

3) **Output Resistance R_o**
To find the output resistance, we remove the (independent) voltage source v_i by short-circuiting it. Then, we can write the output resistance as

$$R_o = R_D||\{(1 + g_m r_o)(R_s||R_S) + r_o\} \overset{R_s \ll R_S}{\rightarrow} R_D||\{(1 + g_m r_o)R_s + r_o\} \overset{R_s, R_D \ll r_o}{\rightarrow} R_D \quad (4.2.17)$$

4) **Maximum Small-Signal Input $v_{s,max}$ for Linear Amplification**
Since $|v_i| = |v_g| = |v_{gs}|$, the maximum small-signal input satisfying the condition (4.2.3) for linear amplification can be determined as

$$v_{i,max} = v_{g,max} = v_{gs,max} = 0.2\left(V_{GS,Q} - V_t\right), \ v_{s,max} = \frac{R_s + R_i}{R_i} v_{i,max} \quad (4.2.18)$$

```
function [VGQ,VSQ,VDQ,IDQ,Av,Ri,Ro,gm,ro,Vom,vsmax] = ...
        FET_CG_analysis(VDD,Rs,R1,R2,RD,RS,RL,Kp,Vt,lambda,Vsm)
% Copyleft: Won Y. Yang, wyyang53@hanmail.net, CAU for academic use only
[VGQ,VSQ,VDQ,VGSQ,VDSQ,IDQ,mode,gm,ro] = ...
        FET_DC_analysis(VDD,R1,R2,RD,RS,Kp,Vt,lambda);
RS1=RS(1); RDL=parallel_comb([RD RL]); Rsg=(ro+RDL)/(1+gm*ro);
if R1<=30, Ri=R2; else Ri=parallel_comb([RS1 Rsg]); end % Eq.(4.2.15)
Ro=parallel_comb([(1+gm*ro)*parallel_comb([Rs RS1])+ro RD]); % Eq.(4.2.17)
Av=RDL/Rsg; Avi=1/(Rs/Ri+1); % Eq.(4.2.16b)
Gv=Avi*Av; % Eq.(4.2.16a)
Vom=Gv*Vsm; vsmax=0.2*(VGSQ-Vt)/Avi; % Eq.(4.2.18)
fprintf(' Ri=%8.2f[kOhm], Ro=%6.0f[Ohm]\n Gv=Ri/(Rs+Ri)xAv =%5.2f x
  %8.2f =%8.2f\n', Ri/1e3,Ro,Avi,Av,Gv,Kp*1e3,gm*1e3,ro/1e3)
if strcmp(mode(1:3),'sat')~=1
 fprintf('/nThis AC analysis is invalid since the FET isn't saturated\n');
 return;
end
if abs(VDSQ)-Av*Vsm<abs(VGSQ-Vt) % Eq.(4.1.9) is not met
 fprintf('\n Possibly violate the ohmic region\n');
end
```

Example 4.8 Analysis of a CG Amplifier

Consider the CG FET amplifier in Figure 4.29(a) where $V_{DD} = 12$ V, $R_s = 2$ kΩ, $R_1 = 2.2$ MΩ, $R_2 = 1.5$ MΩ, $R_D = 22$ kΩ, $R_S = 12$ kΩ, $R_L = 100$ kΩ, $K_p = 0.5$ mA/V^2, $V_t = 1$ V, and $\lambda = 0.01$ V^{-1}.

```
%elec04e08.m
VDD=12; Rs=2e3; R1=22e5; R2=15e5; RD=22e3; RS=12e3; RL=1e5;
Vsm=5e-3;
Kp=0.5e-3; Vt=1; lambda=0.01;
[VGQ,VSQ,VDQ,IDQ,Av1,Ri,Ro,gm,ro,Vom,vsmax] = ...
  FET_CG_analysis(VDD,Rs,R1,R2,RD,RS,RL,Kp,Vt,lambda)
fprintf('vsmax =%10.4fmV\n',vsmax*1e3);
```

To find the values of the amplifier parameters R_i, R_o, $A_v = v_o/v_i$, and $G_v = v_o/v_s$, and determine the maximum small-signal input $v_{s,max}$ allowing linear amplification for the circuit, we run the above MATLAB script "elec04e08.m" to get

```
 VDD   VGQ  VSQ  VDQ    IDQ
12.00 4.86 2.90 6.68 2.42e-04
 in the saturation mode with VGD,Q= -1.82[V] (Vt=1.00)
          gm= 0.501[mS], ro= 429.481[kOhm]
```

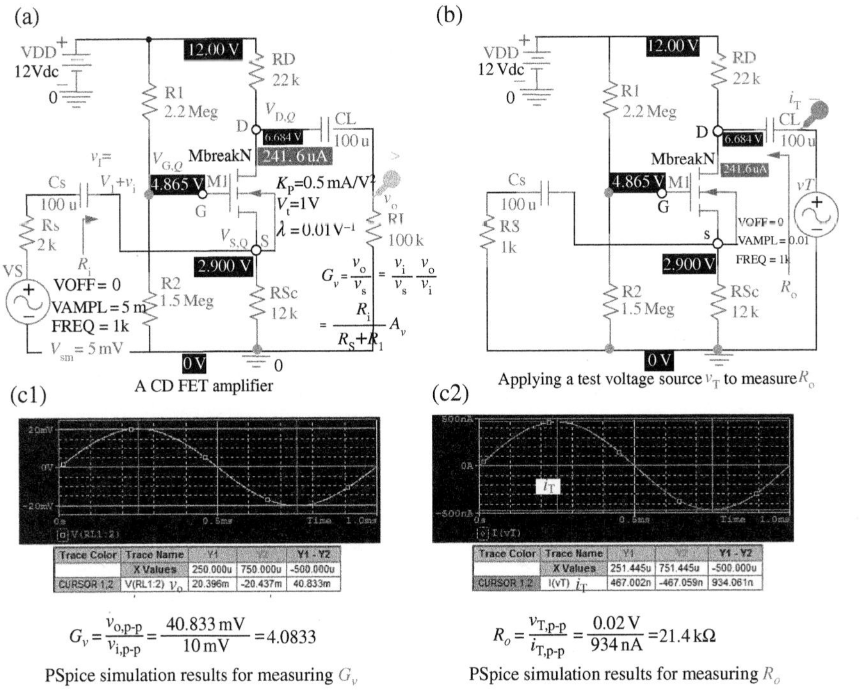

Figure 4.29 A CG amplifier and its PSpice simulation results ("elec04e08.opj").

```
Ri= 1.766[kOhm], Ro= 21411[Ohm]
Gv=Ri/(Rs+Ri)xAvoxRL/(Ro+RL) =0.4690 x ( 10.57x0.8236= 8.707) = 4.083
vsmax = 810.2108mV
```

These DC + AC analysis results are supported by the PSpice simulation results shown in Figure 4.29(a), (c1), and (c2). As can be guessed from Eq. (4.2.15), the input resistance R_i is small so that the overall voltage gain $G_v = v_o/v_s$ can be substantially lower than the terminal voltage gain $A_v = v_o/v_i$.

4.2.4 Common-Source (CS) Amplifier with FET Load

4.2.4.1 CS Amplifier with an Enhancement FET Load

Figure 4.30(a) shows a CS amplifier with an enhancement NMOS driver M_D and an enhancement NMOS load M_L, which is diode connected with its gate and drain short-circuited so that it can act like a nonlinear resistor. The drain (output) characteristic curves of M_D (for several values of v_{GS}) are plotted as solid

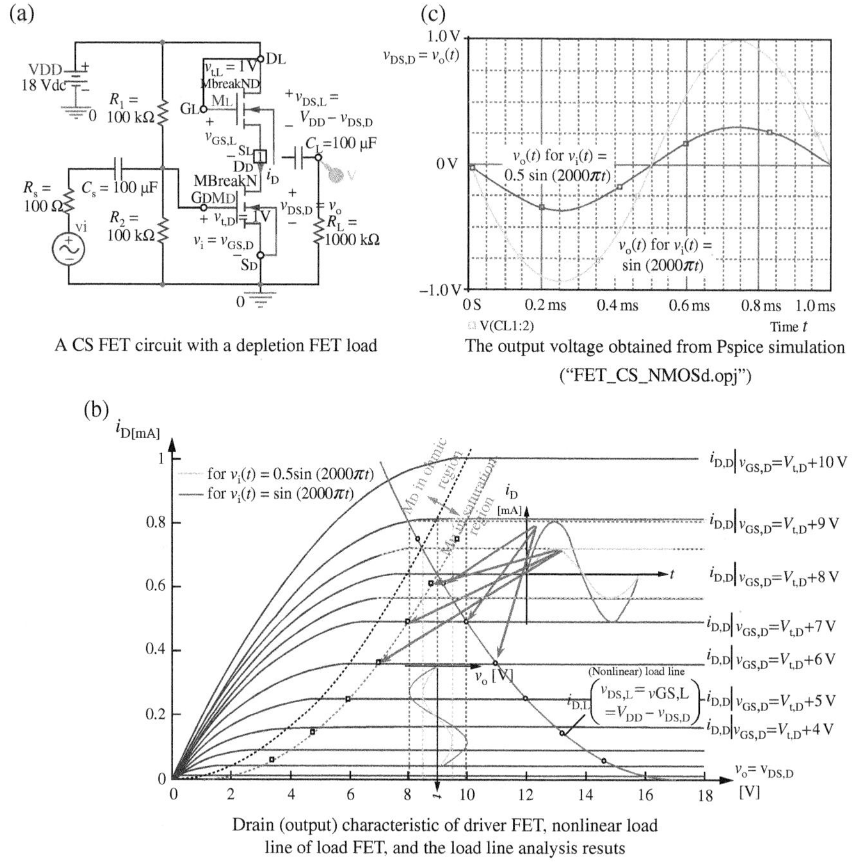

Figure 4.30 A CS amplifier with an enhancement FET load, its load line analysis, and PSpice simulation results.

lines while the $i_{D,L}-v_{DS,L}$ and $i_{D,L}-v_{DS,D}=V_{DD}-v_{DS,L}$ relationships of M_L (with $v_{GS,L} = v_{DS,L}$) are plotted as the lines of square/circle symbols, respectively, in Figure 4.30(b) where the conduction parameter K_p, threshold voltage V_t, and CLM parameter λ of M_D and M_L are

$$K_{p,D} = 20\,\mu A/V^2, K_{p,L} = 20\,\mu A/V^2, V_{t,D} = 1\,V, V_{t,L} = 1\,V, \lambda = 10^{-4}\,V^{-1} \quad (4.2.19)$$

From the locus of dynamic operating points obtained from the intersections of the (pink solid) load line and the characteristic curves of M_D (for several values of $v_i = v_{GS,D}$), the graphs of $i_D(t) = i_{D,L}(t) = i_{D,D}(t)$ and $v_o(t) = v_{DS,D}(t)$ for the sinusoidal input $v_i(t) = V_{sm}$ sin $(2000\pi t)$ (with $V_{sm} = 0.5$ or 1) applied to the

G-S terminals of M_D have been plotted as the green and red lines in Figure 4.30(b).

Note that the intersection of the load line of M_L and the characteristic curve of M_D for a certain value of the input $v_{GS,D} = v_i$ ($> V_{t,D}$) can analytically be obtained as follows:

```
function [VGQ,VSQ,VDQ,IDQ,vo,iD,gm,ro]=
                FET_CS_NMOSe(VDD,Rs,R1,R2,RL,Kp,Vt,lambda,vi)
% Copyleft: Won Y. Yang, wyyang53@hanmail.net, CAU for academic use only
KpD=Kp(1); KpL=Kp(end); VtD=Vt(1); VtL=Vt(end);
if lambda(1)>1, lambda=1./lambda; end
lambdaD=lambda(1); lambdaL=lambda(end);
if nargin<9, vi=0; else vi=[vi 0]; end % 0 for finding the Q point
Ri=R1*R2/(R1+R2); VGQ=VDD*R2/(R1+R2); VSQ=0; VGSQ=VGQ-VSQ;
B=2*(VtL-VDD);
for n=1:length(vi)
  vGSn = VGQ + vi(n)/(Rs/Ri+1);
  C=(VDD-VtL)^2-KpD/KpL*(vGSn-VtD)^2; % Eq.(4.2.20)
  D=B^2-4*C; % Discriminant of Eq.(4.2.20)
  if D<=0 % vGSn<VtD
   fprintf('Driver FET is OFF for vGS(%5.2f)<=VtD(%5.2f)\n', vGSn,VtD);
   iD(n)=0; vo(n)=VDD; continue;
  end
  von=(-B-sqrt(D))/2; iDn=KpD/2*(vGSn-VtD)^2;
  if von<vGSn-VtD % vGD,D=vGS,D-vDS,D=vGSn-von>Vt ?
     A=1+KpD/KpL; B1=B+2*KpD/KpL*(VtD-vGSn); C=(VDD-VtL)^2; % Eq.(4.2.21)
     von=(-B1-sqrt(B1^2-4*A*C))/2/A; iDn=KpL/2*(VDD-von-VtL)^2;
     fprintf('Driver FET is ohmic with vGD(%5.2f)>Vt(%5.2f) ...
          for vi(%4d)=%5.2f\n', vGSn-von,VtD,n,vi(n));
  end
  vo(n)=von; iD(n)=iDn;
end
VDSQ=von; VDQ=VDSQ+VSQ; IDQ=iDn; vo=vo(1:end-1); iD=iD(1:end-1);
gm=KpD*(VGQ-VtD)*(1+lambdaD*VDSQ); ro=1/lambdaD/abs(IDQ); % Eq.(4.1.4)
if VDSQ<sqrt(2*IDQ/KpD)|VDSQ>VDD-VtL
  fprintf('\nLinear AC analysis is invalid for FET isn''t saturated!\n');
end
```

(Step 1) Assuming that both NMOSs are in the saturation region, we find the smaller one of the two roots of the following quadratic equation since the larger one ($> V_{DD}$) must be invalid:

$$\frac{1}{2}K_{p,L}(v_{GS,L} - V_{t,L})^2 \overset{(4.1.13b)\text{ with }\lambda=0}{} \underset{v_{GS,D}=v_i,\, v_{DS,D}=v_o}{\overset{v_{GS,L}=V_{DD}-v_{DS,L}}{\longrightarrow}} \frac{1}{2}K_{p,L}(V_{DD} - v_{DS,D} - V_{t,L})^2 \overset{(4.1.13b)\text{ with }\lambda=0}{=} \frac{1}{2}K_{p,D}(v_{GS,D} - V_{t,D})^2$$

$$;\, v_o^2 + 2(V_{t,L} - V_{DD})v_o + (V_{DD} - V_{t,L})^2 - \frac{K_{p,D}}{K_{p,L}}(v_i - V_{t,D})^2 = 0 \tag{4.2.20}$$

(Step 2) If this (tentative) solution is not in the saturation region of M_D, i.e. $v_{DS,D} < v_{GS,D} - V_{t,D}$ so that $v_{GD,D} = v_{GS,D} - v_{DS,D} > V_{t,D}$, we assume that M_D is in the ohmic region and find the smaller one of the two roots of the following quadratic equation:

$$\frac{1}{2}K_{p,L}(V_{DD} - v_{DS,D} - V_{t,L})^2 \overset{(4.1.13a)}{=} K_{p,D}\left\{(v_{GS,D} - V_{t,D})v_{DS,D} - \frac{1}{2}v_{DS,D}^2\right\} \underset{v_{GS,D}=v_i,\, v_{DS,D}=v_o}{\overset{v_{GS,L}=V_{DD}-v_{DS,D}}{\longrightarrow}}$$

$$K_{p,L}(V_{DD} - v_o - V_{t,L})^2 = K_{p,D}\left(2(v_i - V_{t,D})v_o - v_o^2\right) \rightarrow$$

$$\left(1 + \frac{K_{p,D}}{K_{p,L}}\right)v_o^2 + 2\left\{(V_{t,L} - V_{DD}) + \frac{K_{p,D}}{K_{p,L}}(V_{t,D} - v_i)\right\}v_o + (V_{DD} - V_{t,L})^2 = 0 \tag{4.2.21}$$

This process of analyzing the CS amplifier with an enhanced FET load has been cast into the above MATLAB function '`FET_CS_NMOSe()`'. We can run the following MATLAB script "do_FET_CS_NMOSe.m" to plot the output $v_o(t)$ to the input $v_i(t) = 0.5 \sin(2000\pi t)$ for $t = 0 \sim 1$ ms like the PSpice simulation result shown in Figure 4.30(c).

```
%do_FET_CS_NMOSe.m
VDD=18; Rs=100; R1=1e5; R2=1e5; RL=1e6; Vsm=0.5;
tf=1e-3; t=tf/180*[0:180]; w=2000*pi; vi=Vsm*sin(w*t);
Kp=[2e-5 2e-5]; Vt=[1 1]; lambda=1e-2; % Eq.(4.2.14)
[VGQ,VSQ,VDQ,IDQ,vo,iD]= ...
    FET_CS_NMOSe(VDD,Rs,R1,R2,RL,Kp,Vt,lambda,vi);
subplot(221), plot(t,vi, t,vo-(VDQ-VSQ),'r'), grid on
```

4.2.4.2 CS Amplifier with a Depletion FET Load

Figure 4.31(a) shows a CS amplifier with an enhancement NMOS driver M_D and a depletion NMOS load M_L, which is diode connected with its gate and source short-circuited so that it can act like a current-limited nonlinear resistor. The drain (output) characteristic curves of M_D (for several values of v_{GS}) are plotted as blue lines while the $i_{D,L} - v_{DS,L}$ and $i_{D,L} - v_{DS,D} = V_{DD} - v_{DS,L}$ relationships of M_L (with $v_{GS,L} = 0$) are plotted as the lines of square/circle symbols, respectively, in Figure 4.31(b) where the conduction parameter K_p, threshold voltage V_t, and CLM parameter λ of M_D and M_L are

$$K_{p,D} = 20\,\mu\text{A/V}^2, K_{p,L} = 20\,\mu\text{A/V}^2, V_{t,D} = 1\,\text{V}, V_{t,L} = -10\,\text{V}, \lambda = 10^{-4}\,\text{V}^{-1} \tag{4.2.22}$$

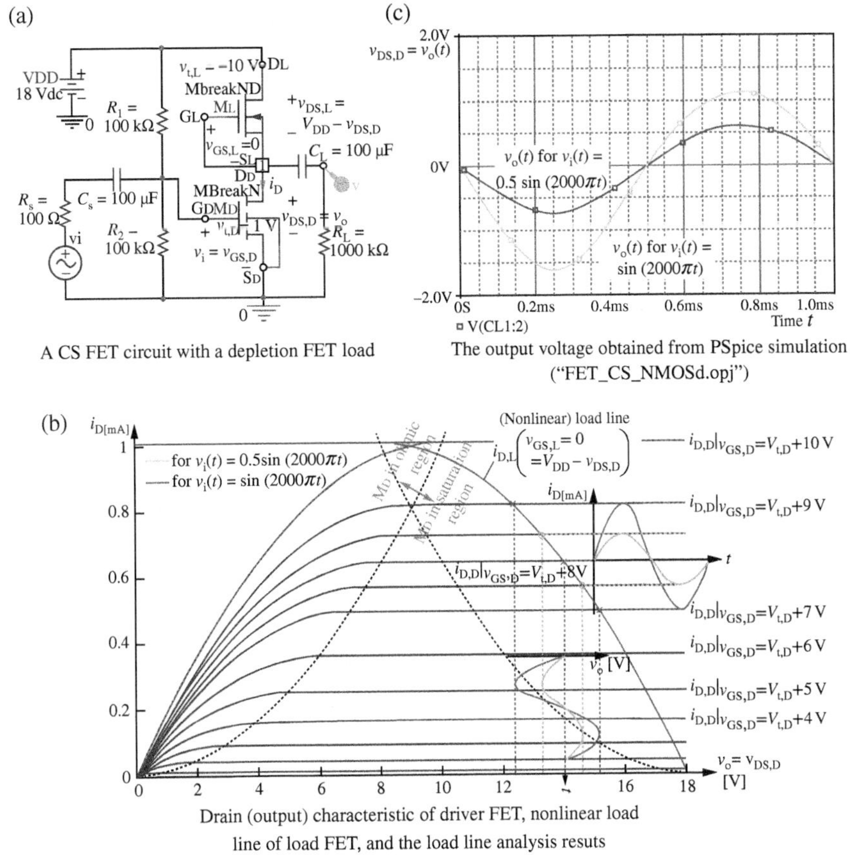

Figure 4.31 A CS amplifier with a depletion FET load, its load line analysis, and PSpice simulation results.

From the locus of dynamic operating points obtained from the intersections of the (pink solid) load line and the characteristic curves of M_D (for several values of $v_i = v_{GS}$), the graphs of $i_D(t) = i_{D,L}(t) = i_{D,D}(t)$ and $v_o(t) = v_{DS,D}(t)$ for the sinusoidal input $v_i(t) = V_{sm} \sin(2000\pi t)$ (with V_{sm} = 0.5 or 1) applied to the G-S terminals of M_D can be plotted as the green and red lines, respectively, in Figure 4.31(b).

Note that the intersection of the load line of M_L and the characteristic curve of M_D for a certain value of the input $v_{GS,D} = v_i$ ($> V_{t,D}$) can analytically be found as follows:

(Step 1) Assuming that both NMOSs are in the saturation region, find the root of the following first-degree polynomial equation:

$$K_{p,L}V_{t,L}^2(1+\lambda_L v_{DS,L}) \overset{(4.1.13b)\text{ with } v_{GS,L}=0}{=} \overset{(4.1.13b)}{K_{p,D}(v_{GS,D}-V_{t,D})^2(1+\lambda_D v_{DS,D})} \underset{v_{GS,D}=v_i,\, v_{DS,D}=v_o}{\overset{v_{GS,L}=0,\, v_{DS,L}=V_{DD}-v_{DS,D}}{\longrightarrow}}$$

$$\left\{\lambda_D K_{p,D}(v_i - V_{t,D})^2 + \lambda_L K_{p,L}V_{t,L}^2\right\}v_o = K_{p,L}V_{t,L}^2(1+\lambda_L V_{DD}) - K_{p,D}(v_i - V_{t,D})^2 \tag{4.2.23}$$

```
function [VGQ,VSQ,VDQ,IDQ,vo,iD,gm,ro] = ...
             FET_CS_NMOSd(VDD,Rs,R1,R2,RL,Kp,Vt,lambda,vi)
% Copyleft: Won Y. Yang, wyyang53@hanmail.net, CAU for academic use only
KpD=Kp(1); KpL=Kp(end); VtD=Vt(1); VtL=Vt(end);
if lambda(1)>1, lambda=1./lambda; end
lambdaD=lambda(1); lambdaL=lambda(end);
Ri=parallel_comb([R1 R2]); % Input resistance
VGQ=VDD*R2/(R1+R2); VSQ=0; VGSQ=VGQ-VSQ;
if nargin<9, vi=0; else vi=[vi 0]; end % 0 for finding the Q point
for n=1:length(vi)
  vGSn = VGQ + vi(n)/(Rs/Ri+1);
  num=VtL^2*(1+lambdaL*VDD)-KpD/KpL*(vGSn-VtD)^2;
  den=lambdaD*KpD/KpL*(vGSn-VtD)^2+lambdaL*VtL^2; % Eq.(4.2.23)
  von=num/den; iDn=KpL/2*VtL^2*(1+lambdaL*(VDD-von));
  if von<vGSn-VtD % vGD,D=vGS,D-vDS,D=vGSn-von>Vt ?
    fprintf('MD is ohmic: vGD(%5.2f)<Vt(%5.2f)\n',vGSn-von,VtD);
    iDn=KpL/2*VtL^2; B=2*(VtD-vGSn); C=iDn/KpD; % Eq.(4.2.24)
    von=(-B-sqrt(B^2-4*C))/2;
  elseif von>VDD+VtL % vGD,L=vSD,L=vDS,D-VDD=von-VDD>VtL ?
    fprintf('ML is ohmic: vGD(%5.2f)>Vt(%5.2f)\n', von-VDD,VtL);
    iDn=KpD/2*(vGSn-VtD)^2;
    B=-2*(VDD+VtL); C=iDn/KpL+VDD*(2*VtL+VDD); % Eq.(4.2.25)
    von=(-B+sqrt(B^2-4*C))/2;
  end
  vo(n)=von; iD(n)=iDn;
end
VDSQ=vo(n); VDQ=VDSQ+VSQ; IDQ=iD(n);
vo=vo(1:end-1); iD=iD(1:end-1);
gm=KpD*(VGQ-VtD)*(1+lambdaD*VDSQ);
ro=1/lambdaD/abs(IDQ); % Eq.(4.1.4)
if VDSQ<VGSQ-VtD
 fprintf(' Linear AC analysis is invalid since driver FET is ohmic\n');
elseif VDSQ>VDD+VtL
 fprintf(' Linear AC analysis is invalid since load FET is ohmic\n');
end
```

(Step 2) If this tentative solution is in the ohmic region of M_D, i.e. $v_o = v_{DS,D} < v_{GS,D} - V_{t,D} = v_i - V_{t,D}$, the following equation should be solved on the assumption that M_D is in the ohmic region:

$$K_{p,L}V_{t,L}^2 = K_{p,D}\left\{2(v_{GS,D} - V_{t,D})v_{DS,D} - v_{DS,D}^2\right\} = K_{p,D}\{2(v_i - V_{t,D})v_o - v_o^2\}$$
$$;K_{p,D}v_o^2 - 2K_{p,D}(v_i - V_{t,D})v_o + K_{p,L}V_{t,L}^2 = 0 \quad (4.2.24)$$

where the smaller one of the two roots should be taken as $v_{o,Q} = V_{DS,D}$ because the larger one must be invalid.

(Step 3) If the tentative solution is in the ohmic region of M_L, i.e. $v_o = v_{DS,D} > V_{DD} + V_{t,L}$ so that $v_{GD,L} = -v_{DS,L} = -(V_{DD} - v_{DS,D}) > V_{t,L}$, the following equation should be solved on the assumption that M_L is in the ohmic region:

$$K_{p,L}\left\{2(v_{GS,L}{}^0 - V_{t,L})v_{DS,L} - v_{DS,L}^2\right\} = K_{p,D}(v_{GS,D} - V_{t,D})^2$$
$$;K_{p,L}\{2(-V_{t,L})(V_{DD} - v_o) - (V_{DD} - v_o)^2\} = K_{p,D}(v_i - V_{t,D})^2$$
$$;K_{p,L}v_o^2 - 2K_{p,L}(V_{DD} + V_{t,L})v_o + K_{p,D}(v_i - V_{t,D})^2 + K_{p,L}V_{DD}(2V_{t,L} + V_{DD}) = 0 \quad (4.2.25)$$

where the larger one of the two roots should be taken as the value of v_o because the smaller one must be invalid.

This process of analyzing the CS amplifier with a depletion FET load has been cast into the above MATLAB function '`FET_CS_NMOSd()`'.

An alternative to analyze the circuit of Figure 4.31(a) is to solve the KCL at node 1:

$$i_{D,L}(0, V_{DD} - v_o) - i_{D,D}(v_{GS,D} + v_i, v_o) = 0 \text{ at node 1} \quad (4.2.26)$$

where $v_{GS,D} = V_{GSQ,D} + v_i$ with $V_{GSQ,D} = V_{DD}R_2/(R_1 + R_2)$ and $i_{Dk}(v_{GSk}, v_{DSk})$ is defined as Eq. (4.1.83). This algorithm can be implemented by replacing the for loop of the above MATLAB function '`FET_CS_NMOSd()`' by the following block of statements:

```
iD_vGS_vDS=@(vGS,vDS,Kp,Vt)Kp/2*(vGS-Vt).^2.*(vGS>=Vt).*(vGS-vDS<Vt)
+ Kp*((vGS-Vt).*vDS-vDS.^2/2).*(vGS>=Vt).*(vGS-vDS>=Vt); % Eq.(4.1.83)
for n=1:length(vi)
  vGSn = VGSQ + vi(n)*Ri/(Rs+Ri); % Eq.(4.2.26)
  eq=@(v)(iD_vGS_vDS(0,VDD-v,KpL,VtL)-iD_vGS_vDS(vGSn,v,KpD,VtD))*1e6;
  if n<2, v0=VDD/2; else v0=vo_(n-1); end
  vo_(n)=fsolve(eq,v0); % Solution of Eq.(4.2.26)
  iD_(n)=iD_vGS_vDS(vGSn,vo_(n),KpD,VtD);
end
discrepancy_vo = norm(vo-vo_)/norm(vo)
```

Isn't it interesting that this simple algorithm of solving just one (seemingly nonlinear) equation can replace the above individual quadratic equation approach?

We can run the following MATLAB script "do_FET_CS_NMOSd.m" to plot the output $v_o(t)$ to the input $v_i(t) = 0.5 \sin(2000\pi t)$ for $t = 0 \sim 1$ ms like the PSpice simulation result shown in Figure 4.31(c).

```
%do_FET_CS_NMOSd.m
VDD=18; Rs=100; R1=1e5; R2=1e5; RL=1e6; Vsm=1;
tf=1e-3; t=tf/180*[0:180]; w=2000*pi; vi=Vsm*sin(w*t);
Kp=[2e-5 2e-5]; Vt=[1 -10]; lambda=[1e-2 1e-2]; % Eq.(4.2.22)
[VGQ,VSQ,VDQ,IDQ,vo,iD]= ...
FET_CS_NMOSd(VDD,Rs,R1,R2,RL,Kp,Vt,lambda,vi);
subplot(222), plot(t,vi, t,vo-(VDQ-VSQ),'r'), grid on
```

4.2.5 Multistage FET Amplifiers

Table 4.3 lists the formulas for finding the input/output resistances, voltage gain, and maximum small-signal input (for linear amplification) of the CS/CD/CG amplifiers. It also shows the conditions to be met by the coupling/bypass capacitors, whose AC impedances have been assumed to be negligibly small (for a frequency range of interest) compared with the equivalent impedance seen from their two terminals for AC analysis (see Section 14.7 of [J-1]).Note that finding the input/output resistance of a CG configuration requires the input/output resistance of the next/previous stage corresponding to its load/source resistance R_L/R_s as implied by Eqs. (4.2.15) and (4.2.17). That is why, for a systematic analysis of a multistage amplifier containing one or more CG configurations, we should find the input/output resistance of each stage, starting from the last/first stage backwards/forwards to the first/last stage where to the last/first stage, the input/output resistance of the next/previous stage is nothing but the load/source resistance.

Each of the formulas listed in Table 4.3 has been coded in MATLAB as follows so that they can be called individually as symbolic expressions whenever and wherever needed. For instance, the formula for the voltage gain Av can be recalled by typing 'Av_CS' at the MATLAB prompt.

Table 4.3 Characteristics of CS/CD/CG amplifiers.

	CS (common source)		CD (common drain) – source follower		CG (common gate)	
R_i	$R_1\|\|R_2$	(4.2.5)	$R_1\|\|R_2$	(4.2.10)	$R_S\|\|\dfrac{r_o+(R_D\|\|R_L)}{1+g_m r_o} \xrightarrow{r_o \gg \max\{1/g_m,(R_D\|\|R_L)\}} R_S\|\|\dfrac{1}{g_m}$	(4.2.15)
R_o	$R_D\|\|\{r_o+(1+g_m r_o)R_{S1}\} \xrightarrow{r_o\to\infty} R_D$	(4.2.7)	$\frac{1}{g_m}\|\|r_o\|\|R_S \xrightarrow{r_o\to\infty} \frac{1}{g_m}\|\|R_S$	(4.2.12)	$R_D\|\|\{r_o+(1+g_m r_o)(R_s\|\|R_S)\} \xrightarrow{R_s, R_D \ll r_o} R_D$	(4.2.17)
A_v	$\dfrac{-g_m r_o(R_D\|\|R_L)}{(R_D\|\|R_L)+r_o+(1+g_m r_o)R_{S1}}$ $\overset{\text{if } r_o >> 1/g_m R_{S1}}{\approx} \dfrac{-g_m(R_D\|\|R_L)}{1+g_m R_{S1}}$	(4.2.6)	$\dfrac{g_m(r_o\|\|R_S\|\|R_L)}{1+g_m(r_o\|\|R_S\|\|R_L)}$ $\xrightarrow[R_s << Ri]{g_m(r_o\|\|R_S\|\|R_L) >> 1} = 1$	(4.2.11)	$\dfrac{(1+g_m r_o)(R_D\|\|R_L)}{r_o+(R_D\|\|R_L)} \xrightarrow{r_o \gg \max\{1/g_m,(R_D\|\|R_L)\}} g_m(R_D\|\|R_L)$	(4.2.16)
G_v			$G_v = \dfrac{R_i}{R_s+R_i}A_v$			
$v_{s,\max}$	$0.2(1+g_m R_{S1})(V_{GS,Q}-V_t) \times \dfrac{R_s+R_i}{R_i}$	(4.2.9)	$0.2\{1+g_m(R_S\|\|R_D)\}(V_{GS,Q}-V_t) \times \dfrac{R_s+R_i}{R_i}$	(4.2.14)	$0.2(V_{GS,Q}-V_t) \times \dfrac{R_s+R_i}{R_i}$	(4.2.18)
C_s	$C_s >> \dfrac{1}{\omega(R_s+R_i)}$ [J-1]		$C_s >> \dfrac{1}{\omega(R_s+R_i)}$		$C_s >> \dfrac{1}{\omega(R_s+R_i)}$	
C_L	$C_L >> \dfrac{1}{\omega(R_L+R_o)}$		$C_L >> \dfrac{1}{\omega(R_L+R_o)}$		$C_L >> \dfrac{1}{\omega(R_L+R_o)}$	
C_S	$C_S >> \dfrac{1}{\omega\{R_{S2}\|\|(R_{S1}+1/g_m)\}}$					

```
function [Av,Avo,Gv,Ri,Ro]=Av_CS(ro_,RS_)
% Put 0 as the 1st input argument ro_ if ro=inf.
% Put 0 as the 2nd input argument RS_ if RS=0.
syms gm ro Rs RG RD RS RL
Ri=RG; % Eq.(4.2.5)
RDL=parallel_comb([RD RL]);
if nargin>0&ro_==0
  Ro=RD; % Eq.(4.2.7)
  if nargin>1&RS_==0, Av=-gm*RDL; % Eq.(4.2.6)
   else Av=-gm*RDL/(1+gm*RS); % Eq.(4.2.6)
  end
 else % if ro is finite
  if nargin>1&RS_==0, gmroRS=ro;
   else gmroRS=ro+(1+gm*ro)*RS;
  end
  Ro=parallel_comb([RD gmroRS]); % Eq.(4.2.7)
  Av=-gm*ro*RDL/(RDL+gmroRS); % Eq.(4.2.6)
end
Avi=1/(Rs/Ri+1); AvL=1/(Ro/RL+1); Avo=Av/AvL;
Gv=Avi*Av;
if nargout<1
  fprintf('\n Av =\n'); pretty(simplify(Av))
  fprintf('\n Avo =\n'); pretty(simplify(Avo))
  fprintf('\n Gv =\n'); pretty(simplify(Gv))
end
```

```
function Ri=Ri_CS()
syms RG
Ri=RG; % Eq.(4.2.5)
```

```
function Ro=Ro_CS(ro_,RS_)
% Put 0 as the 1st input argument ro_ if ro=inf.
% Put 0 as the 2nd input argument RS_ if RS=0.
syms gm ro Rs RG RD RS RL
if nargin>0&ro_==0
 Ro=RD; % Eq.(4.2.7)
 else % if ro is finite
 if nargin>1&RS_==0, gmroRS=ro;
   else gmroRS=ro+(1+gm*ro)*RS;
 end
 Ro=parallel_comb([RD gmroRS]); % Eq.(4.2.7)
end
if nargout<1
  fprintf('\n Ro =\n'); pretty(simplify(Ro))
end
```

```
function Ri=Ri_CD(ro_)
% Put 0 as the 1st input argument ro_ if ro=inf.
syms RG
Ri=RG; % Eq.(4.2.9)
```

```
function Ro=Ro_CD(ro_)
% Put 0 as the 1st input argument ro_ if ro=inf.
syms gm ro Rs RG RD RS RL
if nargin>0&ro_==0, ro=inf; end
Ro=parallel_comb([1/gm RS ro]); % Eq.(4.2.17)
if nargout<1, pretty(simplify(Ro)), end
```

```
function [Av,Avo,Gv,Ri,Ro]=Av_CD(ro_)
% Put 0 as the 1st input argument ro_ if ro=inf.
syms gm ro Rs RG RD RS RL
RSL=parallel_comb([RS RL]);
if nargin>0&ro_==0
 gm1=gm; ro1=inf; % Against the case where RD is nonzero.
 else % if ro is finite
 gm1=gm/(1+RD/ro); ro1=ro+RD;
end
Ro1SL=parallel_comb([ro1 RSL]);
Ro1S=parallel_comb([ro1 RS]);
Ro=parallel_comb([1/gm1 RS ro1]); Ri=RG; % Eq.(4.2.10)
Av=gm1*Ro1SL/(1+gm1*Ro1SL); % Eq.(4.2.11b)
Avi=1/(Rs/Ri+1); AvL=1/(Ro/RL+1);
Avo=Av/AvL; Avo_=gm1*Ro1S/(1+gm1*Ro1S);
discrepancy = simplify(Avo-Avo_)
Gv=Avi*Av; % Eq.(4.2.16a)
if nargout<1
 fprintf('\n Av =\n'); pretty(simplify(Av))
 fprintf('\n Avo =\n'); pretty(simplify(Avo))
 fprintf('\n Gv =\n'); pretty(simplify(Gv))
end
```

```
function [Av,Avo,Gv,Ri,Ro]=Av_CG(ro_)
% Put 0 as the 1st input argument ro_ if ro=inf.
syms gm ro Rs RG RD RS RL
RDL=parallel_comb([RD RL]); RDLo=RD;
if nargin>0&ro_==0
 Rsg=1/gm; Ro=RD; % Eq.(4.2.17)
 else % if ro is finite
 Rsg=(ro+RDL)/(1+gm*ro); Rsgo=(ro+RDLo)/(1+gm*ro);
 Ro=parallel_comb([(1+gm*ro)*parallel_comb([Rs RS])+ro
                  RD]); % Eq.(4.2.17)
end
Ri=parallel_comb([RS Rsg]); % Eq.(4.2.15)
Av=RDL/Rsg; Avo=RDLo/Rsgo; Avi=1/(Rs/Ri+1); % Eq.(4.2.16b)
Gv=Avi*Av; % Eq.(4.2.16a)
if nargout<1
 fprintf('\n Av =\n'); pretty(simplify(Av))
 fprintf('\n Avo =\n'); pretty(simplify(Avo))
 fprintf('\n Gv =\n'); pretty(simplify(Gv))
end
```

```
function Ri=Ri_CG(ro_)
% Put 0 as the 1st input argument ro_ if ro=inf.
syms gm ro RG RD RS RL
RDL=parallel_comb([RD RL]);
if nargin>0&ro_==0, Rsg=1/gm;
 else Rsg=(ro+RDL)/(1+gm*ro); % if ro is finite
end
Ri=parallel_comb([RS Rsg]); % Eq.(4.2.15)
if nargout<1, pretty(simplify(Ri)), end
```

```
function Ro=Ro_CG(ro_)
% Put 0 as the 1st input argument ro_ if ro=inf.
syms gm ro Rs RG RD RS RL
if nargin>0&ro_==0, Ro=RD; % Eq.(4.2.17)
 else % if ro is finite
  Ro=parallel_comb([(1+gm*ro)*parallel_comb([Rs RS])+ro
                     RD]); %Eq.(4.2.17)
end
if nargout<1, pretty(simplify(Ro)), end
```

Example 4.9 Two-Stage (CS-CD) Cascaded Amplifiers with Capacitive or Direct Coupling

Let the device parameters of every FET in Figure 4.32 be $K_p = 0.5\ \text{mA/V}^2$, $V_t = 1$ V, and $\lambda = 10^{-4}\ \text{V}^{-1}$ in common.

a) Find the values of the amplifier parameters R_{i1}, R_{o1}, A_{v1}, R_{i2}, R_{o2}, A_{v2}, and $G_{v1}=v_{o1}/v_s$ for the capacitively coupled two-stage (CS-CD) amplifier in Figure 4.32(a1).

First, set the values of the circuit and device parameters as given:

```
>>Kp=0.5e-3; Vt=1; lambda=1e-4; % Device parameters
  VDD=12; Rs=1e3; RL=100e3; Vsm=1e-3; % Circuit parameters
  R11=2.2e6; R12=1.5e6; RD1=22e3; RS1=12e3; % for Stage 1
  R21=25e6; R22=100e6; RD2=0; RS2=8e3; % for Stage 2
```

Then, we analyze the (last) stage (of CD configuration) with the assumption of $R_{s2} = 0$ (because the output resistance R_{o1} of the previous stage has not yet been determined) to just find its input resistance R_{i2}, which is to be used as the load resistance R_{L1} of the previous stage:

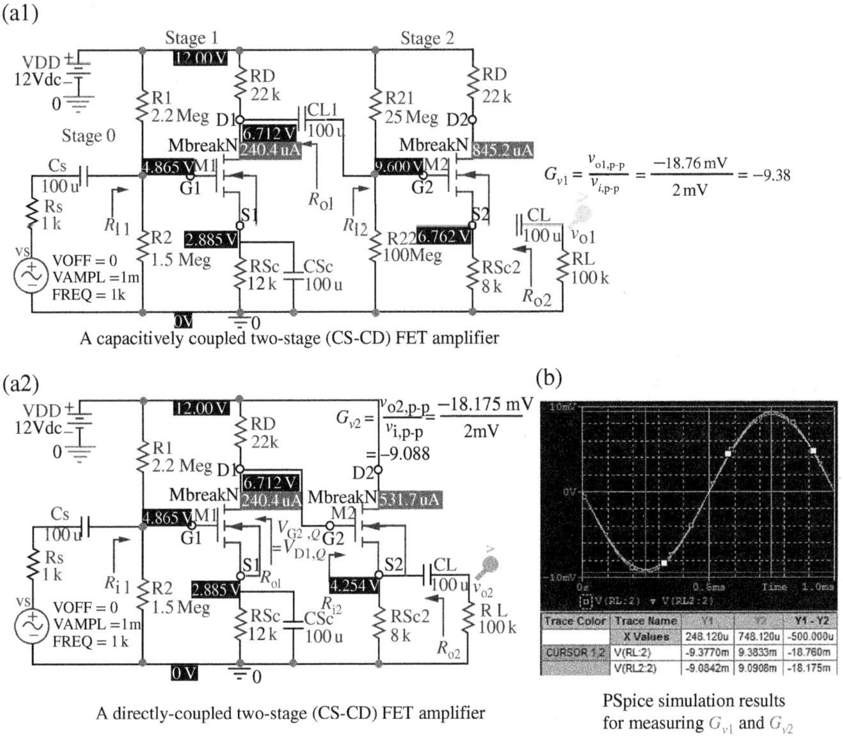

Figure 4.32 Two two-stage (CS-CD) FET amplifiers and their PSpice simulation results ("elec04e09.opj").

```
>>Rs2=0; RL2=RL; % Rs2=Ro1 has not yet been determined.
  [VG2Q,VS2Q,VD2Q,ID2Q,Av2,Ri2,Ro2,gm2,ro2,Vo2m,vs2max]= ...
  FET_CD_analysis(VDD,Rs2,R21,R22,RD2,RS2,RL2,Kp,Vt,lambda);
```

Then, we analyze the previous (first) stage (of CS configuration) with $R_{L1} = R_{i2}$ (obtained here):

```
>>Rs1=Rs; RL1=Ri2;
  [VG1Q,VS1Q,VD1Q,ID1Q,Av1,Ri1,Ro1,gm1,ro1,vs1max]= ...
  FET_CS_analysis(VDD,Rs1,R11,R12,RD1,RS1,RL1,Kp,Vt,lambda);
```

This yields

```
Analysis Results
 VDD  VGQ  VSQ  VDQ   IDQ
12.00 4.86 2.88 6.71 2.40e-04
Ri= 891.89[kOhm], Ro= 21988[Ohm]
Gv=Ri/(Rs+Ri)xAvoxRL/(Ro+RL)=0.999 x ( -10.78x0.9989= -10.77) = -10.76
```

Noting that the first stage has been analyzed with the true values of $R_{s1} = R_s$ and $R_{L1} = R_{i2}$ to yield the proper values of every amplifier parameter, we reanalyze the (last) stage (of CD configuration) with $R_{s2} = R_{o1}$ (obtained just above):

```
>>Rs2=Ro1;
  [VG2Q,VS2Q,VD2Q,ID2Q,Av2,Ri2,Ro2,gm2,ro2,vs2max]= ...
  FET_CD_analysis(VDD,Rs2,R21,R22,RD2,RS2,RL,Kp,Vt,lambda);
```

This yields

```
Analysis Results
 VDD   VGQ  VSQ   VDQ    IDQ
12.00 9.60 6.76 12.00 8.45e-04
Ri=20000.00[kOhm], Ro= 957[Ohm]
Gv=Ri/(Rs+Ri)xAvoxRL/(Ro+RL)=0.999x(0.8803x0.9905=0.8719) = 0.8710
```

Then, we multiply the voltage gains (A_{vn}'s) of every stage including that of the voltage divider at stage 0 (of the input source) to find the overall voltage gain as

```
>>Gv=Ri1/(Rs+Ri1)*Av1*Av2
  Gv =  -9.3806
```

This implies that the overall voltage gain of the CS-CD stage is -9.38 as confirmed by the PSpice simulation result $G_{v,s}$ = -18.76 mV/2 mV = -9.38 in Figure 4.32(b). This is greater than that (-8.66) of the CS stage (in Figure 4.25 (a2)) (see Example 4.6) despite the additional CD stage whose voltage gain is less than one by itself. (Q) Why is that?

b) Find the values of the amplifier parameters R_{i1}, R_{o1}, A_{v1}, R_{i2}, R_{o2}, A_{v2}, and $G_{v2} = v_{o2}/v_s$ for the directly coupled two-stage (CS-CD) amplifier in Figure 4.32(a2).

Since the circuit of (a2) is the same as (a1) except for the nonexistence of C_{L1}, R_{21}, and R_{22}, the values of the circuit and device parameters set for (a1) can be used as they have been set. Noting that due to the direct coupling, the gate voltage of M_2 equals the drain voltage of M_1, i.e. $V_{G2,Q} = V_{D1,Q}$, we should first perform the DC analysis of stage 1 by using 'FET_DC_analysis()' to determine $V_{D1,Q}$:

```
>>[VG1Q,VS1Q,VD1Q,VGS1Q,VDS1Q,ID1Q,mode]= ...
    FET_DC_analysis(VDD,R11,R12,RD1,RS1,Kp,Vt,lambda);
```

Then, we analyze the (last) stage (of CD configuration) with the assumption of $R_{s2} = 0$ (because the output resistance R_{o1} of the previous stage has not yet been determined) to just find its input resistance R_{i2}, which is to be used as the load resistance R_{L1} of the previous stage:

```
>>Rs2=0; vG2=VD1Q; RG2=1e8; RL2=RL; % Rs2=Ro1 is not yet determined.
  [VG2Q,VS2Q,VD2Q,ID2Q,Av2,Ri2,Ro2,gm2,ro2,Vo2m,vs2max] = ...
  FET_CD_analysis(VDD,Rs2,vG2,RG2,RD2,RS2,RL2,Kp,Vt,lambda);
```

Here, we have used '`FET_CD_analysis()`' with the third/fourth input arguments $V_{D1,Q}$ (obtained above)/10^8 in place of R_1/R_2 to analyze stage 2 where 10^8 (a very large value of resistance corresponding to $R_1||R_2 = \infty$) will be assigned as the value of R_{i2} inside '`FET_CD_analysis()`'.

Then, we analyze the previous (first) stage (of CS configuration) with $R_{L1} = R_{i2}$ (obtained just above):

```
>>Rs1=Rs; RL1=Ri2;
  [VG1Q,VS1Q,VD1Q,ID1Q,Av1,Ri1,Ro1,gm1,ro1,vs1max] = ...
  FET_CS_analysis(VDD,Rs1,R11,R12,RD1,RS1,RL1,Kp,Vt,lambda);
```

This yields

```
Analysis Results
  VDD   VGQ   VSQ   VDQ    IDQ
12.00 4.86 2.88 6.71 2.40e-04
Ri= 891.89[kOhm], Ro= 21988[Ohm]
Gv=Ri/(Rs+Ri)xAvoxRL/(Ro+RL)=0.999x(-10.78x0.9998= -10.78) = -10.77
```

Noting that the first stage has been analyzed with the true values of $R_{s1} = R_s$ and $R_{L1} = R_{i2}$ to yield the proper values of every amplifier parameter, we reanalyze the (last) stage (of CD configuration) with $R_{s2} = R_{o1}$ (obtained just above):

```
>>Rs2=Ro1; %vG2=VD1Q; RG2=1e8;
  [VG2Q,VS2Q,VD2Q,ID2Q,Av2,Ri2,Ro2,gm2,ro2,vs2max] = ...
  FET_CD_analysis(VDD,Rs2,vG2,RG2,RD2,RS2,RL,Kp,Vt,lambda);
```

This yields

```
Analysis Results with vG= 6.71[kOhm], RG=100000.00[kOhm]
 VDD    VGQ   VSQ    VDQ    IDQ
12.00 6.71 4.25 12.00 5.32e-04
Ri=100000.00[kOhm], Ro= 1170[Ohm]
Gv=Ri/(Rs+Ri)xAvoxRL/(Ro+RL) =0.9998 x (0.8537x0.9884=0.8438) = 0.8436
```

Then, we multiply the voltage gains (A_{vn}'s) of every stage including that of the voltage divider at stage 0 (of the input source) to find the overall voltage gain as

```
>>Gv=Ri1/(Rs+Ri1)*Av1*Av2
  Gv = -9.0860
```

These MATLAB analysis results conform with the PSpice simulation results shown in Figure 4.32(a2) and (b).

Note that while the direct coupling saves a capacitor and two resistors without impairing the voltage gain, a level shifter (using positive/negative DC voltages) may have to be used to compensate the bias level change that may occur due to the lack of DC isolation between stages that would allow independent design of the biasing circuit of each individual stage.

Example 4.10 Three-Stage (CS-CE-CD) Cascaded Amplifier

Consider the three-stage amplifier of CS-CE-CD configuration in Figure 4.33 where the two FETs are both depletion-type NMOSs with device parameters K_p =10 mA/ V^2, V_t = -1 V, and $\lambda = 10^{-4}\ V^{-1}$ in common and the device parameters of the NPN-BJT are $\beta_F = 150$, $\beta_R = 1$, $\beta_{AC} = 150$, and $V_A = 10^4$ V. Find the values of the amplifier parameters R_{i1}, R_{o1}, A_{v1}, R_{i2}, R_{o2}, A_{v2}, R_{i3}, R_{o3}, A_{v3}, and $G_v = v_o/v_s$.

First, set the values of the circuit and device parameters as given:

```
>>Kp=10e-3; Vt=-1; lambda=1e-4; beta=[150 1 150]; VA=1e4;
  VDD=15; Rs=10e3; RL=10e3; Vsm=1e-5;
  R11=1e10; R12=1e5; RD1=620; RS1=[0 200]; % for Stage 1
  R21=78e3; R22=22e3; RC2=4.7e3; RE2=[0 1.5e3]; % for Stage 2
  R31=4e6; R32=12e6; RD3=0; RS3=10e3; % for Stage 3
```

Then, we analyze the (last) stage (of CD configuration) with the assumption of R_{s3} = 0 (because the output resistance R_{o2} of the previous stage has not yet been determined) to just find its input resistance R_{i3}, which is to be used as the load resistance R_{L2} of the previous stage:

```
>>Rs3=0; RL3=RL; % Rs3=Ro2 has not yet been determined.
  [VG3Q,VS3Q,VD3Q,ID3Q,Av3,Ri3,Ro3,gm3,ro3,Vo3m,vs3max]=...
  FET_CD_analysis(VDD,Rs3,R31,R32,RD3,RS3,RL3,Kp,Vt,lambda);
```

(a)

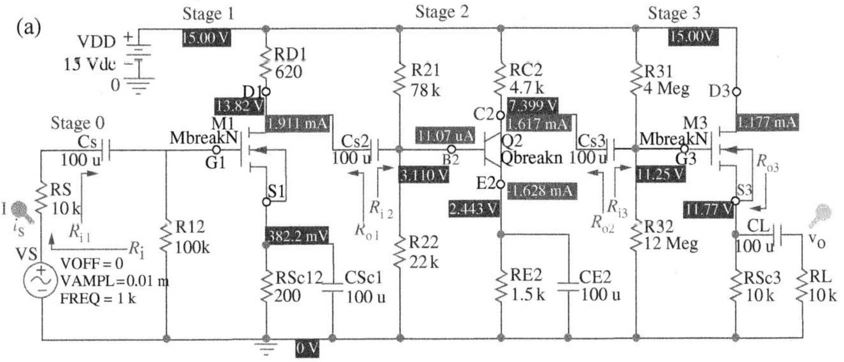

A capacitively coupled (AC coupled) three-stages
(of CS-CE-CD configuration) amplifier

(b)

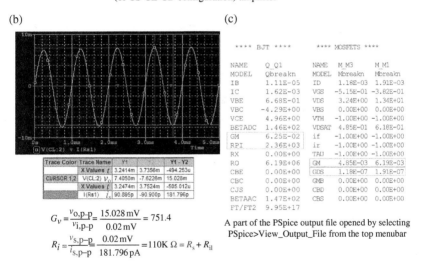

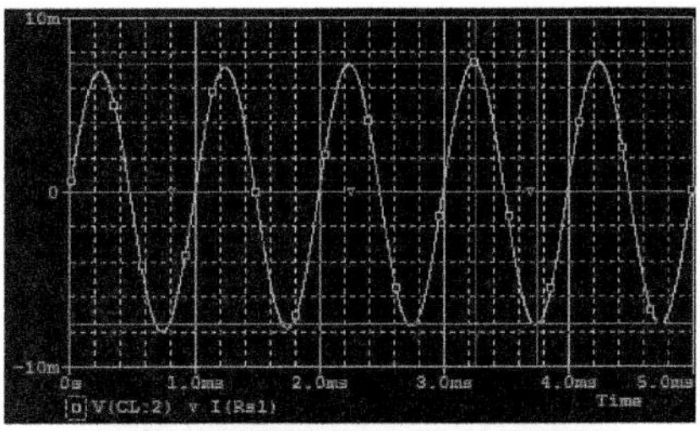

Trace Color	Trace Name	Y1		Y1 - Y2
	X Values t	3.2414m	3.7356m	-494.253u
CURSOR 1,2	V(CL:2) v_o	7.4050m	-7.6226m	15.028m
	X Values t	3.2474m	3.7524m	-505.012u
	I(Rs1) i_s	90.895p	-90.900p	181.796p

$$G_v = \frac{v_{o,p\text{-}p}}{v_{i,p\text{-}p}} = \frac{15.028\,\text{mV}}{0.02\,\text{mV}} = 751.4$$

$$R_i = \frac{v_{s,p\text{-}p}}{i_{s,p\text{-}p}} = \frac{0.02\,\text{mV}}{181.796\,\text{pA}} = 110\text{K}\,\Omega = R_s + R_{i1}$$

PSpice simulation results for measuring G_v and R_i

(c)

**** BJT ****

NAME	Q_Q1
MODEL	Qbreakn
IB	1.11E-05
IC	1.62E-03
VBE	6.68E-01
VBC	-4.29E+00
VCE	4.96E+00
BETADC	1.46E+02
GM	6.25E-02
RPI	2.36E+03
RX	0.00E+00
RO	6.19E+06
CBE	0.00E+00
CBC	0.00E+00
CJS	0.00E+00
BETAAC	1.47E+02
FT/FT2	9.95E+17

**** MOSFETS ****

NAME	M_M3	M_M1
MODEL	Mbreakn	Mbreakn
ID	1.18E-03	1.91E-03
VGS	-5.15E-01	-3.82E-01
VDS	3.24E+00	1.34E+01
VBS	0.00E+00	0.00E+00
VTH	-1.00E+00	-1.00E+00
VDSAT	4.85E-01	6.18E-01
if	-1.00E+00	-1.00E+00
ir	-1.00E+00	-1.00E+00
TAU	-1.00E+00	-1.00E+00
GM	4.85E-03	6.19E-03
GDS	1.18E-07	1.91E-07
GMB	0.00E+00	0.00E+00
CBD	0.00E+00	0.00E+00
CBS	0.00E+00	0.00E+00

A part of the PSpice output file opened by selecting PSpice>View_Output_File from the top menubar

Figure 4.33 A three-stage (CS-CE-CD) FET amplifier and its PSpice simulation result ("elec04e10.opj").

This yields

```
Ri= 3000.00 kOhm, Ro=  202 Ohm
```

Then, we analyze the previous (second) stage (of CE configuration) with $R_{L2} = R_{i3}$ (obtained just above):

```
>>Rs2=0; RL2=Ri3; rb2=0; % Rs2=Ro1 has not yet been determined.
  [VB2Q,VE2Q,VC2Q,IB2Q,IE2Q,IC2Q,Av2,Ai2,Ri2,Ro2,gm2,rbe2,ro2]=...
  BJT_CE_analysis(VDD,rb2,Rs2,R21,R22,RC2,RE2,RL2,beta,Vsm,VA);
```

This yields

```
Ri=  2.12 kOhm, Ro= 4696 Ohm
```

Then, we analyze the previous (first) stage (of CS configuration) with $R_{s1} = R_s$ and $R_{L1} = R_{i2}$ (obtained just above):

```
>>Rs1=Rs; RL1=Ri2;
  [VG1Q,VS1Q,VD1Q,ID1Q,Av(1),Ri(1),Ro(1),gm1,ro1,vs1max]=...
  FET_CS_analysis(VDD,Rs1,R11,R12,RD1,RS1,RL1,Kp,Vt,lambda);
```

This yields

```
  VDD   VGQ   VSQ   VDQ    IDQ
 15.00 0.00 0.38 13.81 1.91e-03
 in the saturation mode with gm= 6.187[mS], ro=5238.819[kOhm]
 Ri= 100.00 kOhm, Ro=  620 Ohm
 Gv=Ri/(Rs+Ri)xAv = 0.91 x -2.97 =  -2.70
```

Noting that the first stage has been analyzed with the true values of $R_{s1} = R_s$ and $R_{L1} = R_{i2}$ to yield the proper values of every amplifier parameter, we reanalyze the (second) stage (of CE configuration) with $R_{s2} = R_{o1}$ (the output resistance of the previous stage obtained just above) and $R_{L2} = R_{i3}$ (the input resistance of the next stage obtained above):

```
>>Rs2=Ro(1); % Now that Rs2=Ro1 has been determined.
  [VB2Q,VE2Q,VC2Q,IB2Q,IE2Q,IC2Q,Av(2),Ri(2),Ro(2)]=...
  BJT_CE_analysis(VDD,rb2,Rs2,R21,R22,RC2,RE2,RL2,beta,Vsm,VA);
```

This yields

```
  VCC   VEE   VBB   VBQ   VEQ   VCQ     IBQ        IEQ       ICQ
 15.00 0.00 3.30 3.12 2.42 7.48 1.07e-05 1.61e-03 1.60e-03
  in the forward-active mode with VCE,Q= 5.06[V]
 where gm= 61.916[mS], rbe= 2423[Ohm], ro=6247.69[kOhm]
  Ri=  2.12 kOhm, Ro= 4696 Ohm
  Gv=Ri/(Rs+Ri)xAv= 0.774 x -290.33 = -224.71
```

Then, we analyze the (last) stage (of CD configuration) with $R_{s3} = R_{o2}$ (the output resistance of the previous stage obtained just above) and $R_{L3} = R_L$:

```
>>Rs3=Ro(2); % Now that Rs3=Ro2 has been determined.
  [VG3Q,VS3Q,VD3Q,ID3Q,Av(3),Ri(3),Ro(3),gm3,ro3,Vo3m,vs3max]=...
  FET_CD_analysis(VDD,Rs3,R31,R32,RD3,RS3,RL3,Kp,Vt,lambda);
```

This yields

```
 VDD   VGQ   VSQ   VDQ   IDQ
15.00 11.25 11.77 15.00 1.18e-03
 in the saturation mode with gm=  4.852 [mS], ro=8502.536 [kOhm]
 Ri= 3000.00 kOhm, Ro=  202 Ohm
 Gv=Ri/(Rs+Ri)xAv = 1.00 x  0.96 =   0.96
```

Lastly, we multiply the voltage gains (A_{vn}'s) of every stage including that of the voltage divider at stage 0 (of the input source) to find the overall voltage gain as

```
>>Gv=Ri(1)/(Rs+Ri(1))*prod(Av)
  Gv =   752.4930
```

This implies that the overall voltage gain of the three-stage amplifier is 752 as confirmed by the PSpice simulation result $G_{v,s} = 15.028\,\text{mV}/0.02\,\text{mV} = 751.4$ in Figure 4.33(b). We can see that the DC analysis results obtained using the MATLAB functions are also close to those obtained from the PSpice simulation results shown on the schematic in Figure 4.33(a) and (c).

Example 4.11 Cascode (CS-CG) Amplifier

Consider the cascode amplifier of CS-CG configuration in Figure 4.34 where the two FETs are both enhancement-type NMOSs with device parameters $K_p = 10\,\text{mA/V}^2$, $V_t = 1\,\text{V}$, and $\lambda = 10^{-4}\,\text{V}^{-1}$ in common. Find the values of the amplifier parameters R_i, R_o, and $G_v = v_o/v_s$.

First, to perform the DC analysis of the circuit, we assume that the two NMOSs operate in the saturation mode and write the KCL equations at nodes 2, 4, and 5:

$$i_{R_D} \underset{\text{at node 5}}{\overset{\text{KCL}}{=}} i_{D2} \underset{\text{with } \lambda=0}{\overset{(4.1.13b)}{=}} \frac{1}{2}K_p(v_{GS2}-V_t)^2 \underset{\text{at node 4}}{\overset{\text{KCL}}{=}} i_{D1} \underset{\text{with } \lambda=0}{\overset{(4.1.13b)}{=}} \frac{1}{2}K_p(v_{GS1}-V_t)^2 \underset{\text{at node 2}}{\overset{\text{KCL}}{=}} i_{RS}$$

$$;\frac{V_{DD}-v_5}{R_D} = \frac{1}{2}K_p(V_3-v_4-V_t)^2 = \frac{1}{2}K_p(V_1-v_2-V_t)^2 = \frac{v_2}{R_S} \tag{E4.11.1}$$

Noting that $V_3 = V_{DD}(R_2 + R_3)/(R_1 + R_2 + R_3) = 12 \times (1+1)/(2+1+1) = 6\,\text{V}$ and $V_1 = V_{DD}R_3/(R_1 + R_2 + R_3) = 12 \times 1/(2+1+1) = 3\,\text{V}$, we can solve the last equality to get v_2 as

$$\frac{1}{2}K_p v_2^2 - \left\{K_p(V_1-V_t) + \frac{1}{R_S}\right\}v_2 + \frac{1}{2}K_p(V_1-V_t)^2 = 0 \tag{E4.11.2}$$

Then, we solve the second equality in Eq. (E4.11.1) to get v_4 as

$$V_3 - v_4 = V_1 - v_2\,;\, v_4 = V_3 - V_1 + v_2 \tag{E4.11.3}$$

Then, we get v_5 from the first equality in Eq. (E4.11.1) as

$$v_5 = V_{DD} - R_D i_D = V_{DD} - R_D\frac{1}{2}K_p(V_3-v_4-V_t)^2 \tag{E4.11.4}$$

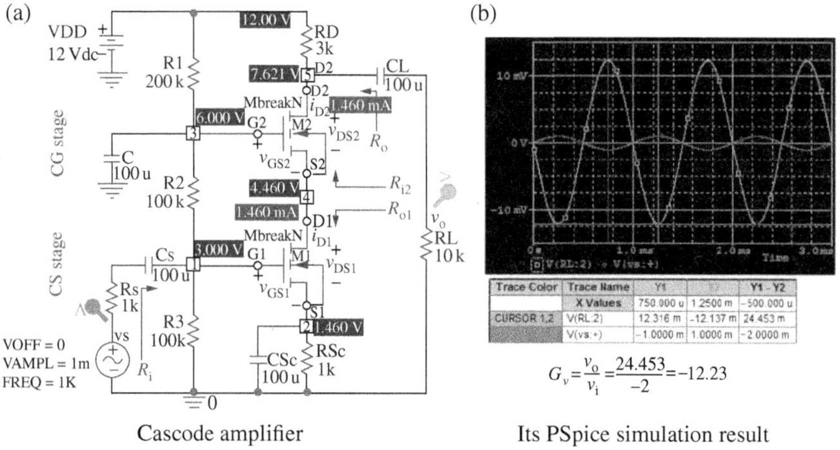

Figure 4.34 A cascode (CS-CG) amplifier circuit and its PSpice simulation result ("elec04e11_cascode.opj").

To find v_2, v_4, and v_5 in this way, we run the following MATLAB statements:

```
>>Kp=10e-3; Vt=1; lambda=1e-4; % Device parameters
  VDD=12; Rs=1e3; RL=10e3; R1=2e5; R2=1e5; R3=1e5; RD=3e3; RS=1e3;
  V3=VDD*(R2+R3)/(R1+R2+R3); V1=VDD*R3/(R1+R2+R3);
   A=Kp/2; B=-(Kp*(V1-Vt)+1/RS); C=Kp/2*(V1-Vt)^2;
   V2=(-B-sqrt(B^2-4*A*C))/2/A; V4=V3-V1+V2;
  IDQ=Kp/2*(V3-V4-Vt)^2; V5=VDD-RD*IDQ; % Eqs.(E4.11.2,3,4)

>>fprintf('V2=%6.2fV, V4=%6.2fV, V5=%6.2fV,
  IDQ=%6.3fmA\n', V2,V4,V5,IDQ*1e3)
```

This yields

```
V2= 1.46V, V4= 4.46V, V5= 7.62V, ID= 1.460mA
```

which conforms with the PSpice simulation results shown in the PSpice schematic of Figure 4.34(a). Here, before going into the AC analysis, we should check if the two NMOSs operate in the saturation:

$$v_{GD_1} = V_1 - V_4 = 3 - 4.46 \le V_t \text{ and } v_{GD_2} = V_3 - V_5 = 6 - 7.62 \le V_t \tag{E4.11.5}$$

Now, to perform the AC analysis, we use Eqs. (4.1.4b,a) to find the transconductances and output resistances of the two FETs:

```
>>VGS1Q=V1-V2; VDS1Q=V4-V2;
  gm1=Kp*(VGS1Q-Vt)*(1+lambda*VDS1Q); ro1=1/lambda/abs(IDQ);
  VGS2Q=V3-V4; VDS2Q=V5-V4;
  gm2=Kp*(VGS2Q-Vt)*(1+lambda*VDS2Q); ro2=1/lambda/abs(IDQ);
  fprintf('gm1=%7.4fmS, ro1=%8.2fkOhm, gm2=%7.4fmS,
  ro2=%8.2fkOhm\n', gm1*1e3,ro1/1e3,gm2*1e3,ro2/1e3)
```

This yields

```
gm1= 5.4047mS, ro1= 6850.78kOhm, gm2= 5.4048mS, ro2= 6850.78kOhm
```

Then, we use Eq. (4.2.15) to find the input resistance of the last stage (stage 2) of CG configuration as

$$R_{i2} \overset{(4.2.15)}{=} R_{S2}\|\frac{1}{g_{m2}} = \infty\|\frac{1}{g_{m2}} = \frac{1}{5.405\text{mS}} = 185\ \Omega \tag{E4.11.6}$$

Then, we use Eq. (4.2.5) to find the input resistance of stage 1 of CS configuration as

$$R_{i1} \overset{(4.2.5)}{=} R_2\|R_3 = 100\ \text{k}\|100\ \text{k} = 50\ \text{k}\Omega \tag{E4.11.7}$$

We also use Eqs. (4.2.7) and (4.2.6b) with $R_{D1} = \infty$, $R_{S1} = 0$, and $R_{L1} = R_{i2}$ to find the output resistance and voltage gain of stage 1 (CS) as

$$R_{o1} \overset{(4.2.7)}{=} R_{D1}\|\{r_{o1} + (1 + g_m r_{o1})R_{S1}\} \overset{R_{D}1=\infty, R_{S1}=0}{=} r_{o1} = 6850.8\text{k}\Omega \tag{E4.11.8}$$

$$A_{v1} \overset{(4.2.6b)}{=} \frac{-g_{m1}r_{o1}(R_{D1}\|R_{L1})}{(R_{D1}\|R_{L1}) + r_{o1} + (1 + g_{m1}r_{o1})R_{S1}} \underset{R_{L1}=R_{i2}}{\overset{R_{D1}=\infty, R_{S1}=0}{=}} \frac{-g_{m1}r_{o1}R_{i2}}{R_{i2} + r_{o1}}$$

$$= \frac{-5.405 \times 6850.8 \times 185}{185 + 6850.8 \times 10^3} \approx -1 \tag{E4.11.9}$$

Then, we use Eqs. (4.2.16b) and (4.2.17) with $R_s = R_{o1}$ to find the voltage gain and output resistance of stage 2 (CG) as

$$A_{v2} \overset{(4.2.16b)}{=} \frac{(1 + g_{m2}r_{o2})(R_{D2}\|R_{L2})}{r_{o2} + (R_{D2}\|R_{L2})}$$

$$\underset{R_{L2}=R_L}{\overset{R_{D2}=R_D}{=}} \frac{(1 + 5.405 \times 6850.8)(3\text{k}\|10\text{k})}{6850.8\text{k} + (3\text{k}\|10\text{k})} = 12.47 \tag{E4.11.10}$$

$$R_{o2} \overset{(4.2.17)}{=} R_D\|\{(1 + g_{m2}r_{o2})(R_s\|R_S) + r_o\}$$

$$\underset{R_S=\infty}{\overset{R_s=R_{o1}}{=}} R_D\|\{(1 + g_{m2}r_{o2})R_{o1} + r_{o2}\} \approx 3\text{k}\Omega \tag{E4.11.11}$$

Thus, the overall voltage gain can be found as

$$G_v = \frac{v_o}{v_s} = \frac{R_{i1}}{R_s + R_{i1}}A_{v1}A_{v2} \overset{(E4.11.7,9,10)}{=} \frac{50}{1 + 50}(-1.00)12.47 \approx \frac{-24.453}{2}(\text{PSpice}) \tag{E4.11.12}$$

Note that if $\lambda = 0$ so that $r_{o1} = \infty$ and $r_{o2} = \infty$, the overall voltage gain will be

$$G_v = \frac{R_{i1}}{R_s + R_{i1}} A_{v1} A_{v2} \overset{(E4.11.7,9,10)}{=} \frac{R_{i1}}{R_s + R_{i1}} (-g_{m1} R_{i2}) g_{m2} (R_D \| R_L)$$
$$\overset{(E4.11.6)}{=} -\frac{R_{i1}}{R_s + R_{i1}} g_{m1} (R_D \| R_L) \tag{E4.11.13}$$

This does not differ from the voltage gain (Eq. (4.2.6) with $R_{S1} = 0$) of a single CS amplifier. Then, what is the CG stage for? It is expected to increase the frequency bandwidth.

This process of analyzing an FET cascode circuit as shown in Figure 4.34(a) has been cast into the following MATLAB function '`FET_cascode()`'.

```
function [VG1Q,VS1Q,VD1Q,VG2Q,VD2Q,IDQ,Av,Ri,Ro,gm,ro,vo,iD] = ...
          FET_cascode(VDD,Rs,R1,R2,R3,RS,RD,RL,Kp,Vt,lambda,vi)
% analyzes an FET cascode circuit of CS-CG like Fig. 4.34(a).
% Copyleft: Won Y. Yang, wyyang53@hanmail.net, CAU for academic use only
if length(VDD)==2, VSS=VDD(2); VDD=VDD(1); else VSS=0; end
if length(RS)==2, RS1=RS(1); RS2=RS(2); else RS1=0; RS2=RS; end
% DC analysis to find the Q-point
VG2Q=VDD*(R2+R3)/(R1+R2+R3); VG1Q=VDD*R3/(R1+R2+R3); RS=sum(RS);
A=Kp/2; B=-(Kp*(VG1Q-Vt)+1/RS); C=Kp/2*(VG1Q-Vt)^2+VSS/RS;
VS1Q=(-B-sqrt(B^2-4*A*C))/2/A; VD1Q=VG2Q-VG1Q+VS1Q;
IDQ=Kp/2*(VG1Q-VS1Q-Vt)^2; VD2Q=VDD-RD*IDQ; % Eqs.(E4.11.2,3,4)
VGS1Q=VG1Q-VS1Q; VGS2Q=VG2Q-VD1Q; VGD1Q=VG1Q-VD1Q; VGD2Q=VG2Q-VD2Q;
% If both FETs are not in saturation, AC analysis is not meaningful.
if VGD1Q>Vt|VGD2Q>Vt
 if VGD1Q>Vt, fprintf('M1 is ohmic: VGD1Q(%5.2f)>Vt\n',VGD1Q); end
 if VGD2Q>Vt, fprintf('M2 is ohmic: VGD2Q(%5.2f)>Vt\n',VGD2Q); end
 fprintf('The linear AC analysis is not meaningful\n');
end
% AC analysis
[gm(1),ro(1)]=gmro_FET(IDQ,VGS1Q,Kp,Vt,lambda);
[gm(2),ro(2)]=gmro_FET(IDQ,VGS2Q,Kp,Vt,lambda);
Ri2=1/gm(2); % Input resistance of stage 2 Eq.(4.2.15)
Ri=parallel_comb([R2 R3]); % Input resistance of stage 1 Eq.(4.2.5)
gmro1RS1=(gm(1)*ro(1)+1)*RS1; RD1=inf;
RDRL1=parallel_comb([RD1 Ri2]);
Ro1=parallel_comb([RD ro(1)+gmro1RS1]); % Eq.(4.2.7)
Av(1)=-gm(1)*ro(1)*RDRL1/(RDRL1+ro(1)+gmro1RS1); % Eq.(4.2.6b)
RDRL2=parallel_comb([RD RL]);
Rs2=Ro1; RS2=inf; RsRS2=parallel_comb([Rs2 RS2]);
Av(2)=(1+gm(2)*ro(2))*RDRL2/(RDRL2+ro(2)); % Eq.(4.2.16b)
Ro=parallel_comb([RD (1+gm(2)*ro(2))*RsRS2+ro(2)]); % Eq.(4.2.17)
Av0=1/(Rs/Ri+1); Gv=Av0*prod(Av); % Overall voltage gain
for n=1:length(vi)
  vGS1n=VGS1Q + vi(n)*Av0;
  iD(n)=Kp/2*(vGS1n-Vt)^2*(1+lambda*(VD1Q-VS1Q)); vo(n)=VD1Q-RD*iD(n);
end
```

```
function [gm,ro]=gmro_FET(ID,VGS,Kp,Vt,lambda)
if abs(Vt)>0.01
 if lambda>=100, lambda=1/lambda; end % lambda must be VA
 gm=2*ID/(VGS-Vt); % Transconductance by Eq.(4.1.4b)
 ro=1/lambda/abs(ID); % Output resistance by Eq.(4.1.4a)
else % gmro_FET(VGS,Kp,Vt,lambda)
 lambda=Vt; Vt=Kp; Kp=VGS; VGS=ID; % If ID is not given,
 if length(VGS)>1, VDS=VGS(2); VGS=VGS(1); else VDS=0; end
 gm=Kp*(VGS-Vt)*(1+lambda*VDS); % Transconductance by Eq.(4.1.4b)
 ro=2/lambda/Kp/(VGS-Vt)^2; % Output resistance by Eq.(4.1.4a)
end
```

Example 4.12 AC Analysis of a CS-CG (Cascode) Amplifier

For the amplifier whose AC equivalent is shown in Figure 4.35(a), find the input/output resistances (R_i/R_o) and the voltage gain $A_v = v_o/v_s$ in terms of g_{m1}, g_{m2}, r_{o1}, r_{o2}, and R_L.

It is obvious that $R_i = \infty$ since no current flows into/from terminal g1 however large v_s may be. To find the voltage gain, we draw the equivalent (with the FETs replaced by the model in Figure 4.3(b)) as depicted in Figure 4.35(b1) and write a set of two node equations (in two unknowns v_{d1} and v_o) as

$$\begin{bmatrix} 1/r_{o1}+1/r_{o2} & -1/r_{o2} \\ -1/r_{o2} & 1/r_{o2}+1/R_L \end{bmatrix} \begin{bmatrix} v_{d1} \\ v_o \end{bmatrix} = \begin{bmatrix} -g_{m1}v_T + g_{m2}v_{gs2} \\ -g_{m2}v_{gs2} \end{bmatrix} = \begin{bmatrix} -g_{m1}v_T - g_{m2}v_{d1} \\ g_{m2}v_{d1} \end{bmatrix}$$

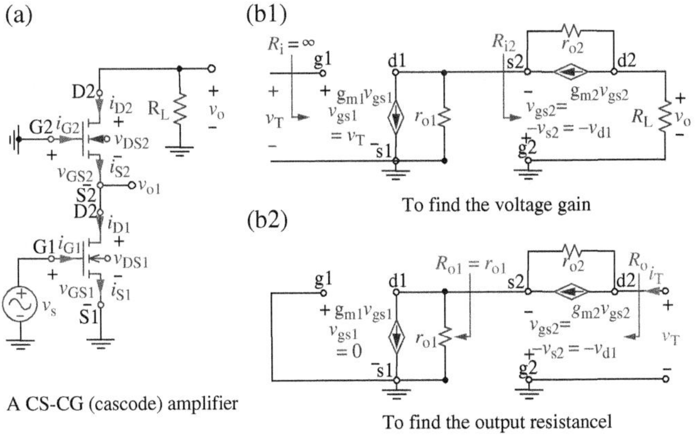

Figure 4.35 A CS-CG (cascode) amplifier and its equivalents to measure the voltage gain and output resistance.

$$\begin{bmatrix} 1/r_{o1}+1/r_{o2}+g_{m2} & -1/r_{o2} \\ -1/r_{o2}-g_{m2} & 1/r_{o2}+1/R_L \end{bmatrix} \begin{bmatrix} v_{d1} \\ v_o \end{bmatrix} = \begin{bmatrix} -g_{m1}v_T \\ 0 \end{bmatrix} \tag{E4.12.1}$$

We can solve this set of equations to find v_{d1}, v_o, and then the voltage gain:

$$\begin{aligned} A_v = \frac{v_o}{v_T} &= \frac{-(1/r_{o2}+g_{m2})g_{m1}}{1/r_{o1}r_{o2}+1/r_{o1}R_L+1/r_{o2}R_L+g_{m2}/R_L} \\ &= \frac{-(1+g_{m2}r_{o2})g_{m1}r_{o1}R_L}{R_L+r_{o1}+r_{o2}+g_{m2}r_{o1}r_{o2}} \end{aligned} \tag{E4.12.2}$$

$$\underset{g_{m2}r_{o2}>>1}{\overset{g_{m2}r_{o1}>>1}{\longrightarrow}} -g_{m1}\{g_{m2}r_{o1}r_{o2}\|R_L\} \overset{(E4.12.6)}{\approx} -g_{m1}\{R_o\|R_L\} \tag{E4.12.3}$$

On the other hand, to find the output resistance, we draw the equivalent as depicted in Figure 4.35(b2) and write a node equation (in unknown v_{d1}) as

$$\left(\frac{1}{r_{o1}}+\frac{1}{r_{o2}}\right)v_{d1} = g_{m2}v_{gs2}+\frac{v_T}{r_{o2}} = g_{m2}(-v_{d1})+\frac{v_T}{r_{o2}} \tag{E4.12.4}$$

We can solve this equation to find v_{d1} and $R_o = v_T/i_T$:

$$\begin{aligned} v_{d1} &= \frac{r_{o1}v_T}{r_{o1}+r_{o2}+g_{m2}r_{o1}r_{o2}} \\ ;\ i_T &= g_{m2}(-v_{d1})+\frac{v_T-v_{d1}}{r_{o2}} = \frac{1}{r_{o1}+r_{o2}+g_{m2}r_{o1}r_{o2}}v_T \end{aligned} \tag{E4.12.5}$$

$$;\ R_o = \frac{v_T}{i_T} = r_{o1}+r_{o2}+g_{m2}r_{o1}r_{o2} \underset{g_{m2}r_{o2}>>1}{\overset{g_{m2}r_{o1}>>1}{\longrightarrow}} g_{m2}r_{o1}r_{o2} \tag{E4.12.6}$$

```
%elec04e12.m
syms gm gm1 gm2 ro ro1 ro2 ro1 Rs RD RS RL
% To find the voltage gain
vT=1; % Test input voltage
Y=[1/ro1+1/ro2+gm2 -1/ro2; -1/ro2-gm2 1/ro2+1/RL];
v=Y\[-gm1*vT; 0]; % solve Eq.(E4.12.1)
vd1=v(1); vo=v(2); Av=vo/vT; % Eq.(E4.12.2)
fprintf('\n Av=\n'); pretty(simplify(Av))
% Output resistance
vd1=vT/ro2/(1/ro1+1/ro2+gm2);
iT=-gm2*vd1+(vT-vd1)/ro2; % Eq.(E4.12.3)
Ro=vT/iT; % Eq.(E4.12.4)
fprintf('\n Ro=\n'); pretty(simplify(Ro))
% Using MATLAB functions
Ri2=subs(Ri_CG,{gm,ro,RD,RS},{gm2,ro2,inf,inf}) % Ri of stage 2
% Output resistance
```

```
Ro1=subs(Ro_CS,{gm,ro,Rs,RD,RS},{gm1,ro1,inf,inf,0})
Roa=subs(Ro_CG,{gm,ro,Rs,RD,RS},{gm2,ro2,Ro1,inf,inf})
discrepancy_Ro=simplify(Ro-Roa)
% Voltage gain
Av1=subs(Av_CS,{gm,ro,Rs,RD,RS,RL},{gm1,ro1,0,inf,0,Ri2});
Av2=subs(Av_CG,{gm,ro,Rs,RD,RS},{gm2,ro2,Ro1,inf,inf});
Ava=Av1*Av2
discrepancy_Av=simplify(Av-Ava)
```

Alternatively, we can use the MATLAB functions like 'Ri_CG()', 'Av_CS()', etc. to obtain the same results by running the above MATLAB script "elec04e12.m".

Note that in determining whether a resistance in a CS/CD/CG amplifier circuit should be set to zero or infinity for removal, it may be helpful to compare the circuit with the corresponding model in Figures 4.24, 4.26, and 4.28, respectively.

Example 4.13 AC Analysis of a CS-CG (Cascode) Amplifier with a Cascode Current Source

For the amplifier whose AC equivalent is shown in Figure 4.36, find the input/output resistances (R_i/R_o) and the voltage gain $A_v = v_o/v_s$ in terms of g_{mk}'s and r_{ok}'s, and R_L on the assumption that $g_{mk}r_{ok} \gg 1$.

It is obvious that $R_i = \infty$. We can use Eq. (E4.12.6) to get the output resistance:

$$R_o = R_{oN} || R_{oP} \approx (g_{m2}\, r_{o1} r_{o2}) || (g_{m3} r_{o3}\, r_{o4}) \tag{E4.13.1}$$

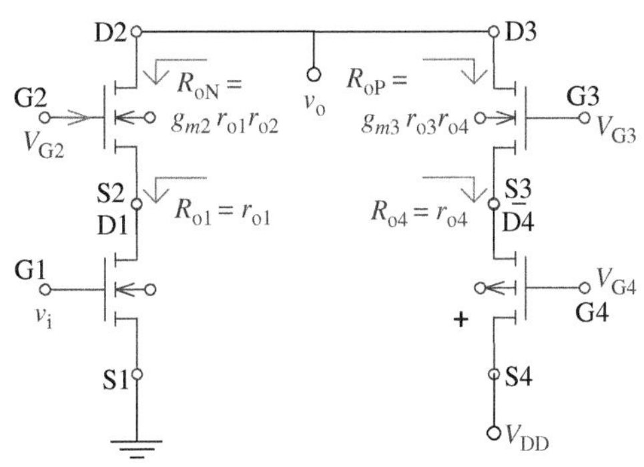

Figure 4.36 A CS-CG (cascode) amplifier with a cascode current-source load (consisting of two PMOSs).

We can use Eq. (E4.12.3) with $R_o = R_{oN}$ and $R_L = R_{oP}$ to get the voltage gain:

$$A_v \overset{(E4.12.3)}{\approx} -g_{m1}\{R_{oN} \| R_{oP}\} = -g_{m1}\,(g_{m2}\, r_{o1} r_{o2}) \| (g_{m3} r_{o3}\, r_{o4}) \tag{E4.13.2}$$

4.3 Design of FET Amplifier

4.3.1 Design of CS Amplifier

This section will show how a CS amplifier (illustrated in Figure 4.37) can be designed to achieve a desired voltage gain $A_{v,d}$ and a desired input resistance $R_{i,d}$, i.e. how the values of resistors constituting the circuit can be determined to make the voltage gain and input resistance as desired.

As with the CE amplifier discussed in Section 3.4.1, to maximize the AC swing of output voltage v_o along the AC load line, it may be good to set the drain current $I_{D,Q}$ and drain-to-source voltage $V_{DS,Q}$ of FET at the operating point Q as half the maximum drain current and about one-third of V_{DD}, respectively:

$$I_{D,Q} = \frac{I_{D,\max}}{2},\ V_{DS,Q} = K_C V_{DD} \text{ with } K_C \approx \frac{1}{3}(\text{design constant}) \tag{4.3.1}$$

The gate-to-source voltage $V_{GS,Q}$ at the operating point can be found from Eq. (4.1.2) or (4.1.13b) with $\lambda = 0$ as

$$I_{D,Q} \underset{\lambda=0}{\overset{(4.1.13b)}{=}} \frac{1}{2} K_p (V_{GS,Q} - V_t)^2;\ V_{GS,Q} = V_t + \text{sign}(I_{D,Q})\sqrt{2\,|I_{D,Q}|/K_p} \tag{4.3.2}$$

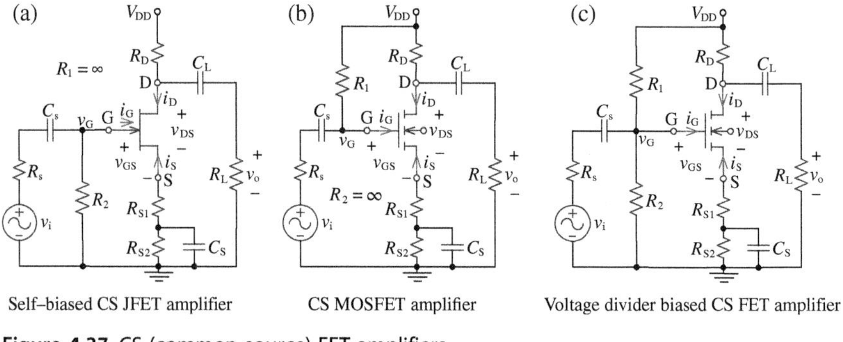
(a)
(b)
(c)
Self–biased CS JFET amplifier
CS MOSFET amplifier
Voltage divider biased CS FET amplifier

Figure 4.37 CS (common-source) FET amplifiers.

Then, R_D and R_S are determined in different ways depending on whether the biasing circuit consists of one resistor (R_2 or R_1 as shown in Figure 4.37(a) or (b)) or two resistors R_1-R_2 (constituting a voltage divider as shown in Figure 4.37(c)):

Case I When the Biasing Circuit Consists of One Resistor R_1 or R_2 (as Shown in Figure 4.37(b) or (a))

Noting that $i_G = 0$ and $v_G = 0/V_{DD}$ for (a)/(b), we apply KVL to the G-S loop to determine R_S as

$$v_G = V_{GS,Q} + R_S I_{D,Q}\,;\; R_S = \frac{v_G - V_{GS,Q}}{I_{D,Q}} \text{ with } v_G = 0 \text{ or } V_{DD} \tag{4.3.3}$$

Then, R_D is determined so that R_D and R_S share the rest of V_{DD}, i.e. $V_{DD} - V_{DS,Q} = (1 - K_C)V_{DD}$:

$$(R_S + R_D)I_{D,Q} = (1 - K_C)V_{DD}\,;\; R_D = (1 - K_C)\frac{V_{DD}}{I_{D,Q}} - R_S \tag{4.3.4}$$

The value of resistor R_1 or R_2 is determined as the desired input resistance.

Case II When the Biasing Circuit Consists of Two Resistors R_1-R_2 (as Shown in Figure 4.37(c))

It will also be good for maximizing the AC swing of output voltage v_o along the AC load line to determine such a value of R_D that the i_D intercept of the AC load line can be about $2I_{D,Q}$:

$$I_{D,Q} + \frac{V_{DS,Q}}{R_{AC} \approx (R_D \| R_L)} = 2I_{D,Q}\,;\; R_{AC} \approx \frac{1}{1/R_D + 1/R_L} = \frac{V_{DS,Q}}{I_{D,Q}}$$
$$;\, R_D = \frac{1}{I_{D,Q}/V_{DS,Q} - 1/R_L} \tag{4.3.5}$$

Here, R_{S1}, which is a part of the AC resistance $R_{AC} = R_{S1} + (R_D||R_L)$, has been neglected because it is presumably much less than $(R_D||R_L)$ and has not yet been determined.

Then, R_S is determined so that R_S and R_D can share the rest of V_{DD}, i.e. $V_{DD} - V_{DS,Q} = (1 - K_C)V_{DD}$:

$$(R_S + R_D)I_{D,Q} = (1 - K_C)V_{DD}\,;\; R_S = (1 - K_C)\frac{V_{DD}}{I_{D,Q}} - R_D \tag{4.3.6}$$

On the condition that the gate voltage does not exceed V_{DD}, i.e.

$$|V_G| = |V_S + V_{GS,Q}| = |R_S I_{D,Q} + V_{GS,Q}| \le V_{DD} \tag{4.3.7}$$

we determine the values of resistances R_1 and R_2 so that their parallel combination equals the desired input resistance $R_{i,d}$:

$$R_1 = \frac{R_1 R_2/(R_1 + R_2)}{R_2/(R_1 + R_2)} = \frac{R_{i,d}}{V_G/V_{DD}}, \quad R_2 = \frac{1}{1/R_{i,d} - 1/R_1} = \frac{R_{i,d}}{(V_{DD} - V_G)/V_{DD}} \tag{4.3.8a,b}$$

However, if the inequality (4.3.7) is not satisfied, we should let $V_G = V_{DD}$ together with

$$R_1 = R_{i,d} \,\text{and}\, R_1 = 10^{10}(\infty) \tag{4.3.9a,b}$$

and adjust $I_{D,Q}$ to the value satisfying Eqs. (4.3.2) and (4.3.7) with $V_G = V_{DD}$, i.e. the (smaller) root of the following quadratic equation:

$$\begin{aligned} &V_{DD} - R_S I_{D,Q} \underset{V_G = V_{DD}}{\overset{(4.3.7)}{=}} V_{GS,Q} \overset{(4.3.2)}{=} V_t + \text{sign}(I_{D,Q})\sqrt{2|I_{D,Q}|/K_p} \\ &; (V_{DD} - R_S I_{D,Q} - V_t)^2 = 2I_{D,Q}/K_p \\ &; R_S^2 I_{D,Q}^2 - 2\{R_S(V_{DD} - V_t) + 1/K_p\} I_{D,Q} + (V_{DD} - V_t)^2 = 0 \end{aligned} \tag{4.3.10}$$

Also, R_S and R_D should be recomputed using Eqs. (4.3.5) and (4.3.6) with the new value of $I_{D,Q}$ obtained as the root of Eq. (4.3.10).

Now, with the output (drain-source) resistance r_o and transconductance g_m:

$$r_o \overset{(4.1.4a)}{\approx} \frac{1}{\lambda I_{D,Q}}, \quad g_m \overset{(4.1.4b)}{\approx} K_p(V_{GS,Q} - V_t) \underset{\lambda=0}{\overset{(4.1.2)}{\approx}} \frac{2I_{D,Q}}{V_{GS,Q} - V_t} \tag{4.3.11a,b}$$

Eq. (4.2.6b) is used to determine the dual source resistance $[R_{S1}, R_{S2}]$ so that the desired voltage gain can be achieved:

$$A_v = \frac{v_o}{v_g} \overset{(4.2.6b)}{=} -\frac{-g_m r_o (R_D \| R_L)}{(R_D \| R_L) + r_o + (1 + g_m r_o) R_{S1}}$$

$$; R_{S1} = \frac{1}{g_m r_o + 1}\left\{\frac{g_m r_o (R_D \| R_L)}{|A_{v,d}|} - (R_D \| R_L) - r_o\right\}, \quad R_{S2} = R_S - R_{S1} \tag{4.3.12a,b}$$

```
function [R1,R2,RD,RS1,RS2,PRs]= ...
           FET_CS_design(VDDSS,Kp,Vt,lambda,Avd,Rid,IDQ,RL,KC)
% Design a CS amp with given Avd & Rid (possibly as [Rid VG] with VG)
% at Q=(VDSQ=KC*VDD,IDQ)
% Set Rid=[Rid VG] to fix VG.
% Copyleft: Won Y. Yang, wyyang53@hanmail.net, CAU for academic use only
if nargin<9, KC=1/3; end % design parameter s.t. VDSQ=KC*VDD;
if length(VDDSS)>1, VSS=VDDSS(2); else VSS=0; end; VDD=VDDSS(1);
VDS=VDD-VSS; VDSQ=KC*VDS; Rid0=Rid;
VGSQ=Vt+sign(IDQ)*sqrt(2*abs(IDQ/Kp)); % Eq.(4.3.2)
if length(Rid)>1
   VG=Rid(2); Rid=Rid(1); RS=(VG-VGSQ)/IDQ; % Eq.(4.3.3)
   RD=max((1-KC)*VDS/IDQ-RS,10); % Eq.(4.3.4)
   K1=(VDD-VG)/VDD; K2=VG/VDD;
   R1=min(Rid/K2,1e10); R2=min(Rid/K1,1e10);
 else
  RD=max(1/(IDQ/VDSQ-1/RL),10); % Eq.(4.3.5)
  RS=max((1-KC)*VDS/IDQ-RD,10); % Eq.(4.3.6)
  VG=RS*IDQ+VGSQ; % Eq.(4.3.7)
  if abs(vG)<abs(VDD)
  R1=Rid*VDD/VG; R2=Rid*VDD/(VDD-VG); % Eq.(4.3.8)
  else
  vG=VDD; R1=Rid; R2=1e10; % Eq.(4.3.9)
  A=RS^2; B=-2*(RS*(VDS-Vt)+1/Kp); C=(VDS-Vt)^2;
  IDQ=(-B-sqrt(B^2-4*A*C))/2/A; % Eq.(4.3.10)
  RD=max(1/(IDQ/VDSQ-1/RL),10); % Eq.(4.3.5)
  RS=max((1-KC)*VDS/IDQ-RD,10); % Eq.(4.3.6)
  end
end
gm=Kp*(VGSQ-Vt)*(1+lambda*VDSQ); ro=1/lambda/abs(IDQ); % Eq.(4.1.4)
mug=gm*ro; RDL=parallel_comb([RD RL]);
RS1=max((mug*RDL/(abs(Avd))-RDL-ro)/(mug+1),0); % Eq.(4.3.12a)
RS2=RS-RS1; % Eq.(4.3.12b)
PR1=(VDD-VG)^2/R1; PR2=VG^2/R2; PRD=RD*IDQ^2;
PRS1=RS1*IDQ^2; PRS2=RS2*IDQ^2; PRs=[PR1 PR2 PRD PRS1 PRS2] % Eq.(4.3.13)
if RS<=10|RD<=10
 if abs(Avd)<mug*RDL/(RDL+ro)
  disp('Try again with smaller/larger values of IDQ/VDD')
  [R1,R2,RD,RS1,RS2,PRs]= ...
    FET_CS_design(VDDSS,Kp,Vt,lambda,Avd,VG,Rid0,0.9*IDQ,RL,KC);
  else
  error('Avd is too large to achieve; try with higher VDD or another TR
         having bigger gm.')
 end
 else
 disp('Design Results')
 disp(' R1 R2 RD RS1 RS2 Avd')
 fprintf('%8.0f %8.0f %8.0f %8.0f %8.0f %8.2f\n', R1,R2,RD,RS1,RS2,Avd)
end
```

Note that the minimum power ratings of R_1, R_2, R_C, R_{S1}, and R_{S2} should be

$$P_{R_1} = \frac{(V_{DD} - V_G)^2}{R_1}, P_{R_2} = \frac{V_G^2}{R_2}, P_{R_D} = R_D I_{D,Q}^2, \text{ and } P_{R_{Si}} = R_{Si} I_{D,Q}^2 \quad (4.3.13)$$

This process of designing a CS amplifier with a specified voltage gain $A_{v,d}$ and a desired input resistance $R_{i,d}$ has been cast into the above MATLAB function '`FET_CS_design()`' where the default value of design constant KC is 1/3.

Remark 4.1 About Fixing the Operating Point of CS FET Amplifier

1) As long as $V_{DS,Q}$ is less than V_{DD}, it seems to be good for reducing the possibility of distortion to increase $V_{DS,Q} = K_C V_{DD}$ so that the operating point can stay away from the saturation and cutoff regions. However, in view of Eq. (4.3.5)/(4.3.6), it should not be so high as to result in negative values of R_D/R_S.
2) The drain current $I_{D,Q}$ at the operating point should not be so small/large as to let (4.3.5)/(4.3.6) yield a negative value of R_D/R_S.

Let us use the MATLAB function '`FET_CS_design()`' to design the four-resistor biasing network for a CS amplifier using an NMOS IRF150 so that it can operate with a desired voltage gain $A_{v,d} = -20$ (for a load resistance of $R_L = 50$ kΩ) and an input resistance $R_{i,d}$=100 kΩ at an operating point Q=($V_{DS,Q}$, $I_{D,Q}$) = (V_{DD}/3,0.3 mA) where the device parameters of the FET are K_p = 3.08 A/V^2 and V_t = 2.831 V and a V_{DD} = 18 V-source is available for biasing the FET. To this end, we run the following script "design_CS_IRF150.m":

```
%design_CS_IRF150.m
clear
VDD=18; RL=5e4; % DC voltage source and load resistance
Avd=20; Rid=1e5; % Design specifications
KC=1/3; IDQ=3e-4; % Design parameters
Kp=3.08; Vt=2.831;
lambda=0.0075; % lambda=1/rds/IDQ=1/444.4e3/3e-4
[R1,R2,RD,RS1,RS2] = ...
   FET_CS_design(VDD,Kp,Vt,lambda,Avd,Rid,IDQ,RL,KC);
Rs=50;
Vsm=0.01; % Resistance/Amplitude of the AC input voltage source
[VGQ,VSQ,VDQ,IDQ,Av,Ri,Ro] = ...
  FET_CS_analysis (VDD,Rs,R1,R2,RD,[RS1 RS2],RL,Kp,Vt,lambda,Vsm);
```

This yields

```
>>design_CS_IRF150
Design Results
    R1       R2       RD      RS1    RS2     Avd
  371520  136830  33333   977   5690   20.00
Analysis Results
  VDD   VGQ  VSQ  VDQ      IDQ
18.00 4.84 2.00 8.00 3.00e-004
 in the saturation mode with VGD,Q= -3.15[V] (Vt=2.83)
 gm= 43.948[mS], ro= 464.368[kOhm]
 Ri= 100.00 kOhm, Ro= 33279 Ohm
Gv=Ri/(Rs+Ri)xAvoxRL/(Ro+RL) = 0.9995x(-33.30x0.6004=-19.99) = -19.98
```

Here, from the PSpice model opened by selecting the part IRF150 and clicking Edit > PSpice_Model from the top menu bar of the PSpice Schematic window or from the PSpice simulation output file (Figure 4.38(b)), we see $K_P = 20.53E{-}06$, $L = 2E{-}06$, $W = 0.3$, $V_{to} = 2.831$, and $R_{DS} = 444.4E{+}03$, which can be interpreted as meaning

$$K_p = K_P\frac{W}{L} = 20.53\,\mu\frac{0.3}{2\,\mu} = 3.08\,[\mathrm{A/V^2}],\ \ V_t = 2.831\,[\mathrm{V}] \tag{4.3.14a,b}$$

$$\lambda \overset{(4.1.4a)}{\approx} \frac{1}{r_o I_{D,Q}} = \frac{1}{r_{ds} I_{D,Q}} = \frac{1}{444.4\,\mathrm{k\Omega} \times 0.3\,\mathrm{mA}} = 0.0075\,\mathrm{V}^{-1} \tag{4.3.15}$$

The above results mean the following values of the resistances of designed CS amplifier:

$$\begin{aligned} &R_1 = 371\,520\,\Omega, R_2 = 136\,830\,\Omega, R_D = 33\,333\,\Omega, \\ &R_{S1} = 977\,\Omega, \text{and}\, R_{S2} = 5690\,\Omega \end{aligned} \tag{4.3.16}$$

The script uses '`FET_CS_analysis()`' (Section 4.2.1) for analyzing the designed circuit to get the DC analysis result:

$$V_{G,Q} = 4.84\,\mathrm{V}, V_{D,Q} = 8\,\mathrm{V}, V_{S,Q} = 2\,\mathrm{V}, \text{and}\, I_{D,Q} = 0.3\,\mathrm{mA} \tag{4.3.17}$$

and the AC analysis result: $G_v = -19.98$, $R_i = 100\ \mathrm{k\Omega}$, and $R_o = 33.3\ \mathrm{k\Omega}$.

Figure 4.38(a) and (b) shows the PSpice schematic of the designed CS amplifier and its simulation results where the overall voltage gain has turned out to be $G_{v,\mathrm{PSpice}} = -1.9887/99.725\ \mathrm{m} \approx -19.94$ as required by the design specification. It is implied that the MATLAB design and analysis functions have worked well.

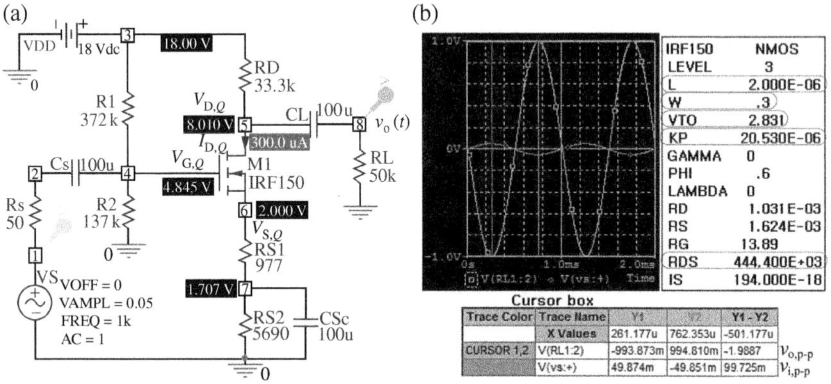

PSpice schematic for a CS circuit

Its simulation result

Figure 4.38 A CS (common-source) circuit and its PSpice simulation ("elec04f38.opj").

Example 4.14 Designing a CS Amplifier

Design a four-resistor biasing network for a CS amplifier using an NMOS whose device parameters are $K_p = 60$ mA/V^2, $V_t = 1.73$ V, and $\lambda = 4.17 \times 10^{-5}$ V^{-1} so that it can operate with a desired voltage gain $A_{v,d} = -20$ (for a load resistance of $R_L = 50$ kΩ) and an input resistance $R_{i,d} = 100$ kΩ at an operating point $Q = (V_{DS,Q}, I_{D,Q}) = (V_{DD}/3, 0.5\text{ mA})$ where a $V_{DD} = 18$ V-source is available for biasing the FET.

To this end, we run the following script "elec04e14.m," which yields

```
>>elec04e14
Design Results
    R1       R2       RD     RS1    RS2    Avd
  301793  149556   15789   471   7740   20.00
Analysis Results
  VDD   VGQ   VSQ   VDQ     IDQ
18.00 5.96 4.11 10.11 5.00e-004
 in the saturation mode with gm= 7.747[mS], ro=47973.445[kOhm]
  Ri= 100.00kOhm, Ro= 15788Ohm
Gv=Ri/(Rs+Ri)xAvoxRL/(Ro+RL) =0.9995x(-26.31x0.7600=-20.00)= -19.99
```

```
%elec04e14.m
VDD=18; RL=5e4; % DC voltage source and load resistance
Rs=50; Vsm=0.01; % Resistance/Amplitude of the AC input voltage source
Avd=20; Rid=1e5; % Design specifications
KC=1/3; IDQ=5e-4; % Design parameters
Kp=0.06; Vt=1.73; lambda=4.17e-5;
[R1,R2,RD,RS1,RS2]= ... FET_CS_design(VDD,Kp,Vt,lambda,Avd,Rid,IDQ,RL,KC);
Rs=50; Vsm=0.1; % Resistance and Amplitude of the AC input voltage source
[VGQ,VSQ,VDQ,IDQ,Av,Ri,Ro]= ...
  FET_CS_analysis(VDD,Rs,R1,R2,RD,[RS1 RS2], RL,Kp,Vt,lambda,Vsm);
```

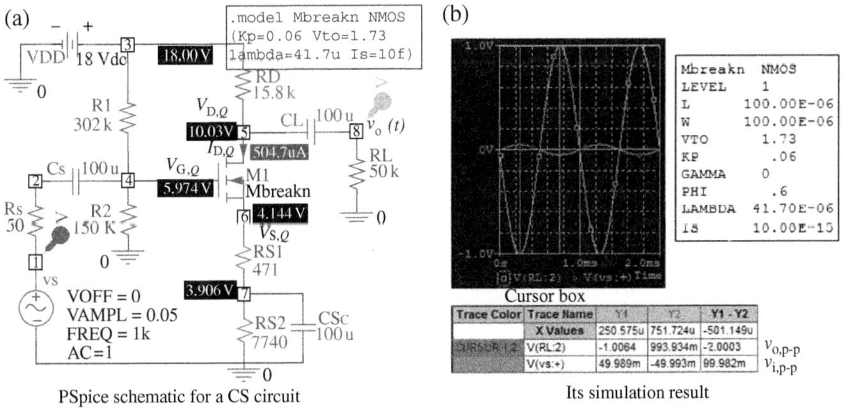

Figure 4.39 A CS (common-source) circuit and its PSpice simulation ("elec04e14.opj").

The above analysis results mean the following values of the resistances of designed CS amplifier:

$$R_1 = 301\,793\,\Omega, R_2 = 149\,556\,\Omega, R_D = 15\,789\,\Omega,$$
$$R_{S1} = 471\,\Omega, \text{ and } R_{S2} = 7740\,\Omega \tag{E4.12.1}$$

The script uses '`FET_CS_analysis()`' for analyzing the designed circuit to get the DC analysis result:

$$V_{G,Q} = 5.96\,\text{V}, V_{D,Q} = 10.11\,\text{V}, V_{S,Q} = 4.11\,\text{V}, I_{D,Q} = 0.5\,\text{mA} \tag{E4.12.2}$$

and the AC analysis result: $G_v = -20.16$, $R_i = 100\ \text{k}\Omega$, and $R_o = 15.8\ \text{k}\Omega$.

Figure 4.39(a) and (b) shows the PSpice schematic of the designed CS amplifier and its simulation results where the overall voltage gain turns out to be $G_{v,\text{PSpice}} = -2.0003/0.1 \approx -20$ as required by the design specification. It is implied that the MATLAB design and analysis functions have worked well.

4.3.2 Design of CD Amplifier

This section will show how a CD amplifier with $R_D = 0$ (as shown in Figure 4.40) can be designed to achieve a desired input resistance $R_{i,d}$, i.e. how the values of resistors constituting the circuit can be determined to yield a desired input resistance.

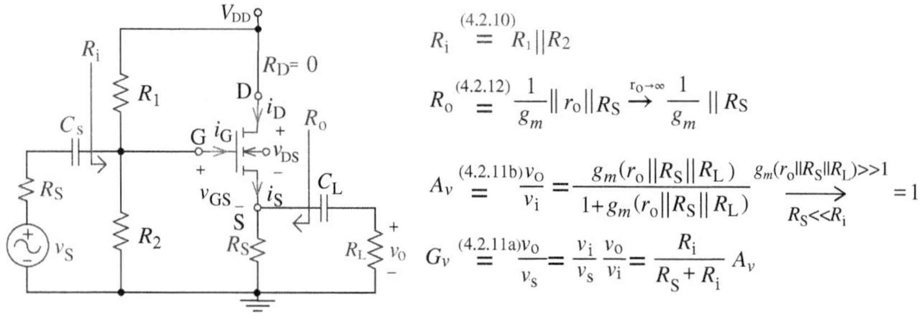

Figure 4.40 A CD (common-drain) FET circuit and its low-frequency AC equivalent.

As with the CS amplifier discussed in Section 4.3.1, to maximize the AC swing of output voltage v_o along the AC load line, it may be good to set the draincurrent $I_{D,Q}$ and drain-to-source voltage $V_{DS,Q}$ of FET at the operating point Q as half the maximum drain current and about one-third of V_{DD}, respectively:

$$I_{D,Q} = \frac{I_{D,max}}{2}, \quad V_{DS,Q} = K_C V_{DD} \text{ with } K_C \approx \frac{1}{3} (\text{design constant}) \tag{4.3.18}$$

The source resistance R_S can be determined as

$$R_S = \frac{V_{DD} - V_{DS,Q}}{I_{D,Q}} \tag{4.3.19}$$

Also, the gate-to-source voltage $V_{GS,Q}$ at the operating point can be determined from Eq. (4.1.1b) or (4.1.13b) as

$$I_{D,Q} \underset{\lambda=0}{\overset{(4.1.13b)}{=}} \frac{1}{2} K_p \left(V_{GS,Q} - V_t\right)^2; \; V_{GS,Q} = V_t + \text{sign}\left(I_{D,Q}\right)\sqrt{\frac{2\,|I_{D,Q}|}{K_p}} \tag{4.3.20}$$

Then, the node voltage $V_{G,Q}$ at the gate terminal can be obtained as

$$V_{G,Q} = R_S I_{D,Q} + V_{GS,Q} \tag{4.3.21}$$

If $V_{G,Q} \le V_{DD}$, the values of resistors R_1 and R_2 can be determined so that their parallel combination equals the desired input resistance $R_{i,d}$:

$$R_1 = \frac{R_1 R_2/(R_1 + R_2)}{R_2/(R_1 + R_2)} = \frac{R_{i,d}}{V_G/V_{DD}} \tag{4.3.22a}$$

$$R_2 = \frac{1}{1/R_{i,d} - 1/R_1} = \frac{R_{i,d}}{(V_{DD} - V_G)/V_{DD}} \tag{4.3.22b}$$

Otherwise, i.e. if $V_{G,Q} > V_{DD}$ (which is impossible), we set $V_{G,Q} = V_{DD}$ and

$$R_1 = R_{i,d}, \; R_2 = \infty \tag{4.3.23a,b}$$

If you want to find the new operating point, the KCL equation at node S,

$$i_D(v_{GS}, v_{DS}) - \frac{v_S}{R_S} = 0;\ v_S - R_S i_D(V_G - v_S, V_{DD} - v_S) = 0 \tag{4.3.24}$$

should be solved for v_S, yielding $V_{S,Q}$ and consequently,

$$I_{D,Q} = i_D\left(V_{G,Q} - V_{S,Q}, V_{DD} - V_{S,Q}\right) \text{ with } V_{G,Q} = V_{DD} \tag{4.3.25}$$

Now, suppose the resulting voltage gain A_v turns out to be intolerably smaller than 1 and/or the resulting output resistance R_o is quite larger than you expected. Should we increase or decrease the drain current $I_{D,Q}$ to improve the values of A_v and R_o? To find the answer to this question, let us recollect the formulas for determining A_v and R_o of a CD amplifier:

$$A_v = \frac{v_o}{v_i} \overset{(4.2.11b)}{=} \frac{g_m}{g_m + 1/(r_o \| R_S \| R_L)} = \frac{g_m(r_o \| R_S \| R_L)}{1 + g_m(r_o \| R_S \| R_L)} = g_m \times \left\{ \frac{1}{g_m} \| r_o \| R_S \| R_L \right\} \tag{4.3.26}$$

$$R_o = \frac{v_T}{i_T} \overset{(4.2.12)}{=} \frac{r_o \| R_S}{1 + g_m(r_o \| R_S)} = \frac{1}{g_m} \| r_o \| R_S \tag{4.3.27}$$

$$g_m = \left.\frac{\partial i_D}{\partial v_{GS}}\right|_Q \overset{(4.1.4b)}{\approx} K_p\left(V_{GS,Q} - V_t\right) \overset{(4.1.2)}{\approx} \frac{2I_{D,Q}}{V_{GS,Q} - V_t} \tag{4.3.28}$$

Noting from Eqs. (4.3.19)?> and (4.3.28) that as $I_{D,Q}$ increases, R_S decreases, and g_m increases, we can tell that a larger value of $I_{D,Q}$ will help to increase A_v (Eq. (4.3.26)) and decrease R_o (Eq. (4.3.27)).

The above process of designing a CD amplifier with a desired input resistance $R_{i,d}$ has been cast into the following MATLAB function '`FET_CD_design()`'.

```
function [R1,R2,RS,PRs,IDQ]=...
       FET_CD_design(VDDSS,Kp,Vt,lambda,Rid,IDQ,RL,KC)
% Design a CD amplifier with RD=0 which has a given value of Rid
%  at Q=(VDSQ=KC*(VDD-VSS),IDQ) where Ri=R1||R2 and Ro=(1/gm)||RS.
% Copyleft: Won Y. Yang, wyyang53@hanmail.net, CAU for academic use only
if nargin<8, KC=1/3; end % Design parameter s.t. VDSQ=KC*(VDD-VSS);
if nargin<7, RL=1e10; end % No load
if length(VDDSS)>1, VSS=VDDSS(2); else VSS=0; end
VDD=VDDSS(1);
VDSQ=KC*(VDD-VSS); % Eq.(4.3.18b)
RS=(VDD-VSS-VDSQ)/IDQ; % Eq.(4.3.19)
VGSQ=Vt+sign(IDQ)*sqrt(2*abs(IDQ/Kp)); % Eq.(4.3.20)
```

```
VG=VSS+RS*IDQ+VGSQ; % Eq.(4.3.21)
if VG<VDD
 R1=Rid*VDD/VG; R2=Rid*VDD/(VDD-VG); % Eq.(4.3.22)
 else
 VG=VDD; R1=Rid; R2=1e10; % Eq.(4.3.23)
 iD=@(vGS,vDS)Kp*(1+lambda*vDS).*((vGS-Vt).^2/2.*(vGS>=Vt).*(vGS-vDS<Vt)
   +((vGS-Vt).*vDS-vDS.^2/2).*(vGS>=Vt).*(vGS-vDS>=Vt)); % Eq.(4.1.13)
 options=optimset('Display','off','Diagnostics','off');
 eq=@(v)v-RS*iD(VG-v,VDD-v)-VSS; % Eq.(4.3.24)
 v0=(VG+VSS)/2;
 VS=fsolve(eq,v0,options);
 IDQ1=iD(VG-VS,VDD-VS); % Eq.(4.3.25)
 VDSQ1=VDD-VS;
 fprintf('\n Q=(IDQ,VDSQ)=(%6.3fmA,%6.3fV) has been adjusted to
          (%6.3fmA,%6.3fV)\n', IDQ,VDSQ,IDQ1,VDSQ1);
end
PRs=[(VDD-VG)^2/R1 VG^2/R2 RS*IDQ^2]; % Power ratings of R1,R2,and RS
fprintf('Design results at Q=(IDQ,VDSQ)=(%6.3fmA,%6.3fV)',
        IDQ*1e3,VDSQ);
disp('   R1       R2      RS')
fprintf('%8.0fOhm %8.0fOhm %6.0fOhm\n', R1,R2,RS)
```

Example 4.15 Designing a CD Amplifier

Design a biasing network for a CD amplifier with $R_D = 0$ (as shown in Figure 4.40) using an NMOS whose device parameters are $K_p = 60\ \text{mA/V}^2$, $V_t = 1.73\ \text{V}$, and $\lambda = 4.17 \times 10^{-5}\ \text{V}^{-1}$ so that it can have a desired input impedance $R_{i,d} = 100\ \text{k}\Omega$ at an operating point $Q = (V_{DS,Q}, I_{D,Q}) = (V_{DD}/3, 0.05\ \text{mA})$ where a $V_{DD} = 18\ \text{V}$-source is available for biasing the FET.

We first run the following MATLAB statements:

```
>>VDD=18; RL=5e3; % DC voltage source and load resistance
  Rid=1e5; KC=1/3; IDQ=5e-5; % Design spec and Design parameters
  Kp=0.06; Vt=1.73; lambda=4.17e-5; % Device parameters
  [R1,R2,RS]=FET_CD_design(VDD,Kp,Vt,lambda,Rid,IDQ,RL,KC);
```

This yields

```
   Design results at Q=(IDQ,VDSQ)=( 0.050mA, 6.000V)
      R1            R2         RS
  130711Ohm  425615Ohm 240000Ohm
```

Based on this design result, we let $R_1 = 131\ \text{k}\Omega$, $R_2 = 426\ \text{k}\Omega$, and $R_S = 240\ \text{k}\Omega$ and use '`FET_CD_analysys()`' to perform the DC + AC analysis of the designed CD amplifier:

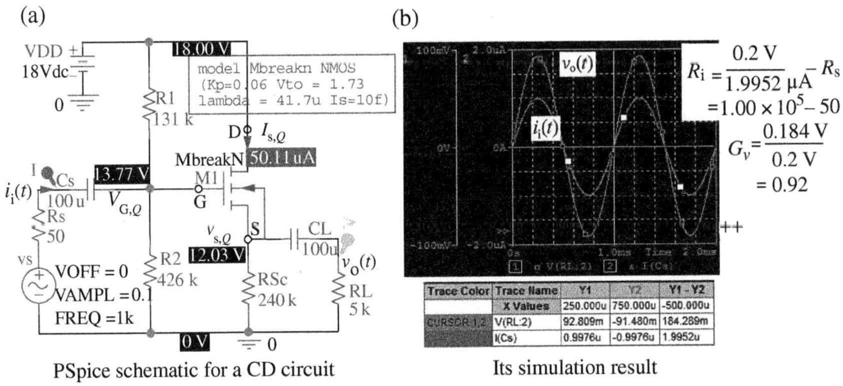

Trace Color	Trace Name	Y1	Y2	Y1 - Y2
	X Values	250.000u	750.000u	-500.000u
CURSOR 1,2	V(RL:2)	92.809m	-91.480m	184.289m
	I(Cs)	0.9976u	-0.9976u	1.9952u

Figure 4.41 A CS (common-source) circuit and its PSpice simulation ("elec04e15.opj").

```
>>Rs=50; R1=131e3; R2=426e3; RD=0;
  [VGQ,VSQ,VDQ,IDQ,Av,Ri,Ro] = ...
     FET_CD_analysis(VDD,Rs,R1,R2,RD,RS,RL,Kp,Vt,lambda);
```

This yields

```
Analysis Results with R1= 131kOhm, R2= 426kOhm, RD= 0Ohm, RS=240000Ohm
 VDD   VGQ   VSQ   VDQ    IDQ
18.00 13.77 12.00 18.00 5.00e-05
 in the saturation mode with VGD,Q= -4.23[V] (Vt=1.73)
   gm=  2.449[mS], ro=479904.587[kOhm]
 Ri= 100.19[kOhm], Ro=  408[Ohm]
Gv=Ri/(Rs+Ri)xAvoxRL/(Ro+RL) =0.9995x(0.9983x0.9246=0.9231)=0.9226
```

This analysis result conforms with the PSpice simulation result shown in Figure 4.41. You can try with a larger value of $I_{D,Q}$, say, 0.5 mA, to increase the voltage gain A_v.

4.4 FET Amplifier Frequency Response

In this section, we will find the transfer function $G(s) = V_o(s)/V_i(s)$ and frequency response $G(j\omega)$ for a CS amplifier, a CD amplifier, and a CG amplifier with the FET replaced by the high-frequency small-signal model shown in Figure 4.3(a).

4.4.1 CS Amplifier

Figure 4.42(a) and (b) shows a CS amplifier circuit and its high-frequency small-signal equivalent, respectively, where one more load capacitor C_{LL}, in addition to the output capacitor C_L, is connected in parallel with the load resistor R_L. For the equivalent circuit shown in Figure 4.42(b), a set of three node equations in $V_1 = V_g$, $V_2 = V_d$, and $V_3 = V_s$ can be set up as

$$\begin{bmatrix} Y_G + sC_{gs} + sC_{gd} & -sC_{gd} & -sC_{gs} \\ -sC_{gd} & sC_{gd} + 1/r_o + sC_{ds} + Y_D & -1/r_o - sC_{ds} \\ -sC_{gs} & -1/r_o - sC_{ds} & sC_{gs} + 1/r_o + sC_{ds} + Y_S \end{bmatrix} \begin{bmatrix} V_1 \\ V_2 \\ V_3 \end{bmatrix} = \begin{bmatrix} I_{Nt} \\ -g_m v_{gs} = -g_m(V_1 - V_3) \\ g_m v_{gs} = g_m(V_1 - V_3) \end{bmatrix} \tag{4.4.1}$$

where

$$Y_s = \frac{1}{Z_s} = \frac{1}{R_s + 1/sC_s},\ I_{Nt} = Y_s V_s,\ Y_G = Y_s + G_G = Y_s + \frac{1}{R_G = (R_1 \| R_2)}$$
$$Y_D = \frac{1}{R_D \| (1/sC_L + Z_L)},\ Y_S = \frac{1}{R_{S1} + (R_{S2} \| 1/sC_S)}, Z_L = R_L \| \frac{1}{sC_{LL}} \tag{4.4.2}$$

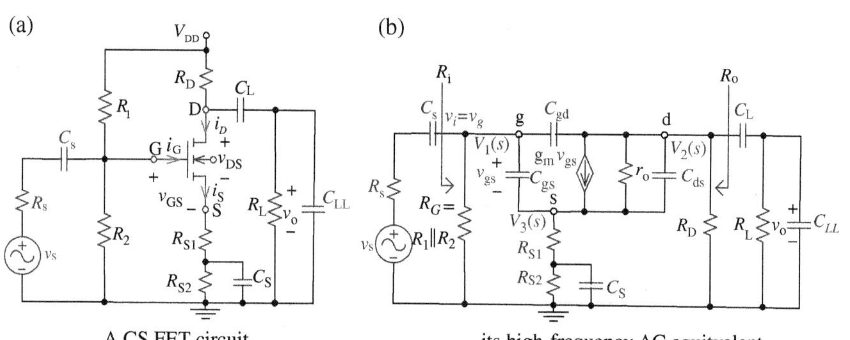

Figure 4.42 A CS (common-source) FET circuit and its high-frequency AC equivalent.

(Q) How about if $R_{S1} + R_{S2} = 0$?

Here, I_{Nt} and Y_B are the values of Norton current source and admittance looking back into the source part from terminals g to 0. This equation can be rearranged into a solvable form with all the unknown/known terms on the LHS/RHS as

$$\begin{bmatrix} Y_G + sC_{gs} + sC_{gd} & -sC_{gd} & -sC_{gs} \\ -sC_{gd} + g_m & sC_{gd} + 1/r_o + sC_{ds} + Y_D & -1/r_o - sC_{ds} - g_m \\ -sC_{gs} - g_m & -1/r_o - sC_{ds} & sC_{gs} + 1/r_o + sC_{ds} + Y_S + g_m \end{bmatrix} \begin{bmatrix} V_1(s) \\ V_2(s) \\ V_3(s) \end{bmatrix} = \begin{bmatrix} I_{Nt} \\ 0 \\ 0 \end{bmatrix} \tag{4.4.3}$$

and solved for V_1, V_2, and V_3. Then, the transfer function and frequency response can be found:

$$G(s) = \frac{V_o(s)}{V_s(s)} \text{ and } G(j\omega) \text{ with } V_o(s) = \frac{R_L \| (1/sC_{LL})}{1/sC_L + (R_L \| 1/sC_{LL})} V_2(s) \tag{4.4.4}$$

This process to find the transfer function and frequency response has been cast into the following MATLAB function '`FET_CS_xfer_ftn()`'.

Note the following about the internal capacitances of an FET, i.e. the gate-to-source capacitance C_{gs}, the gate-to-drain capacitance (sometimes called the overlap capacitance) C_{gd}, and the drain-to-source capacitance C_{ds} ([S-1]):

- C_{ds} can usually be ignored since $C_{ds} \ll C_{gs}$. C_{gs} and C_{gd} can be modeled as voltage-dependent capacitances with values determined by

$$C_{gs} \underset{(10.7\text{-}1)}{\overset{[S\text{-}1]}{=}} \frac{C_{gs0}}{(1 + |V_{GS,Q}|/V_b)^m}, \quad C_{gd} \underset{(10.7\text{-}2)}{\overset{[S\text{-}1]}{=}} \frac{C_{gd0}}{(1 + |V_{GD,Q}|/V_b)^m} \tag{4.4.5}$$

  where C_{gs0}/C_{gd0}[F]: zero-bias gate-source/gate-drain junction capacitances, respectively, $V_{GS,Q}/V_{GD,Q}$[V]: quiescent gate-source/gate-drain voltages, respectively, m(Mj): gate p-n grading coefficient (SPICE default = 0.5), and Vb(Pb): gate junction (barrier) potential, typically 0.6 V (SPICE default = 1 V).

- The zero bias capacitances have dimensions of [F/m]/[F] for MOSFET/JFET, respectively.

Referring to Figure 4.43, four break (pole or corner) frequencies determining roughly the frequency response magnitude can be determined as [R-2]

```
function [Gs,Av,Ri,Ro,Cgs,Cgd,gm,ro,Gw]=...
 FET_CS_xfer_ftn(VDD,Rs,Cs,R1,R2,RD,RS,CS,CL,RLCLL,
                  Kp,Vt,lambda,CVbm,w)
%To find the transfer function/frequency response of a CS amplifier
% CVbm= [Cgs0,Cgd0,Cds,Vb,m]%
with Cgs0|Cgd0: Zero-bias GSJ|GDJ capacitances
% Vb: Gate junction (barrier) potential
% m : Gate p-n grading coefficient
% Copyleft: Won Y. Yang, wyyang53@hanmail.net, CAU for academic use only
Cgs0=CVbm(1); Cgd0=CVbm(2); Cds=CVbm(3); Vb=CVbm(4); m=CVbm(5);
RG=parallel_comb([R1 R2]); RL=RLCLL(1);
if numel(RLCLL)<2, CLL=0; else CLL=RLCLL(2); end
syms s; ZL=parallel_comb([RL 1/s/CLL]); % Eq.(4.4.2f)
if length(RS)==1, RS1=0; RS2=RS;
else RS1=RS(1); RS2=RS(2); end
[VGQ,VSQ,VDQ,IDQ,Av,Ri,Ro,gm,ro]= ...
  FET_CS_analysis(VDD,Rs,R1,R2,RD,RS,RL,Kp,Vt,lambda);
Cgs=Cgs0/(1+abs(VGQ-VSQ)/Vb)^m;
Cgd=Cgd0/(1+abs(VGQ-VDQ)/Vb)^m; % Eq.(4.4.5)
sCs=s*Cs; sCS=s*CS; sCL=s*CL;
sCgs=s*Cgs; sCgd=s*Cgd; sCds=s*Cds;
YD=1/parallel_comb([RD 1/sCL+ZL]);
Ys=1/(Rs+1/sCs); % Eq.(4.4.2d,e)
if sum(RS)>0
 YS=1/(RS1+1/(1/RS2+sCS)); % Admittance at terminal S
 Y=[Ys+1/RG+sCgs+sCgd   -sCgd              -sCgs;
   -sCgd+gm       sCgd+1/ro+sCds+YD    -1/ro-sCds-gm;
   -sCgs-gm -1/ro-sCds sCgs+1/ro+sCds+YS+gm]; % Eq.(4.4.3)
 V=Y\[Ys; 0; 0];
else
 Y=[Ys+1/Ri+sCgs+sCgd -sCgd;
   -sCgd+gm sCgd+1/ro+sCds+YD];
 V=Y\[Ys; 0];
end
Gs=V(2)*ZL/(ZL+1/sCL); % Transfer function Vo(s)/Vs(s)
if nargin>14, Gw=subs(Gs,'s',j*w); % Frequency response
else Gw=0; end
```

$$\omega_{c1} \overset{[R\text{-}2]}{\underset{(8.40)}{=}} \frac{1}{C_s(R_s+R_i)}, \quad \omega_{c2} \overset{[R\text{-}2]}{\underset{(8.41)}{=}} \frac{1}{C_L(R_o+R_L)} \tag{4.4.6a,b}$$

$$\omega_{c3} \overset{[R\text{-}2]}{\underset{(8.57)}{=}} \frac{1}{(C_{gs}+C_m)(R_s\|R_i)}, \quad \omega_{c4} \overset{[R\text{-}2]}{\underset{(8.58)}{=}} \frac{1}{(C_n+C_{LL})(R_o\|R_L)} \tag{4.4.6c,d}$$

Note the following about these four break frequencies:

- ω_{C1}, related with the coupling capacitor C_s, is obtained as the reciprocal of the product of C_s and the equivalent resistance as seen from C_s (Figure 4.43(a)).

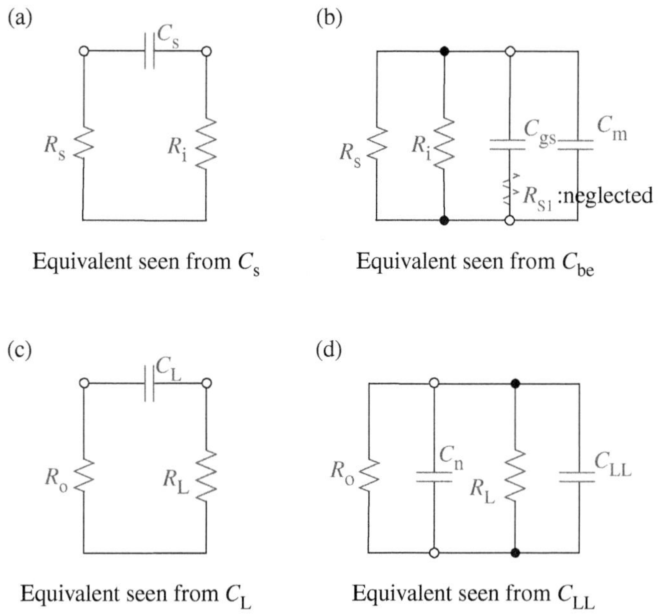

Figure 4.43 (Approximate) equivalent circuits for finding the equivalent resistance seen from each capacitor.

- ω_{C2}, related with the output capacitor C_L, is obtained as the reciprocal of the product of C_L and the equivalent resistance as seen from C_L (Figure 4.43(c)).
- ω_{C3}, related with the internal capacitors $C_{gs}|C_{gd}$, is obtained as the reciprocal of the product of $(C_{gs} + C_m)$ and the equivalent resistance as seen from C_{gs} (Figure 4.43(b)).
- ω_{C4}, related with the internal capacitors $C_{gd}|C_{LL}$, is obtained as the reciprocal of the product of $(C_n + C_{LL})$ and the equivalent resistance as seen from C_{LL} (Figure 4.43(d)).
- The passband of the CS amplifier will be approximately lower-/upper-bounded by the second/third highest break frequencies where not only the values but also the order of the frequencies varies with the values of the related capacitances and resistances.

Note also that R_i/R_o are the input/output resistances of the CS amplifier, respectively, and C_m/C_n are the Miller equivalent capacitances for capacitor C_{gd} seen from the input/output side, respectively (Eq. (1.4.2):

$$C_m = C_{gd}(1 - A_v) \text{ and } C_n = C_{gd}(A_v - 1)/A_v \quad (4.4.7)$$
$$\text{with } A_v = V_2/V_1 \text{ in Figure 4.42(b)}$$

The following MATLAB function 'break_freqs_of_CS()' uses Eqs. (4.4.6a,b,c,d) to find the four break frequencies.

```
function fc=break_freqs_of_CS ...
  (Rs,Cs,CL,RL,CLL,Cgd,Cgs,Cds,Av,Ri,Ro,ro)
% To find the 4 break frequencies of a CS amplifier
RsRi=parallel_comb([Rs Ri]);
RoRL=parallel_comb([Ro RL]);
Cm=Cgd*(1-Av); Cn=Cgd*(1-1/Av); % Eq.(4.4.7a,b)
fc(1)=1/2/pi/Cs/(Rs+Ri); % Eq.(4.4.6a)
fc(2)=1/2/pi/CL/(Ro+RL); % Eq.(4.4.6b)
fc(3)=1/2/pi/(Cgs+Cm)/RsRi; % Eq.(4.4.6c)
fc(4)=1/2/pi/(CLL+Cn)/RoRL; % Eq.(4.4.6d)
```

Example 4.16   Frequency Response of a CS Amplifier

Consider the CS circuit in Figure 4.44(a) where $V_{DD} = 12\,V$, $R_s = 1\,k\Omega$, $C_s = 1\,\mu F$, $R_1 = 2.2\,M\Omega$, $R_2 = 1.5\,M\Omega$, $R_D = 22\,k\Omega$, $R_{S1} = 6\,k\Omega$, $R_{S2} = 6\,k\Omega$, $C_S = 10\,\mu F$, $C_L = 1\,\mu F$, $R_L = 100\,k\Omega$, and $C_{LL} = 0.1\,nF$, and the FET parameters are $K_p = 0.5\,mA/V^2$, $V_t = 1\,V$, $\lambda = 0.01\,V^{-1}$, $C_{gs0} = 10\,pF$, $C_{gd0} = 1\,pF$, $C_{ds} = 0\,F$, $V_b = 1\,V$, and $m = 0.5$. Plot the frequency response for $f = 1\sim100\,MHz$ and see how close it is to the PSpice simulation result. Also estimate the upper 3 dB frequency.

We can use the MATLAB function 'FET_CS_xfer_ftn()' to find the frequency response and plot its magnitude curve (as shown in Figure 4.44(c)) by running the following MATLAB statements:

(a)

PSpice schematic

(b)

Frequency response from PSpice simulation

(c)

Frequency response from MATLAB analysis

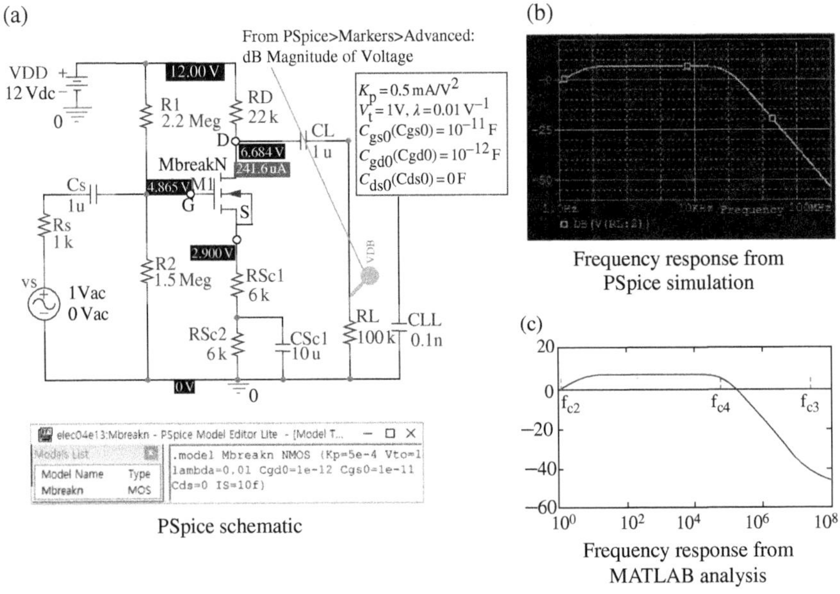

Figure 4.44 A CS amplifier circuit and its frequency response from MATLAB and PSpice ("elec04e16.opj").

```
>>VDD=12; Rs=1e3; R1=22e5; R2=15e5; RD=22e3; RS=[6e3 6e3];
  RL=1e5; Cs=1e-6; CS=1e-5; CL=1e-6; CLL=1e-10; RLCLL=[RL CLL];
  Kp=5e-4; Vt=1; lambda=0.01; % Device parameters
  Cgs0=1e-11; Cgd0=1e-12; Cds=0; Vb=1; m=0.5;
  CVbm=[Cgs0 Cgd0 Cds Vb m];
  f=logspace(0,8,801); w=2*pi*f; % Frequency range
  [Gs,Av,Ri,Ro,Cgs,Cgd,gm,ro]=... FET_CS_xfer_ftn(VDD,
  Rs,Cs,R1,R2,RD,RS,CS,CL,RLCLL,Kp,Vt,lambda,CVbm);
  syms s; Gw=subs(Gs,s,j*w); % Frequency response
  GmagdB=20*log10(abs(Gw)+1e-10); Gmax=max(GmagdB);
  semilogx(f,GmagdB, f([1 end]),[0 0],'k'), hold on
  fc=break_freqs_of_CS(Rs,Cs,CL,RL,CLL,Cgd,Cgs,Cds,Av,Ri,Ro,ro)
  fprintf('\n fc1=%12.3e, fc2=%12.3e, fc3=%12.3e, fc4=%12.3e,
          and fc5=%12.3e\n', fc);
  semilogx(fc(1)*[1 1],[0 Gmax-3],'b:',
  fc(2)*[1 1],[0 Gmax-3],'g:',fc(3)*[1 1],[0 Gmax-3],'r:',
  fc(4)*[1 1],[0 Gmax-3],'m:')
```

This yields Figure 4.44(c) and the values of the four break frequencies as follows:

```
fc1= 1.782e-01, fc2= 1.308e+00, fc3= 2.062e+07, fc4= 8.842e+04
```

from which we can take the third highest frequency f_{c4} = 88.4 kHz as a rough estimate of the upper 3 dB frequency.

4.4.2 CD Amplifier (Source Follower)

Figure 4.45(a) and (b) shows a CD amplifier circuit and its high-frequency small-signal equivalent, respectively, where one more load capacitor C_{LL}, in addition to the output capacitor C_L, is connected in parallel with the load resistor R_L. For the equivalent circuit shown in Figure 4.45(b), a set of three node equations in $V_1 = V_g$, $V_2 = V_s$, and $V_3 = V_d$ can be set up as

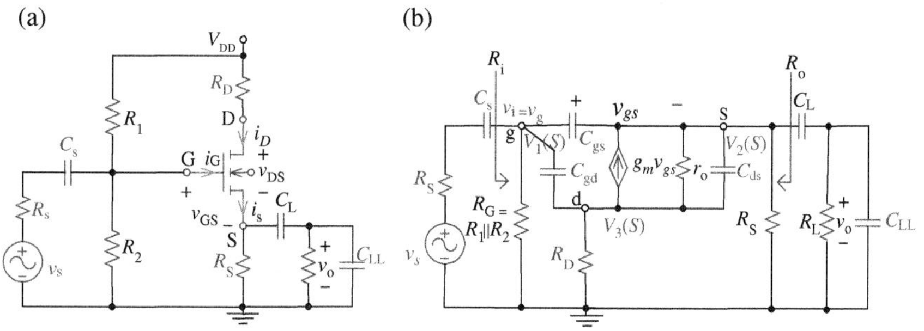

A CD FET circuit | Its high–frequency AC equivalent

Figure 4.45 A CD (common-drain) FET circuit and its high-frequency AC equivalent.

```
function [Gs,Av,Ri,Ro,Cgs,Cgd,gm,ro,Gw]= FET_CD_xfer_ftn ...
  (VDD,Rs,Cs,R1,R2,RD,RS,CS,CL,RLCLL,Kp,Vt,lambda,CVbm,w)
%To find the transfer function/frequency response of a CD amplifier
% CVbm=[Cgs0,Cgd0,Cds,Vb,m] %
with Cgs0|Cgd0: Zero-bias GSJ|GDJ capacitances
% Vb: Gate junction (barrier) potential
% m : Gate p-n grading coefficient
% Copyleft: Won Y. Yang, wyyang53@hanmail.net, CAU for academic use only
Cgs0=CVbm(1); Cgd0=CVbm(2); Cds=CVbm(3);
if numel(CVbm)<4, Vb=1; m=0.5; else Vb=CVbm(4); m=CVbm(5); end
RG=parallel_comb([R1 R2]); RL=RLCLL(1);
if numel(RLCLL)<2, CLL=0; else CLL=RLCLL(2); end
syms s; ZL=parallel_comb([RL 1/s/CLL]); % Eq.(4.4.9e)
if length(RS)==1, RS1=0; RS2=RS; else RS1=RS(1); RS2=RS(2); end
[VGQ,VSQ,VDQ,IDQ,Av,Ri,Ro,gm,ro]= ...
  FET_CD_analysis(VDD,Rs,R1,R2,RD,RS,RL,Kp,Vt,lambda);
Cgs=Cgs0/(1+abs(VGQ-VSQ)/Vb)^m; % Eq.(4.4.5a)
Cgd=Cgd0/(1+abs(VGQ-VDQ)/Vb)^m; % Eq.(4.4.5b)
sCs=s*Cs; sCS=s*CS; sCL=s*CL; sCgs=s*Cgs; sCgd=s*Cgd; sCds=s*Cds;
Ys=1/(Rs+1/sCs); YS=1/parallel_comb([RS ZL]);
if RD>0
 Y=[Ys+1/Ri+sCgs+sCgd     -sCgs                 -sCgd;
   -sCgs-gm    sCgs+1/ro+sCds+YS+gm        -1/ro-sCds;
   -sCgd-gm       -1/ro-sCds-gm  sCgd+1/RD+1/ro+sCds]; % Eq.(4.4.8)
 V=Y\[Ys; 0; 0];
else
 Y=[Ys+1/Ri+sCgs+sCgd  -sCgs;
    -sCgs-gm  sCgs+1/ro+sCds+YS+gm];
 V=Y\[Ys; 0];
end
Gs=V(2)*ZL/(ZL+1/sCL); % Transfer function Vo(s)/Vs(s)
if nargin>14, Gw=subs(Gs,'s',j*w); % Frequency response
 else Gw=0;
end
```

$$\begin{bmatrix} Y_G + sC_{gs} + sC_{gd} & -sC_{gs} & -sC_{gd} \\ -sC_{gs} - g_m & sC_{gs} + 1/r_o + sC_{ds} + Y_S + g_m & -1/r_o - sC_{ds} \\ -sC_{gd} + g_m & -1/r_o - sC_{ds} - g_m & sC_{gd} + 1/R_D + 1/r_o + sC_{ds} \end{bmatrix} \begin{bmatrix} V_1(s) \\ V_2(s) \\ V_3(s) \end{bmatrix} = \begin{bmatrix} I_{Nt} \\ 0 \\ 0 \end{bmatrix} \tag{4.4.8}$$

where

$$Y_s = \frac{1}{Z_s} = \frac{1}{R_s + 1/sC_s},\ I_{Nt} = Y_s V_s,\ Y_G = Y_s + G_G = Y_s + \frac{1}{R_G = (R_1 \| R_2)} \tag{4.4.9}$$

$$Y_S = \frac{1}{R_S \| (1/sC_L + Z_L)},\ Z_L = R_L \| (1/sC_{LL})$$

(Q) How about if $R_D = 0$?

(A) Since $V_3 = V_d = 0$ with $R_D = 0$, Eq. (4.4.8) should be reduced into a 2 × 2 matrix equation in two unknown node voltages V_1 and V_2.

Solving this set of equations, we can find the transfer function and frequency response:

$$G(s) = \frac{V_o(s)}{V_s(s)} \text{ and } G(j\omega) \text{ with } V_o(s) = \frac{R_L \| (1/sC_{LL})}{1/sC_L + (R_L \| 1/sC_{LL})} V_2(s) \quad (4.4.10)$$

This process to find the transfer function and frequency response has been cast into the above MATLAB function '`FET_CD_xfer_ftn()`'.

There are four break (pole or corner) frequencies determining roughly the frequency response magnitude:

$$\omega_{c1} \underset{(8.40)}{\overset{[R\text{-}2]}{=}} \frac{1}{C_s(R_s + R_i)}, \quad \omega_{c2} \underset{(8.41)}{\overset{[R\text{-}2]}{=}} \frac{1}{C_L(R_o + R_L)} \quad (4.4.11a,b)$$

$$\omega_{c3} \underset{(8.57)}{\overset{[R\text{-}2]}{=}} \frac{1}{(C_{gd} + C_m)(R_s \| R_i)}, \quad \omega_{c4} \underset{(8.58)}{\overset{[R\text{-}2]}{=}} \frac{1}{(C_n + C_{LL})(R_o \| R_L)} \quad (4.4.11c,d)$$

where R_i/R_o are the input/output resistances of the CD amplifier, respectively, and C_m/C_n are the Miller equivalent capacitances for capacitor C_{be} seen from the input/output side, respectively (Eq. (1.4.2)):

$$C_m = C_{gs}(1 - A_v) \text{ and } C_n = C_{gs}(A_v - 1)/A_v \quad (4.4.12)$$
$$\text{with } A_v = V_2/V_1 \text{ in Figure 4.45(b)}$$

Note that Eq. (4.4.11) is just like Eq. (4.4.6) with C_{gd} and C_{gs} switched.

The following MATLAB function '`break_freqs_of_CD()`' uses Eqs. (4.4.11a,b,c,d) to find the four break frequencies:

```
function fc=break_freqs_of_CD(Rs,Cs,CL,RL,CLL,Cgd,
                    Cgs,Cds,Av,Ri,Ro)
% To find the 4 break frequencies of a CD amplifier
RsRi=parallel_comb([Rs Ri]);
RoRL=parallel_comb([Ro RL]);
Cm=Cgs*(1-Av); % Eq.(4.4.12a)
Cn=Cgs*(1-1/Av); % Eq.(4.4.12b)
fc(1)=1/2/pi/Cs/(Rs+Ri); % Eq.(4.4.11a)
fc(2)=1/2/pi/CL/(Ro+RL); % Eq.(4.4.11b)
fc(3)=1/2/pi/(Cgd+Cm)/RsRi; % Eq.(4.4.11c)
fc(4)=1/2/pi/(CLL+Cn)/RoRL; % Eq.(4.4.11d)
```

Example 4.17 Frequency Response of a CD Amplifier

Consider the CD circuit in Figure 4.46(a) where $V_{DD}=12\,\text{V}$, $R_s=1\,\text{k}\Omega$, $C_s=1\,\mu\text{F}$, $R_1=2.2\,\text{M}\Omega$, $R_2=1.5\,\text{M}\Omega$, $R_D=22\,\text{k}\Omega$, $R_S=12\,\text{k}\Omega$, $C_L=0.1\,\mu\text{F}$, $R_L=100\,\text{k}\Omega$, and $C_{LL}=0.1\,\text{nF}$, and the FET parameters are $K_p=0.5\,\text{mA/V}^2$, $V_t=1\,\text{V}$, $\lambda=0.01\,\text{V}^{-1}$, $C_{gs0}=1\,\text{F}$, $C_{gd0}=1\,\text{pF}$, $C_{ds}=0\,\text{F}$, $V_b=1\,\text{V}$, and $m=0.5$. Plot the frequency response for $f=1\sim100\,\text{MHz}$ and see how close it is to the PSpice simulation result. Also estimate the upper 3 dB frequency.

We can use the MATLAB function '`FET_CD_xfer_ftn()`' to find the frequency response and plot its magnitude curve (as shown in Figure 4.46(c)) by running the following MATLAB statements:

```
>>Kp=5e-4; Vt=1; lambda=0.01;
  Cgd0=1e-12; Cgs0=1e-11; Cds=0; Vb=1; m=0.5;
  CVbm=[Cgs0 Cgd0 Cds Vb m];
  VDD=12; Rs=1e3; Cs=1e-6; R1=22e5; R2=15e5; RD=0;
  RS=12e3; CS=0;
  CL=1e-7; RL=1e5; CLL=1e-9; RLCLL=[RL CLL];
  f=logspace(0,8,801); w=2*pi*f;
```

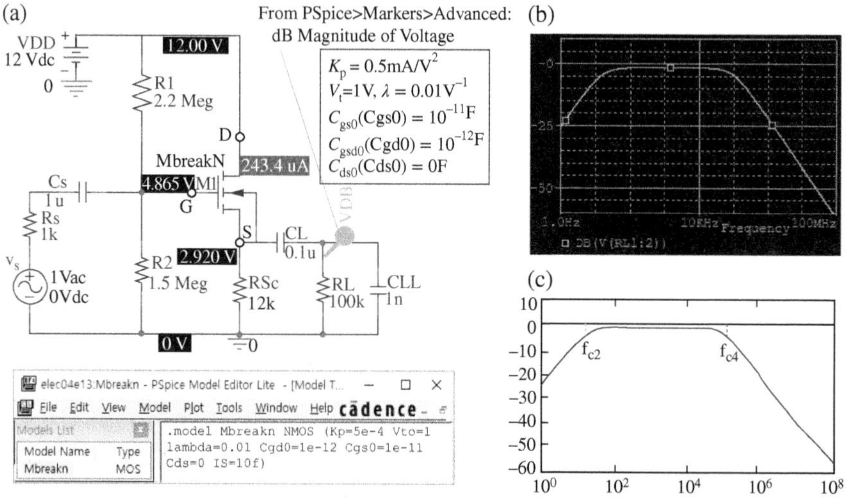

PSpice schematic

Frequency response from MATLAB analysis

Figure 4.46 A CD amplifier circuit and its frequency response from MATLAB and PSpice ("elec04e17.opj").

```
[Gs,Av,Ri,Ro,Cgs,Cgd,gm,ro] = FET_CD_xfer_ftn ...
    (VDD,Rs,Cs,R1,R2,RD,RS,CS,CL,RLCLL,Kp,Vt,lambda,CVbm);
syms s; Gw=subs(Gs,'s',j*w); % Frequency response
GmagdB=20*log10(abs(Gw)+1e-5); Gmax=max(GmagdB);
semilogx(f,GmagdB, f([1 end]),[0 0],'k'), hold on
fc=break_freqs_of_CD(Rs,Cs,CL,RL,CLL,Cgd,Cgs,Cds,Av,Ri,Ro);
fprintf('\n fc1=%12.3e, fc2=%12.3e, fc3=%12.3e, ...
        and fc4=%12.3e\n', fc);
semilogx(fc(1)*[1 1],[0 Gmax-3],'b:', ...
    fc(2)*[1 1],[0 Gmax-3],'g:', fc(3)*[1 1],[0 Gmax-3],'r:', ...
    fc(4)*[1 1],[0 Gmax-3],'m:')
```

This yields Figure 4.46(c) and the values of the four break frequencies as follows:

```
fc1= 1.782e-01, fc2= 1.565e+01, fc3= 1.262e+08, fc4= 9.731e+04
```

from which we can take the third highest frequency $f_{c4} = 97.3$ kHz as a rough estimate of the upper 3 dB frequency.

4.4.3 CG Amplifier

Figure 4.47(a) and (b) shows a CG amplifier circuit and its high-frequency small-signal equivalent, respectively, where one more load capacitor C_{LL}, in addition to the output capacitor C_L, is connected in parallel with the load resistor R_L. Note that if the terminal G(ate) is not AC grounded via a capacitor C_G, we should let $C_G = 0$ as shown in Figure 4.47(b). For the equivalent circuit shown in Figure 4.47(b), a set of three node equations in $V_1 = V_s$, $V_2 = V_d$, and $V_3 = V_g$ can be set up as

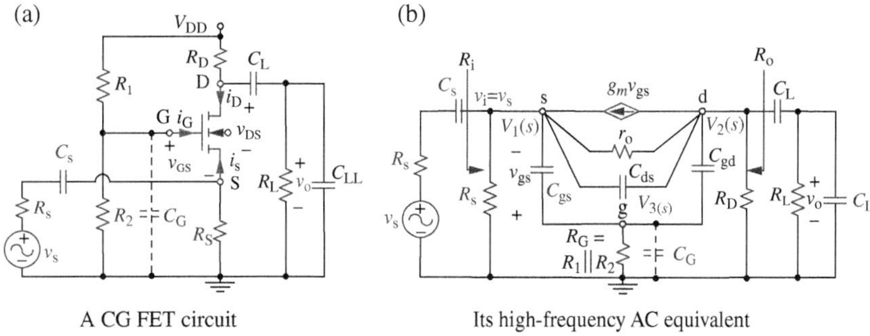
(a)
(b)
A CG FET circuit
Its high-frequency AC equivalent

Figure 4.47 A CG (common-gate) FET circuit and its high-frequency AC equivalent.

```
function [Gs,Av,Ri,Ro,Cgs,Cgd,gm,ro,Gw] = FET_CG_xfer_ftn ...
    (VDD,Rs,Cs,R1,R2,CG,RD,RS,CL,RLCLL,Kp,Vt,lambda,CVbm,w)
%To find the transfer function/frequency response of a CG amplifier
% RLCLL=[RL CLL] with RL and CLL in parallel
% CVbm=[Cgs0,Cgd0,Cds,Vb,m] with Cgs0|Cgd0: Zero-bias GSJ|GDJ capacitances
% Vb: Gate junction (barrier) potential
% m : Gate p-n grading coefficient
% Copyleft: Won Y. Yang, wyyang53@hanmail.net, CAU for academic use only
Cgs0=CVbm(1); Cgd0=CVbm(2); Cds=CVbm(3);
if numel(CVbm)<4, Vb=1; m=0.5; else Vb=CVbm(4); m=CVbm(5); end
RL=RLCLL(1); if numel(RLCLL)<2, CLL=0; else CLL=RLCLL(2); end
syms s; ZL=parallel_comb([RL 1/s/CLL]);
[VGQ,VSQ,VDQ,IDQ,Av,Ri,Ro,gm,ro] = ...
  FET_CG_analysis(VDD,Rs,R1,R2,RD,RS,RL,Kp,Vt,lambda);
Cgs=Cgs0/(1+abs(VGQ-VSQ)/Vb)^m;
Cgd=Cgd0/(1+abs(VGQ-VDQ)/Vb)^m; %Eq.(4.4.5)
sCs=s*Cs; sCG=s*CG; sCL=s*CL; sCgs=s*Cgs; sCgd=s*Cgd; sCds=s*Cds;
RG=parallel_comb([R1 R2]);
Ys=1/(Rs+1/sCs); YS=Ys+1/RS; % Eq.(4.4.14b,c)
YG=1/RG+sCG; YD=1/parallel_comb([RD ZL+1/sCL]); % Eq.(4.4.14d,e)
if RG>0&CG<0.1
 Y=[YS+sCgs+1/ro+sCds+gm  -1/ro-sCds  -sCgs-gm;
   -1/ro-sCds-gm  1/ro+sCds+sCgd+YD  -sCgd+gm;
   -sCgs  -sCgd  sCgs+YG+sCgd]; % Eq.(4.4.13)
 V=Y\[Ys; 0; 0];
else % If RG=0 or CG is so large that RG can be AC-shorted
 Y=[YS+sCgs+1/ro+sCds+gm  -1/ro-sCds;
   -1/ro-sCds-gm  1/ro+sCds+sCgd+YD]; % Eq.(4.4.13)
 V=Y\[Ys; 0];
end
Gs=V(2)*ZL/(ZL+1/sCL); % Transfer function Vo(s)/Vs(s)
if nargin>14, Gw=subs(Gs,'s',j*w); else Gw=0; end
```

$$\begin{bmatrix} Y_S + sC_{gs} + 1/r_o + sC_{ds} + g_m & -1/r_o - sC_{ds} & -sC_{gs} - g_m \\ -1/r_o - sC_{ds} - g_m & 1/r_o + sC_{ds} + sC_{gd} + Y_D & -sC_{gd} + g_m \\ -sC_{gs} & -sC_{gd} & sC_{gs} + Y_G + sC_{gd} \end{bmatrix} \begin{bmatrix} V_1(s) \\ V_2(s) \\ V_3(s) \end{bmatrix} = \begin{bmatrix} I_{Nt} \\ 0 \\ 0 \end{bmatrix} \tag{4.4.13}$$

where

$$Y_S = Y_s + \frac{1}{R_S},\quad I_{Nt} = Y_s V_s,\quad Y_s = \frac{1}{Z_s} = \frac{1}{R_s + 1/sC_s} \tag{4.4.14a,b,c}$$

$$Y_G = \frac{1}{R_G} + sC_G, \text{ and } Y_D = \frac{1}{R_D \| \{1/sC_L + (R_L \| 1/sC_{LL})\}} \tag{4.4.14d,e}$$

(Q) How about if $R_G = 0$ or C_G is so large that R_G can be regarded as AC-shorted?

Solving this set of equations, we can find the transfer function and frequency response:

$$G(s)=\frac{V_o(s)}{V_s(s)} \text{ and } G(j\omega) \text{ with } V_o(s)=\frac{R_L\|(1/sC_{LL})}{1/sC_L+(R_L\|1/sC_{LL})}V_2(s) \quad (4.4.15)$$

This process to find the transfer function and frequency response has been cast into the above MATLAB function '`FET_CG_xfer_ftn()`'.

There are five break (pole or corner) frequencies determining roughly the frequency response magnitude:

$$\omega_{c1} \overset{[R\text{-}2]}{\underset{(8.40)}{=}} \frac{1}{C_s(R_s+R_i)},\ \omega_{c2} \overset{[R\text{-}2]}{\underset{(8.41)}{=}} \frac{1}{C_L(R_o+R_L)} \quad (4.4.16a,b)$$

$$\omega_{c3}=\frac{1}{C_{gs}\{(R_s\|R_S)+R_G\}} \quad (4.4.16c)$$

$$\omega_{c4}=\frac{1}{C_{gd}\{(R_D\|R_L)+R_G\}},\ \omega_{c5}=\frac{1}{C_{LL}(R_o\|R_L)} \quad (4.4.16d,e)$$

where R_i/R_o are the input/output resistances of the CG amplifier, respectively.

The following MATLAB function '`break_freqs_of_CG()`' uses Eqs. (4.4.16a,b,c,d,e) to find the five break frequencies:

```
function fc=break_freqs_of_CG(Rs,Cs,R1,R2,RD,RS,CL,RL,CLL,Cgd,
   Cgs,Ri,Ro)
% To find the 4 break frequencies of a CG amplifier
RG=parallel_comb([R1 R2]);
RsRi=parallel_comb([Rs Ri]); RoRL=parallel_comb([Ro RL]);
fc(1)=1/2/pi/Cs/(Rs+Ri); % Eq.(4.4.16a)
fc(2)=1/2/pi/CL/(Ro+RL); % Eq.(4.4.16b)
fc(3)=1/2/pi/Cgs/(parallel_comb([Rs RS])+RG); % Eq.(4.4.16c)
fc(4)=1/2/pi/Cgd/(parallel_comb([RD RL])+RG); % Eq.(4.4.16d)
fc(5)=1/2/pi/CLL/RoRL; % Eq.(4.4.16e)
```

Example 4.18 Frequency Response of a CG Amplifier

Consider the CG circuit in Figure 4.48(a) where V_{DD}=12 V, R_s=1 kΩ, C_s=1 μF, R_1 = 2.2 MΩ, R_2 = 1.5 MΩ, R_D = 22 kΩ, R_S = 12 kΩ, C_L = 0.1 μF, R_L = 100 kΩ, C_{LL} = 0.1 nF, and the FET parameters are K_p = 0.5 mA/V^2, V_t = 1 V, λ = 0.01 V^{-1}, C_{gs0} = 10 pF, C_{gd0} = 1 pF, C_{ds} = 0 F, V_b = 1 V, and m = 0.5. Plot the frequency response for f = 1∼100 MHz and see how close it is to the PSpice simulation result. Also estimate the lower and upper 3 dB frequencies.

We can use the MATLAB function 'FET_CG_xfer_ftn()' to find the frequency response and plot its magnitude curve (as Figure 4.48(c)) by running the following MATLAB statements:

```
>>Kp=5e-4; Vt=1; lambda=0.01;
  Cgd0=1e-12; Cgs0=1e-11; Cds=0; Vb=1; m=0.5; CCC=[Cgs0 Cgd0 Cds Vb m];
  VDD=12; Rs=2e3; Cs=1e-5; R1=22e5; R2=15e5; RD=22e3; RS=12e3; RL=1e5;
  CG=0; CL=1e-6; RL=1e5; CLL=1e-10; RLCLL=[RL CLL];
  f=logspace(0,8,801); w=2*pi*f;
  [Gs,Av,Ri,Ro,Cgs,Cgd,gm,ro]=...
   FET_CG_xfer_ftn(VDD,Rs,Cs,R1,R2,CG,RD,RS,CL,RLCLL,Kp,Vt,lambda,CCC);
  syms s; Gw=subs(Gs,'s',j*w); % Frequency response
  GmagdB=20*log10(abs(Gw)+1e-5); Gmax=max(GmagdB);
  semilogx(f,GmagdB, f([1 end]),[0 0],'k'), hold on
  fc=break_freqs_of_CG(Rs,Cs,R1,R2,RD,RS,CL,RL,CLL,Cgd,Cgs,Ri,Ro);
  fprintf('\n fc1=%12.3e, fc2=%12.3e, fc3=%12.3e, fc4=%12.3e,
            fc4=%12.3e \n', fc);
```

This yields Figure 4.48(c) and the values of the five break frequencies as follows:

```
fc1= 4.226e+00, fc2= 1.311e+00, fc3= 3.067e+04, fc4= 2.937e+05,
fc5= 9.025e+04
```

from which we can take the second/third highest frequencies f_{c1} = 4.23 Hz and f_{c3} = 30.7 kHz as rough estimates of the lower/upper 3 dB frequencies, respectively.

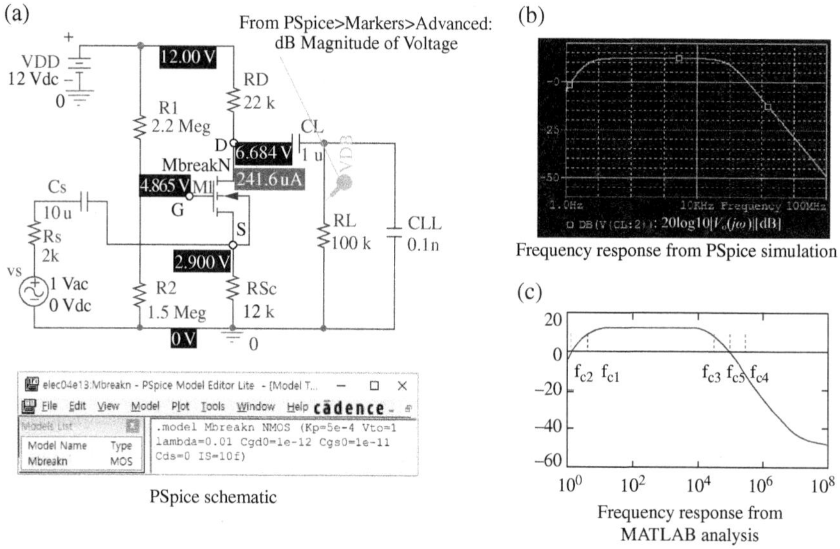

Figure 4.48 A CG amplifier circuit and its frequency response from MATLAB and PSpice ("elec04e18.opj").

4.5 FET Inverter Time Response

In this section, let us see the time response of a basic FET inverter (as shown in Figure 4.16(a) or 4.49(b)) to logic low-to-high and high-to-low input transitions where low-to-high and high-to-low propagation delays due to the internal parasitic capacitances between the terminals of the FET can be observed. Note that the effect of the internal parasitic capacitances becomes conspicuous as the frequency or slope of the input increases, while negligible for a DC or low-frequency input.

Figure 4.49(a) shows a simple DC or large-signal Spice model of an FET where $i_D(v_{GS},v_{DS})$ is defined as Eq. (4.1.83) and

r_g: gate resistance, r_s: source resistance, r_d: drain resistance,
C_{gd}: gate-drain junction capacitance, C_{gs}: gate-source junction capacitance,
C_{bd}: bulk-drain junction capacitance, C_{bs}: bulk-source junction capacitance.

Figure 4.49(c) shows a DC equivalent of the inverter (Figure 4.49(b)) with the FET replaced by its large-signal model (Figure 4.49(a)). Like the MATLAB function 'BJT_inverter_dynamic()' for the BJT inverter (Figure 3.67(b)), a process of solving the FET inverter to find the output $v_o(t)$ to an input $v_i(t)$ has been cast into the following MATLAB function 'FET_inverter_dynamic()'. Note that the drain resistor R_D and the (internal) capacitance $(C_{gd} + C_{bd})$ affect how long it takes for C_{gd} and C_{bd} to be charged up to V_{DD}, via the time constant during the rising period of $v_o(t)$.

```
function [vo,vCgd,vCbd,iD]=FET_inverter_dynamic(Kp,Vt,lambda,
  CVbm,RD,VDD,vi,dt)
% Analyze an FET inverter to find the output vo to an input vi
% which consists of an FET and a resistor RD between VDD and D.
% CVbm=[Cgd0,Cbd0,Vb,m] with Cgd0|Cbd0: Zero-bias GDJ|BDJ capacitances
% Vb: Gate junction (barrier) potential
% m : Gate p-n grading coefficient
% vo: Output voltage(s) for input voltage(s) vi
% Copyleft: Won Y. Yang, wyyang53@hanmail.net, CAU for academic use only
Cgd0=CVbm(1); Cbd0=CVbm(2);
if numel(CVbm)<3, Vb=0.8; m=0.5; else Vb=CVbm(3); m=CVbm(4); end
iDv=@(vGS,vDS)max(Kp*(vGS>=Vt).*(1+lambda*vDS).*((vGS-Vt).^2/2 ...
      .*(vGS-vDS<Vt)+((vGS-Vt).*vDS-vDS.^2/2).*(vGS-vDS>=Vt)), ...
      0); % Eq.(4.1.83)
vCgd(1)=VDD; vCbd(1)=VDD;
for n=1:length(vi)
    vo(n)=vCbd(n); v1(n)=vi(n)+vCgd(n);
    Cgd=Cgd0/(1+abs(vi(n)-vo(n))/Vb)^m; % Eq.(4.4.5)
    Cbd=Cbd0/(1+abs(vo(n))/Vb)^m;
    iD(n)=iDv(vi(n),vo(n));
    iCn=(VDD-vo(n))/RD-iD(n);
    iCgd(n)=iCn*Cgd/(Cgd+Cbd); iCbd(n)=iCn*Cbd/(Cgd+Cbd);
    vCbd(n+1)=vCbd(n)+iCbd(n)*dt/Cbd;
    vCgd(n+1)=vCgd(n)+iCgd(n)*dt/Cgd;
end
vCgd=vCgd(1:n); vCbd=vCbd(1:n); vo=vo(1:n);
```

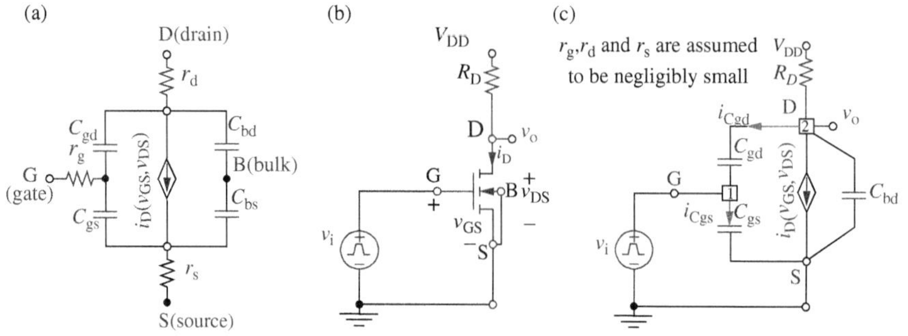

A simple DC (large-signal) model of FET

A basic FET inverter

The FET inverter with the FET replaced by its large-signal model.

Figure 4.49 A simple large-signal Spice model of FET and the equivalent of an FET inverter adopting the model.

Example 4.19 Time Response of an FET Inverter

Consider the FET inverter in Figure 4.50(a) where $V_{DD} = 5$ V, $R_D = 10$ kΩ, and the FET parameters are $K_p = 5$ mA/V^2, $V_t = 1$ V, $\lambda = 10^{-4}$ V^{-1}, $C_{gd0} = 10^{-15}$ F, $C_{bd0} = 10^{-14}$ F, V_b(Pb) $= 0.8$ V, and m(Mj) $= 0.5$. Use the above MATLAB function '`FET_inverter_dynamic()`' to find the output $v_o(t)$ to an input $v_i(t)$ plotted as the dotted line in Figure 4.50(b) or (c) and plot it for $t = 0\sim1$ ns.

To this end, we can run the following MATLAB script "elec04e17.m" to see the plot of $v_i(t)$ and $v_o(t)$ as shown in Figure 4.50(c). The time constant during the rising period of $v_o(t)$ is roughly estimated as

$$R_D\left(C_{gd0} + C_{bd0}\right) = 10^4 \times 11^{-15} = 0.11\,[\text{ns}] \tag{E4.19.1}$$

(Q3) During the falling period of $v_o(t)$, why does it decrease, not exponentially with any time constant, but almost linearly as $v_i(t)$ increases?

(A3) As $v_i(t) = v_{GS}(t)$ increases over $V_t = 1$ V so that the FET M begins to turn on, $C_{gd}||C_{bd}$ gets discharged, not via any resistor, but only through the drain-source junction (DSJ) of M so that $v_o = v_{Cgd} = v_{Cbd}$ can decrease toward 0 V.

```
%elec04e19.m
% To find the dynamic response of an FET amplifier
VDD=5; RD=10000; % Circuit parameters
Kp=5e-3; Vt=1; lambda=1e-4;
Cgd0=1e-15; Cds0=1e-14; % Device parameters
Vb=0.8; m=0.5; % Barrier potential and Gate p-n grading coefficient
CVbm=[Cgd0 Cds0 Vb m];
dt=1e-14; t=[0:100000]*dt; % Time range
ts= [0 2 3 5 6 12 13 15 16 20]/2*1e4*dt;
vis=[0 0 VDD VDD 0 0 VDD VDD 0 0];
vi=interp1(ts,vis,t); % PWL input
[vo,iD]=FET_inverter_dynamic(Kp,Vt,lambda,CVbm,RD,VDD,vi,dt);
plot(t,vi,'r', t,vo)
```

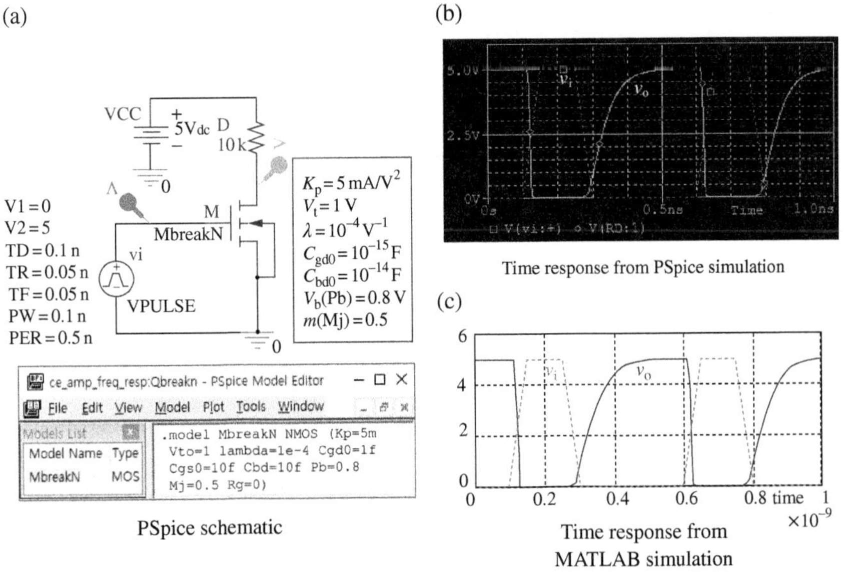

Figure 4.50 A FET inverter and its time response from MATLAB and PSpice ("elec04e19.opj").

Figure 4.51(a) shows three NMOS inverters, one with no load, one with a purely capacitive load, and one with an *RC* load. About Figure 4.51(b) showing their input and outputs, note the following:

- The time constant during the rising period of $v_{o1}(t)$ is much longer than that of $v_o(t)$ because the equivalent capacitance seen from R_D has increased from $C_{gd} + C_{bd} \approx 11[\text{fF}]$ to $C_{gd} + (C_{bd}||C_{L1}) \approx 1 + 10 + 10 = 21[\text{fF}]$ so that the time constant can be made longer.
- The time constant during the rising period of $v_{o2}(t)$ is about half as long as that of $v_{o1}(t)$ because the equivalent resistance seen from $C_{bc}||C_{L2}$ has decreased from $R_D = 10\,\text{k}\Omega$ to $R_{D2}||R_{L2} = 10\text{k}||10\text{k} = 5\,[\text{k}\Omega]$ so that the time constant can halve.

(Q4) During the rising period, why does $v_{o2}(t)$ reach just 2.5 V, which is just half of $V_{DD} = 5\,\text{V}$ unlike $v_o(t)$ or $v_{o1}(t)$?

(Q5) During the falling period, all the inverter outputs decrease almost linearly as $v_i(t)$ increases. Why is that?

(a)

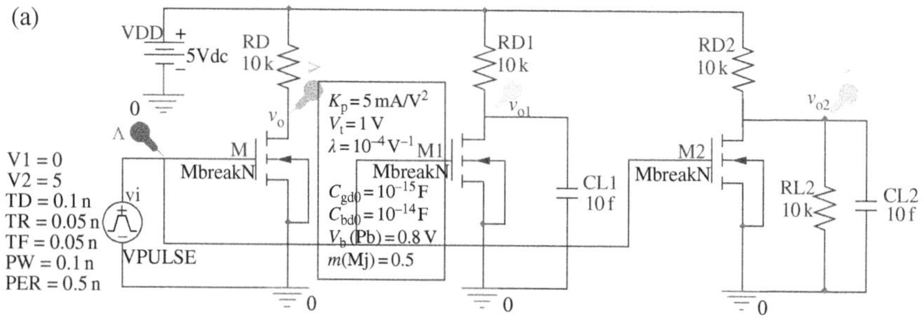

PSpice schematic for NMOS inverters with or without capacitive loads

(b)

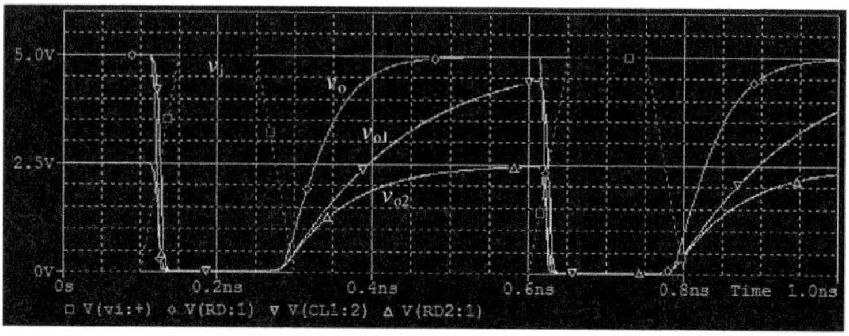

The input and output waveforms of the NMOS inverters from PSpice simulation

Figure 4.51 NMOS inverters with capacitive loads and their time responses from PSpice ("elec04f51.opj").

Figure 4.52(a) shows two CMOS (Complementary MOSFET) inverters, one with no load and one with an *RC* load. About Figure 4.52(b1)/(b2) showing their input and outputs and Figure 4.52(c1)/(c2) showing the currents through Mn and Mp, note the following:

- During the falling period, the output decreases toward 0 almost linearly as $v_i(t)$ increases, almost regardless of the load. Why is that? Because, C_{gd} gets discharged through the NMOS Mn.
- During the rising period, the output increases toward V_{CC} almost linearly as $v_i(t)$ decreases, almost regardless of the load. Why is that? Because, C_{gd} gets charged through the PMOS Mp.
- The current through the CMOS flows only at switching instants when the (internal) capacitors are charged/discharged. Consequently, CMOS logic gates take very little power in a fixed state.

How nice it is of a CMOS inverter to have the ('rail-to-rail') output swing equal to its input swing with almost no propagation delay and very little static power! That is a primary reason why CMOS has become the most used technology to be implemented in Very Large-Scale Integration (VLSI) chips.

(a)

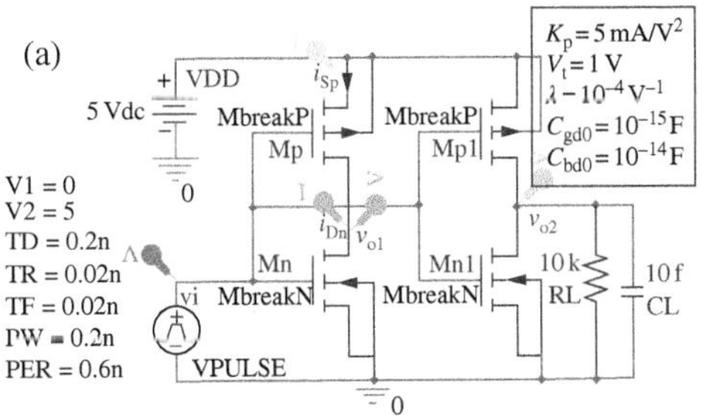

(a1)

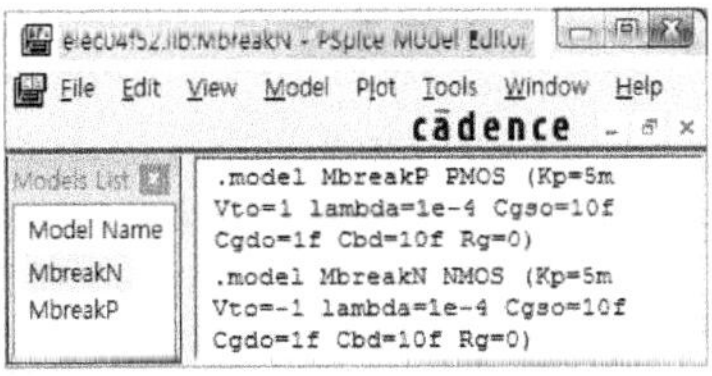

PSpice model editor window

PSpice schematics for CMOS inverters with or without capacitive loads

(b1)

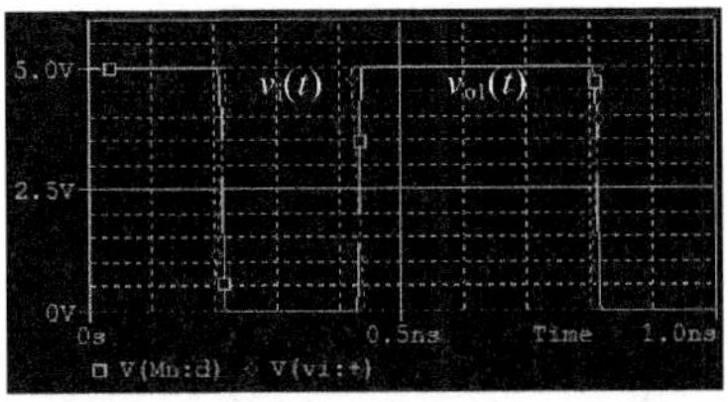

The input and output waveforms with no load

(b2)

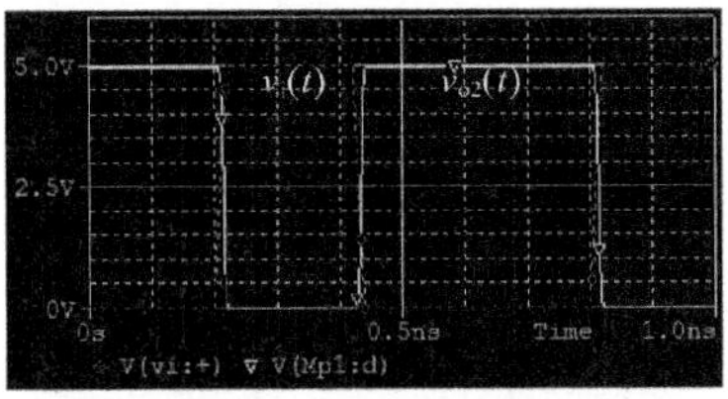

The input and output waveforms with an *RC* load

(c1)

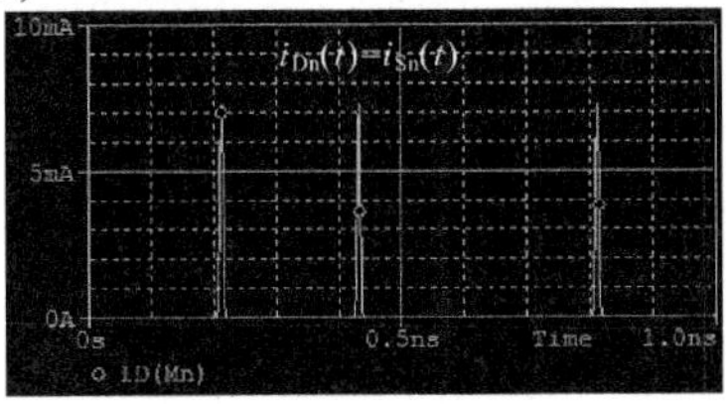

The current waveform $i_D(t)$ of Mn (with no load)

(c2)

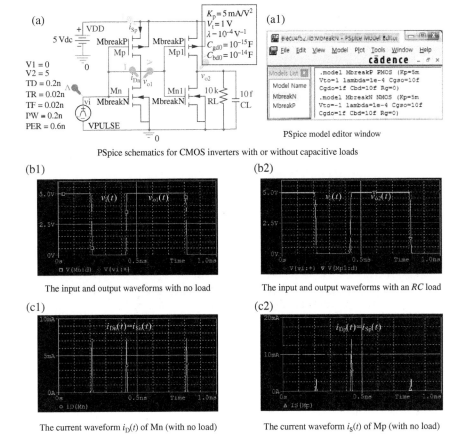

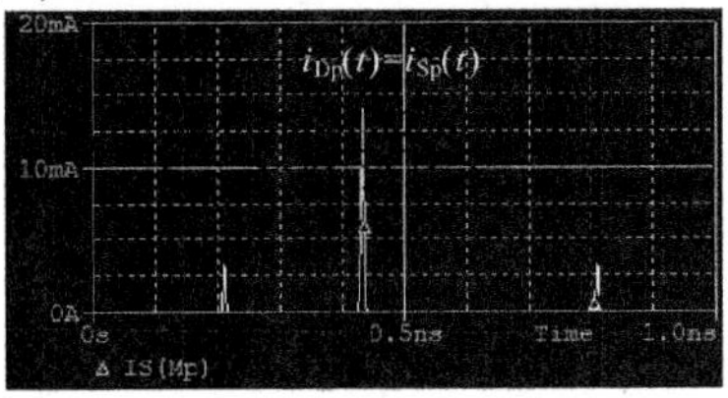

The current waveform $i_S(t)$ of Mp (with no load)

Figure 4.52 CMOS inverter with a capacitive load and its time response from PSpice ("elec04f52.opj").

Problems

4.1 Analysis of JFET Circuits

(a) Consider the JFET circuit in Figure P4.1(a) where the device parameters of the n-channel JFET (NJF) J are $I_{DSS} = 5\ \text{mA/V}^2$, $V_t = -5$ V, and $\lambda = 0$. Find I_D, V_G, V_D, and V_S. Also tell which one of the saturation, triode, and cutoff regions the NJF J operates in.

(Hint)

```
>>VDD=12; VG=0; RG=680e3; RD=2e3; RS=1e3;
  Vt=-5; IDSS=5e-3; Kp=2*IDSS/Vt^2;
  BETA=Kp/2 % SPICE parameter FET_DC_analysis(VDD,VG,RD,RS,Kp,Vt);
```

(b) Consider the JFET circuit in Figure P4.1(b) where the device parameters of the NJF J_2 are $I_{DSS} = 8\ \text{mA/V}^2$, $V_t = -4$ V, and $\lambda = 0$. Find I_D, V_G, V_D, and V_S. Also tell which one of the saturation, triode, and cutoff regions the NJF J_2 operates in.

(Hint)

```
>>VDD=15; R1=1e6; R2=5e5; RD=3e3; RS=1500;
  Vt=-4; IDSS=8e-3; Kp=2*IDSS/Vt^2; BETA=Kp/2
  FET_DC_analysis(VDD,R1,R2,RD,RS,Kp,Vt);
```

Noting that it depends on

$$v_{GS} - V_t \overset{?}{<} v_{DS} : \frac{R_2}{R_1 + R_2} V_{DD} - R_S i_D - V_t \overset{?}{<} V_{DD} - (R_D + R_S) i_D \quad \text{(P4.1.1)}$$

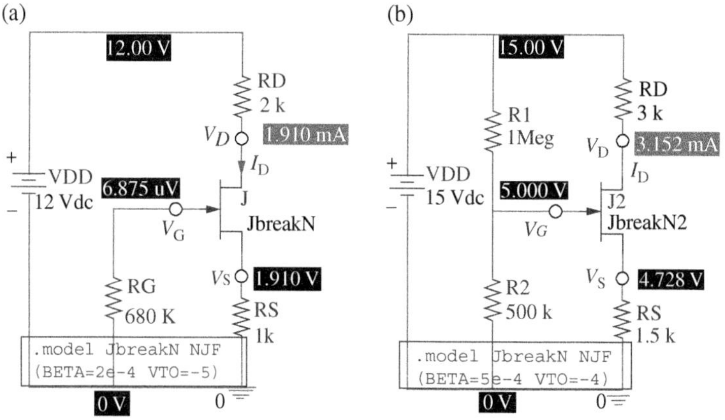

Figure P4.1 JFET circuits for Problem 4.1.

whether the JFET operates in the saturation or triode region, answer the following questions:

(b1) Would you decrease or increase R_2 to push J_2 into the saturation region? With $R_2 = 200\ \mathrm{k\Omega}$ and the other parameter values unchanged, tell which region J_2 operates in.

(b2) Would you increase or decrease R_D to push J_2 into the triode region? With $R_D = 10\ \mathrm{k\Omega}$ and the other parameter values as in (b1), tell which region J_2 operates in.

(b3) Would you increase or decrease R_S to push J_2 into the saturation region? With $R_S = 10\ \mathrm{k\Omega}$ and the other parameter values as in (b2), tell which region J_2 operates in.

(c) Noting that the value of BETA, which is one of the SPICE model parameters for a JFET, is determined from I_{DSS} and V_t as

$$\beta = \frac{K_p}{2} = I_{DSS}/V_t^2 \qquad \text{(P4.1.2)}$$

perform the PSpice simulation (with Bias Point analysis type) for the two JFET circuits to check the validity of the circuit analyses done in (a) and (b).

4.2 Design of a JFET Biasing Circuit

Consider the JFET circuit in Figure P4.2 where the device parameters of the NJF J are $I_{DSS} = 12.3\ \mathrm{mA/V^2}$, $V_t = -3.5\ \mathrm{V}$, and $\lambda = 0$. Determine the values of R_D and R_S so that $I_D = 1\ \mathrm{mA}$ and $V_{DS} = 3\ \mathrm{V}$. Check the validity of your design result by using the MATLAB function 'FET_DC_analysis()' and/or PSpice.

Figure P4.2

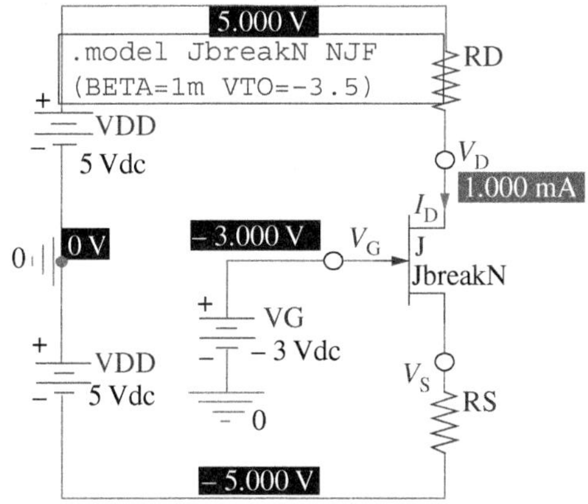

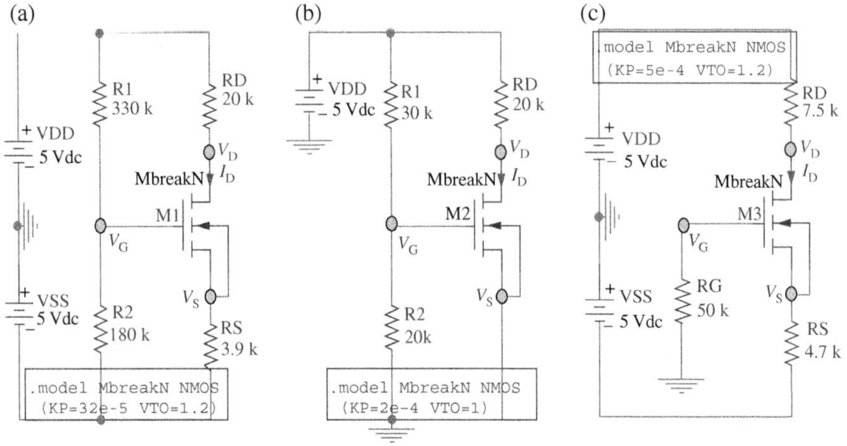

Figure P4.3 NMOS circuits for Problem 4.3.

4.3 Analysis of NMOS Circuits

(a) Consider the NMOS circuit in Figure P4.3(a) where the device parameters of the enhancement NMOS (e-NMOS) M_1 are $K_p = 0.32$ mA/V^2, $V_t = 1.2$ V, and $\lambda = 0$. Find I_D, V_G, V_D, and V_S. Also tell which one of the saturation, triode, and cutoff regions the NMOS M_1 operates in.

(Hint)

```
>>VDD=5; VSS=-5; R1=330e3; R2=180e3; RD=2e4; RS=3900;
  Kp=32e-5; Vt=1.2; lambda=0;
  [VGQ,VSQ,VDQ,VGSQ,VDSQ,IDQ,mode] = ...
  FET_DC_analysis(VDD-VSS,R1,R2,RD,RS,Kp,Vt,lambda);
  fprintf('VG=%6.2fV, VD=%6.2fV, VS=%6.2fV, ID=%8.3fmA\n',...
    VGQ+VSS,VDQ+VSS,VSQ+VSS,IDQ*1e3);
```

(b) Consider the NMOS circuit in Figure P4.3(b) where the device parameters of M_2 are $K_p = 0.2$ mA/V^2, $V_t = 1$ V, and $\lambda = 0$. Find I_D, V_G, V_D, and V_S. Also tell which region M_2 operates in.

(c) Consider the NMOS circuit in Figure P4.3(c) where the device parameters of the e-NMOS M_3 are $K_p = 0.5$ mA/V^2, $V_t = 1.2$ V, and $\lambda = 0$. Find I_D, V_G, V_D, and V_S. Also tell which region the NMOS M_3 operates in. Plot the load line for $i_D = (V_G - V_{SS} - v_{GS})/R_S$ on the i_D-v_{GS} characteristic curve for $i_D = K_p(v_{GS} - V_t)^2/2$ and locate the operating point ($V_{GS,Q}$, $I_{D,Q}$). Also plot the load line for $i_D = (V_{DD} - V_{SS} - v_{DS})/(R_D + R_S)$ on the i_D-v_{DS} characteristic curve (with $v_{GS} = V_{GS,Q}$) and locate the operating point. To those ends, complete and run the following script:

```
%elec04p03c.m
clear,clf
VDD=5; VSS=-5; VG=?; RG=50e3; RD=7500; RS=4700; % Circuit parameters
Kp=0.5e-3; Vt=1.2; lambda=0; % Device parameters
[VGQ,VSQ,VDQ,VGSQ,VDSQ,IDQ,mode] = ...
  FET_DC_analysis([VDD V??],VG,RD,RS,Kp,Vt,lambda);
vGSs=VGSQ+[-2:0.01:2]; vDSs=[0:0.01:VDD-VSS];
% To plot the load line on the iD-vGS characteristic curve
iDs_vGS=Kp/2*(vGSs-Vt).^2.*(vGSs>Vt);
subplot(221), plot(vGSs,iDs_vGS, vGSs,(VGQ-VSS-V???)/R?,'r')
hold on, plot(VGSQ,IDQ,'ro'), axis([vGSs([1 end]) 0 1.2e-3])
% To plot the load line on the iD-vDS characteristic curve
iDs_vDS=iD_NMOS_at_vDS_vGS(vDSs,VGS?,Kp,Vt,lambda);
subplot(222)
plot(vDSs,iDs_vDS, vDSs,(VDD-VSS-v???)/(R?+R?),'r')
hold on, plot(VDSQ,IDQ,'ro')
```

(d) Perform the PSpice simulation (with Bias Point analysis type) for the three NMOS circuits to check the validity of the circuit analyses done in (a), (b), and (c).

4.4 Analysis PMOS Circuits

(a) Consider the PMOS circuit in Figure P4.4(a1) where the device parameters of the enhancement PMOS (e-PMOS) M_1 are $K_p = 0.2\ \text{mA/V}^2$, $V_t = -0.4$ V, and $\lambda = 0$. Find I_D, V_G, V_D, and V_S. Also tell which one of the saturation, triode, and cutoff regions the NMOS M_1 operates in.

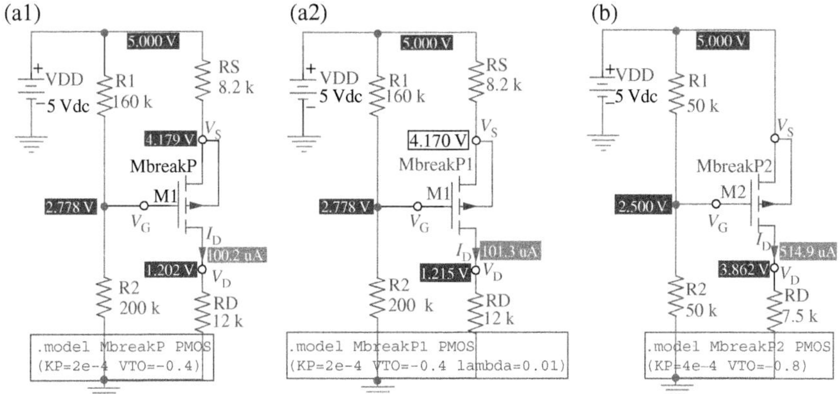

Figure P4.4 PMOS circuits.

(Hint)

```
>>VDD=5; R1=160e3; R2=200e3; RS=8.2e3; RD=12e3;
  Kp=2e-4; Vt=-0.4; lambda=0; % Device parameters
  FET_PMOS_DC_analysis(VDD,R1,R2,RS,RD,Kp,Vt,lambda);
```

(b) Consider the PMOS circuit in Figure P4.4(a2) where the device parameters of the e-PMOS M_1 are $K_p = 0.2\,\text{mA/V}^2$, $V_t = -0.4\,\text{V}$, and $\lambda = 0.01$. Find I_D, V_G, V_D, and V_S. Also tell which region the PMOS M_1 operates in.

(c) Consider the PMOS circuit in Figure P4.4(b) where the device parameters of the e-PMOS M_2 are $K_p = 0.4\,\text{mA/V}^2$, $V_t = -0.8\,\text{V}$, and $\lambda = 0.01$. Find I_D, V_G, V_D, and V_S. Also tell which region the PMOS M_2 operates in.

(d) Perform the PSpice simulation (with Bias Point analysis type) for the three PMOS circuits to check the validity of the circuit analyses done in (a), (b), and (c).

4.5 Design of a MOSFET biasing Circuit

Consider the MOSFET circuit in Figure P4.5 where the device parameters of the NMOS M_1 are $K_p = 1\,\text{mA/V}^2$, $V_t = 1.2\,\text{V}$, and $\lambda = 0$. Determine the values of R_D, R_S, R_1, and R_2 so that $I_D = 0.4\,\text{mA}$, $V_D = 6\,\text{V}$, $V_{DS} = 3.1\,\text{V}$, and $R_i = 100\,\text{k}\Omega$. Check the validity of your design result by using the MATLAB function '`FET_DC_analysis()`' and/or PSpice.

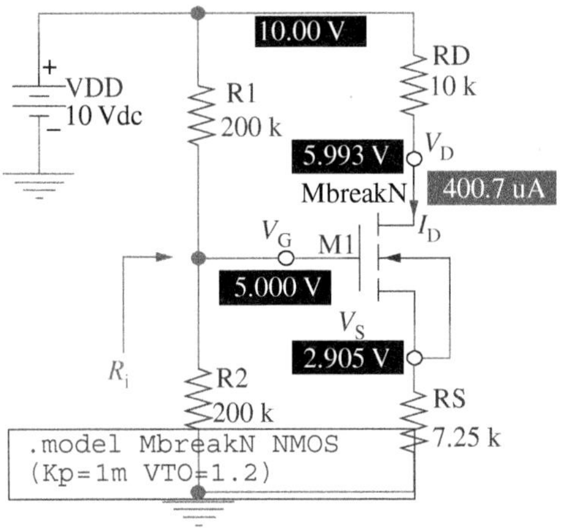

Figure P4.5

4.6 Diode-Connected MOSFETs Used as Nonlinear Resistors

Consider the two NMOS circuits in Figure P4.6 where the device parameters of the enhancement NMOS (e-NMOS) M_1 and the depletion NMOS (d-NMOS) M_2 are $K_{p1} = 0.02\,\text{mA/V}^2$, $V_{t1} = 1\,\text{V}$, $\lambda_1 = 0$ and $K_{p2} = 0.02\,\text{mA/V}^2$, $V_{t2} = -3\,\text{V}$, $\lambda_2 = 0$, respectively.

(a) Perform the load line analysis of finding the operating points Q_1/Q_2 from the intersections between the i_{Dk}-v_{DSk} characteristice curves of M_1/M_2

and the load line for $i_{Dk} = (V_{DD} - v_{DSk})/R_k$ where the load lines for M_1/M_2 are identical since $R_1 = R_2$. If helpful, complete and run the corresponding part of the following MATLAB script "elec04p06.m" to plot the related graphs and find the operating points for M_1/M_2.

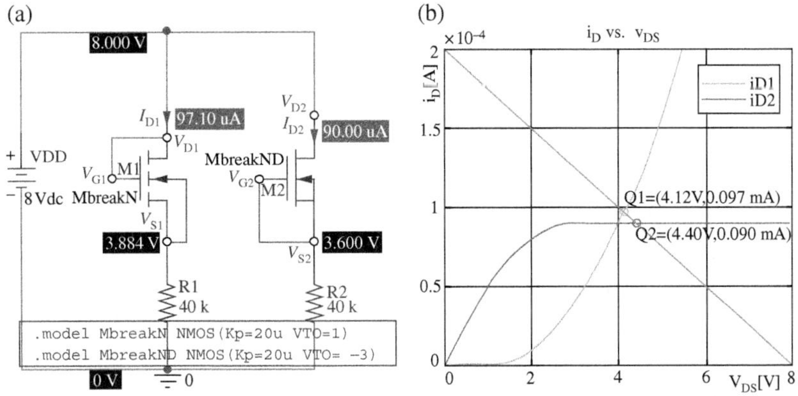

Figure P4.6 Diode-connected MOSFET circuits.

```
%elec04p06.m
vDSs=[0:0.001:8]; % Range of vDS to plot the graphs on
Kp1=0.02e-3; Vt1=1; lam1=0; Kp2=0.02e-3; Vt2=-3; lam2=0;
iDs1=iD_NMOS_at_vDS_vGS(vDSs,v??s,Kp1,Vt1,lam1); % for M1 with vGS=vDS
iDs2=iD_NMOS_at_vDS_vGS(vDSs,?,Kp2,Vt2,lam2); % for M1 with vGS=0
plot(vDSs,iDs1,'g', vDSs,iDs2,'r'), grid on, legend('iD1','iD2')
R1=40e3; R2=40e3;
VDD=8; iDLs=(VDD-vDSs)/R1; % Load line common to M1 and M2
% Load line approach to find the Q-points from the intersections
%  between the two characteristic curves and the identical load line
[fmin1,imin1]=min(abs(iDs1-iDLs)); IDQ1=iDLs(imin1);
VDSQ1=vDSs(imin1);
[fmin2,imin2]=min(abs(iDs2-iDLs)); IDQ2=iDLs(imin2);
VDSQ2=vDSs(imin2);
hold on, plot(vDSs,iDLs, VDSQ1,IDQ1,'go', VDSQ2,IDQ2,'ro')
xlabel('v_{DS}[V]'), ylabel('i_D'), title('i_D vs. v_{DS}')
text(VDSQ1-0.3,IDQ1+7e-6,['Q1=(' num2str(VDSQ1,'%4.2fV') ','
     num2str(IDQ1*1e3,'%6.3fmA') ')'])
text(VDSQ2-0.3,IDQ2-5e-6,['Q2=(' num2str(VDSQ2,'%4.2fV') ','
     num2str(IDQ2*1e3,'%6.3fmA') ')'])
axis([vDSs([1 end]) 0 max(iDLs)])
% Nonlinear equation approach using 'fsolve()'
eq1=@(vDS)VDD-???-R*iD_NMOS_at_vDS_vGS(vDS,???,Kp1,Vt1,lam1);
VDSQ1_=fsolve(eq1,VDD/2);
IDQ1_=iD_NMOS_at_vDS_vGS(VDSQ1,?????,Kp1,Vt1,lam1);
[VDSQ1 VDSQ1_], [IDQ1 IDQ1_]
eq2=@(vDS)VDD-vDS-?*iD_NMOS_at_vDS_vGS(vDS,?,Kp2,Vt2,lam2);
VDSQ2_=fsolve(eq2,VDD/2);
IDQ2_=iD_NMOS_at_vDS_vGS(VDSQ2,?,Kp2,Vt2,lam2);
[VDSQ2 VDSQ2_], [IDQ2 IDQ2_]
```

(b) As a nonlinear approach, use the MATLAB function '`fsolve()`' to solve the following KCL equations:

$$\frac{V_{DD} - v_{DS1}}{R_1} = i_{D1}(v_{GS1}, v_{DS1})$$
$$; V_{DD} - v_{DS1} - R_1 i_{D1}(v_{GS1}, v_{DS1}) = 0 \text{ with } v_{GS1} = v_{DS1} \quad \text{(P4.6.1)}$$

$$\frac{V_{DD} - v_{DS2}}{R_2} = i_{D2}(v_{GS2}, v_{DS2})$$
$$; V_{DD} - v_{DS2} - R_2 i_{D2}(v_{GS2}, v_{DS2}) = 0 \text{ with } v_{GS2} = 0 \quad \text{(P4.6.2)}$$

where $i_{Dk}(v_{GSk}, v_{DSk})$ is defined as Eq. (4.1.83) and implemented by the MATLAB function '`iD_NMOS_at_vDS_vGS()`' (see the end of Section 4.1.1), which is stored in a separate m-file named "iD_NMO-S_at_vDS_vGS.m." If helpful, complete and run the corresponding part of the above MATLAB script "elec04p06.m" to find the operating points for M_1/M_2.

(c) Perform the PSpice simulation (with Bias Point analysis type) for the two MOSFET circuits to check the validity of the operating points found in (a) or (b). Also, perform the PSpice simulation (with DC Sweep analysis type) for the two MOSFETs (with R_1/R_2 short-circuited for removal) to get the i_{Dk}-v_{DSk} characteristice curves of M_1/M_2 (for $V_{DD} = 0:0.01:8$ V) and see if they conform with those obtained in (a) or plotted in Figure 4.13(c1).

4.7 Cascode Current Mirror

Consider the cascode current mirror circuit in Figure P4.7 where the four NMOSs have the device parameters $K_p = 1\text{ mA/V}^2$, $V_t = 1$ V, and $\lambda = 2\text{ mV}^{-1}$ in common.

(a) Noting that $v_{GS1} = v_{DS1} = v_1$ and $v_{GS3} = v_{DS3} = v_3 - v_1$, apply KCL at nodes 1, 2, and 3 to write the node equations:

$$i_{D1}(v_1, \square) = i_{D3}(\boxed{}, v_3 - v_1) = i_D \quad \text{(P4.7.1a)}$$
$$i_{D2}(\square, v_2) = i_{D4}(v_3 - v_2, \boxed{}) = i_o \quad \text{(P4.7.1b)}$$
$$\frac{V_1 - v_3}{R} - i_D = 0;\ V_1 - v_3 - R i_D = 0 \quad \text{(P4.7.1c)}$$

where $i_{Dk}(v_{GSk}, v_{DSk})$ is defined as Eq. (4.1.83) and implemented by the MATLAB function '`iD_NMOS_at_vDS_vGS()`' (see the end of Section 4.1.1). If a current source I is connected into node 3, how will Eq. (P4.7.1c) be changed?

$$i_{D3}(v_3 - v_1, \boxed{}) = i_D = I \quad \text{(P4.7.2)}$$

```
function [io,ID1,v,Ro,Vomin]= ...
        FET4_current_mirror_cascode(Kps,Vt,lambda,R,V12)
% To analyze a cascade 4-FET current mirror like Fig.P4.7(a)
%  with R (resistor) or I (current source)
% If 0<R<1, it will be regarded as a current source I=R with R=inf.
% Copyleft: Won Y. Yang, wyyang53@hanmail.net, CAU for academic use only
V1=V12(1); if length(V12)>1, V2s=V12(2:end); else V2s=V12; end
if length(Kps)<4, Kps=repmat(Kps(1),1,4); end
iDvGS_GDs=@(vGS,Kp,Vt)Kp/2*(vGS-Vt).^2.*(1+lambda*vGS).*(vGS>Vt);
 % This is the iD-(vGS=vDS) characteristic of an FET with G-D shorted.
options=optimset('Display','off','Diagnostics','off');
n0=1; % To determine the index of V2s closest to V1
for n=1:length(V2s)
  V2=V2s(n);
  if n>1&(V2-V1)*(V2s(n-1)-V1)<=0, n0=n; end %Index of V2s closest to V1
  if R>=1|R==0, RL=R;
   eq=@(v) [iDvGS_GDs(v(?)-v(?),Kps(3),Vt)-iDvGS_GDs(v(?),Kps(1),Vt);
        (iD_NMOS_at_vDS_vGS(V2-v(2),v(?)-v(2),Kps(4),Vt,lambda)...
           -iD_NMOS_at_vDS_vGS(v(?),v(1),Kps(2),Vt,lambda));
         V1-R*iDvGS_GDs(v(?)-v(1),Kps(3),Vt)-v(3)]; % Eq.(P4.7.1)
   else I=R; RL=inf;
    eq=@(v) [iDvGS_GDs(v(3)-v(1),Kps(3),Vt)-iDvGS_GDs(v(1),Kps(1),Vt);
         (iD_NMOS_at_vDS_vGS(V2-v(?),v(3)-v(?),Kps(4),Vt,lambda)-...
            iD_NMOS_at_vDS_vGS(v(2),v(?),Kps(2),Vt,lambda));
         I-iDvGS_GDs(v(3)-v(?),Kps(3),Vt)]*1e4; % Eq.(P4.7.2)
  end
  v0=[V1/4 V1/3 2/3*V1]; % Initial guess for v=[v1 v2 v3]
  v=fsolve(eq,v0,options); vs(n,:)=v;
  ID1=iDvGS_GDs(v(?),Kps(1),Vt);
  io(n)=iD_NMOS_at_vDS_vGS(v(2),v(?),Kps(2),Vt,lambda);
end
Io=io(n0); V2=V2s(n0); v=vs(n0,:); % Values of io, V2, vs when V2=V1.
gm2=2*Io/(v(1)-Vt); ro2=1/lambda/Io; % Eqs.(4.1.4b,a) for M2
gm4=2*Io/(v(3)-v(?)-Vt); ro4=1/lambda/Io; % Eqs.(4.1.4b,a) for M4
Ro=ro2+ro4+gm4*ro2*ro4; % Eq.(P4.7.4)
Vomin=v(3)-Vt;
fprintf('Io=%8.4f[mA], ID1=%8.4f[mA],v1=%5.2f[V],v2=%5.2f[V],
  v3=%5.2f[V], Ro=%9.1f[kOhm]\n',Io*1e3,ID1*1e3,v(1),v(2),v(3),
  Ro/1e3);
```

(b) Noting that the small-signal resistance of an e-NMOS with G-D connected is $1/g_m$ (Eq. (4.1.19)), we can draw the small-signal equivalent of the circuit as Figure P4.7(c). To find the output resistance R_o, we apply a test current source i_T into node D_4 and write the KCL equations at nodes D_4 and 2:

$$\begin{bmatrix} 1/r_{o4} & -1/r_{o4} \\ -1/r_{o4} & \boxed{\qquad\qquad} \end{bmatrix} \begin{bmatrix} v_T \\ v_2 \end{bmatrix} = \begin{bmatrix} v_T - g_{m4}v_{gs4} \\ g_{m4}v_{gs4} - g_{m2}v_{gs2} \end{bmatrix} = \begin{bmatrix} i_T + g_{m4}v_2 \\ -g_{m4}v_2 \end{bmatrix}$$

$$;\begin{bmatrix} 1/r_{o4} & -1/r_{o4} - g_{m4} \\ -1/r_{o4} & \boxed{\qquad\qquad} \end{bmatrix} \begin{bmatrix} v_T \\ v_2 \end{bmatrix} = \begin{bmatrix} i_T \\ 0 \end{bmatrix} \tag{P4.7.3}$$

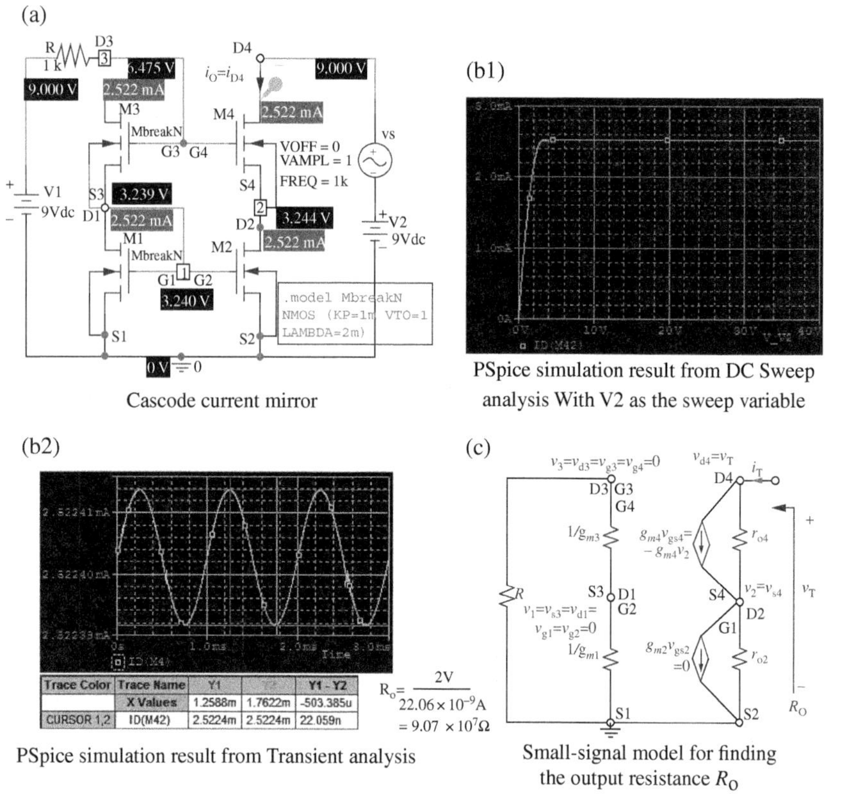

Figure P4.7 Cascode current mirror using FETs ("elec04p07.opj").

We can solve this equation and find the output resistance:

$$R_o = \frac{v_T}{i_T} = \boxed{} \qquad \text{(P4.7.4)}$$

where no current flows in the left part with no source so that $v_3 = v_{d3} = v_{g3} = v_{g4} = 0$ and $v_1 = v_{s3} = v_{d1} = v_{g1} = v_{g2} = 0$.

(Q) In the small-signal equivalent shown in Figure P4.7(c), why have M_1 and M_3 been modeled by just a resistor of $1/g_m$ (with no current source) unlike M_2 and M_4?

(c) Complete the above MATLAB function '`FET4_current_mirror_cascode()`', which implements Eqs. (P4.7.1/2) and (P4.7.4) for analyzing a cascode current mirror (excited by a voltage source V1 or a reference current source I) and finding its output resistance. Use the MATLAB function to find the output current i_o versus V_2 = 0:0.01:40 and plot it. Also, find the output resistance R_o (when $V_2 = 9$ V) of the current mirror (shown in Figure P4.7(a)) two times, once when it is excited by a voltage source V1 of 9 V (with a resistance R of 1 kΩ) and once when it is excited by a current source of $I = 2.5$ mA.

(d) Perform the PSpice simulation of the circuit excited by a voltage source V1 of 9 V (with a resistance R of 1 kΩ) two times, once with DC Sweep analysis to get the plot of the output current i_o versus V_2 = 0:0.01:40 (sweep variable) as depicted in Figure P4.7(b1) and once with Transient analysis to get the plot of i_o versus t = 0 : 1 μs : 3 ms to the AC voltage source v_s (of amplitude 1 V and frequency 1 kHz) as depicted in Figure P4.7(b2). Is the first graph (Figure P4.7(b1)) similar to the plot of i_o versus V_2 obtained in (c)? From the second graph (Figure P4.7 (b2)), find the (small-signal) output resistance:

$$R_o = \frac{V_{s,p\text{-}p}}{I_{o,p\text{-}p}} = \boxed{}\ \Omega \tag{P4.7.5}$$

Is it close to that obtained in (c)?

4.8 CMOS Inverter

Consider the CMOS inverter in Figure 4.19(a) where $V_{DD} = 10$ V and the NMOS/PMOS have the device parameters $K_{pN} = 200\ \mu A/V^2$, $V_{tN} = 2$ V, $K_{pP} = 80\ \mu A/V^2$, $V_{tP} = -2$ V, and $\lambda = 0$.

(a) To determine the point (V_{IL},V_{OH}) with the slope of −1 in Region II (where Mn and Mp are in the saturation and ohmic regions, respectively), we write the KCL equation at the output node (Dp,Dn):

$$K_{pP}\left\{(V_{DD}-v_i+V_{tP})(V_{DD}-v_o)-\frac{1}{2}(V_{DD}-v_o)^2\right\}=\frac{1}{2}K_{pN}(v_i-V_{tN})^2 \tag{P4.8.1a}$$

Differentiating both sides w.r.t. v_i and substituting $dv_o/dv_i = -1$ yields

$$K_{pN}(v_i-V_{tN})=K_{pP}\left\{-(V_{DD}-v_o)-(V_{DD}-v_i+V_{tP})\frac{dv_o}{dv_i}+(V_{DD}-v_o)\frac{dv_o}{dv_i}\right\}$$

$$;v_o=\frac{1}{2}\left\{\frac{K_{pN}}{K_{pP}}(v_i-V_{tN})+V_{DD}+v_i-V_{tP}\right\} \tag{P4.8.1b}$$

Use the MATLAB function ‘`fsolve()`’ to solve this set of two equations and find (V_{IL},V_{OH}) by running the following statements:

```
>>VDD=10; KN=2e-4; KP=8e-5; VtN=2; VtP=-2; Kr2=KN/KP; Kr=sqrt(Kr2)
  Vm=(VDD+VtP+Kr*VtN)/(1+Kr); % Eq.(4.1.65)
  eq_VIL=...
  @(v)[(VDD-v(1)+VtP)*(VDD-v(2))-(VDD-v(2))^2/2-Kr2*(v(1)-VtN)^2/2;
        v(2)-(VDD+v(1)-VtP+Kr2*(v(1)-VtN))/2]; % Eq.(P4.8.1a,b)
  v=fsolve(eq_VIL,[Vm Vm]); VIL=v(1), VOH=v(2)
```

Check if these values conform with those obtained by using Eq. (16.61) or (16.63) of [N-1] depending on $K_{pN}/K_{pP} = 1$ or not:

$$V_{IL} = V_{tN} + \frac{V_{DD} - V_{tN} + V_{tP}}{K_{pN}/K_{pP} - 1}\left(2\sqrt{\frac{K_{pN}/K_{pP}}{K_{pN}/K_{pP} + 3}} - 1\right) \text{ or } V_{tN} + \frac{3}{8}(V_{DD} - V_{tN} + V_{tP}) \tag{P4.8.2}$$

and substituting $v_i = V_{IL}$ into Eq. (P4.8.1b):

```
>>VIL_Neamen=(VtN+(VDD+VtP-VtN)/(Kr2-1)*(2*sqrt(Kr2/(Kr2+3))-1))...
   *(Kr~=1) + (Kr==1)*(VtN+3/8*(VDD+VtP-VtN)); %Eq.(P4.8.2)
  VOH_Neamen=((1+Kr2)*VIL_Neamen+VDD-Kr2*VtN-VtP)/2; % Eq.(P4.8.1b)
  [VIL VIL_Neamen VOH VOH_Neamen] % Do they conform each other?
```

(b) To determine the point (V_{IH}, V_{OL}) with the slope of −1 in Region IV (where Mn and Mp are in the ohmic and saturation regions, respectively), we write the KCL equation at the output node (Dp,Dn):

$$K_{pN}\left\{(v_i - V_{tN})v_o - \frac{1}{2}v_o^2\right\} = \frac{1}{2}K_{pP}(V_{DD} - v_i + V_{tP})^2 \tag{P4.8.3a}$$

Differentiating both sides w.r.t. v_i and substituting $dv_o/dv_i = -1$ yields

$$-K_{pP}(V_{DD} - v_i + V_{tP}) = K_{pN}\left\{v_o + (v_i - V_{tN})\frac{dv_o}{dv_i} - v_o\frac{dv_o}{dv_i}\right\}$$

$$; v_o = \frac{1}{2}\left\{\frac{K_{pP}}{K_{pN}}(v_i - V_{DD} - V_{tP}) + v_i - V_{tN}\right\} \tag{P4.8.3b}$$

We can use the MATLAB function '`fsolve()`' to solve this set of two equations and find (V_{IH}, V_{OL}) by running the following statements:

```
>>eq_VIH=@(v) [Kr2*((v(1)-VtN)*v(2)-v(2)^2/2)-(VDD-v(1)+VtP)^2/2;
    v(2)+((VDD-v(1)+VtP)/Kr2-v(1)+VtN)/2]; % Eq.(P4.8.3a,b)
    v=fsolve(eq_VIH,[Vm Vm]); VIH=v(1), VOL=v(2)
```

Check if these values conform with those obtained by using Eq. (16.66) or (16.69) of [N-1] depending on $K_{pN}/K_{pP} = 1$ or not:

$$V_{IH} = V_{tN} + \frac{V_{DD} - V_{tN} + V_{tP}}{K_{pN}/K_{pP} - 1}\left(\frac{2K_{pN}/K_{pP}}{\sqrt{3K_{pN}/K_{pP} + 1}} - 1\right)$$

$$\text{or } V_{tN} + \frac{5}{8}(V_{DD} - V_{tN} + V_{tP}) \tag{P4.8.4}$$

and substituting $v_i = V_{IH}$ into Eq. (P4.8.3b):

```
>>VIH_Neamen=(VtN+(VDD+VtP-VtN)/(Kr2-1)*(2*Kr2/sqrt(3*Kr2+1)-1))...
   *(Kr~=1) + (Kr==1)*(VtN+5/8*(VDD+VtP-VtN)); %Eq.(P4.8.4)
  VOL_Neamen=(VIH_Neamen*(1+1/Kr2)-VtN-(VDD+VtP)/Kr2)/2; %(P4.8.3b)
  [VIH VIH_Neamen VOL VOL_Neamen] % Do they conform each other?
```

```
%elec04p08.m
% Analyzes a CMOS inverter (Neamen, pp1173-1175)
clear, clf
VDD=10; KN=20e-5; VtN=2; KP=8e-5; VtP=-2;
KNP=[KN K?]; VtNP=[Vt? VtP];
subplot(221)
Kg=2; % To plot the graphs
[VIL,VIH,VOL,VOH,VM,VIT1,VOT1,VIT2,VOT2,VLH,NML,NMH,PDavg]=...
            CMOS_inverter(KNP,????,VDD,Kg),
% Plot the iD-vDS characteristic curve and load line
vis=[VIL VIT2 VM VIT1 VIH VOH]; iDmax=2.5e-3;
vOVNs=vis-VtNP(1); vSGPs=VDD-vis;
vDSs1=[VOH VOT2 VM VOT1 VOL VLH(1)];
iDs1 = iD_NMOS_at_vDS_vGS(VDD-vDSs1,vSGPs,KP,abs(VtP));
subplot(222)
CMOS_inverter_iD_vDS(KNP,VtNP,VDD,vOVNs,iDmax,[vDSs1; iDs1]);
```

```
function CMOS_inverter_iD_vDS(KNP,VtNP,VDD,vOVs,iDmax,vDSiDs)
% To plot the iD-vDS characteristics of CMOS inverter
if nargin<5, iDmax=0.5e-3; end
if numel(KNP)>1, KN=KNP(1); KP=KNP(2); else KN=KNP; KP=KNP; end
if numel(VtNP)>1, VtN=VtNP(1); VtP=VtNP(2); else VtN=VtNP;
VtP=VtNP; end
if nargin<4|isempty(vOVs), vOVs = [0:VDD-VtN]; end
dvDS=VDD/1000; vDSs=[0:dvDS:VDD]; di=iDmax/30;
% Boundaries between triode/saturation regions for Mn and Mp
plot(vDSs,KN/2*vDSs.^2,'b:', vDSs,KP/2*(VDD-vDSs).^2,'r:');
legend('Boundary of Mn between triode/saturation regions', ...
    'Boundary of Mp between triode/saturation regions')
hold on
for i=1:length(vOVs)
  vGS = VtN + vOVs(i);
  iDNs = iD_NMOS_at_vDS_vGS(vDSs,vGS,KN,VtN);
  iDPs = iD_NMOS_at_vDS_vGS(VDD-vDSs,VDD-vGS,KP,abs(VtP));
  plot(vDSs,iDNs, vDSs,iDPs,'r');
  text(VDD/2+1.2,max(iDNs)+di,['v_{GSn}=' num2str(vGS,'%5.2f') 'V'])
  text(VDD/2-2.4,max(iDPs)+di, ...
       ['v_{SGp}=' num2str(VDD-vGS,'%5.2f') 'V'])
end
xlabel('v_{DS}'), ylabel('i_D'); axis([0 VDD 0 iDmax])
if nargin>5&~isempty(vDSiDs) % Operating point, etc
 plot(vDSiDs(1,:),vDSiDs(2,:),'ro','Markersize',5);
end
```

(c) Complete and run the above MATLAB script "elec04p08.m" to plot the VTC together with the critical points as depicted in Figure P4.8(a) and the i_D-v_{DS} characteristic curves (for Mn and Mp) as depicted in Figure P4.8(b).

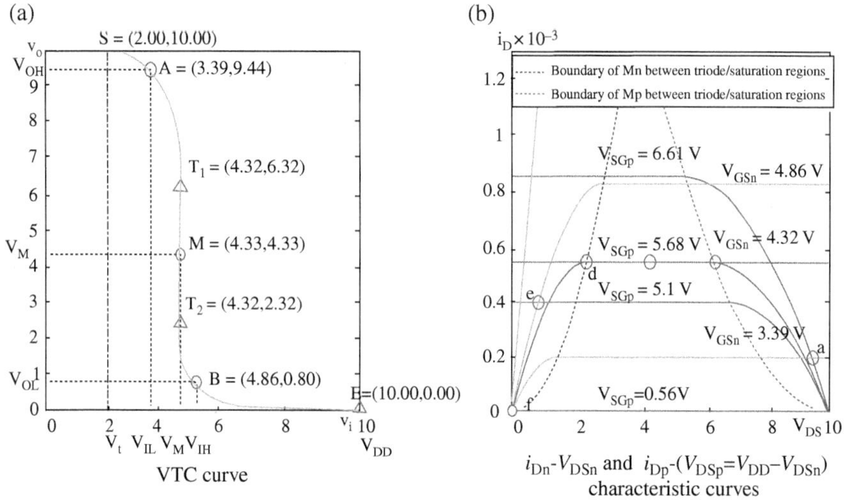

Figure P4.8 VTC curve of a CMOS inverter and the i_D-v_{DS} characteristic curves of NMOS/PMOS.

(d) Find the points on the i_D-v_{DS} characteristic curves (shown in Figure P4.8(b)) corresponding to each of the points A, T_1, M, T_2, and B on the VTC (shown in Figure P4.8(a)). Explain the meaning of each point.

4.9 CMOS Differential (or Source-Coupled) Pair Loaded by a Current Mirror
Consider the CMOS differential pair loaded by a current mirror in Figure P4.9.1(a) where $V_{DD} = 6$ V, $V_{GG} = 4$ V, $V_{SS} = 2$ V, and the NMOS/PMOS have the device parameters $K_{pN} = 2$ mA/V^2, $V_{tN} = 1$ V, $K_{pP} = 2$ mA/V^2, $V_{tP} = -1$ V, and $\lambda = 0.02$ V^{-1}.

(a) Noting that $v_{GS1} = V_{GG} + v_d - v_3$, $v_{DS1} = v_1 - v_3$, $v_{GS2} = V_{GG} - v_3$, $v_{DS2} = v_2 - v_3$, and $v_{SG3} = v_{SD3} = V_{DD} - v_1$, $v_{SG4} = V_{DD} - v_1$, and $v_{SD4} = V_{DD} - v_2$, apply KCL at nodes 1, 2, and 3 to write the node equations:

$$i_{D3}(V_{DD} - v_1, V_{DD} - v_1) = i_{D1}(V_{GG} + v_d - v_3, v_1 - v_3) \tag{P4.9.1a}$$

$$i_{D4}(V_{DD} - v_1, V_{DD} - v_2) = i_{D2}(V_{GG} - v_3, v_2 - v_3) \tag{P4.9.1b}$$

$$i_{D1}(V_{GG} + v_d - v_3, v_1 - v_3) + i_{D2}(V_{GG} - v_3, v_2 - v_3) = i_{D5}(V_{SS}, v_3) \tag{P4.9.1c}$$

where $i_{D1}(v_{GS1}, v_{DS1}) = i_{D2}(v_{GS2}, v_{DS2}) = i_{Dn}(v_{GS}, v_{DS})$ and $i_{D3}(v_{SG3}, v_{SD3}) = i_{D4}(v_{SG4}, v_{SD4}) = i_{Dp}(v_{SG}, v_{SD})$, defined as Eq. (4.1.83), can be implemented by the MATLAB function handles '`iDn()`' and '`iDp()`', respectively, in the MATLAB function '`FET_differential_loaded_ by_current_mirror()`' below.

(b) Noting from Eqs. (3.78) and (3.79) of [C-1] that the output resistance (seen from node 2 and ground) and voltage gain of the differential pair are

```
function [v1s,v2s,v3s,iD1s,iD2s,Av,Ro]=...
      FET_differential_loaded_by_current_mirror(Kp,Vt,lam,VDD,vds)
% Analyze an FET differential pair loaded by a 2 FET current mirror
% (to find v1,v2,v3,iD1, and iD2 to a range vds of differential input vd
%   and plot them vs. vds or t if vds is given as [vds; t].)
% Copyleft: Won Y. Yang, wyyang53@hanmail.net, CAU for academic use only
if numel(Kp)<2, KpN=Kp; KpP=Kp; else KpN=Kp(1); KpP=Kp(2); end
if numel(Vt)<2, VtN=Vt; VtP=-Vt; else VtN=Vt(1); VtP=Vt(2); end
if length(VDD)<3, error('VDD must be given as [VDD VGG VSS]!'); end
VSS=VDD(3); VGG=VDD(2); VDD=VDD(1);
IS=KpN/2*(VSS-VtN^2)^2;
options=optimset('Display','off','Diagnostics','off');
iDn=@(vGS,vDS)KpN/2*(vGS-VtN).^2*(1+lam*vDS).*(vGS>VtN) ...
    .*(vGS-vDS<VtN)+KpN*((vGS-VtN).*vDS-vDS.^2/2)*(1+lam*vDS) ...
    .*(vGS>VtN).*(vGS-vDS>=VtN);
iDp=@(vSG,vSD)KpP/2*(vSG+VtP).^2*(1+lam*vSD).*(vSG>VtPa) ...
    .*(vSG-vSD<VtPa)+KpP*((vSG+VtP).*vSD-vSD.^2/2)*(1+lam*vSD)
    .*(vSG+VtP>0).*(vSG-vSD>=VtPa);
if size(vds,1)==2, t=vds(2,:); vds=vds(1,:); end
lvds=length(vds);
for n=1:lvds
  vd = vds(n);
  eq=@(v)[iDp(VDD-v(?),VDD-v(1))-iDn(VGG+vd-v(3),v(?)-v(3))
       iDp(VDD-v(1),VDD-v(?))-iDn(VGG-v(3),v(2)-v(?)) % Eq.(P4.9.1)
       iDn(VGG+vd-v(3),v(?)-v(3))+iDn(VGG-v(?),v(2)-v(3)) ...
                      -iDn(VSS,v(?))]*1e4;
  if vd<=0
   if ~exist('v'), v0=VDD/6*[4 2 1]; else v0=v; end % Initial guess
  else if vd<=0.02, v0=VDD/6*[4 5.5 1]; else v0=v; end % Initial guess
  end
  v = fsolve(eq,v0,options);
  v1s(n)=v(1); v2s(n)=v(2); v3s(n)=v(3); v12s(n)=v(1)-v(2);
  iD1s(n)=iDn(VGG+vd-v(3),v(1)-v(3));
  iD2s(n)=iDn(VGG-v(3),v(2)-v(3));
  iD5s(n)=iDn(VSS,v(3));
end
im=ceil(lvds/2); ID1=iD1s(im); ID2=iD2s(im);
V1=v1s(im); V2=v2s(im); VSS=v3s(im);
gm1=2*ID1/(VGG-VSS-Vt?); ro2=1/lam/ID2; % Eqs.(4.1.4b,a) for M2
gm2=2*ID2/(VGG-VSS-VtN); ro4=1/lam/ID?; % Eqs.(4.1.4b,a) for M4
gm3=2*ID1/(VDD-V1?Vt?); ro3=1/lam/ID?; % Eqs.(4.1.4b,a) for M3
Ro=1/(1/ro2+1/ro4); % Output resistance Eq.(P4.9.2)
Av=-gm1*Ro; % Voltage gain - Eq.(P4.9.3)
if nargout==0
 if ~exist('t')
  subplot(221), plot(vds,iD1s, vds,iD2s,'r', vds,iD5s,'g:')
  legend('iD1','iD2','iD5'); xlabel('v_d'); ylabel('i_D');
  axis([vds([1 end]) -0.1*IS 1.1*IS]), grid on
```

```
  subplot(222)
  plot(vds,v12s, vds,v1s,'g:', vds,v2s,'r', vds,v3s,'m:')
  legend('v12','v1','v2','v3'); xlabel('v_d'); ylabel('v_o');
  axis([vds([1 end]) min(vos)-0.5 max(v2s)+0.5]), grid on
  if size(vds,1)==2, t=vds(2,:); vds=vds(1,:); end
 else
  subplot(223), plot(t,iD1s, t,iD2s,'r'), legend('iD1(t)','iD2(t)');
  subplot(224)
   plot(t,vds, t,vos,'r', t,v2s-mean(v2s),'r', t,Av*vds,'m:')
  legend('vd(t)','vo(t)','v2(t)-mean(v2(t))');
 end
 fprintf('Ro=%10.2f[kOhm] and Av=%8.2f\n', Ro/1e3,Av);
end
```

```
%elec04p09.m
Kp=2e-3; Vt=1; lambda=0.02; R=100e3;
VDD=6; VGG=4; VSS=2; VDDs=[VDD VGG VSS];
% Please place the average value of vd at the center of vds.
Vdm=1.5; dvd=Vdm/100; vds=[-Vdm:dvd:Vdm]; % Differential input vd
% To plot v1, v2, and v3 vs. vds,
FET_differential_loaded_by_current_mirror(Kp,Vt,lambda,VDDs,vds);
% To plot vd(t) and v2(t) vs. t,
f=1e3; dt=1/f/360; t=[0:360]*dt; Vdm=1e-3;
vds=[Vdm*sin(2*pi*f*t); t];
FET_differential_loaded_by_current_mirror(Kp,Vt,lambda,VDDs,vds);
```

$$R_o \underset{\text{of [C-1]}}{\overset{(3.78)}{=}} r_{o2} \| r_{o4} \overset{(4.1.4a)}{=} \frac{1}{\lambda I_{D2}} \Big\| \frac{1}{\lambda I_{D4}} \overset{I_{D2}=I_{D4}}{=} \frac{r_{o2}}{2} \tag{P4.9.2}$$

$$A_v = \frac{v_o = v_2}{v_d} \underset{\text{of [C-1]}}{\overset{(3.79)}{=}} -g_{m1}(r_{o2} \| r_{o4}) \overset{(4.1.4b)}{=} -\frac{2I_{D1}}{V_{GS1} - V_{t1}} R_o \tag{P4.9.3}$$

complete the above MATLAB function 'FET_differential_loaded_by_current_mirror()', which solves Eq. (P4.9.1) for v_1, v_2, and v_3, and also uses Eqs. (P4.9.2,3) to find the output resistance and voltage gain.

(c) Run the above MATLAB script "elec04p09.m," which uses the MATLAB function 'FET_ differential_loaded_by_current_mirror()' two times, once with the fifth input argument vds = [−1.5 : 0.015 : 1.5] to plot i_{D1}/i_{D2} versus v_d on a graph (as depicted in Figure P4.9.1(b1)) and $v_1/v_2/v_3$ versus v_d on another graph (as depicted in Figure P4.9.1 (b2)), and once the fifth input argument vds = $[v_d(t); t]$ ($t = 0 : 1\ \mu s : 3$ ms) to plot $v_d(t)/A_v v_d(t)/v_o(t) = v_2(t) - \text{mean}(v_2(t))$ versus t on another graph (as depicted in Figure P4.9.2(a1)) where $v_d(t) = \sin(2000\pi t)$ [mV]. You will see not only the related graphs but also the theoretical values of the output resistance and voltage gain.

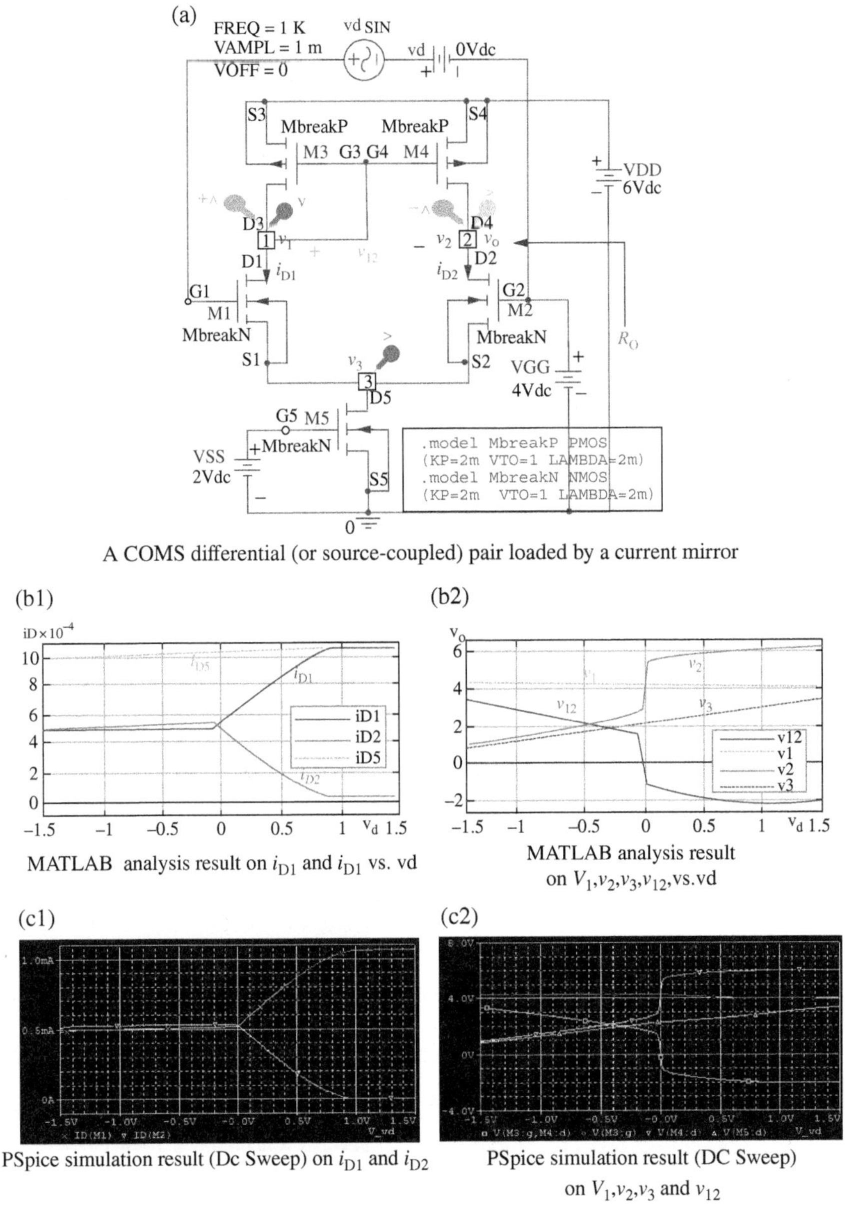

Figure P4.9.1 CMOS differential (source-coupled) pair loaded by a current mirror ("FET_differential_loaded_by_current mirror.opj").

(d) Constructing the schematic as depicted in Figure P4.9.1(a), perform the PSpice simulation of DC Sweep analysis with the sweep variable $v_d = -1.5 : 0.01 : 1.5$ [V] to get the plot of i_{D1}/i_{D1} versus v_d on a graph (as depicted in Figure P4.9.1(c1)) and $v_1/v_2/v_3$ versus v_d on another graph (as depicted in Figure P4.9.1(c2)). Are they similar to the MATLAB analysis results shown in Figure P4.9.1(b1) and (b2)? Also perform the PSpice simulation of Transient analysis on the output voltage v_2 to the sinusoidal input differential voltage source $v_{dSIN} = \sin(2000\pi t)$ [mV] to measure the voltage gain $A_v = v_2/v_{dSIN}$. Is it close to the MATLAB analysis result obtained in (c)?

(e) Constructing the schematic (applying a test current source $i_T = \sin(2000\pi t)$[μA] into the output node 2) as depicted in Figure P4.9.2(b), perform the PSpice simulation of Transient analysis on the voltage v_T across the test current source to measure the output resistance $R_o = v_T/i_T$. Is it close to the MATLAB analysis result obtained in (c)?

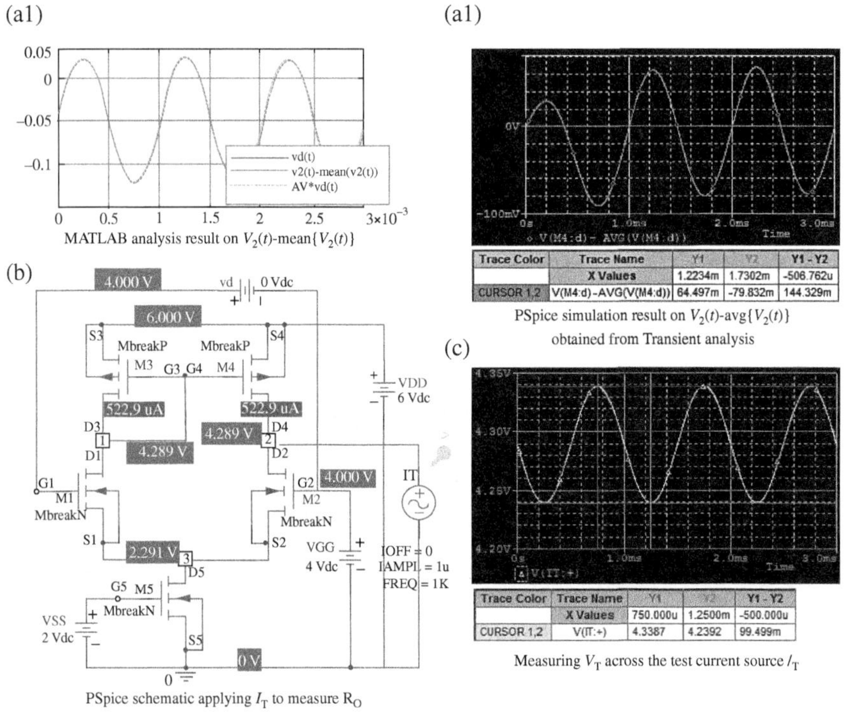

Figure P4.9.2 CMOS differential (or source-coupled) pair loaded by a current mirror.

4.10 Analysis of CS JFET Amplifier Circuit

Consider the JFET circuit in Figure P4.10(a) where the JFET has the device parameters BETA – $K_p/2$ = 0.75 mA/V^2, VTO = V_t = –4 V, and λ – 0.008 V^{-1}.

(a) Complete the following MATLAB script "elec04p10.m" so that it can use the MATLAB function 'FET_CS_analysis()' to analyze the circuit three times, once with R_D = 2.7 kΩ and R_S = 2.7 kΩ (case A), once with R_D = 2.7 kΩ and R_S = 5 kΩ (case B), and once with R_D = 3.5 kΩ and R_S = 5 kΩ (case C). Then, run it to find the overall voltage gain $G_v = v_o/v_i$ for the three cases. Noting that the three output voltage waveforms in Figure P4.10(c) have been obtained from PSpice simulations for the three cases with the small-signal input $v_i(t)$ = 0.1 sin (2000πt), identify which one of $v_{oa}(t)$, $v_{ob}(t)$, and $v_{oc}(t)$ corresponds to the output voltage of three cases A, B, and C, respectively.

(a) PSpice schematic with R_D and R_S unspecified

(c) PSpice simulation results on V_O (t) for (R_D,R_S)=(2.7 kΩ, 2.7 kΩ),(2.7 kΩ, 5 kΩ), and (R_D = 3.5 kΩ, 5 kΩ)

(b1) Q-points on the i_D-V_{GS} characteristic curves

(b2) Q-points on the i_D-V_{DS} characteristics curves

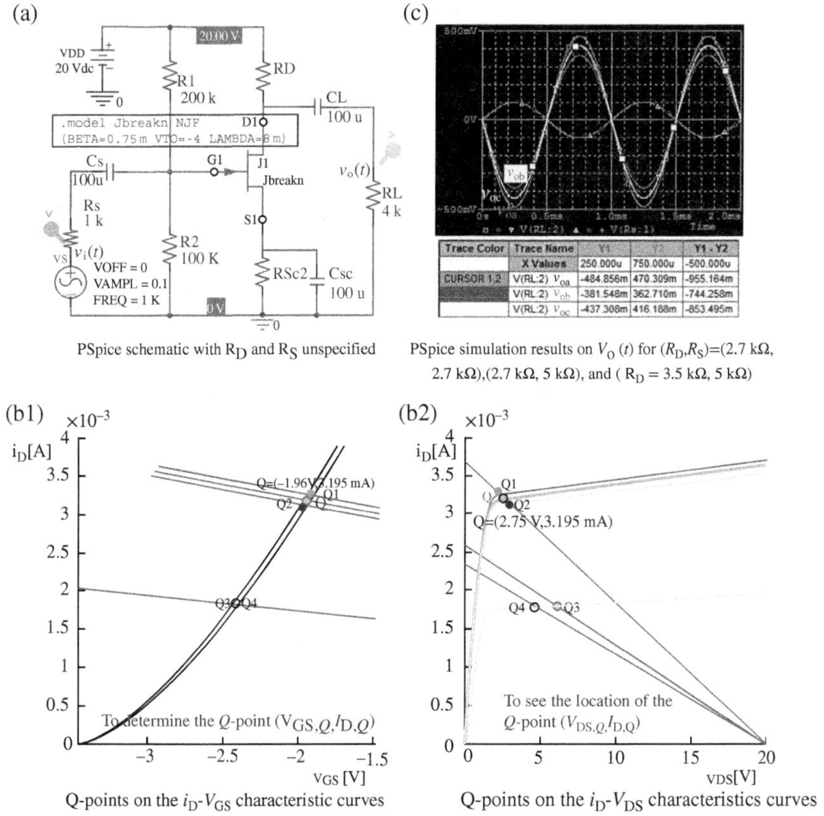

Figure P4.10 A CS JFET amplifier and its analysis based on the operating points.

```
%elec04p10.m
Kp=1.5e-3; Vt=-4; lambda=8e-3; %IDSS=12e-3; Kp=2*IDSS/Vt^2; BETA=Kp/2
disp('With RD=2.7kOhm and RS=2.7kOhm')
VDD=20; Rs=1e3; R1=200e4; R2=100e4; RD=2700; RS=2700; RL=4e3;
FET_CS_analysis(VDD,Rs,R1,R2,RD, [0 ??],RL,Kp,Vt,lambda);
disp('With RD=5kOhm and RS=2.7kOhm')
RS=5000; FET_CS_analysis(VDD,Rs,R1,R2,RD, [? RS],RL,Kp,Vt,lambda);
disp('With RD=3.5kOhm and RS=5kOhm')
RD=3500; FET_CS_analysis(VDD,Rs,R1,R2,RD, [0 RS],RL,Kp,Vt,lambda);
```

(b) Figure P4.10(b1) and (b2) shows the three *Q*-points *Q*, Q_3, and Q_4 on the i_D-v_{GS} and i_D-v_{DS} characteristic curves for the three cases A, B, and C. Based on the slopes of the load lines crossing the *Q*-points, rather than the DC analysis results obtained in (a), identify which one of *Q*, Q_3, and Q_4 corresponds to each of the three cases, respectively, and write the equations for the related load lines. From the locations of the three *Q*-points, tell which case is the most susceptible to the output distortion for an increasing amplitude of the input voltage $v_i(t)$.

(c) To see how easily the output $v_o(t)$ gets distorted by raising the amplitude of the small-signal input increasing $v_i(t)$, two more *Q*-points Q_1 and Q_2 with V_G = 20/3±0.2[V] have been drawn in Figure P4.10(b1) and (b2). Which one of them corresponds to each of V_G = 20/3±0.2[V], respectively?

(d) Considering the above three cases for which the small-signal output voltage waveforms are shown in Figure P4.10(b), can you say that the maximum small-signal output voltage that can be produced with no distortion is proportional to the voltage gain G_v?

(e) Are you going to increase or decrease the value of R_{S1} to push the *Q*-point rightward towards V_{DD}? Will the possibility of output distortion be decreased by moving the *Q*-point rightward, i.e. increasing $V_{DS,Q}$? Generally speaking, if a *Q*-point is too close to $v_{DS} = V_{DD}$, can the output easily get distorted from the lower or upper part?

(f) The output voltage waveforms shown in Figure P4.10(c) have larger/smaller magnitdes when they are negative/positive. To figure out what makes such a difference, find out the relationship among the slope of the i_D-v_{GS} characteristic curve (Figure P4.10(b1)), the transconductance g_m, and the voltage gain $G_v = v_o/v_i$. Based on the relationship and considering the difference between the slopes of the i_D-v_{GS} characteristic curve on the left/right part of a *Q*-point, explain the difference between the magnitudes of $v_o(t)$ during its positive/negative periods.

```
%elec04p11.m
VDD=10; Rs=0; R1=160e3; R2=40e3; RD=9e3; RS=1e3; RL=Inf;
Vt=0.8; lambda=0; Kps=[1.6 2]*1e-3;
% With no bypass capacitor in parallel with RS
for n=1:numel(Kps)
   fprintf('\n With Kp=%6.4fmA/V^2 and Vt=%5.2fV,', [Kps(n)*1e3 Vt]);
   FET_CS_analysis(VDD,Rs,R1,R2,RD,[RS ?],RL,Kps(n),Vt,lambda);
end
% With Kp=2 and a bypass capacitor CSc in parallel with RS
fprintf('\n With Kp=%6.4fmA/V^2, Vt=%5.2fV, and C in parallel with RS',
   [Kps(2)*1e3 Vt]);
FET_CS_analysis(VDD,Rs,R1,R2,RD,[? RS],RL,Kps(2),Vt,lambda);
```

4.11 Analysis of CS MOSFET Amplifier Circuit

Consider the MOSFET circuit in Figure P4.11(a) where the NMOS has the device parameters $K_p = 2\ \text{mA/V}^2$, $V_t = 0.8\ \text{V}$, and $\lambda = 0\ \text{V}^{-1}$.

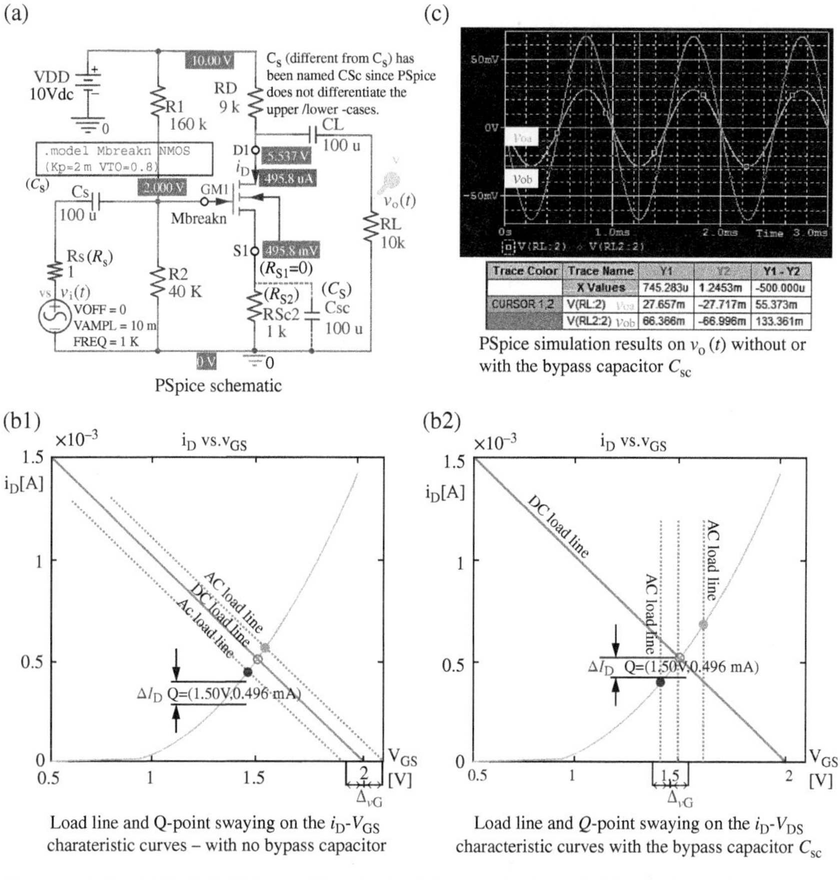

Figure P4.11 A CS MOSFET amplifier, its load line analysis, and PSpice simulation result.

(a) Complete the above MATLAB script "elec04p11.m" so that it can use the MATLAB function 'FET_CS_analysis()' to analyze the circuit three times, once with $K_p = 1.6\ \text{mA/V}^2$ and without a bypass capacitor C_{Sc} in parallel with R_{Sc} (case A), once with $K_p = 2\ \text{mA/V}^2$ and without C_{Sc} (case B), and once with $K_p = 2\ \text{mA/V}^2$ and C_{Sc} in parallel with R_{Sc} (case C). Then, run it to find the voltage gain $G_v = v_o/v_i$ for the three cases.

(b) Which case is the DC analysis result (obtained from the PSpice simulation) displayed on the schematic in Figure P4.11(a) for?

(c) Figure P4.11(b1) and (b2) shows the DC/AC load lines and Q-points on the i_D-v_{GS} characteristic curves for the cases B (without C_{Sc}) and C (with C_{Sc}). Write the equation for the DC load line and find the slopes of the AC load lines for the two cases. From the load lines and corresponding Q-points swaying with the small-signal input voltage $v_i = \Delta v_G$, tell which case will produce a larger small-signal drain current i_d for the same amplitude of v_i, yielding a larger small-signal output voltage like $v_{ob}(t)$ in Figure P4.11(c).

(d) Find the small-signal voltage gains for $v_{oa}(t)$ and $v_{ob}(t)$ in the PSpice simulation result (Figure P4.11(c)) and check if they are close to the MATLAB analysis result obtained in (a).

4.12 Analysis of CD MOSFET Amplifier (Source Follower) Circuit

Consider the MOSFET circuit in Figure P4.12(a) where the NMOS has the device parameters $K_p = 1\ \text{mA/V}^2$, $V_t = 1\ \text{V}$, and $\lambda = 0.025\ \text{V}^{-1}$.

(a) Complete the following MATLAB script "elec04p12.m" so that it can use the MATLAB function 'FET_CD_analysis()' to analyze the circuit three times, once with $R_D = 20\ \text{k}\Omega$ and $R_S = 20\ \text{k}\Omega$ (case A), once with $R_D = 35\ \text{k}\Omega$ and $R_S = 20\ \text{k}\Omega$ (case B), and once with $R_D = 20\ \text{k}\Omega$ and $R_S = 35\ \text{k}\Omega$ (case C). Then, run it to find the voltage gain $G_v = v_o/v_i$ and the input/output resistances for the three cases. Do the changes of the voltage gain and the input/output resistances due to the change of R_D or R_S conform with your expectation based on Eqs. (4.2.11a), (4.2.12), and (4.2.10)?

```
%elec04p12.m
clear, clf
VDD=10; Rs=1e3; R1=200e3; R2=200e3; RL=1e6;
Kp=1e-3; Vt=1; lambda=0.025; % Device parameters
for n=1:3
  if n==1, RD=20e3; RS=20e3; % Case A
  elseif n==2, RD=35e3; RS=20e3; % Case B
  else RD=20e3; RS=35e3; % Case C
  end
  FET_CD_analysis(VDD,Rs,R1,R2,RD,RS,RL,Kp,Vt,lambda);
end
```

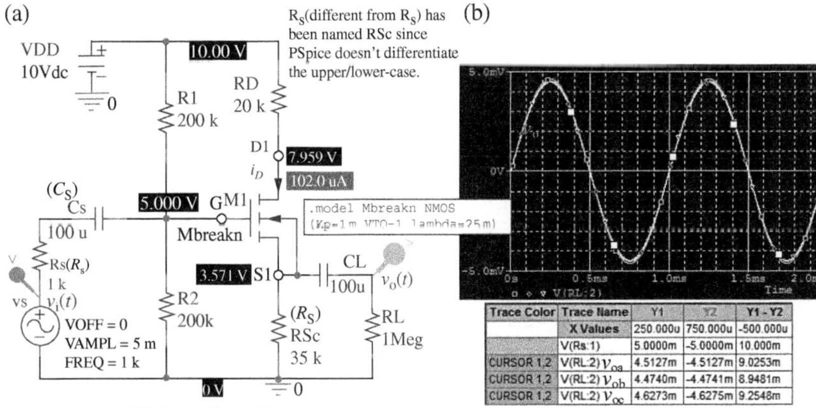

PSpice schematic

PSpice simulation result on $v_o(t)$

Figure P4.12 A CD MOSFET amplifier and its PSpice simulation result.

(b) Check if the PSpice simulation results on the voltage gain $G_v = v_o/v_i$ (shown in Figure P4.12(b)) agree with the MATLAB analysis results.

4.13 Analysis of CG MOSFET Amplifier

Consider the MOSFET circuit in Figure P4.13(a1) or (a2) where the NMOS has the device parameters $K_p = 1\ \text{mA/V}^2$, $V_t = 1$ V, and $\lambda = 0.01\ \text{V}^{-1}$. Complete and run the following MATLAB script "elec04p13.m" to find the voltage gain $G_v = v_o/v_i$ and the input/output resistances. Then see if they conform with the PSpice simulation results shown in Figure P4.13(b1) and (b2).

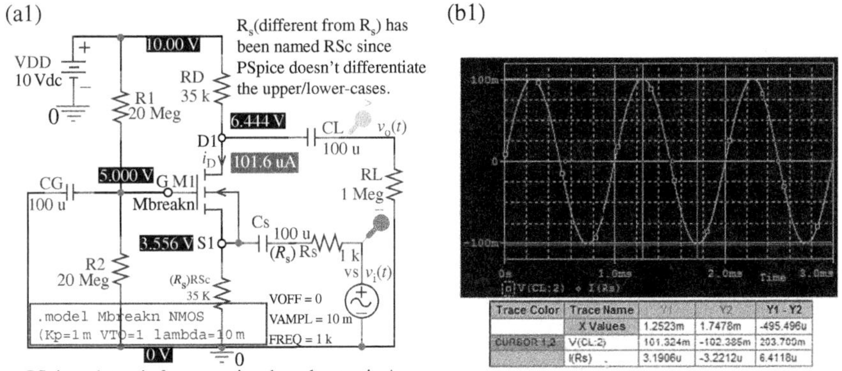

PSpice schematic for measuring the voltage gain A_v and input resistance R_i

PSpice simulation results on $v_o(t)$ and $i_i(t)$

Figure P4.13 A CG MOSFET amplifier and its PSpice simulation result.

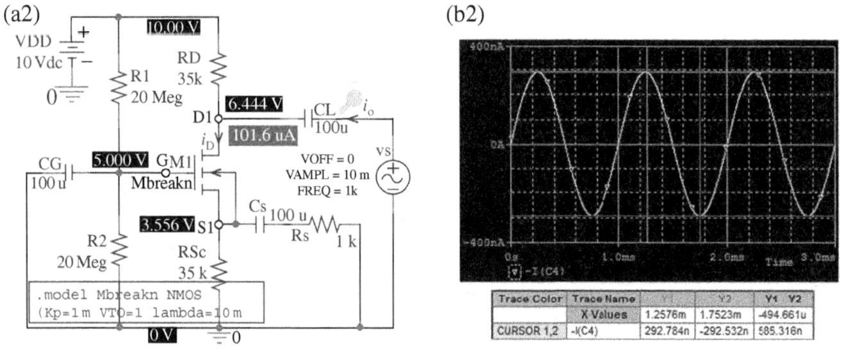

PSpice schematic for measuring the output resistance R_o

PSpice simulation results on i_o (t)

Figure P4.13 (Continued)

```
%elec04p13.m
% For Problem 4.13
% Circuit parameters
VDD=10; Rs=1e3; R1=2e7; R2=2e7; RD=35e3; RS=35e3; RL=1e6;
Kp=1e-3; Vt=1; lambda=0.01; % Device parameters
FET_C?_analysis(VDD,Rs,R1,R2,RD,RS,RL,Kp,Vt,lambda);
```

4.14 Analysis of Multistage MOSFET Amplifier

Consider the two-stage MOSFET circuit in Figure P4.14(a) where the device parameters of the NMOSs M_1 and M_2 are $K_{p1} = 1\ \text{mA/V}^2$, $V_{t1} = 1.2\ \text{V}$, $\lambda_1 = 0\ \text{V}^{-1}$, $K_{p2} = 0.4\ \text{mA/V}^2$, $V_{t2} = 1.2\ \text{V}$, and $\lambda_2 = 0\ \text{V}^{-1}$. Noting that the input resistance R_{i2} of stage 2 is infinite so that stage 1 (consisting of M_1, R_{11}, R_{21}, R_{D1}, and R_{Sc}) can be analyzed first, independently of stage 2 (consisting of M_2 and R_{Sc2}), complete and run the following MATLAB script "elec04p14.m" to find the overall voltage gain $G_v = v_o/v_i$ and the overall input/output resistances. Then see if they conform with the PSpice simulation results shown in Figure P4.14(b).

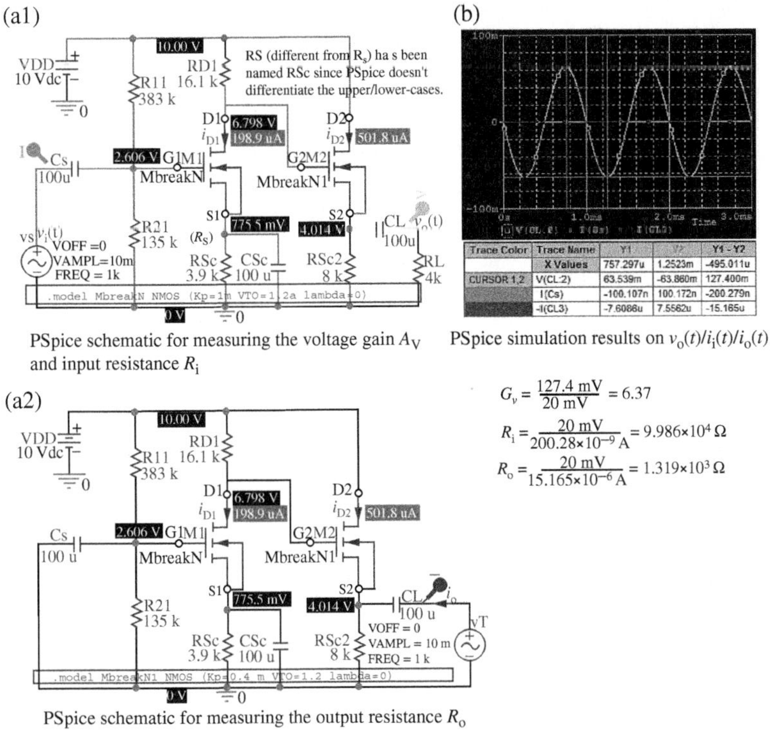

PSpice schematic for measuring the voltage gain A_V and input resistance R_i

PSpice simulation results on $v_o(t)/i_i(t)/i_o(t)$

$$G_v = \frac{127.4\text{ mV}}{20\text{ mV}} = 6.37$$

$$R_i = \frac{20\text{ mV}}{200.28\times10^{-9}\text{ A}} = 9.986\times10^4\,\Omega$$

$$R_o = \frac{20\text{ mV}}{15.165\times10^{-6}\text{ A}} = 1.319\times10^3\,\Omega$$

PSpice schematic for measuring the output resistance R_o

Figure P4.14 A two-stage amplifier and its PSpice simulation result.

```
%elec04p14.m
clear, clf
VDD=10; Rs=4e3; RL=4e3; % DC power supply, Source/Load resistances
Kp1=1e-3; Vt1=1.2; lambda1=0; % Device parameters of M1
Kp2=0.4e-3; Vt2=1.2; lambda2=0; % Device parameters of M2
% Since RL1=Ri2 to Stage 1 is infinite, Stage can be analyzed
% independently of Stage 2.
% Analysis of Stage 1
Rs1=Rs; R11=383e3; R21=135e3; RD1=161e2; RS1=[? ????]; RL1=???;
[VGQ1,VSQ1,VDQ1,IDQ1,Av(1),Ri(1),Ro(1)]= ...
  FET_CS_analysis(VDD,Rs1,R11,R21,RD1,RS1,RL1,Kp1,Vt1,lambda1);
% Analysis of Stage 2
Rs2=Ro(?); VG2=V???; RG2=Inf; RD2=0; RS2=8e3; RL2=RL;
[VGQ2,VSQ2,VDQ2,IDQ2,Av(2),Ri(2),Ro(2)]= ...
  FET_CD_analysis(VDD,Rs2,VG2,RG2,RD2,RS2,RL2,Kp2,Vt2,lambda2);
Gv=Ri(1)/(Rs+Ri(1))*prod(??) % Global voltage gain
RI=Ri(?), RO=Ro(???) % Global input/output resistances
% From PSpice simulation
Gv_PSpice=127.4/20, Ri_PSpice=20/200.28e-6, Ro_PSpice=20/15.165e-3
```

4.15 Analysis of a Two-Stage MOSFET Amplifier

```
%elec04p15.m
% Problem 4.15 : Analysis of a two-stage (CS-CD) amplifier
clear
VDD=10; Rs=0; RL=1e4; % DC power supply, Source/Load resistances
Vsm=0.01; % AC signal source voltage
Kp1=2e-3; Vt1=0.8; lambda1=0; % Device parameters
Kp2=1e-3; Vt2=1; lambda2=0;
% Provisional analysis of Stage 2 to find Ri(2) with Rs2=0 and RL2=RL
Rs2=?; R12=2e5; R22=2e5; RD2=2e4; RS2=35e3; RL2=R?;
[VGQ2,VSQ2,VDQ2,IDQ2,Av2,Ri2,Ro2]=...
  FET_CD_analysis(VDD,Rs2,R12,R22,RD2,RS2,RL2,Kp2,Vt2,lambda2);
% Analysis of Stage 1 with RL1=Ri2
Rs1=??; R11=160e3; R21=40e3; RD1=9e3; RS1=[0 ???]; RL1=R??;
[VGQ1,VSQ1,VDQ1,IDQ1,Av(1),Ri(1),Ro(1)]= ...
   FET_CS_analysis(VDD,Rs1,R11,R21,RD1,RS1,RL1,Kp1,Vt1,...
   lambda1,Vsm);
% Analysis of Stage 2 with Rs2=Ro(1)
Rs2=Ro(?); Vsm2=Av(1)*Vsm;
[VGQ2,VSQ2,VDQ2,IDQ2,Av(2),Ri(2),Ro(2)]=...
   FET_CD_analysis(VDD,Rs2,R12,R22,RD2,RS2,RL2,Kp2,Vt2,...
   lambda2,Vsm2);
RI=Ri(?), RO=Ro(??d) % Overall input/output resistances
Gv_global=R?/(R?+RI)*prod(??) % Global voltage gain
```

Consider the MOSFET circuit in Figure P4.15(a) where the device parameters of the NMOSs M_1 and M_2 are $K_{p1} = 2\ \mathrm{mA/V^2}$, $V_{t1} = 0.8\ \mathrm{V}$, $\lambda_1 = 0\ \mathrm{V^{-1}}$, $K_{p2} = 1\ \mathrm{mA/V^2}$, $V_{t2} = 1\ \mathrm{V}$, and $\lambda_2 = 0\ \mathrm{V^{-1}}$. Noting that in general the analysis of a multistage amplifier needs performing a provisional backward analysis (starting from the last stage with the overall load resistance R_L) to determine the input resistances of each stage and then performing a forward analysis (starting from the first stage with the overall source resistance R_s) to determine the voltage gain and output resistances of each stage, complete the above MATLAB script "elec04p15.m" so that it can use the MATLAB function '`FET_CD_ analysis()`' to analyze stage 2, '`FET_CS_analysis()`' to analyze stage 1, and '`FET_CD_ analysis()`' to analyze stage 2 again. Run it to find the overall voltage gain $G_v = v_o/v_i$ and the overall input/output resistances R_I/R_O. Compare the overall voltage gain G_v with that obtained from the PSpice simulation (shown in Figure P4.15(b)) to see how close they are. Explain how the voltage gain has become larger than that of the single-stage amplifier without the stage 2 (of CD configuration) whose voltage gain is less than 1 (Problem 4.12).

(a)

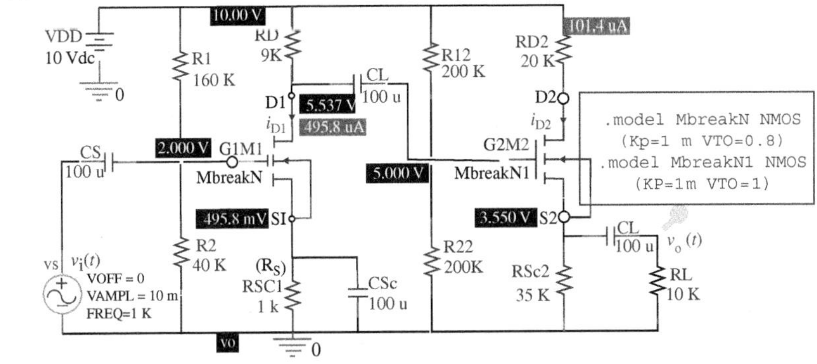

PSpice schematic for measuring the voltage gain G_v

(b)

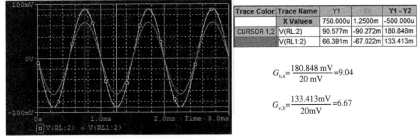

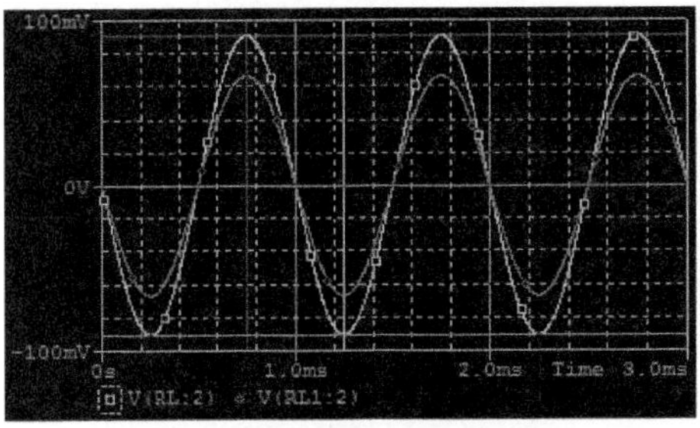

Trace Color	Trace Name	Y1	Y2	Y1 - Y2
	X Values	750.000u	1.2500m	-500.000u
CURSOR 1,2	V(RL:2)	90.577m	-90.272m	180.848m
	V(RL1:2)	66.391m	-67.022m	133.413m

$$G_{v,a}=\frac{180.848\text{ mV}}{20\text{ mV}}=9.04$$

$$G_{v,b}=\frac{133.413\text{mV}}{20\text{mV}}=6.67$$

PSpice simulation results on $v_o(t)$ with the 2nd stage of CD or not

Figure P4.15 A two-stage amplifier and its PSpice simulation result.

4.16 Analysis of a Three-Stage MOSFET Amplifier

Consider the MOSFET circuit in Figure P4.16(a) where the device parameters of the NMOSs M_1/M_3 are $K_p = 1\text{ mA/V}^2$, $V_t = 1\text{ V}$, $\lambda = 0.01\text{ V}^{-1}$ in common, and those of the NPN-BJT Q_2 are $\beta_F = 100$, $\beta_R = 1$, $V_A = 100\text{ V}$, and $I_s = 10^{-16}\text{ V}$. Noting that in general the analysis of a multistage amplifier needs performing a provisional backward analysis (starting from the last stage with the overall load resistance R_L) to determine the input resistances of each stage and then performing a forward analysis (starting from the first stage with the overall source resistance R_s) to determine the voltage gain and output resistances of each stage, complete the following MATLAB script "elec04p16.m" so that it can use the MATLAB function '`FET_CD_analysis()`' to analyze stage 3 (of CD configuration), '`BJT_CE_analysis()`' to analyze stage 2 (of CE configuration), and '`FET_CS_analysis()`'

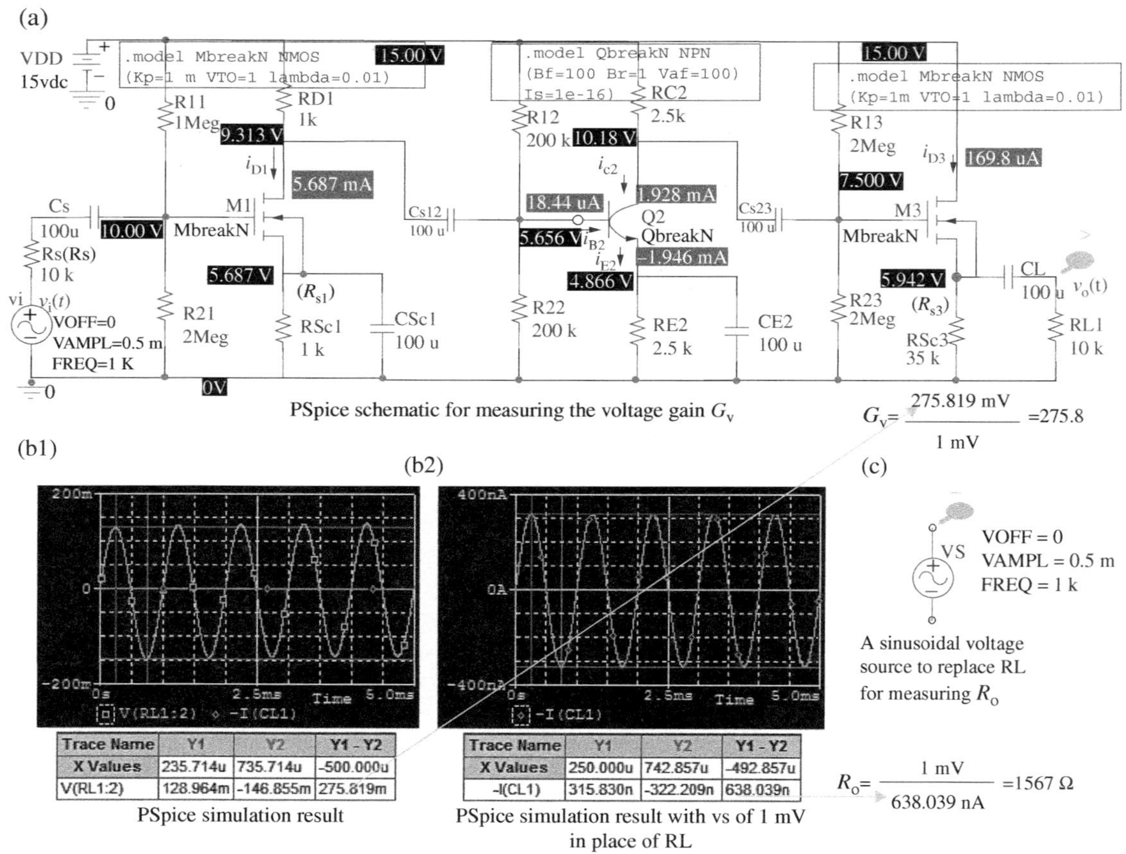

Figure P4.16 A three-stage amplifier and its PSpice simulation result.

```
%elec04p16.m
% Problem 4.16 : Analysis of a 3-stage (CS-CE-CD) amplifier
VDD=15; VCC=VDD; % DC supply voltage sources
Vsm=1e-3; % AC source voltage amplitude
Rs=1e4; RL=1e4; % Source and Load resistances
% Device parameters
Kp=1e-3; Vt=1; lambda=0.01; % Enhancement-type NMOS M1 and M3
BETAF2=100; BETAR2=1; betaAC2=BETAF2; Is=1e-16; VA2=100; rb2=0; % BJT Q2
beta2=[BETAF2,BETAR2,betaAC2,Is];
% Provisional analysis of Stage 3 to find Ri3 with Rs3=0 and RL3=RL
Rs3=0; R13=2e6; R23=2e6; RD3=0; RS3=35e3; RL3=R?;
[VGQ3,VSQ3,VDQ3,IDQ3,Av3,Ri3,Ro3]=...
  FET_CD_analysis(VDD,Rs3,R13,R23,RD3,RS3,RL3,Kp,Vt,lambda);
% Analysis of Stage 2 with Rs2=0 and RL2=Ri3
Rs2=0; rb2=0; R12=2e5; R22=2e5; RC2=2500; RE2=[0 2500]; RL2=Ri?;
[VBQ2,VEQ2,VCQ2,IBQ2,IEQ2,ICQ2,Av2,Ai2,Ri2,Ro2,gm2,rbe2,ro2]=...
  BJT_CE_analysis(VCC,rb2,Rs2,R12,R22,RC2,RE2,RL2,beta2);
% Analysis of Stage 1 with Rs1=Rs and RL=Ri2
Rs1=R?; R11=1e6; R21=2e6; RD1=1e3; RS1=[0 ???]; RL1=Ri?;
[VGQ1,VSQ1,VDQ1,IDQ1,Av(1),Ri(1),Ro(1)]=...
  FET_CS_analysis(VDD,Rs1,R11,R21,RD1,RS1,RL1,Kp,Vt,lambda,Vsm);
% Analysis of Stage 2
Rs2=Ro(?); RL2=Ri?; Vsm2=Av(1)*Vsm;
[VBQ2,VEQ2,VCQ2,IBQ2,IEQ2,ICQ2,Av(2),Ai2,Ri(2),Ro(2)]=...
  BJT_CE_analysis(VCC,rb2,Rs2,R12,R22,RC2,RE2,RL2,beta2,Vsm2,VA2);
% Analysis of Stage 3
Rs3=Ro(?); RL3=R?; Vsm3=prod(Av(1:2))*Vsm;
[VGQ3,VSQ3,VDQ3,IDQ3,Av(3),Ri(3),Ro(3)]=...
  FET_CD_analysis(VDD,Rs3,R13,R23,RD3,RS3,RL3,Kp,Vt,lambda,Vsm3);
RI=Ri(1), RO=Ro(end) % Overall input/output resistances
Gv_global=RI/(Rs+??)*prod(??) % Global voltage gain
```

to analyze stage 1 (of CS configuration), '`BJT_CE_analysis()`' to analyze stage 2, and '`FET_CD_analysis()`' to analyze stage 3. Run it to find the overall voltage gain $G_v = v_o/v_i$ and the overall input/output resistances R_I/R_O. Compare the overall voltage gain G_v with that obtained from the PSpice simulation (shown in Figure P4.16(b1)) to see how close they are.

4.17 Cascode MOSFET Amplifiers

Consider the FET cascode amplifier circuits in Figure P4.17 where the NMOSs have the device parameters $K_p = 1\,\text{mA/V}^2$, $V_t = 1\,\text{V}$, and $\lambda = 10^{-4}\,\text{V}^{-1}$ in common.

(a) For the FET cascode amplifier A (in Figure P4.17(a)) with $R_1 + R_2 + R_3 = 500\,\text{k}\Omega$ and $R_S(R_{Sc}) = 8\,\text{k}\Omega$, determine the values of the resistors R_1, R_2, R_3, and R_D so that $I_{D,Q} = 0.5\,\text{mA}$ and $V_{DS1,Q} = V_{DS2,Q} = 2.5\,\text{V}$ in the saturation mode.

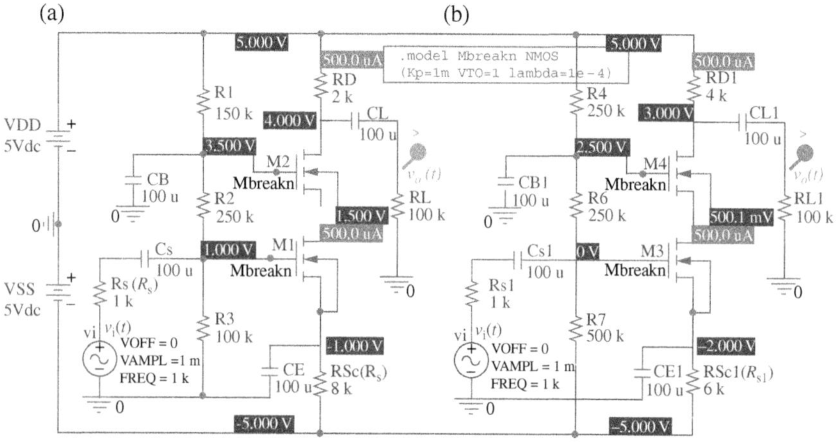

PSpice schematic for cascode amplifier A　　　PSpice schematic for cascode amplifier B

Figure P4.17 PSpice schematics for two cascode amplifier circuits.

(b) Complete the former part of the following MATLAB script "elec04p17.m" so that it can use the MATLAB function '`FET_cascode()`' to analyze the circuit designed in (a). Run it to find the voltage gain $G_v = v_o/v_i$ and the input/output resistances R_i/R_o.

(c) In the light of Eq. (E4.11.13) showing the voltage gain of a cascode amplifier, will a larger or smaller value of R_D be helpful in increasing the voltage gain G_v? Does it increase or decrease the output resistance R_o according to Eq. (E4.11.11)? To make sure that your answer is correct, complete and run the latter part of the MATLAB script "elec04p17.m" to find the voltage gain $G_v = v_o/v_i$ and the input/output resistances R_i/R_o for the circuit with a larger R_D in Figure P4.17(b).

(d) Perform the PSpice simulation to find the voltage gain for the two cascode amplifier circuits and see if the results conform with those obtained from the MATLAB analysis.

```
%elec04p17.m
% Problem 4.17 : Analysis of a cascode amplifier
VDD=5; VSS=-5; % DC supply voltage sources
Kp=1e-3; Vt=1; lambda=1e-4; % Device parameters
Vim=1e-3; t=[0:1e-6:3e-3]; vi=Vim*sin(2*pi*1000*t); % Small-signal input
disp('(a)')
Rs=1e3; R1=???e3; R2=???e3; R3=???e3; RS=8e3; RD=?e3; RL=1e5;
[VG1Q,VS1Q,VD1Q,VG2Q,VD2Q,IDQ,Av,Ri,Ro,gm,ro,vo,iD] = ...
  FET_cascode([VDD VSS],Rs,R1,R2,R3,RS,RD,RL,Kp,Vt,lambda,vi);
Gva=Ri/(Rs+Ri)*prod(Av);
fprintf('Gv=%7.2f, Ri=%8.2sfkOhm, Ro=%7.1fOhm\n', Gva,Ri/1e3,Ro);
```

```
clf, subplot(121), plot(t,vi, t,vo-mean(vo),'g')
disp('(b)')
VDD=VDD-VSS; VSS=0; % This will raise every node voltage by -VSS.
Rs=1e3; R1=???e3; R2=???e3; R3=???e3; RS=?e3; RD=?e3; RL=1e5;
[VG1Q,VS1Q,VD1Q,VG2Q,VD2Q,IDQ,Av,Ri,Ro,gm,ro,vo,iD]= ...
  FET_cascode([VDD VSS],Rs,R1,R2,R3,RS,RD,RL,Kp,Vt,lambda,vi);
Gvb=Ri/(Rs+Ri)*prod(Av);
fprintf('Gv=%7.2f, Ri=%8.2fkOhm, Ro=%7.1fOhm\n', Gvb,Ri/1e3,Ro);
hold on, plot(t,vo-mean(vo),'r')
```

4.18 Design of CS JFET Amplifier Circuit

Consider the (self-biased) CS JFET amplifier circuit of Figure P4.18(a) where the conduction parameter, threshold voltage, and CLM parameter of the JFET are K_P= 20 mA/V^2, V_t= −6.25 V, and λ= 10^{-4}V^{-1}, respectively.

(a) Considering the load resistance R_L = 50 kΩ, determine such resistor values that the voltage gain $A_{v,d}$ = −10 and input resistance $R_{i,d}$ = 100 kΩ can be achieved with the operating point located near Q_1 = $(V_{DS,Q}, I_{D,Q})$ = (6 V,10 mA). Analyze the designed circuit with source/load resistance R_s = 50 Ω/R_L = 50 kΩ to find the voltage gain A_v. Is it close to $A_{v,d}$ = −10?

(Hint) Complete and run the following MATLAB script "elec04p18.m":

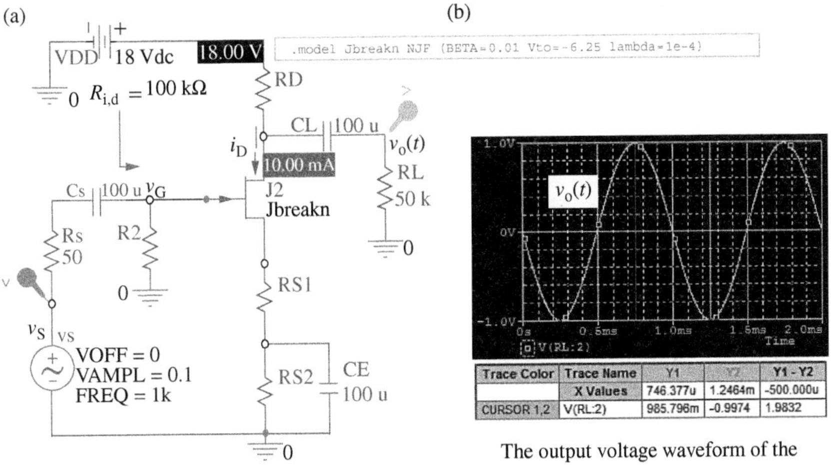

PSpice schematic for a CS JFET amplifier

The output voltage waveform of the CS JFET amplifier

Figure P4.18 A CS JFET circuit and its output voltage waveform ("elec04p18.opj").

```
%elec04p18.m
Avd=??; Kp=0.02; Vt=-6.25; lambda=1e-4; VDD=18;
VDSQd=?; IDQd=????; VG=?; RL=5e4; Rid_VG=[??? VG]; KC=VDSQd/???;
[R1,R2,RD,RS1,RS2]= ...
  FET_CS_design(VDD,Kp,Vt,lambda,Avd,Rid_VG,IDQd,RL,KC);
% Here, the 6th input argument Rid has been set as [Rid VG]
% to fix vG=VG=0 as given in Fig. P4.18(a).
Vsm=0.01; Rs=50; RL=5e4;
[VGQ,VSQ,VDQ,IDQ,Av]=...
  FET_C?_analysis(VDD,Rs,R1,R2,RD,[RS1 RS2],RL,Kp,Vt,lambda,Vsm);
```

(b) Referring to Figure P4.18(a), perform the PSpice simulation of the designed CS JFET amplifier with Time Domain analysis to see the output voltage $v_o(t)$ to an AC input $v_s(t) = V_{sm}\sin(2000\pi t)$ (with $V_{sm} = 0.1$ V) for $t = 0$–2 ms (with the maximum step size 1 us) as shown in Figure P4.18(b). Find the voltage gain and check if it agrees with the analysis result (obtained in (a)) and satisfies the design specification $A_{v,d} = -10$. Note that BETA, one of the PSpice model parameters for JFET, must be given as Kp/2.

(*Note*) If the MATLAB function '`FET_CS_design()`' turns out to work for this problem, it implies that the function can be used to design CS JFET amplifiers as well as CS MOSFET amplifiers.

4.19 Design of CS MOSFET Amplifier Circuit

Consider the CS MOSFET amplifier circuit of Figure P4.19(a) where the conduction parameter, threshold voltage, and CLM parameter of the MOSFET (IRF150) are $K_P = K_P W/L = 20.53\mu \times 0.3/2\mu = 3.08$ A/V^2, $V_t = 2.831$ V, and $\lambda = 2.2$ mV^{-1}, respectively.

```
%elec04p19.m
Avd=10; Kp=3.08; Vt=2.831; lambda=2.2e-3; VDD=18;
IDQd=1e-3; VG=VDD; RL=5e4; Rid_VG=[1e5 ??]; KC=1/7;
[R1,R2,RD,RS1,RS2]= ...
FET_CS_design(VDD,Kp,Vt,lambda,Avd,??????,IDQd,RL,KC);
% Here, the 6th input argument Rid has been set as [Rid VDD]
% to fix vG=VDD as given in Fig. P4.19(a).
Vsm=0.01; Rs=50; RL=5e4;
[VGQ,VSQ,VDQ,IDQ,Av]=...
  FET_C?_analysis(VDD,Rs,R1,R2,RD,[RS1 RS2],RL,Kp,Vt,lambda,Vsm);
```

(a) Considering the load resistance $R_L = 50$ kΩ, determine such resistor values that the voltage gain $A_{v,d} = -10$ and input resistance $R_{i,d} = 100$ kΩ can be achieved with the operating point located near $Q_1 = (V_{DS,Q}, I_{D,Q}) = (18/7\text{ V}, 1\text{ mA})$. Analyze the designed circuit with

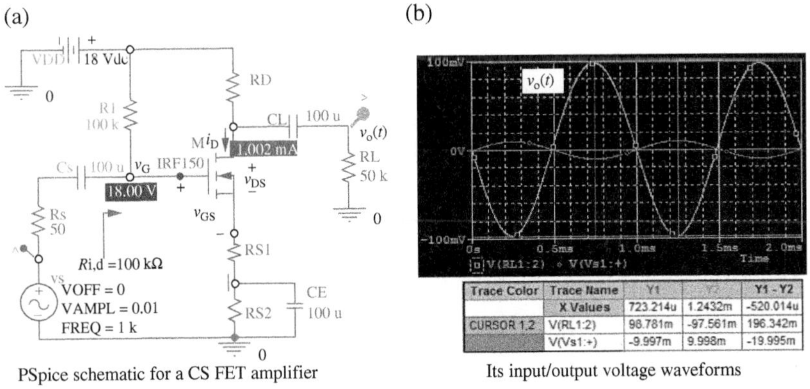

Figure P4.19 A CS MOSFET circuit and its input/output voltage waveforms ("elec04p19.opj").

source/load resistance $R_s = 50\ \Omega/R_L = 50\ \text{k}\Omega$ to find the voltage gain A_v. Is it close to $A_{v,d} = -10$?

(Hint) Complete and run the above MATLAB script "elec04p19.m".

(b) Referring to Figure P4.19(a), perform the PSpice simulation of the designed CS FET amplifier with Time Domain analysis to see the output voltage $v_o(t)$ to an AC input $v_s(t) = V_{sm} \sin(2000\pi t)$ (with $V_{sm} = 0.01$ V) for $t = 0\sim 2$ ms (with the maximum step size 1 μs) as shown in Figure P4.19(b). Find the voltage gain and check if it agrees with the analysis result (obtained in (a)) and satisfies the design specification $A_{v,d} = -10$.

4.20 Design of CS MOSFET Amplifier Circuit

Consider the CS MOSFET amplifier circuit of Figure 4.36(a) where the conduction parameter, threshold voltage, and CLM parameter of the MOSFET (IRF150) are $K_P = K_P W/L = 20.53\mu \times 0.3/2\mu = 3.08\ \text{A/V}^2$, $V_t = 2.831$ V, and $\lambda = 0.0075\ \text{mV}^{-1}$, respectively. Let us see how the location of (static) operating point depending on the design constants such as $I_{D,Q}$ and $V_{DS,Q}$ affects the distortion of the output voltage waveform (to an AC input $v_s(t)=V_{sm}\sin(2000\pi t)$ with $V_{sm}= 0.3$ V) caused by the (dynamic) operating point infringing on the ohmic or cutoff region.

(a) In Section 4.3, a CS FET amplifier (Figure 4.36(a)) with the static operating point at $Q_a = (V_{DS,Q}, I_{D,Q}) = (18/3\ \text{V}, 0.3\ \text{mA})$ was designed so that it could have the voltage gain $A_{v,d} = -20$ and input resistance $R_{i,d} = 100\ \text{k}\Omega$. Perform the PSpice simulation with Time Domain analysis to see the output voltage $v_o(t)$ to an AC input $v_s(t) = V_{sm} \sin(2000\pi t)$ (with

V_{sm} = 0.3 V) as shown in Figure P4.20(a). If $v_o(t)$ has any distortion, could it be predicted from the MATLAB analysis?

(b) Determine the resistor values of the CS FET amplifier with the static operating point at $Q_b = (V_{DS,Q}, I_{D,Q}) = (6\text{ V}, 3\text{ mA})$ so that it can have the same voltage gain $A_{v,d} = -20$ and input resistance $R_{i,d} = 100\text{ k}\Omega$ with $I_{D,Q}$ increased from 0.3 mA to 3 mA. Use the MATLAB function '`FET_CS_analysis()`' to analyze the designed circuit with the source/load resistance of $R_s = 50\ \Omega/R_L = 50\text{ k}\Omega$. Does it give you any warning message of ohmic/cutoff regions to be trespassed on? Perform the PSpice simulation with Time Domain analysis to see the output voltage to an AC input $v_s(t) = V_{sm}\sin(2000\pi t)$ (with $V_{sm} = 0.3$ V) as shown in Figure P4.20(b). Does any distortion of $v_o(t)$ agree with the warning message given by the MATLAB analysis?

(c) Determine the resistor values of the CS FET amplifier with the static operating point at $Q_c = (V_{DS,Q}, I_{D,Q}) = (8\text{ V}, 3\text{ mA})$ so that it can have the same voltage gain $A_{v,d} = -20$ and input resistance $R_{i,d} = 100\text{ k}\Omega$ with $V_{DS,Q}$ increased from 6 V to 8 V. Use the MATLAB function '`FET_CS_analysis()`' to analyze the designed circuit with the source/load resistance of $R_s = 50\ \Omega/R_L = 50\text{ k}\Omega$. Does it give you any warning message of ohmic/cutoff regions to be trespassed on? Perform the PSpice simulation with Time Domain analysis to see the output voltage to an AC input $v_s(t) = V_{sm}\sin(2000\pi t)$ (with $V_{sm} = 0.3$ V) as shown in Figure P4.20(c).

(d) Determine the resistor values of the CS FET amplifier with the static operating point at $Q_d = (V_{DS,Q}, I_{D,Q}) = (4.5\text{ V}, 3\text{ mA})$ so that it can have the same voltage gain $A_{v,d} = -20$ and input resistance $R_{i,d} = 100\text{ k}\Omega$ with $V_{DS,Q}$ decreased from 6 V to 4.5 V. Use the MATLAB function '`FET_CS_analysis()`' to analyze the designed circuit with the source/load resistance of $R_s = 50\ \Omega/R_L = 50\text{ k}\Omega$. Does it give you any warning message of ohmic/cutoff regions to be trespassed on? Perform the PSpice simulation with Time Domain analysis to see the output voltage to an AC input $v_s(t) = V_{sm}\sin(2000\pi t)$ (with $V_{sm} = 0.3$ V) as shown in Figure P4.20(d). Does any distortion of $v_o(t)$ agree with the warning message given by the MATLAB analysis?

(e) Determine the resistor values of the CS FET amplifier with the static operating point at $Q_e = (V_{DS,Q}, I_{D,Q}) = (6\text{ V}, 20\text{ mA})$ so that it can have the same voltage gain $A_{v,d} = -20$ and input resistance $R_{i,d} = 100\text{ k}\Omega$ with $I_{D,Q}$ increased from 0.3 mA to 20 mA. Use the MATLAB function '`FET_CS_analysis()`' to analyze the designed circuit with the source/load resistance of $R_s = 50\ \Omega/R_L = 50\text{ k}\Omega$. Does it give you any warning message of ohmic/cutoff regions to be trespassed on? Perform the PSpice simulation with Time Domain analysis to see the output voltage to an AC input $v_s(t) = V_{sm}\sin(2000\pi t)$ (with $V_{sm} = 0.3$ V) as shown in Figure P4.20(e). Does any distortion of $v_o(t)$ agree with the warning message given by the MATLAB analysis?

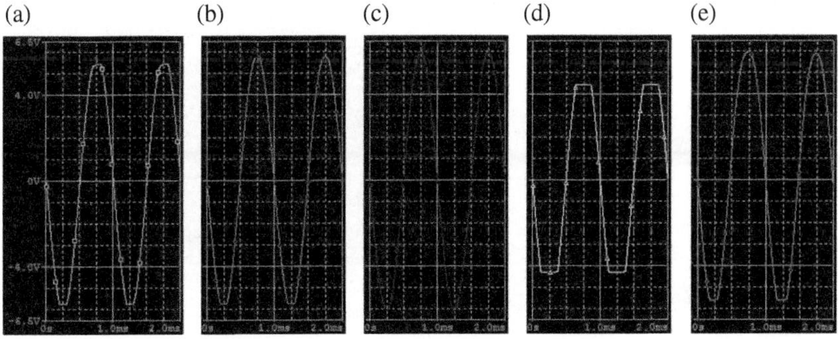

Figure P4.20 Output voltage waveforms of CS FET amplifier with different resistances ("elec04p20.opj").

4.21 Frequency Response of an FET Cascode Amplifier

Consider the FET cascode amplifier with PSpice schematic and simulation result on the frequency response in Figure P4.21(a) and (b), respectively, where $V_{DD} = 12$ V, $R_s = 50\,\Omega$, $C_s = 100\,\mu$F, $R_1 = 200$ kΩ, $R_2 = 100$ kΩ, $R_3 = 100$ kΩ, $R_S = 1$ kΩ, $C_S = 100\,\mu$F, $C_G = 100\,\mu$F, $R_D = 3$ kΩ, $C_L = 100\,\mu$F, $R_L = 100$ kΩ, $C_{LL} = 0.1$ nF, and the parameters of the enhancement type NMOSs are $K_p = 10$ mA/V^2, $V_t = 1$ V, and $\lambda = 10^{-4}$ V^{-1}, $C_{gs0} = 10$ pF, $C_{gd0} = 1$ pF, and $C_{ds} = 0$ F.

(a) Complete the following MATLAB function 'FET_cascode_DC_analysis()', which performs the DC analysis to determine the node voltages and drain currents at the operating point.

```
function [VGS1Q,VGD1Q,ID1Q,VGS2Q,VGD2Q,ID2Q]=...
    FET_cascode_DC_analysis(VDD,R123,RD,RS,Kp,Vt,lambda)
% To do DC analysis of an FET cascode (CS-CG) amplifier in Fig. P4.21(a)
% Note that VDD can be given as [VDD VSS].
% Copyleft: Won Y. Yang, wyyang53@hanmail.net, CAU for academic use only
if numel(VDD)<2, VSS=0; else VSS=VDD(2); VDD=VDD(1); end
Kp1=Kp(1); Kp2=Kp(end); Vt1=Vt(1); Vt2=Vt(end);
lambda1=lambda(1); lambda2=lambda(end);
R1=R123(1); R2=R123(2); R3=R123(3); R_123=R1+R2+R3;
V1=VSS+R3/R_123*(VDD-VSS);
V3=V1+R2/R_123*(VDD-VSS);
options=optimset('Display','off','Diagnostics','off');
iD1=@(vGS,vDS)iD_NMOS_at_vDS_vGS(vDS,vGS,Kp1,Vt1,lambda1); %Eq.(4.1.83)
iD2=@(vGS,vDS)iD_NMOS_at_vDS_vGS(vDS,vGS,Kp2,Vt2,lambda2); %Eq.(4.1.83)
% A set of node equations in v=[V2 V4 V5]:
eq=@(v)[iD1(V1-v(?),v(?)-v(1))-(v(?)-VSS)/RS; % KCL at node 2
    iD2(V3-v(?),v(?)-v(2))-iD1(V1-v(1),v(2)-v(1)); % KCL at node 4
    (VDD-v(?))/RD-iD2(V3-v(2),v(3)-v(2))]*1e3; % KCL at node 5
v0=[V1-Vt1 V3-Vt2 (V3+VDD)/2]; % Initial guess for v=[V2 V4 V5]
v=fsolve(eq,v0,options);
VGS1Q=V1-v(1); VGD1Q=V1-v(2);
VGS2Q=V3-v(2); VGD2Q=V3-v(3);
ID1Q=iD1(V1-v(1),v(2)-v(1));
ID2Q=iD2(V3-v(2),v(3)-v(2));
```

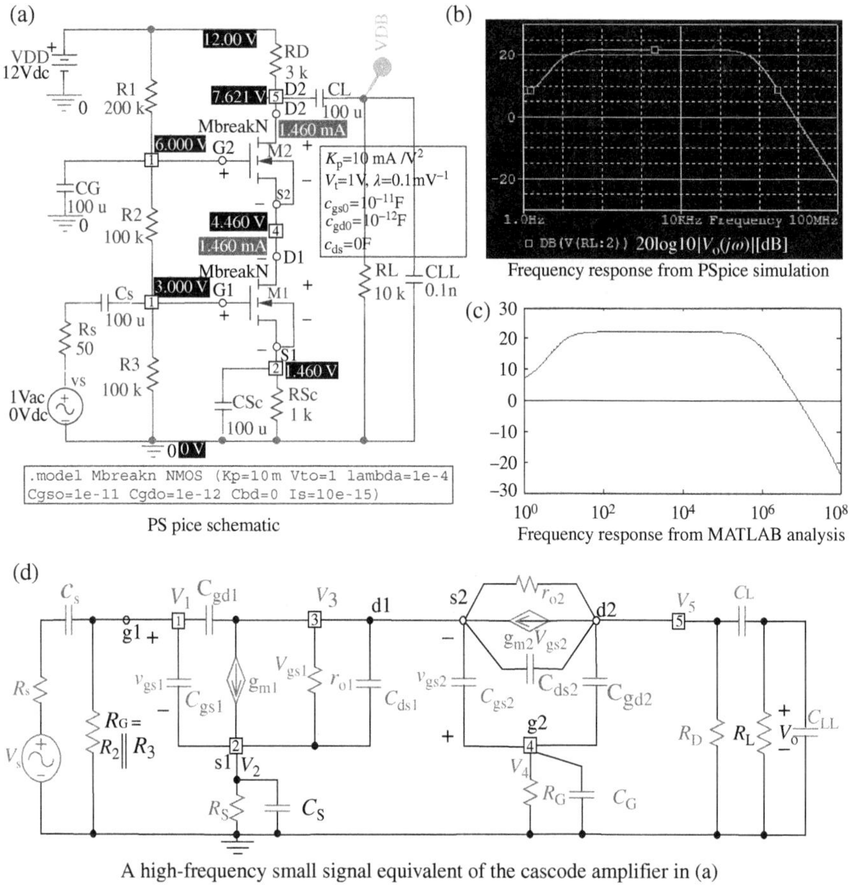

Figure P4.21 Cascode amplifier and its frequency response from PSpice and MATLAB ("elec04p21.opj").

(b) Complete the following MATLAB function '`FET_cascode_xfer_ftn()`', which solves a set of 3, 4, or 5 node equations for the high-frequency small-signal equivalent (Figure P4.21(d)) depending on whether or not nodes 2/3 are AC grounded via capacitors C_{SC}/C_G, respectively, to find the transfer function $G(s) = V_o(s)/V_s(s)$ of the cascade amplifier.

(c) Run the following MATLAB script "elec04p21.m" to perform the DC analysis (using '`FET_cascode_DC_analysis()`'), then based on the DC analysis result, use '`FET_cascode_xfer_ftn()`' to find the transfer function $G(s)$, and plot the frequency response magnitude $20\log_{10}|G(j\omega)|$ [dB] of the cascade amplifier versus $f = 1\sim100$ MHz as shown in Figure P4.21(c).

```
%elec04p21.m
% To find the frequency response of an FET cascode (CS-CG) amplifier
% Circuit parameters
VDD=12; Rs=50; Cs=1e-4; R1=200e3; R2=100e3; R3=100e3; R123=[R1 R2 R3];
RD=3e3; RS=1e3; CS=1e-4; RL=1e4; CG=1e-4; CL=1e-4; RL=1e4; CLL=1e-10;
RLCLL=[RL CLL]; %CG=1; %CL=1; % Without CL
% Device parameters for the two FETs M1 and M2
Kp=0.01; Vt=1; lambda=1e-4; %1e-8;
Cgd0=1e-12; Cgs0=1e-11; Cds=0; CCC=[Cgs0,Cgd0,Cds]; %rds=rds;
f=logspace(0,8,801); w=2*pi*f; % Frequency range
[Gs,fc]=... FET_cascode_xfer_ftn(VDD,Rs,Cs,R123,RD,RS,
CS,CG,CL,RLCLL,Kp,Vt,lambda,CCC);
syms s; Gw=subs(Gs,'s',j*w); % Frequency response
GmagdB=20*log10(abs(Gw)+1e-5); Gmax=max(GmagdB);
semilogx(f,GmagdB, f([1 end]),[0 0],'k'), hold on
semilogx(fc(1)*[1 1],[0 Gmax-3],'b:', ... fc(6)*[1 1],[0 Gmax-3],'k:')
text(fc(1),-10,'f_{c1}'); ... text(fc(6),-10,'f_{c6}');
```

```
function [Gs,fc,gm1,ro1,gm2,ro2]=FET_cascode_xfer_ftn(VDD,Rs,Cs,
   R123,RD,RS,CS,CG,CL,RLCLL,Kp,Vt,lambda,CCC,Vb,m)
%To find the transfer function of an FET cascode amplifier in Fig. 4.21(a)
% R123=[R1 R2 R3]
% RLCLL=[RL CLL] with RL and CLL in parallel
% CCC=[Cgs0,Cgd0,Cds] with Cgs0|Cgd0: Zero-bias GSJ|GDJ capacitances
% Vb: Gate junction (barrier) potential
% m : Gate p-n grading coefficient
% Copyleft: Won Y. Yang, wyyang53@hanmail.net, CAU for academic use only
if nargin<16, m=0.5; end % Gate p-n grading coefficient
if nargin<15, Vb=1; end % Gate junction (barrier) potential
if nargin<14, CCC=[0 0 0]; end
Cgs01=CCC(1,1); Cgd01=CCC(1,2); Cds1=CCC(1,3);
Cgs02=CCC(end,1); Cgd02=CCC(end,2); Cds2=CCC(end,3);
Vt1=Vt(1); Vt2=Vt(end); Kp1=Kp(1); Kp2=Kp(end);
lambda1=lambda(1); lambda2=lambda(end);
RL=RLCLL(1); if numel(RLCLL)<2, CLL=0; else CLL=RLCLL(2); end
R1=R123(1); R2=R123(2); R3=R123(3);
[VGS1Q,VGD1Q,ID1Q,VGS2Q,VGD2Q,ID2Q]= ...
  FET_cascode_DC_analysis(VDD,R123,RD,RS,Kp,Vt,lambda);
[gm1,ro1]=gmro_FET(ID1Q,VGS1Q,Kp1,Vt1,lambda1);
[gm2,ro2]=gmro_FET(ID2Q,VGS2Q,Kp2,Vt2,lambda2);
Cgs1=Cgs01/(1+abs(VGS1Q)/Vb)^m;
Cgd1=Cgd01/(1+abs(VGD1Q)/Vb)^m; %Eq.(4.4.5)
Cgs2=Cgs02/(1+abs(VGS2Q)/Vb)^m;
Cgd2=Cgd02/(1+abs(VGD2Q)/Vb)^m; %Eq.(4.4.5)
syms s; sCs=s*Cs; sCS=s*CS; sCG=s*CG; sCL=s*CL;
sCgs1=s*Cgs1; sCgd1=s*Cgd1; sCds1=s*Cds1;
sCgs2=s*Cgs2; sCgd2=s*Cgd2; sCds2=s*Cds2;
Zs=Rs+1/sCs; Ys=1/Zs; RG=parallel_comb([R2 R3]);
ZL=parallel_comb([RL 1/s/CLL]);
ZD=parallel_comb([RD 1/sCL+ZL]); YD=1/ZD;
YG1=Ys+1/RG; YS=(1/RS+sCS)*(CS<1); % Admittance at terminal S
RG2=parallel_comb([R1 R2+R3]); YG2=(1/RG2+sCG)*(CG<1); %Admittance at G2
if (RS>1&CS<1)&(CG<1)
```

```
Y=[YG1+sCgs1+sCgd1      -sCgs1                        -sCgd1 0 0;
  -sCgs1-gm1    sCgs1+1/ro1+sCds1+YS+gm1     -1/ro1-sCds1 0 0;
  -sCgd1+gm1 -1/ro1-sCds1-gm1 sCgd1+1/ro1+sCds1+sCgs2+1/ro2+sCds2+gm2
                                     -sCgs2-gm2 -1/ro2-sCds2;
   0     0   -sCgs2              sCgs2+sCgd2+YG2          -sCgd2;
   0     0  -1/ro2-sCds2-gm2   -sCgd2+gm2   1/ro2+sCds2+sCgd2+YD];
V=Y\[Ys; 0; 0; 0; 0];
elseif (RS>1&CS<1)&(CG>=1)
Y=[YG1+sCgs1+sCgd1 -sCgs1 -sCgd1 0;
   -sCgs1-gm1 sCgs1+1/ro1+sCds1+YS+gm1 -1/ro1-sCds1  0;
   -sCgd1+gm1 -1/ro1-sCds1-gm1 sCgd1+1/ro1+sCds1+sCgs2+1/ro2+sCds2+gm2
                                                   -1/ro2-sCds2;
    0      0 -1/ro2-sCds2-gm2   1/ro2+sCds2+sCgd2+YD];
V=Y\[Ys; 0; 0; 0];
elseif (RS<=1|CS>=1)&(CG<1)
Y=[YG1+sCgs1+sCgd1 -1/ro1-sCgd1 0 0;
  -sCgd1+gm1 sCgd1+1/ro1+sCds1+sCgs2+1/ro2+sCds2+gm2 -sCgs2-gm2
                                                        -1/ro2-sCds2;
   0          -sCgs2           sCgs2+sCgd2+YG2        -sCgd2;
   0     -1/ro2-sCds2-gm2     -sCgd2+gm2      1/ro2+sCds2+sCgd2+YD];
 V=Y\[Ys; 0; 0; 0];
else %if (RS<=1|CS>=1)&(CG>=1)
Y=[YG1+sCgs1+sCgd1  -1/ro1-sCgd1  0;
  -sCgd1+gm1 sCgd1+1/ro1+sCds1+sCgs2+1/ro2+sCds2+gm2 -1/ro2-sCds2;
   0       -1/ro2-sCds2-gm2   1/ro2+sCds2+sCgd2+YD];
V=Y\[Ys; 0; 0];
end
Gs=V(end)*ZL/(??+1/sCL); %Transfer function G(s)=Vo(s)/Vi(s) with Vi(s)=1
Ri(1)=RG; Ro(1)=ro1; % Eqs.(E4.11.7), (E4.11.8)
RDL=parallel_comb([RD RL]);
Av(2)=(1+gm2*ro2)*RDL/(ro2+RDL); % Eq.(E4.11.10) or (4.2.16b)
Ri(2)=(ro2+RD)/(1+gm2*ro2); % Eq.(4.2.15)
Ro(2)=parallel_comb([RD (1+gm2*ro2)*Ro(1)+ro2]); % Eq.(E4.11.11)
Av(1)=-gm1*ro1*Ri(2)/(Ri(2)+ro1); % Eq.(E4.11.9)
Cgd=[Cgd1 Cgd2]; Cgs=[Cgs1 Cgs2]; Cds=[Cds1 Cds2];
fc=break_freqs_of_FET_cascode(Rs,Cs,CL,RL,CLL,Cgd,Cgs,Cds,Av,Ri,Ro)
```

```
function fc=break_freqs_of_FET_cascode(Rs,Cs,CL,RL,CLL,Cgd,Cgs,
  Cds,Av,Ri,Ro)
% To find the 6 break frequencies of an FET cascode amplifier
RsRi1=parallel_comb([Rs Ri(1)]);
Ro1Ri2=parallel_comb([Ro(1) Ri(2)]);
Ro2RL=parallel_comb([Ro(2) RL]);
Cm1=Cgd(1)*(1-Av(1)); Cn1=Cgd(1)*(1-1/Av(1)); % Eq.(4.4.7a,b)
Cm2=Cds(2)*(1-Av(2)); Cn2=Cds(2)*(1-1/Av(2)); % Eq.(4.4.7a,b)
fc(1)=1/2/pi/Cs/(Rs+Ri(1)); fc(2)=1/2/pi/CL/(Ro(2)+RL);
fc(3)=1/2/pi/(Cgs(1)+Cm1)/RsRi1;
fc(4)=1/2/pi/(Cds(1)+Cgs(1)+Cm2)/Ro1Ri2;
fc(5)=1/2/pi/(Cn2+Cgd(2))/Ro(2); fc(6)=1/2/pi/CLL/Ro2RL;
```

(d) Perform the PSpice simulation (with AC Sweep analysis) to get the frequency response magnitude curve as depicted in Figure P4.21(b). Is the Bias Point analysis result (obtained as a by-product) close to the DC analysis result obtained by using 'FET_cascode_DC_analysis()' in (c)?

4.22 **Time Response of an FET Inverter with another Inverter as a Load**

Consider the FET inverter in Figure P4.22(a) where $V_{DD} = 12\,\text{V}$, $R_D = 10\,\text{k}\Omega$, $R_{D1} = 10\,\text{k}\Omega$, and the FET parameters are $K_p = 5\,\text{mA/V}^2$, $V_t = 1\,\text{V}$, $\lambda = 10^{-4}\,\text{V}^{-1}$, $C_{gs0} = 10\,\text{fF}$, $C_{gd0} = 1\,\text{fF}$, $C_{bd0} = 10\,\text{fF}$ or $20\,\text{fF}$ for M_1, and $K_p = 5\,\text{mA/V}^2$, $V_t = 1\,\text{V}$, $\lambda = 10^{-4}\,\text{V}^{-1}$, $C_{gs0} = 10\,\text{fF}$, $C_{gd0} = 1\,\text{fF}$, $C_{bd0} = 10\,\text{fF}$ for M_2.

Figure P4.22(b1) and (b2) shows the input/output voltage waveforms obtained from PSpice simulation, each with $C_{bd} = 10\,\text{fF}$ and $20\,\text{fF}$ for M_1, respectively. Referring to Section 4.5 and Figure P4.22(b1-b2), explain how a larger value of C_{bd} makes a longer time constant of v_{o1} during its rising period.

(a) PSpice schematic

(a1) (a2)

(b1) Time response with C_{bd}=10pF and C_{gd}=1pF

(b2) Time response with C_{bd}=20pF and C_{gd}=1pF

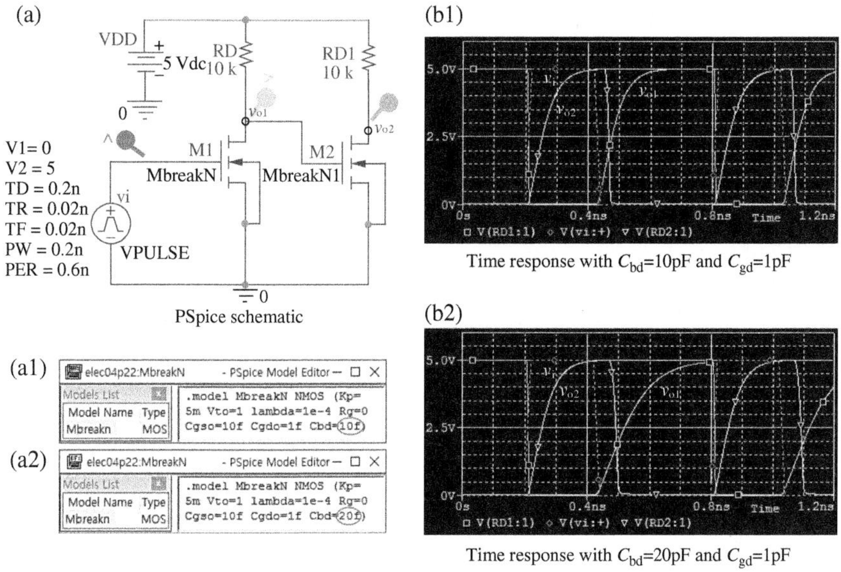

Figure P4.22 FET inverter with another inverter as a load and its time responses from PSpice ("elec04p22.opj").

5

OP Amp Circuits

CHAPTER OUTLINE

The operational amplifier (OP Amp) is probably the most versatile analog chip available. It was originally designed to be used for analog computers that perform mathematical operations such as addition, subtraction, integration, and differentiation. As it was widely used for many other applications, early OP Amps constructed from discrete components (such as vacuum tubes and then transistors and resistors) have been superseded by integrated-circuit (IC) OP Amps made up of a large number of transistors, resistors, and (sometimes) one capacitor. With only a few external components, it can be made to perform a wide variety of analog signal processing tasks. One of the most common and

Electronic Circuits with MATLAB®, PSpice®, and Smith Chart, First Edition. Won Y. Yang, Jaekwon Kim, Kyung W. Park, Donghyun Baek, Sungjoon Lim, Jingon Joung, Suhyun Park, Han L. Lee, Woo June Choi, and Taeho Im.

Companion website: www.wiley.com/go/yang/electroniccircuits

famous OP Amps is the μA741, which was introduced by Fairchild in 1968 and is now available at less than a dollar.

(Note) A handbook of OP Amp applications is available at http://www.focus.ti.com/lit/an/sboa092a/sboa092a.pdf.

5.1 OP Amp Basics [Y-1]

The OP Amp usually comes in the form of an eight-pin DIP (dual in-line package) IC as depicted in Figure 5.1(a) and (b). It is basically a differential amplifier having a large open-loop voltage gain A, a very high input impedance R_I, and a low output impedance R_o. (Impedance is a generalized concept of resistance.) It has an 'inverting' or negative(−) input v_- (through pin 2), a 'non-inverting' or positive(+) input v_+ (through pin 3), and a single output v_o available through pin 6. It is powered by a dual polarity power supply $\pm V_{CC}$ in the range of ±5 V to ±15 V through pins 7 and 4. Though, it is customary to omit the two power supply pins from the OP Amp symbol as depicted in Figure 5.1(d) since we do not need to be concerned about the power supplies in the circuit analysis.

Its major features are as follows:

- Its output voltage v_o is A times as large as the differential input voltage $(v_+ - v_-)$, which is the difference between the two positive/negative input voltages, with the limitation of the upper/lower bounds $\pm V_{om}$ (*maximum output voltage* slightly smaller than the power supply voltages $\pm V_{CC}$), where the *open-loop gain A* of an OP Amp is typically in the order of $10^4 \sim 10^6$. This differential input-output relationship is described by the graph in Figure 5.1(c) and can be written as

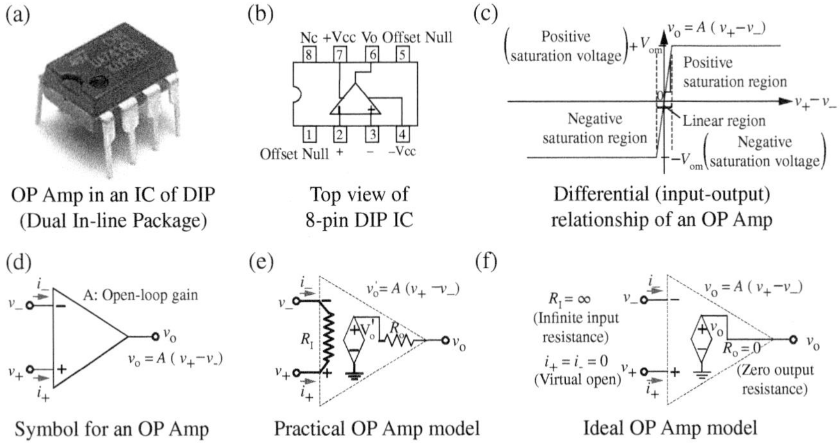

Figure 5.1 Symbol, model, and differential input–output relationship of an operational amplifier (OP Amp).

$$-V_{om} \le v_o = A(v_+ - v_-) \le +V_{om} \tag{5.1.1}$$

or, more specifically,

$$v_o = \begin{cases} +V_{om} & \text{for } v_+ - v_- > (+V_{om}/A) : \text{nonlinear (saturation) region} \\ A(v_+ - v_-) & \text{for } (-V_{om}/A) \le v_+ - v_- \le (+V_{om}/A) : \text{linear region} \\ -V_{om} & \text{for } v_+ - v_- < (-V_{om}/A) : \text{nonlinear (saturation) region} \end{cases} \tag{5.1.2}$$

- It has a very large input resistance $R_I \approx 2\ \text{M}\Omega$ between the two +/− input terminals so that the currents flowing into or out of the input terminals are almost negligible, being normally in the order of μA.
- It has a very low output resistance $R_o \approx 75\ \Omega$ between the output terminal and the ground. (See Figure 5.1(e) for a practical OP Amp model.)

Referring to Figure 5.1(c), the differential input-output relationship (5.1.1) or (5.1.2) indicates that if the magnitude of the differential input voltage, $|v_+ - v_-|$, is larger than V_{om}/A, the OP Amp operates in the positive or negative saturation (nonlinear) region with the output voltage saturated at $+V_{om}$ or $-V_{om}$; otherwise, it operates in the linear region with the output voltage proportional to the differential input as $v_o = A(v_+ - v_-)$. Note that $+V_{om}/-V_{om}$, determining the output voltage swing, are called the *positive/negative saturation* or *maximum output voltages*, respectively.

Here comes a question which could be bothering you: What is the condition for an OP Amp to operate in the linear region and how can it be satisfied? The condition is obtained by dividing both sides of Eq. (5.1.1) by the *open-loop (voltage) gain A* as

$$-\frac{V_{om}}{A} \le v_+ - v_- \le +\frac{V_{om}}{A} \tag{5.1.3}$$

This implies that the two voltages at the +(non-inverting)/−(inverting) input terminals should be 'almost' equal, but it must not be 'exactly' equal. Why? If the two input voltages are exactly equal so that $v_+ - v_- = 0$, the output voltage will be zero:

$$v_o \overset{(5.1.1)}{=} A(v_+ - v_-) = 0$$

In order for the output voltage v_o to be at some nominal value V_o between $-V_{om}$ and $+V_{om}$, the differential input voltage should be $v_+ - v_- = V_o/A$, which is not exactly zero, but close to zero. Appearing not so easy to satisfy, this condition can be satisfied simply by a *negative feedback path,* which is a connection between the output terminal and the negative input terminal, where it does not matter whether the connection is simply a shorted path or via any circuit element.

Another question might arise in your mind: How can the negative feedback make the two voltages at the +/− input terminals so close as to satisfy Eq. (5.1.3)? Suppose the differential input $(v_+ - v_-)$ becomes lower/higher than some nominal value close to zero. Then, the output voltage v_o determined by Eq. (5.1.1) will be lower/higher than its nominal value V_o, which affects the voltage v_- (at the negative input terminal) via the negative feedback path in such a way that v_- becomes lower/higher. This is expected to make $(v_+ - v_-)$ higher/lower so that the differential input and the output can go back to its nominal value. This so-called *stabilization effect* of negative feedback forms the basis of the 'virtual short' or 'imaginary short' principle that the two +/− input terminals of an OP Amp circuit with negative feedback are at almost equal voltage levels as if they were shorted in terms of their voltages. However, note that in order for a negative feedback path to result in 'virtual short' $(v_+ \approx v_-)$, it should not be disturbed by a positive feedback (that is, a connection between the output terminal and the positive input terminal) or any imperative condition forcing $|v_+ - v_-| > V_{om}$.

Together with the infinitely large input resistance and negligibly small output resistance of OP Amp, we can summarize two important properties of an idealized OP Amp model (in Figure 5.1(f)), that are so useful for the analysis of OP Amp circuits.

Remark 5.1 Conditions of the Ideal OP Amp Mode

1) The input resistance between the two (+/−) input terminals and the output resistance between the output terminal and the ground of the ideal OP Amp are assumed to be infinity and zero, respectively. On the assumption of the infinite input resistance, the two input terminals can be regarded as being open in terms of current in the sense that no current flows into or out of them, i.e.

$$i_+ = 0 \text{ and } i_- = 0 \tag{5.1.4}$$

which is referred to as the 'virtual open' principle.

2) The two (+/−) input terminals of an OP Amp with negative feedback path (between the output terminal and the negative input terminal) can be regarded as being short in terms of voltages in the sense that the voltage levels at the two input terminals are almost equal, i.e.

$$v_+ \cong v_- \tag{5.1.5}$$

which is referred to as the 'virtual short' principle. However, this principle may not hold if the OP Amp circuit has also a *positive feedback path* or the absolute nominal value of the output voltage v_o determined by Eq. (5.1.1) exceeds V_{om}. This will be illustrated by an OP Amp circuit (Figure 5.11(a)) with positive feedback as well as negative feedback in Section 5.3.2.1.

5.2 OP Amp Circuits with Resistors [Y-1]

In this section, we discuss the most basic OP Amp configurations, that is, the inverting, non-inverting, and buffer amplifiers with negative feedback, and additionally, the inverting and non-inverting amplifiers with positive feedback. Throughout this chapter, we will be interested only in the terminal behavior without paying much attention to the internal characteristic of the OP Amp.

5.2.1 OP Amp Circuits with Negative Feedback

5.2.1.1 Inverting OP Amp Circuit

Figure 5.2(a) shows an inverting OP Amp circuit where the overall input voltage v_i is applied to the negative (inverting) input terminal through a resistor R_1 and another resistor R_f makes a connection between the output terminal and the negative input terminal, providing a negative feedback path for the OP Amp. As a novice in OP Amp circuits, we replace the OP Amp with the ideal OP Amp model of Figure 5.1(f) to get a common circuit having a dependent voltage source, as depicted in Figure 5.2(b). For this single-mesh circuit, we choose the mesh analysis method, label the mesh current as i, and express the controlling variable $(v_+ - v_-)$ in terms of the mesh current i as

$$v_+ - v_- = 0 - (v_\text{i} - R_1 i) = -v_\text{i} + R_1 i \ (\because v_+ = 0:\ \text{grounded}) \tag{5.2.1}$$

where we have used the fact that the positive input terminal is grounded so that $v_+ = 0$. Then, we set up the mesh equation

$$(R_1 + R_\text{f})i = v_\text{i} - A(v_+ - v_-) \overset{(5.2.1)}{=} v_\text{i} + A(v_\text{i} - R_1 i)$$

which can be solved for the mesh current i as

$$i = \frac{A+1}{AR_1 + (R_1 + R_\text{f})} v_\text{i} \tag{5.2.2}$$

We finally find the output voltage v_o by subtracting the voltage drop across R_1 and R_f from the input voltage v_i as

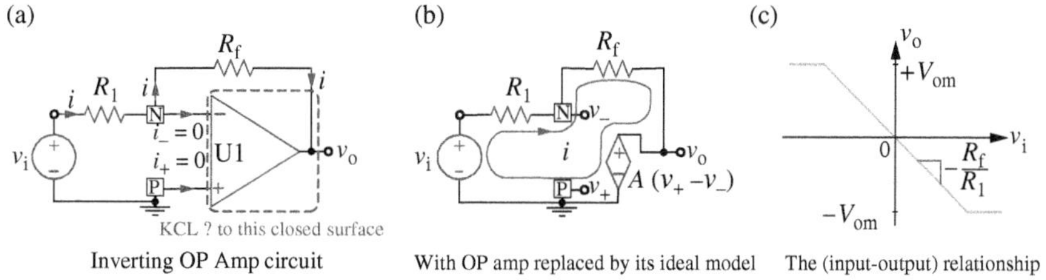

Figure 5.2 Inverting OP Amp circuit.

$$v_o = v_i - (R_1 + R_f)i \overset{(5.2.2)}{=} \frac{AR_1 + (R_1 + R_f) - (A+1)(R_1 + R_f)}{AR_1 + (R_1 + R_f)} v_i$$

$$= \frac{-AR_f}{AR_1 + (R_1 + R_f)} v_i = -\frac{R_f/R_1}{1 + (1 + R_f/R_1)/A} v_i \tag{5.2.3}$$

$$\rightarrow v_o \cong -\frac{R_f}{R_1} v_i \text{ under the assumption that } 1 + \frac{R_f}{R_1} \ll A \tag{5.2.4}$$

where it is reasonable to assume that $1 + R_f/R_1 \ll A$ since the open-loop gain A of an OP Amp is very large. In fact, the value of the open-loop gain A is not certain, in that it varies with temperature and time, as well as from one sample to another. That is why we are so glad to see that A has disappeared in the (approximate) output voltage expression (5.2.4).

Now, as a more practical approach, we are going to use the virtual short principle (Remark 5.1(2)) that can be applied to an OP Amp with negative feedback between the output terminal and the negative input terminal like one contained in our target circuit of Figure 5.2(a). By the virtual short principle, we have

$$v_- \overset{(5.1.5)}{\cong} v_+ \overset{\text{grounded}}{=} 0 \text{ (virtual ground)} \tag{5.2.5}$$

where the negative input terminal is called a virtual ground in the sense that its node voltage is zero, giving the illusion of being grounded. Thus, we can find the current through R_1 to be

$$i = \frac{v_i - v_-}{R_1} \overset{(5.2.5)}{=} \frac{v_i}{R_1} \tag{5.2.6}$$

Noting that all this current flows through R_f to the output terminal since no current flows into or out of the negative input terminal by the virtual open principle (Remark 5.1(1)), we subtract the voltage drop across R_f from the negative input terminal voltage v_- to get the output voltage v_o as

$$v_o = v_- - R_f i \overset{(5.2.5),(5.2.6)}{=} -\frac{R_f}{R_1} v_i \tag{5.2.7}$$

which agrees with Eq. (5.2.4). This input–output relationship is described by the *transfer characteristic curve* in Figure 5.2(c), which implies that the voltage gain, i.e. the ratio of the output voltage to the input voltage of the circuit is

$$\frac{v_o}{v_i} = -\frac{R_f}{R_1} \tag{5.2.8}$$

as long as v_o is not saturated into $\pm V_{om}$. This is referred to as the *closed-loop gain* due to the fact that it is obtained when the feedback path makes a closed loop consisting of the OP Amp and R_f.

Several points are worth mentioning about the OP Amp circuit:

- The closed-loop gain is determined by the external components and can easily be customized to a particular application.
- The OP Amp circuit is called an inverting amplifier because the closed-loop gain is negative and basically, the input voltage is applied to the negative input terminal (via R_1).
- The input impedance is R_1 and the output impedance is $R_o = 0$, as can be obtained in Figure 5.2(b) with the assumption of an ideal OP Amp.

5.2.1.2 Non-Inverting OP Amp Circuit

Figure 5.3(a) shows a non-inverting OP Amp circuit, where the overall input v_i is applied to the positive (non-inverting) input terminal and another resistor R_f makes a connection between the output terminal and the negative input terminal, providing a negative feedback path for the OP Amp. We again replace the OP Amp with the ideal OP Amp model of Figure 5.1(f) to get a common circuit having a dependent voltage source as depicted in Figure 5.3(b).

For this single-mesh circuit, we choose the mesh analysis method, label the mesh current i (in the counterclockwise direction), and express the controlling variable $(v_+ - v_-)$ in terms of the mesh current as

$$v_+ - v_- = v_i - R_1 i \tag{5.2.9}$$

where we have used the fact that the positive input terminal is directly connected to the input voltage source v_i so that $v_+ = v_i$. Then, we set up the mesh equation and solve it as

$$(R_1 + R_f)i = A(v_+ - v_-) \overset{(5.2.9)}{=} A(v_i - R_1 i)$$
$$; i = \frac{A}{AR_1 + (R_1 + R_f)} v_i \tag{5.2.10}$$

We finally find the output voltage v_o by summing the voltage drops across R_1 and R_f as

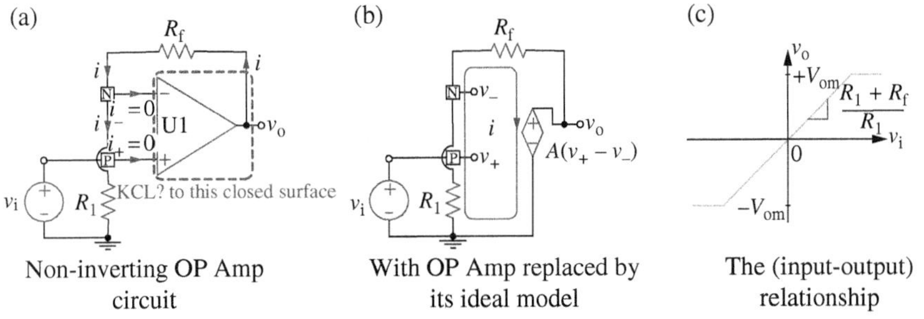

Figure 5.3 Non-inverting OP Amp circuit.

$$v_o = (R_1 + R_f)i \overset{(5.2.10)}{=} \frac{A(R_1 + R_f)}{AR_1 + (R_1 + R_f)} v_i = \frac{(R_1 + R_f)/R_1}{1 + (R_1 + R_f)/R_1/A} v_i \tag{5.2.11}$$

$$\rightarrow v_o \cong \frac{R_1 + R_f}{R_1} v_i \text{ under the assumption that } 1 + \frac{R_f}{R_1} \ll A \tag{5.2.12}$$

where it is reasonable to assume that $1 + R_f/R_1 \ll A$ as mentioned before.

Now, let us take a more practical approach, which is to use the virtual short principle (Remark 5.1(2)). That is, noting that the OP Amp in our target circuit of Figure 5.3(a) has a negative feedback between the output terminal and the negative input terminal and that the positive input terminal is directly connected to the input voltage source v_i so that $v_+ = v_i$, we have

$$v_- \overset{(5.1.5)}{\cong} v_+ = v_i$$

Thus, we can find the current through R_1 to be

$$i = \frac{v_- - 0}{R_1} = \frac{v_i}{R_1} \tag{5.2.13}$$

Noting that all of this current flows through R_f from the output terminal since no current flows into or out of the negative input terminal by the virtual open principle (Remark 5.1(1)), we sum the voltage drops across R_1 and R_f to get the output voltage v_o as

$$v_o = (R_1 + R_f)i \overset{(5.2.13)}{=} \frac{R_1 + R_f}{R_1} v_i \tag{5.2.14}$$

which agrees with Eq. (5.2.12). This input-output relationship is described by the *transfer characteristic curve* in Figure 5.3(c), which implies that the voltage gain, that is, the ratio of the output voltage to the input voltage of the circuit is

$$\frac{v_o}{v_i} = \frac{R_1 + R_f}{R_1} \tag{5.2.15}$$

as long as v_o is not saturated into $\pm V_{om}$. This is referred to as the *closed-loop gain* due to the fact that it is obtained when the feedback path makes a closed loop consisting of the OP Amp and R_f.

Several points are worth mentioning about the OP Amp circuit:

- The closed-loop gain is determined by the external components and can easily be customized to a particular application.
- The OP Amp circuit is called a non-inverting amplifier because the closed-loop gain is positive and basically, the input voltage is applied to the positive input terminal.

- The input impedance is infinitely large (∞) as the input impedance of an ideal OP Amp itself and the output impedance is $R_o = 0$, as can be obtained in Figure 5.3(b) with the assumption of an ideal OP Amp.

(Q) You may have tried to apply KCL to the OP Amp, the group of its two input nodes and output node, or the closed surface denoted by the dotted line in Figure 5.2(a) or 5.3(a) to write the KCL equation as follows. Do they hold?

$$i_+ + i_- + i \stackrel{?}{=} 0, \quad i_+ + i_- - i \stackrel{?}{=} 0$$

(A) No, they do not hold since $i_+ = 0$, $i_- = 0$, and $i = v_i/R_1$. Nevertheless, it cannot be a counterexample which contradicts KCL because you just misapplied KCL. Referring to Figure 5.4, recall that the (dual) power supplies for OP Amps are omitted in the circuit diagram as mentioned in Section 5.1. If you are going to apply KCL to the OP Amp output node or any group of nodes including it, you should restore the omitted power supplies back (Figure 5.4(b)) or replace the OP Amp with its model having a dependent voltage source (Figure 5.1(e) or (f)). In practice, we rarely need to do so for analysis of OP Amp circuits.

Remark 5.2 Practical Analysis Rules of OP Amp Circuits with Negative Feedback

1) Apply the 'virtual open' principle that $i_+ = 0$ and $i_- = 0$, which is mentioned as one of ideal OP Amp conditions in Remark 5.1(1).
2) Apply the 'virtual short' principle that $v_+ \approx v_-$, which is mentioned in Remark 5.1(2). However, this rule may be preempted (invalidated) by positive feedback, if any. Besides, you must apply KCL at each node (individually) to write the node equations in unknown node voltages, except for the output node of an OP Amp or any group of nodes (closed surface) including it.

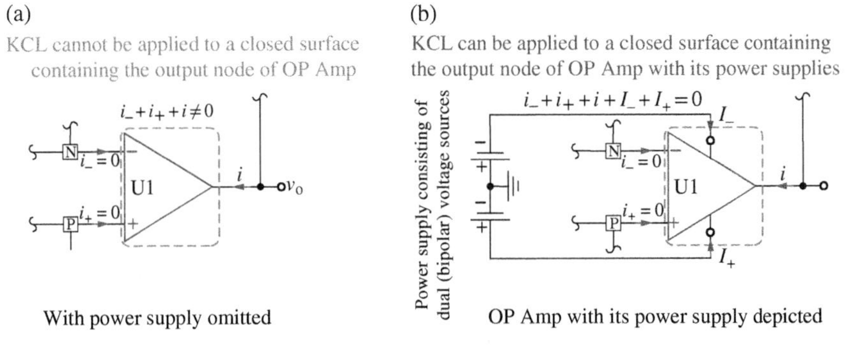

Figure 5.4 Applying KCL to the output node of OP Amp or a closed surface containing it.

5.2.1.3 Voltage Follower

Figure 5.5(a) shows a non-inverting OP Amp circuit with R_1 open-circuited and R_f short-circuited, whose voltage gain (5.2.15) will be unity.

$$\frac{v_o}{v_i} \overset{(5.2.15)}{=} \frac{R_1 + R_f}{R_1} \overset{R_1=\infty, R_f=0}{\longrightarrow} \frac{v_o}{v_i} = 1; \; v_o = v_i \tag{5.2.16}$$

This circuit is called a unity-gain non-inverting amplifier or *voltage follower* in the sense that the output voltage follows the input voltage. Note that its input impedance is (ideally) infinitely large and its output impedance is (ideally) zero as the impedance of a (dependent) voltage source.

The circuit can be used as a voltage buffer for eliminating interstage loading effect since it blockades the flow of current, while presenting a virtual short connection between the + input and output of the OP Amp in terms of the voltage. For example, let us compare the two circuits shown in Figure 5.5(b) and (c) in terms of their voltage gains.

$$\begin{aligned}\text{Figure 5.5b:} \; \frac{v_o}{v_i} &= \frac{R_2 \,\|\, (R_3 + R_4)}{R_1 + (R_2 \,\|\, (R_3 + R_4))} \frac{R_4}{R_3 + R_4} \\ &= \frac{R_2}{R_1 + R_2} \frac{R_4}{\dfrac{R_1 R_2}{R_1 + R_2} + (R_3 + R_4)}\end{aligned} \tag{5.2.17}$$

$$\text{Figure 5.5c:} \; \frac{v_o}{v_i} = \frac{R_2}{R_1 + R_2} \times \frac{R_4}{R_3 + R_4} \tag{5.2.18}$$

This indicates that the voltage gain (Eq. (5.2.17)) of the two-stage voltage divider with no voltage follower in Figure 5.5(b) is smaller than that (Eq. (5.2.18)) of the

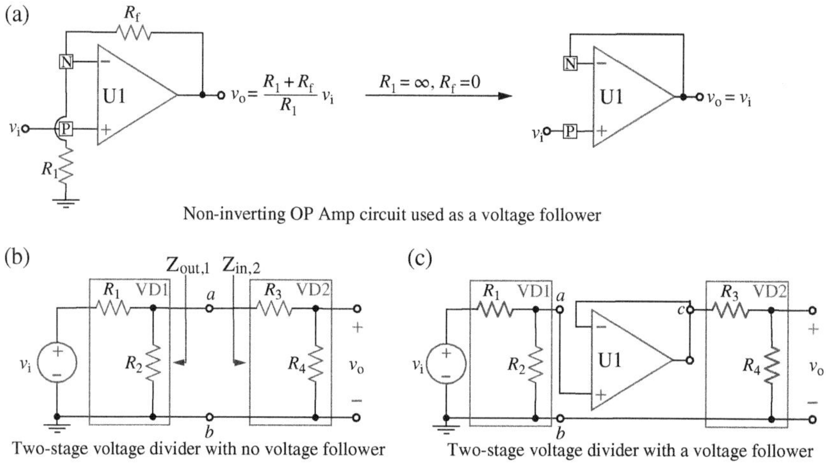

Figure 5.5 Voltage follower to remove or reduce loading effects.

two-stage voltage divider with a voltage follower between the two stages in Figure 5.5(c), as revealed from its larger denominator. What causes this difference in voltage gain? It arises from the *loading effect* of the second-stage voltage divider (VD2) on the first-stage one (VD1) in Figure 5.5(b), while in Figure 5.5(c) such a loading effect is eliminated by the voltage follower having infinitely large input impedance and zero output impedance.

The difference between Eqs. (5.2.17) and (5.2.18) can be made small by making the following inequality satisfied by a wide margin.

$$Z_{\text{out},1} = \frac{R_1 R_2}{R_1 + R_2} \ll R_3 + R_4 = Z_{\text{in},2} \tag{5.2.19}$$

Note that the left-hand side (LHS) is the output impedance of the first voltage divider (on the source side) and the right-hand side (RHS) is the input impedance of the second one (on the load side) seen from the terminals *a*-*b*.

5.2.1.4 Linear Combiner

Let us consider the OP Amp circuits in Figure 5.6 from the analysis and design points of view. For the OP Amp circuit of Figure 5.6(a), we can apply the superposition principle to find $v_P \approx v_N$ (depending on v_{P1} and v_{P2}) by adding v_P with $v_{P2} = 0$ and v_P with $v_{P1} = 0$ as

$$v_P = \frac{R_{P2}}{R_{P1} + R_{P2}} v_{P1} + \frac{R_{P1}}{R_{P1} + R_{P2}} v_{P2} = (R_{P1} \| R_{P2}) \left(\frac{v_{P1}}{R_{P1}} + \frac{v_{P2}}{R_{P2}} \right) \tag{5.2.20}$$

Then, we apply the superposition principle again to find v_o by adding v_o with $v_{N1} = v_{N2} = 0$ and v_o with $v_{P1} = v_{P2} = 0$ as

$$v_o = (R_{P1} \| R_{P2}) \left(\frac{v_{P1}}{R_{P1}} + \frac{v_{P2}}{R_{P2}} \right) \left(1 + \frac{R_f}{R_{N1} \| R_{N2}} \right) - \left(\frac{R_f}{R_{N1}} v_{N1} + \frac{R_f}{R_{N2}} v_{N2} \right) \tag{5.2.21}$$

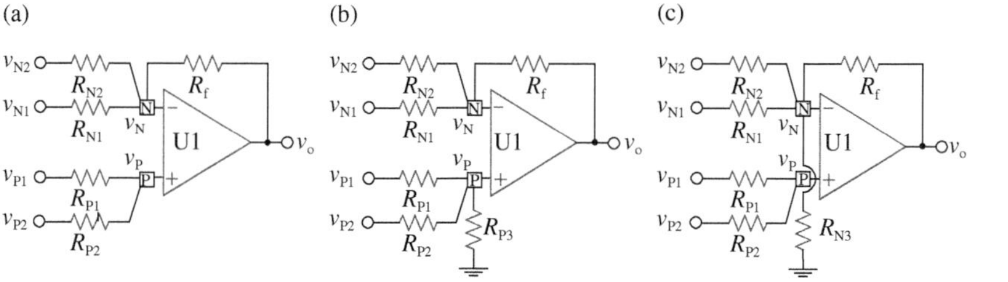

Figure 5.6 Linear combiner using an OP Amp.

Likewise, we can express the output voltage v_o of the OP Amp circuit of Figure 5.6(b) with an additional resistor R_{P3} in terms of the input voltages as

$$v_o = (R_{P1}||R_{P2}||R_{P3})\left(\frac{v_{P1}}{R_{P1}} + \frac{v_{P2}}{R_{P2}}\right)\left(1 + \frac{R_f}{R_{N1}||R_{N2}}\right) - \left(\frac{R_f}{R_{N1}}v_{N1} + \frac{R_f}{R_{N2}}v_{N2}\right) \tag{5.2.22}$$

We compare Eqs. (5.2.21) and (5.2.22) to notice that the additional resistor R_{P3} (connected between node P and GND) decreases the coefficients of v_{P1} and v_{P2}.

Likewise, we can express the output voltage v_o of the OP Amp circuit of Figure 5.9(c) with an additional resistor R_{N3} in terms of the input voltages as

$$v_o = (R_{P1}||R_{P2})\left(\frac{v_{P1}}{R_{P1}} + \frac{v_{P2}}{R_{P2}}\right)\left(1 + \frac{R_f}{R_{N1}||R_{N2}||R_{N3}}\right) - \left(\frac{R_f}{R_{N1}}v_{N1} + \frac{R_f}{R_{N2}}v_{N2}\right) \tag{5.2.23}$$

We compare Eqs. (5.2.21) and (5.2.23) to notice that the additional resistor R_{N3} (connected between node N and GND) increases the coefficients of v_{N1} and v_{N2}.

Now, we design a linear combiner realizing a given linear combination of several input signals, say,

$$v_o = a_1 v_{P1} + a_2 v_{P2} - b_1 v_{N1} - b_2 v_{N2} \quad \text{with } a_1, a_2, b_1, b_2 > 0 \tag{5.2.24}$$

as one of the two configurations in Figure 5.6(b) and (c) by the following procedure.

(Step 1) After choosing an appropriate value of the feedback resistance R_f, determine the values of R_{N1} and R_{N2} to be inversely proportional to the negative coefficients, b_1 and b_2, so that the negative terms can be implemented as Eq. (5.2.21).

(Step 2) Determine the values of R_{P1} and R_{P2} to be inversely proportional to the positive coefficients, a_1 and a_2, so that the positive terms can be implemented as Eq. (5.2.21).

(Step 3) If the coefficients obtained from Eq. (5.2.21) with the chosen values of resistors are larger/smaller than a_1 and a_2, connect another resistor between the input terminal P/N and the ground so that the positive coefficients obtained from Eq. (5.2.22)/(5.2.23) become smaller/larger as required.

This procedure has been cast into the above MATLAB function '`design_combiner()`', which returns the values of the resistances R_{Pi}'s/R_{Ni}'s to be connected to the +/- input terminals for the positive/negative coefficients a's = [a_1 a_2 ...] and b's = [b_1 b_2 ...] given together with the value, Rf, of feedback resistance R_f and the minimum value, RPmin, of R_{Pi}'s.

```
function [RPs,RNs]=design_combiner(as,bs,Rf,RPmin)
% Design a linear combiner to realize vo = as*vPs - bs*vNs
% Input: positive coefficient vector as=[a1 a2..]
%        negative coefficient vector bs=[b1 b2..]
%        Rf: Feedback resistance
%        RPmin: Minimum value among RP1,RP2,...
% Output: resistances to be connected to the +/- input terminal [k ohm]
% Copyleft: Won Y. Yang, wyyang53@hanmail.net, CAU for academic use only
if nargin<4, RPmin= 1; end
if nargin<3, Rf= 1; end
RNs = Rf./bs; amax = max(as); RPs = RPmin*amax./as;
composite_RPs = parallel_comb(RPs);
composite_RNs = parallel_comb(RNs);
as_realized = composite_RPs./RPs*(1+Rf/composite_RNs); % Eq.(5.2.21)
ratio = max(as_realized)/amax;
if ratio>1
 R1= 1/(ratio-1)*composite_RPs; RPs0= RPs; RPs= [RPs R1];
 as_realized=parallel_comb(RPs)./RPs0*(1+Rf/composite_RNs); %Eq.(5.2.22)
elseif ratio<1
 scale= ratio*Rf/((1-ratio)*composite_RNs+Rf);
 R1= scale/(1-scale)*composite_RNs; RNs= [RNS R1];
 as_realized = composite_RPs./RPs*(1+Rf/parallel_comb(RNs)); % Eq.(5.2.23)
end
combiner_coefficients= [as_realized -bs] % Realized
```

For instance, with the feedback resistance R_f and the minimum value of R_{Pi}'s given as 2 kΩ and 1 kΩ, respectively, we can design a linear combiner realizing

$$v_o = 2v_{P1} + v_{P2} - 2v_{N1} - v_{N2} \tag{5.2.25}$$

by running the following MATLAB statements:

```
>>[RPs,RNs]=design_combiner([2 1],[2 1],2,1) % with Rf=2, RPmin=1
   RPs = 1.0000   2.0000   2.0000 % RP1=1kΩ, RP2=2kΩ, RP3=2kΩ
   RNs = 1        2               % RN1=1kΩ, RN2=2kΩ
```

5.2.2 OP Amp Circuits with Positive Feedback

As discussed in Section 5.1, a negative feedback path between the output terminal and the negative input terminal of an OP Amp has the stabilization effect on the differential input voltage ($v_P - v_N$) and the output voltage v_o so that $v_P - v_N = v_o/A \approx 0$. Thus, v_o may stay at some nominal value V_o between the negative/positive maximum output or saturation (limit) voltages $-V_{om}$ and $+V_{om}$. In contrast, a *positive feedback* path between the output terminal and the positive input terminal of an OP Amp has the destabilization effect on the differential input voltage ($v_P - v_N$) and the output voltage v_o. As a result, v_o may diverge till it reaches $-V_{om}$ or $+V_{om}$ to be trapped there.

5.2.2.1 Inverting Positive Feedback OP Amp Circuit

Figure 5.7(a) shows an inverting OP Amp circuit with a positive feedback path (via R_2) between the output terminal and the positive input terminal. Due to the destabilization effect of the positive feedback, the virtual short principle does not apply to this circuit, and besides, the ideal or practical OP Amp model does not work for this circuit. Noting that

- its output voltage will be either $v_o = +V_{om}$ or $-V_{om}$ depending on which one of the two input terminals is of higher voltage, and
- the voltage of the positive input terminal is determined by the voltage divider rule as $bv_o = \pm bV_{om}$ with $b = R_1/(R_1 + R_2)$,

suppose the output voltage at some instant is $v_o = +V_{om}$. Then, the + input terminal of the OP Amp has the node voltage of

$$v_P = +b\,V_{om} \text{ with } b = \frac{R_1}{R_1 + R_2} \tag{5.2.26}$$

and this state will be maintained as long as $v_i = v_N < v_P = +bV_{om}$. If the input voltage v_i somehow rises above $v_P = +bV_{om}$, the voltages at the output and + input terminals will go down to

$$v_o = -V_{om} \text{ and } v_P = -b\,V_{om}$$

respectively, and this state will be maintained as long as $v_i = v_N > v_P = -bV_{om}$. If the input voltage v_i somehow goes below $v_P = -bV_{om}$, the voltages at the output and + input terminals will go up to

$$v_o = +V_{om} \text{ and } v_P = +b\,V_{om}$$

respectively, and this state will be maintained as long as $v_i = v_N < v_P = +bV_{om}$. The output becomes $-V_{om}$ for an input above the upper (higher) threshold value $V_{TH} = +bV_{om}$ and $-V_{om}$ for an input below the lower threshold value

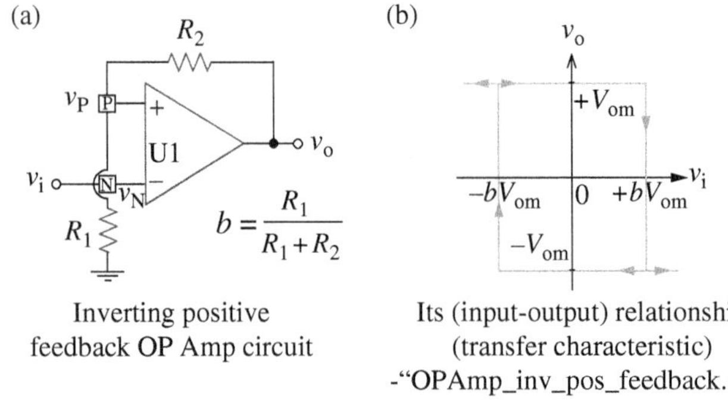

Figure 5.7 Inverting positive feedback OP Amp circuit and its input-output relationship (transfer characteristic).

$V_{TL} = -bV_{om}$, while it stays at the current state for an input between V_{TL} and V_{TH}. This input–output relationship can be described by the following equation and the transfer characteristic curve in Figure 5.7(b).

$$v_o = \begin{cases} +V_{om} & \text{for } v_i < -b\,V_{om} \\ \text{keep the current state} & \text{for } -b\,V_{om} \le v_i \le +b\,V_{om} \text{ with } b = \dfrac{R_1}{R_1+R_2} \\ -V_{om} & \text{for } v_i > +-b\,V_{om} \end{cases} \quad (5.2.27)$$

5.2.2.2 Non-Inverting Positive Feedback OP Amp Circuit

Figure 5.8(a) shows a non-inverting OP Amp circuit with a positive feedback path (via R_2) between the output terminal and the positive input terminal. Note the following:

- Its output voltage will be either $v_o = +V_{om}$ or $-V_{om}$ depending on whether the voltage v_P at the positive input terminal is higher or lower than $v_N = 0$ (grounded) at the negative input terminal.
- The value of the input voltage causing the voltage, v_P, at the positive input terminal to be zero is related with the output voltage v_o as

$$v_i = -\frac{R_1}{R_2}v_o = -b\,v_o \text{ with } b = \frac{R_1}{R_2} \quad (5.2.28)$$

This implies that once $v_o = +V_{om}$, it changes into $-V_{om}$ only when $v_i < -bV_{om}$ and once $v_o = +V_{om}$, it changes into $+V_{om}$ only when $v_i > +bV_{om}$. Thus, the input-output relationship can be described by the following equation and the transfer characteristic curve in Figure 5.8(b).

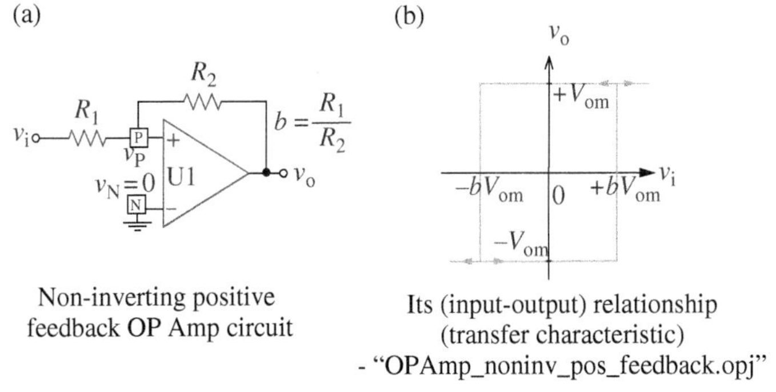

Figure 5.8 Non-inverting positive feedback OP Amp circuit and its input-output relationship (transfer characteristic).

$$v_o = \begin{cases} +V_{om} & \text{for } v_i > +b\,V_{om} \\ \text{keep the current state} & \text{for } -b\,V_{om} \le v_i \le +b\,V_{om} \\ -V_{om} & \text{for } v_i < -b\,V_{om} \end{cases} \quad \text{with } b = \frac{R_1}{R_2} \tag{5.2.29}$$

Interested readers are invited to compose the PSpice schematics of Figures 5.7(a)/5.8(a) (named "OPAmp_inv_pos_feedback.opj"/"OPAmp_noninv_pos_feedback.opj") and run them with the Analysis type of DC Sweep to see the transfer characteristic curves in Figures 5.7(b)/5.8(b).

Remark 5.3 Destabilzation Effect of Positive Feedback and Bistable Multivibrator

1) If an OP Amp has a positive feedback and no negative feedback, changes in the output voltage v_o and the differential input $(v_P - v_N)$ helps each other synergistically so that even the slightest change in v_o instantly results in $v_o = +V_{om}$ or $-V_{om}$. That is why the positive feedback OP Amp circuits operate in saturation with their output voltages at one of the two (extreme) states $\pm V_{om}$ most of the time. This is called the *destabilization effect* of positive feedback.
2) The positive feedback OP Amp circuits in Figures 5.7(a) and 5.8(a) are called 'bistable multi-vibrator' since they have two stable outputs $\pm V_{om}$ (positive or negative saturation voltages) for an input voltage v_i as long as $-bV_{om} \le v_i \le +bV_{om}$. Just like a flip-flop (FF), they are said to have a memory since they keep the previous state, that is, their outputs are either $+V_{om}$ or $-V_{om}$ depending on the (previous) state which they were in, unless they are triggered by any (possibly short duration) input of magnitude greater than bV_{om}. This presents the reason why the circuits are referred to as inverting/non-inverting *Schmitt triggers*.
3) As can be seen from the overall input–output relationships that are described by the transfer characteristic curves in Figure 5.7(b) or 5.8(b), the positive feedback OP Amp circuits have two different output-changing paths since there are two different threshold values of input to change the output depending on whether their inputs are increasing or decreasing. For this reason, they are said to exhibit a *hysteresis* or *deadband* characteristic.
4) The hysteresis or deadband characteristic can be used to reduce the number of contact bounces in an on-off switch for a microprocessor or a temperature control system (see Example 5.1). It also has an important application to periodic wave generation as will be discussed in Section 5.3.2.2.

Example 5.1 Simulation of OP Amp Circuit with Positive Feedback – Schmitt Trigger

Figure 5.9(a1) shows the PSpice schematic for a Schmitt trigger circuit, which is slightly different from the circuit of Figure 5.7(a), in that a DC voltage source of

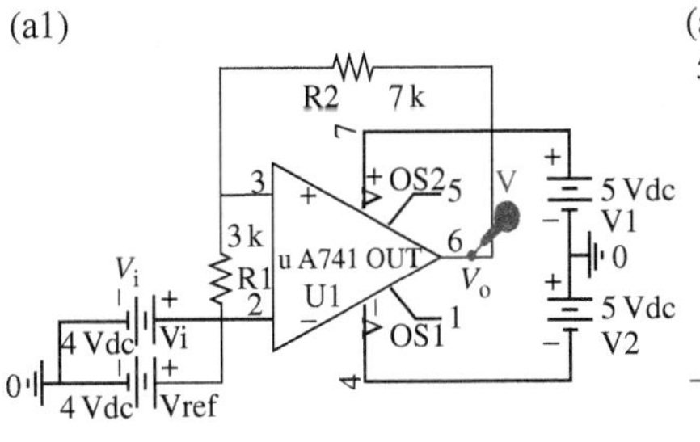

PSpice schematic for a Schmitt trigger circuit – DC Sweep analysis ("elec05e01.opj")

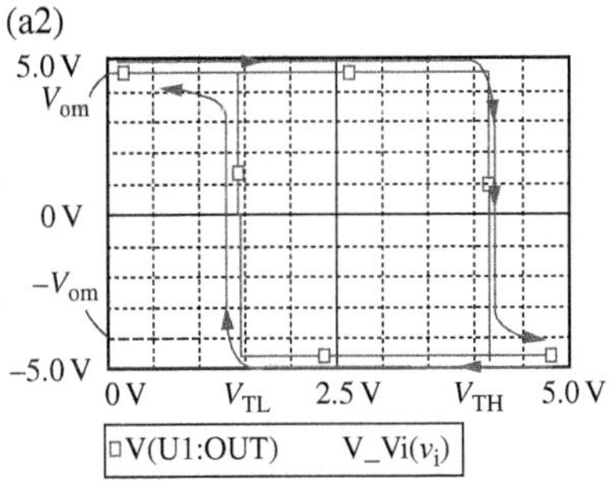

The output with the input swept from 0 to 5 and 5 to 0[V]

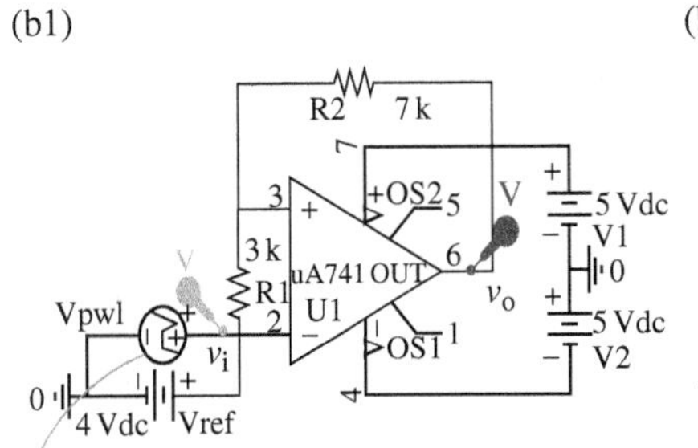

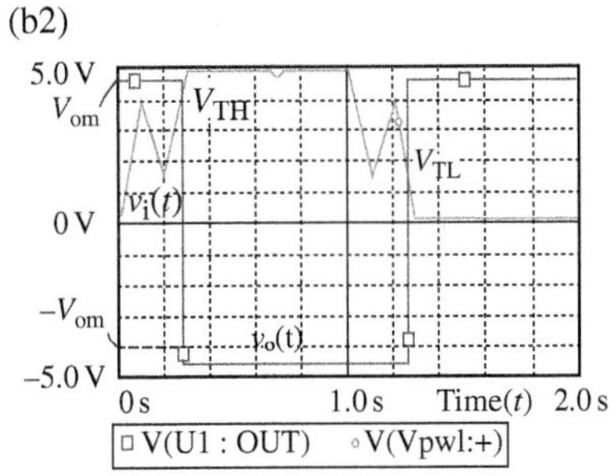

PSpice schematic for a Schmitt trigger circuit – Time Domain (Transient) analysis ("elec05e01_PWL.opj")

The waveforms of input/output voltage

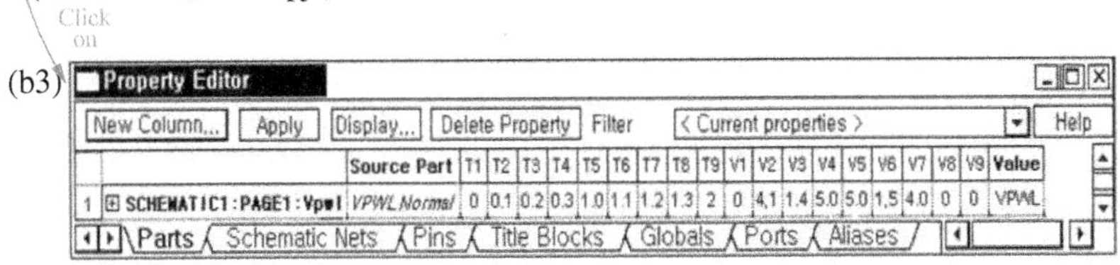

Property Editor spreadsheet for the piece-wise linear voltage source VPWL

Figure 5.9 PSpice schematics and simulation results for Example 5.1.

$V_{ref} = 4$ V is connected in series with R_1. The node voltage at the + input terminal of the OP Amp determining the threshold voltages also differs from Eq. (5.2.26) as

$$v_+ = V_{ref} + b(v_o - V_{ref}) = (1-b)V_{ref} + bv_o \text{ with } b = \frac{R_1}{R_1 + R_2} \tag{E5.1.1}$$

a) Find the two (higher/lower) threshold values of input voltage to reverse the output voltage.

   The higher/lower threshold values are obtained by substituting $v_o = \pm V_{om}$ into Eq. (E5.1.1) as

$$V_{TH}, V_{TL} = (1-b)V_{ref} \pm bV_{om} \tag{E5.1.2}$$

b) With $\pm V_{om} = \pm 4.6$ V, find the values of b and V_{ref} such that the higher/lower threshold values are $V_{TH} = 4.18$ V and $V_{TL} = 1.42$ V, respectively. We solve

$$\begin{aligned} V_{TH} &= (1-b)V_{ref} + 4.6b = 4.18 \\ V_{TL} &= (1-b)V_{ref} - 4.6b = 1.42 \end{aligned} \tag{E5.1.3}$$

to get $b = 0.3$ and $V_{ref} = 4$ V.

c) We take the following steps to perform the PSpice simulation (for DC Sweep analysis) to get the transfer characteristic curve describing the overall input-output relationship shown in Figure 5.9(a2).
   1) Draw the PSpice schematic ("elec05e01.opj") as depicted in Figure 5.9(a1).
   2) Set the Analysis type to DC Sweep, select the sweep variable as Vi, the sweep type as Linear, and the sweep values as (Start value: 0, End value: 5, Increment: 0.01) in the Simulation Settings dialog box, and run the simulation.
   3) Change the sweep values to (Start value: 5, End value: 0, Increment: –0.01) in the Simulation Settings dialog box and run the simulation.

d) We take the following steps to perform the PSpice simulation (for Transient analysis) to get the input and output waveforms as depicted in Figure 5.9(b2), which shows the debouncing feature of the Schmitt trigger due to its deadband characteristic.
   1) Draw the PSpice schematic ("elec05e01_PWL.opj") as shown in Figure 5.9(b1).

```
%elec05e01.m
Vom=4.6; Vref=4; R1=3e3; R2=7e3;
A=1e5; % Open-loop gain of the OP Amp
% (c) DC Sweep Analysis
dv=0.01; vi=[0:499 500:-1:1]*dv; %Increasing/decreasing input voltage
vo(1)=Vom; % Initial value of the OP Amp output voltage
for n=1:length(vi)
   vN(n)=vi(n); vP(n)=(R2*Vref+R1*vo(n))/(R1+R2); % Eq. (E5.1.1)
   vo(n+1)=max(min(A*(vP(n)-vN(n)),Vom),-Vom); % Eq. (5.1.1)
end
subplot(211), plot(vi(1:500),vo(1:500), vi(501:1000),vo(501:1000),'r')
% (d) Transient Analysis with PWL input voltage
TT=[0 0.1 0.2 0.3 1 1.1 1.2 1.3 2]; % Time points for PWL waveform
VV=[0 4.1 1.4 5 5 1.5 4 0 0]; % Voltage points for PWL waveform
T=1e-3; t=0:T:2; vi=interp1(TT,VV,t,'linear'); % PWL input voltage
vo(1)=Vom; % Initial value of the OP Amp output voltage
for n=1:length(t)
   vN(n)=vi(n); vP(n)=(R2*Vref+R1*vo(n))/(R1+R2); % Eq. (E5.1.1)
   vo(n+1)=max(min(A*(vP(n)-vN(n)),Vom),-Vom); % Eq. (5.1.1)
end
subplot(212), plot(t,vi,'g', t,vo(1:end-1),'r')
```

2) Double-click the VPWL (piece-wise linear voltage source) to open the Property Editor spreadsheet and set the values of the parameters (T1,V1), (T2,V2), ... as shown in Figure 5.9(b3). If needed, click the New Column button to create new columns like (T9,V9).
3) Set the Analysis type to Time Domain (Transient), Run_to_time to 2 s, and Maximum step to 0.1 ms in the Simulation Settings dialog box, and run the simulation.

e) We can run the above MATLAB script "elec05e01.m" to get the similar results.

5.3 First-Order OP Amp Circuits [Y-1]

5.3.1 First-Order OP Amp Circuits with Negative Feedback

Let us first consider the *RC* circuit containing an OP Amp in Figure 5.10(a1). Since the OP Amp has a negative feedback path connecting its output terminal to its negative input terminal without having any positive feedback path (connecting its output terminal to its positive input terminal), we can use the virtual short principle (Remark 5.1(2)) to set the node voltage at the negative input terminal (node N) of the OP Amp (almost) equal to that ($v_P = 0$) at the positive terminal that is grounded:

$$v_N(t) \approx 0 \text{ (virtual ground)}$$

Noting also that, by the virtual open principle (Remark 5.1(1)), no current can flow into or out of the input terminals of the (ideal) OP Amp, we apply KCL at node N to write

$$\frac{v_N(t)-v_i(t)}{R}+C\frac{d(v_N(t)-v_o(t))}{dt} \underset{=}{\text{KCL}} 0 \overset{v_N=0}{\rightarrow} \frac{v_i(t)}{R}+C\frac{dv_o(t)}{dt}=0$$

$$; \frac{dv_o(t)}{dt}=-\frac{v_i(t)}{RC}$$

We can solve this first-order differential equation as

$$v_o(t)=-\frac{1}{RC}\int_{-\infty}^{t} v_i(\tau)d\tau=-\frac{1}{C}\int_{-\infty}^{0} i(\tau)d\tau-\frac{1}{RC}\int_{0}^{t} v_i(\tau)d\tau$$

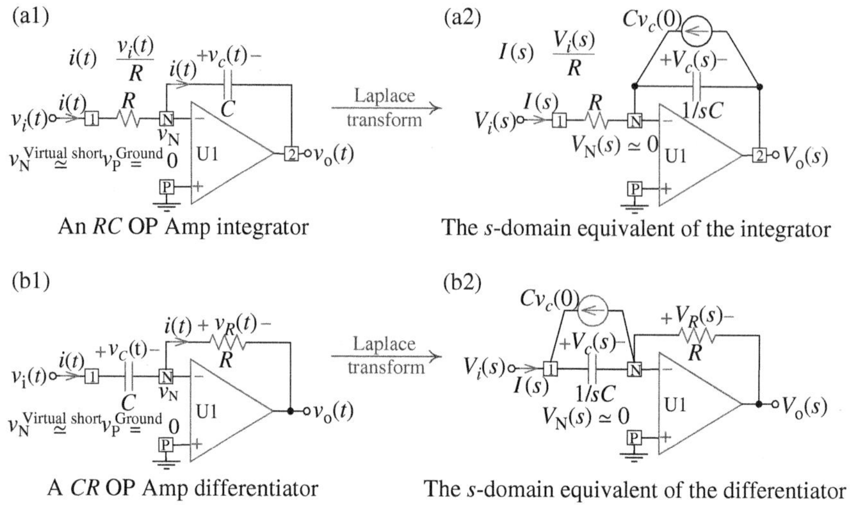
(a1)
Laplace transform
(a2)
An RC OP Amp integrator
The s-domain equivalent of the integrator
(b1)
Laplace transform
(b2)
A CR OP Amp differentiator
The s-domain equivalent of the differentiator

Figure 5.10 First-order circuits containing an OP Amp.

to obtain the input-output relationship as

$$v_o(t) = v_o(0) - \frac{1}{RC}\int_0^t v_i(\tau)d\tau = -v_C(0) - \frac{1}{RC}\int_0^t v_i(\tau)d\tau \tag{5.3.1}$$

$$\text{with } v_C(0) = \frac{1}{C}\int_{-\infty}^{0} i(\tau)d\tau = \frac{1}{C}\int_{-\infty}^{0} \frac{v_i(\tau)}{R}d\tau$$

Since this implies that the output is qualitatively the integral of the input, the *RC* OP Amp circuit is called an inverting *RC* integrator or a *Miller integrator* with the integration time constant *RC*. As an alternative, we can transform this circuit into its *s*-domain equivalent as depicted in Figure 5.10(a2) and apply KCL at node N to obtain the *s*-domain input-output relationship as

$$\frac{0 - V_i(s)}{R} + \frac{0 - V_o(s)}{1/sC} = Cv_C(0);\ V_o(s) = -\frac{1}{s}v_C(0) - \frac{1}{sRC}V_i(s) \tag{5.3.2}$$

which is the Laplace transform of the above *t*-domain input-output relationship (5.3.1).

(cf.) If we omit the initial condition term, this *s*-domain input-output relationship is just like what we can get by substituting R and $1/sC$ for R_1 and R_f into Eq. (5.2.7), which is the input-output relationship of an inverting amplifier consisting of an OP Amp together with two resistors R_1 and R_f.

Now let us consider the *CR* circuit containing an OP Amp in Figure 5.10(b1). Since this circuit has the same structure as the *RC* OP Amp circuit in Figure 5.10 (a1), we apply KCL and the virtual open principle with $v_N(t) \approx 0$ (virtual ground) at node N to obtain the input-output relationship as

$$C\frac{d(v_N(t)-v_i(t))}{dt}+\frac{v_N(t)-v_o(t)}{R}\overset{\text{KCL}}{=}0\overset{v_N=0}{\rightarrow}\frac{v_o(t)}{R}+C\frac{dv_i(t)}{dt}=0$$

$$;v_o(t)=-RC\frac{dv_i(t)}{dt} \tag{5.3.3}$$

Since this implies that the output is qualitatively the derivative of the input, the *CR* OP Amp circuit is called a *CR differentiator*. As an alternative, we can transform this circuit into its *s*-domain equivalent as depicted in Figure 5.10(b2) and apply KCL at node N to obtain the *s*-domain input-output relationship as

$$\frac{0-V_i(s)}{1/sC}+\frac{0-V_o(s)}{R}=-Cv_C(0);\ V_o(s)=RC\,v_C(0)-sRC\ V_i(s) \tag{5.3.4}$$

which is the Laplace transform of the above *t*-domain input-output relationship.

(cf.) If we omit the initial condition term, this *s*-domain input-output relationship is just like what we can get by substituting $1/sC$ and R for R_1 and R_f in Eq. (5.2.7).

5.3.2 First-Order OP Amp Circuits with Positive Feedback

As stated in Remark 5.3(1) (Section 5.2.2.2), a positive feedback path (from the output to the positive input terminal) of an OP Amp destabilizes the output so that the output voltage will have the alternative of $+V_{om}$ (positive saturation voltage $+V_{sat}$) or $-V_{om}$ (negative saturation voltage $-V_{sat}$). As will be illustrated in this section, positive/negative feedbacks can be combined to generate a periodic waveform such as a rectangular or triangular wave, which seems to be a wonderful harmony of two antagonistic enemies each causing instability/stability.

5.3.2.1 Square(Rectangular)-Wave Generator

Consider the OP Amp circuit with positive/negative feedbacks in Figure 5.11(a). Noting that

- its output voltage will be either $v_o = +V_{om}$ or $-V_{om}$ depending on which one of the two input terminals is of higher voltage, and
- the voltage of the positive input terminal is $v_P = bv_o = \pm bV_{om}$ with $b = R_1/(R_1 + R_2)$,

(a)

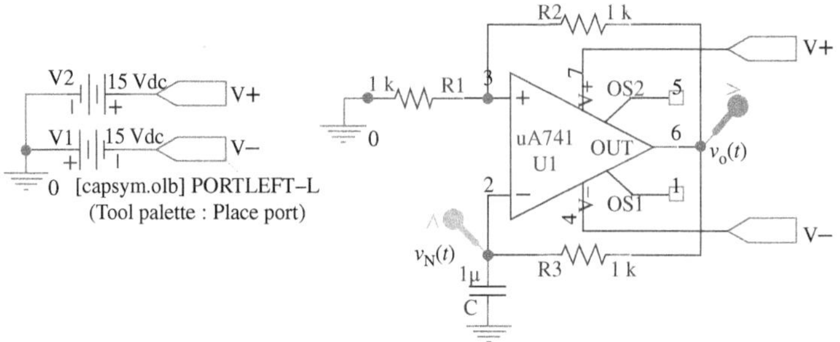

PSpice schematic for a square-wave generator using an OP Amp("OPAmp_RC_squarewave.opj")

(b)

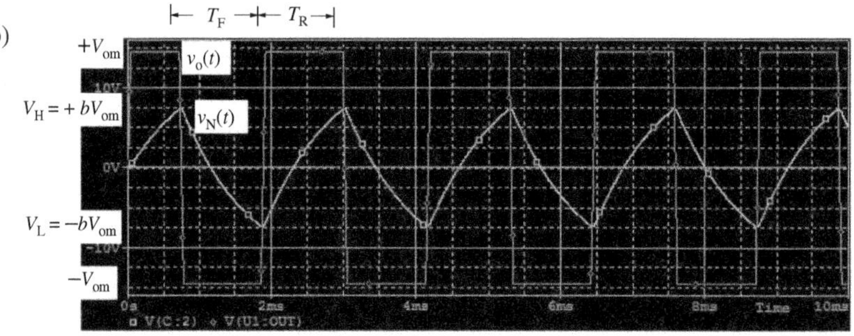

Probe output waveform in the PSpice AD (Probe) window

Figure 5.11 An *RC* OP Amp circuit with positive/negative feedbacks functioning as a square-wave generator.

suppose the output voltage at some instant is $v_o = +V_{om}$. Then, the positive input terminal of the OP Amp has the node voltage of

$$v_P = +b\,V_{om} \text{ with } b = \frac{R_1}{R_1 + R_2}$$

(which is possible only when $v_C = v_N < v_P$) and the voltage $v_C = v_N$ (at the – input terminal) of the capacitor (connected via R_3 to the output terminal with $v_o = +V_{om}$) rises exponentially toward $+V_{om}$ till it catches up with $v_P = +bV_{om}$. As soon as $v_C = v_N$ goes above $v_P = +bV_{om}$ (so that $v_N > v_P$), the voltages at the output and + input terminal will go down to

$$v_o = -V_{om} \text{ and } v_P = -b\,V_{om}$$

respectively. Then, $v_C = v_N$ falls exponentially toward $-V_{om}$ till it touches down at $v_P = -bV_{om}$. As soon as $v_C = v_N$ goes below $v_P = -bV_{om}$ (so that $v_N < v_P$), the voltages at the output and + input terminal will go up to

$$v_o = +V_{om} \text{ and } v_P = +b\,V_{om}$$

respectively. This process repeats itself periodically, generating a rectangular wave at the output as depicted in Figure 5.11(b).

Now, to find the period of the rectangular wave, note that the voltage $v_C = v_N$ (plotted as a sawtooth curve in Figure 5.11(b) will go up and down repetitively in the range limited by the following two threshold values:

$$V_H = +b\,V_{om} \text{ and } V_L = -b\,V_{om} \text{ with } b = \frac{R_1}{R_1 + R_2} \tag{5.3.5}$$

The voltage $v_C = v_N$, as a time-dependent variable of a first-order (RC) circuit, can be written in terms of its initial value $v_N(t_0^+)$, final value $v_N(\infty)$, and the time constant T (Eq. (3.39) of [Y-1]) as

$$v_N(t) \overset{[Y\text{-}1]}{\underset{(3.39)}{=}} \left(v_N(t_0^+) - v_N(\infty)\right) e^{-(t-t_0)/T} + v_N(\infty),\; T = R_3 C \tag{5.3.6}$$

Complying with this formula, the voltage $v_C = v_N$ during the rising/falling interval is described by

$$v_N(t) \overset{(5.3.6)}{=} (V_L - V_{om})\,e^{-(t-t_0)/T} + V_{om} \text{ and } v_N(t) \overset{(5.3.6)}{=} (V_H + V_{om})\,e^{-(t-t_0)/T} - V_{om} \tag{5.3.7a,b}$$

Thus, the times T_R/T_F taken for $v_C = v_N$ to rise/fall from V_L/V_H to V_H/V_L can be obtained as

$$V_H \overset{(5.3.7a)}{=} (V_L - V_{om})\,e^{-T_R/T} + V_{om};\; e^{-T_R/T} = \frac{V_{om} - V_H}{V_{om} - V_L} \overset{(5.3.5)}{=} \frac{1-b}{1+b}$$
$$;\; T_R = -T \ln\frac{1-b}{1+b} = R_3 C \ln\frac{1+b}{1-b} \tag{5.3.8a}$$

$$V_L \overset{(5.3.7b)}{=} (V_H + V_{om})\,e^{-T_F/T} - V_{om};\; e^{-T_F/T} = \frac{V_{om} + V_L}{V_{om} + V_H} \overset{(5.3.5)}{=} \frac{1-b}{1+b}$$
$$;\; T_F = -T \ln\frac{1-b}{1+b} = R_3 C \ln\frac{1+b}{1-b} \tag{5.3.8b}$$

Consequently, the period of the rectangular/sawtooth waves turns out to be

$$P = T_R + T_F \overset{(5.3.8a,b)}{=} 2T \ln\frac{1+b}{1-b} = 2R_3 C \ln\frac{2R_1 + R_2}{R_2} \tag{5.3.9}$$

The PSpice schematic (Figure 5.11(a)) with $R_1 = R_2 = R_3 = 1\,\text{k}\Omega$ and $C = 1\,\mu\text{F}$ has been run with the Analysis type of 'Time Response (Transient)' to yield the waveforms depicted in Figure 5.11(b).

5.3.2.2 Rectangular/Triangular-Wave Generator

Consider the circuit of Figure 5.12(a) in which the left OP Amp U1 with negative feedback makes an inverting integrator (Figure 5.10(a1)) and the right OP Amp U2 with positive feedback forms a non-inverting trigger (Figure 5.8(a)). Noting that

- the output (v_{o1}) of the inverting integrator is applied to the input of the non-inverting trigger,
- the output (v_{o2}) of the non-inverting trigger is fed back into the input of the inverting integrator, and
- the two threshold values at which the non-inverting trigger changes its output voltage v_{o2} are

$$V_H = +bV_{om},\ V_L = -bV_{om} \quad \text{with } b = \frac{R_2}{R_3} \quad \text{(Eq. (5.2.28) or Fig. 5.8)} \qquad (5.3.10)$$

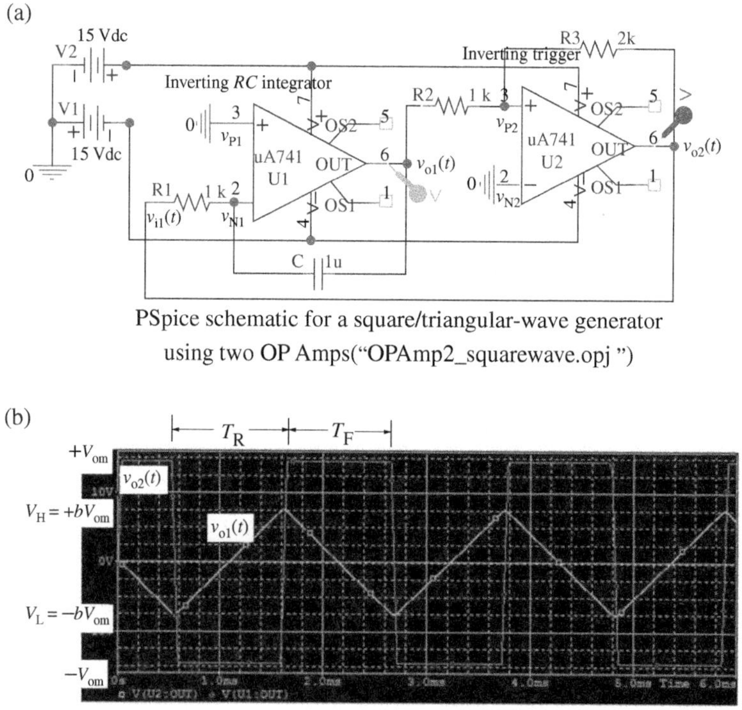

Figure 5.12 Combination of an inverting integrator and a non-inverting positive feedback OP Amp as a trigger.

suppose the output voltage v_{o2} of the OP Amp U2 at some instant is

$$v_{o2} = +V_{om}$$

Then, this constant positive voltage is fed back into the input (v_{i1}) of the inverting integrator to make its output (v_{o1}) decrease linearly till $v_{o1} < -bV_{om}$. When $v_{o1} < -bV_{om}$, the voltage v_{P2} (at the positive input terminal of U2) goes below $v_{N2} = 0$ (at the negative input terminal of U2) as

$$v_{P2} = v_{o1} - R_2\frac{v_{o1} - V_{om}}{R_2 + R_3} = \frac{R_3}{R_2 + R_3}(v_{o1} + bV_{om}) < 0 = v_{N2} \rightarrow v_{P2} - v_{N2} < 0$$

so that the output voltage v_{o2} of U2 changes from $+V_{om}$ to $-V_{om}$:

$$v_{o2} = -V_{om}$$

This constant negative voltage is fed back into the input (v_{i1}) of the inverting integrator to make its output (v_{o1}) increase linearly till $v_{o1} > bV_{om}$. When $v_{o1} > bV_{om}$, the voltage v_{P2} goes above $v_{N2} = 0$ as

$$v_{P2} = v_{o1} - R_2\frac{v_{o1} - (-V_{om})}{R_2 + R_3} = \frac{R_3}{R_2 + R_3}(v_{o1} - bV_{om}) > 0 = v_{N2} \rightarrow v_{P2} - v_{N2} > 0$$

so that the output voltage v_{o2} of U2 changes from $-V_{om}$ back to $+V_{om}$:

$$v_{o2} = +V_{om}$$

This process repeats itself periodically, generating a triangular wave at the output (v_{o1}) of U1 and a rectangular wave at the output (v_{o2}) of U2 as depicted in Figure 5.12(b).

Now, to find the period of the triangular/rectangular waves, let us start from observing that the output voltage v_{o1} of the inverting integrator is described by Eq. (5.3.1) as

$$v_{o1}(t) \overset{(5.3.1)}{=} v_{o1}(t_0) - \frac{1}{R_1C}\int_{t_0}^{t} v_{o2}(\tau)d\tau \tag{5.3.11}$$

Using this equation, the times taken for v_{o1} to rise/fall from V_L/V_H to V_H/V_L with $v_{o2} = -V_{om}/+V_{om}$ can be obtained as

$$V_H = V_L - \frac{1}{R_1C}(-V_{om})T_R; \quad T_R = \frac{R_1C(V_H - V_L)}{V_{om}} \tag{5.3.12a}$$

$$V_L = V_H - \frac{1}{R_1C}(+V_{om})T_F; \quad T_F = \frac{R_1C(V_H - V_L)}{V_{om}} \tag{5.3.12b}$$

Consequently, the period of the triangular/rectangular waves turns out to be

$$P = T_{\rm R} + T_{\rm F} \overset{(5.3.12a,b)}{=} 2R_1C\frac{V_{\rm H} - V_{\rm L}}{V_{\rm om}} \overset{(5.3.10)}{=} 4bR_1C = 4\frac{R_2}{R_3}R_1C \tag{5.3.13}$$

(Note) Figure 5.12(a) is the PSpice schematic of a triangular/rectangular wave generator consisting of an inverting integrator (realized by an *RC* OP Amp circuit with negative feedback) and a non-inverting trigger (realized by a positive feedback OP Amp), where $R_1 = R_2 = 1\ \text{k}\Omega$, $R_3 = 2\ \text{k}\Omega$, and $C = 1\ \mu\text{F}$. This has been run with the Analysis type of 'Time Response (Transient)' to yield the waveforms depicted in Figure 5.12(b) where the square box before SKIPBP (Skip the initial transient bias point calculation) is checked in the Simulation Settings dialog box. To your eyes, is the period of the voltage waveforms close to what you get from Eq. (5.3.13)?

5.3.3 555 Timer Using OP Amp as Comparator

Figure 5.13(a1)/(a2) shows the two circuits which realize astable/monostable circuits, respectively, by using a 555 timer. Note the following about the 555 timer circuit (enclosed by blue-line rectangles):

- Each of the two OP Amps having no feedback functions as a comparator, which exhibits a high/low output for the voltage at its + input terminal higher/lower than that at its − input terminal, respectively.
- The three resistors of equal resistance *R* makes a three-level voltage divider. Thus, the voltage at the − input terminal of the comparator U1 is $2V_{\rm CC}/3$ and the voltage at the + input terminal of the comparator U2 is $V_{\rm CC}/3$ as long as the supply voltage $V_{\rm CC}$ is applied to the terminal +Vcc.
- The FF is set/reset to make its output high (Q = high(logic-1), $\bar{\rm Q}$ = low(logic-0)), or low (Q = low, $\bar{\rm Q}$ = high) by raising the S or R input to high, while its output remains unchanged if the two R and S inputs are both low. It can also be reset by lowering the RESET input (active-low).
- High/low $\bar{\rm Q}$ applies to the base terminal of the NPN-BJT Q_1 to turn it on/off.
- A high output of U1 resets the FF to make Q = low and $\bar{\rm Q}$ = high, while a high output of U2 sets the FF to make Q = high and $\bar{\rm Q}$ = low.

Figure 5.13(a1) shows an application of 555 timer for an *astable* (free-running) circuit, whose output voltage is not stabilized, but repeats periodic transitions between two opposite (high/low) states, appearing as a rectangular-wave generator. To find the period of the rectangular wave, suppose the supply voltage $V_{\rm CC}$ is applied to the terminal +Vcc when the capacitor is uncharged, i.e. $v_C(0) = 0$. Then, U2 with $0 = v_C = v_- < v_+ = V_{\rm CC}/3$ sets the FF to make the output high (Q = high, $\bar{\rm Q}$ = low) and the low $\bar{\rm Q}$ turns off the transistor Q_1 so that the capacitor will be charged (towards $V_{\rm CC}$) through $R_{\rm A}$-$R_{\rm B}$ from $V_{\rm CC}$. This rising (charging) mode

(a1)

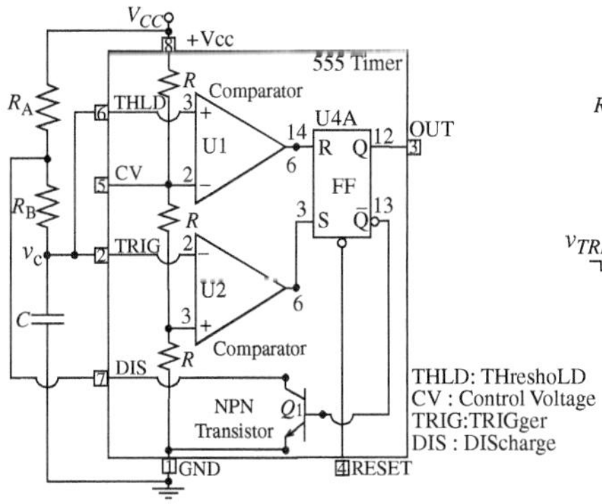

A 555 timer circuit – a stable multi-vibrator

(a2)

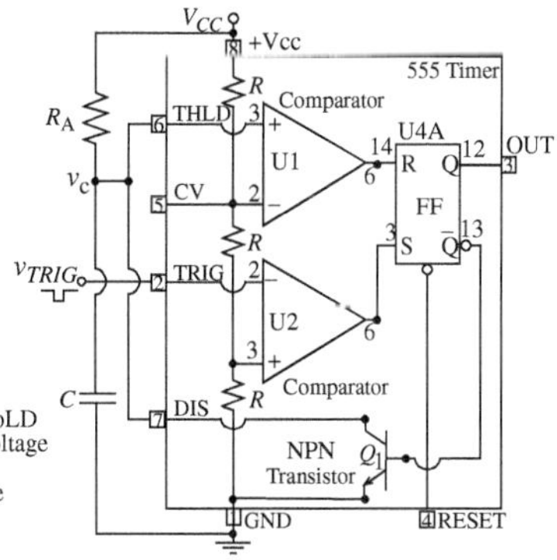

A 555 timer circuit – monostable multi-vibrator

(b1)

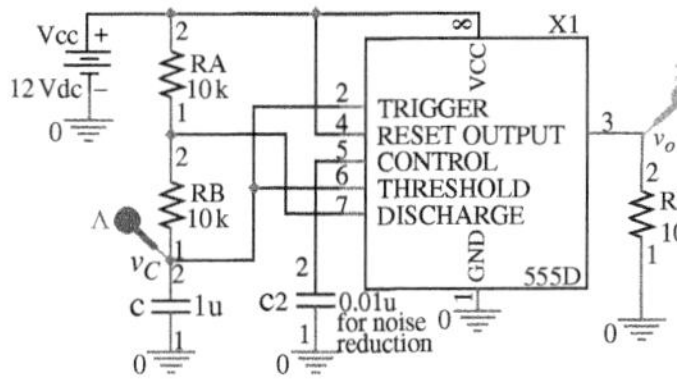

PSpice schematic for (a1) ("timer555.opj")

(b2)

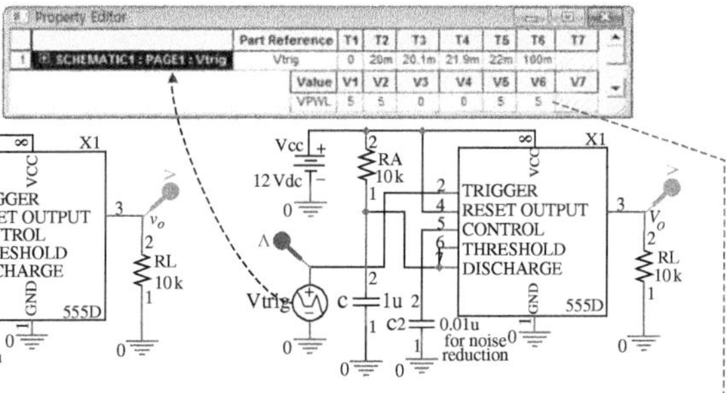

PSpice schematic for (a2) ("timer555-monostable.opj")

(c)

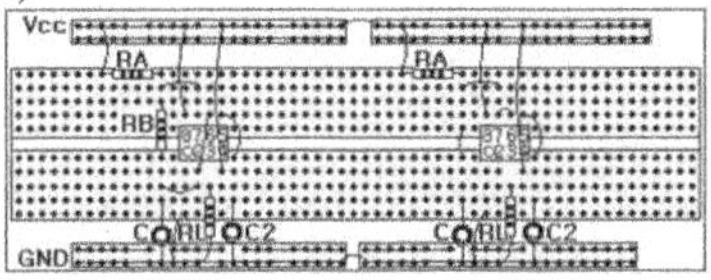

Hardware connection for (a1) and (a2) on the breadboard

(d1)

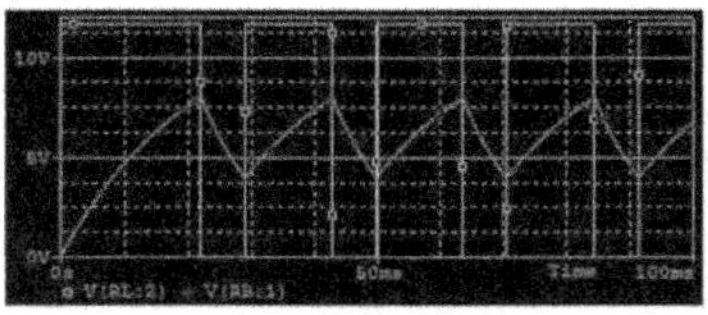

$v_c(t)$ and $v_o(t)$ obtained from PSpice simulation

(d2)

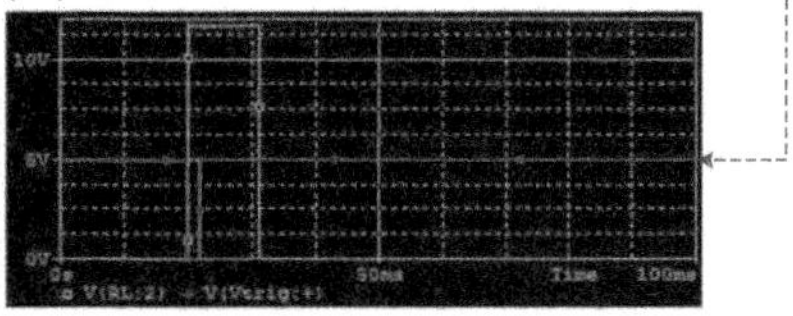

$v_{TRIG}(t)$ and $v_c(t)$ obtained from PSpice simulation

Figure 5.13 Applications of 555 timer.

will continue till v_C goes above $2V_{CC}/3$ (the voltage at the – input terminal of U1) to trigger U1 so that U1's high output resets the FF to make the output low (Q = low, $\bar{Q}$ = high). Then, the high $\bar{Q}$ turns on the transistor Q_1 so that the capacitor will be discharged (towards 0 V) through R_B-Q_1 to the ground (GND). This falling (discharging) mode will continue till v_C goes below $V_{CC}/3$ (the voltage at the + input terminal of U2) to trigger U2 so that U2's high output sets the FF to make the output high again and another rising mode starts. Accordingly, the capacitor voltage v_C will repeat the cycle of going up to $2V_{CC}/3$ and down to $V_{CC}/3$. Noting that the charging/discharging time constants are $(R_A + R_B)C$ and R_BC, respectively, we use Eq. (5.3.6) (Eq. (3.39) of [Y-1]) to get the period as

$$\text{Rising period}: \left(\frac{V_{CC}}{3} - V_{CC}\right)e^{-T_R/(R_A+R_B)C} + V_{CC} = \frac{2V_{CC}}{3}$$

$$; T_R = (R_A + R_B)C\ \ln 2[\mathrm{s}] \quad (5.3.14a)$$

$$\text{Falling period}: \frac{2V_{CC}}{3}e^{-T_F/R_BC} = \frac{V_{CC}}{3};\ T_F = R_B C \ln 2 \quad (5.3.14b)$$

$$\text{Whole oscillation period}: P = T_R + T_F = (R_A + 2R_B)C \ln 2 \quad (5.3.15)$$

Then, the duty cycle of the rectangular wave output turns out to be

$$\text{Duty cycle} = \frac{\text{high(on) time}}{\text{period}} = \frac{T_R}{T_R + T_F} = \frac{R_A + R_B}{R_A + 2R_B} \quad (5.3.16)$$

Figure 5.13(b1) and (d1) shows the PSpice schematic and the simulation result for this circuit, respectively.

Figure 5.13(a2) shows another application of 555 timer for a *mono-stable* (one-shot) circuit, whose output is normally stabilized at the low level, but generates a single positive rectangular pulse of a fixed duration when a short negative (triggering) pulse is applied to the TRIG terminal. To find the duration of the rectangular pulse, suppose the TRIG input is normally high and the FF is reset to make the output low (Q = low, $\bar{Q}$ = high). Then, the transistor Q_1 is turned on and the capacitor C is discharged through Q_1 to the ground (GND) so that $v_C(0) = 0$. This state is stably maintained till U2 is triggered by a low signal at its – input terminal. Let a negative triggering pulse be applied to the terminal TRIG. Then, U2 sets the FF to make the output high (Q = high, $\bar{Q}$ = low) and the low $\bar{Q}$ turns off Q_1 so that C will be charged (towards V_{CC}) through R_A from V_{CC}. This rising (charging) mode will continue till v_C goes above $2V_{CC}/3$ (the voltage of the – input terminal of U1) to trigger U1 so that U1's high output resets the FF to make the output low (Q = low, $\bar{Q}$ = high) again and subsequently, the high $\bar{Q}$ turning on Q_1. Then C will be instantly discharged and the low output level is maintained till another negative triggering pulse is applied to the TRIG terminal. Therefore, the output pulse duration is equal to the time taken for v_C to rise from 0 V to $2V_{CC}/3$, which is obtained by using the formula Eq. (5.3.6) (Eq. (3.39) of [Y-1]) as follows:

$$(0 - V_{CC})e^{-T_R/R_A C} + V_{CC} = \frac{2\,V_{CC}}{3};\quad T_R = R_A C \ln 3\,[\mathrm{s}] \tag{5.3.17}$$

Figure 5.13(b2) and (d2) shows the PSpice schematic and the simulation results, respectively.

5.4 Second-Order OP Amp Circuits [Y-1]

This section deals with two common circuit topologies, MFB (Multi-FeedBack) and Sallen-Key (or VCVS – voltage-controlled voltage source) architectures, that are often used to realize various second-order transfer functions.

5.4.1 MFB (Multi-FeedBack) Topology

Figure 5.14(a) shows the MFB topology. Its transfer function can be found by applying the node analysis. Noting that due to the negative feedback (via Z_5), the virtual short principle (Remark 5.1(2)) gives $V_2(s) \approx 0$, we can write the KCL equations at nodes 1 and 2:

$$\text{Node 1:}\ \frac{V_1(s) - V_i(s)}{Z_1} + \frac{V_1(s)}{Z_2} + \frac{V_1(s) - V_o(s)}{Z_3} + \frac{V_1(s)}{Z_4} = 0 \tag{5.4.1a}$$

$$\text{Node 2:}\ \frac{0 - V_1(s)}{Z_4} + \frac{0 - V_o(s)}{Z_5} = 0 \tag{5.4.1b}$$

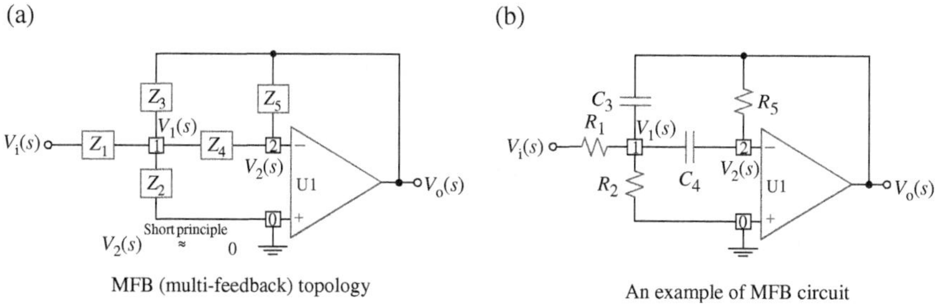

Figure 5.14 MFB (Multi-FeedBack) topology and an example circuit.

and arrange them in matrix-vector form as

$$\begin{bmatrix} Y_1+Y_2+Y_3+Y_4 & -Y_3 \\ Y_4 & Y_5 \end{bmatrix}\begin{bmatrix} V_1(s) \\ V_o(s) \end{bmatrix} = \begin{bmatrix} Y_1 V_i(s) \\ 0 \end{bmatrix} \text{ with } Y_n = \frac{1}{Z_n} \tag{5.4.2}$$

This node equation can be solved to yield the transfer function as

$$G(s) = \frac{V_o(s)}{V_i(s)} = -\frac{Y_1 Y_4}{(Y_1+Y_2+Y_3+Y_4)Y_5 + Y_3 Y_4} \tag{5.4.3}$$

The transfer function of the MFB circuit in Figure 5.14(b) can be written by using this formula with Y_1=1/R_1, Y_2=1/R_2, Y_3=sC_3, Y_4=sC_4, and Y_5=1/R_5 as

$$G(s) = \frac{V_o(s)}{V_i(s)} = \frac{-sR_2R_5C_4}{s^2R_1R_2R_5\,C_3C_4 + sR_1R_2\,(C_3+C_4) + R_1 + R_2} \tag{5.4.4}$$

This transfer function can also be obtained by running the following MATLAB script "elec05f14b.m," which uses the function 'MFB_xfer_ftn()' to solve the node equation (5.4.2).

```
>>elec05f14b

                    C4 G1 s
  - - - - - - - - - - - - - - - - - - - - - - - - -
                                                  2
  G1 G5 + G2 G5 + C3 G5 s + C4 G5 s + C3 C4 s
```

```
function Gs=MFB_xfer_ftn(Y1,Y2,Y3,Y4,Y5)
Y=[Y1+Y2+Y3+Y4 -Y3; Y4 Y5]; % Eq.(5.4.2)
V=Y\[Y1; 0];
Gs=V(2); % Eq.(5.4.3)
```

```
%elec05f14b.m
clear, clf
syms s G1 G2 G5 C3 C4
Gs=MFB_xfer_ftn(G1,G2,s*C3,s*C4,G5);
pretty(simplify(Gs)) % supposed to yield the transfer function (5.4.4)
```

5.4.2 Sallen-Key Topology

Figure 5.15(a) shows the Sallen-Key (or VCVS) topology. Its transfer function can be found by applying the node analysis. Noting that

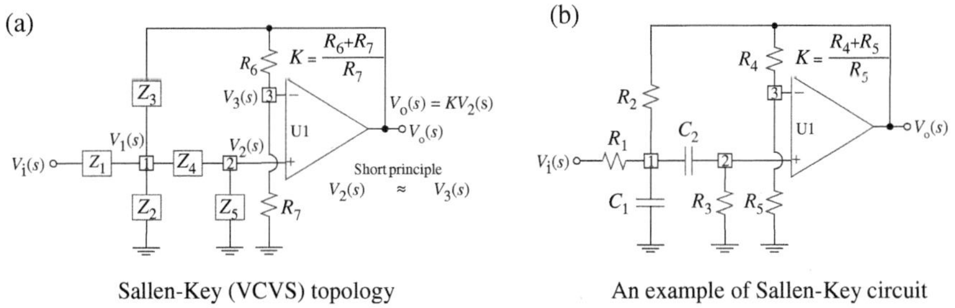

Figure 5.15 Sallen-Key (or VCVS) topology and an example circuit.

- due to the negative feedback (via R_6), the virtual short principle (Remark 5.1 (2)) gives $V_2(s) \approx V_3(s)$, and
- the voltage divider rule applied to the series resistors R_6-R_7 or the KCL applied at node 3 (with no current from/into the input terminals of the OP Amp by the virtual open principle (Remark 5.1(1)) gives

$$V_2(s) \overset{\text{virtual short}}{\approx} V_3(s) \overset{\text{voltage divider}}{=} \frac{R_7}{R_6+R_7} V_o(s)$$
$$; V_o(s) = KV_2(s) \text{ with } K = \frac{R_6+R_7}{R_7} \quad (5.4.5)$$

we can write the KCL equations at nodes 1 and 2:

$$\text{Node 1}: \frac{V_1(s)-V_i(s)}{Z_1} + \frac{V_1(s)}{Z_2} + \frac{V_1(s)-KV_2(s)}{Z_3} + \frac{V_1(s)-V_2(s)}{Z_4} = 0 \quad (5.4.6a)$$

$$\text{Node 2}: \frac{V_2(s)-V_1(s)}{Z_4} + \frac{V_2(s)}{Z_5} = 0 \quad (5.4.6b)$$

and arrange them in matrix-vector form as

$$\begin{bmatrix} Y_1+Y_2+Y_3+Y_4 & -KY_3-Y_4 \\ -Y_4 & Y_4+Y_5 \end{bmatrix} \begin{bmatrix} V_1(s) \\ V_2(s) \end{bmatrix} = \begin{bmatrix} Y_1 V_i(s) \\ 0 \end{bmatrix} \text{ with } Y_n = \frac{1}{Z_n} \quad (5.4.7)$$

This node equation can be solved to yield the transfer function as

$$G(s) = \frac{V_o(s)}{V_i(s)} = \frac{Y_1 Y_4}{(Y_1+Y_2+Y_3)(Y_4+Y_5)+Y_4 Y_5 - KY_3 Y_4} \quad (5.4.8)$$

The transfer function of the Sallen-Key circuit in Figure 5.15(b) can be written by using this formula with $Y_1 = 1/R_1$, $Y_2 = sC_1$, $Y_3 = 1/R_2$, $Y_4 = sC_2$, $Y_5 = 1/R_3$, and $K = (R_4+R_5)/R_5$ as

$$G(s)=\frac{V_o(s)}{V_i(s)}=\frac{KR_2R_3C_2s}{s^2R_1R_2R_3C_1C_2+s((R_1R_2+(1-K)R_1R_3+R_2R_3)C_2+R_1R_2C_1)+R_1+R_2} \tag{5.4.9}$$

This transfer function can also be obtained by running the following MATLAB script "elec05f15b.m," which uses the function '`SallenKey_xfer_ftn()`' to solve the node equation (5.4.7).

```
function Gs=SallenKey_xfer_ftn(Y1,Y2,Y3,Y4,Y5,R6,R7)
K=(R6+R7)/R7;
V=[Y1+Y2+Y3+Y4 -K*Y3-Y4; -Y4 Y4+Y5]\[Y1; 0]; % Eq.(5.4.7)
Gs=K*V(2);
```

```
%elec05f15b.m
syms s G1 G2 G3 R4 R5 C1 C2 K
Gs=SallenKey_xfer_ftn(G1,s*C1,G2,s*C2,G3,(K-1)*R5,R5);
pretty(simplify(Gs))
```

Example 5.2 MATLAB Analysis and PSpice Simulation of a Sallen-Key Circuit

Consider the Sallen-Key OP Amp circuit of Figure 5.15(b) whose input value and parameter values are given as follows:

$$V_i=1\,\text{V}\rightarrow V_i(s)=\frac{1}{s},\ R_1=\frac{1}{5}\text{k}\Omega, R_2=\frac{1}{5}\text{k}\Omega, R_3=\frac{1}{5}\text{k}\Omega, \quad R_4=2.6\,\text{k}\Omega, R_5=1\,\text{k}\Omega, C_1=C_2=1\,\text{mF} \tag{E5.2.1}$$

Substituting these parameter values into Eq. (5.4.9) and taking the inverse Laplace transform yield

$$V_o(s)=G(s)V_i(s)$$

$$\overset{(5.4.9)}{=}\frac{K\,R_2R_3C_2s}{s^2R_1R_2R_3C_1C_2+s((R_1R_2+(1-K)R_1R_3+R_2R_3)C_2+R_1R_2C_1)+R_1+R_2}V_i(s)$$

$$;V_o(s)\underset{K=(R_4+R_5)/R_5}{\overset{(E5.2.1)}{=}}\frac{8s/125}{s^2\times(1/125)+s(4-18/5)/25+2}\times\frac{1}{s}=\frac{18}{s^2+2s+50}$$

$$=\frac{(18/7)\times\omega_d}{(s+1)^2+7^2=(s+\sigma)^2+\omega_d^2} \tag{E5.2.2}$$

$$;v_o(t)=\mathcal{L}^{-1}\{V_o(s)\}\overset{\text{Table A.1(10)}}{=}\frac{18}{7}e^{-t}\sin 7t\,[\text{V}] \tag{E5.2.3}$$

This analytical result indicates that the output voltage of the circuit to a step input $v_i(t)=u_s(t)$ has an oscillation of damped frequency $\omega_d=7$ [rad/s], period

$2\pi/\omega_d = 0.8976$ [s], and the amplitude decreasing exponentially with the time constant of $1/\sigma = 1$ [s].

We run the following MATLAB script "elec05e02a.m" to get the same result together with the plot of $v_o(t)$ (time/step response) depicted in Figure 5.16.1(b).

```
%elec05e02a.m
clear, clf
syms s R1 R2 R3 R4 R5 C1 C2 K Vis
% Transfer function in symbolic expression form
Gs=SallenKey_xfer_ftn(1/R1,s*C1,1/R2,s*C2,1/R3,R4,R5); %Eq.(5.4.9)
disp('G(s)='), pretty(simplify(Gs));
% To substitute the numeric values for the parameters of transfer ftn
% R1=200; R2=200; R3=200; R4=2.6k; R5=1k; C1=1m; C2=1m;
Gs=subs(Gs,{R1,R2,R3,R4,R5,C1,C2},{200,200,200,2600,1e3,1e-3,1e-3})
Vis=1/s; Vos=Gs*Vis % 18/(s^2+2*s+50) : Eq.(E5.2.2)
vo=ilaplace(Vos) % 18/7*exp(-t).*sin(7*t)): Eq.(E5.2.3)
t0=0; tf=3; N=600; dt=(tf-t0)/N; tt=[t0:dt:tf]; % Time vector for [0,3]
% Numeric values of vo(t) for a given time range
for n=1:length(tt), t=tt(n); vot(n)= eval(vo); end
subplot(221), plot(tt,real(vot))
```

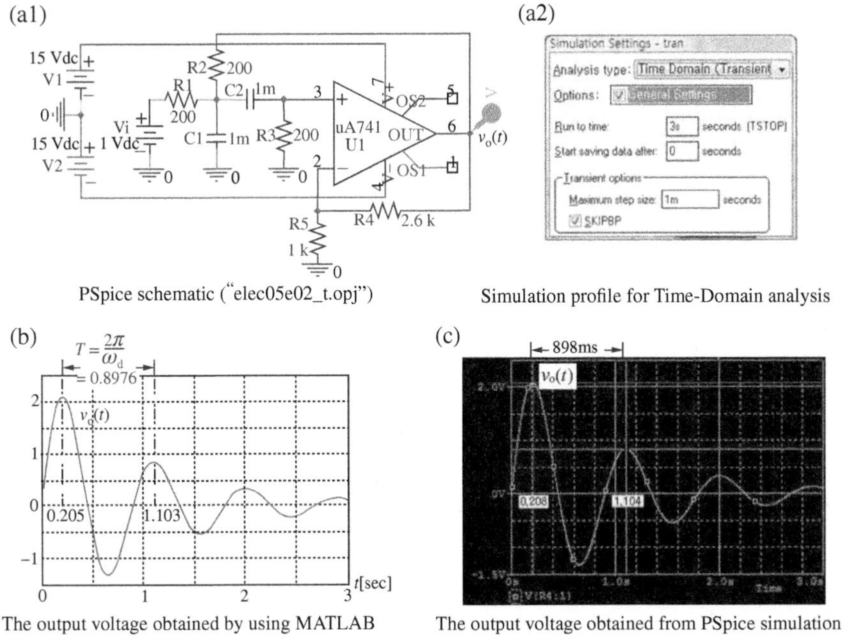

Figure 5.16.1 Simulation of the Sallen-Key circuit in Figure 5.15b for time/step response (Example 5.2).

```
>>elec05e02a
 G(s)=
                          2
   (R4+R5)sC2R3R2/(R5s C1R1R2R3C2
   +(-R3C2R1R4 + R5C1R1R2 + R5R2R3C2 + R5C2R1R2)s + R5R1 + R5R2)
Vos = 144000/(16000*s+400000+8000*s^2) % Eq.(E5.2.2)
vo = 18/7*exp(-t)*sin(7*t) % Eq.(E5.2.3)
```

Figure 5.16.1(a1), (a2), and (c) shows the PSpice schematic, simulation Profile, and the output voltage waveform $v_o(t)$ obtained from the PSpice simulation (with Time Domain analysis type), respectively.

Now, let us find the frequency response of the Sallen-Key OP Amp circuit of Figure 5.15(b), which can be obtained by substituting $s = j\omega$ into the transfer function (5.4.9):

$$G(s) \overset{(5.4.9)}{=} \frac{Ks/R_1C_1}{s^2 + s\dfrac{(R_1R_2 + (1-K)R_1R_3 + R_2R_3)C_2 + R_1R_2C_1}{R_1R_2R_3C_1C_2} + \dfrac{R_1 + R_2}{R_1R_2R_3C_1C_2}}$$

$$= \frac{Ks/R_1C_1}{s^2 + s\omega_b + \omega_p^2} \tag{E5.2.4}$$

$$; G(j\omega) = \frac{j\omega K/R_1C_1}{j\omega\omega_b + \left(\omega_p^2 - \omega^2\right)} \tag{E5.2.5a}$$

$$\text{with } \omega_b = \frac{(R_1R_2 + (1-K)R_1R_3 + R_2R_3)C_2 + R_1R_2C_1}{R_1R_2R_3C_1C_2} \quad \text{and} \quad \omega_p = \sqrt{\frac{R_1 + R_2}{R_1R_2R_3C_1C_2}} \tag{E5.2.5b}$$

```
%elec05e02b.m
% Frequency Response G(jw)
w=logspace(0,2,600); jw=j*w; % Frequency range to plot G(jw) on
Gw=subs(Gs,s,jw); Gw_mag=20*log10(abs(Gw)); % Frequency Response G(jw) [dB]
R1=200; R2=200; R3=200; R4=2600; R5=1000; C1=0.001; C2=0.001; K=(R4+R5)/R5;
wp2=(R1+R2)/(R1*R2*R3*C1*C2); fp=sqrt(wp2)/2/pi; %Eq.(E5.2.5b)
wb=((R1*R2+(1-K)*R1*R3+R2*R3)*C2+R1*R2*C1)/(R1*R2*R3*C1*C2); %Eq.(E5.2.5b)
fdB_u=(wb+sqrt(wb^2+4*wp2))/2/2/pi; % Upper 3dB frequency (Eq.(E5.2.6a))
fdB_l=(-wb+sqrt(wb^2+4*wp2))/2/2/pi; % Lower 3dB frequency (Eq.(E5.2.6b))
subplot(222), semilogx(w,Gw_mag), hold on
plot([wp wp],[0 max(Gw_mag)],'r:', [wdB_l wdB_l],[0 max(Gw_mag)],
                        'k:', [wdB_u wdB_u],[0 max(Gw_mag)],'k:')
```

Note that the system whose transfer function (E5.2.4) has a one-degree term in s as its numerator will be a BPF (Band-Pass Filter) with peak frequency ω_p, bandwidth ω_d, and upper/lower 3 dB frequencies (see Section 6.2.3.1 of [Y-1]):

$$\omega_u, \omega_l \overset{(6.2.11)}{=} \frac{\left(\pm\omega_b + \sqrt{\omega_b^2 + 4\,\omega_p^2}\right)}{2} \tag{E5.2.6}$$

Just after "elec05e02a.m" has been run, the above MATLAB script "elec05e02b.m" can be run to yield the plot of $20\log_{10}|G(\omega)|$ [dB] (frequency response magnitude) depicted in Figure 5.16.2(b). Figure 5.16.2(a1), (a2), and (c) shows the

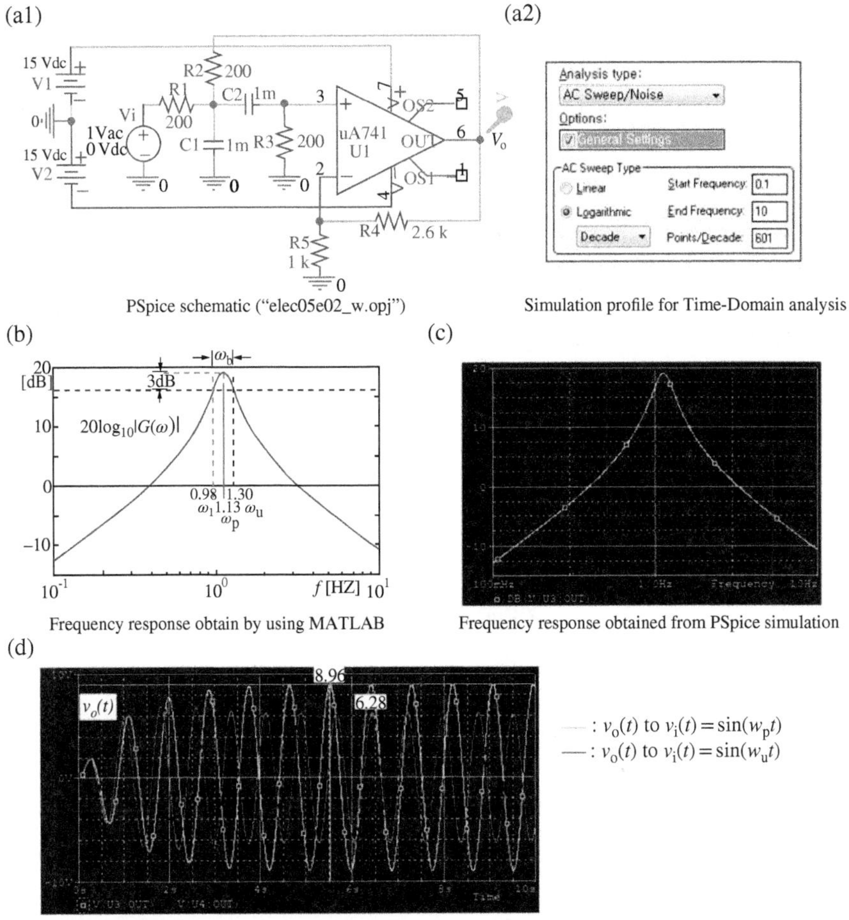

PSpice schematic ("elec05e02_w.opj")

Simulation profile for Time-Domain analysis

Frequency response obtain by using MATLAB

Frequency response obtained from PSpice simulation

Output voltage waveforms to two sinusoidal inputs $v_i(t) = \sin(w_ut)$ and $v_i(t) = \sin(w_pt)$

Figure 5.16.2 Simulation of the Sallen-Key circuit for frequency response in Figure 5.15(b) (Example 5.2).

PSpice schematic (with AC voltage source VAC), Simulation Profile (with AC Sweep analysis type), and the frequency response magnitude [dB] obtained from the PSpice simulation, respectively. Figure 5.16.2(d) shows the outputs $\{v_{o1}(t), v_{o2}(t)\}$ to two sinusoidal inputs $\{v_{i1}(t) = \sin(\omega_p t), v_{i2}(t) = \sin(\omega_u t)\}$ where the amplitude of $v_{o2}(t)$ is $1/\sqrt{2}$ times (3 dB lower than) that of $v_{o1}(t)$.

5.5 Active Filter [Y-1]

5.5.1 First-Order Active Filter

We consider the circuit of Figure 5.17(a) in which the OP Amp has a negative feedback path between the output terminal and the negative input terminal N. By the virtual short principle (Remark 5.1(2)) the voltage at node N is (almost) zero, being almost equal to that at node P (the positive input terminal), which is grounded. Thus, the current flowing from the input $V_i(s)$ to node N (whose voltage is zero) is $I_i(s) = V_i(s)/R_1$ and this current flows through $Z_f(s) = R_2||(1/sC)$ (the parallel combination of R_2 and C) towards the output terminal since no current flows into or out of the negative input terminal of the OP Amp by the virtual open principle (Remark 5.1(1)). Consequently, the output voltage is obtained by subtracting the voltage drop across Z_f from zero (the voltage at node N) as

$$V_o(s) = 0 - Z_f(s)I_i(s) = -Z_f(s)\frac{V_i(s)}{R_1}$$

and the transfer function of the OP Amp circuit is found as

$$G(s) = \frac{V_o(s)}{V_i(s)} = -\frac{Z_f(s)}{R_1} = -\frac{R_2||(1/sC)}{R_1} = -\frac{R_2/sC}{R_1(R_2 + 1/sC)} = -\frac{R_2}{R_1}\frac{1/R_2C}{s + 1/R_2C}$$

$$; G(s) = -K\frac{\omega_c}{s + \omega_c} \text{ with } K = \frac{R_2}{R_1} \text{ and } \omega_c = \frac{1}{R_2C} \tag{5.5.1}$$

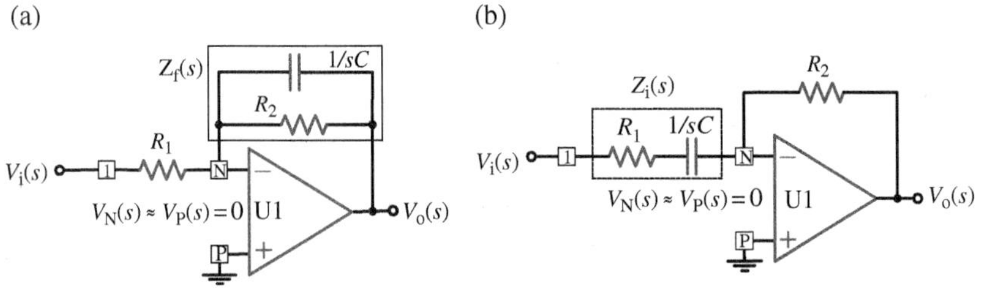

A first-order active low-pass filter (LPF)

A first-order active high-pass filter (HPF)

Figure 5.17 First-order active filters.

This indicates that the OP Amp circuit of Figure 5.17(a) works as a low-pass filter (LPF) with the cutoff frequency of $\omega_c = 1/R_2C$ and the DC gain of $G(s)|_{s=0} = -K = -R_2/R_1$.

The circuit of Figure 5.17(b) is the same as that of Figure 5.17(a) except that R_1 and Z_f are replaced by Z_i and R_2, respectively. Therefore, it has the transfer function as

$$G(s) = \frac{V_o(s)}{V_i(s)} = -\frac{R_2}{Z_i(s)} = -\frac{R_2}{R_1 + 1/sC} = -\frac{R_2}{R_1}\frac{s}{s + 1/R_1C}$$

$$; G(s) = -K\frac{s}{s+\omega_c} \text{ with } K = \frac{R_2}{R_1} \text{ and } \omega_c = \frac{1}{R_1C} \tag{5.5.2}$$

This indicates that the OP Amp circuit in Figure 5.17(b) works as a high-pass filter (HPF) with the cutoff frequency of $\omega_c = 1/R_1C$ and the DC gain of $G(s)|_{s=0} = 0$.

5.5.2 Second-Order Active LPF/HPF

The transfer function of the Sallen-Key circuit in Figure 5.18(a) can be found by using the MATLAB function '`SallenKey_xfer_ftn()`' (introduced in Section 5.4.2) as

```
>>syms s G1 G2 R3 R4 C1 C2 K
>>Gs=SallenKey_xfer_ftn(G1,0,s*C1,G2,s*C2,(K-1)*R3,R3);
  pretty(simplify(Gs))
                          G1 G2 K
  ---------------------------------------------------------
                                                 2
 G1 G2 + C1 G2 s + C2 G1 s + C2 G2 s + C1 C2 s  - C1 G2 K s
```

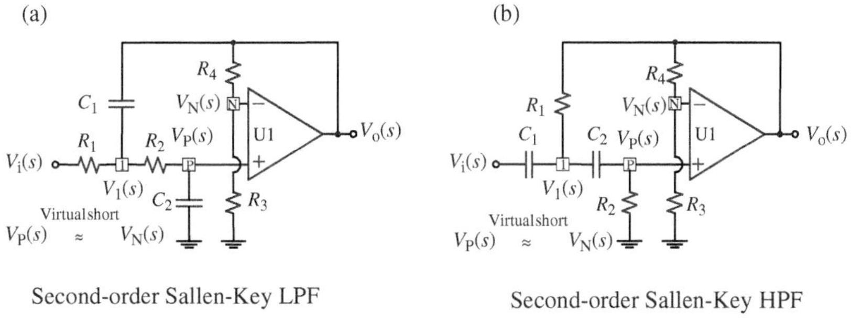

Figure 5.18 Second-order active (Sallen-Key) filters.

This means that the transfer function $G(s)$ is

$$G(s)=\frac{V_o(s)}{V_i(s)}=\frac{KG_1G_2/C_1C_2}{s^2+((G_1+G_2)/C_1+(1-K)G_2/C_2)s+G_1G_2/C_1C_2} \tag{5.5.3}$$

$$\text{with } K=\frac{R_3+R_4}{R_3}$$

This transfer function having only a constant term in the numerator indicates that the circuit will function as a low-pass filter (LPF).

The transfer function of the Sallen-Key circuit in Figure 5.18(b) can also be found by using the MATLAB function '`SallenKey_xfer_ftn()`' as

```
>>syms s G1 G2 R3 R4 C1 C2 K
>>Gs=SallenKey_xfer_ftn(s*C1,0,G1,s*C2,G2,(K-1)*R3,R3);
  pretty(simplify(Gs))
                              2
                     C1 C2 K s
 -----------------------------------------------------------------
                                                  2
 G1 G2 + C1 G2 s + C2 G1 s + C2 G2 s + C1 C2 s  - C2 G1 K s
```

This means that the transfer function $G(s)$ is

$$G(s)=\frac{V_o(s)}{V_i(s)}=\frac{Ks^2}{s^2+(G_2(1/C_1+1/C_2)+(1-K)G_1/C_1)s+G_1G_2/C_1C_2} \tag{5.5.4}$$

$$\text{with } K=\frac{R_3+R_4}{R_3}$$

This transfer function having only a second-degree term in the numerator indicates that the circuit will function as a high-pass filter (HPF).

The transfer function of the MFB (multiple/dual feedback) circuit in Figure 5.19(a) can be found by using the MATLAB function '`MFB_xfer_ftn()`' as

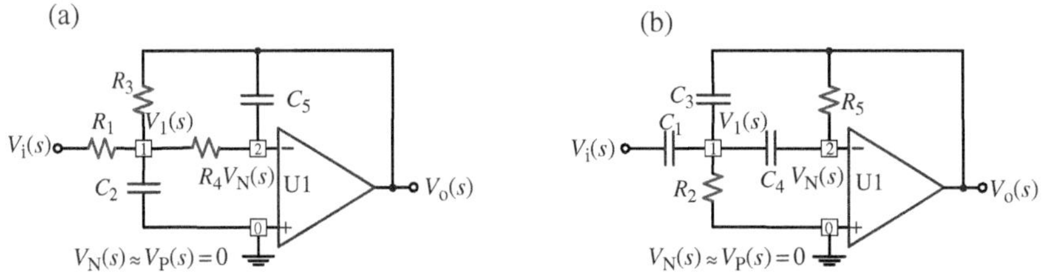

A second-order MFB (multi-feedback) LPF

A second-order MFB (multi-feedback) HPF

Figure 5.19 Second-order active (MFB) filters.

```
>>syms s G1 C2 G3 G4 C5
>>Gs=MFB_xfer_ftn(G1,s*C2,G3,G4,s*C5); pretty(simplify(Gs))

                        G1 G4
  - - - - - - - - - - - - - - - - - - - - -
                                          2
  C5 (G1 s + G3 s + G4 s + C2 s   ) + G3 G4
```

This means that the transfer function $G(s)$ is

$$G(s) = \frac{V_o(s)}{V_i(s)} = \frac{-G_1 G_4}{C_2 C_5 s^2 + C_5(G_1 + G_3 + G_4)s + G_3 G_4} \tag{5.5.5}$$

This transfer function having only a constant term in the numerator indicates that the circuit will function as a low-pass filter (LPF).

The transfer function of the MFB circuit in Figure 5.19(b) can also be found by using the MATLAB function '`MFB_xfer_ftn()`' as

```
>>syms s C1 G2 C3 C4 G5
>>Gs=MFB_xfer_ftn(s*C1,G2,s*C3,s*C4,G5); pretty(simplify(Gs))
                        2
                       s  C4 C1
         - - - - - - - - - - - - - - - - - - - - - - - - -
          2
         s  C3 C4 + (G5 C1 + G5 C3 + G5 C4) s + G5 G2
```

This means that the transfer function $G(s)$ is

$$G(s) = \frac{V_o(s)}{V_i(s)} = \frac{-C_1 C_4 s^2}{C_3 C_4 s^2 + G_5(C_1 + C_3 + C_4)s + G_2 G_5} \tag{5.5.6}$$

This transfer function having only a second-degree term in the numerator indicates that the circuit will function as a high-pass filter (HPF).

5.5.3 Second-Order Active BPF

Noting that the circuit in Figure 5.20(a) is the same as that in Figure 5.19(b) except that C_1 is replaced by R_1, we can replace sC_1 by $G_1=1/R_1$ in Eq. (5.5.6) to write its transfer function as

$$G(s) = \frac{V_o(s)}{V_i(s)} \underset{sC_1 \to G_1}{\overset{(5.5.6)}{=}} \frac{-G_1 C_4 s}{C_3 C_4 s^2 + G_5(C_3 + C_4)s + (G_1 + G_2)G_5}$$

$$= \frac{-(G_1/C_3)s}{s^2 + (G_5(C_3 + C_4)/C_3 C_4)s + (G_1 + G_2)G_5/C_3 C_4} \tag{5.5.7}$$

This transfer function having only a first-degree term in the numerator indicates that the circuit will function as a band-pass filter (BPF).

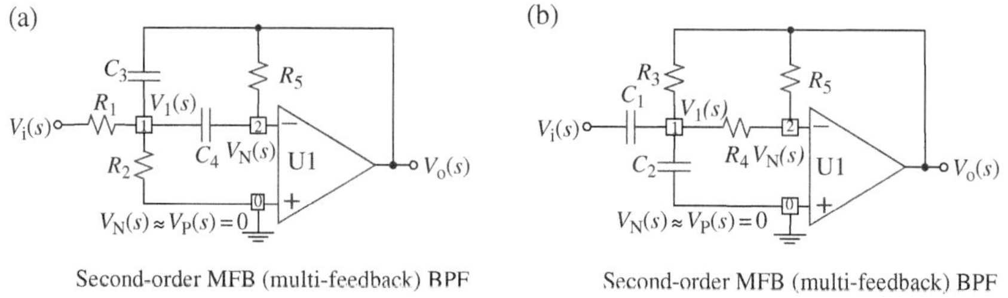

Figure 5.20 Second-order active (MFB) filters.

Now, noting that the circuit in Figure 5.20(b) is the same as that in Figure 5.20 (a) except for that R and C are exchanged, we can switch sC_i with $G_i = 1/R_i$ in Eq. (5.5.7) to write its transfer function as

$$G(s) = \frac{V_o(s)}{V_i(s)} = \frac{-C_1 G_4 s}{(C_1 + C_2)C_5 s^2 + C_5(G_3 + G_4)s + G_3 G_4}$$

$$= \frac{-(C_1 G_4/(C_1 + C_2)C_5)s}{s^2 + ((G_3 + G_4)/(C_1 + C_2))s + G_3 G_4/(C_1 + C_2)C_5} \tag{5.5.8}$$

This transfer function having only a first-degree term in the numerator indicates that the circuit will also function as a band-pass filter (BPF).

The transfer function of the Sallen-Key circuit in Figure 5.21 can be found by using the MATLAB function '`SallenKey_xfer_ftn()`' as

```
>>syms s G1 G2 G3 C1 C2 R4 K
>>Gs=SallenKey_xfer_ftn(G1,0,G2,s*C1,s*C2+G3,(K-1)*R4,R4);
  pretty(Gs)

    C1 G1 K s/ (G1 G3 + G2 G3 + C1 G1 s + C1 G2 s + C2 G1 s + C1 G3 s

                                             2
                    + C2 G2 s + C1 C2 s  - C1 G2 K s)
```

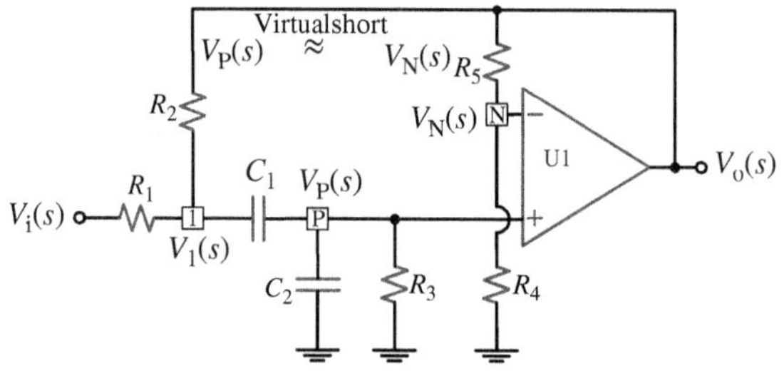

Figure 5.21 Second-order active (Sallen-Key) filter.

This means that the transfer function $G(s)$ is

$$G(s) = \frac{V_o(s)}{V_i(s)} = \frac{KG_1C_1s}{C_1C_2s^2 + s(G_1(C_1 + C_2) + G_2(C_1(1-K) + C_2) + G_3C_1) + G_3(G_1 + G_2)}$$

$$\text{with } K = \frac{R_4 + R_5}{R_4} \tag{5.5.9}$$

5.5.4 Second-Order Active BSF

To analyze the circuit shown in Figure 5.22 ([W-5]), we apply KCL at nodes 1, 2, and 3 to write the node equations

$$\text{Node 1}: \frac{V_1 - V_i}{R} + \frac{V_1 - V_o}{R} + \frac{V_1 - \sigma V_o}{1/2sC} = 0$$

$$\text{Node 2}: \frac{V_2 - V_i}{1/sC} + \frac{V_2 - V_o}{1/sC} + \frac{V_2 - \sigma V_o}{R/2} = 0$$

$$\text{Node 3}: \frac{V_o - V_1}{R} + \frac{V_o - V_2}{1/sC = 0}$$

which can be arranged in compact (matrix-vector) form as

$$\begin{bmatrix} 2sC + 2G & 0 & -2\sigma sC - G \\ 0 & 2sC + 2G & -sC - 2\sigma G \\ -G & -sC & sC + G \end{bmatrix} \begin{bmatrix} V_1 \\ V_2 \\ V_o \end{bmatrix} = \begin{bmatrix} G V_i \\ sC V_i \\ 0 \end{bmatrix}$$

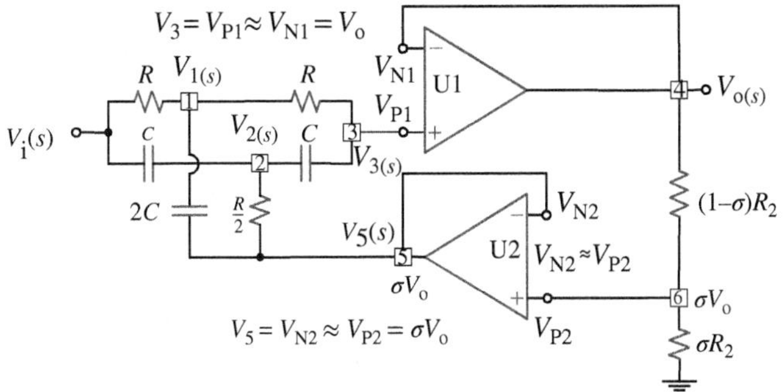

Figure 5.22 Twin-T High Q BSF ("Reproduced by permission of Rod Elliott (https://sound-au.com/articles/dwop a3.htm#s123)").

We can solve this to get V_o and finally the transfer function $G(s)$ as

$$\begin{bmatrix} V_1 \\ V_2 \\ V_o \end{bmatrix} = \frac{1}{\Delta}\begin{bmatrix} C^2s^2+2(2-\sigma)CGs+2G^2 & CGs+2\sigma C^2s^2 & \times \\ CGs+2\sigma G^2 & 2C^2s^2+2(2-\sigma)CGs+G^2 & \times \\ 2CGs+2G^2 & 2C^2s^2+2CGs & \times \end{bmatrix}\begin{bmatrix} GV_i \\ sCV_i \\ 0 \end{bmatrix}$$

$$\text{with } \Delta = 2(sC+G)(C^2s^2+2(2-\sigma)CGs+2G^2)-(2\sigma sC+G)2G(sC+G)$$
$$= 2(sC+G)(C^2s^2+4(1-\sigma)CGs+G^2)$$

$$;G(s) = \frac{V_o(s)}{V_i(s)} = \frac{s^2+1/R^2C^2}{s^2+4(1-\sigma)s/RC+1/R^2C^2} \tag{5.5.10a}$$

$$= \frac{s^2+\omega_0^2}{s^2+\omega_b s+\omega_0^2} \text{ with } \omega_b = \frac{4(1-\sigma)}{RC} \text{ and } \omega_0 = \frac{1}{RC} \tag{5.5.10b}$$

This transfer function having both a second-degree term and a constant term in the numerator indicates that the circuit will function as a band-stop filter (BSF) with rejection frequency ω_0 at which the frequency response is zero. It is noteworthy that this BSF has an adjustable quality factor

$$Q = \frac{\omega_r}{\omega_b} = \frac{\omega_0}{\omega_b} \overset{(5.5.10b)}{=} \frac{1}{4(1-\sigma)} \tag{5.5.11}$$

which can be changed by moving the sliding contact, at node 6, of the variable resistor.

Before going into the next example, there are a couple of things to note, that are summarized in the following remark.

Example 5.3 Tuning of a BPF with MATLAB and Plot of Frequency Response with PSpice

Consider the OP Amp circuit of Figure 5.20(a) with R_2 = 100 Ω and $C_3 = C_4 = C$.

a) Determine the values of R_1, R_5, and $C_3 = C_4 = C$ to make the circuit have the transfer function

$$G(s) \overset{(5.5.7)}{=} \frac{-(G_1/C_3)s}{s^2+(G_5(C_3+C_4)/C_3C_4)s+(G_1+G_2)G_5/C_3C_4} = \frac{-100s}{s^2+100s+100^2} \tag{E5.3.1}$$

so that the bandwidth, the peak frequency, and the lower/upper 3 dB frequencies of the BPF are

$$\omega_b = 100\,[\text{rad/s}],\ \ \omega_p = 100\,[\text{rad/s}]\ (f_p = 15.92\,[\text{Hz}]),\ \text{ and} \tag{E5.3.2}$$

$$\omega_l, \omega_u \overset{(E5.2.6)}{=} \frac{\mp\omega_b+\sqrt{\omega_b^2+4\omega_p^2}}{2} = \mp 50+\sqrt{50^2+100^2} = 61.80,\ 161.80\,[\text{rad/s}] \tag{E5.3.3}$$

$$;f_l = \omega_l/2\pi = 61.80/2\pi = 9.84\,[\text{Hz}] \tag{E5.3.4a}$$

$$f_u = \omega_u/2\pi = 161.80/2\pi = 25.75\,[\text{Hz}] \tag{E5.3.4b}$$

We write a set of nonlinear equations in three unknowns G_1, G_5, and $C_3 = C_4 = C$:

$$G_5\left(\frac{1}{C_3}+\frac{1}{C_4}\right) - 100 = 0,\ \frac{(G_1+G_2)G_5}{C_3C_4} - 10000 = 0,\ \frac{G_1}{C_3} - 100 = 0 \tag{E5.3.5}$$

After saving the MATLAB function 'elec05e03_f()' that describes this set of equations, we run the following MATLAB script "elec05e03.m" to obtain

$$G_1 = 0.01\,\text{S},\ G_5 = 0.005\,\text{S, and } C_3 = C_4 = C = 10^{-4}\,\text{F} \tag{E5.3.6}$$

```
%elec05e03.m
wb=100; wp=100; % Bandwidth and Peak frequency
wlu = ([-wb +wb]+sqrt(wb^2+4*wp^2))/2; flu = wlu/2/pi % Eq.(E5.2.6)
% Indispensable part starts from the next statement.
R2=100; G2=1/R2; % Pre-determined value of R2
G10=0.01; G50=0.01; C0=0.01; x0=[G10 G50 C0]; % Initial guess on G1,G5,C
x= fsolve('elec05e03_f',x0,[],G2) % Solution of nonlinear eq in x
G1=x(1), G5=x(2), C3=x(3), C4=x(3) % Interpret the solution x=[G1 G5 C]
```

```
function fx=elec05e03_f(x,G2)
G1=x(1); G5=x(2); C3=x(3); C4=x(3);
fx= [G5*(1/C3+1/C4)-100; (G1+G2)*G5/C3/C4-10000; G1/C3-100]; %Eq.(E5.3.5)
```

b) We perform the PSpice simulation to get the frequency response of the circuit tuned in (a) and check if the peak frequency and the lower/upper 3 dB frequencies are as expected in (a). To this end, we take the following steps:

- Draw the schematic as depicted in Figure 5.23(a) where a VAC voltage source is placed together with a uA741 OP Amp, three resistors $R_1 = 100\,\Omega$, $R_2 = 100\,\Omega$, and $R_5 = 200\,\Omega$, and two capacitors $C_3 = 10^{-4}$ F and $C_4 = 10^{-4}$ F.
- In the Simulation Settings dialog box, set the Analysis type to 'AC Sweep' and the AC Sweep type as follows:

  Decade (with the frequency plotted in the horizontal log scale)
  Start Frequency: 1, End Frequency: 100, Points/Decade: 500

- Pick up a VDB (dB Magnitude of Voltage) Marker from the menu PSpice > Markers > Advanced and place it at the OP Amp output terminal to measure $10\log_{10}|\mathbf{V}_o(j\omega)|$ [dB].
- Click the Run button on the toolbar to make the PSpice A/D (Probe) window appear on the screen as depicted in Figure 5.23(b).
- To get the peak frequency and the upper/lower 3 dB frequencies from the frequency response magnitude $|\mathbf{V}_o(j\omega)|$, do the following:

(a)

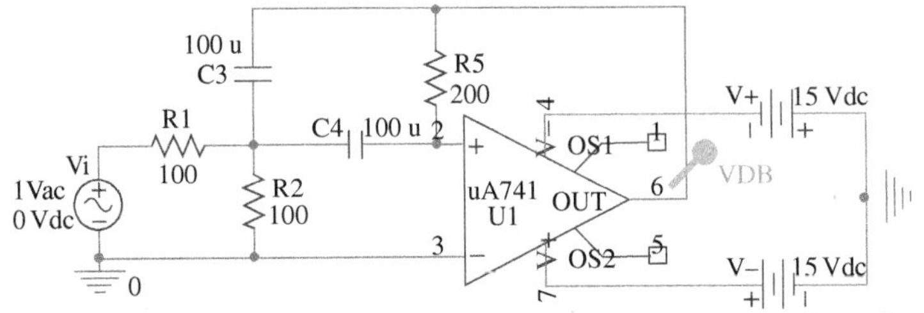

The VDB marker to get the frequency response magnitude curve with is available from the menu PSpice > Markers > Advanced.

PSpice schematic for the circuit in Figure 5.20a ("elec05e03.opj")

(b)

Evaluate	Measurement	Value
☑	CenterFrequency(V(U1:OUT),0.01)	15.919¹3
☑	Max(V(U1:OUT))	999.951/9m
☑	Cutoff_Highpass_3dB(V(U1:OUT))	9.84654
☑	Cutoff_Lowpass_3dB(V(U1:OUT))	25.72490

Click here to evaluate a new measurement...

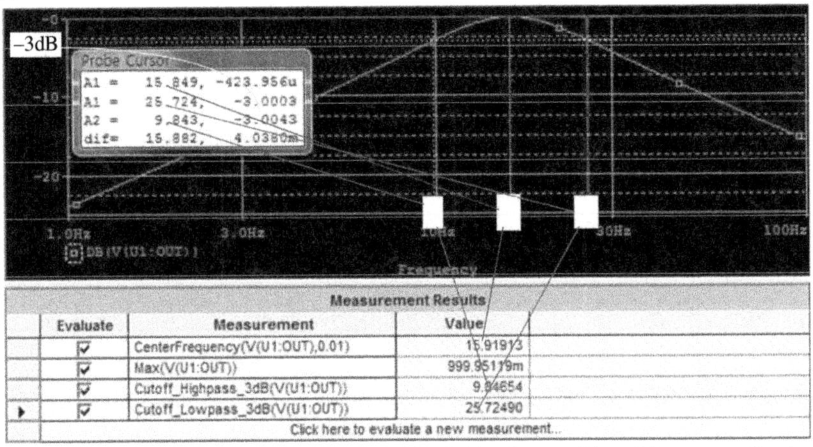

The Measurement Results dialog box can be opened by clicking the menu View > Measurement Results.

Frequency response obtained from PSpice simulation

Figure 5.23 PSpice schematic and simulation results for Example 5.3.

- Click the Toggle Cursor button on the toolbar of the Probe window.
- Click the Cursor Peak button to read the peak frequency 15.849 Hz and the peak magnitude $G_{max} = 1 \approx 0$ dB from the Probe Cursor box.
- Use the left mouse button or the arrow key to move the cursor to the two −3 dB points of the frequency (AC sweep) response curve and read the two 3 dB frequencies, 9.843 Hz and 25.724 Hz, from the Probe Cursor box.
- As an alternative, click the Evaluate Measurement button on the toolbar of the Probe window and type, say,

'Cutoff_Highpass_3dB(V(U1:OUT))', 'Cutoff_Lowpass_3dB(V(U1:OUT))', and 'CenterFrequency(V(U1:OUT),0.1)'

into the Trace Expression field at the bottom of the Evaluate Measurement dialog box.

(Note) For reference, you can see <http://www.ti.com/lit/an/snoa387c/snoa387c.pdf> to see the application note OA-26 for designing active high speed filters. Also, see <http://www.analog.com/media/en/training-seminars/design- handbooks/Basic-Linear-Design/Chapter8.pdf> for various configurations of analog filter.

Problems

5.1 Voltage-to-Current Converter Using OP Amp – Howland Circuit
Consider the OP Amp circuit of Figure P5.1.1(a), which is known as the *Howland circuit.*

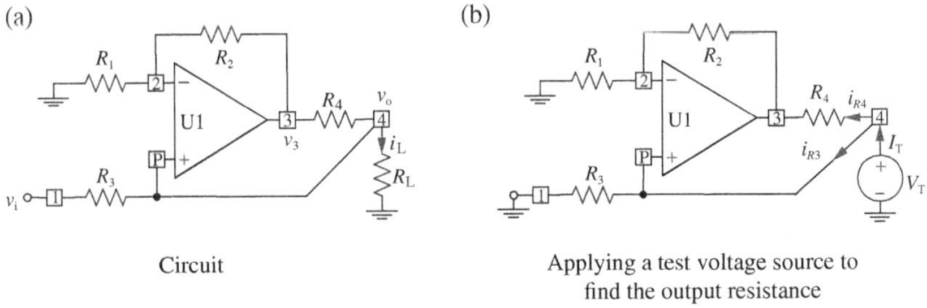

Figure P5.1.1 Voltage-to-current converter for Problem 5.1.

(a) Noting that the circuit has a negative feedback, we can use the virtual short principle and the voltage divider rule to write

$$v_3 = \frac{R_1 + R_2}{R_1} v_- \overset{\text{virtual short}}{=} \frac{R_1 + R_2}{R_1} v_+ = \frac{R_1 + R_2}{R_1} v_o \quad \text{(P5.1.1)}$$

Show that the output voltage v_o can be expressed in terms of the input voltage v_i as

$$v_o = \frac{1}{1 - R_2 R_3 / R_1 R_4 + R_3 / R_L} v_i \quad \text{(P5.1.2)}$$

(b) Referring to Figure P5.1.1(b), show that the output resistance is

$$R_o = \frac{V_T}{I_T} = \frac{R_4}{R_4 / R_3 - R_2 / R_1} \quad \text{(P5.1.3)}$$

(c) Find the condition on which the output resistance R_o is infinitely large. Also, show that if the condition is satisfied, the load current through R_L is

$$i_L = \frac{v_o}{R_L} = \frac{v_i}{R_3} \quad \text{(P5.1.4)}$$

(Note) This load current does not vary with the load resistance R_L and can be controlled by adjusting the input voltage v_i. That is why the circuit is called a voltage-to-current converter.

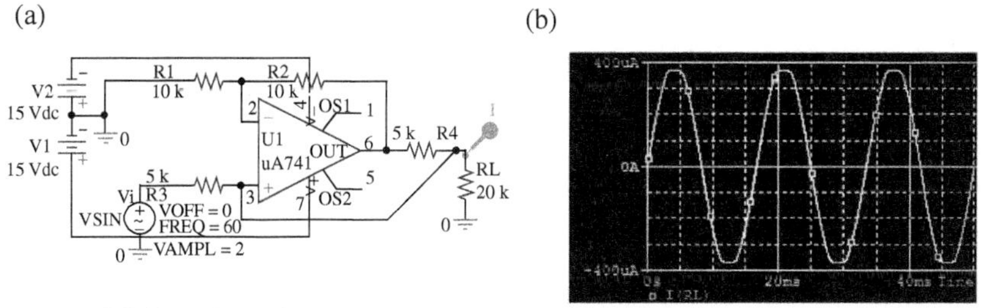

PSpice schematic

Simulation result for VAMPL = 2 and RL = 20 k

Figure P5.1.2 PSpice simulation of a voltage-to-current converter ("V2I_converter.opj").

(d) To see if the load current can really be controlled by adjusting the input voltage v_i regardless of the load resistance R_L, perform the PSpice simulation of the circuit for 0.05 s three times, where $R_1 = R_2 = 10\,\text{k}\Omega$ and $R_3 = R_4 = 5\,\text{k}\Omega$; once with $v_i(t) = \sin 2\pi 60t$ [V] and $R_L = 10\,\text{k}\Omega$, once with $v_i(t) = \sin 2\pi 60t$ [V] and $R_L = 20\,\text{k}\Omega$, and once with $v_i(t) = 2\sin 2\pi 60t$ [V] and $R_L = 10\,\text{k}\Omega$, referring to Figure P5.1.2(a).

(e) Make use of Eq. (P5.1.4) to get the load current for $v_i = 2$ V in the Howland circuit with $R_1 = R_2 = 10\,\text{k}\Omega$, $R_3 = R_4 = 5\,\text{k}\Omega$, and $R_L = 20\,\text{k}\Omega$. Perform the PSpice simulation with $v_i(t) = 2\sin 2\pi 60t$ [V] and $R_L = 20\,\text{k}\Omega$, which will yield the load current waveform depicted in Figure P5.1.2(b). Does the simulation result agree with the analytical result? If not, discuss why $i_L(t)$ is distorted.

<Hint> Find v_3 at the OP Amp output terminal for $v_i = 2$ V and check if it exceeds the maximum output (saturation) voltage V_{om}.

5.2 Realization of *Negative Resistance* Using OP Amp

(a) Show that the input resistance of the circuit of Figure P5.2(a) is

$$R_{in} = \frac{v_i}{i_i} = \frac{v_i}{(v_i - v_o)/R} = -\frac{R_1}{R_2}R \tag{P5.2.1}$$

(b) To see if this negative resistance can really be realized by the OP Amp circuit of Figure P5.2(a), perform the PSpice simulation of the circuit with $R_1 = 10\,\text{k}\Omega$, $R_2 = 20\,\text{k}\Omega$, $R = 2\,\text{k}\Omega$, and $v_i(t) = 4\sin(0.2\pi t)$ [V] for 10 s and have the input voltage and current waveforms plotted in the PSpice A/D (Probe) window as depicted in Figure P5.2(b). Find the ratio of the input voltage to the input current obtained from the simulation and compare it with the resistance obtained from the analytical formula (P5.2.1).

(a)

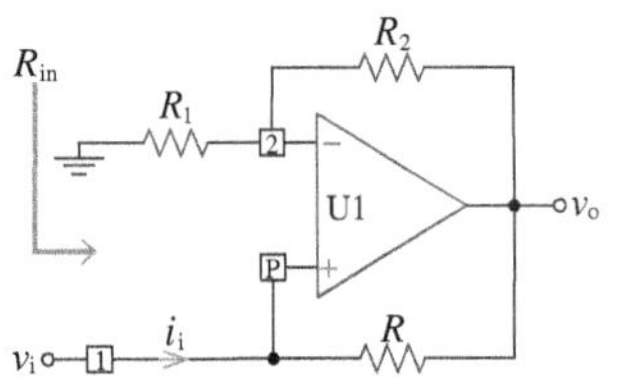

OP Amp circuit to realize a negative resistance

(b)

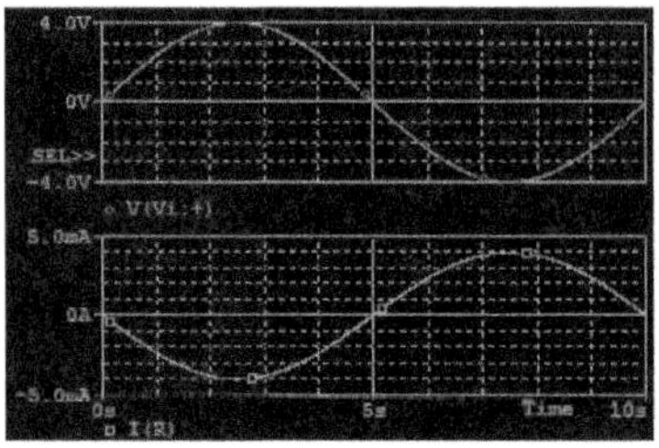

PSpice simulation results: input voltage and current waveforms

Figure P5.2 OP Amp circuit realizing negative resistance for Problem 5.2 ("negative_resistance.opj").

5.3 The Limits on Output Voltage and Current of OP Amp

Referring to Figure P5.3.1(a), perform the PSpice simulation for the DC sweep analysis of a non-inverting OP Amp circuit with $R_1 = 10\ \text{k}\Omega$ and $R_f = 10\ \text{k}\Omega$ to get the plot of v_o vs. v_i in Figure P5.3.1(b1) (with $R_L = 1\ \text{k}\Omega$) and Figure P5.3.1(b2) (with $R_L = 100\ \text{k}\Omega$).

(a)

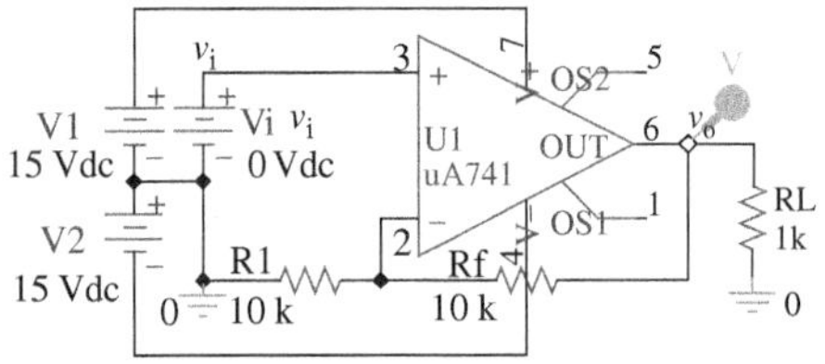

Schematic for a non-inverting OP Amp circuit

(b1)

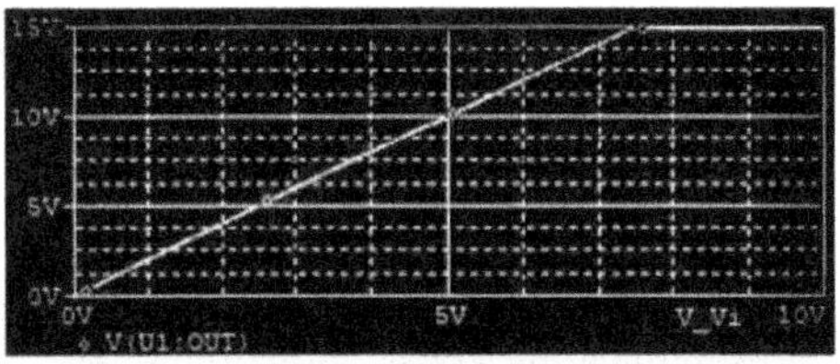

v_o vs. v_i with RL = 1k

(b2)

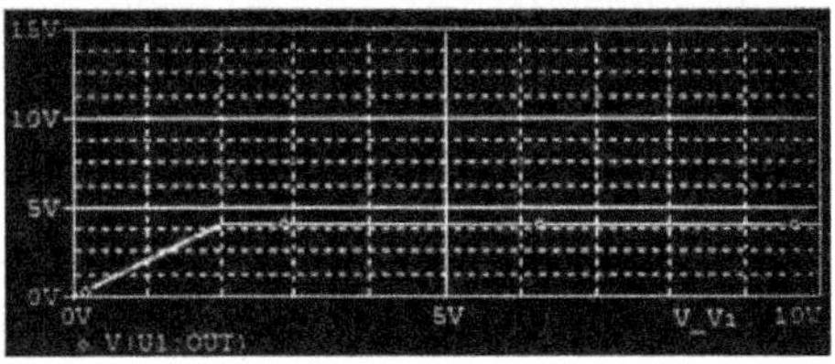

v_o vs. v_i with RL = 100

Figure P5.3.1 OP Amp circuit ("elec05p03a.opj").

(a) Consider Figure P5.3.1(b1). Does the voltage gain of the circuit agree with what you expect from Eq. (5.2.15)? How do you explain the saturation of the output voltage at about 14 V?

<Hint> Read the typical value of the output voltage swing of a μA741 OP Amp denoted by V_{OM} (maximum peak output voltage swing) or Vopp in the datasheet [W-3] or [W-4].

(b) Consider Figure P5.3.1(b2). How do you explain the saturation of the output voltage at about 4 V?

<Hint> Put a current marker at the OP Amp output terminal to measure the output current of the OP Amp and read the maximum value of the (short-circuit) output current of a μA741 OP Amp denoted by Ios (short-circuit output current) in the datasheet [W-3] or [W-4].

(c) Figure P5.3.2(a) shows a non-inverting OP Amp circuit using an NPN-BJT as a current booster. Perform the PSpice simulation two times, once with the feedback from the OP Amp output (A) and once with the feedback from the emitter terminal (B) of the BJT to see that the corresponding input-output relationships are obtained as shown in Figure P5.3.2(b1) and (b2), respectively. Has the overload current problem been remedied by the BJT boosting its base current, which is the OP Amp output current,

Figure P5.3.2 OP Amp circuit ("elec05p03b.opj").

(a)

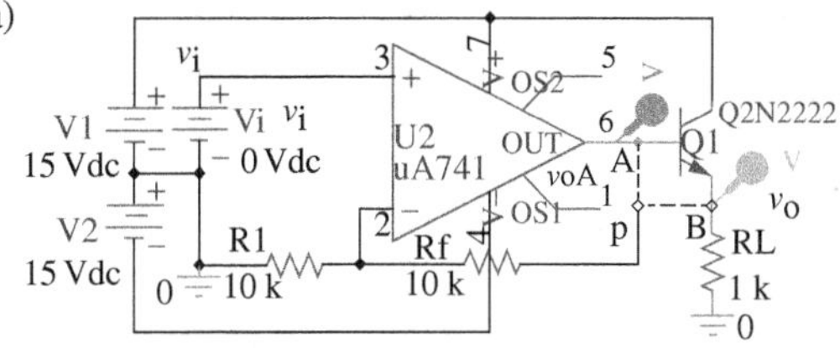

Schematic for a non- inverting OP Amp circuit with BJT

(b1)

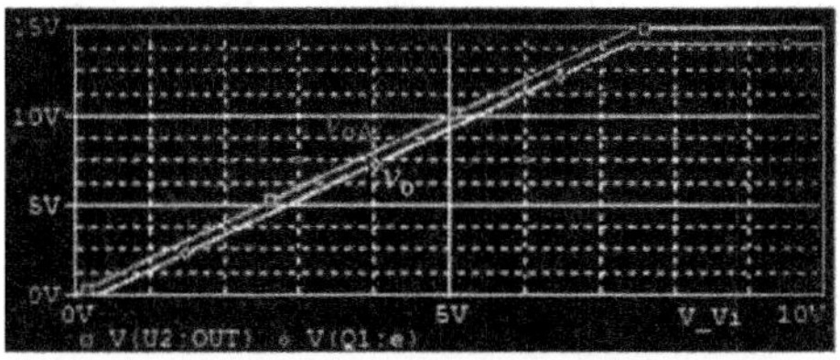

v_o vs. v_i with RL = 1k and P-A or P-B connected

(b2)

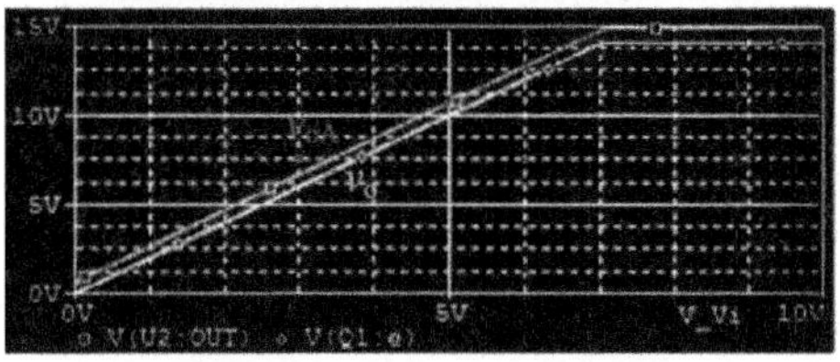

v_o vs. v_i with RL = 100 and P-A or P-B connected

by a factor of ($\beta + 1$)? Which feedback connection is better in realizing the voltage gain of $v_o/v_i = 2$?

(Note) The typical value of the DC current gain β (h_{FE}) is at least 75 for the BJT 2N2222 as can be seen at the website http://www.alldatasheet.com/view.jsp?Searchword=2N2222.

5.4 Applications of 555 Timer/Oscillator [W-2]
Refer to Section 5.3.3 for the internal structure/behavior of the 555 timer.

(a) Let the switch be closed at $t = 0.01$ s in the 555 timer circuit of Figure P5.4 (a), where the switch has been open for a long time before $t = 0.01$ s. Find the duration (pulse-width) of the rectangular pulse $v_o(t)$ appearing at the output terminal. Support your results by PSpice simulation.

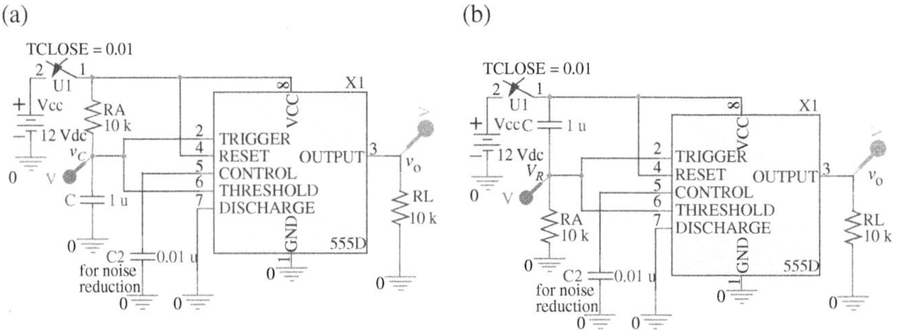

Figure P5.4 Two types of power-on delays realized by using a 555 timer.

(b) Let the switch be closed at $t = 0.01$ s in the 555 timer circuit of Figure P5.4 (b), where the switch has been open for a long time before $t = 0.01$ s. Find the time delay of $v_o(t)$ measured from the switching-on time. Support your results by PSpice simulation.

5.5 Design of a Square-Wave Generator with a Variable Period Depending on a Resistance
Consider the square-wave generator of Figure 5.11(a) in which the values of C and R_2 are $C = 100\ \mu F$ and $R_2 = 10\sim100\ k\Omega$ (variable), respectively.

(a) Determine the values of R_1 and R_3 such that the period of the square wave becomes $P_{max} = 4.5$ s for $R_2 = 10\ k\Omega$ (minimum) and $P_{min} = 0.5$ s for $R_2 = 100\ k\Omega$ (maximum).

(Hint) We can use Eq. (5.3.8) to write the design specifications on the period of the square wave as

$$P_{\max} = 2R_3 C \ln \frac{2R_1 + R_2}{R_2}\bigg|_{R_2 = R_{2,\min} = 10\,\text{k}\Omega} = 4.5\,\text{s} \tag{P5.5.1}$$

$$P_{\min} = 2R_3 C \ln \frac{2R_1 + R_2}{R_2}\bigg|_{R_2 = R_{2,\max} = 100\,\text{k}\Omega} = 0.5\,\text{s} \tag{P5.5.2}$$

These equations can be solved for R_1 by running the following MATLAB script:

```
%elec05p05.m
R2min=1e4; R2max=1e5; C=100e-6;
Pmax=4.5; Pmin=0.5; % desired minimum period of a wiper cycle.
x_0=[10 100]; % Initial guess for 2*R3C and R1
options= optimset('fsolve');
xo=fsolve('elec05p05_f',x_0,options,R2min,R2max,Pmax,Pmin)
R1= xo(2), R3C2= xo(1); R3= R3C2/C/2
R1=standard_value(R1,'R','c',1),
R3=standard_value(R3,'R','c',1)
```

```
function y=elec05p05_f(x,R2min,R2max,Pmax,Pmin)
R3C2=x(1); R1=x(2); % R3C2=2*R3*C
y= [R3C2*log((2*R1+R2min)/R2min)-Pmax;
    R3C2*log((2*R1+R2max)/R2max)-Pmin]; % Eq.(P5.5.1&2)
```

(b) Select the appropriate 1%-tolerance standard resistance values of R_1 and R_3 from Table G.2 (Appendix G) and support your design results by two PSpice simulations, one with R_2 = 10 kΩ for 10 s and one with R_2 = 100 kΩ for 1 s. Set the initial voltage (IC) of the capacitor C to $v_C(0)$ = 0 V in the Property Editor spreadsheet, set the maximum step in the Transient analysis options to 1 ms, and check the square box before 'Skip the initial transient bias point calculation (SKIPBP)' in the Simulation Settings dialog box (Figure H.5(c1)) so that the initial transient bias point calculation will be skipped.

(cf.) If you come across a warning message such as 'Unable to find library file templates. lib; Subcircuit uA741 is undefined' for a device like an OP Amp or a transistor, you can click that device for selection and click Edit/PSpice Model on the menu bar of the Capture window to open the PSpice Model Editor window, press '^s' to save the library file for that device, and just click on x on the PSpice Model Editor window to close it.

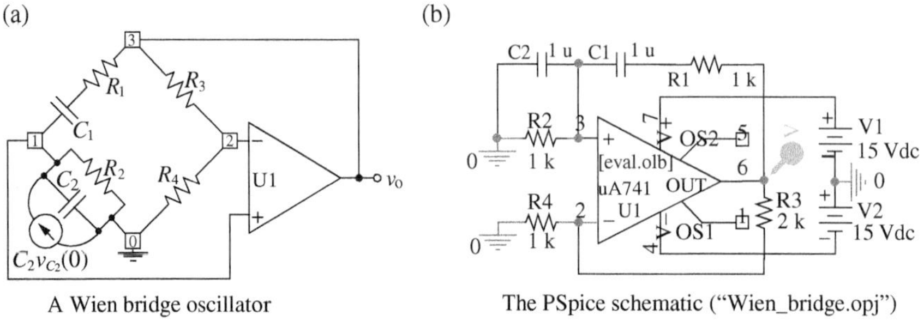

Figure P5.6 Analysis and simulation of a Wien bridge oscillator.

5.6 Wien Bridge Oscillator

Consider the OP Amp circuit of Figure P5.6(a).

(a) For the OP Amp with a negative feedback path connecting the output terminal to the negative input terminal, the virtual short principle (Remark 5.1(2)) says that the voltages at the +/− input terminals are almost equal:

$$v_1 = v_2 = \frac{R_4}{R_3 + R_4} v_o = b v_o \tag{P5.6.1}$$

Let the initial voltages of the capacitors C_1 and C_2 be

$$v_{C_1}(0) = 0 \text{ and } v_{C_2}(0) = V_{20} \tag{P5.6.2}$$

respectively, where the nonzero one of the capacitor C_2 is represented by the current source of $C_2 v_{C2}(0)$ in parallel with C_2 as depicted in Figure P5.6(a). Apply KCL to the node 1 to write the node equation and solve it to find $V_o(s)$ as

$$\left(\frac{b-1}{R_1 + 1/sC_1} + \frac{b}{R_2} + sbC_2\right) V_o(s) = C_2 V_{20} \tag{P5.6.3}$$

$$;\ V_o(s) = \frac{V_{20} R_2 C_2 (1 + sR_1 C_1)}{bR_1 R_2 C_1 C_2 s^2 + s(bR_1 C_1 + bR_2 C_2 + (b-1)R_2 C_1) + b} \tag{P5.6.4}$$

(b) With the initial voltage $v_{C2}(0) = V_{20}$ as a kind of input, we can regard Eq. (P5.5.4) (excluding V_{20}) as a transfer function and set its denominator to zero to obtain the characteristic equation. Verify that the following condition

$$\frac{R_4}{R_3 + R_4} = b = \frac{R_2 C_1}{R_1 C_1 + R_2 C_2 + R_2 C_1} \tag{P5.6.5}$$

guarantees that the characteristic equation has imaginary roots $s = j\omega_r$ so that the output will have an everlasting oscillation of frequency

$$\omega_r = 1/\sqrt{R_1R_2C_1C_2} \tag{P5.6.6}$$

(c) Verify that, with

$$v_{C_2}(0) = V_{20} = 1\,\text{V},\ R_1 = R_2 = R_4 = 1\,\text{k}\Omega,\\ R_3 = 2\,\text{k}\Omega \text{ and } C_1 = C_2 = 1\,\mu\text{F} \tag{P5.6.7}$$

the output voltage is as follows:

$$V_o(s) = \frac{1}{b}\frac{s+\omega_r}{s^2+\omega_r^2} = \frac{3s}{s^2+1000^2} + \frac{3\times 1000}{s^2+1000^2} \tag{P5.6.8}$$

$$;v_o(t) = 3(\cos 1000t + \sin 1000t)u_s(t) = 3\sqrt{2}\sin(1000t + 45^\circ)u_s(t)\ [\text{V}] \tag{P5.6.9}$$

(d) Use MATLAB or equivalent to plot $v_o(t)$ (P5.6.9) for the time interval [0,20 ms].

(e) With reference to the PSpice schematic in Figure P5.6(b), perform the PSpice simulation to get the amplitude and the period of $v_o(t)$.

5.7 Active filters and Their Frequency Responses in Bode Plots

(a) Verify that the circuits in Figure P5.7(a) and (b) have the following transfer functions, respectively:

$$G_a(s) = \frac{V_o(s)}{V_i(s)} = \frac{-s/R_1C_3}{s^2 + s(1/C_3 + 1/C_4)/R_5 + (1/R_1 + 1/R_2)/R_5C_3C_4} \tag{P5.7.1}$$

$$G_b(s) = \frac{Ks^2 + s(K(1/C_3+1/C_4)/R_5 + (K-1)/R_1C_3) + K/R_1R_5C_3C_4}{s^2 + s(1/C_3+1/C_4)/R_5 + 1/R_1R_5C_3C_4} \tag{P5.7.2}$$

$$\text{with } K = \frac{R_6}{R_2 + R_6}$$

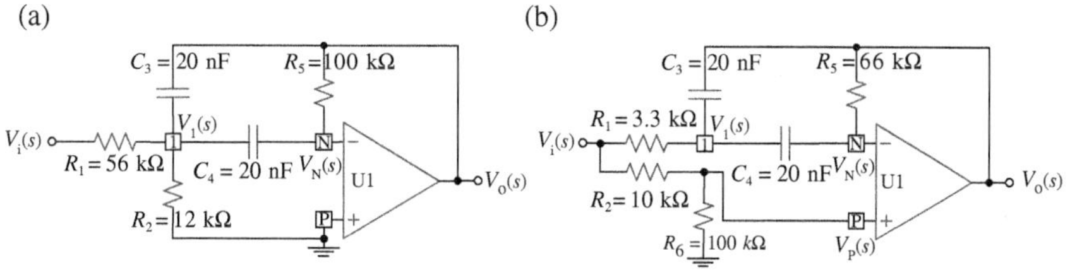

Figure P5.7 Second-order active filters for Problem 5.7.

(b) Using MATLAB or equivalent, plot the magnitude in dB of the frequency response $G_a(j2\pi f)$, that is, $20\log_{10}|G_a(j2\pi f)|$ with $R_1 = 56\ \text{k}\Omega$, $R_2 = 12\ \text{k}\Omega$, $R_5 = 100\ \text{k}\Omega$, $C_3 = 20\ \text{nF}$, and $C_4 = 20\ \text{nF}$ for the 801 log-spaced frequency points between 10 Hz and 1 kHz on a horizontal log scale. Which kind of filter is the circuit of Figure P5.7(a), LPF, HPF, BPF, or BSF? Find the DC gain $|G_a(j0)|$. Find the peak (maximum) magnitude, $G_{a,max}$ [dB], of $|G_a(j2\pi f)|$ and the peak frequency f_p. Find the upper/lower 3 dB frequencies f_u and f_l at which $|G_a(j2\pi f)|=G_{a,\ max}/\sqrt{2}$, or equivalently, $|G_a(j2\pi f)|[\text{dB}]=G_{b,max}-3[\text{dB}]$.

<Hint> You may complete the following script "elec05p07b.m" and run it.

(Note) The MATLAB function [Gmag,Gph] = bode(num,den,w) can be used to find the magnitude (Gmag) and phase(Gph) of the frequency response $G(j\omega)$ for a (radian) frequency vector (w), where the numerator/denominator polynomials of $G(j\omega)$ in $s = j\omega$ are given as the first/second input arguments num and den, respectively. bode(num,den,w) with no output argument yields the Bode plot consisting of the magnitude and phase curves.

(Note) Since we compute the magnitude in dB of frequency response, we do not need to divide the maximum value by $\sqrt{2}$, but just subtract 3 dB from the maximum value in dB when finding the upper/lower 3 dB frequencies. This idea works for the PSpice simulation, too.

```
%elec05p07b.mvb
syms s R1 R2 R5 C3 C4
sC3=s*C3; sC4=s*C4;
% Node admittance matrix
Y = [1/R1+1/R2+sC3+sC4 -sC3; sC4 1/R5];
% Solution of the node equation
Vs = Y\[1/R1; 0];
% Transfer function Eq.(P5.7.1)
Gs=Vs(2)
Gs=subs(Gs,{R1,R2,R5,C3,C4},{56e3,????,1e5,????,2e-8});
pretty(simplify(Gs))
% Numerator and Denominator of the transfer function
[num,den] = numden(Gs);
num=sym2poly(num); den=sym2poly(den);
% Frequency vector from 10^1 to 10^3
f = logspace(1,3,801); w=2*pi*f; jw=j*w;
% Magnitude in dB of the frequency response G(jw)
Gw_mag_dB = 20*log10(abs(polyval(num,jw)./polyval(den,jw)));
Gw_mag_dB1 = 20*log10(abs(freqs(num,den,jw))); % Alternatively
subplot(221), semilogx(f,Gw_mag_dB)
% Just to plot magnitude & phase curves of G(jw) - Bode plot
subplot(222), bode(num,den,w)
% To get the numeric values of the magnitude & phase curves of G(jw)
[Gw_mag,Gw_phase] = bode(num,den,w);
```

```
% the magnitude in dB of G(jw) using bode()
Gw_mag_dB1=20*log10(Gw_mag);
discrepancy-norm(Gw_mag_dB Cw_mag_dB1.')
% Peak frequency fp of the BPF
[GmdB_max,imax]=max(Gw_mag_dB), fp=f(imax)
% DC gain at w=0
Gw0=abs(polyval(num,0)./polyval(den,0))
% Upper/lower 3dB-frequencies
[Gl,il]=min(abs(Gw_mag_dB(1:imax)-(GmdB_max-3))); fl=f(il)
[Gu,iu]=min(abs(Gw_mag_dB(imax:end)-(GmdB_max-3))); fu=f(imax+iu-1)
```

(c) Using MATLAB or equivalent, plot the magnitude in dB of the frequency response $G_b(j2\pi f)$, that is, $20\log_{10}|G_b(j2\pi f)|$ with $R_1 = 3.3\,\text{k}\Omega$, $R_2 = 10\,\text{k}\Omega$, $R_5 = 66\,\text{k}\Omega$, $R_6 = 100\,\text{k}\Omega$, $C_3 = 20\,\text{nF}$, and $C_4 = 20\,\text{nF}$ for the 801 log-spaced frequency points between 10 Hz and 1 kHz on a horizontal log scale. Which kind of filter is the circuit of Figure P5.7(b), LPF, HPF, BPF, or BSF? Find the maximum, $G_{b,max}$, of $|G_b(j2\pi f)|$. Find the minimum, $G_{b,min}$, of $|G_b(j2\pi f)|$ and the rejection frequency f_r. Find the upper/lower 3 dB frequencies f_u and f_l at which $|G_b(j2\pi f)| = G_{b,max}/\sqrt{2}$, or equivalently, $|G_b(j2\pi f)|[\text{dB}] = G_{b,max}-3[\text{dB}]$.

(d) Perform the PSpice simulation to obtain the curves of the frequency response magnitudes $|G_a(j2\pi f)|$ and $|G_b(j2\pi f)|$ for 400 frequency points per decade between 10 Hz and 1 kHz on a horizontal log scale. Based on the PSpice simulations together with the MATLAB analysis results obtained in (b) and (c), fill in the corresponding blanks of Table P5.7 with the appropriate values rounded to three significant digits.

Table P5.7 Results of MATLAB analysis and PSpice simulation.

Figures		MATLAB	PSpice
Figure P5.7(a)	DC gain $\lvert G_a(j0)\rvert$		$10^{-33.1/20} = 0.0221$
	$G_{a,max}$ [dB]	−0.985	
	Peak frequency f_p [Hz]		253
	Lower 3 dB frequency f_l [Hz]	186	
	Upper 3 dB frequency f_u [Hz]		343
Figure P5.7(b)	$G_{b,max}$ [dB]	−0.828	
	$G_{b,min}$ [dB]		−43.2
	Rejection frequency f_r [Hz]	540	
	Lower 3 dB frequency f_l [Hz]		431
	Upper 3 dB frequency f_u [Hz]	672	

(Note) To plot $20\log_{10}|G(j2\pi f)|$ instead of $|G(j2\pi f)|$ in the Probe window, click the ADD Trace button and type the following expression

20*LOG10(ABS(V(U1:OUT)))

into the Trace Expression field at the bottom of the Add Traces dialog box (see Figure H.8(a)).

6

Analog Filter

CHAPTER OUTLINE

6.1 Analog Filter Design

Figure 6.1(a)-(d) shows typical low-pass/band-pass/band-stop/high-pass filter (LPF/BPF/BSF/HPF) specifications on their log magnitude, $20\log_{10}|G(j\omega)|$ [dB], of frequency response. The filter specification can be described as follows:

$$20\log_{10}|G(j\omega_p)| \geq -R_p \text{ [dB] for the passband} \tag{6.1.1a}$$

$$20\log_{10}|G(j\omega_s)| \leq -A_s \text{ [dB] for the stopband} \tag{6.1.1b}$$

Electronic Circuits with MATLAB®, PSpice®, and Smith Chart, First Edition. Won Y. Yang, Jaekwon Kim, Kyung W. Park, Donghyun Baek, Sungjoon Lim, Jingon Joung, Suhyun Park, Han L. Lee, Woo June Choi, and Taeho Im.

Companion website: www.wiley.com/go/yang/electroniccircuits

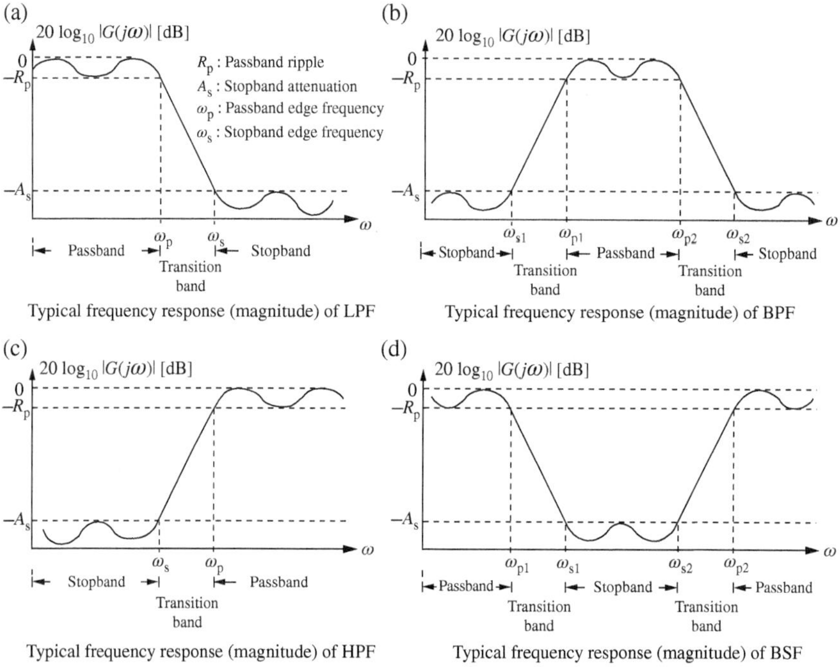

Figure 6.1 Specifications on the log magnitude of analog filter frequency response.

where ω_p, ω_s, R_p, and A_s are referred to as the *passband edge frequency*, the *stopband edge frequency*, the *passband ripple*, and the *stopband attenuation*, respectively. The most commonly used analog filter design techniques are the Butterworth, Chebyshev I, II, and elliptic ones ([K-2], Chapter 8). MATLAB has the built-in functions '`butt()`', '`cheby1()`', '`cheby2()`', and '`ellip()`' for designing the four types of analog/digital filter. As summarized below, '`butt()`' needs the 3 dB cutoff frequency while '`cheby1()`' and '`ellip()`' take the critical passband edge frequency and '`cheby2()`' the critical stopband edge frequency as one of their input arguments. The parametric frequencies together with the filter order can be predetermined using '`buttord()`', '`cheb1ord()`', '`cheb2ord()`', and '`elliord()`'. The frequency input argument should be given in two-dimensional vector for designing BPF or BSF. Also for HPF/BSF, the string `'high'`/`'stop'` should be given as an optional input argument together with `'s'` for analog filter design.

function [N,wc] = **buttord**(wp,ws,Rp,As,opt) % For opt='s', it selects the lowest order N and cutoff frequency wc of analog Butterworth filter. % that has passband ripple<=Rp[dB] and stopband attenuation>=As[dB] % for passband edge frequency wp and stopband edge frequency ws. % Note that for BPF/BSF, the passband edge frequency wp and stopband edge frequency ws % should be given as two-dimensional vectors like [wp1 wp2] and [ws1 ws2].
function [B,A]=**butter**(N,wc,opt) % It designs a digital/analog Butterworth filter, returning the numerator/denominator of system % function where N and wc can be obtained from [N,wc]=buttord(wp,ws,Rp,As,opt). % [B,A]=butter(N,wc,'s') for an analog LPF of order N with cutoff frequency wc[rad/s] % butter(N,[wc1 wc2],'s') for an analog BPF of order 2N with passband wc1<w<wc2[rad/s] % butter(N,[wc1 wc2],'stop','s') for an analog BSF of order 2N with stopband wc1<w<wc2[rad/s] % butter(N,wc,'high','s') for an analog HPF of order N with cutoff frequency wc[rad/s]
function [B,A]=**cheby1**(N,Rp,wpc,opt) % It designs a digital/analog Chebyshev type I filter with passband ripple Rp[dB] % and critical passband edge frequency wpc (Use Rp=0.5 as a starting point, if not sure). % Note that N and wpc can be obtained from [N,wpc]=cheb1ord(wp,ws,Rp,As,opt).
function [B,A]=**cheby2**(N,As,wsc,opt) % It designs a digital/analog Chebyshev type II filter with stopband attenuation As[dB] down % and critical stopband edge frequency wsc (Use As=20 as a starting point, if not sure). % Note that N and wsc can be obtained from [N,wsc]=cheb2ord(wp,ws,Rp,As,opt).
function [B,A]=**ellip**(N,Rp,As,wpc,opt) % It designs a digital/analog Elliptic filter with passband ripple Rp, stopband attenuation As, % and critical passband edge frequency wpc (Use Rp=0.5[dB] & As=20[dB], if unsure). % Note that N and wpc can be obtained from ellipord(wp,ws,Rp,As,opt).

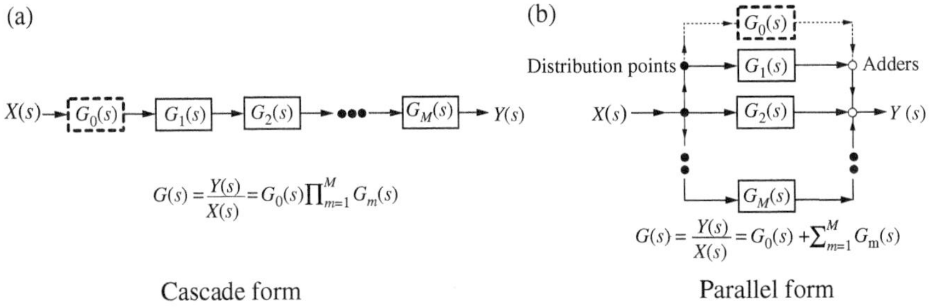

Figure 6.2 Two realizations of analog filter (system or transfer function).

A designed filter transfer function of order N is often factored into the sum or product of SOSs (second-order sections) called biquads (possibly with an additional first-order section in the case of an odd filter order) as

$$G(s) = KG_0(s)\prod_{m=1}^{M} G_m(s) = K\frac{b_{00}s + b_{01}}{s + a_{01}}\prod_{m=1}^{M}\frac{b_{m0}s^2 + b_{m1}s + b_{m2}}{s^2 + a_{i1}s + a_{i2}} \quad \text{with } M = \text{floor}\left(^N/_2\right) \tag{6.1.2a}$$

$$G(s) = G_0(s) + \sum_{m=1}^{M} G_m(s) = \frac{b_{00}s + b_{01}}{s + a_{01}} + \sum_{m=1}^{M}\frac{b_{m0}s^2 + b_{m1}s + b_{m2}}{s^2 + a_{i1}s + a_{i2}} \quad \text{with } M = \text{floor}\left(^N/_2\right) \tag{6.1.2b}$$

and then realized in *cascade* or *parallel* form, respectively, as depicted in Figure 6.2.

Rather than reviewing the design procedures, let us use the MATLAB functions to design a Butterworth low-pass filter, a Chebyshev I band-pass filter, a Chebyshev II BSF, and elliptic HPF in the following example.

Example 6.1 Filter Design Using the MATLAB Routines
Let us find the transfer functions of analog filters meeting the specifications given below.

a) We are going to determine the transfer function of a Butterworth LPF with

$$\omega_p = 2\pi \times 6000\,[\text{rad/s}],\ \ \omega_s = 2\pi \times 15000\,[\text{rad/s}], \quad R_p \le 2\,[\text{dB}], \text{ and } A_s \ge 25\,[\text{dB}] \tag{E6.1.1}$$

First, we use the MATLAB function '`buttord()`' to find the filter order N and the cutoff frequency ω_c at which $20\log_{10}|G(j\omega_c)| = -3$ [dB] by running the following MATLAB statements:

```
>>wp=2*pi*6000; ws=2*pi*15000; Rp=2; As=25;
>>format short e, [N,wc]=buttord(wp,ws,Rp,As,'s')
  N = 4, wc = 4.5914e+004
```

We put these parameter values N and wc into the Butterworth filter design function '`butt()`' as its first and second input arguments where a character 's' is put as the third input argument of '`buttord()`' to specify an analog filter design rather than a digital filter design:

```
>>[Bb,Ab]=butter(N,wc,'s')
   Bb =         0          0          0          0     4.4440e+018
   Ab = 1.0000e+000 1.1998e+005 7.1974e+009 2.5292e+014 4.4440e+018
```

This means that the transfer function of the designed Butterworth filter of order N = 4 is

$$G(s) = \frac{4.444\times 10^{18}}{s^4 + 1.1998\times 10^5 s^3 + 7.1974\times 10^9 s^2 + 2.5292\times 10^{14} s + 4.444\times 10^{18}} \tag{E6.1.2}$$

We can find the cascade and parallel realizations of this transfer function by running the following MATLAB statements:

```
>>[SOS,K]=tf2sos(Bb,Ab); % cascade realization
>>Ns=size(SOS,1); % the number of sections
>>Gm=K^(1/Ns) % the (distributed) gain of each SOS connected in cascade
>>BBc=SOS(:,1:3), AAc=SOS(:,4:6) % the numerator/denominator matrix
  Gm = 2.1081e+009
  BBc = 0   0   1    AAc = 1.0000e+000 3.5141e+004 2.1081e+009
        0   0   1          1.0000e+000 8.4838e+004 2.1081e+009
>>[BBp,AAp]=tf2par_s(Bb,Ab) % parallel realization
 BBp = 0  4.2419e+004  3.5987e+009
       0 -4.2419e+004 -1.4906e+009
 AAp = 1.0000e+000  8.4838e+004 2.1081e+009
       1.0000e+000  3.5141e+004 2.1081e+009
```

This means that the designed transfer function can be realized in cascade and parallel form as

$$G(s) = \frac{2.108\times 10^9}{s^2 + 3.514\times 10^4 s + 2.108\times 10^9} \times \frac{2.108\times 10^9}{s^2 + 8.484\times 10^4 s + 2.108\times 10^9} \tag{E6.1.3a}$$

$$G(s) = \frac{4.242\times 10^4 s + 3.599\times 10^9}{s^2 + 8.484\times 10^4 s + 2.108\times 10^9} - \frac{4.242\times 10^4 s + 1.491\times 10^9}{s^2 + 3.514\times 10^4 s + 2.108\times 10^9} \tag{E6.1.3b}$$

b) We are going to determine the transfer function of a Chebyshev I BPF with

$$\omega_{s1} = 2\pi \times 6000\,[\mathrm{rad/s}],\ \omega_{p1} = 2\pi \times 10000\,[\mathrm{rad/s}],$$
$$\omega_{p2} = 2\pi \times 12000\,[\mathrm{rad/s}],\ \omega_{s2} = 2\pi \times 15000\,[\mathrm{rad/s}]$$
$$R_p \le 2\,[\mathrm{dB}],\ \mathrm{and}\ A_s \ge 25\,[\mathrm{dB}] \tag{E6.1.4}$$

First, we use the MATLAB function 'cheb1ord()' to find the filter order N and the critical passband edge frequencies ω_{pc1} and ω_{pc2} at which the passband ripple condition is closely met, i.e. $20\log_{10}|G(j\omega_{pc})| = -R_p$[dB] by running the following statements:

```
>>ws1=2*pi*6e3; wp1=2*pi*1e4; wp2=2*pi*12e3; ws2=2*pi*15e3;
Rp=2; As=25;
>>[N,wpc]=cheb1ord([wp1 wp2],[ws1 ws2],Rp,As,'s')
   N = 2, wpc = 6.2832e+004 7.5398e+004
```

We put the (half) filter order N, the passband ripple Rp, and the critical passband edge frequency vector wpc = $[\omega_{pc1}\ \omega_{pc2}]$ into the Chebyshev I filter design function 'cheby1()' as

```
>>[Bc1,Ac1]=cheby1(N,Rp,wpc,'s')
   Bc1 =            0            0 1.0324e+008          0           0
   Ac1 = 1.0000e+000 1.0101e+004 9.6048e+009 4.7853e+013 2.2443e+019
```

This means that the transfer function of the designed Chebyshev I filter of order 2N = 4 is

$$G(s) = \frac{1.0324\times10^8 s^2}{s^4 + 10101s^3 + 9.6048\times10^9 s^2 + 4.7853\times10^{13}s + 2.2443\times10^{19}} \tag{E6.1.5}$$

We can find the cascade and parallel realizations of this transfer function by running the following MATLAB statements:

```
>>[SOS,K]=tf2sos(Bc1,Ac1); % cascade realization
>>Ns=size(SOS,1); Gm=K^(1/Ns), BBc=SOS(:,1:3), AAc=SOS(:,4:6)
  Gm = 1.0161e+004
  BBc = 0   0   1   AAc = 1.0000e+000 5.4247e+003 5.4956e+009
        1   0   1         1.0000e+000 4.6763e+003 4.0838e+009
>>[BBp,AAp]=tf2par_s(Bc1,Ac1) % parallel realization
   BBp = 0  1.8390e+002 4.0242e+008
         0 -1.8390e+002 -2.9904e+008
   AAp = 1.0000e+000 5.4247e+003 5.4956e+009
         1.0000e+000 4.6763e+003 4.0838e+009
```

This means that the designed transfer function can be realized in cascade and parallel form as

$$G(s) = \frac{1.0161\times10^4 s}{s^2 + 5.425\times10^3 s + 5.496\times10^9} \times \frac{1.0161\times10^4 s}{s^2 + 4.676\times10^3 s + 4.084\times10^9} \tag{E6.1.6a}$$

$$G(s) = \frac{1.839\times10^2 s + 4.024\times10^8}{s^2 + 5.425\times10^3 s + 5.496\times10^9} - \frac{1.839\times10^2 s + 2.990\times10^8}{s^2 + 4.676\times10^3 s + 4.084\times10^9} \tag{E6.1.6b}$$

c) We are going to determine the transfer function of a Chebyshev II BSF with

$$\omega_{p1} = 2\pi \times 6000\,[\text{rad/s}],\ \omega_{s1} = 2\pi \times 10000\,[\text{rad/s}],$$
$$\omega_{s2} = 2\pi \times 12000\,[\text{rad/s}],\ \omega_{p2} = 2\pi \times 15000\,[\text{rad/s}]$$
$$R_p \le 2\,[\text{dB}], \text{and} A_s \ge 25\,[\text{dB}] \tag{E6.1.7}$$

First, we use the MATLAB function '`cheb2ord()`' to find the filter order N and the critical stopband edge frequencies ω_{sc1} and ω_{sc2} at which the stopband attenuation condition is closely met, i.e. $20\log_{10}|G(j\omega_{sc})| = -A_s$ [dB] by running the following MATLAB statements:

```
>>wp1=2*pi*6000; ws1=2*pi*10000; ws2=2*pi*12000; wp2=2*pi*15000;
>>Rp=2; As=25;
>>[N,wsc]=cheb2ord([wp1 wp2],[ws1 ws2],Rp,As,'s')
   N = 2, wsc = 6.2798e+004 7.5438e+004
```

We put the (half) filter order N, the stopband attenuation As, and the critical stopband edge frequency vector wsc = $[\omega_{sc1}\ \omega_{sc2}]$ into the Chebyshev II filter design function '`cheby2()`' as

```
>>[Bc2,Ac2]=cheby2(N,As,wsc,'stop','s')
  Bc2 = 1.0000e+000 1.0979e-010 9.5547e+009 4.9629e-001 2.2443e+019
  Ac2 = 1.0000e+000 5.1782e+004 1.0895e+010 2.4531e+014 2.2443e+019
```

This means that the transfer function of the designed Chebyshev II filter of order 2N = 4 is

$$G(s) = \frac{s^4 + 9.5547 \times 10^9 s^2 + 4.9629 \times 10^{-1} s + 2.2443 \times 10^{19}}{s^4 + 51782 s^3 + 1.0895 \times 10^{10} s^2 + 2.4531 \times 10^{14} s + 2.2443 \times 10^{19}} \tag{E6.1.8}$$

We can find the cascade and parallel realizations of this transfer function by running the following MATLAB statements:

```
>>[SOS,K]=tf2sos(Bc2,Ac2); % cascade realization
>>Ns=size(SOS,1); Gm=K^(1/Ns), BBc=SOS(:,1:3), AAc=SOS(:,4:6)
  Gm = 1
  BBc = 1.0000e+000 7.7795e-011 5.3938e+009
        1.0000e+000 2.9104e-011 4.1609e+009
  AAc = 1.0000e+000 3.1028e+004 7.0828e+009
        1.0000e+000 2.0754e+004 3.1687e+009
>>[BBp,AAp]=tf2par_s(Bc2,Ac2) % parallel realization
  BBp = 5.0000e-001 -1.5688e+004 3.4426e+009
        5.0000e-001 -1.0204e+004 1.6285e+009
  AAp = 1.0000e+000  3.1028e+004 7.0828e+009
        1.0000e+000  2.0754e+004 3.1687e+009
```

This means that the designed transfer function can be realized in cascade and parallel form as

$$G(s) = \frac{s^2 + 5.394 \times 10^9}{s^2 + 3.103 \times 10^4 s + 7.083 \times 10^9} \times \frac{s^2 + 4.161 \times 10^9}{s^2 + 2.075 \times 10^4 s + 3.169 \times 10^9} \quad \text{(E6.1.9a)}$$

$$G(s) = \frac{0.5s^2 - 1.569 \times 10^4 s + 3.443 \times 10^9}{s^2 + 3.103 \times 10^4 s + 7.083 \times 10^9} + \frac{0.5s^2 - 1.020 \times 10^4 s + 1.6285 \times 10^9}{s^2 + 2.075 \times 10^4 s + 3.169 \times 10^9} \quad \text{(E6.1.9b)}$$

d) We are going to determine the transfer function of an *elliptic* HPF with

$$\omega_s = 2\pi \times 6000\,[\text{rad/s}], \omega_p = 2\pi \times 15000\,[\text{rad/s}], R_p \le 2\,[\text{dB}], \text{ and } A_s \ge 25\,[\text{dB}] \quad \text{(E6.1.10)}$$

First, we use the MATLAB function '`ellipord()`' to find the filter order N and the critical passband edge frequency ω_{pc} at which $20\log_{10}|G(j\omega_{pc})| = -R_p$ [dB] by running the following MATLAB statements:

```
>>ws=2*pi*6000; wp=2*pi*15000; Rp=2; As=25;
>>format short e, [N,wpc]=ellipord(wp,ws,Rp,As,'s')
  N = 3, wpc = 9.4248e+004
```

We put the parameter values N, Rp, As, and wc into the elliptic filter design function '`ellip()`' as

```
>>[Be,Ae]=ellip(N,Rp,As,wpc,'high','s')
   Be = 1.0000e+000 8.9574e-009 3.9429e+009 -5.6429e+002
   Ae = 1.0000e+000 2.3303e+005 1.4972e+010 1.9511e+015
```

This means that the transfer function of the designed elliptic filter of order N = 3 is

$$G(s) = \frac{s^3 + 3.9429 \times 10^9 s - 5.6429 \times 10^2}{s^3 + 2.3303 \times 10^5 s^2 + 1.4972 \times 10^{10} s + 1.9511 \times 10^{15}} \quad \text{(E6.1.11)}$$

We can find the cascade and parallel realizations of this transfer function by running the following MATLAB statements:

```
>>[SOS,K]=tf2sos(Be,Ae); % cascade realization
>>Ns=size(SOS,1); Gm=K^(1/Ns), BBc=SOS(:,1:3), AAc=SOS(:,4:6)
   Gm = 1.0000e+000
   BBc = 1.0000e+000 -1.4311e-007          0
         1.0000e+000  1.5207e-007 3.9429e+009
   AAc = 1.0000e+000  2.0630e+005          0
         1.0000e+000  2.6731e+004 9.4575e+009

>>[BBp,AAp]=tf2par_s(Be,Ae) % parallel realization
   BBp = 5.0000e-001  -1.3365e+004   4.7287e+009
         0             5.0000e-001  -1.0315e+005
   AAp = 1.0000e+000   2.6731e+004   9.4575e+009
                   0   1.0000e+000   2.0630e+005
```

This means that the designed transfer function can be realized in cascade and parallel form as

$$G(s) = \frac{s}{s + 2.063 \times 10^5} \times \frac{s^2 + 3.943 \times 10^9}{s^2 + 2.673 \times 10^4 s + 9.458 \times 10^9} \tag{E6.1.12a}$$

$$G(s) = \frac{0.5s^2 - 1.337 \times 10^4 s + 4.729 \times 10^9}{s^2 + 2.673 \times 10^4 s + 9.458 \times 10^9} + \frac{0.5s - 1.032 \times 10^5}{s + 2.063 \times 10^5} \tag{E6.1.12b}$$

e) We are going to put all the above filter design works into the M-file named "elec06e01.m," which plots the frequency responses of the designed filters so that one can check if the design specifications are satisfied. Figure 6.3, obtained by running the script "elec06e01.m," shows the following points:

- Figure 6.3(a) shows that the cutoff frequency ω_c given as the second argument of '`butter()`' is the frequency at which $20\log_{10}|G(j\omega_c)| = -3$ [dB]. Note that the frequency response magnitude of Butterworth filter is monotonic, i.e. has no ripple.
- Figure 6.3(b) shows that the critical passband edge frequencies ω_{pc1} and ω_{pc2} given as the third input argument wpc = [wpc1 wpc2] of '`cheby1()`' are the frequencies at which the passband ripple condition is closely met, i.e. $20\log_{10}|G(j\omega_{pc})| = -R_p$[dB]. Note that the frequency response magnitude of Chebyshev I filter satisfying the passband ripple condition closely has a ripple in the passband, which is traded off for a narrower transition band than the Butterworth filter (with the same filter order).
- Figure 6.3(c) shows that the critical stopband edge frequencies ω_{sc1} and ω_{sc2} given as the third input argument wsc = [wsc1 wps2] of '`cheby2()`' are the frequencies at which the stopband attenuation condition is closely met, i.e. $20\log_{10}|G(j\omega_{sc})| = -A_s$[dB]. Note that the frequency response magnitude of Chebyshev II filter satisfying the stopband ripple condition closely has a ripple in the stopband.
- Figure 6.3(d) shows that the critical passband edge frequency ω_{pc} given as the fourth input argument wpc of '`ellip()`' is the frequency at which the passband ripple condition is closely met, i.e. $20\log_{10}|G(j\omega_{pc})| = -R_p$[dB]. Note that the frequency response magnitude of elliptic filter has ripples in both the passband and the stopband, yielding a relatively narrow transition band with the smallest filter order N = 3 among the four filters.

```
%elec06e01.m for the filter design and frequency response plot
clear, clf, format short e
disp('(a) Butterworth LPF')
wp=2*pi*6000; ws=2*pi*15000; Rp=2; As=25;
[Nb,wcb]= buttord(wp,ws,Rp,As,'s') % Order of analog BW LPF
[Bb,Ab]= butter(Nb,wcb,'s') % num/den of analog BW LPF transfer ftn
[SOS,K]= tf2sos(Bb,Ab); % cascade realization
Ns=size(SOS,1); Gm=K^(1/Ns), BBc=SOS(:,1:3), AAc=SOS(:,4:6)
[BBp,AAp]= tf2par_s(Bb,Ab) % parallel realization
ww= logspace(4,6,1000); % log frequency vector from 1e4 to 1e6 [rad/s]
subplot(221), semilogx(ww,20*log10(abs(freqs(Bb,Ab,ww))))
title('Butterworth LPF')
disp('(b) Chebyshev I BPF')
ws1=2*pi*6e3; wp1=2*pi*1e4; wp2=2*pi*12e3; ws2=2*pi*15e3; Rp=2; As=25;
[Nc1,wpc]= cheb1ord([wp1 wp2],[ws1 ws2],Rp,As,'s')
[Bc1,Ac1]= cheby1(Nc1,Rp,wpc,'s')
[SOS,K]= tf2sos(Bc1,Ac1); % cascade realization
Ns=size(SOS,1);
Gm=K^(1/Ns), BBc=SOS(:,1:3), AAc=SOS(:,4:6)
[BBp,AAp]= tf2par_s(Bc1,Ac1) % parallel realization
subplot(222), semilogx(ww,20*log10(abs(freqs(Bc1,Ac1,ww))))
title('Chebyshev I BPF')
disp('(c) Chebyshev II BSF')
wp1=2*pi*6e3; ws1=2*pi*1e4; ws2=2*pi*12e3; wp2=2*pi*15e3; Rp=2; As=25;
[Nc2,wsc]= cheb2ord([wp1 wp2],[ws1 ws2],Rp,As,'s')
[Bc2,Ac2]= cheby2(Nc2,As,wsc,'stop','s')
[SOS,K]= tf2sos(Bc2,Ac2); % cascade realization
Ns=size(SOS,1);
Gm=K^(1/Ns), BBc=SOS(:,1:3), AAc=SOS(:,4:6)
[BBp,AAp]= tf2par_s(Bc2,Ac2) % parallel realization
subplot(224), semilogx(ww,20*log10(abs(freqs(Bc2,Ac2,ww))))
title('Chebyshev II BSF')
disp('(d) Elliptic HPF')
ws=2*pi*6000; wp=2*pi*15000; Rp=2; As=25;
[Ne,wpc]= ellipord(wp,ws,Rp,As,'s')
[Be,Ae]= ellip(Ne,Rp,As,wpc,'high','s')
[SOS,K]= tf2sos(Be,Ae); % cascade realization
Ns=size(SOS,1);
Gm=K^(1/Ns), BBc=SOS(:,1:3), AAc=SOS(:,4:6)
[BBp,AAp]= tf2par_s(Be,Ae) % parallel realization
subplot(223), semilogx(ww,20*log10(abs(freqs(Be,Ae,ww))))
```

```
function [BB,AA,K]=tf2par_s(B,A)
% Copyleft: Won Y. Yang, wyyang53@hanmail.net, CAU for academic use only
EPS= 1e-6;
B= B/A(1); A= A/A(1);
I= find(abs(B)>EPS); K= B(I(1)); B= B(I(1):end);
p= roots(A); p= cplxpair(p); Np= length(p);
NB= length(B); N= length(A); M= floor(Np/2);
for m=1:M
   m2= m*2; AA(m,:) = [1 -p(m2-1)-p(m2) p(m2-1)*p(m2)];
end
if Np>2*M
   AA(M+1,:)= [0 1 -p(Np)]; % For a single pole
end
M1= M+(Np>2*M); b= [zeros(1,Np-NB) B]; KM1= K/M1;
% If B(s) and A(s) have the same degree, we let all the coefficients
% of the 2nd-order term in the numerator of each SOS be Bi1=1/M1:
if NB==N, b= b(2:end); end
for m=1:M1
    polynomial = 1; m2=2*m;
    for n=1:M1
       if n~=m, polynomial = conv(polynomial,AA(n,:)); end
    end
    if m<=M
       if M1>M, polynomial = polynomial(2:end); end
       if NB==N, b = b - [polynomial(2:end)*KM1 0 0]; end
       Ac(m2-1,:) = [polynomial 0]; Ac(m2,:) = [0 polynomial];
     else
       if NB==N, b = b - [polynomial(2:end)*KM1 0]; end
       Ac(m2-1,:) = polynomial;
    end
end
Bc = b/Ac; Bc(find(abs(Bc)<EPS)) = 0;
for m=1:M1
   m2= 2*m;
   if m<=M
      BB(m,:) = [0 Bc(m2-1:m2)];
      if NB==N, BB(m,1) = KM1; end
   else
      BB(m,:) = [0 0 Bc(end)];
      if NB==N, BB(m,2) = KM1; end
   end
end
```

6.2 Passive Filter

6.2.1 Low-pass Filter (LPF)

6.2.1.1 Series *LR* Circuit

Figure 6.4(a) shows a series *LR* circuit where the voltage $v_R(t)$ across the resistor R is taken as the output to the input voltage source $v_i(t)$. With $V_R(s) = \mathcal{L}\{v_R(t)\}$

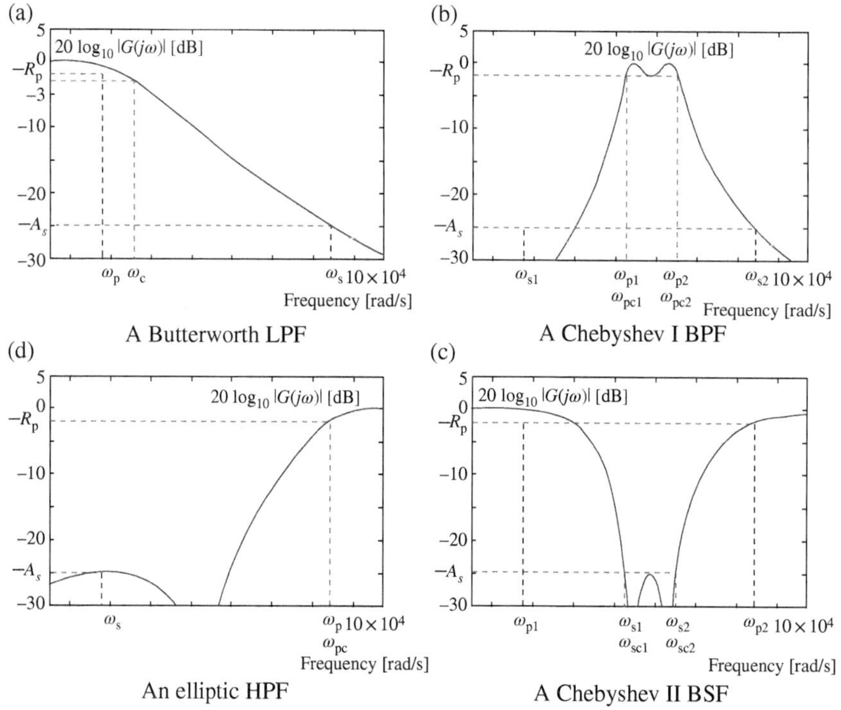

Figure 6.3 Frequency response magnitude curves of the filters designed in Example 8.7.

and $V_i(s)=\mathcal{L}\{v_i(t)\}$, its input-output relationship can be described by the transfer function and the frequency response as

$$G(s)=\frac{V_R(s)}{V_i(s)}=\frac{R}{sL+R}=\frac{R/L}{s+R/L} \tag{6.2.1a}$$

$$;G(j\omega)=\frac{\mathbf{V}_R(\text{phasor transform of } v_R(t))}{\mathbf{V}_i(\text{phasor transform of } v_i(t))}=\frac{R}{j\omega L+R}=\frac{R/L}{j\omega+R/L} \tag{6.2.1b}$$

Since the magnitude $|G(j\omega)|$ of this frequency response becomes smaller as the frequency ω gets higher as depicted in Figure 6.4(b), the circuit is a *lowpass filter* (*LPF*) that prefers to have low-frequency components in its output. Noting that $|G(j\omega)|$ achieves the maximum $G_{max} = G(j0) = 1$ at $\omega = 0$, we can find the cutoff frequency ω_c at which it equals $G_{max}/\sqrt{2} = 1/\sqrt{2}$ as

$$|G(j\omega_c)| \overset{(6.2.1b)}{=} \left.\frac{R/L}{\sqrt{\omega^2+(R/L)^2}}\right|_{\omega=\omega_c} = \frac{1}{\sqrt{2}};\ \omega_c=\frac{R}{L} \tag{6.2.2}$$

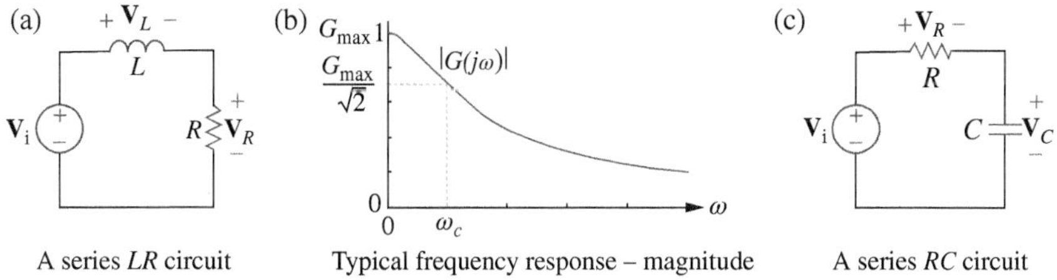

Figure 6.4 Low-pass filters and their typical frequency response.

The *cutoff frequency* has a physical meaning as the boundary frequency between the passband and the stopband. Why do we use $1/\sqrt{2}$ for its definition? In most cases, the transfer function and the frequency response is the ratio of output and input voltages/currents in *s*-frequency domain. Since an electric power is proportional to the squared voltage and current, the ratio of $1/\sqrt{2}$ between voltages/currents corresponds to the ratio of 1/2 between powers. For this reason, the cutoff frequency is also referred to as a *half-power frequency* or a *3dB frequency* on account of that $10\log_{10}(1/2) = -3$ [dB]. In the case of LPF with cutoff frequency ω_c like this circuit, the passband is the range of frequency $[0, \omega_c]$ and its width is the *bandwidth*, where the bandwidth means the width of the range of frequency components that pass the filter better than others.

6.2.1.2 Series *RC* Circuit

Figure 6.4(c) shows a series *RC* circuit where the voltage $v_C(t)$ across the capacitor *C* is taken as the output to the input voltage source $v_i(t)$. With $V_C(s)=\mathcal{L}\{v_C(t)\}$ and $V_i(s)=\mathcal{L}\{v_i(t)\}$, its input-output relationship can be described by the transfer function and the frequency response as

$$G(s) = \frac{V_C(s)}{V_i(s)} = \frac{1/sC}{R+1/sC} = 1/\frac{RC}{s+1/RC}; \tag{6.2.3a}$$

$$G(j\omega) = \frac{\mathbf{V}_C(\text{phasor transform of } v_C(t))}{\mathbf{V}_i(\text{phasor transform of } v_i(t))} = \frac{1/j\omega C}{R+1/j\omega C} = \frac{1/RC}{j\omega+1/RC} \tag{6.2.3b}$$

Everything is the same with the series *LR* circuit in Figure 6.4(a) except for that the cutoff frequency is

$$\omega_c = \frac{1}{RC} \tag{6.2.4}$$

6.2.2 High-pass Filter (HPF)

6.2.2.1 Series *CR* Circuit

Figure 6.5(a) shows a series *CR* circuit where the voltage $v_R(t)$ across the resistor *R* is taken as the output to the input voltage source $v_i(t)$. Its input-output relationship can be described by the transfer function and the frequency response as

$$G(s) = \frac{V_R(s)}{V_i(s)} = \frac{R}{R + 1/sC} = \frac{s}{s + 1/RC} \quad (6.2.5a)$$

$$; G(j\omega) = \frac{\mathbf{V}_R}{\mathbf{V}_i} = \frac{j\omega}{j\omega + 1/RC} \quad (6.2.5b)$$

Since the magnitude $|G(j\omega)|$ of this frequency response becomes larger as the frequency ω gets higher as depicted in Figure 6.5(b), the circuit is a *highpass filter* (*HPF*) that prefers to have high-frequency components in its output. Noting that $|G(j\omega)|$ achieves the maximum $G_{max}=G(j\infty)=1$ at $\omega=\infty$, we can find the cutoff frequency ω_c at which it equals $G_{max}/\sqrt{2} = 1/\sqrt{2}$ as

$$|G(j\omega_c)| = \left.\frac{\omega}{\sqrt{\omega^2 + (1/RC)^2}}\right|_{\omega=\omega_c} = \frac{1}{\sqrt{2}}; \quad \omega_c = \frac{1}{RC} \quad (6.2.6)$$

6.2.2.2 Series *RL* Circuit

Figure 6.5(c) shows a series *RL* circuit where the voltage $v_L(t)$ across the inductor L is taken as the output to the input voltage source $v_i(t)$. Its input-output relationship can be described by the transfer function and the frequency response as

$$G(s) = \frac{V_L(s)}{V_i(s)} = \frac{sL}{R + sL} = \frac{s}{s + R/L} \quad (6.2.7a)$$

$$G(j\omega) = \frac{\mathbf{V}_L}{\mathbf{V}_i} = \frac{j\omega}{j\omega + R/L} \quad (6.2.7b)$$

Everything is the same with the series *CR* circuit in Figure 6.5(a) except for that the cutoff frequency is

$$\omega_c = \frac{R}{L} \quad (6.2.8)$$

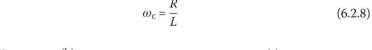

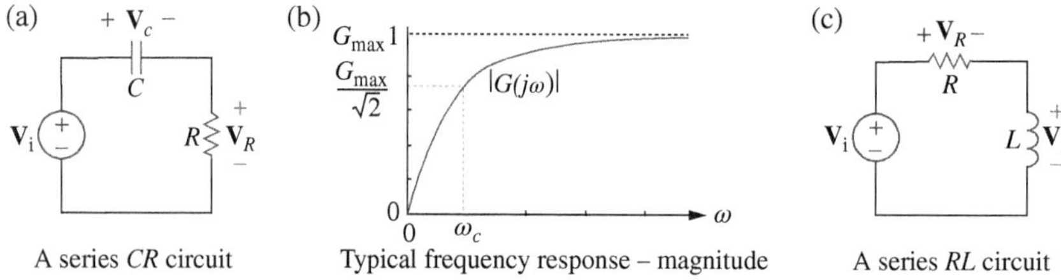

Figure 6.5 High-pass filters and their typical frequency response.

6.2.3 Band-pass Filter (BPF)

6.2.3.1 Series Resistor, an Inductor, and a Capacitor (*RLC*) Circuit and Series Resonance

Let us consider the series resistor, an inductor, and a capacitor (*RLC*) circuit of Figure 6.6(a) in which the voltage $v_R(t)$ across the resistor R is taken as the output to the input voltage source $v_i(t)$. Its input-output relationship can be described by the transfer function and the frequency response as

$$G(s) = \frac{V_R(s)}{V_i(s)} = \frac{R}{sL + R + 1/sC} = \frac{sR/L}{s^2 + sR/L + 1/LC} = \frac{\omega_b s}{s^2 + \omega_b s + \omega_p^2} \tag{6.2.9a}$$

$$G(j\omega) = \frac{\mathbf{V}_R}{\mathbf{V}_i} = \frac{j\omega\omega_b}{\left(\omega_p^2 - \omega^2\right) + j\omega\omega_b} \text{ with } \omega_b = \frac{R}{L} \text{ and } \omega_p = \frac{1}{\sqrt{LC}} \tag{6.2.9b}$$

Referring to the typical magnitude curve of the frequency response of a *band-pass filter* (*BPF*) shown in Figure 6.6(b), we can find the maximum of the magnitude, $|G(j\omega)|$, of the frequency response by setting the derivative of its square $|G(j\omega)|^2$ w.r.t. frequency ω to zero as

$$\frac{d}{d\omega}|G(j\omega)|^2 = \frac{d}{d\omega}\frac{\omega_b^2\omega^2}{\left(\omega_p^2 - \omega^2\right)^2 + \omega_b^2\omega^2} = \frac{2\,\omega_b^2\omega\left(\omega_p^2 + \omega^2\right)\left(\omega_p^2 - \omega^2\right)}{\left(\left(\omega_p^2 - \omega^2\right)^2 + \omega_b^2\omega^2\right)^2} = 0$$

This yields the *peak* or *center frequency*

$$\omega = \omega_p = \frac{1}{\sqrt{LC}} \tag{6.2.10}$$

at which the frequency response achieves the maximum magnitude as

$$G_{max} = |G(j\omega_p)| = \left.\frac{|\omega_b\omega|}{\sqrt{\left(\omega_p^2 - \omega^2\right)^2 + (\omega_b\omega)^2}}\right|_{\omega=\omega_p} = 1$$

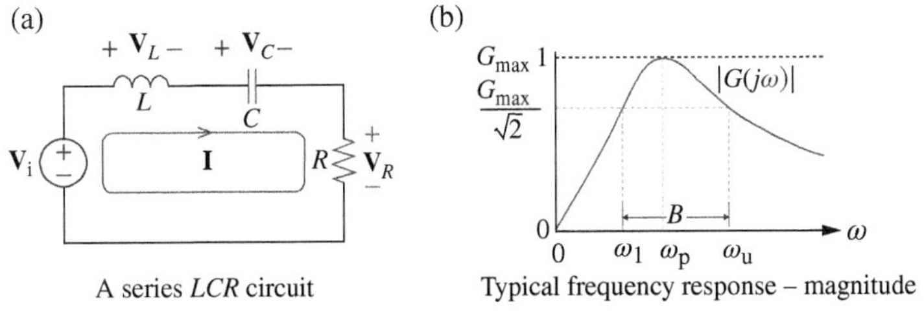

Figure 6.6 A band-pass filter (BPF) and its typical frequency response.

This frequency happens to coincide with the *resonant frequency* ω_r at which the frequency response is real so that the input and the output are in phase at that frequency. We can also find the *lower* and *upper 3dB frequencies* at which the magnitude $|G(j\omega)|$ of this frequency response is $G_{max}/\sqrt{2} = 1/\sqrt{2}$ as

$$|G(j\omega)|^2 = \frac{\omega_b^2\omega^2}{\left(\omega_p^2 - \omega^2\right)^2 + \omega_b^2\omega^2} = \frac{1}{2}; \ \omega_p^2 - \omega^2 = \pm\,\omega_b\,\omega; \ \omega^2 \pm \omega_b\,\omega - \omega_p^2 = 0$$

$$;\omega_l, \omega_u = \frac{\mp\omega_b + \sqrt{\omega_b^2 + 4\,\omega_p^2}}{2} = \mp\frac{R}{2L} + \sqrt{\left(\frac{R}{2L}\right)^2 + \frac{1}{LC}} \tag{6.2.11}$$

Multiplying these two 3 dB frequencies and taking the square root yields

$$\sqrt{\omega_l\,\omega_u} = \omega_p = \omega_r; \ \frac{1}{2}(\log\omega_l + \log\omega_u) = \log\omega_p \tag{6.2.12}$$

This implies that the peak frequency ω_p is the geometrical mean of two 3 dB frequencies ω_l and ω_u and also the arithmetical mean of their logarithmic values that is at the midpoint between the two 3 dB frequency points on the log-frequency ($\log \omega$) axis. That is why ω_p is also called the *center frequency*. The frequency band between the two 3 dB frequencies ω_l and ω_u is the *passband* and its width is the *bandwidth*:

$$B = \omega_u - \omega_l = \frac{R}{L} = \omega_b \tag{6.2.13}$$

Note that the bandwidth for a BPF is the width of the range of frequency components that pass the filter better than others.

For a BPF, we define the *quality factor* or *selectivity* (*sharpness*) as

$$Q = \frac{\omega_p}{B} = \frac{\omega_p}{\omega_u - \omega_l} = \frac{1/\sqrt{LC}}{R/L} = \frac{1}{R}\sqrt{\frac{L}{C}} \tag{6.2.14}$$

This quality factor is referred to as the selectivity since it is a measure of how selectively the filter responds to the peak frequency and frequencies near it within the passband, discriminating against frequencies outside the passband. It is also referred to as the sharpness in the sense that it describes how sharp the magnitude curve of the frequency response is around the peak frequency. It coincides with the *voltage magnification ratio*, that is, the ratio of the magnitude of the voltage across the inductor or the capacitor to that of the input voltage at peak frequency as

$$\frac{|\mathbf{V}_L|}{|\mathbf{V}_i|} = \frac{|\mathbf{V}_C|}{|\mathbf{V}_i|} = \frac{|\,j\omega_p L\,| = |\,1/j\omega_p C\,|}{|\,R + j\omega_p L + 1/j\omega_p C\,|} = \frac{\omega_p L = 1/\omega_p C}{|\,R + j(\omega_p L - 1/\omega_p C)\,|} = \frac{1}{R}\sqrt{\frac{L}{C}} \equiv Q \tag{6.2.15}$$

which may be greater than unity. Isn't it interesting that the voltages across some components of a circuit can be higher than the applied voltage?

Now, let us see the following example showing how to determine the parameters of a series *RLC* circuit so that it can function as a BPF satisfying a design specification on the passband.

Example 6.2 Realization of a BPF with a Series *RLC* Circuit

Consider a series *RLC* circuit with the inductor of inductance $L = 12.9$ mH. Determine the values of the resistance R and capacitance C such that the circuit can function as a BPF with the passband between the two frequencies 691 and 941 Hz.

Noting that the two boundary frequencies are given as the lower/upper 3dB frequencies, we have

$$\omega_\mathrm{l} = 2\pi \times 697 = 4379\ [\mathrm{rad/s}]\ \text{and}\ \omega_\mathrm{u} = 2\pi \times 941 = 5912[\mathrm{rad/s}] \tag{E6.2.1}$$

First, we substitute these upper/lower 3dB frequencies into Eq. (6.2.12) to get the peak frequency as

$$\omega_\mathrm{p} \overset{(6.2.12)}{=} \sqrt{\omega_\mathrm{u}\,\omega_\mathrm{l}} \overset{(E6.2.1)}{=} \sqrt{5912 \times 4379} = 5088\ [\mathrm{rad/s}] \tag{E6.2.2}$$

Then, we use Eq. (6.2.10) for the peak frequency to find the value of C such that the peak frequency of the circuit will be 5088[rad/s]:

$$C \overset{(6.2.10)}{=} \frac{1}{L\,\omega_\mathrm{p}^2} = \frac{1}{0.0129 \times 4379 \times 5912} = 3\ [\mu\mathrm{F}] \tag{E6.2.3}$$

We also use Eq. (6.2.13) for the bandwidth to find the value of R such that the bandwidth of the circuit will be $\omega_\mathrm{u} - \omega_\mathrm{l}$ [rad/s]:

$$R \overset{(6.2.13)}{=} BL = (\omega_u - \omega_l)L = (5912 - 4379) \times 0.0129 = 19.8\ [\Omega] \tag{E6.2.4}$$

6.2.3.2 Parallel *RLC* Circuit and Parallel Resonance

Let us consider the parallel *RLC* circuit of Figure 6.7(a) in which the current $i_R(t)$ through the resistor R is taken as the output to the input current source $i_\mathrm{i}(t)$.

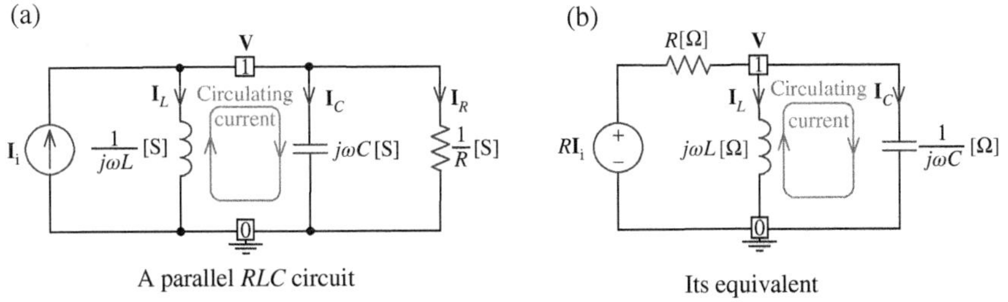

Figure 6.7 A parallel resistor, an inductor, and a capacitor (*RLC*) circuit and its equivalent.

Its input-output relationship can be described by the transfer function and the frequency response as

$$G(s)=\frac{I_R(s)}{I_{\rm i}(s)}=\frac{1/R}{1/R+sC+1/sL}=\frac{s/RC}{s^2+s/RC+1/LC}=\frac{\omega_{\rm b}s}{s^2+\omega_{\rm b}s+\omega_{\rm p}^2} \tag{6.2.16a}$$

$$;G(j\omega)=\frac{\mathbf{I}_R}{\mathbf{I}_{\rm i}}=\frac{j\omega\omega_{\rm b}}{\left(\omega_{\rm p}^2-\omega^2\right)+j\omega\omega_{\rm b}}\ \text{with}\ \omega_{\rm b}=\frac{1}{RC}\ \text{and}\ \omega_{\rm p}=\frac{1}{\sqrt{LC}}. \tag{6.2.16b}$$

Since this frequency response is the same as that, Eq. (6.2.9b), of the series *RLC* circuit in Figure 6.6(a) except for $\omega_{\rm b}$, we can obtain the *peak* or *center frequency*, the *upper/lower 3dB frequencies*, the *bandwidth*, and the *quality factor* as follows:

$$\omega=\omega_{\rm p}=\frac{1}{\sqrt{LC}} \tag{6.2.17a}$$

$$\omega_{\rm l},\omega_{\rm u}=\frac{\mp\omega_{\rm b}+\sqrt{\omega_{\rm b}^2+4\,\omega_{\rm p}^2}}{2}=\mp\frac{1}{2RC}+\sqrt{\left(\frac{1}{2RC}\right)^2+\frac{1}{LC}} \tag{6.2.17b}$$

$$B=\omega_{\rm u}-\omega_{\rm l}=\frac{1}{RC}=\omega_{\rm b} \tag{6.2.17c}$$

$$Q=\frac{\omega_{\rm p}}{B}=\frac{\omega_{\rm p}}{\omega_{\rm u}-\omega_{\rm l}}=\frac{1/\sqrt{LC}}{1/RC}=R\sqrt{\frac{C}{L}}. \tag{6.2.17d}$$

The peak frequency happens to be coinciding with the *resonant frequency* $\omega_{\rm r}$ in the sense that the frequency response is real so that the input and the output are in phase at that frequency. The quality factor coincides with the *current magnification ratio*, that is, the ratio of the amplitude of the current through the inductor or the capacitor to that of the input current at resonant frequency as

$$\frac{|\mathbf{I}_L|}{|\mathbf{I}_{\rm i}|}=\frac{|\mathbf{I}_C|}{|\mathbf{I}_{\rm i}|}=\frac{|1/j\omega_{\rm p}L|=|j\omega_{\rm p}C|}{|1/R+1/j\omega_{\rm p}L+j\omega_{\rm p}C|}=\frac{1/\omega_{\rm p}L=\omega_{\rm p}C}{|1/R+j(\omega_{\rm p}C-1/\omega_{\rm p}L)|}=R\sqrt{\frac{C}{L}}\equiv Q \tag{6.2.18}$$

which may be greater than unity. How strange it is that the currents through some components of a circuit can be larger than the applied current!

Let us now consider *RLC* circuit of Figure 6.7(b) in which the voltage across $L\|C$ (the parallel combination of L and C) is taken as the output to the input voltage source $R\,\mathbf{I}_{\rm i}$. We can use the voltage divider rule to obtain the transfer function as

$$G(s)=\frac{V_{LC}(s)}{RI_{\rm i}(s)}=\frac{(sL\|1/sC)}{R+(sL\|1/sC)}=\frac{(L/C)/(sL+1/sC)}{R+(L/C)/(sL+1/sC)}=\frac{s/RC}{s^2+s/RC+1/LC}$$

This is the same as Eq. (6.2.16a), which is the transfer function of the parallel *RLC* circuit of Figure 6.7(a). In fact, the circuit of Figure 6.7(b) is obtained by transforming the current source $\mathbf{I}_i$ in parallel with R into a voltage source $R\,\mathbf{I}_i$ in series with R in the circuit of Figure 6.7(a).

Remark 6.1 Derivation of Resonance Condition for a BPF

Instead of setting the derivative of the squared magnitude of frequency response to zero, we can derive the *resonance condition* for a BPF by setting the imaginary part of the impedance or admittance to zero.

Example 6.3 Practical Resonance Condition

As suggested in Remark 6.1, the resonance condition of the circuit of Figure 6.8(a) can be obtained by setting the imaginary part of the impedance or admittance to zero. Since the admittance is easier to find than the impedance for this circuit, we set the imaginary part of the admittance to zero so that the frequency response will be real (see the admittance triangle in Figure 6.8(b)).

$$Y(j\omega) = \frac{1}{R+j\omega L} + j\omega C = \frac{R}{R^2+\omega^2L^2} + j\omega\left(C - \frac{L}{R^2+\omega^2L^2}\right) \tag{E6.3.1}$$

$$;\omega\left(C - \frac{L}{R^2+\omega^2L^2}\right) = 0;\; C - \frac{L}{R^2+\omega^2L^2} = 0$$

This yields the resonant frequency as

$$\omega_r = \sqrt{\frac{1}{LC} - \frac{R^2}{L^2}} \tag{E6.3.2}$$

6.2.4 Band-stop Filter (BSF)

6.2.4.1 Series *RLC* Circuit

Let us consider the series *RLC* circuit of Figure 6.9(a) in which the voltage $v_{LC}(t)$ across the series combination of L and C is taken as the output to the input

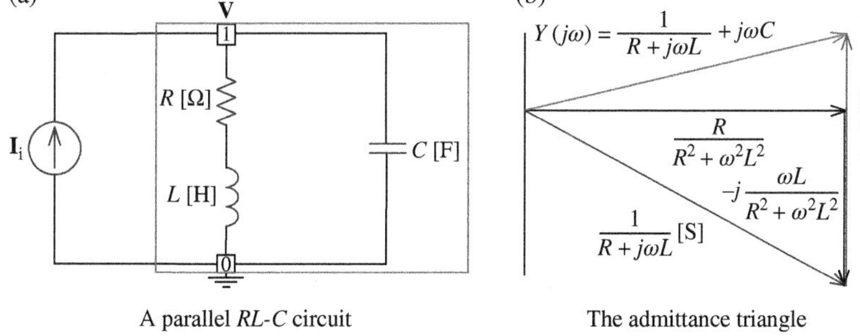

Figure 6.8 A practical parallel resonant circuit and its admittance triangle.

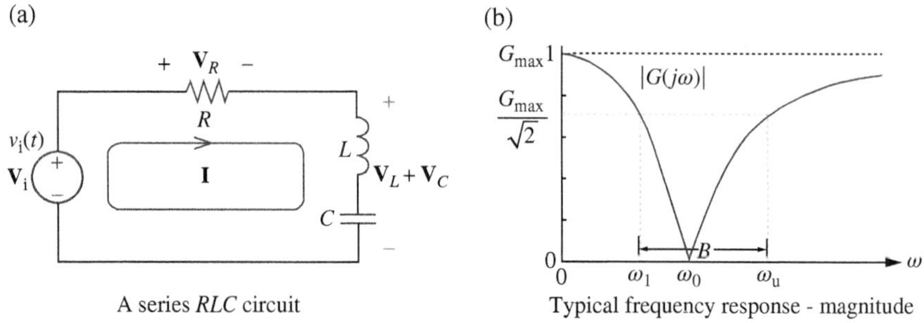

Figure 6.9 A band-stop filter (BSF) and its typical frequency response.

voltage source $v_i(t)$. Its input-output relationship can be described by the transfer function and the frequency response as

$$G(s) = \frac{V_L(s) + V_C(s)}{V_i(s)} = \frac{sL + 1/sC}{sL + R + 1/sC} = \frac{s^2 + 1/LC}{s^2 + sR/L + 1/LC} = \frac{s^2 + \omega_0^2}{s^2 + \omega_b s + \omega_0^2} \quad (6.2.19a)$$

$$;G(j\omega) = \frac{\mathbf{V}_L + \mathbf{V}_C}{\mathbf{V}_i} = \frac{\omega_0^2 - \omega^2}{(\omega_0^2 - \omega^2) + j\omega\omega_b} \text{ with } \omega_b = \frac{R}{L} \text{ and } \omega_0 = \frac{1}{\sqrt{LC}} \quad (6.2.19b)$$

Referring to the typical magnitude curve of the frequency response of a BSF shown in Figure 6.9(b), we can find the minimum of the magnitude of the frequency response by setting the derivative of its square w.r.t. the frequency ω to zero as

$$\frac{d}{d\omega}|G(j\omega)|^2 = \frac{d}{d\omega}\frac{(\omega_0^2 - \omega^2)^2}{(\omega_0^2 - \omega^2)^2 + \omega_b^2\omega^2} = -\frac{2\omega(\omega_0^2 - \omega^2)(\omega_0^2 + \omega^2)\,\omega_b^2}{\left((\omega_0^2 - \omega^2)^2 + \omega_b^2\omega^2\right)^2} = 0$$

This yields the *notch, rejection,* or *center* frequency

$$\omega = \omega_r = \omega_0 = \frac{1}{\sqrt{LC}} \quad (6.2.20)$$

at which the frequency response achieves the minimum magnitude as

$$G_{min} = |G(j\omega_0)| = \left.\frac{|\omega_0^2 - \omega^2|}{\sqrt{(\omega_0^2 - \omega^2)^2 + (\omega_b\omega)^2}}\right|_{\omega = \omega_r = \omega_0} = 0$$

(Note) It is hoped that the readers will enjoy the dual sense of ω_r, as the resonant frequency of BPF or the rejection (notch) frequency of BSF (depending on the context), rather than being confused.

Noting that the magnitude of the frequency response achieves its maximum of unity at another extremum frequency $\omega = 0$ or $\omega = \infty$ as

$$G_{\max} = |G(j0)| = \left.\frac{|\omega_0^2 - \omega^2|}{\sqrt{(\omega_0^2-\omega^2)^2 + (\omega_b\omega)^2}}\right|_{\omega=0} = 1$$

we can find the lower and upper 3 dB frequencies at which the magnitude $|G(j\omega)|$ of this frequency response is $G_{\max}/\sqrt{2} = 1/\sqrt{2}$ as

$$|G(j\omega)|^2 = \frac{(\omega_0^2-\omega^2)^2}{(\omega_0^2-\omega^2)^2 + \omega_b^2\omega^2} = \frac{1}{2};\ \omega_0^2 - \omega^2 = \pm\,\omega_b\,\omega;\ \omega^2 \pm \omega_b\,\omega - \omega_0^2 = 0$$

$$;\omega_l, \omega_u = \frac{\mp\omega_b + \sqrt{\omega_b^2 + 4\,\omega_0^2}}{2} = \mp\frac{R}{2L} + \sqrt{\left(\frac{R}{2L}\right)^2 + \frac{1}{LC}}$$

Multiplying these two 3 dB frequencies and taking the square root yields

$$\sqrt{\omega_l\,\omega_u} = \omega_0 = \omega_r;\quad \frac{1}{2}(\log\omega_l + \log\omega_u) = \log\omega_r \tag{6.2.21}$$

This implies that the notch frequency ω_r is the geometrical mean of two 3 dB frequencies ω_l and ω_u and also the arithmetical mean of their logarithmic values that is at the midpoint between the two 3 dB frequency points on the log-frequency (log ω) axis. The band of frequencies between the two 3 dB frequencies ω_l and ω_u is the stopband and its width is the *bandwidth*:

$$B = \omega_u - \omega_l = \frac{R}{L} = \omega_b \tag{6.2.22}$$

Note that the bandwidth for a BSF is the width of the range of frequency components that are relatively more attenuated or rejected by the filter than other frequencies.

For a BSF, we define the *quality factor* or *selectivity* (*sharpness*) as

$$Q = \frac{\omega_0}{B} = \frac{\omega_0}{\omega_u - \omega_l} = \frac{1/\sqrt{LC}}{R/L} = \frac{1}{R}\sqrt{\frac{L}{C}} \tag{6.2.23}$$

This quality factor is referred to as the selectivity since it is a measure of how selectively the filter rejects the notch frequency and frequencies near it within the stopband, favoring frequencies outside the stopband. It is also referred to as the sharpness in the sense that it describes how sharp the magnitude curve of the frequency response is around the rejection or notch frequency.

Example 6.4 Realization of a BSF with a Series *RLC* Circuit

Consider a series *RLC* circuit with an inductor of inductance L = 13.9 mH. Determine the values of the resistance R and the capacitance C such that the circuit can function as a BSF with the stopband between two frequencies 637 and 1432 Hz.

Noting that the two boundary frequencies are given as the lower/upper 3 dB frequencies, we have

$$\omega_l = 2\pi \times 637 = 4000\ [\text{rad/s}]\,\text{and}\,\omega_u = 2\pi \times 1432 = 9000\ [\text{rad/s}] \tag{E6.4.1}$$

First, we substitute these upper/lower 3 dB frequencies into Eq. (6.2.21) to get the notch frequency as

$$\omega_0 \overset{(6.2.21)}{=} \sqrt{\omega_u \omega_l} = \sqrt{4000 \times 9000} = 6000\ [\text{rad/s}] \tag{E6.4.2}$$

Then, we use Eq. (6.2.20) for the notch frequency to find the value of C such that the notch frequency of the circuit will be 6000 [rad/s]:

$$C \overset{(6.2.20)}{=} \frac{1}{L\,\omega_0^2} = \frac{1}{0.0139 \times 4000 \times 9000} = 2\ [\mu\text{F}] \tag{E6.4.3}$$

We also use Eq. (6.2.22) for the bandwidth to find the value of R such that the bandwidth of the circuit will be ω_u-ω_l[rad/s]:

$$R \overset{(6.2.22)}{=} BL = (\omega_u - \omega_l)L = (9000 - 4000) \times 0.0139 = 69.5\ [\Omega] \tag{E6.4.4}$$

6.2.4.2 Parallel *RLC* Circuit

Let us consider the parallel *RLC* circuit of Figure 6.10(a) in which the current $i_L(t) + i_C(t)$ through the parallel combination of L and C is taken as the output to the input current source $i_i(t)$. Its input-output relationship can be described by the transfer function and the frequency response as

$$G(s) = \frac{I_L(s) + I_C(s)}{I_i(s)} = \frac{1/sL + sC}{1/R + 1/sL + sC} = \frac{s^2 + 1/LC}{s^2 + s/RC + 1/LC} = \frac{s^2 + \omega_0^2}{s^2 + \omega_b s + \omega_0^2} \tag{6.2.24a}$$

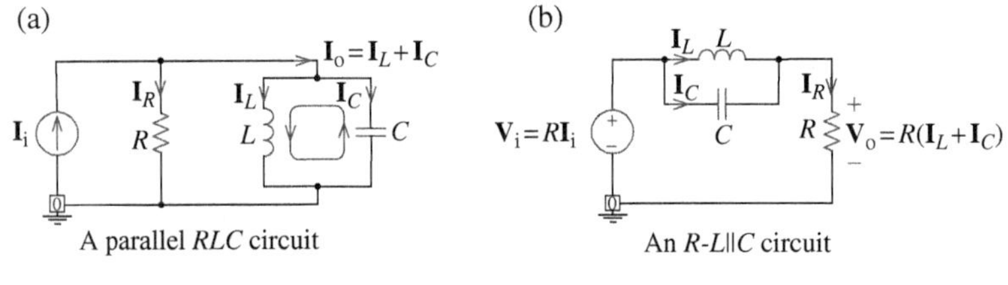

Figure 6.10 Band-stop filter (BSF).

$$G(j\omega) = \frac{\mathbf{I}_R}{\mathbf{I}_{\rm i}} = \frac{\omega_0^2 - \omega^2}{(\omega_0^2 - \omega^2) + j\omega\omega_{\rm b}} \text{ with } \omega_{\rm b} = \frac{1}{RC} \text{ and } \omega_0 = \frac{1}{\sqrt{LC}} \tag{6.2.24b}$$

Since this frequency response is the same as that, Eq. (6.2.19b), of the series *RLC* circuit in Figure 6.9(a) except for $\omega_{\rm b}$, we have the same results for the notch or rejection or center frequency, the upper/lower 3 dB frequencies, the bandwidth, and the quality factor.

Let us now consider the *RLC* circuit of Figure 6.10(b) in which the voltage across R is taken as the output to the input voltage source $R\mathbf{I}_{\rm i}$. We can use the voltage divider rule to obtain the transfer function as

$$G(s) = \frac{V_R(s)}{RI_{\rm i}(s)} = \frac{R}{R + (sL \| 1/sC)} = \frac{s^2 + 1/LC}{s^2 + s/RC + 1/LC}$$

This is exactly the same as Eq. (6.2.24a), which is the transfer function of the parallel *RLC* circuit of Figure 6.10(a). In fact, this circuit is obtained by transforming the current source $\mathbf{I}_{\rm i}$ in parallel with R into a voltage source $R\mathbf{I}_{\rm i}$ in series with R in the circuit of Figure 6.10(a).

6.2.5 Quality Factor

In the previous sections, the quality factors for series and parallel *RLC* circuits were defined as the ratio of the center frequency $\omega_{\rm c}$ to the bandwidth $B = \omega_{\rm u} - \omega_{\rm l}$ where $\omega_{\rm u}$ and $\omega_{\rm l}$ are the upper and lower 3 dB frequencies, respectively:

$$Q \overset{(6.2.14)}{=} \frac{\omega_{\rm c} = \omega_{\rm p} = \sqrt{1/LC}}{\omega_{\rm b} = \omega_{\rm u} - \omega_{\rm l} = R/L} = \frac{\omega_{\rm c} L I^2 = I^2/\omega_{\rm c} C}{RI^2} = \frac{\text{reactive power flowing between } L \text{ and } C}{\text{active power (loss) through } R} \tag{6.2.25a}$$

$$Q \overset{(6.2.17d)}{=} \frac{\omega_{\rm c} = \omega_{\rm p} = \sqrt{1/LC}}{\omega_{\rm b} = \omega_{\rm u} - \omega_{\rm l} = 1/RC} = \frac{\omega_{\rm c} C V^2 = V^2/\omega_{\rm c} L}{V^2/R} = \frac{\text{reactive power flowing between } L \text{ and } C}{\text{active power (loss) through } R} \tag{6.2.25b}$$

Note that the quality factor Q of a circuit can be viewed as the ratio of the reactive power (i.e. the maximum power stored in L or C) to the active power of the circuit.

Now, let us consider the quality factor of an individual reactive element L or C at the center frequency $\omega_{\rm c}$ of the circuit containing the element. It depends on whether the element is modeled as a series or parallel combination with its loss (parasitic) resistance:

$$Q = \frac{|X_{\rm s}|}{R_{\rm s}} \quad \text{or} \quad Q = \frac{R_{\rm p}}{|X_{\rm p}|} \tag{6.2.26a,b}$$

where

$$R_s + jX_s = R_p \| jX_p = \frac{R_p \times jX_p}{R_p + jX_p} = \frac{R_p X_p \left(X_p + jR_p\right)}{R_p^2 + X_p^2} \tag{6.2.27}$$

$$; R_s + jQR_s \overset{(6.2.27)}{\underset{(6.2.26a,b)}{=}} \frac{QX_p^2\left(X_p + jQX_p\right)}{\left(1+Q^2\right)X_p^2}$$

$$\Rightarrow R_s = \frac{QX_p = R_p}{1+Q^2} \text{ and } X_s = QR_s = \frac{Q^2}{1+Q^2}X_p \tag{6.2.28a,b}$$

$$R_p = \left(1+Q^2\right)R_s \overset{Q>>1}{\rightarrow} Q^2 R_s \text{ and } X_p = \frac{R_p}{Q} = \frac{1+Q^2}{Q}R_s = \frac{1+Q^2}{Q^2}X_s \overset{Q>>1}{\rightarrow} X_s \tag{6.2.29a,b}$$

Note that these equivalence conditions between the series and parallel combination models of a reactive element are valid only at a certain frequency since the reactances are frequency dependent. The series-to-parallel and parallel-to-series conversion processes of *L-R* and *C-R* circuits have been cast into the following MATLAB functions:

```
function [Lp,Rp]=ser2par_LR(Ls,Rs,fc)
% Series-to-Parallel conversion of Ls-Rs circuit at frequency fc.
wc=2*pi*fc; Xs=wc*Ls; Q=Xs/Rs; % Quality factor of Ls-Rs by Eq. (6.2.26a)
Xp=(1+Q^2)/Q^2*Xs; Rp=(1+Q^2)*Rs; Lp=Xp/wc; % Eq. (6.2.29a,b)
```

```
function [Ls,Rs]=par2ser_LR(Lp,Rp,fc)
% Parallel-to-Series conversion of Cp|Rp circuit at frequency fc.
wc=2*pi*fc; Xp=wc*Lp; Q=Rp/Xp; % Quality factor of Lp|Rp by Eq. (6.2.26b)
Xs=Q^2/(1+Q^2)*Xp; Rs=Rp/(1+Q^2); Ls=Xs/wc; % Eq. (6.2.28a,b)
```

```
function [Cp,Rp]=ser2par_CR(Cs,Rs,fc)
% Series-to-Parallel conversion of Cs-Rs circuit at frequency fc.
wc=2*pi*fc; Xs=1/wc/Cs; Q=Xs/Rs; % Quality factor of Cs-Rs by Eq. (6.2.26a)
Xp=(1+Q^2)/Q^2*Xs; Rp=(1+Q^2)*Rs; Cp=1/Xp/wc; % Eq. (6.2.29a,b)
```

```
function [Cs,Rs]=par2ser_CR(Cp,Rp,fc)
% Parallel-to-Series conversion of Lp|Rp circuit at frequency fc.
wc=2*pi*fc; Xp=1/wc/Cp; Q=Rp/Xp; % Quality factor of Cp|Rp by Eq. (6.2.26b)
Xs=Q^2/(1+Q^2)*Xp; Rs=Rp/(1+Q^2); Cs=1/Xs/wc; % Eq. (6.2.28a,b)
```

Example 6.5 Series-to-Parallel Conversion of an *L-R* Circuit

Make a series-to-parallel conversion of the *L-R* circuit shown in Figure 6.11(a1) at f_c = 100 MHz and plot the AC impedance (magnitude) curves of the series and parallel *L-R* circuits for comparison.

We can make and run the MATLAB script "elec06e05.m," which yields the following conversion results

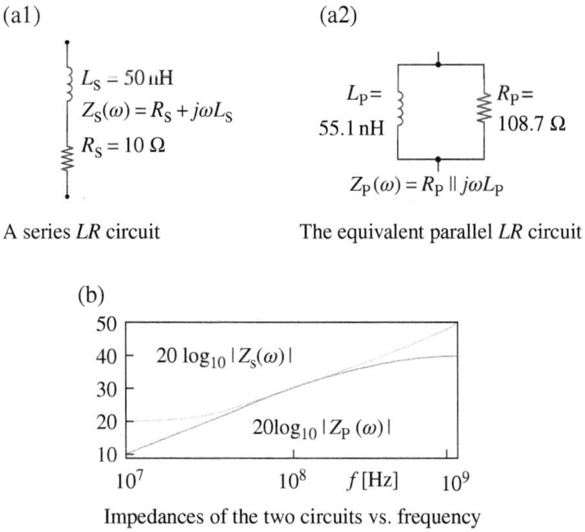

Figure 6.11 Series-to-parallel conversion of an *L-R* circuit.

$$L_s = 50\text{ nH} - R_s = 10\ \Omega \text{ (Series combination)}$$
$$\rightarrow L_p = 55.1\text{ nH} \mid R_p = 108.7\ \Omega \text{ (Parallel combination)}$$

(as shown in Figure 6.11(a2)) and the AC impedance (magnitude) curves shown in Figure 6.11(b).

```
%elec06e05.m
% Series-to-Parallel conversion of an L-R circuit in Example 6.5.
fc=100e6; wc=2*pi*fc; Ls=50e-9; Rs=10;
[Lp,Rp]=ser2par_LR(Ls,Rs,fc)
% To plot the frequency response
ff=logspace(-1,1,601)*fc; w=2*pi*ff;
Zsw=Rs+j*w*Ls; Zpw=Rp*j*w*Lp./(Rp+j*w*Lp);
ZswdB=20*log10(abs(Zsw)); ZpwdB=20*log10(abs(Zpw));
semilogx(ff,ZswdB, ff,ZpwdB,'r')
```

Now, let us consider the (loaded) quality factor [L-1] Q_{LD} of a series *RLC* circuit with the source resistance R_s and the load resistance R_L as shown in Figure 6.12(a). Noting that the circuit is a series connection of (R_s+R+R_L)-L-C, we can use Eq. (6.2.25a) to write the *loaded quality factor* Q_{LD} as

$$Q_{LD} \overset{(6.2.25a)}{=} \frac{X(\text{reactance})}{R_{\text{series, total}}} = \frac{\omega_c L = 1/\omega_c C}{R_s + R + R_L} = \frac{1}{R/\omega_c L + (R_s + R_L)/\omega_c L} = \frac{1}{1/Q_F + 1/Q_E} \tag{6.2.30}$$

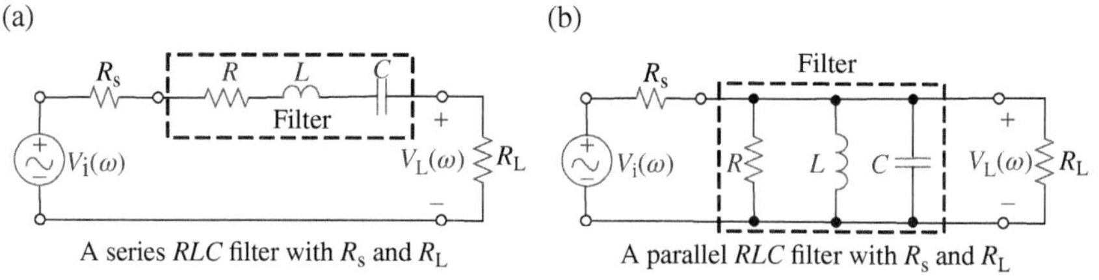

Figure 6.12 Circuits for the definitions of the loaded quality factor Q_{LD}.

where the filter quality factor Q_F and the external quality factor Q_E are as follows:

$$Q_F = \frac{X = \omega_c L = 1/\omega_c C}{R} \quad \text{and} \quad Q_E = \frac{X = \omega_c L = 1/\omega_c C}{R_s + R_L} \tag{6.2.31}$$

Likewise, the *loaded quality factor* Q_{LD} of a parallel RLC circuit with the source resistance R_s and the load resistance R_L as shown in Figure 6.12(b) can be written as

$$\begin{aligned} Q_{LD} &\overset{(6.2.25b)}{=} \frac{R_{\text{parallel,total}}}{X(\text{reactance})} = \frac{\omega_c C = 1/\omega_c L}{1/R + 1/R_s + 1/R_L} \\ &= \frac{1}{1/R/\omega_c C + 1/(R_s \| R_L)/\omega_c C} = \frac{1}{1/Q_F + 1/Q_E} \end{aligned} \tag{6.2.32}$$

where the filter quality factor Q_F and the external quality factor Q_E are as follows:

$$Q_F = \frac{R}{X = \omega_c L = 1/\omega_c C} \quad \text{and} \quad Q_E = \frac{R_s \| R_L}{X = \omega_c L = 1/\omega_c C} \tag{6.2.33}$$

Example 6.6 Design of *LC* Parallel Resonant Filter

Design an LC parallel resonant filter for a transmission line system with the source/load resistance R_s/R_L = 1 kΩ/1 kΩ so that the overall frequency response can have a bandwidth f_B = 10 MHz at center frequency f_c = 100 MHz where the (element) quality factors of the inductor and capacitor (constituting the LC parallel filter) are $Q_{L,p} = R_{L,p}/X = 85$ and $Q_{C,p} = R_{C,p}/X = 0$, i.e. $R_{C,p} = 0$ (lossless), respectively, and $X = \omega_c L = 1/\omega_c C$.

Note that this problem amounts to determining such values of L and C in the circuit of Figure 6.12(b) that can yield the loaded quality factor of

$$\begin{aligned} Q_{LD} &\overset{(6.2.32)}{=} \frac{1}{1/Q_F + 1/Q_E} = \frac{1}{1/Q_{L,p} + X(1/R_s + 1/R_L)} \\ &= \frac{1}{1/85 + X(1/1000 + 1/1000)} = \frac{f_c}{f_B} = \frac{100}{10} = 10 \end{aligned} \tag{E6.6.1}$$

We can solve to get

$$X = 44.1\ \Omega \rightarrow L = \frac{X}{\omega_c} = \frac{X}{2\pi f_c} = \frac{44.1}{2\pi \times 10^8} = 70.2\ \text{nH and } C = \frac{1}{\omega_c X}$$
$$= \frac{1}{2\pi \times 10^8 \times 44.1} = 36.1\ \text{pF} \tag{E6.6.2}$$

and use Eq. (6.2.26b) to get

$$R = R_{L,\text{p}} \overset{(6.2.26b)}{=} Q_{L,\text{p}} X \tag{E6.6.3}$$

6.2.6 Insertion Loss

When designing a filter to be inserted in a telecommunication system as can be modeled by, say, Figure 6.12(a) or (b), it is important to consider the *insertion loss* (*IL*) of the filter, which is defined as the ratio of the power (delivered to the load without the filter) to the power (delivered to the load with the filter inserted):

$$\begin{aligned}\text{IL (Insertion Loss)} &= 10\log_{10}\frac{\{R_\text{L}/(R_\text{s}+R_\text{L})\}^2}{(V_\text{L}(j\omega)/V_\text{i}(j\omega))^2}\\ &= -20\log_{10}\left\{\frac{R_\text{s}+R_\text{L}}{R_\text{L}}\frac{V_\text{L}(j\omega)}{V_\text{i}(j\omega)}\right\}\ [\text{dB}]\end{aligned} \tag{6.2.34}$$

6.2.7 Frequency Scaling and Transformation

Note the following remarks on frequency scaling and frequency transformation:

Remark 6.2 Frequency Response Scaling on Magnitude and Frequency

1) *Magnitude Scaling*: If the frequency response of a circuit is a kind of AC impedance or admittance as a phasor voltage-to-current ratio[V/A] or a phasor current-to-voltage ratio[A/V], we can scale its magnitude curve along the vertical (magnitude) axis with the scale factor k_m by multiplying all resistances/inductances with k_m and all capacitances with $1/k_m$. For example, we can scale the frequency response of a series *RLC* circuit by changing its impedance as

$$Z(j\omega) = R + j\omega L + \frac{1}{j\omega C} \rightarrow k_m R + j\omega k_m L + \frac{1}{j\omega C/k_m} = k_m Z(j\omega) \tag{6.2.35}$$

However, if the frequency response is dimensionless as a phasor voltage-to-voltage ratio[V/V] or a phasor current-to-current ratio[A/A], we cannot scale its magnitude curve along the vertical (magnitude) axis, which implies that all resistances/inductances and capacitances can be multiplied by k_m and $1/k_m$, respectively, without changing the frequency response.

2) *Frequency Scaling*: We can scale the magnitude/phase curve of frequency response along the horizontal (frequency) axis with the scale factor k_f by multiplying all inductances and capacitances with $1/k_f$ and leaving all resistances as they are. This frequency scaling is based on the fact that the reactances of an inductor and a capacitor, ωL and $1/\omega C$, are not changed when we multiply the frequency variable ω by k_f and divide L and C by k_f.

Example 6.7 Magnitude and Frequency Scaling of a Series *RLC* Circuit
Adjust the values of R, L, and C of the RLC filter designed in Example 6.4 by the magnitude scale factor $k_m = 2$ and frequency scale factor $k_f = 10$ and plot the frequency response (magnitude) curves of the filter before/after the scalings.

Referring to Remark 6.2, the values of R, L, and C should be adjusted as follows:

$$R: 69.5 \underset{km=2}{\overset{\times km}{\longrightarrow}} 139\ \Omega,\ L: 13.9 \underset{km=2,kf=10}{\overset{\times km/kf}{\longrightarrow}} 2.78\ \text{mH},\ C: 2 \underset{km=2,kf=10}{\overset{\times 1/km/kf}{\longrightarrow}} 0.1\ \mu F \quad \text{(E6.7.1)}$$

Figure 6.13 shows the frequency response magnitude curves of the filter before/after the magnitude and frequency scalings. Do they agree to your expectation?

Remark 6.3 Transformation of Prototype LPF with Normalized Cutoff Frequency
A prototype LPF with normalized cutoff frequency $\omega_c = 1$ [rad/s] can be transformed into other types like LPF, HPF, BPF, and BSF (with another cutoff frequency) by substituting s as listed in Table 6.1 into its transfer function. The following MATLAB function '`transform_LC()`' finds the values of L's and C's to be determined for such transformations of an LPF.

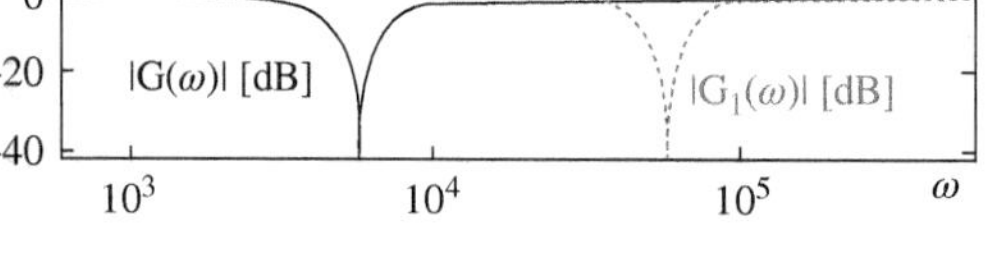

Figure 6.13 Frequency response magnitudes before/after frequency scaling.

```
function [LCs1,LCs1_]=transform_LC(LCs,LCs_,wc,band,Rs)
% Frequency transformations (Remark 6.3 and Table 6.1)
% Input:
%  LCs,LCs_ = [C1 L2 C3],{'C(v)','L(h)','C(v)'}
%  wc       = wc/[wl wu] for (LPF|HPF)/(BPF|BSF)
%  band     = 'LPF'/'HPF'/'BPF'/'BSF'
%  Rs       = Line (Sourece) Resistance
% Output:
%  LCs1,LCs1_ = Frequency transformed results
% Copyleft: Won Y. Yang, wyyang53@hanmail.net, CAU for academic use only
if nargin<5, Rs=1; end
if nargin<4, band='LPF'; end
if nargin<3, wc=1; end
if numel(wc)>1, w0=sqrt(wc(1)*wc(2)); wb=abs(wc(2)-wc(1)); end
if lower(band(1))=='l' % 'LPF'
   LCs1=LCs/wc; LCs1_=LCs_; % Eq.(6.2.36)
   for m=1:numel(LCs_)
     LC_=LCs_{m};
     if LC_(1)=='L', LCs1(m)=LCs1(m)*Rs;
     elseif LC_(1)=='C', LCs1(m)=LCs1(m)/Rs;
    end
  end
else
 LCs1=[]; LCs1_=LCs_;
 for m=1:numel(LCs_)
   LC_=LCs_{m};
   if length(wc)<2 % 'HPF'
    if LC_(1)=='L', LC_(1)='C'; LCs1_{m}=LC_;
      LCs1=[LCs1 1/wc/LCs(m)/Rs]; % Eq.(6.2.37b)
     else LC_(1)='C'; LCs1_{m}=LC_; LCs1=[LCs1 1/wc/LCs(m)*Rs];
    end
   else
    if numel(wc)<2, error('wc must be [wl wu] for BPF or BSF'); end
    if lower(band(1:2))=='bp' % 'BPF'
     if LC_(1)=='L', L=LCs(m);
       LCs1=[LCs1 L/wb*Rs wb/w0^2/L/Rs]; LCs1_{m}='L-C';%(6.2.38b)
      else LC_(1)='C'; C=LCs(m);
       LCs1=[LCs1 wb/w0^2/C*Rs C/wb/Rs]; LCs1_{m}='L|C';%(6.2.38c)
    end
   else % 'BSF'
    if LC_(1)=='L', L=LCs(m);
      LCs1=[LCs1 wb/w0^2*L*Rs 1/wb/L/Rs]; LCs1_{m}='L|C';%(6.2.39b)
     else LC_(1)='C'; C=LCs(m);
      LCs1=[LCs1 1/wb/C*Rs wb/w0^2*C/Rs]; LCs1_{m}='L-C';%(6.2.39c)
     end
   end
  end
 end
end
```

Table 6.1 Frequency transformation.

Frequency transformations	MATLAB function
LPF with cutoff frequency ω_c: $s \to s/\omega_c$ (6.2.36a) This corresponds to substituting $L \to \tilde{L} = L/\omega_c$ and $C \to \tilde{C} = C/\omega_c$. (6.2.36b)	`lp2lp()`
HPF with cutoff frequency ω_c: $s \to \omega_c/s$ (6.2.37a) This corresponds to substituting $L \to \tilde{C} = \frac{1}{\omega_c L}$ and $C \to \tilde{L} = \frac{1}{\omega_c C}$. (6.2.37b)	`lp2hp()`
BPF with upper/lower cutoff frequencies ω_u/ω_l: $s \to \frac{s^2 + \omega_u\omega_l}{s(\omega_u - \omega_l)}$ (6.2.38a) This corresponds to substituting $(\omega_0^2 = \omega_u\omega_l)$ $sL \to s\tilde{L} + \frac{1}{s\tilde{C}} \begin{pmatrix}\text{series connection}\\ \text{of } \tilde{L} \text{ and } \tilde{C}\end{pmatrix} = \frac{sL}{\omega_u - \omega_l} + \frac{1}{s(\omega_u - \omega_l)/\omega_0^2 L}$ (6.2.38b) $sC \to s\tilde{C} + \frac{1}{s\tilde{L}} \begin{pmatrix}\text{parallel connection}\\ \text{of } \tilde{L} \text{ and } \tilde{C}\end{pmatrix} = \frac{sC}{\omega_u - \omega_l} + \frac{1}{s(\omega_u - \omega_l)/\omega_0^2 C}$ (6.2.38c)	`lp2bp()`
BSF with upper/lower cutoff frequencies ω_u/ω_l: $s \to \frac{s(\omega_u - \omega_l)}{s^2 + \omega_u\omega_l}$ (6.2.39a) This corresponds to substituting $(\omega_0^2 = \omega_u\omega_l)$ $sL \to \frac{1}{\frac{1}{s\tilde{L}} + s\tilde{C}} \begin{pmatrix}\text{parallel connection}\\ \text{of } \tilde{L} \text{ and } \tilde{C}\end{pmatrix} = \frac{1}{\frac{\omega_0^2}{s(\omega_u - \omega_l)L} + \frac{s}{(\omega_u - \omega_l)L}}$ (6.2.39b) $sC \to \frac{1}{s\tilde{L} + \frac{1}{s\tilde{C}}} \begin{pmatrix}\text{series connection}\\ \text{of } \tilde{L} \text{ and } \tilde{C}\end{pmatrix} = \frac{1}{\frac{s}{(\omega_u - \omega_l)C} + \frac{\omega_0^2}{sC(\omega_u - \omega_l)}}$ (6.2.39c)	`lp2bs()`

Example 6.8 Frequency Transformation of an LPF

Make a frequency transformation of a series RC circuit with cutoff frequency $\omega_c = RC = 1$[rad/s] into a BSF with $R_s = 69.5[\Omega]$, $\omega_l = 4000$[rad/s], and $\omega_u = 9000$ [rad/s], as specified in Example 6.4.

To this end, we can run the following script "elec06e08.m," which uses the MATLAB function '`transform_LC()`' inside, to get

```
LCs1 =  1.3900e-02  1.9984e-06,   LCs1_ =  'L-C'
```

This means a series connection of $L = 0.0139$ [H] and $C = 2$ [μF], as determined in Example 6.4.

```
%elec06e08.m
wl=4000; wu=9000; wc=sqrt(wl*wu); w=logspace(-1,2,600)*wc; jw=j*w;
Rs=69.5; LCs=[1]; LCs_={'C'}; wc=[wl wu]; band='BSF';
[LCs1,LCs1_]=transform_LC(LCs,LCs_,wc,band,Rs)
syms s; L=LCs1(1); C=LCs1(2); Gs1=(s*L+1/s/C)/(Rs+s*L+1/s/C) %Transfer ftn
[B1,A1]=numden(simplify(Gs1)); B1=sym2poly(B1); A1=sym2poly(A1);
B1=B1/A1(1), A1=A1/A1(1) %Numerator/Denominator of the transfer function
RC=1; b=[0 1]; a=[1 1/RC]; w0=sqrt(wl*wu); wb=wu-wl;
[B,A]=lp2bs(b,a,w0,wb) % Using 'lp2hp()' to get just the transfer ftn
```

6.3 Passive Filter Realization

6.3.1 *LC* Ladder

Figure 6.14.1(a) and (b) shows the Π-type and T-type of *LC* ladder that can be used to realize a prototype LPF where the values of *L*'s, *C*'s, and R_L are listed as $\{g_1, g_2, \dots\}$ in Tables 5.2 (Butterworth), 5.3 (linear-phase), 5.4 (Chebyshev 3 dB), and 5.5 (Chebyshev 0.5 dB) of [L-1] and can also be determined by using the following MATLAB function 'LPF_ladder_coef()'.

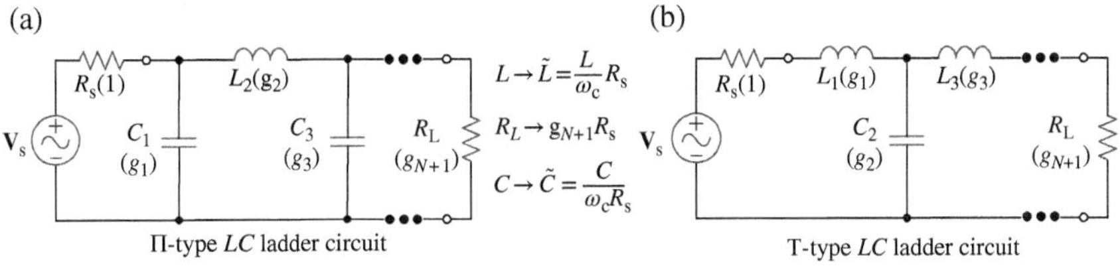

Figure 6.14.1 *LC* ladders realizing low-pass filter (LPF).

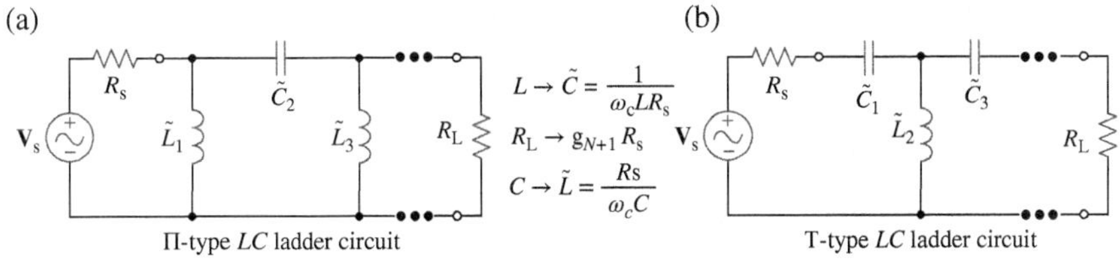

Figure 6.14.2 *LC* ladders realizing high-pass filter (HPF).

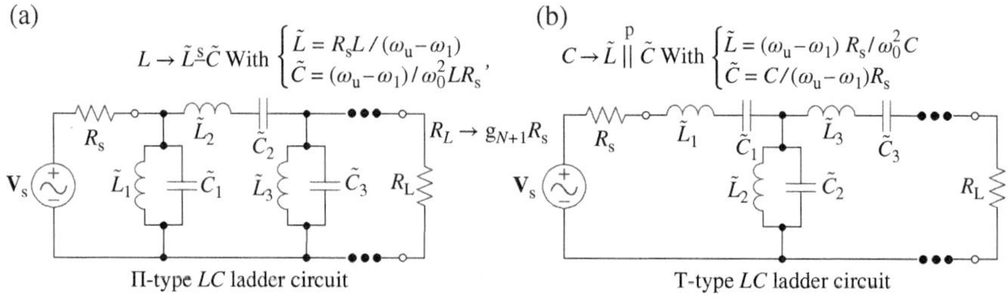

Figure 6.14.3 *LC* ladders realizing BPF.

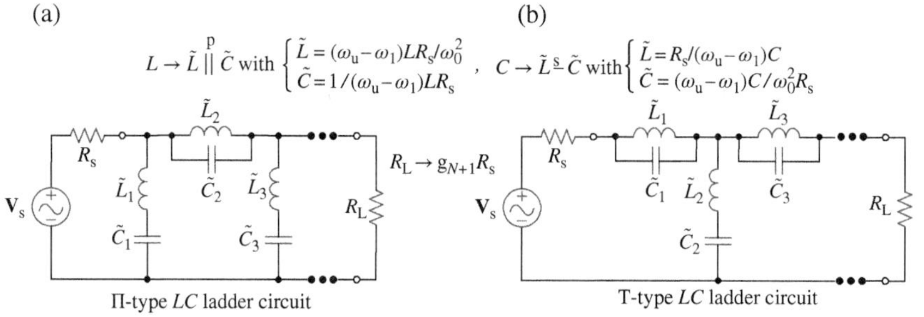

Figure 6.14.4 *LC* ladders realizing BSF.

The formulas listed in Table 6.1 can be used to magnitude/frequency scale the *LC* filter prototypes to make them realize LPF/HPF/BPF/BSF with unnormalized cutoff frequency as depicted in Figures 6.14.1-6.14.4.

Example 6.9 Determining the Values of *L*'s and *C*'s of *LC* Ladder Realizing a Prototype LPF

To determine the values of *L*'s, *C*'s, and R_L of a third-order Butterworth LPF prototype, we can run the following MATLAB statement:

```
>>N=3; gg=LPF_ladder_coef(N,'Bu')
  gg =     1    2    1     1
```

```
function [LCRs,LCRs_,Gs]=LC_ladder(Rs,gg,pT,wc,band)
% Input:
%  Rs = Source resistance
%  gg = Ladder coefficients generated using LPF_ladder_coef()
%  pT = 'p'/'T' or 1/2 for Pi/T-ladder
%  wc = wc/[wl wu] for band=('LPF'|'HPF')/('BPF'|'BSF')
%  band='LPF'/'BPF'/'BSF'/'HPF'
% Output:
% LCRs,LCRs_=[C1 L2 C3 R],{'C(v)','L(h)','C(v)'} for Pi-ladder Fig.6.14.1a
% LCRs,LCRs_=[L1 C2 L3 R],{'L(h)','C(v)','L(h)'} for T-ladder Fig.6.14.1b
% LCRs,LCRs_=[L1 C1 L2 C2 L3 C3 R],{'L-C(v)','L|C(h)','L-C(v)'} for Pi
% LCRs,LCRs_=[L1 C1 L2 C2 L3 C3 R],{'L|C(h)','L-C(v)','L|C(h)'} for T
% Gs = G(s): Transfer function
% Copyleft: Won Y. Yang, wyyang53@hanmail.net, CAU for academic use only
if nargin<5, band='LPF'; end
if nargin<4, wc=1; end
if length(wc)>1, w0=sqrt(wc(1)*wc(2)); wb=wc(2)-wc(1); end
syms s
N1=length(gg); N=N1-1;
if isnumeric(pT), Kc=pT;
 else pT=lower(pT); Kc=(pT(1)=='p')+2*(pT(1)=='t'); % Set Kc=1/2 for pi/T
end
if mod(Kc+N,2)==1, RL=gg(N1)*Rs; % When gg(N1)=RL/Rs
 else RL=Rs/gg(N1); % When gg(N1)=Rs/RL
end
I=1; V=RL; % Set the current through RL to 1 to find the transfer ftn G(s).
LCs=[]; % Set LCs as an empty vector.
if Kc==1 % Pi-ladder
  for n=N:-1:1
    if mod(n,2)==0, [LCsn,LCs_{n},V]=LC_horizontal(Rs,gg(n),wc,band,V,I);
     else [LCsn,LCs_{n},I]=LC_vertical(Rs,gg(n),wc,band,V,I);
    end
    LCs=[LCsn LCs];
  end
else % T-ladder
 for n=N:-1:1
   if mod(n,2)==1, [LCsn,LCs_{n},V]=LC_horizontal(Rs,gg(n),wc,band,V,I);
    else [LCsn,LCs_{n},I]=LC_vertical(Rs,gg(n),wc,band,V,I);
   end
   LCs=[LCsn LCs];
 end
end
LCRs=[LCs RL]; LCRs_=LCs_; LCRs_{N1}='R(v)';
Gs=RL/(V+Rs*I); % Transfer function considering Rs and RL
```

```
function [LCsn,LCs_n,V]=LC_horizontal(Rs,ggn,wc,band,V,I)
syms s
if length(wc)>1, w0=sqrt(wc(1)*wc(2)); wb=wc(2)-wc(1); end
switch upper(band(1:2))
 case 'LP', L=ggn/wc*Rs; sL=s*L; V=V+sL*I; LCsn=L; LCs_n='L(h)';
 case 'HP', C=1/ggn/wc/Rs; sC=s*C; V=V+I/sC; LCsn=C; LCs_n='C(h)';
 case 'BP', L=ggn/wb*Rs; C=1/ggn*wb/w0^2/Rs; sLC=s*L+1/s/C; V=V+sLC*I;
            LCsn=[L C]; LCs_n='L-C(h)'; % Series L-C
 otherwise, L=ggn*wb/w0^2*Rs; C=1/ggn/wb/Rs; sLC=L/C/(s*L+1/s/C);
            V=V+sLC*I; LCsn=[L C]; LCs_n='L|C(h)'; % Shunt L|C
end
```

```
function [LCsn,LCs_n,I]=LC_vertical(Rs,ggn,wc,band,V,I)
syms s
if length(wc)>1, w0=sqrt(wc(1)*wc(2)); wb=wc(2)-wc(1); end
switch upper(band(1:2))
 case 'LP', C=ggn/wc/Rs; sC=s*C; I=I+sC*V; LCsn=C; LCs_n='C(v)';
 case 'HP', L=1/ggn/wc*Rs; sL=s*L; I=I+V/sL; LCsn=L; LCs_n='L(v)';
 case 'BP',  L=1/ggn*wb/w0^2*Rs; C=ggn/wb/Rs; sLC=L/C/(s*L+1/s/C);
             I=I+V/sLC; LCsn=[L C]; LCs_n='L|C(v)'; % Parallel L-C
 otherwise, L=1/ggn/wb*Rs; C=ggn*wb/w0^2/Rs; sLC=s*L+1/s/C; I=I+V/sLC;
            LCsn=[L C]; LCs_n='L-C(v)'; % Series L-C
end
```

```
function gg=LPF_ladder_coef(N,type,Rp)
%returns N+1 L/C values of Butterworth/Linear-phase/Chebyshev 3dB/0.5dB LPF
% depending on the order N and type='Bu'/'Li'/'C1'/'C2'
% L/C values of LC ladder for normalized linear-phase LPF, Table 5.3, [L-1]
if nargin<3, Rp=3; end % Rp is used only for Chebyshev filters
L=[2 1 0 0 0 0 0 0 0 0 0;
  1.5774 0.4226 1 0 0 0 0 0 0 0 0;
  1.255 0.5528 0.1922 1 0 0 0 0 0 0 0;
  1.0598 0.5116 0.3181 0.1104 1 0 0 0 0 0 0;
  0.9303 0.4577 0.3312 0.209 0.0718 1 0 0 0 0 0;
  0.8377 0.4116 0.3158 0.2364 0.248 0.0505 1 0 0 0 0;
  0.7677 0.3744 0.2944 0.2378 0.1778 0.1104 0.0375 1 0 0 0;
  0.7125 0.3446 0.2735 0.2297 0.1867 0.1387 0.0855 0.0289 1 0 0;
  0.6678 0.3203 0.2547 0.2184 0.1859 0.1506 0.1111 0.0682 0.023 1 0;
  0.6305 0.3002 0.2384 0.2066 0.1808 0.1539 0.124 0.0911 0.0557 0.0187 1];
% Data by courtesy of L. Weinberg and Journal of the Franklin Institute.
k=1:N; T=upper(type(1));
if T=='B', gg=[2*sin((2*k-1)*pi/2/N) 1]; % Eq.(6.8) of [C-2]
 elseif T=='L', gg=L(N,1:N+1);
 elseif T=='C'
 beta=log(coth(Rp/17.37)); gamma=sinh(beta/2/N);
 a=sin((2*k-1)*pi/2/N); b=gamma^2+sin(k*pi/N).^2; % Eq.(6.11) of [C-2]
 gg(1)=2*a(1)/gamma;
 for k=2:N, gg(k)=4*a(k-1)*a(k)/b(k-1)/gg(k-1); end % Eq.(6.10) of [C-2]
 if mod(N,2)==1, gg(N+1)=1; else gg(N+1)=coth(beta/4)^2; end
end
```

This result, returned as the output of the above MATLAB function '`LPF_ladder_coef()`', means that the two forms of *LC* ladder depicted in Figure 6.15 can be used as the third-order Butterworth LPF prototype. For reference, the transfer function of the filter in Figure 6.15(a) can be found as

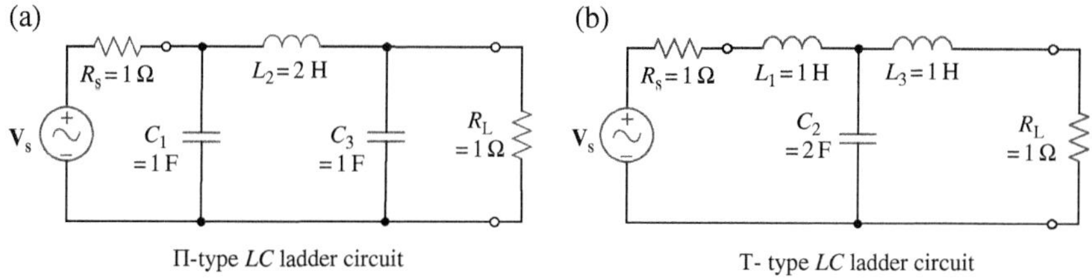

Figure 6.15 A third-order Butterworth LPF prototype realized with Π- and T-type *LC* ladders.

$$G(s) = \frac{(1/sC_1)\|\,[sL_2 + \{(1/sC_3)\|R_L\}]}{R_s + (1/sC_1)\|\,[sL_2 + \{(1/sC_3)\|R_L\}]}\,\frac{(1/sC_3)\|R_L}{sL_2 + \{(1/sC_3)\|R_L\}}$$
$$= \frac{1}{2s^3 + 4s^2 + 4s + 2} \tag{E6.9.1}$$

If you want to get the parameter values for the Π-type *LC* ladder in Figure 6.15(a) more specifically, run the following MATLAB statements:

```
>>Rs=1; pT='Pi'; wc=1; [LCRs,LCRs_,Gs]=LC_ladder(Rs,gg,pT,wc);
  LCRs, LCRs_, pretty(simplify(Gs))
```

This yields not only the parameter values (LCRs) but also the type and vertical/horizontal orientation of each element:

```
LCRs = 1.0000e+00  2.0000e+00  1.0000e+00  1.0000e+00
LCRs_ =   'C(v)'       'L(h)'       'C(v)'       'R(v)'
          1
-------------- % conforms with Eq. (E6.9.1)
   3    2
2 s  + 4 s  + 4 s + 2
```

where `'C(v)'`/`'L(h)'` in the cell LCRs_ mean that their corresponding numbers in the vector LCRs are the values of capacitance/inductance to be placed vertically/horizontally, respectively, in the Π-type *LC* ladder of Figure 6.15(a). If you run the same MATLAB statements, but with `pT = 'T'`,

```
>>Rs=1; pT='T'; wc=1; [LCRs,LCRs_,Gs]=LC_ladder(Rs,gg,pT,wc);
  LCRs, LCRs_, pretty(simplify(Gs))
```

you will get the realization result for the T-type *LC* ladder of Figure 6.15(b):

```
LCRs = 1.0000e+00  2.0000e+00  1.0000e+00  1.0000e+00
LCRs_ =   'L(h)'       'C(v)'       'L(h)'       'R(v)'
          1
------------- % conforms with Eq. (E6.9.1)
   3    2
2 s  + 4 s  + 4 s + 2
```

Example 6.10 Realization of a Fourth-Order Butterworth LPF by an *LC* Ladder

Using a Π-type *LC* ladder circuit, realize a fourth-order Butterworth LPF with cutoff frequency $f_c = 10$ kHz to be connected to a line or source resistance $R_s = 1$ kΩ. Then, plot the frequency response magnitude (relative to that without the filter) for two decades around f_c, i.e. $0.1f_c \le f \le 10f$.

The realization can be determined by running the following MATLAB script "elec06e10.m," which uses '`LPF_ladder_coef()`' to get the *LC* values of the fourth-order Butterworth LPF prototype and then uses '`LC_ladder()`' (with pT = 'pi') or '`transform_LC()`' to find a Π-type *LC* ladder and to plot the magnitude of its frequency response obtained by substituting $s = j\omega$ into the transfer function (obtained as the third output argument Gs of '`LC_ladder()`'), as shown in Figure 6.16:

```
%elec06e10.m
% To realize a Butterworth LPF by an LC ladder in Example 6.10
syms s
N=4; pT='pi'; Rs=1e3; % 4th order, Pi-type
type='B'; band='LPF'; fc=1e4; wc=2*pi*fc; % Butterworth LPF
gg=LPF_ladder_coef(N,type) % LC values of Nth-order B-LPF prototype
[LCRs,LCRs_,Gs]=LC_ladder(Rs,gg,pT,wc,band);
format short e; LCRs, LCRs_, RL=LCRs(end)
% Another way to determine the LC values using transform_LC():
LCs0=gg(1:N); LCs0_={'L(h)','C(v)','L(h)','C(v)'};
[LCs1,LCs1_]=transform_LC(LCs0,LCs0_,wc,band,Rs)
% To plot the frequency response magnitude (considering Rs and RL)
ff=logspace(-1,1,301)*fc; ww=2*pi*ff; % Frequency range
s=j*ww; Gw=eval(Gs); % Frequency response
GwmagdB=20*log10(max(abs(Gw)/(RL/(Rs+RL)),1e-5)); %Freqrespmagnitude
semilogx(ff,GwmagdB)
```

```
>>elec06e10
LCRs = 1.2181e-08 2.9408e-02 2.9408e-08 1.2181e-02 1.0000e+03
LCRs_ = 'C(v)'  'L(h)'  'C(v)'  'L(h)'  'R(v)'
RL = 1000
```

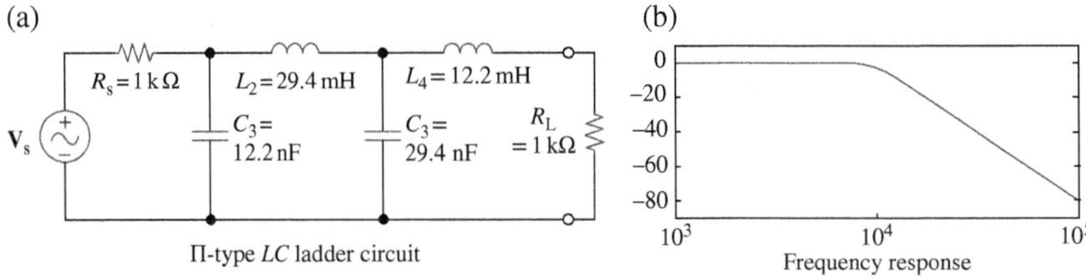

Figure 6.16 Fourth-order Butterworth LPF realized with a Π-type *LC* ladder and its frequency response.

Example 6.11 Realization of a Sixth-Order Chebyshev I BPF with $R_p = 3$ dB

Using a T-type *LC* ladder circuit, realize a sixth-order Chebyshev I BPF with passband ripple $R_p = 3$ dB and lower/upper cutoff frequencies $f_l/f_u = 2.16/2.64$ [Giga-Hz] to be connected to a line or source resistance $R_s = 50\,\Omega$. Plot its *insertion loss* (*IL*), which is defined as the ratio of the power (delivered to the load without the filter) to the power (delivered to the load with the filter inserted):

$$\begin{aligned}\text{IL (Insertion Loss)} &= 10\log_{10}\frac{\{R_L/(R_s+R_L)\}^2}{|G(j\omega)|^2}\ [\text{dB}] \\ &= -20\log_{10}\left\{\frac{R_s+R_L}{R_L}|G(j\omega)|\right\}\ [\text{dB}]\end{aligned} \tag{E6.11.1}$$

The filter realization can be obtained by running the following MATLAB script "elec06e11.m," which uses the MATLAB functions 'LPF_ladder_coef()' (with $N = 6/2 = 3$ for an LPF-to-BPF conversion doubling the order) and 'LC_ladder()' (with pT = 'T') to make the T-type *LC* ladder as shown in Figure 6.17(a) and plot its IL as shown in Figure 6.17(b):

```
%elec06e11.m
% Chebyshev I BPF realization in Example 6.11 (Ex 5.4 of [L-1])
syms s
N=3; pT='T'; Rs=50; type='C'; Rp=3;
fl=2.16e9; fu=2.64e9; wl=2*pi*fl; wu=2*pi*fu; w0=sqrt(wl*wu);
gg=LPF_ladder_coef(N,type,Rp); % Find the LC values
[LCRs,LCRs_,Gs]=LC_ladder(Rs,gg,pT,[wl wu],'BPF');
format short e; LCRs, LCRs_, RL=LCRs(end);
ww=logspace(-1,1,301)*w0; ff=ww/2/pi; % Frequency range
for n=1:length(ww), Gw(n)=subs(Gs,s,j*ww(n)); end
IL_dB=-20*log10(max((Rs+RL)/RL*abs(Gw),1e-5)); % IL by Eq.(E6.11.1)
semilogx(ff,IL_dB)
```

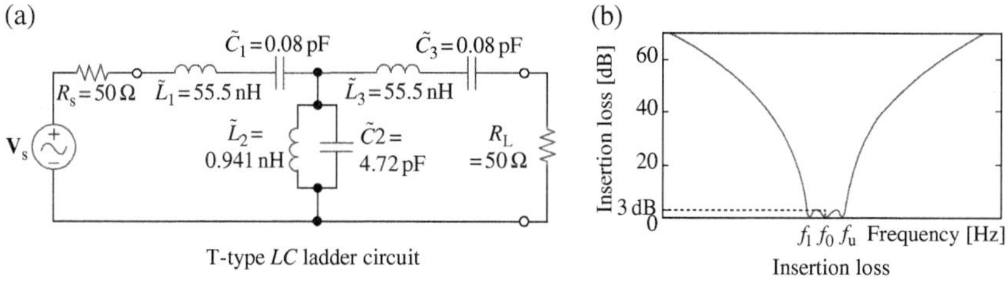

Figure 6.17 Sixth-order Chebyshev BPF realized with a T-type *LC* ladder and its insertion loss.

```
>>elec06e11
 LCRs = 5.5521e-08  8.0007e-14  9.4123e-10  4.7194e-12
        5.5521e-08  8.0007e-14  5.0000e+01
 LCRs_ = 'L-C(h)'   'L|C(v)'    'L-C(h)'   'R(v)'
```

This describes the circuit (excluding the source-line part) of Figure 6.17(a) (identical to Figure 5.30 of [L-1]) where the first element of the cell LCRs_, `'L-C(h)'`, together with the first two elements of the vector LCRs, means the horizontal *LC* series connection with $L = 0.554$ nH and $C = 0.08$ pF, the second element of the cell LCRs_, `'L|C(v)'`, together with the next two elements of the vector LCRs, means the vertical *LC* shunt connection with $L = 0.94$ nH and $C = 4.71$ pF, and so on.

The MATLAB function '`transform_LC()`' can also be used to get the same results:

```
>>LCs=gg(1:N); % From Table 5-4(a) of [L-1] for N=3
  Rs=50; LCs_={'L','C','L'}; % For a T-type filter with L first
  format short e
  [LCs2,LCs2_]=transform_LC(LCs,LCs_,[wl wu],'BPF',Rs) % LC values
  RL=gg(N+1)*Rs % Load resistance
```

Running these MATLAB statements yields

```
LCs2 =
 5.5521e-08 8.0007e-14 9.4123e-10 4.7194e-12 5.5521e-08 8.0007e-14
LCs2_ =   'L-C'   'L|C'   'L-C'
RL =   50
```

This describes the circuit (excluding the source-line part) of Figure 6.17(a) where the first element of the cell LCRs_, `'L-C'`, together with the first two elements of the vector LCRs, means the (horizontal) *LC* series connection with $L = 0.554$ nH and $C = 0.08$ pF, the second element of the cell LCRs_, `'L|C'`, together with the next two elements of the vector LCRs, means the (vertical) LC shunt connection with $L = 0.94$ nH and $C = 4.71$ pF, and so on.

If you want to get the numerator and denominator coefficients of the filter transfer function obtained as the third output argument Gs of '`LC_ladder()`', run the following MATLAB statements:

```
>>[Ns,Ds]=numden(simplify(Gs)); % Extract the numerator/denominator
  B_bpf=sym2poly(Ns); A_bpf=sym2poly(Ds);
  % To normalize the denominator so that its leading coefficient is 1,
  B_bpf=B_bpf/A_bpf(1), A_bpf=A_bpf/A_bpf(1)
```

If you want to get only the numerator and denominator coefficients of the filter transfer function, it is enough to run the following MATLAB statements:

```
>>[B_bpf1,A_bpf1]=cheby1(N,Rp,[wl wu],'s');
  B_bpf1=B_bpf1*RL/(Rs+RL), A_bpf1
```

6.3.2 L-Type Impedance Matcher

Figure 6.18 shows two L-types of LC filters (with $Z_k = jX_k = j\omega L_k$ or $Z_k = jX_k = 1/j\omega C_k$) that can be used to accomplish the impedance matching condition [Y-1], (6.38) $Z_s = Z_L^*$ (*: complex conjugate) between the source impedance $Z_s = R_s + jX_s$ and load impedance $Z_L = R_L + jX_L$. The impedance matching condition for maximum power transfer of L1-type filter (Figure 6.18(a)) can be written as

```
function [X11,X21,X12,X22]=imp_matcher_L(Zs,ZL,fc,type)
% If type is not given, find L1&L2-types of LC impedance matcher(s).
% If type is given as 1|2, find L1|L2-type of LC impedance matcher(s).
% Copyleft: Won Y. Yang, wyyang53@hanmail.net, CAU for academic use only
X11=inf*[1 1]; X21=[0 0]; X12=[0 0]; X22=inf*[1 1];
if nargin<4|type~=2, type=1; end
if type==2, tmp=Zs; Zs=ZL; ZL=tmp; end % switch Zs and ZL
Ys=1/Zs; Gs=real(Ys); Bs=-imag(Ys);
RL=real(ZL); XL=imag(ZL);
tmp=Gs/RL-Gs^2; B=Bs+sqrt(tmp)*[1 -1]; % Eq.(6.3.1a)
if tmp>=0, X11=-1./B; X=RL*(B-Bs)/Gs-XL; X21=X; end % Eq.(6.3.1b)
if type==2, tmp=X11; X11=X21; X21=tmp; end
express_L_filter(X11,X21,2*pi*fc,type);
if nargin<4&type==1
  [X12,X22]=imp_matcher_L(Zs,ZL,fc,2);
end
%-----------------------------------------
function express_L_filter(X1s,X2s,wc,type)
for n=1:length(X1s)
  X1=X1s(n); X2=X2s(n);
  if X1==0|X1==inf|X2==0|X2==inf, continue; end
  if ~(X1==0|X2==0)
   if X1>0
    L1=X1/wc*1e6; fprintf(['L' num2str(type) '-type impedance
                                matcher with L1=%8.3e[uH]'],L1);
   else
     C1=-1e9/X1/wc; fprintf(['L' num2str(type) '-type impedance
                                matcher with C1=%8.3e[nF]'],C1);
   end
   if X2>0, L2=X2/wc*1e6; fprintf(' and L2=%8.3e[uH]\n',L2);
    else C2=-1e9/X2/wc; fprintf(' and C2=%8.3e[nF]\n',C2);
   end
  end
 end
```

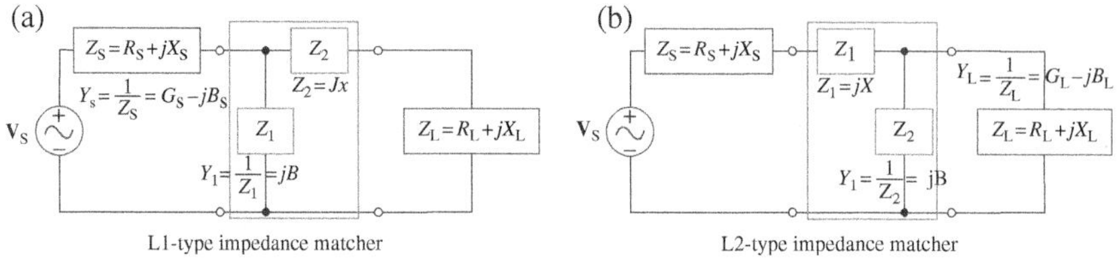

Figure 6.18 L-type impedance matchers.

$$\frac{1}{G_s - jB_s + jB} = (R_L + jX_L + jX)^* \text{ with } G_s = \frac{R_s}{R_s^2 + X_s^2} \text{ and } B_s = \frac{X_s}{R_s^2 + X_s^2}$$

$$;\frac{G_s + j(B_s - B)}{G_s^2 + (B_s - B)^2} = R_L + j(-X - X_L) \rightarrow \begin{cases} B = B_s \pm \sqrt{\dfrac{G_s}{R_L} - G_s^2} \\ X = \dfrac{B - B_s}{(G_s - jB_s)^2 + B^2} - X_L = \dfrac{R_L(B - B_s)}{G_s} - X_L \end{cases} \tag{6.3.1}$$

Likewise, noting that the structure of L2-type filter (Figure 6.18(b)) is just like the reverse of L1-type filter with Z_s and Z_L switched, the impedance matching condition of L2-type filter can be written as

$$\begin{cases} B = B_L \pm \sqrt{\dfrac{G_L}{R_s} - G_L^2} \text{ with } G_L = \dfrac{R_L}{R_L^2 + X_L^2} \text{ and } B_L = \dfrac{X_L}{R_L^2 + X_L^2} \\ X = \dfrac{B - B_L}{(G_L - jB_L)^2 + B^2} - X_L = \dfrac{R_s(B - B_L)}{G_L} - X_s \end{cases} \tag{6.3.2}$$

The above MATLAB function '`imp_matcher_L()`' can be used to find possible L-type filter(s) for impedance matching between given source/load impedances. Note that the function without the fourth input argument 'type' returns all possible L-type *LC* filters while with type = 1/2, it returns just L1-/L2-type *LC* filter(s).

Example 6.12 Design of L-type Impedance Matcher

Design L-type impedance matchers for maximum power transfer from a transmitter with output impedance $Z_s = 50 + 1/j\omega C_s$ ($C_s = 9.362$ pF) $= 50 - j17$ to an antenna with input impedance $Z_L = 40 + j\omega L_L$ ($L_L = 7.958$ nH) $= 40 + j50$ at the center frequency $f_c = 1$ GHz. (Visit http://leleivre.com/rf_lcmatch.html for reference.)

To use the above MATLAB function '`imp_matcher_L()`', we run the MATLAB script "elec06e12.m" below, which uses the MATLAB function '`imp_matcher_L_and_freq_resp()`':

```
%elec06e12.m
% To design L-type impedance matchers in Example 6.12.
clear, clf
fc=1e9; wc=2*pi*fc; ff=logspace(-1,1,601)*fc; % Center frequency
Rs=50; Cs=9.362e-12; RL=40; LL=7.958e-9; % Source and Load impedances
Zs=Rs+1/j/wc/Cs % Output impedance of the transmitter
ZL=RL+j*wc*LL % Input impedance of the load (antenna)
[LC,LC_,ff,Gws]=imp_matcher_L_and_freq_resp(Zs,ZL,fc,ff);
LC, LC_
```

```
function [LC,LC_,ff,Gws]=imp_matcher_L_and_freq_resp(Zs,ZL,fc,ff)
% LC values and Frequency response of L-type impedance matchers
if nargin<4, ff=logspace(-1,1,801)*fc; end
wc=2*pi*fc; w=2*pi*ff; jw=j*w; % Frequency vector
syms s
Rs=real(Zs); Xs=imag(Zs); Zss=Rs+(Xs>=0)*s*Xs/wc-(Xs<0)*wc*Xs/s;
RL=real(ZL); XL=imag(ZL); ZLs=RL+(XL>=0)*s*XL/wc-(XL<0)*wc*XL/s;
[X11,X21,X12,X22]=imp_matcher_L(Zs,ZL,fc);
Gw1s=[]; Gw2s=[]; LC=[]; gss='bgrmkc';
for n=1:numel(X11) % For L1-type,
    syms s; X1=X11(n); X2=X21(n);
    if X1==0|X1==inf|X2==0|X2==inf, continue; end
    % To check if Zs||Z1 and Z2+ZL of L1-type are conjugate,
    [parallel_comb([Zs j*X1]); j*X2+ZL]
    if X1>0, LC(n,1)=X1/wc; LC_{n,1}='L(v)';
    else LC(n,1)=-1/X1/wc; LC_{n,1}='C(v)';
    end
    if X2>0, LC(n,2)=X2/wc; LC_{n,2}='L(h)';
    else LC(n,2)=-1/X2/wc; LC_{n,2}='C(h)';
    end
    Z1s=(X1>=0)*s*X1/wc-(X1<0)*wc*X1/s;
    Z2s=(X2>=0)*s*X2/wc-(X2<0)*wc*X2/s;
    Z12Ls=Z1s*(Z2s+ZLs)/(Z1s+Z2s+ZLs);
    G2s=Z12Ls/(Zss+Z12Ls)*RL/(Z2s+ZLs);
    s=jw; Gw=eval(G2s); Gw1s=[Gw1s; Gw];
    subplot(233), semilogx(ff,abs(Gw).^2/RL,gss(n))
end
m=size(LC,1);
for n=1:numel(X12) % For L2-type,
    syms s; X1=X12(n); X2=X22(n);
    if X1==0|X1==inf|X2==0|X2==inf, continue; end
    % To check if Zs+Z1 and Z2||ZL of L2-type are conjugate,
    [Zs+j*X1; parallel_comb([j*X2 ZL])]
    if X1>0, LC(m+n,1)=X1/wc; LC_{m+n,1}='L(h)';
    else LC(m+n,1)=-1/X1/wc; LC_{m+n,1}='C(h)';
    end
    if X2>0, LC(m+n,2)=X2/wc; LC_{m+n,2}='L(v)';
    else LC(m+n,2)=-1/X2/wc; LC_{m+n,2}='C(v)';
    end
    Z1s=(X1>=0)*s*X1/wc-(X1<0)*wc*X1/s;
    Z2s=(X2>=0)*s*X2/wc-(X2<0)*wc*X2/s;
    Z2Ls=Z2s*ZLs/(Z2s+ZLs);
    G2s=Z2Ls/(Zss+Z1s+Z2Ls)*RL/ZLs;
    s=jw; Gw=eval(G2s); Gw2s=[Gw2s; Gw];
    subplot(236), semilogx(ff,abs(Gw).^2/RL,gss(n))
end
Gws=[Gw1s; Gw2s].';
```

```
>>eles06e12

  LC =                               LC_ =
     8.2200e-13  6.3975e-12             'C(v)'    'C(h)'
     9.1702e-09  2.1185e-12             'L(v)'    'C(h)'
     1.0860e-08  3.5320e-12             'L(h)'    'C(v)'
     4.6487e-12  3.4983e-13             'C(h)'    'C(v)'
```

This result implies that there are four L-type *LC* circuits that can present impedance matching between the source and load as depicted in Figure 6.19. It can be shown that $Z_{1-0} = Z_s^*$ and $Z_{2-0} = Z_L^*$ for all of the *LC* circuits. Say, for the impedance matcher in Figure 6.19(a1), we have

$$Z_{1-0} = Z_1 \| (Z_2 + Z_L) = \left(-j\frac{1}{\omega C_1}\right) \| \left(-j\frac{1}{\omega C_2} + 40 + j50\right) \overset{C_1 = 0.822\times 10^{-12},\; C_2 = 6.397\times 10^{-12}}{\underset{\omega = 2\pi \times 10^9}{=}} 50 + j17 = Z_s^* \tag{E6.12.1}$$

$$Z_{2-0} = (Z_s \| Z_1) + Z_2 = \left\{(50 - j17) \| \left(-j\frac{1}{\omega C_1}\right)\right\} - j\frac{1}{\omega C_2} \overset{C_1 = 0.822\times 10^{-12},\; C_2 = 6.397\times 10^{-12}}{\underset{\omega = 2\pi \times 10^9}{=}} 40 - j50 = Z_L^* \tag{E6.12.2}$$

Figure 6.20(a1), (a2), (b1), and (b2) shows the PSpice schematics for the circuits of Figure 6.19. Figure 6.20(c1) and (c2) shows two plots of power (transferred to and dissipated by the load R_L) versus frequency, one obtained from PSpice simulation and the other obtained from MATLAB analysis, that conform with each other. It can be seen from the power plots that the four L-type circuits commonly perform impedance matching so that the power transfer to the load is maximized at the center frequency f_c = 1 GHz although their quality factors or selectivities differ.

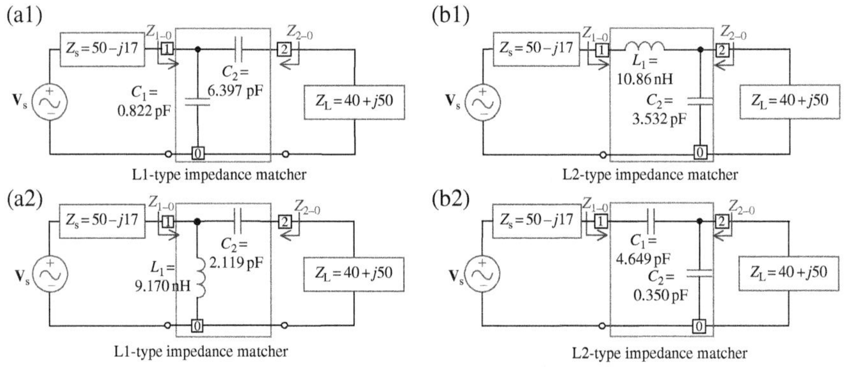

Figure 6.19 L-type filters used for impedance matching in Example 6.12.

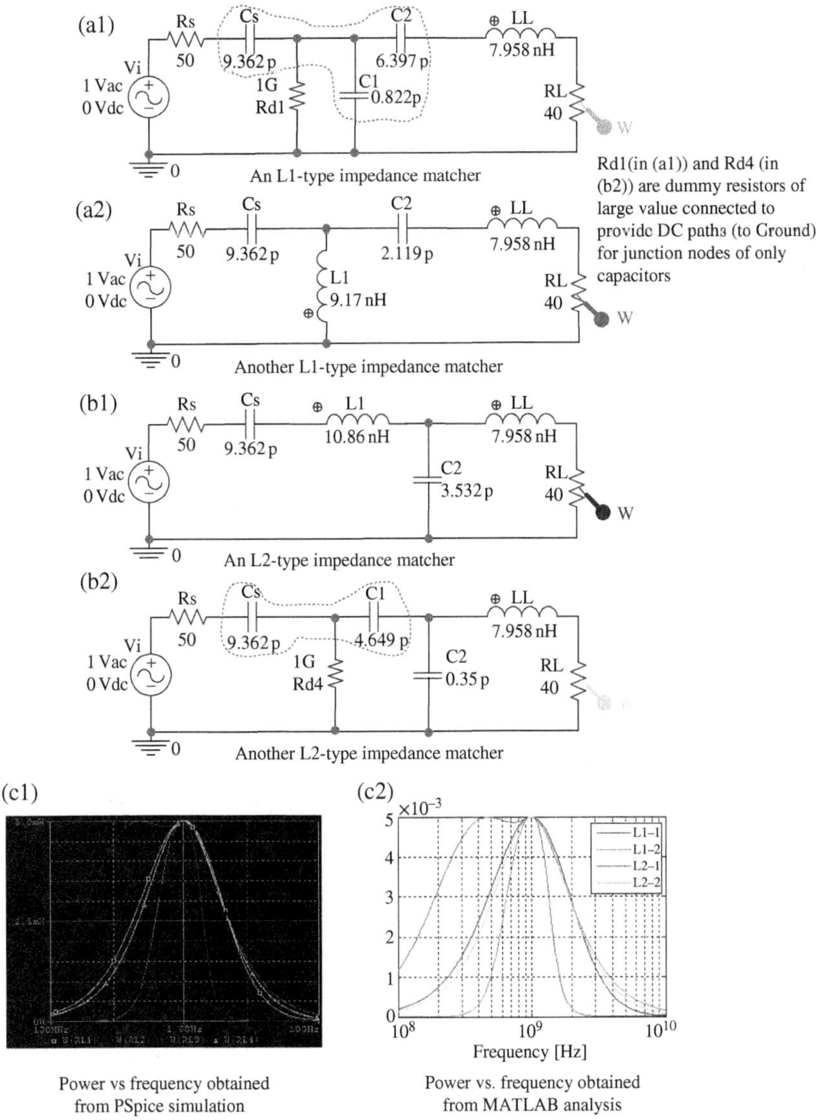

Figure 6.20 PSpice schematics for the circuits of Figure 6.19 and their plots of power P_{RL} versus frequency.

6.3.3 T- and Π-Type Impedance Matchers

Figure 6.21 shows T- and Π-type LC ladders (with $Z_k = jX_k = j\omega L_k$ or $Z_k = jX_k = 1/j\omega C_k$) that can be used to accomplish impedance matching between the source impedance $Z_s = R_s + jX_s$ and load impedance $Z_L = R_L + jX_L$. The

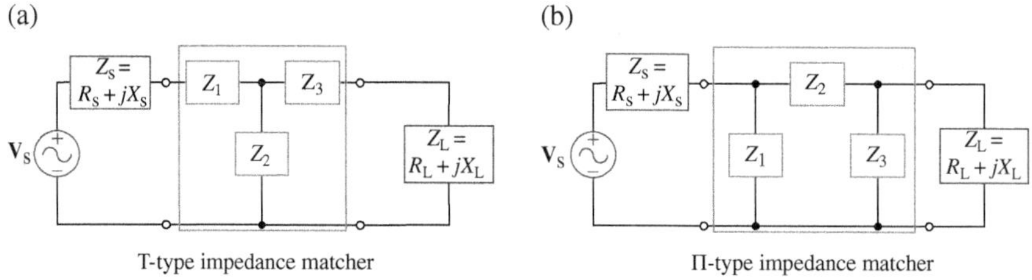

Figure 6.21 T- and Π-type impedance matchers.

conditions for impedance matching of T-type filter (Figure 6.21(a)) with quality factor Q can be formulated into the formulas for computing $Z_3 = jX_3$, $Z_2 = jX_2$, and $Z_1 = jX_1$ as

$$\mathrm{Re}\left\{\frac{1}{R_L + jX_L + jX_3}\right\} = \mathrm{Re}\left\{\frac{1}{R_s + jX'_s}\right\} \text{ with } X'_s \overset{(6.2.26a)}{=} QR_s$$

$$; \frac{R_L}{R_L^2 + (X_L + X_3)^2} = \frac{R_s}{R_s^2 + (QR_s)^2}$$

$$; R_s X_3^2 + 2R_s X_L X_3 + R_s\left(R_L^2 + X_L^2\right) - R_L\left\{R_s^2 + (QR_s)^2\right\} = 0 \rightarrow X_3 \tag{6.3.3a}$$

$$\mathrm{Re}\left\{\frac{(R_L + jX_L + jX_3)jX_2}{(R_L + jX_L + jX_3) + jX_2}\right\} = R_s$$

$$; \mathrm{Re}\left\{\frac{X_2(-(X_L + X_3) + jR_L)(R_L - j(X_L + X_3 + X_2))}{R_L^2 + (X_L + X_3 + X_2)^2}\right\} = R_s$$

$$; \frac{X_2 R_L (X_L + X_3 + X_2 - (X_L + X_3))}{R_L^2 + (X_L + X_3 + X_2)^2} = \frac{R_L X_2^2}{R_L^2 + (X_L + X_3 + X_2)^2} = R_s$$

$$; (R_L - R_s)X_2^2 - 2R_s(X_L + X_3) - R_s\left(R_L^2 + (X_L + X_3)^2\right) = 0 \rightarrow X_2 \tag{6.3.3b}$$

$$X_1 + \mathrm{Im}\left\{\frac{(R_L + jX_L + jX_3)jX_2}{(R_L + jX_L + jX_3) + jX_2}\right\} = -X_s$$

$$; X_1 = -\mathrm{Im}\left\{\frac{(R_L + jX_L + jX_3)jX_2}{(R_L + jX_L + jX_3) + jX_2}\right\} - X_s \tag{6.3.3c}$$

The Y-Δ conversion of a T-type impedance matcher yields its equivalent Π-type impedance matcher (Figure 6.21(b)).

```
function [X1s,X2s,X3s,Zins]=imp_matcher_T(Zs,ZL,Q,f)
% Determine the T-type impedance matcher between Zs and ZL.
% Copyleft: Won Y. Yang, wyyang53@hanmail.net, CAU for academic use only
Rs=real(Zs); Xs=imag(Zs); % Source impedance
RL=real(ZL); XL=imag(ZL); % Load impedance
b=2*XL; c=(RL^2+XL^2)-RL*Rs*(1+Q^2); % Eq.(6.3.3a)
tmp=b^2-4*c;
if tmp<0, error('Rs=real(Zs) must be increased'); end
if tmp<=1e-8; X3=-b/2;
 else X3=(-b+[1 -1]*sqrt(tmp))/2;
end
a=RL-Rs; b=-2*Rs*(XL+X3); c=-Rs*(RL^2+(XL+X3).^2); % Eq.(6.3.3b)
tmp=b.^2-4*a*c; tmp=tmp(1); % Since tmp(1)=tmp(2)
if tmp<0, error('Rin=real(Zin) must be increased'); end
X3s=[]; X2s=[];
for n=1:length(b)
   if abs(a)<eps, X2n=-c/b; X3n=X3(n); else
     if tmp<=1e-8; X2n=-b(n)/2/a; X3n=X3(n);
      else X2n=(-b(n)+[1 -1]*sqrt(tmp))/2/a; X3n=[X3(n) X3(n)];
     end
   end
   X2s=[X2s X2n]; X3s=[X3s X3n];
end
Z23Ls=1./(1./(j*X2s)+1./(j*X3s+ZL));
ids=find(abs(real(Z23Ls)-Rs)<0.1);
X2s=X2s(ids); X3s=X3s(ids); Z23Ls=Z23Ls(ids);
X1s=-Xs-imag(Z23Ls); % Eq.(6.3.3c)
Zins=j*X1s+Z23Ls;
express_T_filter(X1s,X2s,X3s,2*pi*f);
```

```
function express_T_filter(X1s,X2s,X3s,w)
for n=1:length(X1s)
   X1=X1s(n); X2=X2s(n); X3=X3s(n);
   if X1>0
     L1=X1/w*1e6; fprintf('T-type imp matcher: L1=%8.3e[uH]',L1);
    else
     C1=-1e9/X1/w; fprintf('T-type imp matcher: C1=%8.3e[nF]',C1);
   end
   if X2>0, L2=X2/w*1e6; fprintf(', L2=%8.3e[uH]',L2);
    else C2=-1e9/X2/w; fprintf(', C2=%8.3e[nF]',C2);
   end
   if X3>0, L3=X3/w*1e6; fprintf(', L3=%8.3e[uH]\n',L3);
    else C3=-1e9/X3/w; fprintf(', C3=%8.3e[nF]\n',C3);
   end
end
```

```
function [X1s,X2s,X3s,Zins]=imp_matcher_pi(Zs,ZL,Q,f)
% Determine the pi-type impedance matcher between Zs and ZL.
% Copyleft: Won Y. Yang, wyyang53@hanmail.net, CAU for academic use only
[Xas,Xbs,Xcs,Zins]=imp_matcher_T(Zs,ZL,Q,f);
for i=1:length(Xas)
   [X1s(i),X3s(i),X2s(i)]=yd_conversion(Xas(i),Xbs(i),Xcs(i));
end
Y23Ls=1./(j*X2s+1./(1./(j*X3s)+1/ZL));
Zins=1./(1./(j*X1s)+Y23Ls);
express_pi_filter(X1s,X2s,X3s,w);
%--------------------------------
function express_pi_filter(X1s,X2s,X3s,w)
for n=1:length(X1s)
  X1=X1s(n); X2=X2s(n); X3=X3s(n);
  if X1>0
    L1=X1/w*1e6;
    fprintf('pi-type imp matcher: L1=%8.3e[uH]',L1);
   else
    C1=-1e9/X1/w;
    fprintf('pi-type imp matcher: C1=%8.3e[nF]',C1);
  end
  if X2>0, L2=X2/w*1e6; fprintf(', L2=%8.3e[uH]',L2);
    else C2=-1e9/X2/w; fprintf(', C2=%8.3e[nF]',C2);
  end
  if X3>0, L3=X3/w*1e6; fprintf(', L3=%8.3e[uH]\n',L3);
    else C3=-1e9/X3/w; fprintf(', C3=%8.3e[nF]\n',C3);
  end
end
```

```
function [Zab,Zbc,Zca]=yd_conversion(Za,Zb,Zc)
% Wye-Delta Transformation
temp = Za*Zb+Zb*Zc+Zc*Za;
Zab=temp/Zc; Zbc=temp/Za; Zca=temp/Zb;
```

```
function [Za,Zb,Zc]=dy_conversion(Zab,Zbc,Zca)
% Delta-Wye Transformation
temp = Zab+Zbc+Zca;
Za=Zca*Zab/temp; Zb=Zab*Zbc/temp; Zc=Zbc*Zca/temp;
```

The above MATLAB functions 'imp_matcher_T()'/'imp_matcher_pi()' can be used to find possible T/Π-type *LC* filter(s) (with given quality factor *Q*) tuned for impedance matching between given source and load impedances, respectively.

Example 6.13 Design of T-type Impedance Matcher

Make as many T-type ladders as possible, with quality factor $Q = 3$ for maximum power transfer from a transmitter with output impedance $Z_s = 10 - j/\omega C_s$ ($C_s = 7.958$ pF) to an antenna with input impedance $Z_L = 60 - j/\omega_c C_L$ ($C_L = 5.305$ pF) at the center frequency $f_c = 1$ GHz.

We run the following MATLAB statements:

```
>>fc=1e9; wc=2*pi*fc; Q=3;
  Rs=10; Cs=7.958e-12; Zs=Rs-j/(wc*Cs) % Zs=10-20i
  RL=60; CL=5.305e-12; ZL=RL-j/(wc*CL) % ZL=60-30i
  [X1s,X2s,X3s,Zins_T]=imp_matcher_T(Zs,ZL,Q,fc);
```

This yields

```
T-type imp matcher: C1=1.591e-02[nF], L2=7.289e-03[uH], L3=1.257e-02[uH]
T-type imp matcher: L1=7.958e-03[uH], C2=6.074e-03[nF], L3=1.257e-02[uH]
T-type imp matcher: C1=1.591e-02[nF], L2=4.170e-03[uH], C3=8.381e-03[nF]
T-type imp matcher: L1=7.958e-03[uH], C2=3.475e-03[nF], C3=8.381e-03[nF]
```

This result implies that there are four T-types of LC circuits that can present impedance matching between the source and load as depicted in Figure 6.22.

If you want to check if each ladder satisfies the impedance matching condition, it is enough to see that the input impedance (seen from nodes 1-0) obtained as the fourth output argument Zins_T of '`imp_matcher_pi()`' is equal to Z_s^*. If you are still not sure, check if $Z_{1-0} = Z_s^*$ by running the following statements:

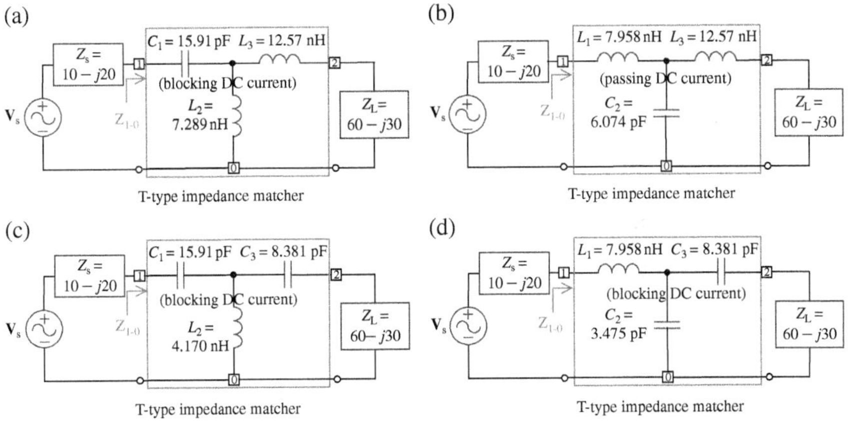

Figure 6.22 T-type ladder circuits used for impedance matching in Example 6.13.

```
>>for n=1:numel(X1s)
   Z10=j*X1s(n)+parallel_comb([j*X2s(n) j*X3s(n)+ZL]) %Z1+{Z2||(Z3+ZL)}
   % This is expected to be equal to the conjugate of Zs.
  end
```

This yields

```
Z10 = 10.0000 +19.9994i
Z10 = 10.0000 +19.9994i
Z10 = 10.0000 +19.9994i
Z10 = 10.0000 +19.9994i
```

That is, for the ladder in Figure 6.22(a), we have

$$Z_{1-0} = Z_1 + \{Z_2 \| (Z_3 + Z_L)\} = -j\frac{1}{\omega C_1} + \{j\omega L_2 \| (j\omega L_3 + Z_L)\} \overset{\substack{C_1 = 15.91\times 10^{-12}\\ L_2 = 7.289\times 10^{-9}\\ L_3 = 12.57\times 10^{-9}}}{\underset{\omega = 2\pi \times 10^9}{=}} 10 + j20 \tag{E6.13.1}$$

Also, for the ladder in Figure 6.22(b), we have

$$Z_{1-0} = Z_1 + \{Z_2 \| (Z_3 + Z_L)\} = j\omega L_1 + \left\{\frac{1}{j\omega C_2} \| (j\omega L_3 + Z_L)\right\} \overset{\substack{L_1 = 7.958\times 10^{-9}\\ C_2 = 6.074\times 10^{-12}\\ L_3 = 12.57\times 10^{-9}}}{\underset{\omega = 2\pi \times 10^9}{=}} 10 + j20 \tag{E6.13.2}$$

Example 6.14 Design of Π-type Impedance Matcher

Make as many Π-type LC ladders as possible, with quality factor $Q = 2$ for maximum power transfer from a transmitter with output impedance $Z_s = 20 - j/\omega C_s$ ($C_s = 1.658$ pF) to an antenna with input impedance $Z_L = 10 - j/\omega_c C_L$ ($C_L = 6.632$ pF) at the center frequency $f_c = 2.4$ GHz.

We can run the following MATLAB statements:

```
>>fc=2.4e9; wc=2*pi*fc; Q=2;
  Rs=20; Cs=1.658e-12; Zs=Rs-j/(wc*Cs) % Zs=20-40i
  RL=10; CL=6.632e-12; ZL=RL-j/(wc*CL) % ZL=10-10i
  [X1s,X2s,X3s,Zins_pi]=imp_matcher_pi(Zs,ZL,Q,fc);
```

This yields

```
pi-type imp matcher: C1=6.632e-04[nF], L2=2.652e-03[uH], L3=8.083e+01[uH]
pi-type imp matcher: L1=1.326e-03[uH], C2=1.658e-03[nF], L3=6.631e-04[uH]
pi-type imp matcher: L1=1.326e-03[uH], C2=3.316e-03[nF], L3=8.084e+00[uH]
pi-type imp matcher: C1=3.014e-04[nF], L2=2.918e-03[uH], L3=3.647e-03[uH]
```

This result implies that there are four Π-type LC ladders that can present impedance matching between the source and load as depicted in Figure 6.23.

If you want to check if each ladder satisfies the impedance matching condition, it is enough to see that the input impedance (seen from nodes 1-0) obtained as the

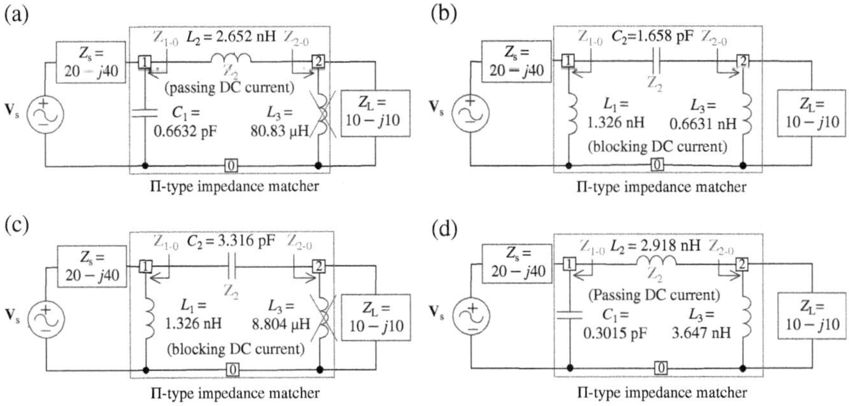

Figure 6.23 Π-type ladder circuits used for impedance matching in Example 6.14.

fourth output argument Zins_pi of '`imp_matcher_pi()`' is equal to Z_s^*. If you are still not sure, check if $Z_{1\text{-}0} = (Z_2 + Z_{2\text{-}0})^*$:

```
>>for n=1:numel(X1s)
     Z10=parallel_comb([Zs j*X1s(n)])
     Z2=j*X2s(n); Z20=parallel_comb([j*X3s(n) ZL]); Z2+Z20
  end
```

However, the values of $X_3 = \omega_c L_3$ of Figure 6.23(a) and (c) are overwhelmingly larger than $|Z_L| = 10\sqrt{2}\ \Omega$, the parallel L_3's are not meaningful as can be seen by typing 'X3s' at the MATLAB prompt:

```
>>X3s
  X3s = 1.0e+06 * 1.21895  0.00001  0.12191  0.00005
```

This implies that Figure 6.23(a) and (c) is virtually L-type impedance matcher without L_3.

6.3.4 Tapped-C Impedance Matchers

Figure 6.24(a) shows a tapped-C(capacitor) impedance matching circuit for a desired quality factor of $Q/2$ between R_1 and R_2. Figure 6.24(b) shows the intermediate equivalent circuit where the parallel $R_1|C_2$ part (with a target quality factor Q_p to be determined) has been transformed into its series equivalent consisting of R_{1s}-C_{2s} and the parallel $L|R_2$ part (with a quality factor Q given) has been transformed into its series equivalent consisting of L_s-R_{2s} by applying Eq. (6.2.28a). Figure 6.24(c) shows the final equivalent circuit with the series C_1-C_{2s} part replaced by its series combination where

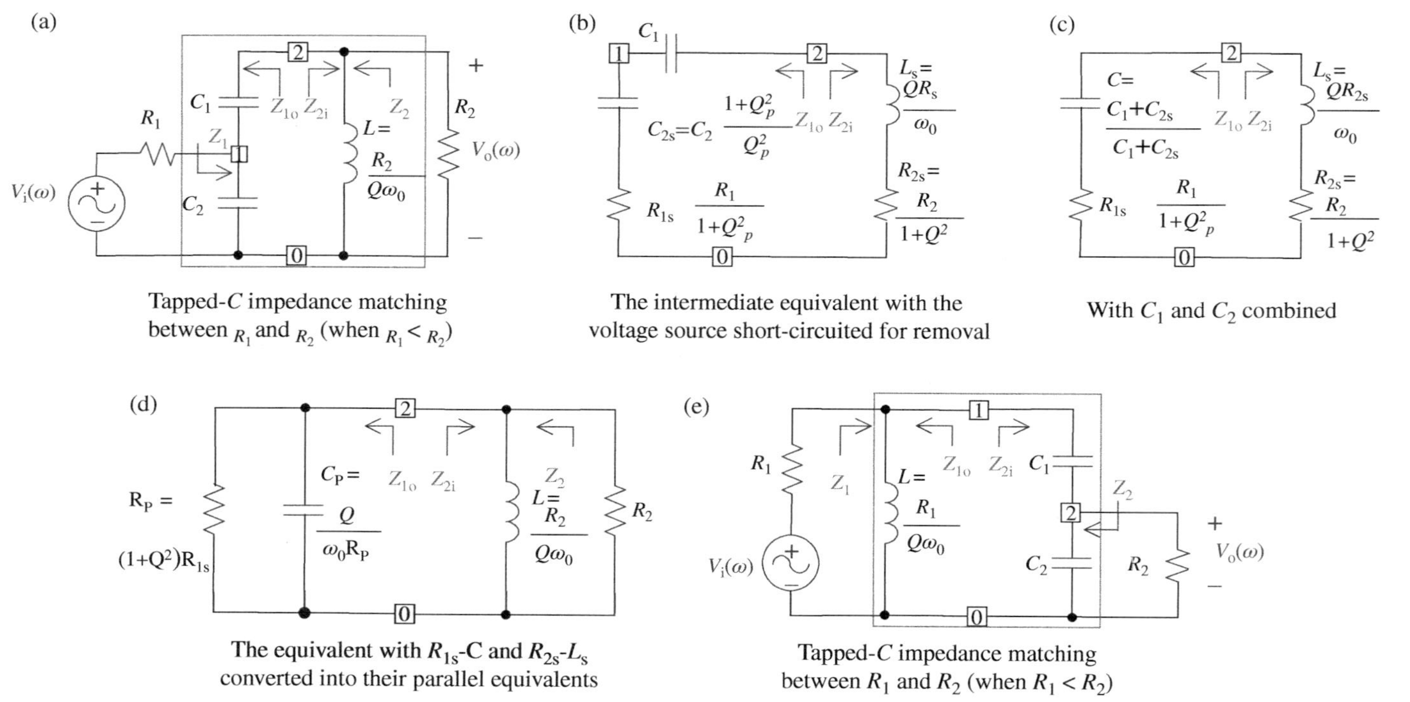

Figure 6.24 Tapped-*C* impedance matching circuit and its equivalents.

$$R_{1s} \overset{(6.2.28a)}{=} \frac{R_1}{1+Q_p^2},\ C_2 \overset{(6.2.26b)}{=} \frac{Q_p}{\omega_0 R_1},\ C_{2s} \overset{(6.2.28b)}{=} C_2\frac{1+Q_p^2}{Q_p^2},\ C = \frac{C_1 C_{2s}}{C_1 + C_{2s}} \tag{6.3.4a,b,c,d}$$

$$R_{2s} \overset{(6.2.28a)}{=} \frac{R_2}{1+Q^2},\ L_s \overset{(6.2.28b)}{=} \frac{QR_{2s}}{\omega_0} \tag{6.3.5a,b}$$

Noting that if the quality factors of the source and load parts from node 2 are equally Q, their composite quality factor will be $1/(1/Q + 1/Q) = Q/2$, we write the condition for the quality factor of the load ($L|R_2$) part (i.e. the parallel combination of L and R_2) to be Q as

$$X_L = \omega_0 L \overset{(6.2.26b)}{=} \frac{R_2}{Q};\ L = \frac{R_2}{\omega_0 Q} \tag{6.3.6}$$

Also, the impedance matching conditions to make the source and load impedances, Z_1 and Z_2, a complex conjugate pair can be written as

$$R_{1s} - j\frac{1}{\omega_0 C} = (R_{2s} + j\omega_0 L_s)^* \rightarrow \begin{cases} R_{1s} \overset{(6.3.4a)}{=} \dfrac{R_1}{1+Q_p^2} = \dfrac{R_2}{1+Q^2} \overset{(6.3.5a)}{=} R_{2s};\ Q_p = \sqrt{\dfrac{R_1}{R_2}(1+Q^2)-1} & (6.3.7) \\ \dfrac{1}{\omega_0 C} = \omega_0 L_s \overset{(6.3.5b)}{=} QR_{2s};\ C = \dfrac{C_1 C_{2s}}{C_1 + C_{2s}} = \dfrac{1}{\omega_0 R_{2s} Q} & (6.3.8) \end{cases}$$

```
function [L,C1,C2,Qr,Qp,Rp,Z1,Z2]=imp_matcher_tapped_C(R1,R2,f0,Q)
% Q : Desired quality factor.
w0=2*pi*f0; Q2=2*Q; X2=R2/Q2; % Eq.(6.3.6a)
R2s=R2/(1+Q2^2); L=X2/w0; % Eqs.(6.3.5a),(6.3.6b)
Qp = sqrt(R1/R2*(1+Q2^2)-1); % Eq.(6.3.7)
R1s = R1/(1+Qp^2); % Eq.(6.3.4a)
C2 = Qp/w0/R1; C2s = C2*(1+Qp^2)/Qp^2; % Eq.(6.3.4b,c)
C = 1/w0/R2s/Q2; % Eq.(6.3.8)
C1 = C*C2s/(C2s-C); % Eq.(6.3.9)
if C1<0, error('\n How about switching R1 and R2?\n'); end
Rp=(1+Q2^2)*R1s; % Eq.(6.2.29a): Series-to-parallel conversion of R1s
Qr = parallel_comb([Rp R2])/X2; % Resulting value of Quality factor
% To get the impedance Z1(w) that is expected to be matched to R1,
jw0=j*w0; ZC1LRL = 1/jw0/C1 + parallel_comb([jw0*L R2]);
Z1 = parallel_comb([1/jw0/C2 ZC1LRL]); % = R1 ?
fprintf('\n Z1=%5.0f + j%5.0f matches to R1=%5.0f?',real(Z1),imag(Z1),R1)
% To get the impedance Z2(w) that is expected to be matched to R2,
ZC1C2Rs = 1/jw0/C1 + parallel_comb([1/jw0/C2 R1]);
Z2 = parallel_comb([ZC1C2Rs jw0*L]); % = R2 ?
fprintf('\n Z2=%5.0f + j%8.0f matches to R2=%5.0f?\n',real(Z2),imag(Z2),R2)
fprintf('\n Design results for given R1=%5.0f [Ohm], R2=%5.0f [Ohm], and
                                                   Q=%5.0f, ',R1,R2,Q)
fprintf('\n C1=%11.3e [mF], C2=%11.3e [mF], L=%11.3e [mH]\n',
                                          C1*1e3,C2*1e3,L*1e3)
```

Once the series combination of C_1-C_{2s} part, that is C, is determined from Eq. (6.3.8), we can use the determined value of C_{2s} (Eq. (6.3.4c)) together with Eq. (6.3.8) to find C_1:

$$C = \frac{C_1 C_{2s}}{C_1 + C_{2s}}; \; C_1 = \frac{C\,C_{2s}}{C_{2s} - C} \tag{6.3.9}$$

This process of designing a tapped-C impedance matcher has been cast into the above MATLAB function '`imp_matcher_tapped_C()`', in which the resulting value of quality factor Q, in the variable name of '`Qr`', is computed as

$$Q_r \overset{(6.2.26b)}{=} \frac{R_p \| R_2}{X_2 = \omega_0 L = 1/\omega_0 C_p = X_1} = \frac{1}{X_1/R_p + X_2/R_2} \overset{(6.2.26b)}{=} \frac{1}{1/Q + 1/Q} \tag{6.3.10a}$$

$$\text{with } R_p \overset{(6.2.29a)}{=} \left(1 + Q^2\right) R_{1s} \text{ and } C_p \overset{(6.2.29b)}{=} \frac{Q}{\omega_0 R_p}, \tag{6.3.10b}$$

since if the series R_{1s}-C part (in Figure 6.24(c)) is converted to its parallel equivalent $R_p|C_p$ while the series R_{2s}-L_s part (in Figure 6.24(c)) is converted back to its parallel equivalent (as shown in Figure 6.24(d)), the resistance in parallel with L and C_p is not just R_2 but $R_p \| R_2$.

Example 6.15 Design of Tapped-C Impedance Matcher

Design a tapped-C impedance matcher for source/load resistances $R_1/R_2 = 5/50$ [Ω] at $f_0 = 150$ MHz such that the loaded quality factor is $Q = 10$, find the equivalent impedances Z_1/Z_2 (see Figure 6.25(a)) to check if they match to R_1/R_2, respectively, and plot its frequency response magnitude $G(\omega) = V_o(\omega)/V_i(\omega)$ to see the resulting value of $Q = f_p/(f_u - f_l)$ where f_p is the peak frequency that is expected to conform with the resonant frequency for most BPFs.

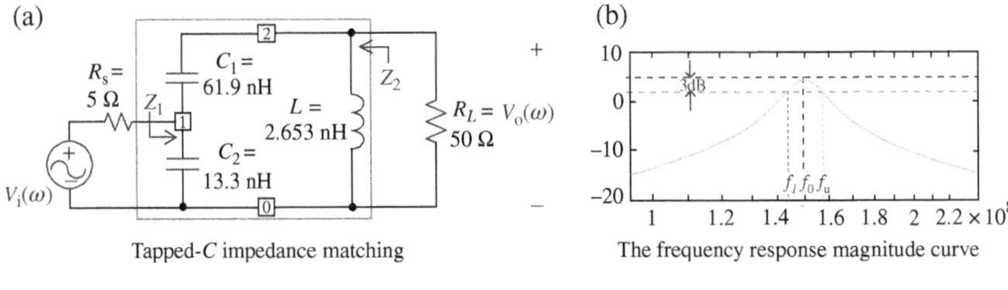

Figure 6.25 Tapped-C impedance matching circuit for Example 6.15.

```
%elec06e15.m
% Try the tapped-C impedance matching.
clear, clf
Rs=5; RL=50; f0=150e6; w0=2*pi*f0; jw0=j*w0; Q=10;
[L,C1,C2,Q_resulting,Qp,Rp,Z1,Z2]=...
                    imp_matcher_tapped_C(Rs,RL,f0,Q);
% To get the frequency response,
f=logspace(-0.2,0.2,600)*f0; w=2*pi*f; % Frequency range
jw=j*w;
ZLRLw = 1./(1./jw/L+1/RL);
ZC1LRLw = 1./jw/C1 + ZLRLw;
ZC1C2LRLw = 1./(jw*C2 + 1./ZC1LRLw);
Gw=ZC1C2LRLw./(Rs+ZC1C2LRLw).*ZLRLw./ZC1LRLw/(RL/(Rs+RL));
GwmagdB = 20*log10(abs(Gw)); % Frequency response magnitude
semilogx(f,GwmagdB), hold on
[peak,idx]=max(GwmagdB); f0=f(idx) % Peak frequency
% To find the lower/upper 3dB frequencies
[peaks,idx]=findpeaks(-abs(GwmagdB-peak+3));
if numel(idx)>1
 fl=f(idx(1)), fu=f(idx(2)), Q_loaded_measured = f0/(fu-fl)
else
 fprintf('\n Q should be larger to get such a concave-down freq.
        resp. magnitude curve\n that has two 3dB frequencies
        fl and fu.\n')
end
```

To this end, we can run the above script "elec06e15.m," which uses the MATLAB function 'imp_matcher_tapped_C()' inside, to get

```
>>elec06e15
  Z1=  5 + j  0 matches to R1=  5?
  Z2= 50 + j  0 matches to R2= 50?
  Design results for given R1=  5[Ohm], R2= 50[Ohm], and Q= 10
  C1= 6.190e-07[mF], C2= 1.327e-06[mF], L= 2.653e-06[mH]
  f0 = 1.5012e+08, fl = 1.4269e+08,  fu = 1.5769e+08
  Q_loaded_measured =  1.0009e+01
```

This means the tapped-C impedance matching circuit shown in Figure 6.25(a) where the impedances Z_1 and Z_2 match $R_1 = 5$ [Ω] and $R_2 = 50$ [Ω], respectively, and the loaded $Q = f_p/(f_u - f_l)$ has turned out to be 10 as required by the design specifications. Figure 6.25(b) shows the frequency response magnitude curve together with the lower and upper 3 dB frequencies.

Note that in case where $R_s > R_L$, the MATLAB function 'imp_matcher_tapped_C()' should be used with R_s and R_L switched and also the tapped-C impedance matcher should be flipped horizontally.

6.4 Active Filter Realization

This section introduces some MATLAB functions that can be used to determine the parameters of the circuits depicted in Figures 5.17-5.20 so that they can implement a given (designed) transfer function.

```
function [R1,CR2,Gs]=LPF_RC_OPAmp_design(B,A,R2C,KC)
% Design a 1st-order LPF with the circuit in Fig. 5.17(a)
%           R2    1/R2C    B  A        B
% G(s) = - -- --------- = - ---- = - ---          (5.5.1)
%           R1  s + 1/R2C   A  s+A      s+A
if nargin<4, KC=1; end
if KC==1 % Find R1 and C for given R2.
 R2=R2C; C=1/(A*R2); R1=1/(B*C); CR2=C;
 else % Find R1 and R2 for given C.
 C=R2C; R2=1/(A*C); R1=1/(B*C); CR2=R2;
end
syms s; Gs = B/(s+A);
```

```
function [R1,CR2,Gs]=HPF_RC_OPAmp_design(B,A,R2C,KC)
% Design a 1st-order HPF with the circuit in Fig. 5.17(b)
%           R2     s           s
% G(s) = - -- --------- = - B ---          (5.5.2)
%          R1  s + 1/R1C      s+A
if nargin<4, KC=1; end
if KC==1 % Find R1 and C for given R2.
 R2=R2C; C=1/(A*R1); R1=R2/B; CR2=C;
 else % Find R1 and R2 for given C.
 C=R2C; R1=1/(A*C); R2=R1*B; CR2=R2;
end
syms s; Gs = B*s/(s+A);
```

```
function [CR1,CR2,Gs]=LPF_Sallen_design(A2,A3,K,RC1,RC2,KC)
% Design an LPF with the circuit in Fig. 5.18(a)
%                        KG1G2/C1C2                        B3=K*A3
%G(s) = ------------------------------------ = ----------- (5.5.3)
%       s^2 + ((G1+G2)/C1+(1-K)G2/C2)*s +G1G2/C1C2   s^2 + A2*s + A3
if K<1, fprintf('Let K=1 since we must have K=(R3+R4)/R3>=1!'); K=1; end
if nargin<6, KC=1; end
if KC==1 % Find C1 and C2 for given K, R1, and R2.
  R1= RC1; R2= RC2; G1= 1/R1; G2= 1/R2;
  a= G1+G2; b= -(K-1)*G2; c= A2; d= A3/G1/G2; tmp = c^2-4*a*b*d;
 C1= 2*a/(c + sqrt(tmp)); C2= 1/d/C1; CR1= C1; CR2= C2;
 else % Find R1 and R2 for given K, C1, and C2.
  C1= RC1; C2= RC2;
  a= 1/C1; b= 1/C1 - (K-1)/C2; c= A2; d= A3*C1*C2; tmp = c^2-4*a*b*d;
 if tmp<0, error('Increase C1 and K, or decrease C2'); end
 G1= (c + sqrt(tmp))/2/a; G2= d/G1; R1= 1/G1; R2= 1/G2; CR1= R1; CR2= R2;
end
B3= K*A3; A2= (G1+G2)/C1 + (1-K)*G2/C2; A3= G1*G2/C1/C2;
syms s; Gs = B3/(s^2+A2*s+A3);
```

```
function [CR1,CR2,Gs]=HPF_Sallen_design(A2,A3,K,RC1,RC2,KC)
% Design an HPF with the circuit in Fig. 5.18(b)
%                             K*s^2                            K*s^2
%G(s) = - - - - - - - - - - - - - - - - - - - - - - - - - - - - = - - - - - - - - - - (5.5.4)
%        s^2 +(G2(1/C1+1/C2)-(K-1)G1/C1)s +G1G2/C1C2    s^2 + A2*s + A3
if K<1, fprintf('Let K=1 since we must have K=(R3+R4)/R3>=1!'); K=1; end
if nargin<6, KC=1; end
if KC==1 % Find C1 and C2 for given K, R1, and R2.
  R1= RC1; R2= RC2; G1= 1/R1; G2= 1/R2;
  a= G2+(1-K)*G1; b= G2; c= A2; d= A3/G1/G2; tmp= c^2-4*a*b*d;
  if tmp<0, error('Try with smaller/greater values of R1/K'); end
  C1= 2*a/(c + sign(a)*sqrt(tmp)); C2= 1/d/C1; CR1= C1; CR2= C2;
 else % Find R1 and R2 for given K, C1, and C2.
  C1=RC1; C2=RC2;
  a=(1-K)/C1; b=1/C1+1/C2; c=A2; d=A3*C1*C2; tmp=c^2-4*a*b*d;
  if tmp<0, error('Try with smaller/greater values of C2/K'); end
  if abs(a)<eps, G2= A2/b; G1= d/G2;
   else G1= (c + sign(a)*sqrt(tmp))/2/a; G2= d/G1;
  end
  R1= 1/G1; R2= 1/G2; CR1= R1; CR2= R2;
end
B1= K; A2= G2*(1/C1+1/C2) - (K-1)*G1/C1; A3= G1*G2/C1/C2;
syms s; Gs = B1*s^2/(s^2+A2*s+A3);
```

```
function [R1,C2R3,C5R4,Gs]=LPF_MFB_design(B3,A2,A3,R3C2,R4C5,KC)
% Design an LPF with the circuit in Fig. 5.19(a)
%                    -G1G4/C2C5                      -B3
%G(s) = - - - - - - - - - - - - - - - - - - - - - - = - - - - - - - - - -    (5.5.5)
%        s^2 + (G1+G3+G4)/C2*s + G3G4/C2C5   s^2 + A2*s + A3
if nargin<6, KC=1; end
if KC==1 % Find R1, C2 and C5 for given R3 and R4.
 R3= R3C2; R4= R4C5; G3= 1/R3; G4= 1/R4;
 G1=G3*B3/A3; C2=(G1+G3+G4)/A2; C5=G3*G4/C2/A3; R1=1/G1;
                                                  C2R3=C2; C5R4=C5;
 else % Find R1, R3 and R4 for given C2 and C5.
 C2=R3C2; C5=R4C5; a=1+B3/A3; b=1; c=A2*C2; d=A3*C2*C5;
                                               tmp = c^2-4*a*b*d;
 if tmp<0, error('Try with greater/smaller values of C2/C5'); end
 G3= (c + sign(a)*sqrt(tmp))/2/a; G4= d/G3;
 G1= B3/A3*G3; R3= 1/G3; R4= 1/G4; R1=1/G1; C2R3= R3; C5R4= R4;
end
B3= G1*G4/C2/C5; A2= (G1+G3+G4)/C2; A3= G3*G4/C2/C5;
syms s; Gs = -B3/(s^2+A2*s+A3);
```

```
function [C1,C3R2,C4R5,Gs]=HPF_MFB_design(B1,A2,A3,R2C3,R5C4,KC)
% Design an HPF with the circuit in Fig. 5.19(b)
%                 -(C1/C3)*s^2                           -B1*s^2
%G(s) = ----------------------------------------- = ---------- (5.5.6)
%       s^2 + G5(C1+C3+C4)/C3/C4*s + G2G5/C3C4   s^2 + A2*s + A3
if nargin<6, KC=1; end
if KC==1 % Find C1, C3 and C4 for given R2 and R5.
  R2= R2C3; R5= R5C4; G2= 1/R2; G5= 1/R5;
  a= 1; b= 1+B1; c= A2/G5; d= A3/G2/G5; tmp = c^2-4*a*b*d;
 if tmp<0, error('Try with smaller/greater values of R2/R5'); end
 C3= 2*a/(c + sqrt(tmp)); C4= 1/d/C3; C1= B1*C3; C3R2= C3; C4R5= C4;
else % Find C1, R2 and R5 for given C3 and C4.
 C3= R2C3; C4= R5C4;
 C1 = B1*C3; G5= A2/(C1+C3+C4)*C3*C4; G2= A3*C3*C4/G5;
 R2= 1/G2; R5= 1/G5; C3R2= R2; C4R5= R5;
end
B1= C1/C3; A2= G5*(C1+C3+C4)/C3/C4; A3= G2*G5/C3/C4;
syms s; Gs = -B1*s^2/(s^2+A2*s+A3);
```

```
function [C3R1,C4R2,R5,Gs]=BPF_MFBa_design(B2,A2,A3,R1C3,R2C4,KC)
% Design a BPF with the circuit in Fig. 5.20(a)
%                       -(G1/C3)*s                          -B2*s
%G(s) = --------------------------------------------- = ----------      (5.5.7)
%       s^2 + G5(1/C3+1/C4)*s + (G1+G2)G5/C3C4    s^2 + A2*s + A3
if nargin<6, KC=(R1C3+R2C4>1)+2*(R1C3+R2C4<1); end
if KC==1 % Find C3, C4, and R5 for given R1 and R2.
  R1=R1C3; R2=R2C4; G1=1/R1; G2=1/R2; C3=G1/B2;
  G5=(A2-A3*C3/(G1+G2))*C3;
 if G5<0, error('Try with smaller values of R2'); end
  C4= G5*(G1+G2)/C3/A3; R5=1/G5;
  C3R1=C3; C4R2=C4;
 elseif KC==2 % Find R1, R2 and R5 for given C3 and C4.
 C3=R1C3; C4=R2C4; G1=B2*C3; G5=A2/(1/C3+1/C4); G2=A3*C3*C4/G5-G1;
 R5=1/G5; R1=1/G1; R2=1/G2; C3R1=R1; C4R2=R2;
 elseif KC==3 % Find R1, R5, and C3=C4=C for given R2 and C3=C4.
 R2=R1C3; G2=1/R2;
 C=A2*G2/(2*A3-A2*B2); G5=A2/2*C;
 C3=C; C4=C; G1=B2*C; R1=1/G1; R5=1/G5;
 C3R1=C3; C4R2=R1;
 else % Find C3=C4, R5 for given R1 and R2 (not caring about B2).
 R1=R1C3; R2=R2C4; G1=1/R1; G2=1/R2;
  C=(G1+G2)*A2/2/A3; C3=C; C4=C; G5=A2/2*C; R5=1/G5;
  C3R1=C; C4R2=C;
end
fprintf('\n R1=%8.0f, R2=%8.0f, R5=%8.0f, C3=%10.3e, C4=%10.3e\n',
R1,R2,R5,C3,C4)
B2=G1/C3; A2=G5*(1/C3+1/C4); A3=(G1+G2)*G5/C3/C4;
syms s; Gs = -B2*s/(s^2+A2*s+A3);
```

```
function [C1,C2R3,C5R4,Gs]=BPF_MFBb_design(B2,A2,A3,R3C5,R4C5,KC)
% Design a BPF with the circuit in Fig. 5.20(b)
%                (C1G4/(C1+C2)C5)*s                      -B2*s
%G(s)=-----------------------------------=------------ (5.5.8)
%     s^2 + ((G3+G4)/(C1+C2))*s + G3G4/(C1+C2)C5     s^2 + A2*s + A3
if nargin<6, KC=1; end
if KC==1 % Find C1, C2 and C5 for given R3 and R4.
  R3=R3C5; R4=R4C5; G3=1/R3; G4=1/R4;
  C1pC2= (G3+G4)/A2; C5= G3*G4/A3/C1pC2; C1=B2*C1pC2*C5/G4; %=B2*G3/A3
  C2= C1pC2 - C1;
  if C2<0, error('Try with greater/smaller values of R3/R4'); end
  C2R3= C2; C5R4= C5;
 elseif KC==2 % Find C1, R3 and R4 for given C5 and C1=C2.
  C5=R3C5; G4= 2*C5*B2; G3_2C= A3/G4*C5; %=A3/2/B2: not adjustable
  C= G4/2/(A2-G3_2C); C1=C; C2=C;
  if C<0, error('It may work with greater values of B2/A2 and smaller
                                                   value of A3');
  end
  G3= G3_2C*2*C; R3= 1/G3; R4= 1/G4; C2R3= R3; C5R4= R4;
 else % Find C1=C2, R3=R4 for given C5 (not caring about B2).
  C5=R3C5; R=A2/2/C5/A3; R3=R; R4=R3; G3=1/R; G4=G3;
  C=1/R/A2; C1=C; C2=C; C2R3=R; C5R4=C5;
end
fprintf('R3=%8.0f, R4=%8.0f, C1=%10.3e, C2=%10.3e, C5=%10.3e\n', ...
    R3,R4,C1,C2,C5)
B2= C1*G4/(C1+C2)/C5; A2= (G3+G4)/(C1+C2); A3= G3*G4/(C1+C2)/C5;
syms s; Gs = -B2*s/(s^2+A2*s+A3);
```

```
function [R1,R2,R3,Gs]=BPF_Sallen_design(B2,A2,A3,K,C1,C2)
% Design a BPF with the Sallen-Key circuit in Fig. 5.21
% by determining R1, R2, and R3 for given K, C1, and C2.
%                                  (KG1/C2) s
% G(s) = -------------------------------------------------------
%       s^2 +(G1(1/C1+1/C2)+G2((1-K)/C2+1/C1)+G3/C2)*s +G3(G1+G2)/C1/C2
if K<=1, fprintf('Let K=1 since we must have K=(R4+R5)/R4>=1!'); K=1; end
G1=B2*C2/K;
if G1>0.1, error('Try with smaller C2!'); end
D=K-1-C2/C1; E=C2*(A2-G1/C1)-G1;
b=(D*G1+E)/D; c=-A3*C1*C2/D; tmp = b^2-4*c;
if tmp<0, error('Try with larger K,C1 and/or smaller C2'); end
G2 = (-b+sqrt(tmp))/2; G3=D*G2+E;
if G3>0.1, error('Try with smaller C1!'); end
R1=1/G1; R2=1/G2; R3=1/G3; R4=1e4; R5=(K-1)*R4;
B2=K*G1/C2; A2=G1*(1/C1+1/C2)+G2*((1-K)/C2+1/C1)+G3/C2;
A3=G3*(G1+G2)/C1/C2;
syms s;
Gs = B2*s/(s^2+A2*s+A3)
fprintf('R1=%8.0f, R2=%8.0f, R3=%8.0f, R4=%8.0f, R5=%8.0f, C1=%10.3e,
                                C2=%10.3e\n', R1,R2,R3,R4,R5,C1,C2)
%draw_BPF_Sallen(R1,R2,R3,C1,C2,R4,R5)
%K=10; C1=1e-11; C2=1e-11; wr=2e7*pi; Q=0.707; A2=wr/Q; A3=wr^2; B2=A2;
%BPF_Sallen_design(B2,A2,A3,K,C1,C2);
```

For example, we can use the MATLAB function 'BPF_MFBa_design()' to tune the parameters of the MFB (multiple feedback) circuit of Figure 5.20(a) so that the circuit realizes the BPF transfer function (E5.3.1) tuned in Example 5.3:

$$G(s) \overset{(5.5.7)}{=} \frac{-(G_1/C_3)s}{s^2 + (G_5(C_3 + C_4)/C_3C_4)s + (G_1 + G_2)G_5/C_3C_4} \overset{(E5.3.1)}{=} \frac{-100s}{s^2 + 100s + 100^2} \tag{6.4.1}$$

To this end, we have only to run the following statements:

```
>>B2=100; A2=100; A3=10000; % Desired transfer function B2*s/(s^2+A2*s+A3)
>>R2=1e2; KC=3; % With R2=100 and C3=C4 given
>> [C3,R1,R5,Gs]=BPF_MFBa_design(B2,A2,A3,R2,0,KC);

  R1= 100, R2= 100, R5= 200, C3= 1.000e-004, C4= 1.000e-004
  Gs = -100*s/(s^2+100*s+10000)
```

For another example, we can use the function 'LPF_Sallen_design()' to tune the parameters of the Sallen-Key circuit of Figure 5.18(a) so that it realizes the following LPF transfer function:

$$G(s) \overset{(5.5.3)}{=} \frac{KG_1G_2/C_1C_2}{s^2 + ((G_1 + G_2)/C_1 + (1-K)G_2/C_2)s + G_1G_2/C_1C_2} = \frac{K\,\omega_r^2}{s^2 + \omega_b s + \omega_r^2} \tag{6.4.2}$$

More specifically, suppose we need to determine the values of R_1 and R_2 of the Sallen-Key circuit of Figure 5.18(a) with the predetermined values of capacitances $C_1 = C_2 = 100$ pF so that it realizes a second-order LPF with the DC gain $K = 1.5$, the corner frequency $\omega_r = 2\pi \times 10^7$ [rad/s], and the quality factor $Q = 0.707$ (for $\omega_b \overset{(6.2.14)}{=} \omega_r/Q$). To this end, we have only to run the following statements:

```
>>K=1.5; C1=1e-10; C2=1e-10; wr=2e7*pi; Q=0.707;
  % The coefficients of denominator of desired transfer ftn
>>A2=wr/Q; A3=wr^2; % G(s)=K*A3/(s^2+A2*s+A3)
>>KC=2; [R1,R2,Gs]=LPF_Sallen_design(A2,A3,K,C1,C2,KC);

   R1= 221, R2= 115, R3= 10000, R4= 5000, C1= 1.000e-010, C2= 1.000e-010
   Gs=5921762640653615/(s^2+5964037174912491/67108864*s+7895683520871487/2)
```

Example 6.16 Fifth-Order Butterworth LPF Design

```
%elec06e16.m
N=5; fc=1e4; wc=2*pi*fc; % Order and Cutoff frequency of the LPF
[B,A]=butter(N,wc,'s') % Butterworth LPF transfer function G(s)=B(s)/A(s)
disp('Cascade realization of 5th-order Butterworth LPF')
[SOS,K0]= tf2sos(B,A); % Cascade realization
BBc=SOS(:,1:3); AAc=SOS(:,4:6); % Numerator/Denominator of each SOS
KC=1; K=1; R1=1e4; R2=1e4; % Predetermined values of R1, R2
B1=BBc(1,2)*K0; % Numerator of the 1st-order section initialized
for n=2:ceil(N/2)
   A2=AAc(n,2); A3=AAc(n,3); B1=B1/A3;
   LPF_Sallen_design(A2,A3,K,R1,R2,KC);
end
KC=2; C=1e-9;
LPF_RC_OPAmp_design(B1,AAc(1,2),C,KC); % 1st-order section
```

Find the cascade realization of the fifth-order Butterworth LPF with cutoff frequency f_c = 10 kHz using two Sallen-Key circuits of Figure 5.18(a) and one first-order LPF circuit of Figure 5.17(a). To this end, we compose the above MATLAB script "elec06e16.m" and run it to get the C values of the two-stage Sallen-Key LPFs (with all the R values fixed at 10 kΩ) and the R values of the first-order LPF (with C fixed at 1 nF) as follows:

```
R1= 10000, R2= 10000, R3= 10000, R4=  0, C1=5.150e-009, C2=4.918e-010
R1= 10000, R2= 10000, R3= 10000, R4=  0, C1=1.967e-009, C2=1.288e-009
R1= 15915, R2= 15915, C=1.000e-009
```

Note that the voltage dividers consisting of R_3 = 10 kΩ and R_4 = 0 Ω for the two Sallen-Key circuits (with $K = v_o/v_- = (R_3 + R_4)/R_3 = 1$) are not needed and instead, the negative input terminals of the OP Amps should be directly connected (R_4 = 0 Ω) to their output terminals with no connection ($R_3 = \infty$ Ω) to the ground as depicted in the PSpice schematic of Figure 6.26(a). Note also that the numerator of the overall transfer function has been divided by the numerator of each SOS (equal to the constant term of its denominator) to yield the numerator of the first-order section. Figure 6.26(b) shows the frequency response magnitude curve obtained from PSpice simulation (with the analysis type of AC Sweep) where the voltage divider bias (VDB) voltage marker measuring the dB magnitude of the output voltage has been fetched from PSpice > Markers > Advanced > dB magnitude voltage in the top menu bar of the Schematic window.

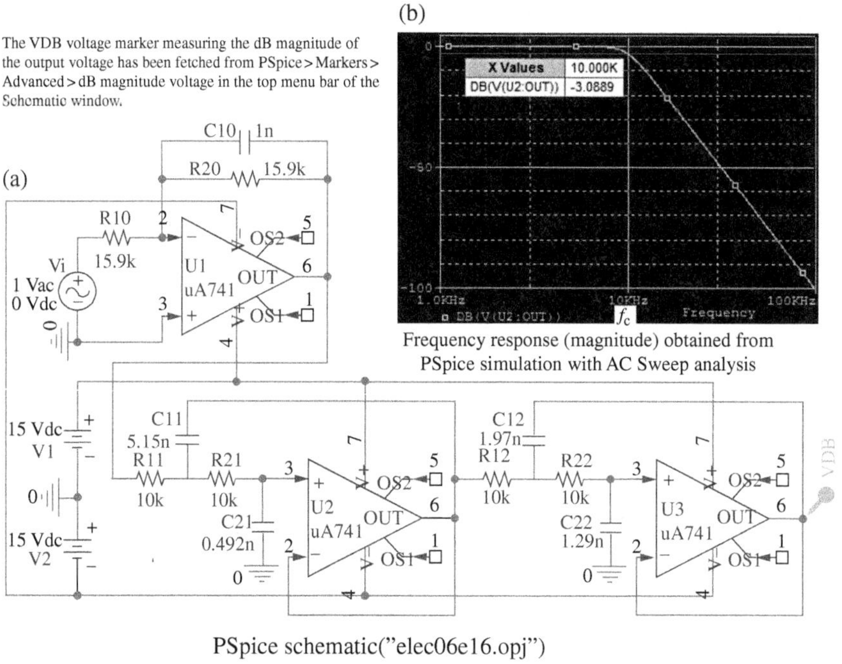
(b)
X Values 10.000K
DB(V(U2:OUT)) -3.0889
fc
Frequency
Frequency response (magnitude) obtained from PSpice simulation with AC Sweep analysis
The VDB voltage marker measuring the dB magnitude of the output voltage has been fetched from PSpice>Markers> Advanced>dB magnitude voltage in the top menu bar of the Schematic window.
(a)
PSpice schematic("elec06e16.opj")

Figure 6.26 A fifth-order Butterworth LPF implemented with one first-order section and two Sallen-Key LPFs.

Example 6.17 Chebyshev I BPF Design with MFB Circuit

Find the cascade realization of the Chebyshev I BPF using the MFB circuit of Figure 5.20(a) so that it can satisfy the following design specifications:

$$\omega_{s1} = 2\pi \times 6000,\ \omega_{p1} = 2\pi \times 10000,\ \omega_{p2} = 2\pi \times 12000,\ \omega_{s2} = 2\pi \times 15000\ [\text{rad/s}],$$
$$R_p = 3\,[\text{dB}],\ \text{and}\ A_s = 25\,[\text{dB}] \tag{E6.17.1}$$

To this end, we compose the following MATLAB script "elec06e17.m" and run it to get the R values of the two-stage MFB BPFs (with all the C values fixed at 10 nF) as follows:

```
R1=11241, R2=40, R5=46095, C3=1.000e-008, C4=1.000e-008
R1=11241, R2=46, R5=53119, C3=1.000e-008, C4=1.000e-008
```

```
%elec06e17.m
fs1=6e3; fp1=10e3; fp2=12e3; fs2=15e3; Rp=3; As=25; fp=sqrt(fp1*fp2);
ws1=2*pi*fs1; wp1=2*pi*fp1; wp2=2*pi*fp2; ws2=2*pi*fs2;
[N,wpc]=cheb1ord([wp1 wp2],[ws1 ws2],Rp,As,'s');
[B,A]=cheby1(N,Rp,wpc,'s');
f=logspace(-1,1,600)*fp; Gw=freqs(B,A,2*pi*f); % Frequency response
subplot(221), semilogx(f,20*log10(abs(Gw))), hold on
[SOS,K]= tf2sos(B,A);
Ns=size(SOS,1); % Number of sections
Gm=K^(1/Ns), BBc=SOS(:,1:3), AAc=SOS(:,4:6)
KC=2; % With all the C values fixed
for n=1:Ns
   B2=Gm; A2 = AAc(n,2); A3 = AAc(n,3); C3=1e-8; C4=1e-8;
   subplot(222+n), BPF_MFBa_design(B2,A2,A3,C3,C4,KC);
end
```

Figure 6.27(a) shows the PSpice schematic with all the parameter values set as suggested by the MATLAB realization functions. Figure 6.27(b1) and (b2) shows the frequency response magnitude curves obtained from PSpice simulation and

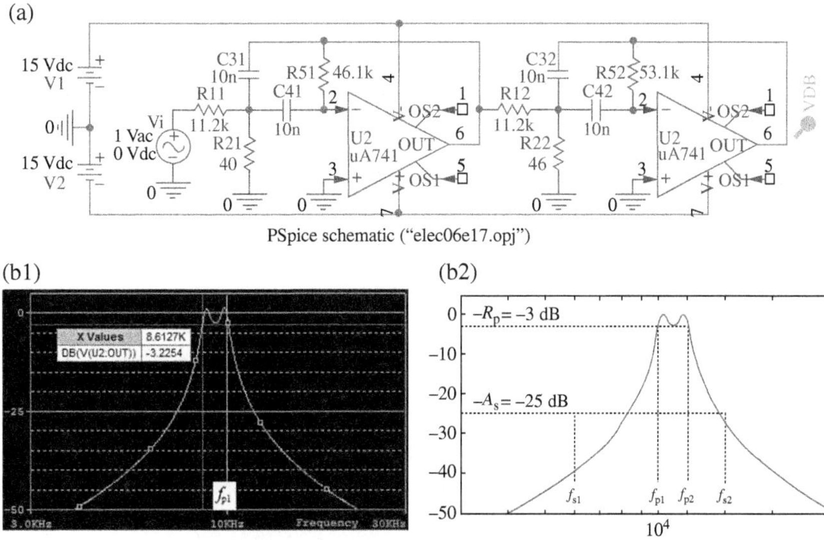

Frequency response (magnitude) obtained from PSpice simulation with AC Sweep analysis

Frequency response (magnitude) obtained from MATLAB analysis

Figure 6.27 A fourth-order Chebyshev I BPF implemented with two cascade-connected MFB BPFs.

MATLAB analysis, respectively. Oh, my God! The former seems to have been shifted by −2 kHz compared with the design specification (depicted in Figure 6.27(b2)) although their shapes of having ripples only in the passband look similar. Why is that? The discrepancy may have been caused by the load effect of the second stage on the first stage that was not considered here. Anyway, to make up for this (unexpected) frequency shift, you can try the frequency scaling of the designed filter by a factor of about 6/5 (see Problem 6.9).

(Note) Multiplying all the resistances/capacitances by the same constant does not change the transfer function and frequency response. This implies that if you want to scale up/down the tuned capacitances/resistances without affecting the transfer function and frequency response, you can scale up/down the predetermined values of resistances/capacitances by the same factor (see Remark 6.2(1)).

(Note) See http://www.analog.com/library/analogdialogue/archives/43-09/EDCh%208%20filter.pdf; for various configurations of analog filter.

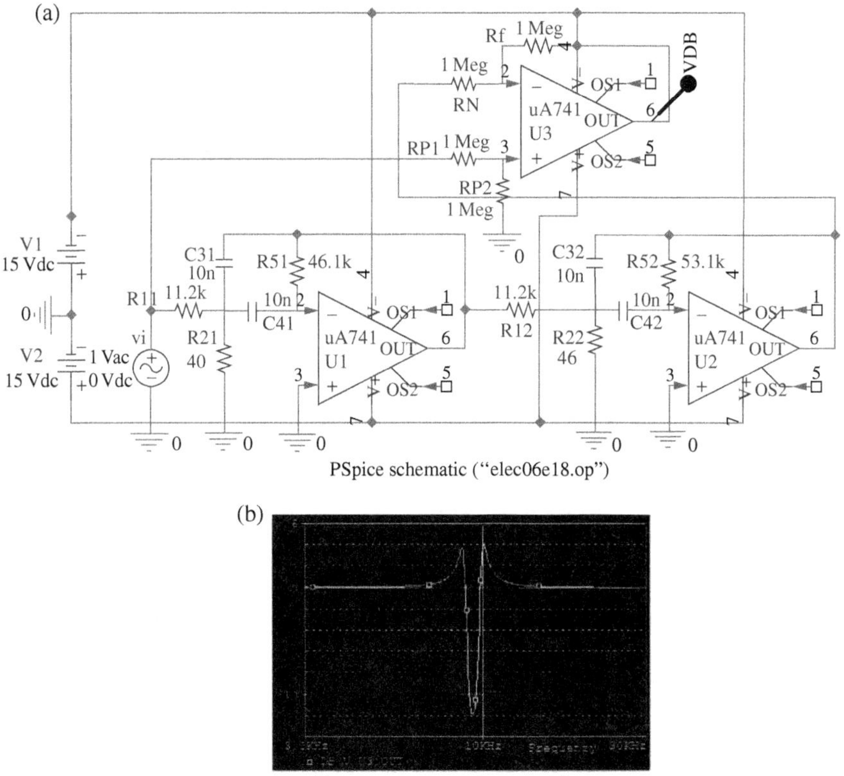

Figure 6.28 A BSF realized by using the BPF of Figure 6.27 (in Example 6.17) and a difference amplifier.

One may wonder why we have no example of BSF design and realization. It is because we may design a BSF satisfying a given frequency response specification as illustrated in Example 6.1(c), but the transfer functions of the SOSs as shown in Eqs. (E6.1.9a,b) do not fit Eq. (5.5.10), which is the transfer function of the BSF in Figure 5.22. However, a BSF can be realized by the difference amplifier having the input signal and a BPF output as its two inputs as will be illustrated in the next example. Likewise, an HPF can be realized by the difference amplifier having the input signal and an LPF output as its two inputs.

Example 6.18 Realization of a BSF Using a BPF and a Difference Amplifier

Figure 6.28(a) shows the PSpice schematic of a BSF realized by connecting the BPF (in Figure 6.27) and a difference amplifier so that the overall output can be the difference between the input signal and the BPF output:

$$v_o(t) = v_i(t) - v_{o,\mathrm{BPF}}(t) \tag{E6.18.1}$$

Figure 6.28(b) shows the frequency response magnitude obtained from the PSpice simulation. Regrettably, it is not a typical shape of BSF frequency response, being contrary to our expectation.

Problems

6.1 Design of Passive Filter in the form of LC Ladder Circuit

(a) Design a Butterworth LPF with cutoff frequency $f_c = 2$ GHz in the form of $N = 5$-stage Π-type LC ladder circuit (depicted in Figure 6.14.1(a)) where the source impedance is assumed to be 50 Ω. Plot the frequency response magnitude [dB] (relative to that without the filter) to check the validity of the designed filter.

```
%elec06p01a.m
clear, clf
%(a) Butterworth (maximally flat) LPF (Example 8.4 of [P-1] by Pozar)
syms s
N=5; pT='??'; Rs=50; type='?';
fc=2e9; wc=2*pi*fc;
gg=LPF_ladder_coef(N,type);
[LCRs,LCRs_,Gs]=LC_ladder(Rs,gg,pT,wc,'???');
format short e; LCRs, LCRs_, RL=LCRs(end)
ff=logspace(-1,1,401)*fc; ww=2*pi*ff; % Frequency range
s=j*ww; Gw=eval(Gs/(RL/(Rs+RL))); % Frequency response
GwmagdB=20*log10(max(abs(Gw),1e-5));
subplot(222), semilogx(ff,GwmagdB), hold on
plot([ff(1) fc],[-3 -3],'k:', [fc fc],[0 -3],'k:')
xlabel('Frequency[Hz]'), ylabel('Relative frequency response[dB]')
```

(b) Design a Chebyshev BPF with passband ripple $R_p = 0.5$ dB, peak frequency $f_p = 1$ GHz, and bandwidth 0.1 GHz in the form of $N = 3$-stage T-type LC ladder circuit (depicted in Figure 6.14.1(b)) where the source impedance is assumed to be 50 Ω. Plot the frequency response magnitude or its reciprocal (called the insertion loss) [dB] to check the validity of the designed filter.

```
%elec06p01b.m
% Chebyshev 0.5dB BPF (Example 8.5 of [P-1] by Pozar)
syms s
N=3; pT='?'; Rs=50; type='?'; Rp=0.5;
fp=1e9; wp=2*pi*fp; wb=0.1*wp; sqBw=sqrt(wb^2/4+wp^2); %wb=Bandwidth
w12=[-1 1]*wb/2+sqw; fl=w12(1)/2/pi; fu=w12(2)/2/pi;
gg=LPF_ladder_coef(N,type,Rp);
[LCRs,LCRs_,Gs]=LC_ladder(Rs,gg,pT,w12,'???');
format short e; LCRs, LCRs_, RL=LCRs(end)
ff=logspace(-0.3,0.3,401)*fp; ww=2*pi*ff; % Frequency range
s=j*ww; Gw=eval(Gs/(RL/(Rs+RL))); % Frequency response
GwmagdB=20*log10(max(abs(Gw),1e-5));
subplot(224), semilogx(ff,GwmagdB), hold on
plot([fl fl],[-100 -Rp],'k:', [fu fu],[-100 -3],'k:')
```

6.2 Design of Passive Filters in the form of *LC* Ladder Circuit

(a) Design a Butterworth LPF with cutoff frequency f_c = 10 MHz, passband ripple R_p – 3 dB, and stopband attenuation A_s = 30 dB at f_s = 31.62 MHz where the source and load impedances are equally 50 Ω. Note that the order of the Butterworth LPF can be determined by using the MATLAB function 'buttord()' as

```
>>wp=2*pi*7e6; ws=2*pi*31.62e6; Rp=3; As=30;
  [N,wc]=buttord(wp,ws,Rp,As,'s'), fc=wc/2/pi
```

```
%elec06p02a.m
%(a) Butterworth LPF (Example 8.6 of [M-2] by Misra)
syms s
N=3; Rs=50; %% Filter order 3 and Source impedance 50[Ohm]
type='B'; band='LPF';
fc=1e7; wc=2*pi*fc; % Butterworth LPF
gg=LPF_ladder_coef(N,type) %LC values of Nth-order B-LPF prototype
ff=logspace(-1,1,401)*fc; ww=2*pi*ff; % Frequency range
for m=1:2
   if m==1, pT='T'; else pT='p'; end
   [LCRs,LCRs_,Gs]=LC_ladder(Rs,gg,pT,wc,????);
   format short e; LCRs, LCRs_, RL=LCRs(end)
   subplot(2,2,2*m-1), draw_ladder(LCRs,LCRs_)
   % To plot the frequency response magnitude (considering Rs,RL)
   s=j*ww; Gw=eval(Gs); % Frequency response
   GwmagdB=20*log10(max(abs(Gw)/(RL/(Rs+RL)),1e-5));
   subplot(2,2,2*m), semilogx(ff,GwmagdB)
end
```

(b) Design a Chebyshev LPF with cutoff frequency f_c = 100 MHz, passband ripple R_p = 0.01 dB, and stopband attenuation A_s = 5 dB at f_s = 40 MHz where the source and load impedances are equally 75 Ω. Note that the minimum order of the Chebyshev LPF can be determined by using the MATLAB function 'cheb1ord()' as

```
>>wp=2*pi*1e8; ws=2*pi*4e8; Rp=0.01; As=5;
  [N,wc]=cheb1ord(wp,ws,Rp,As,'s')
```

```
%elec06p02b.m
% Chebyshev LPF (Example 8.7 of [M-2] by Misra)
clear, clf
syms s
N=3; Rs=75; % Filter order 3 and Source impedance 75 [Ohm]
type='C'; band='LPF';
Rp=0.01; fc=1e8; wc=2*pi*fc; % Chebyshev LPF
gg=LPF_ladder_coef(N,type,??) %LCs of Nth-order C-LPF prototype
ff=logspace(-1,1,401)*fc; ww=2*pi*ff; % Frequency range
for m=1:2
  if m==1, pT='T'; else pT='p'; end
  [LCRs,LCRs_,Gs]=LC_ladder(??,gg,pT,wc,band);
  format short e; LCRs, LCRs_, RL=LCRs(end)
  subplot(2,2,2*m-1)
  draw_ladder(LCRs,LCRs_)
  s=j*ww; Gw=eval(Gs); % Frequency response
  GwmagdB=20*log10(max(abs(Gw)/(RL/(Rs+RL)),1e-5));
  subplot(2,2,2*m), semilogx(ff,GwmagdB)
end
```

(c) Design a Chebyshev HPF with cutoff frequency f_c = 100 MHz, passband ripple R_p = 0.01 dB, and stopband attenuation A_s = 5 dB at f_s = 25 MHz where the source and load impedances are equally 75 Ω. Note that the minimum order of the Chebyshev HPF can be determined by using the MATLAB function 'cheb1ord()' as

```
>>wp=2*pi*1e8; ws=2*pi*25e6; Rp=0.01; As=5;
  [N,wc]=cheb1ord(wp,ws,Rp,As,'s')
```

```
%elec06p02c.m
% Chebyshev HPF (Example 8.9 of [M-2] by Misra)
N=3; Rs=75; % 3rd order
type='C'; band='HPF'; Rp=0.01; fc=1e8; wc=2*pi*fc; % Chebyshev HPF
gg=LPF_ladder_coef(?,type,Rp) % LCs of Nth-order C-LPF prototype
ff=logspace(-1,1,401)*fc; ww=2*pi*ff; % Frequency range
for m=1:2
  if m==1, pT='T'; else pT='p'; end
  [LCRs,LCRs_,Gs]=LC_ladder(Rs,gg,pT,??,band);
  format short e; LCRs, LCRs_, RL=LCRs(end)
  subplot(2,2,2*m-1), draw_ladder(LCRs,LCRs_)
  s=j*ww; Gw=eval(Gs); % Frequency response
  GwmagdB=20*log10(max(abs(Gw)/(RL/(Rs+RL)),1e-5));
  subplot(2,2,2*m), semilogx(ff,GwmagdB, [fc fc],[-100 -Rp],'r:')
end
```

(d) Design a sixth-order Chebyshev BPF with cutoff frequencies $\{f_{c1}$ = 10 MHz, f_{c2} = 40 MHz$\}$ and passband ripple R_p = 0.01 dB, where the source and load impedances are equally 75 Ω.

```
%elec06p02d.m
% Chebyshev BPF (Example 8.10 of [M-2] by Misra)
N=3; Rs=75; % N=6/2=3rd-order LPF prototype
type='C'; band='BPF'; % Chebyshev BPF
Rp=0.01; fb=[10 40]*1e6; wb=2*pi*fb; fc=sqrt(fb(1)*fb(2));
gg=LPF_ladder_coef(N,????,Rp) % LCs of 2Nth-order C-BPF prototype
ff=logspace(-1,1,401)*fc; ww=2*pi*ff;
for m=1:2
   if m==1, pT='T'; else pT='p'; end
   [LCRs,LCRs_,Gs]=LC_ladder(Rs,gg,??,wb,band);
   format short e; LCRs, LCRs_, RL=LCRs(end)
   subplot(2,2,2*m-1), draw_ladder(LCRs,LCRs_)
   s=j*ww; Gw=eval(Gs); % Frequency response
   GwmagdB=20*log10(max(abs(Gw)/(RL/(Rs+RL)),1e-5));
   subplot(2,2,2*m), semilogx(ff,GwmagdB), hold on
   semilogx(fb(1)*[1 1],[-100 -Rp],'r:', fb(2)*[1 1],[-100 -Rp],'r:')
end
```

(e) Design a sixth-order Butterworth BSF with cutoff frequencies $\{f_{c1} = 10$ MHz, $f_{c2} = 40$ MHz$\}$ and passband ripple $R_p = 0.1$ dB, where the source and load impedances are equally 75 Ω.

```
%elec06p02e.m
% Butterworth BSF (Example 8.11 of [M-2] by Misra)
N=3; Rs=75; % N=6/2=3rd-order LPF prototype
type='B'; band='BSF'; % Butterworth BSF
Rp=0.1; fb=[10 40]*1e6; wb=2*pi*fb; fc=sqrt(fb(1)*fb(2));
gg=LPF_ladder_coef(?,type,??) % LCs of 2Nth-order B-LPF prototype
ff=logspace(-1,1,401)*fc; ww=2*pi*ff; % Frequency range
for m=1:2
   if m==1, pT='T'; else pT='p'; end
   [LCRs,LCRs_,Gs]=LC_ladder(Rs,??,pT,wb,band);
   format short e; LCRs, LCRs_, RL=LCRs(end)
   subplot(2,2,2*m-1), draw_ladder(LCRs,LCRs_)
   s=j*ww; Gw=eval(Gs); % Frequency response
   GwmagdB=20*log10(max(abs(Gw)/(RL/(Rs+RL)),1e-5));
   subplot(2,2,2*m)
   semilogx(ff,GwmagdB,[fc fc],[-100 -3],'r:')
end
```

6.3 Design of L-Type Impedance Matcher for Maximum Power Transfer
Design as many L-type impedance matchers as possible for maximum power transfer from a transmitter with output impedance $Z_s = 50\,\Omega$ to an antenna with input impedance $Z_L = R_L + 1/j\omega C_L$ ($R_L = 80\,\Omega$, $C_L = 2.653$ pF) at the center frequency $f_c = 1$ GHz. Plot the power delivered to the load R_L as shown in Figure 6.20 twice, once by using the MATLAB function '`imp_matcher_L_and_freq_resp()`' and once by using PSpice (with AC Sweep analysis type

and a power marker on R_L in the schematic). Which one of the L-type filters are you going to use for impedance matching after viewing the power curves?

6.4 Design of L Type Impedance Matcher

(a) Make as many L-type impedance matchers as possible to match a load of $Z_L = 600\ \Omega$ to a source of $Z_s = 50\ \Omega$ at a frequency $f_c = 400$ MHz and plot their frequency responses for some frequency band around f_c.

```
%elec06p04a.m
% To design L-type impedance matchers (Example 5.6 of [M-2])
fc=4e8; wc=2*pi*fc; % Center frequency
Zs=50; ZL=600; % Source and Load impedances
ff=logspace(-1,1,801)*fc; % Frequency range
[LC,LC_]=imp_matcher_L_and_freq_resp(Zs,??,fc,ff);
```

(b) Make as many two-stage impedance matchers as possible to match a load of $Z_L = 600\ \Omega$ to a source of $Z_s = 50\ \Omega$ at a frequency $f_c = 400$ MHz where the second stage matches $Z_L = 600$ to $173.2\ \Omega$ and the first stage matches 173.2 to $Z_s = 50\ \Omega$. Use the following MATLAB function '`ladder_xfer_ftn()`' to find the frequency responses (for two decades around f_c) and the input impedances ($Z_i(j2\pi f)$'s) of the impedance matchers (with the load impedance $Z_L = 600\ \Omega$) at f_c. Plot the frequency responses to see if they have peaks at f_c and check if $Z_i(j2\pi f)$'s match the source impedance $Z_s = 50\ \Omega$ at f_c.

```
function [Gs,Zis]=ladder_xfer_ftn(Rs,LCRs,LCRs_,w)
% Input:
%  Zs = Input impedance
%  LCRs_ = [C1 L2 C3 RL] for Pi-ladder like Fig. 6.14.1a
%  LCRs_ = {'C(v)','L(h)','C(v)' 'Z(v)'}
%  w = (Operation) Frequency[rad/s]
% Output:
%  Gs = G(s) or G(jw) if w is given.
%  Zis = Input impedance Zi(s) or Zi(jw) if w is given.
% Copyleft: Won Y. Yang, wyyang53@hanmail.net, CAU for academic use only
if nargin>3, s=j*w; else syms s; end
[M,N1]=size(LCRs); [M_,N1_]=size(LCRs_); N=N1-1;
if M~=M_|N1~=N1, error('LCR and LCR_ are not of the same size!'); end
for m=1:M
   ZL=LCRs(m,N1);
   I=1; V=ZL; % Set the current through RL to 1 to find G(s)
   for n=N:-1:1
      LC=LCRs_{m,n}(1); hv=LCRs_{m,n}(3);
      if LC=='L', Zs=s*LCRs(m,n); else Zs=1/s/LCRs(m,n); end
      if hv=='v', I=I+V./Zs; else V=V+Zs.*I; end % Parallel/Series
   end
   Gs(m,:)=ZL./(V+Rs.*I); % Transfer function
   Zis(m,:)=V./I; % Input impedance
end
```

(c) Make as many L-type filters as possible to match a load of $Y_L = 8 - j12$ [mS] to a 50 Ω line at a frequency $f_c = 1$ GHz. Use the above MATLAB function 'ladder_xfer_ftn()' to find the input impedances of the impedance matchers (with the load impedance $Z_L = 1/Y_L$) and check if they match the source impedance $Z_s = 50\,\Omega$ at $f_c = 1$ GHz.

```
%elec06p04b.m
% To design L-type impedance matchers (Example 5.7 of [M-2])
fc=4e8; wc=2*pi*fc; % Center frequency
Zs=50; ZL=600; % Source and Load impedances
ff=logspace(-1,1,801)*fc; % Frequency range
[LC2,LC2_]=imp_matcher_L_and_freq_resp(?????,ZL,fc,ff); % 2nd stage
[LC1,LC1_]=imp_matcher_L_and_freq_resp(??,173.2,fc,ff); % 1st stage
LC11=[LC1(1,:) LC2(1,:) ZL], LC11_=[LC1_(1,:) LC2_(?,:) 'RL(v)']
[Gss(1,:),Ziws(1)]=ladder_xfer_ftn(Zs,LC11,LC11_); % G(s) and Zi(jw)
LC12=[LC1(1,:) LC2(?,:) ZL], LC12_=[LC1_(1,:) LC2_(2,:) 'RL(v)']
[Gss(2,:),Ziws(2)]=ladder_xfer_ftn(Zs,LC12,LC12_); % G(s) and Zi(jw)
LC21=[LC1(?,:) LC2(1,:) ZL], LC21_=[LC1_(2,:) LC2_(1,:) 'RL(v)']
[Gss(3,:),Ziws(3)]=ladder_xfer_ftn(Zs,LC21,LC21_); % G(s) and Zi(jw)
LC22=[LC1(2,:) LC2(?,:) ZL], LC22_=[LC1_(2,:) LC2_(2,:) 'RL(v)']
[Gss(4,:),Ziws(4)]=ladder_xfer_ftn(Zs,LC22,LC22_); % G(s) and Zi(jw)
% To check if the input impedance of each ladder is matched to Zs
s=j*wc; eval(Ziws) % Input AC impedances at w=wc
% To get the frequency response
s=j*2*pi*ff;
for m=1:4, Gws(:,m)=eval(Gss(m)); end
GmagdB=20*log10(abs(Gws)); G3dB=max(GmagdB)-3;
semilogx(ff,GmagdB); hold on
semilogx(ff([1 end]),[1;1]*G3dB,'r:', [fc fc],[-100 0],'r:')
legend('Gw1','Gw2','Gw3','Gw4')
```

```
%elec06p04c.m
% To design L-type impedance matchers (Example 5.9 of [M-1])
clear, clf
fc=1e9; wc=2*pi*fc; % Center frequency
Zs=50; ZL=1e3/(8-12i); % Source and Load impedances
ff=logspace(-1,1,801)*fc; % Frequency range
[LCs,LCs_,ff,Gws]=imp_matcher_L_and_freq_resp(Zs,ZL,fc,ff);
% To check if ZL has been matched to Zs through the LC ladders,
[M,N]=size(LCs); N1=N+1;
LCRs=[LCs repmat(ZL,M,1)]; LCRs_=LCs_;
for m=1:M, LCRs_{m,N1}='Z(v)'; end
[Gss,Ziws]=ladder_xfer_ftn(Zs,LCRs,?????); % G(s)'s and Zi(jw)'s
s=j*wc; eval(Ziws) % Input AC impedances at w=wc
```

6.5 Design of T-Type Impedance Matcher

(a) Use the MATLAB function '`imp_matcher_T()`' to make as many T-type filter(s) as possible matching a load of $Z_L = 225\ \Omega$ to a source of $Z_s = 15 + j15\ \Omega$ with quality factor $Q = 5$ at a frequency $f_c = 30$MHz. You can visit the website https://www.eeweb.com/tools/t-match to find two of the designs, one with DC current passed and one with DC current blocked.

(b) Use the MATLAB function '`imp_matcher_T()`' to determine just the reactances constituting T-type filter(s) to match a load of $Z_L = 50\ \Omega$ to a source of $Z_s = 10\ \Omega$ with quality factor $Q = 10$. (See Example 4.5 of [B-2].)

6.6 Design of Π-Type Impedance Matcher

Determine just the reactances constituting Π-type filter(s) to match a load of $Z_L = 1000\ \Omega$ to a source of $Z_s = 100\ \Omega$ with quality factor $Q = 15$. (See Example 4.4 of [B-2].)

6.7 Design of Tapped-*C* Impedance Matcher

Make a tapped-*C* impedance matcher to match a load of $Z_L = 5\ \Omega$ to a source of $Z_s = 50\ \Omega$ with quality factor $Q = 5$ at $f_0 = 150$ MHz. Plot the frequency response magnitude curve as shown in Figure 6.25(b).

```
%elec06p07.m
Rs=50; RL=5; f0=150e6; w0=2*pi*f0; jw0=j*w0; Q=5;
[L,C1,C2,Qr,Qp,Rp1,Z1,Z2]=imp_matcher_tapped_C(RL,R?,??,Q);
% To get the frequency response,
f=logspace(-0.2,0.2,600)*f0; w=2*pi*f; jw=j*w; % Frequency range
ZC2RLw = 1./(jw*C2+1/RL);
ZC12RLw = 1./jw/C1 + ZC2RLw;
ZLC12RLw = 1./(1./jw/L + 1./ZC12RLw);
Gw = ZLC12RLw./(Rs+ZLC12RLw).*ZC2RLw./ZC12RLw; % Frequency response
GwmagdB = 20*log10(abs(Gw));
%........
```

6.8 Design, Realization, and Implementation of High-Order Butterworth Lowpass Filter (LPF)

Given the order, say, $N = 4$ and cutoff frequency, say, $\omega_c = 2\pi f_c = 2\pi \times 10^4$ [rad/s] of an LPF, we can use the following MATLAB statements:

```
>>format short e; N=4; fc=1e4; wc=2*pi*fc;
  [B,A]=butter(N,wc,'s')
```

to find the numerator (B) and denominator (A) polynomials in *s* of the transfer function of the desired fourth-order Butterworth LPF as

```
B =           0           0           0           0 1.5585e+019
A = 1.0000e+000 1.6419e+005 1.3479e+010 6.4819e+014 1.5585e+019
```

This means the following transfer function:

$$G(s) = \frac{1.5585 \times 10^{19}}{s^4 + 1.6419 \times 10^5 s^3 + 1.3479 \times 10^{10} s^2 + 6.4819 \times 10^{14} s + 1.5585 \times 10^{19}} \tag{P6.8.1}$$

(a) Plot the frequency response magnitude $20\log_{10}|G(j\omega)|$ [dB] of the LPF (for the frequency range of two decades around $f_c = 10^4$ [Hz]) and show that the transfer function can be realized in cascade form, i.e. as a product of second-order transfer functions as

$$G(s) = G_{C1}(s)G_{C2}(s) = \frac{39478 \times 10^5}{s^2 + 116100s + 39478 \times 10^5} \frac{39478 \times 10^5}{s^2 + 48089s + 39478 \times 1010^5} \tag{P6.8.2}$$

```
%elec06p08a.m
clear
format short e
% To design a 4th-order Butterworth LPF with fc=1e4
N=4; fc=1e4; wc=2*pi*fc;
[B,A]=butter(?,??,'s')
% To plot the frequency response of G(s)=B(s)/A(s)
f=logspace(-1,1,800)*fc; w=2*pi*f; % Frequency of 2 decades around wc
Gw=freqs(B,A,?); % Frequency response of a system with G(s)=B(s)/A(s)
subplot(332), semilogx(f,20*log10(abs(Gw)), fc*[1 1],[0 -3])
axis([f(1) f(end) -80 10]), hold on
title('Designed frequency response')
% To find a cascade realization of G(s)=B(s)/A(s)
fprintf('\n\nCascade form of realization of G(s):')
[SOS,Kc]=tf2sos(B,A); % Cascade realization
Ns=size(SOS,1); % Number of sections
Gm=??^(1/??); % (Distributed) gain of each SOS in cascade realization
BBc=Gm*SOS(:,1:3), AAc=SOS(:,4:6) % the numerator/denominator matrix
% To find a parallel realization of G(s)=B(s)/A(s)
fprintf('\n\nParallel form of realization of G(s):')
[BBp,AAp]=tf2par_s(B,A) % the numerator/denominator matrix
```

```
%elec06p08b.m
Gw_cas=1; % Initialize the frequency response of cascade realization.
R3=1e10; R4=0; K=1+R4/R3; % R3=inf;
for m=1:Ns
   A2=AAc(m,2); A3=AAc(m,3); R1=1e4; R2=1e4; KC=1;
   subplot(333+m), [C3,C4,Gs]=LPF_Sallen_design(A2,A3,K,R1,R2,KC);
   Gw_cas = Gw_cas.?freqs(BBc(m,:),AAc(m,:),w);
end
semilogx(f,20*log10(abs(Gw_cas)),'r:')
```

```
%elec06p08c.m
Gw_par=0; % Initialize the frequency response of cascade realization.
L=0.01; Rf=BBp(1,2)*L % Common to numerators of SOS 1/2 in Eq.(P6.8.5)
for m-1:Ns
  A2=AAp(m,2); A3=AAp(m,3); % Denominators of SOS 1/2 in Eq.(P6.8.5)
  C=1/A?/L; R=1/A?/C; %Sol of 2 simulataneous eqs: 1/R/C=A2, 1/L/C=A3
  mc=num2str(m);
  fprintf(['\nR' mc '=%10.4e, L' mc '=%10.4e, C' mc '=%10.4e\n'],R,L,C)
  Gw_par = Gw_par ? freqs(BBp(m,:),AAp(m,:),w);
end
semilogx(f,20*log10(abs(Gw_par)),'g:')
```

or in parallel form, i.e. as a sum of second-order transfer functions as

$$G(s) = G_{P1}(s) + G_{P2}(s) = \frac{58049(s+116100)}{s^2+116100\,s+39478\times 10^5} - \frac{58049(s+48089)}{s^2+48089s+39478\times 10^5} \tag{P6.8.3}$$

In fact, this implies that the transfer function can be realized as a cascade connection of the two SOSs (second-order sections), $G_{C1}(s)$ and $G_{C2}(s)$, or a parallel connection of the two SOSs, $G_{P1}(s)$ and $G_{P2}(s)$. You can complete and run the above MATLAB script "elec06p08a.m" to find the cascade and parallel realizations.

(b) Noting that each SOS of the circuit in Figure P6.8(a) has the following transfer function

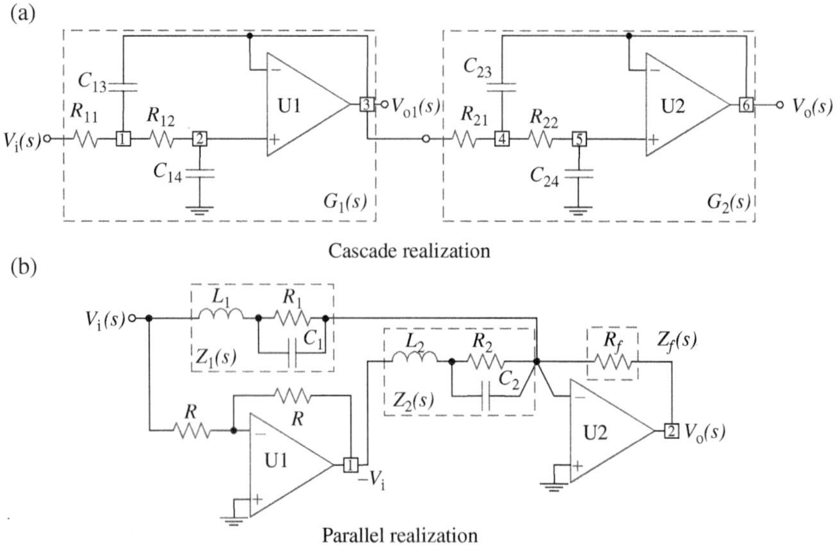

Figure P6.8 Two realizations of a fourth-order Butterworth LPF.

$$G(s) \overset{(5.5.3)}{\underset{K=1}{=}} \frac{1/R_1R_2C_3C_4}{s^2 + s(1/R_1 + 1/R_2)/C_3 + 1/R_1R_2C_3C_4} \quad \text{(P6.8.4)}$$

let $R_{11} = R_{12} = R_{21} = R_{22} = 10\ \text{k}\Omega$ and find the values of C_{13}, C_{14}, C_{23}, and C_{24} such that the desired transfer function will be implemented. Plot the frequency response of the cascade realization (P6.8.2) to see if it conforms with that of $G(s)$ obtained in (a).

(c) Noting that the circuit of Figure P6.8(b) has the following transfer function

$$G(s) = -\frac{Z_f(s)}{Z_1(s)} + \frac{Z_f(s)}{Z_2(s)} = -\frac{(R_f/L_1)(s + 1/R_1C_1)}{s^2 + s/R_1C_1 + 1/L_1C_1} + \frac{(R_f/L_2)(s + 1/R_2C_2)}{s^2 + s/R_2C_2 + 1/L_2C_2} \quad \text{(P6.8.5)}$$

let $L_1 = L_2 = L = 0.01$ H and find the values of R_1, C_1, R_2, C_2, and R_f such that the desired transfer function will be implemented. Plot the frequency response of the parallel realization (P6.8.3) to see if it conforms with that of $G(s)$ obtained in (a).

(d) Perform the PSpice simulation to see that the cutoff frequency of the two circuits, one realized in cascade form and the other in parallel form, is $f_c = 10$ kHz at which the magnitude of the frequency response is $1/\sqrt{2} = 0.7071 = -3$ dB. You can use the VDB marker (from PSpice > Markers > Advanced) in the Schematic window to measure the frequency response in dB.

6.9 Frequency Scaling

Referring to Remark 6.2(2), try the frequency scaling for Example 6.17 to reduce the discrepancy between the two frequency responses, one obtained from PSpice simulation and the other obtained from MATLAB analysis. Using PSpice, plot the frequency response magnitude of the modified filter to show that you worked properly.

6.10 Design and Realization of High-Order Chebyshev I Bandpass Filter (BPF)

Given the filter order, say, $N = 4$ and the lower/upper 3 dB frequencies, say, $\omega_l = 2\pi \times 8000$ and $\omega_u = 2\pi \times 12500$ [rad/s] of a BPF, we can run the following MATLAB statements to find the numerator (B) and denominator (A) polynomials in s of the transfer function of the desired fourth-order *Chebyshev* I BPF.

```
>>format short e; N=4; Rp=3;
  fl=8000; fu=12500; wl=2*pi*fl; wu=2*pi*fu;
  [B,A]=cheby1(N/2,Rp,[wl wu],'s')
```

This yields

```
B =           0            0 4.0067e+008            0           0
A = 1.0000e+000 1.8234e+004 8.4616e+009 7.1985e+013 1.5585e+19
```

meaning the following transfer function:

$$G(s) = \frac{4.0067 \times 10^8 s^2}{s^4 + 1.8234 \times 10^4 s^3 + 8.4616 \times 10^9 s^2 + 7.1985 \times 10^{13} s + 1.5585 \times 10^{19}} \tag{P6.10.1}$$

This transfer function can be expressed in the form of product of two second-order transfer functions as

$$G(s) = G_{C1}(s)G_{C2}(s) = \frac{20000s}{s^2 + 7543s + 27852 \times 10^5} \frac{20000s}{s^2 + 10691s + 55958 \times 10^5} \tag{P6.10.2}$$

In fact, this implies that the transfer function can be realized as a cascade connection of the two SOSs, $G_{C1}(s)$ and $G_{C2}(s)$.

```
%elec06p10.m
clear, clf
f=logspace(3,5,801); w=2*pi*f; jw=j*w; % Frequency range
N=4; % Filter order
fl=8000; fu=12500; wl=2*pi*fl; wu=2*pi*fu; % Lower/upper 3dB freqs
Rp=3; % Passband ripple
[B,A]= cheby1(N/2,Rp, [wl wu],'s') % 4th-order Chebyshev I filter
[SOS,K0]= tf2sos(B,A); BBc=SOS(:,1:3); AAc=SOS(:,4:6);
K=1; C3=1e-8; C4=1e-8; KC=2; num=1; den=1; N2=floor(N/2);
for n=1:N2
   B2 = K0^(1/N2); A2 = AAc(n,2); A3 = AAc(n,3);
   num = conv(num, [B2 0]); den = conv(den, [1 A2 A3]);
   BPF_MFBa_design(B2,??,A3,??,C4,??);
end
Gwd=freqs(B,A,w); Gwd_magdB=20*log10(abs(Gwd));
Gw=freqs(num,den,w); Gw_magdB=20*log10(abs(Gw));
subplot(2,N2,N2+1)
semilogx(f,Gw_magdB, f,Gwd_magdB,'r:'), hold on
plot([fl fl],[0 -Rp],'k:', [fu fu],[0 -Rp],'k:')
plot(f([1 end]),[-Rp -Rp],'k:'), axis([f(1) f(end) -60 5])
```

(a) Noting that each SOS of the circuit in Figure P6.10.1 has the following transfer function

$$G(s) \overset{(5.5.7)}{=} \frac{-(G_1/C_3)s}{s^2 + (G_5(C_3 + C_4)/C_3C_4)s + (G_1 + G_2)G_5/C_3C_4} \tag{P6.10.3}$$

let $C_{13} = C_{14} = C_{23} = C_{24} = 10$ nF and find the values of R_{11}, R_{12}, R_{15}, R_{21}, R_{22}, and R_{25} such that the desired transfer function will be realized. You can complete and run the above script "elec06p10.m," which uses the MATLAB function 'BPF_MFBa_design()'.

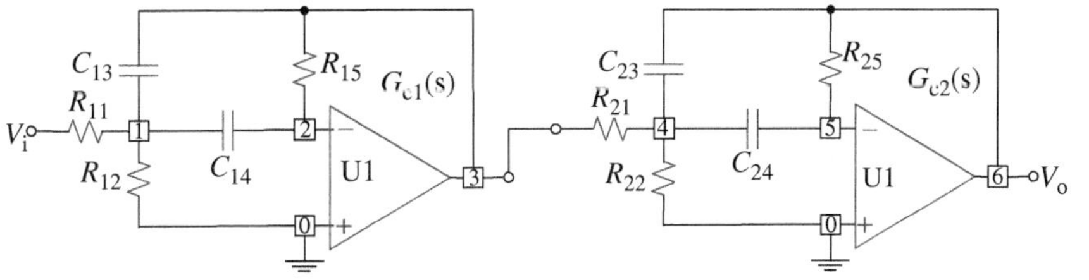

Figure P6.10.1 Cascade connection of two second-order MFB BPFs.

(b) As depicted in Figure P6.10.2(a), plot the magnitude curve of the frequency response of the filter having the transfer function (P6.10.1) for $f = 1–100$ kHz. Does it have any ripple in the passband or the stopband? Then perform the PSpice simulation to get the frequency response of the BPF circuit that has been designed in part (a) (as depicted in Figure P6.10.2(b)). Check if the lower and upper 3 dB frequencies of the circuit are close to $f_{l,3\,\text{dB}} = 8000$ Hz and $f_{u,3\,\text{dB}} = 12500$ Hz, respectively, at which the magnitudes of the frequency response are $1/\sqrt{2} = 0.7071 = -3$ dB.

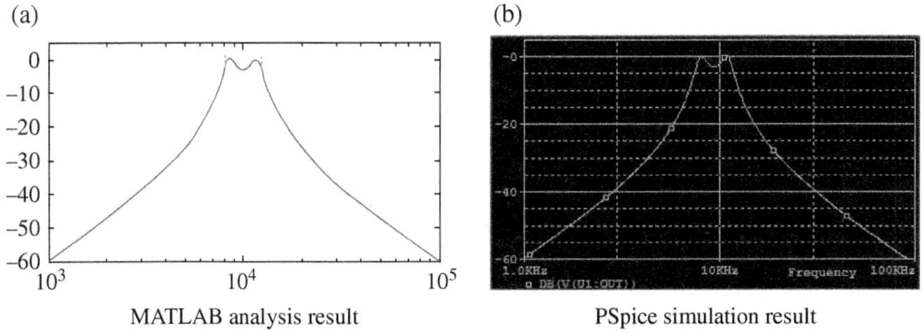

Figure P6.10.2 Frequency response of the circuit in Figure P6.10.1.

(c) If the PSpice simulation result tells you that the upper/lower 3 dB frequencies are considerably deviated from the expected ones, scale the capacitance values by an appropriate factor for frequency scaling (Remark 6.2(1)) and perform the PSpice simulation again to see if the upper/lower 3 dB frequencies get closer to the expected ones.

6.11 Fourier Series Solution of an Active Second-Order OP Amp Circuit Excited by a Triangular Wave

Consider the active second-order OP Amp circuit in Figure P6.11(a), which is excited by a triangular wave voltage source of period 0.0004π as depicted in

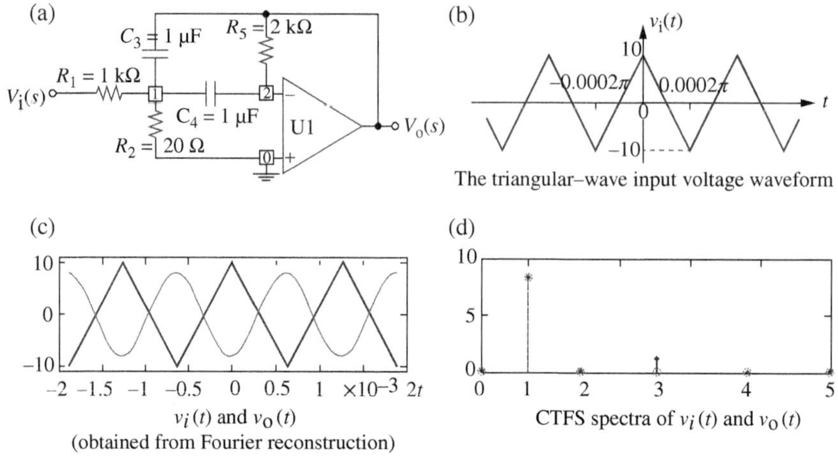

Figure P6.11

Figure P6.11(b). Find the magnitudes of the leading three frequency components (up to three significant digits) of the input $v_i(t)$ and output $v_o(t)$ in two ways, one using MATLAB and one using PSpice, and fill in Table P6.11.

Table P6.11 Magnitudes of the leading three frequency components of $v_i(t)$ and $v_o(t)$.

	$k = 1$		$k = 3$		$k = 5$	
Order of harmonics	**MATLAB**	**PSpice**	**MATLAB**	**PSpice**	**MATLAB**	**PSpice**
Input $v_i(t)$	8.11			0.903	0.324	
Output $v_o(t)$		8.16	0.0675			0.0135

(a) Complete the following MATLAB script "elec06p11.m," which uses the MATLAB function '`Fourier_analysis()`' introduced in Section 1.5.2, together with the MATLAB function '`elec06p11_f()`' (to be saved in a separate M-file named "elec06p11_f.m") and run it to get the waveforms and Fourier spectra of $v_i(t)$ and $v_o(t)$ as shown in Figure P6.11(c) and (d).

(b) Find the peak (center) frequency ω_p and bandwidth ω_b of the circuit from its transfer function or frequency response. How is ω_p related with the frequency component of the input passing through the filter that can be seen in the input and output spectra shown in Figure P6.11(d)?

(c) Use the PSpice simulation where the parameters of the periodic voltage source VPULS can be set as follows:

$$V1 = 10, V2 = -10, TD = 0, TR = 0.000628,\ PW = 0.000001, PER = 0.001257$$

In the Simulation Settings dialog box, set the Analysis type, Run_to_time, and Maximum step as Time Domain (Transient), 20 ms, and 1 u, respectively, and click the Output File Options button to open the Transient Output File Options dialog box, in which the parameters are to be set as

$$\text{Center frequency} = 796\text{ Hz, Number harmonics} = 5$$
$$\text{Output variables} = V(R1:1)\ V(U1:OUT)$$

```
%elec06p11.m : to solve Problem 6.11
% Perform Fourier analysis for a BPF with a triangular wave input
clear
format short e
global T D Vm
T=0.0004*pi, D=T/2, Vm=10; % Period, Duration, Amplitude
N=5; kk=0:N; % Frequency indices to analyze using Fourier analysis
x='elec06p11_f'; % Name of an M-file defining triangular wave function
R1=1e3; R2=20; C3=1e-6; C4=1e-6; R5=???;
n= - [1/R1/C3 ?] % Numerator of transfer function G(s) Eq.(5.5.7)
d=[? (C3+C4)/R5/C3/C4 ????????????????????] %Denominator of Eq.(5.5.7)
[Y,X,THDy,THDx] = ???????????????(n,d,x,T,N);
THDx, THDy
Magnitude_and_phase_of_input_and_output_spectrum= ...
  [kk; abs(X); phase(X)*180/pi; abs(Y); angle(Y)*180/pi].'
```

```
function y=elec06p11_f(t)
% defines a triangular wave with period T, duration D, amplitude Vm
global T D Vm
t=mod(t,T); y=Vm*((t<=D).*(1-(2/D)*t) + (t>D).*(-1+(2/D)*(t-D)));
```

7

Smith Chart and Impedance Matching

CHAPTER OUTLINE

This chapter introduces the concept of the transmission line (TL) theory, impedance matching, and how the Smith Chart can be used for impedance matching.

7.1 Transmission Line

The sinusoidal steady-state voltage and current of a TL (as depicted in Figure 7.1) at point of distance d from the load end can be written in cosine-based phasor form as[P-1]

$$\mathbf{V}(d) = \mathbf{V}_o^+ e^{\gamma d} + \mathbf{V}_o^- e^{-\gamma d} \tag{7.1.1}$$

$$\mathbf{I}(d) = \mathbf{I}_o^+ e^{\gamma d} + \mathbf{I}_o^- e^{-\gamma d} = \frac{1}{Z_0}\left(\mathbf{V}_o^+ e^{\gamma d} - \mathbf{V}_o^- e^{-\gamma d}\right) \tag{7.1.2}$$

Electronic Circuits with MATLAB®, PSpice®, and Smith Chart, First Edition. Won Y. Yang, Jaekwon Kim, Kyung W. Park, Donghyun Baek, Sungjoon Lim, Jingon Joung, Suhyun Park, Han L. Lee, Woo June Choi, and Taeho Im.

Companion website: www.wiley.com/go/yang/electroniccircuits

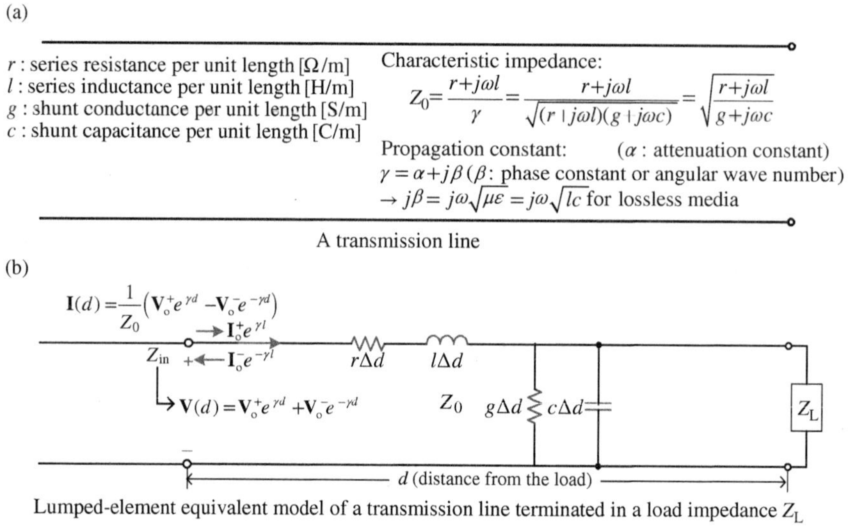

Figure 7.1 A transmission line (TL) terminated in a load impedance Z_L.

where $\mathbf{V}_o^+/\mathbf{V}_o^-$ and $\mathbf{I}_o^+/\mathbf{I}_o^-$ are called the forward(incident)/reflected voltages and currents (each propagating from the source/load to the load/source), respectively. The characteristic impedance Z_0 and the propagation constant $\gamma = \alpha + j\beta$ (α: attenuation constant, β: phase constant) are defined as

$$\text{Characteristic impedance}: Z_0 = \frac{r + j\omega l}{\gamma} = \frac{r + j\omega l}{\sqrt{(r + j\omega l)(g + j\omega c)}} = \sqrt{\frac{r + j\omega l}{g + j\omega c}} \quad (7.1.3)$$

$$\text{Propagation constant}: \gamma = \alpha + j\beta = \alpha + j\frac{2\pi}{\lambda(\text{wavelength})} = \sqrt{(r + j\omega l)(g + j\omega c)} \quad (7.1.4)$$

(α: attenuation constant, β: phase constant, angular wave number, or spatial frequency)

with

$$\begin{aligned} &r: \text{series resistance per unit length } [\Omega/\text{m}] \\ &l: \text{series inductance per unit length } [\text{H/m}] \\ &g: \text{shunt conductance per unit length } [\text{S/m}] \\ &c: \text{shunt capacitance per unit length } [\text{C/m}] \end{aligned} \quad (7.1.5)$$

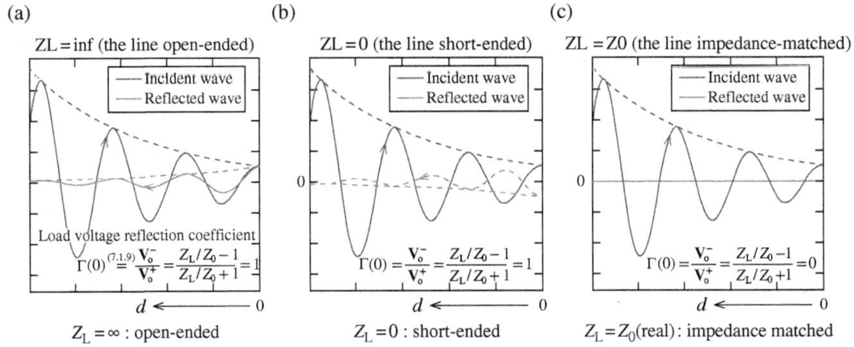

Figure 7.2 The shapes of forward (incident) and reflected waves along the distance d from the load end.

If the forward (incident) and reflected voltage phasors are

$$\mathbf{V}_o^+ = V^+ e^{j\phi} \text{ and } \mathbf{V}_o^- = V^- e^{j\phi} \tag{7.1.6}$$

the line voltage (with angular frequency ω) at point of distance d from the load can be expressed as

$$v(d,t) = \text{Re}\{\mathbf{V}(d)e^{j\omega t}\} \underset{(7.1.6)}{\overset{(7.1.1)}{=}} \text{Re}\{V^+ e^{j\phi} e^{\gamma d} e^{j\omega t} + V^- e^{j\phi} e^{-\gamma d} e^{j\omega t}\}$$

$$\overset{\gamma = \alpha + j\beta}{=} V^+ e^{\alpha d} \cos(\omega t + \beta d + \phi) + V^- e^{-\alpha d} \cos(\omega t - \beta d + \phi) \tag{7.1.7}$$

Figure 7.2(a), (b), and (c) show the shapes of forward (incident) and reflected waves when the TL is open-ended, short-ended, and impedance-matched, respectively.

We divide $\mathbf{V}(d)$ (Eq. (7.1.1)) by $\mathbf{I}(d)$ (Eq. (7.1.2)) to get the input impedance Z_{in} of the TL at point of distance d from the load as

$$Z_{\text{in}}(d) = \frac{\mathbf{V}(d)}{\mathbf{I}(d)} \underset{(7.1.2)}{\overset{(7.1.1)}{=}} \frac{\mathbf{V}_o^+ e^{\gamma d} + \mathbf{V}_o^- e^{-\gamma d}}{(\mathbf{V}_o^+ e^{\gamma d} - \mathbf{V}_o^- e^{-\gamma d})/Z_0} \overset{(7.1.9)}{=} \frac{1 + \Gamma(d)}{1 - \Gamma(d)} Z_0 \tag{7.1.8}$$

where the *(input) voltage reflection coefficient* $\Gamma(d)$ is defined as the ratio of reflected voltage phasor to incident voltage phasor at a distance d from the load:

$$\Gamma(d) = \frac{\mathbf{V}_o^- e^{-\gamma d}}{\mathbf{V}_o^+ e^{\gamma d}} = \Gamma(0) e^{-2\gamma d} \tag{7.1.9}$$

$$\text{with } \Gamma(0) = \frac{\mathbf{V}_o^-}{\mathbf{V}_o^+} \text{(Load voltage reflection coefficient)}$$

The voltage and current phasors at the load ($d = 0$) are related by the load impedance as

$$Z_L = Z_{in}(0) = \frac{\mathbf{V}(0)}{\mathbf{I}(0)} \overset{(7.1.8)}{=} \frac{\mathbf{V}_o^+ + \mathbf{V}_o^-}{\mathbf{V}_o^+ - \mathbf{V}_o^-} Z_0 = \frac{1 + \mathbf{V}_o^- / \mathbf{V}_o^+}{1 - \mathbf{V}_o^- / \mathbf{V}_o^+} Z_0 \overset{(7.1.9)}{=} \frac{1 + \Gamma(0)}{1 - \Gamma(0)} Z_0 \quad (7.1.10)$$

This can be solved for $\Gamma(0)$ (the voltage reflection coefficient at the load, called the *load [voltage] reflection coefficient*) as

$$(1 - \Gamma(0)) Z_L \overset{(7.1.10)}{=} (1 + \Gamma(0)) Z_0 \Gamma(0) \underset{d=0}{\overset{(7.1.9)}{=}} \frac{\mathbf{V}_o^-}{\mathbf{V}_o^+} = \frac{Z_L - Z_0}{Z_L + Z_0} = \frac{Z_L / Z_0 - 1}{Z_L / Z_0 + 1} = \frac{\bar{Z}_L - 1}{\bar{Z}_L + 1} \quad (7.1.11)$$

where the *normalized load* and *input impedances* are defined as

$$\bar{Z}_L = \frac{Z_L}{Z_0} \text{ and } \bar{Z}_{in}(d) = \frac{Z_{in}(d)}{Z_0} \overset{(7.1.8)}{=} \frac{1 + \Gamma(d)}{1 - \Gamma(d)} \quad (7.1.12a,b)$$

As a measure of the mismatch of a lossless TL with $\gamma = j\beta$ (β: phase constant), we define the *standing wave ratio* (*SWR*) and *attenuation ratio* (*ATR*) as

$$\text{SWR} = \frac{V_{max}}{V_{min}} = \frac{1 + |\Gamma|}{1 - |\Gamma|} \text{ for } \Gamma < 0 \text{ (negative reflection)} \quad (7.1.13a)$$

$$\text{ATR} = \frac{1}{|\Gamma|} \text{ for } \Gamma > 0 \text{ (positive reflection)} \quad (7.1.13b)$$

where

$$|\Gamma| = |\Gamma(d)| = |\Gamma(0)e^{-2\gamma d}| = |\Gamma(0)e^{-j2\beta d}| = |\Gamma(0)| \quad (7.1.14)$$

(irrespective of the distance d)

Remark 7.1 Reflection Coefficient and SWR (Standing Wave Ratio)

1) Since $|\Gamma| < 1$, the SWR is always greater than or equal to unity and it will be closer to $1/\infty$ for better/worse impedance matching.
2) If $Z_L = \infty$ (the TL is open-ended), we have $\Gamma(0) = 1$ so that SWR $= \infty$, implying a complete positive reflection (see Figure 7.2(a)).
3) If $Z_L = 0$ (the TL is short-ended), we have $\Gamma(0) = -1$ so that SWR $= \infty$, implying a complete negative reflection (see Figure 7.2(b)).
4) If $Z_L = Z_0$ (the TL is perfectly matched), we have $\Gamma(0) = 0$ so that SWR $= 1$, implying no reflection, i.e. perfect absorption by the load (see Figure 7.2(c)).

We can substitute Eq. (7.1.9) into Eq. (7.1.12) to rewrite the (normalized) load impedance as

$$\bar{Z}_{\text{in}}(d)=\frac{Z_{\text{in}}(d)}{Z_0}\overset{(7.1.12\text{b})}{=}\frac{1+\Gamma(d)}{1-\Gamma(d)}\overset{(7.1.9)}{=}\frac{1+\Gamma(0)e^{-2\gamma d}}{1-\Gamma(0)e^{-2\gamma d}}\overset{(7.1.11)}{=}\frac{1+\dfrac{\bar{Z}_{\text{L}}-1}{\bar{Z}_{\text{L}}+1}e^{-2\gamma d}}{1-\dfrac{\bar{Z}_{\text{L}}-1}{\bar{Z}_{\text{L}}+1}e^{-2\gamma d}}$$

$$=\frac{\bar{Z}_{\text{L}}+1+(\bar{Z}_{\text{L}}-1)e^{-2\gamma d}}{\bar{Z}_{\text{L}}+1-(\bar{Z}_{\text{L}}-1)e^{-2\gamma d}}=\frac{\bar{Z}_{\text{L}}(1+e^{-2\gamma d})+(1-e^{-2\gamma d})}{\bar{Z}_{\text{L}}(1-e^{-2\gamma d})+(1+e^{-2\gamma d})}=\frac{\bar{Z}_{\text{L}}+\tanh\gamma d}{1+\bar{Z}_{\text{L}}\tanh\gamma d} \tag{7.1.15}$$

or

$$Z_{\text{in}}(d)=Z_0\frac{\bar{Z}_{\text{L}}+\tanh\gamma d}{1+\bar{Z}_{\text{L}}\tanh\gamma d}=Z_0\frac{Z_{\text{L}}+Z_0\tanh\gamma d}{Z_0+Z_{\text{L}}\tanh\gamma d} \tag{7.1.16}$$

This can be written for a lossless TL (with an imaginary propagation constant $\gamma=j\beta=j2\pi/\lambda$) terminated with a load impedance Z_{L} as

$$Z_{\text{in}}(d)\underset{\gamma=j\beta}{\overset{(7.1.16)}{=}}Z_0\frac{\bar{Z}_{\text{L}}+\tanh(j\beta d)}{1+\bar{Z}_{\text{L}}\tanh(j\beta d)}=Z_0\frac{Z_{\text{L}}+jZ_0\tan\beta d}{Z_0+jZ_{\text{L}}\tan\beta d} \tag{7.1.17}$$

$$\text{where }\tanh j\theta=\frac{e^{j\theta}-e^{-j\theta}}{e^{j\theta}+e^{-j\theta}}=j\tan\theta$$

This implies that the input impedance has the same value at every one-half wave length ($\lambda/2$), i.e. at $d\pm n\lambda/2$ (n: an integer) on a lossless TL terminated with a load impedance Z_{L} since

$$Z_{\text{in}}\left(d\pm n\frac{\lambda}{2}\right)\underset{\beta=2\pi/\lambda}{\overset{(7.1.17)}{=}}Z_0\frac{Z_{\text{L}}+jZ_0\tan\{(2\pi/\lambda)(d\pm n\lambda/2)\}}{Z_0+jZ_{\text{L}}\tan\{(2\pi/\lambda)(d\pm n\lambda/2)\}}$$

$$=Z_0\frac{Z_{\text{L}}+jZ_0\tan(2\pi d/\lambda)}{Z_0+jZ_{\text{L}}\tan(2\pi d/\lambda)}\underset{\beta=2\pi/\lambda}{\overset{(7.1.17)}{=}}Z_{\text{in}}(d)$$

At points of distance $d=n\lambda/2$ (n: an integer), the input impedance of a lossless TL terminated with a load impedance Z_{L} is the same as the load impedance:

$$Z_{\text{in}}\left(d=n\frac{\lambda}{2}\right)\underset{d=n\lambda/2}{\overset{(7.1.17)}{=}}Z_0\frac{Z_{\text{L}}+jZ_0\tan\beta d}{Z_0+jZ_{\text{L}}\tan\beta d}=Z_0\frac{Z_{\text{L}}+jZ_0\tan n\pi}{Z_0+jZ_{\text{L}}\tan n\pi}=Z_{\text{L}} \tag{7.1.18}$$

Especially, at point of distance $d=\lambda/4$, the input impedance of a lossless TL terminated in a load impedance Z_{L} turns out to be

$$Z_{\text{in}}\left(d=\frac{\lambda}{4}\right)\underset{d=\lambda/4}{\overset{(7.1.17)}{=}}Z_0\frac{Z_{\text{L}}+jZ_0\tan\pi/2}{Z_0+jZ_{\text{L}}\tan\pi/2}=\frac{Z_0^2}{Z_{\text{L}}} \tag{7.1.19}$$

This can be used to determine the characteristic impedance of a $\lambda/4$-length lossless line, called a *quarter-wave* ($\lambda/4$) (*impedance*) *transformer*, with a desired input impedance $Z_{\text{in,d}}$ as

$$Z_0 = \sqrt{Z_{\text{in,d}} Z_L} \tag{7.1.20}$$

In addition to the SWR, as another measure of discontinuity at the load in a TL, the *return loss* (*RL*) is defined as the ratio of the incident power to (negatively) reflected power if $\Gamma(0)$ is real-valued and negative, i.e. $\Gamma(0) < 0$:

$$\text{RL(Return Loss)} = 10\log_{10}\frac{(\mathbf{V}_o^+)^2}{(\mathbf{V}_o^-)^2} \overset{(7.1.11)}{=} 20\log_{10}\frac{1}{|\Gamma(0)|}\ [\text{dB}] \tag{7.1.21a}$$

If $\Gamma(0)$ is real-valued and positive, i.e. $\Gamma(0) > 0$, the *reflection loss* (*RFL*) is defined as

$$\text{RFL(Reflection Loss)} = 10\log_{10}\left\{\frac{1}{1-\Gamma(0)^2}\right\}\ [\text{dB}] \tag{7.1.21b}$$

If $\Gamma(0) = 0$ (perfect impedance matching between the line and the load), we have RL = ∞ and RFL = 0. Figure 7.3 shows the lower part of Smith chart giving the SWR, RL/RFL, etc., where the values of the SWRs and RLs for $\Gamma = -0.5$ and $\Gamma = 0.5$ can be read as

$$\text{SWR} \underset{\Gamma<0}{\overset{(7.1.13a)}{=}} \frac{1+|\Gamma|}{1-|\Gamma|} = \frac{1+0.5}{1-0.5} = 3 \tag{7.1.22a}$$

$$\text{ATR} \underset{\Gamma>0}{\overset{(7.1.13b)}{=}} 10\log_{10}\frac{1}{\Gamma} = 10\log_{10}\frac{1}{0.5} \approx 3\ [\text{dB}] \tag{7.1.22b}$$

$$\text{RL} \underset{\Gamma<0}{\overset{(7.1.21a)}{=}} 20\log_{10}\frac{1}{|\Gamma| = 0.5} \approx 6[\text{dB}] \tag{7.1.23a}$$

$$\text{RFL} \underset{\Gamma>0}{\overset{(7.1.21b)}{=}} 10\log_{10}\frac{1}{1-\Gamma^2} \overset{\Gamma=0.5}{\approx} 1.25\ [\text{dB}] \tag{7.1.23b}$$

To see how the voltage reflection coefficient is related with the power flow along the TL, let us rewrite Eqs. (7.1.1) and (7.1.2) for the total voltage and current as

$$\mathbf{V}(d) \overset{(7.1.1)}{=} \mathbf{V}_o^+ e^{j\beta d} + \mathbf{V}_o^- e^{-j\beta d} \overset{(7.1.9)}{=} \mathbf{V}_o^+ \{e^{j\beta d} + \Gamma(d) e^{-j\beta d}\} \tag{7.1.24}$$

$$\mathbf{I}(d) \overset{(7.1.2)}{=} \frac{1}{Z_0}\{\mathbf{V}_o^+ e^{\gamma d} - \mathbf{V}_o^- e^{-\gamma d}\} \overset{(7.1.9)}{=} \frac{\mathbf{V}_o^+}{Z_0}\{e^{j\beta d} - \Gamma(d) e^{-j\beta d}\} \tag{7.1.25}$$

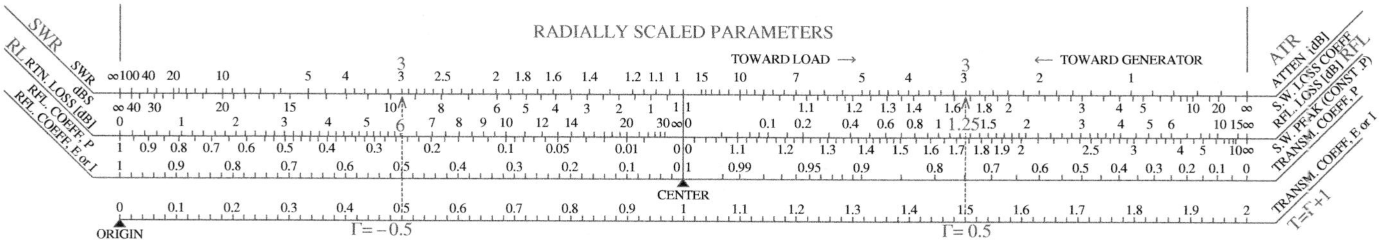

Figure 7.3 Lower part of Smith chart giving the Standing Wave Ratio (SWR), return/reflection loss, and reflection/transmission coefficients. (A part of the Smith chart from https://leleivre.com/rf_smith.html)

Then, the time-average power flow along the TL at a point of distance d from the load can be written as

$$P_{av} = \frac{1}{2}\text{Re}\{\mathbf{V}(d)\mathbf{I}^*(d)\} \underset{(7.1.25)}{\overset{(7.1.24)}{=}} \frac{1}{2}\frac{|\mathbf{V}_o^+|^2}{Z_0} e^{2\alpha d}\{1+\text{Re}\{\Gamma(d)-\Gamma^*(d)\}-|\Gamma(d)|^2\}$$

$$\underset{\text{make imaginary}}{\overset{\text{the 2nd and 3rd terms}}{=}} \frac{1}{2}\frac{|\mathbf{V}_o^+|^2}{Z_0} e^{2\alpha d}\{1-|\Gamma(d)|^2\} \underset{\text{at the load}}{\overset{d=0}{\rightarrow}} \frac{1}{2}\frac{|\mathbf{V}_o^+|^2}{Z_0}\{1-|\Gamma(0)|^2\} \tag{7.1.26}$$

Here, $|\Gamma(0)|^2$ may well be called the power reflection coefficient.

7.2 Smith Chart

The Smith chart was developed by P. W. Smith as a graphical tool to analyze and design TL circuits, in 1939. It can be used for the conversions between reflection coefficients and normalized impedances based on Eq. (7.1.12):

$$\bar{Z} = \bar{R} + j\bar{X} \overset{(7.1.12)}{=} \frac{1+\Gamma}{1-\Gamma} = \frac{1+\Gamma_R+j\Gamma_I}{1-\Gamma_R-j\Gamma_I} = \frac{1-\Gamma_R^2-\Gamma_I^2+j2\Gamma_I}{(1-\Gamma_R)^2+\Gamma_I^2} \tag{7.2.1}$$

This can be reduced to two real equations each corresponding to the real and imaginary parts:

$$\bar{R} = \frac{1-\Gamma_R^2-\Gamma_I^2}{(1-\Gamma_R)^2+\Gamma_I^2}\ ;\ \Gamma_R^2 - \frac{2\bar{R}}{1+\bar{R}}\Gamma_R + \frac{\bar{R}-1}{1+\bar{R}} + \Gamma_I^2 = 0$$
$$;\left(\Gamma_R - \frac{\bar{R}}{1+\bar{R}}\right)^2 + \Gamma_I^2 = \left(\frac{1}{1+\bar{R}}\right)^2 \tag{7.2.2a}$$

$$\bar{X} = \frac{2\Gamma_I}{(1-\Gamma_R)^2+\Gamma_I^2}\ ;\ \Gamma_R^2 - 2\Gamma_R + \Gamma_I^2 - \frac{2\Gamma_I}{\bar{X}} = 0$$
$$;(\Gamma_R-1)^2 + \left(\Gamma_I - \frac{1}{\bar{X}}\right)^2 \Gamma_I^2 = \left(\frac{1}{\bar{X}}\right)^2 \tag{7.2.2b}$$

where

$$\Gamma = \frac{\bar{Z}-1}{\bar{Z}+1} = |\Gamma|e^{j\theta};\ \bar{Z} = \frac{Z}{Z_0} = \frac{1+|\Gamma|e^{j\theta}}{1-|\Gamma|e^{j\theta}} : \text{Normalized load impedance} \tag{7.2.3}$$

Eqs. (7.2.2a,b) represent two families of circles, each called resistance circles for $\bar{R}$ = constant and reactance circles for $\bar{X}$ = constant, respectively, that constitute the Smith chart as plotted in Figure 7.4. Note that the center $(\bar{R}/(\bar{R}+1), 0)$ of

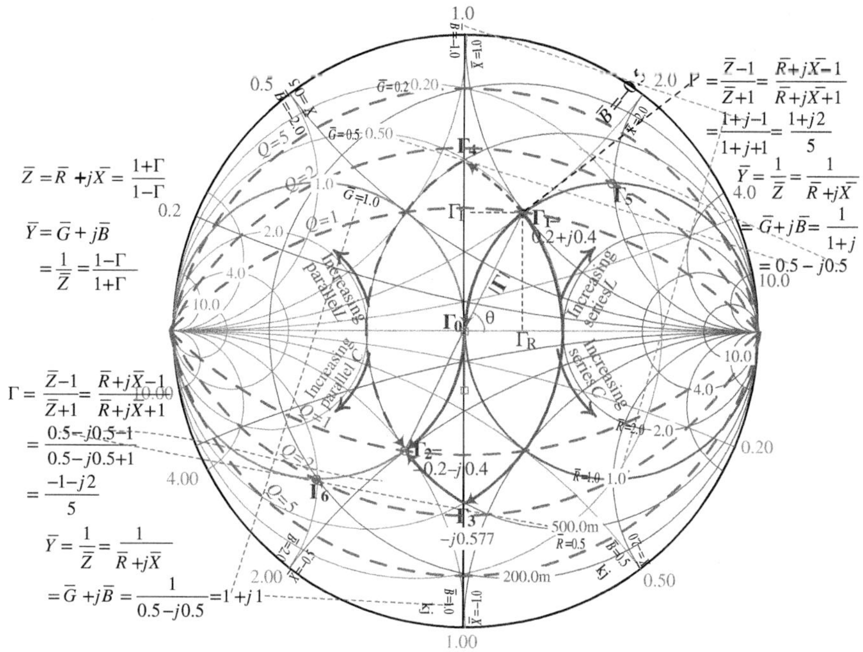

Figure 7.4 The Smith chart.

resistance circle gets closer to (1,0) as $\bar{R} \to \infty$ and the center $(1,1/\bar{X})$ of reactance circle moves up or down (depending on the sign of $\bar{X}$) as $|\bar{X}| \to 0$.

Likewise, the normalized admittance is related with the reflection coefficient as

$$\bar{Y} = \frac{Y = G + jB}{Y_0 = 1/Z_0} = \bar{G} + j\bar{B} = \frac{1 - \Gamma}{1 + \Gamma} \tag{7.2.4}$$

only with the reversed sign of Γ in comparison with Eq. (7.2.1). This implies that conductance circles for $\bar{G}$ = constant and susceptance circles for $\bar{B}$ = constant can be obtained by making symmetric transposes of the resistance and reactance circles w.r.t. the origin ($\Gamma = 0$) or rotating them by 180° so that the sign of Γ can be reversed. In the Smith chart in Figure 7.4, conductance and susceptance circles share the point (−1,0) while the resistance and reactance circles share the point (1,0).

Note that a point on the Γ-plane (in the Smith chart) represents the (input) voltage reflection coefficient at a distance d from the load:

$$\Gamma(d) \overset{(7.1.9)}{=} \Gamma(0)e^{-2\gamma d} \overset{\gamma=\alpha+j\beta}{=} \rho_L e^{-2\alpha d} e^{j(\theta-2\beta d)} \tag{7.2.5}$$

$$\text{with } \Gamma(0) = \rho_L e^{j\theta} \begin{pmatrix}\text{Load voltage} \\ \text{reflection coefficient}\end{pmatrix}$$

Remark 7.2 Some Observations about the Smith Chart

1) In view of Eq. (7.2.5), as d increases, i.e. an observation point (on a TL) moves away from the load toward the source (generator) so that the WTG (wavelength toward generator) increases, the corresponding point on the Γ-plane (in the Smith chart) representing the (input) voltage reflection coefficient at the observation point makes a clockwise decreasing spiral movement toward the point corresponding to Z_0 (the characteristic impedance of the line), which is not always, but often, the origin on the Γ-plane if the impedances are normalized to Z_0 as usually. Especially, the effect of increasing the length of a lossless TL (with $\alpha = 0$) by d [m] can be described by rotating the observation point on the Γ-plane clockwise by $2\beta d = 4\pi(d/\lambda)$ [rad] on the circle of radius $|\Gamma|$ centered at the point corresponding to Z_0.
2) Since $\lambda = c/f$ (c: light speed, f: frequency), increasing the frequency f has the same effect (on $\beta d = 2\pi(d/\lambda)$) with increasing the length d of TL.
3) In view of Eq. (7.2.5), as an observation point moves away from the load by 0.25λ (quarter WTG), the corresponding point on the Γ-plane rotates clockwise by π[rad]. It makes a complete revolution as the observation point moves by each half-wavelength, i.e. $0.5n\lambda$ (n: an integer) and that is why the circumference of the (unit) circle is said to be $\lambda/2$ regardless of its radius.
4) In view of Eqs. (7.2.1)/(7.2.4), the points $\Gamma = -1$ and $\Gamma = 1$ corresponds to $\bar{Z} = 0$ (short-circuit) and $\bar{Y} = 0$ (open-circuit), respectively.
5) Comparing Eqs. (7.1.13a) and (7.2.1), we can see that

$$\bar{Z} = (1+\Gamma_R)/(1-\Gamma_R) = \text{SWR if } \Gamma_R > 0 \text{ and } \Gamma_I = 0$$
$$\bar{Z} = (1+\Gamma_R)/(1-\Gamma_R) = 1/\text{SWR if } \Gamma_R < 0 \text{ and } \Gamma_I = 0$$

6) If $Z_L = Z_0$, i.e. the load impedance is perfectly matched to the characteristic impedance of the TL, the corresponding point on the Γ-plane is at the origin $\Gamma = 0$ so that SWR = 1, implying no reflection, i.e. perfect absorption by the load (see Figure 7.2(c)).
7) In view of Eqs. (7.1.13) and (7.1.21), the constant SWR and constant RL (return loss) lines on the Smith chart are circles of radius $|\Gamma|$ centered at the origin on the Γ-plane.

8) The normalized impedance represented by point $\Gamma_1 = 0.2 + j0.4$ can be found as $1 + j1$ from the values of the (blue) resistance and reactance circles crossing the point and can also be computed using Eq. (7.2.1). The normalized admittance represented by the point can be found as $0.5 - j0.5$ from the values of the (red) conductance and susceptance circles crossing the point and can also be computed using Eq. (7.2.4), which is just the reciprocal of Eq. (7.2.1).
9) The sign of reactance values for all the resistance curves above/below the real axis on the Γ-plane is +/−, respectively, and the sign of susceptance values for all the conductance curves above/below the real axis on the Γ-plane is −/+, respectively.
10) Connecting a capacitor or an inductor in shunt with a load rotates the point (representing the impedance) along a conductance circle clockwise or counterclockwise, respectively. Connecting an inductor or a capacitor in series with a load rotates the point (representing the impedance) along a resistance circle clockwise or counterclockwise, respectively.
11) The six dotted arcs on the Smith chart in Figure 7.4 are Q-circles (also called constant Q circles) of quality factor $Q = |X|/R = 1, 2,$ and 5, whose equations can be derived by dividing Eq. (7.2.2b) by Eq. (7.2.2a):

$$Q = \pm\frac{\bar{X}}{\bar{R}} \underset{(7.2.2a)}{\overset{(7.2.2b)}{=}} \frac{\pm 2\Gamma_I}{1-\Gamma_R^2-\Gamma_I^2};\ 1-\Gamma_R^2-\Gamma_I^2 = \pm\frac{2\Gamma_I}{Q}$$

$$;\Gamma_R^2 + \left(\Gamma_I \pm \frac{1}{Q}\right)^2 = \left(\sqrt{1+\frac{1}{Q^2}}\right)^2 \qquad (7.2.6)$$

Example 7.1 Design of L-type Impedance Transformers Using the Smith Chart

Find L-type LC circuits transforming the normalized impedance $Z_1 = 1 + j1\ [\Omega]$ or admittance $Y_1 = 1/Z_1 = 0.5 - j0.5[\text{S}]$ represented by point Γ_1 into the normalized impedance $Z_2 = 0.5 - j0.5[\Omega]$ or admittance $Y_2 = 1/Z_2 = 1 + j1[\text{S}]$ represented by point Γ_2 on the Smith chart of Figure 7.4.

a) There are three paths along resistance/conductance curves from point Γ_1 to Γ_2, each via Γ_3, Γ_4, and Γ_0. The first path consists of a section from Γ_1 to Γ_3 (corresponding to $Z_3 = 0.5 - j0.866[\Omega]$ or admittance $Y_3 = 1/Z_3 = 0.5 + j0.866[\text{S}]$) and one from Γ_3 to Γ_2 where

- the former section clockwise along a conductance circle corresponds to the shunt connection of a capacitor with susceptance $B_{13} = \omega C_{13} = 0.866 - (-0.5) = 1.366\ \Omega$,

- and the latter section clockwise along a resistance circle corresponds to the series connection of an inductor with reactance $X_{32} = \omega L_{32} = -0.5 - (-0.866) = 0.366\ \Omega$.

The resulting normalized impedance corresponds to Γ_2 as desired:

$$Z = \frac{1}{Y_1 + jB_{13}} + jX_{32} = \frac{1}{0.5 - j0.5 + j1.366} + j0.366 = 0.5 - j0.5\ [\Omega] \tag{E7.1.1}$$

The second path consists of a section from Γ_1 to Γ_4 (corresponding to $Z_4 = 0.5 + j0.866[\Omega]$ or admittance $Y_4 = 1/Z_4 = 0.5 - j0.866[S]$) and one from Γ_4 to Γ_2 where

- the former section counterclockwise along a conductance circle corresponds to the shunt connection of an inductor with susceptance $B_{14} = -1/\omega L_{14} = -0.866 - (-0.5) = -0.366\ \Omega$,
- and the latter section counterclockwise along a resistance circle corresponds to the series connection of a capacitor with reactance $X_{42} = -1/\omega C_{42} = -0.5 - 0.866 = -1.366\ \Omega$.

The resulting normalized impedance corresponds to Γ_2 as desired:

$$Z = \frac{1}{Y_1 + jB_{14}} + jX_{42} = \frac{1}{0.5 - j0.5 - j0.366} - j1.366 = 0.5 - j0.5\ [\Omega] \tag{E7.1.2}$$

The third path consists of a section from Γ_1 to Γ_0 and one from Γ_0 to Γ_2 where

- the former section counterclockwise along a resistance circle corresponds to the series connection of a capacitor with reactance $X_{10} = -1/\omega C_{10} = 0 - 1 = -1\ \Omega$,
- and the latter section clockwise along a conductance circle corresponds to the shunt connection of a capacitor with susceptance $B_{02} = \omega C_{02} = 1 - 0 = 1\ \Omega$.

The resulting normalized admittance and impedance correspond to Γ_2 as desired:

$$Y = \frac{1}{Z_1 + jX_{10}} + jB_{02} = \frac{1}{1 + j1 - j1} + j1 = 1 + j\ [S] \tag{E7.1.3a}$$

$$;Z = \frac{1}{Y} = 0.5 - j0.5\ [\Omega] \tag{E7.1.3b}$$

b) Noting that this problem is to find L-type impedance matchers between load impedance of $Z_1 = 1 + j1[\Omega]$ and source impedance $Z_2^* = 0.5 + j0.5[\Omega]$, we can use the MATLAB function '`imp_matcher_L()`' (Section 6.3.2) as follows:

```
>>Zs=0.5+0.5i; ZL=1+i; % Source/Load impedances
  fc=1/2/pi; % Faked carrier frequency
  [X11,X21,X12,X22]=imp_matcher_L(Zs,ZL,fc)
```

This results in

```
L1-type impedance matcher with C1=1.000e+09[nF] and C2=1.000e+09[nF] %1
L1-type impedance matcher with C1=1.000e+09[nF] and C2=1.000e+09[nF] %2
L2-type impedance matcher with L1=3.660e+05[uH] and C2=1.366e+09[nF] %3
L2-type impedance matcher with C1=7.321e+08[nF] and L2=2.732e+06[uH] %4
X11 =   -1   -1 %1
X21 =   -1   -1 %2
X12 =    0.366025403784439  -1.366025403784439 %3
X22 =   -0.732050807568877   2.732050807568878 %4
```

Here, we set the frequency as $f_c = 1/2\pi$ so that $\omega_c = 2\pi f_c = 1$ and consequently, the reactance values (ωL) of inductors and the susceptance values (ωC) of capacitors can be equal to their inductance and capacitance values, respectively, for simplicity. Noting that $1/0.7321 = 1.366$ and $1/2.732 = 0.366$, we see that the first and second results happened to be identical, corresponding to the third case in (a), the third one corresponds to the first case in (a), and the forth one corresponds to the second case in (a).

c) Let us use the Smith Chart software 'Smith V4.1' (developed by Professor Fritz Dellsperger) to find the LC filter implied by the path from Γ_1 to Γ_2 via Γ_3. To this end, take the following steps:

1) Start 'Smith V4.1' to open the main window as shown in Figure 7.5(a).
2) From the top menu bar, select Tools > Settings to open the Settings dialog box (Figure 7.5(b)), set the values of Z_0 (characteristic impedance) and frequency to $1[\Omega]$ and $1/2\pi = 0.159155[\text{Hz}]$, respectively, and click OK button to close the dialog box.
3) From the toolbar (below the menu bar), select 'Mouse' and click on Γ_1 (the intersection of the $1\,\Omega$-circle and the 1 S-circle). Alternatively, select Keyboard to open Data Point dialog box and insert the impedance $Z_1 = 1 + j1[\Omega]$ as a data point (Figure 7.5(c)).
4) Click OK button to close the Data Point dialog box. Then DP 1 will appear at Γ_1 on the Smith chart and its impedance value will appear in the Data Points pane at the right side as shown in Figure 7.6.
5) From the tool bar, click on 'Insert parallel capacitor' button and move the cross-type mouse cursor to click on the point Γ_3 (the intersection of the $500\,\text{m}\Omega$-circle and 0.5 S-circle). Then, TP 2 will appear at Γ_3 on the Smith chart and its impedance value will appear in the Data Points pane at the right side as shown in Figure 7.6(a).

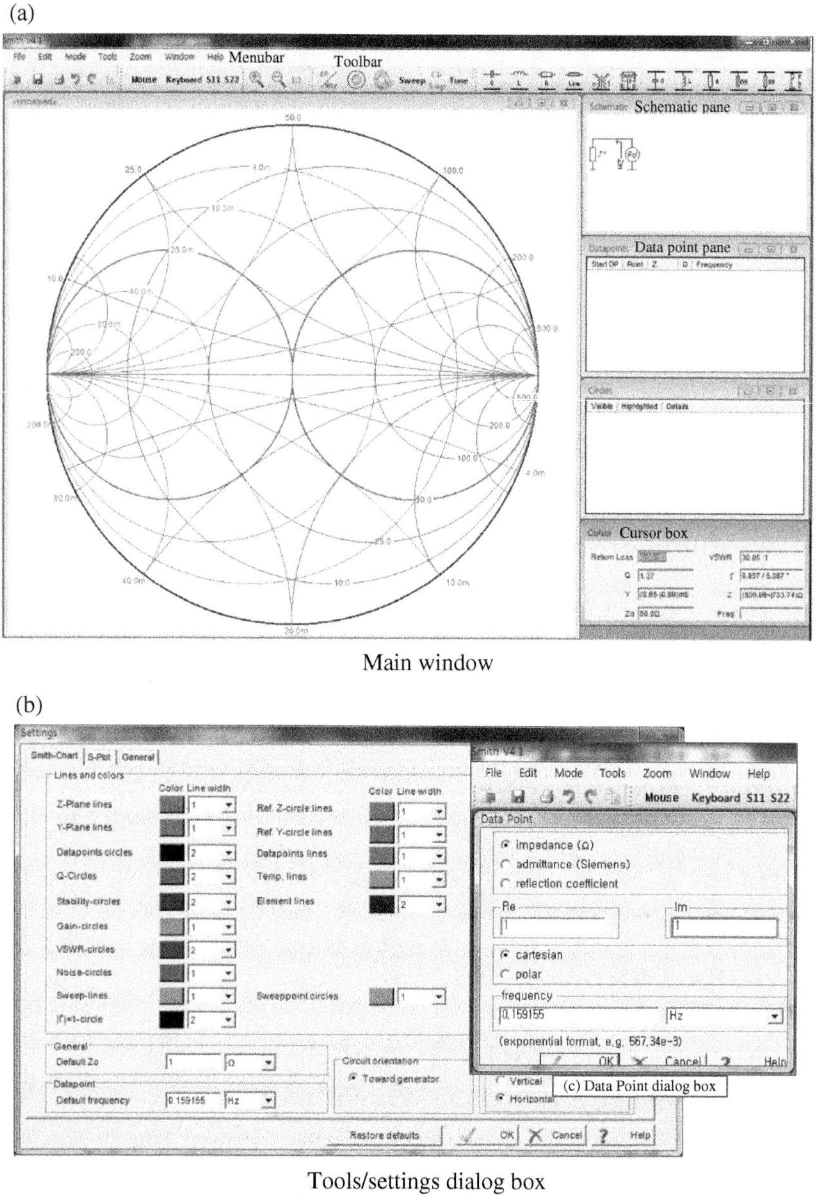

Figure 7.5 The main window, Tools/Setting dialog box, and Data Point dialog box of 'Smith 4.1'.

(a)

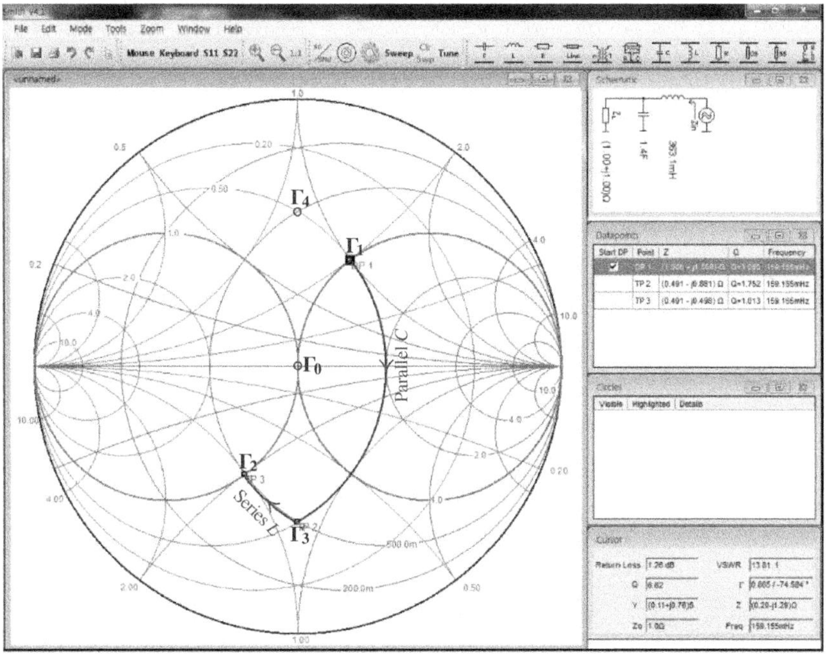

Main window after an L-type impedance matcher has been determined

(b)

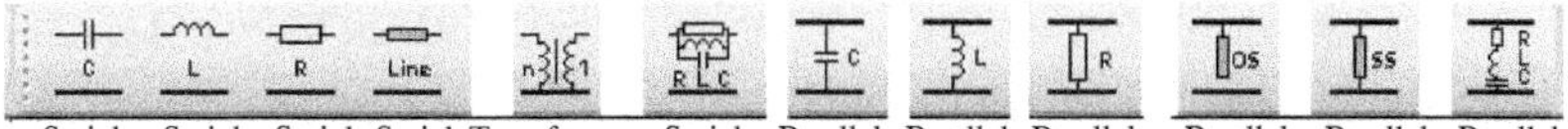

Serial capacitor | Serial inductor | Serial resistor | Serial line | Transformer | Serial *RLC* | Parallel capacitor | Parallel inductor | Parallel resistor | Parallel open stub | Parallel short stub | Parallel *RLC*

Toolbar below the top menubar in the Main window

Figure 7.6 Main window with a design result of 'Smith 4.1'.

6) From the tool bar, click on 'Insert series inductor' button and move the mouse cursor to click on the point Γ_2 (the intersection of the $R = 500$ mΩ-circle, the $G= 1$ S-circle, the $B = 1$ S-circle, and also the $X = -0.5$ Ω-circle) representing $Z_2 = 0.5 - j0.5$ [Ω], or equivalently, $Y_2 = 1 + j1$ [S]. Then, TP 3 will appear at Γ_2 on the Smith chart and its impedance value will appear in the Data Points pane at the right side as shown in Figure 7.6(a).
7) The designed network with the parameter values $C = 1.4$ F and $L = 363.1$ mH is shown in the Schematic pane at the upper-right part of the window. It conforms with the first one ($\omega C_{13} = 1.366\ \Omega$ and $\omega L_{32} = 0.366\ \Omega$ with $\omega = 1$) of the three design results obtained in (a).

8) If you want to read more digits of a numeric parameter value or change it, double-click on it to open the Edit Element dialog box.

It would be a good practice if you try the other two paths from Γ_1 to Γ_2, one via Γ_4 and one via Γ_0, and check if the resulting parameter values are fairly close to those obtained in (a).

Figure 7.6(b) shows several insert functions (like connecting a resistor/inductor/capacitor, an open stub, or a short(-circuited) stub in series or parallel with the load side) that can be activated by clicking on the corresponding button in the toolbar.

7.3 Impedance Matching Using Smith Chart

7.3.1 Reactance Effect of a Lossless Line

From Eq. (7.1.17), we see that connecting a lossless TL (with an imaginary propagation constant $\gamma = j\beta = j2\pi/\lambda$ and characteristic impedance Z_0) of length l in series with a load of impedance Z_L changes the impedance as

$$Z_{in}(l) \underset{d=l}{\overset{(7.1.17)}{=}} Z_0 \frac{Z_L + jZ_0 \tan \beta l}{Z_0 + jZ_L \tan \beta l} \tag{7.3.1}$$

How about a short-circuited stub (SS) with $Z_L = 0$ and an open-circuited stub (OS) with $Z_L = \infty$? Their impedances will be

$$Z_{ss}(l) \underset{Z_L=0}{\overset{(7.3.1)}{=}} jZ_0 \tan\beta l, \quad Z_{os}(l) \underset{Z_L=\infty}{\overset{(7.3.1)}{=}} -j\frac{Z_0}{\tan\beta l} = -jZ_0 \cot\beta l \tag{7.3.2a,b}$$

like those of an inductor or a capacitor, respectively. That is why a short-circuited or an open-circuited stub (line) is used for impedance matching where stubs can easily be fabricated in microstrip or stripline form compared with the 'equivalent' lumped element. To feel how connecting a stub line of length l in series with an impedance Z_L affects the impedance in terms of its reflection coefficient Γ (on the Smith chart), let us run the following MATLAB script to get Figure 7.7. It shows that as the length of the stub increases, the reflection coefficient Γ rotates clockwise around the point Γ_{01} corresponding to the characteristic impedance Z_{01} of the stub line, which is not necessarily equal to the characteristic impedance Z_0 for normalization. Figure 7.7 also shows the effect of connecting a lumped element of reactance X in series with an impedance $Z_L = R_L + jX_L$ is to rotate the reflection coefficient Γ clockwise/counterclockwise along the R_L-circle depending on the sign of the reactance as $|X|$ increases where $\Gamma = 1$ for $X = \pm\infty$ (see Eq. (7.2.3) with $\bar{Z} = (R + jX)/Z_0 = \pm j\infty$ where $Z_0 = R_0 > 0$).

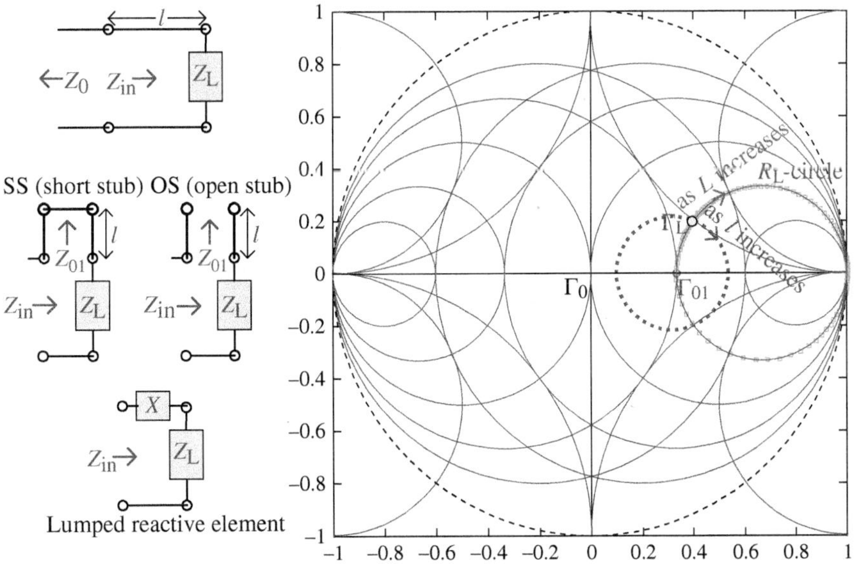

Figure 7.7 The effects of a stub and a lumped reactive element on the impedance.

```
%plot_line_imp.m
clear, clf
Z0=50; % Characteristic impedance for normalization
Z01=100; % Characteristic impedance of the line
b=2*pi; % Phase constant
ZL=100+50i; % Load impedance
Z=@(l)Z01*(ZL+j*Z01*tan(b*l))./(Z01+j*ZL*tan(b*l)); % Eq.(7.3.1)
GammaZ=@(Z)(Z/Z0-1)./(Z/Z0+1); % Eq.(7.2.3)
% To see the effect of increasing the length of stub
ds=0:0.01:0.5; % Lengths of the stub
plot(GammaZ(ZL)+j*eps,'o'), hold on
axis('equal'), axis([-1 1 -1 1])
for n=1:length(ds)
    plot(GammaZ(Z(ds(n))),'r.'), pause(0.05)
end
% To see the effect of changing series reactance of lumped element
Xs=0.5*3.^[0:0.1:20]; % Values of the reactance
Xs=[-fliplr(Xs) 0 Xs]; %-1000:0.1:1000;
for n=1:length(Xs)
    plot(GammaZ(ZL+j*Xs(n))), pause(0.01)
end
% plot the characteristic impedance of the line
plot(GammaZ(Z01)+j*eps,'ro')
% To plot the Smith chart as a background
Rs=[12.5 25 50 100 200]; % Resistance values for Smith chart
Xs=[-50 -25 0 25 50]*2; % Reactance values for Smith chart
Gs=1./[12.5 25 50 100 200]; % Conductance values for Smith chart
Bs=1./[-50 -25 0 25 50]; % Susceptance values for Smith chart
plot_Smith_chart(Z0,Rs,Xs,Gs,Bs)
```

```
function plot_Smith_chart(Z0,Rs,Xs,Gs,Bs)
% plot the Smith chart.
ths=pi/180*[0:360];
if ~isempty(Rs)
 for n=1:length(Rs), [c,r]=Rcircle(Rs(n),Z0); plot(c+r*exp(j*ths)); end
end
if nargin>1&~isempty(Xs)
 for n=1:length(Xs), [c,r]=Xcircle(Xs(n),Z0); plot(c+r*exp(j*ths)); end
end
if nargin>2&~isempty(Gs)
 for n=1:length(Gs), [c,r]=Gcircle(Gs(n),Z0); plot(c+r*exp(j*ths)); end
end
if nargin>3&~isempty(Bs)
 for n=1:length(Bs), [c,r]=Bcircle(Bs(n),Z0); plot(c+r*exp(j*ths)); end
end
plot(exp(j*ths),'k:'); plot([-1 1],[0 0],'k', [0 0],[-1 1],'k')
```

```
function [c,r]=Rcircle(R,Z0)
% To find the radius r and center c (in complex value) of an R-circle
R0=R/Z0; c=R0/(1+R0)*exp(j*eps); r=1/(1+R0); % Eq.(7.2.2a)
```

```
function [c,r]=Xcircle(X,Z0)
% To find the radius r and center c (in complex value) of an X-circle
X0=X/Z0; c=1+i/X0; r=1/abs(X0); % Eq.(7.2.2b)
```

```
function [c,r]=Gcircle(G,Z0)
% To find the radius r and center c (in complex value) of a G-circle
G0=G*Z0; c=G0/(1+G0)*exp(j*pi); r=1/(1+G0);
```

```
function [c,r]=Bcircle(B,Z0)
% To find the radius r and center c (in complex value) of a B-circle
B0=B*Z0; c=-1-i/B0; r=1/abs(B0);
```

7.3.2 Single-Stub Impedance Matching

The *single-stub impedance matching (tuning)* is to connect a single open-circuited or short-circuited lossless stub line in shunt (parallel) or series with an existing lossless TL for matching a (possibly complex-valued) load impedance Z_L to the TL of characteristic impedance R_0 as shown in Figure 7.8. The problem is how distant the stub should be from the load and how long it should be.

7.3.2.1 Shunt-Connected Single Stub

To consider how to determine the distance d from the load and the length l of the shunt stub in Figure 7.8(a), let us write the normalized impedance and admittance of the load as

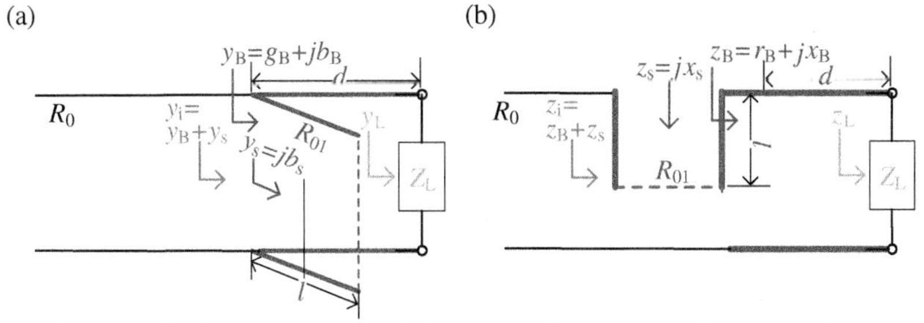

Shunt (parallel) connection of a stub   Series connection of a stub

Figure 7.8 Single-stub impedance matching.

$$z_L = \frac{Z_L}{R_0} = \frac{R_L + jX_L}{R_0} = r_L + jx_L = \frac{1}{R_0 Y_L} = \frac{1}{y_L} \tag{7.3.3a}$$

$$y_L = \frac{Y_L = G_L + jB_L}{Y_0} = \frac{Y_L}{G_0} = R_0 Y_L = g_L + jb_L \tag{7.3.3b}$$

where R_0 and $G_0 = 1/R_0$ are the characteristic impedance and admittance of the TL, respectively, that are used to normalize the impedances and admittances. Noting that

- the impedance matching is achieved when the (parallel) combination of the line-load and the shunt stub results in a unity normalized admittance/impedance corresponding to G_0/R_0, i.e.

$$y_i = y_B + y_s = 1 \tag{7.3.4}$$

- and the stub (whose admittance y_s will be added to the admittance y_B of the line-load to yield the input admittance y_i), short-circuited or open-circuited, is purely reactive/susceptive,

we should determine the distance d of the stub from the load so that the real part (conductance) of the normalized line-load admittance can be unity:

$$g_B = \text{Re}\{y_B\} = 1 \tag{7.3.5}$$

where

$$y_B(d) = \frac{1}{z(d)} = \frac{Z_0}{Z(d)} \overset{(7.1.17)}{=} \frac{Z_0 + Z_L \tanh\gamma d}{Z_L + Z_0 \tanh\gamma d} \underset{\gamma=j\beta}{\overset{Z_0=R_0}{=}} \frac{R_0 + jZ_L \tan\beta d}{Z_L + jR_0 \tan\beta d} \tag{7.3.6}$$

If the stub has its own characteristic impedance $R_{01} \neq R_0$, y_B should be multiplied with R_{01}/R_0. Now we determine the length l of the stub such that the susceptance can be $b_s = -\text{Im}\{y_B\}$ to make the imaginary part of the input admittance $y_i = y_B + y_s$ zero, satisfying Eq. (7.3.4):

$$\text{Im}\{z_s(l)\} \overset{(7.3.2a)}{=} \tan\beta l = \frac{1}{\text{Im}\{y_B\}}$$

$$;\, l = \frac{1}{\beta}\cot^{-1}\{\text{Im}\{y_B\}\} \text{ for a short-circuited stub} \tag{7.3.7a}$$

$$\text{Im}\{z_s(l)\} \overset{(7.3.2b)}{=} -\cot\beta l = \frac{1}{\text{Im}\{y_B\}}$$

$$;\, l = \frac{1}{\beta}\tan^{-1}\{-\text{Im}\{y_B\}\} \text{ for an open-circuited stub} \tag{7.3.7b}$$

Note that OSs (open-circuited stubs) are preferable for easy fabrication in microstrip or stripline form while SSs (short-circuited stubs) are preferable for coax or waveguide.

```
function [ds,ls,Bs,Gammas]=imp_match_1stub_shunt(Z0,ZL,r,os)
% Input:  Z0 = Characteristic impedances [Z0,Z01] of the TL and stub
%         ZL = Load impedance
%          r = a+bi = Propagation constant
%         os = 'o'/'s' for open/short(-circuited) stub
% Output: ds = Distance of the stub from load
%         ls = Length of the stub
if nargin<4, os='s'; end
if length(Z0)>1, Z01=Z0(2); Z0=Z0(1); else Z01=Z0; end
a=real(r); beta=imag(r); lambda=2*pi/beta;
yB=@(ZL,Z0,d)(Z0+ZL*tanh(r*d))./(ZL+Z0*tanh(r*d)); %Eq.(7.3.1)normalized
eq=@(d)real(yB(ZL,Z0,d))-1; % Eq.(7.3.5)
options=optimoptions('fsolve','TolX',1e-20,'TolFun',1e-20);
ds=real([fsolve(eq,0,options) fsolve(eq,0.25*lambda,options)]);
ds=mod(ds*beta,pi)/beta; % Possibly two solutions for d
bB=imag(yB(ZL,Z0,ds))*Z01/Z0; bs=-bB; Bs=bs/Z0; % Susceptances
ys=1+j*bs; Gammas=(1-ys)./(1+ys); % Intermediate points
if lower(os(1))=='s', ls=acot(bB)/beta; % Eq.(7.3.7a)
 else ls=atan(-bB)/beta; % Eq.(7.3.7b)
end
ls=mod(ls*b,pi)/beta;
% Usage: Z0=50; ZL=35-47.5i; lambda=1; r=2*pi*i/lambda;
% [ds,ls]=imp_match_1stub_shunt(Z0,ZL,r,'s')
```

Example 7.2 Design of Shunt Single Short(-Circuited) Stubs for Impedance Matching

Determine the distances (from the load) and lengths of shunt single short-circuited stubs in the unit of wavelength so that they can match a load impedance $Z_L = 30 - j40[\Omega]$ to the lossless transmission (or feed) line of characteristic impedance $Z_0 = 50\,\Omega$.

a) We can easily get the solution by using the MATLAB function '`imp_matcher_1stub_shunt()`' (listed above) as

```
>>Z0=50; ZL=30-40i; lambda=1; beta=2*pi/lambda; r=i*beta;
  [ds,ls]=imp_match_1stub_shunt(Z0,ZL,r,'s') % for SS
```

This yields

```
ds =   0.0417   0.2083
ls =   0.1136   0.3864
```

which means two solutions, one of length 0.1136λ and distance 0.0417λ from the load and the other of length 0.3864λ and distance 0.2083λ from the load.

b) To find the solutions using the Smith chart, we first should recall the following:

- Our target point $\Gamma_3 = 0$ (corresponding to both R_0 and G_0) is on the $G = G_0 = 1/R_0$-circle.
- A lossless TL of length d connected to the load rotates the Γ-point clockwise by $2\beta d = 4\pi(d/\lambda)$ [rad] along the $|\Gamma_L|$-circle, i.e. the circle of radius $|\Gamma_L|$ centered at the origin $\Gamma = 0$ where $\Gamma_L \overset{(7.1.11)}{=} (Z_L - 1)/(Z_L + 1)$ (Remark 7.2(1)).
- A shunt reactive element, a lumped one or a stub, can change only the susceptance, not the conductance, so that the Γ-point can be rotated along a G-circle.

That is why we first should rotate the Γ-point clockwise along the $|\Gamma_L|$-circle (passing through $Z_L = 30 - j40[\Omega]$) to some point (like Γ_2 or Γ_4 in Figure 7.9) on the $G = G_0 = 1/R_0 = 20$ mS-circle. Here, the distance d of a stub (to be shunt-connected) from the load is to be determined from the angles by which the Γ-point has been rotated. Then, we should determine the length l of the parallel (shunt) short-circuited stub, which rotates the Γ-point counterclockwise along the G-circle to the target point $\Gamma = 0$. With this strategy in mind, let us use 'Smith V4.1' by taking the following steps:

1) From the top menu bar, select Tools > Settings to open the Settings dialog box (Figure 7.5(b)), set the values of Z_0 (characteristic impedance) and frequency to $50[\Omega]$ and $1[\text{GHz}]$, respectively, and click OK button to close the dialog box. Here, the frequency does not matter because we will determine d and l just in the unit of wavelength.
2) From the toolbar (below the menu bar), select Keyboard to open Data Point dialog box, insert $Z_1 = 30 - j40[\Omega]$ as a data point, and click OK button to close the dialog box.
3) From the tool bar, click on 'Insert serial line' button and move the cross-type mouse cursor to click on the point Γ_2 (see Figure 7.9).
4) From the tool bar, click on 'Insert parallel line shorted' button (denoted by SS: shorted stub) and move the mouse cursor to click on the point $\Gamma_3 = 0$.

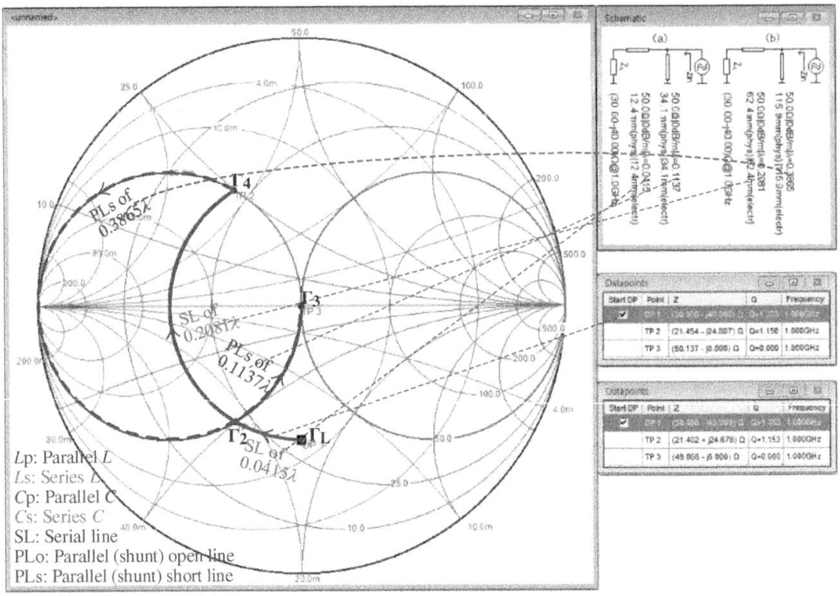

Figure 7.9 Using 'Smith 4.1' to design shunt single short-stub impedance matchers in Example 7.2.

5) Then, the designed network with $d_1 = 0.0415\lambda$ and $l_1 = 0.1137\lambda$ will be shown in the Schematic pane at the upper-right part of the window. It conforms with the results obtained using the MATLAB function '`imp_matcher_1stub_shunt()`' in (a).

7.3.2.2 Series-Connected Single Stub

Similarly to the case of shunt-connected single-stub matching, note that

- the impedance matching is achieved when the (series) combination of the line-load and the series stub (Figure 7.8(b)) results in a unity normalized impedance corresponding to R_0, i.e.

$$z_i = z_B + z_s = 1 \tag{7.3.8}$$

- and the stub (whose impedance z_s will be added to the impedance z_B of the line-load to yield the input impedance z_i), short-circuited or open-circuited, is purely reactive.

We should determine the distance d of the stub from the load so that the real part (resistance) of the normalized line-load impedance can be unity:

$$r_B = \text{Re}\{z_B\} = 1 \tag{7.3.9}$$

where

$$z_B(d) = \frac{Z(d)}{Z_0} \overset{(7.1.17)}{=} \frac{Z_L + Z_0 \tanh \gamma d}{Z_0 + Z_L \tanh \gamma d} \underset{\gamma = j\beta}{\overset{Z_0 = R_0}{=}} \frac{Z_L + jR_0 \tan \beta d}{R_0 + jZ_L \tan \beta d} \tag{7.3.10}$$

If the stub has a characteristic impedance $R_{01} \neq R_0$, z_B should be multiplied with R_0/R_{01}.

Now we determine the length l of the stub such that the reactance can be $x_s = -\text{Im}\{z_B\}$ to make the imaginary part of the input impedance $z_i = z_B + z_s$ zero, satisfying Eq. (7.3.8):

$$\text{Im}\{z_s(l)\} \overset{(7.3.2a)}{=} \tan \beta l = -\text{Im}\{z_B\}$$

$$; l = \frac{1}{\beta} \tan^{-1}\{-\text{Im}\{z_B\}\} \quad \text{for a short-circuited stub} \tag{7.3.11a}$$

$$\text{Im}\{z_s(l)\} \overset{(7.3.2b)}{=} -\cot \beta l = -\text{Im}\{z_B\}$$

$$; l = \frac{1}{\beta} \cot^{-1}\{\text{Im}\{z_B\}\} \text{for an open-circuited stub} \tag{7.3.11b}$$

```
function [ds,ls,Xs,Gammas]=imp_match_1stub_series(Z0,ZL,r,os)
% Input:  Z0 = Characteristic impedances [Z0,Z01] of the TL and stub
%         ZL = Load impedance
%          r = alpha+beta*i = Propagation constant
%         os = 'o'/'s' for open/short(-circuited) stub
% Output: ds = Distances of the stub from load
%         ls = Lengths of the stub
if nargin<4, os='s'; end
if length(Z0)>1, Z01=Z0(2); Z0=Z0(1);
 else Z01=Z0;
end
beta=imag(r); lambda=2*pi/beta;
zB=@(ZL,Z0,d) (ZL+Z0*tanh(r*d))./(Z0+ZL*tanh(r*d)); %Eq.(7.3.1)normalized
eq=@(d)real(zB(ZL,Z0,d))-1; % Eq.(7.3.9,10)
options=optimoptions('fsolve', 'TolX',1e-20, 'TolFun',1e-20);
ds=real([fsolve(eq,0,options) fsolve(eq,0.25*lambda,options)]);
ds=mod(ds*b,pi)/beta;
xB=imag(zB(ZL,Z0,ds))*Z0/Z01; xs=-xB; Xs=xs*Z0; % Reactances
zs=1+j*xs; Xs=xs*Z0; Gammas=(zs-1)./(zs+1); % Intermediate points
if lower(os(1))=='o', ls=acot(xB)/beta; % Eq.(7.3.11b)
 else ls=atan(-xB)/beta; % Eq.(7.3.11a)
end
ls=mod(ls*b,pi)/beta;
% Usage: Z0=50; ZL=35-47.5i; lambda=1; r=2*pi*i/lambda;
% [ds,ls]=imp_match_1stub_series(Z0,ZL,r,'o')
```

Example 7.3 Design of Series Single Stubs for Impedance Matching
Determine the distances (from the load) and lengths of series single open-/short-circuited stubs in the unit of wavelength so that they can match a load impedance $Z_L = 30 - j40[\Omega]$ to the lossless transmission (or feed) line of characteristic impedance $Z_0 = 50\ \Omega$.

a) We can easily get the solutions by using the MATLAB function '`imp_matcher_1stub_shunt()`' as

```
>>Z0=50; ZL=30-40i; lambda=1; beta=2*pi/lambda; r=i*beta;
  [dso,lso]=imp_match_1stub_series(Z0,ZL,r,'o') % for OSs
  [dss, lss]=imp_match_1stub_series(Z0,ZL,r,'s') % for SSs
```

This yields

```
dso = 0.4583 0.2917   lso = 0.3864 0.1136
dss = 0.4583 0.2917   lss = 0.1364 0.3636
```

which means two open stubs, one of length 0.3864λ and distance 0.4583λ (from the load) and the other of length 0.1136λ and distance 0.2917λ, and also two short stubs, one of length 0.1364λ and distance 0.4583λ (from the load) and the other of length 0.3636λ and distance 0.2917λ.

There are some observations to make about the comparison of the open and short stubs:

- It does not matter to the distance d (i.e. the point of attachment) of the stub whether the stub is open-circuited or short-circuited.
- Even the difference in the stub length is just 0.25λ (a quarter wavelength), corresponding to halfway round the Smith chart (by $2\beta l = 4\pi/\lambda \times 0.25\lambda = \pi$[rad]), which is not surprising.
- If either short(-circuited) stub or open(-circuited) stub can be freely chosen, we can keep the stub length in the range of $0 \sim 0.25\lambda$. In general, open stubs are preferred on microstrip for constructional reasons while short stubs are preferred on coax line or parallel wire line because they have less radiation from the ends.

b) To find the solutions using the Smith chart, we first should recall the following:

- Our target point $\Gamma_3 = 0$ (corresponding to both R_0 and G_0) is on the $R = R_0$-circle.
- A lossless TL of length d connected to the load rotates the Γ-point clockwise by $2\beta d = 4\pi(d/\lambda)$ [rad] along the circle of radius $|\Gamma_L|$ centered at the origin $\Gamma = 0$ (Remark 7.2(1)).
- A series reactive element, a lumped one or a stub, can change only the reactance, not the resistance, so that the Γ-point can be rotated along an R-circle.

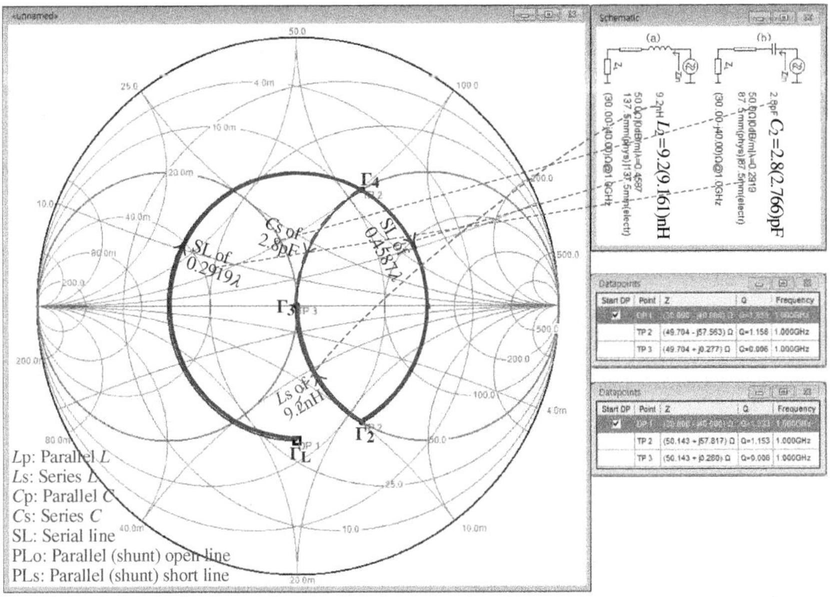

Figure 7.10 Using 'Smith 4.1' to design series single open-stub impedance matchers in Example 7.3.

That is why we first should rotate the Γ-point clockwise along the $|\Gamma_L|$-circle (passing through $Z_L = 30 - j40$ [Ω]) to some point (like Γ_2 or Γ_4 in Figure 7.10) on the $R = R_0 = 50\ \Omega$-circle. Here, the distance d of a stub (to be series-connected) from the load is to be determined from the angles by which the Γ-point has been rotated. Then, we should determine the length l of the open-circuited stub, which rotates the Γ-point clockwise along the R-circle to the target point $\Gamma_3 = 0$). With this strategy in mind, let us use 'Smith V4.1' by taking the following steps:

1) Take the first three steps as done in Example 7.2.
2) With the path Γ_L-Γ_2-Γ_3 in mind, click on 'Insert serial line' button from the tool bar and move the cross-type mouse cursor to click on the point Γ_2, which is one of the intersections between the $|\Gamma|$-circle crossing Γ_L and the $R = R_0 = 50\ \Omega$-circle (see Figure 7.10).
3) To realize the section Γ_2-Γ_3, you may look for 'Insert serial line open' button in the tool bar, but in vain, because there is nothing like that. Then, how can we move the Γ-point to $\Gamma_3(\Gamma = 0)$? Instead of a series stub, a lumped element of inductor can be connected in series with the load to rotate the Γ-point clockwise along the R-circle and that is why we select the 'Insert serial inductor' button and move the mouse cursor to click on the point $\Gamma_3 = 0$).

4) Then, the designed network with $d_1 = 0.4587\lambda$ and $L_1 = 9.2(\approx 9.161)$ nH will be shown in the Schematic pane. Can we check if this value of inductance conforms with the open-stub length $l_1 = 0.3864\lambda$ obtained using MATLAB in (a)? Yes, we can be sure of the conformity by computing the reactance value of this inductor as $X_1 = \omega_c L_1 = 2\pi f_c L_1 = 2\pi \times 10^9 \times 9.161 \times 10^{-9} = 57.56\ \Omega$ and also using Eq. (7.3.2b) to find the reactance of the open stub of length 0.3864λ as

$$X(l_1) \overset{(7.3.2b)}{=} -R_0 \cot \beta l_1 = -50 \cot(2\pi \times 0.3864) = 57.73\ \Omega \qquad \text{(E7.3.1)}$$

5) To find another series open-stub impedance matcher corresponding to another path Γ_L-Γ_4-Γ_3, right-click anywhere on the Smith chart two times to delete the existing two data points Γ_3 and Γ_2 (in LIFO order) except Γ_L, click on 'Insert serial line' button, and move the mouse cursor to click on the point Γ_4, which is another intersection between the $|\Gamma|$-circle crossing Γ_L and the $R = R_0 = 50\ \Omega$-circle (see Figure 7.10).
6) To rotate the Γ-point to $\Gamma_3 = 0$ counterclockwise along the $R = R_0 = 50\ \Omega$-circle, we select the 'Insert serial capacitor' button and move the mouse cursor to click on the point $\Gamma_3 = 0$.
7) Then, the designed network with $d_2 = 0.2919\lambda$ and $C_2 = 2.8(\approx 2.766)$ [pF] will be shown in the Schematic pane. To check if this value of capacitance conforms with the open-stub length $l_2 = 0.1136\lambda$ obtained using MATLAB in (a), we compute the reactance value of this capacitor as $X_2 = -1/\omega_c C_2 = -1/2\pi f_c C_1 = -1/(2\pi \times 10^9 \times 2.766 \times 10^{-12}) = -57.54\ \Omega$, use Eq. (7.3.2b) to find the reactance of the open stub of length 0.1136λ as

$$X(l_2) \overset{(7.3.2b)}{=} -R_0 \cot \beta l_2 = -50 \cot(2\pi \times 0.1136) = -57.73\ \Omega \qquad \text{(E7.3.2)}$$

and see that these two reactance values are close enough.

7.3.3 Double-Stub Impedance Matching

The single-stub matching introduced in the previous section requires the stub to be attached to the main line at a specific point of precise distance from the load, presenting practical difficulties because the specified point may be undesirable to attach a stub and it is very difficult to build a variable-length coaxial line with a constant characteristic impedance. An alternative is to attach two short-circuited stubs at fixed positions, as shown in Figure 7.11 where the distance d_0 (of stub A from a load) and distance d (of stub B from stub A) can be arbitrarily chosen and fixed (say, as $\lambda/16$, $\lambda/8$, $3\lambda/8$, etc.), and the lengths (l_A and l_B) of the two stubs are determined to match the load impedance Z_{L0} to the characteristic impedance R_0 of the main lossless line. This scheme is the so-called *double-stub matching (tuning)*.

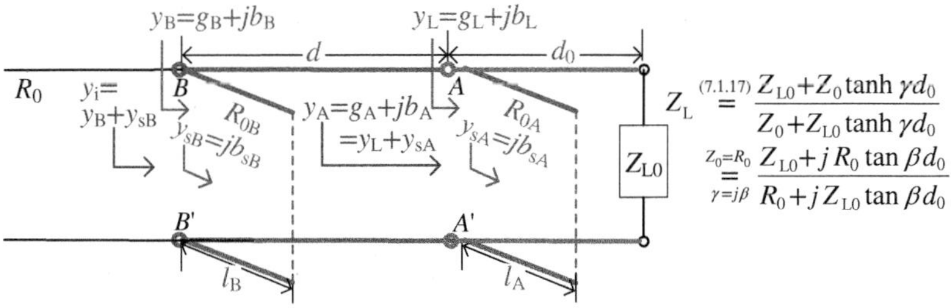

Figure 7.11 Double-stub impedance matching.

For convenience, let the serial line of length d_0 be a part of the load so that the virtual load impedance can be obtained by using Eq. (7.1.17) as

$$Z_L \overset{(7.1.17)}{=} R_0 \frac{Z_{L0} + jR_0 \tan\beta d_0}{R_0 + jZ_{L0} \tan\beta d_0} \quad \text{with } R_0 \text{ : the original load impedance} \tag{7.3.12}$$

Also, let us write the normalized impedance and admittance of the load as

$$z_L = \frac{Z_L}{R_0} = \frac{R_L + jX_L}{R_0} = r_L + jx_L = \frac{1}{R_0 Y_L} = \frac{1}{y_L} \tag{7.3.13a}$$

$$y_L = \frac{Y_L = G_L + jB_L}{Y_0} = \frac{Y_L}{G_0} = R_0 Y_L = g_L + jb_L \tag{7.3.13b}$$

where the lowercase/uppercase letters stand for normalized/unnormalized values. Then, to satisfy the impedance matching condition

$$y_i = y_B + y_{sB} = y_B + jb_{sB} = 1 \tag{7.3.14}$$

the real part (conductance) of y_B (the input admittance at terminals B-B' looking into the load excluding stub B) should be 1 since $y_{sB} = jb_{sB}$ (the admittance of stub B) has only an imaginary part:

$$g_B = \text{Re}\{y_B\} = 1 \tag{7.3.15}$$

where

$$y_B = \frac{1}{z_B} = \frac{Z_0}{Z_B} \overset{(7.1.17)}{=} \frac{R_0 + jZ_A \tan\beta d}{Z_A + jR_0 \tan\beta d} = \frac{1 + jz_A \tan\beta d}{z_A + j \tan\beta d}$$

$$= \frac{y_A + j\tan\beta d}{1 + jy_A \tan\beta d} = \frac{g_A + jb_A + j\tan\beta d}{1 + j(g_A + jb_A)\tan\beta d} = \frac{g_L + jb_A + j\tan\beta d}{1 + j(g_L + jb_A)\tan\beta d} \tag{7.3.16a}$$

```
function [lA,lB,bsA,bsB,bA]=imp_match_2stub(Z0,ZL,r,d)
% Input:  Z0 = Characteristic impedance
%         ZL = Load impedance
%          r = a+bi = Propagation constant
%          d = Distance of the stub B from A (load)
%   where d may be given as [d d0] with d0: Distance of A from load
% Output: lA = Length of stub A at (d0 from) the load
%         lB = Length of stub B at d from stub A
if numel(d)>1
  d0=d(2); d=d(1);
  ZL=Z0*(ZL+Z0*tanh(r*d0))/(Z0+ZL*tanh(r*d0)); % Eq.(7.1.16)/(7.3.12a)
end
if imag(r)<eps, r=r*i; end % Since a real-valued r must be beta.
RL=real(ZL); a=real(r); b=imag(r); lambda=2*pi/b;
yBgb=@(G,B)(G+j*B+j*tan(b*d))./(1+j*(G+j*B)*tan(b*d)); % Eq.(7.3.16a)
YL=1/ZL; yL=YL*Z0; gL=real(yL); bL=imag(yL);
eq=@(bA)real(yBgb(gL,bA))-1; % Eq.(7.3.15)
options=optimoptions('fsolve','TolX',1e-20, 'TolFun',1e-20);
bA=real([fsolve(eq,0,options) fsolve(eq,1,options)])
bsA=bA-bL; lA=acot(-bsA)/b; % Eq.(7.3.17a), Eq.(7.3.7a)
yB=yBgb(gL,bA); % Eq.(7.3.16a)
bsB=imag(yB); lB=acot(bsB)/b; % Eq.(7.3.17b), Eq.(7.3.7a)
lA=mod(lA*b,pi)/b; lB=mod(lB*b,pi)/b;
```

$$y_A = g_A + jb_A = y_L + jy_{sA} = g_L + j(b_L + b_{sA}) \tag{7.3.16b}$$

and $y_{sB} = jb_{sB}$ and $y_{sA} = jb_{sA}$ denote the admittances of stubs B and A, respectively, that depend on their lengths as described by Eq. (7.3.2). Thus, we should first find the value of b_A such that Eq. (7.3.15) can be satisfied. Then, we can determine the susceptances $\{b_{sA}, b_{sB}\}$ of stubs A and B as

$$b_{sA} = b_A - b_L \text{ and } b_{sB} = -\text{Im}\{y_B\} \tag{7.3.17a,b}$$

so that the susceptance value of y_B can be really b_{sA} and that of the overall input admittance can be cancelled out to be zero. Then, the stub lengths l_A and l_B are determined from Eq. (7.3.7). This solution process has been cast into the above MATLAB function '`imp_match_2stub()`'.

One of the drawbacks of the double-stub matching is that it can be used for a certain range of load impedances satisfying[M-1]

$$0 \le g_L < \csc(\beta d) = \frac{1}{\sin\beta d} \overset{(7.2.4)}{\rightarrow} \frac{1-\Gamma_r^2-\Gamma_i^2}{(1+\Gamma_r)^2+\Gamma_i^2} < \frac{1}{\sin\beta d} \text{ on the Smith chart} \tag{7.3.18}$$

Still it is not so fatal because it is easy to make this condition satisfied by using a serial line (of a certain length d_0) between the load and the first stub as long as it is allowed.

Example 7.4 Design of Double Stubs for Impedance Matching

Determine the lengths of double-stub matchers with $d = \lambda/8$ and $d_0 = 0$ in the unit of wavelength λ so that they can match a load impedance $Z_L = 60 - j80$ [Ω] to the lossless transmission (or feed) line of characteristic impedance $Z_0 = 50\ \Omega$. Note that $z_L = Z_L/Z_0 = 1.2 - j1.6[\Omega]$ and $y_L = 1/z_L = 0.3 + j0.4$[S].

a) We can easily get the solutions by using the MATLAB function '`imp_matcher_2stub()`' as

```
>>Z0=50; ZL=60-80i; lambda=1; r=2*pi*i/lambda; d=1/8; d0=0;
  [lA,lB]=imp_match_2stub(Z0,ZL,r,[d d0]) % for SSs
```

This yields

```
lA = 0.2319    0.3965
lB = 0.0998    0.4542
```

which means two double short stubs, $\{l_A = 0.2319\lambda,\ l_B = 0.0998\lambda\}$ and $\{l_A = 0.3965,\ l_B = 0.4542\lambda\}$. If you want open stubs, add/subtract 0.25λ depending on whether the stub is shorter/longer than 0.25λ.

b) To find the solutions using the Smith chart, we first should recall the following:

- Our target point $\Gamma = 0$ (corresponding to both R_0 and G_0) is on the $R = R_0$-circle.
- As a data point (just before the target point $\Gamma = 0$), we need a point, say, Γ_B, corresponding to y_B (the input admittance at terminals B–B' looking into the load excluding stub B) on the $G = G_0 = 20$ [mS]-circle or $g = g_0 = 1$[S]-circle (along which the Γ-point rotates to reach $\Gamma = 0$ by the effect of the shunt stub B).
- The data point Γ_B should be the $-2\beta d = -4\pi/\lambda \times \lambda/8 = -\pi/2$ [rad]-rotated version (due to the serial line of length $d = \lambda/8$ according to Remark 7.2(1)) of another data point, say, Γ_A, corresponding to y_A (the input admittance at A-A'), which should be on the $G = G_L = 6$ [mS]-circle or $g = g_L = 0.3$ [S]-circle because the shunt stub A changes only the susceptance, not the conductance.
- It is easier to rotate the $G = G_0 = 20$ [mS]-circle (by $+\pi/2$ [rad] (counterclockwise)) than to rotate the $G = G_L = 6$ [mS]-circle (by $-\pi/2$ [rad] (clockwise)).

Thus, we are going to rotate the $G = G_0 = 20$[mS]-circle by $+\pi/2$[rad] (counterclockwise), find its intersection with the $G = G_L$=6 [mS]-circle as Γ_A, and rotate Γ_A back by $-\pi/2$[rad] (clockwise) to get Γ_B. Then, we can get the length l_A of stub A from the difference between the susceptances b_A (corresponding to Γ_A) and b_L (of the load). Also, we can get the length l_B of stub B from the susceptance b_B

(corresponding to Γ_B). This idea is implemented in the following MATLAB script "elec07e04.m," which has been run to yield the same result as obtained in (a):

```
GamA = -0.4382 - 0.7407i    0.4675 - 0.3227i
lA = 0.3965    0.2319
GamB = -0.7407 + 0.4382i  -0.3227 - 0.4675i
lB = 0.4542    0.0998
```

```
%elec07e04.m
Z0=50; Y0=1/Z0; % Characteristic impedance/admittance
R0=real(Z0); G0=real(Y0);
ZL=60-80i; % Load impedance
zL=ZL/Z0; rL=real(zL); xL=imag(zL);
yL=1/zL; gL=real(yL); bL=imag(yL);
lambda=1; b=2*pi/lambda; d=1/8;
% To use the Smith chart
[cG0,rG0]=Gcircle(G0,Z0); % G0-circle
[cGL,rGL]=Gcircle(gL,1); % GL-circle
cG0_r=cG0*exp(j*2*b*d); % Center of the rotated G0-circle
% To find the intersections of the GL-circle and the rotated G0-circle.
cG0_r_R=real(cG0_r); cG0_r_I=imag(cG0_r);
[GamA_R,GamA_I]=circcirc(cG0_r_R,cG0_r_I,rG0,real(cGL),imag(cGL),rGL);
GamA=GamA_R+i*GamA_I % Two GammaA points
G2Z=@(G,Z0)Z0*(1+G)./(1-G); % Eq.(7.1.12)
yA=Z0./G2Z(GamA,Z0); % Normalized admittance at A
bsA=imag(yA)-bL; % Eq.(7.3.17a): Susceptance of stub A
lA=acot(-bsA)/b; lA=mod(lA*b,pi)/b % Eq.(7.3.7a)
GamB=GamA*exp(-j*2*b*d) % Two GammaB points with the d[m]-serial line
yB=Z0./G2Z(GamB,Z0); % Normalized admittance at B
bsB=-imag(yB); % Eq.(7.3.17b): Susceptance of stub B
lB=acot(-bsB)/b; lB=mod(lB*b,pi)/b % Eq.(7.3.7a)
```

Additionally, we have got two intermediate points $\Gamma_A = 0.4675 - j0.3227$ and $\Gamma_B = -0.3227 - j0.4675$, that can be of a help for working with 'Smith V4.1'.

Now, let us work with 'Smith V4.1' taking the following steps:

1) From the top menu bar, select Tools > Settings to open the Settings dialog box (Figure 7.5(b)), set the values of Z_0 (characteristic impedance) and frequency to 50[Ω] and 1[GHz], respectively, and click OK button.
2) From the toolbar (below the menu bar), select Keyboard to open Data Point dialog box, insert $Z_L = 60 - j80$ [Ω] as a data point, and click OK button to close the Data Point dialog box.
3) Select Keyboard again to open Data Point dialog box, insert $\Gamma_A = 0.4675 - j0.3227$ as a data point, and click OK button to close the Data Point dialog box. It seems to be difficult to draw auxiliary circles like a rotated version of standard constant-parameter circle.

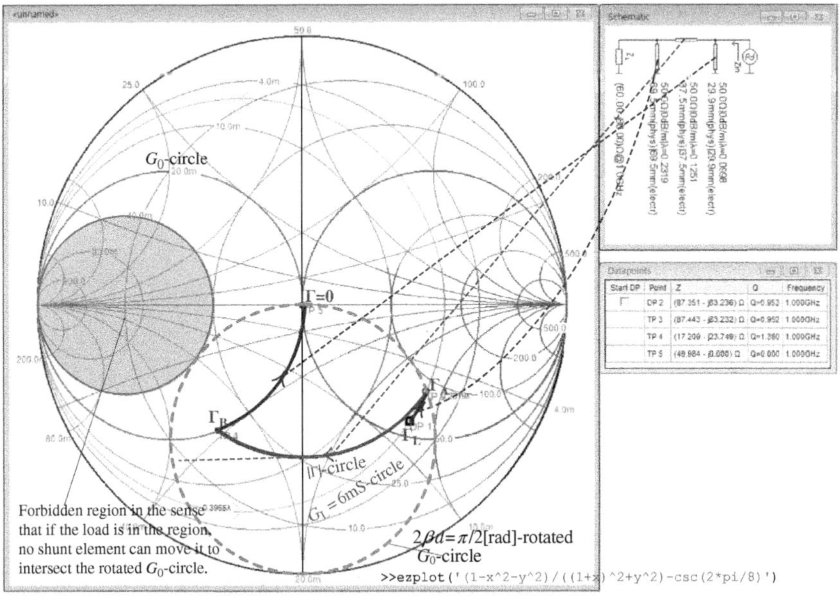

Figure 7.12 Using 'Smith 4.1' to design a double-stub tuner (impedance matcher) in Example 7.4.

4) From the tool bar, click on 'Insert parallel line shorted' button (denoted by SS: short stub) and move the cross-type mouse cursor to click on the point Γ_A (DP2) (see Figure 7.12).
5) From the tool bar, click on 'Insert serial line' button and move the mouse cursor along the automatically appearing $|\Gamma|$-circle by $d = \lambda/8$ to click on an intersection point Γ_B with the $G = G_0 = 20$ [mS]-circle.
6) From the tool bar, click on 'Insert parallel line shorted' button and move the mouse cursor to click on $\Gamma = 0$.
7) Then, the designed double-stub circuit with $l_A = 0.2319\lambda$ and $l_B = 0.0998\lambda$ will be shown in the Schematic pane.

Note that there is some *forbidden region* (Eq. (7.3.18)) in the sense that if the reflective coefficient of a load is in the region, no shunt element can move it to intersect the rotated G_0-circle.

7.3.4 The Quarter-Wave Transformer

As mentioned in Section 7.1, a real load impedance R_L can be matched to a feed line of characteristic impedance Z_0 by using a $\lambda/4$-length lossless line (Figure 7.13(a)), called a *QWT* (*quarter-wave* ($\lambda/4$) (*impedance*) *transformer*), with characteristic impedance Z_1

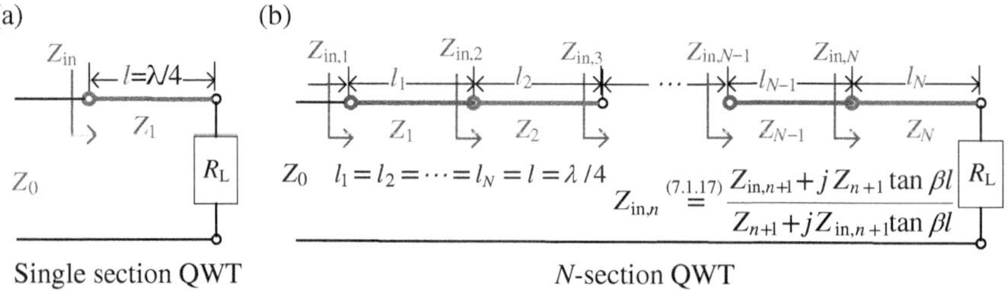

Figure 7.13 An *N*-section impedance transformer.

$$Z_1 = \sqrt{Z_0 R_L} \tag{7.3.19}$$

which transforms the load impedance Z_L into

$$\begin{aligned} Z_{in}(l) &\overset{(7.3.1)}{\underset{(7.1.17)}{=}} Z_1 \frac{R_L + jZ_1 \tan(\beta l)}{Z_1 + jR_L \tan(\beta l)} \overset{\beta=2\pi/\lambda}{\underset{l=\lambda/4}{=}} Z_1 \frac{R_L + jZ_1 \tan \pi/2}{Z_1 + jR_L \tan \pi/2} \\ &= \frac{Z_1^2}{R_L} \overset{(7.3.19)}{=} \frac{\left(\sqrt{Z_0 R_L}\right)^2}{R_L} = Z_0 \end{aligned} \tag{7.3.20}$$

With this QWT, the reflection coefficient will be zero at $\beta_0 = 2\pi/\lambda_0$ (for a certain wavelength λ_0) so that there is no reflected wave beyond the point distant from the load by $\lambda_0/4$. However, at other frequency with $\beta l \neq \pi/2$, the *reflection coefficient response* will be

$$\Gamma_{in}(\beta l) \overset{(7.1.11)}{=} \frac{Z_{in} - Z_0}{Z_{in} + Z_0} \overset{(7.1.17)}{=} \frac{Z_1 \dfrac{R_L + jZ_1 \tan\beta l}{Z_1 + jR_L \tan\beta l} - Z_0}{Z_1 \dfrac{R_L + jZ_1 \tan\beta l}{Z_1 + jR_L \tan\beta l} + Z_0} \overset{(7.3.19)}{=} \frac{R_L - Z_0}{R_L + Z_0 + j2\sqrt{Z_0 R_L}\tan\beta l} \tag{7.3.21a}$$

$$\rho_{in}(\beta l) = |\Gamma_{in}|(\beta l) \overset{(7.3.21a)}{=} \frac{R_L - Z_0}{\sqrt{(R_L + Z_0)^2 + 4Z_0 R_L \tan^2 \beta l}} (\beta l: \text{Electrical length}) \tag{7.3.21b}$$

Now, we consider a multisection transformer connected between a TL of characteristic impedance Z_0 and a load R_L as shown in Figure 7.13 where Z_n denotes the characteristic impedance of each section n. The impedance matching occurs if

$$Z_{in,1} = Z_0 \tag{7.3.22}$$

where $Z_{in,n}$ is the input impedance of the nth section that can be obtained from Eq. (7.1.17) with the input impedance of the $(n + 1)$th section as the load impedance as

$$Z_{\text{in},n} \overset{(7.1.17)}{=} \frac{Z_{\text{in},n+1} + jZ_n \tan\beta l}{Z_n + jZ_{\text{in},n+1} \tan\beta l} Z_n \text{ with } Z_{\text{in},N+1} = R_L \text{ for } n = 1, \ldots, N \tag{7.3.23}$$

There are two classical N-section transformers, the binomial transformer and the Chebyshev transformer where the reflection coefficient of the former over the frequency βd is flat while that of the latter has equiripples as a result of trying to minimize the maximum deviation for better response.

The following MATLAB function 'multi_line()' returns the input impedance and reflection coefficient response of a QWT with characteristic impedance vector, etc., given as input arguments.

```
function [Zin,Gam,bb]=multi_line(Z0s,ls,Z0,ZL,b)
% To find the input reflection coefficient and impedance of a multi-line
% consisting of multisections each of impedance Z0s(m) and length ls(m).
% Input:
%  Z0s = Vector of characteristic impedances of each section of line
%  ls = Length(s) of each section, which is a scalar for equal length.
%  Z0 = Characteristic impedance of transmission line
%  ZL = Load resistance
%  b = Phase constant (a number or a vector with 2*pi*fc/c as 1st entry)
%      where fc=cutoff frequency, c=light velocity
% Output:
%  Zin = Input impedance, Gam = Reflection coefficient
N=numel(Z0s);
if numel(ls)<2, ls=repmat(ls,1,N); end
if numel(b)>1, bb=b(2:end); b=b(1); else bb=pi/90*[0:360]; end
Zl=@(ZL,Z0,bl)(ZL+j*Z0*tan(bl))./(Z0+j*ZL.*tan(bl))*Z0;
Zi(N+1)=ZL; Zibl=ZL;
for n=N:-1:1
  Zi(n)=Zl(Zi(n+1),Z0s(n),b*ls(n)); % Eq.(7.3.23)
  Z0s_(n)=sqrt(Zi(n)*Zi(n+1)); % Eq.(7.3.19)
  Zibl=Zl(Zibl,Z0s(n),bb*ls(n)); % Cascaded reflection response
end
Zin=Zi(1); Gam=(Zibl-Z0)./(Zibl+Z0); % Eq.(7.3.21b) Ref coef response
Mismatching = norm(Z0s-Z0s_)
```

7.3.4.1 Binomial Multisection QWT

According to the design method of binomial N-section QWT introduced in [P-1], the characteristic impedance of each section can be approximately determined in turn from the first one as

$$\ln Z_n = \ln Z_{n-1} + 2^{-N} C_{n-1}^N \ln\frac{Z_L}{Z_0} \text{ with } C_n^N = \binom{N}{n} = \frac{N!}{(N-n)!n!} \tag{7.3.24}$$

This design process has been cast into the following MATLAB function 'binomial_QWT()'.

```
function Z0s=binomial_QWT(Z0,ZL,N)
% To determine characteristic impedances of Nth-order binomial QWT
% Input : Z0 = Characteristic impedance of the transmission line
%         RL = Load resistance, N = Number of sections
% Output: Z0s = Characteristic impedances of each section of binomial QWT
A=2^-N*(ZL-Z0)/(ZL+Z0); Zp=Z0;
for n=N:-1:1
  logZn=log(Zp)+2^-N*nchoosek(N,N-n)*log(ZL/Z0); % Eq.(7.3.24)
  Zp=real(exp(logZn)); Z0s(N+1-n)=Zp;
end
```

7.3.4.2 Chebyshev Multisection QWT

The following MATLAB function 'Cheby_QWT()' uses the functions 'chebtr2()' and 'chebtr3()' implementing the Collin/Riblet algorithm (presented in [O-1]) to design Chebyshev QWTs. It regards the fourth input argument as a desired value of (minimum) fractional bandwidth or a required (minimum) attenuation [dB] in the reflectionless band (relative to unmatched case) depending on whether its value is less than 1 or not.

```
function Z0s=Cheby_QWT(Z0,ZL,N,A_fB)
% To determine the characteristic impedance of Nth-order Chebyshev QWT
% Input:
%  Z0   = Characteristic impedance of the transmission line
%  RL   = Load resistance
%  N    = Number of sections
%  A_fb = Attenuation or Fractional bandwidth
%         depending on whether A_fb >=1 or not.
% Output:
%  Z0s = Characteristic impedances of each section of Chebyshev QWT
% This uses the MATLAB functions chebtr2() and chebtr3() presented in
%  "Electromagnetic Waves and Antennas" by Sophocles J. Orfanidis,
%  which is available at <http://www.ece.rutgers.edu/~orfanidi/ewa/>,
%  much appreciated.
if A_fB<1, Y0s=chebtr2(1/Z0,1/ZL,N,A_fb); % with A_fb=fb/f0 (fractional BW)
   else Y0s=chebtr3(1/Z0,1/ZL,N,A_fb); % with A_fb=As (attenuation)
end
Z0s=1./Y0s(2:end-1);
```

We can run the following MATLAB script "test_QWT.m" to use 'binomial_QWT()' and 'Cheby_QWT()' to design 1,2,3,4-section binomial and Chebyshev QWTs and use 'multi_line()' to plot their reflection coefficient responses as shown in Figure 7.14, which shows the following:

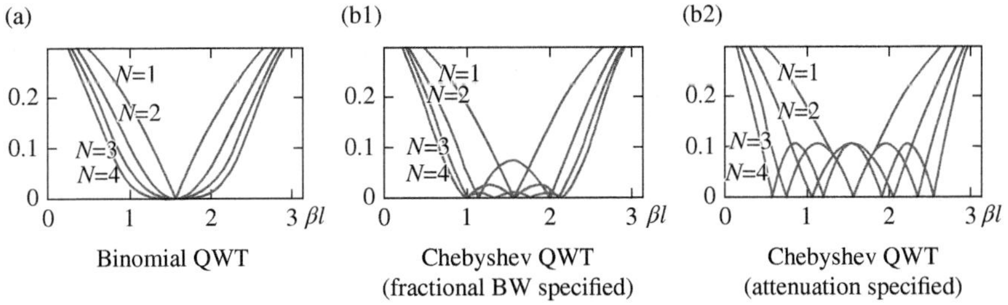

Figure 7.14 Reflection coefficient responses of binomial and Chebyshev QWTs.

- Chebyshev QWTs ((b1) and (b2)) have wider bandwidth than binomial QWTs ((a)).
- Chebyshev QWTs with the fractional bandwidth (BW) specified have smaller ripples as the number N of sections increases.
- Chebyshev QWTs with the attenuation specified have larger attenuations paying the cost of larger equiripples.

```
%test_QWT.m
Z0=50; ZL=100; % Line characteristic impedance and Load impedance
fb=0.8; % Required fractional bandwidth (f2-f1)/fc where fc=(f1+f2)/2
A=10; %Attenuation[dB] of reflectionless band (relative to unmatched case)
b=2*pi; l=1/4; % Phase constant and Length of each section
for N=1:5
  Z0sb=binomial_QWT(Z0,ZL,N) % Binomial QWT
  [Zin,Gam,bb]=multi_line(Z0sb,l,Z0,ZL,b);
  subplot(231), plot(bb*l,abs(Gam)), hold on
  Z0sCf=Cheby_QWT(Z0,ZL,N,fb) % Chebyshev QWT with fractional BW specified
  [Zin,Gam,bb]=multi_line(Z0sCf,l,Z0,ZL,b);
  subplot(232), plot(bb*l,abs(Gam)), hold on
  Z0sCa=Cheby_QWT(Z0,ZL,N,A) % Chebyshev QWT with attenuation specified
  [Zin,Gam,bb]=multi_line(Z0sCa,l,Z0,ZL,b);
  subplot(233), plot(bb*l,abs(Gam)), hold on
end
```

7.3.5 Filter Implementation Using Stubs[P-1]

The LC ladder realization using lumped elements discussed in Section 6.3.1 has two problems if they are going to be used at microwave frequencies. First, lumped elements such as inductors and capacitors are available only for a limited range of values. Second, the distances between components is not negligible at microwave frequencies. For these reasons, distributed components like short- or open-circuited stubs (TL sections) are used. To implement an LC ladder using stubs, we use Richard's transformation to convert lumped elements to stubs and then use Kuroda's identities to transform series/parallel stubs into parallel/series stubs.

<Richard's Transformation>

To understand the Richard's transformation, we need to recall Eqs. (7.3.2a,b) that the impedances of short- and open-circuited stubs with characteristic impedance Z_0 are

$$Z_{ss}(l) \underset{Z_L=0}{\overset{(7.3.1)}{=}} jZ_0 \tan \beta l \quad \text{and} \quad Z_{os}(l) \underset{Z_L=\infty}{\overset{(7.3.1)}{=}} -j\frac{Z_0}{\tan \beta l} \tag{7.3.25a,b}$$

and therefore, with characteristic impedance

$$Z_0 = L \quad \text{and} \quad Z_0 = 1/C \tag{7.3.26a,b}$$

a short-circuited stub (SS) and an open-circuited stub (OS) of length $l = \lambda_0/8$ ($\lambda_0 = v_p/f_0$) can act as an inductor of impedance $j\Omega L$ and a capacitor of impedance $1/j\Omega C$, that is,

$$Z_{ss}(\Omega) = j\Omega L = jL \tan \beta l \quad \text{and} \quad Z_{os}(\Omega) = \frac{1}{j\Omega C} = \frac{1}{jC \tan \beta l} \tag{7.3.27a,b}$$

respectively, at a frequency

$$\Omega = \tan \beta l = \tan\left(\frac{2\pi}{\lambda}\frac{\lambda_0}{8}\right) = \tan\left(\frac{\pi}{4\lambda}\frac{v_p}{f_0}\right) = \tan\left(\frac{\pi f}{4f_0}\right) \tag{7.3.28}$$

Here v_p is the phase velocity of signal propagating on the line and f_0 is some reference frequency, which is often the cutoff frequency f_c of a filter implemented using the SSs and OSs, corresponding to

$$\Omega = \tan\left(\frac{\pi f}{4f_0}\right) \overset{f=f_0}{=} 1 \tag{7.3.29}$$

just like the unity cutoff frequency of a low-pass filter (LPF) prototype. Since the tangent function is periodic in its argument with period π, the impedances of the SSs and OSs are periodic in f with period $4f_0$. Here, replacing inductance L and capacitance C by SS (with characteristic impedance $Z_0 = L$) and OS (with characteristic impedance $Z_0 = 1/C$), respectively, through Eq. (7.3.26) based on Eq. (7.3.28) (see Figure 7.15) is referred to as the *Richard's transformation* where their impedances are regarded as Eq. (7.3.27a,b).

Note that in principle, the inductors and capacitors of a lumped-element filter can be replaced with SSs and OSs, respectively, and if all the stubs are of equal length, they are said to be *commensurate*.

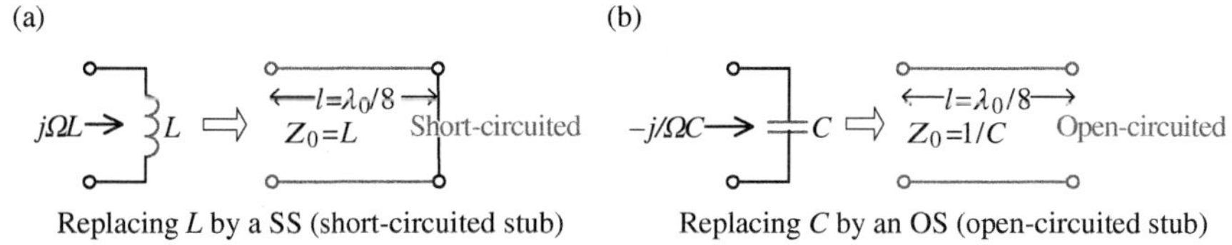

Figure 7.15 Richard's transformation.

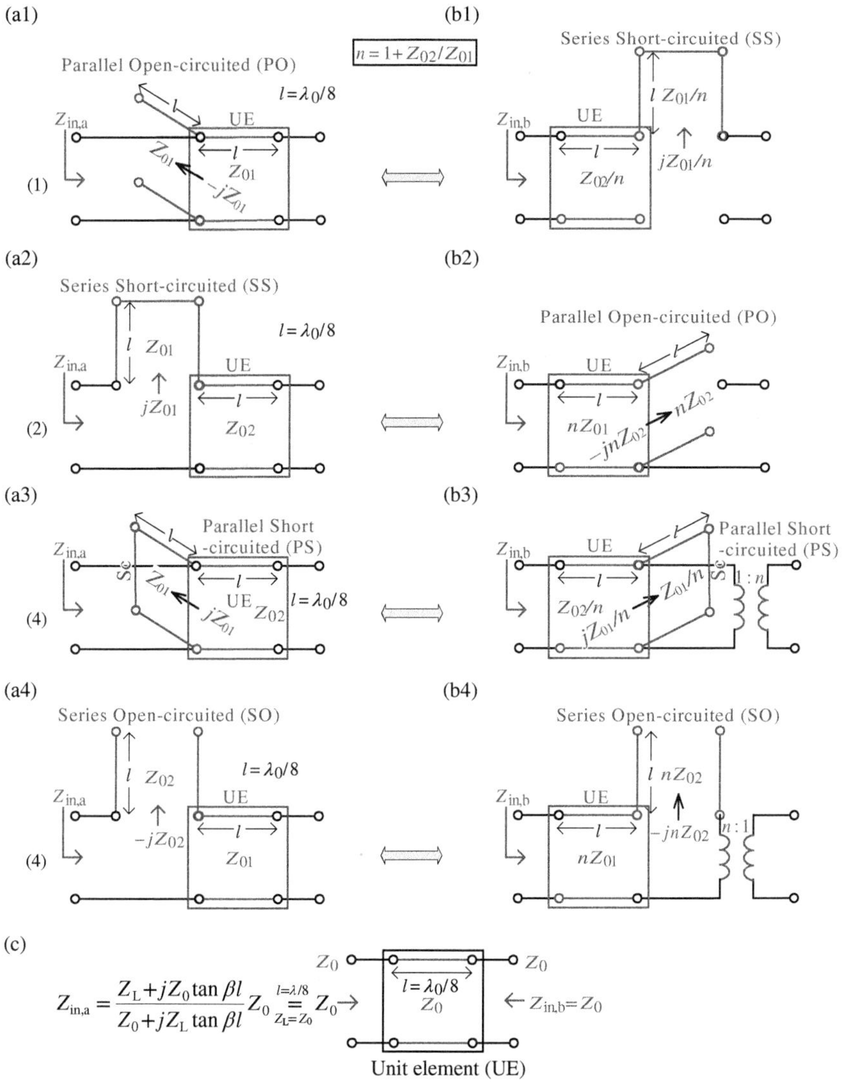

$$Z_{\text{in,a}} = \frac{Z_L + jZ_0 \tan \beta l}{Z_0 + jZ_L \tan \beta l} Z_0 \underset{Z_L = Z_0}{\overset{l=\lambda/8}{=}} Z_0$$

Figure 7.16 Kuroda's identities.

<Kuroda's Identities>

Figure 7.16(a1)-(b4) shows the four Kuroda's identities that can be used to transform series/parallel stubs into parallel/series stubs, physically separate stubs, and change impractical characteristic impedances into more realizable ones without changing any two-port network property such as the input and output impedances. To understand and get familiar with them, let us show the equivalence of each pair of two-port networks in terms of the input impedance by running the following MATLAB script "show_Kuroda.m" and checking if $Z_{\text{in,a}} = Z_{\text{in,b}}$ for the four identities where the ideal transformer with turns ratio of 1: n in Figure 7.16(b3) makes the load impedance Z_L (on the secondary [load] side) reflected onto the primary (source) side as Z_L/n^2:

$$Z_{\text{in,a}} \overset{f=f_0}{=} (-jZ_{02})\|\left(Z_{01}\frac{Z_L+jZ_{01}}{Z_{01}+jZ_L}\right) \overset{?}{\equiv} Z_{\text{in,b}} \overset{f=f_0}{=} \left(\frac{Z_{02}}{n}\right)\frac{j(Z_{01}/n)+Z_L+j(Z_{02}/n)}{Z_{02}/n+j\{j(Z_{01}/n)+Z_L\}} \tag{7.3.30a,b}$$

$$Z_{\text{in,a}} \overset{f=f_0}{=} jZ_{01}+Z_{02}\frac{Z_L+jZ_{02}}{Z_{02}+jZ_L} \overset{?}{\equiv} Z_{\text{in,b}} \overset{f=f_0}{=} nZ_{01}\frac{(-jnZ_{02})\|Z_L+jnZ_{01}}{nZ_{01}+j\{(-jnZ_{02})\|Z_L\}} \tag{7.3.31a,b}$$

$$Z_{\text{in,a}} \overset{f=f_0}{=} (jZ_{01})\|\left(Z_{02}\frac{Z_L+jZ_{02}}{Z_{02}+jZ_L}\right) \overset{?}{\equiv} Z_{\text{in,b}} \overset{f=f_0}{=} \left(\frac{Z_{02}}{n}\right)\frac{(jZ_{01}/n)\|(Z_L/n^2)+j(Z_{02}/n)}{Z_{02}/n+j\{j(jZ_{01}/n)\|(Z_L/n^2)\}} \tag{7.3.32a,b}$$

$$Z_{\text{in,a}} \overset{f=f_0}{=} -jZ_{02}+Z_{01}\frac{Z_L+jZ_{01}}{Z_{01}+jZ_L} \overset{?}{\equiv} Z_{\text{in,b}} \overset{f=f_0}{=} nZ_{01}\frac{-jnZ_{02}+n^2Z_L+jnZ_{01}}{nZ_{01}+j\{-jnZ_{02}+n^2Z_L\}} \tag{7.3.33a,b}$$

```
%show_Kuroda.m
%Z1=@(ZL,Z0,bl)(ZL+j*Z0*tan(bl))./(Z0+j*ZL.*tan(bl))*Z0;
Zu=@(ZL,Z0)(ZL+j*Z0)./(Z0+j*ZL)*Z0; % with bl=pi/4
syms Z1 Z2 ZL; n=1+Z2/Z1; %
% Kuroda Identity 1 (Table 8.7 of [P-1])
Zina=parallel_comb([-j*Z2 Zu(ZL,Z1)]) % Fig.7.16(a1)
Zinb=Zu(j*Z1/n+ZL,Z2/n), simplify(Zina-Zinb) % Fig.7.16(b1)
% Kuroda Identity 2 (Table 8.7 of [P-1])
Zina=j*Z1+Zu(ZL,Z2) % Fig.7.16(a2)
Zinb=Zu(parallel_comb([-j*n*Z2 ZL]),n*Z1) % Fig.7.16(b2)
simplify(Zina-Zinb)
% Kuroda Identity 3 (Table 8.7 of [P-1])
Zina=parallel_comb([j*Z1 Zu(ZL,Z2)]) % Fig.7.16(a3)
Zinb=Zu(parallel_comb([j*Z1/n ZL/n^2]),Z2/n) % Fig.7.16(b3)
simplify(Zina-Zinb)
% Kuroda Identity 4 (Table 8.7 of [P-1])
Zina=-j*Z2+Zu(ZL,Z1) % Fig.7.16(a4)
Zinb=Zu(-j*n*Z2+n^2*ZL,n*Z1), simplify(Zina-Zinb) % Fig.7.16(b4)
```

```
function [Z0,Z2,SPso]=Kuroda(Z1,SPso,Z0)
% SPso='Ss' for Series-to-Parallel of Z1(L) or Z1(short)
% SPso='So' for Series-to-Parallel of Z1(C) or Z1(open)
% SPso='Ps' for Parallel-to-Series of Z1(L) or Z1(short)
% SPso='Po' for Parallel-to-Series of Z1(C) or Z1(open)
% where Z0 is the characteristic impedance of a UE (unit element)
% Input:  Z1 = Impedance (Reactance) of the other element than UE
%         Z0 = Characteristic impedance of UE
% Output: Z0 = Characteristic impedance of UE
%         Z2 = Impedance (Reactance) of the other element than UE
if nargin<3|~isnumeric(Z0)|Z0==0, Z0=1; end
if nargin<2, SPsp='Ss'; end
SPso=upper(SPso);
if SPso(1)=='S' % Series-to-Parallel conversion
 if SPso(2)=='S', SPso='Po'; n=1+Z0/Z1; Z2=-n*Z0; Z0=n*Z1; % Kuroda 2
  else SPso='Po'; Z1=abs(Z1); n=1+Z1/Z0; Z0=n*Z0; Z2=[-n*Z1 n]; % K-4
 end
 % The +/- sign of Z2 means L(short stub)/C(open stub), respectively.
else % Parallel-to-Series conversion
 if SPso(2)=='S', SPso='Ss'; n=1+Z0/Z1; Z0=Z0/n; Z2=[Z1/n 1/n]; % K-3
  else SPso='Ss'; Z1=abs(Z1); n=1+Z1/Z0; Z2=Z0/n; Z0=Z1/n; % K-1
 end
end
```

Figure 7.16(c) shows a unit element (UE) of characteristic impedance Z_0 and length $l = \lambda_0/8$ ($\lambda_0 = v_p/f_0$) or equivalently, electrical (phase) length $\beta l = \pi/4$ that has input impedances at ports a and b as

$$Z_{\text{in,a}} = \frac{Z_{\text{L,b}} + jZ_0 \tan\beta l}{Z_0 + jZ_{\text{L,b}} \tan\beta l} Z_0 \underset{Z_{\text{L,b}}=Z_0}{\overset{l=\lambda/8}{=}} Z_0 \text{ and } Z_{\text{in,b}} = \frac{Z_{\text{L,a}} + jZ_0 \tan\beta l}{Z_0 + jZ_{\text{L,a}} \tan\beta l} Z_0 \underset{Z_{\text{L,a}}=Z_0}{\overset{l=\lambda/8}{=}} Z_0 \tag{7.3.34a,b}$$

This implies that the frequency characteristic of a TL is not affected at frequency $f_0 = v_p/\lambda_0$ by the (cascade) insertion of UE (a $\lambda_0/8$-long TL) if only its characteristic impedance is equal to that of the TL. That is why UEs are inserted between a TL and a distributed or lumped element as a preparation for applying some Kuroda's identity.

Example 7.5 Design of LPF Using Stubs [P-1]

With source/load impedances $R_s = 50\ \Omega/Z_L = 50\ \Omega$, design a third-order Chebyshev (equiripple) LPF with passband ripple $R_p = 3$ dB and $f_c = 4$ GHz for fabrication using microstrip lines with shunt stubs only.

To design such an LPF, let us take the following steps:

(Step 1) Determine the normalized LPF prototype element values by running the following MATLAB statements:

```
>>Rs=50; % Source (line) characteristic impedance
  N=3; type='C'; Rp=3; % for 3rd order Chebyshev with Rp=3dB
  gg=LPF_ladder_coef(N,type,Rp)
```

This yields

```
gg = 3.3489e+00 7.1167e-01 3.3489e+00 1.0000e+00
```

(Step 2) Determine the normalized *LC* values of the prototype LPF by running the following MATLAB statements:

```
>>Rs0=1; pT='T'; wc=1; % for a T-type prototype filter
  [LCRs,LCRs_,Gs]=LC_ladder(Rs0,gg,pT,wc);
  LCRs, LCRs_, pretty(simplify(Gs))
```

This yields

```
LCRs  = 3.3489 0.7117 3.3489 1.0000
LCRs_ = 'L(h)' 'C(v)' 'L(h)' 'R(v)'
570899077082383952423314387779798054553098 6496/
                                                3
(45566839819273848699938746290718052889107427673 s
                                                2
+ 27212766570123614952226532668306611436833996800 s
+ 42300936562233083587497427629177459394738126848 s
+ 11417981541647679048466287755595961091061972992)
```

This, with 'h' = horizontal and 'v' = vertical, means the filter of Figure 7.17.1.

(Step 3) Convert the *L*/*C* values into the impedances of short-/open-circuited stubs by running

```
>>[Zs,SPsos]=LC2Z(LCRs,LCRs_)
```

where the following MATLAB function 'LC2Z()' is supposed to be saved on the computer.

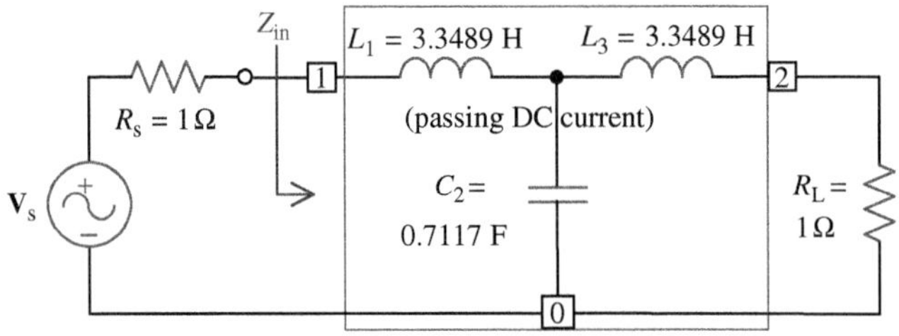

Figure 7.17.1 Normalized prototype low-pass filter (LPF) of order $N = 3$.

```
function [Zs,SPsos]=LC2Z(LCs,LCs_)
% To convert the L/C's into short/open-circuited stubs
for m=1:numel(LCs)
 if LCs_{m}(1)=='L'
   Zs(m)=LCs(m); % Eq.(7.3.26a)
   if LCs_{m}(3)=='h', SPsos{m}='Ss'; else SPsos{m}='Ps'; end
  elseif LCs_{m}(1)=='C'
   Zs(m)=-1/LCs(m); % Eq.(7.3.26b)
   if LCs_{m}(3)=='h', SPsos{m}='So'; else SPsos{m}='Po'; end
  else break;
 end
end
```

This yields the characteristic impedances of the Ss (series short-circuited stub), Po (parallel open-circuited stub), and Ss constituting the microstrip filter in Figure 7.17.2(a):

```
Zs = 3.3489  -1.4052  3.3489,  SPsos = 'Ss'  'Po'  'Ss'
```

where the +/− signs stand for L/C or short/open-circuited stubs, respectively.

(Step 4) Now, to use Kuroda identity 2 to convert the two SSs (series short-circuited stubs) Z_1 and Z_3 into POs (parallel open-circuited stubs), we insert UEs between Z_1/Z_2 and the TL (as depicted in Figure 7.17.2(b)) and run the following statements:

```
>>[Z0s(1),Zs(1),SPsos{1}]=Kuroda(Zs(1),SPsos{1});
  [Z0s(3),Zs(3),SPsos{3}]=Kuroda(Zs(3),SPsos{3});
  Z0s, Zs, SPsos
```

This yields the characteristic impedances of the UEs and stubs

```
Z0s =  4.3489        0     4.3489
Zs = -1.2986  -1.4052   -1.2986
SPsos = 'Po'     'Po'      'Po'
```

With the +/− signs of Zs values standing for L/C or short-/open-circuited stubs, respectively, this implies the microstrip filter in Figure 7.17.2(c) where, say, Zs(1) = −1.2986 together with SPsos(1) = Po means that the first stub of characteristic impedance 1.2986 Ω is a parallel open-circuited one and, say, Z0s(1) = 4.3489 means that the characteristic impedance of the UE next to the first stub is 4.3489 Ω.

(Step 5) Denormalize the normalized characteristic impedances by multiplying them with R_s = 50 Ω:

```
>>Z0s=Z0s*Rs, Zs=Zs*Rs
```

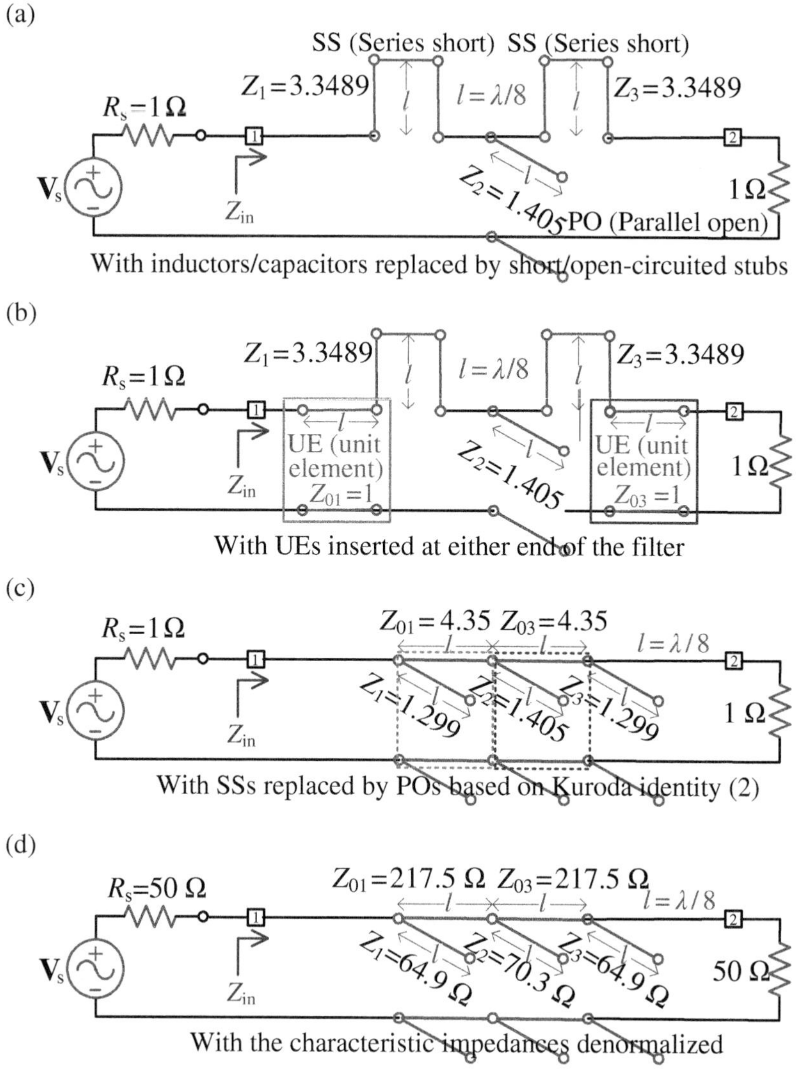

Figure 7.17.2 Microstrip filter design for Example 7.5.

This yields

```
Z0s = 217.4466       0  217.4466
Zs =  -64.9301  -70.2575  -64.9301
```

This implies the microstrip filter in Figure 7.17.2(d).

(Step 6) Frequency scale the circuit by determining the line/stub lengths as $l = \lambda/8 = v_p/f_c/8$[m] where v_p[m/s] is the phase velocity of signal propagating on the line and f_c is the cutoff frequency given as a design specification.

(Step 7) To plot the frequency responses of the two filters, one (called a lumped-element filter) consisting of lumped elements L/C (Figure 7.17.1) and the other (called a distributed-element filter) consisting of microstrip lines/stubs (Figure 7.17.2(d)), we do the following:

1) We first construct a vector of the characteristic impedances of each stub/line:

```
>>Z0strip=[Zs(1) Z0s(1) Zs(2) Z0s(3) Zs(3)];
```

and the corresponding cell

```
>>SPsos={SPsos{1} 'Tl' SPsos{2} 'Tl' SPsos{3}};
```

which contains one of the five strings {'Po', 'Tl', 'Po', 'Tl', 'Po'} for each stub/line depending on whether it is a parallel-/series-connected stub or a section of TL and also whether it is short- or open-circuited if it is a stub.

2) Referring to Eqs. (7.3.27) and (7.3.28), set the frequency range which to plot the frequency response for:

```
>>l=1/8; fc=4e9; % Stub length and Cutoff frequency
  fN=linspace(0,pi/4,1000); % Normalized frequency range
  w=fN*2*pi*fc; % Real frequency range
  bl=2*pi*w*l/fc; % Range of bl (bl=2*pi*l at f=fc=4e9[Hz])
```

3) To get the overall frequency response considering R_s, we use the MATLAB function 'GZ_strip()' to get the frequency response $G(\omega)$ (Gw) and input impedance $Z_{in}(\omega)$ (Zinw) of the filter (excluding R_s) for the frequency range set above:

```
>>ZL=Rs; [Gw,Zinw]=GZ_strip(Z0strip,SPsos,ZL,bl);
```

and we multiply $G(\omega)$ with the voltage divider ratio $Z_{in}(\omega)/(R_s+Z_{in}(\omega))$ on the source side:

```
>>Gw=Zinw./(Rs+Zinw).*Gw; % the overall frequency response
```

Here, the below MATLAB function 'GZ_strip()' is supposed to be saved on the computer.

```
function [Gw,Zinw]=GZ_strip(Z0strip,SPsos,ZL,bl)
% To get the frequency response Gw and input impedance Zinw
% Input:
% Z0strip = Impedances of stub, line, stub, line, ...
% SPsos   = e.g. {'Ss','Po',..} for Series-short, Parallel-open, ..
% ZL      = Load impedance
% bl      = Range of electrical length
% Output:
% Gw      = Frequency response
% Zinw    = Input AC impedance
Zl=@(ZL,Z0,bl)(ZL+j*Z0*tan(bl))./(Z0+j*ZL.*tan(bl))*Z0;
% Impedance of a line of characteristic impedance Z0 and
% electrical length bl with load ZL
Zls=@(Z0,bl)j*Z0*tan(bl); % Impedance of a an SS
Zlo=@(Z0,bl)-j*Z0./tan(max([bl; eps*ones(1,numel(bl))]));
% Impedance of an open-circuited stub
Zlso=@(Z0,bl,so)(so=='s')*Zls(Z0,bl)+(so=='o')*Zlo(Z0,bl);
Z_par=@(Z1,Z2)Z1.*Z2./(Z1+Z2); % Parallel(shunt) combination
Z0strip=abs(Z0strip); Zinw=ZL; I=1;
for m=numel(Z0strip):-1:1
  if isempty(SPsos{m})|Z0strip(m)<eps, continue; end
  SP=upper(SPsos{m}(1)); so=lower(SPsos{m}(2));
  if SP=='P'
   Zp=Z_par(Zlso(Z0strip(m),bl,so),Zinw); % Parallel combination
   I=I.*Zinw./Zp; Zinw=Zp; % Sum of shunt currents.
  elseif SP=='S'
   Zinw=Zinw+Zlso(Z0strip(m),bl,so); % Series combination
  elseif SP=='T', Zinw=Zl(Zinw,Z0strip(m),bl); % Serial line
  end
end
Gw=ZL./(Zinw.*I); % Ratio of voltage across ZL to Input voltage
```

4) To plot the magnitudes [dB] of the overall frequency responses considering R_s (with maximum of 0 dB) for the two filters, we run the following MATLAB statements:

```
>>Gwmag=20*log10(max(abs(Gw),1e-5));
  clf, plot(w,Gwmag-max(Gwmag)), hold on
  s=j*w/fc; Gw0=eval(Gs); Gwmag0=20*log10(max(abs(Gw0),1e-5));
  plot(w,Gwmag0-max(Gwmag0),'r') % To make the maximum 0dB
```

where Gs is the transfer function obtained in Step 2. This yields Figure 7.17.3.

An easy alternative to implement an *LC* ladder using stubs is a *stepped-impedance* or hi-*Z*, low-*Z* filter [P-1] obtained by replacing the normalized values of inductance *L* and capacitance *C* with TL sections of high/low characteristic impedances Z_h/Z_l and electrical lengths

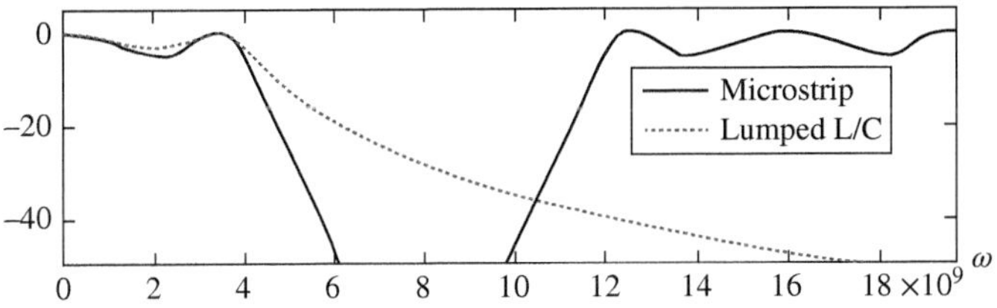

Figure 7.17.3 Frequency responses of the lumped-element filter and distributed-element filter.

$$\beta l_L = \frac{LR_0}{Z_\mathrm{h}} \text{ for normalized inductance } L \text{ and} \tag{7.3.35a}$$

$$\beta l_C = \frac{CZ_\mathrm{l}}{R_0} \text{ for normalized capacitance } L \tag{7.3.35b}$$

respectively.

Example 7.6 Design of Stepped-Impedance Filter [P-1]
For line/load impedances $R_0 = 50\ [\Omega]/Z_\mathrm{L} = 50\ [\Omega]$, design a sixth-order Butterworth LPF prototype with passband ripple $R_\mathrm{p} = 3$ dB and implement it as a stepped-impedance filter with $Z_\mathrm{h} = 150\ [\Omega]$ and $Z_\mathrm{l} = 10\ [\Omega]$.

To design such an LPF, we run the following MATLAB script "elec7e06.m"

```
%elec07e06.m: Butterworth LPF design in Example 8.7 of [P-1]
N=6; Rp=3; % Order and Passband ripple of Butterworth LPF
type='B'; % Butterworth
gg=LPF_ladder_coef(N,type,Rp)
% To find the L & C values of the prototype Pi filter
R0=50; Rs=R0; pT='pi'; wc=1; % for a prototype filter
[LCRs,LCRs_,Gs]=LC_ladder(Rs,gg,pT,wc);
disp('Values of the lumped elements L, C, and RL')
format short e; LCRs, LCRs_
LCs=LCRs(1:N); LCs_=LCRs_(1:N);
Zh=150; Zl=10; % Highest/lowest practical line impedances
bl=@(g,LC_)(LC_=='L')*g*R0/Zh+(LC_=='C')*g*Zl/R0; % Eq.(7.3.35a,b)
Z0=@(LC_)(LC_=='L')*Zh+(LC_=='C')*Zl;
% Find the electrical lengths of the hi/low-Z transmission line sections
% to replace the series inductors and shunt capacitors.
for m=1:N
  bls(m)=bl(gg(m),LCs_{m}(1)); Z0s(m)=Z0(LCs_{m}(1));
end
disp('Electrical lengths, Characteristic impedances of microstrip lines')
Z0s, Els=rad2deg(bls) % Characteristic impedances, Electrical lengths
```

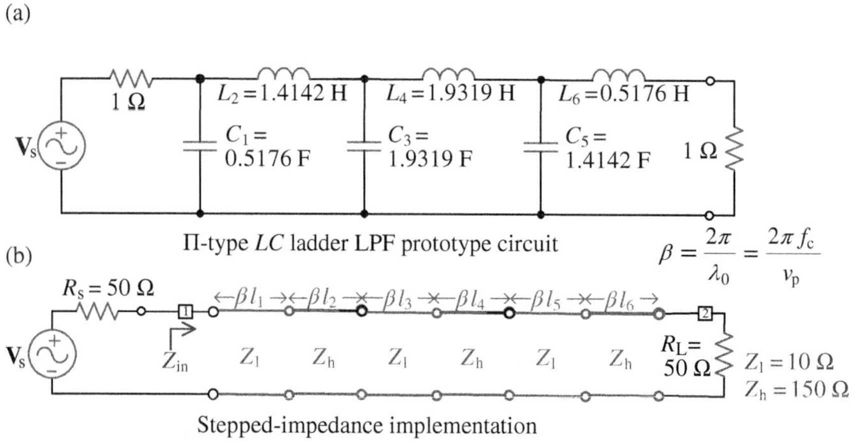

Figure 7.18 Sixth-order Butterworth 3 dB LPF realized with a Π-type *LC* ladder and implemented as a stepped-impedance filter.

to get the electrical lengths and characteristic impedances of each TL section as

```
Z0s =    10      150      10      150      10      150
ELs = 5.9317  2.7009  2.2137  3.6896  1.6206  9.8862
```

This means the stepped-impedance LPF of Figure 7.18(b).

7.3.6 Impedance Matching with Lumped Elements

This section presents several examples illustrating how some types of impedance matchers consisting of lumped elements can be designed by using the Smith chart.

Example 7.7 Design of L-type Impedance Transformers Using the Smith Chart

Find L-type *LC* circuits transforming the impedance $Z_1 = 40 + j50[\Omega]$ or admittance $Y_1 = 1/Z_1 = 0.009756 - j0.012195$ [S] represented by point Γ_1 into the impedance $Z_2 = 50 + j17$ [Ω] or $Y_2 = 1/Z_2 = 0.017928 - j0.060954$ [S] (so that it can be matched to $Z_s = 50 - j17$ [Ω]) by using the Smith chart and compare the results with those obtained in Example 6.12.

Noting that there are four paths along resistance/conductance curves from point Γ_1 (representing $Z_1 = 40 + j50$ [Ω]) to Γ_2 (representing $Z_2 = 50 + j17$ [Ω]), each via Γ_3, Γ_5, Γ_6, and Γ_7 in Figure 7.13, let us work with 'Smith V4.1' taking the following steps:

1) From the top menu bar, select Tools > Settings to open the Settings dialog box, set the values of Z_0 (characteristic impedance) and frequency to 50 [Ω] (default) and 1 [GHz], respectively, and click OK button. In fact, as long as the design objective is not to match the input impedance with the characteristic impedance, it does not matter whatever value Z_0 is set to, because 'Smith V4.1' accommodates even unnormalized impedances/admittances as they are.
2) From the toolbar (below the menu bar), select Keyboard to open Data Point dialog box, insert the impedance $Z_1 = 40 + j50$ [Ω] as a (start) point (DP1), and click OK button. Also, insert $Z_2 = 50 + j17$ [Ω] as a (target) point (DP2). If you want to cancel (undo) any work or delete the data points starting from the last one (in LIFO order), just right-click anywhere on the Smith chart.
3) To design the L1-type *C*-*C* impedance matcher as shown in Figure 6.19(a1) corresponding to the path from Γ_1 to Γ_2 via Γ_3, click on 'Insert series capacitor' button and move the mouse cursor to click on Γ_3 (the intersection of the $R = 40$ Ω-circle and $G = 0.0179$ S-circle) at which the value of $G = \text{Re}\{Y\}$ is close to 0.0179 S = 17.9 mS as can be read in the Cursor pane. Then, TP 3 will appear at Γ_3 and its impedance value will appear in the Data Points pane at the right side as shown in Figure 7.19.

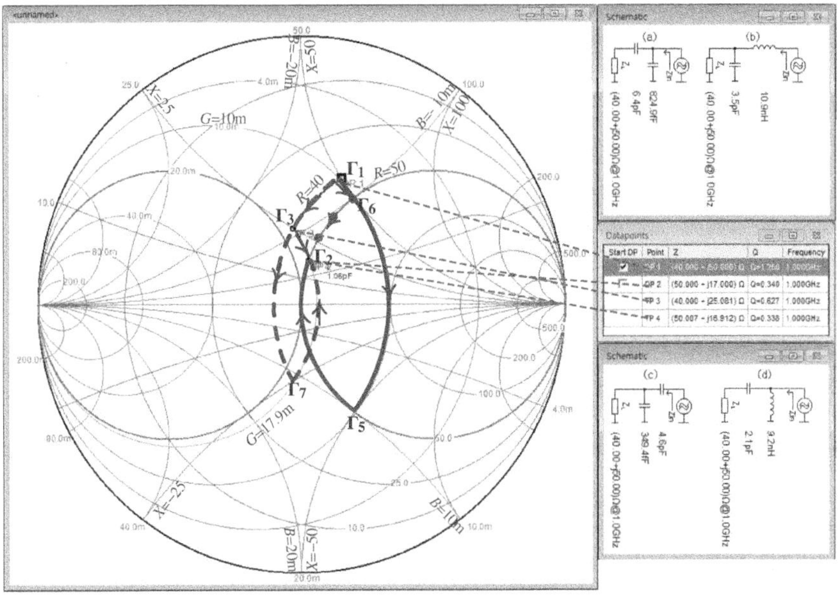

Figure 7.19 Using 'Smith 4.1' to design L-type impedance matchers in Example 7.7 or 6.12.

4) From the tool bar, click on 'Insert parallel capacitor' button and move the mouse cursor to click on Γ_2 (the intersection of the R = 50 Ω-circle and the G = 17.9 mS-circle) representing Z_2 = 50 + j17 [Ω]. Then, TP 4 will appear at Γ_2 and its impedance value will appear in the Data Points pane.
5) Then, the designed network with the parameter values C_2 = 6.4 pF and C_1 = 824.9 fH will be shown in the Schematic pane at the upper-right part of the window. It conforms with Figure 6.19(a1).
6) If you want to read more digits of a numeric value or change it, double-click on it to open the Edit Element dialog box.
7) Right-click anywhere on the Smith chart twice to delete the last two data points.
8) To design the L2-type L-C impedance matcher as shown in Figure 6.19(b1) corresponding to the path from Γ_1 to Γ_2 via Γ_5, click on 'Insert parallel capacitor' button (from the tool bar) and move the mouse cursor to click on Γ_5 (the intersection of the G=0.009756 S ≈ 0.01 S-circle and the R=50 Ω-circle) at which the value of R=Re{Z} is close to 50 Ω as can be read in the Cursor pane.
9) From the tool bar, click on 'Insert series inductor' button and move the mouse cursor to click on Γ_2 (the intersection of the R=50 Ω-circle and the X=17 Ω-circle) representing Z_2=50 + j17 [Ω].
10) Then, the designed network with the parameter values C_2 = 3.5 pF and L_1 = 10.9 nH will be shown in the Schematic pane at the upper-right part of the window. It conforms with Figure 6.19(b1).

Example 7.8 Design of T-type Impedance Transformers Using the Smith Chart

Make T-type ladders with quality factor Q = 3 for transforming Z_1 = 60 – j30 [Ω] to Z_2 = 10 + j20 [Ω] at frequency f_c = 1 GHz.

Note that, with $Q \leq 3$, there are four paths along resistance/conductance curves from point Γ_1 (representing Z_1=60 – j30[Ω]) to Γ_2 (representing Z_2=10+j20[Ω]), each via Γ_5-Γ_3, Γ_5-Γ_4, Γ_6-Γ_4, and Γ_6-Γ_3, in Figure 7.20 where Γ_3 and Γ_4 have been determined as the intersections between 10 Ω-resistance circle and Q= 3-circles. Let us work with 'Smith V4.1' taking the following steps:

1) From the top menu bar, select Tools > Circles to open the Circles dialog box, insert 3 as a value of Q, and click OK to close the dialog box so that two Q=3-circles appear above/below the real axis.
2) From the top menu bar, select Tools > Settings to open the Settings dialog box, set the values of Z_0 and frequency to 50[Ω] (default) and 1[GHz], respectively, and click OK.
3) From the toolbar, select Keyboard to open Data Point dialog box, insert Z_1 = 60 – j30[Ω] as a (start) point (DP1), and click OK. Also, insert Z_2=10 + j20[Ω] as a (target) point (DP2). See Figure 7.20.

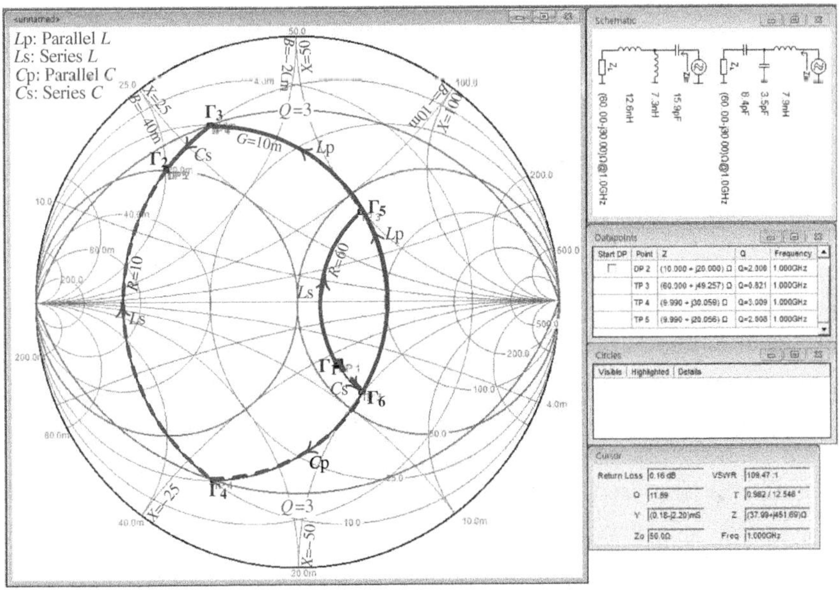

Figure 7.20 Using 'Smith 4.1' to design T-type impedance matchers in Example 7.8 or 6.13.

4) To designate Γ_3 and Γ_4 as intermediate points just before the target point Γ_2, click on the 'Mouse' button, find two points with $R = R_2 = 10[\Omega]$ and $Q = 3$, one by one, on the two Q=3-circles (keeping your eye on the values of Q and $R = \text{Re}\{Z\}$ in the Cursor pane), and click on them to insert as data points. In fact, noting that the R-value of (the resistance circle through) Γ_2 is 10 [Ω], we can compute the X-value of (the reactance circle through) Γ_3 with $Q = 3$ as

$$X_3 \underset{X_3-L}{\overset{(6.2.26\text{a})}{=}} QR_2 = 3 \times 10 = 30\,\Omega \rightarrow Z_3 = R_3 + jX_3 = 10 + j30\,[\Omega], \qquad Y_3 = \frac{1}{Z_3} = 0.01 - j0.03\,[\text{S}] \tag{E7.8.1}$$

5) Noting that the number of data points is not allowed to exceed 5, designate Γ_5 as an intermediate point by clicking on a point with $G_5 = G_3 = 10$ [mS] (equal to the conductance value of Γ_3) and $R_5 = R_1 = 60$ [Ω] (equal to the resistance value of Γ_1) (keeping your eye on the values of $R = \text{Re}\{Y\}$ and $G = \text{Re}\{Y\}$ in the Cursor pane) to insert it. Why is that? Because we are going to connect the reactive elements

(each corresponding to the Γ_1-Γ_5 and Γ_5-Γ_3 sections, respectively) in series and in parallel, respectively. In fact, from $R_5 = 60\ \Omega$ and $G_5 = 0.01$ S, we can compute the X_5 for Γ_5 as

$$Y_5 = G_5 + jB_5 = 0.01 + jB_5 = \frac{1}{Z_5} = \frac{1}{R_5 + jX_5} = \frac{1}{60 + jX_5} = \frac{60 - jX_5}{60^2 + X_5^2}$$

$$;\ 0.01 = \frac{60}{60^2 + X_5^2} \quad \rightarrow X_5 = \sqrt{60/0.01 - 60^2} = 49\ [\Omega] \tag{E7.8.2}$$

or

```
>>R5=60; G5=0.01; [Z5,Y5]=RG2ZY(R5,G5) % see Problem 7.9
```

6) To design the T-type C-L-L impedance matcher as shown in Figure 6.22(a) corresponding to the path from Γ_1 to Γ_2 via Γ_5-Γ_3, click on 'Insert series inductor' and move the mouse cursor to click on Γ_5 (corresponding to Z_5=60+j49[Ω]). Then, click on 'Insert parallel inductor' and move the mouse cursor to click on Γ_3 (corresponding to Z_3 = 10 + j30 [Ω]). Then, click on 'Insert series capacitor' and move the mouse cursor to click on Γ_2 (corresponding to Z_2=60−j30[Ω]). Then, the designed network appears in the Schematic pane as shown in Figure 7.20 where L_3 = 12.61 nH, L_2 = 7.306 nH, and C_1 = 15.91 pF.
7) To design the T-type L-C-C impedance matcher as shown in Figure 6.22(d) corresponding to the path from Γ_1 to Γ_2 via Γ_6-Γ_4, click on 'Insert series capacitor' and move the mouse cursor to click on Γ_6 (corresponding to Z_6=60−j49[Ω]). Then, click on 'Insert parallel capacitor' and move the mouse cursor to click on Γ_4 (corresponding to Z_4 = 10 − j30 [Ω]). Then, click on 'Insert series inductor' and move the mouse cursor to click on Γ_2. Then, the designed network appears in the Schematic pane (Figure 7.20) where C_3 = 8.383 pF, C_2 = 3.488 pF, and L_1 = 7.945 nH.

Example 7.9 Design of Π-type Impedance Transformers Using the Smith Chart

Make Π-type ladders with quality factor $Q = 2$ for transforming $Z_1 = 10 - j10$ [Ω] to Z_2=20+j40[Ω] at frequency f_c = 2.4 GHz.

Note that, with $Q \le 2$, there is a path along resistance/conductance curves from point Γ_1 (representing Z_1 = 10 − j10 [Ω] or equivalently Y_1 = 50 + j50 [mS]) to Γ_2 (representing Z_2=20+j40[Ω] or equivalently Y_2=10+j20[mS]) via Γ_3-Γ_4 in Figure 7.21 where Γ_4 has been determined as the intersection between G = 10 mS-circle and the lower Q = 2-circle. Let us work with 'Smith V4.1' taking the following steps:

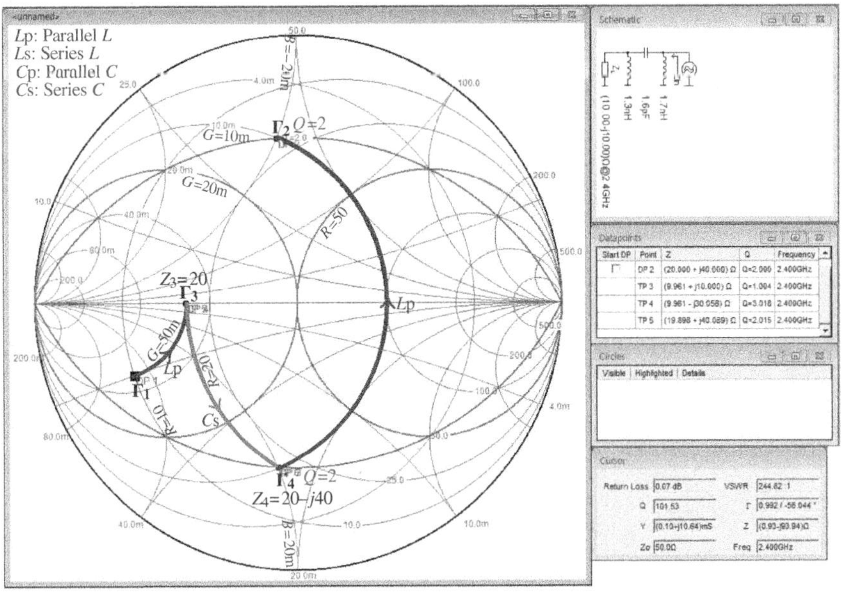

Figure 7.21 Using 'Smith 4.1' to design Π-type impedance matchers in Example 7.9 or 6.14.

1) From the top menu bar, select Tools > Circles to open the Circles dialog box, insert 2 as a value of Q, and click OK to close the dialog box so that two Q = 2-circles appear above/below the real axis.
2) From the top menu bar, select Tools > Settings to open the Settings dialog box, set the values of Z_0 and frequency to 50[Ω] (default) and 2.4[GHz], respectively, and click OK.
3) From the toolbar, select Keyboard to open Data Point dialog box, insert $Z_1=10-j10[\Omega]$ as a (start) point (DP1), and click OK. Also, insert $Z_2 = 20 + j40$ [Ω] as a (target) point (DP2). See Figure 7.21.
4) To set an intermediate point Γ_4 just before the target point Γ_2, click on the 'Mouse' button, find a point with $G = G_2 = 10$[mS] and $Q = 2$ on the lower Q = 2-circle (keeping your eye on the values of Q and $G = \text{Re}\{Y\}$ in the Cursor pane), and click on it to insert as data points. Why not on the upper Q=2-circle? Because Γ_2 is on the upper Q=2-circle. In fact, noting that the G-value of (the G-circle through) Γ_2 is 0.01[S], we can compute the B-value of (B-circle through) Γ_4 with $Q = 2$ as

$$B_4 \underset{B_4\|L}{\overset{(6.2.26b)}{=}} QG_2 = 2 \times 0.01 = 0.02\,\text{S}$$

This implies that the admittance and impedance for Γ_4 are

$$Y_4 = G_4 + jB_4 = G_2 + jB_4 = 0.01 + j0.02,\ Z_4 = \frac{1}{Y_4} = \frac{1}{0.01 + j0.02} = 20 - j40 \quad \text{(E7.9.1)}$$

Why have we set $G_4 = G_2$? Because we are going to connect the reactive element (corresponding to the Γ_4-Γ_2 section) in parallel with Z_2, as denoted by '*Lp*'.

5) To determine another intermediate point Γ_3 (between Γ_1 and Γ_4), we set the conductance and resistance to 50 mS (that for Γ_1) and 20 Ω (that for Γ_4), respectively, because we are going to connect the reactive elements (each corresponding to the Γ_1-Γ_3 and Γ_3-Γ_4 sections, respectively,) in parallel and in series, respectively:

$$Y_3 = G_3 + jB_3 = 0.05 + jB_3 = \frac{1}{Z_3} = \frac{1}{R_3 + jX_3} = \frac{1}{20 + jX_3} = \frac{20 - jX_3}{20^2 + X_3^2}$$
$$;\ 0.05 = \frac{20}{20^2 + X_3^2} \rightarrow X_3 = \sqrt{20/0.05 - 20^2} = 0\,[\Omega] \quad \text{(E7.9.2)}$$

or

```
>>R3=20; G3=0.05; [Z3,Y3]=RG2ZY(R3,G3)
```

If you somehow have got the numeric value of the *Z*, *Y*, or Γ of a data point to be inserted, you can select Keyboard to open the Data Point dialog box and enter the value.

6) To design the Π-type *L-C-L* impedance matcher as shown in Figure 6.23(b) corresponding to the path from Γ_1 to Γ_2 via Γ_3-Γ_4, click on 'Insert parallel inductor' and move the mouse cursor to click on Γ_3 (corresponding to $Z_3 = 20 + j0$ [Ω]). Then, click on 'Insert series capacitor' and move the mouse cursor to click on Γ_4 (corresponding to $Z_4 = 20 - j40$[Ω]). Then, click on 'Insert parallel inductor' and move the mouse cursor to click on Γ_2 (corresponding to $Z_2 = 20 + j40$ [Ω]). Then, the designed network appears in the Schematic pane as shown in Figure 7.22 where

$$L_3 = 1.326\ \text{nH}, C_2 = 1.634\ \text{pF, and}\ L_1 = 1.666\ \text{nH} \quad \text{(E7.9.3)}$$

Regrettably, this design result is quite different from that of Figure 6.23(b). To figure out the difference between the design result using the Smith chart and that using the MATLAB function '`imp_matcher_pi()`', we have drawn the latter on the Smith chart in Figure 7.22, which shows that some intermediate points such as Γ_4 and Γ_6 do not satisfy the upper bound condition for *Q*. It implies that the design result using the Smith chart satisfies the *Q*-condition at every stage while that using MATLAB may not satisfy the *Q*-condition at some stage even if both design results satisfy the condition as a whole.

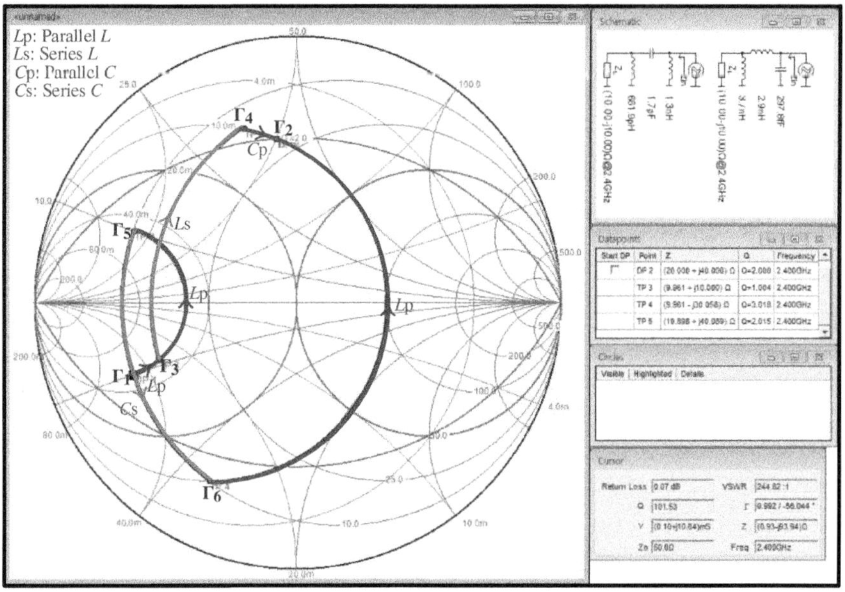

Figure 7.22 The design results of Figure 6.23(b) and (d) (Example 6.14) drawn on the Smith chart.

Example 7.10 Impedance Matching with an *LC* Ladder

Make *LC* ladder-type impedance matchers (with quality factor $Q \leq 1$ at every stage) for transforming Z_1=12.5+j10[Ω] to 50 Ω at frequency f_c=100 MHz.

a) Let us work with 'Smith V4.1' taking the following steps to make an LPF *LC* ladder-type impedance matcher:

   1) From the top menu bar, select Tools > Circles to open the Circles dialog box, insert 1 as a value of Q and click OK to close the dialog box so that two Q=1-circles appear above/below the real axis.
   2) From the top menu bar, select Tools > Settings to open the Settings dialog box, set the values of Z_0 and frequency to 50 [Ω] (default) and 100 [MHz], respectively, and click OK.
   3) From the toolbar, click on 'Keyboard' to open Data Point dialog box, insert the impedance Z_1 = 12.5 + j10 [Ω] or Y_1 = 48.8 + j39 [mS] as a (start) point (DP1), and click OK. The target point Γ = 0, i.e. the origin corresponding to Z_0 = 50 [Ω], is so apparent that it does not have to be designated as a data point.

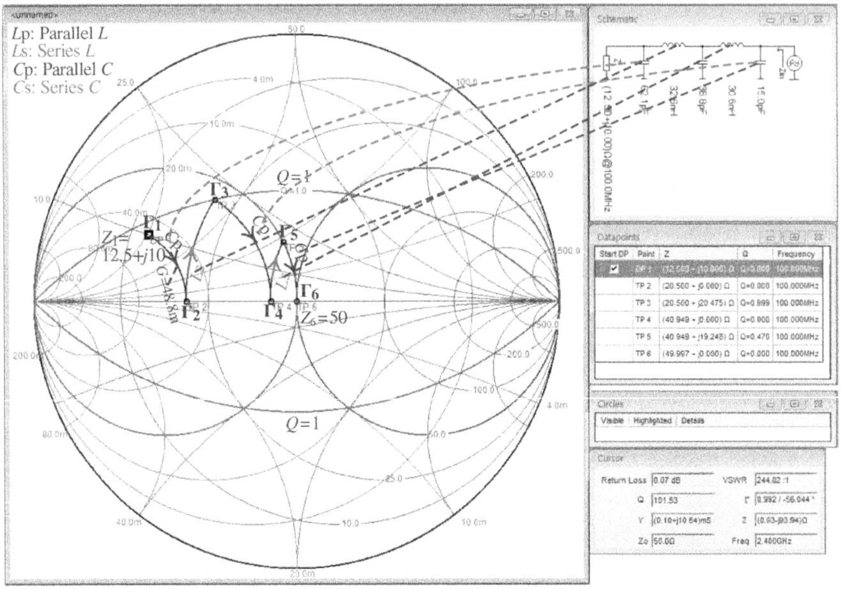

Figure 7.23 Using 'Smith 4.1' for impedance matching between $Z_1 = 12.5 + j10$ and $Z_6 = 50$ in Example 7.10(a).

4) To design an LC ladder-type impedance matcher as an LPF, take the following steps, referring to Figure 7.23:

- Click on 'Insert parallel capacitor' and move the mouse cursor to click on Γ_2 (the intersection of $G = G_1 = 48.8$ [mS]-circle and the horizontal axis).
- Click on 'Insert series inductor' and move the mouse cursor to click on Γ_3 (the intersection of $R = R_2 = 1/G_1 = 20.5[\Omega]$-circle and the upper $Q = 1$-circle corresponding to $Y_3 = G_3 + jB_3$[S]).
- Click on 'Insert parallel capacitor' and move the mouse cursor to click on Γ_4 (the intersection of $G = G_3$-circle and the horizontal axis).
- Click on 'Insert series inductor' and move the mouse cursor to click on Γ_5 (the intersection of $R = R_4 = 1/G_3$-circle and the $G = 1/R_0$-circle).
- Click on 'Insert parallel capacitor' and move the mouse cursor to click on $\Gamma_6 = 0$ (i.e. the origin corresponding to $Z_0 = 50[\Omega]$).

5) Then, the designed LC ladder appears in the Schematic pane in Figure 7.23.

```
function Xs=imp_matcher_LC_LPF1(R0,ZL,fc,Qm)
% Design an LC ladder LPF with Q<=Qm to make impedance matching with R0
% when RL<R0
% Copyleft: Won Y. Yang, wyyang53@hanmail.net, CAU for academic use only
EPS=1e-4;
if abs(ZL-R0)<EPS, Xs=[]; return; end
RL=real(ZL); XL=imag(ZL); YL=1/ZL; GL=real(YL); BL=imag(YL);
if abs(XL)<EPS
 M1=abs(Qm); M2=sqrt(R0/RL-1);
 % Does the R-circle cross G=1/R0-circle or Q-circle first?
 [X,imin]=min([M1 M2]); X=X*RL;
 Z=RL+j*X; Y=1/Z; B=imag(Y);
 if abs(M1-M2)<EPS|imin>1 % If it crosses G=1/R0-circle first
  Xs=[1/B X]; % Series L & Parallel C from Load
 else % If it crosses Q-circle first
  RL=1/real(Y); Xsp=imp_matcher_LC_LPF1(R0,RL,fc,Qm);
  Xs=[Xsp 1/B X]; % Series L & Parallel C
 end
elseif XL>0 % First, C in parallel with ZL for LPF
 RL=1/GL;
 Xsp=imp_matcher_LC_LPF1(R0,RL,fc,Qm);
 Xs=[Xsp 1/BL]; % Parallel C
else % if XL<0
 % Move the point from the 3rd quadrant into the 2nd one by Ls (series L).
 M1=abs(Qm); M2=sqrt(R0/RL-1);
 [X,imin]=min([M1 M2]); X=X*RL;
 Z=RL+j*X; Y=1/Z; B=imag(Y);
 if abs(M1-M2)<EPS|imin>1
  Xs=[1/B (X-XL)]; % Series L and Parallel C from Load
 else
  R=1/real(Y); Xsp=imp_matcher_LC_LPF1(R0,R,fc,Qm);
  Xs=[Xsp 1/B (X-XL)]; % Series L and Parallel C
 end
end
```

```
function [LCs,LCs_]=L_or_C_from_X(Xs,fc,vh)
% vh = 'v' or 'h' for vertical/parallel or horizontal/series first
% Find L or C from given values of reactance.
% Copyleft: Won Y. Yang, wyyang53@hanmail.net, CAU for academic use only
wc=2*pi*fc;
for n=1:length(Xs)
  X=Xs(n);
  if abs(X)<eps|abs(X)>realmax, LCs(n)=0; LCs_{n}='????';
   else
    if X>0, LCs(n)=X/wc; LCs_n='L'; unit='H';
     else LCs(n)=-1/X/wc; LCs_n='C'; unit='F';
    end
    if nargin>2
     LCs_n=[LCs_n '(' vh ')'];
     if vh=='v', vh='h';
      else vh='v';
     end % To alternate 'L' and 'C'
    else fprintf([LCs_n '(%d)=%10.3e[' unit ']\n'],n,LCs(n));
    end
    LCs_{n}=LCs_n;
   end
end
```

This LPF *LC* ladder-type impedance matcher (with quality factor $Q \leq 1$ at every stage) can also be designed by using the above MATLAB function 'imp_matcher_LC_LPF1()' as follows:

```
>>fc=1e8; Qm=1; % Frequency and Maximum quality factor
  Xs=imp_matcher_LC_LPF1(50,12.5+10i,fc,Qm)
  [LCs,LCs_]=L_or_C_from_X(Xs,fc,'v')
  LCs  = 1.4913e-11 3.0573e-08 3.8818e-11 3.2627e-08 6.2109e-11
  LCs_= 'C(v)'      'L(h)'     'C(v)'     'L(h)'     'C(v)'
```

where the third input argument of 'L_or_C_from_X()' has been given as 'v' (meaning 'vertical') since we know that the last section of the path from Γ_1 to Γ_2 leads to a parallel *C* and consequently the first element of the *LC* ladder (from the source) is to be connected vertically. This design result conforms with that obtained using 'Smith 4.1' and shown in the Schematic pane in Figure 7.23.

b) Now, to design an *LC* ladder-type impedance matcher as a HPF, take the following steps:

   1) Right-click anywhere on the Smith chart four times to delete the existing four data points Γ_2, Γ_3, Γ_4, and Γ_5 (in LIFO order) except Γ_1.
   2) From the top menu bar, select Tools > Circles to open the Circles dialog box, insert 1 as a value of Q and click OK to close the dialog box so that two $Q = 1$-circles appear above/below the real axis.
   3) From the top menu bar, select Tools > Settings to open the Settings dialog box, set the values of Z_0 and frequency to 50[Ω] (default) and 100 [MHz], respectively, and click OK.
   4) From the toolbar, click on 'Keyboard' to open the Data Point dialog box, insert Z_1=12.5+j10[Ω] as a (start) point (DP1), and click OK.
   5) To design an *LC* ladder-type impedance matcher, do the following, referring to Figure 7.24:

      - Click on 'Insert series capacitor' (from the tool bar) and move the mouse cursor to click on the point Γ_2 (the intersection of R=R_1=12.5[Ω]-circle and the <u>lower</u> Q=1-circle) corresponding to Y_2=40+j40[mS].
      - Click on 'Insert parallel inductor' and move the mouse cursor to click on the point Γ_3 (the intersection of G=G_2=40[mS]-circle and the horizontal axis).
      - Click on 'Insert series capacitor' and move the mouse cursor to click on the point Γ_4 (the intersection of $R = 1/G_3 = 1/0.04$=25 Ω-circle and the G=1/R_0-circle, which coincidentally happens to be on the lower Q = 1-circle), corresponding to Y_4=20+j20 [mS].

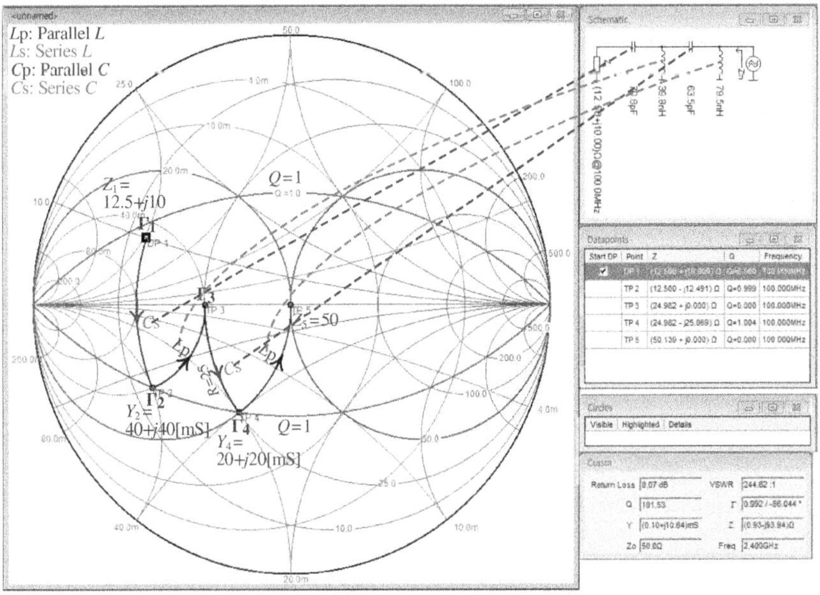

Figure 7.24 Using 'Smith 4.1' for impedance matching between $Z_1 = 12.5 + j10$ and $Z_5 = 50$ in Example 7.10(b).

- Click on 'Insert parallel inductor' and move the mouse cursor to click on the point $\Gamma_5 = 0$ (i.e. the origin corresponding to $Z_0 = 50$ [Ω]).

6) Then, the designed *LC* ladder appears in the Schematic pane in Figure 7.24.

This HPF *LC* ladder-type impedance matcher (with quality factor $Q \leq 1$ at every stage) can also be designed by using the above MATLAB function '`imp_matcher_LC_HPF1()`' as follows:

```
>>fc=1e8; Qm=1; % Frequency and Maximum quality factor
  format short e; Xs=imp_matcher_LC_HPF1(50,12.5+10i,fc,Qm)
  [LCs,LCs_]=L_or_C_from_X(Xs,fc,'v')
```

where the third input argument of '`L_or_C_from_X()`' has been given as '`v`' (meaning 'vertical') since we know that the last section of the path from Γ_1 to Γ_2 leads to a parallel *L* and consequently, the first element of the *LC* ladder (from the source) is to be connected vertically. Running these statements yields the following result:

```
LCs  =  7.9577e-08  6.3662e-11   3.9789e-08   7.0736e-11
LCs_=  'L(v)'       'C(h)'       'L(v)'       'C(h)'
```

This design result conforms with that obtained using 'Smith 4.1' and shown in the Schematic pane in Figure 7.24.

```
function Xs=imp_matcher_LC_HPF1(R0,ZL,fc,Qm)
% Design an LC ladder HPF with Q<=Qm to make impedance matching with R0
% when RL<R0
% Copyleft: Won Y. Yang, wyyang53@hanmail.net, CAU for academic use only
EPS=1e-4;
if abs(ZL-R0)<EPS, Xs=[]; return; end
RL=real(ZL); XL=imag(ZL); YL=1/ZL; GL=real(YL); BL=imag(YL);
if abs(XL)<EPS
 M1=abs(Qm); M2=sqrt(R0/RL-1);
 % Does the R-circle cross G=1/R0-circle or Q-circle first?
 [X,imin]=min([M1 M2]); X=X*RL;
 Z=RL+j*X; Y=1/Z; B=imag(Y);
 if abs(M1-M2)<EPS|imin>1 % If it crosses G=1/R0-circle first
  Xs=[-1/B -X]; % Series C and Parallel L from Load
 else % If it crosses Q-circle first
  RL=1/real(Y); Xsp=imp_matcher_LC_HPF1(R0,RL,fc,Qm);
  Xs=[Xsp -1/B -X]; % Series C and Parallel L from Load
 end
elseif XL<0 % First, L in parallel with ZL for HPF
 RL=1/GL; Xsp=imp_matcher_LC_HPF1(R0,RL,fc,Qm);
 Xs=[Xsp 1/BL]; % Parallel L
else % if XL>0
 % Move the point from the 2nd quadrant into the 3rd one by Ls (series C).
 M1=abs(Qm); M2=sqrt(R0/RL-1);
 [X,imin]=min([M1 M2]); X=-X*RL;
 Z=RL+j*X; Y=1/Z; B=imag(Y);
 if abs(M1-M2)<EPS|imin>1 % If it crosses G=1/R0-circle first
  Xs=[1/B (X-XL)]; % Series C and Parallel L from Load
 else % If it crosses Q-circle first
  R=1/real(Y); Xsp=imp_matcher_LC_HPF1(R0,R,fc,Qm);
  Xs=[Xsp 1/B (X-XL)]; % Series C and parallel L
 end
end
%Usage:
%fc=1e8; Qm=1; Xs=imp_matcher_LC_HPF1(50,10+5i,fc,Qm);
%[LCs,LCs_]=L_or_C_from_X(Xs,fc,'v')
```

Example 7.11 Impedance Matching with Serial Line

Use a serial transmission line of characteristic impedance $Z_0 = 50\ \Omega$ and a parallel capacitor to make an impedance matching between $Z_L = 30 - j40[\Omega]$ and $50\ \Omega$ at frequency $f_c = 430$ MHz.

Having started 'Smith V4.1' to open the main window, take the following steps:

1) From the top menu bar, select Tools > Settings to open the Settings dialog box, set the values of Z_0 and frequency to 50[Ω](default) and 430[MHz], respectively, and click OK.
2) From the toolbar, click on 'Keyboard' to open Data Point dialog box, insert the impedance $Z_L=30-j40[\Omega]$ as a (start) point Γ_L (DP1), and click OK. The target point $\Gamma_3 = 0$, i.e. the origin corresponding to $Z_0 = 50$ [Ω], is so apparent that it does not have to be designated as a data point.
3) To design an impedance matcher consisting of a serial line and a reactive element C or L, you need to visualize a $|\Gamma|$-circle centered at the origin and crossing Γ_L (which shows the effect of a lossless SL [serial line] on the reflection coefficient Γ of the load impedance as described by Eq. (7.2.5)) and the $G = 1/R_0$-circle along which a Γ point can be moved by a parallel C or L to reach the target point $\Gamma = 0$, as have been shown in Figure 7.25. From Figure 7.25, notice that there are two intersections between the two circles, Γ_2 and Γ_4, each of which can be an intermediate point. Therefore, there are two paths from Γ_L to the target point $\Gamma_3 = 0$, one via Γ_2 and one via Γ_4.

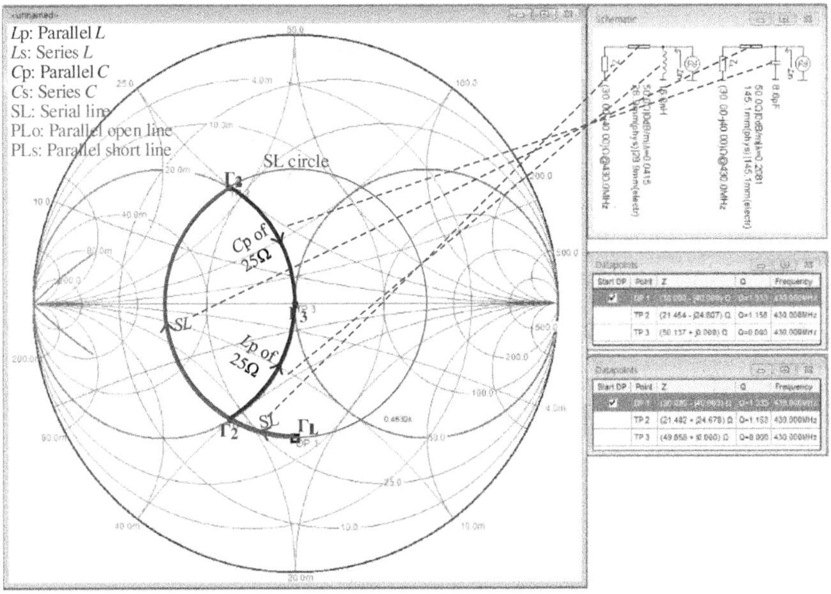

Figure 7.25 Using 'Smith 4.1' for impedance matching between $Z_L = 30 - j40$ and $Z_0 = 50$ in Example 7.11.

First, to take the first path Γ_L-Γ_2-Γ_3, do the following, referring to Figure 7.25:

- Click on 'Insert serial line' (from the tool bar) and move the mouse cursor to click on the point Γ_2 (one intersection of the G = 1/R0-circle and the $|\Gamma|$-circle crossing Γ_L).
- Click on 'Insert parallel inductor' and move the mouse cursor upward to click on the point $\Gamma_3 = 0$ (i.e. the origin corresponding to $Z_0 = 50[\Omega]$).
- Then, the designed impedance matcher consisting of SL (serial line) and Lp (parallel inductor) appears in the Schematic pane as shown in Figure 7.25.

Second, to take the second path Γ_L-Γ_4-Γ_3, do the following:

- Click on 'Insert serial line' (from the tool bar) and move the mouse cursor to click on the point Γ_4 (another intersection of the G = 1/R0-circle and the $|\Gamma|$-circle).
- Click on 'Insert parallel capacitor' and move the mouse cursor downward to click on $\Gamma_2 = 0$.
- Then, the designed impedance matcher consisting of SL (serial line) and Cp (parallel capacitor) appears in the Schematic pane as shown in Figure 7.25.

As an alternative, we can compose the following MATLAB script "elec07e11.m" and run it to get

```
>>elec07e11
    ds =   0.0417   0.2083 % Transmission length(s)
    Xs =  24.7436  -24.7436 % Reactances to be adjusted
    L(1) = 1.603e-08 [H]   C(2) = 8.548e-12
```

These results conform with those obtained using 'Smith 4.1'.

```
%elec07e11.m
clear, clf
Z0=50; b=2*pi; fc=430e6; ZL=30-40i; G0=1/Z0;
[c,r]=Gcircle(G0,Z0); % Center and radius of the G-circle
% To determine the length(s) of line to move from Gamma 1 to Gamma 2|4
Zd=@(d)Z0*(ZL+j*Z0*tan(b*d))./(Z0+j*ZL*tan(b*d)); % Eq.(7.1.17)
GammaZ=@(Z)(Z-Z0)./(Z+Z0); % Eq.(7.1.11)
f=@(d)abs(GammaZ(Zd(d))-c)-r; % To be solved for the line length d
ds=[fsolve(f,0) fsolve(f,0.25)];
ds=mod(ds,0.5) % Zd is periodic with period 0.5 in d[wavelength]
Xs=1./(imag(1./Zd(ds)-1/Z0)) % Reactances to be adjusted for imp matching
LCs=L_or_C_from_X(Xs,fc);
% To plot the G-circle, the Gamma-circle, and their intersections
dth=0.01; ths=-pi+dth:dth:pi; ucircle=exp(j*ths);
dd=[0:1e-4:0.5];
plot(c+r*ucircle,'b'), hold on, plot(GammaZ(Zd(dd))+j*eps,'k')
legend(['G=' num2str(G0,'%6.3e') '-circle'],'Line impedance effect')
plot(GammaZ(Zd(ds)),'o'), axis('equal')
```

Problems

7.1 Skinship with the Smith Chart
To get familiar with the Smith chart, do the following:

(a) Compose the following MATLAB functions and save them as M-files each named "G2Z.m" and "Z2G.m" if you don't have such similar M-files.

```
function Z=G2Z(Gam,Z0)
if nargin<2, Z0=1; end
Z=Z0*(1+Gam)./(1-Gam); %Eq.(7.2.1)
```

```
function Gam=Z2G(Z,Z0)
if nargin<2, Z0=1; end
Gam=(Z-Z0)./(Z+Z0); %Eq.(7.2.3)
```

(b) For the point $\Gamma_1 = 0.2 + j0.4$ (with $Z_0 = 50$) on the Smith chart in Figure 7.4, find the corresponding impedance/admittance and their normalized values, hopefully, by using one of the above MATLAB functions. Then, plot the point Γ_1 and the *R-/X-/G-/B*-circles passing through Γ_1 by completing and running the following MATLAB script "elec07p01b.m".

```
%elec07p01b.m
clear, clf
Z0=50; Gam1=0.2+0.4i;
Z1=???(Gam1,Z0), Y1=1/Z1 % Impedance and Admittance
Z1n=Z1/Z0, Y1n=Y1*?? % Normalized impedance and admittance
ths=pi/180*[0:360];
plot(Gam1,'ro'), hold on
R1=real(Z1); [cR1,rR1]=Rcircle(R1,Z0);
plot(cR1+rR1*exp(j*ths)); % plot the R1-circle
X1=imag(Z1); [cX1,rX1]=Xcircle(X1,Z0);
plot(cX1+rX1*exp(j*ths),':'); % plot the X1-circle
G1=real(Y1); [cG1,rG1]=Gcircle(G1,Z0);
plot(cG1+rG1*exp(j*ths),'r'); % plot the G1-circle
B1=imag(Y1); [cB1,rB1]=Bcircle(B1,Z0);
plot(cB1+rB1*exp(j*ths),'r:'); % plot the B1-circle
plot(exp(j*ths)); % plot the unit circle as background
plot([-1 1],[0 0],'k', [0 0],[-1 1],'k') % Horizontal/vertical axes
axis('equal'), axis([-1 1 -1 1])
```

(c) For the point $\Gamma_2 = -0.2 - j0.4$ (with $Z_0 = 50$) on the Smith chart in Figure 7.4, find the corresponding impedance/admittance and their normalized values, hopefully, by using one of the above MATLAB functions. Then, plot the point Γ_2 and the *R-/X-/G-/B*-circles passing through Γ_2 by completing and running the following MATLAB script "elec07p01c.m".

```
%elec07p01c.m
Z0=50; Gam2=-0.2-0.4i;
Z2=G2Z(Gam2,??), Y2=1/Z2 % Impedance and Admittance
Z2n=Z2/??, Y2n=Y2*Z0 % Normalized impedance and admittance
plot(Gam2,'ro'), hold on
R2=real(Z2); [cR2,rR2]=Rcircle(R2,Z0);
plot(cR2+rR2*exp(j*ths)); % plot the R1-circle
X2=imag(Z2); [cX2,rX2]=Xcircle(X2,Z0);
plot(cX2+rX2*exp(j*ths),':'); % plot the X1-circle
G2=real(Y2); [cG2,rG2]=Gcircle(G2,Z0);
plot(cG2+rG2*exp(j*ths),'r'); % plot the G1-circle
B2=imag(Y2); [cB2,rB2]=Bcircle(B2,Z0);
plot(cB2+rB2*exp(j*ths),'r:'); % plot the B1-circle
```

(d) For the points $\Gamma_1 = 0.2 + j0.4$ and $\Gamma_2 = -0.2 - j0.4$ (with $Z_0 = 50$), find the corresponding impedance/ admittance and their normalized values by using the Smith Chart software 'Smith V4.1'

(e) Which one of the circles drawn in (b) can the Γ-point move from Γ_1 to Γ=0 along? Which one of the circles drawn in (c) can the Γ-point move from $\Gamma = 0$ to Γ_2 along?

(f) Find the Γ-points for $Z_3 = 25 + j25$ (with $Z_0 = 50$) and $Y_4 = 0.01 + j0.01$ (with $Z_0 = 50$).

7.2 Using the Smith Chart

In connection with Example 7.1, find two L-type *LC* circuits transforming the normalized impedance Z_1=1+j1[Ω] or admittance Y_1=1/Z_1=0.5 – j0.5 [S] represented by point Γ_1 into the normalized impedance Z_2 = 0.5 – j0.5 [Ω] represented by point Γ_2, one via Γ_4 and the other via $\Gamma_0 = 0$ on the Smith chart in Figure 7.4.

7.3 Single-Stub Shunt Impedance Matching

Consider the single-stub shunt impedance matching discussed in Section 7.3.2.1 to find the distance (from the load) and length of shunt single stub so that they can match a given load impedance Z_L to the lossless transmission (or feed) line of characteristic impedance Z_0.

(a) Misra [M-2] presents the following formulas for determining the distance *d* and length *l* of the shunt single stub:

$$d \underset{(5.1.2)}{\overset{[\text{M-2}]}{=}} \frac{1}{\beta}\tan^{-1}\left\{\frac{b_L \pm \sqrt{b_L^2 - A(1-g_L)}}{A}\right\} \text{ with } A = g_L(g_L-1) + b_L^2 \quad \text{(P7.3.1a)}$$

$$l \underset{(5.1.5,6)}{\overset{[M-2]}{=}} \begin{cases} \tan^{-1}(b)/\beta & \text{for an open-circuited stub} \\ \cot^{-1}(-b)/\beta & \text{for a short-circuited stub} \end{cases} \tag{P7.3.1b}$$

where

$$y_L = \frac{Z_0}{Z_L} = g_L + jb_L \tag{P7.3.2}$$

$$b \underset{(5.1.3)}{\overset{[M-2]}{=}} \frac{\{b_L + \tan(\beta d)\}\{1 - b_L \tan(\beta d)\} - g_L^2 \tan(\beta d)}{\{g_L \tan(\beta d)\}^2 + \{1 - b_L \tan(\beta d)\}^2} \tag{P7.3.3}$$

```
function [ds,ls]=imp_match_1stub_shunt_Misra(Z0,ZL,r,os)
% Input: Z0 = Characteristic impedance, ZL = Load impedance
%         r = a+bi = Propagation constant
%         os = 'o'/'s' for open/short stub
% Output: ds = Distances of the stub from load
%          ls = Lengths of the stub
if nargin<4, os='s'; end
if length(Z0)>1, Z01=Z0(2); Z0=Z0(1); else Z01=Z0; end
a=real(r); b=imag(r); %lambda=2*pi/b;
YL=1/ZL; yL=YL*Z0; gL=real(yL); bL=imag(yL);
A=gL*(gL-1)+bL^2;
ds=atan((bL+[? ??]*sqrt(bL^2-A*(1-gL)))/A)/b; % Eq.(5.1.2) of [M-2]
tbd=tan(b*ds); den=(gL*tbd).^2+(1-bL*tbd).^2;
bs=-((bL+tbd).*(1-bL*tbd)-gL^2*tbd)./den; % Eq.(5.1.3,4) of [M-2]
ds=mod(ds*b,pi)/b;
if lower(os(1))=='s', ls=acot(-bs)/b; % Eq.(5.1.5) of [M-2]
 else ls=atan(bs)/b; % Eq.(5.1.6) of [M-2]
end
ls=mod(ls*b,pi)/b;
```

Complete the above MATLAB function '`imp_match_1stub_ shunt_ Misra()`' so that it uses the above formulas to compute the distances and lengths of two shunt single stubs for impedance matching.

(b) Use the MATLAB function '`imp_match_1stub_shunt()`' (presented in Section 7.3.2.1) and/or the above function '`imp_match_1 stub_-shunt_Misra()`' to determine the distances (from the load) and lengths of single shunt short-circuited stubs in the unit of wavelength so that they can match a load impedance $Z_L = 50 - j75[\Omega]$ to the lossless transmission (or feed) line of characteristic impedance $Z_0 = 100\,\Omega$.

(c) Use 'Smith 4.1' to get the solution for the problem dealt with in (b).

7.4 Single-Stub Series Impedance Matching

Consider the single-stub series impedance matching discussed in Section 7.3.2.2 to find the distance (from the load) and length of series single stub so that they can match a given load impedance Z_L to the lossless transmission (or feed) line of characteristic impedance Z_0.

(a) Misra [M-2] presents the following formulas for determining the distance d and length l of the series single stub:

$$d \underset{(5.1.8)}{\overset{[M\text{-}2]}{=}} \frac{1}{\beta}\tan^{-1}\left\{\frac{x_L \pm \sqrt{x_L^2 - A(1-r_L)}}{A}\right\} \text{ with } A = r_L(r_L-1) + x_L^2 \quad \text{(P7.4.1a)}$$

$$l \underset{(5.1.12,13)}{\overset{[M\text{-}2]}{=}} \begin{cases} \tan^{-1}(x)/\beta & \text{for a short-circuited stub} \\ \cot^{-1}(-x)/\beta & \text{for an open-circuited stub} \end{cases} \quad \text{(P7.4.1b)}$$

where

$$z_L = \frac{Z_L}{Z_0} = r_L + jx_L \text{ and}$$

$$x \underset{(5.1.9)}{\overset{[M\text{-}2]}{=}} \frac{\{x_L + \tan(\beta d)\}\{1 - x_L\tan(\beta d)\} - r_L^2\tan(\beta d)}{\{r_L\tan(\beta d)\}^2 + \{1 - x_L\tan(\beta d)\}^2} \quad \text{(P7.4.2,3)}$$

Complete the following MATLAB function '`imp_match_1stub_series_Misra()`' so that it uses the above formulas to compute the distances and lengths of two series single stubs for impedance matching.

```
function [ds,ls]=imp_match_1stub_series_Misra(Z0,ZL,r,os)
% Input: Z0 = Characteristic impedance of the TL and stub
%         ZL = Load impedance
%          r = a+bi = Propagation constant
%         os = 'o'/'s' for open/short stub
% Output: ds = Distance of the stub from the load
%         ls = Length of the stub
if nargin<4, os='s'; end
a=real(r); b=imag(r); lambda=2*pi/b;
zL=ZL/Z0; rL=real(zL); xL=imag(zL);
A=rL*(rL-1)+xL^2;
ds=atan((xL+[? ??]*sqrt(xL^2-A*(1-rL)))/A)/b; % Eq.(5.1.8) of [M-1]
tbd=tan(b*ds); den=(rL*tbd).^2+(1-xL*tbd).^2;
xs=-((xL+tbd).*(1-xL*tbd)-rL^2*tbd)./den; % Eq.(5.1.9) of [M-2]
if lower(os(1))=='o', ls=acot(-xs)/b; % Eq.(5.1.13) of [M-2]
 else ls=atan(xs)/b; % Eq.(5.1.12) of [M-2]
end
ds=mod(ds*b,pi)/b; ls=mod(ls*b,pi)/b;
```

(b) Use the MATLAB function '`imp_match_1stub_series()`' (presented in Section 7.3.2.2) and/or the above function '`imp_match_1stub_series_Misra()`' to determine the distances (from the load) and lengths of single series short-circuited stubs in the unit of wavelength so that they can match a load impedance $Z_L = 100 + j80$ [Ω] to the lossless transmission (or feed) line of characteristic impedance $Z_0 = 50\,\Omega$.

(c) Use 'Smith 4.1' to get the solution for the problem dealt with in (b).

7.5 Double-Stub Impedance Matching

Consider the double-stub impedance matching discussed in Section 7.3.2.3 to find the lengths of stub A (at d_0 from load) and stub B (at d from stub A) so that they can match a given load impedance Z_L to the lossless transmission (or feed) line of characteristic impedance Z_0.

(a) Misra [M-2] presents the following formulas for determining the distance d and length l of the shunt double stub:

$$l_A \underset{(5.2.7)}{\overset{[M\text{-}2]}{=}} \frac{1}{\beta}\cot^{-1}(-b_A) \text{ and } l_B \underset{(5.2.8)}{\overset{[M\text{-}2]}{=}} \frac{1}{\beta}\cot^{-1}(-b_B) \tag{P7.5.1a,b}$$

where

$$z_L = \frac{Z_L}{Z_0} = r_L + jx_L,\ y \underset{(5.2.2)}{\overset{[M\text{-}2]}{=}} \frac{1 + jz_L \tan(\beta d_0)}{z_L + j\tan(\beta d_0)} = g + jb \tag{P7.5.2}$$

$$b_A \underset{(5.2.5)}{\overset{[M\text{-}2]}{=}} \cot(\beta d)\left[1 - b\tan(\beta d) \pm \sqrt{g\sec^2(\beta d) - \{g\tan(\beta d)\}^2}\right] \tag{P7.5.3a}$$

$$b_B \underset{(5.2.6)}{\overset{[M\text{-}2]}{=}} \frac{g^2\tan(\beta d) - \{b + b_A + \tan(\beta d)\}\{1 - (b + b_A)\tan(\beta d)\}}{\{g\tan(\beta d)\}^2 + \{1 - (b + b_A)\tan(\beta d)\}^2} \tag{P7.5.3b}$$

```
function [lA,lB]=imp_match_2stub_Misra(Z0,ZL,r,d)
% Input:  Z0 = Characteristic impedance
%         ZL = Load impedance
%          r = a+bi = Propagation constant
%          d = Distance of the stub B from A (load)
%    where d may be given as [d d0] with d0: Distance of A from load
% Output: lA = Length of the stub A at (d0 from) load
%         lB = Length of the stub B at d from stub A
if numel(d)>1
 d0=d(2); d=d(1);
 ZL=Z0*(ZL+Z0*tanh(r*d0))/(Z0+ZL*tanh(r*d0)); % Eq.(7.1.16)
end
if imag(r)<eps, r=r*i; end % Since a real-valued r must be beta.
RL=real(ZL); a=real(r); b=imag(r); lambda=2*pi/b;
YL=1/ZL; yL=YL*Z0; gL=real(yL);
if gL>csc(b*d)^2 % Eq.(5.2.4) of [M-2]
 error('Double stub matching is impossible since gL>csc(b*d)^2!');
end
bd=b*d; tbd=tan(bd);
sgd=sqrt(gL*sec(bd)^2-(gL*tbd)^2);
bA=cot(bd)*(1-bL*tbd+[? ??]*sgd); % Eq.(P7.5.3a), Eq.(5.2.5) of [M-2]
bb1tbd=1-(bL+bA)*tbd;
sqgt=(gL*tbd)^2+bb1tbd.^2;
bB=(gL^2*tbd-(bL+bA+tbd).*bb1tbd)./sqgt; % Eq.(5.2.6) of [M-2]
lA=acot(-bA)/b; lB=acot(-bB)/b; % Eq.(P7.5.1a,b)
lA=mod(lA*b,pi)/b; lB=mod(lB*b,pi)/b;
```

Complete the above MATLAB function '`imp_match_2stub_Misra()`' so that it uses the above formulas to compute the lengths of two stubs for impedance matching.

(b) Use the MATLAB function '`imp_match_2stub()`' (presented in Section 7.3.2.3) and/or the above function '`imp_match_2stub_Misra()`' to determine the lengths of two stub A (at $d_0 = \lambda/8$ from load) and stub B (at $d = \lambda/8$ from stub A) in the unit of wavelength so that they can match a load impedance $Z_L = 100 + j50$ [Ω] to the lossless transmission (or feed) line of characteristic impedance $Z_0 = 50\,\Omega$.

(c) Noting that with the distance $d_0 = \lambda/8$ of stub A from the load, the virtual load impedance can be regraded as

$$Z_L' \underset{\gamma=j\beta}{\overset{(7.1.16)}{=}} Z_0 \frac{Z_L + Z_0 \tanh(j\beta d_0)}{Z_0 + Z_L \tanh(j\beta d_0)} = Z_0 \frac{Z_L + jZ_0 \tan(\beta d_0)}{Z_0 + jZ_L \tan(\beta d_0)} \tag{P7.5.4}$$

modify the MATLAB script "elec07e04.m" into "elec07p05.m" so that it returns the solution to this problem.

(d) Use 'Smith 4.1' to get the solution for the problem dealt with in (b) where you may use some transit points (that can be found in (c)) to move the Γ-point from Γ_L to $\Gamma_0 = 0$ as in Example 7.4.

7.6 Quarter-Wave Transformer (QWT)
Consider the MATLAB function '`chebtr()`' presented in [O-1] to design the Chebyshev QWTs (Section 7.3.4.2) satisfying the requirements for attenuation and fractional bandwidth where the attenuation is specified indirectly through reflection coefficient or SWR (standing wave ratio) as

$$A \underset{(13.3.8)}{\overset{[O-1]}{=}} 20\log_{10}\frac{|\Gamma_L|}{|\Gamma_1|_{max}} \overset{(7.1.13a)}{=} 20\log_{10}\left(\frac{SWR_L - 1}{SWR_L + 1}\frac{SWR_{max} - 1}{SWR_{max} + 1}\right) \quad (P7.6.1)$$

Here, $|\Gamma_1|$ is the reflection coefficient from the first section of the several section constituting the QWT and $|\Gamma_L|$ is the reflection coefficient from the load. Note that the usage of '`chebtr()`' is as follows:

```
[Y,Az,Bz]=chebtr(Ys,YL,A,fb)
```

where
Input: `Ys` = source (line) admittance Y_s, `YL` = load admittance Y_L
`A` = attenuation defined by Eq. (P7.6.1)

$$\texttt{fb} = \text{fractional bandwidth } f_b = \frac{f_2 - f_1}{f_0} \text{ with } f_0 = \frac{f_1 + f_2}{2} \quad (P7.6.2)$$

and $f_0/f_1/f_2$: center/l ower(left)/upper(right) band edge frequencies

Output: `Y` = $[Y_s\ Y_1\ \dots\ Y_N\ Y_L]$ consisting of Y_s, characteristic admittances of each section, and Y_L
`Az/Bz` = Denominator/Numerator polynomials of reflection response in $z = \exp(j\pi f/f_c)$

(a) Design a Chebyshev QWT that matches a $Z_L = 200\,\Omega$-load to a Z_0=50 Ω-line with SWR_{max}=1.25 over the frequency band [50, 150] MHz where the center frequency $f_0 = 100$ MHz. You can complete and run the following script "elec07p06a.m", which uses '`chebtr()`' to design such a QWT. Noting that the characteristic admittance vector Y returned as the first output argument of '`chebtr()`' contains Y_s/Y_L, as the first/last entry, tell the number of sections and characteristic impedances of each section of the designed QWT.

```
%elec07p06a.m
clear, clf
Z0=50; ZL=200; % Line characteristic impedance and Load impedance
GL = Z2G(ZL,Z0); SWRmax = 1.25;
f1 = 50; f2 = 150; % Band edge frequencies
f0 = (f1+f2)/2; % Operating frequency
fb = (f2-f1)/f0; % Required fractional bandwidth
As = 20*log10(GL*(SWRmax+1)/(SWRmax-1)); % Attenuation of
% reflectionless band (relative to unmatched case)
[Y,a,b] = chebtr(1/Z0,1/ZL,As,fb); % Chebyshev transformer design
% where b/a = Numerator/Denominator polynomials of reflection response
Z = 1./? % Characteristic impedances of each section of designed QWT
N=length(Z)-2; % Number of sections
Zr=Z(2:N+1); % Real characteristic impedance vector of designed QWT
```

(b) To see if the above design Zr satisfies the following QWT conditions:

$$Z_{\text{in},1} \overset{(7.3.22)}{=} Z_0 \text{ where } Z_{\text{in},n} \overset{(7.3.23)}{=} \frac{Z_{\text{in},n+1} + jZ_{\text{r},n}\tan\beta l}{Z_{\text{r},n} + jZ_{\text{in},n+1}\tan\beta l} Z_n \tag{P7.6.3}$$

$$\text{with } Z_{\text{in},N+1} = Z_{\text{L}} \text{ for } n = 1,\ldots,N$$

$$Z_{\text{r},n} = \sqrt{Z_{\text{out},n}Z_{\text{in},n+1}} = \sqrt{Z_{\text{in},n}Z_{\text{in},n+1}},\; Z_{\text{out},n} = Z_{\text{in},n} \text{ for } n = 1,\ldots,N \tag{P7.6.4}$$

$$Z_{\text{out},N+1} = Z_{\text{L}} \text{ where } Z_{\text{out},n+1} \overset{(7.3.23)}{=} \frac{Z_{\text{out},n} + jZ_{\text{r},n}\tan\beta l}{Z_{\text{r},n} + jZ_{\text{out},n}\tan\beta l} Z_n \tag{P7.6.5}$$

$$\text{with } Z_{\text{out},1} = Z_0 \text{ for } n = 1,\ldots,N$$

complete/run the above MATLAB script "elec07p06b.m" and discuss the result (see Figure P7.6.1).

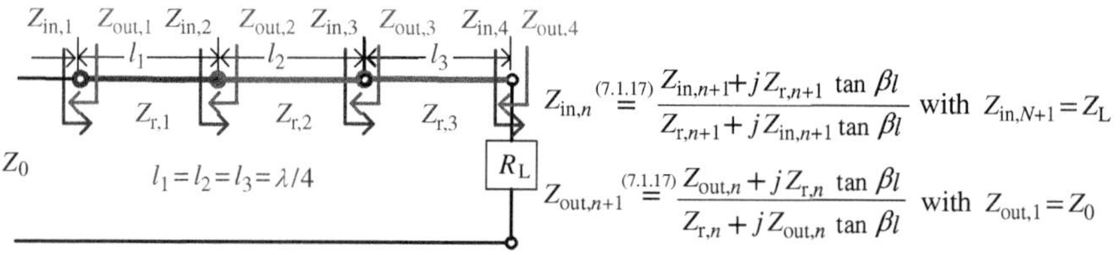

Figure P7.6.1 An *N*-section impedance transformer.

```
%elec07p06b.m
Zl=@(ZL,Z0,bl) (ZL+j*Z0*tan(bl))./(Z0+j*ZL.*tan(bl))*Z0;
Zo(1)=Z0; Zi(N+1)=ZL; bl=pi/2;
for n=1:N, Zo(n+1)=Zl(Zo(n),Z?(n),bl); end % Eq. (P7.6.3b)
for n=N:-1:1
  Zi(n)=Zl(Z?(n+1),Zr(n),bl); % Eq. (P7.6.3b)
  Zr1(n)=sqrt(Zo(n)*Zi(n+1)); % Eq. (7.3.19) at each stage
end
Mismatching=norm([Zi(1)-Z0 Zr-Zr1 Zo-Zi Zo(N+1)-ZL])
```

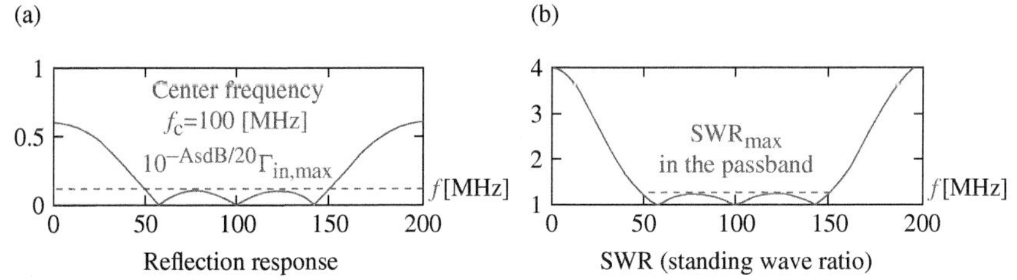

Figure P7.6.2 Reflection response and SWR.

(c) Noting that the fifth input argument of '`multi_line()`' (presented in Section 7.3.4) should have the normalized phase constant $\beta = 2\pi$ as the first entry, use the MATLAB function to get the reflection response of the QWT designed in (a) and plot it together with SWR (Eq. (7.1.13a)) for the frequency range 0~$2f_c$. To this end, complete and run the following script "elec07p06c.m" to get the graphs as shown in Figure P7.6.2.

```
%elec07p06c.m
f = linspace(0,2*f0,401); % Frequency range [0,200]MHz
ls = ones(1,N)/4; % Quarter-wave lengths
bb = 2*pi*[1 f/f0]; % Normalized frequency vector
%  with phase constant beta as its 1st entry
[Zin,Gam,bb]=multi_line(Z?,ls,Z0,ZL,bb);
SWR = (1+abs(Gam))./(1-abs(Gam)); % SWR by Eq.(7.1.13a)
subplot(221), plot(f,abs(Gam)), hold on
  plot(f([1 end]),max(Gam)*10^(-As/20)*[1 1],'r:')
subplot(222), plot(f,SWR, [f1 f2],SWRmax*[1 1],'r:')
```

(d) To realize the meanings of the second/third output arguments of '`chebtr()`' as the denominator/numerator polynomials of the reflection response in $z = \exp(j\pi f/f_0)$, use them to compute the reflection response

$$\Gamma[z] = \frac{b_1 z^{M-1} + b_2 z^{M-2} + \cdots + b_{M-1} z + b_M}{a_1 z^{N-1} + a_2 z^{N-2} + \cdots + a_{N-1} z + a_N} \text{ with } z = e^{j\pi f/f_0} \quad \text{(P7.6.5)}$$

and see if it conforms with that obtained as the second output argument of '`multi_line()`' in (c) by running the following MATLAB script "elec07p06d.m".

```
%elec07p06d.m
w=pi*f/fc; Gam1=freqz(b,a,w); % Alternative way of computing Gam
Gam1_=polyval(b,exp(j*w))./polyval(a,exp(j*w)); % Eq.(P7.6.5)
discrepancy=norm([Gam-Gam1 Gam-Gam1_])
```

7.7 Design of LPF Using Stubs [L-1]

With source/load impedances R_s = 50 [Ω]/Z_L = 50 [Ω], design a Chebyshev I LPF with passband ripple R_p = 0.5 dB for $f \le f_p$ (passband edge frequency) = 3 GHz and stopband attenuation A_s = 40 dB for $f \ge f_s$ (stopband edge frequency) = 6 GHz for fabrication using microstrip lines with shunt stubs only.

Referring to Example 7.5, take the following steps to design such an LPF:

(Step 0) Determine the order N and cutoff frequency ω_c [rad/s] of the LPF to satisfy the given frequency response specifications by running the following MATLAB statements:

```
>>wp=2*pi*3e9; ws=2*pi*6e9; Rp=0.5; As=40
  [N,wc]=cheb1ord(wp,ws,Rp,As,'s')
```

(Step 1) Determine the normalized LPF prototype element values by running

```
>>Rs=50; % Source (line) characteristic impedance
  type='C'; % for the Nth order Chebyshev with Rp[dB]
  gg=LPF_ladder_coef(N,type,Rp)
```

(Step 2) Determine the normalized LC values of the prototype LPF by running

```
>>Rs0=1; pT='T'; wc=1; % for a T-type prototype filter
  [LCRs,LCRs_,Gs]=LC_ladder(Rs0,gg,pT,wc);
  LCRs, LCRs_
```

With the results, fill in the parameter values of the circuit in Figure P7.7.1.

(Step 3) Convert the L/C values into the impedances of short-/open-circuited stubs by running the following statement:

```
>>[Zs,SPsos]=LC2Z(LCRs,LCRs_); Zs_0=Zs, SPsos_0=SPsos
```

With the results, fill in the parameter values in Figure P7.7.2(a).

(Step 4) Noting that even if we are going to convert every stub into parallel (shunt) stubs, we provisionally should convert the two Po's (parallel open-circuited stubs) Z_1 and Z_5 (at either ends) into Ss's (series

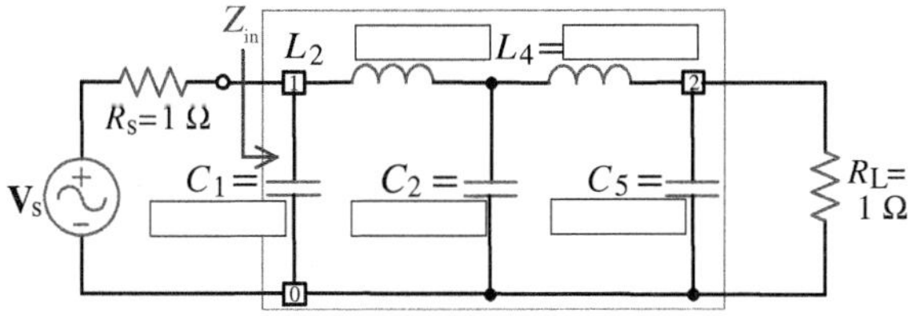

Figure P7.7.1 Normalized prototype LPF of order $N = 3$.

short-circuited stubs) so that the Ss's Z_2 and Z_4 can have UEs of $Z_0 = 1$ (needed for conversion), let us use Kuroda identity 1 to convert Z_1 and Z_5 into Ss's. Thus, we insert UEs between Z_1/Z_5 and the transmission line (as Figure P7.7.2(b)) and run the following statements:

```
>> [Z0s(2),Zs(1),SPsos{1}]=Kuroda(Zs(1),SPsos{1});
   [Z0s(4),Zs(5),SPsos{5}]=Kuroda(Zs(5),SPsos{5});
   Z0s, Zs, SPsos
```

With the results, fill in the parameter values in Figure P7.7.2(c).

(Step 5) Noting that in Figure P7.7.2(c) the two Ss's Z_1 and Z_5 (at either ends) have no UE needed for conversion (while Z_2 and Z_4 have UEs), let us insert UEs of Z_0=1 between them and TL as shown in Figure P7.7.2(d).

```
>>Z0s(1)=1; Z0s(5)=1;
```

(Step 6) Now that every Ss has got its UEs for conversion, let us use Kuroda identity 2 to convert the Ss's Z_1, Z_2, Z_4, and Z_5 into Po's by running

```
>>for m=[1 2 4 5]
      [Z0s(m),Zs(m),SPsos{m}]=Kuroda(Zs(m),SPsos{m},Z0s(m));
  end
  Z0s, Zs, SPsos
```

With the results, fill in the parameter values in Figure P7.7.2(e).

Alternatively, the process from Step 4 to Step 6 can be performed by using the MATLAB function '`S2P_stubs()`' (listed below) as follows:

```
>>Zs=Zs_0; Z0s=zeros(size(Zs)); SPsos=SPsos_0;
  [Z0sf,Zsf,SPsosf]=S2P_stubs(Z0s(1:2),Zs(1:2),SPsos(1:2));
  [Z0sb,Zsb,SPsosb]= ...
       S2P_stubs(Z0s(end:-1:4),Zs(end:-1:4),SPsos(end:-1:4));
  Z0s_=[Z0sf Z0s(3) Z0sb(end:-1:1)]; Zs_=[Zsf Zs(3)
  Zsb(end:-1:1)]
  SPsos_=[SPsosf SPsos(3) SPsosb(end:-1:1)]
```

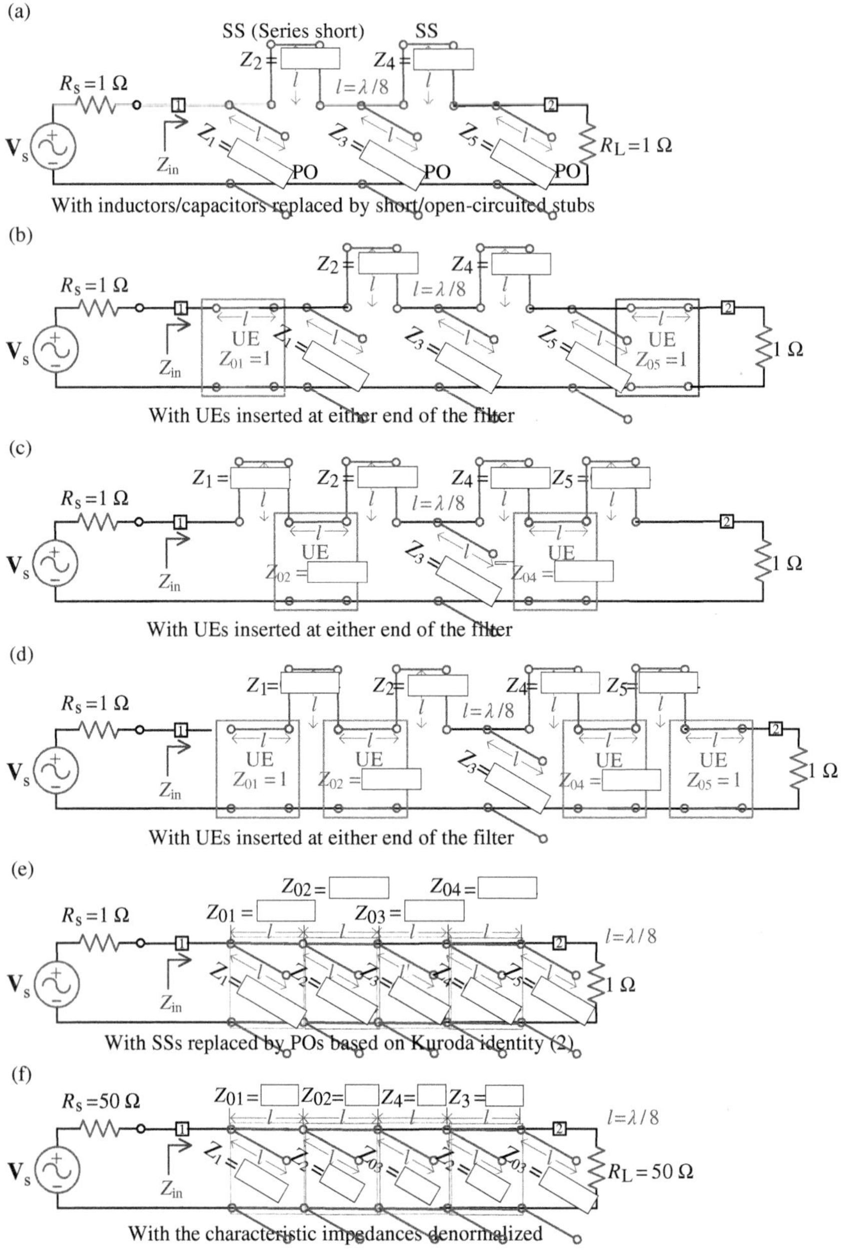

Figure P7.7.2 Microstrip filter design for Problem 7.7.

(Step 7) Denormalize the normalized characteristic impedances by multiplying them with $R_s = 50\,\Omega$:

```
>>Zs=Zs*Rs, Z0s=Z0s*Rs
```

With the results, fill in the parameter values in Figure P7.7.2(f).

```
function [Z0s,Zs,SPsos]=S2P_stubs(Z0s,Zs,SPsos)
% To convert all series stubs (SS) into parallel stubs (PS)
% Input:
%  Z0s = Characteristic impedances of UEs with 0 for no UE
%  Zs  = Stub impedances
%  SPsos = {'Po','Ss', .. } for parallel-open, series-short stubs
% Output:
%  Z0s, Zs, SPsos = Results with all stubs parallel
% where any -ve sign of Zs implies that it is an open-circuited stub.
N=numel(Zs); nS=N+1;
for n=1:N % To find the position of the 1st SS
   if upper(SPsos{n}(1))=='S', nS=n; break; end
end
if nS<=N % If any stub is a SS
  for n=1:nS-1 % All PSs into SSs before the 1st SS
    [Z0s(n+1),Zs(n),SPsos{n}]=Kuroda(Zs(n),SPsos{n},Z0s(n));
  end
  for n=1:nS % All the SSs into PSs up to the 1st SS
    [Z0s(n),Zs(n),SPsos{n}]=Kuroda(Zs(n),SPsos{n},Z0s(n));
  end
[Z0s,Zs,SPsos]=S2P_stubs(Z0s,Zs,SPsos); % Self-calling recursively
end
```

(Step 8) Frequency scale the circuit by determining the line/stub lengths as $l = \lambda/8 = v_p/f_c/8$ [m] where v_p [m/s] is the phase velocity of signal propagating on the line and f_c is the cutoff frequency given as a design specification.

(Step 9) To plot the frequency responses of the two filters, one (called a lumped-element filter) consisting of lumped elements L/C (Figure P7.7.1) and the other (called a distributed-element filter) consisting of microstrip lines/stubs (Figure P7.7.2(f)), complete and run the following MATLAB script "elec07p07b.m" to get the frequency responses as shown in Figure P7.7.3:

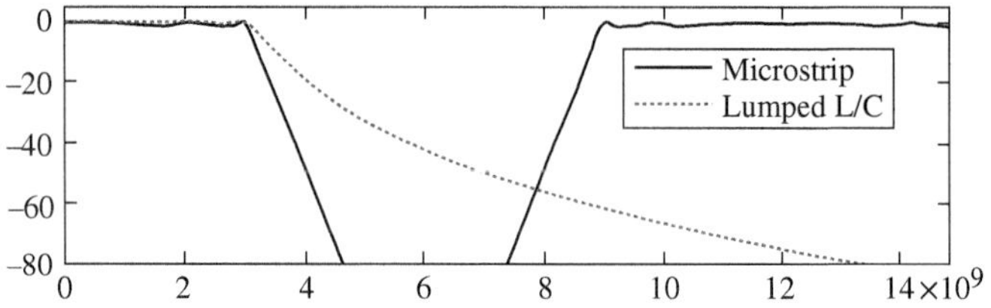

Figure P7.7.3 Frequency responses of the lumped-element filter and distributed-element filter.

```
%elec07p07b.m
Z0strip=[Zs(1) Z0?(1) Zs(2) Z0s(?) Zs(3) Z0s(4) ??(4) Z0s(5) Zs(5)];
SPsos={SPsos{1} 'Tl' SPsos{2} '??' SPsos{3} 'Tl' SPsos{?} 'Tl' SPsos{5}};
l=1/8; fc=3e9; % Stub length and Cutoff frequency
fN=linspace(0,pi/4,1000); % Normalized frequency range
w=fN*2*pi*fc; % Real frequency range
bl=2*pi*w*l/fc; % Range of bl (bl=2*pi*l at f=fc=4e9[Hz])
% To get the frequency response and input impedance
ZL=Rs; [Gw,Zinw]=GZ_strip(Z0?????,??sos,ZL,bl);
Gw=Zinw./(Rs+Zinw).*Gw; % Overall frequency response
Gwmag=20*log10(max(abs(Gw),1e-5)); % Frequency response magnitude in dB
plot(w,Gwmag-max(Gwmag),'b'), hold on % with the maximum 0dB
s=j*w/fc; Gw0=eval(Gs); % Frequency response from the transfer function
Gwmag0=20*log10(max(abs(Gw0),1e-5));
plot(w,Gwmag0-max(Gwmag0),'r:') % To make the maximum 0dB
legend('Microstrip','Lumped L/C'), axis([w([1 end]) -80 5])
```

Note that the whole process from Step 1 to Step 7 can be performed by using the MATLAB function '`ustrip_LPF()`' (listed below) as follows:

```
>>Rs=50; type='C'; wp=2*pi*3e9; ws=2*pi*6e9; Rp=0.5; As=40
  [Z0strip,SPsos]=ustrip_LPF(Rs,type,[Rp As],[wp ws],pT)
```

7.8 Design of T-type Impedance Transformers Using the Smith Chart
Consider making T-type ladders with quality factor $Q = 2$ for transforming $Z_L = 20 - j15[\Omega]$ to $Z_s = 50[\Omega]$ at frequency $f_c = 2$ GHz.

```
function [Z0strip,SPsos,LCRs,LCRs_]= ...
                    ustrip_LPF(Rs,type,RpAs,wps,pT)
% Input:
%  Rs = Line characteristic impedance
%  type - 'B'/'C1'/'C2'/'E' for B/Chebyshev I/II/Elliptic filter
%  RpAs = [Rp As] Passband ripple/Stopband attenuation[dB]
%  wps = [wp ws] = Passband/Stopband edge frequencies[rad]
%  pT = 'pi'/'T' for pi-/T-form
% This function can be called as ustrip_LPF(Rs,type,Rp,N,pT).
% Copyleft: Won Y. Yang, wyyang53@hanmail.net, CAU for academic use only
if nargin<5, pT='pi'; end % Pi-form by default
if numel(wps)<2 % When called as ustrip_LPF(Rs,type,Rp,N,pT)
  N=wps; Rp=RpAs;
 else
 Rp=RpAs(1); As=RpAs(2); wp=wps(1); ws=wps(2);
 if lower(type(1))=='b', [N,wc]=buttord(wp,ws,Rp,As,'s');
  elseif lower(type(1))=='c'
   if numel(type)<2|type(2)=='1', [N,wc]=cheb1ord(wp,ws,Rp,As,'s');
    else [N,wc]=cheb2ord(wp,ws,Rp,As,'s');
   end
  elseif lower(type(1))=='e', [N,wc]=ellipord(wp,ws,Rp,As,'s');
  else error('What type of filter do you want? (inside ustrip_LPF)');
 end
end
gg=LPF_ladder_coef(N,type,Rp);
% To find the L & C values of the prototype pi filter
Rs0=1;
[LCRs,LCRs_,Gs]=LC_ladder(Rs0,gg,pT,1);
format short e;
LCs=LCRs(1:end-1);
LCs_=LCRs_(1:end-1);
% To convert the L/C values into the characteristic impedance
% of short/open-circuited stub
[Zs,SPsos]=LC2Z(LCRs,LCRs_);
% To convert all SSs (series stubs) into PSs (parallel stubs)
M=ceil(N/2);
if SPsos{M}(1)~='P', M=M+1; end
M1=M+1; Z0s=zeros(1,N);
Z0s_f=Z0s(1:M-1); Zs_f=Zs(1:M-1); SPsos_f=SPsos(1:M-1);
Z0s_b=Z0s(end:-1:M1); Zs_b=Zs(end:-1:M1); SPsos_b=SPsos(end:-1:M1);
[Z0s_f,Zs_f,SPsos_f]=S2P_stubs(Z0s_f,Zs_f,SPsos_f );
[Z0s_b,Zs_b,SPsos_b]=S2P_stubs(Z0s_b,Zs_b,SPsos_b);
Z0s=[Z0s_f Z0s_b(end:-1:1)];
Zs=[Zs_f Zs(M) Zs_b(end:-1:1)];
SPsos=[SPsos_f SPsos(M) SPsos_b(end:-1:1)];
% Denormalize the normalized characteristic impedances
% by multiplying them with Rs.
Z0s=Z0s*Rs, Zs=Zs*Rs
Z0strip=reshape([Zs; [Z0s 0]],1,2*N); Z0strip=Z0strip(1:end-1);
SPsos=[SPsos; repmat({'Tl'},1,N)];
SPsos=reshape(SPsos,1,2*N); SPsos=SPsos(1:end-1);
```

(a) Complete and run the following MATLAB script "elec07p08.m" to make such T-type ladders.

```
%elec07p08.m
Zs=50; ZL=20-15i; Q=2; fc=2e9;
[LCs1,LCs1_,ff1,Gws1]=imp_matcher_T(??,ZL,-Q,fc);
[LCs2,LCs2_,ff2,Gws2]=imp_matcher_T(Zs,??,Q,fc);
```

(b) Referring to Figure P7.8, note that, with $Q \leq 2$, there are four paths along resistance/conductance curves from point Γ_L (representing $Z_L = 20 - j15$ [Ω]) to Γ_s (representing $Z_s = 50$ [Ω]), each via Γ_2-Γ_3, Γ_4-Γ_3, Γ_2-Γ_5, and Γ_4-Γ_5 where Γ_2 and Γ_4 are on the lower and upper $Q = 2$-circles, respectively. Use 'Smith V4.1' to solve the same design problem as (b).

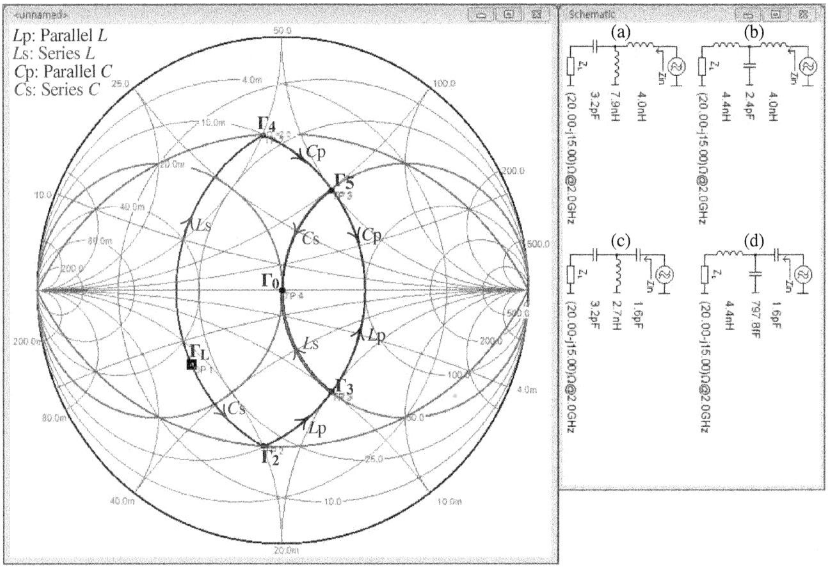

Figure P7.8 Using 'Smith 4.1' to design T-type impedance matchers in Problem 7.8

7.9 Impedance/Admittance from Resistance/Conductance
Complete the following MATLAB function 'RG2ZY()' so that it can return the impedance Z and admittance Y with given resistance R and conductance G. Use it for $R_5 = 60\,\Omega$ and $G_5 = 0.01$ S (Eq. (E7.8.2)) to check if it works properly.

```
function [Z,Y]=RG2ZY(R,G)
X=sqrt(R/G-?^2); Z=R+j*X; % Reactance and Impedance
B=sqrt(?/R-G^2); Y=G+j*B; % Susceptance and Admittance
```

8

Two-Port Network and Parameters

CHAPTER OUTLINE

This chapter introduces two-port parameters such as immittance parameters, hybrid parameters, and transmission (ABCD) parameters, and scattering parameter.

8.1 Two-Port Parameters [Y-1]

You might feel like calling any pair of terminals in a circuit a port if only an external element can be connected between the two terminals (nodes). However, in order for a pair of terminals to be a *port*, it should satisfy the *port condition* or *current requirement* that the current flowing into the circuit through one terminal must equal the current flowing out of the circuit through the other terminal. In fact, for one-port (two-terminal) networks, we do not

Electronic Circuits with MATLAB®, PSpice®, and Smith Chart, First Edition. Won Y. Yang, Jaekwon Kim, Kyung W. Park, Donghyun Baek, Sungjoon Lim, Jingon Joung, Suhyun Park, Han L. Lee, Woo June Choi, and Taeho Im.

Companion website: www.wiley.com/go/yang/electroniccircuits

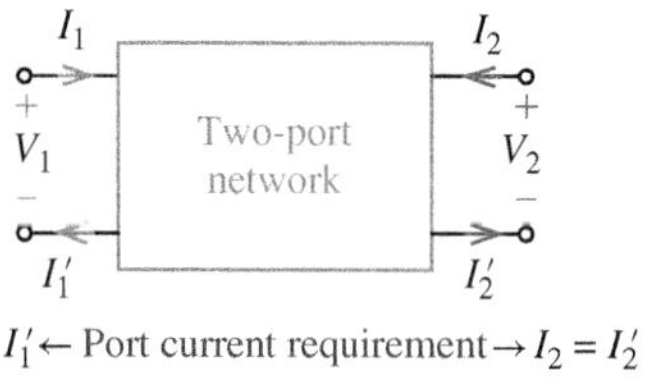

Figure 8.1 A two-port network.

have to bother about the port condition since Kirchhoff's current law (KCL) does not allow the port condition to be violated for any pair of terminals. This is no longer true for two-port networks.

A circuit is called a *two-port network* if it has two ports available for external connection, each of which is supposed to satisfy the port condition. A two-port network is often depicted as in Figure 8.1, where the left/right port is somewhat arbitrarily referred to as the input(source)/output(load) port or the *primary/secondary port*, respectively, and the voltage/current are defined at each port.

8.1.1 Definitions and Examples of Two-Port Parameters

There are six sets of parameters characterizing a two-port network defined as follows.

<*Immittance* (Impedance/Admittance) or *z/y-parameters*>

$$\begin{bmatrix} V_1 \\ V_2 \end{bmatrix} = \begin{bmatrix} z_{11} & z_{12} \\ z_{21} & z_{22} \end{bmatrix} \begin{bmatrix} I_1 \\ I_2 \end{bmatrix} \text{ with } \begin{matrix} z_{11} = \left.\dfrac{V_1}{I_1}\right|_{I_2=0}, & z_{12} = \left.\dfrac{V_1}{I_2}\right|_{I_1=0} \\ z_{21} = \left.\dfrac{V_2}{I_1}\right|_{I_2=0}, & z_{22} = \left.\dfrac{V_2}{I_2}\right|_{I_1=0} \end{matrix} \tag{8.1.1}$$

$$\begin{bmatrix} I_1 \\ I_2 \end{bmatrix} = \begin{bmatrix} y_{11} & y_{12} \\ y_{21} & y_{22} \end{bmatrix} \begin{bmatrix} V_1 \\ V_2 \end{bmatrix} \text{ with } \begin{matrix} y_{11} = \left.\dfrac{I_1}{V_1}\right|_{V_2=0}, & y_{12} = \left.\dfrac{I_1}{V_2}\right|_{V_1=0} \\ y_{21} = \left.\dfrac{I_2}{V_1}\right|_{V_2=0}, & y_{22} = \left.\dfrac{I_2}{V_2}\right|_{V_1=0} \end{matrix} \tag{8.1.2}$$

<*Transmission* (ABCD) or *a/b-parameters*>

$$\begin{bmatrix} V_1 \\ I_1 \end{bmatrix} = \begin{bmatrix} a_{11} & a_{12} \\ a_{21} & a_{22} \end{bmatrix} \begin{bmatrix} V_2 \\ -I_2 \end{bmatrix} \text{ with } \begin{matrix} a_{11} = \left.\dfrac{V_1}{V_2}\right|_{I_2=0}, & a_{12} = \left.\dfrac{V_1}{-I_2}\right|_{V_2=0} \\ a_{21} = \left.\dfrac{I_1}{V_2}\right|_{I_2=0}, & a_{22} = \left.\dfrac{I_1}{-I_2}\right|_{V_2=0} \end{matrix} \tag{8.1.3}$$

$$\begin{bmatrix} V_2 \\ I_2 \end{bmatrix} = \begin{bmatrix} b_{11} & b_{12} \\ b_{21} & b_{22} \end{bmatrix} \begin{bmatrix} V_1 \\ -I_1 \end{bmatrix} \text{ with } \begin{matrix} b_{11} = \left.\dfrac{V_2}{V_1}\right|_{I_1=0}, & b_{12} = \left.\dfrac{V_2}{-I_1}\right|_{V_1=0} \\ b_{21} = \left.\dfrac{I_2}{V_1}\right|_{I_1=0}, & b_{22} = \left.\dfrac{I_2}{-I_1}\right|_{V_1=0} \end{matrix} \quad (8.1.4)$$

<*Hybrid* or *h/g-parameters*>

$$\begin{bmatrix} V_1 \\ I_2 \end{bmatrix} = \begin{bmatrix} h_{11} & h_{12} \\ h_{21} & h_{22} \end{bmatrix} \begin{bmatrix} I_1 \\ V_2 \end{bmatrix} \text{ with } \begin{matrix} h_{11} = \left.\dfrac{V_1}{I_1}\right|_{V_2=0}, & h_{12} = \left.\dfrac{V_1}{V_2}\right|_{I_1=0} \\ h_{21} = \left.\dfrac{I_2}{I_1}\right|_{V_2=0}, & h_{22} = \left.\dfrac{I_2}{V_2}\right|_{I_1=0} \end{matrix} \quad (8.1.5)$$

$$\begin{bmatrix} I_1 \\ V_2 \end{bmatrix} = \begin{bmatrix} g_{11} & g_{12} \\ g_{21} & g_{22} \end{bmatrix} \begin{bmatrix} V_1 \\ I_2 \end{bmatrix} \text{ with } \begin{matrix} g_{11} = \left.\dfrac{I_1}{V_1}\right|_{I_2=0}, & g_{12} = \left.\dfrac{I_1}{I_2}\right|_{V_1=0} \\ g_{21} = \left.\dfrac{V_2}{V_1}\right|_{I_2=0}, & g_{22} = \left.\dfrac{V_2}{I_2}\right|_{V_1=0} \end{matrix} \quad (8.1.6)$$

Note that, while the z- and y-parameter matrices make the inverse of each other and so do the h- and g-parameter matrices, the a- and b-parameter matrices do not have such an inverse relationship.

Example 8.1 ***z*-parameters**

We can use Eq. (8.1.1) to find the z-parameters of the two-port network depicted in Figure 8.2(a) as

$$\begin{aligned} z_{11} &= \left.\frac{V_1}{I_1}\right|_{I_2=0} = \frac{11}{3} \| \left(\frac{11}{2} + 11\right) = \frac{11/3 \times (11/2 + 11)}{11/3 + (11/2 + 11)} = 3 \\ z_{12} &= \left.\frac{V_1}{I_2}\right|_{I_1=0} = \frac{11}{(11/3 + 11/2) + 11} \times \frac{11}{3} = 2 \\ z_{21} &= \left.\frac{V_2}{I_1}\right|_{I_2=0} = \frac{11/3}{11/3 + (11/2 + 11)} \times 11 = 2 \\ z_{22} &= \left.\frac{V_2}{I_2}\right|_{I_1=0} = \left(\frac{11}{2} + \frac{11}{3}\right) \| 11 = \frac{(11/3 + 11/2) \times 11}{(11/3 + 11/2) + 11} = 5 \end{aligned} \quad \text{(E8.1.1)}$$

Alternatively, noting that the two-port network is of Π-type, we can suppose two current sources I_1 and I_2 are connected at each port as depicted in Figure 8.2(b), set up the node equation, and solve it for V_1 and V_2 as

$$\begin{bmatrix} 3/11 + 2/11 & -2/11 \\ -2/11 & 1/11 + 2/11 \end{bmatrix} \begin{bmatrix} V_1 \\ V_2 \end{bmatrix} = \begin{bmatrix} I_1 \\ I_2 \end{bmatrix}; \; Y\,V = I; \; V = Y^{-1} I \quad \text{(E8.1.2)}$$

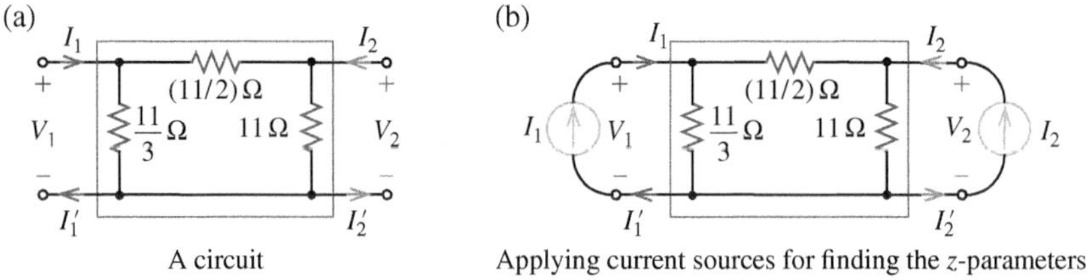

Figure 8.2 A circuit and applying current sources to it for finding its z-parameters.

$$\begin{bmatrix} V_1 \\ V_2 \end{bmatrix} = \frac{1}{15/11^2 - 4/11^2} \begin{bmatrix} 3/11 & 2/11 \\ 2/11 & 5/11 \end{bmatrix} \begin{bmatrix} I_1 \\ I_2 \end{bmatrix} = \begin{bmatrix} 3 & 2 \\ 2 & 5 \end{bmatrix} \begin{bmatrix} I_1 \\ I_2 \end{bmatrix} \tag{E8.1.3}$$

Then, we match this equation with Eq. (8.1.1) to get the same z-parameters as the above in matrix form.

$$\begin{bmatrix} z_{11} & z_{12} \\ z_{21} & z_{22} \end{bmatrix} = \begin{bmatrix} 3 & 2 \\ 2 & 5 \end{bmatrix} \tag{E8.1.4}$$

Example 8.2 ***y*-parameters**

We can use Eq. (8.1.2) to find the y-parameters of the two-port network depicted in Figure 8.3(a).

$$\begin{aligned} y_{11} &= \left.\frac{I_1}{V_1}\right|_{V_2=0} = \frac{1}{1+(2\|3)} = \frac{1}{1+6/5} = \frac{5}{11} \\ y_{12} &= \left.\frac{I_1}{V_2}\right|_{V_1=0} = -\frac{1}{3+(1\|2)} \times \frac{2}{1+2} = -\frac{2}{11} \\ y_{21} &= \left.\frac{I_2}{V_1}\right|_{V_2=0} = -\frac{1}{1+(2\|3)} \times \frac{2}{2+3} = -\frac{2}{11} \\ y_{22} &= \left.\frac{I_2}{V_2}\right|_{V_1=0} = \frac{1}{3+(1\|2)} = \frac{3}{11} \end{aligned} \tag{E8.2.1}$$

Alternatively, noting that the two-port network is of T-type, we suppose two voltage sources V_1 and V_2 are connected at each port as depicted in Figure 8.3 (b), set up the mesh equation in I_1 and I_2, and solve it as

$$\begin{bmatrix} 1+2 & 2 \\ 2 & 2+3 \end{bmatrix} \begin{bmatrix} I_1 \\ I_2 \end{bmatrix} = \begin{bmatrix} V_1 \\ V_2 \end{bmatrix}; \; ZI = V; \; I = Z^{-1}V \tag{E8.2.2}$$

$$; \begin{bmatrix} I_1 \\ I_2 \end{bmatrix} = \frac{1}{3\times 5 - 2\times 2} \begin{bmatrix} 5 & -2 \\ -2 & 3 \end{bmatrix} \begin{bmatrix} V_1 \\ V_2 \end{bmatrix} = \begin{bmatrix} 5/11 & -2/11 \\ -2/11 & 3/11 \end{bmatrix} \begin{bmatrix} V_1 \\ V_2 \end{bmatrix} \tag{E8.2.3}$$

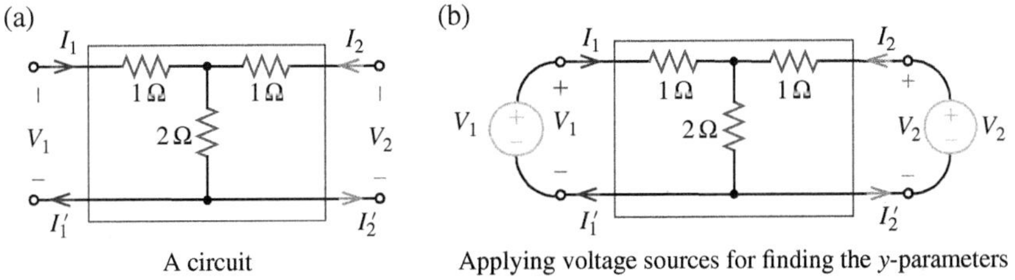

Figure 8.3 A circuit and applying voltage sources to it for finding its *y*-parameters.

Then, we match this equation with Eq. (8.1.2) to get the same *y*-parameters as the above in matrix form:

$$\begin{bmatrix} y_{11} & y_{12} \\ y_{21} & y_{22} \end{bmatrix} = \begin{bmatrix} 5/11 & -2/11 \\ -2/11 & 3/11 \end{bmatrix} \tag{E8.2.4}$$

Example 8.3 ***a*-parameters**

We can use Eq. (8.1.3) to find the *a*-parameters of the two-port network depicted in Figure 8.4(a) as

$$\begin{aligned} a_{11} &= \left.\frac{V_1}{V_2}\right|_{I_2=0} = 1, \quad a_{12} = \left.\frac{V_1}{-I_2}\right|_{V_2=0} = 0 \\ a_{21} &= \left.\frac{I_1}{V_2}\right|_{I_2=0} = Y_a, \quad a_{22} = \left.\frac{I_1}{-I_2}\right|_{V_2=0} = 1 \end{aligned} ; \quad \begin{bmatrix} a_{11} & a_{12} \\ a_{21} & a_{22} \end{bmatrix} = \begin{bmatrix} 1 & 0 \\ Y_a & 1 \end{bmatrix} \tag{E8.3.1}$$

The *a*-parameters of the two-port network depicted in Figure 8.4(b) can be obtained as

$$\begin{aligned} a_{11} &= \left.\frac{V_1}{V_2}\right|_{I_2=0} = 1, \quad a_{12} = \left.\frac{V_1}{-I_2}\right|_{V_2=0} = Z_b \\ a_{21} &= \left.\frac{I_1}{V_2}\right|_{I_2=0} = 0, \quad a_{22} = \left.\frac{I_1}{-I_2}\right|_{V_2=0} = 1 \end{aligned} ; \quad \begin{bmatrix} a_{11} & a_{12} \\ a_{21} & a_{22} \end{bmatrix} = \begin{bmatrix} 1 & Z_b \\ 0 & 1 \end{bmatrix} \tag{E8.3.2}$$

The *a*-parameters of the two-port network depicted in Figure 8.4(c) can be obtained as

$$a_{11} = \left.\frac{V_1}{V_2}\right|_{I_2=0} = 1, \quad a_{12} = \left.\frac{V_1}{-I_2}\right|_{V_2=0} = Z_b$$

$$a_{21} = \left.\frac{I_1}{V_2}\right|_{I_2=0} = Y_a, \quad a_{22} = \left.\frac{I_1}{-I_2}\right|_{V_2=0} = \frac{Y_a + Y_b}{Y_b} = 1 + Y_a Z_b$$

$$; \quad \begin{bmatrix} a_{11} & a_{12} \\ a_{21} & a_{22} \end{bmatrix} = \begin{bmatrix} 1 & Z_b \\ Y_a & 1 + Y_a Z_b \end{bmatrix} \tag{E8.3.3}$$

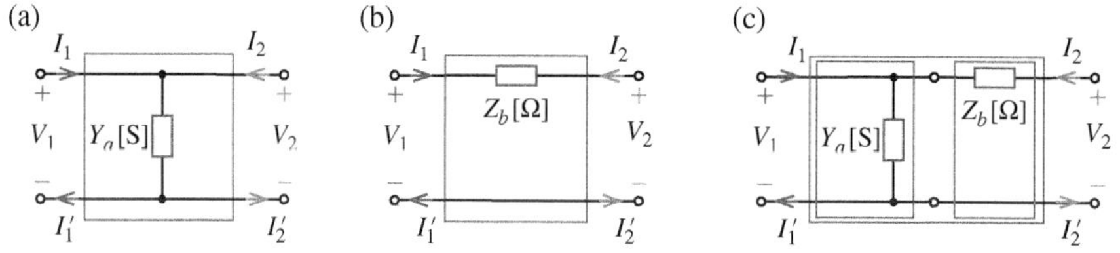

Figure 8.4 Circuits for Example 8.3 to find the a-parameters.

(cf.) The two-port network depicted in Figure 8.4(a) has the z-parameters, but its y-parameters do not exist:

$$z=\begin{bmatrix} Z_a & Z_a \\ Z_a & Z_a \end{bmatrix};\ y=z^{-1}=? \tag{E8.3.4}$$

On the other hand, the two-port network depicted in Figure 8.4(b) has the y-parameters, but its z-parameters do not exist:

$$y=\begin{bmatrix} Y_b & -Y_b \\ -Y_b & Y_b \end{bmatrix};\ z=y^{-1}=? \tag{E8.3.5}$$

Example 8.4 a-parameters of an Ideal Transformer

The a-parameters of an ideal transformer with turns ratio n depicted in Figure 8.5 are obtained from its voltage-current relationship as follows:

$$\begin{matrix} V_1(s) \overset{(5.21a)\text{ of [Y-1]}}{=} n\,V_2(s) \\ I_1(s) \overset{(5.21b)\text{ of [Y-1]}}{=} -\dfrac{1}{n}I_2(s) \end{matrix};\quad \begin{bmatrix} V_1 \\ I_1 \end{bmatrix}=\begin{bmatrix} n & 0 \\ 0 & 1/n \end{bmatrix}\begin{bmatrix} V_2 \\ -I_2 \end{bmatrix};\ A=\begin{bmatrix} n & 0 \\ 0 & 1/n \end{bmatrix} \tag{E8.4}$$

(cf.) This two-port network has neither z-parameters nor y-parameters.

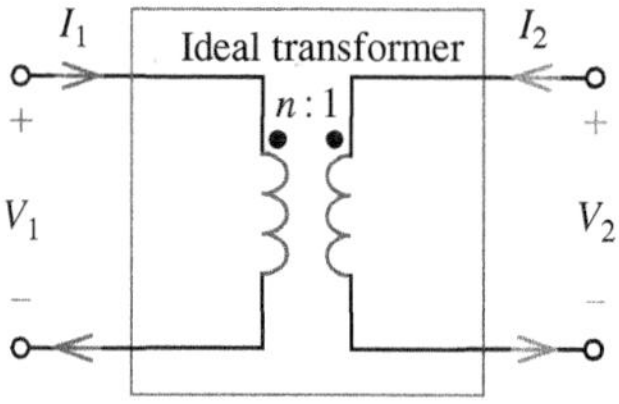

Figure 8.5 Ideal transformer circuit for Example 8.4.

Example 8.5 *a*-parameters of a Transmission Line

To find the *a*-parameters of the transmission line (TL) of characteristic impedance Z_0 and length l as shown in Figure 8.6, recall Eqs. (7.1.1) and (7.1.2) for the total phasor voltage and current at a point of distance d from the load at port 2:

$$\mathbf{V}(d) \overset{(7.1.1)}{=} \mathbf{V}_o^+ e^{\gamma d} + \mathbf{V}_o^- e^{-\gamma d} \tag{E8.5.1a}$$

$$\mathbf{I}(d) \overset{(7.1.2)}{=} \mathbf{I}_o^+ e^{\gamma d} + \mathbf{I}_o^- e^{-\gamma d} = \frac{1}{Z_0}\left(\mathbf{V}_o^+ e^{\gamma d} - \mathbf{V}_o^- e^{-\gamma d}\right) \tag{E8.5.1b}$$

According to these equations, we can write the voltage and current at port 2 (with $d = 0$) as

$$\begin{aligned} \mathbf{V}_2 &= \mathbf{V}(0) \underset{d=0}{\overset{(\text{E8.5.1a})}{=}} \mathbf{V}_o^+ + \mathbf{V}_o^- \\ \mathbf{I}_2 &= -\mathbf{I}(0) \underset{d=0}{\overset{(\text{E8.5.1b})}{=}} \frac{1}{Z_0}\left(-\mathbf{V}_o^+ + \mathbf{V}_o^-\right) \end{aligned} \tag{E8.5.2}$$

and solve this set of equations for $\mathbf{V}_o^+$ and $\mathbf{V}_o^-$ to get

$$\begin{bmatrix} \mathbf{V}_o^+ \\ \mathbf{V}_o^- \end{bmatrix} = \begin{bmatrix} 1 & 1 \\ -1/Z_0 & 1/Z_0 \end{bmatrix}^{-1} \begin{bmatrix} \mathbf{V}_2 \\ \mathbf{I}_2 \end{bmatrix} = \begin{bmatrix} 1/2 & -Z_0/2 \\ 1/2 & Z_0/2 \end{bmatrix} \begin{bmatrix} \mathbf{V}_2 \\ \mathbf{I}_2 \end{bmatrix} \tag{E8.5.3}$$

Also, we can use Eqs. (E8.5.1a,b) to write the voltage and current at port 1 (with $d = l$) as

$$\begin{aligned} \mathbf{V}_1 &= \mathbf{V}(l) \underset{d=l}{\overset{(\text{E8.5.1a})}{=}} \mathbf{V}_o^+ e^{\gamma l} + \mathbf{V}_o^- e^{-\gamma l} \\ \mathbf{I}_1 &= \mathbf{I}(l) \underset{d=l}{\overset{(\text{E8.5.1b})}{=}} \frac{1}{Z_0}\left(\mathbf{V}_o^+ e^{\gamma l} - \mathbf{V}_o^- e^{-\gamma l}\right) \end{aligned} \tag{E8.5.4}$$

Plugging Eq. (E8.5.3) into this equation yields

$$\begin{aligned} \mathbf{V}_1 &\underset{(\text{E8.5.3})}{\overset{(\text{E8.5.4a})}{=}} \frac{\mathbf{V}_2 - Z_0\mathbf{I}_2}{2} e^{\gamma l} + \frac{\mathbf{V}_2 + Z_0\mathbf{I}_2}{2} e^{-\gamma l} = \mathbf{V}_2 \cosh\gamma l - Z_0\mathbf{I}_2 \sinh\gamma l \\ \mathbf{I}_1 &\underset{(\text{E8.5.3})}{\overset{(\text{E8.5.4b})}{=}} \frac{\mathbf{V}_2 - Z_0\mathbf{I}_2}{2Z_0} e^{\gamma l} - \frac{\mathbf{V}_2 + Z_0\mathbf{I}_2}{2Z_0} e^{-\gamma l} = \mathbf{V}_2 \frac{\sinh\gamma l}{Z_0} - \mathbf{I}_2 \cosh\gamma l \end{aligned} \tag{E8.5.5}$$

Matching this equation with Eq. (8.1.3) yields

$$\begin{bmatrix} a_{11} & a_{12} \\ a_{21} & a_{22} \end{bmatrix} = \begin{bmatrix} \cosh\gamma l & Z_0 \sinh\gamma l \\ \sinh\gamma l / Z_0 & \cosh\gamma l \end{bmatrix} \tag{E8.5.6}$$

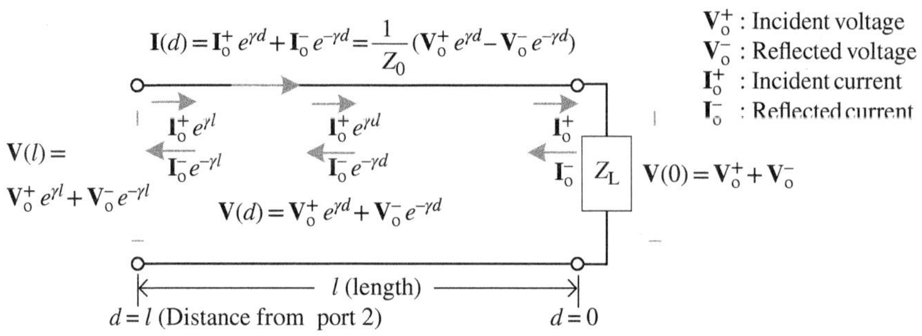

Figure 8.6 A transmission line (TL) of characteristic impedance Z_0 and length l.

Example 8.6 ***h*-parameters**

Let us find the h-parameters of the two-port network depicted in Figure 8.7(a). Instead of using Eq. (8.1.5) directly, we will find an expression of V_1 and I_2 in terms of I_1 and V_2. With this objective in mind, we connect a current source I_1 and a voltage source V_2 at the primary (left) and secondary (right) ports, respectively, as shown in Figure 8.7(a). Noting that this circuit has more current sources than voltage sources, we convert the voltage source into a current source as shown in Figure 8.7(b), set up the node equation, and then solve it as

$$\begin{bmatrix} \frac{1}{r_b} & -\frac{1}{r_b} \\ -\frac{1}{r_b} & \frac{1}{r_b} + \frac{1}{r_c} + \frac{1}{r_e} \end{bmatrix} \begin{bmatrix} V_1 \\ V_d \end{bmatrix} = \begin{bmatrix} I_1 \\ \beta I_1 + \frac{V_2}{r_c} \end{bmatrix} \tag{E8.6.1}$$

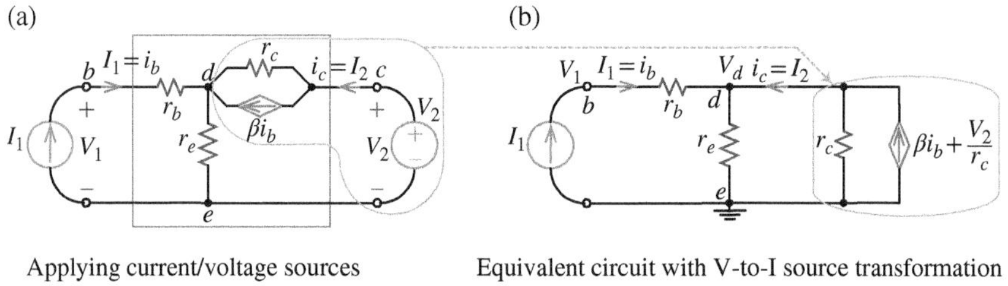

Figure 8.7 Applying current/voltage sources to find the h-parameters of common-emitter transistor circuit

$$\begin{bmatrix} V_1 \\ V_d \end{bmatrix} = \frac{r_b r_c r_e}{r_c + r_e} \begin{bmatrix} \frac{1}{r_b} + \frac{1}{r_c} + \frac{1}{r_e} & \frac{1}{r_b} \\ \frac{1}{r_b} & \frac{1}{r_b} \end{bmatrix} \begin{bmatrix} I_1 \\ \beta I_1 + \frac{V_2}{r_c} \end{bmatrix} = \begin{bmatrix} \frac{r_c r_e(\beta+1)}{r_c + r_e} + r_b & \frac{r_e}{r_c + r_e} \\ \frac{r_c r_e(\beta+1)}{r_c + r_e} & \frac{r_e}{r_c + r_e} \end{bmatrix} \begin{bmatrix} I_1 \\ V_2 \end{bmatrix} \quad \text{(E8.6.2)}$$

This gives the expression of V_1 in terms of I_1 and V_2. One more thing to get is the expression of $I_2 = i_c$ in terms of I_1 and V_2, which is obtained by summing up the current through r_c and the value of the dependent current source $\beta i_b = \beta I_1$ (in the original circuit of Figure 8.7(a)) as follows:

$$I_2 = \beta I_1 + \frac{1}{r_c}(V_2 - V_d) \overset{\text{(E8.6.2b)}}{=} \beta I_1 + \frac{1}{r_c}\left(V_2 - \frac{r_c r_e(\beta+1)}{r_c + r_e} I_1 - \frac{r_e}{r_c + r_e} V_2 \right)$$

$$= \frac{\beta r_c - r_e}{r_c + r_e} I_1 + \frac{1}{r_c + r_e} V_2 \quad \text{(E8.6.3)}$$

Combining this equation with Eq. (E8.6.2a) yields

$$\begin{bmatrix} V_1 \\ I_2 \end{bmatrix} = \begin{bmatrix} \frac{r_c r_e(\beta+1)}{r_c + r_e} + r_b & \frac{r_e}{r_c + r_e} \\ \frac{\beta r_c - r_e}{r_c + r_e} & \frac{1}{r_c + r_e} \end{bmatrix} \begin{bmatrix} I_1 \\ V_2 \end{bmatrix}$$

$$\overset{(8.1.5)}{\rightarrow} \begin{bmatrix} h_{11} & h_{12} \\ h_{21} & h_{22} \end{bmatrix} = \begin{bmatrix} \frac{r_c r_e(\beta+1)}{r_c + r_e} + r_b & \frac{r_e}{r_c + r_e} \\ \frac{\beta r_c - r_e}{r_c + r_e} & \frac{1}{r_c + r_e} \end{bmatrix} \quad \text{(E8.6.4)}$$

8.1.2 Relationships Among Two-Port Parameters

Let two or more sets of two-port parameters describe the relationship among the input/output voltages and currents of a two-port network in common. Then, they must be interrelated with each other. Accordingly, if two sets of two-port parameters exist, we may convert one set to the other set by using the formulas in Table 8.1 or equivalently, the MATLAB function '`port_conversion()`' listed below.

Table 8.1 Conversion among two-port parameters.

	z	**y**	**a**	**b**	**h**	**g**
z	$\begin{bmatrix} z_{11} & z_{12} \\ z_{21} & z_{22} \end{bmatrix}$	$\begin{bmatrix} \frac{y_{22}}{\Delta y} & -\frac{y_{12}}{\Delta y} \\ -\frac{y_{21}}{\Delta y} & \frac{y_{11}}{\Delta y} \end{bmatrix}$	$\begin{bmatrix} \frac{a_{11}}{a_{21}} & \frac{\Delta a}{a_{21}} \\ \frac{1}{a_{21}} & \frac{a_{22}}{a_{21}} \end{bmatrix}$	$\begin{bmatrix} \frac{b_{22}}{b_{21}} & \frac{1}{b_{21}} \\ \frac{\Delta b}{b_{21}} & \frac{b_{11}}{b_{21}} \end{bmatrix}$	$\begin{bmatrix} \frac{\Delta h}{h_{22}} & \frac{h_{12}}{h_{22}} \\ -\frac{h_{21}}{h_{22}} & \frac{1}{h_{22}} \end{bmatrix}$	$\begin{bmatrix} \frac{1}{g_{11}} & -\frac{g_{12}}{g_{11}} \\ \frac{g_{21}}{g_{11}} & \frac{\Delta g}{g_{11}} \end{bmatrix}$
y	$\begin{bmatrix} \frac{z_{22}}{\Delta z} & -\frac{z_{12}}{\Delta z} \\ -\frac{z_{21}}{\Delta z} & \frac{z_{11}}{\Delta z} \end{bmatrix}$	$\begin{bmatrix} y_{11} & y_{12} \\ y_{21} & y_{22} \end{bmatrix}$	$\begin{bmatrix} \frac{a_{22}}{a_{12}} & -\frac{\Delta a}{a_{12}} \\ -\frac{1}{a_{12}} & \frac{a_{11}}{a_{12}} \end{bmatrix}$	$\begin{bmatrix} \frac{b_{11}}{b_{12}} & -\frac{1}{b_{12}} \\ -\frac{\Delta b}{b_{12}} & \frac{b_{22}}{b_{12}} \end{bmatrix}$	$\begin{bmatrix} \frac{1}{h_{11}} & -\frac{h_{12}}{h_{11}} \\ \frac{h_{21}}{h_{11}} & \frac{\Delta h}{h_{11}} \end{bmatrix}$	$\begin{bmatrix} \frac{\Delta g}{g_{22}} & \frac{g_{12}}{g_{22}} \\ -\frac{g_{21}}{g_{22}} & \frac{1}{g_{22}} \end{bmatrix}$
a	$\begin{bmatrix} \frac{z_{11}}{z_{21}} & \frac{\Delta z}{z_{21}} \\ \frac{1}{z_{21}} & \frac{z_{22}}{z_{21}} \end{bmatrix}$	$\begin{bmatrix} -\frac{y_{22}}{y_{21}} & -\frac{1}{y_{21}} \\ -\frac{\Delta y}{y_{21}} & -\frac{y_{11}}{y_{21}} \end{bmatrix}$	$\begin{bmatrix} a_{11} & a_{12} \\ a_{21} & a_{22} \end{bmatrix}$	$\begin{bmatrix} \frac{b_{22}}{\Delta b} & \frac{b_{12}}{\Delta b} \\ \frac{b_{21}}{\Delta b} & \frac{b_{11}}{\Delta b} \end{bmatrix}$	$\begin{bmatrix} -\frac{\Delta h}{h_{21}} & -\frac{h_{11}}{h_{21}} \\ -\frac{h_{22}}{h_{21}} & -\frac{1}{h_{21}} \end{bmatrix}$	$\begin{bmatrix} \frac{1}{g_{21}} & \frac{g_{22}}{g_{21}} \\ \frac{g_{11}}{g_{21}} & \frac{\Delta g}{g_{21}} \end{bmatrix}$
b	$\begin{bmatrix} \frac{z_{22}}{z_{12}} & \frac{\Delta z}{z_{12}} \\ \frac{1}{z_{12}} & \frac{z_{11}}{z_{12}} \end{bmatrix}$	$\begin{bmatrix} -\frac{y_{11}}{y_{12}} & \frac{1}{y_{12}} \\ -\frac{\Delta y}{y_{12}} & \frac{y_{22}}{y_{12}} \end{bmatrix}$	$\begin{bmatrix} \frac{a_{22}}{\Delta a} & \frac{a_{12}}{\Delta a} \\ \frac{a_{21}}{\Delta a} & \frac{a_{11}}{\Delta a} \end{bmatrix}$	$\begin{bmatrix} b_{11} & b_{12} \\ b_{21} & b_{22} \end{bmatrix}$	$\begin{bmatrix} \frac{1}{h_{12}} & \frac{h_{11}}{h_{12}} \\ \frac{h_{22}}{h_{12}} & \frac{\Delta h}{h_{12}} \end{bmatrix}$	$\begin{bmatrix} -\frac{\Delta g}{g_{12}} & -\frac{g_{22}}{g_{12}} \\ -\frac{g_{11}}{g_{12}} & -\frac{1}{g_{12}} \end{bmatrix}$
h	$\begin{bmatrix} \frac{\Delta z}{z_{22}} & \frac{z_{12}}{z_{22}} \\ -\frac{z_{21}}{z_{22}} & \frac{1}{z_{22}} \end{bmatrix}$	$\begin{bmatrix} \frac{1}{y_{11}} & -\frac{y_{12}}{y_{11}} \\ \frac{y_{21}}{y_{11}} & \frac{\Delta y}{y_{11}} \end{bmatrix}$	$\begin{bmatrix} \frac{a_{12}}{a_{22}} & \frac{\Delta a}{a_{22}} \\ -\frac{1}{a_{22}} & \frac{a_{21}}{a_{22}} \end{bmatrix}$	$\begin{bmatrix} \frac{b_{12}}{b_{11}} & \frac{1}{b_{11}} \\ \frac{\Delta b}{b_{11}} & \frac{b_{21}}{b_{11}} \end{bmatrix}$	$\begin{bmatrix} h_{11} & h_{12} \\ h_{21} & h_{22} \end{bmatrix}$	$\begin{bmatrix} \frac{g_{22}}{\Delta g} & -\frac{g_{12}}{\Delta g} \\ -\frac{g_{21}}{\Delta g} & \frac{g_{11}}{\Delta g} \end{bmatrix}$
g	$\begin{bmatrix} \frac{1}{z_{11}} & -\frac{z_{12}}{z_{11}} \\ \frac{z_{21}}{z_{11}} & \frac{\Delta z}{z_{11}} \end{bmatrix}$	$\begin{bmatrix} \frac{\Delta y}{y_{22}} & \frac{y_{12}}{y_{22}} \\ -\frac{y_{21}}{y_{22}} & \frac{1}{y_{22}} \end{bmatrix}$	$\begin{bmatrix} \frac{a_{21}}{a_{11}} & -\frac{\Delta a}{a_{11}} \\ \frac{1}{a_{11}} & \frac{a_{12}}{a_{11}} \end{bmatrix}$	$\begin{bmatrix} \frac{b_{21}}{b_{22}} & -\frac{1}{b_{22}} \\ \frac{\Delta b}{b_{22}} & \frac{b_{12}}{b_{22}} \end{bmatrix}$	$\begin{bmatrix} \frac{h_{22}}{\Delta h} & -\frac{h_{12}}{\Delta h} \\ -\frac{h_{21}}{\Delta h} & \frac{h_{11}}{\Delta h} \end{bmatrix}$	$\begin{bmatrix} g_{11} & g_{12} \\ g_{21} & g_{22} \end{bmatrix}$

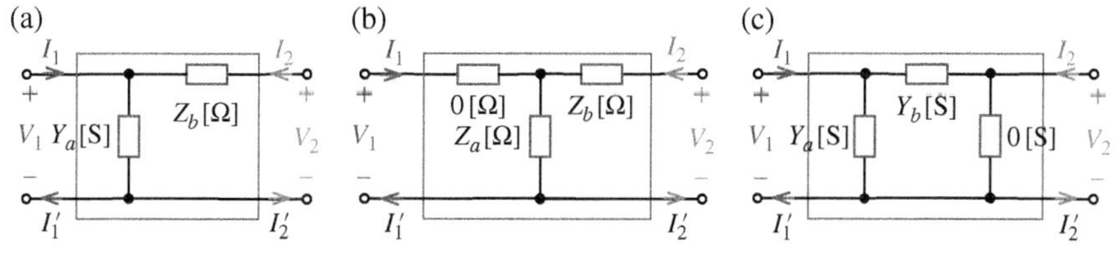

Figure 8.8 Circuits for Example 8.7 to find the *a*-parameters.

Example 8.7 *a*-to-*z* Parameter Conversion

Let us find the *z*-parameters of the two-port network depicted in Figure 8.8(a) by applying the *a*-to-*z* parameter conversion for the *a*-parameters (Eq. (E8.3.3)) obtained in Example 8.3.

We can use the MATLAB function 'port_conversion()' as

```
>>syms Za Zb; Ya=1/Za; Yb=1/Zb;
>>a=[1 Zb; Ya 1+Ya*Zb]; % Eq.(E8.3.3)
>>z=port_conversion(a,'a2z')
```

This yields

```
z = [ Za,              Za]
    [ Za, Za*(Zb/Za + 1)]
```

This can also be obtained by regarding the network as the T-network of Figure 8.8(b) or the Π-network of Figure 8.8(c) and determining the *z*-parameter matrix or the inverse of the *y*-parameter matrix, respectively.

Remark 8.1 Reciprocity and Symmetry of a Two-Port Network

A two-port network is said to be *reciprocal* if the transfer impedance/admittance from port 1 to port 2 and those from port 2 to port 1 are the same, or equivalently, if one of the following relations among the parameters holds:

$$z_{12} = z_{21},\ \ y_{12} = y_{21},\ \ h_{12} = -h_{21},\ \ g_{12} = -g_{21},\ \ \Delta a = 1,\ \ \Delta b = 1 \tag{8.1.7}$$

In general, the networks composed of two-terminal elements such as resistors, inductors, and capacitors with no dependent (controlled) source turn out to be reciprocal.

A reciprocal two-port network is also said to be *symmetrical* if interchanging the input/output ports does not alter the voltages/currents at the two ports, or equivalently, if one of the following relations among the parameters holds:

$$z_{11} = z_{22},\ \ y_{11} = y_{22},\ \ a_{11} = a_{22},\ \ b_{11} = b_{22},\ \ \Delta h = 1,\ \ \Delta g = 1 \tag{8.1.8}$$

```
function P2=port_conversion(P1,which)
% Conversion Formulas among Two-port network parameters (Table 8.1)
% Copyleft: Won Y. Yang, wyyang53@hanmail.net, CAU for academic use only
switch lower(which)
  case 'z2y', P2=P1^-1;
  case 'z2a', P2=[P1(1,1) det(P1); 1 P1(2,2)]/P1(2,1);
  case 'z2b', P2=[P1(2,2) det(P1); 1 P1(1,1)]/P1(1,2);
  case 'z2h', P2=[det(P1) P1(1,2); -P1(2,1) 1]/P1(2,2);
  case 'z2g', P2=[1 -P1(1,2); P1(2,1) det(P1)]/P1(1,1);
  case 'y2z', P2=P1^-1;
  case 'y2a', P2=-[P1(2,2) 1; det(P1) P1(1,1)]/P1(2,1);
  case 'y2b', P2=[-P1(1,1) 1; -det(P1) P1(2,2)]/P1(1,2);
  case 'y2h', P2=[1 -P1(1,2); P1(2,1) det(P1)]/P1(1,1);
  case 'y2g', P2=[det(P1) P1(1,2); -P1(2,1) 1]/P1(2,2);
  case 'a2z', P2=[P1(1,1) det(P1);1 P1(2,2)]/P1(2,1);
  case 'a2y', P2=[P1(2,2) -det(P1); -1 P1(1,1)]/P1(1,2);
  case 'a2b', P2=[P1(2,2) P1(1,2); P1(2,1) P1(1,1)]/det(P1);
  case 'a2h', P2=[P1(1,2) det(P1); -1 P1(2,1)]/P1(2,2);
  case 'a2g', P2=[P1(2,1) -det(P1); 1 P1(1,2)]/P1(1,1);
  case 'b2z', P2=[P1(2,2) 1; det(P1) P1(1,1)]/P1(2,1);
  case 'b2y', P2=[P1(1,1) -1; -det(P1) P1(2,2)]/P1(1,2);
  case 'b2a', P2=[P1(2,2) P1(1,2); P1(2,1) P1(1,1)]/det(P1);
  case 'b2h', P2=[P1(1,2) 1; det(P1) P1(2,1)]/P1(1,1);
  case 'b2g', P2=[P1(2,1) -1; det(P1) P1(1,2)]/P1(2,2);
  case 'h2z', P2=[det(P1) P1(1,2); -P1(2,1) 1]/P1(2,2);
  case 'h2y', P2=[1 -P1(1,2); P1(2,1) det(P1)]/P1(1,1);
  case 'h2a', P2=-[det(P1) P1(1,1); P1(2,2) 1]/P1(2,1);
  case 'h2b', P2= [1 P1(1,1); P1(2,2) det(P1)]/P1(1,2);
  case 'h2g', P2=P1^-1;
  case 'g2z', P2=[1 -P1(1,2); P1(2,1) det(P1)]/P1(1,1);
  case 'g2y', P2=[det(P1) P1(1,2); -P1(2,1) 1]/P1(2,2);
  case 'g2a', P2=[1 P1(2,2); P1(1,1) det(P1)]/P1(2,1);
  case 'g2b', P2=-[det(P1) P1(2,2); P1(1,1) 1]/P1(1,2);
  case 'g2h', P2=P1^-1;
  otherwise error('What do you want to do by port_conversion()?')
end
```

Example 8.8 A Reciprocal Network

In order to test for the reciprocity of the two-port network of Figure 8.9, we will find its z-parameters by applying two current sources I_1/I_2 to port 1/2, respectively, setting up the node equations in the unknowns V_1, V_2, and V_3, and obtaining the expressions of V_1/V_2 in terms of I_1/I_2.

$$\begin{bmatrix} 1+s & -1 & -s \\ -1 & 1+1/s & -1/s \\ -s & -1/s & s+1/s+2 \end{bmatrix} \begin{bmatrix} V_1 \\ V_2 \\ V_3 \end{bmatrix} = \begin{bmatrix} I_1 \\ I_2 \\ 0 \end{bmatrix} \tag{E8.8.1}$$

Figure 8.9 A reciprocal two-port network for Example 8.8.

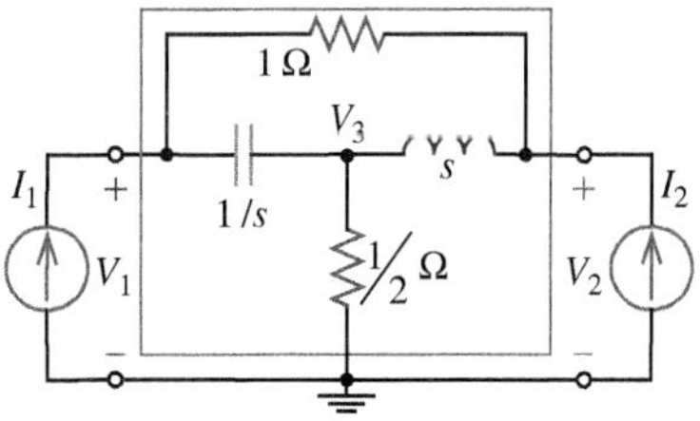

Since this seems to be a bit complicated to solve, let us compose and run the following MATLAB script "elec08e08.m," which performs the necessary symbolic computations for us.

```
%elec08e08.m: Symbolic computation to solve a 3x3 node equation
clear
syms s
Y=[1+s -1 -s;-1 1+1/s -1/s;-s -1/s s+1/s+2]; % Node admittance matrix
V1=Y\[1 0 0].'; % Node voltage vector with I1=1 and I2=0
V2=Y\[0 1 0].'; % Node voltage vector with I1=0 and I2=1
z=[V1(1:2) V2(1:2)] % z-parameter matrix
y=z^-1 % y-parameter matrix if necessary
```

```
>>elec08e08
 z = [ 1/2*(3*s+3+s^2)/(s^2+1+s),   1/2*(s^2+1+3*s)/(s^2+1+s)   ]
     [ 1/2*(s^2+1+3*s)/(s^2+1+s), 1/2*(3*s^2+1+3*s)/(s^2+1+s) ]
 y = [ (3*s^2+1+3*s)/(s^2+1+2*s),    -(s^2+1+3*s)/(s^2+1+2*s)   ]
     [ -(s^2+1+3*s)/(s^2+1+2*s),      (3*s+3+s^2)/(s^2+1+2*s)   ]
```

From this result of running "elec08e08.m," we get the z-/y-parameters as

$$\begin{bmatrix} z_{11} & z_{12} \\ z_{21} & z_{22} \end{bmatrix} = \begin{bmatrix} \dfrac{s^2+3s+3}{2(s^2+s+1)} & \dfrac{s^2+3s+1}{2(s^2+s+1)} \\ \dfrac{s^2+3s+1}{2(s^2+s+1)} & \dfrac{3s^2+3s+1}{2(s^2+s+1)} \end{bmatrix}, \quad \begin{bmatrix} y_{11} & y_{12} \\ y_{21} & y_{22} \end{bmatrix} = \begin{bmatrix} \dfrac{3s^2+3s+1}{(s+1)^2} & -\dfrac{s^2+3s+1}{(s+1)^2} \\ -\dfrac{s^2+3s+1}{(s+1)^2} & \dfrac{s^2+3s+3}{(s+1)^2} \end{bmatrix} \tag{E8.8.2}$$

where the y-parameter matrix has been obtained from the inverse of the z-parameter matrix. In view of Eq. (8.1.7), the symmetry of this z-/y-parameter matrices ($z_{12} = z_{21}$, $y_{12} = y_{21}$) implies that the two-port network is reciprocal, from which we can guess that it is composed of two-terminal elements such as resistors, inductors, and capacitors with no dependent source.

8.1.3 Interconnection of Two-Port Networks

For resistors connected in series/parallel, it is convenient to use the impedance/admittance for obtaining their equivalent parameters. Likewise, depending on the type of connection of two 2-port networks, there is some preferable set of parameters that is more convenient to work with than other sets of parameters for obtaining their equivalent parameters.

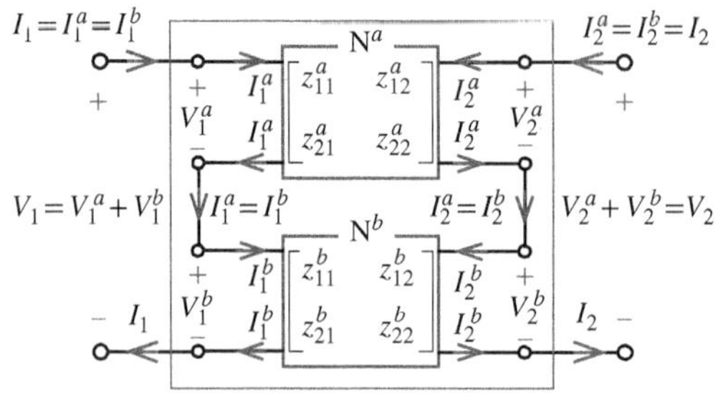

Figure 8.10 Series connection of two 2-port networks.

8.1.3.1 Series Connection and z-parameters

Consider the connection of two 2-port networks shown in Figure 8.10. The currents of two 2-port networks are the same ($I_1^a = I_1^b = I_1, I_2^a = I_2^b = I_2$) and their voltages are added to yield the voltages of the overall network ($V_1^a + V_1^b = V_1$, $V_2^a + V_2^b = V_2$). Thus, it is called the *series connection* and its overall z-parameter matrix equals the sum of the z-parameter matrices of two networks.

$$\begin{bmatrix} V_1 \\ V_2 \end{bmatrix} = \begin{bmatrix} V_1^a \\ V_2^a \end{bmatrix} + \begin{bmatrix} V_1^b \\ V_2^b \end{bmatrix} = \begin{bmatrix} z_{11}^a & z_{12}^a \\ z_{21}^a & z_{22}^a \end{bmatrix} \begin{bmatrix} I_1^a \\ I_2^a \end{bmatrix} + \begin{bmatrix} z_{11}^b & z_{12}^b \\ z_{21}^b & z_{22}^b \end{bmatrix} \begin{bmatrix} I_1^b \\ I_2^b \end{bmatrix}$$
$$= \begin{bmatrix} z_{11}^a + z_{11}^b & z_{12}^a + z_{12}^b \\ z_{21}^a + z_{21}^b & z_{22}^a + z_{22}^b \end{bmatrix} \begin{bmatrix} I_1 \\ I_2 \end{bmatrix} = \begin{bmatrix} z_{11} & z_{12} \\ z_{21} & z_{22} \end{bmatrix} \begin{bmatrix} I_1 \\ I_2 \end{bmatrix} \tag{8.1.9}$$

8.1.3.2 Parallel (Shunt) Connection and y-parameters

Consider the connection of two 2-port networks shown in Figure 8.11. The voltages of two 2-port networks are the same ($V_1^a = V_1^b = V_1$, $V_2^a = V_2^b = V_2$) and their currents are added to yield the currents of the overall network ($I_1^a + I_1^b = I_1, I_2^a + I_2^b = I_2$). Thus, it is called the *parallel connection* and its overall y-parameter matrix equals the sum of the y-parameter matrices of two networks.

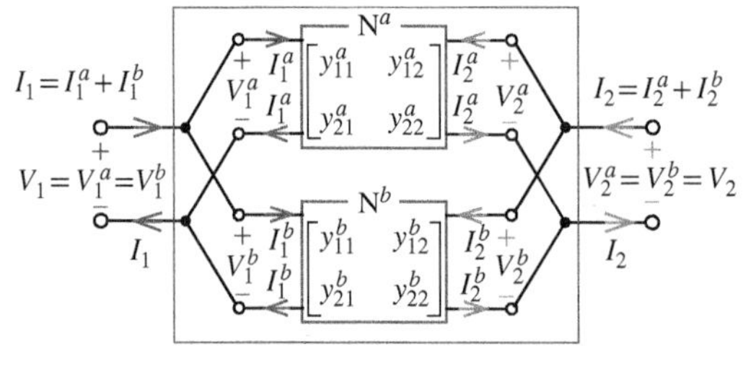

Figure 8.11 Parallel connection of two 2-port networks.

$$\begin{bmatrix} I_1 \\ I_2 \end{bmatrix} = \begin{bmatrix} I_1^a \\ I_2^a \end{bmatrix} + \begin{bmatrix} I_1^b \\ I_2^b \end{bmatrix} = \begin{bmatrix} y_{11}^a & y_{12}^a \\ y_{21}^a & y_{22}^a \end{bmatrix} \begin{bmatrix} V_1^a \\ V_2^a \end{bmatrix} + \begin{bmatrix} y_{11}^b & y_{12}^b \\ y_{21}^b & y_{22}^b \end{bmatrix} \begin{bmatrix} V_1^b \\ V_2^b \end{bmatrix}$$
$$= \begin{bmatrix} y_{11}^a + y_{11}^b & y_{12}^a + y_{12}^b \\ y_{21}^a + y_{21}^b & y_{22}^a + y_{22}^b \end{bmatrix} \begin{bmatrix} V_1 \\ V_2 \end{bmatrix} = \begin{bmatrix} y_{11} & y_{12} \\ y_{21} & y_{22} \end{bmatrix} \begin{bmatrix} V_1 \\ V_2 \end{bmatrix} \tag{8.1.10}$$

8.1.3.3 Series-Parallel(Shunt) Connection and *h*-parameters

Consider the connection of two 2-port networks in Figure 8.12. The input ports of two networks are connected in series ($I_1^a = I_1^b = I_1$, $V_1^a + V_1^b = V_1$) and their output ports are connected in parallel ($V_2^a = V_2^b = V_2$, $I_2^a + I_2^b = I_2$). Thus, it is called the *series-parallel connection* and its overall *h*-parameter matrix equals the sum of the *h*-parameter matrices of two networks.

$$\begin{bmatrix} V_1 \\ I_2 \end{bmatrix} = \begin{bmatrix} V_1^a \\ I_2^a \end{bmatrix} + \begin{bmatrix} V_1^b \\ I_2^b \end{bmatrix} = \begin{bmatrix} h_{11}^a & h_{12}^a \\ h_{21}^a & h_{22}^a \end{bmatrix} \begin{bmatrix} I_1^a \\ V_2^a \end{bmatrix} + \begin{bmatrix} h_{11}^b & h_{12}^b \\ h_{21}^b & h_{22}^b \end{bmatrix} \begin{bmatrix} I_1^b \\ V_2^b \end{bmatrix}$$
$$= \begin{bmatrix} h_{11}^a + h_{11}^b & h_{12}^a + h_{12}^b \\ h_{21}^a + h_{21}^b & h_{22}^a + h_{22}^b \end{bmatrix} \begin{bmatrix} I_1 \\ V_2 \end{bmatrix} = \begin{bmatrix} h_{11} & h_{12} \\ h_{21} & h_{22} \end{bmatrix} \begin{bmatrix} I_1 \\ V_2 \end{bmatrix} \tag{8.1.11}$$

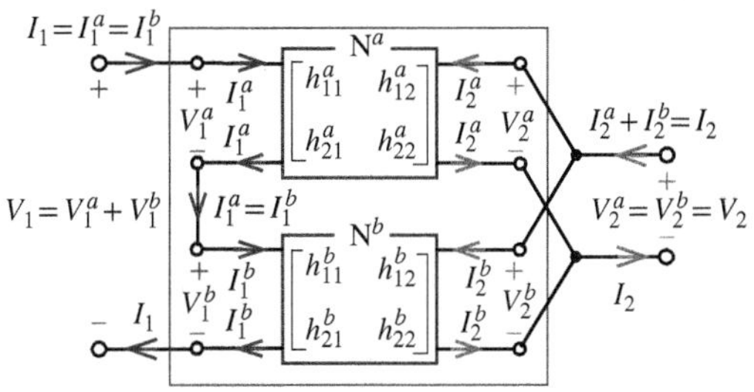

Figure 8.12 Series-parallel connection of two 2-port networks.

8.1.3.4 Parallel(Shunt)-Series Connection and *g*-parameters

Consider the connection of two 2-port networks in Figure 8.13. The input ports of two networks are connected in parallel ($V_1^a = V_1^b = V_1$, $I_1^a + I_1^b = I_1$) and their output ports are connected in series ($V_2^a + V_2^b = V_2$, $I_2^a = I_2^b = I_2$). Thus, it is called the *parallel-series connection* and its overall *g*-parameter matrix equals the sum of the *g*-parameter matrices of two networks.

$$\begin{bmatrix} I_1 \\ V_2 \end{bmatrix} = \begin{bmatrix} I_1^a \\ V_2^a \end{bmatrix} + \begin{bmatrix} I_1^b \\ V_2^b \end{bmatrix} = \begin{bmatrix} g_{11}^a & g_{12}^a \\ g_{21}^a & g_{22}^a \end{bmatrix} \begin{bmatrix} V_1^a \\ I_2^a \end{bmatrix} + \begin{bmatrix} g_{11}^b & g_{12}^b \\ g_{21}^b & g_{22}^b \end{bmatrix} \begin{bmatrix} V_1^b \\ I_2^b \end{bmatrix}$$
$$= \begin{bmatrix} g_{11}^a + g_{11}^b & g_{12}^a + g_{12}^b \\ g_{21}^a + g_{21}^b & g_{22}^a + g_{22}^b \end{bmatrix} \begin{bmatrix} V_1 \\ I_2 \end{bmatrix} = \begin{bmatrix} g_{11} & g_{12} \\ g_{21} & g_{22} \end{bmatrix} \begin{bmatrix} V_1 \\ I_2 \end{bmatrix} \tag{8.1.12}$$

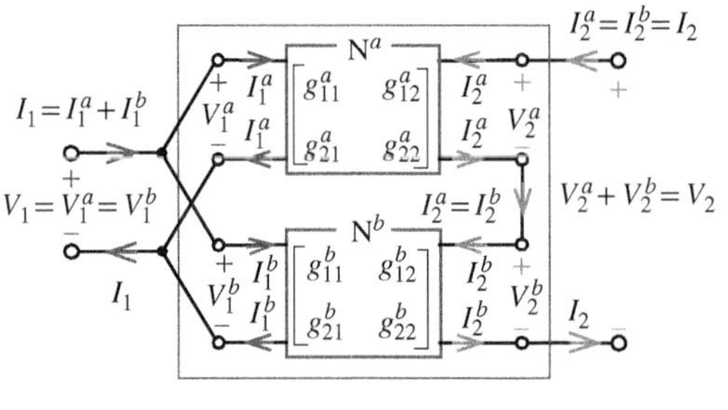

Figure 8.13 Parallel-series connection of two 2-port networks.

8.1.3.5 Cascade Connection and a-parameters

Consider the *cascade* connection of two 2-port networks in Figure 8.14, where the output port of a network N^a is connected to the input port of another network N^b. Since the voltage and current at the input port of the network N^b are the same as those at the output port of the network N^a $(-I_2^a = I_1^b,\ V_2^a = V_1^b)$, its overall a-parameter matrix equals the product of the a-parameter matrices of the two networks.

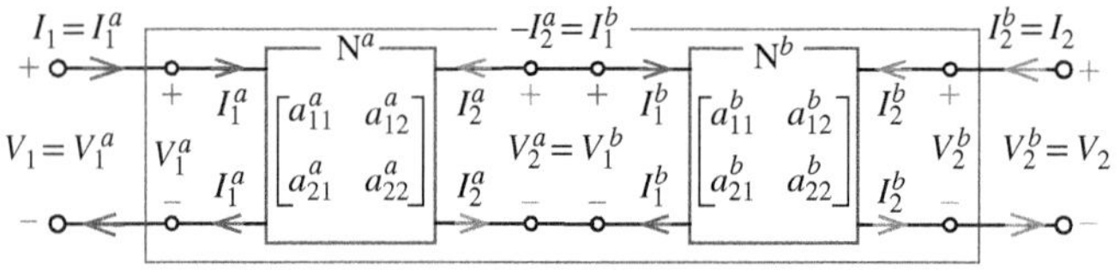

Figure 8.14 Cascade connection of two 2-port networks.

$$\begin{bmatrix} V_1 \\ I_1 \end{bmatrix} = \begin{bmatrix} V_1^a \\ I_1^a \end{bmatrix} = \begin{bmatrix} a_{11}^a & a_{12}^a \\ a_{21}^a & a_{22}^a \end{bmatrix} \begin{bmatrix} V_2^a \\ -I_2^a \end{bmatrix} = \begin{bmatrix} a_{11}^a & a_{12}^a \\ a_{21}^a & a_{22}^a \end{bmatrix} \begin{bmatrix} V_1^b \\ I_1^b \end{bmatrix}$$
$$= \begin{bmatrix} a_{11}^a & a_{12}^a \\ a_{21}^a & a_{22}^a \end{bmatrix} \begin{bmatrix} a_{11}^b & a_{12}^b \\ a_{21}^b & a_{22}^b \end{bmatrix} \begin{bmatrix} V_2^b \\ -I_2^b \end{bmatrix} = \begin{bmatrix} a_{11} & a_{12} \\ a_{21} & a_{22} \end{bmatrix} \begin{bmatrix} V_2 \\ -I_2 \end{bmatrix} \quad (8.1.13)$$

8.1.4 Curse of Port Condition

In order for the overall parameters of an interconnection of two-port networks to be obtained by using one of the above formulas (Eqs. (8.1.9)-(8.1.13)), each port of the two-port networks should satisfy the *port condition* requiring that the current flowing into the network through one terminal should equal the current flowing out of the network through the other terminal.

Although the port condition is satisfied for each of two 2-port networks alone, it may not be satisfied after the two 2-port networks are interconnected. Let us see the following example:

Example 8.9 The Curse of Port Condition Violated by Interconnection

a) Consider the two-port network shown in Figure 8.15(a). We will find its y-parameters by applying two current sources I_1 and I_2 to port 1 and 2, respectively, obtaining the expression of V_1 and $V_2 = V_c - V_d$ in terms of I_1 and I_2 to get the z-parameter matrix, and then taking its inverse. According to this scheme, we write the node equation and solve it as

$$\begin{bmatrix} 1+1/2 & -1 & -1/2 \\ -1 & 1+1/3 & 0 \\ -1/2 & 0 & 1/2+1/4 \end{bmatrix} \begin{bmatrix} V_1 \\ V_c \\ V_d \end{bmatrix} = \begin{bmatrix} I_1 \\ I_2 \\ -I_2 \end{bmatrix} \tag{E8.9.1}$$

$$; \begin{bmatrix} V_1 \\ V_c \\ V_d \end{bmatrix} = \begin{bmatrix} 12/5 & 9/5 & 8/5 \\ 9/5 & 21/10 & 6/5 \\ 8/5 & 6/5 & 12/5 \end{bmatrix} \begin{bmatrix} I_1 \\ I_2 \\ -I_2 \end{bmatrix} = \begin{bmatrix} 12/5 I_1 + 1/5 I_2 \\ 9/5 I_1 + 9/10 I_2 \\ 8/5 I_1 - 6/5 I_2 \end{bmatrix}$$

$$; \begin{bmatrix} V_1 \\ V_2 \end{bmatrix} = \begin{bmatrix} V_1 \\ V_c - V_d \end{bmatrix} = \begin{bmatrix} 12/5 & 1/5 \\ 1/5 & 21/10 \end{bmatrix} \begin{bmatrix} I_1 \\ I_2 \end{bmatrix} \tag{E8.9.2}$$

Thus, we get the z-parameter matrix and take its inverse to get the y-parameter matrix as

$$z = \begin{bmatrix} 12/5 & 1/5 \\ 1/5 & 21/10 \end{bmatrix}; \; y = z^{-1} = \frac{50}{12 \times 21 - 2} \begin{bmatrix} 21/10 & -1/5 \\ -1/5 & 12/5 \end{bmatrix} = \begin{bmatrix} 21/50 & -1/25 \\ -1/25 & 12/25 \end{bmatrix} \tag{E8.9.3}$$

You can run the following MATLAB script 'elec08e09.m' to get this result:

```
%elec08e09.m: Symbolic computation to solve a 3x3 node equation
format rat
Y=[1+1/2 -1 -1/2;-1 1+1/3 0;-1/2 0 1/2+1/4]; % Node admittance matrix
Y^-1
V1=Y\[1 0 0]'; V2=Y\[0 1 -1]'; % Node Voltage vectors for I1 and I2
z=[V1(1) V2(1); V1(2)-V1(3) V2(2)-V2(3)]; % z-parameter matrix
y=z^-1, format short % y-parameter matrix
```

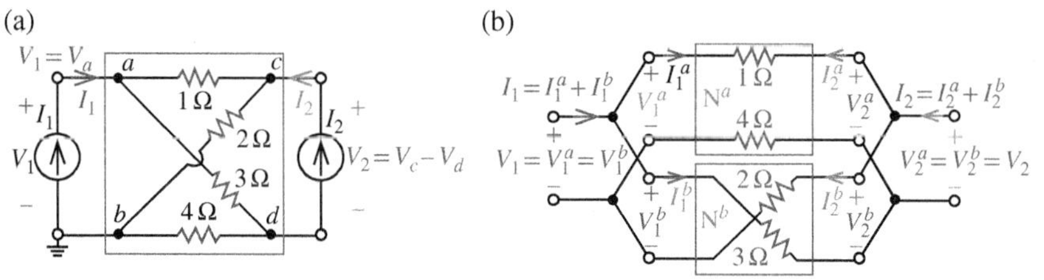

A 2-port network with two current sources to find the z-parameter

The same network regarded as a parallel connection of two 2-port subnetworks

Figure 8.15 An example of two-port networks violating the port current condition after connection.

Now, regarding the two-port network as the parallel interconnection of two subnetworks as depicted in Figure 8.15(b), let us add the y-parameter matrices of the subnetworks to get the overall y-parameter matrix as

$$\begin{bmatrix} I_1^a \\ I_2^a \end{bmatrix} = \begin{bmatrix} 1/(1+4) & -1/(1+4) \\ -1/(1+4) & 1/(1+4) \end{bmatrix} \begin{bmatrix} V_1^a \\ V_2^a \end{bmatrix}, \begin{bmatrix} I_1^b \\ I_2^b \end{bmatrix} = \begin{bmatrix} 1/(2+3) & 1/(2+3) \\ 1/(2+3) & 1/(2+3) \end{bmatrix} \begin{bmatrix} V_1^b \\ V_2^b \end{bmatrix} \tag{E8.9.4}$$

$$;y \overset{?}{=} y^a + y^b = \begin{bmatrix} 1/5 & -1/5 \\ -1/5 & 1/5 \end{bmatrix} + \begin{bmatrix} 1/5 & 1/5 \\ 1/5 & 1/5 \end{bmatrix} = \begin{bmatrix} 2/5 & 0 \\ 0 & 2/5 \end{bmatrix} \tag{E8.9.5}$$

Surprisingly, this does not agree with the previous result (E8.9.3). What is happening? This is the case where the subnetworks violate the port condition after interconnection, which can be verified by finding the currents flowing through the resistors 1 and 4 Ω (belonging to subnetwork N^a):

$$I_{1\Omega} = \frac{V_1 - V_c}{1} \overset{(E8.9.2)}{=} \frac{3}{5}I_1 - \frac{7}{10}I_2 \neq I_{4\Omega} = \frac{V_d - 0}{4} \overset{(E8.9.2)}{=} \frac{2}{5}I_1 - \frac{3}{10}I_2 \tag{E8.9.6}$$

b) Consider the two-port network shown in Figure 8.16(a). First, regarding the network as the series connection of two subnetworks, we would add the z-parameter matrices of the two subnetworks to get its (overall) z-parameter matrix. Second, we might simplify the whole network as depicted in Figure 8.16(b) and find its z-parameter matrix as

$$z^a + z^b = \begin{bmatrix} 1+1+1 & 1 \\ 1 & 1+1+1 \end{bmatrix} + \begin{bmatrix} 1+1 & 1 \\ 1 & 1+1 \end{bmatrix} \neq z = \begin{bmatrix} 4 & 3 \\ 3 & 4 \end{bmatrix} \tag{E8.9.7}$$

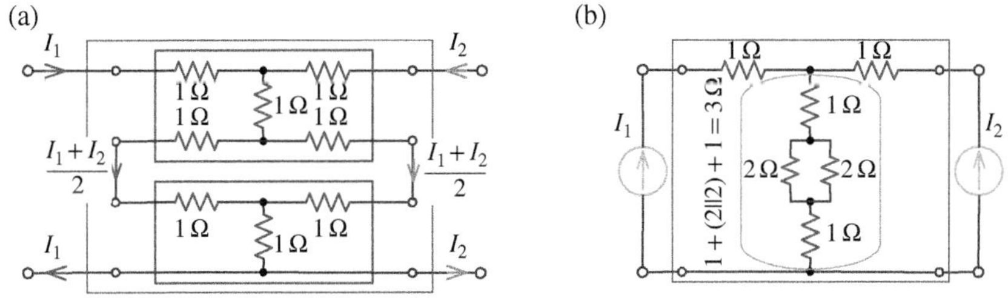

A 2-port network regarded as a series connection of two 2-port subnetworks

The equivalent network with two current sources to find the z-parameter

Figure 8.16 An example of two-port networks violating the port current condition after interconnection.

Why are these results different? It is because the two subnetworks in the series interconnection (Figure 8.16(a)) do not satisfy the port condition, that is, the currents flowing into each subnetwork through one terminal of the ports are not the same as the currents flowing out of it through the other terminal of the ports ($I_1 \neq (I_1 + I_2)/2 \neq I_2$) as long as $I_1 \neq I_2$.

(cf.) This example implies that we should be careful in regarding a two-port network as an interconnection of subnetworks for the purpose of finding its parameters more easily.

Example 8.10 Examples of Interconnection Not Irritating the Port Condition

a) Consider the two-port network shown in Figure 8.17(a). We will find its z-parameters by regarding it as a series connection of two subnetworks as shown in Figure 8.17(b) and adding the z-parameter matrices of the two subnetworks

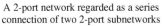

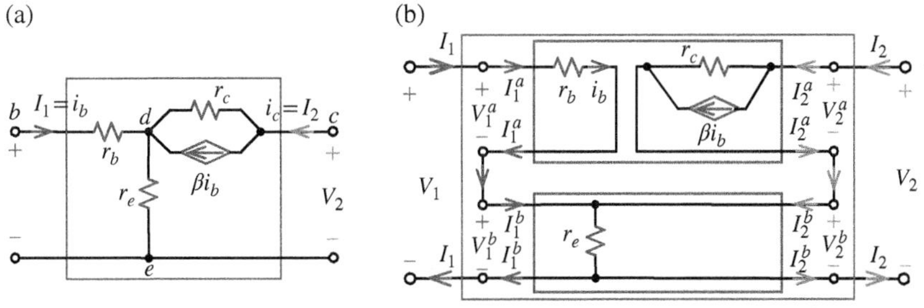

A 2-port network

The same network regarded as a series connection of two 2-port subnetworks

Figure 8.17 A two-port network and its series decomposition.

(see Eq. (8.1.9)). Noting that $I_1^a = i_b$, the input/output port voltages are expressed in terms of the input/output port currents as

$$\begin{bmatrix} V_1^a \\ V_2^a \end{bmatrix} = \begin{bmatrix} r_b & 0 \\ -\beta r_c & r_c \end{bmatrix} \begin{bmatrix} I_1^a \\ I_2^a \end{bmatrix}, \quad \begin{bmatrix} V_1^b \\ V_2^b \end{bmatrix} = \begin{bmatrix} r_e & r_e \\ r_e & r_e \end{bmatrix} \begin{bmatrix} I_1^b \\ I_2^b \end{bmatrix} \qquad \text{(E8.10.1)}$$

Thus, we add the two z-parameter matrices to get the overall z-parameter matrix as

$$z = z^a + z^b = \begin{bmatrix} r_b & 0 \\ -\beta r_c & r_c \end{bmatrix} + \begin{bmatrix} r_e & r_e \\ r_e & r_e \end{bmatrix} = \begin{bmatrix} r_b + r_e & r_e \\ -\beta r_c + r_e & r_c + r_e \end{bmatrix} \qquad \text{(E8.10.2)}$$

To check if this is right, we convert this z-parameter into the h-parameter by using the MATLAB function 'port_conversion()':

```
>>syms rb rc re beta
>>z=[rb+re re; re-beta*rc rc+re]; % Eq.(E8.10.2)
>>h=port_conversion(z,'z2h') % convert z-param to h-param
  h= [(rb*rc+rb*re+re*rc+re*beta*rc)/(rc+re),    re/(rc+re)]
     [ (-re+beta*rc)/(rc+re),                     1/(rc+re) ]
```

This agrees with Eq. (E8.6.4), which was obtained in Example 8.6.

b) Consider the two-port network shown in Figure 8.18(a). We will find its y-parameters by regarding it as a parallel connection of two subnetworks as depicted in Figure 8.18(b) and adding the y-parameter matrices of the two subnetworks (see Eq. (8.1.10)). From Eq. (E8.3.5), the y-parameter matrix of the upper subnetwork can be obtained as

$$y^a = \begin{bmatrix} 1 & -1 \\ -1 & 1 \end{bmatrix} \qquad \text{(E8.10.3)}$$

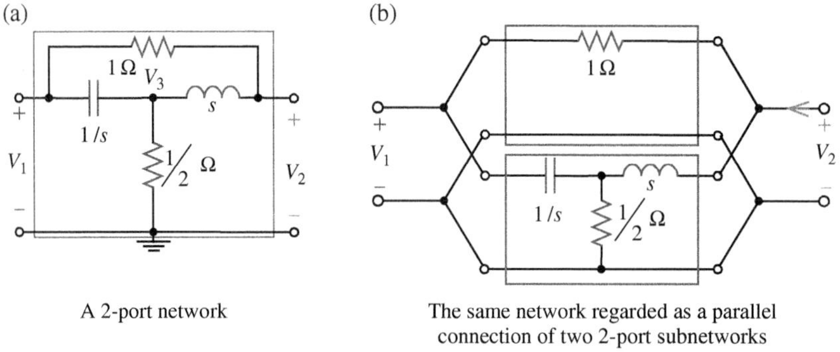

(a)
1Ω
V_3
1/s
s
1/2 Ω
V_1
V_2
A 2-port network
(b)
1Ω
1/s
s
1/2 Ω
V_1
V_2
The same network regarded as a parallel connection of two 2-port subnetworks

Figure 8.18 A two-port network and its parallel decomposition.

and that of the lower subnetwork can be obtained by taking the inverse of the z-parameter matrix as

$$y^b = \left[z^b\right]^{-1} = \begin{bmatrix} 1/s + 1/2 & 1/2 \\ 1/2 & s + 1/2 \end{bmatrix}^{-1} = \frac{2}{2 + s + 1/s}\begin{bmatrix} s + 1/2 & -1/2 \\ -1/2 & 1/s + 1/2 \end{bmatrix}$$

$$= \frac{1}{(s+1)^2}\begin{bmatrix} s(2s+1) & -s \\ -s & s+2 \end{bmatrix} \tag{E8.10.4}$$

We add these two y-parameter matrices to get the y-parameters of the overall network as

$$y = y^a + y^b = \begin{bmatrix} \dfrac{3s^2 + 3s + 1}{(s+1)^2} & -\dfrac{s^2 + 3s + 1}{(s+1)^2} \\ -\dfrac{s^2 + 3s + 1}{(s+1)^2} & \dfrac{s^2 + 3s + 3}{(s+1)^2} \end{bmatrix} \tag{E8.10.5}$$

```
>>syms s; ya=[1 -1; -1 1]; yb=[1/s+1/2 1/2; 1/2 s+1/2]^-1;
>>y=ya+yb; pretty(simplify(y))
```

This agrees with Eq. (E8.8.2), which is the result obtained in Example 8.8.

c) Consider the two-port network shown in Figure 8.4(c). Regarding it as a cascade connection of the two 2-port networks depicted in Figure 8.4(a) and (b) and using Eq. (8.1.13), we multiply the a-parameter matrices of the two networks, i.e. Eqs. (E8.3.1) and (E8.3.2), to get its overall a-parameters as

$$\begin{bmatrix} a_{11} & a_{12} \\ a_{21} & a_{22} \end{bmatrix} = \begin{bmatrix} 1 & 0 \\ Y_a & 1 \end{bmatrix}\begin{bmatrix} 1 & Z_b \\ 0 & 1 \end{bmatrix} = \begin{bmatrix} 1 & Z_b \\ Y_a & 1 + Y_a Z_b \end{bmatrix} \tag{E8.10.6}$$

This agrees with Eq. (E8.3.3), which is the result obtained in Example 8.3.

8.1.5 Circuit Models with Given Parameters

So far we have discussed how to find the parameters for a given two-port network. In this section, we will see how to construct a two-port network model with a given parameter.

8.1.5.1 Circuit Model with Given *z*-parameters

If a given z-parameter matrix is symmetric, the T-model depicted in Figure 8.19 (a) suffices; otherwise, we should resort to the model having two CCVSs (current-controlled voltage sources) as shown in Figure 8.19(b).

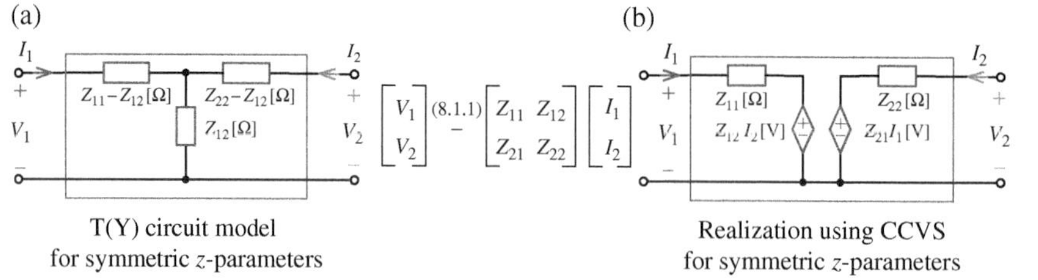

Figure 8.19 Two-port network models with a given z-parameter matrix.

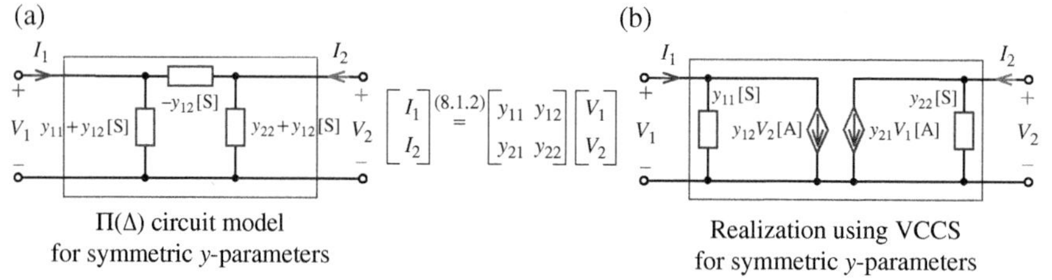

Figure 8.20 Two-port network models with a given y-parameter matrix.

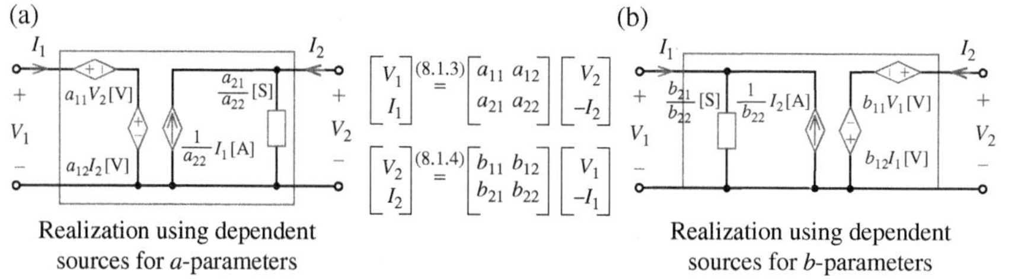

Figure 8.21 Two-port network models with a given a/b-parameter matrix.

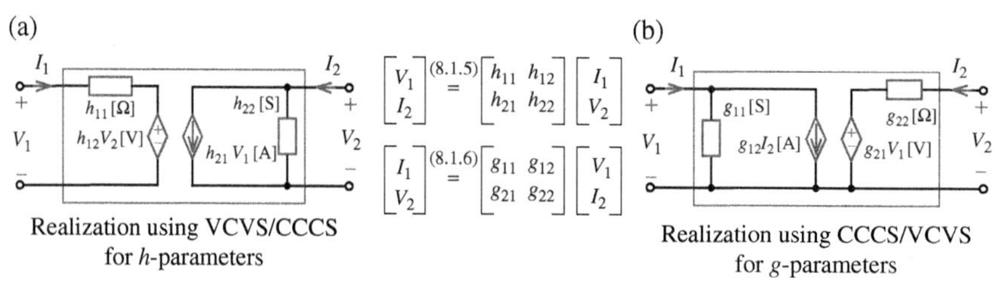

Figure 8.22 Two-port network models with a given h/g-parameter matrix.

8.1.5.2 Circuit Model with Given *y*-parameters

If a given *y*-parameter matrix is symmetric, the Π-model depicted in Figure 8.20 (a) suffices; otherwise, we should resort to the model having two VCCSs (voltage-controlled current sources) as shown in Figure 8.20(b).

8.1.5.3 Circuit Model with Given *a*/*b*-parameters

For given *a*/*b*-parameters, one choice is the model having a VCVS (voltage-controlled voltage source), a CCVS, and a CCCS (current-controlled current source) as shown in Figure 8.21(a) and (b).

8.1.5.4 Circuit Model with Given *h*/*g*-parameters

For given *h*/*g*-parameters, one choice is the model having a VCVS and a CCCS as depicted in Figure 8.22(a) and (b).

Example 8.11 T/Π-Models for a Transmission Line

Let us find the T-/Π-models (T-/Π-equivalents) of the TL whose *a*-parameter matrix was found in Example 8.5 as

$$\begin{bmatrix} a_{11} & a_{12} \\ a_{21} & a_{22} \end{bmatrix} \overset{(E8.5.6)}{=} \begin{bmatrix} \cosh\gamma l & Z_0\sinh\gamma l \\ \sinh\gamma l/Z_0 & \cosh\gamma l \end{bmatrix} \tag{E8.11.1}$$

First, to find the T-model, we convert this *a*-parameter matrix into the *z*-parameter matrix:

```
>>syms rl Z0;
  a=[cosh(rl) Z0*sinh(rl); sinh(rl)/Z0 cosh(rl)]; % Eq.(E8.11.1)
  z=port_conversion(a,'a2z'); pretty(simplify(z))
```

This yields

```
/ Z0 cosh(rl)          Z0        \
| -----------,      ---------    |
|   sinh(rl)         sinh(rl)    |
|                                |
|     Z0          Z0 cosh(rl)    |
|  --------,      -----------    |
\  sinh(rl)        sinh(rl)      /
```

$$; \begin{bmatrix} z_{11} & z_{12} \\ z_{21} & z_{22} \end{bmatrix} = \begin{bmatrix} Z_0\coth\gamma l & Z_0/\sinh\gamma l \\ Z_0/\sinh\gamma l & Z_0\coth\gamma l \end{bmatrix} \tag{E8.11.2}$$

Second, to find the Π-model, we convert this *a*-parameter matrix into the *y*-parameter matrix:

```
>>y=port_conversion(a,'a2y'); pretty(simplify(y))
```

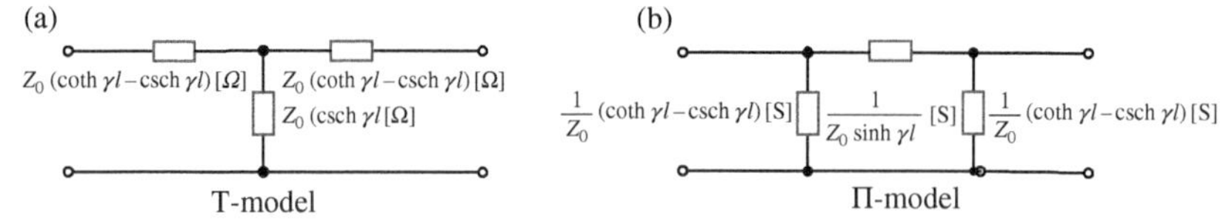

Figure 8.23 T- and Π-models for a TL.

This yields

```
/     cosh(rl)             1        \
|  -----------,   - -----------     |
|  Z0 sinh(rl)      Z0 sinh(rl)     |
|                                   |
|        1          cosh(rl)        |
| - -----------,    -----------     |
\   Z0 sinh(rl)     Z0 sinh(rl)     /
```

$$;\begin{bmatrix} y_{11} & y_{12} \\ y_{21} & y_{22} \end{bmatrix} = \begin{bmatrix} \coth\gamma l/Z_0 & -1/Z_0\sinh\gamma l \\ -1/Z_0\sinh\gamma l & \coth\gamma l/Z_0 \end{bmatrix} \quad \text{(E8.11.3)}$$

In view of Figures 8.19 and 8.20, the T- and Π-models can be drawn as Figure 8.23 (a) and (b), respectively, where the z-parameter and y-parameter matrices can be written as.

$$\begin{bmatrix} z_{11} & z_{12} \\ z_{21} & z_{22} \end{bmatrix} = \begin{bmatrix} Z_0(\coth\gamma l - \operatorname{csch}\gamma l + \operatorname{csch}\gamma l) & Z_0\operatorname{csch}\gamma l \\ Z_0\operatorname{csch}\gamma l & Z_0(\coth\gamma l - \operatorname{csch}\gamma l + \operatorname{csch}\gamma l) \end{bmatrix} \quad \text{(E8.11.4)}$$

$$\begin{bmatrix} y_{11} & y_{12} \\ y_{21} & y_{22} \end{bmatrix} = \begin{bmatrix} (\coth\gamma l - \operatorname{csch}\gamma l + \operatorname{csch}\gamma l)/Z_0 & -\operatorname{csch}\gamma l/Z_0 \\ -\operatorname{csch}\gamma l/Z_0 & (\coth\gamma l - \operatorname{csch}\gamma l + \operatorname{csch}\gamma l)/Z_0 \end{bmatrix} \quad \text{(E8.11.5)}$$

8.1.6 Properties of Two-Port Networks with Source/Load

In most applications, a two-port network is driven by a voltage source V_s with source impedance Z_s at its input port and drives a load impedance Z_L at its output port (as shown in Figure 8.24) and then, it is called a *terminated two-port network*. Table 8.2 lists the formulas to find the input impedance Z_{in} (looking into the two-port network at port 1), the voltage gain $A_v = V_2/V_1$, the current gain $A_i = -I_2/I_1$, and the output impedance Z_{out} (looking into the two-port network at port 2), most of which have been derived in [Y-1]. The table has been cast into the following MATLAB function '`port_property()`'.

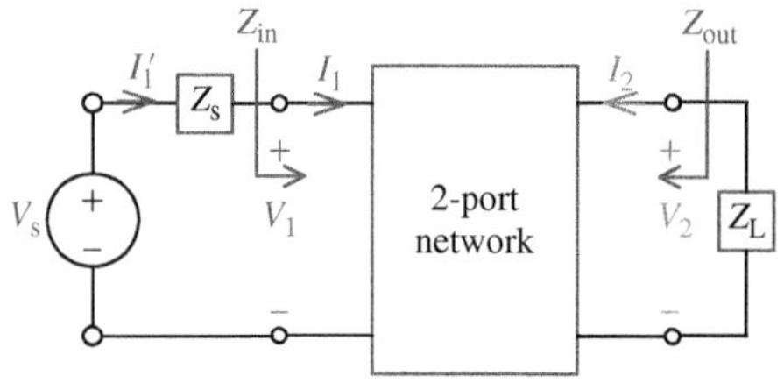

Figure 8.24 Two-port network with source/load at port 1/2.

Table 8.2 Properties of a two-port network having a source/load in terms of its parameters.

	z	y	a	b	h	g
Z_{in} (Z_L at port 2)	$\frac{\Delta z + z_{11}Z_L}{z_{22}+Z_L}$	$\frac{y_{22}Z_L+1}{\Delta y Z_L + y_{11}}$	$\frac{a_{11}Z_L+a_{12}}{a_{21}Z_L+a_{22}}$	$\frac{b_{22}Z_L+b_{12}}{b_{21}Z_L+b_{11}}$	$\Delta\frac{hZ_L+h_{11}}{h_{22}Z_L+1}$	$\frac{g_{22}+Z_L}{\Delta g + g_{11}Z_L}$
$A_v = V_2/V_1$ (Z_L at port 2)	$\frac{z_{21}Z_L}{\Delta z + z_{11}Z_L}$	$\frac{-y_{21}Z_L}{y_{22}Z_L+1}$	$\frac{Z_L}{a_{11}Z_L+a_{12}}$	$\frac{\Delta b Z_L}{b_{22}Z_L+b_{12}}$	$\frac{-h_{21}Z_L}{\Delta h Z_L + h_{11}}$	$\frac{g_{21}Z_L}{g_{22}+Z_L}$
$A_i = -I_2/I_1$ (Z_L at port 2)	$\frac{z_{21}}{z_{22}+Z_L}$	$\frac{-y_{21}}{\Delta y Z_L + y_{11}}$	$\frac{1}{a_{22}+a_{21}Z_L}$	$\frac{\Delta b}{b_{11}+b_{21}Z_L}$	$\frac{-h_{21}}{h_{22}Z_L+1}$	$\frac{g_{21}}{\Delta g + g_{11}Z_L}$
$Z_{out} = Z_{Th}$ (Z_s at port 1)	$\frac{\Delta z + z_{22}Z_s}{z_{11}+Z_s}$	$\frac{y_{11}Z_s+1}{\Delta y Z_s + y_{22}}$	$\frac{a_{12}+a_{22}Z_s}{a_{11}+a_{21}Z_s}$	$\frac{b_{12}+b_{11}Z_s}{b_{22}+b_{21}Z_s}$	$\frac{h_{11}+Z_s}{\Delta h + h_{22}Z_s}$	$\frac{\Delta g Z_s + g_{22}}{g_{11}Z_s+1}$

(cf) Δz, Determinant of the z-parameter matrix.

```
function p=port_property(P,Z,which) % Table 8.2
%Input: P= Base parameter matrix
%      Z= ZL(load impedance) or Zs(source impedance)
%      which= specifies the property and the base parameter
%       e.g., to find the z-parameter-based current gain(Ai),
%        you put 'Aiz' as the 3rd input argument 'which'.
%Output: p=Zi(input impedance)/Av(voltage gain)/Ai(current gain),
%        Zo(output impedance) depending on the 1st two chs of which
%      based on the parameter specified by the 3rd character
% Copyleft: Won Y. Yang, wyyang53@hanmail.net, CAU for academic use only
switch lower(which)
  case 'ziz', Y=1/Z; p=(det(P)*Y+P(1,1))/(P(2,2)*Y+1);
  case 'ziy', Y=1/Z; p=(P(2,2)+Y)/(det(P)+P(1,1)*Y);
  case 'zia', Y=1/Z; p=(P(1,1)+P(1,2)*Y)/(P(2,1)+P(2,2)*Y);
  case 'zib', Y=1/Z; p=(P(2,2)+P(1,2)*Y)/(P(2,1)+P(1,1)*Y);
  case 'zih', Y=1/Z; p=(det(P)+P(1,1)*Y)/(P(2,2)+Y);
  case 'zig', Y=1/Z; p=(P(2,2)*Y+1)/(det(P)*Y+P(1,1));
  case 'avz', Y=1/Z; p=P(2,1)/(det(P)*Y+P(1,1));
  case 'avy', Y=1/Z; p=-P(2,1)/(P(2,2)+Y);
  case 'ava', Y=1/Z; p=1/(P(1,1)+P(1,2)*Y);
  case 'avb', Y=1/Z; p=det(P)/(P(2,2)+P(1,2)*Y);
  case 'avh', Y=1/Z; p=-P(2,1)/(det(P)+P(1,1)*Y);
  case 'avg', Y=1/Z; p=P(2,1)/(P(2,2)*Y+1);
  case 'aiz', p=P(2,1)/(P(2,2)+Z);
  case 'aiy', p=-P(2,1)/(det(P)*Z+P(1,1));
  case 'aia', p=1/(P(2,2)+P(2,1)*Z);
  case 'aib', p=det(P)/(P(1,1)+P(2,1)*Z);
  case 'aih', p=-P(2,1)/(P(2,2)*Z+1);
  case 'aig', p=P(2,1)/(det(P)+P(1,1)*Z);
  case 'zoz', p=(det(P)+P(2,2)*Z)/(P(1,1)+Z);
  case 'zoy', p=(P(1,1)*Z+1)/(det(P)*Z+P(2,2));
  case 'zoa', p=(P(1,2)+P(2,2)*Z)/(P(1,1)+P(2,1)*Z);
  case 'zob', p=(P(1,2)+P(1,1)*Z)/(P(2,2)+P(2,1)*Z);
  case 'zoh', p=(P(1,1)+Z)/(det(P)+P(2,2)*Z);
  case 'zog', p=(det(P)*Z+P(2,2))/(P(1,1)*Z+1);
  otherwise error('What do you want to do by using port_property()?')
end
```

Example 8.12 Input Impedances of a Transmission Line
Find the input impedance, voltage gain, and current gain of a two-port network, which consists of a TL of characteristic impedance Z_0, propagation constant γ, and length l, with a load impedance Z_L at port 2 where the a parameter matrix of the TL was found in Example 8.5 as

$$\begin{bmatrix} a_{11} & a_{12} \\ a_{21} & a_{22} \end{bmatrix} \overset{\text{(E8.5.6)}}{=} \begin{bmatrix} \cosh\gamma l & Z_0\sinh\gamma l \\ \sinh\gamma l/Z_0 & \cosh\gamma l \end{bmatrix} \tag{E8.12.1}$$

To find the input impedance, we run the following MATLAB statements:

```
>>syms Z0 r l ZL;
  a=[cosh(r*l) Z0*sinh(r*l); sinh(r*l)/Z0 cosh(r*l)]; % Eq.(E8.12.1)
  Zin=port_property(a,ZL,'Zia'); pretty(simplify(Zin))
```

which yields

```
Z0 (ZL cosh(l r) + Z0 sinh(l r))
--------------------------------
  Z0 cosh(l r) + ZL sinh(l r)
```

$$; \ Z_{in} = \frac{Z_0(Z_L\cosh\gamma l + Z_0\sinh\gamma l)}{Z_0\cosh\gamma l + Z_L\sinh\gamma l} \tag{E8.12.2}$$

This turns out to be conformed with Eq. (7.1.16).

To find the voltage and current gains, we run the following MATLAB statements:

```
>>Av=port_property(a,ZL,'Ava'); pretty(simplify(Av))
  Ai=port_property(a,ZL,'Aia'); pretty(simplify(Ai))
```

which yields

```
              1
---------------------------
             Z0 sinh(l r)
cosh(l r) +  ------------
                  ZL
```

$$; \ A_v = \frac{1}{\cosh\gamma l + \dfrac{Z_0\sinh\gamma l}{Z_L}} \tag{E8.12.3a}$$

```
              1
---------------------------
             ZL sinh(l r)
cosh(l r) +  ------------
                  Z0
```

$$; \ A_i = \frac{1}{\cosh\gamma l + \dfrac{Z_L\sinh\gamma l}{Z_0}} \tag{E8.12.3b}$$

To get comfortable about these results and also to reflect on their physical meanings, consider two special situations, one is a short-ended TL and the other is an open-ended TL. First, for a short-ended TL with $Z_L = 0$, the voltage and current gains are

$$A_v = \frac{\mathbf{V}_2}{\mathbf{V}_1} \overset{\text{(E8.12.3a)}}{=} \frac{1}{\cosh\gamma l + (Z_0/Z_L)\sinh\gamma l} \overset{Z_L=0}{=} 0 \tag{E8.12.4a}$$

$$A_i = \frac{-\mathbf{I}_2}{\mathbf{I}_1} \overset{\text{(E8.12.3b)}}{=} \frac{1}{\cosh\gamma l + (Z_L/Z_0)\sinh\gamma l} \overset{Z_L=0}{=} \frac{1}{\cosh\gamma l} \tag{E8.12.4b}$$

Apart from this, we can use Eq. (E8.5.3) with $\mathbf{V}_2 = 0$ (short-ended) to write the incident/reflected voltage on/from port 2 as

$$\begin{bmatrix} \mathbf{V}_o^+ \\ \mathbf{V}_o^- \end{bmatrix} \overset{\text{(E8.5.3)}}{=} \begin{bmatrix} 1/2 & -Z_0/2 \\ 1/2 & Z_0/2 \end{bmatrix} \begin{bmatrix} \mathbf{V}_2 \\ \mathbf{I}_2 \end{bmatrix} \overset{\mathbf{V}_2=0}{=} \begin{bmatrix} -Z_0/2 \\ Z_0/2 \end{bmatrix} \mathbf{I}_2 \tag{E8.12.5}$$

which implies that $\mathbf{V}_o^- = -\mathbf{V}_o^+$, i.e. the reflected voltage at port 2 (short-ended) is 180° out of phase w.r.t. the incident voltage (see Figure 7.2(b)) so that the load voltage reflection coefficient (Eq. (7.1.9)) is −1. We can substitute this into Eq. (7.1.2) to write the phase current at port 1 as

$$\mathbf{I}_1 \underset{d=l}{\overset{(7.1.2)}{=}} \mathbf{I}_o^+ e^{\gamma l} + \mathbf{I}_o^- e^{-\gamma l} = \frac{1}{Z_0}\left(\mathbf{V}_o^+ e^{\gamma l} - \mathbf{V}_o^- e^{-\gamma l}\right)$$

$$\overset{\text{(E8.12.5)}}{=} -\frac{\mathbf{I}_2}{2}\left(e^{\gamma l} + e^{-\gamma l}\right) = -\mathbf{I}_2\cosh\gamma l \tag{E8.12.6}$$

From this, we can get the same current gain as Eq. (E8.12.4b).

Second, for an open-ended TL with $Z_L = \infty$, the voltage and current gains are

$$A_v = \frac{\mathbf{V}_2}{\mathbf{V}_1} \overset{\text{(E8.12.3a)}}{=} \frac{1}{\cosh\gamma l + (Z_0/Z_L)\sinh\gamma l} \overset{Z_L=\infty}{=} \frac{1}{\cosh\gamma l} \tag{E8.12.7a}$$

$$A_i = \frac{-\mathbf{I}_2}{\mathbf{I}_1} \overset{\text{(E8.12.3b)}}{=} \frac{1}{\cosh\gamma l + (Z_L/Z_0)\sinh\gamma l} \overset{Z_L=\infty}{=} 0 \tag{E8.12.7b}$$

Apart from this, we can use Eq. (E8.5.3) with $\mathbf{I}_2 = 0$ (open-ended) to write the incident/reflected voltage on/from port 2 as

$$\begin{bmatrix} \mathbf{V}_o^+ \\ \mathbf{V}_o^- \end{bmatrix} \overset{\text{(E8.5.3)}}{=} \begin{bmatrix} 1/2 & -Z_0/2 \\ 1/2 & Z_0/2 \end{bmatrix} \begin{bmatrix} \mathbf{V}_2 \\ \mathbf{I}_2 \end{bmatrix} \overset{\mathbf{I}_2=0}{=} \begin{bmatrix} 1/2 \\ 1/2 \end{bmatrix} \mathbf{V}_2 \tag{E8.12.8}$$

which implies that $\mathbf{V}_o^- = \mathbf{V}_o^+$, i.e. the reflected voltage from port 2 (open-ended) is equal to the incident voltage (see Figure 7.2(a)) so that the load voltage reflection coefficient (Eq. (7.1.9)) is 1. We can substitute this into Eq. (7.1.1) to write the phasor voltage at port 1 as

$$\mathbf{V}_1 \underset{d=l}{\overset{(7.1.1)}{=}} \mathbf{V}_o^+ e^{\gamma l} + \mathbf{V}_o^- e^{-\gamma l} \overset{(E8.12.8)}{=} \frac{\mathbf{V}_2}{2}\left(e^{\gamma l} + e^{-\gamma l}\right) = \mathbf{V}_2 \cosh \gamma l \tag{E8.12.9}$$

From this, we can get the same voltage gain as Eq. (E8.12.7a).

Example 8.13 Making Use of '`port_conversion()`' and '`port_property()`'

Figure 8.25 shows a feedback amplifier circuit which can be regarded as a parallel-parallel (shunt) connection of two 2-port subnetworks. Note that the y-parameters of the upper subnetwork can be written from Eq. (E8.3.5) as

$$y^a \overset{(E8.3.5)}{=} \begin{bmatrix} 1/R_f & -1/R_f \\ -1/R_f & 1/R_f \end{bmatrix} \tag{E8.13.1}$$

The voltage-current relations of the lower subnetwork can be written as

$$\begin{bmatrix} V_1^b \\ I_2^b \end{bmatrix} = \begin{bmatrix} r_{be} I_1^b \\ g_m v_{be} + V_2^b/R_C \end{bmatrix} = \begin{bmatrix} r_{be} I_1^b \\ g_m r_{be} I_1^b + V_2^b/R_C \end{bmatrix} = \begin{bmatrix} r_{be} & 0 \\ g_m r_{be} & 1/R_C \end{bmatrix} \begin{bmatrix} I_1^b \\ V_2^b \end{bmatrix} \tag{E8.13.2}$$

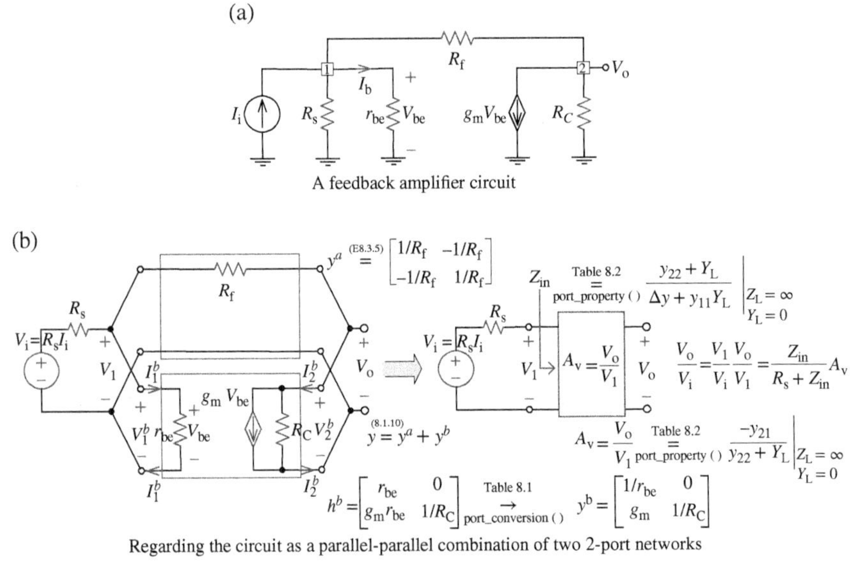

Figure 8.25 Circuit regarded as a parallel-parallel interconnection of two 2-port subnetworks.

We can match this with Eq. (8.1.5) to get its h-parameters and then use Table 8.1 to make an h-to-y conversion as

$$h^b \overset{(E8.13.2)}{\underset{(8.1.5)}{=}} \begin{bmatrix} r_{be} & 0 \\ g_m r_{be} & 1/R_C \end{bmatrix} \underset{\text{port\_conversion()}}{\overset{\text{Table 8.1}}{\longrightarrow}} y^b = \begin{bmatrix} 1/r_{be} & 0 \\ g_m & 1/R_C \end{bmatrix} \tag{E8.13.3}$$

This conversion can be done by using the MATLAB function '`port_conversion()`' as

```
>>syms Rf rbe gm RC
  h=[rbe 0; gm*rbe 1/RC]; yb=port_conversion(h,'h2y')
```

which yields the y-parameter matrix of the lower subnetwork:

```
yb = [ 1/rbe,      0]
     [    gm, 1/RC]
```

We can add these two y-parameter matrices y^a and y^b to get the (overall) y-parameters, the input impedance Z_{in}, and the voltage gain A_v of the composite 2-port network (excluding the source part consisting of V_s and R_s) as follows:

$$y \overset{(8.1.10)}{=} y^a + y^b \overset{(E8.13.1)}{\underset{(E8.13.4)}{=}} \begin{bmatrix} 1/R_f + 1/r_{be} & -1/R_f \\ -1/R_f + g_m & 1/R_f + 1/R_C \end{bmatrix} \tag{E8.13.5}$$

$$Z_{in} \overset{\text{Table 8.2}}{=} \left. \frac{y_{22}Z_L + 1}{\Delta y Z_L + y_{11}} \right|_{\substack{Z_L=\infty \\ Y_L=0}} = \frac{y_{22}}{\Delta y} \overset{(E8.13.5)}{=} \frac{(1/R_f + 1/R_C)\, r_{be} R_f R_C}{r_{be} + R_f + R_C + g_m r_{be} R_C} \tag{E8.13.6}$$

$$A_v = \frac{V_o}{V_1} \underset{\text{port\_property()}}{\overset{\text{Table 8.2}}{=}} \left. \frac{-y_{21}Z_L}{y_{22}Z_L + 1} \right|_{\substack{Z_L=\infty \\ Y_L=0}} = \frac{-y_{21}}{y_{22}} \overset{(E8.13.5)}{=} -\frac{g_m - 1/R_f}{1/R_f + 1/R_C} \tag{E8.13.7}$$

Now, we can obtain the overall voltage gain of the whole two-port network including the source part consisting of V_s and R_s as

$$\frac{V_o}{V_i} = \frac{V_1}{V_i}\frac{V_o}{V_1} = \frac{Z_{in}}{R_s + Z_{in}} A_v$$

$$\overset{(E8.13.6,7)}{=} \frac{(1 - g_m R_f)\, r_{be} R_C}{R_s r_{be} + R_s R_f + R_s R_C + g_m R_s r_{be} R_C + r_{be} R_C + r_{be} R_f} \tag{E8.13.8}$$

All these computations have been listed in the following MATLAB script 'elec08e13.m' so that it can be run at a time.

```
%elec08e13.m
syms Rs rbe Rf gm Rc ZL Ii
y1=[1/Rf -1/Rf; -1/Rf 1/Rf]; % Eq.(E8.13.1)
h2=[rbe 0; gm*rbe 1/Rc]; % Eq.(E8.13.3)
y2=port_conversion(h2,'h2y') % Eq.(E8.13.4)
y=y1+y2 % Eq.(E8.13.5)
ZL=inf; % No load
Zin=port_property(y,ZL,'ZiY'); % Eq.(E8.13.6)
Av=port_property(y,ZL,'AvY') % Eq.(E8.13.7)
% To get the overall voltage gain: Eq.(E8.13.8)
Avo=Av*Zin/(Rs+Zin); pretty(simplify(Avo))
% Node Analysis for crosscheck
Y=[1/Rs+1/rbe+1/Rf -1/Rf;
   gm-1/Rf  1/Rf+1/Rc]; % Node admittance matrix
Is=[1/Rs; 0]; % RHS vector of the node equation
V=Y\Is; % Is equivalent to Vs=1 in series with Rs
pretty(V(2)) % Vo to a unity voltage source
```

Example 8.14 Analysis of a Two-Port Network with Source and Load
Consider the two-port network (enclosed by the rectangle) with a source and a load depicted in Figure 8.26(a).

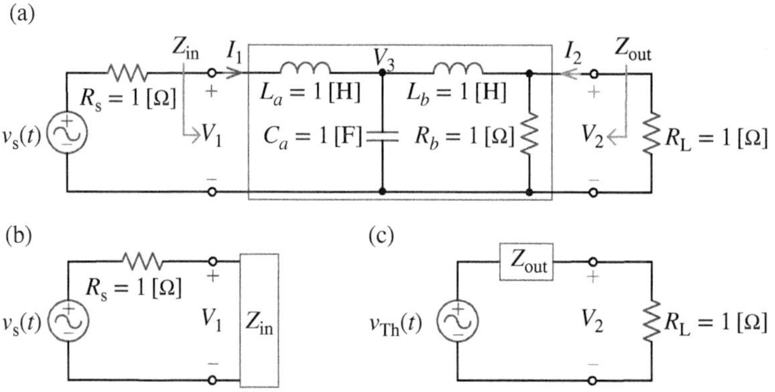

Figure 8.26 Two-port network with a source and a load.

a) Find the a-parameters of the two-port network.
Regarding the circuit as a four-stage cascade connection, we can use Eq. (8.1.13) together with Eqs. (E8.3.1) and (E8.3.2) to get the overall a-parameters as

$$\begin{bmatrix} a_{11} & a_{12} \\ a_{21} & a_{22} \end{bmatrix} \overset{(8.1.13)}{\underset{(E8.3.1,2)}{=}} \begin{bmatrix} 1 & sL_a \\ 0 & 1 \end{bmatrix} \begin{bmatrix} 1 & 0 \\ sC_a & 1 \end{bmatrix} \begin{bmatrix} 1 & sL_b \\ 0 & 1 \end{bmatrix} \begin{bmatrix} 1 & 0 \\ 1/R_b & 1 \end{bmatrix} = \begin{bmatrix} 1 & s \\ 0 & 1 \end{bmatrix} \begin{bmatrix} 1 & 0 \\ s & 1 \end{bmatrix} \begin{bmatrix} 1 & s \\ 0 & 1 \end{bmatrix} \begin{bmatrix} 1 & 0 \\ 1 & 1 \end{bmatrix}$$

$$= \begin{bmatrix} s^2+1 & s \\ s & 1 \end{bmatrix} \begin{bmatrix} s+1 & s \\ 1 & 1 \end{bmatrix} = \begin{bmatrix} s^3+s^2+2s+1 & s^3+2s \\ s^2+s+1 & s^2+1 \end{bmatrix} \tag{E8.14.1}$$

b) Find the input impedance Z_{in} of the two-port network with port 2 open ($R_L=\infty[\Omega]$).

$$Z_{\text{in,o}} \overset{\text{Table 8.2}}{=} \frac{a_{11}Z_L + a_{12}}{a_{21}Z_L + a_{22}} \underset{Z_L = R_L = \infty}{\overset{(E8.14.1)}{=}} \frac{a_{11}}{a_{21}} = \frac{s^3+s^2+2s+1}{s^2+s+1} \tag{E8.14.2}$$

c) Referring to the equivalent circuit depicted in Figure 8.26(b), find $V_1(s)/V_s(s)$ with port 2 open.

$$\left.\frac{V_1(s)}{V_s(s)}\right|_{\text{port 2 open}} = \frac{Z_{\text{in,o}}}{R_s + Z_{\text{in,o}}} \underset{Z_L=\infty}{\overset{(E8.14.2)}{=}} \frac{(s^3+s^2+2s+1)/(s^2+s+1)}{1+(s^3+s^2+2s+1)/(s^2+s+1)}$$
$$= \frac{s^3+s^2+2s+1}{s^3+2s^2+3s+2} \tag{E8.14.3}$$

d) Find the voltage gain $V_2(s)/V_1(s)$ with port 2 open.

$$A_{v,o} = \frac{V_2(s)}{V_1(s)} \overset{\text{Table 8.2}}{=} \frac{1}{a_{11}+a_{12}Y_L} \underset{Y_L=0}{\overset{(E8.14.1)}{=}} \frac{1}{s^3+s^2+2s+1} \tag{E8.14.4}$$

e) Find the overall voltage gain $V_2(s)/V_s(s)$ with port 2 open.

$$A_{v,o,ov} = \frac{V_2(s)}{V_s(s)} = \frac{V_1(s)}{V_s(s)}\frac{V_2(s)}{V_1(s)} \underset{(E8.14.4)}{\overset{(E8.14.3)}{=}} \frac{s^3+s^2+2s+1}{s^3+2s^2+3s+2}\frac{1}{s^3+s^2+2s+1}$$
$$= \frac{1}{s^3+2s^2+3s+2} \tag{E8.14.5}$$

f) Find the input impedance of the two-port network with the load resistance R_L = 1 Ω connected.

$$Z_{\text{in}} \overset{\text{Table 8.2}}{=} \frac{a_{11}Z_L + a_{12}}{a_{21}Z_L + a_{22}} \underset{Z_L=R_L=1}{\overset{(E8.14.1)}{=}} \frac{(s^3+s^2+2s+1)+(s^3+2s)}{(s^2+s+1)+(s^2+1)}$$
$$= \frac{2s^3+s^2+4s+1}{2s^2+s+2} \tag{E8.14.6}$$

g) Referring to the equivalent circuit depicted in Figure 8.26(b), find $V_1(s)/V_s(s)$ with R_L=1 Ω connected.

$$\frac{V_1(s)}{V_s(s)} = \frac{Z_{\text{in}}}{R_s+Z_{\text{in}}} \overset{(E8.14.6)}{=} \frac{(2s^3+s^2+4s+1)/(2s^2+s+2)}{1+(2s^3+s^2+4s+1)/(2s^2+s+2)}$$
$$= \frac{2s^3+s^2+4s+1}{2s^3+3s^2+5s+3} \tag{E8.14.7}$$

h) Find the voltage gain $V_2(s)/V_1(s)$ with $R_L = 1\,\Omega$ connected.

$$A_v = \frac{V_2(s)}{V_1(s)} \overset{\text{Table 8.2}}{=} \frac{Z_L}{a_{11}Z_L + a_{12}} \underset{Z_L = R_L = 1}{\overset{(E8.14.1)}{=}} \frac{1}{(s^3 + s^2 + 2s + 1) + (s^3 + 2s)}$$
$$= \frac{1}{2s^3 + s^2 + 4s + 1} \quad \text{(E8.14.8)}$$

i) Find the overall voltage gain $V_2(s)/V_s(s)$ with $R_L=1\,\Omega$ connected.

$$A_{v,ov} = \frac{V_2(s)}{V_s(s)} = \frac{V_1(s)}{V_s(s)}\frac{V_2(s)}{V_1(s)} \underset{(E8.14.8)}{\overset{(E8.14.7)}{=}} \frac{2s^3 + s^2 + 4s + 1}{2s^3 + 3s^2 + 5s + 3}\frac{1}{2s^3 + s^2 + 4s + 1}$$
$$= \frac{1}{2s^3 + 3s^2 + 5s + 3} \quad \text{(E8.14.9)}$$

j) Find the current gain $-I_2(s)/I_1(s)$ with $R_L=1\,\Omega$ connected.

$$A_i = -\frac{I_2(s)}{I_1(s)} \overset{\text{Table 8.2}}{=} \frac{1}{a_{22} + a_{21}Z_L} \underset{Z_L = R_L = 1}{\overset{(E8.14.1)}{=}} \frac{1}{(s^2 + 1) + (s^2 + s + 1)}$$
$$= \frac{1}{2s^2 + s + 2} \quad \text{(E8.14.10)}$$

k) Find the output impedance of the two-port network with the source resistance $R_s=1\,\Omega$ connected.

$$Z_{out} \overset{\text{Table 8.2}}{=} \frac{a_{12} + a_{22}Z_s}{a_{11} + a_{21}Z_s} \underset{Z_s = 1}{\overset{(E8.14.1)}{=}} \frac{(s^3 + 2s) + (s^2 + 1)}{(s^3 + s^2 + 2s + 1) + (s^2 + s + 1)}$$
$$= \frac{s^3 + s^2 + 2s + 1}{s^3 + 2s^2 + 3s + 2} \quad \text{(E8.14.11)}$$

l) Find the Thevenin equivalent source of the circuit seen from port 2.

$$V_{Th} = V_{2,\text{port 2 open}} = A_{v,o}V_s(s) \overset{(E8.14.5)}{=} \frac{1}{s^3 + 2s^2 + 3s + 2}V_s(s) \quad \text{(E8.14.12)}$$

m) Referring to the equivalent circuit depicted in Figure 8.26(c), find $V_2(s)/V_s(s)$ with $R_L=1\,\Omega$ connected.

$$\frac{V_2(s)}{V_s(s)} = \frac{V_{Th}(s)}{V_s(s)}\frac{R_L}{Z_{out} + R_L} \underset{(E8.14.12)}{\overset{(E8.14.11)}{=}} \frac{1}{s^3 + 2s^2 + 3s + 2}\frac{1}{\dfrac{s^3 + s^2 + 2s + 1}{s^3 + 2s^2 + 3s + 2} + 1}$$
$$= \frac{1}{2s^3 + 3s^2 + 5s + 3} \quad \text{(E8.14.13)}$$

n) Apply the node analysis to find $V_2(s)/V_s(s)$ with R_L=1 Ω connected.
We can set up the node equation and solve it to get the same result as Eq. (E8.14.13):

$$\begin{bmatrix} 1/Rs+1/sL_a & -1/sL_a & 0 \\ -1/sL_a & 1/sL_a+sC_a+1/sL_b & -1/sL_b \\ 0 & -1/sL_b & 1/sL_b+1/R_b+1/RL \end{bmatrix} \begin{bmatrix} V_1 \\ V_3 \\ V_2 \end{bmatrix} = \begin{bmatrix} V_s/Rs \\ 0 \\ 0 \end{bmatrix}$$

$$; \begin{bmatrix} 1/s+1 & -1/s & 0 \\ -1/s & s+2/s & -1/s \\ 0 & -1/s & 1/s+2 \end{bmatrix} \begin{bmatrix} V_1 \\ V_3 \\ V_2 \end{bmatrix} = \begin{bmatrix} V_s \\ 0 \\ 0 \end{bmatrix}; \begin{bmatrix} s+1 & -1 & 0 \\ -1 & s^2+2 & -1 \\ 0 & -1 & s+2 \end{bmatrix} \begin{bmatrix} V_1 \\ V_3 \\ V_2 \end{bmatrix} = \begin{bmatrix} sV_s \\ 0 \\ 0 \end{bmatrix}$$

$$; \begin{bmatrix} V_1 \\ V_3 \\ V_2 \end{bmatrix} = \frac{1}{2s^3+3s^2+5s+3} \begin{bmatrix} 2s^3+s^2+4s+1 & \times & \times \\ 2s+1 & \times & \times \\ 1 & \times & \times \end{bmatrix} \begin{bmatrix} V_s \\ 0 \\ 0 \end{bmatrix} \tag{E8.14.14}$$

All the computations for (a)-(m) have been listed in the following MATLAB script 'elec08e14.m' so that it can be run at a time.

```
%elec08e14.m
syms s Vs %Rs La Ca Lb Rb RL
Rs=1; La=1; Ca=1; Lb=1; Rb=1; RL=1; sLa=s*La; sCa=s*Ca; sLb=s*Lb;
a=[1 sLa; 0 1]*[1 0; sCa 1]*[1 sLb; 0 1]*[1 0; 1/Rb 1]; % Eq.(E8.14.1)
Zino=port_property(a,inf,'ZiA'); pretty(Zino) % Eq.(E8.14.2)
pretty(simplify(Zino/(Rs+Zino))) % Eq.(E8.14.3)
Avo=port_property(a,inf,'AvA'); % Eq.(E8.14.4): Open-loop voltage gain
Avoov=Zino/(Rs+Zino)*Avo; %Eq.(E8.14.5): Overall open-loop voltage gain
Zin=port_property(a,RL,'ZiA'); % Eq.(E8.14.6): Input impedance
pretty(Zin)
pretty(simplify(Zin/(Rs+Zin))) % Eq.(E8.14.7)
Av=port_property(a,RL,'AvA'); %Av=limit(Av,ZL,inf) % Eq.(E8.14.8)
pretty(Av)
Avov=Zin/(Rs+Zin)*Av; %Av=limit(Av,ZL,inf) % Eq.(E8.14.9)
pretty(simplify(Avov))
%.................
```

8.2 Scattering Parameters

8.2.1 Definition of Scattering Parameters

Figure 8.27 shows a two-port network where the total sinusoidal steady-state voltages and currents at ports 1 and 2 can be written in cosine-based (maximum) phasor form as

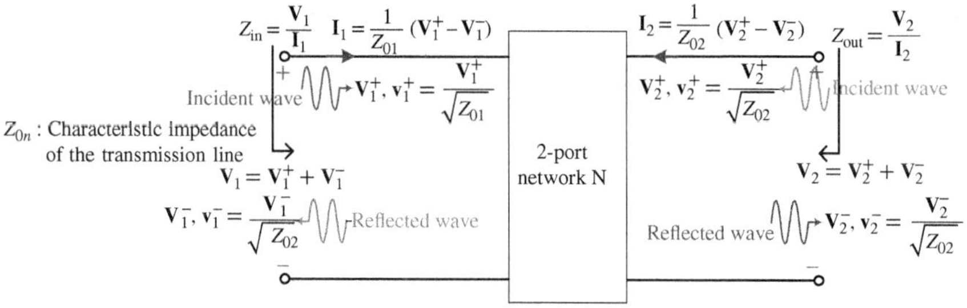

Figure 8.27 Two-port network.

$$\mathbf{V}_1 = \mathbf{V}_1^+ + \mathbf{V}_1^- = \sqrt{Z_{01}}\left(\mathbf{v}_1^+ + \mathbf{v}_1^-\right),\quad \mathbf{V}_2 = \mathbf{V}_2^+ + \mathbf{V}_2^- = \sqrt{Z_{02}}\left(\mathbf{v}_2^+ + \mathbf{v}_2^-\right) \tag{8.2.1a,b}$$

$$\mathbf{I}_1 = \frac{1}{Z_{01}}\left(\mathbf{V}_1^+ - \mathbf{V}_1^-\right) = \frac{1}{\sqrt{Z_{01}}}\left(\mathbf{v}_1^+ - \mathbf{v}_1^-\right),\quad \mathbf{I}_2 = \frac{1}{Z_{02}}\left(\mathbf{V}_2^+ - \mathbf{V}_2^-\right) = \frac{1}{\sqrt{Z_{02}}}\left(\mathbf{v}_2^+ - \mathbf{v}_2^-\right) \tag{8.2.2a,b}$$

where Z_{0n} and $\mathbf{V}_n^+/\mathbf{V}_n^-$ are the characteristic impedances at port n and the incident/reflected voltage into/from port n, respectively, and $\mathbf{v}_n^+/\mathbf{v}_n^-$ are the normalized representations of $\mathbf{V}_n^+/\mathbf{V}_n^-$:

$$\mathbf{v}_n^+ = \frac{\mathbf{V}_n^+}{\sqrt{Z_{0n}}} \text{ and } \mathbf{v}_n^- = \frac{\mathbf{V}_n^-}{\sqrt{Z_{0n}}} \tag{8.2.3a,b}$$

Note that $\mathbf{V}_2^+/\mathbf{V}_2^-$ are backward-/forward-propagating waves, respectively, unlike $\mathbf{V}_o^+/\mathbf{V}_o^-$ in Eqs. (7.1.1) and (7.1.2) while $\mathbf{V}_1^+/\mathbf{V}_1^-$ are forward-/backward-propagating waves, respectively.

Substituting these equations into the definition of the a-parameter matrix

$$\begin{bmatrix} \mathbf{V}_1 \\ \mathbf{I}_1 \end{bmatrix} \overset{(8.1.3)}{=} \begin{bmatrix} a_{11} & a_{12} \\ a_{21} & a_{22} \end{bmatrix} \begin{bmatrix} \mathbf{V}_2 \\ -\mathbf{I}_2 \end{bmatrix} \tag{8.2.4}$$

yields

$$\begin{bmatrix} \sqrt{Z_{01}}\left(\mathbf{v}_1^+ + \mathbf{v}_1^-\right) \\ \left(\mathbf{v}_1^+ - \mathbf{v}_1^-\right)/\sqrt{Z_{01}} \end{bmatrix} = \begin{bmatrix} a_{11} & a_{12} \\ a_{21} & a_{22} \end{bmatrix} \begin{bmatrix} \sqrt{Z_{02}}\left(\mathbf{v}_2^+ + \mathbf{v}_2^-\right) \\ -\left(\mathbf{v}_2^+ - \mathbf{v}_2^-\right)/\sqrt{Z_{02}} \end{bmatrix}$$

$$; \begin{bmatrix} \sqrt{Z_{01}}\mathbf{v}_1^- - \left(a_{11}\sqrt{Z_{02}} + a_{12}/\sqrt{Z_{02}}\right)\mathbf{v}_2^- \\ -\mathbf{v}_1^-/\sqrt{Z_{01}} - \left(a_{21}\sqrt{Z_{02}} + a_{22}/\sqrt{Z_{02}}\right)\mathbf{v}_2^- \end{bmatrix} = \begin{bmatrix} a_{11} & a_{12} \\ a_{21} & a_{22} \end{bmatrix} \begin{bmatrix} \sqrt{Z_{02}}\mathbf{v}_2^+ \\ -\mathbf{v}_2^+/\sqrt{Z_{02}} \end{bmatrix} - \begin{bmatrix} \sqrt{Z_{01}} \\ 1/\sqrt{Z_{01}} \end{bmatrix} \mathbf{v}_1^+$$

$$; \begin{bmatrix} \sqrt{Z_{01}} & -\left(a_{11}\sqrt{Z_{02}} + a_{12}/\sqrt{Z_{02}}\right) \\ -1/\sqrt{Z_{01}} & -\left(a_{21}\sqrt{Z_{02}} + a_{22}/\sqrt{Z_{02}}\right) \end{bmatrix} \begin{bmatrix} \mathbf{v}_1^- \\ \mathbf{v}_2^- \end{bmatrix} = \begin{bmatrix} -\sqrt{Z_{01}} & a_{11}\sqrt{Z_{02}} - a_{12}/\sqrt{Z_{02}} \\ -1/\sqrt{Z_{01}} & a_{21}\sqrt{Z_{02}} - a_{22}/\sqrt{Z_{02}} \end{bmatrix} \begin{bmatrix} \mathbf{v}_1^+ \\ \mathbf{v}_2^+ \end{bmatrix} \tag{8.2.5}$$

We can solve this equation to find the expressions of the normalized reflected waves $\mathbf{v}_1^+$ and $\mathbf{v}_2^+$ in terms of the normalized incident waves $\mathbf{v}_1^+$ and $\mathbf{v}_2^+$ as

$$\begin{bmatrix} \mathbf{v}_1^- \\ \mathbf{v}_2^- \end{bmatrix} = \begin{bmatrix} s_{11} & s_{12} \\ s_{21} & s_{22} \end{bmatrix} \begin{bmatrix} \mathbf{v}_1^+ \\ \mathbf{v}_2^+ \end{bmatrix} \text{ or } \mathbf{v}^- = S\mathbf{v}^+ \text{ with } \begin{array}{l} s_{11}: \text{Input reflection gain with } \mathbf{v}_2^+ = 0 \\ s_{12}: \text{Reverse transmission gain with } \mathbf{v}_1^+ = 0 \\ s_{21}: \text{Forward transmission gain with } \mathbf{v}_2^+ = 0 \\ s_{22}: \text{Output reflection gain with } \mathbf{v}_1^+ = 0 \end{array} \tag{8.2.6}$$

where S is called the *scattering* (*s-parameter*) *matrix* and can be obtained from the a-parameter (transmission) matrix as

$$S = \begin{bmatrix} s_{11} & s_{12} \\ s_{21} & s_{22} \end{bmatrix} = \begin{bmatrix} \sqrt{Z_{01}} & -\left(a_{11}\sqrt{Z_{02}} + a_{12}/\sqrt{Z_{02}}\right) \\ -1/\sqrt{Z_{01}} & -\left(a_{21}\sqrt{Z_{02}} + a_{22}/\sqrt{Z_{02}}\right) \end{bmatrix}^{-1} \begin{bmatrix} -\sqrt{Z_{01}} & a_{11}\sqrt{Z_{02}} - a_{12}/\sqrt{Z_{02}} \\ -1/\sqrt{Z_{01}} & a_{21}\sqrt{Z_{02}} - a_{22}/\sqrt{Z_{02}} \end{bmatrix}$$

$$= \frac{1}{a_{11}Z_{02} + a_{12} + a_{21}Z_{01}Z_{02} + a_{22}Z_{01}} \begin{bmatrix} a_{11}Z_{02} + a_{12} + a_{21}Z_{01}Z_{02} + a_{22}Z_{01} & 2(a_{11}a_{22} - a_{12}a_{21})\sqrt{Z_{01}Z_{02}} \\ 2\sqrt{Z_{01}Z_{02}} & -a_{11}Z_{02} + a_{12} - a_{21}Z_{01}Z_{02} + a_{22}Z_{01} \end{bmatrix} \tag{8.2.7}$$

```
>>syms sZ01 sZ02 a11 a12 a21 a22
  Ma=[sZ01 -(a11*sZ02+a12/sZ02); -1/sZ01 -(a21*sZ02+a22/sZ02)];
  Mb=[-sZ01 a11*sZ02-a12/sZ02; -1/sZ01 a21*sZ02-a22/sZ02];
  S=Ma\Mb; pretty(simplify(S)) % Eq.(8.2.7)
```

Note that in view of the definition (8.2.6) of scattering matrix S, its (m, n)th entry can be determined as

$$s_{mn} = \left.\frac{\mathbf{v}_n^-}{\mathbf{v}_m^+}\right|_{\mathbf{v}_k^+ = 0 \text{ for } k \neq m} \overset{(8.2.3a,b)}{=} \left.\frac{\mathbf{V}_n^-/\sqrt{Z_{0m}}}{\mathbf{V}_m^+/\sqrt{Z_{0n}}}\right|_{\mathbf{v}_k^+ = 0 \text{ for } k \neq m} \tag{8.2.8}$$

This is the ratio of the reflected wave amplitude from port n to the incident wave amplitude on port m where the incident waves on all ports other than port m are set to zero, which is made possible by terminating the ports with matched loads to remove reflections.

On the other hand, we can solve Eqs. (8.2.1a) and 8.2.2a for $\mathbf{v}_1^+$ and $\mathbf{v}_1^-$ to write

$$\mathbf{v}_1^+ = \frac{1}{2}\left(\frac{1}{\sqrt{Z_{01}}}\mathbf{V}_1 + \sqrt{Z_{01}}\mathbf{I}_1\right),\quad \mathbf{v}_1^- = \frac{1}{2}\left(\frac{1}{\sqrt{Z_{01}}}\mathbf{V}_1 - \sqrt{Z_{01}}\mathbf{I}_1\right) \tag{8.2.9a,b}$$

and also solve Eqs. (8.2.1b) and (8.2.2b) for $\mathbf{v}_2^+$ and $\mathbf{v}_2^-$ to write

$$\mathbf{v}_2^+ = \frac{1}{2}\left(\frac{1}{\sqrt{Z_{02}}}\mathbf{V}_2 + \sqrt{Z_{02}}\mathbf{I}_2\right),\quad \mathbf{v}_2^- = \frac{1}{2}\left(\frac{1}{\sqrt{Z_{02}}}\mathbf{V}_2 - \sqrt{Z_{02}}\mathbf{I}_2\right) \tag{8.2.10a,b}$$

Substituting these equations into Eq. (8.2.6) yields

$$\begin{bmatrix} \frac{1}{\sqrt{Z_{01}}}\mathbf{V}_1 - \sqrt{Z_{01}}\mathbf{I}_1 \\ \frac{1}{\sqrt{Z_{02}}}\mathbf{V}_2 - \sqrt{Z_{02}}\mathbf{I}_2 \end{bmatrix} = \begin{bmatrix} s_{11} & s_{12} \\ s_{21} & s_{22} \end{bmatrix} \begin{bmatrix} \frac{1}{\sqrt{Z_{01}}}\mathbf{V}_1 + \sqrt{Z_{01}}\mathbf{I}_1 \\ \frac{1}{\sqrt{Z_{02}}}\mathbf{V}_2 + \sqrt{Z_{02}}\mathbf{I}_2 \end{bmatrix}$$

$$; \begin{bmatrix} (1-s_{11})/\sqrt{Z_{01}} & -(1+s_{11})\sqrt{Z_{01}} \\ -s_{21}/\sqrt{Z_{01}} & -s_{21}\sqrt{Z_{01}} \end{bmatrix} \begin{bmatrix} \mathbf{V}_1 \\ \mathbf{I}_1 \end{bmatrix} = \begin{bmatrix} s_{12}/\sqrt{Z_{02}} & -s_{12}\sqrt{Z_{02}} \\ (-1+s_{22})/\sqrt{Z_{02}} & -(1+s_{22})\sqrt{Z_{02}} \end{bmatrix} \begin{bmatrix} \mathbf{V}_2 \\ -\mathbf{I}_2 \end{bmatrix} \tag{8.2.11}$$

In view of Eq. (8.1.3) defining the a-parameter, this implies that the a-parameter (transmission) matrix can be obtained from the s-parameter matrix as

$$A = \begin{bmatrix} a_{11} & a_{12} \\ a_{21} & a_{22} \end{bmatrix} = \begin{bmatrix} (1-s_{11})/\sqrt{Z_{01}} & -(1+s_{11})\sqrt{Z_{01}} \\ -s_{21}/\sqrt{Z_{01}} & -s_{21}\sqrt{Z_{01}} \end{bmatrix}^{-1} \begin{bmatrix} s_{12}/\sqrt{Z_{02}} & -s_{12}\sqrt{Z_{02}} \\ (-1+s_{22})/\sqrt{Z_{02}} & -(1+s_{22})\sqrt{Z_{02}} \end{bmatrix}$$

$$= \frac{1}{2s_{21}\sqrt{Z_{01}Z_{02}}} \begin{bmatrix} (1+s_{11})(1-s_{22})Z_{01} + s_{12}s_{21}Z_{01} & \{(1+s_{11})(1+s_{22}) - s_{12}s_{21}\}Z_{01}Z_{02} \\ (1-s_{11})(1-s_{22}) - s_{12}s_{21} & (1-s_{11})(1+s_{22})Z_{02} + s_{12}s_{21}Z_{02} \end{bmatrix} \tag{8.2.12}$$

```
>>syms sZ01 sZ02 s11 s12 s21 s22
  Z01=sZ01^2; Z02=sZ02^2;
  Ma=[(1-s11)/sZ01 -(1+s11)*sZ01; -s21/sZ01 -s21*sZ01];
  Mb=[s12/sZ02 -s12*sZ02; (-1+s22)/sZ02 -(1+s22)*sZ02];
  A=Ma\Mb; pretty(simplify(A)) % Eq.(8.2.12)
```

The a-to-s and s-to-a conversion processes have been cast into the following MATLAB functions '`a2s()`' and '`s2a()`', respectively. Since the conversion of s-parameters into other parameters than a-parameters or that of other parameters than a-parameters into s-parameters can be done indirectly via the s-parameters, let us not derive the conversion formulas between s-parameters and other parameters than a-parameters.

```
function S=a2s(A,Z01,Z02)
% A-to-S parameter conversion
if nargin<3, Z02=Z01; end
sZ01=sqrt(Z01); sZ02=sqrt(Z02);
D=A(1,1)*Z02+A(1,2)+A(2,1)*Z01*Z02+A(2,2)*Z01;
S(1,1)=(A(1,1)*Z02+A(1,2)-A(2,1)*Z01*Z02-A(2,2)*Z01)/D; % Eq.(8.2.7)
S(1,2)=(2*(A(1,1)*A(2,2)-A(1,2)*A(2,1))*sZ01*sZ02)/D;
S(2,:)=[2*sZ01*sZ02 -A(1,1)*Z02+A(1,2)-A(2,1)*Z01*Z02+A(2,2)*Z01]/D;
```

```
function A=s2a(S,Z01,Z02)
% S-to-A parameter conversion
if nargin<3, Z02=Z01; end
sZ01=sqrt(Z01); sZ02=sqrt(Z02);
D=2*S(2,1)*sZ01*sZ02;
A(1,1)=((Z01+S(1,1)*Z01)*(1-S(2,2))+S(1,2)*S(2,1)*Z01)/D;
A(1,2)=((Z01+S(1,1)*Z01)*(Z02+S(2,2)*Z02)-S(1,2)*S(2,1)*Z01*Z02)/D;
A(2,1)=((1-S(1,1))*(1-S(2,2))-S(1,2)*S(2,1))/D;
A(2,2)=((1-S(1,1))*(Z02+S(2,2)*Z02)+S(1,2)*S(2,1)*Z02)/D; % Eq. (8.2.12)
```

With the relationships (8.2.1) and (8.2.2) between the total voltage/currents ($\mathbf{V}_n/\mathbf{I}_n$) and traveling incident/reflected waves ($\mathbf{v}_n^+/\mathbf{v}_n^-$), we can write the power loss in the two-port network N as the difference between the power entering N through port 1 and the power leaving N through port 2:

$$
\begin{aligned}
P_{\text{loss}} &= \frac{1}{2}\text{Re}\{\mathbf{V}_1\mathbf{I}_1^*\} - \frac{1}{2}\text{Re}\{\mathbf{V}_2(-\mathbf{I}_2^*)\} \\
&= \frac{1}{2}\text{Re}\left\{\sqrt{Z_0}(\mathbf{v}_1^+ + \mathbf{v}_1^-)\frac{1}{\sqrt{Z_0}}(\mathbf{v}_1^{+*} - \mathbf{v}_1^{-*}) + \sqrt{Z_0}(\mathbf{v}_2^+ + \mathbf{v}_2^-)\frac{1}{\sqrt{Z_0}}(\mathbf{v}_2^{+*} - \mathbf{v}_2^{-*})\right\} \\
&= \frac{1}{2}\text{Re}\left\{|\mathbf{v}_1^+|^2 + |\mathbf{v}_2^+|^2 - |\mathbf{v}_1^-|^2 - |\mathbf{v}_2^-|^2\right\} + \frac{1}{2}\text{Re}\{-\mathbf{v}_1^+\mathbf{v}_1^{-*} + \mathbf{v}_1^{+*}\mathbf{v}_1^-\} + \frac{1}{2}\text{Re}\{-\mathbf{v}_2^+\mathbf{v}_2^{-*} + \mathbf{v}_2^{+*}\mathbf{v}_2^-\} \\
&= \frac{1}{2}\begin{bmatrix}\mathbf{v}_1^{+*} & \mathbf{v}_2^{+*}\end{bmatrix}\begin{bmatrix}\mathbf{v}_1^+ \\ \mathbf{v}_2^+\end{bmatrix} - \frac{1}{2}\begin{bmatrix}\mathbf{v}_1^{-*} & \mathbf{v}_2^{-*}\end{bmatrix}\begin{bmatrix}\mathbf{v}_1^- \\ \mathbf{v}_2^-\end{bmatrix} = \frac{1}{2}\mathbf{v}^{+*\text{T}}\mathbf{v}^+ - \frac{1}{2}\mathbf{v}^{-*\text{T}}\mathbf{v}^- = (P_{1,\text{in}} + P_{2,\text{in}}) - (P_{1,\text{r}} + P_{2,\text{r}}) \\
&\overset{(8.2.6)}{=} \frac{1}{2}\mathbf{v}^{+*\text{T}}\mathbf{v}^+ - \frac{1}{2}\mathbf{v}^{+*\text{T}}S^{*\text{T}}S\mathbf{v}^+ = \frac{1}{2}\mathbf{v}^{+*\text{T}}[I - S^{*\text{T}}S]\,\mathbf{v}^+
\end{aligned}
\tag{8.2.13}
$$

If the S-matrix is unitary, i.e. $S^{*\text{T}}S = I$, we have $P_{\text{loss}} = 0$, implying that the two-port network is lossless.

Table 8.3 shows the conditions for a two-port network to be reciprocal, symmetrical, and lossless (see Remark 8.1).

Table 8.3 The conditions for a two-port network to be reciprocal, symmetrical, and lossless.

Property \ Parameter	*Z*	*Y*	*A*	*B*	*H*	*G*	*S*
Reciprocal	$z_{12} = z_{21}$	$y_{12} = y_{21}$	$\Delta a = 0$	$\Delta b = 0$	$h_{12} = -h_{21}$	$g_{12} = -g_{21}$	$s_{12}/Z_{01} = s_{21}/Z_{02}$
Symmetrical	$z_{11} = z_{22}$	$y_{11} = y_{22}$	$a_{11} = a_{22}$	$b_{11} = b_{22}$	$\Delta h = 1$	$\Delta g = 1$	$s_{11}/Z_{01} = s_{22}/Z_{02}$
Lossless	$\text{Im}\{Z\} = 0$	$\text{Im}\{Y\} = 0$					$S^{*\text{T}}S = I$

(cf) Δa: Determinant of the a-parameter matrix.

8.2.2 Two-Port Network with Source/Load

In this section, we consider a two-port network, which is connected to a voltage source (in series with a source impedance Z_s) at port 1 and a load impedance Z_L at port 2 as depicted in Figure 8.28. We can write the voltage-current relationships of the network at ports 1 and 2 as

$$\mathbf{V}_1 = Z_{in}\mathbf{I}_1, \qquad \mathbf{V}_2 = -Z_L\mathbf{I}_2 \tag{8.2.14a,b}$$

where Z_{in} and Z_L are the input impedance (of the network including the load) and load impedance, respectively. Assuming that the characteristic impedances of the TLs connected to ports 1 and 2 are equal, i.e. $Z_{01} = Z_{02} = Z_0$), we substitute Eq. (8.2.14b) into Eqs. (8.2.10a) and (8.2.10b) to write the traveling waves at port 2 as

$$\mathbf{v}_2^+ \overset{(8.2.10a)}{=} \frac{1}{2}\left(\frac{1}{\sqrt{Z_0}}\mathbf{V}_2 + \sqrt{Z_0}\mathbf{I}_2\right) \overset{(8.2.14b)}{=} \frac{1}{2\sqrt{Z_0}}(-Z_L + Z_0)\mathbf{I}_2 \tag{8.2.15}$$

$$\mathbf{v}_2^- \overset{(8.2.10b)}{=} \frac{1}{2}\left(\frac{1}{\sqrt{Z_0}}\mathbf{V}_2 - \sqrt{Z_0}\mathbf{I}_2\right) \overset{(8.2.14b)}{=} \frac{1}{2\sqrt{Z_0}}(-Z_L - Z_0)\mathbf{I}_2 \tag{8.2.16}$$

which yields the reflection coefficient at the load:

$$\Gamma_L = \frac{\mathbf{v}_2^+ \text{ (reflected wave amplitude from load)}}{\mathbf{v}_2^- \text{ (incident wave amplitude on load)}} \overset{(8.2.15)}{\underset{(8.2.16)}{=}} \frac{Z_L - Z_0}{Z_L + Z_0} \tag{8.2.17}$$

Also, we can substitute Eq. (8.2.14a) into Eqs. (8.2.9a) and (8.2.9b) to write the traveling waves at port 1 as

$$\mathbf{v}_1^+ \overset{(8.2.9a)}{=} \frac{1}{2}\left(\frac{1}{\sqrt{Z_0}}\mathbf{V}_1 + \sqrt{Z_0}\mathbf{I}_1\right) \overset{(8.2.14a)}{=} \frac{1}{2\sqrt{Z_0}}(Z_{in} + Z_0)\mathbf{I}_1 \tag{8.2.18}$$

$$\mathbf{v}_1^- \overset{(8.2.9b)}{=} \frac{1}{2}\left(\frac{1}{\sqrt{Z_0}}\mathbf{V}_1 - \sqrt{Z_0}\mathbf{I}_1\right) \overset{(8.2.14a)}{=} \frac{1}{2\sqrt{Z_0}}(Z_{in} - Z_0)\mathbf{I}_1 \tag{8.2.19}$$

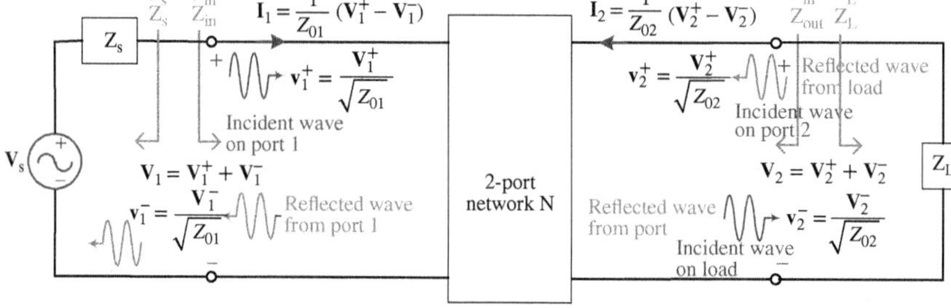

Figure 8.28 Two-port network connected to a voltage source (in series with a source impedance and a load).

which yields the reflection coefficient at port 1:

$$\Gamma_{in} = \frac{\mathbf{v}_1^- \text{ (reflected wave amplitude from port 1)}}{\mathbf{v}_1^+ \text{ (incident wave amplitude on port 1)}} = \frac{Z_{in} - Z_0}{Z_{in} + Z_0} \tag{8.2.20}$$

On the other hand, we can use the definition (8.2.6) of the S-parameters to write

$$\mathbf{v}_2^- \overset{(8.2.6b)}{=} s_{21}\mathbf{v}_1^+ + s_{22}\mathbf{v}_2^+ \overset{(8.2.17)}{=} s_{21}\mathbf{v}_1^+ + s_{22}\Gamma_L\mathbf{v}_2^- \rightarrow \mathbf{v}_2^- = \frac{s_{21}}{1 - s_{22}\Gamma_L}\mathbf{v}_1^+ \tag{8.2.21}$$

$$\mathbf{v}_1^- \overset{(8.2.6a)}{=} s_{11}\mathbf{v}_1^+ + s_{12}\mathbf{v}_2^+ \overset{(8.2.17)}{=} s_{11}\mathbf{v}_1^+ + s_{12}\Gamma_L\mathbf{v}_2^- \overset{(8.2.21)}{=} \left(s_{11} + \frac{s_{12}s_{21}\Gamma_L}{1 - s_{22}\Gamma_L}\right)\mathbf{v}_1^+ \tag{8.2.22}$$

which yields the relationship between the reflection coefficient at port 1 and that at the load as

$$\Gamma_{in} = \frac{\mathbf{v}_1^-}{\mathbf{v}_1^+} \overset{(8.2.22)}{=} s_{11} + \frac{s_{12}s_{21}\Gamma_L}{1 - s_{22}\Gamma_L} \overset{(8.2.20)}{=} \frac{Z_{in} - Z_0}{Z_{in} + Z_0} \text{ with } \Gamma_L = \frac{\mathbf{v}_2^+}{\mathbf{v}_2^-} \overset{(8.2.17)}{=} \frac{Z_L - Z_0}{Z_L + Z_0} \tag{8.2.23}$$

Matching conditions at port 1 and at the load can be obtained as

$$Z_{in} - Z_0 = 0 \rightarrow \Gamma_{in} = 0, \quad Z_L - Z_0 = 0 \rightarrow \Gamma_L = 0 \tag{8.2.24}$$

each implying that the reflection coefficients at port 1 and at the load are zero, respectively.

Reversing the roles of voltage source and load will yield

$$\Gamma_{out} = \frac{\mathbf{v}_2^-}{\mathbf{v}_2^+} = s_{22} + \frac{s_{21}s_{12}\Gamma_s}{1 - s_{11}\Gamma_s} = \frac{Z_{out} + Z_0}{Z_{out} - Z_0} \text{ with } \Gamma_s = \frac{\mathbf{v}_1^+}{\mathbf{v}_1^-} = \frac{Z_s + Z_0}{Z_s - Z_0} \tag{8.2.25}$$

Remark 8.2 Effect of Shifting Reference Planes on Scattering Parameters

In the two-port network shown in Figure 8.29, the incident/reflected waves at plane A are delayed/advanced in phase by βl_1 compared with those at plane A':

$$\mathbf{v}_1^+ = \mathbf{v}_1^{+'} e^{-j\beta l_1}, \quad \mathbf{v}_1^- = \mathbf{v}_1^{-'} e^{j\beta l_1} \tag{8.2.26}$$

and the incident/reflected waves at plane B are delayed/advanced in phase by βl_2 compared with those at plane B':

$$\mathbf{v}_2^+ = \mathbf{v}_2^{+'} e^{-j\beta l_2}, \quad \mathbf{v}_2^- = \mathbf{v}_2^{-'} e^{j\beta l_2} \tag{8.2.27}$$

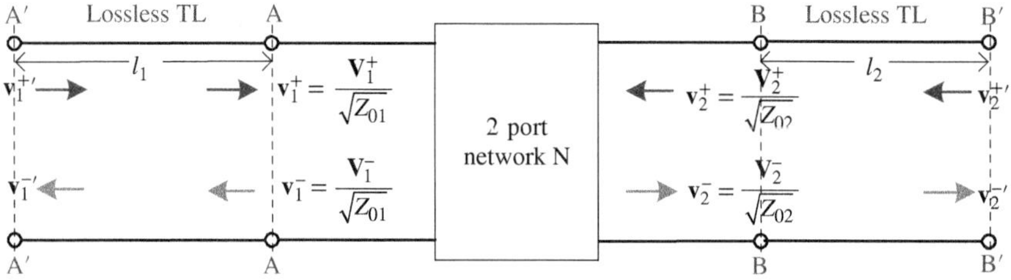

Figure 8.29 Shifting reference planes for a two-port network.

Substituting these relations into Eq. (8.2.6) yields

$$\begin{bmatrix} \mathbf{v}_1^- \\ \mathbf{v}_2^- \end{bmatrix} \overset{(8.2.6)}{=} \begin{bmatrix} s_{11} & s_{12} \\ s_{21} & s_{22} \end{bmatrix} \begin{bmatrix} \mathbf{v}_1^+ \\ \mathbf{v}_2^+ \end{bmatrix} \overset{(8.2.26,27)}{\Rightarrow} \begin{bmatrix} e^{j\beta l_1} & 0 \\ 0 & e^{j\beta l_2} \end{bmatrix} \begin{bmatrix} \mathbf{v}_1^{-\prime} \\ \mathbf{v}_2^{-\prime} \end{bmatrix} = \begin{bmatrix} s_{11} & s_{12} \\ s_{21} & s_{22} \end{bmatrix} \begin{bmatrix} e^{-j\beta l_1} & 0 \\ 0 & e^{-j\beta l_2} \end{bmatrix} \begin{bmatrix} \mathbf{v}_1^{+\prime} \\ \mathbf{v}_2^{+\prime} \end{bmatrix}$$

$$; \begin{bmatrix} \mathbf{v}_1^{-\prime} \\ \mathbf{v}_2^{-\prime} \end{bmatrix} = \begin{bmatrix} e^{-j\beta l_1} & 0 \\ 0 & e^{-j\beta l_2} \end{bmatrix} \begin{bmatrix} s_{11} & s_{12} \\ s_{21} & s_{22} \end{bmatrix} \begin{bmatrix} e^{-j\beta l_1} & 0 \\ 0 & e^{-j\beta l_2} \end{bmatrix} \begin{bmatrix} \mathbf{v}_1^{+\prime} \\ \mathbf{v}_2^{+\prime} \end{bmatrix} = \begin{bmatrix} s'_{11} & s'_{12} \\ s'_{21} & s'_{22} \end{bmatrix} \begin{bmatrix} \mathbf{v}_1^{+\prime} \\ \mathbf{v}_2^{+\prime} \end{bmatrix} \tag{8.2.28}$$

where

$$\begin{bmatrix} s'_{11} & s'_{12} \\ s'_{21} & s'_{22} \end{bmatrix} = \begin{bmatrix} e^{-j\beta l_1} & 0 \\ 0 & e^{-j\beta l_2} \end{bmatrix} \begin{bmatrix} s_{11} & s_{12} \\ s_{21} & s_{22} \end{bmatrix} \begin{bmatrix} e^{-j\beta l_1} & 0 \\ 0 & e^{-j\beta l_2} \end{bmatrix} = \begin{bmatrix} s_{11}e^{-j2\beta l_1} & s_{12}e^{-j\beta(l_1+l_2)} \\ s_{21}e^{-j\beta(l_1+l_2)} & s_{22}e^{-j2\beta l_2} \end{bmatrix} \tag{8.2.29}$$

Example 8.15 Scattering Matrix of a Two-Port Network
Consider the two-port network with TLs of characteristic impedance $Z_{01} = Z_{02} = 50\ \Omega$ at both ports shown in Figure 8.30.

a) Find the s-parameters of the two-port network and check if the S-matrix is unitary, i.e. $S^{*T}S = I$.

   Instead of directly using Eq. (8.2.8), let us find the a-parameters and make the a-to-s conversion. The circuit can be regarded as a cascade connection of L and C so that we can use Eq. (8.1.13) together with Eqs. (E8.3.2) and (E8.3.1) to determine the a-parameters as

$$A \overset{(8.1.13)}{=} A_L A_C \underset{(E8.3.1)}{\overset{(E8.3.2)}{=}} \begin{bmatrix} 1 & j50 \\ 0 & 1 \end{bmatrix} \begin{bmatrix} 1 & 0 \\ j/50 & 1 \end{bmatrix} = \begin{bmatrix} 0 & j50 \\ j/50 & 1 \end{bmatrix} \tag{E8.15.1}$$

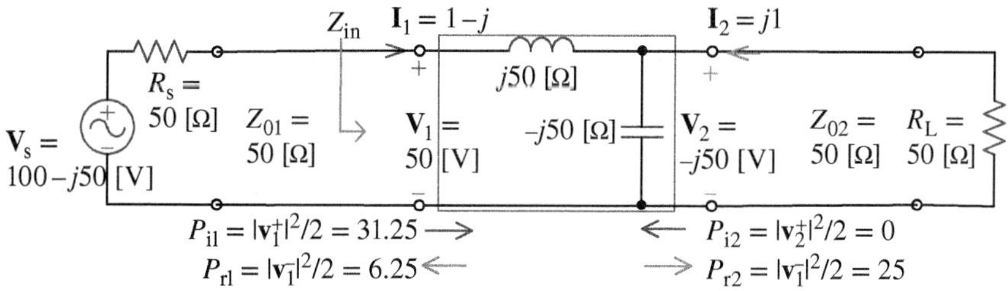

Figure 8.30 Two-port network for Example 8.15.

Thus, we convert this *a*-parameters into the *s*-parameters by running the following MATLAB statements:

```
>>Z01=50; Z02=50; % Characteristic impedances
  A=[1 50i; 0 1]*[1 0; i/50 1]; % a-parameter matrix
  S=a2s(A,Z01,Z02) % a-to-s conversion
```

which yields the *S*-parameter matrix:

```
S  =  -0.2000  +  0.4000i    0.4000  -
0.8000i
```

$$; \ S=\frac{1}{5}\begin{bmatrix}-1+j2 & 2-j4\\ 2-j4 & 1-j2\end{bmatrix} \tag{E8.15.2}$$

To get the polar form, i.e. the magnitudes/phases of each entry of the *S*-parameter matrix, we run

```
>>Smag=abs(S), Sth=rad2deg(angle(S)) % Magnitudes, Angles in degree
```

which yields

```
Smag =  0.4472   0.8944     Sth =  116.5651  -63.4349
        0.8944   0.4472            -63.4349  -63.4349
```

To check if the *S*-matrix is unitary, we run

```
>>S'*S
```

which yields an identity matrix, meaning that the *S*-matrix is unitary.

```
ans =  1.0000   0.0000
       0.0000   1.0000
```

This implies that the two-port network is lossless (Table 8.3), which is natural since it is made of only reactive elements such as L and C.

b) With the voltage source $\mathbf{V}_s = 100 - j50$ [V] and load impedance $Z_L = 50$ [Ω] connected at the input and output ports, respectively, find $\mathbf{V}_1$, $\mathbf{V}_2$, $\mathbf{I}_1$, and $\mathbf{I}_2$. Then, use Eqs. (8.2.9a,b), (8.2.10a,b), (8.2.23), and (8.2.25) to find $\mathbf{v}_1^+, \mathbf{v}_1^-, \mathbf{v}_2^+, \mathbf{v}_2^-, \Gamma_{in}$, and Γ_L.

We set up a set of node equations and solve it as

$$\begin{bmatrix} 1/50 + 1/j50 & -1/j50 \\ -1/j50 & 1/j50 - 1/j50 + 1/50 \end{bmatrix} \begin{bmatrix} \mathbf{V}_1 \\ \mathbf{V}_2 \end{bmatrix} = \begin{bmatrix} (100 - j50)/50) \\ 0 \end{bmatrix}$$

$$; \begin{bmatrix} \mathbf{V}_1 \\ \mathbf{V}_2 \end{bmatrix} = \begin{bmatrix} 1-j & j \\ j & 1 \end{bmatrix}^{-1} \begin{bmatrix} 100 - j50 \\ 0 \end{bmatrix} = \frac{1}{2-j} \begin{bmatrix} 1 & -j \\ -j & 1-j \end{bmatrix} \begin{bmatrix} 100 - j50 \\ 0 \end{bmatrix} = \begin{bmatrix} 50 \\ -j50 \end{bmatrix} \tag{E8.15.3}$$

Thus, we get the currents

$$\mathbf{I}_1 = \frac{\mathbf{V}_1 - \mathbf{V}_2}{j50} = \frac{50 - (-j50)}{j50} = 1 - j, \quad \mathbf{I}_2 = -\frac{\mathbf{V}_2}{R_L} = \frac{j50}{50} = j \tag{E8.15.4}$$

Then, we substitute the voltage/current values into Eqs. (8.2.9a,b) and (8.2.10a,b) to get the normalized incident and reflected waves:

$$\mathbf{v}_1^+ \overset{(8.2.9a)}{=} \frac{1}{2}\left(\frac{1}{\sqrt{Z_{01}}}\mathbf{V}_1 + \sqrt{Z_{01}}\mathbf{I}_1\right) \overset{(E8.15.3,4)}{=} \frac{1}{2}\left(\frac{1}{\sqrt{50}}50 + \sqrt{50}(1-j)\right)$$
$$= 7.0711 - j3.5355 \tag{E8.15.5a}$$

$$\mathbf{v}_1^- \overset{(8.2.9b)}{=} \frac{1}{2}\left(\frac{1}{\sqrt{Z_{01}}}\mathbf{V}_1 - \sqrt{Z_{01}}\mathbf{I}_1\right) \overset{(E8.15.3,4)}{=} \frac{1}{2}\left(\frac{1}{\sqrt{50}}50 - \sqrt{50}(1-j)\right)$$
$$= j3.5355 \tag{E8.15.5b}$$

$$\mathbf{v}_2^+ \overset{(8.2.10a)}{=} \frac{1}{2}\left(\frac{1}{\sqrt{Z_{02}}}\mathbf{V}_2 + \sqrt{Z_{02}}\mathbf{I}_2\right) \overset{(E8.15.3,4)}{=} \frac{1}{2}\left(\frac{1}{\sqrt{50}}(-j50) + \sqrt{50}j\right) = 0 \tag{E8.15.6a}$$

$$\mathbf{v}_2^- = {}^{(8.2.10b)} \frac{1}{2}\left(\frac{1}{\sqrt{Z_{02}}}\mathbf{V}_2 - \sqrt{Z_{02}}\mathbf{I}_2\right) \overset{(E8.15.3,4)}{=} \frac{1}{2}\left(\frac{1}{\sqrt{50}}(-j50) - \sqrt{50}j\right)$$
$$= -j7.0711 \tag{E8.15.6b}$$

```
>>Y=[1/50-i/50 i/50; i/50 1/50]; V=Y\[(100-50i)/50; 0];% Eq.(E8.15.3)
  I=[(V(1)-V(2))/50i; -V(2)/50]; sZ0=sqrt([Z01; Z02]); % Eq.(E8.15.4)
  vp=(V./sZ0+sZ0.*I)/2, vn=(V./sZ0-sZ0.*I)/2 % Eq.(E8.15.5)
```

Then, we use Eqs. (8.2.17) and (8.2.23) to get the reflection coefficients at the load and at port 1:

$$\Gamma_{\mathrm{L}} \overset{(8.2.17)}{=} \frac{\mathbf{v}_2^+ \text{ (reflected wave amplitude from load)}}{\mathbf{v}_2^- \text{ (incident wave amplitude on load)}} \overset{\substack{(E8.15.6a)\\(E8.15.6b)}}{=} 0 \tag{E8.15.7}$$

$$\Gamma_{\mathrm{in}} \overset{(8.2.23)}{=} \frac{\mathbf{v}_1^- \text{ (reflected wave amplitude from port 1)}}{\mathbf{v}_1^+ \text{ (incident wave amplitude on port 1)}} \overset{\substack{(E8.15.5b)\\(E8.15.5a)}}{=} -0.2 + j0.4$$

$$= 0.4472\angle 116.6^\circ \tag{E8.15.8}$$

c) With the analysis results obtained in (b), find the incident/reflected powers at port 1 and port 2.

Since $\mathbf{v}_n = \mathbf{V}_n/\sqrt{Z_0}$ where $\mathbf{V}_n$'s are supposed to be maximum phasor representations of sinusoidal voltages, we can compute the powers as

```
>>abs([vp(1) vn(1)]).^2/2, abs([vp(2) vn(2)]).^2/2
```

$$P_{1,\mathrm{in}} = \frac{1}{2}|\mathbf{v}_1^+|^2 = 31.25[\mathrm{W}],\; P_{1,\mathrm{r}} = \frac{1}{2}|\mathbf{v}_1^-|^2 = 6.25[\mathrm{W}] \tag{E8.15.9a}$$

$$P_{2,\mathrm{in}} = \frac{1}{2}|\mathbf{v}_2^+|^2 = 0\,[\mathrm{W}],\; P_{2,\mathrm{r}} = \frac{1}{2}|\mathbf{v}_2^-|^2 = 25[\mathrm{W}] \tag{E8.15.9b}$$

Through this example, we make some observations about the physical meaning of s-parameters and (normalized) incident/reflected waves:

Remark 8.3 Observations About Physical Meaning of s-parameters

1) Halving the squared absolute normalized incident/reflected waves on/from a port yields the incident/reflected active (average or real) power on/from the port:

$$P_{n,\mathrm{in}} = \frac{|\mathbf{v}_n^+|^2}{2} \quad \text{and} \quad P_{n,\mathrm{r}} = \frac{|\mathbf{v}_n^-|^2}{2} \tag{8.2.30}$$

since $\mathbf{v}_n = \mathbf{V}_n/\sqrt{Z_0}$ where $\mathbf{V}_n$'s are supposed to be not root-mean-square (r.m.s.) phasor but maximum phasor representations of sinusoidal voltages at port n.

2) At port 2, the reflected power (incident to the load) is $|\mathbf{v}_2^-|^2/2 = 25$ [W] while the incident power $|\mathbf{v}_2^+|^2/2$ (reflected from the load) is 0 [W] since that port is terminated in a matched load ($Z_{\mathrm{L}} = Z_0 = 50\,\Omega$) so that $\Gamma_{\mathrm{L}} = 0$ (see Eq. (8.2.23)).

3) The reflection coefficient Γ_{in} at port 1 is related with the s-parameters and Γ_{L} by Eq. (8.2.23):

$$\Gamma_{1,\mathrm{in}} = \frac{\mathbf{v}_1^-}{\mathbf{v}_1^+} \overset{(8.2.23)}{=} s_{11} + \frac{s_{12}s_{21}\Gamma_{\mathrm{L}}}{1 - s_{22}\Gamma_{\mathrm{L}}} \overset{(8.2.20)}{=} \frac{Z_{\mathrm{in}} - Z_0}{Z_{\mathrm{in}} + Z_0} \quad \text{with}\, \Gamma_{\mathrm{L}} = \frac{\mathbf{v}_2^+}{\mathbf{v}_2^-} \overset{(8.2.17)}{=} \frac{Z_{\mathrm{L}} - Z_0}{Z_{\mathrm{L}} + Z_0} \tag{8.2.31}$$

where the input impedance Z_{in}, if necessary, can be obtained from the a-parameters as

$$\text{>> ZL = 50; Zin = port\_property(A, ZL, 'ZiA')} \rightarrow Z_{\text{in}} = 25 + j25\,[\Omega] \tag{8.2.32}$$

Especially for Example 8.15 with port 2 matched so that $\Gamma_L = 0$, we have $\Gamma_{1,\text{in}} = s_{11}$ (see Eq. (8.2.8) for the definition of s-parameters) and thus, the input power at port 1 is

$$P_1 \overset{(7.1.26)}{=} \frac{1}{2}\frac{|\mathbf{V}_1^+|^2}{Z_0}\{1 - |\Gamma_{1,\text{in}}|^2\} \overset{(8.2.3)}{=} \frac{1}{2}|\mathbf{v}_1^+|^2\{1 - |s_{11}|^2\}$$

$$\underset{(\text{E}8.15.2)}{\overset{(\text{E}8.15.5\text{a})}{=}} \frac{1}{2}62.5(1 - 0.2) = 25[\text{W}]$$

where the complex power (having the active power P and reactive power Q as real and imaginary parts, respectively) at port 1 is computed as a half of the product of the voltage (maximum) phasor and the conjugated current (maximum) phasor:

$$S_1 \underset{(6.28)}{\overset{[\text{Y-1}]}{=}} \frac{1}{2}\mathbf{V}_1\mathbf{I}_1^* \overset{(\text{E}8.15.3\text{a},4\text{a})}{=} \frac{1}{2}50(1 + j) = 25 + j25 = P_1[\text{W}] + jQ_1[\text{VAR}]$$

Note that the input reflection coefficient $\Gamma_{n,\text{in}}$ at port n of a two-port network is not generally equal to s_{nn} except for the case where all other ports are terminated in matched loads.

4) At port 1, the reflected power $P_{1,\text{r}} = |\mathbf{v}_1^-|^2/2 = 6.25$ [W] can be obtained by multiplying the incident power $P_{1,\text{in}} = |\mathbf{v}_1^+|^2/2 = 31.25$ [W] with $|s_{11}|^2 = 0.2$. Also, multiplying the incident power $P_{1,\text{in}}$ by 2 times the imaginary part of s_{11}, i.e. $2\text{Im}\{s_{11} = -0.2 + j0.4\} = 0.8$, yields the incident reactive power.
5) Multiplying the incident power $P_{1,\text{in}}$ by $|s_{21}|^2 = 0.8$ yields the reflected power $P_{2,\text{r}} = |\mathbf{v}_2^-|^2/2 = 25[\text{W}]$ from port 2 where port 2 is matched so that $P_{2,\text{in}} = 0$ [W] (see Eq. (8.2.8) for the definition of s-parameters).
6) With port 2 open-/short-circuited so that $\Gamma_{\text{L,in}} = +1/-1$, the reflection coefficient Γ_{in} at port 1 is

$$\Gamma_{\text{in}} = \frac{\mathbf{v}_1^-}{\mathbf{v}_1^+} \overset{(8.2.23)}{=} s_{11} + \frac{s_{12}s_{21}\Gamma_L}{1 - s_{22}\Gamma_L} = s_{11} + \frac{s_{12}s_{21}}{1 - s_{22}} \text{ or } s_{11} - \frac{s_{12}s_{21}}{1 + s_{22}} \tag{8.2.33}$$

7) Do multiplying the incident power $P_{2,\text{in}}$ by $|s_{22}|^2 = 0.2$ and $|s_{12}|^2 = 0.8$ yield the reflected powers $P_{2,\text{r}} = |\mathbf{v}_2^-|^2/2 = 25[\text{W}]$ from port 2 and $P_{1,\text{r}} = |\mathbf{v}_1^-|^2/2 = 6.25[\text{W}]$ from port 1, respectively? If not, why?

Example 8.16 Reciprocity and Symmetry of a Two-Port Network
To reflect on the reciprocity and symmetry, reconsider the two-port network given in Example 8.15.

a) Referring to Table 8.3, judge the reciprocity and symmetry of the network.
b) Find the current $\mathbf{I}_2$ with port 2 short-circuited where the network is excited by a voltage source of $\mathbf{V}_s = 50[\mathrm{V}]$ at port 1 (Figure 8.31(a1)) and the current $\mathbf{I}_1$ with port 1 short-circuited where the network is excited by a voltage source of $\mathbf{V}_s = 50[\mathrm{V}]$ at port 2 (Figure 8.31(b1)). Are the two currents equal?

$$\mathbf{I}_2 = -\frac{50}{j50} = j \text{ in Figure 8.31(a1)} \tag{E8.16.1a}$$

$$\mathbf{I}_1 = -\frac{50}{j50} = j \text{ in Figure 8.31(b1)} \tag{E8.16.1b}$$

They are equal as meant by the reciprocity.
c) Find the voltages $\mathbf{V}_1$ and $\mathbf{V}_2$ with port 2 open-circuited where the network is excited by a current source of $\mathbf{I}_s = 1[\mathrm{A}]$ at port 1 (Figure 8.31(a2)) and the voltages $\mathbf{V}_1$ and $\mathbf{V}_2$ with port 1 open-circuited where the network is excited by a current source of $\mathbf{I}_s = 1[\mathrm{A}]$ at port 2 (Figure 8.31(b2)).

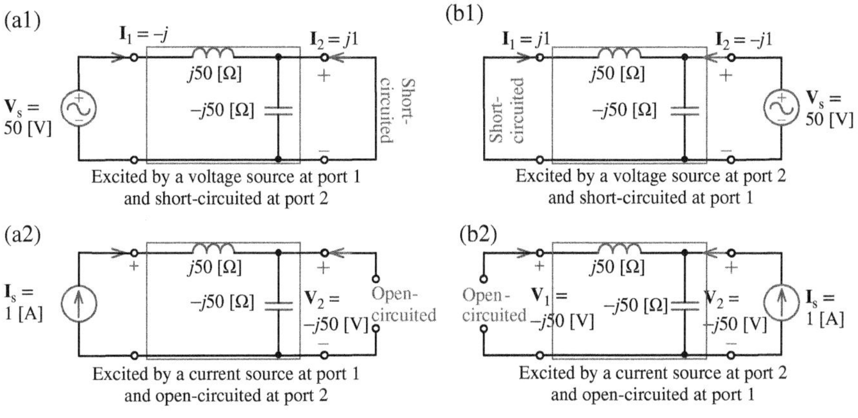

Figure 8.31 Two-port network for Example 8.16.

$$\mathbf{V}_1 = 0,\ \mathbf{V}_2 = -j50[\text{V}] \text{ in Figure 8.31(a2)} \tag{E8.16.2a}$$

$$\mathbf{V}_1 = \mathbf{V}_2 = -j50[\text{V}] \text{ in Figure 8.31(b2)} \tag{E8.16.2b}$$

We have $\mathbf{V}_2$(in (a2)) = $\mathbf{V}_1$(in (b2)), implying the reciprocity. If $\mathbf{V}_1$(in (a2)) = $\mathbf{V}_2$(in (b2)), the network can be said to be symmetrical. Is it symmetrical?

Example 8.17 Scattering Matrix of an Ideal Transformer

Find the s-parameters of an ideal transformer with turns ratio n:1 (Figure 8.5), whose A-matrix was found in Example 8.4 as

$$A \overset{\text{(E8.4)}}{=} \begin{bmatrix} n & 0 \\ 0 & 1/n \end{bmatrix} \tag{E8.17.1}$$

We convert this A-matrix into the S-matrix by running the following MATLAB statements:

```
>>syms n Z0; % Turns ratio, Characteristic impedance
  A=[n 0; 0 1/n]; % A-parameter matrix (E8.17.1)
  S=a2s(A,Z0); simplify(S) % To make a-to-s conversion
```

This yields the S-matrix:

```
S = [ (n^2 - 1)/(n^2 + 1),     (2*n)/(n^2 + 1)]
    [   (2*n)/(n^2 + 1), -(n^2 - 1)/(n^2 + 1)]
```

that is,

$$S = \frac{1}{n^2+1}\begin{bmatrix} n^2-1 & 2n \\ 2n & -(n^2-1) \end{bmatrix} \tag{E8.17.2}$$

Example 8.18 Scattering Matrix of a Transmission Line

Find the s-parameters of a TL with characteristic impedance Z_0 (Figure 8.6), whose A-matrix was found in Example 8.5 as

$$A \overset{\text{(E8.5.6)}}{=} \begin{bmatrix} \cosh\gamma l & Z_0\sinh\gamma l \\ \sinh\gamma l/Z_0 & \cosh\gamma l \end{bmatrix} \tag{E8.18.1}$$

Also, using Eq. (8.2.23), determine the input reflection coefficient Γ_{in} looking into the network at port 1 when port 2 is open-circuited, short-circuited, and matched, i.e. terminated in Z_0.

We convert this A-matrix into the S-matrix by running the following MATLAB statements:

```
>>syms rl Z0; % Turns ratio, Characteristic impedance
  A=[cosh(rl) Z0*sinh(rl); sinh(rl)/Z0 cosh(rl)]; % A-matrix Eq. (E8.18.1)
  S=a2s(A,Z0); simplify(S) % To make a-to-s conversion
```

This yields the S-matrix:

```
S = [          0, exp(-rl)]
    [ exp(-rl),          0]
```

$$; S=\begin{bmatrix} 0 & e^{-\gamma l} \\ e^{-\gamma l} & 0 \end{bmatrix} \tag{E8.18.2}$$

When port 2 is open-circuited/short-circuited/matched so that $\Gamma_L = +1/-1/0$, we use Eq. (8.2.23) to compute the reflection coefficient Γ_{in} at port 1 as

$$\Gamma_{in} = \frac{\mathbf{V}_1^-}{\mathbf{V}_1^+} \overset{(8.2.23)}{=} s_{11} + \frac{s_{12}s_{21}\Gamma_L}{1-s_{22}\Gamma_L} \begin{cases} \overset{\Gamma_L=1}{=} s_{11} + \dfrac{s_{12}s_{21}}{1-s_{22}} \overset{(E8.18.2)}{=} e^{-2\gamma l} \overset{\text{lossless}}{\longrightarrow} e^{-j2\beta l} \\ \overset{\Gamma_L=-1}{=} s_{11} - \dfrac{s_{12}s_{21}}{1+s_{22}} \overset{(E8.18.2)}{=} -e^{-2\gamma l} \overset{\text{lossless}}{\longrightarrow} -e^{-j2\beta l} \\ \underset{(E8.18.2)}{\overset{\Gamma_L=0}{=}} s_{11} - \dfrac{s_{12}s_{21}}{1+s_{22}} \overset{(E8.18.2)}{=} 0 \end{cases}$$

(E8.18.3a,b,c)

If the TL is lossless, the input power, computed by Eq. (7.1.26), is zero (since $|e^{-j\beta l}| = 1$), implying that the TL is receiving no power, when port 2 is open-circuited or short-circuited. When port 2 is matched, the input power equals the incident power since the TL delivers all the incident power to the load.

8.3 Gain and Stability

8.3.1 Two-Port Power Gains [L-1, P-1]

Pozar [P-1] defines three types of power gain, i.e. (operating) power gain G, transducer power gain G_T, and available power gain G_A, in terms of the s-parameters of a two-port network excited by a voltage source $\mathbf{V}_s$ and driving a load impedance Z_L as shown in Figure 8.32, where the characteristic impedances at ports 1 and 2 are equally Z_0. To examine them, let us recall Eq. (7.1.11) to write the reflection coefficients Γ_s/Γ_{in} looking into the source/network at port 1 and that looking to the load at port 2 as

$$\Gamma_s = \frac{Z_s - Z_0}{Z_s + Z_0}, \quad \Gamma_{in} = \frac{\mathbf{V}_1^-}{\mathbf{V}_1^+} = \frac{Z_{in} - Z_0}{Z_{in} + Z_0}, \quad \text{and } \Gamma_L = \frac{\mathbf{V}_2^+}{\mathbf{V}_2^-} = \frac{Z_L - Z_0}{Z_L + Z_0} \tag{8.3.1a,b,c}$$

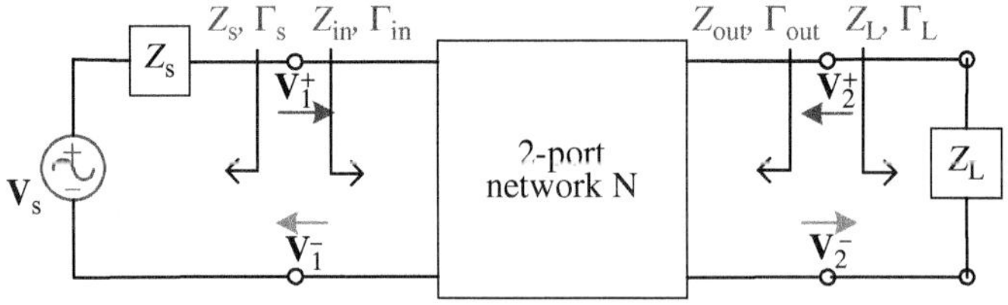

Figure 8.32 A two-port network with a source and a load.

From the first two of these equations, we can get

$$Z_s = Z_0 \frac{1+\Gamma_s}{1-\Gamma_s}, \quad Z_{in} = Z_0 \frac{1+\Gamma_{in}}{1-\Gamma_{in}} \tag{8.3.2a,b}$$

Also recall the expressions (8.2.23)/(8.2.25) of the reflection coefficients Γ_{in}/Γ_{out} looking into the network at port 1/2 as

$$\Gamma_{in} = \frac{\mathbf{V}_1^-}{\mathbf{V}_1^+} \overset{(8.2.3)}{=} \frac{\mathbf{v}_1^-}{\mathbf{v}_1^+} \overset{(8.2.23)}{=} s_{11} + \frac{s_{12}s_{21}\Gamma_L}{1-s_{22}\Gamma_L} \overset{(8.2.20)}{=} \frac{Z_{in}-Z_0}{Z_{in}+Z_0} \tag{8.3.3a}$$

$$\Gamma_{out} = \frac{\mathbf{V}_2^-}{\mathbf{V}_2^+} \overset{(8.2.3)}{=} \frac{\mathbf{v}_2^-}{\mathbf{v}_2^+} \overset{(8.2.25)}{=} s_{22} + \frac{s_{21}s_{12}\Gamma_s}{1-s_{11}\Gamma_s} = \frac{Z_{out}+Z_0}{Z_{out}-Z_0} \tag{8.3.3b}$$

where Eq. (8.2.6) can be rewritten by multiplying both sides with $\sqrt{Z_0}$ as

$$\begin{bmatrix} \mathbf{V}_1^- \\ \mathbf{V}_2^- \end{bmatrix} = \begin{bmatrix} s_{11} & s_{12} \\ s_{21} & s_{22} \end{bmatrix} \begin{bmatrix} \mathbf{V}_1^+ \\ \mathbf{V}_2^+ \end{bmatrix}; \quad \begin{matrix} \mathbf{V}_1^- = s_{11}\mathbf{V}_1^+ + s_{12}\mathbf{V}_2^+ \\ \mathbf{V}_2^- = s_{21}\mathbf{V}_1^+ + s_{22}\mathbf{V}_2^+ \end{matrix} \tag{8.3.4a,b}$$

Now, for the voltage divider stage on the source side, we can write the total voltage $\mathbf{V}_1$ at port 1 in terms of the source voltage $\mathbf{V}_s$ as

$$\mathbf{V}_1 = \frac{Z_{in}}{Z_s + Z_{in}}\mathbf{V}_s \overset{(8.2.1a)}{=} \mathbf{V}_1^+ + \mathbf{V}_1^- \overset{(8.3.1b)}{=} \mathbf{V}_1^+(1+\Gamma_{in}) \tag{8.3.5}$$

From this together with Eqs. (8.3.2a) and (8.3.2b), we can write the expression for $\mathbf{V}_1^+$ as

$$\begin{aligned} \mathbf{V}_1^+ &\overset{(8.3.5)}{=} \frac{1}{1+\Gamma_{in}} \frac{Z_{in}}{Z_s + Z_{in}} \mathbf{V}_s \\ &\overset{(8.3.2a,b)}{=} \frac{1}{1+\Gamma_{in}} \frac{Z_0(1+\Gamma_{in})/(1-\Gamma_{in})}{Z_0(1+\Gamma_s)/(1-\Gamma_s) + Z_0(1+\Gamma_{in})/(1-\Gamma_{in})} \mathbf{V}_s = \frac{1-\Gamma_s}{2(1-\Gamma_s\Gamma_{in})} \mathbf{V}_s \end{aligned} \tag{8.3.6}$$

Thus, referring to Eq. (7.1.26), we can express the average input power delivered to the network as

$$P_{\text{in}} \overset{(7.1.26)}{=} \frac{1}{2}\frac{|\mathbf{V}_1^+|^2}{Z_0}\left(1-|\Gamma_{\text{in}}|^2\right) \overset{(8.3.6)}{=} \frac{|\mathbf{V}_s|^2}{8Z_0}\left(1-|\Gamma_{\text{in}}|^2\right)\frac{|1-\Gamma_s|^2}{|1-\Gamma_s\Gamma_{\text{in}}|^2} \tag{8.3.7}$$

Likewise, noting that substituting Eq. (8.3.1c) for $\mathbf{V}_2^+$ into Eq. (8.3.4b) yields

$$\mathbf{V}_2^- \overset{(8.3.4b)}{=} s_{21}\mathbf{V}_1^+ + s_{22}\mathbf{V}_2^+ \overset{(8.3.1c)}{=} s_{21}\mathbf{V}_1^+ + s_{22}\Gamma_L\mathbf{V}_2^-;\ \mathbf{V}_2^- = \frac{s_{21}}{1-s_{22}\Gamma_L}\mathbf{V}_1^+ \tag{8.3.8}$$

we can write the average power delivered to the load as

$$P_L \overset{(7.1.26)}{=} \frac{1}{2}\frac{|\mathbf{V}_2^-|^2}{Z_0}\left(1-|\Gamma_L|^2\right) \overset{(8.3.8)}{=} \frac{|\mathbf{V}_1^+|^2}{2Z_0}\frac{|s_{21}|^2\left(1-|\Gamma_L|^2\right)}{|1-s_{22}\Gamma_L|^2} \overset{(8.3.6)}{=} \frac{|\mathbf{V}_s|^2}{8Z_0}\frac{|s_{21}|^2\left(1-|\Gamma_L|^2\right)|1-\Gamma_s|^2}{|1-s_{22}\Gamma_L|^2|1-\Gamma_s\Gamma_{\text{in}}|^2} \tag{8.3.9}$$

Here, we define the (*operating*) *power gain* as

$$G_p = \frac{P_L(\text{power delivered to load})}{P_{\text{in}}(\text{power supplied to network})} \overset{(8.3.9)}{\underset{(8.3.7)}{=}} \frac{|s_{21}|^2\left(1-|\Gamma_L|^2\right)}{|1-s_{22}\Gamma_L|^2\left(1-|\Gamma_{\text{in}}|^2\right)}$$
$$\overset{(8.3.3a)}{=} \frac{|s_{21}|^2\left(1-|\Gamma_L|^2\right)}{|1-s_{22}\Gamma_L|^2-|s_{11}-\Delta s\Gamma_L|^2} \tag{8.3.10}$$

Apart from this, note that the available power through port 1 from the source is the maximum power that is delivered to the network when the maximum power transfer or impedance (conjugate) matching condition at port 1 is satisfied, i.e. $Z_{\text{in}} = Z_s^*$ so that $\Gamma_{\text{in}} = \Gamma_s^*$:

$$P_{\text{av1}} \overset{(8.3.7)}{=} \frac{|\mathbf{V}_s|^2}{8Z_0}\left(1-|\Gamma_{\text{in}}|^2\right)\frac{|1-\Gamma_s|^2}{|1-\Gamma_s\Gamma_{\text{in}}|^2} \overset{\Gamma_{\text{in}}=\Gamma_s^*}{=} \frac{|\mathbf{V}_s|^2}{8Z_0}\frac{|1-\Gamma_s|^2}{1-|\Gamma_s|^2} \tag{8.3.11}$$

Here, we define the *transducer gain* as

$$G_t = \frac{P_L(\text{power delivered to load})}{P_{\text{av1}}(\text{power available to network})} \overset{(8.3.9)}{\underset{(8.3.11)}{=}} \frac{\left(1-|\Gamma_s|^2\right)|s_{21}|^2\left(1-|\Gamma_L|^2\right)}{|1-\Gamma_s\Gamma_{\text{in}}|^2|1-s_{22}\Gamma_L|^2} \tag{8.3.12a}$$

which was shown to be equal to [L-1]

$$G_t \overset{[L\text{-}1]}{\underset{(9.11)}{=}} \frac{\left(1-|\Gamma_s|^2\right)|s_{21}|^2\left(1-|\Gamma_L|^2\right)}{|1-s_{11}\Gamma_s|^2|1-\Gamma_L\Gamma_{\text{out}}|^2} \tag{8.3.12b}$$

Also the *available gain* was defined and shown to be [L-1, P-1]

$$G_a = \frac{P_{av2}(\text{power available to load})}{P_{av1}(\text{power available to network})} = G_t\big|_{\Gamma_L = \Gamma_{out}^*}$$

$$\underset{(11.15)\text{ of [P-1]}}{\overset{(9.13)\text{ of [L-1]}}{=}} \frac{|s_{21}|^2(1-|\Gamma_s|^2)}{|1-s_{11}\Gamma_s|^2(1-|\Gamma_{out}|^2)} \underset{(10.3.17)\text{ of [L-1]}}{\overset{(8.3.3b)}{=}} \frac{|s_{21}|^2(1-|\Gamma_s|^2)}{|1-s_{11}\Gamma_s|^2 - |s_{22}-\Delta s\Gamma_s|^2} \tag{8.3.13}$$

where the numerator is the maximum average power that can be delivered to the load, which is realized when the impedance matching condition at port 2 is satisfied, i.e. $Z_L = Z_{out}^*$ so that $\Gamma_L = \Gamma_{out}^*$:

$$P_L \underset{\Gamma_L = \Gamma_{out}^*}{\overset{(8.3.9)}{=}} \frac{|\mathbf{V}_s|^2}{8Z_0} \frac{|s_{21}|^2(1-|\Gamma_L|^2)|1-\Gamma_s|^2}{|1-s_{22}\Gamma_L|^2|1-\Gamma_s\Gamma_{in}|^2} \underset{(11.14)}{\overset{[P-1]}{=}} \frac{|\mathbf{V}_s|^2}{8Z_0} \frac{|s_{21}|^2(1-|\Gamma_s|^2)}{|1-s_{11}\Gamma_s|^2|1-\Gamma_L\Gamma_{out}|^2} \tag{8.3.14}$$

If s_{12} is negligibly small, i.e. $s_{12} \approx 0$ so that $\Gamma_{in} \approx s_{11}$, the transducer gain G_t can be approximated as

$$G_{tu} \underset{(8.3.3a)\text{ with } s_{12}=0}{\overset{(8.3.12a)}{=}} \frac{1-|\Gamma_s|^2}{|1-s_{11}\Gamma_s|^2}|s_{21}|^2 \frac{1-|\Gamma_L|^2}{|1-s_{22}\Gamma_L|^2} \tag{8.3.15}$$

which is called the *unilateral transducer gain*. If $\Gamma_L = \Gamma_s = 0$, i.e. both the input and output are matched for zero reflection (different from conjugate matching), the transducer gain (8.3.12) becomes

$$G_t = |s_{21}|^2 \tag{8.3.16}$$

Note that G_t depends on both Γ_s and Γ_L while G_p is not dependent on Γ_s (or Z_s) and G_a is not dependent on Γ_L (or Z_L).

Here are the formulas for determining the centers and radii of constant operating/available power gain circles derived in [M-2] or [L-1]:

$$c_p \underset{(10.3.11)}{\overset{[M-2]}{=}} \frac{g_p(s_{22}-s_{11}^*\Delta s)^*}{1+g_p(|s_{22}|^2-|\Delta s|^2)},\ r_p \underset{(10.3.12)}{\overset{[M-2]}{=}} \frac{\sqrt{1-2k|s_{12}s_{21}|g_p + |s_{12}s_{21}|^2 g_p^2}}{1+g_p(|s_{22}|^2-|\Delta s|^2)} \text{ with } g_p = \frac{G_p}{|s_{21}|^2} \tag{8.3.17}$$

$$c_a \underset{(10.3.14)}{\overset{[M-2]}{=}} \frac{g_a(s_{11}-s_{22}^*\Delta s)^*}{1+g_a(|s_{11}|^2-|\Delta s|^2)},\ r_a \underset{(10.3.15)}{\overset{[M-2]}{=}} \frac{\sqrt{1-2k|s_{12}s_{21}|g_a + |s_{12}s_{21}|^2 g_a^2}}{1+g_a(|s_{11}|^2-|\Delta s|^2)} \text{ with } g_a = \frac{G_a}{|s_{21}|^2} \tag{8.3.18}$$

where

$$k = \frac{1 - |s_{11}|^2 - |s_{22}|^2 + |\Delta s|^2}{2\,|s_{12}s_{21}|} \quad \text{and } \Delta s = |S| = s_{11}s_{22} - s_{12}s_{21} \tag{8.3.19}$$

```
function [c,r,k,aDS]=gain_circle(S,GdB,oa,Kg)
% Given a S-matrix and a value of operating/available power gain GdB,
% it returns the center c and radius r, drawing the constant gain circle
% depending on whether the 3rd input argument oa is 'o' or 'a'.
if nargin<4, Kg=0; end
if nargin<3, oa='o'; end
D=det(S); aDS=abs(D); % Eq.(8.3.19)
s11=S(1,1); s12=S(1,2); s21=S(2,1); s22=S(2,2);
k=(1-abs(s11)^2-abs(s22)^2+aDS^2)/2/abs(s12*s21); % Eq.(8.3.19)
% Center/Radius of gain circle
oa=lower(oa(1)); % Operating/Available power gain
g=10^(GdB/10)/abs(s21)^2; % gp or ga
if oa=='o'|oa=='p', SL='L'; str='Operating power gain circle of ';
 C=s22-s11'*D; den=1+g*(abs(s22)^2-aDS^2); % Eq.(8.3.17)
 else SL='S'; str='Available power gain circle of ';
 C=s11-s22'*D; den=1+g*(abs(s11)^2-aDS^2); % Eq.(8.3.18)
end
c=g*C'/den; % Eq.(8.3.17/18a)
r=sqrt(1-2*k*abs(s12*s21)*g+abs(s12*s21)^2*g^2)/den; % Eq.(8.3.17/18b)

if Kg==0, return; end
% To plot the operating/available gain circle on GamL/GamS-plane
ths=pi/180*[0:360];
if Kg>0, figure(Kg), smithchart, hold on; end % A blank Smith chart
plot(c+r*exp(j*ths),'c'), axis('off'), set(gcf,'color','white')
plot([-1 1],[0 0],'k', [0 0],[-1 1],'k'), axis('equal'),
title([str num2str(GdB,'%6.2f') ' [dB] on \Gamma_' SL '-plane'])
```

Example 8.19 Constant-Gain Circles from Scattering Matrix

Suppose a two-port network with the reference impedance of $Z_0 = 50\ \Omega$ and the following S-matrix:

$$S = \begin{bmatrix} 0.45\angle 150^\circ & 0.01\angle -10^\circ \\ 2.05\angle 10^\circ & 0.4\angle -150^\circ \end{bmatrix} \tag{E8.19.1}$$

is excited by a voltage source with source impedance $Z_s = 20\ \Omega$ and loaded by a load impedance of $Z_L = 30\ \Omega$. Compute its operating power gain, available power gain, and transducer power gain. Then, plot the constant operating gain circle crossing Γ_L and the constant available gain circle crossing Γ_s.

```
%elec08e19.m
S=[0.45*exp(j*deg2rad(150)) 0.01*exp(j*deg2rad(-10));
   2.05*exp(j*deg2rad(10)) 0.4*exp(j*deg2rad(-150))]; %Eq.(E8.19.1)
Zs=20; ZL=30; Z0=50;
[Gp,Ga,Gt,Gami,Gamo,GamS,GamL]=gain_S(S,Zs,ZL,Z0)
GpdB=10*log10(max([Gp 1e-10])), GadB=10*log10(max([Ga 1e-10]))
Kg=1; [cp,rp,k,aDS]=gain_circle(S,GpdB,'o',Kg), plot(GamL+j*eps,'mo')
Kg=2; [ca,ra,k,aDS]=gain_circle(S,GadB,'a',Kg), plot(GamS+j*eps,'mo')
```

We can run the above MATLAB script "elec08e19.m" to get Figure 8.33(a1) and (a2). Also, we can use 'Smith 4.1' to plot the same constant-gain circles as shown in Figure 8.33(b), where the values of the *s*-parameters and power gain should be given through the Circles (Gain) dialog box that is opened by clicking on the menu Tools > Circles and selecting the Gain tab.

8.3.2 Stability [E-1, L-1, P-1]

Edwards and Sinsky [E-1] define two types of stability:

1) Unconditional stability: The network is *unconditionally stable* if $|\Gamma_{in}| < 1$ and $|\Gamma_{out}| < 1$ for any passive source/load impedances (with $|\Gamma_s| < 1$ and $|\Gamma_L| < 1$).
2) Conditional stability: The network is *conditionally stable* or *potentially unstable* if $|\Gamma_{in}| < 1$ and $|\Gamma_{out}| < 1$ only for a certain range of passive source/load impedances.

Note that the stability condition of a network is frequency dependent and consequently, a network may be stable at its design frequency but unstable at other frequencies.

We can use Eqs. (8.2.23) and (8.2.25) to write the output and input conditions for unconditional stability in terms of Γ_L and Γ_s as

$$\begin{matrix}\text{Output stability}\\ \text{condition}\end{matrix} \text{ on } \Gamma_L : |\Gamma_{in}| \overset{(8.2.23)}{=} \left| s_{11} + \frac{s_{12}s_{21}\Gamma_L}{1-s_{22}\Gamma_L} \right| = \left| \frac{s_{11} - \Delta s \Gamma_L}{1-s_{22}\Gamma_L} \right| < 1 \quad (8.3.20a)$$

$$\begin{matrix}\text{Input stability}\\ \text{condition}\end{matrix} \text{ on } \Gamma_s : |\Gamma_{out}| \overset{(8.2.25)}{=} \left| s_{22} + \frac{s_{21}s_{12}\Gamma_s}{1-s_{11}\Gamma_s} \right| = \left| \frac{s_{22} - \Delta s \Gamma_s}{1-s_{11}\Gamma_s} \right| < 1 \quad (8.3.20b)$$

If the network is unilateral, i.e. $s_{12} = 0$, these conditions reduce to

$$|\Gamma_{in}| \underset{s_{12}=0}{\overset{(8.2.23)}{=}} |s_{11}| < 1 \text{ and } |\Gamma_{out}| \underset{s_{12}=0}{\overset{(8.2.25)}{=}} |s_{22}| < 1 \quad (8.3.21a,b)$$

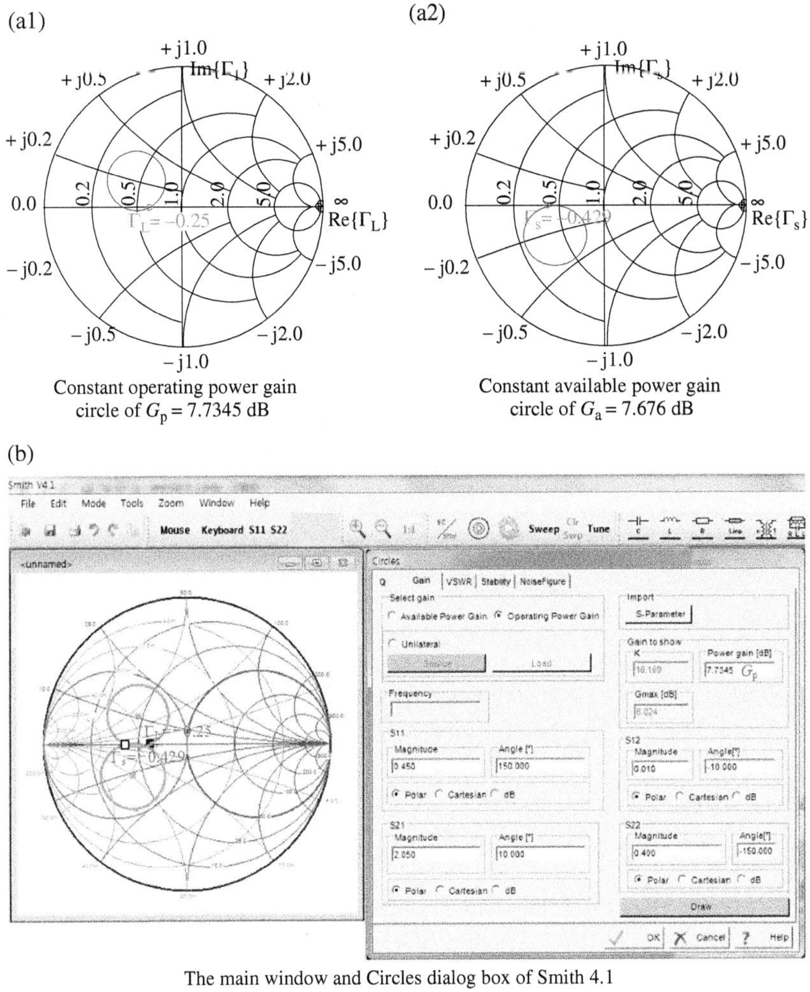

Figure 8.33 Examples of plotting constant operating/available power gain circles.

Otherwise, the input and output stability conditions can be described by certain regions on the Γ_s-plane and Γ_L-plane, respectively, as illustrated in Figure 8.34, where the equations for the input and output stability circles are [L-1]

$$(\mathrm{Re}\{\Gamma_s\}-\mathrm{Re}\{c_i\})^2+(\mathrm{Im}\{\Gamma_s\}-\mathrm{Im}\{c_i\})^2=r_i^2 \tag{8.3.22a}$$

$$(\mathrm{Re}\{\Gamma_L\}-\mathrm{Re}\{c_o\})^2+(\mathrm{Im}\{\Gamma_L\}-\mathrm{Im}\{c_o\})^2=r_o^2 \tag{8.3.22b}$$

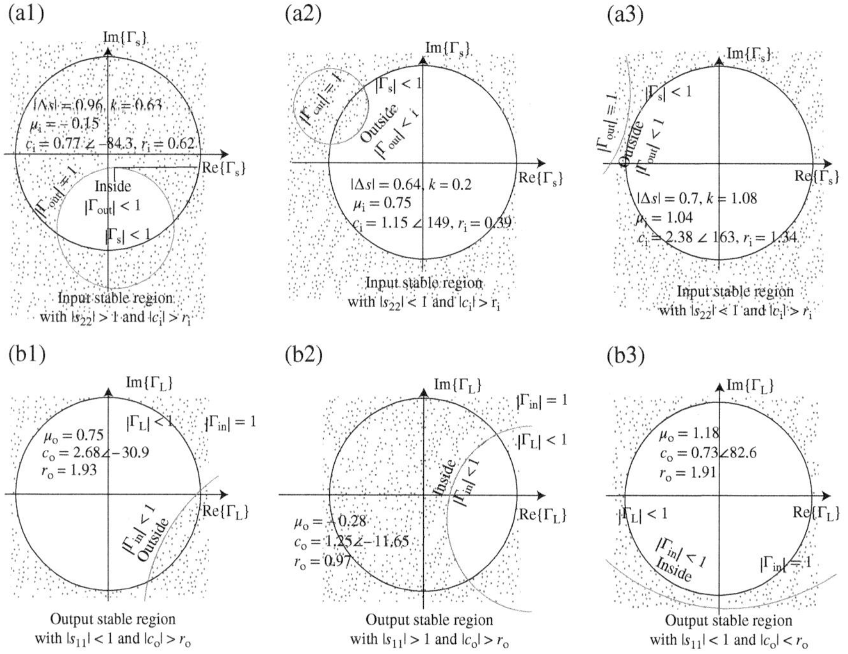

Figure 8.34 Examples of Input/output stable regions on Γ_s-plane and Γ_L-plane.

where, with $\Delta s = |S|$ (determinant of the scattering matrix) $= s_{11}s_{22} - s_{12}s_{12}$, their centers and radii are

$$c_i = \frac{\left(s_{11} - s_{22}^{*}\Delta s\right)^{*}}{|s_{11}|^2 - |\Delta s|^2} \text{ and } r_i = \frac{|\, s_{12}s_{21} \,|}{\left||s_{11}|^2 - |\Delta s|^2\right|} \tag{8.3.23}$$

$$c_o = \frac{\left(s_{22} - s_{11}^{*}\Delta s\right)^{*}}{|s_{22}|^2 - |\Delta s|^2} \text{ and } r_o = \frac{|\, s_{12}s_{21} \,|}{\left||s_{22}|^2 - |\Delta s|^2\right|} \tag{8.3.24}$$

Remark 8.4 Comments on Stability Condition

If $|s_{11}| > 1$ or $|s_{22}| > 1$, the network cannot be unconditionally stable since the condition (8.3.20a/b) is not satisfied at the origin ($\Gamma_L = 0/\Gamma_s = 0$) on the Smith chart that is reachable just by the load/source impedance equal to Z_0 (Eqs. (8.3.1c)/(8.3.1a)). See Figure 8.33(a1) and (b2).

1) An alternative (simpler) condition for unconditional stability, called '*k*-*Δ test*' [R-5], is

$$k = \frac{1 - |s_{11}|^2 - |s_{22}|^2 + |\Delta s|^2}{2\,|s_{12}s_{21}|} > 1 \text{ and } |\Delta s| < 1\;(\Delta s = |S|) \tag{8.3.25}$$

where k is called the *Rollett stability factor.*

2) Another alternative stability condition, referred to as 'μ-parameter test' [E-1], is

$$\mu_i = \frac{1 - |s_{22}|^2}{|s_{11} - s_{22}^{*}\Delta s| + |s_{21}s_{12}|} > 1 \text{ for input stability} \tag{8.3.26a}$$

$$;\mu_o = \frac{1 - |s_{11}|^2}{|s_{22} - s_{11}^{*}\Delta s| + |s_{21}s_{12}|} > 1 \text{ for output stability} \tag{8.3.26b}$$

where μ_i/μ_o, called the *geometric input/output stability factor,* are the minimum distances from the origin of the Smith chart to the input/output stability circles, respectively, and therefore, a larger value of μ_i/μ_o implies a greater input/output stability on the premise that $|s_{22}| < 1/|s_{11}| < 1$ (see Figure 8.33(a2) and (b1)). Negative values (due to $|s_{22}| > 1/|s_{11}| > 1$) indicate potential instability.

The following MATLAB function '`stability_S()`', given a *S*-matrix as the first input argument, uses Eqs. (8.3.25) and (8.3.26) to compute the stability factors, plots the input/output stability circles of centers and radii specified by Eqs. (8.3.23a,b)/(8.3.24a,b), and then denotes the unstable region by marking many values of Γ_s/Γ_L violating the stability conditions (8.3.20b/a) with dots.

```
function [aDS,k,mui,muo,ci,ri,co,ro]=stability_S(S,Kg)
% Given a S-matrix, it finds stability factors and plots the stable region.
if nargin<2|Kg>3, Kg=0; end % Kg specifies which section to plot the graph.
D=det(S); aDS=abs(D); % Eq.(8.3.25b) < 1 for stability
s11=S(1,1); s12=S(1,2); s21=S(2,1); s22=S(2,2);
k=(1-abs(s11)^2-abs(s22)^2+aDS^2)/2/abs(s12*s21); %Eq.(8.3.25a)>1: stability
mui=(1-abs(s22)^2)/(abs(s11-s22'*D)+abs(s12*s21)); %Eq.(8.3.26)>1: stability
muo=(1-abs(s11)^2)/(abs(s22-s11'*D)+abs(s12*s21)); %Eq.(8.3.26)>1: stability
% To plot the stable regions on GamS-plane and GamL-plane
if Kg<1, return; end
if numel(Kg)>2, m1=Kg(1); m2=Kg(2); m=Kg(3); else m1=1; m2=1; m=1; end
subplot(m1,m2,m)
Kp=2; [ci,ri]=stability_circle(S,'i',Kp) % Input stable region
subplot(m1,m2,(m2>1)*m2+m+(m2<2))
[co,ro]=stability_circle(S,'o',Kp) % Output stable region
```

```
function [c,r]=stability_circle(S,io,Kp)
% finds the center and radius of input/output stability circle
% and plots the input/output stable regions on the GamS/GamL plane.
ths=pi/180*[0:360]; D=det(S); aDS=abs(D); s12=S(1,2); s21=S(2,1);
if io(1)=='i', s1=S(1,1); s2=S(2,2); den=abs(s1)^2-aDS^2; gs='b';
  c=(s1-D*s2')'/den; r=abs(s12*s21)/abs(den); % Eq.(8.3.23)
  str='Input stability region on \Gamma_S-plane';
 else s1=S(2,2); s2=S(1,1); den=abs(s1)^2-aDS^2; gs='r';
  c=(s1-D*s2')'/den; r=abs(s12*s21)/abs(den); % Eq.(8.3.24)
  str='Output stability region on \Gamma_L-plane';
end
if nargin<3|Kp<1, return; end
plot(exp(j*ths),'k'), hold on, plot(c+r*exp(j*ths),gs)
plot([-1.5 1.5],[0 0],'k', [0 0],[-1.5 1.5],'k')
axis('equal'), axis('off'), title(str); axis([-1.5 1.5 -1.5 1.5])
N=2e3; Gams=random('unif',-1,1,[1,N])+i*random('unif',-1,1,[1,N]);
for n=1:N
   Gam=Gams(n); absG=abs(Gam); sG=(s2-Gam*D)/(1-s1*Gam); % Eq.(8.3.20)
   if absG>=1|abs(sG)>=1, plot(Gam,'.','Markersize',2); end
end
```

Figure 8.34 shows the input/output stable regions for the following three S-matrices

$$S_1 = \begin{bmatrix} 0.52\angle 88^o & 0.80\angle 26.0^o \\ 0.51\angle -93^o & 1.07\angle 29.1^o \end{bmatrix}, \; S_2 = \begin{bmatrix} 1.09\angle -147^o & 0.51\angle -62.9^o \\ 0.61\angle 109^o & 0.31\angle -0.27^o \end{bmatrix},$$
$$S_3 = \begin{bmatrix} 0.81\angle -170^o & 0.43\angle -121^o \\ 0.49\angle 158^o & 0.61\angle 48^o \end{bmatrix} \tag{8.3.27}$$

that have been obtained by running the MATLAB script "plot_stability_region.m" listed below, which uses the MATLAB function '`stability_S()`'.

There are some observations to make about the stability circles and stable regions:

Remark 8.5 Some Observations About Stability Circles and Stable Regions

1) How can we decide which side of the input/output stability circle is stable? In view of Eq. (8.3.20b), the region containing $\Gamma_s = 0$ becomes the input stable or unstable region depending on whether $|s_{22}| < 1$ or $|s_{22}| > 1$. In view of Eq. (8.3.20a), the region containing $\Gamma_L = 0$ becomes the output stable or unstable region depending on whether $|s_{11}| < 1$ or $|s_{11}| > 1$.
2) The input/output stable regions depicted in Figure 8.34(a3) and b3 are the whole disks inside $|\Gamma_s| = 1/|\Gamma_L| = 1$, corresponding to unconditional

stability, as can also be seen from $\mu_i = 1.04/\mu_o = 1.18$. In Figure 8.34(a3), the stable region required by $|\Gamma_{out}| < 1$ is outside the stability circle $|\Gamma_{out}| = 1$ (of center c_i and radius r_i), not affecting the whole disk inside $|\Gamma_s| = 1$ since $|c_i| > r_i + 1$ so that the stability circle is outside the circle $|\Gamma_s| = 1$. Also, in Figure 8.34(b3), the stable region required by $|\Gamma_{in}| < 1$ is inside the stability circle $|\Gamma_{in}| = 1$ (of center c_o and radius r_o), not affecting the whole disk inside $|\Gamma_L| = 1$ since $|c_o| < r_o - 1$ so that the stability circle includes the circle $|\Gamma_L| = 1$.

```
%plot_stability_region.m
clear, clf
Ss(:,:,1)=[0.5196*exp(j*deg2rad(88.4)) 0.8047*exp(j*deg2rad(26))
          0.5074*exp(j*deg2rad(-93.2)) 1.066*exp(j*deg2rad(29.1))];
Ss(:,:,2)=[1.0928*exp(j*deg2rad(-147.5)) 0.5051*exp(j*deg2rad(-62.9))
        0.6086*exp(j*deg2rad(109.2)) 0.3126*exp(j*deg2rad(-0.275))];
Ss(:,:,3)=[0.8056*exp(j*deg2rad(-170)) 0.4371*exp(j*deg2rad(-120.6))
  0.4923*exp(j*deg2rad(157.6)) 0.613*exp(j*deg2rad(48))]; % Eq.(8.3.27)
for n=1:3
  S=Ss(:,:,n), Kg=[2 3 n];
  [D,k,mui,muo,ci,ri,co,ro]=stability_S(S,Kg);
  fprintf('|det(S)|=%6.2f, k=%6.2f, mui=%6.2f, muo=%6.2f',
          D,k,mui,muo)
 fprintf('\n ci=%6.2f<%6.2f, ri=%5.2f, co=%6.2f<%6.2f, ro=%5.2f\n', ...
    abs(ci),rad2deg(angle(ci)),ri, abs(co),rad2deg(angle(co)),ro);
end
```

8.3.3 Design for Maximum Gain [M-2, P-1]

The simultaneous (conjugate) impedance matching conditions between source/amplifier and amplifier/load, that is,

$$\begin{matrix}\text{Source-amplifier}\\ \text{impedance matching}\end{matrix}: \Gamma_{in} \overset{(8.2.23)}{=} s_{11} + \frac{s_{12}s_{21}\Gamma_L}{1 - s_{22}\Gamma_L} = \Gamma_s^* \tag{8.3.28a}$$

$$\begin{matrix}\text{Amplifier-load}\\ \text{impedance matching}\end{matrix}: \Gamma_{out} \overset{(8.2.25)}{=} s_{22} + \frac{s_{21}s_{12}\Gamma_s}{1 - s_{11}\Gamma_s} = \Gamma_L^* \tag{8.3.28b}$$

lead to the matched source/load reflection coefficients:

$$\begin{matrix}\text{Matched source}\\ \text{reflection coefficient}\end{matrix}: \Gamma_{sm} \underset{(10.2.5)}{\overset{[M-2]}{=}} \frac{B_1 \pm \sqrt{B_1^2 - 4|C_1|^2}}{2C_1} \text{ such that } |\Gamma_{sm}| < 1 \tag{8.3.29a}$$

$$\begin{array}{l}\text{Matched load}\\\text{reflection coefficient}\end{array}:\ \Gamma_{Lm} \overset{[M\text{-}2]}{\underset{(10.2.6)}{=}} \frac{B_2 \pm \sqrt{B_2^2 - 4|C_2|^2}}{2C_2} \text{ such that } |\Gamma_{Lm}|<1 \quad (8.3.29b)$$

where

$$B_1 = 1 + |s_{11}|^2 - |s_{22}|^2 - |\Delta s|^2, C_1 = s_{11} - (\Delta s)s_{22}^* \quad (8.3.30a)$$

$$B_2 = 1 + |s_{22}|^2 - |s_{11}|^2 - |\Delta s|^2, C_2 = s_{22} - (\Delta s)s_{11}^* \quad (8.3.30b)$$

With these matched source/load reflection coefficients, the maximized available/transducer power gains (MTGs) equally become

$$G_t \overset{(8.3.12a)}{=} \frac{(1-|\Gamma_s|^2)|s_{21}|^2(1-|\Gamma_L|^2)}{|1-\Gamma_s\Gamma_{in}|^2|1-s_{22}\Gamma_L|^2}$$

$$\overset{\Gamma_{in}=\Gamma_s^*}{\rightarrow} G_{t,max} = \frac{1}{1-|\Gamma_s|^2}|s_{21}|^2\frac{1-|\Gamma_L|^2}{|1-s_{22}\Gamma_L|^2} \overset{[M\text{-}2]}{\underset{(10.2.7)}{=}} \frac{|s_{21}|}{|s_{12}|}\left(k-\sqrt{k^2-1}\right) \quad (8.3.31)$$

Noting that $k > 1$ for unconditional stability (Eq. (8.3.25)), this value of gain with $k = 1$ (on the verge of instability) results in the *maximum stable gain* (*MSG*):

$$\text{MSG}: G_{MS} \overset{(8.3.31)}{\underset{k=1}{=}} \frac{|s_{21}|}{|s_{12}|} \quad (8.3.32)$$

The following MATLAB function '`max_gain()`' tests the unconditional stability based on the criterion (8.3.25), uses the maximum gain formulas (8.3.29a,b) to determine the source/load reflection coefficients Γ_s/Γ_L, respectively, and uses Eqs. (8.3.31) and (8.3.32) to find the resulting maximum transducer gain and MSG, respectively. It also checks if $\Gamma_{in} = \Gamma_s^*$ and $\Gamma_{out} = \Gamma_L^*$ are in the input/output stable regions, respectively.

Example 8.20 Maximum Gain Design Using Conjugate Impedance Matching

Consider a two-port network whose S-matrix (with reference impedance Z_0 = 50 Ω) is

$$S = \begin{bmatrix} 0.72\angle-89^\circ & 0.03\angle57^\circ \\ 2.6\angle76^\circ & 0.73\angle-54^\circ \end{bmatrix} \quad (E8.20.1)$$

where the source and load impedances Z_s/Z_L are matched with Z_0 = 50 Ω so that $\Gamma_s = 0$ and $\Gamma_L = 0$.

a) Determine the values of source/load impedances Z_s/Z_L such that the transducer gain can be maximized.

```
function [GamS,GamL,Gtmax,Gamax]=max_gain(S,GpdB)
% To determine GamS and GamL for the maximum gain design
% based on the simultaneous conjugate matching condition (8.3.28).
% Output:
%  GamS/GamL = Source/Load reflection coefficients
%  Gtmax = Maximized transducer power gain
%  Gamax = Maximized available power gain
D=det(S); aDS=abs(D); EPS=1e-6;
s11=S(1,1); s12=S(1,2); s21=S(2,1); s22=S(2,2);
k=(1-abs(s11)^2-abs(s22)^2+aDS^2)/2/abs(s12*s21); % Eq.(8.3.25)>1
% To test if the network is unconditionally stable.
fprintf(' With stability factor k=%6.2f or |det(S)|=%5.2f,', k,aDS);
if k1-EPS|aDS>1-EPS % with some margin considering numerical errors
  fprintf(' the network is potentially unstable!\n')
 else fprintf(' the network is unconditionally stable!\n');
end
% Simultaneous conjugate impedance matching
B1=1+abs(s11)^2-abs(s22)^2-aDS^2; C1=s11-D*s22'; % Eq.(8.3.30a)
B2=1+abs(s22)^2-abs(s11)^2-aDS^2; C2=s22-D*s11'; % Eq.(8.3.30b)
GamS=(B1-sqrt(B1^2-4*abs(C1)^2))/2/C1; % Eq.(8.3.29a)
GamS=GamS(find(abs(GamS)<1)); aGamS=abs(GamS);
if numel(GamS)>1 % If B1^2-4*abs(C1)^2)<0 so that sqrt() is imaginary
  GamS=GamS(1); aGamS=abs(GamS);
end
GamL=(B2+[1 -1]*sqrt(B2^2-4*abs(C2)^2))/2/C2; % Eq.(8.3.29b)
GamL=GamL(find(abs(GamL)
if numel(GamL)>1 % If B2^2-4*abs(C2)^2)<0 so that sqrt() is imaginary
  GamL=GamL(1); aGamL=abs(GamL);
end
if nargin>1, Kg=1; [c,r,k,aDS]=gain_circle(S,GpdB,'o',Kg); end
% Gain analysis
if isempty(GamS)|isempty(GamL)
 fprintf('\nGain analysis is impossible since at least one of GamS=Gami*
 and GamL=Gamo* is in the unstable region!\n'); Gtmax=[]; Gamax=[];
else
 GS=1/(1-aGamS^2); G0=abs(s21)^2;
 GL=(1-aGamL^2)/abs(1-s22*GamL)^2;
 GS_G0_GLdB=10*log10([GS G0 GL])
 Gtmax=GS*G0*GL; % Maximum Transducer Gain (MTG) - Eq.(8.3.31)
 Gms=abs(s21)/abs(s12); % Maximum Stable Gain (MSG) - Eq.(8.3.32)
 Gamax=Gms*(k-sqrt(k^2-1)); % Eq.(8.3.31b), Eq.(10.2.7) of [M-2]
 fprintf('Maximum Transducer Gain(MTG)=%7.2f[dB]', 10*log10(Gamax));
 fprintf(' where Maximum Stable Gain(MSG)=%7.2f[dB]', 10*log10(Gms));
 % To see if GamL and GamS are in the stable region
 Gami=s11+s12*s21*GamL/(1-s22*GamL); % (Zi-Z0)/(Zi+Z0) % Eq.(8.2.23)
 Gamo=s22+s21*s12*GamS/(1-s11*GamS); % (Zo+Z0)/(Zo-Z0) % Eq.(8.2.25)
 if abs(Gami)>1-1e-6 % with some margin considering numerical errors
  fprintf('Gami=%6.2f<%5.1f is out of the stable region!\n', abs
   (Gami),rad2deg(angle(Gami)));
 end
 if abs(Gamo)>1-EPS % with some margin considering numerical errors
  fprintf('Gamo=%6.2f<%5.1f is out of the stable region!\n', abs
   (Gamo),rad2deg(angle(Gamo)));
 end
 plot(GamS+j*eps,'o'), plot(GamL+j*eps,'mo')
end
```

To this end, we compose and run the following MATLAB script "elec08e20a.m":

```
%elec08e20a.m
S=[0.72*exp(j*deg2rad(-116)) 0.03*exp(j*deg2rad(57));
   2.6*exp(j*deg2rad(76)) 0.73*exp(j*deg2rad(-54))]; % Eq.(E8.20.1)
[GamS,GamL,Gtmax,Gamax]=max_gain(S);
fprintf('\nGamS=%6.3f<%5.1f, GamL=%6.3f<%5.1f\n', ...
 abs(GamS),rad2deg(angle(GamS)),abs(GamL),rad2deg(angle(GamL)));
```

to get

```
With stability factor k= 1.19 or |det(S)|= 0.49,
  the network is unconditionally stable!
Maximum Transducer Gain=16.71[dB]
where Maximum Stable Gain(MSG)=19.38[dB]
GamS= 0.872<123.4, GamL= 0.876< 61.0
```

Then, we can convert these source/load reflection coefficients into the source/load impedances Z_s/Z_L by running the following statements:

```
>>Z0=50; Zsd=G2Z(GamS,Z0), ZLd=G2Z(GamL,Z0)
```

where the MATLAB function 'G2Z()' was listed in Problem 7.1. This yields desired values of source/load impedances Z_s/Z_L:

```
Zsd =  4.4125 +26.7554i, ZLd =  12.6298 +83.4260i
```

b) Supposing that a single stub and a TL section are used for input and output matching as shown in Figure 8.35, determine the lengths of the open-circuited stubs and TL sections to realize the desired source and load impedances (obtained in (a)) for maximizing the (transducer) power gain.

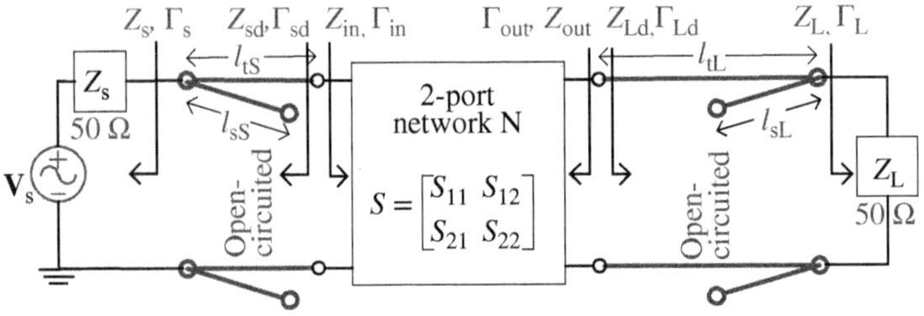

Figure 8.35 An radio-frequency (RF) network using single stubs for input and output impedance matching for maximum gain.

Referring to Eqs. (7.3.1) and (7.3.2b), we can write the input/output matching conditions as

$$Z_l(\beta l_{sS},\beta l_{tS}) \underset{Z_L=Z_s\|Z_{os}(\beta l_{sS})}{\overset{(7.3.1)}{=}} Z_0\frac{(Z_s\|Z_{os}(\beta l_{sS}))+jZ_0\tan\beta l_{tS}}{Z_0+j(Z_s\|Z_{os}(\beta l_{sS}))\tan\beta l_{tS}}=Z_{sd} \tag{E8.20.2a}$$

$$Z_l(\beta l_{sL},\beta l_{tL}) \underset{Z_L=Z_L\|Z_{os}(\beta l_{sL})}{\overset{(7.3.1)}{=}} Z_0\frac{(Z_L\|Z_{os}(\beta l_{sL}))+jZ_0\tan\beta l_{tL}}{Z_0+j(Z_L\|Z_{os}(\beta l_{sL}))\tan\beta l_{tL}}=Z_{Ld} \tag{E8.20.2b}$$

$$\text{with } Z_{os}(\beta l)\overset{(7.3.2b)}{=}-jZ_0\cot\beta l \tag{E8.20.3}$$

Note that a complex equation amounts two real equations each for the real and imaginary parts. To use the nonlinear equation solver '`fsolve()`' for solving these nonlinear equations, let us compose and run the following MATLAB script "elec08e20b.m" where the MATLAB function '`imp_making_1stub()`' is supposed to be stored on the computer:

```
%elec08e20b.m
Zs=Z0; ZL=Z0;
[blsS,bltS]=imp_making_1stub(Zsd,Zs,Z0,'o'); % Input matcher design
lsS=blsS/2/pi, ltS=bltS/2/pi % Lengths of stub/TL for input matching
[blsL,bltL]=imp_making_1stub(ZLd,ZL,Z0,'o'); % Output matcher design
lsL=blsL/2/pi, ltL=bltL/2/pi % Lengths of stub/TL for output matching
```

```
function [bls,blt]=imp_making_1stub(Zd,Z,Z0,os)
% determines the (b*l)s of a shunt open/short-circuited stub and
% a lossless TL section of characteristic impedance Z0
% to make a desired impedance Zd from Z.
% Input:
%  Zd, Z = Desired and Current impedances
%  Z0 = Characteristic impedance of the TL
%  os = 'o'/'s' for open/short-circuited stub
% Output:
%  bls, blt = beta*lengths of the stub and TL section
if nargin<4, os='o'; end
os=lower(os);
Zl=@(Z,bl) (Z+j*Z0*tan(bl))./(Z0+j*Z.*tan(bl))*Z0; %LHS of Eq. (E8.20.2)
if os(1)=='o', Zs=@(bl)-j*Z0*cot(bl); % Eq. (7.3.2a)
 else Zs=@(bl)j*Z0*tan(bl); % Eq. (7.3.2b)
end
eq=@(bl) [real(Zl(parallel_comb([Z Zs(bl(1))]),bl(2))-Zd);
          imag(Zl(parallel_comb([Z Zs(bl(1))]),bl(2))-Zd)]; %Eq.(E8.20.2)
bl=fsolve(eq,0.1*[1 1]); bl1=mod(bl,pi);
bls=bl1(1); blt=bl1(2);
% Try with a larger initial guess.
bl=fsolve(eq,[1 1]); bl2=mod(bl,pi);
if norm(bl2-bl1)>0.01
   bls=[bls bl2(1)]; blt=[blt bl2(2)];
end
```

This yields

```
lsS = 0.2936   0.2064, ltS = 0.0379   0.1193
lsL = 0.2927   0.2073, ltL = 0.1252   0.2052
```

which means the lengths of the stubs and TL sections as

$$\{l_{sS} = 0.2936\lambda,\ l_{tS} = 0.0379\lambda\}$$
$$\text{or } \{l_{sS} = 0.2064\lambda,\ l_{tS} = 0.1193\lambda\} \text{ for input matching} \tag{E8.20.4a}$$

$$\{l_{sL} = 0.2927\lambda,\ l_{tL} = 0.1252\lambda\}$$
$$\text{or } \{l_{sL} = 0.2073\lambda,\ l_{tL} = 0.2052\lambda\} \text{ for output matching} \tag{E8.20.4b}$$

c) To check if the impedance matchers designed in (b) satisfy the input and output impedance matching conditions, convert the s-parameters into the (equivalent) a-parameters and find the input/output impedances of the network equipped with the desired input/output matching sections.
To this end, we compose and run the following MATLAB script "elec08e20c.m":

```
%elec08e20c.m
A=s2a(S,Z0);
Zi=port_property(A,ZLd,'ZiA'), Zsd
Zo=port_property(A,Zsd,'ZoA'), ZLd
```

to get

```
Zi =  4.4125  -26.7554i,   Zsd =  4.4125  +26.7554i
Zo = 12.6298  -83.4260i,   ZLd = 12.6298  +83.4260i
```

These results imply that the conjugate impedance matching conditions have been satisfied.

d) Let use 'Smith 4.1' to design the input/output impedance matchers, take the following steps:

1) From the top menu bar, select Tools > Settings to open the Settings dialog box (Figure 7.5(b)), set the values of Z_0 (characteristic impedance) and frequency to 50 [Ω] and 1 [GHz], respectively, and click OK button. Here, the frequency does not matter because we will determine the lengths of stubs and TL sections in the unit of wavelength.
2) From the toolbar (below the menu bar), select Keyboard to open Data Point dialog box, insert $\Gamma_1 = 0\angle 0^\circ$ (the origin corresponding to $Z_s = Z_0 = 50\ \Omega$) and $\Gamma_2 = \Gamma_{sd} = 0.872\angle 123.4^\circ$ as starting and target data points, respectively, and click OK button to close the Data Point dialog box.

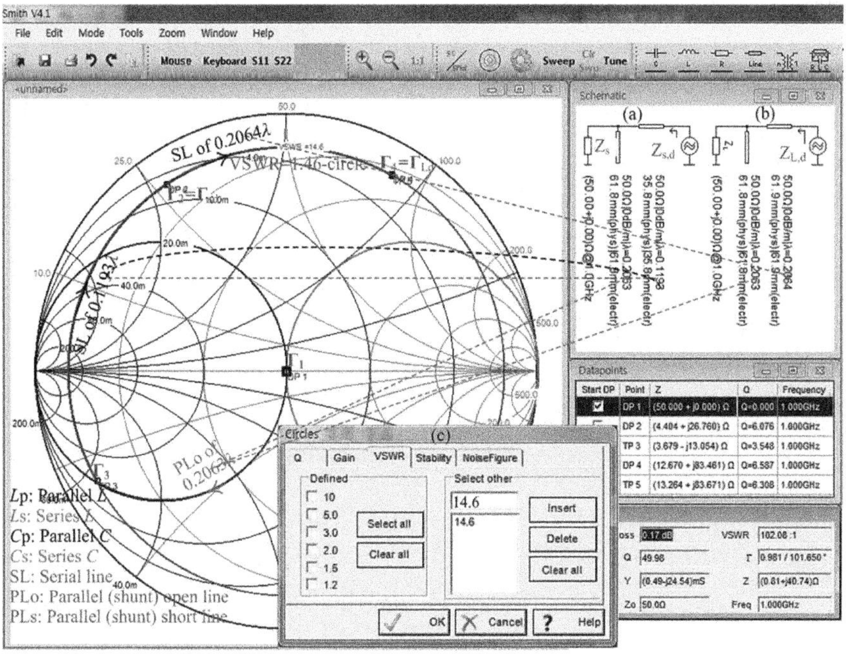

Figure 8.36 Using 'Smith 4.1' to design the input/output impedance matchers for Example 8.20.

3) Noting that the Γ point will be moved along an SWR circle by the TL section and $VSWR_{sd}=(1+|\Gamma_{sd}|)/(1-|\Gamma_{sd}|)=14.6$, select the Tools > Circles menu to open the Circles dialog box, insert the circle of VSWR = 14.6, and click OK button to close the dialog box (Figure 8.36(c)).
4) From the tool bar, click on 'Insert parallel line open' button (denoted by OS: open stub) and move the mouse cursor to click on the point Γ_3, which is an intersection of the $G = 1/50$ [S]-circle and the VSWR = 14.6-circle.
5) From the tool bar, click on Line ('Insert serial line') button and move the cross-type mouse cursor to click on the point Γ_2 (see Figure 8.36). Then, you will see the schematic (a) corresponding to $l_{sS} = 0.2063\lambda$ and $l_{tS} = 0.1193\lambda$.
6) Right-click on any point to delete the last data point and from the tool bar, select Keyboard to open Data Point dialog box, insert $\Gamma_4 = \Gamma_{Ld} = 0.876\angle 61^\circ$ as a target data point, and click OK button to close the Data Point dialog box.
7) From the tool bar, click on 'Insert parallel line open (OS)' button and move the mouse cursor to click on the point Γ_4 (see Figure 8.36). Then you will see the schematic (b) corresponding to $l_{sL} = 0.2063\lambda$ and $l_{tL} = 0.2064\lambda$. Here, we

did not have to draw another VSWR-circle because $VSWR_{Ld} = (1 + |\Gamma_{Ld}|)/(1 - |\Gamma_{Ld}|) = 14.6$ is almost identical to $VSWR_{sd}$. Besides, the transit Γ_3 (obtained as an intersection of the $G = 1/50$[S]-circle and the same VSWR-circle) is also common to the input/output impedance matching.

8.3.4 Design for Specified Gain [M-2, P-1]

If a desired value $G_{p,d}$ of the operating power gain for a network to achieve is specified, the design process is as follows:

1) Draw the input/output stability circles and the constant operating power gain circle of $G_{p,d}$.
2) Determine a desired value $\Gamma_{L,d}$ of load reflection coefficient at a position on the constant-gain circle (inside the input stable region with some margin, i.e. not too close to the input stability circle) such that the desired operating gain $G_{p,d}$ can be realized.
3) Use Eq. (8.2.23a) to find the value of Γ_{in} (corresponding to $\Gamma_{L,d}$) and its conjugate as a desired value of source reflection coefficient Γ_s.

$$\text{Source-amplifier impedance matching}: \Gamma_{s,d} = \Gamma^*_{in,d} \overset{(8.2.23a)}{=} \left(s_{11} + \frac{s_{12}s_{21}\Gamma_{L,d}}{1 - s_{22}\Gamma_{L,d}}\right)^* \tag{8.3.33}$$

4) Check to see if $\Gamma_{s,d}$ is in the output stable region. If so, find the corresponding values of $Z_{s,d}$ and $Z_{L,d}$. Otherwise, make a compromise with the conjugate impedance matching condition (8.3.33) or go to Step 2 and repeat the iteration with another value of $\Gamma_{L,d}$.

Example 8.21 Specified Gain Design

Consider a two-port network whose S-matrix (with reference impedance $Z_0 = 50\,\Omega$) is

$$S = \begin{bmatrix} 0.5\angle -180^\circ & 0.08\angle 30^\circ \\ 2.5\angle 70^\circ & 0.8\angle -100^\circ \end{bmatrix} \tag{E8.21.1}$$

```
%elec08e21a.m
S=[0.5*exp(j*deg2rad(-180)) 0.08*exp(j*deg2rad(30));
   2.5*exp(j*deg2rad(70)) 0.8*exp(j*deg2rad(-100))]; % Eq.(E8.21.1)
smithchart; hold on
Kp=1; stability_circle(S,'i',Kp); stability_circle(S,'o',Kp);
oa='o'; GdB=10; Kg=-1; [c,r,k,aDS]=gain_circle(S,GdB,oa,Kg)
```

a) Plot the input and output stable regions together with the constant operating power gain circle of G_p = 10 dB on the Smith chart. Also determine the stability.
   To this end, we compose and run the above MATLAB script "elec08e21a.m" to get Figure 8.37, the stability factor k, and $|\Delta s|$:

```
k =  0.3991,  aDS = 0.2228
```

   Since $k = 0.3991 < 1$ (violating the stability condition (8.3.25a)), the network is potentially unstable.

b) Use 'Smith 4.1' to draw the input/output stability circles and the constant operating power gain circle of G_p = 10 dB as depicted in Figure 8.38.2. Then, referring to Figures 8.37 and 8.38.2, determine such a value $\Gamma_{L,d}$ of load reflection coefficient (inside the input stable region with some margin, i.e. at a position not to close the input stability circle) that achieves the operating gain of 10 dB and that can be realized by inserting a shunt open-circuited stub and a TL section, in addition to the current load of Z_L = 50 Ω. Then, use Eq. (8.3.33) to find a desired value $\Gamma_{s,d}$ of source reflection coefficient. Then, check to see if $\Gamma_{s,d}$ is in the output stable region. If not, repeat the iteration on $\Gamma_{L,d}/\Gamma_{s,d}$ or compromise.
   To this end, let us run 'Smith V4.1' and take the following steps:

   1) From the top menu bar, select Tools > Settings to open the Settings dialog box (Figure 7.5(b)), set the values of Z_0 (characteristic impedance) and frequency to 50 [Ω] and 1 [GHz], respectively, and click OK button. Here, the frequency does not matter because we will determine the lengths of stubs and TL sections in the unit of wavelength.

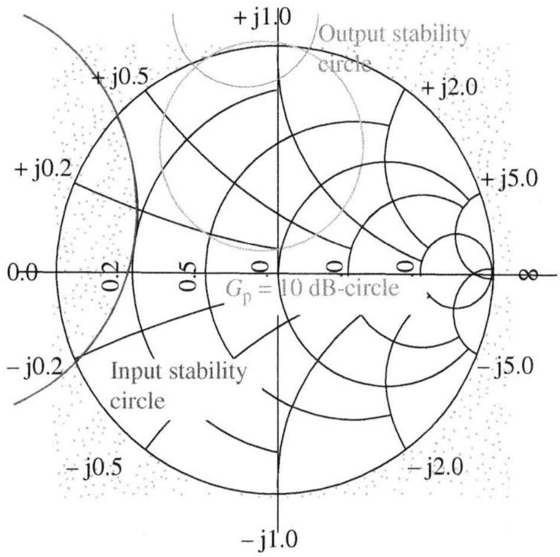

Figure 8.37 Input/output stable regions and G_p = 10 dB-circle for Example 8.21.

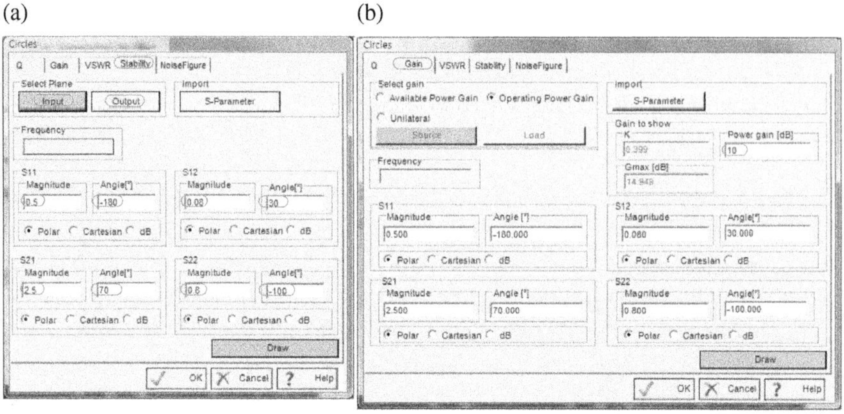

The Stability tab in the Circles dialog box — The Gain tab in the Circles dialog box

Figure 8.38.1 To input the parameter values for drawing the stability circles and constant-gain circle in 'Smith 4.1' for Example 8.21.

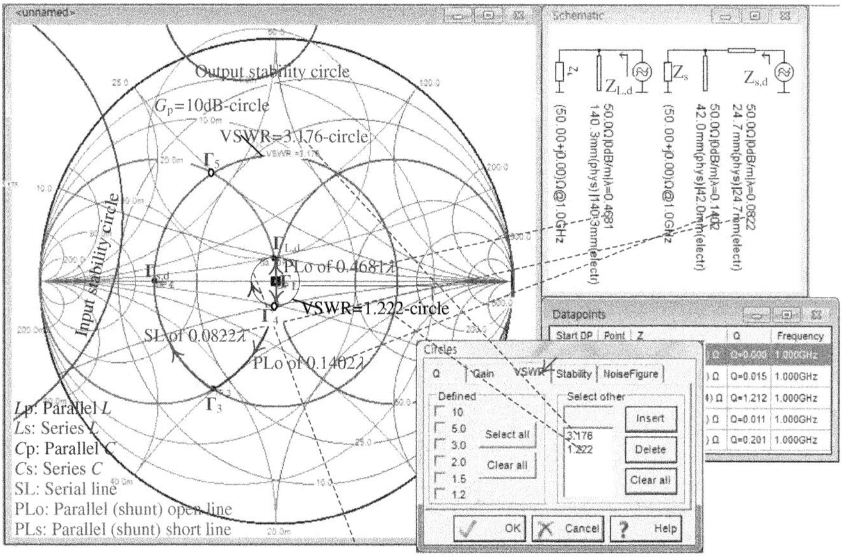

Figure 8.38.2 Using 'Smith 4.1' to design the input/output stages for Example 8.21.

2) From the toolbar (below the menu bar), select Keyboard to open Data Point dialog box, insert $\Gamma_1 = 0\angle 0^\circ$ (the origin corresponding to $Z_s = Z_0 = 50\ \Omega$) as starting data point, and click OK button to close the Data Point dialog box.

3) From the top menu bar, select Tools > Circles to open the Circles dialog box, select the Stability tab, click on Input in the Select Plane box, type in the *s*-parameter values (given by Eq. (E8.21.1)) as shown in Figure 8.38.1(a), and click the Draw button. Then, click on Output in the Select Plane box and click the Draw button.
4) Staying in the Circles dialog box, select the Gain tab, check the radio button before operating power gain in the Select gain box, type in the value 10 of Power gain in the 'Gain to show' box, type in the *s*-parameter values as shown in Figure 8.38.1(b), click the Draw button, and then click OK to close the Circles dialog box.
5) Noting the constant gain G_p = 10 dB-circle and the constant conductance $G = 1/Z_{L,0}$ = 20 mS-circle have two intersections, let us pick one of them, $\Gamma_{L,d} = 0.1\angle 95^\circ$ (as seen in the Cursor pane), which is farther from the input stable circle (as target point) such that we need just a single shunt stub to move from $\Gamma_{L,0} = 0$ to that point along the G = 20 mS-circle.
6) We use Eq. (8.3.33) to find a desired value of source reflection coefficient Γ_s corresponding to $\Gamma_{L,d}$ as

$$\Gamma_{s,d} = \Gamma^*_{in,d} \overset{(8.3.33)}{=} \left(s_{11} + \frac{s_{12}s_{21}\Gamma_{L,d}}{1 - s_{22}\Gamma_{L,d}}\right)^* \overset{\Gamma_{L,d}=0.1\angle 95^\circ}{=} 0.521\angle 179.4^\circ \quad \text{(E8.21.2)}$$

This can be computed by running the following MATLAB statements:

```
>>GamLd=0.1*exp(j*deg2rad(95)); GamSd=conj(Gami_L(GamLd,S))
  abs(GamSd), rad2deg(angle(GamSd)) % conversion into polar form
```

where the following MATLAB function '`Gami_L()`' is supposed to have been stored on the computer.

```
function Gami=Gami_L(GamL,S)
Gami=S(1,1)+S(1,2)*S(2,1)*GamL/(1-S(2,2)*GamL);
     % Eq.(8.3.33 or 28a)
```

c) Referring to Figure 8.35, find the lengths of the (open-circuited) stubs and TL sections to realize the desired source/load reflection coefficients $\Gamma_{s,d}/\Gamma_{L,d}$ on the source/load stages.

1) From the toolbar in the main window of 'Smith 4.1', select Keyboard to open Data Point dialog box, insert $\Gamma_{s,d} = 0.521\angle 179.4^\circ$ (Eq. (E8.21.2)) as a data point, click OK button to close the Data Point dialog box, and check to see if the new data point $\Gamma_{s,d}$ is in the output stable region (see Figure 8.38.2).
2) Now that $\Gamma_{L,d}/\Gamma_{s,d}$ have been determined and that in the input/output stable regions, respectively, we use Eq. (7.1.10) or '`G2Z()`' to convert $\Gamma_{L,d}/\Gamma_{s,d}$ into the desired load/source impedances $Z_{L,d}/Z_{s,d}$:

```
>>Z0=50; ZLd=G2Z(GamLd,Z0), Zsd=G2Z(GamSd,Z0)
   ZLd = 48.1784 + 9.6960i, Zsd = 15.7440 + 0.2362i
```

This means that the desired load/source impedances $Z_{L,d}/Z_{s,d}$ for realizing the gain G_p = 10 dB are

$$Z_{L,d} = 48.1785 + j9.6960, Z_{s,d} = 15.744 + j0.2362 \tag{E8.21.3}$$

3) Now we can use the MATLAB function '`imp_making_1stub()`' (introduced in Example 8.20) to determine the (electrical) lengths of shunt stubs and TL sections to connect with Z_s/Z_L so that the resulting source/load impedances can be $Z_{s,d}/Z_{L,d}$ as desired:

```
>>Zs=Z0; ZL=Z0; pi2=2*pi;
  [blsS,bltS]=imp_making_1stub(Zsd,Zs,Z0,'o'); % Input stage design
  lsS=blsS/pi2, ltS=bltS/pi2 % Lengths of stub/TL for input stage
  [blsL,bltL]=imp_making_1stub(ZLd,ZL,Z0,'o'); % Output stage design
  lsL=blsL/pi2, ltL=bltL/pi2 % Lengths of stub/TL for output stage
```

Running these statements yields

```
lsS =    0.1408,  ltS =    0.0822
lsL =    0.4684   0.0316,    ltL =    0.0010   0.2351
```

which means the lengths of the stubs and TL sections as

$$l_{sS} = 0.1408\lambda,\quad l_{tS} = 0.0822\lambda \text{ for input stage} \tag{E8.21.4a}$$

$$l_{sL} = 0.4684\lambda,\quad l_{tL} = 0.001\lambda \text{ or } l_{sL} = 0.0316\lambda,\ l_{tL} = 0.2351\lambda \text{ for output stage} \tag{E8.21.4b}$$

4) For practice and crosscheck, we use Smith 4.1 to get the microstrip realizations of the input and output stages. First, let us find a path from $\Gamma_1 = 0\angle 0^\circ$ to $\Gamma_{s,d} = 0.521\angle 179.4^\circ$ consisting of two sections, one covered by a shunt stub and one covered by a serial TL. To this end, we draw the constant VSWR circle crossing $\Gamma_{s,d}$, i.e. the VSWR = $(1 + |\Gamma_{s,d}|)/(1 - |\Gamma_{s,d}|) = (1 + 0.521)/(1 - 0.521) = 3.176$-circle (along which the Γ-point moves to reach $\Gamma_{s,d}$ through the effect of a TL section) by selecting the menu Tools > Circles to open the Circles dialog box, selecting VSWR tab, inserting 3.176, and clicking Insert and OK buttons in turns.
5) With $\Gamma_1 = 0\angle 0^\circ$ (the origin corresponding to $Z_s = Z_0 = 50\ \Omega$) as starting point, click on OS (Insert parallel line open) from the toolbar, move the mouse cursor to Γ_3 (the intersection of the VSWR = 3.176-curve and the constant conductance G = 20 mS-circle), and click on it. Then, an open-circuited (floating) stub of length 0.1402λ will appear in the Schematic pane (see Figure 8.38.2).

6) Then, click on Line (Insert serial line) from the toolbar, move the mouse cursor from Γ_3 to $\Gamma_{s,d}$ (a desired source reflection coefficient) along a part of the VSWR curve, and click on $\Gamma_{s,d}$. Then, a serial line of length 0.0822λ will appear in the Schematic pane (see Figure 8.38.2), completing an input stage design. This result conforms with Eq. (E8.21.4a) obtained using the MATLAB function '`imp_making_1stub()`'.
7) Second, let us find a path from $\Gamma_1 = 0\angle 0^\circ$ to $\Gamma_{L,d} = 0.1\angle 95^\circ$ consisting of only a shunt stub, which is possible because the two points are on the same constant conductance circle, i.e. $G = 20$ mS-circle. To this end, right-click on anywhere on the Smith chart twice to delete the last two data points (excluding Γ_1). Then, click on OS from the toolbar, move the mouse cursor to $\Gamma_{L,d}$, and click on it. Then, an open-circuited (floating) stub of length 0.4681λ will appear in the Schematic pane (see Figure 8.38.2), completing an output stage design. This result conforms with one of two designs obtained using '`imp_making_1stub()`' and given by Eq. (E8.21.4b).

Note the following:

- The other of two output designs given by Eq. (E8.21.4b) can be obtained by moving the Γ-point (designated by the mouse cursor) from Γ_1 to $\Gamma_{L,d}$ via Γ_4 that is an intersection of the G = 20 mS-circle and the VSWR = 1.222-circle (crossing $\Gamma_{L,d}$).
- Another input stage design like $\{l_{sS} = 0.3592\lambda, l_{tS} = 0.4195\lambda\}$ can be obtained by moving the Γ-point from Γ_1 to $\Gamma_{s,d}$ via Γ_5 that is another intersection of the G = 20 mS-circle and the VSWR = 3.176-circle (crossing $\Gamma_{s,d}$). It could have been obtained using '`imp_making_1stub()`' with different initial guess, say, 0 instead of 0.1, inside it.

Problems

8.1 Transfer Function of LC Ladder Circuits

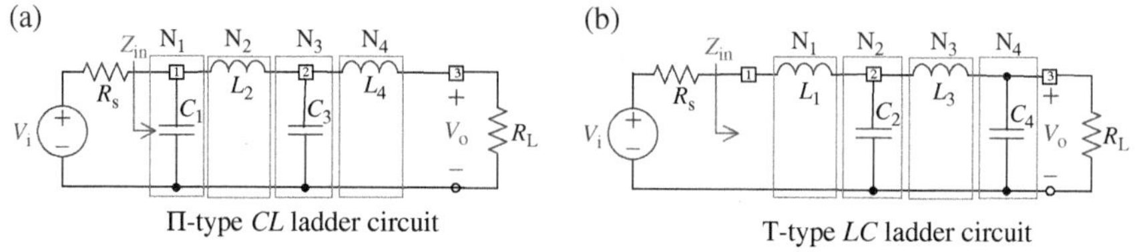

Figure P8.1 *LC* ladder circuits.

(a) Noting that the CL ladder of Figure P8.1(a) can be regarded as a cascade connection of the four 2-port networks N_1, N_2, N_3, and N_4 so that its a-parameter matrix A can be obtained as

$$A = \begin{bmatrix} 1 & R_s \\ 0 & 1 \end{bmatrix} \begin{bmatrix} 1 & 0 \\ sC_1 & 1 \end{bmatrix} \begin{bmatrix} 1 & sL_2 \\ 0 & 1 \end{bmatrix} \begin{bmatrix} 1 & 0 \\ sC_3 & 1 \end{bmatrix} \begin{bmatrix} 1 & sL_4 \\ 0 & 1 \end{bmatrix} \tag{P8.1.1}$$

complete the following MATLAB function '`ladder_xfer_ftn_CL()`' so that it can find the overall A-matrix and then get the transfer function $G(s) = V_o(s)/V_i(s)$ from the A-matrix and the load impedance R_L by using the MATLAB function '`port_property()`'.

```
function [Gs,A]=ladder_xfer_ftn_CL(Rs,CL,RL)
% CL : [C1 L2 C3 L4 ..]
% Gs=Vo(s)/Vi(s) : Transfer function of Rs-CL-RL filter
syms s; A=[1 Rs; 0 1];
for n=1:length(CL)
   if mod(n,2)==1, A=A*[1 0; s*????? 1]; % Y=sC
   else A=A*[1 s*CL(n); 0 1]; % Z=sL
   end
end
Gs=port_property(A,RL,'???'); %Gs_=1/(A(1,1)+A(1,2)/RL); Table 8.2
```

To check the validity of this function, complete and run the following MATLAB script "elec08p01a.m," which uses the node analysis to find the transfer function $G(s)$ and compare it with that obtained using the above function '`ladder_xfer_ftn_CL()`'. Discuss the running results.

```
%elec08p01a.m
syms s Rs RL C1 L2 C3 L4
Y=[1/Rs+s*C1+1/s/L2    -1/s/L2 0;
 -1/s/L2   1/s/L2+s*C3+1/s/L4    -1/s/L4;
   0                 -1/s/L4     1/s/L4+1/RL];
V=Y\[1/Rs; 0; 0];
Gs=V(?)
CL=[C1 L2 C3 L4]; Gs1=ladder_xfer_ftn_CL(Rs,??,RL)
Gs_=subs(Gs,{Rs,RL,C1,L2,C3,L4},{1 2 3 4 5 6})
Gs1_=subs(Gs1,{Rs,RL,C1,L2,C3,L4},{1 2 3 4 5 6})
pretty(simplify(Gs_)), pretty(simplify(Gs1_))
```

```
function [Gs,A]=ladder_xfer_ftn_LC(Rs,LC,RL)
% LC : [L1 C2 L3 C4 L5 ..]
% Gs=Vo(s)/Vi(s) : Transfer function of Rs-LC-RL filter
syms s; A=[1 Rs; 0 1];
for n=1:length(LC)
   if mod(n,2)==1, A=A*[1 s*LC(n); 0 1]; % Z=sL
   else A=A*[1 0; s*????? 1]; % Y=sC
   end
end
Gs=port_property(A,RL,'Ava'); %Gs_=1/(A(1,1)+A(1,2)/RL); Table 8.2
```

(b) Noting that the *LC* ladder of Figure P8.1(b) can be regarded as a cascade connection of the four 2-port networks N_1, N_2, N_3, and N_4, complete the above MATLAB function '`ladder_xfer_ftn_ LC()`' so that it can find the overall *A*-matrix and then get the voltage gain or transfer function $G(s) = V_o(s)/V_i(s)$ from the *A*-matrix and the load impedance R_L by using the MATLAB function '`port_property()`'. Also, to check the validity of the function, complete and run the following MATLAB script "elec08p01b.m," which uses the node analysis to find the transfer function $G(s)$ and compare it with that obtained using the above function '`ladder_xfer_ftn_LC()`'. Discuss the running results.

```
%elec08p01b.m
syms s Rs RL L1 C2 L3 C4
Y=[1/Rs+1/s/L1      -1/s/L1              0;
   -1/s/L1     1/s/L1+s*C2+1/s/L3 -1/s/L3;
     0        -1/s/L3                 1/s/L3+s*C4+1/RL];
V=Y\[?/Rs; 0; 0];
Gs=V(3)
LC=[L1 C2 L3 C4]; Gs1=ladder_xfer_ftn_LC(Rs,??,RL)
Gs_=subs(Gs,{Rs,RL,L1,C2,L3,C4},{1 2 3 4 5 6})
Gs1_=subs(Gs1,{Rs,RL,L1,C2,L3,C4},{1 2 3 4 5 6})
pretty(simplify(Gs_)), pretty(simplify(Gs1_))
```

8.2 Hybrid Transistor Model

(a) Show that the h-parameter matrix of the two-port network in Figure P8.2.1(a) is

$$H = \frac{1}{1 + sr_{\text{be}}(C_{\text{bc}} + C_{\text{be}})} \begin{bmatrix} r_{\text{be}} & sr_{\text{be}}C_{\text{bc}} \\ r_{\text{be}}(g_m - sC_{\text{bc}}) & s^2 r_{\text{be}} r_{\text{ce}} C_{\text{bc}} C_{\text{be}} + s\{r_{\text{be}}(C_{\text{bc}}+C_{\text{be}}) + r_{\text{ce}}C_{\text{bc}}(1+g_m r_{\text{be}})\}+1 \end{bmatrix} \quad \text{(P8.2.1)}$$

To that end, applying a current source $I_1 = 1$ [A] at port 1 and short-circuiting port 2 so that $V_2 = 0$ [V], find V_1 and I_2 to get the h-parameters h_{11} and h_{21}. Then, open-circuiting port 1 so that $I_1 = 0$ [A] and applying a voltage source $V_2 = 1$ [V] at port 2, find V_1 and I_2 to get the h-parameters h_{12} and h_{22}. To this end, you can complete and run the following MATLAB script "elec08p02a.m."

```
%elec08p02a.m
syms s rbe Cbe Cbc gm rce C4
V1=1/(1/rbe+s*(C??+Cbc)); % with I1=1 and V2=0 from KCL at node 1
I2=??*V1-s*Cbc*V1; % with I1=1 and V2=0 from KCL at node 2
H(1,1)=V1; H(2,1)=I?;
Zbe=1/(1/rbe+s*Cbe);
V1=Zbe/(1/s/C??+Zbe); % with I1=0 and V2=1 from KCL at node 1
I2=s*C??*(1-V1)+gm*V1+1/rce; % with I1=0 and V2=1 from KCL at node 2
H(1,2)=V?; H(2,2)=I2;
```

(b) As an alternative to get the H-matrix of the two-port network in Figure P8.2.1 (a), regard the network as a parallel–parallel (or shunt-shunt) connection of the two subnetworks shown in Figure P8.2.1(b1) and (b2) and combine the parameters of the two subnetworks to find the H-matrix. To this end, you can complete and run the following MATLAB script "elec08p02b.m."

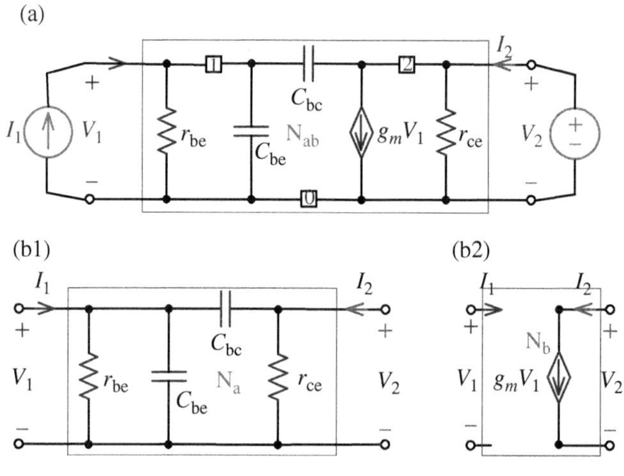

Figure P8.2.1 Hybrid transistor model.

```
%elec08p02b.m
% Regarding the circuit as a shunt-shunt connection of two 2-port NWs
Ya=[1/rbe+s*(Cbe+C??)          -s*Cbc;
    -s*C??                      s*C??+1/r??]; % Y-matrix of subnetwork a
Yb=[0 0; ?? 0]; % Y-matrix of subnetwork b
Yab=Ya?Yb; Hab=port_conversion(Y,'y2?');
pretty(simplify(Hab))
simplify(H-Hab)
```

(c) Consider the circuit shown in Figure P8.2.2(a) where the subnetwork N_{ab} is the circuit of Figure P8.2.1(a) whose H-matrix is given by Eq. (P8.2.1). Noting that the network N_2 in the central part can be regarded as a parallel-parallel connection of N_{ab} and N_c in Figure P8.2.2(b) and the cascade connection of the three subnetworks N_1, N_2, and N_3 makes the whole circuit, plot the overall current gain $A_i(s)|_{s=j\omega} = -I_o(j\omega)/I_i(j\omega)$ (with port 2 short-circuited) for $\omega = 2\pi f$ (with $f = 10^4$–10^9[Hz]) and $R = 200\,\Omega$, $300\,\Omega$, $500\,\Omega$, $1\,k\Omega$, and $10\,k\Omega$ as depicted in Figure P8.2.3 where the parameter values are as follows:

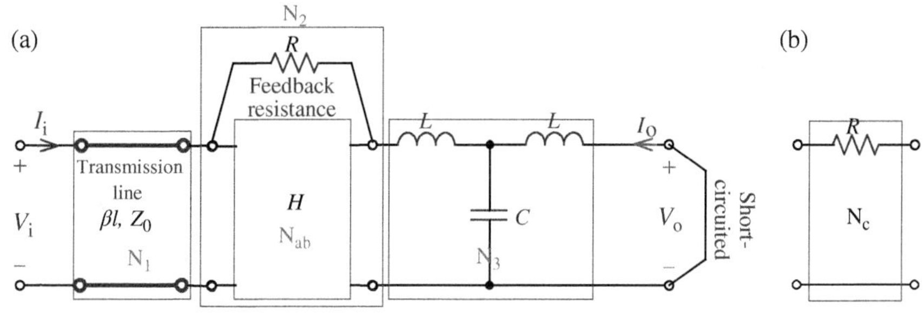

Figure P8.2.2 Microwave amplifier circuit viewed as a connection of several subnetworks.

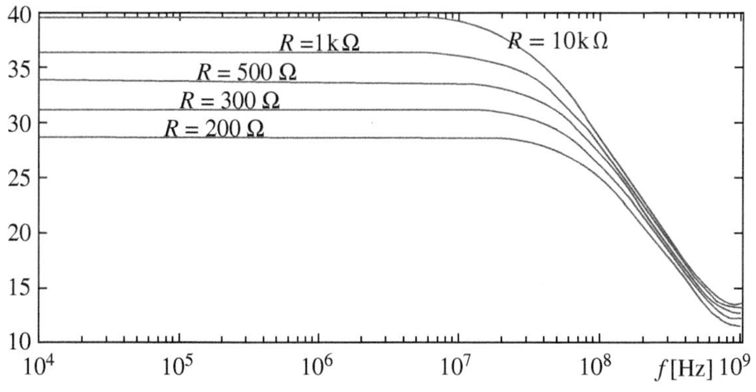

Figure P8.2.3 Current gain of the microwave amplifier shown in Figure P8.2.2a.

$$\beta = 2\pi/\lambda\,[\text{rad/m}],\ \lambda = 500\,[\text{nm}] = 5\times10^{-7}\,[\text{m}],\ l = 0.05\,[\text{m}],$$
$$Z_0 = 50\,[\Omega],\ r_{be} = 520\,[\Omega],\ r_{ce} = 80\,[\text{k}\Omega],\ C_{be} = 10\,[\text{pF}],$$
$$C_{bc} = 1\,[\text{pF}],\ g_m = 0.192\ [\text{S}],\ L = 1\ [\text{nF}],\ C = 10\,[\text{pF}]$$

```
%elec08p02c.m
Yab=port_conversion(H,'???');
Yc=[1/R -1/R; -1/R 1/R];
Y2=Yab?Yc;
A2=port_conversion(Y2,'???');
Z0=50; lambda=500e-9; b=2*pi/lambda; l=0.05;
A1=[cosh(j*b*l)     Z0*sinh(j*b*l);
    sinh(j*b*l)/Z0 cosh(j*b*l)]; % Eq.(E8.5.6)
Z3=[s*L+1/s/C 1/s/C; 1/s/C s*L+1/s/C];
A3=port_conversion(Z3,'???')
A=A1?A2?A3; % Overall A-matrix
A=subs(A,{rbe,rce,Cbe,Cbc,gm,L,C},...
   {520,8e4,10e-12,1e-12,0.192,1e-9,10e-12});
RL=0; % With port 2 short-circuited
Ais=port_property(A,RL,'???') % Current gain
f=logspace(4,9,1000); % Frequency range to plot the current gain on
s=j*2*pi*f;
Ai=eval(Ais); % Current gain evaluated for the frequency range
Rs=[200 300 500 1e3 10e3]; % Several values of feedback resistance R
for m=1:numel(Rs)
  Aim=subs(Ai,R,Rs(m)); % Current gain with R specified
  AimdB=20*log10(abs(Aim));
  semilogx(f,AimdB), hold on
end
```

8.3 Transistor Model

Consider the transistor model shown in Figure P8.3.

(a) Find the current gain $A_i(s) = -I_2(s)/I_1(s)$ (with port 2 short-circuited) by using the node analysis to find $-I_2(s) = V_3(s)/(r_c + sL_c)$ with $I_1(s) = 1$. You can complete and run the following MATLAB script "elec08p03a.m."

```
%elec08p03a.m
clear
syms s rb re rbe ro rc gm Lb Lc Le Cbe Cbc
% Using the node analysis
Y=[1/(s*Lb+rb)    -1/(s*??+rb)                 0           0;
  -1/(s*Lb+??) 1/(s*??+rb)+s*Cbe+1/rbe+s*C?? -s*Cbc  -s*Cbe-1/rbe;
   0          ??-s*Cbc     s*Cbc+1/??+1/(??+s*Lc)     -??-1/ro;
   0      -??-s*Cbe-1/rbe -1/ro  gm+s*Cbe+1/rbe+1/ro+1/(re+s*??)];
V=Y\[1; 0; 0; 0];
Ais=V(3)/(rc+s*Lc);
```

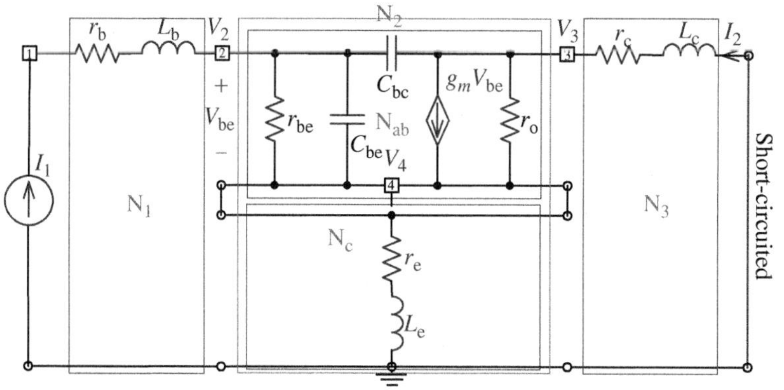

Figure P8.3 Transistor model.

```
%elec08p03b.m
% Combination approach of several 2-port subnetworks
Ya=[1/rbe+s*(Cbe+Cbc) -s*Cbc; -s*Cbc s*Cbc+1/ro]; % Y-matrix of Na
Yb=[0 0; gm 0]; % Y-matrix of Nb
Yab=Ya+Yb;
Zab=port_conversion(Yab,'y2?');
Zc=[re+s*Le re+s*??; ??+s*Le re+s*Le];
Z2=Zab+Zc;
A1=[1 s*Lb+??; 0 1]; A3=[1 rc+s*??; 0 1];
A=A1*port_conversion(Z2,'???')*??; % Overall A-matrix
RL=0; % With port 2 short-circuited
Ais1=port_property(A,RL,'???')
simplify(Ais-Ais1)
```

(b) To take the approach of combining several two-port networks, note the following:
- The two-port subnetwork N_{ab} in the central part of Figure P8.3 is just like the two-port network of Figure P8.2.1(a) with r_o instead of r_{ce} so that its H-matrix can be obtained by substituting r_o for r_{ce} into Eq. (P8.2.1).
- The series-series connection of N_{ab} and N_c makes N_2.
- The cascade connection of N_1, N_2, and N_3 makes the whole network.
- Then, combine the appropriate parameters of the subnetworks in appropriate ways to find the overall two-port parameters and use them together with Table 8.2 or the MATLAB function '`port_property()`' to find the current gain $A_i(s) = -I_2(s)/I_1(s)$ (with port 2 short-circuited). To this end, you can complete and run the above MATLAB script "elec08p03b.m."

8.4 Two-Port Networks in Various Type of Connection

(a) Regarding the two-port network of Figure P8.4(a) as appropriate connections of subnetworks N_1, N_2, and N_3, find the overall a-parameters, the input impedance Z_{in} with R_L taken into consideration, the output impedance Z_{out} with R_1 taken into consideration, and the overall voltage gain $A_v = V_o/V_s$.

```
%elec08p04a.m
R1=1e3; R2=2e3; RI=1e6; Ro=100; mu=1e5; RL=1e4;
A1= [1 R1; 0 1]; Y2=[1 -1; -1 1]/R2;
G3=[1/RI 0; -mu Ro];
Y3= port_conversion(G3,'???');
A23=port_conversion(Y2+Y3,'???'); % y-parameter of Parallel
combination
A123= A1*A23 % Overall a-parameter
Zi= port_property(A23,RL,'???') % Input impedance
Zo= port_property(A23,R1,'???') % Output impedance
Av= port_property(A123,RL,'???') % Overall voltage gain
```

(a)

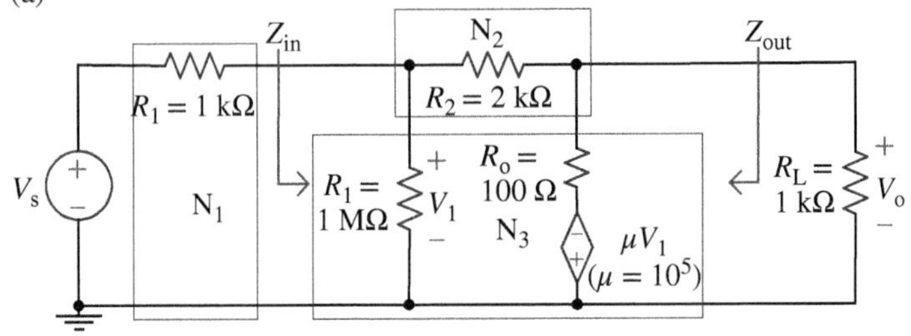

(b)

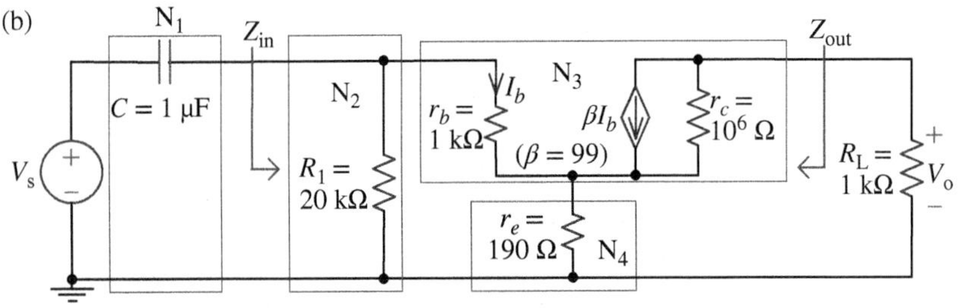

Figure P8.4

(b) Regarding the two-port network of Figure P8.4(b) as appropriate connections of subnetworks N_1, N_2, N_3, and N_4, find the input impedance Z_{in} and the overall voltage gain $A_v = V_o/V_s$.

```
%elec08p04b.m
syms s;
C=1e-6; R1= 2e4; rb=1000; re=190; b=99; rc=1e6; RL=1000;
Y1=C*[s -s; -s s]; A1=port_conversion(Y1,'y2a');
Z2=R1*[1 1; 1 1]; H3=[rb ?; b 1/??]; Z4= re*[1 1; 1 1];
Z3= port_conversion(H3,'???'); % z-parameter of N3
A234= port_conversion(Z2,'???')*port_conversion(Z3?Z4,'z2a');
A1234= A1*A234 % Overall a-parameter
Zi= port_property(A234,RL,'???') % Input impedance
```

8.5 Two-Port Networks in Cascade Connection and Impedance Matching Design

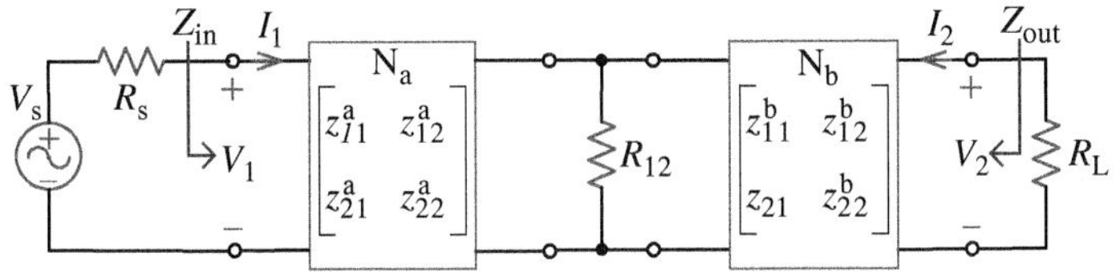

Figure P8.5 Two-port networks in cascade interconnection.

Consider the circuit of Figure P8.5 in which R_s=75 Ω, R_{12}=2 kΩ, R_L=16 Ω, and the z-parameter matrices of the two 2-port subnetworks are

$$\begin{bmatrix} z_{11}^a & z_{12}^a \\ z_{21}^a & z_{22}^a \end{bmatrix} = \begin{bmatrix} 350 & 2.667 \\ -10^6 & 6667 \end{bmatrix} \text{ and } \begin{bmatrix} z_{11}^b & z_{12}^b \\ z_{21}^b & z_{22}^b \end{bmatrix} = \begin{bmatrix} 1.0262\times10^6 & 6791 \\ 1.0258\times10^6 & 6794 \end{bmatrix} \quad \text{(P8.5.1)}$$

respectively.

```
%elec08p05a.m
Rs=75; R12=2000; RL=16;
Za=[350 2.667; -1e6 6667]; Zc=[R12 R12; R12 R12];
Zb=[1.0262e6 6791; 1.0258e6 6794]; Yd= [1/Rs -1/Rs;-1/?? 1/Rs];
disp('(a)')
Ad=port_conversion(Yd,'???'); %A1=[1 Rs; 0 1]
Aa=port_conversion(Za,'z2a'); % z-to-a parameter conversion
Ac=port_conversion(Zc,'???');
Ab=port_conversion(Zb,'z2a');
A = Aa*Ac*Ab; % The a-parameter of a, c, and b
Zin=port_property(A,RL,'Zia') %Input impedance based on a-parameter
Zout=port_property(A,Rs,'???'); %Output impedance based on a-parameter
Zout1=(A(1,2)+A(2,2)*Rs)/(A(1,1)+A(2,1)*Rs); % Table 3.4
[Zout Zout1] % Output impedance
Av= port_property(A,RL,'???'); % Voltage gain V2/V1
A1 = Ad*A; % Overall a-parameter matrix
Av1= port_property(A1,RL,'???'); Av1_1= Zin/(Rs+Zin)*Av; % V2/Vs
[Av1 Av1_1] % Overall voltage gain with Rs and RL considered
```

(a) Find the overall a-parameters for the cascade interconnection of N_a, R_{12}, and N_b. Then, find the input impedance Z_{in} with R_L taken into consideration, the output impedance Z_{out} with R_s taken into consideration, and the overall voltage gain $A_v = V_2/V_s$.

(b) Does the output impedance Z_{out} match the load impedance R_L? If not, find the new value of the resistance R_{12} such that Z_{out} matches R_L.

```
%elec08p05b.m
disp('(b)')
syms R12
% Overall a-parameter with R12 as a symbolic variable
A=Aa*port_conversion([R12 R12; R12 R12],'???')*Ab;
Zout=port_property(A,Rs,'???'); % Output impedance based on a-parameters
pretty(simplify(Zout))
R12=eval(solve(Zout-RL,'R12')) % Solve Zout(R12)=RL for R12
A=Aa*port_conversion([R12 R12; R12 R12],'???')*Ab; %Overall a (numeric)
Zout=port_property(A,Rs,'???') % To check the output impedance
```

8.6 Conversion Between S-parameters and Z-parameters

(a) According to [D-1], the conversion formulas between the s-parameters and z-parameters are

$$S = F(Z - G^*)(Z + G)^{-1}F^{-1} \text{ and } Z = F^{-1}(I - S)^{-1}(SG + G^*)F \quad \text{(P8.6.1)}$$

where

$$F = \begin{bmatrix} 1/2\sqrt{Z_{01}} & 0 \\ 0 & 1/2\sqrt{Z_{02}} \end{bmatrix} \text{ and } G = \begin{bmatrix} Z_{01} & 0 \\ 0 & Z_{02} \end{bmatrix} \quad \text{(P8.6.2)}$$

Complete the following two MATLAB functions 'z2s_my()' and 's2z_my()' to perform the z-to-s and s-to-z parameter conversions, respectively, and use them for an arbitrary z- or s-parameter matrix to compare with the MATLAB built-in functions 'z2s()' and 's2z()'.

```
function S=z2s_my(Z,Z01,Z02)
% Z-to-S parameter conversion
if nargin<3, Z02=Z01; end
Z0s=[Z01 Z02]; F=diag(2*sqrt(Z0s)); G=diag(Z0s);
S=F*(Z-G')*(Z+G)^-1*?^-1; % [D-1]
```

```
function Z=s2z_my(S,Z01,Z02)
% S-to-Z parameter conversion
if nargin<3, Z02=Z01; end
Z0s=[Z01 Z02]; F=diag(2*sqrt(Z0s)); G=diag(Z0s);
Z=?\(eye(2)-S)^-1*(S*G+G')*F; % [D-1]
```

(b) According to [F-2], the conversion formulas between the s-parameters and h-parameters are

$$S = \frac{1}{(h_{11} + Z_{01})(h_{22}Z_{02} + 1) - h_{12}h_{21}Z_{02}} \begin{bmatrix} (h_{11} - Z_{01}^*)(h_{22}Z_{02} + 1) - h_{12}h_{21}Z_{02} & 2h_{12}\sqrt{Z_{01}Z_{02}} \\ -2h_{21}\sqrt{Z_{01}Z_{02}} & (h_{11} + Z_{01})(1 - h_{22}Z_{02}^*) + h_{12}h_{21}Z_{02}^* \end{bmatrix} \tag{P8.6.3}$$

$$H = \frac{1}{(1 - s_{11})(Z_{02}^* + s_{22}Z_{02}) + s_{12}s_{21}Z_{02}} \begin{bmatrix} (Z_{01}^* + s_{11}Z_{01})(Z_{02}^* + s_{22}Z_{02}) - s_{12}s_{21}Z_{01}Z_{02} & 2s_{12}\sqrt{Z_{01}Z_{02}} \\ -2s_{21}\sqrt{Z_{01}Z_{02}} & (1 - s_{11})(1 - s_{22}) - s_{12}s_{21} \end{bmatrix} \tag{P8.6.4}$$

Complete the following two MATLAB functions '`h2s_my()`' and '`s2h_my()`' to perform the h-to-s and s-to-h parameter conversions, respectively, and use them for an arbitrary h- or s-parameter matrix to compare with the MATLAB built-in functions '`h2s()`' and '`s2h()`'.

```
function S=h2s_my(H,Z01,Z02)
% H-to-S parameter conversion ([D-1])
if nargin<3, Z02=Z01; end
D=(H(1,1)+Z01)*(H(2,2)*Z02+1)-H(1,2)*H(2,1)*Z02; sZ12=sqrt(Z01*Z02);
S=[(H(1,1)-Z01')*(H(2,2)*Z02+1)-H(1,2)*H(2,1)*Z02 2*H(1,2)*sZ12;
  -2*H(2,1)*sZ12 (H(1,1)+Z01)*(1-H(2,2)*Z02')+H(1,2)*H(2,1)*Z02']/?;
```

```
function H=s2h_my(S,Z01,Z02)
% S-to-H parameter conversion ([D-1])
if nargin<3, Z02=Z01; end
Z12=Z01*Z02; sZ12=sqrt(Z12); s12=S(1,2); s21=S(2,1);
D=(1-S(1,1))*(Z02'+S(2,2)*Z02)+s12*s21*Z02;
H=[(Z01'+S(1,1)*Z01)*(Z02'+S(2,2)*Z02)-s12*s21*Z12 2*s12*sZ12;
   -2*s21*sZ12 ((1-S(1,1))*(1-S(2,2))-s12*s21)]/?;
```

(cf.) See [D-1] (Appendix D) or [F-2] for conversion formulas for several two-port parameters.

8.7 Stabilization of a Two-Port Network
Consider a two-port network whose S-matrix is

$$\begin{bmatrix} s_{11} & s_{12} \\ s_{21} & s_{22} \end{bmatrix} = \begin{bmatrix} 0.56\angle -78^{\circ} & 8.64\angle 122^{\circ} \\ 0.05\angle 33^{\circ} & 0.66\angle -42^{\circ} \end{bmatrix} \tag{P8.7.1}$$

(a) Draw the input and output stable regions by completing and running the following MATLAB script "elec08p07a.m":

```
%elec08p07a.m
S=[0.56*exp(j*deg2rad(-78))  8.64*exp(j*deg2rad(122))
   0.05*exp(j*deg2rad(33))    0.66*exp(j*deg2rad(-??))];
Kg=1; [D,k,mui,muo,ci,ri,co,ro]=stability_S(S,Kg);
```

(b) To estimate the limit of source resistance R_s such that the input stability is kept, run the following MATLAB script "elec08p07b.m" and find the maximum or minimum value of R_s assuring the input stability (see Figure P8.7(a)):

```
%elec08p07b.m
ths=[0:360]*(pi/180);
Rs=50;
while Rs>0
 smithchart, hold on; [c,r]=Rcircle(??,Z0);
 plot(ci+ri*exp(j*ths),'b'); hold on, plot(c+r*exp(j*ths),'r:')
 Rs_str=num2str(Rs,'%8.2f [Ohm]'); text(0,0.9,Rs_str)
 Rs=input(['New value of Rs?(Rs=' Rs_str ')']);
  if Rs>0, plot(c+r*exp(j*ths),'w'); end
end
```

(c) To estimate the limit of source resistance G_s such that the input stability is kept, complete and run the following MATLAB script "elec08p07c.m" and find the maximum or minimum value of G_s assuring the input stability (see Figure P8.7(a)):

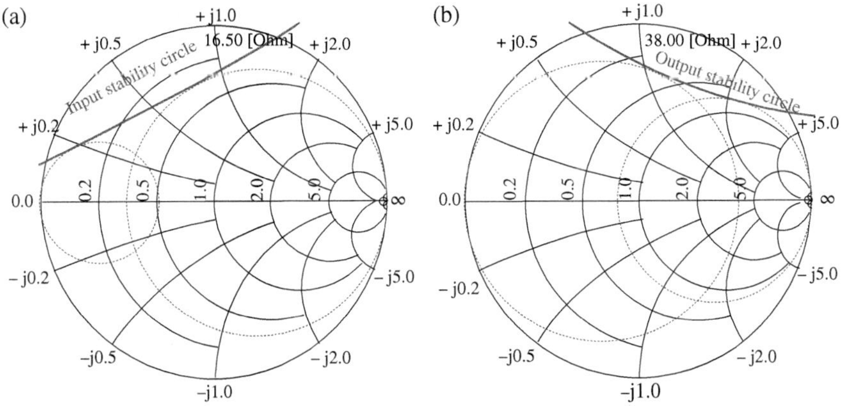

Limits of R_s and G_s considering the input stability

Limits of R_L and G_L considering the input stability

Figure P8.7 Finding the limits of R_s, G_s, R_L, and G_L in terms of the input/output stability for Problem 8.7.

```
%elec08p07c.m
Gs=1/50;
while Gs>0
 smithchart, hold on; [c,r]=Gcircle(??,Z0);
 plot(c?+r?*exp(j*ths),'b'); hold on, plot(c+r*exp(j*ths),'r:')
 Gs_str=num2str(Gs*1e3,'%8.2f[mS]'); text(0,0.9,Gs_str)
 Gs=input(['New value of Gs?(Gs=' Gs_str ')']);
 if Gs>0, plot(c+r*exp(j*ths),'w'); end
end
```

(d) To estimate the limit of load resistance R_L such that the output stability is kept, complete and run the following MATLAB script "elec08p07d.m" and find the maximum or minimum value of R_L assuring the output stability (see Figure P8.7(b)):

```
%elec08p07d.m
RL=50;
while RL>0
 smithchart, hold on; [c,r]=?circle(RL,Z0);
 plot(co+ro*exp(j*ths),'b'); hold on, plot(c+r*exp(j*ths),'r:')
 RL_str=num2str(RL,'%8.2f[Ohm]'); text(0,0.9,RL_str)
 RL=input(['New value of RL?(RL=' RL_str ')']);
 if RL>0, plot(c+r*exp(j*ths),'w'); end
end
```

(e) To estimate the limit of load resistance G_L such that the output stability is kept, complete and run the following MATLAB script "elec08p07e.m" and find the maximum or minimum value of G_L assuring the output stability (see Figure P8.7(b)):

```
%elec08p07e.m
GL=1/50;
while GL>0
 smithchart, hold on; [c,r]=?circle(GL,Z0);
 plot(c?+r?*exp(j*ths),'b'); hold on, plot(c+r*exp(j*ths),'r:')
 GL_str=num2str(GL*1e3,'%8.2f[mS]'); text(0,0.9,GL_str)
 GL=input(['New value of GL?(GL=' GL_str ')']);
  if GL>0, plot(c+r*exp(j*ths),'w'); end
end
```

8.8 Maximum Gain Design and Specified Gain Design

Consider a two-port network whose S-matrix (with reference impedance $Z_0 = 50\,\Omega$) is

$$\begin{bmatrix} s_{11} & s_{12} \\ s_{21} & s_{22} \end{bmatrix} = \begin{bmatrix} 0.614\angle -167.4^{\circ} & 0.046\angle 65^{\circ} \\ 2.187\angle 32.4^{\circ} & 0.716\angle -83^{\circ} \end{bmatrix} \tag{P8.8.1}$$

where the source and load impedances Z_s/Z_L are matched with $Z_0 = 50\,\Omega$ so that $\Gamma_s = 0$ and $\Gamma_L = 0$.

(a) Determine desired values of source/load impedances Z_s/Z_L such that the transducer gain can be maximized. Find the operating gain with those values of Z_s/Z_L.

(b) Referring to Figure P8.8, determine desired values of source/load impedances Z_s/Z_L such that the operating gain will be 10 dB and $|\Gamma_L| = 0.2$ by taking the following steps:

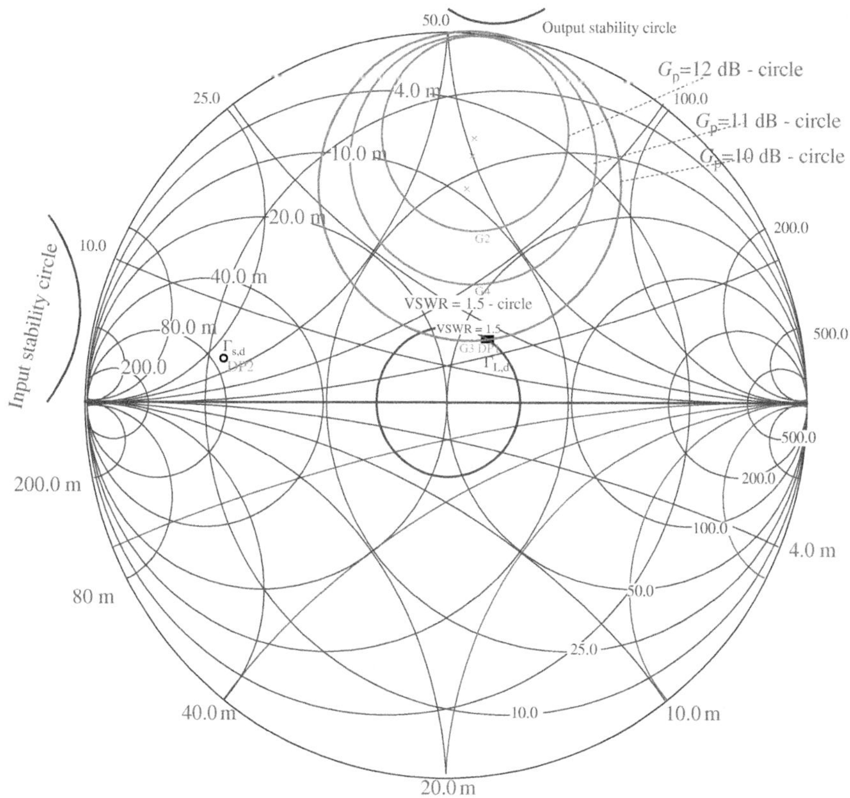

Figure P8.8 Using 'Smith 4.1/ to design the input/output impedances for Problem 8.8.

1) To make $G_p = 10$ dB and $|\Gamma_L| = 0.2$, locate the load reflection coefficient Γ_L at an intersection of the G_p=10 dB-circle and the VSWR = 1.5-circle (corresponding to $|\Gamma_L| = 0.2$) and make the Γ-to-Z conversion to determine $Z_{L,d}$.
2) Use Eq. (8.3.33) or MATLAB function '`Gami_L()`' to get the input reflection coefficient Γ_{in}. Then, set its conjugate as a desired value of the source reflection coefficient Γ_s and make the Γ-to-Z conversion to determine $Z_{s,d}$.

Appendix A

Laplace Transform

A.1 Definition of Laplace Transform

The (unilateral or one-sided) Laplace transform is defined for a function $x(t)$ of a real variable t (often meaning the time) as

$$X(s) = \mathcal{L}\{x(t)\} = \int_{0-}^{\infty} x(t)\, e^{-st} dt \tag{A.1}$$

where s is a complex variable, the lower limit, t_{-}, of integration interval is the instant just before $t = 0$, and $x(t)$ is often assumed to be causal in the sense that it is zero for all $t < 0$.

A.2 Inverse Laplace Transform

Suppose an s-function $X(s)$ is given in the form of a rational function, i.e. the ratio of an Mth-degree polynomial $Q(s)$ to an Nth-degree polynomial $P(s)$ in s and it is expanded into the partial fraction form as

$$X(s) = \frac{Q(s)}{P(s)} = \frac{b_M s^M + \cdots + b_1 s + b_0}{a_N s^N + \cdots + a_1 s + a_0} \; (M \le N) \tag{A.2}$$

$$= \left(\sum_{n=1}^{N-L} \frac{r_n}{s - p_n}\right) + \frac{r_{N-L+1}}{s-p} + \cdots + \frac{r_N}{(s-p)^L} + K \tag{A.3}$$

where

$$r_n = (s - p_n)\frac{Q(s)}{P(s)}\bigg|_{s=p_n}, \quad n = 1, 2, \ldots, N - L \tag{A.4a}$$

Electronic Circuits with MATLAB®, PSpice®, and Smith Chart, First Edition. Won Y. Yang, Jaekwon Kim, Kyung W. Park, Donghyun Baek, Sungjoon Lim, Jingon Joung, Suhyun Park, Han L. Lee, Woo June Choi, and Taeho Im.

Companion website: www.wiley.com/go/yang/electroniccircuits

$$r_{N-l} = \frac{1}{l!}\frac{d^l}{ds^l}\left\{(s-p)^L\frac{Q(s)}{P(s)}\right\}\bigg|_{s=p}, \quad l = 0,1,\ldots,L-1 \tag{A.4b}$$

Then the inverse Laplace transform of $X(s)$ can be obtained as

$$x(t) = \left\{\sum\nolimits_{n=1}^{N-L} r_n e^{p_n t} + r_{N-L+1}e^{pt} + r_{N-L+2}\,t\,e^{pt} + \cdots + \frac{r_N}{(L-1)!}t^{L-1}e^{pt}\right\}u_s(t) + K\delta(t) \tag{A.5}$$

Example A.1 Inverse Laplace Transform

Let us find the inverse Laplace transform of the following s-functions.

a)

$$X(s) = \frac{3s^2+11s+11}{s^3+4s^2+5s+2} = \frac{3s^2+11s+11}{(s+1)^2(s+2)} = \frac{r_1}{s+2} + \frac{r_2}{s+1} + \frac{r_3}{(s+1)^2} \tag{1}$$
(with a double pole at $s = -1$)

We can use Formula (A.4a) to find the simple-pole coefficient r_1 as

$$r_1 = (s+2)X(s)|_{s=-2} \overset{(1)}{=} (s+2)\frac{3s^2+11s+11}{(s+1)^2(s+2)}\bigg|_{s=-2} = 1 \tag{2}$$

and can also use Formula (A.4b) to find the multiple-pole coefficient r_2 and r_3 as

$$r_3 = (s+1)^2X(s)|_{s=-1} \overset{(1)}{=} \frac{3s^2+11s+11}{s+2}\bigg|_{s=-1} = 3 \tag{3}$$

$$r_2 = \frac{d}{ds}(s+1)^2X(s)\bigg|_{s=-1} = \frac{d}{ds}\frac{3s^2+11s+11}{s+2}\bigg|_{s=-1}$$

$$\overset{(F.31)}{=} \frac{(6s+11)(s+2)-(3s^2+11s+11)}{(s+2)^2}\bigg|_{s=-1} = 2. \tag{4}$$

Thus, the inverse Laplace transform can be written as

$$x(t) \underset{(A.5)}{\overset{(1)}{=}} \left(r_1 e^{-2t} + r_2 e^{-t} + r_3 t e^{-t}\right)u_s(t) = \left(e^{-2t} + 2e^{-t} + 3te^{-t}\right)u_s(t) \tag{5}$$

```
>>Ns=[3 11 11]; Ds=[1 4 5 2];
  [r,p,k]=residue(Ns,Ds), [A,B]=residue(r,p,k)
```

b) $X(s) = \frac{s}{s^2 + 2s + 2} = \frac{s}{(s+1)^2 + 1^2}$ (with complex conjugate poles at $s = -1 \pm j$) (6)

We may use Formula (A.4a) to find the coefficients of the partial fraction expansion form

$$X(s) = \frac{s}{s^2 + 2s + 2} = \frac{r_1}{s + 1 - j} + \frac{r_2}{s + 1 + j} \tag{7}$$

as

$$r_1 \overset{(A.4a)}{=} (s + 1 - j) X(s)\big|_{s=-1+j} \overset{(6)}{=} \frac{s}{s + 1 + j}\bigg|_{s=-1+j} = 0.5(1 + j)$$
$$r_2 \overset{(A.4a)}{=} (s + 1 + j) X(s)\big|_{s=-1-j} \overset{(6)}{=} \frac{s}{s + 1 - j}\bigg|_{s=-1-j} = 0.5(1 - j) \tag{8}$$

Thus, we can write the inverse Laplace transform as

$$x(t) \underset{(A.5)}{\overset{(7)}{=}} \left(0.5(1 + j) e^{(-1+j)t} + 0.5(1 - j) e^{(-1-j)t}\right) u_s(t)$$
$$\overset{(F.20)}{=} e^{-t}(\cos t - \sin t) u_s(t) \tag{9}$$

In the case of complex conjugate poles, it may be simpler to equate $X(s)$ with the following form

$$X(s) = \frac{s}{s^2 + 2s + 2} = \frac{A(s+1)}{(s+1)^2 + 1^2} + \frac{B \times 1}{(s+1)^2 + 1^2} \overset{\text{common denominator}}{=} \frac{As + (A + B)}{(s+1)^2 + 1^2} \tag{10}$$

and get the coefficients as $A = 1$ and $B = -1$. Then we can directly find each term of the inverse Laplace transform from the Laplace transform Table A.1 and write the inverse Laplace transform as

$$X(s) \overset{(10)}{=} \frac{s + 1}{(s+1)^2 + 1^2} + \frac{(-1) \times 1}{(s+1)^2 + 1^2}$$
$$\overset{\text{Table A.1(9),(10)}}{\longrightarrow} x(t) = (e^{-t}\cos t - e^{-t}\sin t) u_s(t) \tag{11}$$

c) Use of MATLAB for Partial Fraction Expansion and Inverse Laplace Transform

We can use the MATLAB command "`residue()`" to get the partial fraction expansion and "`ilaplace()`" to obtain the whole inverse Laplace transform. It should, however, be noted that "`ilaplace()`" might not work properly for high-degree rational functions.

Table A.1 Laplace transforms of basic functions.

x(t)	X(s)	x(t)	X(s)	x(t)	X(s)
(1) $\delta(t)$	1	(5) $e^{-at}u_s(t)$	$\frac{1}{s+a}$	(9) $e^{at}\sin\omega t\ u_s(t)$	$\frac{\omega}{(s+a)^2+\omega^2}$
(2) $\delta(t-t_1)$	$e^{-t_1 s}$	(6) $t^m e^{-at}u_s(t)$	$\frac{m!}{(s+a)^{m+1}}$	(10) $e^{-at}\cos\omega t\ u_s(t)$	$\frac{s+a}{(s+a)^2+\omega^2}$
(3) $u_s(t)$	$\frac{1}{s}$	(7) $\sin\omega t\ u_s(t)$	$\frac{\omega}{s^2+\omega^2}$		
(4) $t^m u_s(t)$	$\frac{m!}{s^{m+1}}$	(8) $\cos\omega t\ u_s(t)$	$\frac{s}{s^2+\omega^2}$		

Table A.2 Properties of Laplace transform.

(0) Definition	$X(s)=\mathcal{L}\{x(t)\} = \int_0^\infty x(t)\,e^{-st}\,dt;\ x(t)\leftrightarrow X(s)$
(1) Linearity	$\alpha\,x(t)+\beta\,y(t)\leftrightarrow\ \alpha\,X(s)+\beta\,Y(s)$
(2) Time shifting (Real translation)	$x(t-t_1)u_s(t-t_1),t_1>0\leftrightarrow e^{-st_1}\left\{X(s)+\int_{-t_1}^{0}x(\tau)e^{-s\tau}d\tau\right\}$ $=e^{-st_1}X(s)$ for $x(t)=0\ \forall t<0$
(3) Frequency shifting (Complex translation)	$e^{s_1 t}x(t)\ \leftrightarrow\ X(s-s_1)$
(4) Real convolution	$g(t)*x(t)\ \leftrightarrow\ G(s)\,X(s)$
(5) Time derivative (Differentiation property)	$x'(t)=\frac{d}{dt}x(t)\leftrightarrow sX(s)-x(0)$
(6) Time integral (Integration property)	$\int_{-\infty}^{t}x(\tau)\,d\tau\ \leftrightarrow\ \frac{1}{s}X(s)+\frac{1}{s}\int_{-\infty}^{0}x(\tau)\,d\tau$
(7) Complex derivative	$t\,x(t)\ \leftrightarrow\ -\frac{d}{dsX(s)}$
(8) Complex convolution	$x(t)y(t)\ \leftrightarrow\ \frac{1}{2\pi j}\int_{\sigma_0-\infty}^{\sigma_0+\infty}X(v)Y(s-v)dv$
(9) Initial value theorem	$x(0)\ =\ \lim_{s\to\infty}sX(s)$
(10) Final value theorem	$x(\infty)\ =\ \lim_{s\to 0}sX(s)$

```
>>Ns=[3 11 11]; Ds=[1 4 5 2]; [r,p,k]=residue(Ns,Ds); [r p],k % (1)

 r = 1.0000   p = -2.0000    % (2) 1/(s-(-2))
     2.0000       -1.0000    % (4) 2/(s-(-1))
     3.0000       -1.0000    % (3) 3/(s-(-1))^2
 k = []
```

```
>>syms s, x= ilaplace((3*s^2+11*s+11)/(s^3+4*s^2+5*s+2))
   x = exp(-2*t)+3*t*exp(-t)+2*exp(-t)   % (5)
>>Ns=[1 0]; Ds=[1 2 2]; [r,p,k]=residue(Ns,Ds); [r p],k % (6)
   r = 0.5000 + 0.5000i  p = -1.0000 + 1.0000i % (8) (0.5+0.5i)/(s+1-i)
       0.5000 - 0.5000i  p = -1.0000-1.0000i % (8) (0.5-0.5i)/(s+1+i)
   k = []
>>syms s, x= ilaplace(s/(s^2+2*s+2))
   x = exp(-t)*cos(t)-exp(-t)*sin(t)   % (9) or (11)
>>ilaplace(s/(s^4+10*s^3+9*s^2+6*s+1)) %?? Ns=[1 0]; Ds=[1 10 9 6 1];
```

A.3 Laplace Transform of Electric Circuits

In order to transform a circuit into the *s*-domain, we need the *s-domain* (*transformed*) *equivalent* circuits for a resistor, an inductor, and a capacitor, as shown in Figure A.1. These circuits are obtained by taking the Laplace transform of their voltage-current relationships as follows:

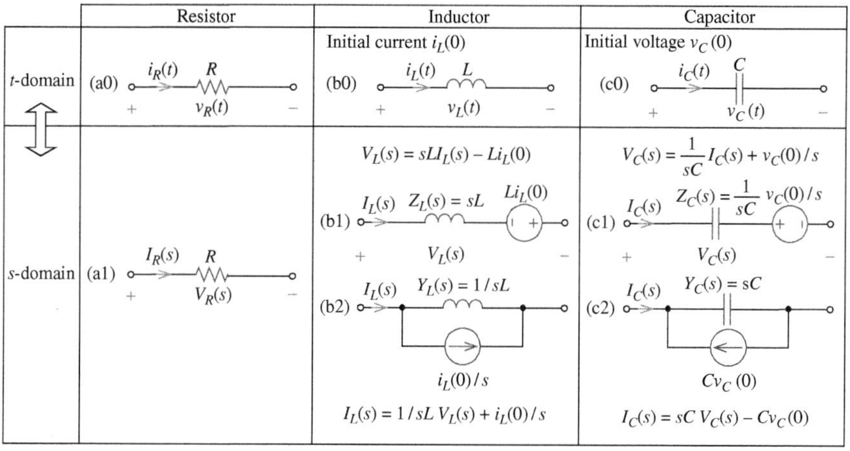

Figure A.1 The transformed (*s*-domain) equivalent circuits of *R*, *L*, and *C*.

Appendix B

Matrix Operations with MATLAB

B.1 Addition and Subtraction

$$A+B=\begin{bmatrix} a_{11} & a_{12} & \cdot & a_{1N} \\ a_{21} & a_{22} & \cdot & a_{2N} \\ \cdot & \cdot & \cdot & \cdot \\ a_{M1} & a_{M2} & \cdot & a_{MN} \end{bmatrix}+\begin{bmatrix} b_{11} & b_{12} & \cdot & b_{1N} \\ b_{21} & b_{22} & \cdot & b_{2N} \\ \cdot & \cdot & \cdot & \cdot \\ b_{M1} & b_{M2} & \cdot & b_{MN} \end{bmatrix}=\begin{bmatrix} c_{11} & c_{12} & \cdot & c_{1N} \\ c_{21} & c_{22} & \cdot & c_{2N} \\ \cdot & \cdot & \cdot & \cdot \\ c_{M1} & c_{M2} & \cdot & c_{MN} \end{bmatrix}=C \tag{B.1a}$$

$$\text{with } a_{mn}+b_{mn}=c_{mn} \tag{B.1b}$$

B.2 Multiplication

$$AB=\begin{bmatrix} a_{11} & a_{12} & \cdot & a_{1K} \\ a_{21} & a_{22} & \cdot & a_{2K} \\ \cdot & \cdot & \cdot & \cdot \\ a_{M1} & a_{M2} & \cdot & a_{MK} \end{bmatrix}\begin{bmatrix} b_{11} & b_{12} & \cdot & b_{1N} \\ b_{21} & b_{22} & \cdot & b_{2N} \\ \cdot & \cdot & \cdot & \cdot \\ b_{K1} & b_{K2} & \cdot & b_{KN} \end{bmatrix}=\begin{bmatrix} c_{11} & c_{12} & \cdot & c_{1N} \\ c_{21} & c_{22} & \cdot & c_{2N} \\ \cdot & \cdot & \cdot & \cdot \\ c_{M1} & c_{M2} & \cdot & c_{MN} \end{bmatrix}=C \tag{B.2a}$$

$$\text{with } c_{mn}=\sum_{k=1}^{K} a_{mk}b_{kn} \tag{B.2b}$$

(cf.) For this multiplication to be done, the number of columns of A must equal the number of rows of B.

(cf.) Note that the commutative law does not hold for the matrix multiplication, i.e. $AB \neq BA$.

Electronic Circuits with MATLAB®, PSpice®, and Smith Chart, First Edition. Won Y. Yang, Jaekwon Kim, Kyung W. Park, Donghyun Baek, Sungjoon Lim, Jingon Joung, Suhyun Park, Han L. Lee, Woo June Choi, and Taeho Im.

Companion website: www.wiley.com/go/yang/electroniccircuits

B.3 Determinant

The determinant of a $K \times K$ (square) matrix $A = [a_{mn}]$ is defined by

$$\det(A) = |A| = \sum_{k=0}^{K} a_{kn}(-1)^{k+n} M_{kn} \text{ or } \sum_{k=0}^{K} a_{mk}(-1)^{m+k} M_{mk} \quad \text{(B.3)}$$
$$\text{for any fixed } 1 \le n \le K \text{ or } 1 \le m \le K$$

where the minor M_{kn} is the determinant of the $(K-1) \times (K-1)$ (minor) matrix formed by removing the kth row and the nth column from A and $c_{kn} = (-1)^{k+n} M_{kn}$ is called the cofactor of a_{kn}.

Especially, the determinants of a 2×2 matrix $A_{2\times2}$ and a 3×3 matrix $A_{3\times3}$ are

$$\det(A_{2\times2}) = \begin{vmatrix} a_{11} & a_{12} \\ a_{21} & a_{22} \end{vmatrix} = \sum_{k=0}^{2} a_{kn}(-1)^{k+n} M_{kn} = a_{11}a_{22} - a_{12}a_{21} \quad \text{(B.4a)}$$

$$\det(A_{3\times3}) = \begin{vmatrix} a_{11} & a_{12} & a_{13} \\ a_{21} & a_{22} & a_{23} \\ a_{31} & a_{32} & a_{33} \end{vmatrix} = a_{11}\begin{vmatrix} a_{22} & a_{23} \\ a_{32} & a_{33} \end{vmatrix} - a_{12}\begin{vmatrix} a_{21} & a_{23} \\ a_{31} & a_{33} \end{vmatrix} + a_{13}\begin{vmatrix} a_{21} & a_{22} \\ a_{31} & a_{32} \end{vmatrix}$$
$$= a_{11}(a_{22}a_{33} - a_{23}a_{32}) - a_{12}(a_{21}a_{33} - a_{23}a_{31}) + a_{13}(a_{21}a_{32} - a_{22}a_{31}) \quad \text{(B.4b)}$$

Note the following properties.

- If the determinant of a matrix is zero, the matrix is singular.
- The determinant of a matrix equals the product of the eigenvalues of a matrix.
- If A is upper/lower triangular having only zeros below/above the diagonal in each column, its determinant is the product of the diagonal elements.
- $\det(A^T) = \det(A)$; $\det(AB) = \det(A)\det(B)$; $\det(A^{-1}) = 1/\det(A)$

B.4 Inverse Matrix

The inverse matrix of a $K \times K$ (square) matrix $A = [a_{mn}]$ is denoted by A^{-1} and defined to be a matrix which is pre-multiplied/post-multiplied by A to form an identity matrix i.e. satisfies

$$AA^{-1} = A^{-1}A = I \quad \text{(B.5)}$$

An element of the inverse matrix $A^{-1} = [\alpha_{mn}]$ can be computed as

$$\alpha_{mn} = \frac{1}{\det(A)} c_{mn} = \frac{1}{|A|}(-1)^{m+n} M_{nm} \quad \text{(B.6)}$$

where M_{nm} is the minor of a_{nm} and $c_{mn} = (-1)^{m+n} M_{nm}$ is the cofactor of a_{mn}.

Note that a square matrix A is invertible/nonsingular if and only if

- no eigenvalue of A is zero or equivalently,
- the rows (and the columns) of A are linearly independent or equivalently, and
- the determinant of A is nonzero.

Example B.1 Matrix Inversion
Let us find the inverse of a 3 × 3 matrix

$$A = \begin{bmatrix} 3 & -2 & -1 \\ -2 & 5 & -3 \\ -1 & -3 & -8 \end{bmatrix}, \tag{1}$$

where the cofactors are

$$c_{11} = (-1)^{1+1}M_{11} = \begin{vmatrix} 5 & -3 \\ -3 & 8 \end{vmatrix} = 5 \times 8 - (-3)(-3) = 31$$

$$c_{12} = (-1)^{1+2}M_{21} = -\begin{vmatrix} -2 & -1 \\ -3 & 8 \end{vmatrix} = -\{-2 \times 8 - (-1)(-3)\} = 19$$

$$c_{13} = (-1)^{1+3}M_{31} = \begin{vmatrix} -2 & -1 \\ 5 & -3 \end{vmatrix} = (-2) \times (-3) - (-1) \times 5 = 11$$

$$c_{21} = (-1)^{2+1}M_{12} = -\begin{vmatrix} -2 & -3 \\ -1 & 8 \end{vmatrix} = -(-2 \times 8 - (-3)(-1)) = 19$$

$$c_{22} = (-1)^{2+2}M_{22} = \begin{vmatrix} 3 & -1 \\ -1 & 8 \end{vmatrix} = 3 \times 8 - (-1)(-1) = 23$$

$$c_{23} = (-1)^{2+3}M_{32} = -\begin{vmatrix} 3 & -1 \\ -2 & -3 \end{vmatrix} = -(3 \times (-3) - (-1)(-2)) = 11$$

$$c_{31} = (-1)^{3+1}M_{13} = \begin{vmatrix} -2 & 5 \\ -1 & -3 \end{vmatrix} = -2 \times (-3) - 5(-1) = 11$$

$$c_{32} = (-1)^{3+2}M_{23} = -\begin{vmatrix} 3 & -2 \\ -1 & -3 \end{vmatrix} = -\{3 \times (-3) - (-2)(-1)\} = 11$$

$$c_{33} = (-1)^{3+3}M_{33} = \begin{vmatrix} 3 & -2 \\ -2 & 5 \end{vmatrix} = 3 \times 5 - (-2)(-2) = 11$$

We can use Eqs. (B.3) and (B.6) to get the determinant and the inverse matrix as

$$|A| = a_{11}c_{11} + a_{12}c_{12} + a_{13}c_{13} = 3\times 31 - 2\times 19 - 1\times 11 = 93 - 38 - 11 = 44 \quad (2)$$

$$A^{-1} = \frac{1}{|A|}\begin{bmatrix} c_{11} & c_{21} & c_{31} \\ c_{12} & c_{22} & c_{32} \\ c_{13} & c_{23} & c_{33} \end{bmatrix} = \frac{1}{44}\begin{bmatrix} 31 & 19 & 11 \\ 19 & 23 & 11 \\ 11 & 11 & 11 \end{bmatrix} \quad (3)$$

B.5 Solution of a Set of Linear Equations Using Inverse Matrix

Let us consider the following set of linear equations in three unknown variables x_1, x_2, and x_3:

$$\begin{aligned} 3x_1 - 2x_2 - x_3 &= -4 \\ -2x_1 + 5x_2 - 3x_3 &= -1 \\ -x_1 - 3x_2 + 8x_3 &= 17 \end{aligned}$$

This can be formulated in the matrix-vector form

$$\begin{bmatrix} 3 & -2 & -1 \\ -2 & 5 & -3 \\ -1 & -3 & 8 \end{bmatrix}\begin{bmatrix} x_1 \\ x_2 \\ x_3 \end{bmatrix} = \begin{bmatrix} -4 \\ -1 \\ 17 \end{bmatrix}; \quad Ax = b$$

so that it can be solved as

$$x = A^{-1}b = \frac{1}{44}\begin{bmatrix} 31 & 19 & 11 \\ 19 & 23 & 11 \\ 11 & 11 & 11 \end{bmatrix}\begin{bmatrix} -4 \\ -1 \\ 17 \end{bmatrix} = \frac{1}{44}\begin{bmatrix} 31\times(-4) + 19\times(-1) + 11\times 17 \\ 19\times(-4) + 23\times(-1) + 11\times 17 \\ 11\times(-4) + 11\times(-1) + 11\times 17 \end{bmatrix} = \begin{bmatrix} 1 \\ 2 \\ 3 \end{bmatrix}$$

B.6 Operations on Matrices and Vectors Using MATLAB

The following statements and their running results illustrate the powerful usage of MATLAB in dealing with matrices and vectors.

```
>>a= [-2 2 3] % a 1x3 matrix (row vector)
  a = -2  2  3
>>b= [-2; 2; 3] % 3x1 matrix (column vector)
  B =  -2
        2
        3
>>b= a.' % transpose

>>A= [1 -1 2; 0 1 0; -1 5 1] % a 3x3 matrix
```

$$A = \begin{bmatrix} 1 & -1 & 2 \\ 0 & 1 & 0 \\ -1 & 5 & 1 \end{bmatrix}$$

```
>>A(1,2) % will return -1
>>A(:,1) % will return the 1st column of the matrix A
 ans = -1
        0
       -1
>>A(:,2:3) % will return the 2nd and 3rd columns of the matrix A
  ans = -1  2
         1  0
         5  1
>>c= a*A % vector-matrix multiplication
 C= -5 19 -1
>>A*b
 ans =  2
        2
       15
>>A*a % not permissible for matrices with incompatible dimensions
 ??? Error using ==> mtimes
 Inner matrix dimensions must agree
>>a*c' % Inner product : multiply a with the conjugate transpose of c
  ans = 45
>>a.*c % (termwise) multiplication element by element
  ans = 10 38 -3
>>a./c % (termwise) division element by element
  ans = 0.4000  0.1053  -3.0000
>>det(A)  % determinant of matrix A
  ans = 3
>>inv(A)  % inverse of matrix A
  ans =  0.3333    3.6667  -0.6667
         0         1.0000   0
         0.3333   -1.3333   0.3333
>>[V,E]= eig(A) % eigenvector and eigenvalue of matrix A
   V = 0.8165             0.8165              0.9759
       0                  0                   0.1952
       0 + 0.5774i        0 - 0.5774i         0.0976
   E = 1.0000 + 1.4142i   0                   0
       0                  1.0000 - 1.4142i    0
       0                  0                   1.0000
>>I= eye(3) % a 3x3 identity matrix
>>O= zeros(size(I)) % a zero matrix of the same size as I
>>A= sym('[a b c; d e f]') % a matrix consisting of (non-numeric) symbols
 A =    [ a, b, c]
        [ d, e, f]
>>B= [1 0; 0 1; 1 1]
 B =    1   0
        0   1
        1   1
>>A*B
 ans =  [a+c,  b+c]
        [d+f,  e+f]
```

Appendix C

Complex Number Operations with MATLAB

C.1 Addition

$$(a_1 + jb_1) + (a_2 + jb_2) = (a_1 + a_2) + j(b_1 + b_2) \tag{C.1}$$

C.2 Multiplication

$$\text{Rectangular form: } (a_1 + jb_1) \times (a_2 + jb_2) = (a_1 a_2 - b_1 b_2) + j(a_1 b_2 + b_1 a_2) \tag{C.2a}$$

$$\text{Polar form: } r_1 \angle \theta_1 \times r_2 \angle \theta_2 = r_1 e^{j\theta_1} r_2 e^{j\theta_2} = r_1 r_2 e^{j(\theta_1 + \theta_2)} = r_1 r_2 \angle (\theta_1 + \theta_2) \tag{C.2b}$$

C.3 Division

$$\text{Rectangular form: } \frac{a_2 + jb_2}{a_1 + jb_1} = \frac{a_2 + jb_2}{a_1 + jb_1} \times \frac{a_1 - jb_1}{a_1 - jb_1} = \frac{a_1 a_2 + b_1 b_2}{a_1^2 + b_1^2} + j\frac{a_1 b_2 - a_2 b_1}{a_1^2 + b_1^2} \tag{C.3a}$$

$$\text{Polar form: } \frac{r_2 \angle \theta_2}{r_1 \angle \theta_1} = \frac{r_2 e^{j\theta_2}}{r_1 e^{j\theta_1}} = \frac{r_2}{r_1} e^{j(\theta_2 - \theta_1)} = \frac{r_2}{r_1} \angle (\theta_2 - \theta_1) \tag{C.3b}$$

Electronic Circuits with MATLAB®, PSpice®, and Smith Chart, First Edition. Won Y. Yang, Jaekwon Kim, Kyung W. Park, Donghyun Baek, Sungjoon Lim, Jingon Joung, Suhyun Park, Han L. Lee, Woo June Choi, and Taeho Im.

Companion website: www.wiley.com/go/yang/electroniccircuits

C.4 Conversion between Rectangular Form and Polar/Exponential Form

$$a_r + ja_i = r\angle\theta = re^{j\theta} \text{ with } r = |a_r + ja_i| = \sqrt{a_r^2 + a_i^2} \text{ and } \theta = \tan^{-1}\frac{a_i}{a_r} \quad (C.4)$$

Here, r and θ are referred to as the absolute value and argument or phase angle of complex number a_r+ja_i, respectively, and j is the unit imaginary number $\sqrt{-1}$.

C.5 Operations on Complex Numbers Using MATLAB

If we do not use `i` and `j` for any other purpose, they represent the basic imaginary unit $\sqrt{-1}$ by default. Try typing the following statements into the MATLAB Command Window.

```
>>c1= 1+2i; c2= 3-4i;
>>c1*c2 % multiplication of complex numbers
>>c1/c2 % division of complex numbers
>>r=abs(c2) % absolute value of the complex number c2
>>sqrt(real(c2)^2+imag(c2)^2) % equivalent to the absolute value
>>th=angle(c2) % phase angle of the complex number c2 in radians
>>atan2(imag(c2),real(c2)) % equivalent to the phase angle
>>imag(log(c2)) % equivalent to the phase angle
>>th*180/pi % radian-to-degree conversion
>>r*exp(j*th) % polar-to-rectangular conversion
  ans = 3.0000 - 4.0000i
>>C= [1+i 1-2i; -1+3i -1-4i] % a complex matrix
  C = 1.0000 + 1.0000i  1.0000 - 2.0000i
     -1.0000 + 3.0000i -1.0000 - 4.0000i
>>C1= C' % conjugate transpose
  C1 = 1.0000 - 1.0000i -1.0000 - 3.0000i
       1.0000 + 2.0000i -1.0000 + 4.0000i
```

Appendix D

Nonlinear/Differential Equations with MATLAB

D.1 Nonlinear Equation Solver <fsolve>

MATLAB has the built-in function '`fsolve(f,x0)`', which can give us a solution for a system of (nonlinear) equations. Suppose we have the following system of nonlinear equations to solve.

$$\begin{aligned} x_1^2 + 4x_2^2 &= 5 \\ 2x_1^2 - 2x_1 - 3x_2 &= 2.5 \end{aligned} \tag{D.1}$$

To solve this system of equations, we should rewrite as

$$\begin{aligned} f_1(x_1, x_2) &= x_1^2 + 4x_2^2 - 5 = 0 \\ f_2(x_1, x_2) &= 2x_1^2 - 2x_1 - 3x_2 - 2.5 = 0 \end{aligned} \tag{D.2}$$

and convert it into a MATLAB function defined in an M-file, say, '`f_d02.m`' as follows.

```
function y=f_d02(x)
y(1)= x(1)*x(1)+4*x(2)*x(2) -5;
y(2)=2*x(1)*x(1)-2*x(1)-3*x(2) -2.5;
```

Then we type the following statements into the MATLAB Command Window.

```
>>x0=[0.8 0.2]; % Initial guess [0.8 0.2]
>>x=fsolve('f_d02',x0,optimset('fsolve')) % with default parameters
  x = 2.0000   0.5000
```

If you see some warning message with the MATLAB solution obtained by using '`fsolve()`' like this, you can change the value(s) of the optional parameters such as

Electronic Circuits with MATLAB®, PSpice®, and Smith Chart, First Edition. Won Y. Yang, Jaekwon Kim, Kyung W. Park, Donghyun Baek, Sungjoon Lim, Jingon Joung, Suhyun Park, Han L. Lee, Woo June Choi, and Taeho Im.

Companion website: www.wiley.com/go/yang/electroniccircuits

- MaxFunEvals: Maximum number of function evaluations allowed
- MaxIter: Maximum number of iterations allowed
- TolFun: Termination tolerance on the function value
- TolX: Termination tolerance on x

by using the MATLAB command 'optimset( )' as follows.

```
options = optimset('param1',value1,'param2',value2,...)
```

For example, if you feel it contributive to better solution to set MaxFunEvals and TolX to 1000 and 1e-8, respectively, you need to type the following MATLAB statements.

```
>>options = optimset('MaxFunEvals',1000,'TolX',1e-8);
>>x=fsolve('f_d02',x0,options) % with new values of parameters
  x = 2.0000   0.5000
```

If you feel like doing cross-check, you might use another MATLAB routine 'Newtons( )' which is defined in the next page.

```
>>x=Newtons('f_d02',x0) % Alternatively,
  x = 2.0000   0.5000
```

```
function [x,fx,xx]=Newtons(f,x0,TolX,MaxIter,varargin)
% Newtons.m to solve a set of nonlinear eqs. f1(x)=0, f2(x)=0,..
%input:  f = 1st-order vector function equivalent to a set of equations
%         x0 = Initial guess of the solution
%       TolX = Upper limit of |x(k)-x(k-1)|
%    MaxIter= Maximum # of iteration
%output:  x = Point which the algorithm has reached
%         fx = f(x(last))
%         xx = History of x
% Copyleft: Won Y. Yang, wyyang53@hanmail.net, CAU for academic use only
h=1e-4; TolFun=eps; EPS=1e-6;
fx= feval(f,x0,varargin{:}); Nf=length(fx); Nx=length(x0);
if Nf~=Nx, error('Incompatible dimensions of f and x0!'); end
if nargin<4, MaxIter=100; end
if nargin<3, TolX=EPS; end
xx(1,:)=x0(:).'; % Initialize the solution as the initial row vector
for k=1: MaxIter
   dx= -jacob(f,xx(k,:),h,varargin{:})\fx(:);/; %-[dfdx]^-1*fx
   xx(k+1,:)= xx(k,:)+dx.';
   fx= feval(f,xx(k+1,:),varargin{:}); fxn=norm(fx);
   if fxn<TolFun|norm(dx)<TolX, break; end
end
x= xx(k+1,:);
if k==MaxIter, fprintf('The best in %d iterations\n', MaxIter), end
```

```
function g=jacob(f,x,h,varargin) % Jacobian of f(x)
if nargin<3, h=1e-4; end
h2= 2*h; N=length(x); x=x(:).'; I=eye(N);
for n=1:N
  g(:,n)=(feval(f,x+I(n,:)*h,varargin{:}) ...
         -feval(f,x-I(n,:)*h,varargin{:}))'/h2;
end
```

D.2 Differential Equation Solver <ode45>

Suppose we have a third-order differential equation

$$x^{(3)}(t) + 7x^{(2)}(t) + 14x'(t) + 8x(t) = \sin(t) \quad \text{with} \quad x(0) = 1, x'(0) = 2, x^{(2)}(0) = 3 \quad \text{(D.3)}$$

We can rewrite this in a first-order vector differential equation called a 'state equation' as

$$\begin{bmatrix} x_1'(t) \\ x_2'(t) \\ x_3'(t) \end{bmatrix} = \begin{bmatrix} 0 & 1 & 0 \\ 0 & 0 & 1 \\ -8 & -14 & -7 \end{bmatrix} \begin{bmatrix} x_1(t) \\ x_2(t) \\ x_3(t) \end{bmatrix} + \begin{bmatrix} 0 \\ 0 \\ 1 \end{bmatrix} \sin(t) \text{ with } \begin{bmatrix} x_1(0) \\ x_2(0) \\ x_3(0) \end{bmatrix} = \begin{bmatrix} x(0) \\ x'(0) \\ x^{(2)}(0) \end{bmatrix} = \begin{bmatrix} 1 \\ 2 \\ 3 \end{bmatrix} \quad \text{(D.4a)}$$

$$x(t) = x_1(t) \quad \text{(D.4b)}$$

With this equation cast into a MATLAB function and saved as an M-file named, say, "df_apd.m,"

```
function dx=df_apd(t,x)
dx= zeros(size(x)); dx(1)=x(2); dx(2)=x(3);
dx(3)= -8*x(1) -14*x(2) - 7*x(3) + sin(t); % (D.3)
```

we can type the following statements into the MATLAB Command Window to solve it:

```
>>[t,x]=ode45(@df_apd, [0 10], [1 2 3]); plot(t,x(:,1))
  % for time span [0,10]
```

Appendix E

Symbolic Computations with MATLAB

E.1 How to Declare Symbolic Variables and Handle Symbolic Expressions

To declare any variable(s) as a symbolic variable, you should use the sym or syms command as follows:

```
>>a=sym('a'); x=sym('x'); y=sym('y'); t=sym('t'); n=sym('n');
>>syms a x y t n %or, equivalently and more efficiently
```

Once the variables have been declared as symbolic, they can be used in expressions and as arguments to many functions without being evaluated as numeric.

```
>>f=x^2/(1+tan(x)^2);
>>ezplot(f,-pi,pi) % easy plot of f for [-π,+π]
>>simplify(cos(x)^2+sin(x)^2) % simplify an expression
  ans= 1
>>simplify(cos(x)^2-sin(x)^2) % simplify an expression
  ans= 2*cos(x)^2-1
>>simplify(cos(x)^2-sin(x)^2) % simple expression
  ans= cos(2*x)
>>simplify(cos(x)+i*sin(x)) % simple expression
  ans= exp(i*x)
>>eq1=expand((x+y)^3-(x+y)^2) % expand
  eq1= x^3+3*x^2*y+3*x*y^2+y^3-x^2-2*x*y-y^2
>>collect(eq1,y) % collect similar terms in descending order w.r.t. y
  ans= y^3+(3*x-1)*y^2+(3*x^2-2*x)*y+x^3-x^2
>>factor(eq1) % factorize
  ans= (x+y-1)*(x+y)^2
>>horner(eq1) % nested multiplication form
  ans= (-1+y)*y^2+((-2+3*y)*y+(-1+3*y+x)*x)*x
>>pretty(ans) % pretty form
            2
  (-1 + y) y  + ((-2 + 3 y) y + (-1 + 3 y + x) x) x
```

Electronic Circuits with MATLAB®, PSpice®, and Smith Chart, First Edition. Won Y. Yang, Jaekwon Kim, Kyung W. Park, Donghyun Baek, Sungjoon Lim, Jingon Joung, Suhyun Park, Han L. Lee, Woo June Choi, and Taeho Im.

Companion website: www.wiley.com/go/yang/electroniccircuits

If you need to substitute numeric values or other expressions for some symbolic variables in an expression or take the limit of an expression, you can use subs( ) and limit( ) as follows:

```
>>subs(eq1,x,0) % substitute numeric value x=0 into eq1=(x+y-1)*(x+y)^2
   ans= -y^2+y^3
>>subs(eq1,{x,y},{0,x-1}) % substitute numeric values x=0 and y=x-1
   ans= (x-1)^3-(x-1)^2
>>limit((1+x/n)^n,n,inf) % lim_{n→∞}(1+x/n)^n = e^x
   ans= exp(x)
```

E.2 Solving Algebraic Equations

We can use the backslash(\) operator to solve a set of linear equations written in a matrix-vector form.

```
>>syms R11 R12 R21 R22 b1 b2
>>R=[R11 R12; R21 R22]; b=[b1; b2];
>>x=R\b % or R^-1*b or inv(R)*b
  x= [ (R12*b2-b1*R22)/(-R11*R22+R21*R12)]
     [ (-R11*b2+R21*b1)/(-R11*R22+R21*R12)]
```

MATLAB has many commands and functions that can be very helpful in dealing with complex analytic(symbolic) expressions and equations as well as in getting numerical solutions. One of them is '`solve()`', which can be used for obtaining the symbolic or numeric roots of equations. According to what we could see by typing '`help solve`' into the MATLAB Command Window, its usages are as follows:

```
>>syms a b c x
>>fx=a*x^2+b*x+c;
>>solve(fx) % Formula for roots of 2nd-order polynomial eq
  ans= [ 1/2/a*(-b+(b^2-4*a*c)^(1/2))]
       [ 1/2/a*(-b-(b^2-4*a*c)^(1/2))]
>>syms x1 x2 b1 b2
>>fx1=x1+x2-b1; fx2=x1+2*x2-b2; % a system of simultaneous algebraic eq.
>>[x1o,x2o]=solve(fx1,fx2) %
   x1o= 2*b1-b2
   x2o= -b1+b2
>>solve('p*sin(x)=r') %regarding x as an unknown variable and p as a
  parameter
   ans = asin(r/p) %sin^-1(r/p)
>>[x1,x2]=solve('x1^2+4*x2^2-5=0','2*x1^2-2*x1-3*x2-2.5=0')
   x1 = [                 2.]   x2 = [                  0.500000]
        [          -1.206459]        [                  0.941336]
        [0.603229 -0.392630*i]       [-1.095668 -0.540415e-1*i]
        [0.603229 +0.392630*i]       [-1.095668 +0.540415e-1*i]
```

```
>>S=solve('x^3-y^3=2','x=y') %returns the solution in a structure.
  S = x: [3x1 sym]
  y: [3x1 sym]
>>S.x
  ans = [                   1]
        [ -1/2+1/2*i*3^(1/2)]
        [ -1/2-1/2*i*3^(1/2)]
>>S.y
  ans = [                 -1]
        [ 1/2-1/2*i*3^(1/2)]
        [ 1/2+1/2*i*3^(1/2)]
```

Appendix F

Useful Formulas

$\sin(A \pm 90°) = \pm\cos A$ (F.1)	$\cos(A \pm 90°) = \mp\sin A$ (F.2)
$\sin(A \pm 180°) = -\sin A$ (F.3)	$\cos(A \pm 180°) = \cos A$ (F.4)
$\sin(A \pm B) = \sin A\cos B \pm \cos A\sin B$ (F.5) $\cos(A \pm B) = \cos A\cos B \mp \sin A\sin B$ (F.6)	$\tan(A \pm B) = \dfrac{\tan A \pm \tan B}{1 \mp \tan A\tan B}$ (F.7)
$\sin A\sin B = \frac{1}{2}\{\cos(A-B) - \cos(A+B)\}$ (F.8)	$\sin A\cos B = \frac{1}{2}\{\sin(A+B) + \sin(A-B)\}$ (F.9)
$\cos A\sin B = \frac{1}{2}\{\sin(A+B) - \sin(A-B)\}$ (F.10)	$\cos A\cos B = \frac{1}{2}\{\cos(A+B) + \cos(A-B)\}$ (F.11)
$\sin A + \sin B = 2\sin\left(\frac{A+B}{2}\right)\cos\left(\frac{A-B}{2}\right)$ (F.12)	$\cos A + \cos B = 2\cos\left(\frac{A+B}{2}\right)\cos\left(\frac{A-B}{2}\right)$ (F.13)
$\sin^2 A = \frac{1}{2}(1 - \cos 2A)$ (F.14)	$\cos^2 A = \frac{1}{2}(1 + \cos 2A)$ (F.15)
$\sin 2A = 2\sin A\cos A$ (F.16)	$\cos 2A = \cos^2 A - \sin^2 A = 1 - 2\sin^2 A = 2\cos^2 A - 1$ (F.17)
$a\cos A - b\sin A = \sqrt{a^2+b^2}\cos(A+\theta),\ \theta = \tan^{-1}\left(\frac{b}{a}\right)$ (F.18)	
$a\sin A + b\cos A = \sqrt{a^2+b^2}\sin(A+\theta),\ \theta = \tan^{-1}\left(\frac{b}{a}\right)$ (F.19)	
Euler identity: $e^{\pm j\theta} = \cos\theta \pm j\sin\theta$ (F.20)	$e^{j\theta} + e^{-j\theta} = 2\cos\theta$ (F.21) $e^{j\theta} - e^{-j\theta} = j2\sin\theta$ (F.22)

Electronic Circuits with MATLAB®, PSpice®, and Smith Chart, First Edition. Won Y. Yang, Jaekwon Kim, Kyung W. Park, Donghyun Baek, Sungjoon Lim, Jingon Joung, Suhyun Park, Han L. Lee, Woo June Choi, and Taeho Im.

Companion website: www.wiley.com/go/yang/electroniccircuits

For a complex number z, $\cosh z = \dfrac{e^z + e^{-z}}{2}$ (F.23a) $\sinh z = \dfrac{e^z - e^{-z}}{2}$ (F.23b) $\tanh z = \dfrac{\sinh z}{\cosh z}$ (F.23c)	For a complex number z, $\cosh jz = \cos z$ (F.24a) $\sinh jz = j \sin z$ (F.24b) $\tanh jz = j \tan z$ (F.24c)
$e^x = \sum_{n=0}^{\infty} \dfrac{1}{n!} x^n = 1 + \dfrac{1}{1!}x + \dfrac{1}{2!}x^2 + \dfrac{1}{3!}x^3 + \cdots \overset{x \to 0}{\rightarrow} e^x = 1 + x$ (F.25)	
$\dfrac{d}{dt} t^n = n t^{n-1}$ (F.26)	$\int t^n \, dt = \dfrac{1}{n+1} t^{n+1}$ for $n \neq -1$ (F.32)
$\dfrac{d}{dt} e^{at} = a e^{at}$ (F.27)	$\int e^{at} \, dt = \dfrac{1}{a} e^{at}$ (F.33)
$\dfrac{d}{dt} \cos \omega t = -\omega \sin \omega t$ (F.28)	$\int \cos \omega t \, dt = \dfrac{1}{\omega} \sin \omega t$ (F.34)
$\dfrac{d}{dt} \sin \omega t = \omega \cos \omega t$ (F.29)	$\int \sin \omega t \, dt = -\dfrac{1}{\omega} \cos \omega t$ (F.35)
$\dfrac{d}{dt}(uv) = u \dfrac{d}{dt} v + v \dfrac{d}{dt} u$ (F.30)	$\int u \dfrac{dv}{dt} dt = uv - \int v \dfrac{du}{dt} dt$ (Partial integral) (F.36)
$\dfrac{d}{dt}\left(\dfrac{v}{u}\right) = \dfrac{u(dv/dt) - v(du/dt)}{u^2}$ (F.31)	

Appendix G

Standard Values of Resistors, Capacitors, and Inductors

G.1 Color Code of Resistors

Except for wire-wound/cermet/high-power/precision resistors, most common resistors do not have their resistance value printed on them, but rather have a color code representing their resistance value as illustrated in Figure G.1. Table G.1 shows the numerical value or tolerance (manufacturer's reliability rating) represented by each color. For example, the resistance value of a resistor with the four-color band of yellow-violet-red-silver is

$$R = (10a + b) \times 10^m \pm \% = (10 \times 4 + 7) \times 10^2 \pm 10\% = 4700 \pm 470\,[\Omega] \tag{G.1}$$

and that of a resistor with the five-color band of orange–black–white–gold–gold is

$$\begin{aligned} R &= (100a + 10b + c) \times 10^m \pm \% \\ &= (100 \times 3 + 10 \times 0 + 1) \times 10^{-1} \pm 5\% = 30.1 \pm 0.15\,[\Omega] \end{aligned} \tag{G.2}$$

(cf.) Visit the web site http://xtronics.com/kits/rcode.htm for more details about the color code.

G.2 Standard Values of Resistors

Discrete resistors are commercially available only in standard values depending on their tolerance as listed in Table G.2. Consequently, when the designed value of a resistor is 3.1 kΩ, we should use $30 \times 10^2 \pm 5\%[\Omega]$ or $309 \times 10^1 \pm 1\%[\Omega]$ unless we somehow have a resistor of 3.1 kΩ fabricated.

Electronic Circuits with MATLAB®, PSpice®, and Smith Chart, First Edition. Won Y. Yang, Jaekwon Kim, Kyung W. Park, Donghyun Baek, Sungjoon Lim, Jingon Joung, Suhyun Park, Han L. Lee, Woo June Choi, and Taeho Im.

Companion website: www.wiley.com/go/yang/electroniccircuits

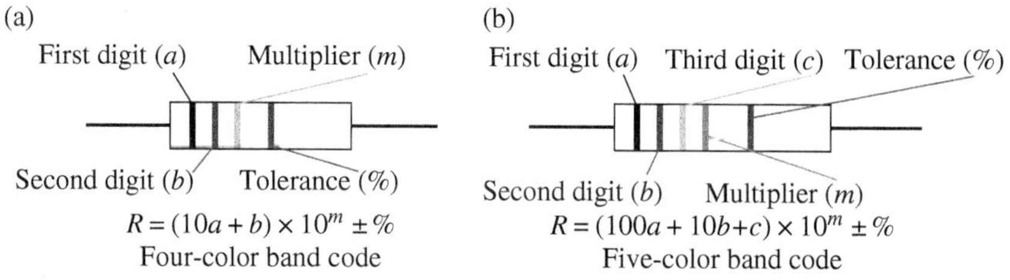

Figure G.1 Color code for resistors. (*See insert for color representation of the figure.*)

Table G.1 Color code of resistors.

Color	Digit	Color	Digit	Color	Tolerance (%)
Black	0	Blue	6	Brown	1
Brown	1	Violet	7	Red	2
Red	2	Gray	8	Gold	5
Orange	3	White	9	Silver	10
Yellow	4	Gold	-1 (applied only to multiplier m)	None	20
Green	5	Silver	-2 (applied only to multiplier m)		

(cf.) The commercially available resistors have the power (wattage) ratings of 1/8, 1/4, 1/2, 1, and 2W depending on their physical sizes.

(cf.) Some resistors may exhibit parasitic effects of series inductance and parallel capacitance especially if the circuit containing them operates at very high frequency.

G.3 Standard Values of Capacitors

Discrete capacitors are commercially available only in standard values depending on their physical material/shape as listed in Tables G.3.1 and G.3.2. Table G.3.3 shows the letter tolerance code of capacitors. Most of them have their value (like 22 μF) printed on their body together with their breakdown voltage, while the capacitance value of ceramic condenser is printed as, say, 104, which means the capacity of

$$10 \times 10^4\,[\text{pF}] = 10^5 \times 10^{-12}\,[\text{F}] = 0.1 \times 10^{-6}\,[\text{F}] = 0.1\,[\mu\text{F}]$$

Table G.2 Standard values of resistors.

1% tolerance												5% tolerance				10% tolerance		20% tolerance
100	**121**	**147**	**178**	**215**	**261**	**316**	**383**	**464**	**562**	**681**	**825**	**10**	**18**	**33**	**56**	**10**	**33**	**10**
102	124	150	182	221	267	324	392	475	576	698	845	11	20	36	62	12	39	15
105	127	154	187	226	274	332	402	487	590	715	866	12	22	39	68	15	47	22
107	130	158	191	232	280	340	412	499	604	732	887	13	24	43	75	18	56	33
110	133	162	196	237	287	348	422	511	619	750	909	15	27	47	82	22	68	47
113	137	165	200	243	294	357	432	523	634	768	931	16	30	51	91	27	82	68
115	140	169	205	249	301	365	442	536	649	787	953							
118	143	174	210	255	309	374	453	549	665	806	976							

Table G.3.1 Standard values of electrolytic capacitors [μF].

Maximum voltage 10 V			Maximum voltage 25 V			Maximum voltage 50 V				
100	1000	10 000	10	100	1000	0.1	1.0	10	100	1000
22	220	2200	22	220	2200	0.22	2.2	22	220	2200
33	330	3300	33	330	3300	0.33	3.3	33	330	
47	470	4700	47	470	4700	0.47	4.7	47	470	
		6800								

Table G.3.2 Standard values of ceramic and mylar polyester capacitors.

Ceramic disc capacitors [pF] with maximum voltage 200 V				Mylar polyester capacitors [μF] with maximum voltage 100 V			
10	100	1000	10 000	0.001	0.01	0.1	0.33
15	150	1500	15 000	0.0015	0.015	0.12	0.39
22	220	2200		0.0022	0.022	0.15	0.47
33	330	3300		0.0033	0.033	0.18	0.56
47	470	4700		0.0047	0.047	0.22	0.68
68	680	6800		0.0068	0.068	0.27	0.82
				0.0082	0.082		1.0

Table G.3.3 Letter tolerance code of capacitors.

A	B	C	D	E	F	G	H	J	K	M	N	P	Z
+0.05 pF	+0.1 pF	+0.25 pF	+0.5 pF	+0.5%	+1%	+2%	+3%	+5%	+10%	+20%	+30%	+100%	+80%
−0.05 pF	−0.1 pF	−0.25 pF	−0.5 pF	−0.5%	−1%	−2%	−3%	−5%	−10%	−20%	−30%	−0%	−20%

If the tolerance is missing, it can be assumed to be ±20%. Note that electrolytic capacitors have positive/negative terminals with the positive one having longer leg, and if we are not careful to connect them in accord with the polarity they will leak or may be destroyed.

(cf.) The commercially available capacitors have the working voltages of about 3∼1000 V.

(cf.) Some capacitors may exhibit parasitic effects of parallel/series resistance and series inductance.

```
function std_val=standard_value(val,RLC,glc,tol)
%Input: R= the value of desired Resistance/Inductance/Capacitance
%    RLC= 'R'/'L'/'C'
%    glc='g'/'l'/'c' for greater than or equal/less than or equal/close
%    tol=1/5/10 for error tolerance
if nargin<4, tol=1; end
if nargin<3, glc='c'; end
if nargin<2, RLC='R'; end
Rs1=[100 102 105 107 110 113 115 118 121 124 127 130 133 137 140 143 147];
Rs1=[Rs1 150 154 158 162 165 169 174 178 182 187 191 196 200 205 210 215];
Rs1=[Rs1 221 226 232 237 243 249 255 261 267 274 280 287 294 301 309 316];
Rs1=[Rs1 324 332 340 348 357 365 374 383 392 402 412 422 432 442 453 464];
Rs1=[Rs1 475 487 499 511 523 536 549 562 576 590 604 619 634 649 665 681];
Rs1=[698 715 732 750 768 787 806 825 845 866 887 909 931 953 976];
Multipliers=10.^[-2:9]; % Table G.1
Rs1=Multipliers'*Rs1; Rs1=Rs1(:)'; % 1%-tolerance (Table G.2)
Rs=[10 11 12 13 15 16 18 20 22 24 27 30 33 36 39 43 47 51 56 62 68 75 82 91];
Rs5=Multipliers'*Rs5; Rs5=Rs5(:)'; % 5%-tolerance (Table G.2)
Rs10=[10 12 15 18 22 27 33 39 47 56 68 82];
Rs10=Multipliers'*Rs10; Rs10=Rs10(:)'; % 10%-tolerance (Table G.2)
Cs1=[10 15 22 33 47 68 100 150 220 330 470 680 1000 1500 2200 3300 4700 6800];
Cs1=[Cs1 10000 15000]*1e-12;
Cs2=[0.001 0.0015 0.0022 0.0033 0.0047 0.0068 0.0082 0.01 0.015 0.022 0.033];
Cs2=[Cs2 0.047 0.068 0.082 0.1 0.12 0.15 0.18 0.22 0.27 0.33 0.39 0.47 0.56];
Cs2=[0.68 0.82 1 2.2 3.3 4.7 10 22 33 47 100 220 330 470 1000 2200 3300 4700];
Cs2=[Cs2 6800 10000]*1e-6; Cs = [Cs1 Cs2];
Ls0=[1 1.1 1.2 1.3 1.5 1.6 1.8 2 2.2 2.4 2.7 3 3.3 3.6 3.9 4.3 4.7 5.1 5.6 6.2];
Ls0 = [Ls0 6.8 7.5 8.2 8.7 9.1];
Ls = [Ls0*1e-6 Ls0*1e-9]; Ls=[1; 10; 100]*Ls; Ls=[Ls(:)' Ls0*1e-3];
switch lower(RLC)
  case 'r' % Resistor
   if tol==5
    if lower(glc(1))=='g', ind=find(Rs5>=val); list=Rs5(ind);
     elseif lower(glc(1))=='l', ind=find(Rs5<=val); list=Rs5(ind);
     else list=Rs5;
    end
   elseif tol==10
    if lower(glc(1))=='g', ind=find(Rs10>=val); list=Rs10(ind);
    elseif lower(glc(1))=='l', ind=find(Rs10<=val); list=Rs10(ind);
     else list=Rs10;
    end
   else % if tol=1%
    if lower(glc(1))=='g', ind=find(Rs1>=val); list=Rs1(ind);
     elseif lower(glc(1))=='l', ind=find(Rs1<=val); list=Rs1(ind);
     else list=Rs1;
    end
   end
  case 'l' % Inductor
   if lower(glc(1))=='g', ind=find(Ls>=val); list=Ls(ind);
    elseif lower(glc(1))=='l', ind=find(Ls<=val); list=Ls(ind);
    else list=Ls;
   end
  case 'c' % Capacitor
   if lower(glc(1))=='g', ind=find(Cs>=val); list=Cs(ind);
    elseif lower(glc(1))=='l', ind=find(Cs<=val); list=Cs(ind);
    else list=Cs;
   end
  otherwise error('What do you want, neither R nor L nor C?')
end
[em,im] = min(abs(list-val)); std_val = list(im);
```

G.4 Standard Values of Inductors

Discrete inductors are commercially available only in standard values (with the tolerance of +5%, +10%, and ±20%) listed in Table G.4.

(cf.) Most inductors exhibit parasitic effects of series resistance (due to the large number of turns) and parallel capacitance.
(cf.) Visit the web site http://www.rfcafe.com/references/electrical/inductor-values.htm for more details about the standard values of inductors.

For the desired values of the resistances, capacitances, and inductances, the standard values can easily be chosen from Tables G.2, G.3.1, G.3.2, and G.4 by using the following MATLAB routine 'standard_value(val,RLC,glc,tol)' where the input arguments are supposed to be given as follows:

val: the desired value of a resistance, a capacitance, or an inductance;
RLC: 'R', 'C', or 'L' for resistance, a capacitance, or an inductance;
glc: 'g', 'l', or 'c' to look for the standard value slightly greater than, less than, or just close to the desired value; and
tol: 1, 5, or 10 for 1%-/5%-/10%-tolerance standard resistance value.

Table G.4 Standard values of inductors.

1.0 1.1 1.2 1.3 1.5 1.6 1.8 2.0 2.2 2.4 2.7 3.0 3.3 3.6 3.9 4.3 4.7 5.1 5.6 6.2 6.8 7.5 8.2 8.7 9.1 nH, μH
1.0 1.1 1.2 1.3 1.5 1.6 1.8 2.0 2.2 2.4 2.7 3.0 3.3 3.6 3.9 4.3 4.7 5.1 5.6 6.2 6.8 7.5 8.2 8.7 9.1×10 nH, μH
1.0 1.1 1.2 1.3 1.5 1.6 1.8 2.0 2.2 2.4 2.7 3.0 3.3 3.6 3.9 4.3 4.7 5.1 5.6 6.2 6.8 7.5 8.2 8.7 9.1×10^2nH, μH
1.0 1.1 1.2 1.3 1.5 1.6 1.8 2.0 2.2 2.4 2.7 3.0 3.3 3.6 3.9 4.3 4.7 5.1 5.6 6.2 6.8 7.5 8.2 8.7 9.1×10^3nH, μH

G.5 Standard Values of Zener Diode Voltage

Table G.5 Standard values of Zener diode voltages.

2.4 2.5 2.7 2.8 3.0 3.3 3.6 3.9 4.3 4.7 5.1 5.6 6.0 6.2 6.8 7.5 8.2 8.7 9.1
10 11 12 13 14 15 16 17 18 19 20 22
24 25 27 28 30 33 36 39 43 47 51 56 60 62 68 75 82 87 91
100 110 120 130 140 150 160 170 180 190 200

http://www.logwell.com/tech/components/zener.html

Appendix H

OrCAD/PSpice®

OrCAD PSpice®, formerly MicroSim PSpice® (Professional Simulation Program with Integrated Circuit Emphasis), is a powerful circuit simulation program with graphic interface to analyze electric and electronic circuits. You can download the OrCAD demo software together with the OrCAD flow tutorial at the web site http://www.cadence.com/products.

H.1 Starting Capture Component Information System (CIS) Session

Once you have installed the OrCAD software, you may start the Capture Component Information System (CIS) Demo session by clicking the corresponding icon to open the Capture CIS Window (Figure H.1), in which you can click File>New>Project on the menu bar to open the New Project dialog box (Figure H.1(b)). Here, you are supposed to do the following in the New Project dialog box:

- In the Name field, type, say, 'test' (with no quotation mark) as a new project name.
- Check the radio button of Analog or Mixed A/D.
- In the Location field, type the pathname of the subdirectory in which you want the new project to be saved. You can use the Browse button to select the location for the project files and click OK.
- Then click OK to close the New Project dialog box, which will make the Create PSpice Project dialog box pop-up.
- Check the radio button of Create a blank project and click OK in the Create PSpice Project box (Figure H.1(c)).

Then you will see the activated Capture (CIS) Window with the menu bar together with the tool bar, which has the Project Manager Window containing

Electronic Circuits with MATLAB®, PSpice®, and Smith Chart, First Edition. Won Y. Yang, Jaekwon Kim, Kyung W. Park, Donghyun Baek, Sungjoon Lim, Jingon Joung, Suhyun Park, Han L. Lee, Woo June Choi, and Taeho Im.

Companion website: www.wiley.com/go/yang/electroniccircuits

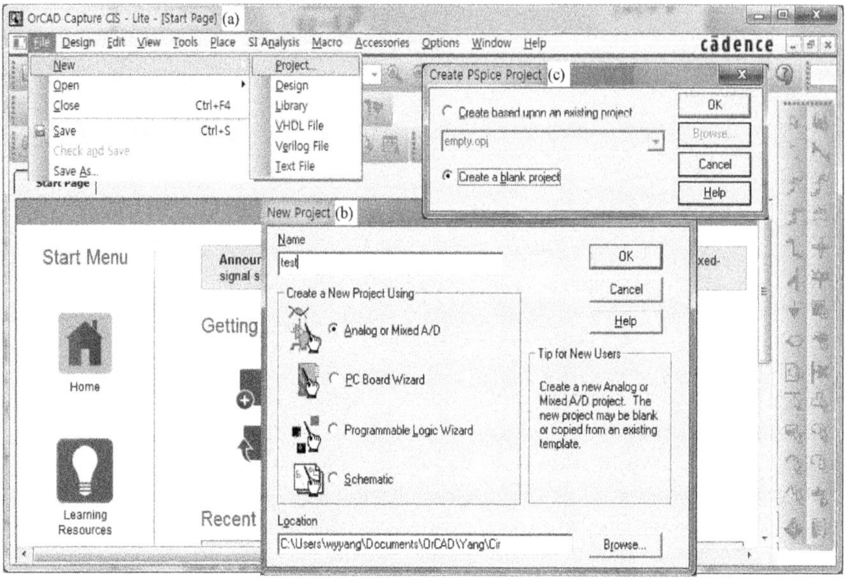

Figure H.1 New Project dialog box opened by clicking File>New>Project in the Capture Component Information System (CIS) Window.

the design file name 'test.dsn' under the subdirectory named 'Design Resources' on the left side (Figure H.2). You can click the design file name (like 'test.dsn') and successively, SCHEMATIC1 and PAGE1 to open the Schematic Editor Window, in which you can draw the schematic for the circuit which you would like to analyze.

H.2 Drawing Schematic

H.2.1 Fetching Parts (Place>Part or 'p')

Click Place/Part on the menu bar or the Place Part button on the tool palette, or just press the 'p' key on the keyboard to open the Place Part dialog box (Figure H.2), in which you can either scroll through the part list and select (click) the needed part or just type the part name into the Part field in the upper part of the dialog box. It may be helpful for you to know the part names representing some often-used device types listed in Table H.1. In case you cannot find the device (component) you want to place from the current part list, you need to search for it after clicking the Part Search button and then add the library (like \tools\capture\library\pspice\'analog.olb', 'source.olb', 'eval.olb', 'breakout.olb', 'special.olb', 'sourcstm.olb' etc.) containing the device.

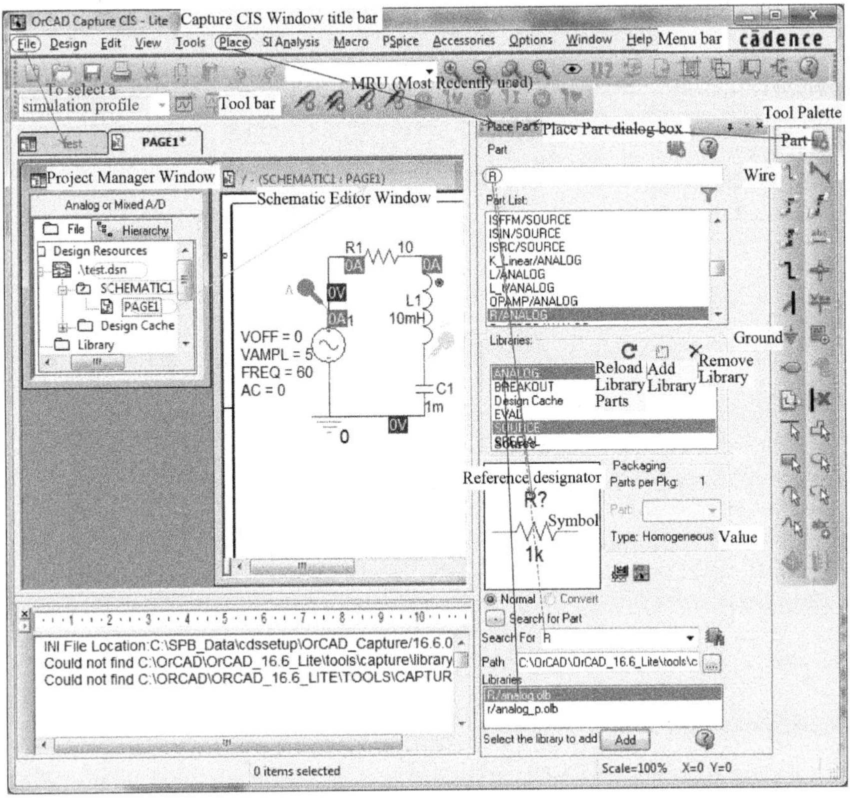

Figure H.2 Capture window, Project Manager window, Schematic Editor Window, and Place Part dialog box opened by clicking the Place Part button in the Tool Palette.

Table H.1 Device types and PSpice part names.

Device type	Part name	Device type	Part name
Capacitor	C	Inductor	L
Diode	D	Bipolar junction transistor (BJT)	Q
Voltage-controlled voltage source (VCVS)	E	Resistor	R
Current-controlled current source (CCCS)	F	Switch to be closed at specified time	Sw_tClose

(Continued)

Table H.1 (Continued)

Device type	Part name	Device type	Part name
Voltage controlled current source (VCCS)	G	Switch to be opened at specified time	Sw_tOpen
Current-controlled voltage source (CCVS)	H	OP Amp	uA741
Independent current source	I	Independent voltage source	V
Independent AC current source	IAC	Independent AC voltage source	VAC
Independent DC current source	IDC	Independent DC voltage source	VDC
Independent periodic current source	IPULSE	Independent periodic voltage source	VPULSE
Independent piecewise linear current source	IPWL	Independent piecewise linear voltage source	VPWL
Sinusoidal current source	ISIN	Sinusoidal voltage source	VSIN
		Transformer	XFRM_LINEAR

H.2.2 Placing Parts

Once you have got the wanted part as can be inspected from the graphic symbol displayed in the preview box at the bottom-right corner of the Place Part dialog box, click OK to close the dialog box. Then the selected part attaches to the mouse pointer so that we can move and place it anywhere on the schematic page by clicking. The device will remain attached to the mouse pointer, so you may use it as many times as needed. To detach the device from the mouse pointer, press ESC key or right-click and click (select) End Mode in the pop-up menu. You can have one of the recently fetched devices attached to the mouse pointer by clicking (selecting) it among the MRU list in the center part of the tool bar just below the top menu bar. You can select any part on the schematic by clicking it so that it gets pink-hot and then rotate it counterclockwise by pressing the 'r' key or mirror it horizontally/vertically by pressing the 'h'/'v' key. You can also select any area of the schematic by clicking one corner of the area and dragging (pressing/holding the left mouse button) to the opposite corner and then move(click and drag), cut('^x'), paste('^v'), delete(Del), rotate('r'), or mirror horizontally('h')/vertically('v') the selected (pink-hot) area, where ^ indicates 'with the CTRL key held down' You can zoom in/out the whole schematic by pressing the 'i'/'o' key. After placing all the parts at the appropriate locations, you should click the GND (ground) button on the tool palette, click (select) 0/SOURCE in the Place Ground dialog box, click OK to close the dialog box, and move the mouse pointer (with the Ground symbol attached to it) to click at an appropriate location to place the ground symbol at that location.

Remark H.1 Difference Between VSIN and VAC
The two parts VSIN/ISIN (sine wave V/I-source) and VAC/IAC (AC V/I-source) are represented by the graphic symbols of the same shape, but with different parameters. Specifically, compared with VAC, VSIN has an additional parameter, which is the frequency in Hz, while the frequency range of VAC is specified in the simulation profile. In fact, VAC is used for AC sweep (steady-state) analysis to find the frequency response, while VSIN is for time-domain (transient) analysis.

H.2.3 Defining(Assigning)/Changing Part Values

You can double-click the value or reference designator of a part to define/change it in the Display Properties dialog box, or double-click the part to open its Property Editor spreadsheet, define or change the values of the parameters, and even create a new parameter (New Column) and set its value. Figure H.3 shows the waveforms of sine/pulse/piecewise-linear/digital sources and their Property Editor spreadsheets, in which the parameters are defined and changed. Sometimes, you should set the value of initial condition (IC) for inductors/capacitors, where IC denotes the initial current/voltage for inductor/capacitor, respectively. Table H.2 shows magnitude identifiers used by PSpice.

H.2.4 Wiring (Place>Wire or 'w')

In order to connect the parts here and there in the Schematic page, you click the menu Place/Wire on the menu bar or the Place Wire button on the tool palette, or just press the 'w' key on the keyboard to change the arrow-type pointer into a cross-type pointer, which you can use like a pencil. Then click the one point of the connection you want to make, click successively the intermediate points, and either double-click the other point or right-click the other point and click (select) End Wire in the pop-up box. You can draw a slant line by holding down the SHIFT key and clicking the connection points. To stop wiring, press ESC key or right-click and click (select) End Wire in the pop-up box.

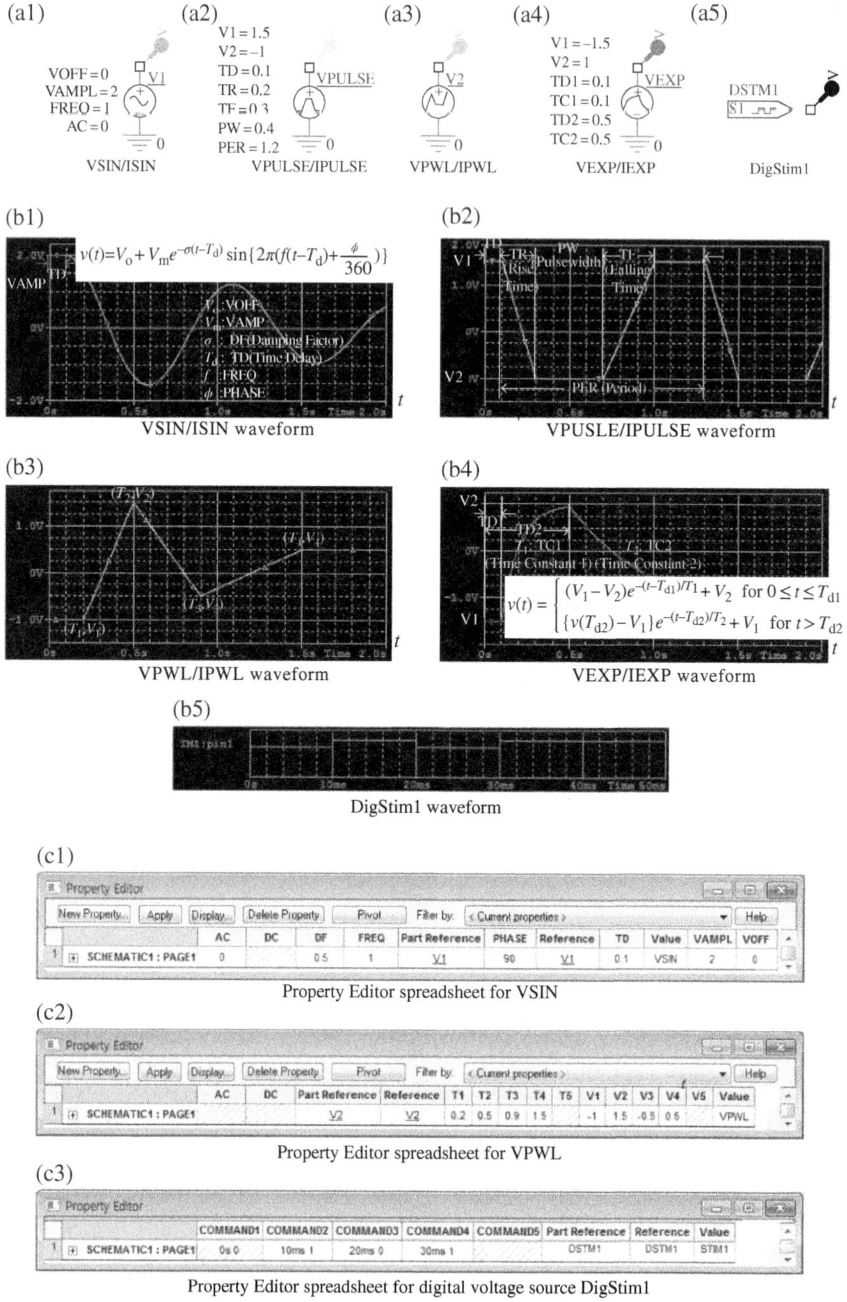

Figure H.3 Waveforms and Property Editor spreadsheets of the sources VSIN/VPULSE/VPWL/VEXP/STIM1.

Table H.2 Magnitude identifiers used by PSpice (Capture).

Unit identifier	F or f (femto)	P or p (pico)	N or n (nano)	U or u (micro)	M or m (milli)	K or k (kilo)	MEG or meg (mega)	G or g (giga)	T or t (tera)
Magnitude	10^{-15}	10^{-12}	10^{-9}	10^{-6}	10^{-3}	10^{3}	10^{6}	10^{9}	10^{12}

(cf.) PSpice does not require the units in specifying the part values; but if the unit is shown, there must be no space between the value and the unit. In fact, PSpice does not differentiate the uppercase and lowercase letters.

Remark H.2 Connections Rejected by PSpice
Figure H.4 shows some examples of connection that are rejected by PSpice, together with the corresponding countermeasures:

1) PSpice insists that every node should have a DC path to ground for a proper analysis, but the top node of the circuit in Figure H.4(a1) does not satisfy the requirement, which is the case of missing DC path to ground. It can be circumvented by connecting a very large-valued resistor of, say, 1 GΩ between the node and the ground node as depicted in Figure H.4(a2).
2) The secondary side of the transformer in the circuit of Figure H.4(b1) has no DC path to ground, dissatisfying the above requirement. It can also be circumvented by connecting a very large-valued dummy resistor between the two nodes of primary and secondary coils as depicted in Figure H.4(b2).
3) PSpice does not accept any loop consisting of only voltage source(s) and/or inductor(s) that may result in a DC short. This condition is violated by the circuit in Figure H.4(c1). This case of voltage source and/or inductor loop can be avoided by inserting a very small-valued dummy resistor of, say, 1 μΩ into the loop as depicted in Figure H.4(c2).

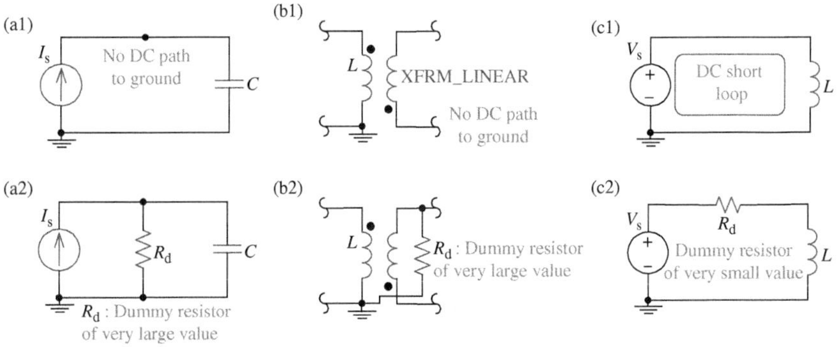

Figure H.4 Examples of connections rejected by PSpice and the corresponding measures.

H.3 Setting Simulation Conditions

H.3.1 Creating Simulation Profile

To set the simulation conditions, click PSpice/New Simulation Profile on the menu bar or the corresponding button on the tool bar to open the New Simulation dialog box, in which you can type a profile name, say, 'tran' (with no quotation mark) into the Name field and click the Create button (Figure H.5(a)).

(a)

New Simulation dialog box opened by clicking New Simulation Profile button from the tool bar

(b)

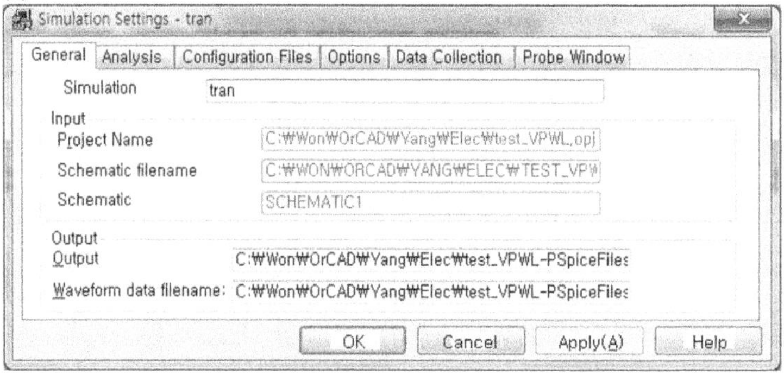

General tab of Simulation Settings dialog box to set the output directory and filename in

(c1)

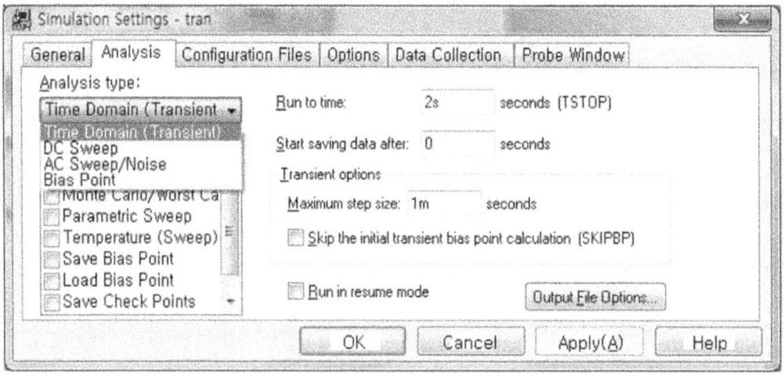

Analysis tab of Simulation Settings dialog box to set Anaysis type (as Transient) and parameters in

Figure H.5 Setting the simulation conditions in the Simulation Settings dialog box.

(c2)

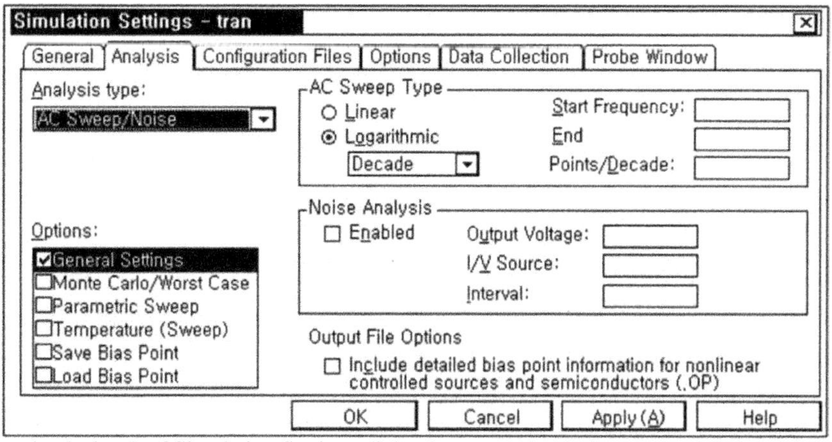

Analysis tab of Simulation Settings dialog box to set Analysis type (as AC Sweep) and parameters in

(d)

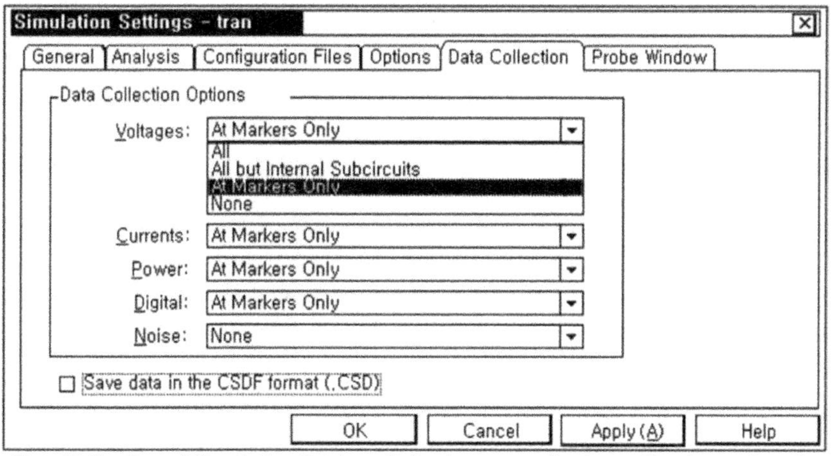

Data Collection of Simulation Settings dialog box to set the data collection options in

(e)

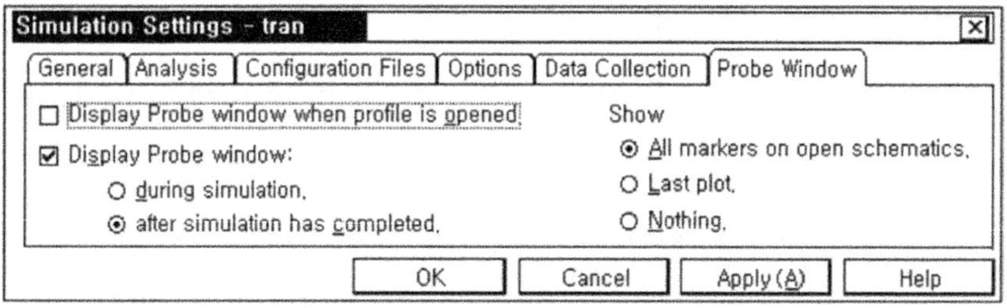

Probe Window tab of Simulation Settings dialog box to set PSpice A/D (Probe) window options in

Figure H.5 (Continued)

Then the Simulation Settings dialog box is opened, in which you can do the following:

- Click on the General tab to set the output directory and waveform data filename (Figure H.5(b))
- Click the Analysis tab to set the Analysis type to one of {Time Domain (Transient), DC Sweep, AC Sweep, and Bias Point}, select the Options, and set the simulation time/frequency range, etc. (Figure H.5(c1) or (c2)). Especially in the case of transient analysis, you can click the Output File Options button to make the results of Fourier analysis and/or operating bias point analysis printed into the output file.
- Click the Data Collection tab to select Data Collection Options (Figure H.5(d)).
- Click the Probe Window tab to set the conditions on PSpice A/D (Probe) window (Figure H.5(e)).
- Most often, you are supposed to do only the second thing about the Analysis type and the related parameters.

H.3.2 Placing Voltage/Current Markers (Probes) – PSpice>Markers

To obtain the voltage/current/power waveform, you can click PSpice>Markers on the menu bar and select one of {Voltage Level, Voltage Differential, Current, Power Dissipation, Advanced (dB, Phase, Real, Imaginary)} from the submenu or click one of {V, Vd, I, W}-Marker buttons on the tool bar to have the probe pins attached to the mouse and then put one at each target point/device on the schematic. The probe pin will remain attached to the mouse pointer, so you may use it as many times as needed. To detach the probe pin from the mouse pointer, press ESC key or right-click and click (select) End Mode in the pop-up menu. Alternatively, you can use the Add Trace function in the Probe window in order to make measurements of any voltage/current/power without using schematic markers. Specifically, you can click Trace>Add_Trace on the menu bar or the Add Trace button on the tool bar of the Probe window to open the Add Traces dialog box, select the variables from the Simulation Output Variables list, and click OK. On the other hand, you can have all the values of voltage/currents (at the operating point) displayed/hidden on the schematic by clicking the Enable Bias Voltage(V)/Current(I)/Power(W) Display button on the tool bar of the Schematic Editor Window.

H.3.3 Editing Simulation Profile

To modify the simulation conditions, either click PSpice>Edit Simulation Profile on the menu bar or the corresponding button on the tool bar in the Capture window, or click Simulation/Edit Profile on the menu bar or the corresponding button on the (left) tool bar in the Probe window to reopen the Simulation

Settings dialog box, in which you can change the simulation conditions and the related parameters. In case there is any existing Simulation Profile, you can select it from the Simulation Profile list on the left side of the tool bar in the Capture window.

H.4 Running PSpice Simulation and Observing the Results

To perform the simulation of the circuit you have drawn in the schematic page, either click PSpice>Run on the menu bar or the corresponding button (of right arrow shape) on the tool bar in the Capture window, or click Simulation/Run on the menu bar or the corresponding button on the right side of the tool bar in the Probe window. Pressing the 'F11' key is another way to perform the simulation. Then you will see the waveforms of the voltages/currents/powers which you have chosen as the measurement outputs by placing the corresponding Markers. Besides, you can find their peaks, troughs, minimum, maximum, etc. as if you had a digital storage oscilloscope.

H.4.1 Analyzing Waveforms – Trace>Cursor

Suppose we have drawn the schematic of an *RLC* circuit as Figure H.6(a) and performed the simulation to get the result as Figure H.6(b) in the PSpice A/D (Probe) window. Then you can click Trace/Cursor on the menu bar or the Toggle Cursor button on the tool bar in the Probe window to have two cross-type cursors appear on a waveform, each of which can be moved either by left-/right-clicking at any desired position or by pressing the left or right Arrow/Shift+Arrow key, where the coordinates of the cursor points on the waveform together with their pairwise differences can be seen in the Probe Cursor box. In case you have several waveforms in the Probe window, you can select the waveform to be traced using each cursor by left-/right-clicking on the graphic symbol next to the name of the variable (in the legend just below the waveform graph) which you want to trace. Once you have moved a cursor, you can find the peak|trough|minimum|maximum|... of the waveform associated with that cursor by clicking Peak|Trough|Min|Max|... on the menu bar or the corresponding button on the toolbar. You can also click Trace>Evaluate_Measurement on the menu bar or the Evaluate Measurement button on the toolbar to open the Evaluate Measurement dialog box and select a desired measurement (such as the maximum, the minimum, the center frequency, the bandwidth, etc.) and a simulation output variable so that the target trace expression will appear in the Trace Expression field at the bottom of the Evaluate Measurement dialog box as illustrated in Figure H.7. (You may directly type the target trace expression into the Trace Expression field.) Then you can click OK to have the desired measurement results appear at the bottom of the Probe window as illustrated in Figure H.6(b).

(a)

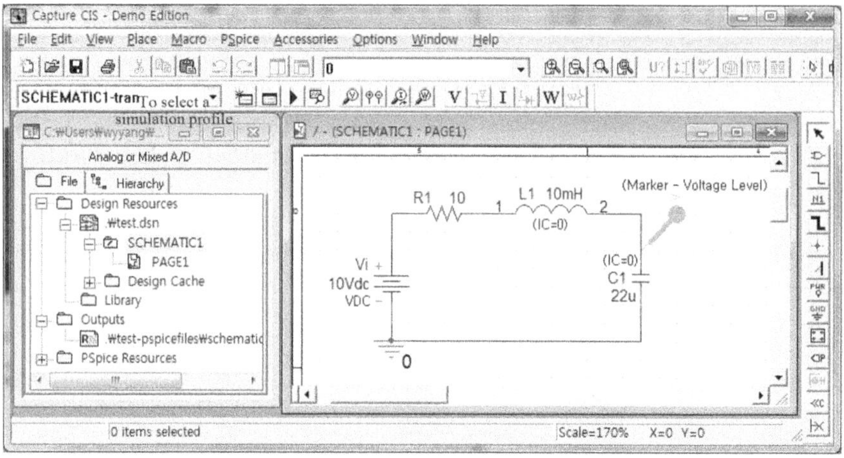

A schematic in the Schematic Editor window

(b)

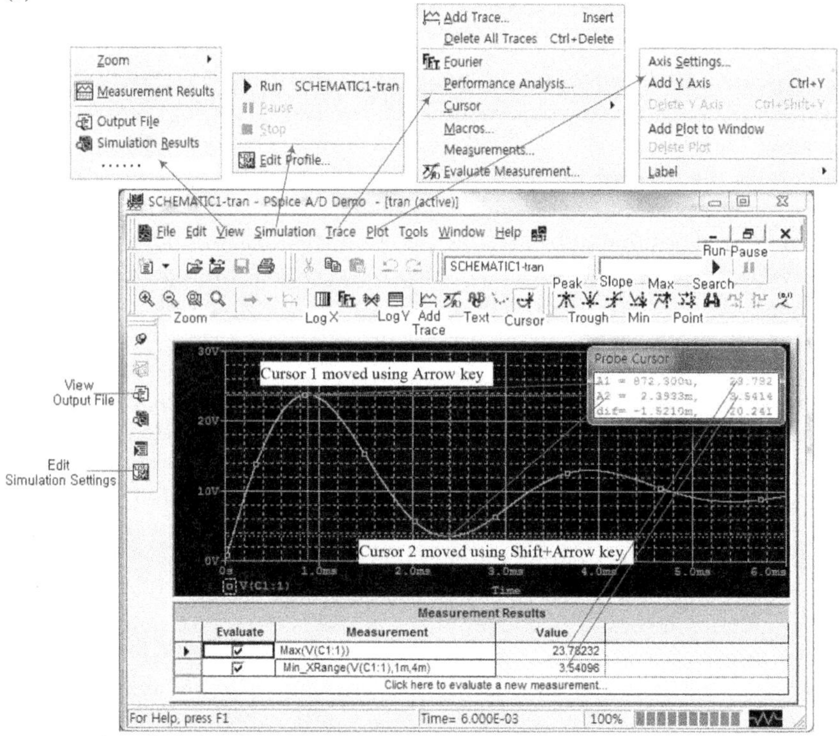

A typical signal waveform obtained from Transient Analysis and seen through the Probe window

Figure H.6 A typical schematic and its simulation result.

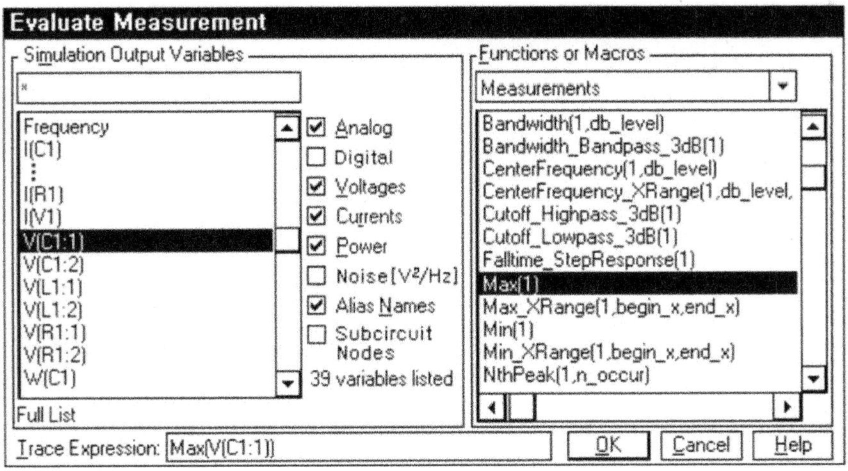

Figure H.7 Evaluate Measurement dialog box.

H.4.2 Adding/Removing Waveforms – Trace>Add_Trace or 'Ins'/Del key

To see the waveform of another variable (Figure H.8(b)), you can click the Voltage/Current/Power Marker button on the toolbar of the Schematic Editor Window and put the probe pin at the target node or element. Alternatively, you can just press the 'Ins' key or click Trace>Add_Trace on the menu bar or the Add Trace button on the toolbar (of the Probe window) to open the Add Traces dialog box (Figure H.8(a)) and enter the target trace expression into the Trace Expression field (at the bottom) directly or by selecting related variable(s) from the Simulation Output Variable list (on the left side) and/or operator/function from the Function or Macro list (on the right side). You can delete any existing waveform by clicking the variable name below the waveform graph and pressing the 'Del' key.

In case the plotted variables differ in the order of magnitude and/or the unit so that you feel like having another graph on which to plot one or some of them, you can click Plot>Add_Plot_to_ Window on the menu bar (of the Probe window) to add another graph as illustrated in Figure H.8(c), press the 'Ins' key to open the Add Traces dialog box, and then enter the trace expression to plot in the new graph. Note that you might have to increase the vertical size of the Probe window to make it accommodate additional graph(s) to be created. After getting more than one graph, you can select one of them by clicking it so that you can modify it, where the selected graph is denoted by 'SEL>>' If you want to delete a graph, click the graph for selection and then click Plot>Delete_Plot on the menu bar (of the Probe window) for deletion.

If you have got any data file on the same-named variable for the same range in another simulation, you can plot it in the same graph by clicking File>Append_Waveform(.DAT) on the menu bar or the Append File button and then selecting the data file.

(a)

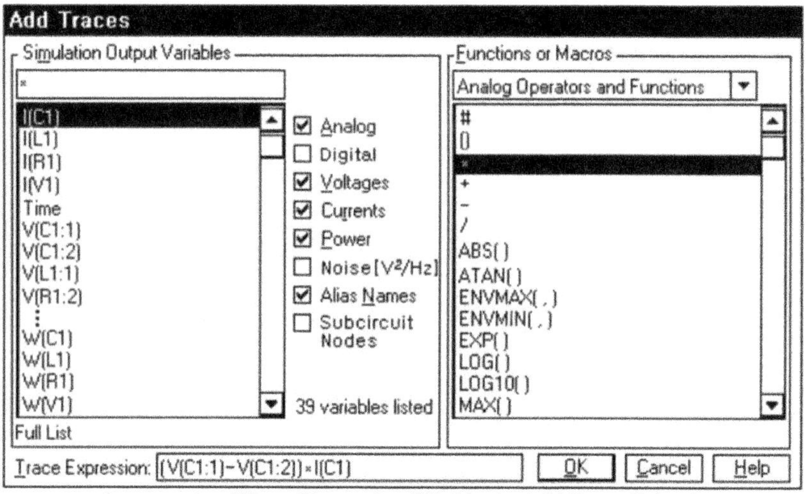

An example of Trace Expression filled in the Add Trace dialog box

(b)

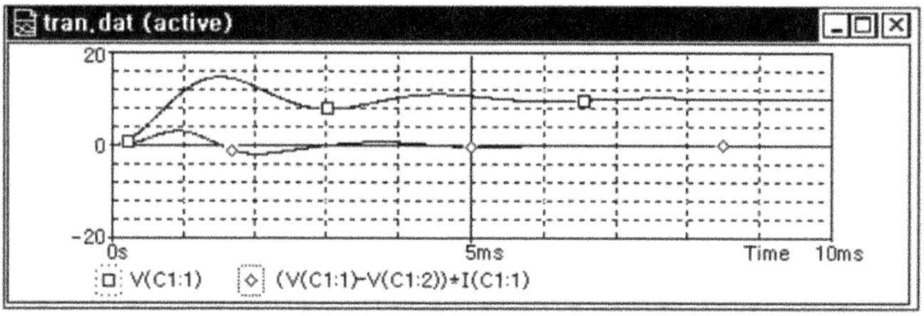

Waveforms seen through the PSpice A/D (Probe) window

(c)

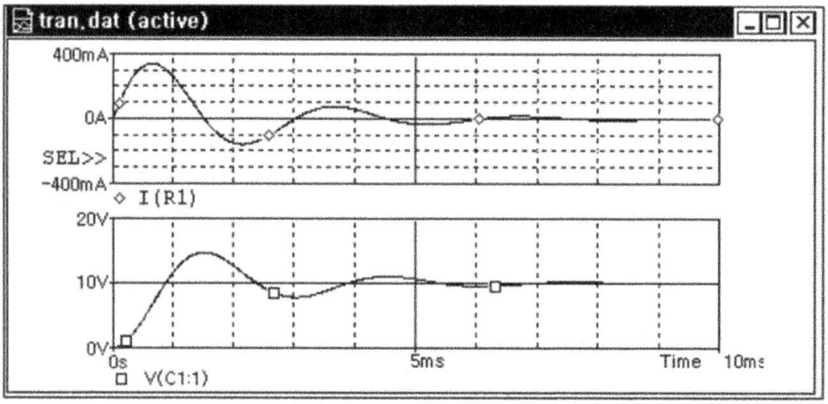

Graph added by using Plot > Add Plot and Trace > Add Trace

Figure H.8 Adding waveform(s) (using Add Trace and Add Plot).

H.4.3 Processing Waveforms/Graph

You can scale up/down the graph by either pressing the '^i'|'^o' key or clicking View>Zoom> (In|Out) on the menu bar or the Zoom_(In|Out) button on the toolbar, and then clicking at any point around which the graph will be zoomed in|out by a factor of 2. You can scale up/down a partial area of the graph by either pressing the '^a' key or clicking View>Zoom>Area on the menu bar or the Zoom_Area button on the toolbar, and then dragging the mouse pointer from a corner point to the opposite one to get the area selected. You can click View>Zoom>Fit on the menu bar or the Zoom_Fit button on the toolbar to have the whole waveform(s) fit the graph .

You can also click the Plot>Axis_Settings menu or double-click the x- or y-axis to open the Axis Settings dialog box (Figure H.9) and select a tab of {X-Axis, Y-Axis,X-Grid,Y-Grid} to make a change of linear/log scales, ranges, and grid

(a)

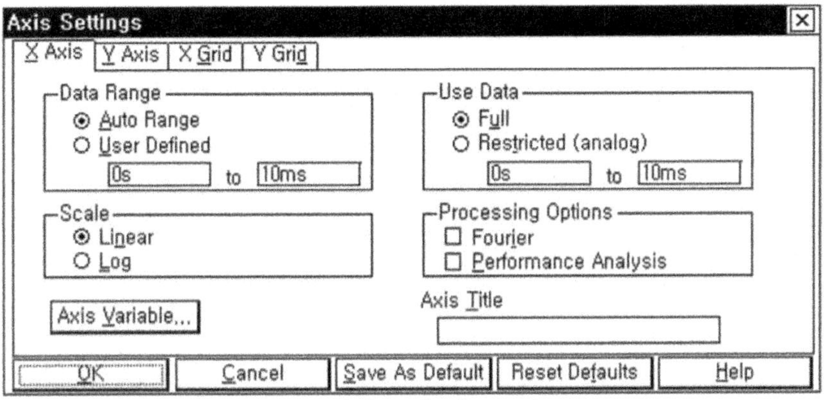

Axis setting for x-axis

(b)

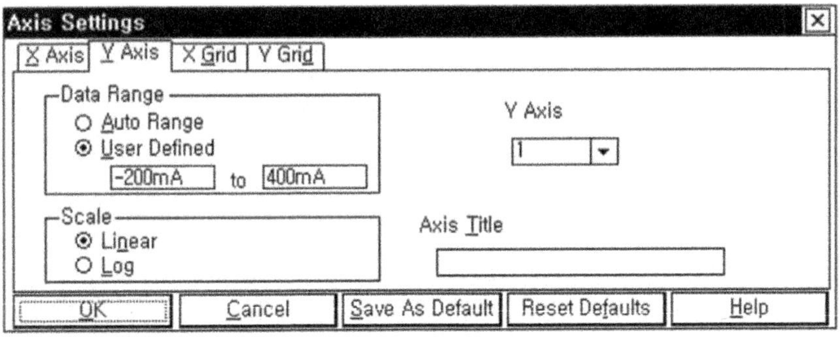

Axis setting for y-axis

Figure H.9 Axis Settings dialog box

spacings (intervals) of x-axis an y-axis. Especially, you can draw an x–y plot instead of a t–y plot by clicking the Axis Variable button (Figure H.9(a)) and then changing the x-axis variable from Time to another variable in the Axis Settings dialog box for the x-axis.

You can label the x-/y-axis by entering a title string into the Axis Title field in the Axis Settings dialog box for the x-/y-axis (Figure H.9(a)/(b)). You can also write a text string (name) and draw a line/arrow/box/circle on the graph by clicking Plot>Label>(Text|Line|Arrow|Box|Circle). You can see the Fourier spectra of signals plotted versus frequency in the Probe window just by clicking Trace>Fourier on the menu bar or the Fourier button on the toolbar.

H.4.4 Initial Transient Bias Point

The *bias point,* also referred to as the operating *point,* means the DC steady-state values of voltages and currents obtained with every inductor/capacitor short-/open-circuited and every AC source removed. PSpice finds it not only in the Bias Point analysis but also in the Transient (Time Domain) analysis (in the name of the *initial transient bias point*) as long as the SKIPBP box (in the corresponding Simulation Settings dialog box) is unchecked (as default) and the ICs of inductors and capacitors are not specified. Hence, if you want the transient analysis (of a circuit containing inductors/capacitors) to start from some other ICs than the DC steady-state values for inductor currents and capacitor voltages, check the SKIPBP box and specify the ICs in the spreadsheets of the inductors/capacitors if they are not zeros (see Exercise H.1).

To have the bias point analysis results displayed/removed on the schematic, you should click PSpice>Bias_Points>Enable_Bias_(Voltage|Current|Power)_Display on the menu bar or the corresponding button on the toolbar of the Capture window. To have the bias point analysis results printed in the output file, you should check the square box before 'Include detailed bias point information for nonlinear controlled sources and semiconductors (/OP)' in the Simulation Settings dialog box for bias point analysis or in the Transient Output File Options dialog box opened by clicking the Output File Options button in the Simulation Settings dialog box for transient analysis.

H.4.5 Viewing Output File and Netlist File

To see the simulation output file (Figure H.10), you should click PSpice>View_Output_File on the menu bar of the Capture window or the View Simulation Output File button on the left side of the Probe window (Figure H.6(b)). The output file contains the contents of the netlist file named 'schematic1.net', which shows the detailed description of the circuit. You can also see the netlist file by clicking PSpice>View_Netlist_File on the menu bar of the Capture window.

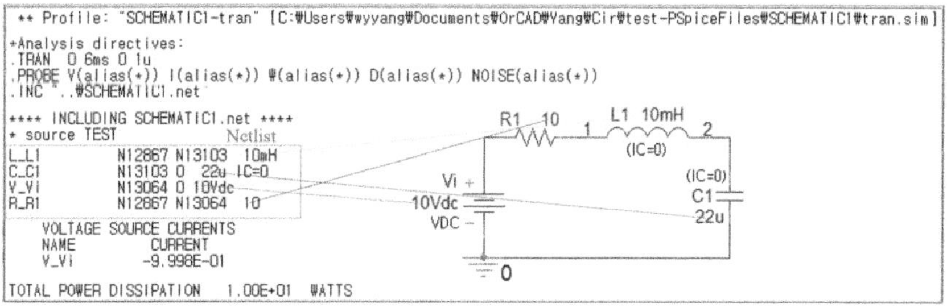

Figure H.10 An example of PSpice Simulation Output file.

H.5 Circuit Analysis Using OrCAD/Capture

H.5.1 DC Sweep Analysis

To practice the DC sweep analysis, let us get the common-emitter (CE) collector characteristic curves of a bipolar junction transistor (BJT) by simulating the circuit of Figure H.11(a). For this job, we should draw the schematic and set the parameters in the Simulation Settings dialog box as depicted in Figure H.11(c) and (d). Then we will obtain the CE collector characteristic curves of the collector current i_C versus the collector-emitter junction voltage v_{CE} for several values of the base current i_B as depicted in Figure H.11(b).

As another application example of DC sweep analysis, we can draw the schematic of a CE transistor circuit as Figure H.12(a) and run it to get the curves of the B-E junction voltage v_{BE}, the base current i_B, the collector current i_C, the collector-emitter junction voltage v_{CE}, and the DC current gain h_{FE} versus V_{BB} as depicted in Figure H.12(b1)-(b5). Instead of Voltage/Current Markers, we have clicked the Add Trace button on the tool bar of the Probe window to open the Add Traces dialog box and typed each target expression like 'V(Q1:b)-V(Q1:e)' or 'IC(Q1)/IB(Q1)' (with no quotation mark) into the Trace Expression field to obtain these plots in the Probe window.

H.5.2 Transient Analysis

Figure H.13(a),(b), and (c) shows the schematic of an OP Amp circuit, the Simulation Settings dialog box for the time-domain (transient) analysis, and the output waveforms obtained from the simulation, respectively. We may read the period of the output waveform either by using the cursor or by clicking the Evaluate Measurement button on the toolbar in the PSpice A/D (Probe) window to open the Evaluate Measurement window and by filling out the Trace Expression field as depicted in Figure H.13(d) to get the measurement result (Figure H.13(e)).

(a)

[eval.olb]
Q2N2222

0Adc i_B
(IDC)
IB

I

i_C

C +

v_{CE}

Q1

B

E −

(VDC)
+ VCE

0 Vdc

0(GND)

BJT circuit to measure the CE
collector characteristic

(b)

40 mA

i_C

$i_B = 200$ uA

$i_B = 160$ uA

20 mA

$i_B = 120$ uA

$i_B = 80$ uA

$i_B = 40$ uA

$i_B = 0$ uA

0 A

0 V

2.5 V

v_{CE} 5 V

□ IC(Q1) i_C

V_VCE

The CE collector characteristic

(c)

Simulation Settings – dc_sweep ☒

General | Analysis | Configuration Files | Options | Data Collection | Probe Window

Analysis type:
DC Sweep ▼

Options:
☑ Primary Sweep ▲
☐ Secondary Sweep
☐ Monte Carlo/Worst Case
☐ Parametric Sweep
☐ Temperature (Sweep)
☐ Save Bias Point
☐ Load Bias Point ▼

Sweep variable
⦿ Voltage source
○ Current source
○ Global parameter
○ Model parameter
○ Temperature

Name: VCE v_{CE}
Model type: ▼
Model
Parameter

Sweep type
⦿ Linear
○ Logarithmic Decade ▼
○ Value list

Start value: 0
End value: 5
Increment: 0.1

OK | Cancel | Apply(A) | Help

The Simulation Settings dialog box for the primary sweep variable VCE

(d)

Simulation Settings – dc_sweep ☒

General | Analysis | Configuration Files | Options | Data Collection | Probe Window

Analysis type:
DC Sweep ▼

Options:
☐ Primary Sweep ▲
☑ Secondary Sweep
☐ Monte Carlo/Worst Case
☐ Parametric Sweep
☐ Temperature (Sweep)
☐ Save Bias Point
☐ Load Bias Point ▼

Sweep variable
○ Voltage source
⦿ Current source
○ Global parameter
○ Model parameter
○ Temperature

Name: IB i_B
Model type: ▼
Model
Parameter

Sweep type
⦿ Linear
○ Logarithmic Decade ▼
○ Value list

Start value: 0
End value: 200u
Increment: 40u

OK | Cancel | Apply(A) | Help

The Simulation Settings dialog box for the secondary sweep variable IB

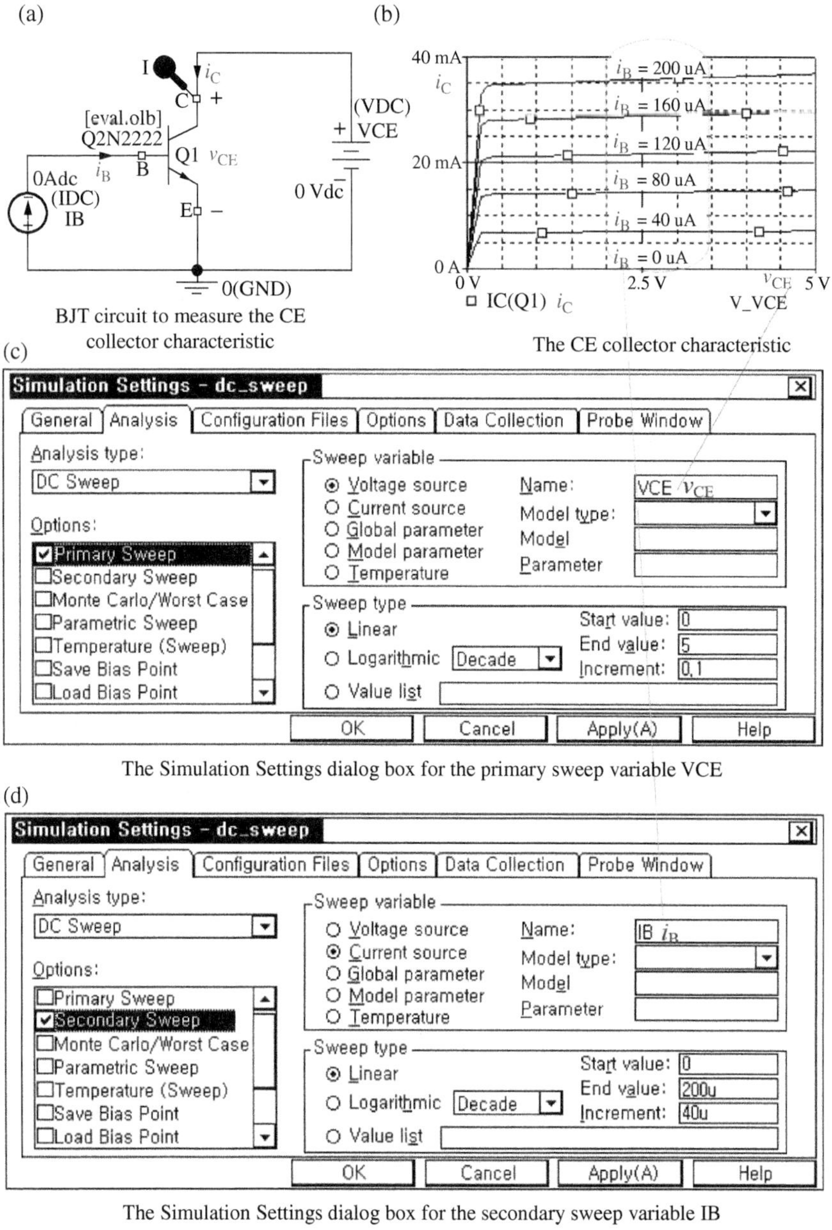

Figure H.11 Obtaining the common-emitter (CE) collector characteristic curves
("BJT_CE0.opj").

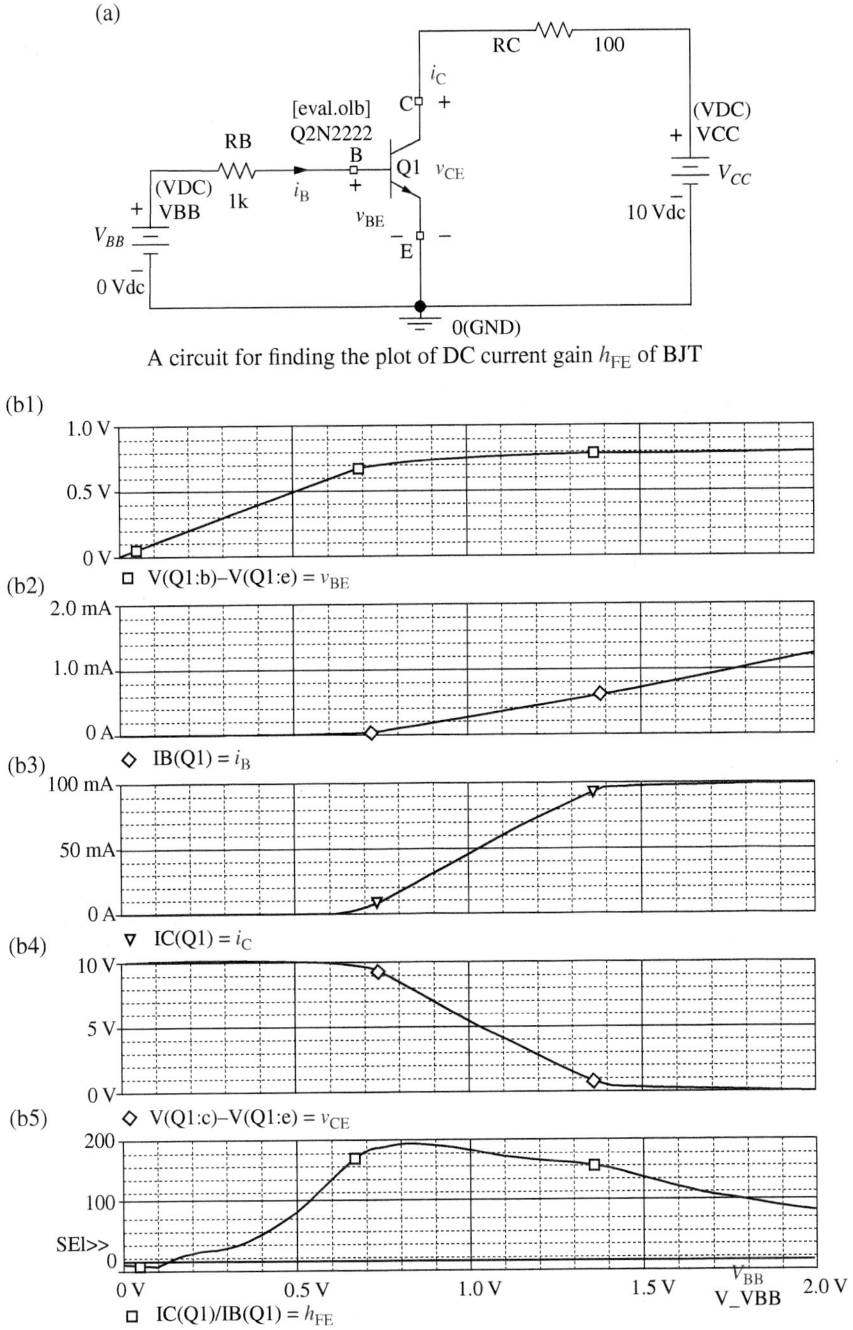

Figure H.12 Plots of v_{BE}, i_B, i_C, v_{CE}, and h_{FE} versus V_{BB} ("BJT_hFE.opj").

(a)

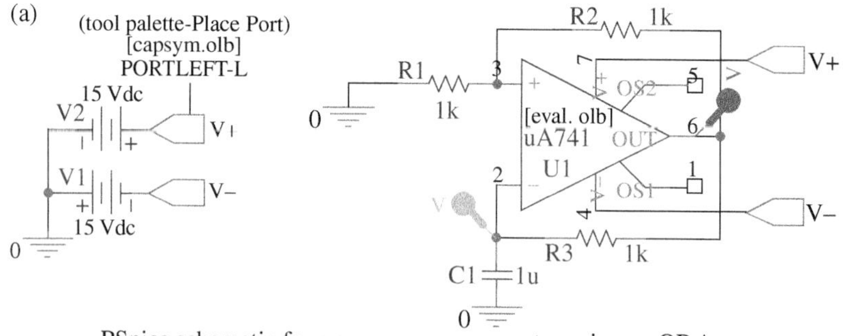

PSpice schematic for a square-wave generator using an OP Amp

(b)

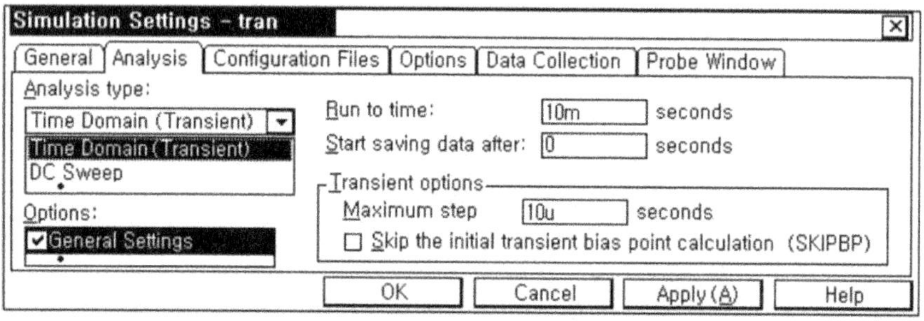

Simulation Settings dialog box for time-domain (transient) analysis

(c)

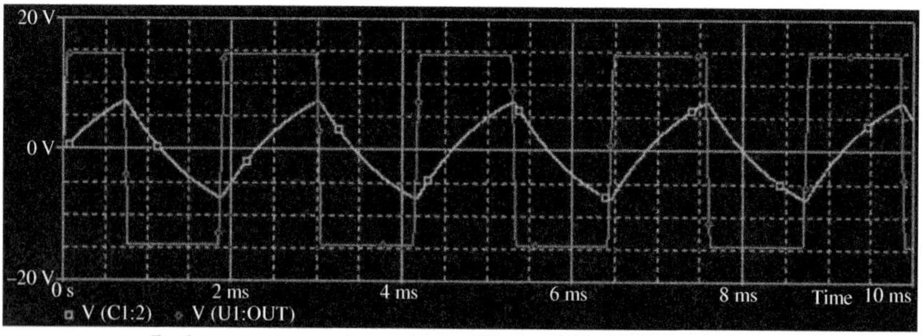

Probe output waveforms in the PSpice A/D (Probe) window

(d)

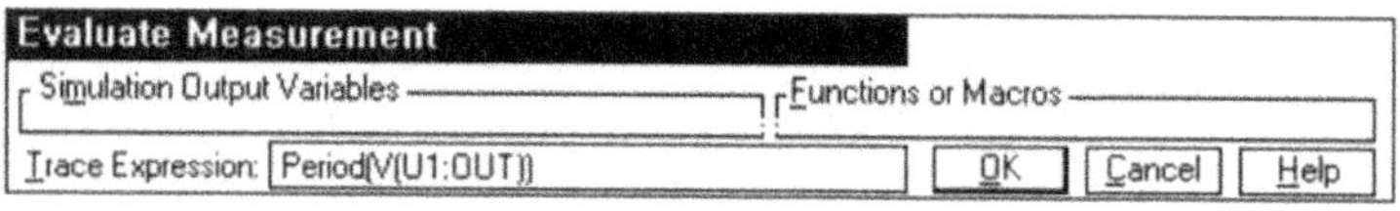

Evaluate Measurement dialog box

(e)

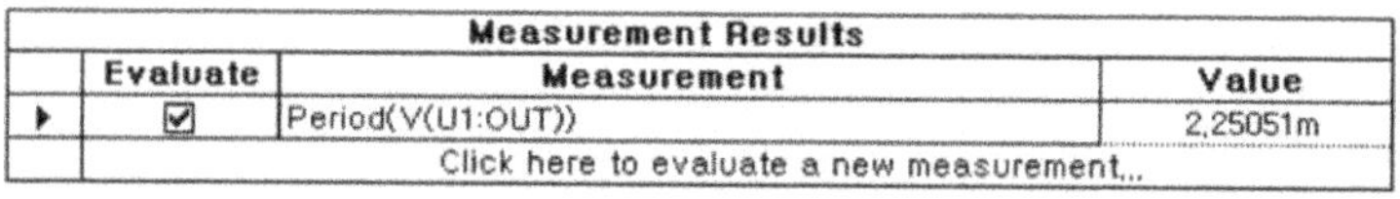

Measurement Results box

Figure H.13 PSpice simulation of an OP Amp circuit ("OPAmp_pos_feedback.opj").

Figure H.14(a) shows a logic circuit for one-bit full adder, which gets three inputs A_n, B_n, and C_{n-1} (carry from bit $n-1$) and computes two outputs S_n (sum) and C_n (carry) as

$$\begin{aligned} S_n &= (\bar{A}_n\bar{B}_n + A_nB_n)C_{n-1} + (\bar{A}_nB_n + A_n\bar{B}_n)\bar{C}_{n-1} \\ &= \overline{\mathrm{XOR}(A_n,B_n)}\,C_{n-1} + \mathrm{XOR}(A_n,B_n)\bar{C}_{n-1} \\ &= \mathrm{XOR}(\mathrm{XOR}(A_n,B_n),C_{n-1}) \end{aligned} \tag{H.1}$$

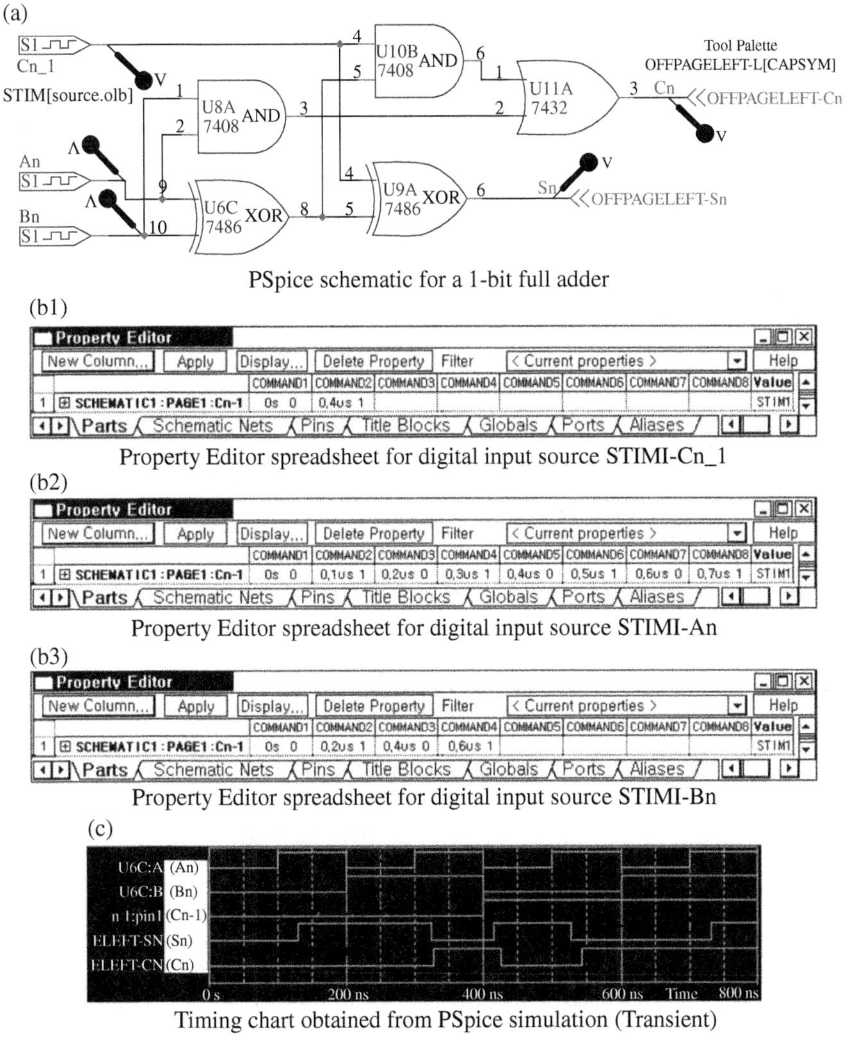

Figure H.14 PSpice simulation of a 1-bit full adder ("full_adder.opj").

$$C_n = A_n B_n (C_{n-1} + \bar{C}_{n-1}) + (\bar{A}_n B_n + A_n \bar{B}_n) C_{n-1} \\ = A_n B_n + \mathrm{XOR}(A_n, B_n) C_{n-1} \tag{H.2}$$

Figure H.14(b1)-(b3) and (c) shows the Property Editor spreadsheets for each of the three inputs and the timing chart obtained from the transient analysis and seen in the PSpice A/D (Probe) window, respectively.

H.5.3 AC Sweep Analysis

Figure H.15(a), (b), and (c) shows the schematic of an OP Amp circuit, the Simulation Settings dialog box for the frequency-domain (AC sweep) analysis, and the frequency response curve obtained from the simulation. We may read the

(a)

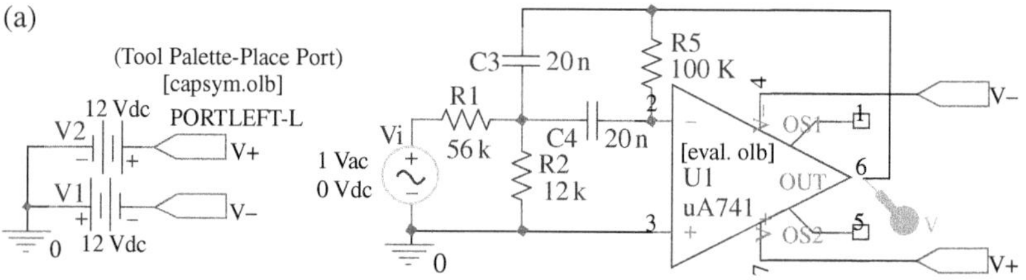

PSpice schematic for an active 2nd-order BPF

(b)

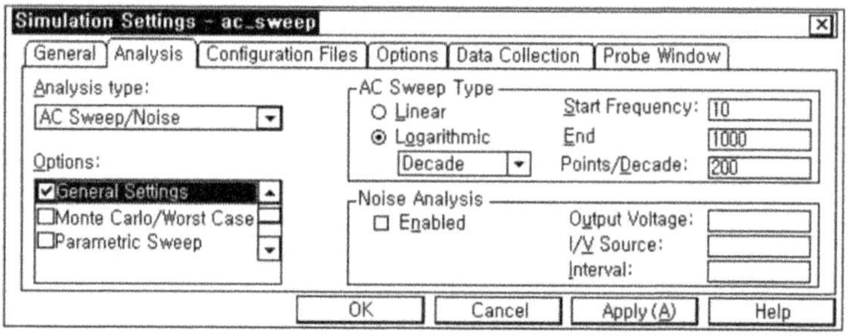

Simulation Settings dialog box for AC Sweep analysis to get the frequency response

(c)

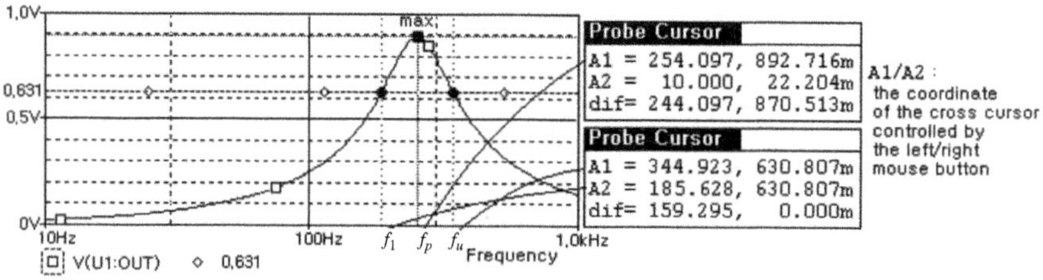

Magnitude curve of the frequency response in the PSpice A/D (Probe) window

Figure H.15 PSpice simulation to get a frequency response ("OPAmp_filter_ac_sweep.opj").

(d)

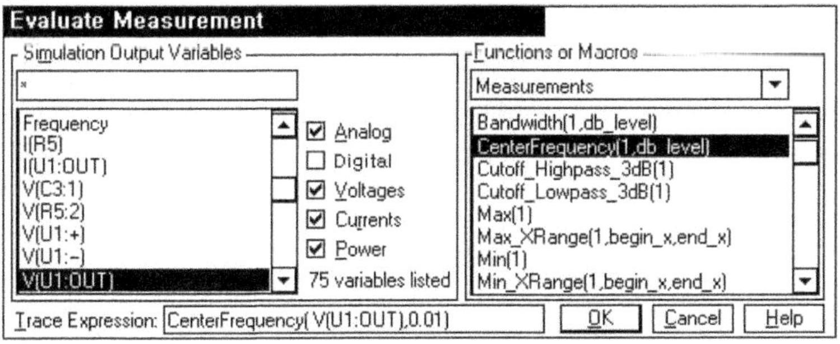

Evaluate Measurement dialog box

(e)

Measurement Results			
	Evaluate	Measurement	Value
	☑	Cutoff_Lowpass_3dB(V(U1:OUT))	344,53009
	☑	Bandwidth_Bandpass_3dB(V(U1:OUT))	158,68715
	☑	CenterFrequency(V(U1:OUT),0,01)	253,19116
▶	☑	Cutoff_Highpass_3dB(V(U1:OUT))	185,84294
	Click here to evaluate a new measurement...		

Measurement Results box

Figure H.15 (Continued)

peak or center frequency f_p and the lower/upper 3dB frequencies f_l and f_u of the band-pass filter (BPF) by using the cursor or the Evaluate Measurement function as follows.

1) To find the peak or center frequency f_p at which the magnitude of the frequency response achieves its maximum, click the Toggle Cursor button to activate the cursor and then click the Cursor Max button on the toolbar of the Probe window, which will show you the cursor at the maximum together with the coordinates $(f_p, |G(j2\pi f_p)|)$=(254.097, 892.716m) in the Probe Cursor box (Figure H.15(c)).
2) To find the two 3dB frequencies f_l and f_u at which the magnitude of the frequency response is $(1/\sqrt{2})$ times its maximum, that is, $0.893/\sqrt{2} = 0.631$, let us draw a horizontal line of $y = 0.631$ by clicking the Add Trace button to open the Add Traces dialog box and typing '0.631' (with no quotation mark) into the Trace Expression field. Then position the cursor at the two intersection points of the frequency response curve and the horizontal line of $y = 0.631$ by left-/right-clicking at the target points or pressing the (left or right) arrow/Shift + arrow key and read the coordinates of the two points from the Probe Cursor box, which will be

$$f_l \approx 185.6\,\text{Hz and } f_u \approx 344.9\,\text{Hz}$$

respectively (see Figure H.15(c)). It is also implied that the bandwidth is $B = f_u - f_u \approx 159.3$ Hz.

3) Once we have got the frequency response in the Probe window, we can easily find the center frequency, the 3 dB frequencies, and the bandwidth by using the Evaluate Measurement function. That is, you can click the Evaluate Measurement button to open the Evaluate Measurement dialog box and do the following:

   - Fill out the Trace Expression field with

Cutoff_Highpass_3dB(V(U1:OUT))

     and click OK. Then you will see the value of the lower 3dB frequency in the Measurement Results box appearing (below the graph) at the bottom of the Probe window. Consecutively, click the bottom part of the Measurement Results box saying that

Click here to evaluate a new measurement ..

     which will reopen the Evaluate Measurement dialog box. Then in the same way, you can type

Center Frequency(V(U1:OUT),0.01)
Bandwidth_Bandpass_3dB(V(U1:OUT))
Cutoff_Lowpass_3dB(V(U1:OUT))

     one at a time, which will make the center frequency, the bandwidth, and the upper 3 dB frequency appear in the Measurement dialog box as illustrated in Figure H.15(e).

H.5.4 Parametric Sweep Analysis

While DC Sweep (Section H.5.1) and AC Sweep (Section H.5.3) analyses aim at observing the variation of the output versus the value of a DC source or the frequency of an AC source, Parametric Sweep analysis aims at observing the variation of the output versus the value of *R*, *L*, *C*, a model parameter, temperature, etc. Let us observe how the frequency response of the BPF in Figure H.16(a) varies with the value of $R_5 = 50$, 100, and 150 kΩ by taking the following steps:

1) Click the value of R_5 (to be varied) and set it to {Rvar} (including the braces) in the schematic (Figure H.16(a)).
2) Click the Place Part button on the tool palette of the Schematic window to open the Place Part dialog box, get PARAM (contained in the library 'special.olb'), and click OK to place it somewhere in the schematic (Figure H.16(a)).

(a)

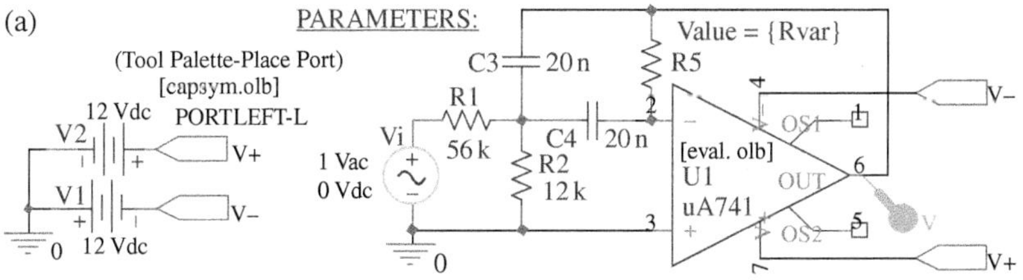

PSpice schemetic for an active 2nd-order BPF

(b)

Property Editor

New Column... | Apply | Display... | Delete Property | Filter < Current properties > | Help

		Color	Primitive	Reference	Source Library	Source Part	Value
1	⊞ SCHEMATIC1 : PAGE1 : 1	Default	DEFAULT	1	C:/ORCAD/...	PARAM.Normal	PARAM

Parts / Schematic Nets / Pins / Title Blocks / Globals / Ports / Aliases

Property Editor spreadsheet for PARAM-Parametric Sweep

(c)

Add New Column

Name:
Rvar
Value:
100
Enter a name and click Apply or OK to add a column/row to the property editor and optionally the current filter (but not the <Current properties> filter).
No properties will be added to selected objects until you enter a value here or in the newly created cells in the property editor spreadsheet.
☐ Always show this column/row

Add New Column dialog box popped out of the above Property Editor spreadsheet

(d)

		Color	Primitive	Reference	Rvar	Source Library	Source Part	Value
1	⊞ SCHEMATIC1 : PAGE1 : 1	Default	DEFAULT	1	100	C:/ORCAD/...	PARAM.Normal	PARAM

Property Editor spreadsheet for PARAM changed by adding a new column

Figure H.16 Multiple AC Sweeps using Parametric Sweep ("OPAmp_filter_par.opj").

3) Double-click the PARAMETERS placed in the schematic to open the Property Editor spreadsheet (Figure H.16(b)) and click the New Column button to open the Add New Column box (Figure H.16(c)), in which you can type Rvar and 100 into the Name and Value fields, respectively, where the numerical value entered as 100 here does not matter.
4) Click the New Simulation Profile button to create a new simulation profile named, say, 'AC_sweep', click the Edit Simulation Profile button to open the Simulation Settings dialog box, and fill it out as depicted in Figure H.16(e), where the menu of Options is selected as Parametric Sweep, the Sweep variable is chosen to be Global parameter named 'Rvar', and the Sweep type

(e)

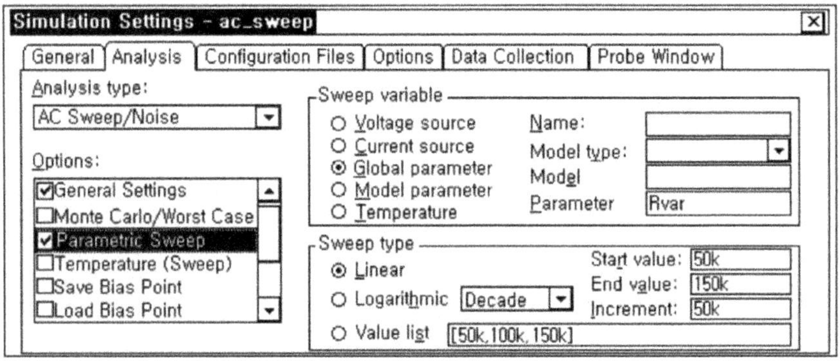

Simulation Settings dialog box

(f)

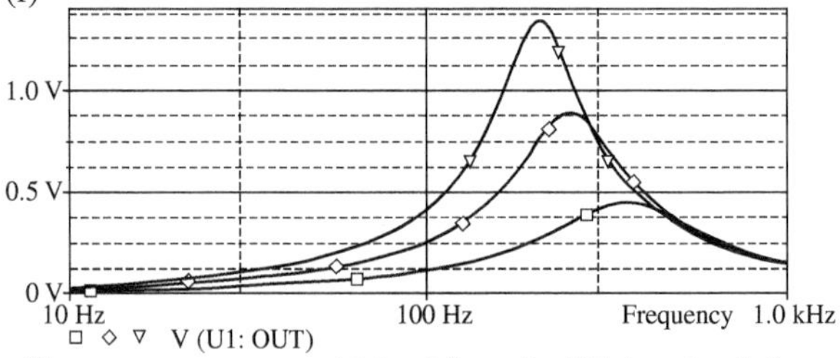

Frequency responses obtained from the PSpice simulation

(g)

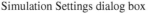

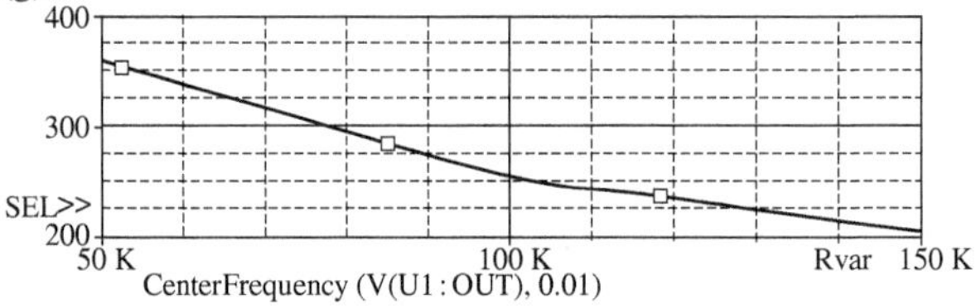

Result of Trace/Performance Analysis w.r.t. the center frequency

Figure H.16 (Continued)

is set to Linear with the Start value 50k, the End value 150k, and the Increment 50k. If the values of parameters do not form an arithmetic progression (with constant difference), you may type the list of the parameter values into the Value list field as illustrated in Figure H.16(e). Then click OK to close the Simulation Settings dialog box.

5) Click the Voltage/Level Marker button to put the voltage probe pin at the OP Amp output node, click the Run button to perform the simulation, and see the multiple frequency curves as illustrated in Figure H.16(f).

6) To see how the center frequency f_p varies with the value of R_5 = {Rvar}, click Trace/Performance Analysis on the menu bar of the PSpice A/D (Probe) window to create an additional graph, click the Add Trace button to open the Add Traces dialog box, and fill out the Trace Expression field with

```
CenterFrequency(V(U1:OUT),0.01)
```

Then click OK to see the plot of f_p versus R_5 as depicted in Figure H.16(g).

H.5.5 Transient/Fourier Analysis

Figure H.17(a) shows the PSpice schematic of an *RC* circuit excited by a rectangular (square) wave voltage source of height $\pm V_m = \pm\pi$, period $T = 2$ [s], and duration (pulse width) $D = 1$ [s], where the voltage across the capacitor is taken

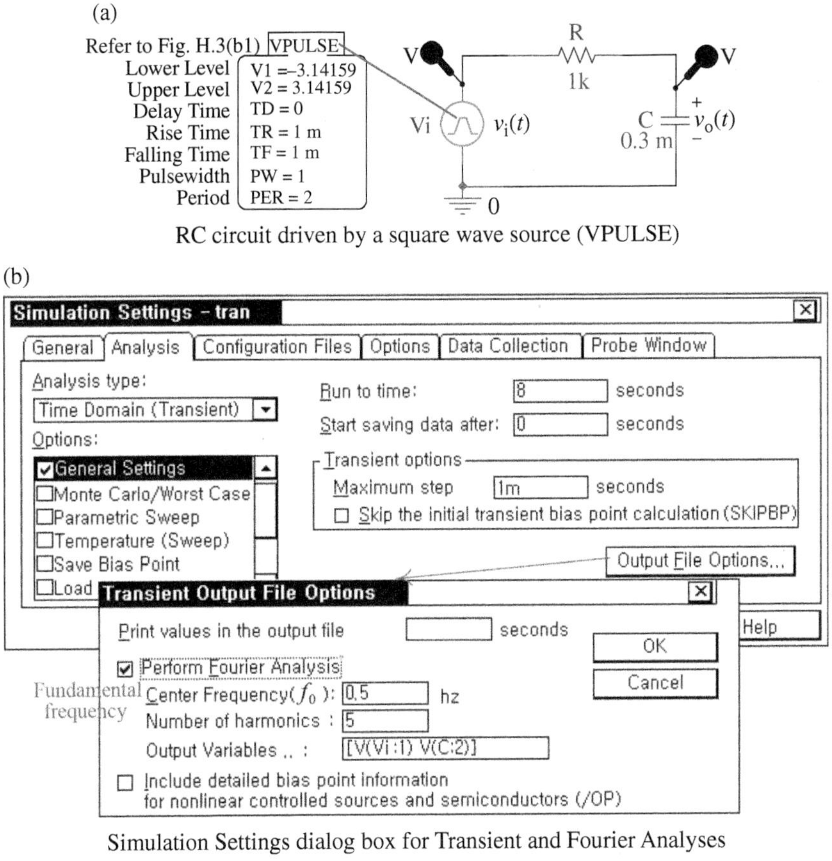

Figure H.17 Example of Fourier analysis using PSpice ("RC_pulse_Fourier.opj").

(c)

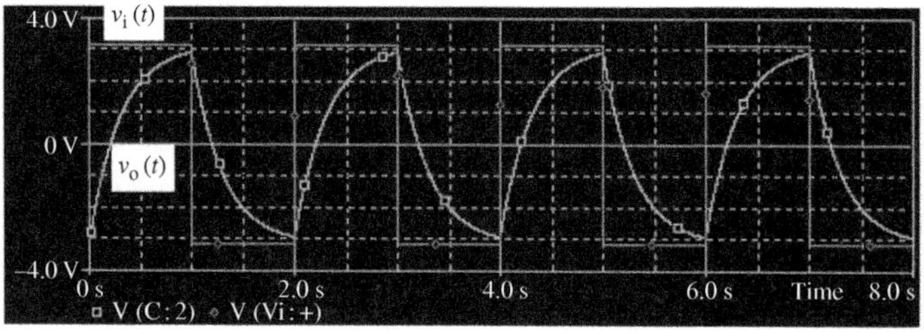

The voltage generated by the square wave voltage source Vi and the voltage across the capacitor C

(d)

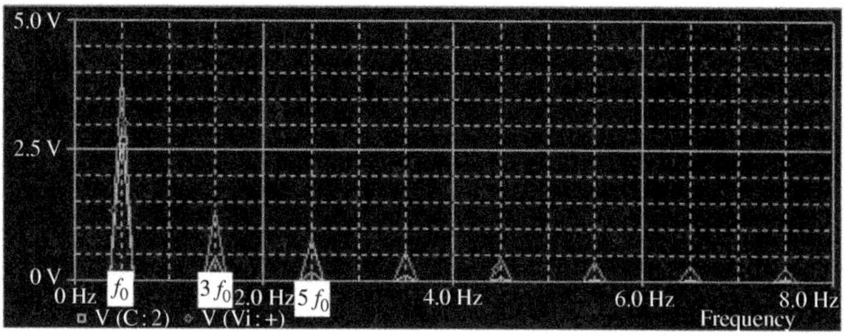

Fourier spectra of Vi and V(C) obtained by clicking on the FFT button in the tool bar

(e)

tran.out

FOURIER COMPONENTS OF TRANSIENT RESPONSE V(V_Vi)

DC COMPONENT = -3.110485E-02 (d_0)

HARMONIC NO (k)	FREQUENCY (HZ)	FOURIER COMPONENT (d_k)	NORMALIZED COMPONENT	PHASE (DEG) $(\;_k)$	NORMALIZED PHASE (DEG)
1	5.000E-01 (f_0)	4.000E+00	1.000E+00	-8.911E-01	0.000E+00
2	1.000E+00	6.224E-02	1.556E-02	-9.178E+01	-9.000E+01
3	1.500E+00 $(3f_0)$	1.334E+00	3.334E-01	-2.673E+00	-3.002E-09
4	2.000E+00	6.233E-02	1.558E-02	-9.356E+01	-9.000E+01
5	2.500E+00 $(5f_0)$	8.008E-01	2.002E-01	-4.455E+00	-1.487E-08

TOTAL HARMONIC DISTORTION = 3.895449E+01%

FOURIER COMPONENTS OF TRANSIENT RESPONSE V(N00580)

DC COMPONENT = 2.539615E-03 (d_0)

HARMONIC NO (k)	FREQUENCY (HZ)	FOURIER COMPONENT (d_k)	NORMALIZED COMPONENT	PHASE (DEG) $(\;_k)$	NORMALIZED PHASE (DEG)
1	5.000E-01 (f_0)	2.923E+00	1.000E+00	-4.400E+01	0.000E+00
2	1.000E+00	2.537E-03	8.681E-04	1.408E+01	1.021E+02
3	1.500E+00 $(3f_0)$	4.318E-01	1.477E-01	-7.236E+01	5.965E+01
4	2.000E+00	1.623E-03	5.553E-04	-1.428E+01	1.617E+02
5	2.500E+00 $(5f_0)$	1.588E-01	5.435E-02	-7.732E+01	1.427E+02

TOTAL HARMONIC DISTORTION = 1.574085E+01%

Fourier analysis result listed in the output file

Figure H.17 (Continued)

as the output. To have the Fourier analysis results printed in the output file, we set the simulation profile and click the Output File Options button to open the Transient Output File Options dialog box, in which you are supposed to enter the fundamental frequency $f_0 = 1/T = 1/2 = 0.5$ Hz, the number of harmonics which you want to observe, and the list of output variable(s) for Fourier analysis into the corresponding fields, say, as illustrated in Figure H.17(b). Now, we put the Voltage/Level probe pins at the input/output nodes, click the Run button to perform the simulation, and see the input/output waveforms in the Probe window as illustrated in Figure H.17(c). Whenever you want to see the spectra of the signals (Figure H.17(d)) whose waveforms are shown in the Probe window, you just click the (toggle) fast Fourier transform (FFT) button on the toolbar. You can also view the output file as shown in Figure H.17(e) by clicking (selecting) PSpice/View_Output_File on the top menu bar and pull-down menu in the Capture window or View/Output_File on the menu bar or the View_Simulation_Output_File button on the left toolbar of the Probe window.

Note that based on the Fourier analysis results printed in the output file, we can write the Fourier series representation of the input and output signals as

$$\begin{aligned} v_i(t) &= \sum_{k=0}^{\infty} d_k \sin(2\pi k f_0 t + \phi_k) \ (f_0 = 0.5\,\text{Hz}) \\ &\cong -0.031 + 4\sin(\pi t - 0.89^\circ) + 0.062\sin(2\pi t - 91.78^\circ) \\ &\quad + 1.334\sin(3\pi t - 2.67^\circ) + \ldots \end{aligned} \tag{H.3}$$

$$v_o(t) \cong 2.92\sin(\pi t - 44^\circ) - 0.43\sin(3\pi t - 72^\circ) + 0.16\sin(5\pi t - 77^\circ) + \cdots \tag{H.4}$$

and the total harmonic distortion (THD) is defined as

$$THD \overset{(1.5.11)}{=} \sqrt{\frac{\sum_{k=2}^{\infty} d_k^2}{d_1}} \tag{H.5}$$

H.5.6 Bias Point/DC Sensitivity and Transfer Function Analysis

Figure H.18(a1) and (a2) shows two interface circuits each of which takes on a duty of transferring the DC voltage of 9 V from the 12 V source (with the source impedance of 1.5 kΩ) to a 9 kΩ resistor. Figure H.18(b) shows the Simulation Settings dialog box to be set for DC Sensitivity analysis of the output voltage w.r.t. the load resistance and Transfer Function analysis to find the DC gain and the input/output resistances, where 'V(RL2)' should be typed into the Output and To_Output variable fields instead of 'V(RL1)' for the analysis of circuit (a2). Figure H.18(c1)/(c2) and (d1)/(d2) shows the results of the DC Sensitivity and Transfer Function analyses for the two circuits (a1)/(a2), that can be seen from the output file.

(a1)

Rs1   RS
1.5 k   1.5 k

12 Vdc +
Vs1 ⎓

RL1
9 k $v_o(t)$

0

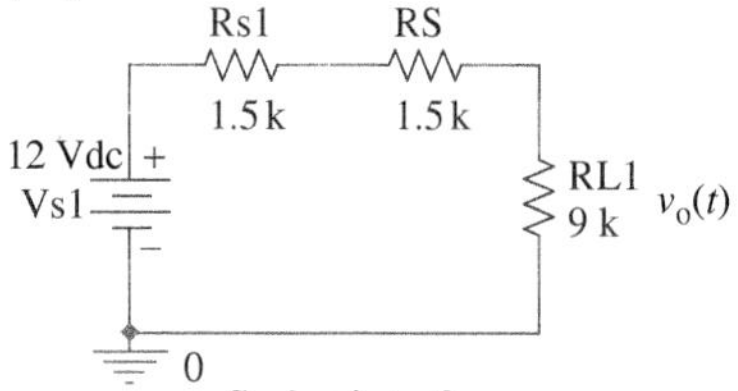

Series interface

(a2)

Rs2
1.5 k

12 Vdc +
Vs1 ⎓

RP
9 k

RL2
9 k $v_o(t)$

0

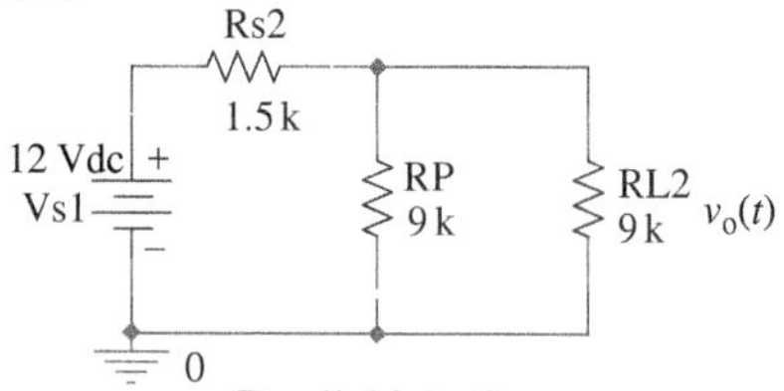

Parallel interface

(b)

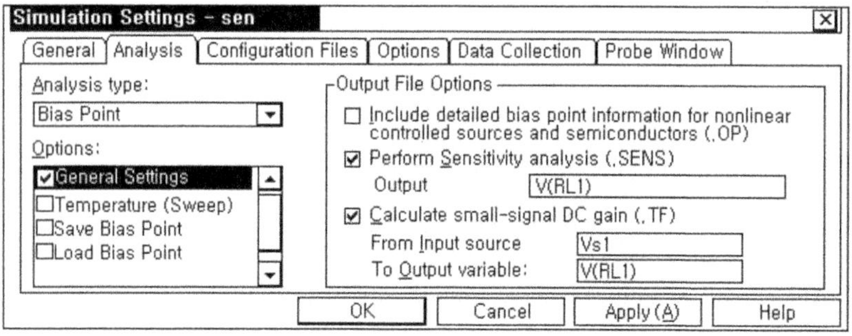

Simulation Settings – sen   ☒

| General | Analysis | Configuration Files | Options | Data Collection | Probe Window |

Analysis type:

Bias Point ▼

Options:

☑ General Settings ▲
☐ Temperature (Sweep)
☐ Save Bias Point
☐ Load Bias Point ▼

Output File Options

☐ Include detailed bias point information for nonlinear controlled sources and semiconductors (.OP)

☑ Perform Sensitivity analysis (.SENS)

Output   V(RL1)

☑ Calculate small-signal DC gain (.TF)

From Input source   Vs1
To Output variable: V(RL1)

| OK | Cancel | Apply (A) | Help |

Simulation Settings dialog box for DC senitivity and Transfer Function analyses

(c1)

DC SENSITIVITIES OF OUTPUT V(R_RL1)

ELEMENT NAME	ELEMENT VALUE	ELEMENT SENSITIVITY (VOLTS/UNIT)	NORMALIZED ENSITIVITY (VOLTS/PERCENT)
R_Rs1	1.500E+03	-7.500E-04	-1.125E-02
R_RS	1.500E+03	-7.500E-04	-1.125E-02
R_RL1	9.000E+03	2.500E-04	2.250E-02

The result of DC Sensitivity analysis for the circuit in (a1)

(c2)

DC SENSITIVITIES OF OUTPUT V(R_RL2)

ELEMENT NAME	ELEMENT VALUE	ELEMENT SENSITIVITY (VOLTS/UNIT)	NORMALIZED SENSITIVITY (VOLTS/PERCENT)
R_Rs2	1.500E+03	1.500E-03	2.250E-02
R_RP	9.000E+03	1.250E-04	1.125E-02
R_RL2	9.000E+03	1.250E-04	1.125E-02

The result of DC Sensitivity analysis for the circuit in (a2)

(d1)

```
****     SMALL-SIGNAL CHARACTERISTICS

V(R_RL2)/V_Vs1 =  7.500E-01
INPUT RESISTANCE AT V_Vs1 =  1.200E+04
OUTPUT RESISTANCE AT V(R_RL1) =  2.250E+03
```
The result of Transfer Function analysis for the circuit in (a1)

(d2)

```
****     SMALL-SIGNAL CHARACTERISTICS

V(R_RL2)/V_Vs1 =  7.500E-01
INPUT RESISTANCE AT V_Vs1 =  6.000E+03
OUTPUT RESISTANCE AT V(R_RL2) =  1.125E+03
```
The result of Transfer Function analysis for the circuit in (a2)

Figure H.18 DC Sensitivity and Transfer Function analyses ("test_sensitivity.opj").

In order to understand and verify these analysis results, let us find the sensitivities, DC gains, and the input/output impedances by using an analytical method as follows:

a) For the circuit (a), we have

$$v_o(R_{L1}) = \frac{R_{L1}}{R_{s1} + R_S + R_{L1}} V_{s1} = \frac{R_{L1}}{3 + R_{L1}} 12$$

$$\text{Sensitivity of } v_o(R_{L1}) \text{ w.r.t. } R_{L1}\,|_{R_{L1}=9\text{k}\Omega} = \left.\frac{dv(R_{L1})}{dR_{L1}}\right|_{R_{L1}=9\text{k}\Omega}$$

$$= \left.\frac{3V_{s1}}{(3 + R_{L1})^2}\right|_{R_{L1}=9\text{k}\Omega} = \frac{1}{4\text{k} = 4000} = 2.5\text{E-}4$$

$$\text{DC gain} = \left.\frac{v(R_{L1})}{V_{s1}}\right|_{R_{L1}=9\text{k}\Omega} = \left.\frac{R_{L1}}{3 + R_{L1}}\right|_{R_{L1}=9\text{k}\Omega} = \frac{3}{4} = 7.5\text{E-}1$$

$$\text{Input resistance} = R_{s1} + R_S + R_{L1} = 1.5 + 1.5 + 9 = 12\,\text{k}\Omega = 1.2\text{E+}4$$

$$\text{Output resistance} = (R_{s1} + R_S)\|R_{L1} = (1.5 + 1.5)\|9 = \frac{3 \times 9}{3 + 9}$$

$$= 2.25\text{k}\Omega = 2.25\text{E+}3$$

b) For the circuit (b), we have

$$v_o(R_{L2}) = \frac{R_P \| R_{L2}}{R_{s2} + (R_P \| R_{L2})} V_s = \frac{9 \times R_{L2}/(9 + R_{L2})}{1.5 + 9 \times R_{L2}/(9 + R_{L2})} 12 = \frac{9R_{L2}}{13.5 + 10.5R_{L2}} 12$$

$$\text{Sensitivity of } v_o(R_{L2}) \text{ w.r.t. } R_{L2}\,|_{R_{L2}=9\text{k}\Omega} = \left.\frac{dv(R_{L2})}{dR_{L2}}\right|_{R_{L2}=9\text{k}\Omega}$$

$$= \left.\frac{9 \times 13.5V_{s1}}{(13.5 + 10.5R_{L2})^2}\right|_{R_{L2}=9\text{k}\Omega} = 1.25\text{E-}4$$

$$\text{DC gain} = \left.\frac{v(R_{L2})}{V_{s1}}\right|_{R_{L2}=9\text{k}\Omega} = \left.\frac{9R_{L2}}{13.5 + 10.5R_{L2}}\right|_{R_{L2}=9\text{k}\Omega} = \frac{3}{4} = 7.5\text{E-}1$$

$$\text{Input resistance} = R_{s2} + (R_P\|R_{L2}) = 1.5 + \frac{9 \times 9}{9 + 9} = 6\,\text{k}\Omega = 6\text{E+}3$$

$$\text{Output resistance} = R_{s2}\|R_P\|R_{L2} = \frac{1}{1/1.5 + 1/9 + 1/9} = 1.125\,\text{k}\Omega = 1.125\text{E+}3$$

H.5.7 Monte Carlo Analysis

Figure H.19(a) shows the Capture window, which has the schematic of a CE BJT amplifier circuit. To see how the output voltage varies with the ±20% change of the load resistance R_L(RL) and the ±50% change of the CE forward current gain β_F, let us take the following steps of Monte Carlo/Worst-Case analysis presented by PSpice.

1) Create a PSpice project, say, named 'BJT_CE_Amp1_mon.opj' and draw a schematic in the Schematic Editor Window as depicted in Figure H.19(a).
2) Click the BJT(Q1) for selection so that it will be pink-hot and boxed by a dotted-line rectangle. Then click Edit/PSpice Model on the menu bar to open the PSpice Model Editor Window (Figure H.19(b)), in which you can type 'Dev=50%' (including the first blank space, but without the single quotation marks) just after 'Bf=255.9' in the model statement. Then press the '^s' key to save the PSpice model for the device Q2N2222 placed in the schematic and click on x to close the Editor Window.
3) Click the load resistor RL for selection and press the 'Del' key to delete it. Then click the Place Part button to open the Place Part dialog box, search/select Rbreak[breakout.olb], and click OK to place it in place of RL.
4) Click the load resistor RL (having the part name 'Rbreak') for selection, click Edit/PSpice Model on the menu bar to open the PSpice Model Editor Window (Figure H.19(b)), in which you can change the model name from Rbreak for RMonte1, and type 'Dev=20%' just after 'R=1' (the default value of the resistance) in the model statement. Then press the '^s' key to save the PSpice model for the device RMonte1 placed in the schematic and click on x to close the Editor Window.
5) Click the New Simulation Profile button on the toolbar to create a new simulation profile named, say, 'tran' and then click the Edit Simulation Settings button to open the Simulation Settings dialog box, in which you can do the following:

   - Set the Analysis type and Run_to_time to 'Time Domain (Transient)' and '2ms' respectively.
   - Select (click) Monte Carlo/Worst Case from the Options menu, select Monte Carlo from the upper-right menu, type 'V(RL:2)' into the Output variable field, and type '5' into the Number of Runs field, ... as depicted in Figure H.19(c).
   - Then click OK to close the Simulation Settings dialog box.

6) Click the Voltage/Level Marker button to place two voltage probe pins, one at the upper node of RL and one at the + terminal of the sinusoidal input voltage source VSIN.
7) Click the Run button to perform the simulation to see the input/output waveforms in the Probe window as depicted in Figure H.19(d). You can also click the View Simulation Output File button to view the output file as illustrated in Figure H.19(e).

(a)

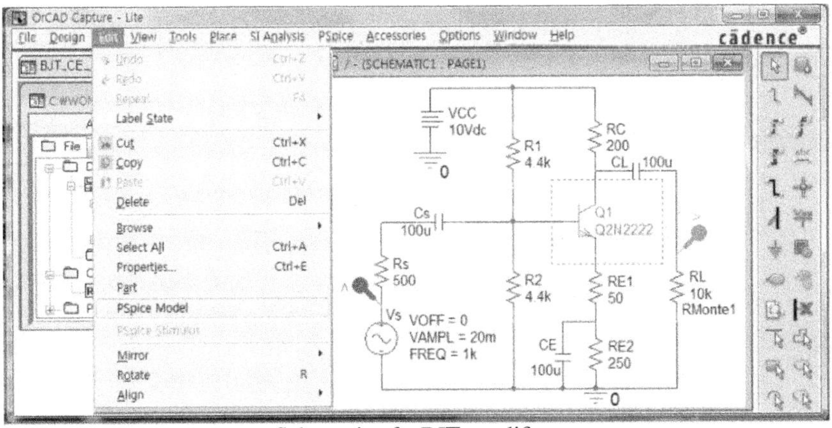

Schematic of a BJT amplifier

(b)

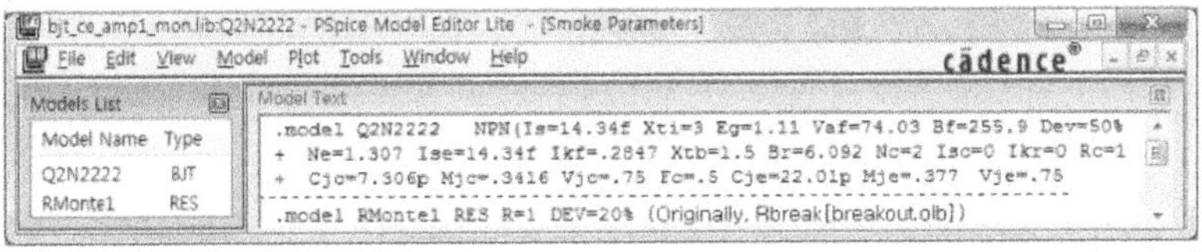

Model Editor window for setting the deviation (Dev) of forward current gain β_F of a BJT

(c)

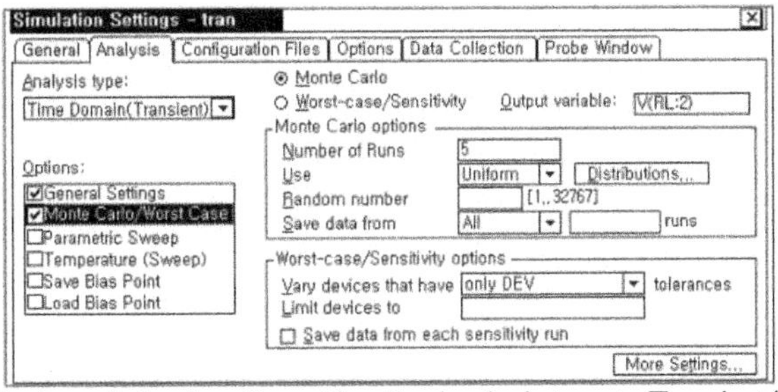

Simulation Settings dialog box for setting the Analysis type to Transient/Monte Carlo

(d)

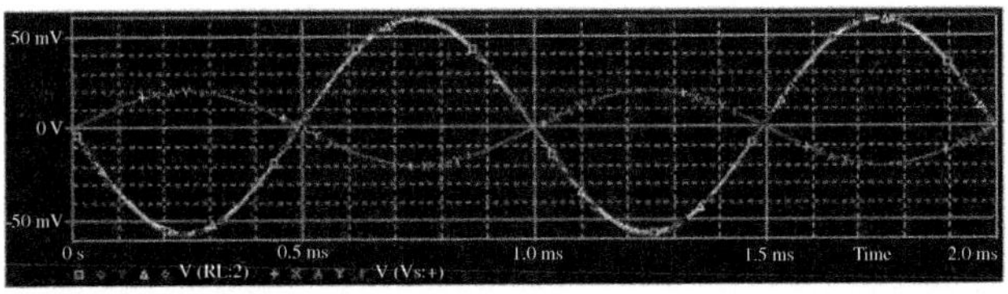

Monte Carlo simulation result for 50% deviation of β_F of the BJT

Figure H.19 Example of Monte Carlo simulation ("BJT_CE_amp1_mon.opj").

(e)

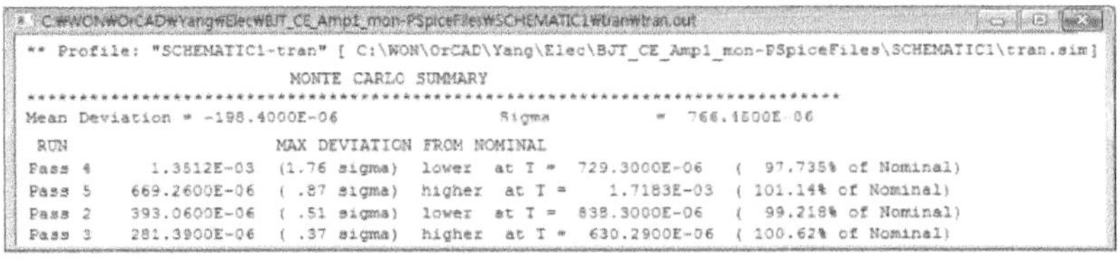

```
** Profile: "SCHEMATIC1-tran" [ C:\WON\OrCAD\Yang\Elec\BJT_CE_Amp1_mon-PSpiceFiles\SCHEMATIC1\tran.sim]
                    MONTE CARLO SUMMARY
******************************************************************************
Mean Deviation = -198.4000E-06                 Sigma          = 766.1500E-06
 RUN              MAX DEVIATION FROM NOMINAL
Pass 4    1.3512E-03  (1.76 sigma)  lower  at T = 729.3000E-06  (  97.735% of Nominal)
Pass 5  669.2600E-06  ( .87 sigma)  higher  at T =   1.7183E-03  ( 101.14% of Nominal)
Pass 2  393.0600E-06  ( .51 sigma)  lower  at T = 838.3000E-06  (  99.218% of Nominal)
Pass 3  281.3900E-06  ( .37 sigma)  higher  at T = 630.2900E-06  ( 100.62% of Nominal)
```

Monte Carlo simulation result printed in the output file

(f)

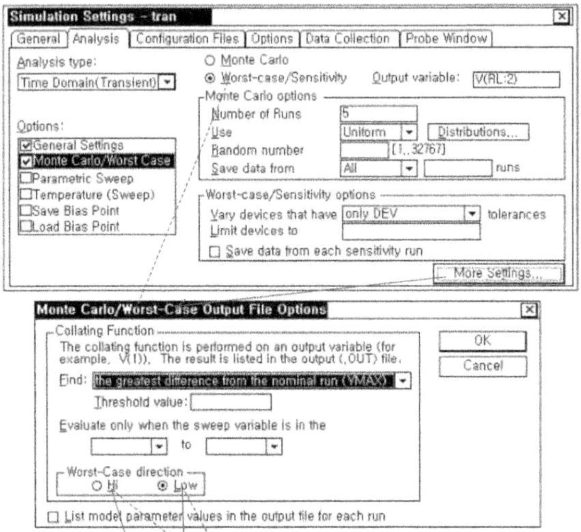

Simulation Settings dialog box for Monte Carlo/Worst-Case analysis

(g)

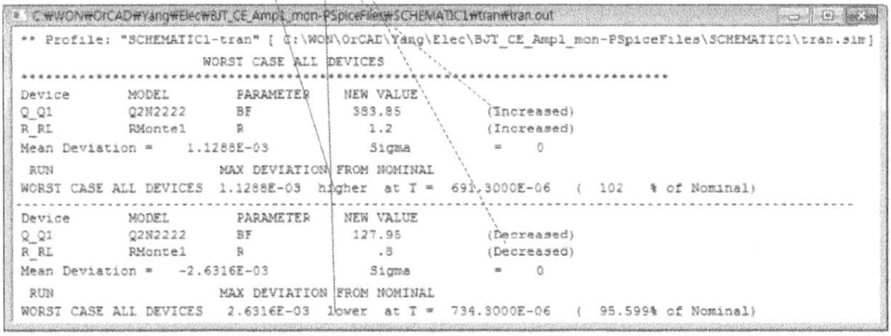

```
** Profile: "SCHEMATIC1-tran" [ C:\WON\OrCAD\Yang\Elec\BJT_CE_Amp1_mon-PSpiceFiles\SCHEMATIC1\tran.sim]
                 WORST CASE ALL DEVICES
******************************************************************************
Device      MODEL       PARAMETER    NEW VALUE
Q_Q1        Q2N2222     BF           383.85          (Increased)
R_RL        RMonte1     R              1.2           (Increased)
Mean Deviation =    1.1288E-03       Sigma          =   0
 RUN                 MAX DEVIATION FROM NOMINAL
WORST CASE ALL DEVICES  1.1288E-03  higher  at T =  691.3000E-06  (  102   % of Nominal)
------------------------------------------------------------------------------
Device      MODEL       PARAMETER    NEW VALUE
Q_Q1        Q2N2222     BF           127.95          (Decreased)
R_RL        RMonte1     R               .8           (Decreased)
Mean Deviation =   -2.6316E-03       Sigma          =   0
 RUN                 MAX DEVIATION FROM NOMINAL
WORST CASE ALL DEVICES   2.6316E-03  lower  at T =  734.3000E-06  (  95.599% of Nominal)
```

Worst Case simulation result printed in the output file

Figure H.19 (Continued)

8) Click the Edit Simulation Settings button to reopen the Simulation Settings dialog box (Figure H.19(f)), in which you can do the following:
   - Select Monte Carlo/Worst Case in the Options menu, select Worst-Case/Sensitivity from the upper-right menu, and select 'only DEV' in the tolerance field of the Worst-Case/Sensitivity options box.
   - Click the More Settings button to open the Monte Carlo/Worst-Case Output File Options dialog box, in which you can select High or Low in the Worst-Case direction box.
   - Click OK to close the Monte Carlo/Worst-Case Output File Options dialog box and then click OK to close the Simulation Settings dialog box.
9) Click Run to perform the simulation and see the input/output waveforms in the Probe window. You can also click the View Simulation Output File button to view the output file as illustrated in Figure H.19(g), the contents of which depends on which of High and Low has been selected as the Worst-Case direction.

H.5.8 AC Sweep/Noise Analysis

Figure H.20(a) and (b) shows the PSpice schematic of a CE BJT amplifier circuit and the Simulation Settings dialog box for AC Sweep/Noise analysis, respectively. With the simulation profile made by setting the Simulation Settings dialog box as illustrated in Figure H.20(b) and the Voltage/Level probe pins attached at the upper nodes of RL and Vs, we can run the PSpice simulation to get the input and output noises for the specified frequency range of 10 Hz∼10 MHz as depicted in Figure H.20(c), where the output noise is the root-mean-square (rms) sum of all the device contributions propagated to a specified output and the input noise is the equivalent noise that would be needed at the input source to generate the calculated output noise in an ideal (noiseless) circuit. You can also view the output file (Figure H.20(d)) containing every device noise, that is the noise contribution propagated to the specified output from every resistor and semiconductor device in the circuit. Note that the three frequencies $f = 10, 10^4$, and 10^7 Hz at which the noise analysis has been done are extracted with the interval of 300 points from the total 601 frequency points (100 points per decade) from 10 Hz to 10 MHz as specified in the Simulation Settings dialog box (Figure H.20(b)).

Exercise H.1 Initial Transient Bias Point and SKIPBP
To see how the PSpice simulation result may differ depending on whether the SKIPBP box has been checked or unchecked in the Simulation Settings dialog box for transient analysis and also whether the ICs of inductor L and capacitor C are specified or not, do the following:

(a)

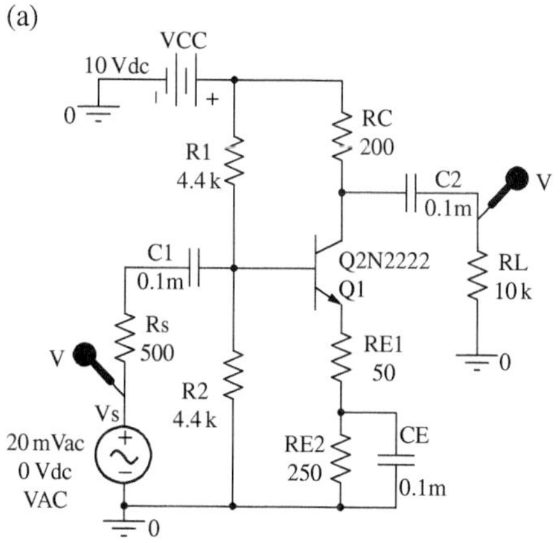

Schematic of a BJT amplifier

(b)

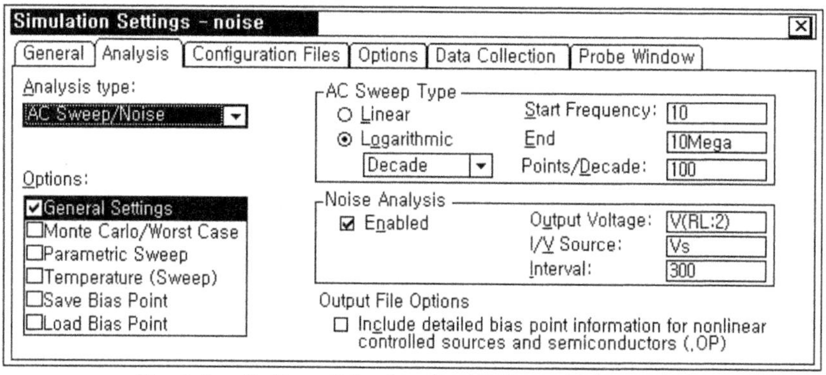

Simulation Settings dialog box for AC Sweep/Noise analysis

(c)

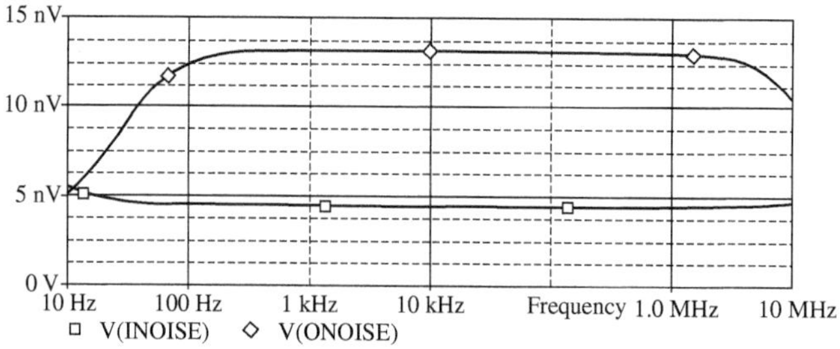

The result of Noise analysis

Figure H.20 Example of noise analysis ("BJT_amp1_noise.opj").

(d)

```
**** 08/07/05 17:41:12 ************** PSpice Lite (Jan 2003) *****************
** Profile: "SCHEMATIC1-noise" [C:\OrCAD\Yang\bjt_ampl_noise-pspicefiles\schematic1\noise.sim]
****      NOISE ANALYSIS                      TEMPERATURE =   27.000 DEG C
******************************************************************************

   FREQUENCY =   1.000E+01 HZ

 **** SQUARED NOISE VOLTAGES (SQ V/HZ)

 TRANSISTOR Q_Q1            RESISTOR
 RB        2.172E-19        R_R1      8.993E-19
 RC        1.770E-26        R_RC      3.180E-18
 RE        0.000E+00        R_RE1     1.088E-18    TOTAL OUTPUT NOISE VOLTAGE   =  2.622E-17 SQ V/HZ
 IBSN      1.093E-17        R_R2      8.993E-19                                 =  5.120E-09 V/RT HZ
 IC        1.464E-19        R_Rs      7.186E-18    TRANSFER FUNCTION VALUE:
 IBFN      0.000E+00        R_RL      1.040E-19     V(N09196)/V_Vs              =  9.312E-01
                            R_RE2     1.569E-18
 TOTAL     1.129E-17                               EQUIVALENT INPUT NOISE AT V_Vs = 5.499E-09 V/RT HZ
****

   FREQUENCY =   1.000E+04 HZ

 **** SQUARED NOISE VOLTAGES (SQ V/HZ)

 TRANSISTOR Q_Q1            RESISTOR
 RB        2.131E-18        R_R1      8.037E-18
 RC        1.410E-25        R_RC      3.168E-18
 RE        0.000E+00        R_RE1     1.067E-17    TOTAL OUTPUT NOISE VOLTAGE   =  1.717E-16 SQ V/HZ
 IBSN      6.774E-17        R_R2      8.037E-18                                 =  1.310E-08 V/RT HZ
 IC        1.135E-18        R_Rs      7.073E-17    TRANSFER FUNCTION VALUE:
 IBFN      0.000E+00        R_RL      6.336E-20     V(N09196)/V_Vs              =  2.921E+00
                            R_RE2     2.162E-23
 TOTAL     7.101E-17                               EQUIVALENT INPUT NOISE AT V_Vs = 4.486E-09 V/RT HZ
****

   FREQUENCY =   1.000E+07 HZ

 **** SQUARED NOISE VOLTAGES (SQ V/HZ)

 TRANSISTOR Q_Q1            RESISTOR
 RB        1.260E-18        R_R1      4.755E-18
 RC        2.224E-21        R_RC      2.127E-18
 RE        0.000E+00        R_RE1     6.348E-18    TOTAL OUTPUT NOISE VOLTAGE   =  1.089E-16 SQ V/HZ
 IBSN      4.003E-17        R_R2      4.755E-18                                 =  1.043E-08 V/RT HZ
 IC        7.712E-18        R_Rs      4.184E-17    TRANSFER FUNCTION VALUE:
 IBFN      0.000E+00        R_RL      4.253E-20     V(N09196)/V_Vs              =  2.247E+00
                            R_RE2     1.286E-29
 TOTAL     4.901E-17                               EQUIVALENT INPUT NOISE AT V_Vs = 4.644E-09 V/RT HZ
```

The output file obtained from the PSpice simulation for Noise analysis

Figure H.20 (Continued)

(a) Create a project named, say, "test_RLC.opj" and draw the schematic of an *RLC* series circuit (as shown in Figure H.21(a)) with Vi = 1 V, R1 = 1 Ω, L1 = 10 mH, and C1 = 1 mF.

(b) Click on the New Simulation Profile button to create a Simulation Profile named, say, 'tran' and click on Create to open a Simulation Settings dialog box (like Figure H.5(c1)) where you set the Analysis type to Time Domain (Transient), Run_to_time=0.1s, Maximum step=0.1ms, uncheck the SKIPBP box (as default), and click OK to close the dialog box.

(c) Place a current marker and a voltage marker as depicted in Figure H.21(a).

(d) Click on the Run button to see PSpice simulation result having shown up in the PSpice A/D window. Which one of Figure H.21(b1), (b2), and (b3) do you see?

(e) Click on the Edit Simulation Profile button to reopen the Simulation Settings dialog box where you check the SKIPBP box and click OK to close the dialog box. Then click on the Run button to see PSpice simulation result having shown up in the PSpice A/D window. Which one of Figure H.21(b1), (b2), and (b3) do you see?

(f) Double-click on the inductor L to open its Property Editor spreadsheet (as shown in Figure H.21(c1)), set its IC value to 1, and click on x to close the spreadsheet. Also, double-click on the capacitor C to open its Property

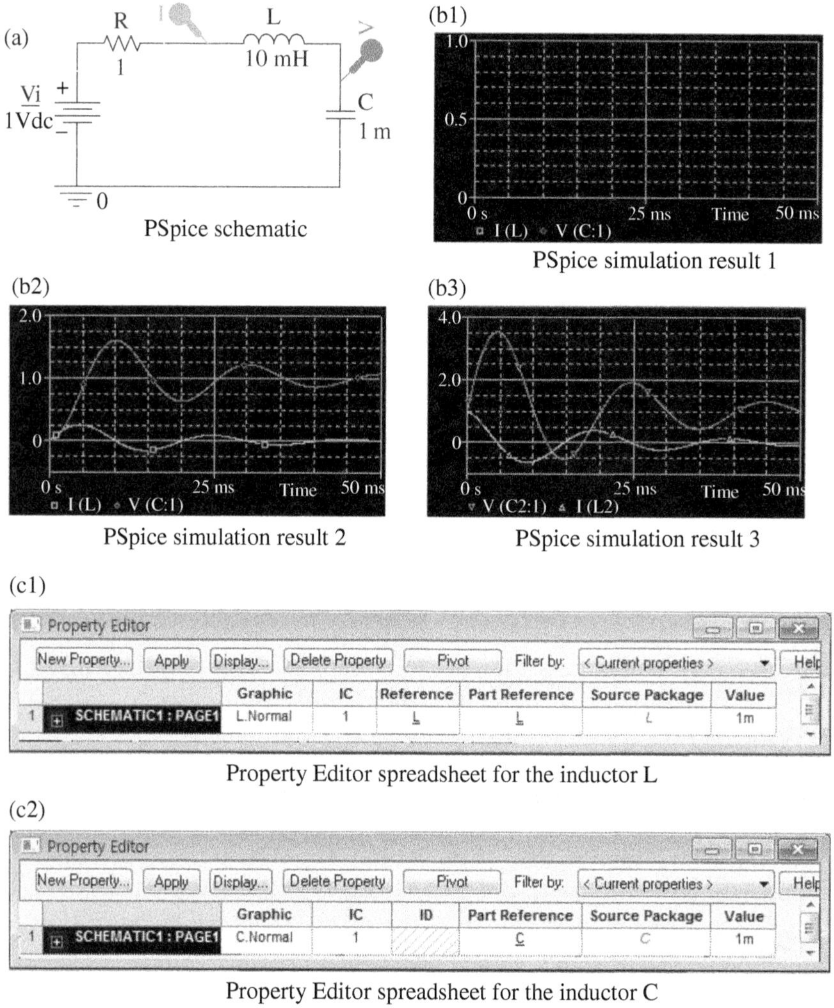

Figure H.21 Exercise H.1 to know about SKIPBP.

Editor spreadsheet (as shown in Figure H.21(c2)), set its IC value to 1 [V], and click on x to close the spreadsheet. Then click on the Run button to see PSpice simulation result having shown up in the PSpice A/D window. Which one of Figure H.21(b1), (b2), and (b3) do you see?

(g) In light of Remark 3.9, find the DC steady steady-state (final) values of $i_L(t)$ and $v_C(t)$, as also can be seen from Figure H.21(b1), (b2), or (b3). How are they related with their initial values $i_L(0)$ and $v_C(0)$ in the PSpice simulation result obtained with the SKIPBP box unchecked?

(h) Complete and run the following MATLAB script "Q_SKIPBP.m" so that it can use the Laplace transform to find $i_L(t)$ and $v_C(t)$ with the ICs $v_C(0) = 1$ V and $i_L(0) = 1$ A and plot them.

```
%Q_SKIPBP.m
syms s
R=1; L=0.01; C=1e-3; Vis=1/s; vC0=1; iL0=1;
Is=((s*Vis-vC0)/L+iL0*s)/(s^2+s*R/L+1/L/C);
VCs=Is/s/C+vC0/s;
iL=ilaplace(Is), vC=ilaplace(VCs)
t=0:1e-4:0.1; iLt=eval(iL); vCt=eval(vC);
plot(t,iLt, t,vCt)
```

H.6 How to Save a Project as another Project

To save a current (existing) project as another project (with a different file name), click on the Project Manager tab to activate (highlight) the corresponding Project Manager window, select File>Save_Project_As from the top menu bar to open the Save Project dialog box as depicted in Figure H.22, specify a location, enter a new project name, and click OK.

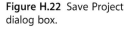

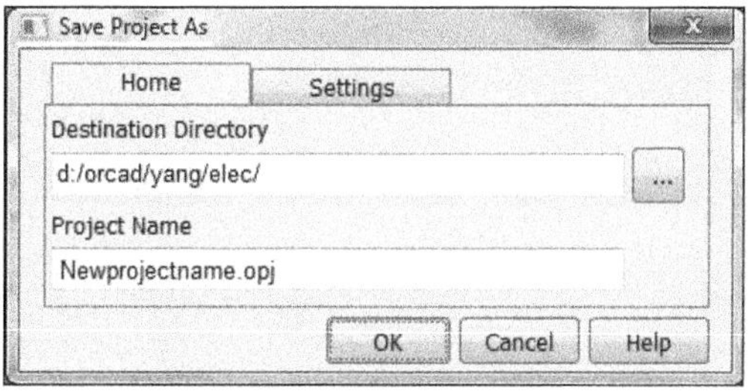

Figure H.22 Save Project dialog box.

Appendix I

MATLAB® Introduction

On a Windows-based computer with MATLAB® software installed, you can double-click on the MATLAB icon (like the one on the right side) to start MATLAB. Once MATLAB has been started, you will see the MATLAB desktop as shown in Figure I.1, which has the MATLAB Command window in the lower middle where a blinking cursor appears to the right of the MATLAB prompt '>>' waiting for you to type in MATLAB commands/statements. It also contains other windows each labeled Current Folder, Editor, and Workspace (showing the contents of MATLAB memory).

How do we work with the MATLAB Command window?

- By clicking on 'New'/'Open' in the groups of functionalities under HOME tab and then selecting an item in their dropdown menus, you can create/edit any file with the MATLAB editor.
- By typing any MATLAB commands/statements in the MATLAB Command Window, you can use various powerful mathematic/graphic functions of MATLAB.
- If you have an M-file which contains a series of commands/statements for performing a target procedure, you can type in the filename (without the extension '.m') to run it.
- By clicking on 'Set Path' in the right part of the toolstrip, you can make the MATLAB search path list include or exclude the paths containing the files you want to or not to be run.
- By typing a search term like a keyword or a function name into the Search Documentation box at the top-right part, you can search help documentation for what you want to know about.
- To undock, i.e. separate the Editor Window from the MATLAB desktop, click on the Editor Actions button (down-arrow-shaped) in the top-right part of the Editor Window and select 'Undock' in its drop-down menu. To dock

Electronic Circuits with MATLAB®, PSpice®, and Smith Chart, First Edition. Won Y. Yang, Jaekwon Kim, Kyung W. Park, Donghyun Baek, Sungjoon Lim, Jingon Joung, Suhyun Park, Han L. Lee, Woo June Choi, and Taeho Im.

Companion website: www.wiley.com/go/yang/electroniccircuits

the Editor Window back to the desktop, click on the Editor Actions button and select 'Dock Editor' in its drop-down menu.

Some of mathematical functions and special reserved constants/variables defined in MATLAB are listed in Table I.1. Table I.2 shows the graphic line specifications used in the plot( ) command.

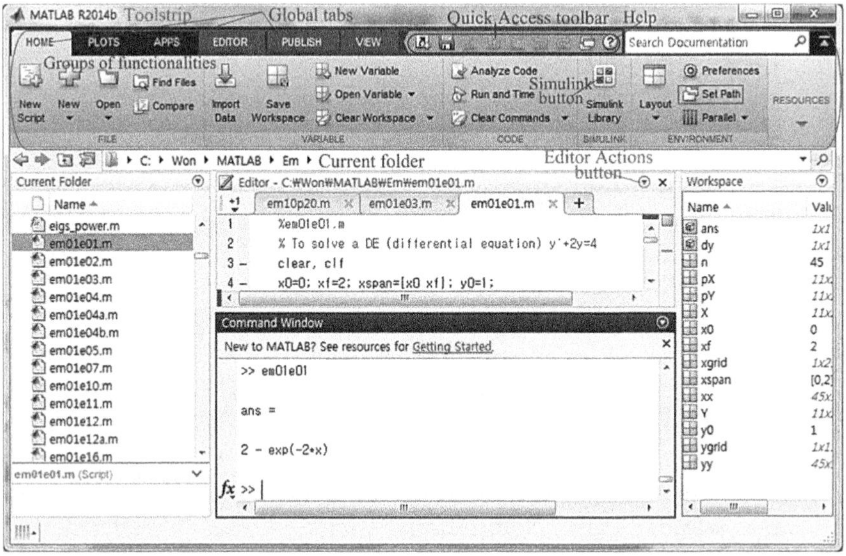

Figure I.1 The MATLAB desktop with Command/Editor/Current folder/Workspace windows.

Table I.1 Functions and variables in MATLAB.

Function	Remark	Function	Remark
`cos(x)`		`exp(x)`	Exponential function
`sin(x)`		`log(x)`	Natural logarithm
`tan(x)`		`log10(x)`	Common logarithm
`acos(x)`	$\cos^{-1}(x)$	`abs(x)`	Absolute value
`asin(x)`	$\sin^{-1}(x)$	`angle(x)`	Phase of a complex number [rad]
`atan(x)`	$-\pi/2 \le \tan^{-1}(x) \le \pi/2$	`sqrt(x)`	Square root

Table I.1 (Continued)

Function	Remark	Function	Remark
`atan2(y,x)`	$-\pi \le \tan^{-1}(y/x) \le \pi$	`real(x)`	Real part
`cosh(x)`	$(e^x + e^{-x})/2$	`imag(x)`	Imaginary part
`sinh(x)`	$(e^x - e^{-x})/2$	`conj(x)`	Complex conjugate
`tanh(x)`	$(e^x - e^{-x})/(e^x + e^{-x})$	`round(x)`	The nearest integer (round-off)
`acosh(x)`	$\cos \mathrm{h}^{-1}(x)$	`fix(x)`	The nearest integer towards 0
`asinh(x)`	$\sin \mathrm{h}^{-1}(x)$	`floor(x)`	The greatest integer $\le x$
`atanh(x)`	$\tan \mathrm{h}^{-1}(x)$	`ceil(x)`	The smallest integer $\ge x$
`max`	Maximum and its index	`sign(x)`	1(positive)/0/−1 (negative)
`min`	Minimum and its index	`mod(y,x)`	Remainder of y/x
`sum`	Sum	`rem(y,x)`	Remainder of y/x
`prod`	Product	`eval(f )`	Evaluate an expression
`norm`	Norm	`feval(f,a)`	Function evaluation
`sort`	Sort in the ascending order	`polyval`	Value of a polynomial function
`clock`	Present time	`poly`	Polynomial with given roots
`find`	Index of element(s)	`roots`	Roots of polynomial
`tic`	Start a stopwatch timer	`toc`	Read the stopwatch timer (elapsed time from tic)
`date`	Present date		
Reserved variables with special meaning			
`i, j`	$\sqrt{-1}$	`pi`	π
`eps`	Machine epsilon	`Inf, inf`	Largest number(∞)
realmax, realmin	Largest/smallest positive number	`NaN`	Not_a_Number (undetermined)
`end`	The end of for loop or if, while	`break`	Exit while/for loop
`nargin`	Number of input arguments	`nargout`	Number of output arguments
`varargin`	Variable input argument list	`varargout`	Variable output argument list

Table I.2 Graphic line specifications used in the plot( ) command.

Line type	Point type (Marker symbol)			Color	
- solid line	.(dot)	+(plus)	*(asterisk)	r: red	m: magenta
: dotted line	^: Δ	>: >	o(circle)	g: green	y: yellow
-- dashed line	p: ☆	v: ∇	x: x-mark	b: blue	c: cyan(sky blue)
-. dash-dot	d: ◇	<: <	s: □	k: black	

Appendix J

Diode/BJT/FET

J.1 Diode

The junction (depletion) and diffusion capacitances are

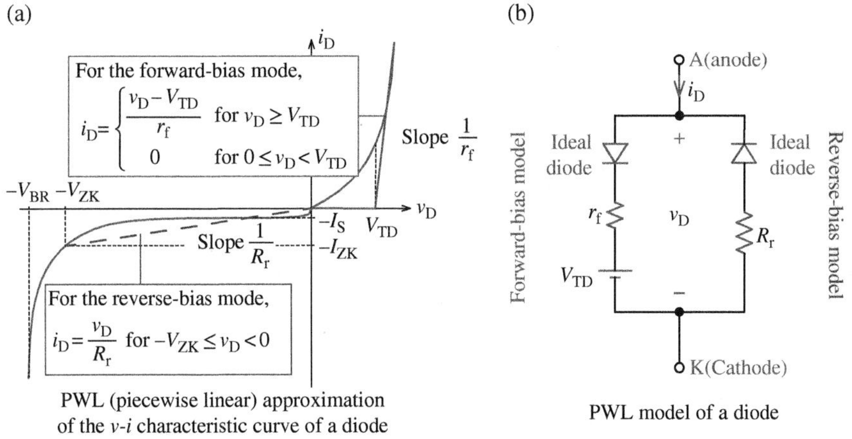

Figure J.1 Piecewise linear current (PWL) approximation of the *v-i* characteristic curve of a diode and the corresponding model.

Electronic Circuits with MATLAB®, PSpice®, and Smith Chart, First Edition. Won Y. Yang, Jaekwon Kim, Kyung W. Park, Donghyun Baek, Sungjoon Lim, Jingon Joung, Suhyun Park, Han L. Lee, Woo June Choi, and Taeho Im.

Companion website: www.wiley.com/go/yang/electroniccircuits

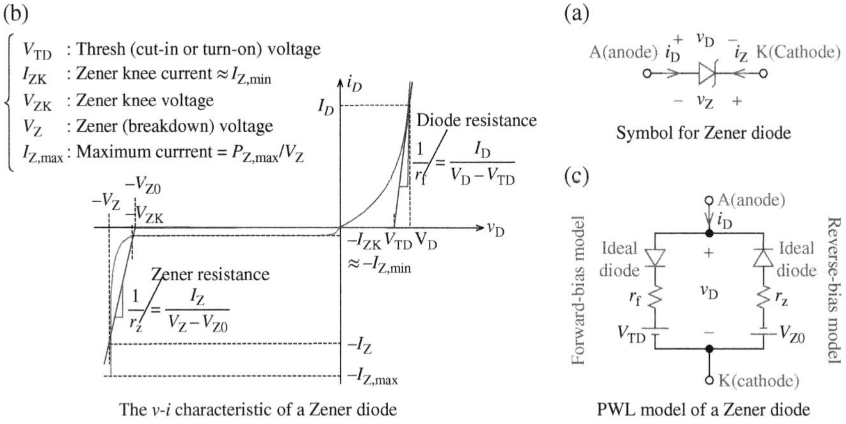

Figure J.2 Symbol, *v-i* characteristic, and PWL model of a Zener diode.

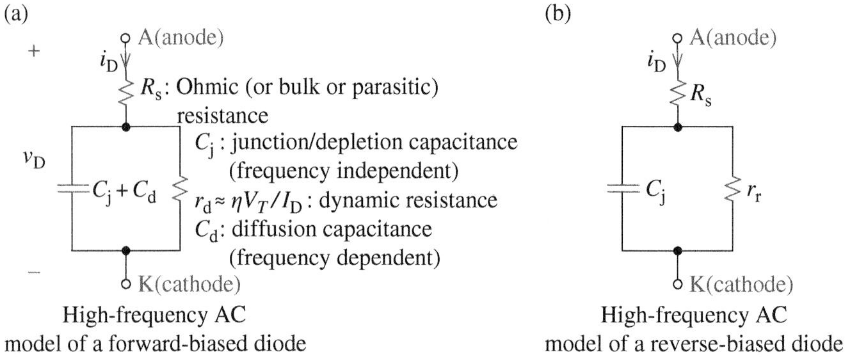

Figure J.3 High-frequency AC (small-signal) model of a diode.

Table J.1 PSpice model parameters of diode.

SPICE name	Model parameter	Default value
IS	Transport saturation current I_s	10^{-14}
RS	Ohmic (series) resistance R_s	0 [Ω]
N	Emission (ideality) coefficient η	1
CJO	Zero-bias junction capacitance C_{j0}	0
VJ	PN junction (built-in) potential φ_B	1 [V]
MJ	PN grading coefficient m	0.5
TT	Transit time τ_t	0 [s]
BV	Reverse breakdown voltage V_Z	∞ [V]
IBV	Reverse breakdown current I_Z	10^{-3}

$$C_j = \frac{C_{j0}}{\left(1 - V_{D,Q}/\phi_B\right)^m}, \quad C_d = \frac{\tau_t(\text{transit time})}{r_d\,(\text{dynamic resistance})} \tag{J.1}$$

J.2 BJT (Bipolar Junction Transistor)

g_m: *transconductance (gain)*

$$g_m = \left.\frac{\partial i_C}{\partial v_{BE}}\right|_Q \overset{(3.1.1b)}{\approx} \left.\frac{\partial}{\partial v_{BE}}\left\{\pm I_s\left(e^{\pm v_{BE}/V_T} - 1\right)\right\}\right|_Q = \frac{1}{V_T} I_s e^{\pm v_{BE,Q}/V_T} \overset{(3.1.1b)}{\approx} \frac{|I_{C,Q}|}{V_T} \tag{J.2}$$

$$r_{be} = r_\pi = \beta/g_m : \text{incremental resistance of forward-biased B-E junction } [\Omega] \tag{J.3}$$

C_{be}/C_{bc}: zero-bias B-E/B-C junction capacitances

$$C_{be} = \frac{C_{be0}}{\left(1 - V_{BE,Q}/\phi_{be}\right)^{m_{be}}}, \quad C_{bc} = \frac{C_{bc0}}{\left(1 - V_{BC,Q}/\phi_{bc}\right)^{m_{bc}}} \tag{J.4}$$

C_{be0}(CJE)/C_{bc0}(CJC): zero-bias B-E/B-C junction capacitances
$V_{BE,Q}/V_{BC,Q}$: quiescent B-E/B-C voltages [V]
m_{be}(MJE)/m_{bc}(MJC): B-E/B-C grading coefficient
ϕ_{be}(VJE)/ϕ_{bc}(VJC): B-E/B-C built-in potential

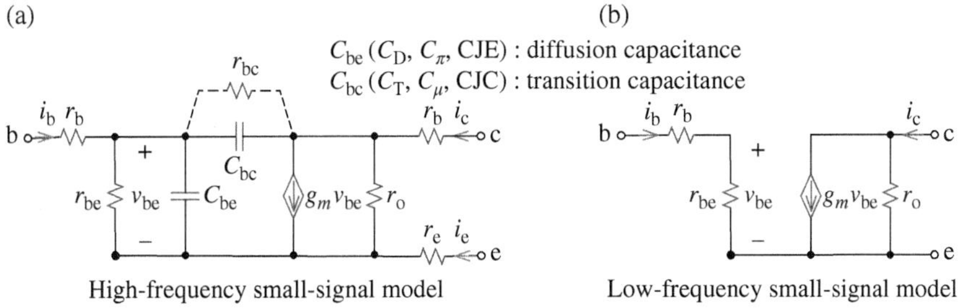

Figure J.4 Hybrid-π small-signal models of NPN-BJT with or without frequency dependence.

Table J.2 PSpice model parameters of BJT.

SPICE name	Model parameter	Default value
BF	Forward active current gain β_F	100
BR	Reverse active current gain β_R	1
IS	Transport saturation current I_s	10^{-16}
ISC	B-C leakage saturation current I_{sc}	0
CJE	B-E zero-bias junction (depletion) capacitance C_{be0}	0
CJC	B-C zero-bias junction (depletion) capacitance C_{bc0}	0
VJE	B-E built-in potential φ_{be}	0.75 [V]
VJC	B-C built-in potential φ_{bc}	0.75 [V]
VAF	Forward mode Early voltage	∞ [V]
VAR	Reverse mode Early voltage	∞ [V]
NF	Forward ideality factor (current emission coefficient)	1
NR	Reverse ideality factor (current emission coefficient)	1
MJE	B-E capacitance exponent m_{be}	0.33
MJC	B-C capacitance exponent m_{bc}	0.33
RB	Base resistance r_b	0 [Ω]
RE	Emitter resistance r_e	0 [Ω]
RC	Collector resistance r_c	0 [Ω]

Table J.3 Characteristics of CE/CC/CB amplifiers ($\beta = g_m r_{be} = (I_{C,Q}/V_T) r_{be}$).

	CE		CC		CB	
R_i	$R_B \| \| \{r_b + r_{be} + (\beta+1)R_{E1}\}$	(3.2.1)	$R_B \| \| \{r_b + r_{be} + (\beta+1)(R_E \| \| R_L)\}$	(3.2.5)	$R_E \| \| \frac{r_b + r_{be} + R_B}{\beta+1}$	(3.2.9)
R_o	$R_C \| \| r_o \approx R_C$	(3.2.4)	$R_E \| \| \frac{(R_s \| R_B) + r_b + r_{be}}{\beta+1}$	(3.2.8)	$R_C \| \| r_{o1}$	(3.2.12)
A_v	$-\frac{\beta(R_C \| R_L)}{r_b + r_{be} + (\beta+1)R_{E1}}$	(3.2.3)	$\frac{(\beta+1)(R_E \| R_L)}{r_b + r_{be} + (\beta+1)(R_E \| R_L)}$	(3.2.7)	$\frac{\beta(R_C \| R_L)}{r_b + r_{be} + R_B}$	(3.2.11)
A_i	$-\frac{\beta R_C}{R_C + R_L}$	(3.2.2)	$\frac{(\beta+1)R_B R_E}{(R_B + r_b + r_{be})(R_E + R_L) + (\beta+1)R_E R_L}$		$\frac{R_C R_E}{R_C + R_L} \frac{\beta}{r_b + r_{be} + R_B + (\beta+1)R_E}$	

```
function [gm,rbe,re,ro]=gmrbero_BJT(ICQ,beta,VA,VT)
if VT>0.1, VT=(273+VT)/11605; end % considering VT as T
gm=abs(ICQ)/VT; % Transconductance gain
rbe=beta/gm; re=rbe/(beta+1);
ro=VA/abs(ICQ); % Eqs.(3.1.26-28)
```

$$i_C(v_{BE}, v_{BC}) \underset{V_A=\infty}{\overset{(3.1.7a)}{=}} \left\{ I_S\left(e^{v_{BE}/V_T}-1\right) - I_{SC}\left(e^{v_{BC}/V_T}-1\right)\right\}\left(1+\frac{v_{CE}}{V_A}\right)$$
$$\text{with } V_T = \frac{273+T\,[^\circ\text{C}]}{11\,605}\,[\text{V}] \tag{J.5a}$$

$$i_B(v_{BE}, v_{BC}) \underset{V_A=\infty}{\overset{(3.1.7b)}{=}} \left\{ \frac{I_S}{\beta_F}\left(e^{v_{BE}/V_T}-1\right) + \frac{I_{SC}}{\beta_R+1}\left(e^{v_{BC}/V_T}-1\right)\right\}\left(1+\frac{v_{CE}}{V_A}\right) \tag{J.5b}$$

$$i_E = i_B + i_C \tag{J.5c}$$

```
function [iC,iB,iE]=iCiBiE(vBE,vCE,Is,Isc,bF,bR,T,VA)
if nargin<8, VA=1e8; end
if nargin<7, T=27; end
VT=(T+273)/11605; vBC=vBE-vCE;
iCf=@(vBE,vCE)(Is*(exp(vBE/VT)-1)-Isc*(exp(vBC/VT)-1)) ...
             .*(1+vCE/VA); % Eq.(J.5a)
iBf=@(vBE,vCE)(Is/bF*(exp(vBE/VT)-1)-Isc/(bR+1) ...
*(exp(vBC/VT)-1)).*(1+vCE/VA); % Eq.(J.5b)
iEf=@(vBE,vCE)iBf(vBE,vCE)+iCf(vBE,vCE);
iB=iBf(vBE,vCE); iC=iCf(vBE,vCE); iE=iEf(vBE,vCE);
```

J.3 FET (Field Effect Transistor)

The body effect or substrate sensitivity[J-1] of the threshold voltage V_t is described by

$$V_t = V_{t0} + \gamma\left(\sqrt{|\,2\phi + v_{SB}\,|} - \sqrt{|\,2\phi\,|}\right) \text{with } v_{SB}: \text{the source-bulk voltage } v_{SB} \tag{J.6}$$

The gate-source and gate-drain capacitances are modeled as voltage-dependent capacitances:

$$C_{gs} \underset{(10.7\text{-}1)}{\overset{[S\text{-}1]}{=}} \frac{C_{gs0}}{(1+|V_{GS,Q}|/V_b)^m}, \quad C_{gd} \underset{(10.7\text{-}2)}{\overset{[S\text{-}1]}{=}} \frac{C_{gd0}}{(1+|V_{GD,Q}|/V_b)^m} \tag{J.7}$$

where C_{gs0}/C_{gd0}: zero-bias gate-source/gate-drain junction capacitances, respectively, $V_{GS,Q}/V_{GD,Q}$ [V]: quiescent gate-source/gate-drain voltages, respectively, m(MJ): gate p-n grading coefficient (SPICE default = 0.5), and V_b(PB): gate junction (barrier) potential, typically 0.6 V (SPICE default = 1 V).

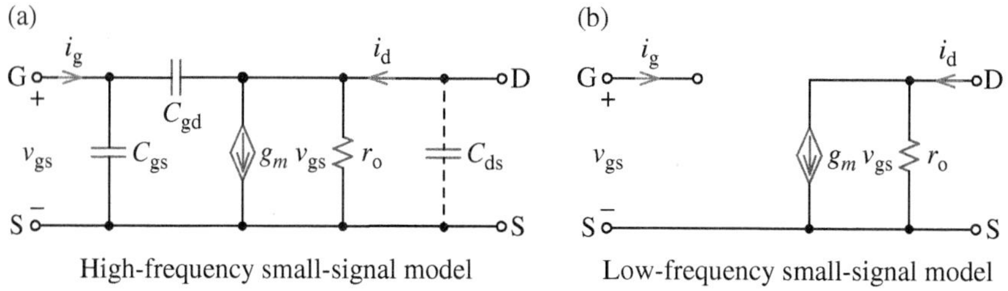

Figure J.5 Small-signal models of FET.

Table J.4 PSpice model parameters of FET.

SPICE name	Model parameter	Default value
VTO	Zero-bias threshold voltage V_{t0}	0 [V]
KP	Transconductance parameter $K_p = \mu_p C_{OX}(W/L)$	2×10^{-5} [A/V^2]
GAMMA	Bulk threshold parameter γ	0 [V$^{1/2}$]
PHI	Surface (bulk) potential φ	0.6 [V]
LAMBDA	Channel length modulation parameter λ	0 [V^{-1}]
PB	Built-in potential for the bulk (substrate) junction V_b	0.8 [V]
CGSO	Gate-source overlap capacitance per meter channel length C_{gs0}	0 [F/m]
CGDO	Gate-drain overlap capacitance per meter channel length C_{gd0}	0 [F/m]
CGBO	Gate-bulk overlap capacitance per meter channel length C_{gb0}	0 [F/m]
MJ	Bulk junction bottom grading coefficient m	0.5

Table J.5 Circuit symbols and *i-v* relationships of JFET and MOSFET.

FET type	*n*-Channel			*p*-Channel		
	JFET	Enhancement MOSFET	Depletion MOSFET	JFET	Enhancement MOSFET	Depletion MOSFET
Circuit symbols	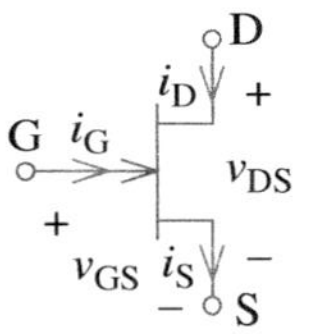	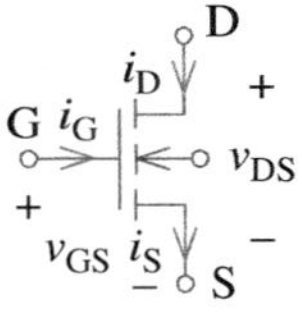	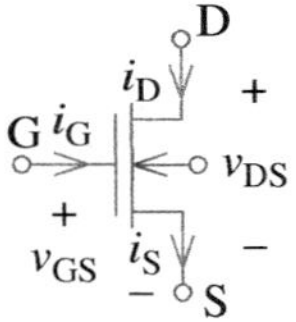	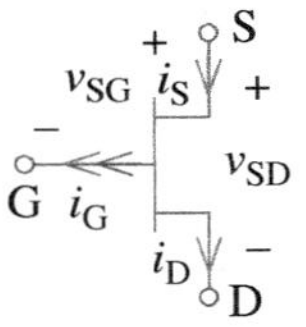	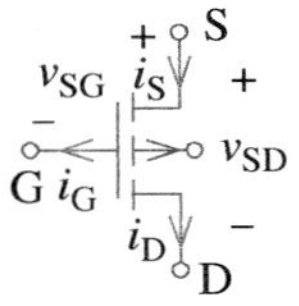	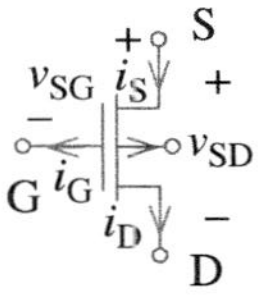
Threshold voltage V_t	–	+	–	+	–	–
Conduction constant K_p	$\frac{2I_{DSS}}{V_t^2}$	Process conduction parameter $K_p = \boxed{\mu_n C_{OX}} \frac{W}{L}$		$\frac{2I_{DSS}}{V_t^2}$	Process conduction parameter $K_p = \boxed{\mu_p C_{OX}} = \frac{W}{L}$	
Turn-on condition	$v_{GS} > V_t$ and $v_{DS} > 0$			$v_{SG} > \|V_t\|$ and $v_{SD} > 0$		
Triode region (Ohmic mode)	$v_{GD} = v_G - v_D > V_t > 0$			$v_{DG} = v_D - v_G > \|V_t\|$		
	$i_D \cong K_p\{v_{GS} - V_t)v_{DS} - v_{DS}^2/2\}$ with overdrive voltage $v_{OV} = v_{GS} - V_t$			$i_D \cong K_p\{(v_{SG} - \|V_t\|)v_{SD} - V_{SD}^2/2\}$ with overdrive voltage $v_{OV} = v_{SG} - \|V_t\|$ (4.1.13a)		
Saturation region (Pinch-off mode)	$v_{GD} = v_G - v_D \leq V_t$			$v_{DG} = v_D - v_G \leq \|V_t\|$		
	$i_D \cong K_p(v_{GS} - V_t)^2/2$			$i_D \cong K_p(v_{SG} - \|V_t\|)^2/2$ (4.1.13b)		

Table J.6 Characteristics of CS/CD/CG amplifiers.

	CS (common source)		CD (common drain) – source follower		CG (common gate)	
R_i	$R_1 \| R_2$	(4.2.5)	$R_1 \| R_2$	(4.2.10)	$R_S \| \dfrac{r_o + (R_D \| R_L)}{1 + g_m r_o} \xrightarrow{r_o >> \max\{1/g_m, (R_D \| R_L)\}} R_S \| \dfrac{1}{g_m}$	(4.2.15)
R_o	$R_D \| \{r_o + (1 + g_m r_o) R_{S1}\} \xrightarrow{r_o \to \infty} R_D$	(4.2.7)	$\dfrac{1}{g_m} \| r_o \| R_S \xrightarrow{r_o \to \infty} \dfrac{1}{g_m} \| R_S$	(4.2.12)	$R_D \| \{r_o + (1 + g_m r_o)(R_s \| R_S)\} \xrightarrow{R_s, R_D << r_o} R_D$	(4.2.17)
A_v	$\dfrac{-g_m r_o (R_D \| R_L)}{(R_D \| R_L) + r_o + (1 + g_m r_o) R_{S1}}$ if $r_o >> 1/g_m R_{S1}$ $\approx \dfrac{-g_m (R_D \| R_L)}{1 + g_m R_{S1}}$	(4.2.6)	$\dfrac{g_m (r_o \| R_S \| R_L)}{1 + g_m (r_o \| R_S \| R_L)} \xrightarrow[R_s << Ri]{g_m (r_o \| R_S \| R_L) >> 1} = 1$	(4.2.11)	$\dfrac{(1 + g_m r_o)(R_D \| R_L)}{r_o + (R_D \| R_L)} \xrightarrow{r_o >> \max\{1/g_m, (R_D \| R_L)\}} g_m (R_D \| R_L)$	(4.2.16)

```
function [iD,mode]=iD_NMOS_at_vDS_vGS(vDS,vGS,Kp,Vt,lambda)
if nargin<5, lambda=0; end
% Kp=k'(W/L)-mu*Cox*W/L
vGD=vGS-vDS; ON=(vGS>Vt)&(vDS>0); SAT=(vGD<=Vt)&ON; TRI=
(vGD>Vt)&ON;
iD = Kp*(1+lambda*vDS).*((vGS-Vt).^2/2.*SAT + ...
    ((vGS-Vt).*vDS-vDS.^2/2).*TRI); %Eq.(4.1.12,13)
iD = max(iD,0);
if SAT>0, mode='saturation';
elseif TRI>0, mode='ohmic';
 else mode='cutoff';
end
```

Bibliography

[B-1] Baker, R.J. (2010). *CMOS: Circuit Design, Layout, and Simulation*, 3e. New Jersey: IEEE Press and Wiley.

[B-2] Bowick, C.J. (2007). *RF Circuit Design*, 2e. Elsevier (Newnes).

[C-1] Carusone, T.C., Johns, D.A., and Martin, K.W. (2012). *Analog Integrated Circuit Design*. New Jersey: Wiley.

[C-2] Chang, K., Bahl, I., and Nair, V. (2002). *RF and Microwave Circuit and Component Design for Wireless Systems*. New Jersey: Wiley.

[C-3] Colclaser, R.A., Neamen, D.A., and Hawkins, C.F. (1984). *Electronic Circuit Analysis: Basic Principles*. New Jersey: Wiley.

[D-1] Davis, W.A. and Agarwal, K. (2001). *Radio Frequency Circuit Design*. New Jersey: Wiley.

[E-1] Edwards, M.L. and Sinsky, J.H. (December 1992). A new criteria for linear 2-port stability using a single geometrically derived parameter. *IEEE Trans. Microwave Theory Tech.* MTT-40: 2803–2811.

[E-2] Elliot, Rod, Precision Rectifiers. http://sound.westhost.com/appnotes/an001.htm

[F-1] Franco, S. (1995). *Electric Circuits Fundamentals*. Philadelphia, Pennsylvania: Saunders College Publishing.

[F-2] Frickey, D.A. (1994). Conversions between S, Z, Y, h, ABCD, and T parameters which are valid for complex source and load impedances. *IEEE Trans. Microwave Theory Tech.* 42 (2).

[H-1] Herrick, R.J. (2003). *DC/AC Circuits and Electronics: Principles & Applications*. New York: Thomson Delmar Learning, Inc.

[H-2] Hu, C.C. (2009). *Modern Semiconductor Devices for Integrated Circuits*. Pearson Education.

[H-3] Huelsman, L.P. (1998). *Basic Circuit Theory*, 3e. New Jersey: Prentice Hall, Inc.

Electronic Circuits with MATLAB®, PSpice®, and Smith Chart, First Edition. Won Y. Yang, Jaekwon Kim, Kyung W. Park, Donghyun Baek, Sungjoon Lim, Jingon Joung, Suhyun Park, Han L. Lee, Woo June Choi, and Taeho Im.

Companion website: www.wiley.com/go/yang/electroniccircuits

[I-1] Irwin, J.D. and Chwan-Hwa, W. (1999). *Basic Engineering Circuit Analysis*, 6e. New Jersey: Prentice Hall, Inc.

[J-1] Jaeger, R.C. and Blalock, T.N. (2011). *MicroElectronic Circuit Design*, 4e. New York: McGraw-Hill, Inc.

[J-2] Jones, D. and Stitt, M. *Burr-Brown SBOA068 – Precision Absolute Value Circuits*. Texas Instruments http://www.ti.com/lit/an/sboa068/sboa068.pdf.

[K-1] Keown, J. (2001). *OrCAD PSpice and Circuit Analysis*, 4e. New Jersey: Prentice Hall, Inc.

[K-2] Kuc, R. (1988). *Introduction to Digital Signal Processing*. New York: McGraw-Hill.

[L-1] Ludwig, R. and Bogdanov, G. (1998). *RF Circuit Design: Theory and Applications*, 2e. Englewood Cliffs, New Jersey: Prentice Hall.

[M-1] Millman, J. and Grabel, A. (1987). *Microelectronics*, 2e. New York: McGraw-Hill Book Company, Inc.

[M-2] Misra, D.K. (2001). *Radio-Frequency and Microwave Communication Circuits*. New Jersey: Wiley.

[N-1] Neamen, D.A. (2007). *MicroElectronics – Circuit Analysis and Design*, 3e. New York: McGraw-Hill, Inc.

[O-1] Orfanidis, S.J. *Electromagnetic Waves and Antennas*. ECE Dept. of Rutgers University http://www.ece.rutgers.edu/~orfanidi/ewa.

[P-1] Pozar, D.M. (1998). *Microwave Engineering*. New York: Wiley.

[R-1] Radmanesh, M.M. (1999). *RF and Microwave Design Essentials: Engineering Design and Analysis from DC to Microwaves*. Boston, Massachusetts: PWS Publishing Company.

[R-2] Rashid, M.H. (2007). *Microelectronic Circuits: Analysis and Design*. Bloomington, Indiana: Authorhouse.

[R-3] Rashid, M.H. (2004). *Introduction to PSpice Using OrCAD for Circuits and Electronics*, 3e. New Jersey: Pearson Education, Inc.

[R-4] Rizzi, P.A. (1988). *Microwave Engineering: Passive Circuits*. Englewood Cliffs, New Jersey: Prentice Hall.

[R-5] Rollett, J.M. (March 1962). Stability and power gain invariants of linear two-ports. *IRE Trans. Circuit Theory* CT-9: 29–32.

[S-1] Schubert, T.F. Jr. and Kim, E.M. (2016). *Fundamentals of Electronics: Book 3: Active Filters and Amplifier Frequency Response*. Morgan & Claypool Publishers.

[S-2] Sedra, A.S. and Smith, K.C. (2016). *Microelectronic Circuits*, 7e. Philadelphia: Saunders College Publishing.

[T-1] Thomas, R.E. and Rosa, A.J. (1998). *The Analysis and Design of Linear Circuits*, 2e. New Jersey: Wiley.

[T-2] Thomas, R.E. and Rosa, A.J. (1984). *Circuits and Signals: An Introduction to Linear and Interface Circuit*. New Jersey: Wiley.

[W-1] Website. www.mathworks.com

[W-2] Website. https://www.jameco.com/Jameco/workshop/TechTip/555-timer-tutorial.html

[W-3] Website. www.datasheetcatalog.com

[W-4] Website. www.alldatasheet.com

[W-5] Website. http://sound.westhost.com/articles/active-filters.htm#s6

[W-6] Website. http://www.analog.com/library/analogdialogue/archives/43-9/EDCh%208%20filter.pdf

[W-7] Wilamowski, B.M. and Irwin, J.D. (2011). *Fundamentals of Industrial Electronics*. CRC Press.

[W-8] Wong, James, Analog Devices, Application Briefs, AB-109. http://www.analog.com/media/en/technical-documentation/application-notes/130445851AB109.pdf

[Y-1] Yang, W.Y., Lee, S.C., Kim, J. et al. (2012). *Circuit System with MATLAB and PSpice*. Korea: Hongrung Publishing Company.

Index

~ refers the main-entry of the corresponding sub-entry

Electronic Circuits with MATLAB®, PSpice®, and Smith Chart, First Edition. Won Y. Yang, Jaekwon Kim, Kyung W. Park, Donghyun Baek, Sungjoon Lim, Jingon Joung, Suhyun Park, Han L. Lee, Woo June Choi, and Taeho Im.

Companion website: www.wiley.com/go/yang/electroniccircuits

MATLAB

(*Continued*)

(*Continued*)

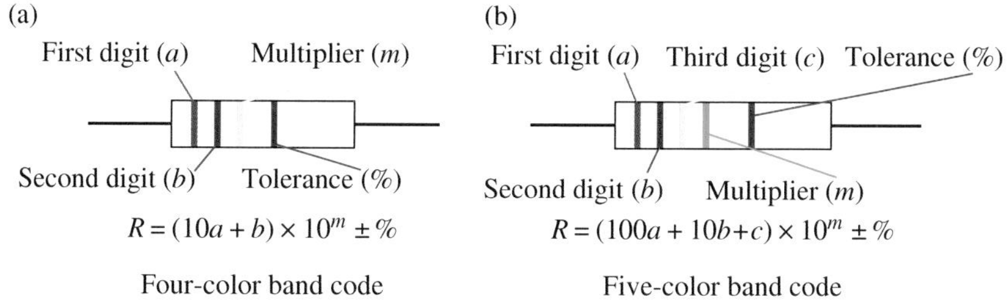

Figure G.1 Color code for resistors.

Electronic Circuits with MATLAB®, PSpice®, and Smith Chart, First Edition. Won Y. Yang, Jaekwon Kim, Kyung W. Park, Donghyun Baek, Sungjoon Lim, Jingon Joung, Suhyun Park, Han L. Lee, Woo June Choi, and Taeho Im.

Companion website: www.wiley.com/go/yang/electroniccircuits